TO METHODS

For complete Subject Index see page 959.

Fundamentals of

CLINICAL CHEMISTRY

Edited by

NORBERT W. TIETZ, Ph.D.

Director of Clinical Chemistry, Mount Sinai Hospital Medical Center, and Professor of Clinical Chemistry and Biochemistry, The University of Health Sciences/The Chicago Medical School, Chicago, Illinois.

with contributions by

Sheldon Berger
Edward W. Bermes, Jr.
Robert V. Blanke
Wendell T. Caraway
Sati C. Chattoraj
Robert L. Dryer
Merle A. Evenson
Willard R. Faulkner
Ermalinda A. Fiereck
Donald T. Forman
Esther F. Freier
Martin E. Hanke
John F. Kachmar
John W. King
Joseph I. Routh
Hyman J. Zimmerman

W. B. Saunders Company

Philadelphia · London · Toronto

W. B. Saunders Company: West Washington Square
Philadelphia, Pa. 19105

12 Dyott Street
London, WC1A 1DB

833 Oxford Street
Toronto 18, Ontario

Listed here is the latest translated edition of this book together with the language of the translation and the publisher.

Spanish *(1st Edition)*–Nueva Editorial Interamericana, S.A., de C.V.,
Mexico City

Fundamentals of Clinical Chemistry ISBN 0-7216-8865-9

Print No.: 9 8 7 6 5

In Thankfulness Dedicated to
MY WIFE GERTRUD
and to my children
MARGARET ANN, KURT RICHARD,
ANNETTE MARIE *and* MICHAEL GERHARD

CONTRIBUTORS

SHELDON BERGER, M.D.

Associate Professor of Medicine, Northwestern University Medical School. Attending Physician, Department of Medicine, Passavant Memorial Hospital, Chicago, Illinois.

EDWARD W. BERMES, JR., Ph.D.

Associate Professor of Biochemistry, Departments of Biochemistry and Pathology, Stritch School of Medicine, Loyola University, Maywood, Illinois. Director of Clinical Chemistry, Loyola University Hospital, Maywood, Illinois.

ROBERT V. BLANKE, Ph.D.

Associate Professor of Legal Medicine, Medical College of Virginia, Health Sciences Division, Virginia Commonwealth University. Chief Toxicologist, Office of Chief Medical Examiner, Virginia. Consultant in Toxicology, McGuire Veterans Administration Hospital, Richmond, Virginia.

WENDELL T. CARAWAY, Ph.D.

Biochemist, McLaren General Hospital, St. Joseph Hospital, and the Flint Medical Laboratory, Flint, Michigan.

SATI C. CHATTORAJ, Ph.D.

Assistant Professor, Department of Biochemistry and Department of Obstetrics and Gynecology, Boston University School of Medicine, Boston, Massachusetts. Director, Research Laboratories, Department of Obstetrics and Gynecology, Boston University School of Medicine, Boston, Massachusetts.

ROBERT L. DRYER, Ph.D.

Professor of Biochemistry, College of Medicine, University of Iowa, Iowa City.

MERLE A. EVENSON, Ph.D.

Assistant Professor of Medicine, University of Wisconsin. Director of Clinical Chemistry Laboratories, University of Wisconsin Hospitals, Madison, Wisconsin.

WILLARD R. FAULKNER, Ph.D.

Associate Professor of Biochemistry and Director of Clinical Chemistry Laboratories, Vanderbilt University Medical Center, Nashville, Tennessee.

ERMALINDA A. FIERECK, M.S.

Associate Chemist, Mount Sinai Hospital Medical Center, Chicago, Illinois.

DONALD T. FORMAN, Ph.D.

Assistant Professor of Biochemistry, Northwestern University Medical School, Chicago, Illinois. Director of Clinical Biochemistry, Evanston Hospital, Evanston, Illinois.

ESTHER F. FREIER, M.S., M.T.(ASCP)

Professor and Hospital Chemist, Department of Laboratory Medicine, University of Minnesota, Minneapolis, Minnesota.

MARTIN E. HANKE, Ph.D.

Professor Emeritus of Biochemistry, University of Chicago, Chicago, Illinois. Formerly Director of Clinical Chemistry Laboratories, University of Chicago Hospitals and Clinics, Chicago, Illinois.

JOHN F. KACHMAR, Ph.D.

Assistant Professor of Biochemistry and Director, Clinical Chemistry Laboratory, Department of Biochemistry, Rush-Presbyterian-St. Luke's Medical Center, Chicago, Illinois.

JOHN W. KING, M.D., Ph.D.

Head, Department of Clinical Pathology, The Cleveland Clinic Foundation, Cleveland, Ohio.

JOSEPH I. ROUTH, Ph.D.

Professor of Biochemistry, University of Iowa College of Medicine. Consultant in Clinical Chemistry, Veterans Administration Hospital, Iowa City, Iowa.

NORBERT W. TIETZ, Ph.D.

Professor of Clinical Chemistry, Department of Pathology; Professor of Biochemistry, Department of Biochemistry, The University of Health Sciences/The Chicago Medical School. Director of Clinical Chemistry, Mount Sinai Hospital Medical Center, Chicago, Illinois.

HYMAN J. ZIMMERMAN, M.D.

Professor of Medicine, Boston University School of Medicine, Boston, Massachusetts. Chief, Medical Service, Veterans Administration Hospital, Boston, Massachusetts.

FOREWORD

By Michael Somogyi, Ph.D.

The clinical chemistry laboratory a few decades ago was a modest handmaiden supplying a few chemical tests that appeared serviceable in diagnostic procedures. These tests were developed in Otto Folin's laboratory at Harvard University, in D. D. Van Slyke's laboratory in the hospital of the Rockefeller Institute, and in a number of biochemistry departments, which were themselves new in some medical schools. These methods, which filled small booklets, were in most instances like cookbook recipes, and were frequently in the hands of semiskilled persons with no great competence in any branch of chemistry. A great change was inaugurated when a few progressive hospitals appointed to their laboratories highly trained professional chemists, men who were qualified to carry on investigative work in addition to supplying reliable data for diagnostic use. These chemists faced great challenges in refining existing methods and in the development of a rapidly increasing number of new tests and techniques. Today, in place of small "cookbooks," the chemist is confronted with a burgeoning amount of literature on clinical chemistry, the "fundamentals" of which are presented in this volume by the cooperative and coordinated effort of a group of competent and erudite clinical chemists.

With the well organized, rich content of this volume before me, and with the picture of instrumentation and equipment of an up-to-date clinical laboratory in mind, I cannot help but hark back to the humble beginnings in this field. In 1926, when I moved from academic work to the newly constructed Jewish Hospital of St. Louis, this was the first hospital in this city (priding itself on two medical schools!) to place a chemist in charge of its clinical laboratory. The space originally allotted to chemistry consisted of a single room of moderate size, with $1200 available for equipment. As my technical assistant I trained a young, small-town school teacher with no previous experience in chemistry. True, the routine work comprised at that time but a dozen or so analytical procedures, involving exclusively simple manual operations. Thus we found time for some investigative and educational activities, which year after year attracted the interest of the interns, residents, and younger members of the medical staff. They learned to use and appreciate the services of the laboratory so that the erstwhile modest handmaiden has developed into "laboratory medicine." My participation in this process, my collaboration with progressive, spirited, mostly youthful, clinicians was the most rewarding experience in my professional life. I am confident that these progressive young clinicians will find the study of the *Fundamentals of Clinical Chemistry* as useful as will workers in the laboratory. By so doing they will narrow the gap between medical science and medical practice.

Michael Somogyi

PREFACE

The editor and the authors of this volume are engaged in teaching clinical chemistry to the great variety of persons interested in this field, namely students of biochemistry, medicine, and medical technology, as well as laboratory technicians and new trainees. Clinical chemistry by its nature is a discipline that draws upon all fields of chemistry (inorganic, analytical, organic, physical, and biological). In addition, it requires a certain familiarity with many aspects of related fields, such as biology, physiology, hematology, immunology, physics, instrumentation, and electronics. The editor, authors, and their students believe that there is a need for a book that conveys the fundamentals of clinical chemistry and guides the novice in initial studies of this complex and rapidly growing field. The auxiliary subjects are included only to the extent that they are necessary to explain and understand the clinical chemical aspects. The text includes not only analytical procedures, but also the chemical principles upon which these methods are based. It is our hope that this will help in training individuals who are prepared to stand up to the challenge of the rapidly growing number of manual and automated analytical techniques for the detection and quantitation of, at times, complex metabolites or other compounds of interest to contemporary medicine and physiological chemistry. The concerted efforts of chemists, pathologists, and public health officials to upgrade the quality of analytical work produced in clinical laboratories cannot succeed without such a group of well trained laboratory technicians.

The sections related to the clinical significance of laboratory tests, although limited, are of value not only to the clinician but also to the technician because they provide the latter with a realization of the use of his data and of the importance of accurate laboratory work. For a comprehensive coverage of these aspects, the student is referred to appropriate texts, professional journals, and references.

Occasionally the material presented goes beyond the scope of basic teaching programs and thus provides information for advanced students; an effort was made to present such material in a form understandable to the undergraduate student.

A list of references and reading assignments are included at the end of each chapter. The references pertaining to methods given in detail are listed directly after the procedure.

The work on this book started several years ago. Throughout this time the International Union of Biochemistry, various scientific organizations, and various individuals have expended a great amount of effort to standardize various units related to Clinical Chemistry so that they may be more meaningful and also consistent throughout the world. The use of nm. instead of mμ. for expressing wavelength, the use of mEq./L. instead of degree acidity for expressing the acidity of gastric content, and the use of Absorbance (A) instead of Optical Density (O.D.) in spectrophotometric measurements are examples of this trend. These new units have been used throughout the book. Although the use of mmoles/L. instead of mM/L. has also been recommended, this recommendation has not been accepted as readily and the term mM./L. is still widely used. Consequently, in this book some contributing authors have continued to use the term mM./L. In a future edition, uniform use of the term mmoles/L. will be enforced.

Each of the chapters presented in this textbook has been reviewed by a number of the other contributors. Thus, it was hoped to incorporate the viewpoint of several individuals and thereby to improve the quality of the material and promote continuity and intellectual uniformity in this multiauthor book. If the authors have failed in certain objectives they would greatly appreciate receiving comments from instructors and students so that these suggestions may be incorporated in any future revision of the text.

NORBERT W. TIETZ, Ph.D.

ACKNOWLEDGMENTS

Preparation of a multiauthor book requires the close cooperation not only of the contributors with the editor, but also among the contributors themselves. I am most grateful to all contributors for submitting so willingly to this principle in preparing their manuscripts and for cooperating so well in the review of the manuscripts of other contributors. Thus, this book is truly a joint effort of the entire staff. My special thanks, however, go to Drs. W. T. Caraway, M. Friedman, M. E. Hanke, and J. F. Kachmar, as well as to Miss E. F. Freier, for their significant support in editing the various manuscripts. Dr. H. J. Zimmerman reviewed all chapters in regard to clinical application and made many helpful suggestions, for which I am most appreciative.

The cooperation of the staff of W. B. Saunders Company and especially of John Dusseau, Vice-President and Editor in Chief, Robert Rowan, Medical Editor, Mr. Grant Lashbrook, Head of the Illustration Department, Deborah A. Roseman and Doris Holmes of the editorial staff is also gratefully acknowledged.

NORBERT W. TIETZ, Ph.D.

CONTENTS

Chapter 1 / BASIC LABORATORY PRINCIPLES AND PROCEDURES

by Edward W. Bermes, Jr., Ph.D., and Donald T. Forman, Ph.D.

The function of the clinical chemistry laboratory is to perform qualitative and quantitative analyses of body fluids, such as blood, urine, and spinal fluid, as well as feces, calculi, and other materials. If the results of these tests are to be useful to the physician in the diagnosis and treatment of disease, they must be done as accurately as possible. This involves using sound analytical methods and good instrumentation. Thus, the persons performing these analyses should not only be thoroughly familiar with the technique involved in the test being done, but they should also be well grounded in general methods of analytical chemistry. A good background in instrumentation is also desirable and, indeed, is becoming essential. In addition to a complete knowledge of the technical aspects of the tests involved, the technologist should know the principles of the method. This involves a knowledge of the chemical reactions and the effect of physical variables on them and the purpose of each reagent used. An understanding of normal values and the extent of deviation from these normal values, which are found in health and disease, is also necessary.

Before a new procedure is introduced it must be evaluated in the laboratory. The variables of the method must be thoroughly explored. For a colorimetric technique, adherence to Beer's law should be investigated. The time for maximum color development, stability of color, as well as reagent and sample stability, must be known. The effect of various anticoagulants should be known if plasma or whole blood is to be used in the procedure. Finally, the accuracy and precision of the procedure have to be evaluated using statistical techniques (see section on statistics). In the years to come the clinical chemistry laboratory will rely even more heavily on both manual and automated types of instrumentation. A thorough knowledge of the theoretical aspects of the equipment is needed as well as an understanding of the practical operational principles required for its day to day use and, to some extent, its maintenance.

GENERAL LABORATORY SUPPLIES

Clinical chemistry laboratories of today are still served in the main by the classic material glass in spite of the modern introduction of plastics and stainless steel. The advantages as well as the limitations imposed by the properties of these materials should be appreciated in light of their possible effects upon analysis.

Glassware

It was first discovered that glassware might be attacked by its contents about the middle of the eighteenth century by Lavoisier. Stas,[27] also found that his glassware was attacked by reagents to such an extent that there was interference with his atomic weight determinations. Many of these difficulties were associated with the use of glasses that had a high alkali content. This problem was decidedly improved after World War I when borosilicate glasses became available.

Clinical laboratory glassware can be divided into five general types of glass:[20] (1) high thermal resistance glass, (2) high silica glass, (3) glass with high resistance to alkali, (4) low actinic glassware, and (5) standard flint glass.

TABLE 1-1. *Thermal Durability of Borosilicate Glass*

Strain point*	510°C.
Annealing point†	555°C.
Softening point	820°C.

* Temperature at which deformation of glass may result due to heat stress.

† Temperature to which glass must be heated to relieve strains.

High thermal resistant glass. This type is usually a borosilicate glass that has a low alkali content. It is free from magnesia-lime-zinc group elements, heavy metals, arsenic, and antimony. The borosilicate glass resists heat, corrosion, and thermal shock.[2] Since its dimensions change very little with temperature (relatively low coefficient of expansion), this type of glassware should be used whenever heating or sterilization by heat is employed. Some borosilicate glassware, if properly supported and not under internal pressures, can be heated to about 600°C. for a relatively short period of time. If the glass is cooled too quickly, however, it will acquire strains that may affect its future serviceability. The highest safe operating temperature for this glass is its strain point. Table 1-1 shows the thermal durability of borosilicate glass at varying temperatures.

Pyrex and *Kimax* brand glass are the most common thermal resistant borosilicate glassware found in the laboratory. Laboratory apparatus such as beakers, flasks, and pipets are usually made of this type of borosilicate glass. Because it contains so few elements, contamination of liquids by the vessel is minimal, even when liquids are hot. *Exax* brand glassware is a lower grade borosilicate glass and may be used when it is not necessary or desirable that high quality borosilicate glass be used.

Several years ago a special alumina-silicate glass was developed that is at least six times stronger than borosilicate glass.[23] This *Corex* brand laboratory glassware has been strengthened chemically rather than thermally. Corex pipets have a typical impact strength of 30,000 p.s.i. compared to a rating of 2000 to 5000 p.s.i. for borosilicate pipets. Corex laboratory glassware is harder than conventional borosilicates and better able to resist clouding due to alkali and scratching. This glass is also used in higher temperature thermometers, graduated cylinders, and centrifuge tubes.

Vycor brand laboratory glassware is recommended for use in applications involving high temperatures, drastic heat shock, and extreme chemical treatment with acids and dilute alkalies. This transparent glassware is resistant to attack by all acids except hydrofluoric and even in the upper temperature range more resistant to alkalies than borosilicate glasses. Vycor ware is used primarily in ashing and ignition techniques. It can be heated to 900°C. and can withstand downshocks from 900°C. to ice water.

High silica glass. The high silica content of this glass (over 96 per cent) makes it comparable to fused quartz in its thermal endurance, chemical stability, and electrical characteristics. The glass is made by removing almost all elements except silica from borosilicate glass. It is radiation resistant and has good optical qualities and temperature capabilities; it is used for high precision analytical work and can also be used for optical reflectors and mirrors.

Glass with high resistance to alkali. This boron-free glassware was developed particularly for use with strong alkaline solutions. Its thermal resistance is much less than that of borosilicate glass and therefore must be heated and cooled very carefully. Its primary use should be whenever solutions or digestions with strong alkali are made. This glass is often referred to as soft glass.

Low actinic glassware. This glassware contains materials that usually impart an amber or red color to the glass and reduce the amount of light passing through to the substance within the glassware. It was developed to provide a highly protective laboratory glassware for handling materials sensitive to light in the 3000 to 5000 Å range (i.e., bilirubin, carotene, and vitamin A).

Standard flint glass. This is a soda-lime glass that will be used at or near room temperature only. It is the glass of choice for weighing bottles since it is less prone to develop static surface charges. Since this glass is relatively easy to melt and shape, it has been used for bottles and disposable laboratory glassware.

The following is a list of special glasses:

Colored and opal glasses. These are made by adding small amounts of coloring materials to glass batches. These glasses are used in light filters, lamp bulbs, and lighting lenses.

Coated glass. This glass has a thin, metallic oxide permanently fire bonded to the surface of glass. The film, unlike glass, can conduct electricity. It has electronic applications as heat shields to protect against infrared light and as electrostatic shields to carry off charges.

Optical glass. It is made of soda-lime, lead, and borosilicate and is of high optical purity. It is used in making prisms, lenses, and optical mirrors.

Glass-ceramics (Pyroceram). This glassware has high thermal resistance, chemical stability, and corrosion resistance (like borosilicate glass). This material is useful in hot plates, table tops, and heat exchangers.

Radiation absorbing glass. This is made of soda-lime and lead and is useful in preventing transmission of high energy radiation (i.e., gamma ray, x-ray).

TABLE 1-2. *Physical and Chemical Properties of Laboratory Plastics*

Material	*Special Properties or Uses*	*Clarity*	*Heat Resistance °C.**	*Effect of* *Weak Acids* / *Strong Acids*	*Weak Alkalies* / *Strong Alkalies*	*Organic Solvents* / *Sunlight*
Polystyrene (styrene)	Excellent optical properties, hard, biologically inert	Transparent	65–80	None Attacked by oxidizing acids	None None	Soluble in aromatic and chlorinated hydrocarbons Yellows
High-impact polystyrene	For molding where impact resistance important	Translucent to opaque	60–80	None Attacked by oxidizing acids	None None	Soluble in aromatic and chlorinated hydrocarbons Some strength loss
Polyethylene regular (PE)	Tough, chemically and biologically inert, pliable	Translucent to opaque	80–99	Resistant Attacked by oxidizing acids	Resistant Resistant	Resistant below 60°C. Surface cracking
Polyethylene high density	Tough, chemically and biologically inert, high heat resistance, harder than conventional polyethylene	Translucent to opaque	105–115	Very resistant Attacked by oxidizing acids	Very resistant Very resistant	Resistant below 80°C. Surface cracking
Polypropylene (PP)	Very tough, heat resistant autoclavable, inert	Transparent to opaque	135–160	Very resistant Attacked by oxidizing acids	None Very resistant	Resistant below 80°C. Surface cracking

Methyl Methacrylate (Plexiglass)	Good for casting, fabricating, machining, polishing, fairly hard, tough, good optical properties	Transparent	60–90	Practically none Attacked by oxidizing acids	Practically nil Practically nil	Soluble in ketones, esters, aromatic hydrocarbons Very slight
Cellulose acetate (acetate)	Clear, easily fabricated (sheet), quite tough	Translucent to opaque	60–105	Slight Decomposes	Slight Decomposes	Soluble in ketones and esters, softens in alcohol Slight
Polyamide (nylon)	Tough, heat resistant machinable, autoclavable	Transparent to opaque	130–150	Resistant Attacked	None None	Resistant to common solvents Slight discoloration
Rigid vinyl (PVC)	Tough, clear, ideal for vacuum forming, plasticized for tubing	Transparent to opaque	50–73	None None to slight	None None	Soluble in ketones, esters, swells in aromatic hydrocarbons Slight
Polytetrafluoroethylene (Teflon)	Extremely high heat resistance, waxy slippery surface, machinable, chemically inert	Opaque	255	None None	None None	None None
Polycarbonate	Very tough, heat resistant, biologically inert	Transparent	135–160	Resistant Attacked by oxidizing agents	Slight Attacked	Soluble in aromatic and chlorinated hydrocarbons Surface cracking

* Heat resistance (continuous) = maximum temperature at which no distortion occurs during continuous exposure (not under tension).

Plasticware[25]

The introduction of plasticware to clinical chemistry laboratories has greatly enhanced laboratory analysis. Beakers, bottles, flasks, graduated cylinders, funnels, centrifuge tubes, tubing, and pipets now have unique qualities that make them ideal for use when high corrosion resistance and unusual impact and tensile strength are required. Table 1-2 describes the physical and chemical properties of various resins that are used in the preparation of laboratory ware.

Polyolefins (polyethylene, polypropylene). These are a unique group of resins with relatively inert chemical properties. Generally, the polyolefins are unaffected by acids (however, concentrated sulfuric acid slowly attacks polyethylene at room temperature), alkalies, salt solutions, and most aqueous solutions; aromatic, aliphatic, and chlorinated hydrocarbons cause moderate swelling at room temperature; organic acids, essential oils, and halogens slowly penetrate these plastics. Strong oxidizing agents attack this group of resins at elevated temperatures only. Polypropylene plasticware is slightly more vulnerable to attack by oxidizing agents, but it has the advantage of withstanding higher temperatures. Polyethylene and polypropylene are used primarily to fabricate bottles, beakers, jars, carboys, jugs, funnels, pipet jars, pipet baskets, tanks, buret covers, check valves, disconnect valves, twistcock connectors, needle valves, hollow stoppers, dropping pipets, hydrometer jars, stirring rods, tubing and reagent dispensers. Polyethylene is less expensive than polypropylene and is used in most disposable plasticware. Polypropylene has a distinct advantage in that it can be sterilized; however, it absorbs pigments and tends to become discolored.

Polycarbonate resin. It is twice as strong as polypropylene and may be used at temperatures ranging from −100 to +160°C.; however, its chemical resistance is not as wide as that of the polyolefins. This resin is unsuitable for use with bases such as amines, ammonia, and alkalies, as well as oxidizing agents. It is dissolved by chlorinated aliphatic and aromatic hydrocarbons. Polycarbonate resin is insoluble in aliphatic hydrocarbons, some alcohols, and dilute aqueous acids and salts. Since labware molded from this resin is glass-clear and shatterproof it is used extensively in centrifuge tubes and graduated cylinders.

Tygon. A nontoxic, clear plastic of modified plasticized polyvinyl chloride, it is used extensively for the manufacture of tubing (i.e., tubing used in AutoAnalyzers). It is very flexible and can be used to handle most chemicals, although it should not be subjected to prolonged immersion in aliphatic or aromatic hydrocarbons, ketones, and esters. It is flexible at −30°C. (brittle at −45°C.), resists dry heat to 95°C., and can be steam autoclaved or chemically sterilized. This tubing is soft and flexible and quickly slips over tubulatures gripping tightly on glass or metal.

Teflon fluorocarbon resins. They have unique qualities that make them almost chemically inert and ideal when high corrosion resistance at extreme temperatures is essential. Bottles and beakers made of Teflon resist severe temperatures ranging from −270 to +255°C., which allows use in cryogenic experiments or work at high temperatures over extended periods.

This labware is pure translucent white and inert to such corrosive reagents as boiling aqua regia, nitric and sulfuric acids, boiling hydrocarbons, ketones, esters, and alcohols. Because of its unique antiadhesive properties and its nonwettable surface, Teflon is used for self-lubricating stopcocks, stirring bars, bottle cap liners, and tubing. It is also quite easy to clean and is fast drying, but it can be scratched and misshaped.

VOLUMETRIC EQUIPMENT AND ITS CALIBRATION[21]

Most clinical chemistry procedures require accurate measurements of volumes. In precise work it is never safe to assume that the volume contained or delivered by any piece of equipment is exactly that amount indicated by the graduation mark. Ideally all volumetric apparatus should be either purchased with a calibration certificate or calibrated by the analyst.

Pipets

In general, two main types of pipets are used in clinical chemistry (Fig. 1-1). The *volumetric* or *transfer pipet* (Fig. 1*A*) is designed to deliver a fixed volume of liquid and consists of a cylindrical bulb joined at both ends to narrower glass tubing. A calibration mark is etched around the upper suction tube and the lower delivery tube is drawn out to a fine tip. The more important requirements of the volumetric pipet are that the calibration mark should not be too close to the top of the suction tube, that the bulb should merge gradually into the narrower tubes, and that the tip should have a gradual taper. The orifice should be such that the outflow of liquid is not too rapid in order to reduce drainage errors to negligible proportions.[8]

Volumetric or transfer pipets are calibrated to deliver the volume specified. The most commonly used sizes are 1, 2, 3, 4, 5, 10, 25, 50, and 100 ml. Less frequently used sizes, delivering 6, 8, and so on, ml. can also be obtained. They are used for accurate measurements of aliquots of nonviscous samples, filtrates, and standard solutions. These pipets should be of Kimax or Pyrex grade. The reliability of the calibration of the volumetric pipet decreases with a decrease in size and, therefore, special micropipets have been developed for chemical microanalysis.

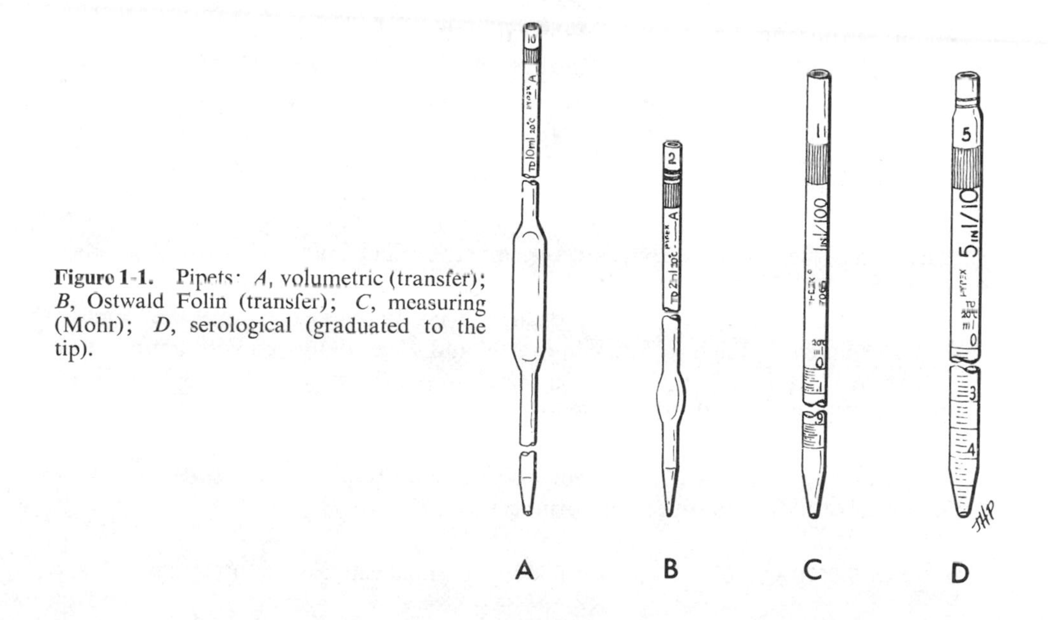

Figure 1-1. Pipets: *A*, volumetric (transfer); *B*, Ostwald Folin (transfer); *C*, measuring (Mohr); *D*, serological (graduated to the tip).

Ostwald-Folin pipets (Fig. 1*B*) are similar to volumetric pipets, but have their bulb closer to the delivery tip. These pipets are found in 0.5, 1.0, 2.0, and 3.0 ml. sizes. They are used for measuring viscous fluids such as blood or serum. The Ostwald-Folin pipet has an etched ring near the mouthpiece, indicating that it is a blow out pipet. The pipet is not blown out, however, until the blood or serum has drained to the last drop in the delivery tip. With opaque fluids such as blood, the top of the meniscus must be read when filling, and controlled slow drainage is required in order that no residual film of blood be left on the walls of the pipet. All the previously mentioned pipets have been calibrated to deliver (TD) a specific volume.

The second type of pipet is the *graduated* or *measuring pipet* (Fig. 1*C*). This is a plain, narrow tube drawn out to a tip and graduated uniformly along its length. Two types calibrated for delivery are available; one is calibrated between two marks on the stem (*Mohr*) and the other has graduation marks down to the tip (*serological*). These pipets are intended for the delivery of predetermined volumes. The serological type of pipet (Fig. 1*D*) must be blown out to deliver the entire volume of the pipet. It has an etched ring (or pair of rings) near the mouth end of the pipet signifying that it is a blow out pipet. Measuring pipets are commonly used in 0.1, 0.2, 0.5, 1.0, 5.0, and 10.0 ml. sizes. They are principally used for the measurement of reagents and are not generally considered accurate enough for measuring samples and standards. The serological pipet has a larger orifice than the measuring (Mohr) pipet and thus drains faster.

Calibration of to deliver pipets is usually performed by measuring the amount of water delivered by the pipet. This measurement may be made by weighing the water delivered and calculating the volume from its density. It is important to note the flow out time and to maintain the pipet and receiving vessel in the same position for a 10 second drainage period. The receiving vessel is stoppered and weighed and the equilibrated water and air temperature recorded. The weight of water delivered from the to deliver pipet should be corrected for temperature to obtain the true delivery volume. Water is commonly used as a calibration liquid because it is readily available and because it is similar in viscosity and speed of drainage to the dilute solutions ordinarily employed in clinical chemical analysis.

The following formula is used for the calibration of a to deliver pipet:

$$\text{Apparent weight of water in gm. at } t^\circ\text{C. (equilibrated temperature)} + \text{stated capacity of pipet in ml.} \times \text{correction at } t^\circ\text{C.} = \text{actual capacity in ml. at } 20^\circ\text{C.}$$

One determines the actual capacity of a volumetric pipet made of borosilicate glass and marked to deliver 10 ml. at 20°C. in the following manner:

1. First determine the weight of the empty receiving vessel, e.g., 22.0391 gm.

2. Determine the weight of receiving vessel and water delivered with temperature of water at 24.1°C., e.g., 31.9961 gm. In this example the weight of water, therefore, is $31.9961 - 22.0391 = 9.9570$ gm. Correcting per ml. of stated capacity, using a table that corrects for determining true capacities of borosilicate glass vessels from weight of water in air at 24.1°C., we get 0.00369 (24.1°C.).

3. Substituting in the preceding formula:

$$9.9570 + (10 \times 0.00369) = 9.9939 \text{ ml. actual capacity at } 20^\circ\text{C.}$$

Finally:

$$\frac{0.0061}{10.0000} \times 100 = 0.061\% \text{ error}$$

This small error requires no correction. A correction need not be applied in general analysis if it is 0.1 per cent or less.[12]

In microwork, the remaining volume left in a pipet can cause a significant error. For this reason most micropipets are calibrated to contain (TC) the stated volume rather than to deliver it.[1] They are generally available in small sizes such as 0.02, 0.05, 0.1, and 0.2 ml. The contents of these pipets are expelled into the recipient solution and then the pipet is rinsed several times by drawing up the solution to the calibration mark and expelling it back into the receiving vessel. Care must be taken not to exceed the calibration mark or a positive error will occur. Hemoglobin pipets are an example of this type of pipet.

Inaccurate calibration of to contain micropipets constitutes an important source of error in clinical chemistry. A gravimetric procedure employing mercury as the substance delivered or contained by the micropipet obviates this problem. In calibrating with mercury the level may be adjusted so that the bottom of the meniscus is set against the graduation line being tested. Linderstrom-Lang micropipets have a constriction above the bulb so that the pipet is self-leveling at the stated volume.

For clinical chemistry work it should not be necessary to calibrate pipets manufactured to Class A capacity tolerance as prescribed by the National Bureau of Standards. When using Class B pipets the analyst should be aware that their tolerances are about twice as large as those for Class A pipets.

Burets

Burets (Fig. 1-2) are calibrated by filling them to a point just above the zero line with freshly boiled and cooled (to room temperature) distilled water and carefully adjusting the meniscus to the zero line. The drop of water adhering to the buret tip is removed by touching the tip to the inside of a glass vessel. A tared vessel is placed beneath the tip and the buret is fully opened. Delivery should proceed freely with the buret tip not in contact with the vessel wall until the meniscus is about 1 cm. above the graduation line to be tested. The buret tip should be touched to the wall of the container and the content allowed to drain into the solution, until the meniscus reaches the graduation mark. The tared vessel is then stoppered and reweighed again, noting the equilibrated water and air temperature. The test should be repeated in triplicate and the final weight of water corrected to true delivery volumes.

As in the case with graduated pipets, the precision of calibration using water decreases with a decrease in buret size.

Burets used in macroanalysis have major graduation marks completely around the long cylindrical tube and minor graduation marks at least halfway around. By this means, errors of parallax in reading the meniscus can be avoided. The standard buret described is found in sizes varying from 1 to 100 ml.; a buret of 50 ml. capacity is subdivided at 0.1 ml. intervals. Burets having a capacity of 10 ml. or less are classified as microburets.

The outflow of liquid from the buret is usually controlled by an all-glass or all Teflon stopcock. The latter type of stopcock does not require any lubricant and is especially useful when the titrant is an alkali. The all-glass stopcock should be lightly

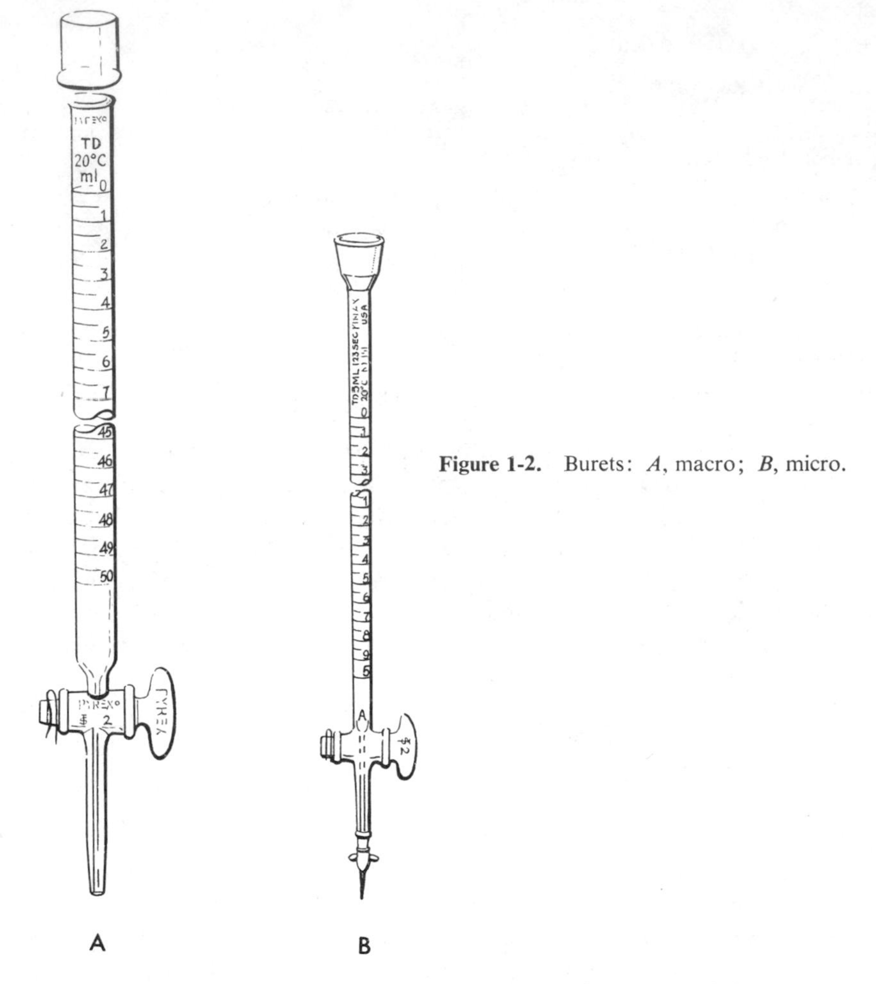

Figure 1-2. Burets: *A*, macro; *B*, micro.

greased with pure petrolatum-like lubricants. Silicone-containing lubricants are not recommended since they "creep" along the length of the buret with subsequent contamination of the walls.[17]

Volumetric Flasks

Volumetric flasks (Fig. 1-3) should be thoroughly cleaned and dried with purified ethanol or acetone before calibration. The flask is weighed and then filled with carbon dioxide free distilled water until just above the graduation mark. The neck of the flask just above the water level should be kept free of water. The meniscus mark is set at the graduation line by removing excess water and the flask is reweighed. The final weight is corrected for the equilibrated water and air temperature to obtain the volume of the flask.

Volumetric flasks are found in the following sizes: 1, 2, 5, 10, 25, 50, 100, 200, 250, 500, 1000, 2000, and 4000 ml. They are primarily used in preparing solutions of known concentration.

An important factor in the use of volumetric apparatus is the need for accurate setting or reading of the meniscus level. A small piece of card that is half-black and

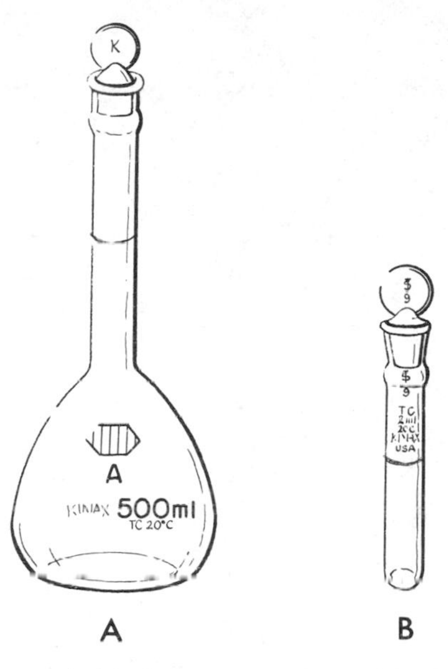

Figure 1-3. Volumetric flasks: *A*, macro; *B*, micro.

half-white is most useful. The card is placed 1 cm. behind the apparatus with the white half uppermost and the top of the black area about 1 mm. below the meniscus. The meniscus then appears as a clearly defined thin black line. This device is also useful in reading the meniscus of a buret.

Volumetric equipment should be used with solutions equilibrated at room temperature. Solutions diluted in volumetric flasks should be repeatedly mixed during dilution so that the contents are quite homogeneous before making up to volume. In this way, errors due to expansion or contraction of liquids in mixing are negligible.

For accurate work with pipets and burets, they should be rinsed out and drained twice with some of the liquid to be measured. The filling of pipets with poisonous, corrosive, or volatile liquids should be done using a rubber bulb or other pipet aids.

Separatory Funnels

Separatory funnels (Fig. 1-4) are employed in the clinical chemistry laboratory for simple extraction procedures. This involves the bringing of a given volume of solution into contact with a given volume of solvent (immiscible in the solution) by vigorous shaking until equilibrium has been attained, followed by separation of the liquid layers. If necessary, the procedure may be repeated after the addition of fresh solvent. This type of extraction gives rapid, simple, and clean separations. When extracting from one liquid to a lighter solvent, it is necessary to remove the lower phase from the funnel after each extraction before removing the extraction solvent, as in ethyl ether extractions of aqueous solutions. Most separatory funnels taper off into a narrow bottom, which contains a sealed stopcock. Thus, it is relatively easy to separate two phases for further analysis.

Automatic Diluting Systems

Semiautomatic versions of *pipets* and burets are obtainable in sizes ranging from 0.02 to 100 ml. and are very useful in the routine clinical laboratory. Figure 1-5 shows

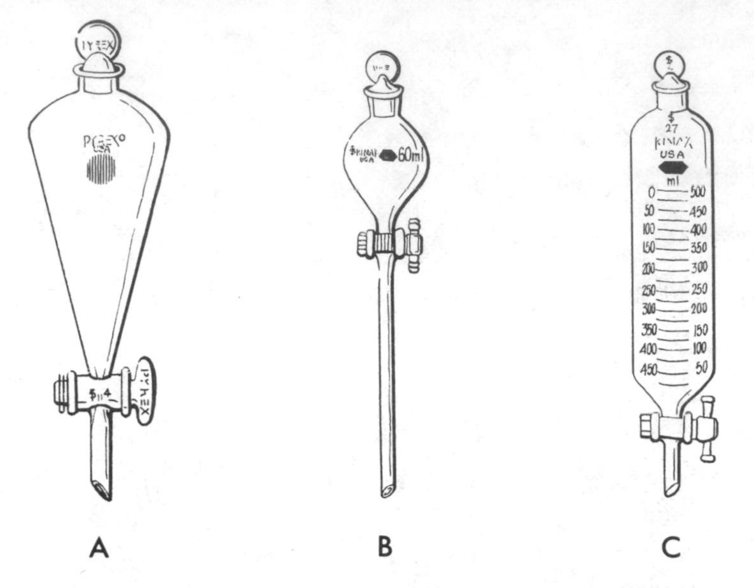

Figure 1-4. Separatory funnels: *A*, Pear shaped; *B*, globe shaped, *C*, cylindrical, graduated.

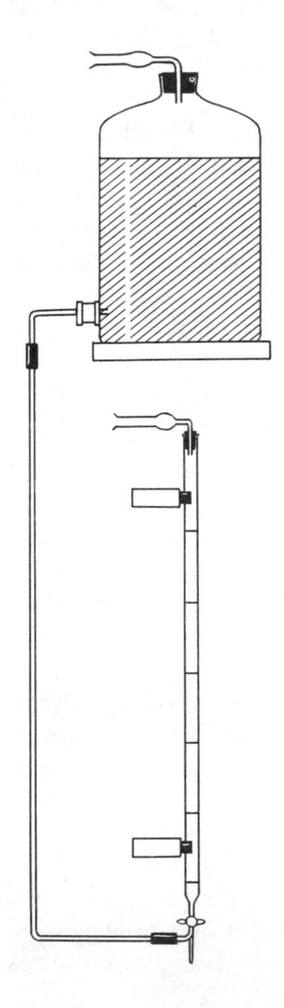

Figure 1-5. Semiautomatic buret, buret-stock bottle arrangement.

an older system that has a large stock bottle of solution conveniently attached to a buret. By incorporating a sidearm tap just above the buret tap, or by using a buret fitted with a three-way stopcock, which is used for filling and delivering, solutions can be rapidly delivered to the waiting vessel. Gravity feed to the buret is the most common arrangement.

Syringe-type, continuous *pipets* (Fig. 1-6*A*) allow the automatic delivery of measured, repeated doses. These glass syringes can be calibrated and adjusted easily. All metal parts of this pipettor, except for the stainless steel plunger springs, are made of chromeplated brass. This pipet can accommodate 1, 5, and 10 ml. standard syringes and has been reported to be accurate to ± 1.5 per cent with a reproducibility of ± 0.5 per cent. It can be set to deliver any volume from 0.2 ml. to 10 ml.

Another automatic pipettor is shown in Figure 1-6*B*. Made of glass, Teflon, and plastic, only the glass and Teflon come in direct contact with the reagent. The reproducibility of this device has been reported to be ± 0.5 per cent.

A manual *syringe type dilutor* is shown in Figure 1-7*A*. This one piece, borosilicate glass system of syringes and valves aspirates and dispenses sample and diluent.

Figure 1-7*B* shows a dilutor that is a semiautomatic sampling, diluting, and dispensing apparatus. It measures and dispenses preset volumes of solutions by means of two motor-driven syringes—one for metering the sample and one for metering the diluent. The dilutor can be adjusted to accept as little as 10 μl. of sample and deliver it with as much as 10 ml. of diluent. The reproducibility of the metered pumps has been claimed to be $+0.5$ per cent.

These automatic and semiautomatic devices must be correctly calibrated by the analyst before they can be accepted as a useful tool. The volume dispensed by the

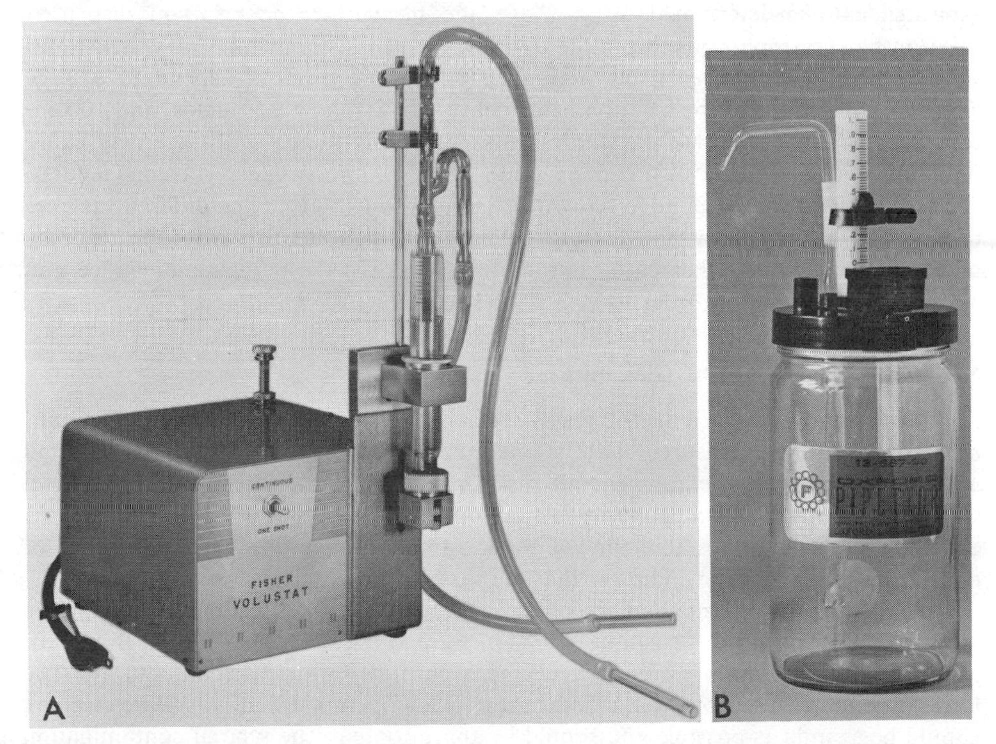

Figure 1-6. Automatic pipettors: *A*, motor driven all glass syringe type; *B*, manual teflon glass syringe type.

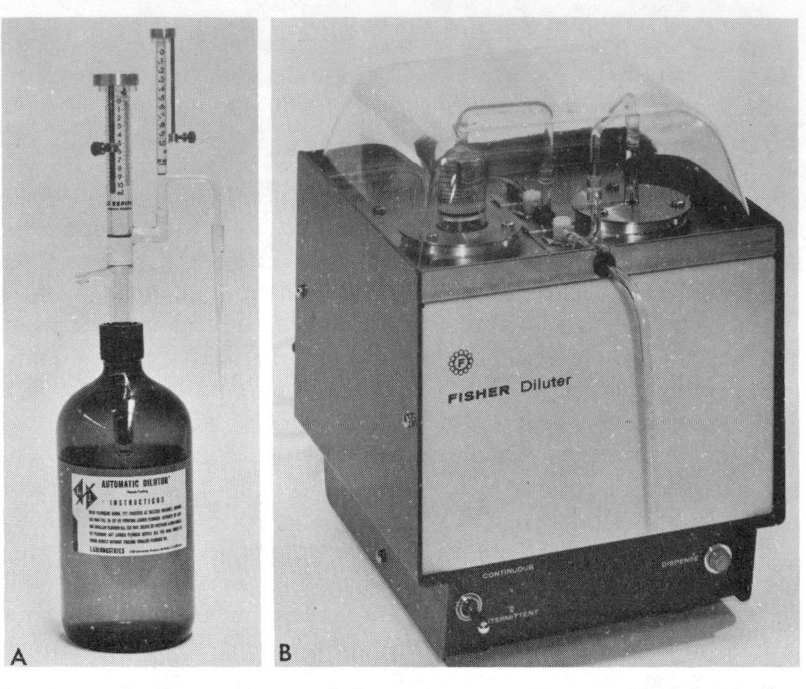

Figure 1-7. Automatic dilutor: *A*, manual glass syringe type; *B*, motor driven teflon-glass syringe type.

diluent syringe can be determined by gravimetric calibration and the volume of sample aspirated can be determined using a dye and measuring the resultant dispensed solution by spectrophotometry.

Problems encountered using automatic pipets and dilutors depend to a large degree on the nature of the solution being delivered. Strong bases, acids, and 100 per cent organic solvents cause some problems. Reagents (acids or bases) may cause bubbling and reproducibility can be quite poor. Some organic solvents tend to "creep" in the system and adversely affect the reproducibility. The different degrees of viscosity of samples and the degrees of washout of samples also affect the accuracy of these devices. In either case, autodilutors must receive daily maintenance and periodic recalibration in order to insure their reproducibility and accuracy.

Maintenance and Care of Glassware

It is essential that volumetric glassware and glass apparatus be absolutely clean, otherwise volumes measured will be inaccurate and chemical reactions affected adversely. One gross method generally used to test for cleanliness is to fill the vessel with distilled water and then empty it and examine the walls to see whether they are covered by a continuous thin film of water. Imperfect wetting or the presence of discrete droplets of water indicates that the vessel is not sufficiently clean.

A wide variety of methods has been suggested for the cleaning of volumetric glassware. Of the various cleaning agents in common usage, the National Bureau of Standards prefers fuming sulfuric acid and a chromic-sulfuric acid mixture.[16] The latter is primarily used by most laboratories. It is imperative that glassware cleaning should be as mild as possible and should be appropriate to the type of contamination present. Fats and grease are the most frequent causes of severe contamination and it is advisable to dissolve these contaminants by a lipid solvent (water-miscible organic solvent) followed by water washing before removing the last traces with an oxidizing agent (chromic acid). The most widely used oxidant is a solution of sodium dichromate in concentrated sulfuric acid. Because of its oxidizing power, the solution, particularly when hot, removes grease quickly and completely. Cleaning solution, as the mixture is called, is not a general solvent for cleaning all apparatus, only for cleaning borosilicate glassware, including volumetric ware. Glassware is generally left in contact with the mixture for 1 to 24 hours, depending upon the amount of grease or lipid present. After removal of the acid and draining, the glassware should be washed out at least four times with tap water and then three times with distilled water. Glassware contaminated with chromic acid is unsuitable for enzyme analyses.

PREPARATION OF CHROMIC ACID CLEANING SOLUTION

Stir about 20 gm. of powdered technical sodium dichromate with just enough water to make a thick paste. Slowly and carefully add 300 ml. of technical grade concentrated sulfuric acid, stirring well. This preparation is best carried out in a sink. Store in a glass-stoppered bottle or covered glass jar. Clear supernatant solution should be decanted from the bottle each time it is used. The solution may be used repeatedly until the reddish color of dichromate has been replaced by the green color of the chromic ion. Do not allow this cleaning solution to come into contact with you or your clothing because it will burn the skin severely and destroy clothing. If any acid is spilled on the floor or bench top, neutralize it immediately with commercial grade sodium bicarbonate; then wash completely with water. If it is spilled on the skin, wash under running water as soon as possible.

Alternative methods of cleaning. These include the use of fuming nitric acid, 10 per cent alcoholic potassium hydroxide, hydrochloric-nitric acid (1:1), 50 per cent potash solution, and acid, neutral or alkaline permanganate solutions. In case of the latter, manganese dioxide separates at the grease sites and its subsequent removal by concentrated hydrochloric acid causes the evolution of chlorine, which destroys the grease. Precipitated material may also be removed by aqua regia, fuming sulfuric acid, or an ultrasonic device. When chemical cleaning is inadequate, mechanical cleaning using warm soapy water or synthetic alkaline detergents can be used to good effect. Care is necessary to insure complete removal of these cleaning agents, since even low concentrations can result not only in a chemical contamination, but also in a marked lowering of the surface tension of aqueous solutions with consequent change in meniscus shape. In extreme cases, steaming out glassware has been suggested, but the thermal retardation effect on the glass can cause measurable changes in its volume.

Borosilicate laboratory glassware is extremely resistant to acidic cleaning solutions, with the exception of hydrofluoric acid; however, strongly alkaline solutions will attack any glass over a period of time. Therefore, borosilicate glassware should generally be washed in a cleaning solution that is neutral or acid in reaction. It is important that scratching of glass be avoided since scratching diminishes the thermal shock resistance of glassware and can lead to breakage.

CARE OF ABSORPTION CELLS (CUVETS)

Absorption cells must be scrupulously clean. Optical surfaces should not be touched as grease smudges are difficult to remove. As soon as possible after use, absorption cells should be rinsed and soaked in distilled water. When cleaning cells, a mild detergent should be used. Stubborn contaminants can be removed by soaking the cells in dilute sulfuric acid. Absorption cells should never be allowed to soak in hot concentrated acids, alkalies, or other agents that may etch the optical surfaces. When drying cuvets, high temperatures and unclean air should be avoided. A low to medium temperature oven (not to exceed 100°C.) or vacuum or a combination of the two can be used to rapidly dry cuvets.

CLEANING PIPETS

Pipets should be placed in a vertical position with the tips up in a jar of cleaning solution. A pad of glass wool is placed at the bottom of the jar to prevent breakage. After soaking for several hours, the pipets are drained and rinsed with tap water until all traces of cleaning solution are removed. The pipets are then soaked in distilled water for at least an hour. A gross test for cleanliness is made by filling with water, allowing the pipets to empty, and observing whether drops form on the side within the graduated portion. Formation of drops indicates greasy surfaces. After the final distilled water rinse, the pipets are dried in an oven at not more than 110°C. Most laboratories that use large numbers of pipets daily use a convenient automatic pipet washer. These devices are made of metal or polyethylene and can be connected directly to hot and cold water supplies. Polyethylene baskets and jars may be used for soaking and rinsing pipets in chromic acid cleaning solution.

CLEANING AND CARE OF BURETS

Inspect the stopcock plug before cleaning. If it is well greased, the plug will turn easily and the surface between the plug and barrel will appear transparent. If the plug needs greasing, remove and wipe it clean, also wipe out the inside of the barrel. Both parts must be dry before greasing. Apply a thin layer of good grade stopcock grease. Heat about 100 ml. of chromic acid cleaning solution to a temperature of 60 to 70°C. After clamping the buret in an inverted position with the opening reaching nearly to the bottom of the beaker, draw the cleaning solution into the buret by suction until the level is slightly past the final graduation mark. Do not allow the cleaning solution to reach the stopcock area where it will remove the grease. After closing the stopcock, allow the filled buret to stand 3 to 5 minutes. The stopcock is then opened and the buret raised above the liquid level and allowed to drain thoroughly. The buret should then be flushed out well with tap water and finally rinsed at least three times with distilled water. When all traces of cleaning solution have been removed and the buret has been properly rinsed, drops of water will not adhere to the inner wall surface. The buret should be filled with clean distilled water and left in this state until needed. If the buret is left empty, it will quickly become contaminated with a film of grease.

CLEANING FLASKS, BEAKERS, CYLINDERS, OTHER GLASSWARE, AND VOLUMETRIC EQUIPMENT

Pour warm cleaning solution into each vessel and stopper or cover carefully. Each vessel should be manipulated so that all portions of the wall are repeatedly brought into contact with the solution. This procedure should be followed for at least 5 minutes. The cleaning solution can be poured from one vessel to another and then returned to its original container. The vessels should then be rinsed repeatedly with tap water (four rinses) and finally rinsed three times with distilled water. It is important that the necks of volumetric flasks above the graduation mark be clean because, when solutions are diluted in the flask, drops of water may adhere to an unclean wall and may invalidate the measurement of volume.

FILTER PAPERS

A properly stocked clinical chemistry laboratory must have on hand filter papers in a variety of sizes and characteristics. These differ in thickness, porosity, and wet strength.

Filter papers can be classified into various categories. Within each category there are papers with slow, medium, and rapid filtering speeds, which correspond to fine, medium, and coarse retentions. A grade of filter paper is selected on the basis of precipitate size and method of filtration (gravity or suction). The success or failure of a separation depends a great deal on the method of handling the filtration. The majority of analytical filtrations are made with a 60° funnel using a gravity filtration. Many analysts have their own individual methods of folding filter paper and, in general, paper for use in a funnel should be just large enough to reach the edge of the funnel or slightly below and should never extend over the top. Filter papers with diameters of 9.0 and 12.5 cm. are in common use in the clinical laboratory because they fit funnels generally used for preparation of blood and urine filtrates. Precipitates are poured into the paper with the aid of a stirring rod to avoid splashing and to keep the precipitate near the apex of the filter paper cone. Suction and other filter aids (accelerators, ashless floc, ashless clipping) may be used to enhance the filtration process.

Table 1-3 describes the grade, flow rate, and retention characteristics of commonly used filter papers. Retention refers to the type of precipitate the grade will retain; speed refers to relative mean flow rates.

Protein precipitates are best removed on a smooth surfaced, medium speed filter paper. Gelatinous precipitates such as the hydroxides of many multivalent metals are quantitatively retained on a rapid filtering paper. The granular crystalline precipitates of metallic halides and phosphates are best collected on a slower paper of

TABLE 1-3. *Selection of Useful Filter Papers—Properties of the Various Grades*

S and S Grade	*Whatman No.*	*Retention*	*Speed**	*Thickness (inches)*	*Surface*
595	1	Medium	75	0.005	Smooth
597	2	Fine	110	0.006	Smooth
604	4	Coarse	23	0.007	Smooth
602	5	Very fine	275	0.007	Rough
588	12	Whatman No. 2 ready folded			
402	30	Medium	95	0.007	Rough
589-white	40	Medium	95	0.006	Rough
589-black	41	Coarse	22	0.007	Smooth
589-1H	41H	Coarse	17	0.005	Smooth
589-blue	42	Very fine	300	0.007	Rough
589-red	44	Very fine	250	0.006	Rough

* Relative flow values; the larger the number, the slower the filtration rate.

a medium texture. Finely divided precipitates such as calcium oxalate or barium sulfate should be filtered on a very slow, fine pore paper.

Filter paper circles are made in several diameters, of which the most common are 5.5, 7, 9, 11, 12.5, and 15 cm. The size taken for a particular filtration depends on the bulk of the precipitate to be caught, and the volume of the solution to be filtered. The total insoluble matter after filtration should occupy less than one third of the volume of the paper cone.

A recent filtering aid has been the development of Whatman *Phase Separating Paper* No. 1 PS. This paper is a water repellent cellulose acetate that acts as a phase separator in place of conventional separatory funnels. It separates aqueous solutions from water immiscible solutions, quickly and conveniently. Because this paper is hydrophobic, it retains the aqueous layer but allows the organic layer to go through. It also filters out any solid that may be present in the system at the time of filtration. This phase separating paper can be used with organic solvents that are lighter or heavier than water. If the organic solvent is lighter than water, it will gradually migrate to the edge of the water meniscus and pass through the paper, leaving the water phase behind. If heavier than water, it will pass directly through the paper, and the water will be retained.

Glass fiber filters are also available for clinical techniques. These filters are produced entirely from borosilicate fibers and when used in a Gooch, Büchner, or similar filtering apparatus they give a combination of fine retention with extremely rapid filtering speed not usually found in any cellulose grade. Glass fiber filters are well suited for filtration of heavy viscous solutions or gels since they do not clog as quickly as open cellulose papers and yet can retain very fine particles.

Filter Papers for Special Applications

Cellulose products for chromatography and electrophoresis can be separated into four general classes: (1) papers, (2) powders, (3) ion exchange celluloses, and (4) specialties.

Papers. The papers for chromatography and electrophoresis vary with respect to flow rate, loading capacity, and adsorption. The flow rates are relative and are classified as fast, medium, and slow. In general, the loading capacity is related to the thickness of the paper, and the thicker the grade, the more heavily it can be loaded. Adsorption on the cellulose fibers of one or more of the constituents being separated occurs occasionally because of the nature of the cellulose itself. The thin grades like Whatman No. 1 and No. 44 and the hardened grades show the least adsorption.

Powders. Cellulose powders for column chromatography meet the demand for a cellulose material capable of larger scale separations than is possible on paper. The powders are fibrous in nature and, in this respect, differ from many of the chromatographic adsorbents, which are powdered inorganic compounds. The grades can be of a coarse material with no consistently measurable particle size and of a powder ground to pass a 200 mesh sieve and give a closely packed, slow running column.

Ion exchange celluloses. These materials include phosphorylated cellulose, carboxymethyl cellulose, DEAE (diethylaminoethyl) cellulose, ECTEOLA (epichlorohydrin triethanolamine), and other celluloses.

Specialties. These include the many individual precut papers and membranes of special design useful for chromatographic and electrophoretic techniques. In this group are barium-impregnated strips for the removal of bilirubin in the estimation of urobilinogen and cation and anion exchange paper useful for chromatography of amino acids.

Cellulose acetate membranes made of homogenous cellulose acetate with uniform pores of less than 2 μ diameter are used in electrophoresis systems to separate proteins, hemoglobins, immunoglobulins, and various isoenzymes. The porosity of these membranes when carefully controlled permits a precisely predictable flow of buffer solution between the cell reservoirs.

The Millipore filters (Millipore Corp., Bedford, Mass.) are cellulose acetate porous membranes ranging in pore size from 8 μ downward to 10 mμ. These filters enable the analyst to concentrate efficiently any particles by pore size. The Millipore filter is free of biological inhibitors and provides an optimum collection environment for microorganisms.

Centriflo membrane filter cones (Amicon Corp., Lexington, Mass.) have recently been introduced in ultrafiltration techniques for concentration and purification of multiple small samples of biological fluids. The Centriflo cones retain molecules greater than 50,000 (M.W.) and are intended essentially for protein concentration.

MEASURES AND WEIGHTS

Ultimately, most of the calculations in clinical chemistry are concerned with measurements of volume and mass. Every measurement includes both a number and a unit. The unit identifies the kind of dimension and the number indicates how many of the reference units are contained in the quantity being measured.

The metric system is used primarily in clinical chemistry. It is frequently called the centimeter-gram-second (or cgs) system after its three reference units: the centimeter (cm.) of length, the gram (gm.) of mass, and the second (sec.) of time. Other units in this system are derived from these three reference units. For example, the unit for volume in the cgs system is the cubic centimeter (cm^3), since volume = length $\times$ length $\times$ length = cm. $\times$ cm. $\times$ cm. = cm^3; the unit for density is the unit for mass divided by the unit for volume, or the gram per cm^3 (gm./cm^3) since density = mass/volume = gm./cm^3.

Length

The metric standard of length is 1 meter (m.). It is divided into 1000 equal parts called millimeters (mm.). The millimeter in turn is divided into 1000 equal parts called micrometers (μm.). Micro means one millionth, and a micrometer is one millionth of a meter. The angstrom (Å) is used in atomic measurements and the micrometer equals 10,000 angstroms.

1 meter (m.) = 100 centimeters (cm.) = 1000 millimeters (mm.)

1 kilometer (km.) = 1000 m.,	1 micrometer (μm.) = 10^{-3} mm.
1 meter (m.) = 100 cm.,	1 nanometer (nm.) = 10^{-6} mm.
1 centimeter (cm.) = 10 mm.	1 angstrom (Å) = 10^{-8} cm.

All nonmetric units of length are easily converted to metric units, remembering the interrelation:

1 inch = 2.540005 cm., or simply, 2.54 cm.
e.g., 1 foot = 12 inches = 12 × 2.54 = 30.48 cm.

Conversion factors for length

1 inch (in.) = 2.540 cm.,	i.e., cm. = in. × 2.540
1 centimeter = 0.3937 in.,	i.e., in. = cm. × 0.3937
1 meter = 3.281 ft.,	i.e., ft. = m. × 3.281

WAVELENGTH (λ)

When discussing the wavelength of radiant energy, one refers to the international standard measure of 1 meter. One thousandth of a meter is the millimeter (mm.), and one millionth of the meter or one thousandth of a millimeter is micrometer (μm.), formerly called micron (μ). One thousandth of a micrometer is the nanometer (nm.), formerly designated millimicron (mμ). The wavelengths of the visible part of the spectrum extend from approximately 400 to 700 nanometers (nm.). The angstrom unit (Å) is equal to one tenth of a nanometer. Therefore, 500 nm. is equivalent to 5000 Å.

Infrared measurements are usually made beyond 1000 nm. (one micrometer). It is in the infrared region from 2 to 12 micrometers that the most common organic compounds are identified.

Volume

The volumetric unit of capacity most commonly used in clinical chemistry when liquids are measured is standardized from a unit mass (weight) of water. Thus, 1 *liter* is defined as the volume of 1 kilogram of water at its temperature of maximum density, 4°C. It is supposedly identical with 1000 cubic centimeters (cm^3); however, after the kilogram of mass was established, very precise measurements showed that it was too large by about 27 parts per million—1 kilogram of water at 4°C. measured 1000.027 cm^3 instead of the intended exact 1000. Thus, the cgs unit of capacity, the milliliter (ml.) equals 1.000027 cm^3. This difference has no practical significance in elementary chemical problems and 1 *liter is now defined as exactly equal to* 1 *cubic decimeter or* 1000 *cm*^{3} (CGPM,* 1964).[9]

1 liter (L.) = 1000 milliliters (ml.) = 1000 cubic centimeters (cm^3)

The *cubic foot* (cu ft. or ft^3) unit of volume is equal to 12 in. × 12 in. × 12 in. = 1728 cubic inches (cu in. or in^3). A frequent unit of liquid measure is the *U.S. gallon*, which is defined as 231 cubic inches or 3785 ml. Either of these measures can be converted to metric equivalents by remembering the linear factor of 2.54 cm. per inch, as just explained.

Conversion factors for volume

1 liter = 1.057 quarts	1 cubic foot = 28.32 liters
1 quart = 0.9463 liter	1 cubic inch = 16.39 cm^3 or ml.
1 gallon = 3.785 liters	1 fluid ounce = 29.57 cm^3 or ml.

* Conférence Générale des Poids et Mésures.

For volumes smaller than 1 ml., the preferred unit is the microliter (μl.), which is the one-thousandth part of a milliliter or one millionth of a liter. Another unit used to indicate a microliter is the term cubic millimeter (mm^3). The volume of ultramicropipets may be written in any of the following ways: 0.05 ml., 0.05 cm^3, 50 μl., and 50 mm^3.

Weight (Mass)

Weight is a measure of quantity of matter and the total weight of any system undergoing changes remains constant. The cgs unit of weight is the gram (gm.), which is the mass of 1 ml. of water at 4°C. (the temperature of maximum density of water). At this temperature, 1 gm. of water occupies 1 ml., or exactly 1 cm^3. A larger unit of weight is the kilogram (kg.), which is equal to 1000 grams. A smaller unit is the milligram (mg.), which is one thousandth of a gram (0.001 gm.). A still smaller unit is the microgram (μg.), which is one millionth of a gram (10^{-6} gm.). One finds 1 microgram often written as 10^{-3} mg. or 10^{-6} gm. One nanogram (ng.) is equal to 10^{-9} gm. and 10^{-12} gm. is equivalent to 1 picogram (pg.).

Since 1 μg. is a one millionth part of 1 gram, industrial laboratories will often report micrograms per milliliter or per gram as parts per million (ppm). Thus, 20 μg. per ml. will be reported as 20 ppm. If a serum contains 2.5 μg. of copper in 1 ml., it may be expressed in the following ways: 0.0025 mg./ml., 2.5 μg./ml., 2.5 ppm, 2.5×10^{-6} gm./ml., 2.5×10^{-3} mg./ml.

Conversion of weight units

1 pound = 453.6 gm. 1 kilogram = 2.205 lb.
1 ounce = 28.35 gm. 1 gram = 15.43 grains
1 grain = 65 mg.

Gas Concentration and Pressure

The common units for gas concentrations are volume percent (v/v) or, preferably, mmole/L. (or m*M*./L.). For example, the CO_2 content of blood is expressed in mmole/L. If all the carbon dioxide in the blood were in the form of bicarbonate, mEq./L. of bicarbonate and mmole/L. of carbon dioxide would be identical. Since this is not the case (carbon dioxide in the blood is in the form of free CO_2, HCO_3^-, and H_2CO_3), it is best to report CO_2 levels in mmole/L. (mEq./L. is no longer preferred, CGPM, 1964). Until recently the common unit of expression for partial pressure of a gas as carbon dioxide was mm. Hg; however, the CGPM in 1964 recommended the substitution of "millibar" (mbar) for the conventional millimeter of mercury (1 mm. Hg = 1.33322 mbar). As of now, this recommendation has not been followed in the United States and mm. Hg. is still widely used.

Ionic Strength

The concentration of buffer solutions is often given in terms of ionic strength since the effect of a buffer solution is a function of the concentration and the charge of all ions present. The ionic strength is the sum of the numbers obtained by multiplying the concentration (moles/liter) of each ion by the square of its valence and dividing the sum by 2.

Example:

1. 0.2 mole/L. NaCl

$$\text{ionic strength} = \frac{0.2 \times (1)^2 + 0.2 \times (1)^2}{2}$$

$$\text{ionic strength} = 0.2$$

2. 0.5 mole/L. Na_2SO_4

$$\text{ionic strength} = \frac{0.5 \times 2 \times (1)^2 + 0.5 \times (2)^2}{2}$$

$$\text{ionic strength} = 1.5$$

NOMENCLATURE FOR PREFIXES

Prefixes approved by the CGPM, 1964, and the International Congress of Clinical Chemistry, 1966, for use in clinical chemistry that denote the approved decimal factors are shown in Table 1-4. It is recommended that the clinical laboratory have the book by Dybkaer and Jorgensen[9] as an available reference.

TABLE 1-4. *Prefixes Denoting Decimal Factors*

Factor	*Name*	*Symbol*	*Factor*	*Name*	*Symbol*
			10^{-3}	milli	m
			10^{-6}	micro	μ
10^{12}	tera	T	10^{-9}	nano	n
10^{9}	giga	G	10^{-12}	pico	p
10^{6}	mega	M	10^{-15}	femto	f
10^{3}	kilo	k	10^{-18}	atto	a
10^{2}	hecto	h	10^{-1}	deci	d
10^{1}	deca	da	10^{-2}	centi	c

CHEMICALS

Laboratory chemicals are supplied in about six grades. The differentiation of these is quite unofficial and there is no agreement among manufacturers concerning the designation of the various degrees of purity. The most common designations follow.

Primary Standards

These highly purified chemicals may be weighed out directly for the preparation of solutions of selected concentration or for the standardization of solutions of unknown strength. They are supplied with an analysis for contaminating elements and an assay for each lot. The assay should not be less than 99.95 per cent. The specifications for primary standards have been prepared by the Committee on Analytical Reagents of the American Chemical Society.[24] These chemicals must be stable substances of definite composition, which can be dried, preferably at 104°C. to 110°C., without a change in composition. They must not be hygroscopic, so that water is not adsorbed during weighing (see discussion on preparation of standard solutions).

Secondary standards are solutions whose concentration cannot be determined directly from the weight of solute and volume of solution. The concentration of secondary standards is usually determined by analysis of an aliquot of the solution by an acceptable reference method, using a primary standard.

Reagent Grade or Analytical Reagent Grade (A.R.)

This degree of purity belongs to several hundred chemicals called reagents; these chemicals meet specifications designed to permit use in quantitative and qualitative analyses. These are found in two forms: (1) *lot analyzed reagents*, in which each individual lot is analyzed and the actual amount of impurity reported (e.g., arsenic—0.0005 per cent) and (2) *maximum impurities reagents*, for which maximum impurities are listed (e.g., arsenic—maximum 0.001 per cent). In the latter instance, the arsenic may only be 0.0004 per cent, but, of course, the analyst has no tangible figure to put his finger on other than the guaranteed maximum limit of 0.001 per cent.

In this reagent group the specifications have been prepared by the Committee on Analytical Reagents of the American Chemical Society for over 187 of the most commonly used items.[24] Establishment of the A.C.S. specifications marked a turning point in the development of chemical purity. Manufacturers of "A.C.S." chemicals check each lot in a control laboratory and only place A.C.S. on the labels of those lot chemicals that meet the Society's published specifications. These reagent grade chemicals are of very high purity and are recommended for trace metal analyses and for standardization of reference methods.

Chemically Pure Grade (C.P.)

The degree of purity of materials bearing this label is shared by the terms "highest purity" and "chemically pure." The designation chemically pure fails to reveal what limits of impurities are tolerated, and the practice followed by different manufacturers in the use of this designation is not uniform. Chemicals in this category are probably not dependable for research and various clinical chemical techniques unless the chemist has analyzed the materials to assure himself of the absence of impurities that may cause trouble. The term highest purity is used by manufacturers for organic chemicals that they have purified to as great a degree as they find practical. The purity is usually determined by measurement of melting points or boiling points. Again, for research purposes, this grade of purity may be unsatisfactory without further purification. This group of chemicals may be used in clinical chemical analysis when higher purity biochemicals are not available.

U.S.P. and N.F. Grade

These chemicals are produced to meet specifications set out in the *United States Pharmacopeia* (U.S.P.) or the *National Formulary* (N.F.). These designations are of primary concern to the pharmaceutical chemist, and the tolerances specified for impurities are those that will not be injurious to health. In many cases these compounds may be very pure and can be used in chemical analysis and in the preparation of various reagents, but this cannot be assumed to be the case in all instances. The important thing to remember is that in these categories chemical purity is only incidental.

Purified, Practical, or Pure Grade

These chemicals can be used as starting materials for laboratory synthesis of other chemicals of greater purity, but probably require purification and analysis before they can be used as analytical reagents. For certain analyses when reagents are not available, practical chemicals can be used if a blank is also run. In general, these chemicals should not be used in clinical chemical analysis.

Technical or Commercial Grade

These chemicals are generally used only in manufacturing. The degree of purity varies widely and depends on the ease with which contaminants can be removed.

Reference Standards

The purity of some biochemicals used in clinical chemical analysis has been assured by the availability of reliable preparations. It is presently possible to obtain reference standards from the National Bureau of Standards for cholesterol, creatinine, urea and uric acid.

Spectroanalyzed solvents, chromatographically pure reagents (i.e., amino acids), and calibrating reagents (for bilirubin, cholesterol, and protein assays) are now also commercially available.

Purity of Organic Reagents

The purity of commercially obtained organic reagents for clinical chemistry purposes is generally inferior to that of inorganic reagents. The majority of impurities in these compounds will have been introduced in their synthesis either with the starting materials or as by-products, and these are presumably more difficult or more expensive to remove than impurities in inorganic substances. In addition, some organic compounds oxidize or decompose on standing and the amount of impurities from this cause will depend on how long the bottle of reagent has been opened or stored. A well known example of this sort of deterioration is provided by solutions of the chloride indicator s-diphenylcarbazone. This orange-yellow solution must be protected from light, since it turns orange-red when exposed to daylight and cannot be used thereafter. Its stability is much improved when it is stored in amber bottles and refrigerated; however, phenols and amines oxidize on standing and tend to darken even when refrigerated. Sugars have been shown to be quite hygroscopic and absorb moisture rapidly unless they are properly stored.

The presence of impurities in an organic reagent may be a source of difficulty in its use. If the contaminant does not react with the substance being determined, interference will not occur as long as there are no interfering side reactions and there is enough of the original reactant remaining. If a reagent is impure, the net final color developed by a reaction may be considerably less than ideal because of a high blank due to the impurity. The existence of isomers and their presence in a particular lot of an organic reagent may be a source of difficulty, since in the rather specific geometrical requirements of a chelate ring only certain isomers may produce the desired colored complex[18] or in enzyme reactions, only one of the isomers may be suitable as substrate.

DESICCANTS AND USE OF DESICCATORS

Most published information concerning the comparative efficiency of drying agents is based upon experiments that measure the amount of moisture absorbed from air flowing through a bed of desiccant.[5] The limited information that is available concerning the behavior of drying reagents in a desiccator suggests that their comparative usefulness for this purpose may not be the same as in flow measurements. Every effort should be made to avoid desiccants that produce dust when used in desiccators (granular calcium chloride frequently carries a large amount of dusty "fines" when fresh). Drying agents that incorporate cobalt chloride or other indicators to show when they are exhausted are much preferred to those that do not. Silica gel and anhydrous calcium sulfate (Drierite) are sold in indicating forms.

TABLE 1-5. *Chemistry and Activity of Desiccants*

Drying Agent	*Activity**	*Capacity*	*Deliquescence*	*Easy Regeneration*	*Chemical Reaction*
Phosphorus pentoxide	0.02	v low	yes	no	acidic
Barium oxide	0.6–0.8	moderate	no	no	alkaline
Alumina	0.8–1.2	low	no	yes	neutral
Magnesium perchlorate (anhydrous)	1.6–2.4	high	yes	no	neutral
Calcium sulfate (Drierite)	4–6	moderate	no	yes	neutral
Silica gel	2–10	low	no	yes	neutral
Potassium hydroxide (stick)	10–17	moderate	yes	no	alkaline
Calcium chloride (anhydrous)	330–380	high	yes	no	neutral

* Micrograms residual water per liter of air at 30°C.

Desiccators should be opened carefully. Ordinary desiccators may contain air at less than atmospheric pressure; this is a result of heating of the air when hot specimens are inserted and then cooling of the air with the lid on. If the desiccator is not opened slowly, the inrush of air may create draughts sufficient to dislodge materials from open vessels or to stir up dust particles from the drying agent that may subsequently settle in the vessels that are being stored. Vacuum desiccators should be provided with a curved inlet tube to deflect incoming air against the lid, and the stopcocks on these should be opened very carefully when restoring the internal pressure to that of the atmosphere.

Table 1-5 describes various desiccants. From this table it is apparent that several are distinctly alkaline and one is strongly acidic. The choice of the drying reagent required for the quantitative absorption of moisture depends on the composition of the gases or materials to be dried, convenience, efficiency, and sometimes cost. Magnesium perchlorate quantitatively absorbs ammonia gas, and anhydrous calcium chloride (technical grade) absorbs carbon dioxide and ammonia. These facts should be kept in mind when choosing a desiccant intended for the quantitative removal of water from gases that may also contain ammonia, carbon dioxide, or other reactive substances. Certain drying reagents are deliquescent, and, when liquefaction of the drying agent occurs, a decline in drying efficiency results. Calcium chloride and magnesium perchlorate are examples; however, both have a considerable capacity before deliquescence sets in. Phosphorus pentoxide is one of the most powerful drying agents in use, but its effective capacity is rapidly reduced by formation of metaphosphoric acid. Drying agents prepared with moisture-sensitive salts, which indicate exhaustion of drying capacity by a change in color (such as silica gel, activated alumina, and anhydrous calcium sulfate), are advantageous but considerably more costly.

Some drying agents can easily be regenerated, and this is an important consideration when comparing costs. Silica gel can be regenerated by heating in a drying oven at 120°C., but anhydrous calcium sulfate and activated alumina require temperatures of 275°C. and 175°C., respectively. Magnesium perchlorate can be regenerated by heating to 240°C. in a partial vacuum.

DISTILLED AND DEIONIZED WATER

Distilled or deionized water is necessary for the preparation of all reagents and solutions in the laboratory. Even such water is not entirely pure, however, for it is contaminated by dissolved gases and by material dissolved from the container in which it has been stored. The dissolved gases may be removed by boiling the water for a short time. Occasionally distilled water is found to be contaminated by nonvolatile impurities that have been carried over by steam, in the form of a spray. The substances most likely to be carried over by spurting or spray are sodium, potassium, manganese, carbonates, and sulfates. The kinds and types of impurities introduced into the distilled water from the distilling apparatus also depend on the material used in the construction of the equipment. The most common impurities of this type are copper and glass products. Also, a still may froth while in operation and badly contaminate the distillate.

When water of the highest purity is required, distilled water is redistilled from an alkaline permanganate solution, in a silica or block tin apparatus. The permanganate solution oxidizes nitrogenous matter present. Redistilled water prepared in this manner is known as conductivity water. Many distillation apparatus are now equipped with deionization resin beds, which feed deionized water into the still and thereby make the redistillation from an alkaline permanganate solution unnecessary.

The storage vessel employed may have a marked effect on the purity of water and reagents. Solutions stored in soda glass vessels are much more easily contaminated by trace metals than those stored in borosilicate glass vessels. Thus, when reagents or water are to be stored for long periods of time, it is desirable to use Pyrex bottles.

The simplest overall check on the quality of distilled and deionized water is the measurement of specific conductance. Some laboratories endeavor to keep the specific conductance at around 1.5 to 2.0 micromhos,[24] with a resistance over 1 million ohms; however, the conductance does not measure the nonionized substances (organic contaminants) that may be present in the water and the conductivity of a given sample may in part be due to dissolved gases ($CO_2 \rightleftharpoons H_2CO_3$).

The following properties meet the general analytical requirements of distilled or deionized water:[24]

Property	*Requirement*
Residue after evaporation	Not more than 1 mg./L. based on drying 1 hour at 105°C.
Chloride content	Not more than 0.1 mg./L.
Ammonia content	Not more than 0.1 mg./L.
Heavy metals (e.g., lead)	Not more than 0.01 mg./L.
Consumption of permanganate	When 0.03 ml. of 0.1 *N* potassium permanganate is added to 500 ml. of water containing 1 ml. of concentrated sulfuric acid, the color should not completely disappear on standing for 1 hour at room temperature.

Deionized water may be prepared by using a commercial deionizer. Deionized water is water from which mineral salts have been removed by a process of ion exchange. Mixed-bed synthetic anionic and cationic exchangers are used as the filtering beds through which water flows. Generally this water is not free from organic matter, bacteria or other nonionizable material, but may be prefiltered through a charcoal bed before deionization, which reduces the degree of such contamination. Deionized water can be substituted for distilled water in most laboratory operations; however, in fluorometric and enzyme analyses, double or triple distilled water is recommended in place of deionized water.

PREPARATION AND CONCENTRATION OF SOLUTIONS

Solute and solvent. In a solution of one substance in another substance, the dissolved substance is called the solute. The substance in which the solute is dissolved is called the solvent. When the relative amount of one substance in a solution is much greater than that of the other, the substance present in a greater amount is generally regarded as the solvent. When the relative amounts of the two substances are of the same order of magnitude, it becomes difficult, in fact arbitrary, to specify which substance is the solvent.

Dilute, Saturated, Concentrated, and Standard Solutions

A *dilute* solution contains a relatively small proportion of solute, and a *concentrated* solution contains a relatively large proportion of solute. Concentrated solutions are possible only when the solute is very soluble.

A *saturated* solution exists when the molecules of the solute in solution are in equilibrium with the excess undissolved molecules. Since temperature affects solubility, the exact temperature of the solution should be specified.

A *supersaturated* solution exists when there is more solute in solution than is present in a saturated solution of the same substance at the same temperature and pressure. Supersaturated solutions are unstable and cannot exist in equilibrium with the solid phase.

Standard solutions are an integral part of every quantitative analysis in the clinical chemistry laboratory and are used during the assay of an unknown sample. This is a solution whose precise concentration is known. The process of determining or adjusting the concentration of the standard solution is known as *standardization*, which may be carried out in one of three ways:

1. Direct preparation of the standard solution by dissolving a weighted amount of a pure, dry chemical and diluting the solution to an exactly known volume. For example, a 1.0 molar sodium chloride solution may be prepared by dissolving 58.50 gm. of dry NaCl and diluting the solution in a volumetric flask to exactly 1 L.

2. Titration of a solution of a weighed portion of a pure, dry chemical by the solution to be standardized. The weighed material used for this purpose is known as a primary standard. If a solution of sodium hydroxide is standardized by determining the volume of the solution required to react with a known weight of pure, dry sulfamic acid ($NH_2 \cdot SO_3H$), the latter serves as a primary standard. The equation of the reaction is:

$$NH_2 \cdot SO_3H + NaOH = NH_2SO_3Na + H_2O$$

3. Titration against a primary standard such as hydrochloric acid that has been made up from constant boiling hydrochloric acid. The solution thus standardized is known as a secondary standard. The concentration of the base to be standardized in this titration cannot be known to any greater degree of accuracy than that of the standard hydrochloric acid solution.

A well equipped clinical chemistry laboratory will maintain a supply of primary reference standard solutions for use, when necessary, in the standardization of other secondary standard solutions (working standards). The solutions used as reference standards must be stable on storage; i.e., their concentrations must not change appreciably with time. The solution should not react with glass or with constituents of the atmosphere and it should not be affected by light. Many standard solutions that can be used safely within a few days of their preparation are not sufficiently stable for use as permanent reference standards.

EXPRESSING CONCENTRATIONS IN PHYSICAL UNITS

The concentrations of solutions are generally expressed in the following ways:

1. By the weight of solute per unit volume of solution (w/v), i.e., 20 gm. of KCl per liter (gm./L.) of solution (2.0 per cent, w/v).

2. By the weight percentage composition, or the number of grams of solute per each 100 grams of solution (w/w), i.e., a 10 per cent (w/w) NaCl solution contains 10 gm. of NaCl per 100 gm. of solution (10 gm. of NaCl are dissolved in 90 gm. of water to form 100 gm. of solution).

3. By the weight of solute per weight of *solvent*, i.e., 10 gm. of NaCl in 100 gm. of water (10 gm. NaCl/110 gm. solution).

4. By volume of solute per volume of solvent (v/v). The volume-per-volume solutions are the least accurate and are used when the solute is a liquid as in alcohol solutions. A 70 per cent solution of alcohol would be prepared by adding 70 ml. of alcohol to water and diluting to a final volume of 100 ml.

5. By parts per million (ppm) i.e., μg. of solute per gram of solution. Since 1 ml. of water weighs 1 gm., ppm is equal to μg./ml. or mg./L.

EXPRESSING CONCENTRATION IN CHEMICAL UNITS

1. A *molar* solution contains 1 mole of the solute in 1 liter of solution, i.e., a 1 molar (*M*) solution of H_2SO_4 contains 98.08 gm. of H_2SO_4 per liter of solution, since the molecular weight of H_2SO_4 is 98.08. A 0.50 *M* solution contains 0.50×98.08 gm. = 49.04 gm. H_2SO_4 per liter of solution.

2. A *normal* solution contains 1 gram equivalent weight of the solute in 1 liter of solution, i.e., 1 mole of HCl, 0.5 mole of H_2SO_4, and 0.333 mole of H_3PO_4, each in 1 liter of solution, give normal (*N*) solutions of these substances. A normal (*N*) solution of HCl is also a molar (*M*) solution. A normal (*N*) solution of H_2SO_4 is also a 0.5 molar (*M*) solution.

3. A *molal* solution (*m*) contains 1 mole of solute in 1000 gm. of solvent. In comparison, a molar solution has a final volume of 1000 ml., whereas the molal solution has a volume that exceeds 1000 ml. The following equations define the expressions of concentration:

$$\text{Molarity of a solution} = \frac{\text{number of moles of solute}}{\text{number of liters of solution}} \tag{1}$$

$$\text{moles} = \frac{\text{weight (grams)}}{\text{gram molecular weight}}$$

$$\text{Normality of a solution} = \frac{\text{number of gram-equivalents of solute}}{\text{number of liters of solution}} = \frac{\text{number of milligram equivalents of solute}}{\text{number of milliliters of solution}} \tag{2}$$

$$\text{Number of milligram equivalents (milliequivalents)} = \text{number of milliliters} \times \text{normality}$$

$$\text{Normality (in oxidation-reduction reaction)} = \text{molarity} \times \text{oxidation state change}$$

$$\text{gram equivalent weight (as oxidant or reductant)} = \frac{\text{formula weight (grams)}}{\text{oxidation state change}}$$

$$\text{Molality of a solution} = \frac{\text{number of moles of solute}}{\text{number of kilograms of solvent}} \tag{3}$$

The molarity and normality expressions are useful in clinical chemistry when the amount of solute in a given portion of solution is related to the measured volume of solution and when comparing the relative volumes required for two solutions to react chemically with each other. A limitation of the normality scale is that a given solution may have more than one normality, depending on the reaction for which it is used. The molarity of a solution, however, is a fixed number since there is only one molecular weight for any substance.

A milligram equivalent of a substance is its equivalent weight expressed in milligrams. The equivalent weight of H_2SO_4 is 49.04 gm. Then 1 gram equivalent of H_2SO_4 = 49.04 gm. H_2SO_4, and 1 milligram equivalent of H_2SO_4 = 49.04 mg. H_2SO_4. Since substances may react on the basis of their valence, 1 mole of calcium (atomic weight = 40), which is bivalent, has twice the combining power of 1 mole of sodium. Therefore, 1 mole of calcium has the combining power of two equivalents or 2 moles of sodium, or 40 gm. of Ca is equivalent to 2 times 23 gm. of Na.

The unit of measure commonly used is the milliequivalent (mEq.), which is 1/1000 of an equivalent.

$$\text{milliequivalents (mEq.)} = \frac{\text{weight (grams)}}{\text{milliequivalent weight (grams)}}$$

Milligrams per 100 ml. can be converted to mEq. per liter using the following formula:

$$\text{mEq./L.} = \frac{\text{mg./100 ml.} \times 10 \times \text{valence}}{\text{atomic weight}}$$

Example:
If serum sodium is 322 mg./100 ml., then there are 3220 mg./L. The equivalent weight of sodium is 23, and the valence is 1, therefore:

$$\text{mEq./L.} = \frac{322 \times 10 \times 1}{23} = 140$$

DILUTION PROBLEMS

All of the solutions considered thus far are volumetric solutions that contain a definite amount of solute in a fixed volume of solution. In percentage, molar, normal, or molal solutions, the amount (mass) of solute contained in a given volume of solution is equal to the product of the volume times the concentration. Whenever a solution is diluted its volume is increased and its concentration decreased, but the total amount of solute remains unchanged. Hence, two solutions of different concentrations but containing the same amounts of solute will be related to each other as follows:

$$\text{amount of solute}_1 = \text{volume}_1 \times \text{concentration}_1$$
$$\text{amount of solute}_2 = \text{volume}_2 \times \text{concentration}_2$$

Since amount of solute_1 = amount of solute_2, then the $\text{volume}_1 \times \text{concentration}_1 = \text{volume}_2 \times \text{concentration}_2$.

The volume and concentration on both sides of the equation must be expressed in the same units.

Examples:

$$\text{ml}_1 \times N_1 = \text{ml}_2 \times N_2$$
$$\text{liters}_1 \times M_1 = \text{liters}_2 \times M_2$$
$$\text{ml}_1 \times \text{per cent}_1 = \text{ml}_2 \times \text{per cent}_2$$

Example:
It is desired to make 500 ml. of a 0.12 *N* solution of HCl from a stock solution of 1.00 *N*. Substituting into the equation:

$$\text{ml}_1 \times N_1 = \text{ml}_2 \times N_2$$
$$500 \times 0.12 = X \times 1.00$$
$$X = \frac{500 \times 0.12}{1.00} = 60$$

Diluting 60 ml. of 1.00 *N* HCl to 500 ml. will give a solution of 0.12 *N* HCl.

Buffer Solutions and Their Action

Buffers are defined as substances that resist changes in the pH of a system. All weak acids or bases, in the presence of their salts, form buffer systems. The action of buffers and their role in maintaining the pH of a solution can best be explained with the aid of the Henderson-Hasselbalch equation.

$$pH = pK_a + \log \frac{[salt]}{[acid]}$$

where [salt] = $[A^-]$ = dissociated salt
and [acid] = [HA] = undissociated acid

Consequently, the pH of the system is determined by the pK of the acid and the ratio of $[A^-]$ to [HA]. The buffer has its greatest buffer capacity at the point where the $[A^-]$ = [HA] and the pH = pK_a.*

This entered into the preceding equation gives

$$pH = pK_a + \log 1$$
$$pH = pK_a + 0$$

The capacity of the buffer decreases as the ratio deviates from 1. If the ratio is beyond 50/1 or 1/50, the system is considered to have lost its buffer capacity. This point is approximately 1.7 pH units to either side of the pK of the acid since

$$pH = pK_a + \log 50/1$$
$$pH = pK_a + 1.7$$

Table 1-6 demonstrates the relationship between pH and the ratio of CH_3COONa to CH_3COOH (A^-:HA).

TABLE 1-6. *Salt-Acid Ratio and* pH

Concentration CH_3COONa *Molar*	CH_3COOH *Molar*	*Ratio Salt/Acid*	pH
0.00	0.20	0.00	2.7
0.01	0.20	0.05	3.4
0.05	0.20	0.25	4.1
0.10	0.20	0.50	4.4
0.20	0.20	1.00	4.7
0.40	0.20	2.00	5.0
1.00	0.20	5.00	5.4
2.00	0.20	10.00	5.7

The chemical mechanisms by which buffers exert their effect may be seen by considering the reactions involved upon the addition of base to a buffer solution containing acetate ions, CH_3COO^-, and acetic acid molecules, CH_3COOH.

On addition of NaOH: CH_3COOH / CH_3COONa $+ Na^+OH^- \rightarrow 2CH_3COONa + H_2O$ (OH^- is removed)

The addition of alkali decreases the CH_3COOH in the buffer and increases the CH_3COONa. The pH of the solution increases in proportion to the change in ratio of salt to acid in the buffer solution.

* The pK_a is defined as the negative logarithm of the dissociation constant of the acid.

On addition of HCl:

$$\left.\begin{array}{l} CH_3COOH \\ CH_3COONa \end{array}\right\} + H^+Cl^- \rightarrow 2CH_3COOH + NaCl$$

H^+ is removed (the hydrogen ion has combined with acetate to form poorly dissociated acetic acid.)

In this case the HCl reacts to decrease CH_3COONa and increase CH_3COOH in the buffer. The pH of the solution falls in proportion to the change in ratio of salt to acid in the solution; however, since the pH is related to the logarithm of the A^-/HA ratio only a small change in pH occurs.

TABLE 1-7. *Phosphate Buffer* (0.1 *M*)

pH range: 5.29 to 8.04; 20°C. (Sorenson); 10 ml. mixtures of *X* ml. of 0.1 *M* Na_2HPO_4 (14.2 gm./L.) and *Y* ml. of 0.1 *M* KH_2PO_4 (13.6 gm./L.). The pH values are about 0.03 pH unit less at 37°C.

X ml. Na_2HPO_4	*Y* ml. KH_2PO_4	pH *at* 20°C.
0.25	9.75	5.29
0.5	9.5	5.59
1.0	9.0	5.91
2.0	8.0	6.24
3.0	7.0	6.47
4.0	6.0	6.64
5.0	5.0	6.81
6.0	4.0	6.98
7.0	3.0	7.17
8.0	2.0	7.38
9.0	1.0	7.73
9.5	0.5	8.04

TABLE 1-8. *Tris(hydroxymethyl)aminomethane Buffer*

pH range: 7.20 to 9.10; 23°C. 0.5057 gm. of tris-(hydroxymethyl)aminomethane dissolved in 50 ml. of distilled water and mixed with the indicated amounts of *X* ml. of 0.1 *N* HCl and diluted to 100 ml. give the pH values shown in the table. The pH values are approximately 0.15 pH unit lower at 37°C.

X ml. 0.1 *N* HCl *Added*	pH *at* 23°C.
5.0	9.10
7.5	8.92
10.0	8.74
12.5	8.62
15.0	8.50
17.5	8.40
20.0	8.32
22.5	8.23
25.0	8.14
27.5	8.05
30.0	7.96
32.5	7.87
35.0	7.77
37.5	7.66
40.0	7.54
42.5	7.36
45.0	7.20

TABLE 1-9. *Carbonate-Bicarbonate Buffer*

pH range: 9.1 to 10.6 at 25°C. (Delory and King); 10 ml. mixtures of *X* ml. of 0.1 *M* Na_2CO_3 (10.6 gm./L.) and *Y* ml. of 0.1 *M* $NaHCO_3$ (8.4 gm./L.). The pH values are about 0.1 pH unit less at 37°C.

X ml. Na_2CO_3	*Y* ml. $NaHCO_3$	pH *at* 25°C.
1.1	8.9	9.1
1.4	8.6	9.2
2.2	7.8	9.4
2.7	7.3	9.5
3.9	6.2	9.7
5.1	4.9	9.9
6.4	3.6	10.1
7.4	2.5	10.3
7.9	2.1	10.4
8.3	1.6	10.5
8.8	1.2	10.6

TABLE 1-10. *Acetic Acid–Sodium Acetate Buffer*

pH range: 3.6 to 5.8; 25°C. Mixtures of *X* ml. of 0.2 *N* CH_3COOH (dilute 11.5 ml. A.R. grade glacial acetic acid to liter) and *Y* ml. of 0.2 *N* CH_3COONa (27.2 gm./L.). The pH values are approximately 0.05 pH unit lower at 37°C.

X ml. CH_3COOH	*Y* ml. CH_3COONa	pH *at* 25°C.
92.5	7.5	3.6
88.0	12.0	3.8
82.0	18.0	4.0
73.5	26.5	4.2
63.0	37.0	4.4
52.0	48.0	4.6
41.0	59.0	4.8
30.0	70.0	5.0
21.0	79.0	5.2
14.0	86.0	5.4
9.0	91.0	5.6
6.0	94.0	5.8

Other buffers commonly used and important in clinical chemistry are phosphate (Sorenson), citrate (Sorenson), carbonate-bicarbonate (Delory-King), acetate, phthalate, boric acid–borate (Palitzsch), veronal-sodium veronal, glycine-glycinate, and tris(hydroxymethyl)aminomethane.

Tables 1-7 to 1-10 describe the preparation of commonly used buffers in clinical chemistry laboratories.

Preparation of Various Solutions

STANDARD ACIDS AND BASES

Accurately prepared standard solutions of hydrochloric and sulfuric acids, potassium hydrogen phthalate, and sodium hydroxide are necessary in a clinical chemistry laboratory. These standard solutions are used to establish the normality of all acids and alkalies in the laboratory. The standard solutions are usually not prepared directly, but are made approximately and standardized by titration. The final concentration most frequently required is 0.1 N or less. Table 1-11 shows the strengths of various concentrated solutions of acids and alkalies.

TABLE 1-11. *Strengths of Concentrated Solutions of Acids and Alkalies*

Acid or Alkali	*Specific Gravity*	*% by Weight*	*Grams/Liter*	*Approximate Normality*	*Milliliters Required to Make 1 L. of 1 N Solution*
Hydrochloric acid (HCl)	1.19	37	440	12.1	83
Sulfuric acid (H_2SO_4)	1.84	96	1730	36	28
Nitric acid (HNO_3)	1.42	70	990	15.7	64
Acetic acid (CH_3COOH)	1.06	99.5	1060	17.4	57
Ammonium hydroxide (NH_4OH)	0.880	29	250	15–17	57–67
Sodium hydroxide (saturated solution) (NaOH)	1.50–53	about 50	600 to 700	15–18	57–67
Potassium hydroxide* (saturated solution) (KOH)	1.55	about 50	800	14	70

* Saturated solutions made from the usual C.P. potassium hydroxide will vary in strength, chiefly because of the variable amount of carbonate that such solutions contain.

Potassium hydrogen phthalate. A primary standard for acidimetry, this chemical may be obtained from the National Bureau of Standards in very pure form. An accurate normality can be obtained by carefully weighing out the desired amount on an analytical balance. The molecular weight of potassium hydrogen phthalate ($HKC_8H_4O_4$) is 204.22. Since there is one dissociable hydrogen, the molecular weight is also the equivalent weight. For a 0.1 N solution, weigh 20.422 gm. of the dry substance and dissolve in distilled water diluting volumetrically to 1 L. *Alkali* can be standardized against this or other acidimetry standards, and other acids standardized by titration against the *alkali*. It must be remembered that the normality of alkaline solutions may change as a result of absorption of CO_2 or reaction with the glass container; hence, they must be restandardized often. Acids are relatively stable.

Hydrochloric acid. Prepare 0.1000 N HCl from constant-boiling HCl. This constant-boiling HCl is prepared by distillation and has a constant concentration of 20.22 per cent of HCl by weight, and is used to prepare the dilute primary standard (0.1000 $N \pm 0.0004$) acid by diluting the proper weight (not volume) of the acid to the required volume. Further standardization is unnecessary. Acid prepared in this manner can be relied on for the standardization of alkali and other reagents.

Sulfuric acid. Prepare an approximately 1 N solution by running 30 ml. of concentrated H_2SO_4 (Sp. gr., 1.84) slowly with mixing into about 500 ml. of distilled water. Dilute to a volume of 1 L. and standardize this solution against borax. As a suitable primary standard, borax (sodium tetraborate, $Na_2B_4O_7{\cdot}10H_2O$) has a very constant composition and is as accurate as anhydrous sodium carbonate or tris (hydroxy-methyl) aminomethane. Dissolve 4 gm. of solid borax in warm distilled water and bring to a volume of 100 ml. Add a few drops of methyl red indicator and titrate the solution with the H_2SO_4 solution from a 50 ml. buret until the yellow color changes to red. The titration should be carried out in triplicate.

The equation of the reaction is:

$$B_4O_7^{=} + 2H^+ \longrightarrow H_2B_4O_7 \xrightarrow{5H_2O} 4H_3BO_3$$

One equivalent weight of borax is 381.4/2 = 190.7. Thus, X gm. of borax represents:

$$\frac{X \times 1000}{190.7} \text{ milliequivalents}$$

If the titration required Y ml. of the acid:

$$Y \times \text{normality of acid} = \frac{X \cdot 1000}{190.7}$$

$$\text{normality} = \frac{X \cdot 1000}{Y \cdot 190.7} = 5.24 \frac{X}{Y}$$

The average of the triplicate results is the normality of the H_2SO_4. The acid is now diluted to make a 0.1 N solution using the formula:

$$\text{normality}_1 \times \text{volume}_1 = \text{normality}_2 \times \text{volume}_2$$

where N_1 and V_1 are the initial normality and volume of the acid and N_2 and V_2 are the final normality and volume after dilution.

Sodium hydroxide (concentrated). Add 100 gm. of NaOH sticks (sticks being purer than the pellets) to 100 ml. of water warmed to 55°C. Mix until the NaOH is dissolved and allow to stand overnight in a covered container. The NaOH solution will contain carbonate that will settle out and the supernatant can be removed by decanting or centrifugation. This concentrated solution is 16 to 18 N and can be diluted to any strength desired. Its normality can be established by titration against standard potassium acid phthalate, standard HCl, or standard sulfamic acid. The NaOH standard should be stored in a plastic bottle or a paraffin lined container. The standard should be well protected from CO_2 of the air. Rubber or plastic stoppers should be used with alkaline solutions since alkali will cause a glass stopper to freeze. It is advantageous to use a soda-lime trap in the container in order to prevent atmospheric CO_2 from entering and weakening the solution.

As a general rule one should not risk contamination of a standard by introducing a pipet into the container. It is better technique to pour off the necessary amount into a beaker or tube and carefully wipe off the pouring edge. If there is a dried residue present on the mouth of the container, it should be rinsed off with a gauze moistened with distilled water and wiped with a dry gauze before pouring any standard. Condensed moisture that forms on the upper inside of the standard container on storage should also be swirled back into the solution before removing any standard solutions.

pH REFERENCE SOLUTIONS

For the standardization of pH meters, the following standards are useful.

1. *Potassium hydrogen phthalate*, 0.05 M. It is made by dissolving 10.2 gm. per liter in water, and has pH values of 4.001 at 20°C. and 4.025 at 37°C.

2. *Equimolar phosphate buffer*, contains 0.025 M dihydrogen phosphate and 0.025 M monohydrogen phosphate. It is made by dissolving 3.402 gm. of anhydrous potassium dihydrogen phosphate (KH_2PO_4) and 3.549 gm. of anhydrous disodium hydrogen phosphate (Na_2HPO_4) per liter in water. The pH values are 6.88 at 20°C. and 6.81 at 37°C.

3. Borax (sodium tetraborate) 0.05 M. It is made by dissolving 19.07 gm. of $Na_2B_4O_7 \cdot 10H_2O$ per liter in water. It has pH values of 9.22 at 20°C. and 9.08 at 37°C.

LABORATORY MATHEMATICS

Exponents

Any number may be expressed as an integral power of 10, or as a product of two numbers, one of which is an integral power of ten.

Examples:

$$100 = 10^2$$
$$600 = 6 \times 10^2$$
$$460 = 4.6 \times 10^2$$
$$0.46 = 4.6 \times 10^{-1}$$

In multiplication, exponents of like bases are added and in division exponents of like bases are subtracted.

Examples:

$$20 \times 600 = 2 \times 10^1 \times 6 \times 10^2 = (6 \times 2) \times 10^{1+2} = 12 \times 10^3 = 1.2 \times 10^4 = 12{,}000 \quad (1)$$

$$\frac{20}{400} = \frac{2 \times 10^1}{4 \times 10^2} = \frac{2}{4} \times 10^{1-2} = 0.5 \times 10^{-1} = 5 \times 10^{-2} = 0.05 \quad (2)$$

Logarithms

Logarithms may be used in the laboratory to carry out multiplications and divisions, to extract square roots, square numbers, and so on, but their value in biochemistry is more than merely a convenience. A number of basic phenomena are described by equations or formulas that utilize logarithms. Two obvious examples are pH and absorption of light.

Because their use is necessary, a brief review will be given of the system of logarithms to the base 10 (called the common or Briggsian system). The "common" logarithm of a number is the exponent or power to which 10 must be raised to give that number.

Examples:

$$\text{Log } 10 = 1, \text{ since } 10^1 = 10$$
$$\text{Log } 100 = 2.0 \text{ since } 10^2 = 100$$
$$\text{Log } 2 = 0.301, \text{ since } 10^{0.301} = 2$$

A common logarithm always consists of two parts, an integer, called the characteristic, and a decimal number, called the mantissa.

The mantissa of the log of the number is found in logarithm tables or from a slide rule and is independent of the position of the decimal point in the original number. Thus 352, 0.352, and 0.00352 all have the same mantissa.

The characteristic indicates the position of the decimal point in the set of figures represented by the mantissa and is determined by inspection of the original number according to the following rules:

1. For a number greater than 1, the characteristic is positive and is one less than the number of digits to the left of the decimal point.

Examples:

The characteristic for 167.0 = 2
16.7 = 1
1.67 = 0

2. For a number less than 1, the characteristic is negative and is numerically one more than the number of zeros *immediately* following the decimal point. The negative sign of the characteristic may be expressed in two ways:

a. by placing a bar above the characteristic as $\bar{1}$, $\bar{2}$, and so on

b. as 9. −10, 8. −10, and so on.

Examples:

The characteristic for 0.0741 is $\bar{2}$ or 8. −10

0.00067 is $\bar{4}$ or 6. −10

The antilogarithm is the number corresponding to a given logarithm. Thus, "the antilog of 2" really means "the number whose logarithm is 2," which in this case would be 100. Suppose it is required to find the antilog of 1.5020. The characteristic is 1 and the mantissa is 0.5020, which would be looked up in a table and found to be 3177. Since the characteristic is 1, there are two digits to the left of the decimal point and the number becomes 31.77.

MULTIPLICATION

When multiplying two numbers, the logarithm of each number is determined, the logarithms are added, and the antilogarithm of the sum is found.

Example:

$$\begin{aligned} &2.2 \times 4.8 = \\ &\text{Log of } 2.2 = 0.3424 \\ &\text{Log of } 4.8 = 0.6812 \\ &\text{Sum} \quad 1.0236 \\ &\text{Antilog of } 1.0236 = 10.56 \text{ (answer)} \end{aligned}$$

DIVISION

When carrying out divisions, the logarithm of each number is determined and then the logarithm of the divisor is subtracted from the logarithm of the dividend.

Example:

$$\begin{aligned} &\text{Log of } 4.8 = 0.6812 \\ &\text{Log of } 2.2 = 0.3424 \\ &\text{Difference} \quad 0.3388 \\ &\text{Antilog of } 0.3388 = 2.18 \text{ (answer)} \end{aligned}$$

Logarithms are very useful when carrying out multiple multiplications or when solving problems involving both multiplication and division.

Example:

$$\begin{aligned} &\frac{4.8}{2.2} \times 2.5 = \\ &\text{Log } 4.8 = 0.6812 \\ &\text{Log } 2.2 = 0.3424 \\ &\text{Difference } 0.3388 \\ &\text{Log } 2.5 = 0.3979 \\ &\text{Sum } 0.7367 \\ &\text{Antilog of } 0.7367 = 5.45 \end{aligned}$$

The logarithmic relationship used daily in all clinical chemistry laboratories is that of pH, which may be defined as the negative logarithm of the hydrogen ion concentration.

$$\text{pH} = \log \frac{1}{[\text{H}^+]} = -\log [\text{H}^+]$$

In the living organism, concentration of acids and bases is very dilute. For example the hydrogen ion concentration of the living organism is about 0.000,000,05 *M*. Sorenson believed that laboratory workers would have difficulty in expressing $[H^+]$ of acid and basic solutions and devised the term "pH." The relationship is very simple if the $[H^+]$ is exactly $1 \times 10^{-\text{power}}$ because the pH is then equal to the value of the minus power of 10.

For example, in water

$$K_w = 10^{-14}$$

and

$$[H^+] \times [OH^-] = 10^{-14};$$

since

$$[H^+] = [OH^-]$$

therefore

$$[H^+] = 10^{-7},$$

or

$$pH = 7$$

To calculate the pH of solutions with a (H^+) more complex than $1 \times 10^{-\text{power}}$, logarithms must be used. Thus the pH of a solution with a $[H^+]$ of 0.004 may be calculated as follows:

$$0.004 \text{ may be expressed as } 4 \times 10^{-3}$$

then

$$\begin{aligned} pH &= -\log(4 \times 10^{-3}) = -(\log 4 - 3) \\ &= -\log 4 + 3 \\ &= 3 - 0.602 = 2.398 \end{aligned}$$

A valuable working relationship between the pH of a buffer solution and the dissociation constant of the weak acid in the presence of its salt is given by the Henderson-Hasselbalch equation, which may be stated as follows:

$$pH = pK + \log \frac{[\text{salt}]}{[\text{acid}]}$$

The pK is defined as $-\log K$ and may be found in tables such as the table in the appendix or in suitable handbooks. The concentration of salt and acid is commonly expressed as moles or equivalents per liter. From this equation it can be seen that if the pK of a buffer pair is known, the amount of salt and acid needed to prepare a buffer of a given pH and concentration may be calculated.

Example:

It is desired to prepare 1L. of a 0.1 *M* acetate buffer with a pH of 4.9 The pK is 4.76. Substituting in the equation

$$pH = pK + \log \frac{A^-}{HA}$$

$$4.9 = 4.76 + \log \frac{A^-}{HA}$$

$$\log \frac{A^-}{HA} = 0.14$$

or

$$\frac{A^-}{HA} = \text{antilog } 0.14 = 1.38$$

If moles/L. of salt + moles/L. of HA = 0.1, then moles/L. of A^- = (0.1 − moles/L. of HA) or

$$A^- = 0.10 - HA$$

Since

$$\frac{\text{salt}}{\text{acid}} = \frac{A^-}{HA} = 1.38$$

we may now substitute into this equation

$$\frac{0.1 - HA}{HA} = 1.38$$

$$0.1 - HA = 1.38 \times HA$$

$$0.1 = HA + 1.38 \times HA$$

$$0.1 = 2.38\ HA$$

$$HA = \frac{0.1}{2.38} = 0.042$$

$$A = 0.1 - HA$$

$$A = 0.1 - 0.042 = 0.058$$

The buffer then should consist of

0.042 moles acetic acid per liter
and 0.058 moles sodium acetate per liter

To prepare 1 liter of this buffer, the following amounts are used:

60 gm. (molecular weight of acetic acid in grams)
× 0.042
2.52 grams of acetic acid

and

82 gm. (molecular weight of sodium acetate in grams)
× 0.058
4.75 grams of sodium acetate

These amounts dissolved and diluted to 1 liter will yield a 0.1 *M* acetate buffer with a pH of 4.9.

Expressing and Calculating Dilutions

Dilution of solutions, which involves making a weaker solution from a stronger one, is a frequent laboratory procedure. The following are some of the most common reasons for dilutions: (1) if the concentration of the material in solution (usually the specimen to be analyzed) is too great to be accurately determined; (2) if in the removal of undesirable substances (e.g., proteins), solutions are added to precipitate the proteins and a dilution has taken place; and (3) in the preparation of working solutions from stock solutions.

Dilutions are usually expressed as a ratio, such as 1:10. This refers to 1 unit of the original solution diluted to a final volume of 10 units. This would yield a solution which is 1/10 the concentration of the original solution. To calculate the strength of the dilute solution, multiply the concentration of the original solution by the dilution, expressed as a fraction.

Example:

A 500 mg./100 ml. solution is diluted 1:25. The concentration of the final solution is: 500 × 1/25 = 20 mg./100 ml.

If more than one dilution is made with a given solution, the concentration of the final solution is obtained by multiplying the original concentration by the product of the dilutions.

Example:

A 1000 mg./100 ml. solution is diluted 1:10 and then this diluted solution is further diluted 1:100. The concentration of the final solution would be: $1000 \times 1/10 \times 1/100 =$ 1 mg./100 ml.

Large dilutions can conveniently be made in two steps. For example, if a 1:1000 dilution is required but only 100 ml. is needed or the diluent is in short supply, this dilution could be accomplished, as in the previous example, without having to resort to micropipets.

The systematic dilution and redilution of a solution is called "serial dilution." This technique is commonly used in serology when the test may actually be reported in terms of the number of times a solution had to be serially diluted to still give a positive or negative reaction. In serial dilutions, to find the concentration in any one tube, the dilution in that tube is multiplied by each of the preceding dilutions, including the original tube.

Example:

Into each of 8 tubes is placed 0.2 ml. of diluent. Then 0.2 ml. of serum is added to the first tube and mixed. Next 0.2 ml. of the mixture is removed and placed in the second tube and mixed. Then 0.2 ml. is removed and placed in the third tube and so on until tube eight when the 0.2 ml. removed is discarded. The dilutions are calculated as follows:

Tube	*Dilution*	
1	$\frac{0.2}{0.2 + 0.2} = \frac{0.2}{0.4}$	= 1:2
2	$1/2 \times 1/2$	= 1:4
3	$1/4 \times 1/2$	= 1:8
4	$1/8 \times 1/2$	= 1:16
5	$1/16 \times 1/2$	= 1:32
6	$1/32 \times 1/2$	= 1:64
7	$1/64 \times 1/2$	= 1:128
8	$1/128 \times 1/2$	= 1:256

In the preparation of a so-called protein-free filtrate, protein precipitating reagents are added and thus a dilution has taken place. If an aqueous standard is used in the procedure, this standard is sometimes used directly since deproteinization is unnecessary. In this case the dilution of the serum must be calculated so that the standard may be appropriately diluted.

If the standard is always used this way, it is usually made up to contain an amount that is less, by a factor of the dilution, than the amount it is equivalent to in the procedure. Its true concentration is thus less (by a known factor) than the serum concentration it represents.

Example:

Serum is diluted 1:10 in the preparation of a protein-free filtrate. Then 1 ml. of this filtrate is added to a cuvet, followed by the appropriate reagents for colorimetric analysis. Also, 1 ml. of standard is added to another cuvet. If the standard is to be equivalent to a 100 mg./100 ml. concentration of the substance, it is obvious that the standard cannot contain 100 mg./100 ml. since the serum was diluted and the standard was not diluted. Thus, this standard must be diluted 1:10 before use or more practically it would be made up to contain $100 \times 1/10 = 10$ mg./100 ml. The standard therefore would contain 10 mg./100 ml., but would be equivalent to a serum concentration of 100 mg./100 ml. in this particular test.

Significant Figures

Some degree of error is involved in all chemical determinations and, indeed, in all measurements taken whether reading a spectrophotometer, making a weighing, or measuring a distance. None of these measurements would be absolutely correct and the accuracy would be limited by the reliability of the measuring instrument. In a chemical determination the accuracy would be limited by all the steps involved, such as pipetting. In reporting results some indication of the reliability of the measurement is given by the number of significant figures that are given in the result. A significant figure is one that is known to be reasonably reliable. As an example, consider that the length of an object is recorded as 12.8 cm. By convention, this means that the length was measured to the nearest tenth of a centimeter, and that its exact value lies between 12.75 and 12.85 cm. If this measurement were exact to the nearest hundredth of a centimeter, it would have been reported as 12.80 cm. Thus in the clinical laboratory a result reported as 14.5 means that the result is accurate to the nearest tenth and that the exact value lies between 14.45 and 14.55. If this result were reliable to the nearest hundredth it would be reported as 14.50. The figure 14.5 contains three significant figures while 14.50 contains four significant figures. The table below gives the implied maximal limits of a result reported to various significant figures.

Reported Result	*Number of Significant Figures*	*Implied Limits*
14	2	13.5–14.5
14.5	3	14.45–14.55
14.50	4	14.495–14.505

Zeros to the right of the decimal point are always significant if they follow the digits and not significant if they precede a digit and the total number is less than one. Thus, 0.072 contains two significant figures, 0.720 contains three significant figures and 1.072 contains four significant figures. Zeros to the left of the decimal point may or may not be significant. Thus, a report of 220 does not indicate whether the measurement was to the nearest tens or to the nearest ones. If there are digits on both sides of the zero, as 202, it is a significant figure. If the zero is followed by a decimal point and digits or zeros, as 220.1, the zero is a significant figure.

As was mentioned earlier a result reported as 14.5 implies that the result is accurate to the nearest tenth and that the exact value lies between 14.45 and 14.55. Although the result may not be that accurate, three significant figures may still be used if the result is significantly more accurate than would be indicated by only two significant figures. In other words, if the result is not as good as between 14.45 and 14.55 but much better than 13.5 to 14.5 there is good reason to report the result as 14.5.

The dropping of one or more digits to the right to give the desired number of figures is called "rounding off." When the first digit dropped is less than 5, the last digit retained remains unchanged; i.e., 4.571 = 4.57. When it is more than 5 or is followed by digits greater than 0, the last digit is increased by 1; i.e., 3.788 = 3.79. When it is exactly 5, the last digit is increased by 1 if that digit is odd, otherwise the digit remains unchanged; in other words, it is rounded to the nearest even number. Thus, 6.85 becomes 6.8 and 6.75 also becomes 6.8.

COLLECTION AND HANDLING OF SPECIMENS

The preservation of the chemical integrity of the specimen from the time of collection to the time of testing is of utmost importance if the results are to be meaningful. All tubes and syringes must be chemically clean. In general, tubes for chemical analysis do not have to be sterile, but one should not assume that a sterile tube is chemically clean. When blood is drawn from a patient by a syringe, it is immediately transferred to a clean dry tube after the needle has been removed. The blood is then allowed to clot for at least 10 to 15 minutes at room temperature and longer if allowed to stand in a refrigerator. The clot may adhere to the wall of the tube so that "ringing" should be performed before centrifugation. This is conveniently done by making a gentle sweep around the inside walls of the tube with a wooden applicator stick. Excessive ringing is unnecessary and can produce hemolysis. By allowing the clot to retract for a longer period of time, hemolysis is minimized and the yield of serum is greater; however, during this time glycolysis takes place and there can be a shift of substances from cells to serum. Thus, the time allowed for clot retraction is dependent upon the procedure to be done.

An alternative procedure is to draw blood into a vacuum tube (Vacutainer). Vacuum tubes may be siliconized to minimize hemolysis and prevent the clot from adhering to the wall of the tube.

After clotting, the tube is centrifuged and the supernatant serum removed. The individual handling of the serum will now depend on the analysis that is to be done and the time that will elapse before the analysis is started. The serum may be allowed to remain either at room temperature, refrigerated and protected from the light, or frozen, depending on circumstances. The stability of individual components of serum will be discussed later in the respective chapters.

Anticoagulants

If whole blood or plasma is desired, an anticoagulant must be added to the specimen immediately after it is drawn or placed into the tube into which the blood is collected. In the case of heparin it is more advantageous to coat the walls of the syringe with the anticoagulant.

The technologist must be certain that the anticoagulant used does not affect the chemical analysis. This may happen in a number of ways. If the anticoagulant is present as the sodium or potassium salt and electrolytes are being analyzed, a significant error can occur. In this case, use of the lithium or ammonium salt would obviate the problem. The anticoagulant may also remove the constituent to be measured, as in the case of oxalate, which removes calcium from the serum by forming an insoluble salt. The action of anticoagulants on enzymes is a function of anticoagulants as well as enzymes. Oxalate has been reported to inhibit lactic dehydrogenase, acid phosphatase, and amylase. Fluoride inhibits urease, but activates amylase.[7]

Heparin is the anticoagulant that least interferes with clinical chemical tests. It is present in most of the tissues of the body, but in concentrations less than that required to prevent the clotting of blood. Heparin exists in the highest concentrations in the liver and the lungs. It is a mucoitin polysulfuric acid and is available as the sodium, potassium, and ammonium salt. This anticoagulant is believed to act as an antithrombin, preventing the transformation of prothrombin into thrombin, and thus preventing the formation of fibrin from fibrinogen. Several workers have shown that this action of heparin needs the presence of a co-factor that seems to be associated with the albumin fraction of plasma. Heparin has also been shown to possess antithromboplastin activity and to inhibit the lysis of platelets.

Usually about 20 units of heparin are used per milliliter of blood. Since heparin is not readily soluble, it is frequently used as a solution. If it is to be dried, it is better to dry it on the walls of the tube so that when the blood is added, solution may be as rapid as possible. Some disadvantages of heparin are high cost, temporary action, and the fact that it gives a blue background on a blood smear stained with Wright's stain.

Oxalates such as sodium, potassium, ammonium, or lithium oxalate inhibit blood coagulation by forming rather insoluble complexes with calcium ions, which are necessary for coagulation. Of these, potassium oxalate at a concentration of about 1 to 2 mg./ml. of blood is the most widely used. If the laboratory prepares its own tubes, the potassium oxalate may be added as a 30 per cent solution and dried in an oven. The clotting of 15 ml. of blood may be prevented by 30 mg. (0.1 ml.) of potassium oxalate. Temperatures in excess of 150°C. should be avoided during the drying since at elevated temperatures the oxalate will be converted to carbonate, which has no anticoagulant activity.

Sodium fluoride, although usually considered as a preservative for blood glucose determinations, also acts as a weak anticoagulant. When used as a preservative along with an anticoagulant such as potassium oxalate, it is effective in a concentration of about 2 mg./ml. of blood. It exerts its action by inhibiting the enzyme system involved in glycolysis. When used as an anticoagulant the concentration must be much greater, i.e., 6 to 10 mg./ml. of blood. As a general rule fluoride should not be used when collecting specimens for enzyme determinations or when using an enzyme in a test, e.g., the urease method for the determination of urea.

Ethylenediaminetetraacetic acid (*EDTA*) is a chelating agent that is particularly useful for hematologic examinations since it preserves the cellular constituents of the blood. It is used as the disodium or dipotassium salt, the latter being more soluble. It is effective in a final concentration of 1 to 2 mg./ml. of blood. This chelating agent derives its anticoagulant activity from the fact that it binds calcium, which is essential for the clotting mechanism.

Specimen Variables

In general, blood for chemical analysis should be drawn while the patient is in the postabsorptive state. An overnight fast is the usual procedure although a six hour fast is ample. During this time there is no need to restrict water intake. The common procedures that are affected most significantly by eating are glucose (elevated), inorganic phosphorus (decreased), thymol turbidity (increased), and triglycerides (increased). In addition, lipemia, which is caused by a transient rise in chylomicrons following a meal containing fat, causes interference with a large number of chemical analyses because of turbidity.

There is diurnal variation (not related to eating) on certain constituents such as iron and corticosteroids. Dietary habits influence constituents such as uric acid and lipids. A proper glucose tolerance test is performed when the patient has had an adequate carbohydrate intake (250 gm. per day) for 3 days before the test.

At present our knowledge of the effect of medications on chemical tests is rather fragmentary, although articles devoted to this subject are beginning to appear.[10] Some drug manufacturers are now beginning to supply information on the effect of new medications on chemical analysis. When a medication is suspected of giving a false elevation or depression it should be removed and the test repeated in a few days.[7]

Hemolysis may interfere with a number of chemical procedures and should be avoided. Several constituents, such as glutamic oxalacetic transaminase, lactic dehydrogenase, acid phosphatase, and potassium, are present in large amounts in erythrocytes so that hemolysis will significantly elevate the values for these substances in serum. Hemoglobin may directly interfere in a chemical determination by inhibiting an enzyme such as lipase, by interfering with a reaction such as the diazotization of bilirubin, or by yielding a significant amount of color and thus interfering with a colorimetric analysis. This is particularly true when the reading is taken in the blue portion of the spectrum. In grossly hemolyzed serum a dilution of the serum components occurs if the concentration of the metabolite in the red cell is less than in plasma; thus, the sodium and chloride concentration of serum would be falsely low in a grossly hemolyzed serum. Table 1-12 shows the concentration of some common constituents in erythrocytes and plasma.

TABLE 1-12. *Concentration of Substances in Erythrocytes and Plasma**

Substance	*Erythrocytes*	*Plasma*	*Erythrocytes/Plasma*
Glucose, mg./100 ml.	74.0	90.0	0.82
Nonsugar-reducing substances, mg./100 ml.	40.0	8.0	5.00
Nonprotein N, mg./100 ml.	44.0	25.0	1.76
Urea N, mg./100 ml.	14.0	16.0	0.88
Creatinine (Jaffe), mg./100 ml.	1.8	1.1	1.63
Uric acid, mg./100 ml.	2.5	4.6	0.55
Total cholesterol, mg./100 ml.	139.0	194.0	0.72
Cholesterol esters, mg./100 ml.	0.0	129.0	0.00
Sodium, mEq./L.	16.0	140.0	0.11
Potassium, mEq./L.	100.0	4.4	22.70
Chloride, mEq./L.	52.0	104.0	0.50
Bicarbonate, m*M*./L.	19.0	26.0	0.73
Calcium, mEq./L.	0.5	5.0	0.10
Inorganic P, mg./100 ml.	2.5	3.2	0.78
Acid β-glycerophosphatase, units	3.0	0.25	12.00
Acid phenylphosphatase, units	200.0	3.0	67.00
Lactic dehydrogenase, units	58,000.0	360.0	160.00
Transaminase, GO, units	500	25	20.00
Transaminase, GP, units	150	30	5.00

* After Caraway, W. T.: Am. J. Clin. Path., *37*: 445, 1962.

Some common causes of hemolysis are moisture in the syringe and the mechanical destruction of cells by forcing the blood from the syringe into a tube without removing the needle. The blood should be allowed to slowly run down the side of the tube after the needle has been removed from the syringe. The use of vacuum tubes has largely taken care of these problems. Too vigorous mixing of the tube after drawing the blood may also lead to hemolysis.

Cerebrospinal Fluid

Cerebrospinal fluid is usually collected in three or four sterile tubes. The first and second tubes of spinal fluid are frequently contaminated with blood, so that these tubes are sometimes not suitable for chemical analysis. Even a small amount of plasma may cause a significant error in the protein analysis. If an analysis for glucose is to be performed on the spinal fluid, the filtrate should be prepared immediately after collection in order to avoid glycolysis.

Urine Collection

Many of the chemical analyses performed on urine specimens must be carried out on 24 hour urine collections. One of the reasons for this is that many constituents, including most of the hormones, exhibit a diurnal variation. In order to eliminate the variability of these peak excretion times, it is much more definitive to perform the analysis on a 24 hour specimen. Collection of this specimen requires the utmost cooperation of the patient, nursing staff, and laboratory. At the beginning of the collecting period (usually when the patient wakens) the bladder should be emptied, that specimen discarded and the time noted. All urine specimens passed thereafter are collected in an appropriate container. At the end of the 24 hour collecting period, the bladder is emptied and this urine specimen added to those already collected.

Inadequate preservative, loss of voided specimens, or the inclusion of two morning specimens in a 24 hour period are common errors encountered in the collection of a 24 hour urine specimen. Determination of total creatinine excretion in the 24 hour period may be used as a guide to the adequacy of the 24 hour collection. This is particularly useful if several 24 hour urine specimens are collected from the same person, since the 24 hour creatinine excretion, largely a function of muscle mass, is relatively constant from day to day in the same individual.

It is the responsibility of the clinical chemical laboratory to provide an appropriate chemically clean container, containing the proper preservative, for the collection of a 24 hour urine specimen. The container should hold at least 3 to 4 liters. One-gallon bottles are convenient. These should be properly labeled. The label should have space for the name and room number of the patient and the test desired as well as the time the collection was started and the time the collection was finished. In addition the bottle should contain a warning "Do Not Discard," since this is a 24 hour urine collection. If at all possible, all 24 hour urine collection bottles should be refrigerated during collection. If this is possible, the label on the bottle should also state this information.

The *preservative* for a 24 hour urine specimen will depend not only on the procedure to be carried out, but to a certain degree on the methodology used. In general, no preservatives are used when a bioassay is to be performed on the specimen. Since catecholamines and VMA (3-methoxy-4-hydroxymandelic acid) are stable only in acidic solution, 10 to 15 ml. of concentrated hydrochloric acid is the most common preservative. For the chemical analysis of the urinary steroids, in general, refrigeration is adequate. If the specimen is to be mailed to a reference laboratory, they will supply information about the proper preservative to be added (such as hydrochloric or boric acid). This preservative must not only insure stability during collection but also during the transit of the specimen to the reference laboratory. The directions of the reference laboratory should be followed carefully. Specimens for certain analyses (e.g., renin) must be transported frozen. Styrofoam containers are available and when these are used with dry ice they will keep the specimen frozen for 48 hours, which is long enough for most cases. Specimens should be mailed early enough during the week so that they will not be in transit over a weekend.

Porphyrins are in a class by themselves because, in general, they are most stable in alkaline urine. This alkalinization is usually accomplished by adding 5 gm. of sodium carbonate to the collection bottle. Two additional precautions are the use of a brown bottle and the addition of 100 ml. of petroleum ether to retard oxidation by formation of a protective layer.

Collection of Blood and Serum

The stability of various constituents will be discussed in detail in later chapters; however, a few general comments are appropriate at this time. Serum is usually preferred to blood when analyzing for constituents that are relatively evenly distributed between the intracellular and extracellular compartments. The serum or plasma should be removed from the clot or cells soon after collection. Glucose changes most rapidly of all the commonly measured chemical constituents when left in contact with the cells. Since glycolysis is an enzyme reaction it is very sensitive to temperature. Many laboratories collect the blood for glucose analysis in a tube containing fluoride, which inhibits the enzyme enolase involved in glycolysis. Glucose in serum or plasma (which is essentially free of cells that contain the glycolytic enzymes) is quite stable, especially in the refrigerator. The stability of glucose in serum at room temperature has been reported to be variable, but it is much greater than it is in whole blood.

Most enzymes are stable in serum for at least 24 hours under refrigeration and longer if frozen. The inorganic ions are stable for at least 8 hours at room temperature and for days under refrigeration. Bilirubin (particularly unconjugated) is very sensitive to light so that the serum must be assayed immediately or protected from direct light. Urea nitrogen, creatinine, and uric acid are stable for at least 24 hours without refrigeration, and longer if refrigerated.

STATISTICS IN THE CLINICAL CHEMISTRY LABORATORY

Clinical chemistry laboratories produce data that are used as a basis for important clinical decisions; therefore, one must be able to evaluate the reliability of data by objective tests. Repeat analyses on the same or on different aliquots of the same sample will not, as a rule, give identical values for any given constituent. Thus, the analyst, should be able to know how large the variation may be and still be within the bounds of *random* experimental error for the given analytical procedure. Often, a series of extended analyses must be made in order to obtain this information. Replicate analyses reduce the risk of the first analysis being erroneous because of technical error or the presence of an unsuspected variable; unfortunately, replicate analyses are often impractical in a busy laboratory.

Variations

There are many effects that can produce variations in laboratory analyses. Often some components of an apparatus are affected by temperature or humidity or both. "Aging" phenomena are important causes of systematic errors. For example, biological specimens may alter with time, parts of an apparatus may become excessively worn or corroded, or the operator may suffer fatigue as the day proceeds. Chemicals during storage may decompose or be gradually altered in composition by distillation, crystallization, or microbial attack. Traces of glass containers can slowly dissolve in alkaline solutions. Impurities can be absorbed from the air (i.e., carbon dioxide). Chemicals and specimens can also lose weight through water loss. Deviations in results may occur due to mechanical changes of apparatus or periodic variations in some variable such as temperature, humidity, sunlight or vibration. Another type of variation may result from the fact that different technicians usually will not carry out a procedure in the exact same way. This technician variation may be even greater if a new or less skilled operator performs the test. Combinations of such sources of variations may occur and usually take place independently of one another. A system is said to be in a state of *statistical control* when all the variables that have significant effects are known and kept under control, i.e., known and random sources of variability and their degree of interaction have been evaluated and are known to be within acceptable limits. This does not necessarily eliminate all variability in data because there will be a very large number of random and nonrandom sources of variation that individually have small effects and will produce only a minor scatter in the results.

SOURCES OF VARIATION

The following sources of variations in analytical results can be distinguished:

1. *Sampling error.* Error inherent in the method of sampling; i.e., the constituent of the specimen being sampled is affected by an anticoagulant, by drug interference, and so on.

2. *Experimental random error.* This reflects variations in the results of an analytical method obtained when successive analyses of the same sample are made in exactly the same way by the same analyst; i.e., if the analyst made two duplicate analyses of serum calcium, his results might be 9.5 and 10.0 mg./100 ml.

3. *Experimental systematic error* due to the bias of the method. A given method yields results lower or higher than another analytical procedure. This may be due to variations in methods based on the same principle (e.g., glucose by Somogyi or Folin-Wu) or methods based on different principles but measuring the same constituent (e.g., glucose by *o*-toluidine or glucose oxidase).

4. *Personal bias of the analyst* (analyst's systematic error). In this error, the analyst unconsciously introduces a constant variation into his results; however, this variation differs from one that might be introduced, equally unconsciously, by another analyst using an identical procedure in the same laboratory. For example, one analyst may use the meniscus mark on a pipet consistently lower than another.

5. *Bias of the laboratory.* This is due to differences arising from such factors as reagents, apparatus, environment, and methods. These differences can result in either an upward or downward bias relative to results that would be obtained in a different laboratory by the same analyst using similar techniques and equipment.

Among other sources of variability are:

a. sample-to-sample variation within a patient (hour-to-hour, day-to-day and seasonal).

b. patient-to-patient variation (race, age, sex, status of health).

Frequency Distributions

If serum urea nitrogen values were determined on a large series of apparently healthy individuals, and the results arranged in order of magnitude, the list would

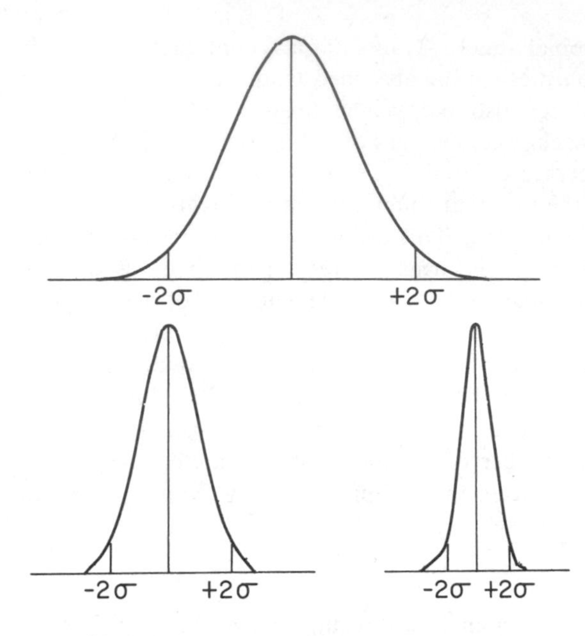

Figure 1-8. Possible normal frequency distributions. In all distributions $\pm 2\sigma$ represents approximately 95 per cent of the population.

most likely be symmetrical in the sense that the average or (arithmetic) mean of the series would be close to the median (the central value). The number of values that would exceed the mean by some given amount would roughly equal in number those that would fall short of it by the same amount. Results close to the mean will also occur more frequently than results further removed from it. The distribution of these results can be succinctly expressed in the form of curves such as those shown in Figure 1-8. These are theoretical frequency distributions, in which each numerical value (variable) is plotted against the number of times (frequency) it occurs in a given population. The scatter of results is also shown graphically in Figure 1-9. Experience has shown that frequency plots of analytical data generally follow a symmetrical, bell-shaped curve, the so-called Gaussian curve.[26] If the total area under the curve includes all the values plotted, then the area from -1σ to $+1\sigma$ will include 68.27 per cent of the values. The range from -2σ to $+2\sigma$ will include 95.45 per cent of the values and that from -3σ to $+3\sigma$ will include 99.93 per cent of the values. The parameter, σ or s,* is called the standard deviation (S.D.). The standard deviation of

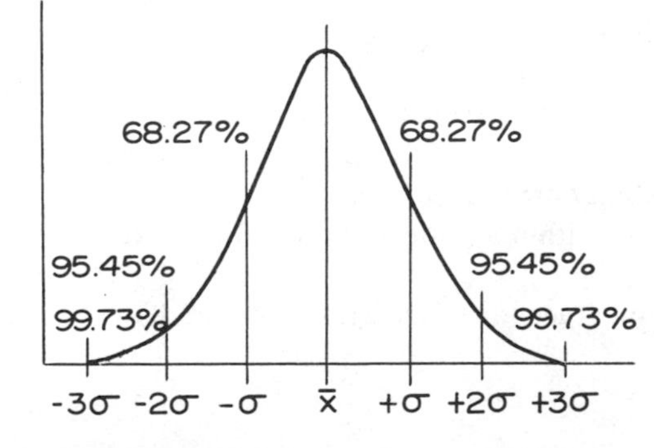

Figure 1-9. Normal (gaussian) frequency curve showing distribution with 1σ, 2σ, and 3σ.

* One differentiates between the universal population, that of all possible values of the variable, and a segment of this population. The symbol σ applies to the former, and s to the latter. The corresponding symbols for the means are μ and $\bar{x}$.

a set of measurements of a given variable is a measure of the scatter of values around the mean and can be calculated for any set of values as follows:

$$s = \sqrt{\frac{\sum (x - \bar{x})^2}{N - 1}}$$

where

$\sum$ = "sum of"

$\bar{x}$ = mean or arithmetic average $\frac{\sum x}{N}$

x = observed values of variable

$(x - \bar{x})$ = deviation (d) of a value from the mean ($\bar{x}$)

N = number of observations in the set of variables

In case of a normal distribution curve, the previously listed percentages of the total number of values of a test series should fall approximately within a particular standard deviation value. For example, when 20 or more determinations of glucose are performed on the same specimen and the standard deviation is calculated, the values would be distributed approximately as predicted by the following percentages:

±0.5 S.D.	38.30 per cent	
±1.0 S.D.	68.27 per cent	
±1.5 S.D.	86.64 per cent	
±2.0 S.D.	95.45 per cent	
±2.5 S.D.	98.76 per cent	
±3.0 S.D.	99.73 per cent	
±3.5 S.D.	99.96 per cent	
±4.0 S.D.	100.00 per cent	(99.99 per cent)

Confidence Limits and Confidence Interval

Confidence limits refer to the lower and upper values of the range (interval) within which random variations are acceptable. When one refers to two standard deviations as a confidence interval for some variable, this is equivalent to saying that if a single result is drawn at random from a series of analyses of the same sample population, the odds are about 95 per cent to 5 per cent (20:1) that this value will not be further removed from the mean than twice the standard deviation. In other words, the probability is 0.05 or 1/20 ($P = 0.05$) that the result may lie outside these limits.

The "probability of unity" indicates absolute certainty that the predicted value will be obtained, a "probability of zero" an absolute certainty that it will not. All intermediate probabilities are expressed by fractions, so that $P = 0.50$, represents a so-called 50–50 chance that a predicted value will be obtained.[26] Customarily, probability is expressed as percentages. $P = 0.50$ is equivalent to $P = 50$ per cent.

Standard Error of Sample Mean (S.E.$_{\bar{x}}$)

Two sets of populations of the same large population need not give the same mean or average. The statistical tool needed for the comparison of mean values is called the standard error or the standard deviation of the mean. This factor is calculated from the standard deviation by the equation:

$$\text{S.E.}_{\bar{x}} = \frac{s}{\sqrt{N}}$$

where s is the standard deviation and N is the number of individual results. Thus, S.E. is a measure of expected variation in the sample mean. The larger the number of values, the smaller the standard error, and the more reliable the value of the mean; i.e., it will more closely approximate the true mean of the population. By using the standard error we can measure the variation of the means of repeated samplings, calculate what are known as "t" values, and establish an interval of confidence for the mean.[26] The quantity "t" is the deviation of the estimated mean from that of the population, measured in terms of $s/\sqrt{N}$ as the unit. The t distribution is symmetrical about the mean and is calculated

$$t = \frac{\bar{x} - \mu}{\frac{s}{\sqrt{N}}}$$

where μ is the true mean. The true mean will be in the interval:

$$\bar{x} - \frac{t_s}{\sqrt{N}} < \mu < \bar{x} + \frac{t_s}{\sqrt{N}}$$

For example, in a random sample of 100 white males ages 20 to 29 years, the mean urinary nitrogen excretion was found to be 8.0 gm./24 hr. with a standard deviation of 1.6 gm. On the basis of figures from a single sampling, what can be said about the true mean of the population from which this sample was obtained? First, the value 8.0 gm. is considered to be the best estimate of the true mean. Second, since the standard deviation of the sample is the best estimate of the standard deviation in the population, the standard error is calculated by use of the formula:

$$S_{\bar{x}} = \text{S.E.}_{\cdot\bar{x}} = \frac{s}{\sqrt{N}} = \frac{1.6}{\sqrt{100}} = 0.16 \text{ gm.}$$

For the 95 per cent confidence limits of the mean, a range included in $\bar{x} \pm 2$ S.E.$_{\cdot\bar{x}}$ is $8.0 \pm 2 \times (0.16)$ or 7.68 gm. to 8.32 gm./24 hr. In 95 out of 100 such studies the mean value found will lie between 7.68 and 8.32 gm. For any single sample one cannot infer whether the interval does in truth bracket the true mean; however, in repeated samples it is to be expected that 95 per cent of the sample means will fall within the confidence intervals calculated. For 99.7 per cent confidence limits the range would be $\bar{x} = \pm 3$ S.E.$_{\cdot\bar{x}}$. If in a given problem the estimate of the range of confidence, determined on the basis of any given probability, is too great to be practically useful, this range can be reduced by increasing the number of samples on which means are determined.

Coefficient of Variation

Frequently it is necessary to compare the relative variability between various sets of data and methods. This cannot be done merely by examining the means and standard deviations of the two groups for they may be of different magnitudes or of different units. To make such comparisons, the standard deviations (s) are expressed as percentages of the means. This gives a measure of relative variability and is called the coefficient of variation (C.V.). It is calculated as

$$\text{C.V.} = \frac{s}{\bar{x}} \times 100$$

As an example, suppose that for a group of individuals the mean glucose value determined ($\bar{x}$) by a reducing sugar methodology was 98.5 mg./100 ml. with a standard deviation (s) of 2.5 mg./100 ml. and the mean glucose by an enzymatic true glucose procedure was 78 mg./100 ml. with a standard deviation of 0.9. The relative variability for the reducing sugar method would be:

$$\frac{2.5 \times 100}{98.5} = 2.6 \text{ per cent}$$

and that for the enzymatic glucose method would be:

$$\frac{0.9 \times 100}{78} = 1.2 \text{ per cent}$$

Thus, the variation inherent in the reducing sugar method is more than twice as great as that in the enzymatic procedure.

Example:

The following illustrates calculation of s, S.E.$_{\bar{x}}$ and C.V.

Analyses for blood glucose in 10 normal fasting individuals gave these results:

Blood Glucose (mg./100 ml.)	*Deviation from Mean*	*Deviation Squared*
(x)	$(x - x)$	$(x - \bar{x})^2$
90	+3	9
89	+2	4
85	−2	4
92	+5	25
88	+1	1
83	−4	16
88	+1	1
87	0	0
86	−1	1
82	−5	25
870	0	86

$$\text{Mean} = \bar{x} = \frac{\sum x}{N} = \frac{870}{10} = 87 \text{ mg./100 ml.}$$

$$\text{Standard Deviation } (s) = \sqrt{V^2} = \sqrt{\frac{\sum (x - \bar{x})^2}{N - 1}}$$

$$= \sqrt{\tfrac{86}{9}} = \sqrt{9.55} = 3.09 \text{ mg./100 ml.}$$

$$\text{Standard Error (S.E.}_{\bar{x}}) = \frac{s}{\sqrt{N}} = \frac{3.09}{\sqrt{10}} = \frac{3.09}{3.16} = 0.97 \text{ mg./100 ml.}$$

$$\text{Coefficient of Variation (C.V.)} = \frac{s}{\bar{x}} \times 100 = \tfrac{309}{87} = 3.6 \text{ per cent}$$

The preceding example described a convenient arithmetical process for calculating the mean, the standard deviation, the standard error of the sample mean, and the coefficient of variation. From this example it can be seen that the standard deviation is a statistical measurement of the actual limits of the variation of the individual values about a sample mean. The greater the standard deviation, the greater the differences between individual determinations and the less precise the analytical method. The standard error of the mean is inversely proportional to the square root of the number of observations contributing to the mean. In other words, the standard error of the mean of four observations is exactly one half of the mean of a pair of observations and the standard error of a mean of nine, is one third of the mean of a pair of observations.

OTHER MEASURES OF VARIATION

1. *Range.* This measure is used most frequently by the individual who has little knowledge of statistical methods. It is of limited value for it states merely what are the highest and lowest values. This can easily be seen by examining the following series of numbers:

	Series	
	21	
	24	
	45	
	46	
	48	
	48	
	49	
	50	Median(*Md*)
	50	
	51	
	52	
	52	
	54	
	55	
	55	
	100	
	800	
$\bar{x}$ =	50	

The sum of the series is 800 and since there are 16 observations in the series the mean is $\frac{800}{16} = 50$. The median for the series is also 50 and the range is 21 to 100 with a span of 79. This system is not as meaningful as others, and the variability of the data is better described by other constants that give more information about the distribution.

2. *Median.* If the distribution of data is asymmetrical or skewed, the arithmetic mean will be higher or lower than the point of most frequent values. For distributions such as this, the second centering constant or measure of central tendency, the median, is the one of choice. This constant, as its name implies, is the value in the middle of the distribution or the value that divides the number of observations into two equal parts. Thus, when the series is arranged in order from lowest to highest, the value that divides the preceding series in half is 50.

3. *Average deviation.* This measure of dispersion or variation is used to show the average variation of the measurements from either the mean or median. It is most commonly used as a measure of variation about the median and, therefore, is the parameter of choice in asymmetrical distributions. It is calculated by taking the average of the absolute values (without regard to sign) of the deviations of the individual measurements from the median. In the next series of numbers, the average

deviation from the median would be calculated as follows:

Serum Urea Nitrogen (mg./100 ml.)(x)	*x-Median*	*Difference*
13	13 − 16	−3
14	14 − 16	−2
median 16	16 − 16	0
18	18 − 16	+2
19	19 − 16	+3
		10:5 = average deviation

d_a = average deviation = 2 mg./100 ml.

The average deviation of 2 indicates that three of five samples or 60 per cent will be within 2 mg./100 ml. of the median value. The average deviation (d_a) is a biased (too low) measure of the dispersion but

$$1.25 \times d_a = s \text{ (approximately).}$$

d_a is not as precise as s, but can be calculated rapidly, especially for small samples, and thus is convenient to use.

4. *Variance.* The variance (V, s^2) is obtained in the course of calculating s.

$$V = s^2 = \frac{(x - \bar{x})^2}{N - 1}$$

This is the only true, unbiased measure of variation. Variance is used to determine whether a significant difference in precision exists between two different methods or two analysts using the same method.

The "F" test is used to compare measured variances of two sets of related measurements and to ascertain if they can occur by chance error. F relates the variances in the following equation:

$$F = \frac{\text{larger variance}}{\text{smaller variance}}$$

To obtain the limiting value of F for any particular probability of significance a Table of F is used to correct for "degrees of freedom." If the F value exceeds the tabular value, the precision of the data corresponding to the numerator for any given confidence limits is significantly poorer than that corresponding to the denominator.

Accuracy, Precision, and Reliability

It is appropriate at this point to explain the terms "accuracy," "precision," and "reliability." These words, which are often used loosely, almost as synonyms, have distinct and different meanings. *Accuracy* refers to the extent to which measurements agree with or approach the true value of the quantity being measured. Determination of accuracy consists of comparing observed results with actual "true" values. In the laboratory the true value is unknown (except for reference standards) and one usually compares measurements made by the method in question with those obtained by an acceptable reference method, employing control specimens whenever possible. When there is a constant systematic error in a clinical method, such as that due to air oxidation of a colorimetric reaction, causing the mean of a long series of determinations to vary from the actual true value, the method is inaccurate. Weighings made on a platform balance will not be precise, but they can be reasonably accurate; if made on a fine analytical balance that is not zeroed properly, they can be very precise but inaccurate. Various methods for determining glucose may give different results because of differences in specificity. In the Folin-Wu method for determining blood glucose, nonglucose reducing substances normally present in blood (ascorbic acid, glutathione, ergothioneine, and so on) are determined along with glucose. Such a test has poor accuracy, but this does not negate the value of the test if one becomes accustomed to the normal values associated with the procedure.

Precision refers to the magnitude of the random errors and demonstrates the reproducibility of the measurements. The precision of a clinical method is measured by its variance or standard deviation (s). The smaller the value of s, the greater the precision, and if two methods are being compared, the method with the smaller s is the more precise.

The *reliability* of a method is measured by its capacity to maintain both accuracy and precision. If a procedure has maintained a steady state of accuracy and precision over a considerable period, the method can be considered reliable. Reliability of a procedure can only be established by periodic checks of the method, using appropriate primary standards and control specimens.

Controlling Laboratory Error

The primary concern of the analyst is to ascertain that his results are as accurate as is possible with the use of a particular method. The correct procedure in such a case is to compare the differences between the values obtained by the particular method with the values derived from a reference method having acceptable accuracy.

Depending on the procedure, precalibrated standard curves or standards included in the run, or both, may be used for the calculation of results; however, curves and standards fall short of controlling the complete procedure. They frequently control only some step of a test, and although the value of a standard and the instrument calibration may be correct, it is possible for the final value of the unknown sample to be considerably in error.

In order to control all the steps of the procedure it is necessary to use a specimen, the analysis of which is known and which is similar in composition to the unknown sample. This type of preparation is then carried through the entire test procedure in parallel with the unknown and is affected by any and all variables that affect the unknown. This is used as the "control sample" and serves as a check on procedure, techniques, reagents, and instrument calibration.

The control serum usually does not meet the requirements of a primary standard. No control serum can be assigned a definite value for all constituents since even a "weighed in" constituent of a control sample is, as a rule, only a part of the total of that component. Furthermore, the assay values for both residual level and final levels of any constituent must be, eventually, based on pure aqueous reference standards.

Additional Techniques

A discussion of statistics in clinical chemistry would not be complete without mentioning the following three techniques useful in the evaluation of the quality of clinical chemistry procedures:

1. Statistical evaluation of the significance of the difference between two means
2. Tests of significance and analysis of variance
3. Regression analysis

These more advanced techniques for determining the magnitude of individual and multiple variables in clinical chemistry procedures enable the analyst to evaluate properly the chemical, mechanical, and subjective aspects of analysis. It is beyond the scope of this discussion to consider sampling, which, if inefficiently or improperly carried out, can completely invalidate the analyst's work.

QUALITY CONTROL

As the field of clinical chemistry has developed, there has been a corresponding increase in the number of tests requested per patient. Many hospitals are now performing chemical tests on almost all admissions and there is greater use of the "profile" tests for diagnostic purposes. The net result is that chemistry laboratories are faced with ever increasing workloads. Another factor causing an increase in the number of tests being requested per patient has been the development of micromethods, which make it possible to perform more analyses on the same aliquot of blood. Although many established test methods are undergoing automation and other modifications that make them simpler, the advent of newer and more complex tests, such as the measurement of various enzyme activities and the trend toward analyses of trace substances, has more than counteracted the modifications of established methods. These developments have further increased the need for effective quality control programs in the clinical chemistry laboratory.

Quality control is performed in the laboratory to assure that accuracy and precision are maintained in the day-to-day performance of clinical chemistry analyses.

During the past 15 years clinical biochemists and clinical pathologists have become increasingly aware of the lack of agreement seen in assays performed by different laboratories, as indicated by surveys.[4,29] In these surveys gross and shocking differences were observed in the values obtained from different laboratories. This proves that the data produced in many clinical laboratories are less accurate than is generally assumed, and, indeed, results may not be sufficiently precise to serve the purpose for which they are required. The following section describes an effective quality control system.

Outline of a Quality Control Program

The accuracy of an analytical result depends to a large extent on:

1. The ability to maintain proper analytical skill from start to finish.
2. The reduction of the number of manipulations to a minimum and the simplification of such manipulations, if this results in an increase in accuracy.
3. The ability to maintain the required quality of reagents.
4. The ability to maintain the desired performance of equipment.
5. The selection of the most accurate method best suited for the needs of the particular clinical chemistry laboratory.
6. The routine use of primary standards, quality control, reference sera, and statistical methods to evaluate results obtained.
7. The availability of conscientious and well trained technologists.

The Use of Quality Control Specimens (Pools)

A constructive and economical way to study and control the variations of methods is the use of pooled, preanalyzed sera, with each run of unknowns followed by the tabulation of data obtained on control charts and statistical evaluation of the data.

The procedures necessary for this phase of an effective quality control program can be summarized as follows:

1. Use of serum pools or commercial control sera with each run of unknowns.
2. Determination of mean values for each component of the control specimen, by a current laboratory method.
3. Determination of standard deviations for each component of the control specimen.
4. Setting up quality control charts.
5. Choosing acceptable limits of variation.
6. Setting up specific trouble shooting procedures for "out of control" results for each method.
7. Post-graduate education of technical staff in regard to quality control and advances in clinical chemistry, in order to increase skills and technical ability.

PREPARATION OF CONTROLS FROM POOLED SERUM

1. Pool volume

A pool should be sufficient for 3 to 4 months supply. Less than this needlessly increases the necessary work to establish means; more than this introduces the possibility of deterioration in the freezer.

2. Sources of serum

a. The most practical source is the salvaging of usable *excess serum* at the end of each day.

b. *Commercial preparations.* (These vary considerably in method of preparation, analysis, and properties.)

c. *Blood bank plasma.* (Concentrated by dialysis or by freezing in a separatory funnel and taking off the lower portion.) This approach is a tedious method and may permanently alter the characteristics of some of the materials contained. (Blood bank plasma is also contaminated with citrate.)

3. Collection of pool

Each day collect only clear serum samples free of hemolysis, lipemia, jaundice, or dyes. Place in a plastic bottle in the deep freeze (−20°C.). Subsequent daily samples are added directly to the frozen mixture until an appropriate amount (4 to 5 L.) is collected for processing.

4. Preparation of pool

Allow frozen serum to thaw completely at room temperature; mix thoroughly. A magnetic stirrer used for 1 hour will accomplish complete mixing without producing any foam. Centrifuge at 3000 rpm for 30 minutes to eliminate fibrin and other debris. (Thrombin may be added to enhance fibrin formation.) Although filtration through filter paper has been used, it is relatively slow and bacterial growth and consequent alteration of ingredients may occur. If possible, filtration through a Seitz or Millipore filter is recommended. Mix well again.

5. Aliquoting

a. *Size of aliquots.* The aliquots may be large enough to provide for the needs of a complete day's work (e.g., 10 ml.) or aliquots may contain smaller amounts to provide for individual tests or groups of tests.

b. *Storage.* Place tubes in deep freeze and keep them stoppered or covered with Parafilm, until use.

6. *Adjustment of serum values* to levels meeting the needs of the laboratory

Pool values are best set at critical levels of clinical importance, such as bilirubin values at 1 mg. and 20 mg. and glucose at 100 mg. Examples of adding supplemental amounts of pure materials follow:

a. *Bicarbonate.* Add 84 mg. of $NaHCO_3$/L. for each desired millimolar increase in concentration of CO_2/L.

b. *Glucose.* If the raw pool contains 60 mg./100 ml. and the desired value is 100 mg./100 ml., add 40 mg. of glucose/100 ml. of pool.

c. *Bilirubin.* If the raw pool contains 0.4 mg./100 ml. and the desired value is 1.0 mg./100 ml., add 0.6 mg. of bilirubin/each 100 ml. of pool.

Dissolve crystalline bilirubin in 0.1 *N* NaOH (approximately 1 ml. 0.1 *N* NaOH for 10 mg. of bilirubin). *Minimize exposure to light!* Dilute to 3 to 5 ml. with distilled water and add slowly with gentle stirring to the serum pool. The pH of the pool should be readjusted to pH 7.40 using 0.1 *N* HCl.

PREPARATION OF POOL SAMPLE FROM SALVAGED PLASMA OF OUTDATED BLOOD

Control plasma may be prepared from outdated blood bank blood. Plasma to be used is carefully chosen, selecting only the plasma showing no hemolysis and containing a minimum amount of chyle. The salvaged plasma is centrifuged to remove cells and clots. Since the blood bank plasma has been diluted approximately 20 per cent by the addition of anticoagulants and preservatives, it is restored to normal concentration by dialysis. Smaller molecules and ions lost during the dialyzing process may be restored by adding a calculated amount of these materials to the plasma after dialysis.

Dialysis is carried out in a cellulose casing suspended in a 25 per cent solution of polyvinylpyrrolidone (P.V.P.) in barbital buffer at pH 8.6. The barbital buffer is prepared by dissolving 2.76 gm. of diethylbarbituric acid and 15.4 gm. of sodium barbital in 1 L. of distilled water.

To concentrate 2 L. of plasma, three lengths of this casing (each $1\frac{1}{2}$ to 2 yards long) are used. One end of the casing is securely fastened, and the plasma is poured inside. The other end of the casing is similarly closed, and then it is laid out in a large vessel containing 1 to $1\frac{1}{2}$ L. of the dialyzing agent (P.V.P.) The time usually required to concentrate 2 L. of plasma is 5 to 6 hours.

After dialysis is completed, the contents are drained into a flask. The plasma is then mixed and the concentration of the constituents determined. If the total protein and cholesterol values are too low, further dialysis is carried out; if the values are too high, they are adjusted by a calculated dilution with distilled water. After this volume adjustment has been made, the chloride and uric acid concentrations are determined. Chloride in the form of sodium chloride is added to bring the value to approximately 100 mEq./L. Uric acid content is increased by the addition of a calculated amount of a standard uric acid solution containing 1 mg./ml.

PREPARATION OF PROTEIN-BOUND IODINE POOL

This pool must be prepared separately in order to avoid contamination from sera containing iodides or mercury compounds. After the unknown sera have been analyzed for protein-bound iodine, the remainder of all sera with values lying between 3.5 and 8 μg./100 ml. are pooled until about 250 ml. have been accumulated. Keep pooled sera in a flask in the refrigerator and proceed as for the regular pool, aliquoting 2.5 ml. per tube.

PREPARATION OF ENZYME POOLS

A pool for amylase, transaminase, lactic acid dehydrogenase, or other enzymes should be prepared separately, since it is desirable to have a high level of these enzymes to check properly the activity of the reagents.

Transaminase. Pool the remainder of all sera with transaminase values at least a threefold above normal, i.e., about 150 Karmen units. *Do not freeze.* When 35 ml. have been collected, the pool is assayed. If it is lower than 150 Karmen units save and add only very high transaminases. If it is higher, calculate the amount of normal serum needed to reduce its level to 140 to 180 Karmen units; readjust it until it does have this value. Filter through a Seitz or Millipore filter, and place small aliquots in (5 ml. capacity) screw-top vials. Place vials in freezer. Defrost 1 vial each week, and keep in the refrigerator. It will be found that freezing lowers the transaminase value of the pool considerably, but once it has dropped, it remains stable.

Amylase. Prepare in exactly the same manner as described for transaminase, and pool the remainder of all sera with amylase values above the upper limit of normal.

Establishing Acceptable Limits

To obtain data for statistical treatment and to establish limits of acceptability,

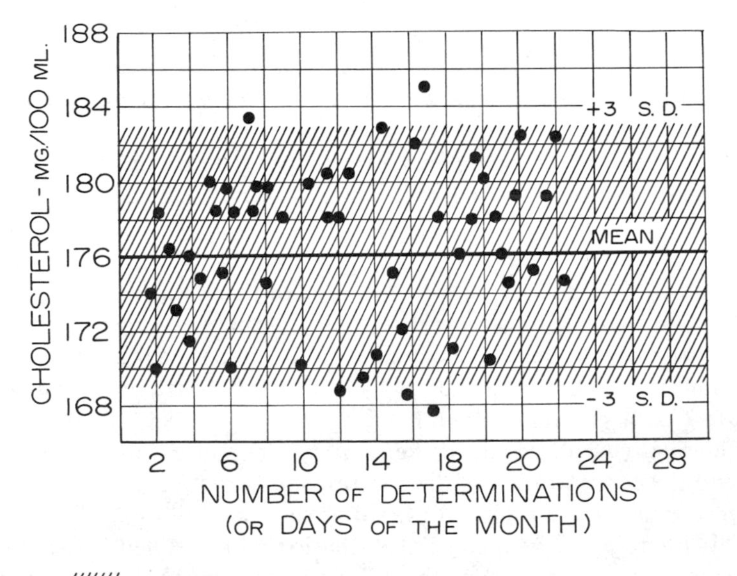

Figure 1-10. Quality control chart for cholesterol. Each point represents a cholesterol value for the control serum. On some days several batches of cholesterols were run, and a control was run with each batch.

TABLE 1-13. *Calculation of Standard Deviation for a Series of Cholesterol Determinations (Standard Equation)*

Number of Tests	A *values* (x)	B *Deviation* ($x - \bar{x}$)	C *(Deviation)*^{2} ($x - \bar{x}$)^2
1	173	3	9
2	177	1	1
3	176	0	0
4	178	2	4
5	168	8	64
6	174	2	4
7	175	1	1
8	177	1	1
9	174	2	4
10	174	2	4
11	177	1	1
12	175	1	1
13	173	3	9
14	173	3	9
15	175	1	1
16	175	1	1
17	180	4	16
18	180	4	16
19	178	2	4
20	178	2	4
	3510		154

$$\bar{x} = \frac{\Sigma x}{N} = \frac{3510}{20} = 175.5 \text{ mean* (or 176)} \qquad (1)$$

$$\text{Standard Deviation} = \sqrt{\frac{\text{sum of squared differences (Column C) from mean}}{\text{No. of values} - 1}} \qquad (2)$$

$$\text{or} \quad s = \sqrt{\frac{\Sigma(x - \bar{x})^2}{N - 1}}$$

$$\text{S.D.} = \sqrt{\frac{154}{19}} \quad \text{or} \quad \sqrt{8.10}$$

$$\text{S.D.} = \sqrt{8.10} \qquad (3)$$

$$\text{S.D.} = 2.84^*$$

* Round off to nearest significant number.

a minimum of 20 analyses should be performed and the mean, standard deviation, and acceptable limits calculated. An example of the calculation is shown in Table 1-13 (see also the section on statistics). The control chart, as shown in Figure 1-10, is the simplest way to utilize the daily control data.[13] The control chart is a sheet of rectangular coordinate graph paper marked by days. The units of the vertical axis are in terms of the assay concerned and "limits of acceptable error" are drawn as heavy lines. The daily value is plotted on the control chart and, thus, any deviation beyond established limits can easily be seen. The control chart shown (Fig. 1-10) is for cholesterol with limits between 169 and 183 mg./100 ml. and a mean of 176 mg./100 ml. As can be seen, the data of several days should have been the subject of close scrutiny since the control value on those days is beyond the limit of acceptability.

In the example shown in Table 1-13, the cholesterol control yields a mean value of 176 mg./100 ml. and a standard deviation of 2.8 mg./100 ml. Thus, the range for

1 S.D. is 173 to 179 mg./100 ml., for 2 S.D., 170 to 182 mg./100 ml., and for 3 S.D. 167 to 185 mg./100 ml.

Establishing Ranges

As control values are analyzed, it may be noted that in the case of certain constituents the range of the control value is as great as the range of normal values; however, the clinical usefulness of such a procedure would be open to question. Thus, there is a need for a guideline that the laboratory may use to indicate the permissible range by which the obtained results may vary from the mean and still remain useful.

The expression used to indicate the permissible range by which the obtained results may vary from the correct result is "acceptable limit of error." Repeated analysis of the control or an unknown specimen should give results within this limit. If a result outside of these limits is obtained, the result is considered to be "out of control."

In a control program the use of 1 standard deviation as "acceptable limit of error" would be of little value since about one third of the replicate values would fall by chance above or below the established range. Use of 2 standard deviations is helpful, (and is widely used), but still 5 out of every 100 assays might fall beyond the limits of acceptability and still be only due to random error. Assays that fall outside of our limits should be examined in order to determine whether they are random errors. Use of 3 standard deviations would eliminate any possibility of a value being beyond the acceptable range without overt error, but the use of 3 standard deviations is discouraged, since it gives a wide range that, in some cases, may extend even beyond the biological normal range.

Although most laboratories use 2 standard deviations in their control program, its use has definite drawbacks since a laboratory with a low degree of precision will have a figure for ± 2 standard deviations that is rather large when compared with a laboratory with a high degree of precision. Therefore, the laboratory that performs with a low degree of precision would consider a much larger range of values as acceptable. In general, this is not desirable since values obtained from a hospital laboratory should be equally reliable, regardless of the laboratory submitting them.[28]

TABLE 1-14. *Acceptable Limits of Error (In Per Cent)*[28]

Test	*Tonks*	*Other Institutions*
Albumin	8.5	6.6, 7.8, 4.2, 5.0
Bilirubin	10	18.3, 13.3, 3.0
Calcium	5	3.3, 6.6, 2.0, 3.0
Chloride	2	2.1, 3.0, 2.5
Cholesterol, total	10	2.6, 8.6, 10.0, 6.0
Creatinine	10	11.1, 10.0, 10.1
Glucose	10	5.4, 7.1, 8.0, 5.0
Iron	10	10.0, 9.1
Phosphorus	10	5.6, 7.0, 5.0
Potassium	10	2.8, 9.0, 4.0
Protein, total	7.5	3.3, 5.2, 4.0
Sodium	2.3	1.5, 2.9, 2.0
Thymol turbidity	10	11.7, 12.0, 10.0
Urea nitrogen	10	10.0, 10.9, 15.0, 8.0
Uric acid	10	4.6, 7.0, 6.0

Another approach is the calculation of the acceptable limits of error by the following formula suggested by Tonks:[29]

$$\frac{1/4 \text{ of the normal range}}{\text{mean of the normal range}} \times 100 = \text{acceptable limits of error in per cent}$$

If the result is higher than 10 per cent, the figure should be reduced to 10 per cent. But even this system has deficiencies, since some of the calculated values give a too wide range of allowable limits of error. Thus, at present, there is no definite system that can be used to establish the desired figures. Respective laboratories will have to set their own limits based on their own experience and needs.[28,30] The values published in the literature can be of considerable help. Table 1-14 is a summary of the acceptable limits used at some institutions.

Interpretation of Quality Control Charts

Any analytical procedure, regardless of how carefully and accurately performed, is subject to a certain degree of error since it is physically impossible to duplicate exact conditions from day to day. These errors that occur in an otherwise well controlled procedure are referred to as "*experimental errors.*"

Experimental random errors are errors over which the analyst has limited or no control (see statistics). The magnitude of the experimental error is directly related to the accuracy with which reagents are prepared, the number and complexity of steps in the procedure, and the skill of the analyst performing the test. It is these uncontrollable variables that determine how large or small the total "experimental error" (standard deviation) is for a particular procedure.

In a quality control program one measures the magnitude of the variation expected (s) and from these data plots the acceptable limits on a quality control chart (see under establishing ranges).

TRENDS

A trend in a quality control chart is formed by the values for the control that continued either to increase or decrease over a period of 6 consecutive days. (Fig. 1-11.)

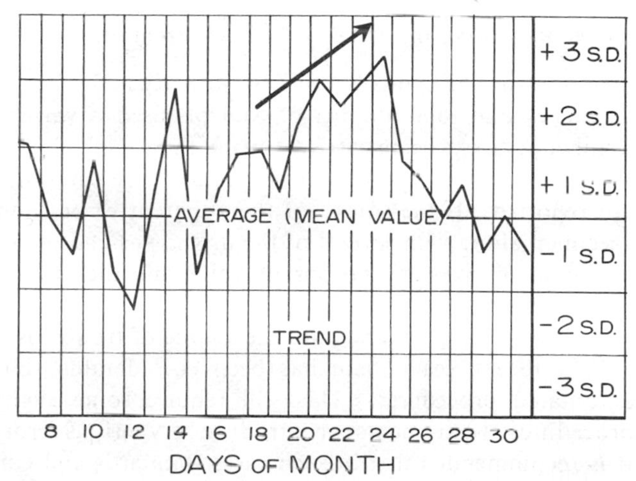

Figure 1-11. Trend in quality control chart. The arrow denotes the upward trend.

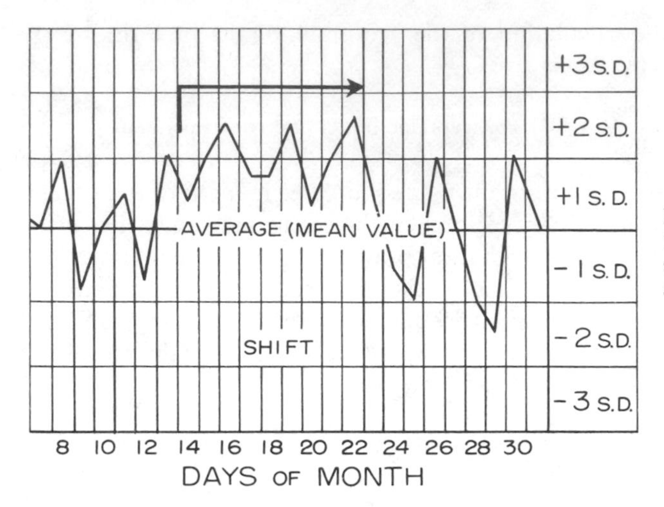

Figure 1-12. Shift in quality control chart. The arrow denotes the shift in the distribution of values away from the mean.

Trends may be the result of deterioration of one or more of the reagents, changes in standards, incomplete protein precipitation, and so on.

SHIFTS

When six or more consecutive daily values distribute themselves on one side of the mean value line, but are maintaining a constant level, the chart is said to have taken a shift (Fig. 1-12). An upward shift might indicate, for example, that a standard has deteriorated but is maintaining a constant level, a new standard has been prepared at a lower concentration than that required for a test, or a reagent has shifted to a new level of sensitivity. In test procedures in which solutions are boiled, an upward shift could indicate a prolonged boiling time due to an inaccurate timer, or an indicator that has lost sensitivity, thereby prolonging the end point reading. As a rule, downward shifts are caused by conditions that are opposite to those causing upward shifts.

The distinguishing feature between a shift and a trend is that the values in a shift do not continue to rise, but instead the distribution of these values is away from the mean on one side or the other. In both cases, it should become immediately evident that a problem exists.

Notes Regarding the Quality Control Program

A series of controls are run with each day's group of determinations. The frequency and total number of controls used is variable, depending upon the total number of assays being carried out and the complexity of the procedures. If the values are outside the acceptable range, the data are carefully evaluated before any results are reported. If the source of the apparent error cannot be found, a new control specimen along with several of the unknowns chosen at random is reassayed. If the new control gives an appropriate value and the unknowns check the previous run, the values are reported; if not, the data is held back and the reagents and procedural steps checked one by one until the source of trouble is determined.

In recent years there has been considerable shift from manual methods to automated procedures. This will require some slight change in quality control procedures. Instruments can introduce very serious errors if not sufficiently supervised. It is recommended that in such cases standards and controls be run not only at the beginning of the test run but also among the unknowns and at the end of the run.[28] In this way, any changes in instrument performance, such as change in light source or weakening of photocells and changes in environment (temperature), can be detected.

Evaluation of a New Procedure

A good analyst is aware of the importance of his results and his *procedures* will be based on common sense and good scientific practice. It is good practice to spend some time analyzing sets of standard solutions, until one can define over what range the absorbance is *linear with the concentration.* Such repeated runs will also show whether the results are reproducible from batch to batch of reagents and from day to day. After this, a number of specimens from normal subjects should be analyzed to confirm or establish the normal range of the method. In most instances, it is also advisable to add known quantities of a constituent to specimens at various steps of the procedure, perform the analyses, and calculate the "recovery" to show that significant quantities are not being lost in protein precipitation or at some other stage of the method. Finally, if the method is intended to replace one already in use in the laboratory, it is necessary to make a careful comparison of the results produced by the old and new methods and to accept the newer method only if the results are *statistically acceptable.* If the new method is superior to the old, but the results are significantly different from those obtained with the old method, the reasons for serious discrepancies should be sought. It will then be necessary to inform the clinical staff of the change in method and the difference in the normal values.

NORMAL VALUES

A normal value for a given constituent of clinical interest is considered to be that amount of the constituent of interest which is found in the body fluid or excretions of a group of clinically normal (apparently healthy) persons.* As will be discussed later, these normal values are arbitrarily defined as the range of values that would encompass 95 per cent of a population of these clinically normal persons. Values thus obtained (and processed as will be explained) are considered as "normal data"; however, it is important to differentiate between these "normal data" and those that are physiologically "desirable" or "ideal" for any given constituent. Let us consider that we determined the serum cholesterol values and the body weight of 20 clinically healthy American businessmen between the ages of 40 and 50 years. There would be general agreement that the "normal range" obtained after statistical evaluation of these data would be higher than that which we would like to consider as "ideal." The Metropolitan Life Insurance Company uses terminology that rather clearly points out this distinction. In 1912 its standards of weights for a given body height were called "average" weights. The term "average" was altered in 1942–43 to "ideal" and

* The term "clinically normal" or "apparently healthy" is not easily defined, however, it is generally considered to be a group of individuals who are considered to be in a state free of any obvious or overt abnormalities. It should be emphasized that a possible biochemical abnormality may not have progressed far enough to be clinically detectable so that the subject is still an "apparently" healthy person.

again in 1959 to "desirable." The weights listed in the respective tables are considerably below the present average weight for Americans.[3]

When establishing normal values for chemical constituents of body fluids and excretions, one should take into consideration physiologic variations, which may include diurnal variation, true day-to-day variation, and the effect of the environment. One should also consider that there may be differences due to race, age, sex, weight, nutritional and absorptive states, degree of physical activity, position of the body during blood sampling, stage of the menstrual cycle, ovarian status, emotional state, geographic location, and the time of day at which the sample was taken.[3]

Establishment of Normal Values

The clinical chemistry laboratory is frequently confronted with the need to establish normal values for a body constituent of clinical interest. Such situations may arise when a new method has been developed, (e.g., for a new enzyme) or if the laboratory merely wishes to compare its range of normal values with that found in the literature (an approach that is certainly recommended). Also, automated equipment has generated a massive amount of data that needs to be evaluated in terms of trends of values and present normal values at hand.

In practice, when establishing normal values, it is customary to analyze the blood of a group of apparently healthy individuals, as previously defined, for the level of the constituent of interest. Values thus obtained have to be treated in a way that will result in a "meaningful" normal range. The procedure to be selected will depend on the number of data available, on the type of data obtained and their relative distribution, and on the "type" of normal range that is desired (see the following).

Any type of statistical analysis is most valuable if a large population was analyzed. The same principle applies to normal values, when an infinite number of analyses would be considered ideal. In practice, it is usually considered sufficient to obtain approximately 100 data. Frequently, however, even such a number is not available and, consequently, the type of statistical method applied to treat the obtained data may have to be selected accordingly (see examples to be listed). The type of statistical procedure used to determine the normal values will also depend on the type of data. Data distributed according to a symmetrical bell-shaped pattern would be treated differently than those that are skewed in one or the other direction.

Finally, it is important to decide on the type of normal values desired. For most constituents tested in clinical chemistry laboratories there is an overlap of values obtained from the normal and abnormal populations. If the normal range is set in such a way that it would include the entire normal population, some of the abnormal values would be in the "normal" range and go undetected. If, on the other hand, a narrower normal range is used, a number of normal individuals would have an "abnormal" laboratory result. In the first case, a disease state may be overlooked, and, in the latter case, undue concern over a normal individual could develop. It has therefore been proposed that two ranges be used.[30] The first range would include 98 per cent of the apparently normal population and any value outside this range would be considered as definitely abnormal. The second range would include 80 per cent of the apparently normal population. All values that fall into this range would be considered as definitely normal. All values falling within the range of 80–98 per cent constitute a "gray zone" and these values should be viewed with suspicion. The concept of two normal values is appealing but impractical and, hence, has not found widespread use. The range most frequently used today is the so-called 95 per cent confidence limits, i.e., the range into which 95 per cent of the apparently normal population will fall. The important point is that these are statistically obtained ranges of "normal values" and that 1 out of 20 individuals will have values outside of this range. Similarly, a value just inside the normal range should not be viewed as absolutely normal.

Thus, in summary, there appears to be no procedure for the establishment of normal values that can be applied uniformly to all situations. Before any one of the following listed (or any other) procedures is selected, the obtained data should be thoroughly inspected and the desired goals should be clearly set.

DETERMINATION OF NORMAL VALUES BASED ON A NORMAL GAUSSIAN DISTRIBUTION

Many analytical data will follow the so-called "normal" Gaussian frequency distribution. This was first described by the French mathematician Abraham de Moivre in his treatise of 1733 and further developed by the astronomer-mathematician, Karl Friedrich Gauss in the nineteenth century. To obtain a Gaussian distribution curve, the values obtained on normal individuals are plotted in bar graph (histogram) form. In some cases, the distribution curve obtained will be a symmetrical bell-shaped pattern, but in many cases the curve is skewed to the left or more frequently to the right. If the distribution curve is symmetrical, with the necessary experience and knowledge, it may be possible to arrive at a normal range after visual inspection of the curve. A mathematical approach, however, will yield a result that is statistically more precise; that is, the degree of confidence for the normal range is greater. To obtain such data, the mean ($\bar{x}$) and the standard deviation (s) are calculated. All values falling between $\pm 2s$ would constitute the "normal range" and would include 95 per cent of all (presumably) normals.*

$$\text{Normal range} \quad = \bar{x} \pm 2s \tag{1}$$

(95% confidence limits)

The accuracy of such a range would be affected by the number of data analyzed, as previously mentioned. Since, in practice, the sample size is usually relatively small, it has been suggested[12] that the so-called "tolerance limits" be calculated using the following expression:

$$\bar{x} \pm K \times s \tag{2}$$

The K factor in the equation has been taken from Bowker (see Table 1-15) who gives complete tables of K for populations up to 1000 and for additional confidence coefficients (other than $\gamma = 0.90$) and additional confidence limits (other than 95 per cent).[6] Condensed tables giving K values also appear elsewhere.[12,14] The arithmetic mean is represented by $\bar{x}$ and s represents the standard deviation calculated from the data obtained.

As an example, let us assume that glucose determinations have been performed on the plasma of 10 fasting normal individuals and that the arithmetic mean ($\bar{x}$) equals

* Calculation of the "95 per cent confidence limits" based on the calculation of $2s$ from the arithmetic mean ($\bar{x}$) is considered acceptable only if the distribution is symmetrical.

TABLE 1-15. K *Factors for* 0.90 *Confidence Coefficient*, γ, *to Obtain* 95 *per cent Limits of the Normal Range**

N	*K Factor*
5	4.152
6	3.723
7	3.452
8	3.264
9	3.125
10	3.018
15	2.713
20	2.564
25	2.474
30	2.413
40	2.334
50	2.284
60	2.248
70	2.222
80	2.202
90	2.185
100	2.172

* From Bowker, A. H.: *In* Selected Techniques of Statistical Analysis. C. Eisenhart, M. W. Hastay, and W. A. Wallis, Eds. New York, McGraw-Hill Book Co., Inc., 1947.

90 mg./100 ml. and $s = 7.5$. Using equation (2), the normal range with "95 per cent confidence limits" is:

$$90 \pm 3.02 \times 7.5 \qquad (3)$$

or

$$90 \pm 22.7$$

Thus, the normal range for plasma glucose based on our sampling is 90 ± 22.7 or 67.3 to 112.7 mg./100 ml. If, on the other hand, we had studied 100 normal individuals and found a mean of 90 mg./100 ml. and s of 7.5, then the normal range would be:

$$90 \pm 2.17 \times 7.5 \qquad (4)$$

or

$$90 \pm 16.3$$

In this second example, the normal range for plasma glucose would be 73.7 mg./100 ml. to 106.3 mg./100 ml.

The numbers for K listed in Table 1-15 show that the larger the population sampled, the closer K approaches 2. In other words, if a sufficiently large population is analyzed, data obtained with equation (1) would be identical to data obtained with equation (2). It should be pointed out again that data obtained with either equation (1) or (2) include only 19 out of 20 individuals and thus are subject to the limitations previously discussed.

The use of the K factors for the so-called tolerance limits in place of the more conventional 2 may seem to make the range unnecessarily wide. Although this is not in common usage, Henry feels that it is a statistically correct approach.[12] Some workers have suggested 3 standard deviations when working with a limited number.

The use of ± 3 S.D. is discouraged, since this range includes not only most of the normal population, but also a relatively large number of the "abnormal" population.

DETERMINATION OF NORMAL VALUES FROM DATA THAT DO NOT FOLLOW A "NORMAL" GAUSSIAN DISTRIBUTION

The previous discussion dealt with the handling of data that were distributed according to a "symmetrical" (normal) Gaussian distribution curve. In reality, much biological data does not follow such a distribution and calculations performed using equation (1) or (2) may be erroneous. In fact, such approaches could lead to normal ranges extending to below zero.[11] In other situations, in which the distribution curve is skewed to the right, ($\bar{x} > Md$), the calculated normal range would extend too far to the high side.

TABLE 1-16

Thymol Units	*Log Thymol Units*	*N.E.D. (for 20 samples)*
0.4	−0.40	−1.960
0.6	−0.22	−1.440
0.6	−0.22	−1.150
0.8	−0.10	−0.935
0.8	−0.10	−0.755
1.1	0.04	−0.598
1.3	0.11	−0.454
1.4	0.15	−0.319
1.6	0.20	−0.189
1.6	0.20	−0.063
1.8	0.25	0.063
2.0	0.30	0.189
2.0	0.30	0.319
2.0	0.30	0.454
2.3	0.36	0.598
2.3	0.36	0.755
2.6	0.41	0.935
3.4	0.53	1.150
3.9	0.59	1.440
4.3	0.63	1.960

Inspection of a histogram will generally reveal whether the distribution is symmetrical or skewed; however, if a small number of samples are analyzed, it may not be obvious whether the data follow a true Gaussian distribution or a log normal or other distribution. In these cases, Moore[19] has suggested the use of a graphic method called "the normal equivalent deviate" (N.E.D.). This method requires that the values obtained be arranged in order of increasing magnitude. Each number is then plotted on the horizontal axis of a linear graph paper against the normal equivalent deviate (N.E.D.) on the vertical axis. The N.E.D. values for samples from 1 to 1000 can be taken from the article by Moore from which the data listed in Table 1-16 were taken. N.E.D. tables for limited sets of data also appear elsewhere.[12,14] After plotting all values, a straight line is drawn representing all points. If the data obtained represent a curve rather than a straight line, a log transformation of all data is made and these values are plotted against the N.E.D. If there is a random distribution of these values from the straight line, the distribution of values is considered to be normal and the "normal range" and the "mean" (in this case the geometric mean) may be determined directly from the graph, as described by Henry and Dryer.[14] If the deviation from the straight line is systematic, the data probably did not come from a homogeneous population. Examples of the most common variables to be considered here include differences due to sex, race, age, and postabsorptive state.

EXAMPLE OF HANDLING A SMALL SET OF DATA, THE DISTRIBUTION OF WHICH IS NOT APPARENT

Occasionally a normal range must be derived from a relatively small set of data. It should be emphasized that this should only be considered a "temporary" or "working" normal range until more data become available. Frequently with a small set of data the distribution is not readily apparent even when the data are presented in histogram form. As mentioned earlier, when large data are available the normal range may be arrived at visually upon inspection of the curve, but with limited data mathematical methods must be used. The following example shows an approach to deriving a "working" normal range from a limited set of data.

Figure 1-13 gives the data for thymol turbidity from 20 healthy adults. This histogram shows the scatter of the data but from direct observation it is difficult to decide if this set of data could better fit a normal or log normal distribution. This can be decided by subjecting the data to the N.E.D. graphic test of linearity. The first step is to arrange a frequency table listing the data in increasing order of magnitude and to associate each value with its N.E.D. The log of each figure is also taken. This data is shown in Table 1-16. Figures 1-14 and 1-15 show the graph of N.E.D. plotted versus the thymol units and the log of the thymol units, respectively. From these graphs it is evident that only the log normal distribution gives a straight line.

Once it has been ascertained that the distribution follows a log normal distribution, a mathematical approach may be used to determine the mean and standard deviation. The log of each value for thymol turbidity is taken and the logarithms are

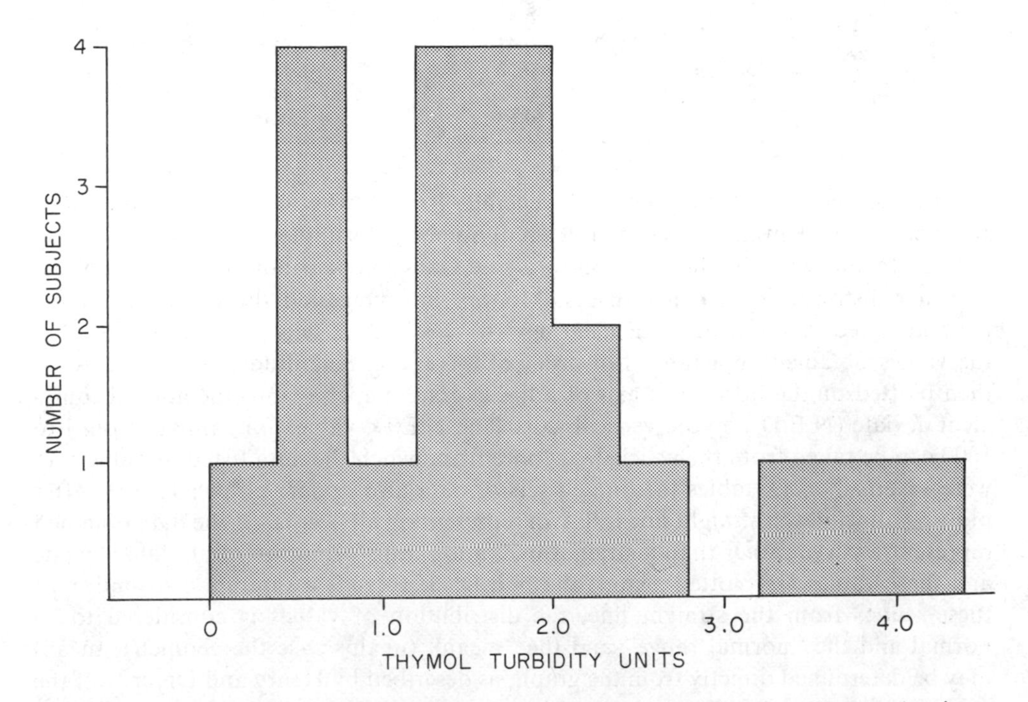

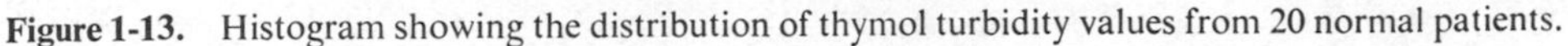
Figure 1-13. Histogram showing the distribution of thymol turbidity values from 20 normal patients.

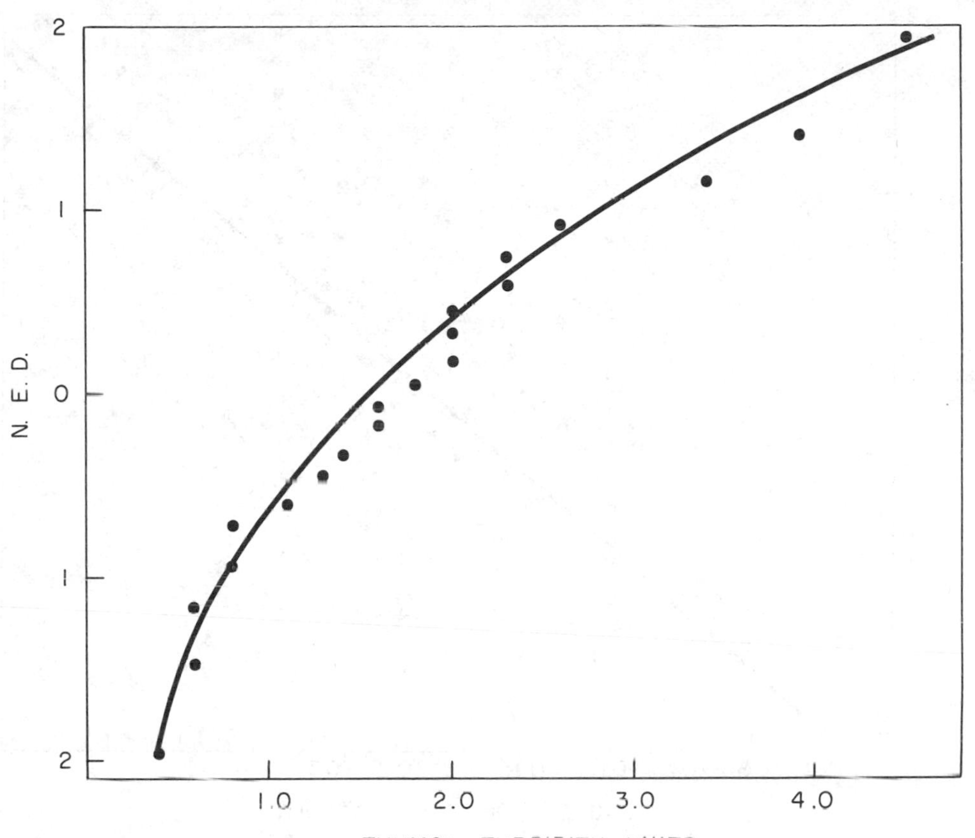

Figure 1-14. N.E.D. plot of thymol turbidity data from Table 1-15 on a linear scale.

now handled exactly as the original number for the determination of standard deviation.

From Table 1-17 it is found that:

$$\bar{x} = 0.18 \quad \text{and} \quad \sum(\bar{x} - x)^2 = 1.49$$

Substituting into the equation for standard deviation:

$$s = \sqrt{\frac{\sum(\bar{x} - x)^2}{N - 1}}$$

$$s = \sqrt{\frac{1.49}{19}} = \sqrt{0.078} = 0.28$$

This now represents the standard deviation expressed as a log. This data is now substituted into the formula used for establishing the range (interval).

$$\bar{x} \pm 2s$$
$$0.18 \pm 2\ (0.28)$$
$$0.18 \pm 0.56$$

The normal range (still expressed as logs) is -0.38 to 0.74 with a mean of 0.18. Taking the antilogs, the mean now becomes 1.5 units, with a normal range of 0.4 to 5.5 units.

If the data for thymol turbidity had been handled as if it followed a simple

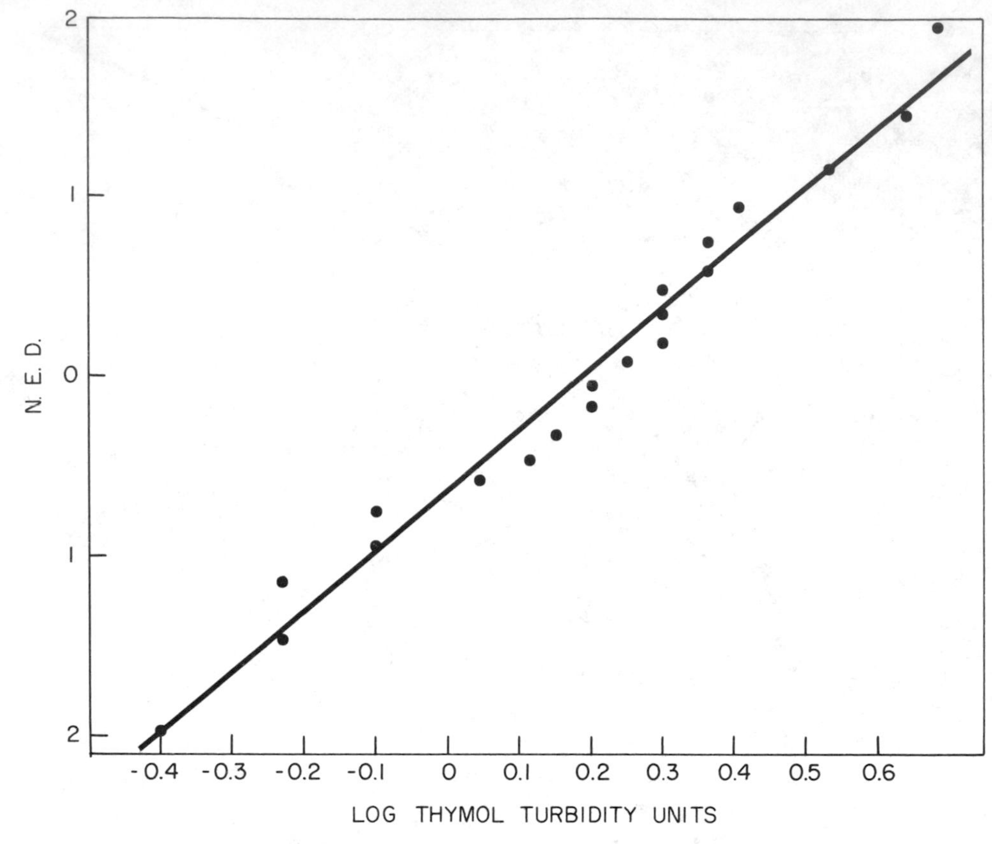

Figure 1-15. N.E.D. plot of thymol turbidity data from Table 1-15 on a log scale.

TABLE 1-17

Thymol Units	*Log Thymol Units* (x)	*Deviation* $\bar{x} - x$	*(Deviation)*^{2} $(\bar{x} - x)^2$
0.4	−0.40	0.58	0.34
0.6	−0.22	0.40	0.16
0.6	−0.22	0.40	0.16
0.8	−0.10	0.28	0.08
0.8	−0.10	0.28	0.08
1.1	0.04	0.14	0.02
1.3	0.11	0.07	0.01
1.4	0.15	0.03	0.00
1.6	0.20	0.02	0.00
1.6	0.20	−0.02	0.00
1.8	0.25	−0.07	0.01
2.0	0.30	−0.12	0.01
2.0	0.30	−0.12	0.01
2.0	0.30	−0.12	0.01
2.3	0.36	−0.18	0.03
2.3	0.36	−0.18	0.03
2.6	0.41	−0.23	0.05
3.4	0.53	−0.35	0.12
3.9	0.59	−0.41	0.17
4.3	0.63	−0.45	0.20
	sum = 3.69		sum = 1.49

$$\bar{x} = \frac{\Sigma x}{N} = \frac{3.69}{20} = 0.18$$

Gaussian distribution, we would have had

$$\bar{x} = 1.8$$
$$s = 1.2$$

Thus, based on $\bar{x} \pm 2s$, the range for thymol turbidity would have been $1.8 \pm 2x\,(1.2)$ or -0.6 to 4.2 units. This illustrates what was mentioned earlier, namely, that the formula $\bar{x} \pm 2s$, without log transformation, may only be used when the distribution approaches Gaussian.

Another technique that may be helpful in the establishment of a normal range is *probability paper;* however, this may be used only when dealing with a normal distribution. A single straight line (obtained by plotting cumulative per cent versus values) indicates that the distribution is normal. Some workers have felt that probability paper can be used to determine a normal range from a mixed population (normal and "sick").[15] There are usually two problems associated with this use. First, the data for the "sick" population are not clearly separable from the normal population without resorting to mathematical devices much more complicated than probability paper.[22] Second, the distribution of data from the normal population is frequently not Gaussian.

To illustrate the use of probability paper the thymol turbidity data will again be used; however, ordinarily much more data should be available when using probability paper. The data are arranged in order of increasing magnitude and the cumulative frequency and cumulative per cent then determined. In Figure 1-16 the values for

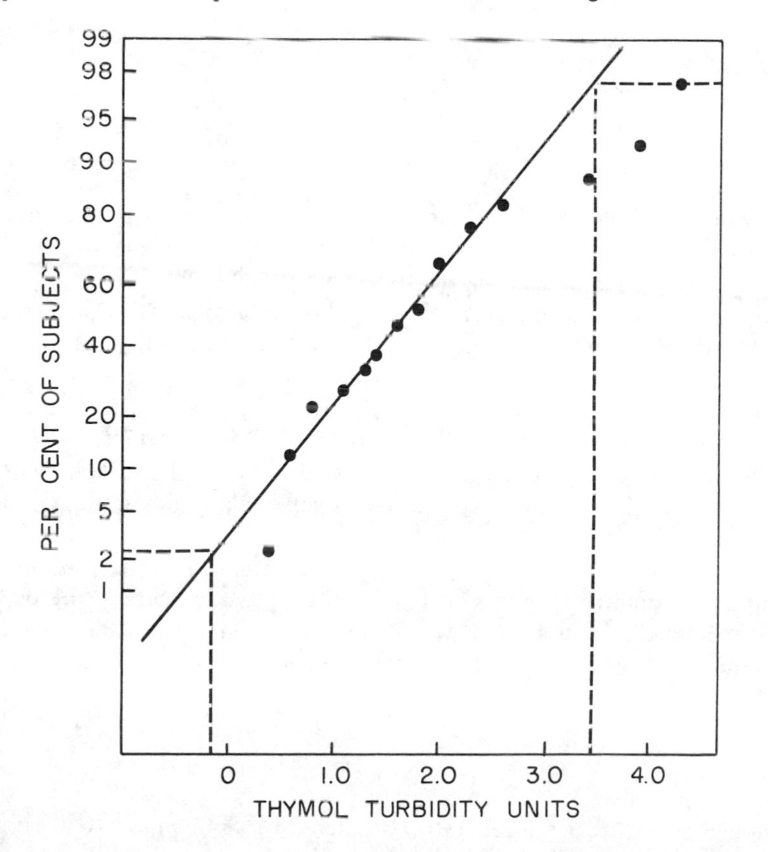

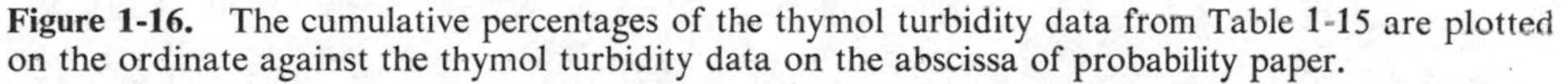
Figure 1-16. The cumulative percentages of the thymol turbidity data from Table 1-15 are plotted on the ordinate against the thymol turbidity data on the abscissa of probability paper.

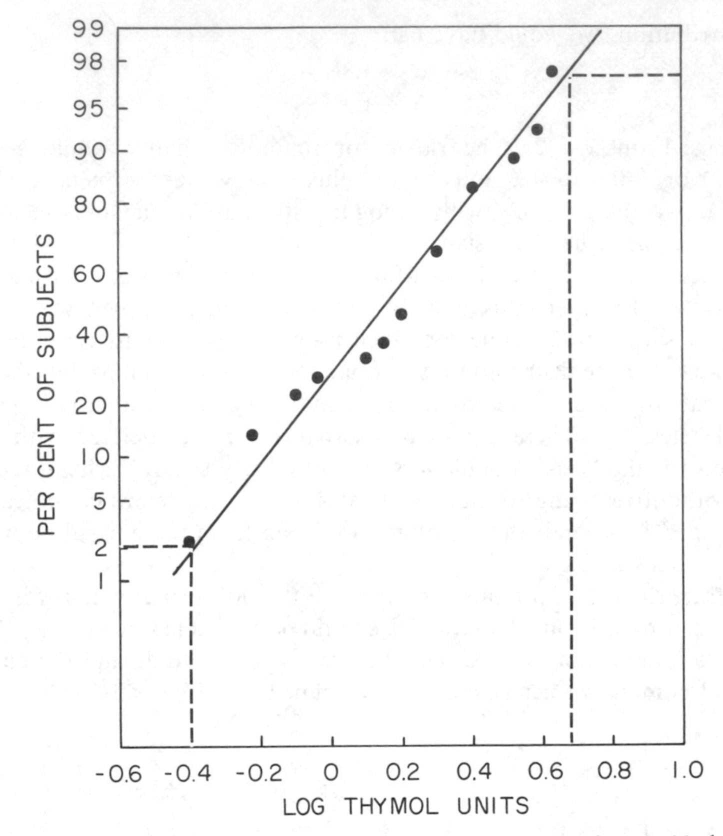

Figure 1-17. The cumulative percentages of the thymol turbidity data from Table 1-15 are plotted on the ordinate against the log of the thymol turbidity on the abscissa of probability paper.

thymol turbidity are plotted on the abscissa and the cumulative percentages on the ordinate on probability paper. The best straight line is drawn through the points. Horizontal lines are drawn from the 2.5 and 97.5 per cent points on the abscissa to the straight line. Vertical lines are dropped from these intersections to the ordinate.

The 95 per cent limits from the ordinate are −0.2 to 3.7 thymol units. This is fairly close to the normal range of −0.6 to 4.2 obtained assuming a simple Gaussian distribution, and points out the fact that probability paper cannot be used without log transformation if a distribution is not Gaussian. In Figure 1-17 the log of thymol units is used as the ordinate and the best line again drawn. The 95 per cent range is now shown to be −0.4 to 0.68 (expressed as logs). Taking the antilogs, the normal range becomes 0.4 to 4.8.

Aside from the selection of the subjects, the single most important point in determining a normal range for a given constituent is ascertaining the distribution pattern of the data. Presentation of the data in histogram form will frequently reveal the distribution as Gaussian or not. A more frequent distribution of clinical chemistry normal data is a log normal distribution.

SAFETY

Although laboratories are generally considered very safe places to work, it is only because necessary safety rules and procedures have been incorporated into the daily activities of the workers. In general, safety rules are similar from one laboratory to another, and should be learned early in the training of a chemist or technologist. Future practices in a laboratory are then governed almost automatically by these safety habits.

Electrical Hazards

With the ever increasing trend toward instrumentation, it must be remembered that wherever there are electrical wires or connections, there is a potential shock or fire hazard. Worn wires should be replaced immediately on all electrical equipment; all apparatus should be grounded using either three prong plugs or pigtail adapters. If grounded receptacles are not available, check with an electrician for proper grounding techniques.

Electrical equipment and connections should not be handled with wet hands. Electrical apparatus should not be used after liquid has been spilled on it. It must be turned off immediately and dried thoroughly; a fan or hair dryer will speed up the drying process. In the case of a wet or malfunctioning electrical instrument, pulling out the plug is not enough if the instrument is used by several people; a note cautioning co-workers not to use it should be left on the instrument.

Electrical apparatus, especially motors, that are to be operated in an area where there are flammable vapors should be explosion proof and air driven stirrers should be considered. Induction driven motors are well suited for these areas. The laboratory, particularly the bench top, is no place for numerous extension cords.

Hazards from Chemicals

The storage and use of chemicals may be attended by a variety of dangers such as burns, explosions, fires, and toxic fumes. A knowledge of the properties of these substances and proper handling procedures will virtually eliminate any danger. Bottles of chemicals and solutions should always be handled carefully, and a cart used to transport two or more containers or one heavy container from one area to another. An excellent manual on the handling of chemicals is available.*

Spattering from acids, caustic materials, and strong oxidizing agents probably represents the greatest hazard to clothing and eyes and is a potential source of chemical burns. Bottles should never be grasped by the neck, but firmly around the body with one or both hands, depending on the size of the bottle. When diluting acids, always add acid slowly to the water with mixing; water should never be added to concentrated acid. When working with acid or alkali solutions, particularly when making up reagents in large amounts, safety glasses should be worn. Thought should be given to the possibility of breakage, and if possible this mixing should be done in a sink, which would provide water for cooling as well as for confinement of the reagent in the event the flask or bottle were to break.

All bottles containing reagents should be properly labeled. Before adding the reagent, it is good practice to label the container; in this way the possibility of having an unlabeled reagent is avoided. The label should bear the initials of the person who made up the reagent and the date the reagent was prepared. If appropriate, the

---

* Safety in Handling Hazardous Chemicals. Matheson, Coleman, & Bell, Norwood, Ohio 45212 (available without charge).

expiration date should also be included. The labeling may be color coded to designate specific storage instructions, such as the need for refrigeration or special storage. All reagents found in unlabeled bottles should be discarded. The reagent should be flushed down the drain with large amounts of water. As a general rule, do not spare the water, when pouring reagents down the drain. In the case of an acid, it must be diluted sufficiently in order not to harm the plumbing. Other reagents that are not harmful in the drain in themselves may present a real hazard if followed by an acid. An example of such a chemical is cyanide, which results in the release of hydrogen cyanide if brought in contact with acid. Large amounts of water must be used after each reagent that presents a danger, either actual or potential, to health.

Strong *acids*, *caustic* materials, and strong *oxidizing* agents should, whenever possible, be dispensed with automatic dispensing devices. When it is necessary to pipet these materials there are a number of excellent safety devices commercially available to eliminate the necessity of pipetting by mouth.

Most of the general precautions just mentioned also hold for *automatic chemistry equipment* such as AutoAnalyzers. With this equipment the disposing of reagents should be considered carefully in view of what was mentioned earlier. In certain methods rather strong acids are pumped into the drain and should be accompanied by a steady flow of water from the faucet. Many of these instruments pump hazardous reagents. The use of safety glasses is recommended when inspecting a plugged tube or a malfunctioning manifold when these reagents are involved.

Perchloric acid, because of its ability to explode when it comes in contact with organic materials, requires careful handling procedures. Do not use perchloric acid around wooden bench tops. Bottles of this acid should be stored on a glass tray. Disposal may be accomplished by slowly adding the acid to at least 10 volumes of cold water and pouring down the sink with large amounts of additional cold water. Special perchloric acid hoods with special wash-down facilities are available when one is using large amounts of this acid.

Special care is needed in dealing with *mercury*, a rather common element in a clinical laboratory. Many think of mercury as a material used to fill a gasometer and forget that it is also a potential hazard. Small drops of mercury on bench tops and floors may poison the atmosphere in a poorly ventilated room. Mercury toxicity is cumulative and the element's ability to amalgamate with a number of metals is well known. After an accidental spillage of mercury, the area should be gone over carefully until there are no globules remaining. All containers of mercury should be kept well stoppered.

A chemical *fume hood* is a necessity for every clinical chemistry laboratory. Any container with a material giving off a harmful vapor should be opened only with caution in the fume hood. Reagent preparation that results in fume production should also be done in a hood. The heating of all flammable solvents should be done in a hood with a sealed electric hot plate, water bath, or heating mantle. In the event an unexpected explosion or fire occurs the hood may be closed and the fire confined.

Safety Equipment

There are a large number of safety items available for the laboratory. Most of these are shown in a manual on laboratory safety.* A few will be mentioned briefly

---

* Manual of Laboratory Safety. Fisher Scientific Co., Pittsburgh Pa., 15219 (available without charge).

and some of these should be considered in areas where they are appropriate. *Eyewashes* or facewashes should be available in every chemistry laboratory. Some of these may merely be connected to existing plumbing. *Asbestos gloves* should be available for handling hot glassware and for handling dry ice. Safety goggles, glasses, and visors are available in many sizes and shapes. Some fit conveniently over regular eye glasses. Shatterproof laminated safety shields should be used in front of systems that represent a potential danger because of vacuum collapse or pressure explosions. Desiccator guards should be used with vacuum desiccators. Hot beakers should be handled with *tongs*. Floor standing gas cylinders should be strapped to the bench or wall with special *cylinder supports*. Inexpensive polyethylene pumps are available to pump acids from large bottles.

Fire Hazards

Every laboratory should have the necessary equipment for general fire fighting to put out or confine the fire to a given area, as well as to put out the fire on the clothing of an individual. Easy access to *safety showers*, which may be near the door of the laboratory, is desirable. These showers should have either a pull chain attached to the wall at a convenient height or a large ring attached to the chain so that the shower may easily be activated even with the eyes closed. *Fire blankets* in an easily accessible wall mounted case should be available for smothering clothing blazes. By taking hold of the rope that is attached to the blanket and turning the body around, the blanket is unrolled from the case and rolled around the body, thus smothering the flames.

There are various types of *fire extinguishers* available. Their use depends on the type of fire. Almost any type of fire extinguisher may be used on a fire where only wood, paper, textiles, and similar materials are involved. For fires involving grease or oils, foam, dry chemical or vaporizing liquid extinguishers should be used. In fires involving electrical equipment or in areas where live electricity is present, water or soda type extinguishers are not to be used. These materials will conduct electricity and will lead to a serious electrical hazard in addition to the existing fire. For these fires carbon dioxide, dry chemical, or vaporizing liquid extinguishers should be used. It is rather impractical to have several types of extinguishers present in every area. Dry chemical fire extinguishers are among the best all purpose extinguishers for laboratory areas. They should be provided near every laboratory door and in a large laboratory also at the opposite end of the room. The assumption should not be made that all the laboratory personnel know how to use this equipment. Everyone in the laboratory should be instructed in the use of these extinguishers and any other fire fighting equipment. All dry chemical extinguishers should be tested twice a year by qualified personnel.

Isotopes

Radioisotopes as used in the clinical chemistry laboratory are usually obtained in very dilute solutions and used for such tests as T_3 and blood volume. Since the actual amount of radiation present is usually very small there are relatively few, if any, special precautions needed. When special precautions are needed they will be stated by the supplier of the isotopes and should be followed exactly.

Signs must be available so that all radioactive material is thus labeled. Careful observance of good laboratory technique is essential in an area where radioisotopes are handled. More than anywhere else carelessness must be avoided in this area.

REFERENCES

1. Anderson, H. H.: Anal. Chem., *20*:1241, 1948.
2. American Society for Testing Materials: Standards. Bull. #C-225-45 T, pt. 3, Philadelphia, 1949, p. 354.
3. Beeler, M. F.: Postgrad. Med., *43*:67, 1968.
4. Belk, W. P., and Sunderman, F. W.: Am. J. Clin. Path., *17*:853, 1947.
5. Bower, J. H.: J. Res. Nat. Bur. Standards, *33*:199, 1944.
6. Bowker, A. H.: *In* Selected Techniques of Statistical Analysis. C. Eisenhart, M. W. Hastay, and W. A. Wallis, Eds. New York, McGraw-Hill Book Co., Inc., 1947, p. 97.
7. Caraway, W. T.: Am. J. Clin. Path., *37*:445, 1962.
8. Dean, G. A., and Herringshaw, J. F.: Analyst, *86*:434, 1961.
9. Dybkaer, R., and Jorgensen, K.: Quantities and Units in Clinical Chemistry. Baltimore, The Williams & Wilkins Co., 1967.
10. Elking, M. P., and Kabat, H. J.: Am. J. Hosp. Pharm., *25*:485, 1968.
11. Henry, R. J.: Am. J. Clin. Path., *34*:326, 1960.
12. Henry, R. J.: Clinical Chemistry: Principles and Techniques. New York, Harper & Row, Publishers, 1964.
13. Henry, R. J.: Clin. Chem., *5*:309, 1959.
14. Henry, R. J., and Dryer, R. L.: *In* Standard Methods of Clinical Chemistry. New York, Academic Press, Inc., 1963, vol. 4, p. 205.
15. Hoffman, R. G.: J.A.M.A., *185*:865, 1963.
16. Horst, F. W.: Chem. Ztg., *45*:604, 1921.
17. Kirsten, W.: Anal. Chem., *25*:1137, 1953.
18. Martell, A. E., and Calvin, M.: Chemistry of the Metal Chelate Compounds. Englewood Cliffs, N.J., Prentice-Hall, Inc., 1952.
19. Moore, F. J., Cramer, F. B., and Knowles, R. G.: Statistics for Medical Students and Investigators in the Clinical and Biological Sciences., New York, Blakiston, 1951.
20. Morey, G. W.: The Properties of Glass. American Chemical Society Monograph, New York, Reinhold Publishing Co, 1954.
21. National Bureau of Standards, Circular C-602, 1959.
22. Neumann, G. J.: Clin. Chem., *14*:979, 1968.
23. Properties of Selected Commercial Glasses: Pyrex, Corning Vycor, B-83, Corning, N.Y., Corning Glass Works, 1959.
24. Reagent Chemicals. The American Chemical Society, Washington, D.C., 1955.
25. Simonds, H. R., Weith, A. J., and Bigelow, M. H.: Handbook of Plastics. 2nd. ed. New York, D. Van Nostrand Co., Inc., 1949.
26. Snedecor, G., and Cochran, J.: Statistical Methods. 6th ed. Ames, Iowa, Iowa State Press, 1967.
27. Stas, J. S.: Chem. News, *17*:1, 1868.
28. Tietz, N. W.: Hosp. Prog., *45*:140, 1964.
29. Tonks, D.: Clin. Chem., *9*:217, 1963.
30. Wootton, I. D. P., and King, E. J.: Lancet, *1*:470, 1953.

ADDITIONAL READINGS

Bancroft, H.: Introduction to Biostatistics. New York, Harper Bros., 1959.

Davidsohn, I., and Henry, J. B. (Eds.): Todd-Sanford Clinical Diagnosis by Laboratory Methods. 14th ed. Philadelphia, W. B. Saunders Co., 1969.

Henry, R. J.: Clinical Chemistry, Principles and Technics. New York, Harper & Row, Publishers, 1965.

MacFate, R. B.: Introduction to the Clinical Laboratory. Chicago, Year Book Medical Publishers, Inc., 1961.

Chapter 2

ANALYTICAL PROCEDURES AND INSTRUMENTATION

by Merle A. Evenson, Ph.D.

The purpose of this chapter is to provide a brief explanation of the general principles involved in the physical and chemical measurements commonly made in clinical chemistry laboratories. The discussion will develop only simple mathematical equations based upon physical laws of chemistry, and it will only moderately involve descriptive material on each of the methods. Entire books are devoted to each of the sections discussed here; consequently, the subject cannot be covered in detail in one short chapter. The references listed at the end of the chapter are organized by subject matter and should be consulted for further details.

Analytical procedures in the modern clinical chemistry laboratory are numerous and vary greatly in complexity. A need exists for laboratory personnel to be well acquainted with a number of disciplines. Basic electronics, physics, inorganic, organic, physical, and analytical chemistry, biochemistry, and statistics are all important subjects to be blended in teaching prospective personnel. Knowledge or information from all of these subjects will be needed to understand the analytical methodology and instrumentation used in the clinical chemistry laboratory. Details of the actual measurements will be found in the respective chapters of this book where the methods of analysis are described.

Gravimetric Analysis

Gravimetric analysis involves the separation of the substance of interest from other components in the matrix and the weighing of that pure substance or its derivative to obtain the amount present in the original sample. The earlier reliance upon gravimetric analysis is shown by the units still employed in reporting results, e.g., gm./100 ml., mg./100 ml., μg./100 ml.

Determination of total lipid in serum and feces is one of the few procedures still in use in clinical chemistry that requires direct weighing of the isolated material.

Before flame photometry was introduced, sodium and potassium were determined by precipitation of the uranium and platinum salts, respectively, followed by washing, drying, and weighing of the precipitates. Gravimetric procedures not only require a large sample, but also the separation of substances is difficult and time-consuming; consequently, other more sensitive, more rapid, and more convenient analytical approaches have almost entirely superseded gravimetric methods of analysis.

Although few analytical procedures performed in the clinical chemistry laboratory today involve the direct use of the analytical balance, the weight of a purified standard is almost always the reference point upon which the method of analysis is based. The analytical balance should, therefore, be considered as an important instrument, in the clinical laboratory. The absence of spectacular operating modules like flashing lights, flames, and noisy read-out devices contributes to the limited attention generally paid to the analytical balance.

Balances

Balances used in clinical chemistry laboratories are generally of three types: the double pan (trip) balance, the suspended single pan triple beam balance, and the analytical balance. The trip balance is used to weigh objects up to 2 kg., the triple beam balance is usually used to weigh objects up to 200 gm., and the analytical balance is used to weigh objects up to 100 gm. The trip balance will generally weigh with an accuracy of 100 mg., and the analytical balance will weigh accurately to 0.1 mg. Automatic single pan balances with tare corrections are available that fit into the three categories just mentioned.

The most common problem associated with the use of an analytical balance is carelessness by the operator. Fingerprints either on the weighing bottle or on other glass containers as a result of handling the vessels directly produces errors and should be avoided. The handling of weights with fingers must also be avoided. The weight of a fingerprint is measurable and the chemical reaction that the fingerprints will initiate will limit the useful life of the weight set.

The balance of choice for clinical chemistry laboratories today is an automatic single pan analytical balance. The balance weighs by substitution and operates by the removal of calibrated weights from above the arm of the balance holding the pan. The total of the weights removed corresponds to the weight of the object on the pan. The advantages of such a balance are that the weights are enclosed in the balance and are not openly exposed to the atmosphere. The object to be weighed is also physically isolated from the weights with this type of balance. Since the weights are never touched by the user, the weights and the balance do not need calibration as frequently as a conventional double pan analytical balance. In summary, the balance should be treated carefully and the following few rules should be followed strictly.

1. Do not place hot objects on the balance pans.
2. Do not place chemicals directly on the pan. Parchment paper, weighing paper, weighing bottles, or other glassware should be used on the pans.
3. Do not overload the balance. A 50 to 100 gm. load is a safe limit for most conventional analytical balances.
4. Do not handle the weights, the objects to be weighed, or the internal parts of the balance with your fingers, as fingerprints add weight and will initiate corrosion.
5. Keep the balance clean and covered when it is not in use.

6. Corrosive materials should only be weighed in closed containers. This applies especially to iodine crystals and cyanide compounds.

Excellent discussions on the use of balances are available in several textbooks of college chemistry[4] and quantitative analysis.[1] MacFate[3] and Boutwell[2] have sections devoted to balances and should be referred to for additional information.

Volumetric Analysis

Volumetric analysis involves the determination of a substance by reacting it with a measured volume of a known standardized solution. This process is commonly called titration.

The principle of volumetric analysis is illustrated with the formula:

$$V_1 \times C_1 = V_2 \times C_2$$

where V_1 = measured volume (ml.) of titrant required to reach the end point

C_1 = known concentration of titrant expressed in equivalents/L. or mEq./L.

V_2 = measured volume (ml.) of unknown to be titrated

C_2 = calculated concentration of unknown in equivalents/L. or mEq./L.

Another way of considering the preceding equation is to look at the conventional units of each quantity in the formula. V_1 and V_2 are usually measured in milliliters (ml.) of solution, and C_1 and C_2 are usually in equivalents/L. or mEq./ml. (normality). When the volume is expressed in identical volume units (ml. or L.), and then multiplied by the concentration per unit volume, the volume units cancel. Thus, at the equivalence point, the amount of titrant is chemically equivalent to the amount of the substance being measured.

The common types of glassware used for volumetric analysis are pipets, burets, and volumetric flasks, of which there are many kinds and sizes (see Chapter 1 for a discussion of glassware).

In general, to perform a titration, a measured volume of standard solution is made to react with a substance whose concentration is to be measured. The chemical reaction then proceeds in a stoichiometric or controlled manner in the presence of an indicator. The indicator possesses the property of producing a physical change in the solution when the substance of interest has quantitatively reacted. This physical change may be a color change, as is frequently the case in acid-base titrations, a change in the electrical conductance of the solution, a change in the electrical potential of the solution, or other changes in the physical characteristics of the solution. Changes in absorbance of indicators may also indicate an end point in the titration. In the Schales and Schales chloride titration, free mercuric ions form a purple complex with diphenylcarbazone after all the chloride in solution has been complexed as non-ionized mercuric chloride. A color change in the presence of free mercuric ions signals the equivalence point. Details of the chloride measurement are found in Chapter 10.

In discussing titrations, it is appropriate to illustrate stages of a titration by the

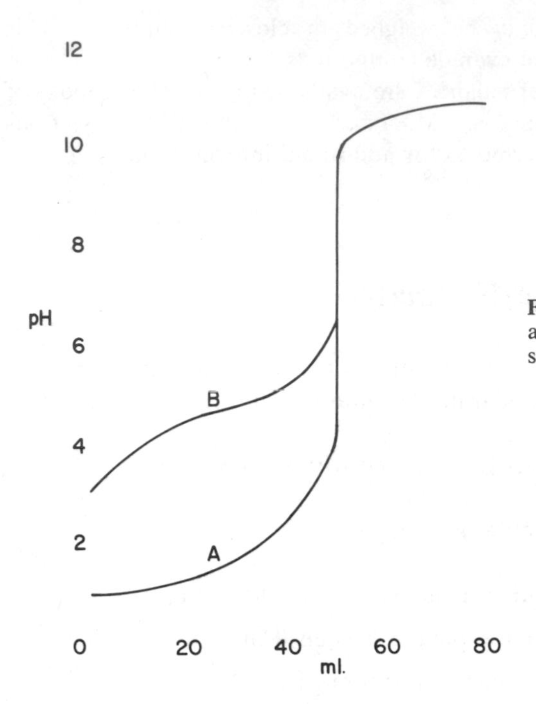

Figure 2-1. Comparison of titration curves of a strong acid (A) and a weak acid (B) with a strong base.

following example. In the titration of a strong acid with a strong base, the choice of indicator does not significantly affect the end point. In Figure 2-1, curve A illustrates the titration of 0.1 *N* HCl with 0.1 *N* NaOH. At pH 4 the residual concentration of acid is 10^{-4} *N* and at pH 10 the excess concentration of base is 10^{-4} *N*. At this point in the titration, less than 1 drop of 0.1 *N* NaOH is required to change the pH from 4 to 10. Consequently, any indicator that changes color between pH 4 and 10 will give the same end point. Curve B illustrates the titration of a weak acid (0.1 *N* acetic acid) with NaOH. As the titration proceeds, a buffer system is set up between the acetic acid and the acetate ions formed. The steep inflection does not occur until pH 6.8, when 99 per cent of the acetic acid has been neutralized. At this point, the solution's buffer capacity is greatly decreased and, therefore, the pH increases rapidly. The indicator of choice for this titration is one that changes color between pH 7 and 10.

Sometimes a titrant may serve as its own indicator. In the Clark-Collip method for the determination of calcium, the precipitated calcium oxalate is dissolved in dilute sulfuric acid and the oxalic acid liberated is titrated with potassium permanganate. In this redox reaction, oxalic acid is oxidized and permanganate is reduced. When the reaction is complete, purple MnO_4^- appears in solution and signals the end point of the titration.

The normality of an oxidizing or reducing substance is determined by the number of electrons gained or lost by a reactant. In the preceding example, the (Mn^{+7}) in the MnO_4^- (oxidizing agent) is reduced to manganous ion (Mn^{+2}). In this reaction MnO_4^- gains five electrons from the oxalic acid (reducing agent). The equivalent weight of potassium permanganate is therefore one fifth of the molecular weight for this chemical reaction:

$$MnO_4^- + 8H^+ + 5e^- \rightleftharpoons 4H_2O + Mn^{+2}$$

Electroanalytical Chemistry

Measurement or separation techniques, or both, based on electroanalytical methods have been well known and carefully studied for a number of years, but have not been extensively used in routine clinical chemistry laboratories. One of the reasons may be that the matrix, containing the substance to be measured, is not well defined for biological samples. The effects of proteins, lipids, and other materials found in serum and urine are not easily measured and controlled for certain types of electroanalytical measurements.

Before discussing specific kinds of electrochemical measurements, certain definitions of terms and conditions are necessary. Oxidation is defined as a process that results in the release of electrons and occurs at the electrode called the anode. Reduction is the reverse process of oxidation and involves the absorption of electrons at an electrode. This electrode is called the cathode. By definition, oxidation always occurs at the anode, and reduction always occurs at the cathode. These definitions and conventions follow the recommendations adopted at the International Union of Pure and Applied Chemistry meeting in Stockholm, July, 1953.

To understand elementary electrochemical definitions, notations, and the operation of a galvanic cell, a description of such a cell is in order. A simple galvanic cell may be built by placing a strip of zinc metal in a beaker of 1.0 *M* zinc chloride and a strip of copper metal in another beaker of 1.0 *M* copper chloride. The zinc and copper metal strips are connected with a wire and the circuit is completed by connecting the solutions in the two beakers with an inverted U-tube filled with saturated potassium chloride solution, called a salt bridge. If it is bubble free, ions will flow through it to maintain a charge balance between the two beakers while electrons flow through the wire from the anode to the cathode. The zinc metal ionizes more easily than the copper. Thus, the zinc electrode becomes oxidized and forms the anode. The copper ions from the solution are plated out onto the copper electrode and the copper electrode becomes the cathode.

A compact way of schematically representing the galvanic cell just discussed is:

$$\mathrm{Cu};\quad 1\,M\,\mathrm{CuCl_2}\,\|\,1\,M\,\mathrm{ZnCl_2};\quad \mathrm{Zn}$$

The semicolons are used to separate the electrode materials from solutions. If more than one ion occurs in solution, a comma is used to indicate this, and the parallel sign is used to indicate the salt bridge. In this notation, the composition of the cathode is usually written on the left hand portion of the cell, and the composition of the anode is written on the right hand portion.

The recommended convention is to write all half-cell reactions as reductions. For example, the half-cell reactions for zinc and copper would be written as:

$$\mathrm{Zn^{+2}} + 2e^- \rightarrow \mathrm{Zn}_{(s)} \qquad E_0 = -0.763 \text{ volts}$$
$$\mathrm{Cu^{+2}} + 2e^- \rightarrow \mathrm{Cu}_{(s)} \qquad E_0 = +0.337 \text{ volts}$$

E_0 is the symbol for the standard reduction potential and is measured against a standard hydrogen reference electrode. Elements with positive E_0 values for their half-cell reactions, such as copper, show a greater tendency to go to the reduced state as exemplified in the half-cell reaction. Elements with negative E_0 values for their half-cell reactions, such as zinc, show a greater tendency to go to the oxidized state, i.e., in a direction opposite to that shown in the half-cell reaction. For a given galvanic cell using the preceding sign convention, the more negative half-cell becomes the anode and assumes a positive charge and the more positive half-cell the cathode. The net voltage is the algebraic difference between the two electrodes. In the preceding example, the net voltage of the cell is $+0.337 - (-0.763)$ or 1.100 volts.

Tables are available that provide standard electrode reduction potentials (E_0) from which one can predict whether reactions will occur spontaneously. Some older tables list standard electrode *oxidation* potentials, in which case the signs are opposite of the standard reduction potentials. Oxidation potentials can be used in the same manner as reduction potentials. Care must be used in predicting the direction in which a reaction will spontaneously go because of confusion over the sign of the voltage of a reaction. This information can then be used to determine if a system can be designed to measure the substance of interest. The kinds and numbers of analyses performed by electroanalytical methods are certain to expand at a rapid rate in clinical chemistry laboratories. Examples of electrochemical measurements in current use will be presented later in this chapter.

POTENTIOMETRY

The measurement of the potential between two electrodes in solution forms the basis for a variety of measurements that can be used to quantitate concentrations of the substance of interest.

To measure the potential (E) or the electromotive force (EMF) directly, both an indicator and a reference electrode are necessary. The EMF of the indicator electrode can be made to respond proportionally to the concentration of the substance of interest, but the reference electrode must maintain a constant voltage under controlled conditions for a significant length of time. The most frequently used reference electrode in electroanalytical measurements is the saturated calomel electrode (SCE). Figure 2-2 shows the schematic diagram of the components of an SCE reference electrode. A porous plug shown at the bottom of the figure stoppers the outside glass jacket and yet allows solution contact with the inner part of the calomel electrode. A small

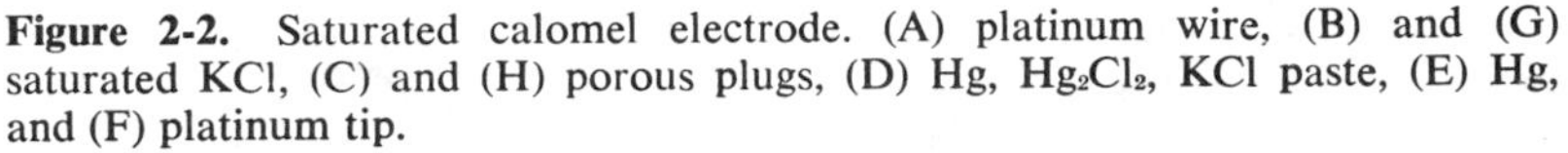

Figure 2-2. Saturated calomel electrode. (A) platinum wire, (B) and (G) saturated KCl, (C) and (H) porous plugs, (D) Hg, Hg_2Cl_2, KCl paste, (E) Hg, and (F) platinum tip.

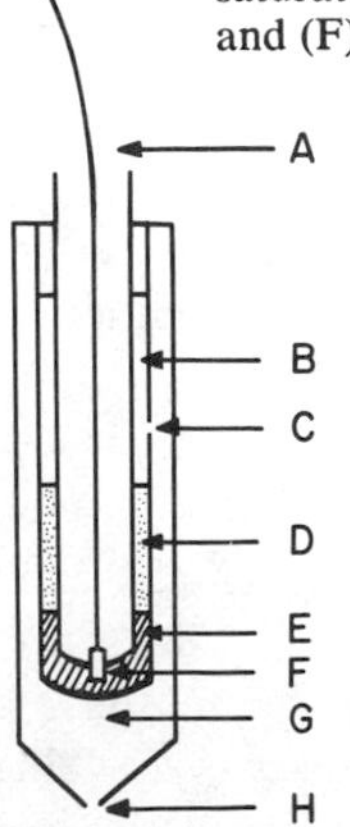

platinum wire is sealed into the end of the inner glass jacket. Mercury metal, mercurous chloride, and a paste made of mercury, saturated mercurous chloride and potassium chloride are inserted into the center glass jacket. Finally, electrical contact is made to the mercury- mercurous chloride interface and the platinum tip by a wire to the measuring device. The standard reduction potential (E_0) of the calomel electrode versus a normal hydrogen electrode is −0.242 volts.

A silver-silver chloride electrode is another common reference electrode and may be easily constructed using a low voltage (less than 6 volts) battery, a platinum wire, and a silver wire. The silver wire is usually coiled, connected to the positive pole of the battery, and submerged in a dilute chloride solution. The platinum wire is connected to the negative pole of the battery, and then submerged in the same solution containing the coil of silver wire. Hydrogen will be liberated at the platinum cathode and silver chloride will be deposited on the silver anode when connected to the battery.

The silver wire will develop a rose to purple color in a few minutes. The silver-silver chloride wire is then placed in a standard chloride solution that has been saturated with silver ions to decrease the solubility of silver chloride and maintain a stable reference potential. A silver-silver chloride electrode has the advantage over the calomel electrode of being less sensitive to temperature changes.

The normal hydrogen electrode (NHE) consists of a platinized platinum electrode in a 1.228 *N* HCl solution with hydrogen at 760 mm. Hg bubbled over the platinum surface. This reference electrode has an assigned E_0 of 0.000 volts. Because of difficult maintenance problems, the NHE is not frequently used in routine measurements.

There are several types of indicator electrodes that can be used with reference electrodes like those just described. Indicator electrodes can be a platinum wire, a planar surface of almost any metal, a carbon rod, or a thin stream of mercury flowing into the solution where a measurement will be made.

In potentiometry the EMF or potential (E) is measured and the relationship between the measured voltage and the sought for concentration is shown by the Nernst equation:

$$E = E_0 + \frac{0.059}{n} \log \frac{[C_{ox}]}{[C_{red}]} \qquad (1)$$

where E = the potential measured at 25°C.

E_0 = the standard reduction potential

n = the number of electrons involved in the reaction

C_{ox} = the molar concentration of the oxidized reaction form

C_{red} = the molar concentration of the reduced reaction form

The measurement of potential may be made either by using an instrument called a potentiometer or a vacuum tube volt meter (VTVM). The VTVM is an electron tube device that is capable of measuring very small amounts of current and, because of a high input impedance, the measurement does not change the system being measured. Schematic diagrams of potentiometers and VTVM's and how each operates may be found in many elementary physics or chemistry textbooks and will not be discussed here.

MEASUREMENT OF pH

The pH of a solution is defined by the equation:

$$\text{pH} = -\log\,[\text{H}^+] \tag{2}$$

where $[\text{H}^+]$ = hydrogen ion concentration in moles/L.

In buffer solutions, the pH is related to the concentrations of the undissociated acid and its corresponding anion according to the following equation:

$$\text{pH} = \text{pK}_a + \log \frac{[\text{A}^-]}{[\text{HA}]} \tag{3}$$

where $[\text{A}^-]$ = the molar concentration of anion
$[\text{HA}]$ = the molar concentration of acid
K_a = the acid dissociation constant
$\text{pK}_a = -\log \text{K}_a$

Notice the similarities and relationships between the Nernst equation (1), the definition of pH (2), and the buffer equation (3). A single potential measurement under proper conditions can be directly related to the H^+ concentration in solution.

A pH measurement is usually made with the aid of a glass electrode as the indicator electrode. One type consists of a bulb of special glass filled with 0.1 *N* HCl in contact with a suitable metallic electrode. When immersed in solution, a potential difference develops between the solution inside the glass electrode and the solution being measured for H^+ concentration. The magnitude depends upon the hydrogen ion concentration of the solution. This potential difference is measured by combining the glass electrode with some standard reference electrode, such as the saturated calomel electrode, and measuring the voltage of the system.

Calibration is achieved by using a known solution that has a pH value assigned (pH_s) by the National Bureau of Standards. The pH of an unknown solution is compared to the known solution by potential measurements using a pH meter. A pH meter simply measures the potential produced in a solution using the electrodes just described.

The electrode arrangement for a pH measurement may be considered as a special type of concentration cell. A modification of equation (1) that can be used for a concentration cell is:

$$\Delta E = \frac{RT}{nF} \ln \frac{C_1}{C_2} \tag{4}$$

where ΔE = measured change in potential
R = gas constant
T = temperature in degrees Kelvin
n = number of electrons in electrochemical reaction
F = value of the Faraday constant
C_1 = concentration of unknown (outside of glass electrode)
C_2 = concentration of known (inside glass electrode)

At 25°C. for a one electron reaction and with C_2 equal to 1 *M*, equation (4) becomes $\Delta E = 0.059$ pH. In other words, a 59 mv. change will occur when the pH changes 1 unit.

A schematic diagram for a pH meter is shown in Fig. 2-3. While making the pH measurement, it is important that the amount of current drawn from the measuring

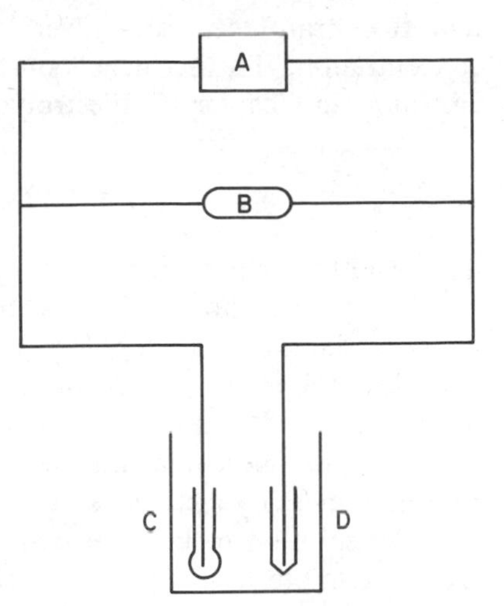

Figure 2-3. Schematic diagram of a pH meter. In making pH measurements one adjusts the potentiometer (A) until the current across the glass electrode (C) and the reference electrode (D) is zero, the voltage is indicated by the meter (B).

electrodes is very small. Generally 10^{-10} to 10^{-12} amp or less are drawn. The principle behind the pH measurement involves the adjustment of the potentiometer potential to be equal and in the opposite direction to the potential of the measuring electrodes. At balance, the pH reading is taken off the pH calibrated scale connected to the slide wire of the potentiometer. A description of a potentiometer and how it operates may be found in many general chemistry or physics reference books.

In pH instruments of recent design, the potentiometer is replaced by a direct reading meter calibrated in pH units. In other pH meters, the potential is electronically measured, amplified, and fed to a direct readout device.

COULOMETRIC MEASUREMENTS

Coulometry involves the measurement of the quantity of electricity (in coulombs) at a fixed potential:

$$Q = I \times T \tag{5}$$

where Q = coulombs of electricity
I = the current in amperes
T = the time in seconds that the current is flowing

Two approaches can be used in coulometric measurements. When the current is kept constant, the elapsed time is proportional to the total coulombs consumed. Alternatively, the current may be changed in a known manner for a fixed time and the area beneath the current curve integrated with respect to time to obtain the number of coulombs. A coulomb is equal to a current flow of 1 amp sec. A Faraday is defined as 96,500 coulombs and corresponds to the electrical charge carried by 1 gram equivalent of substance. One equivalent is equal to 1 mole if only one electron is involved in the electrochemical reaction. Thus, the number of coulombs consumed can be related directly to the concentration of the unknown.

In the chloride determination using the Cotlove titrator, a constant current is applied across silver electrodes, which liberate silver ions into the solution at a constant rate. When all of the chloride ions in solution have been complexed by the liberated silver ions, a pair of indicator electrodes senses the excess silver ions, activates a relay that shuts off the timer, and stops the titration. The length of time that the titrator generates silver ions is directly proportional to the chloride ion concentration. Further details on the principle of operation of the chloride titrator are found in Chapter 10, Electrolytes.

POLAROGRAPHY

Another amperometric technique frequently used in electroanalytical chemistry is called polarography. The technique involves making a current measurement as the potential is varied. A polarographic measurement in its simplest form involves a two electrode system, consisting of a reference electrode and an indicator electrode. Many kinds and combinations of indicator and reference electrodes may be used for polarographic measurements. The indicator electrode frequently consists of a thin ribbon of mercury metal flowing slowly through a small glass capillary and dropping into the solution to be measured. Generally, linearly increasing cathodic voltage is applied to the dropping mercury electrode (DME) and the current that flows between the two electrodes is measured and is proportional to the concentration of the substance of interest.

The current measured is called the diffusion current (i_d) because it is the maximum amount of current that can flow in the system. The i_d is limited by the rate of diffusion of the electroactive species to the surface of the DME. The quantitative relationship between the i_d and the concentration is given by the Ilkovic equation:

$$i_d = 607\, nD^{1/2}M^{2/3}t^{1/6}C \tag{6}$$

where n = number of Faradays per mole of substance reduced
D = diffusion coefficient in $cm^2\ sec^{-1}$
M = rate of flow of mercury in mg./sec.
t = drop time in seconds
C = millimolar concentration of substance reduced

For a given substance, n and D will be constant and, under experimental conditions, M and t are kept constant. Hence, from equation (6), it is apparent that the concentration of the electroactive species is directly proportional to the diffusion current (i_d).

Figure 2-4 shows a typical polarogram similar to one that would be obtained

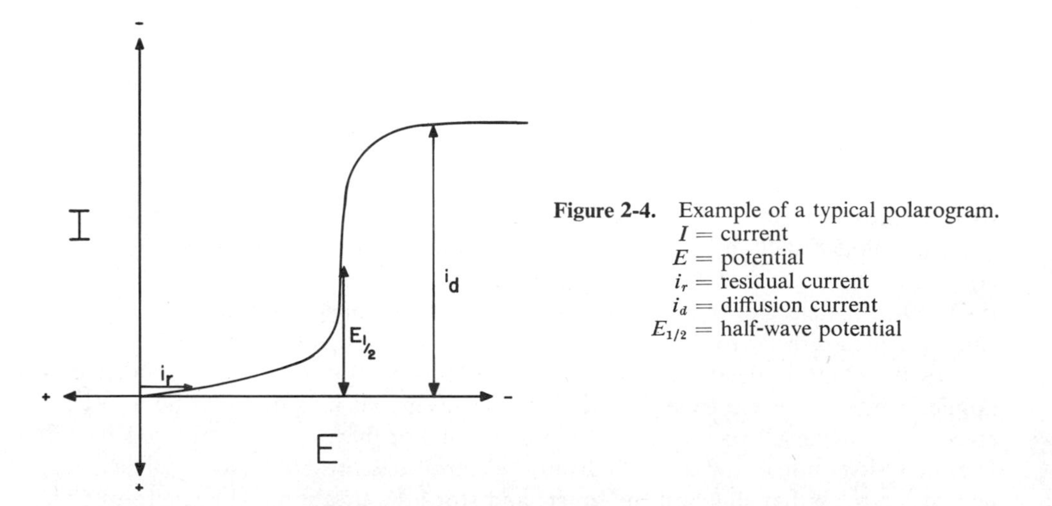

Figure 2-4. Example of a typical polarogram.
I = current
E = potential
i_r = residual current
i_d = diffusion current
$E_{1/2}$ = half-wave potential

if the amount of a reducible substance were being measured. In Figure 2-4 notice that little current flows (i_r) until the applied cathodic potential becomes large enough to reduce the substance of interest at the mercury surface. When sufficient potential is obtained, the amount of current measured increases rapidly until the plateau, where the current no longer increases with increased cathodic potential applied. The current measured at this plateau is the i_d for the system under study and is proportional to the concentration of the species being reduced.

The half-wave potential ($E_{1/2}$) as shown in Figure 2-4 may be used to identify the electroactive species in solution. Different substances frequently are reduced at various potentials. Once this potential is known or established and if substances having overlapping potentials are avoided, the $E_{1/2}$ can be used to identify certain electroactive substances.

Polarographic measurements are used in clinical chemistry laboratories to measure trace metals, oxygen, vitamin C, and amino acid concentrations. As with other electroanalytical methods, polarography is not widely used at present in clinical chemistry laboratories. The reason is probably because of the complicated, uncontrolled interferences in biological samples.

SPECTROPHOTOMETRIC AND PHOTOMETRIC MEASUREMENTS

The term photometric measurement was originally defined as making a measurement of light intensity independent of wavelength; spectrophotometric measurements refer to making light intensity measurements in a narrow wavelength range of the spectrum selected by the instrument. It has recently become common usage to refer to instruments that use filters for isolation of part of the spectrum as photometers, and to refer to instruments that use gratings or prisms as spectrophotometers. Definitions of these terms will be clarified later in this section.

Electromagnetic radiation (EMR) includes radiant energy from short wavelength gamma rays to long wavelength radio waves. Frequently, a white light source or light containing a mixture of various wavelengths is used by photometers. The wavelength of light is defined as the distance between peaks as the light travels in a wavelike manner. The distance between peaks in the ultraviolet (UV) and visible ranges is measured either in Angstroms (Å), nanometers (nm.) or millimicrons (mμ). There are 10^7 Å, 10^6 nm. or 10^6 mμ in 1 mm. A recent trend is to use nanometer (nm.) for expressing the wavelength of light. There are 10 Å/nm., and a nm. numerically equals a mμ. In addition to possessing wavelength characteristics, light also has properties that indicate it is composed of discrete energy packets called photons. The relationship between the energy of photons and their frequency is given by the equation:

$$E = h\nu \tag{7}$$

E refers to the energy in ergs when the frequency ν is given in cycles per second and h, Planck's constant, is given as 6.62×10^{-27} erg seconds. The frequency of light is related to the wavelength by an equation:

$$\nu = \frac{c}{\lambda} \tag{8}$$

where ν is the frequency in cycles per second, c is the speed of light in a vacuum

(3×10^{10} cm./sec.) and λ the wavelength in centimeters. By looking at equations (7) and (8), we can readily see that as the frequency of light increases, so does the energy. If we substitute the value of ν from equation (8) into equation (7), we obtain:

$$E = \frac{hc}{\lambda} \tag{9}$$

This equation shows that the energy of light is inversely proportional to the wavelength. For example, UV radiation at 200 nm. possesses greater energy than infrared radiation at 750 nm.

TABLE 2-1. *Characteristics of EMR Spectrum (Numbers indicate approximate lower limit for each type of radiation)*

Name	*Gamma Rays*	*X-rays*	*UV*	*Visible*	*IR*	*Micro-waves*	*Radio Waves*
Wavelength							
Angstrom (Å)	1	10	1800	3800	7000		
Nanometers (nm.)			180	380	700		
(Millimicrons) (mμ)							
Microns (μ)					0.7	400	
Centimeters (cm.)						0.04	25

Table 2-1 shows the relationship of the wavelength to the name assigned to certain areas in the electromagnetic radiation spectrum and also the relationships between the various units that are used in the measurement of wavelength. Table 2-2

TABLE 2-2. *UV-Visible Spectrum Characteristics*

Wavelength (nm.)	*Region Name*	*Color Absorbed*	*Complementary Color*
180–220	short UV	not visible	—
220–380	UV	not visible	—
380–430	visible	violet	yellow-green
430–475	visible	blue	yellow
475–495	visible	green-blue	orange
495–505	visible	blue-green	red
505–555	visible	green	purple
555–575	visible	yellow-green	violet
575–600	visible	yellow	blue
600–620	visible	orange	green-blue
620–700	visible	red	blue-green

shows similar relationships except that it is limited to the UV and visible range of the spectrum. The areas are classified as to the name of the region, its wavelength, its color, and its complementary color. If a solution absorbs all light between 435 to 480 nm. (blue), it will appear yellow to the eye. Therefore, yellow is the complementary color of blue. Similarly, if all the green color is absorbed, the solution will appear purple. The human eye responds to radiation only between about 400 and 720 nm., but modern instrumentation permits measurements at both shorter wavelength—ultraviolet (UV)—and longer wavelength—infrared (IR)—portions of the spectrum.

PRISMS FOR WAVELENGTH ISOLATION

In spectrophotometry, generally the wavelength under consideration is isolated by means of gratings or prisms.

Prisms are wedge shaped pieces of glass, quartz, sodium chloride, or some other material that allows transmission of light. Because of the variation of the refractive index with wavelength, the light that enters the prism is dispersed to varying degrees, depending upon the wavelength of the light. The red end of the spectrum is bent least by the prism, and the blue or violet end of the spectrum is bent the most as it passes through the prism.

GRATINGS FOR WAVELENGTH ISOLATION

The grating is a device that has small grooves cut into it at such an angle that each groove behaves like a very small prism. Light is reflected or transmitted from or through the grating in such a manner that light is again separated into its various components throughout the spectrum. A grating may have 3000 or more small grooves per millimeter cut into the grating surface.

SELECTION OF WAVELENGTH FOR MEASUREMENT

When a measurement is made in a spectrophotometer, the color of light that shows maximum or near maximum absorption should be passed through the solution to obtain maximum sensitivity. A blue solution absorbs red strongly, therefore, a wavelength in the red portion of the spectrum would be chosen for measurements of blue solutions. Occasionally an absorption measurement will intentionally be made at a wavelength off the absorption maximum. Although less than ideal, a reduction in sensitivity is achieved. This reduction in sensitivity will frequently linearize or extend the linear portion of a working curve. Some glucose methodologies use this approach to reduce nonlinearity.

Most analyses performed in the clinical chemistry laboratory today depend upon making measurements of the amount of light absorbed for each of the particular substances being measured. Most of the measurements are made in the visible range of the spectrum, some in the ultraviolet range, and even fewer in the infrared region; however, new instruments and improved methodology are making more and more applications practical in the field of infrared spectrophotometry. In the laboratories, infrared instruments are used to determine the composition of kidney stones and gallstones (see Chapter 17) and to analyze for toxicological substances, their metabolites, steroids, and other hormones found in biological specimens.

BEER'S LAW

Beer's law states that the concentration of a substance is directly proportional to the amount of light absorbed or inversely proportional to the logarithm of the transmitted light.

The mathematical relationship between absorption of radiant energy and the concentration of a solution is shown by Beer's law:

$$A = abc = \log\left(\frac{100}{\%T}\right) = 2 - \log \%T$$

where A = absorbance
a = absorptivity
b = light path of the solution in centimeters
c = concentration of the substance of interest
$\%T$ = per cent transmittance

This relationship is the basis for all spectrophotometric absorption measurements and results from contributions by several individuals. Lambert, Bouguer, and others independently contributed to this formula, commonly called "Beer's law."

Beer's law is an ideal mathematical relationship and therefore contains several limitations in practice. Deviations from Beer's law can result from the following conditions: (1) simultaneous absorption at multiple wavelengths; (2) absorption of light by other species; (3) measurements throughout extreme concentration ranges; and (4) transmission of light by other mechanisms. Strictly speaking, the absorptivity (a) is different for each wavelength of light. Unless the absorptivity is constant over the range of wavelengths being used, Beer's law will not be followed.

If two or more chemical species are absorbing the wavelength of light being used, each with a different absorptivity, Beer's law will not be followed. Deviations from Beer's law also occur when a very wide range of concentrations is measured. The range of concentrations that are linear with absorbance vary with each substance. Finally, if the absorption of a fluorescent solution is being measured, Beer's law will not be followed.

Figure 2-5 shows a plot of per cent transmittance versus concentration, illustrating that $\%T$ is inversely and logarithmically related to concentration. Also shown is a plot of absorbance versus concentration, in which the absorbance is directly and linearly related to the concentration of interest. From the figure, notice that the absorbance decreases by 50 per cent when the concentration (or light path) decreases by 50 per cent, but $\%T$ shows a nonlinear relationship to the same condition. Since the per cent transmittance has a reciprocal log relationship to concentration, a decrease in concentration produces a logarithmic increase in the per cent transmittance. Unfortunately, most laboratory instruments produce an electrical signal that is proportional to $\%T$. If one wants to take advantage of the linear relationship between absorbance and concentration, the $\%T$ values have to be converted to absorbance either electronically or with the aid of logarithmic scales or tables.

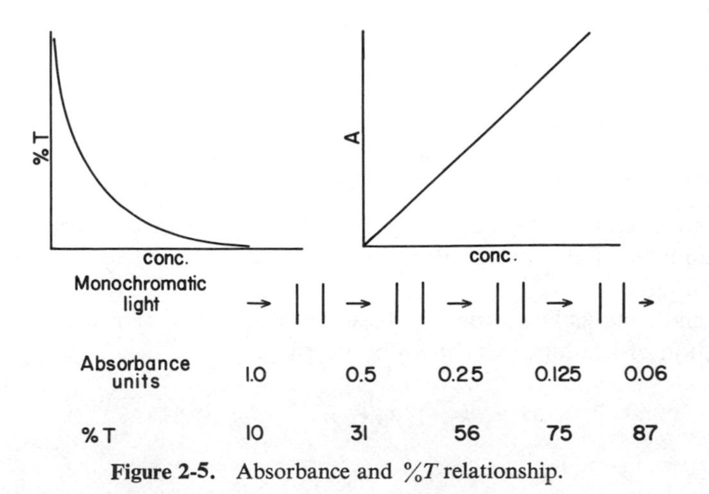

Figure 2-5. Absorbance and $\%T$ relationship.

In summary, Beer's law will only be followed if the incident radiation on the substance of interest is monochromatic, if the solvent absorption is insignificant compared to the solute absorbance, if the solute concentration is within given limits, and if a chemical reaction does not occur between the molecule of interest and another solute or solvent molecule.

Spectrophotometric measurements have gained significantly in popularity in recent years. The principle advantages of spectrophotometric measurements are relatively high levels of sensitivity, ease at which rapid measurements can be made, and a relatively high degree of specificity. Specificity is obtained by reacting the substance of interest with the proper reagents, thus producing different colors. Spectral isolation of interferences is achieved easily by the monochromators. In addition, spectrophotometric methods are widely applicable in both qualitative and quantitative analysis. Nearly all substances of interest will either absorb energy of a specific wavelength themselves or they can be chemically converted to form colored compounds that can then be measured. The specificity of many spectrophotometric procedures used in measurement for clinical chemistry is not always adequate and continuous efforts to improve methods are being made.

Components of Spectrophotometers

The top half of Figure 2-6 shows the basic components of all spectrophotometers. All spectrophotometers need a light source and an entrance slit so that the light that enters the monochromator will come from a common, well defined point. The monochromator consists of a system of prisms or gratings by which light is resolved into the various wavelengths. The exit slit is used to control the size of the beam (incident light) that passes to the analytical cell or cuvet. The analytical cell or cuvet (to be discussed later in the chapter) is a glass, quartz, or plastic container that holds the solution whose absorption is to be measured. The detector is the module that measures the intensity of light that passes through the cuvet (emergent light). The output of the detector is related to the concentration of the substance of interest.

THE LIGHT SOURCE

The function of the light source is to provide radiant energy in the form of visible or nonvisible light that may be passed through the monochromator to be separated into discrete wavelengths. The light of the proper wavelength is then made to be incident on the analytical cell that is holding the solution whose absorption is to be measured.

In the conventional spectrophotometer commercially available 15 to 20 years

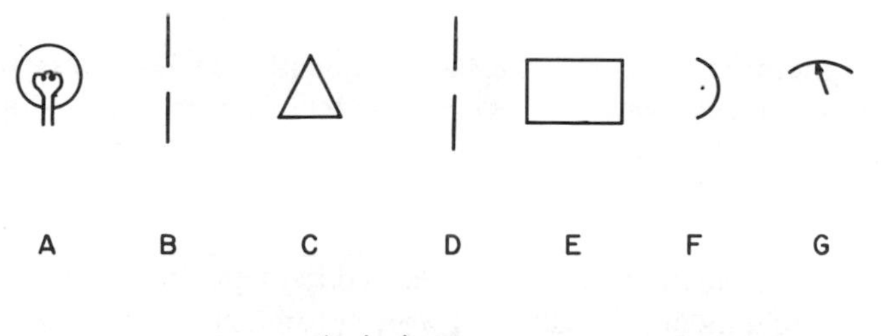

Figure 2-6. Components of single beam spectrophotometers. (A) Light source, (B) entrance slit, (C) monochromator, (D) exit slit, (E) cuvet, (F) detector, and (G) meter.

ago, a tungsten light bulb was usually the only source of visible light. Today other sources are commonly used and these will be mentioned later in this section. The tungsten bulb was and is acceptable for making measurements in moderately dilute solutions in which the difference in color intensity varies significantly with small changes in concentration. A common disadvantage with some early photometers is that a considerable amount of electrical energy is necessary to maintain a constant high energy output from such a light bulb. As a result, significant heat is generated by the light source and may cause measurement problems. The generated heat may change the geometry of the optical system as well as change the sensitivity of the detectors. The thermal change can shift the optics (lenses) so that a different wavelength of light is incident on the cuvet between the standardization and analysis steps. This wavelength change or the sensitivity change of the detectors may produce significant errors.

A tungsten light source does not supply sufficient radiant energy for measurements in the ultraviolet region below 320 nm. For this purpose a mercury arc light is suitable since the amount of energy supplied is high over the entire ultraviolet range. Recently a deuterium light source has come into use and although it does not possess as much intensity as the mercury arc, it has approximately three times as much intensity as a hydrogen lamp.

Quartz iodide sources are now frequently used light sources for visible and UV measurements and are mounted in a quartz envelope so that they may also be used in the ultraviolet region of the spectrum. These sources are high intensity and long lasting. They will frequently operate 2000 to 3000 hours before replacement is necessary.

THE ENTRANCE SLIT

The function of the entrance slit is to minimize stray light and prevent scattered light from entering the monochromator system. Stray light must be excluded from passing through the cuvet, otherwise Beer's law will not be followed and significant errors may be introduced, depending upon how compensations for such design deficiencies are handled.

MONOCHROMATORS

The function of the monochromator in spectrophotometers is to isolate specific wavelengths of light emitted by the source. This is achieved by the use of prisms or gratings, or both, in spectrophotometry. The prism or grating may be tilted or rotated in the light beam in order to permit the proper wavelength to be incident upon the cuvet and the detector. Still other spectrophotometers, like the Coleman, obtain the proper wavelength of light incident on the cuvet by moving the light source. In visible spectrophotometry, glass prisms are frequently used, but quartz is required for ultraviolet region measurements. Prisms made of sodium chloride or potassium bromide were frequently used in infrared spectrophotometers before gratings became popular. Some spectrophotometers designed years ago, such as the Beckman DU, contain prisms and certain high quality, high performance UV-visible recording spectrophotometers, such as the Cary, also contain prisms today. In many of the medium price range, medium quality spectrophotometers, such as the Beckman DBG and the Bausch & Lomb 505, the prisms have been replaced with gratings. Detailed descriptions of how gratings are made are available elsewhere in the literature;

therefore only a brief discussion will be included here. Generally, a master grating is made by carefully controlled etching or machining of grooves into a metal surface. A thin layer of aluminum is then deposited on the master grating and an optically flat surface is glued to the aluminized grooves deposited in the master grating. When hardened, the replica grating is removed from the master grating and performs nearly as well as an original. Such a grating frequently costs 1/100 or less than the master grating and performs adequately. Gratings are inexpensive and can increase the amount of energy available by using reflection instead of transmittance techniques. Light must always pass through a prism, but a grating may be used in either reflectance or transmittance modes of operation.

INTERFERENCE FILTERS

Filter photometers use conventional filters (see under Gelatin Filters) or *interference filters* instead of prisms or gratings to obtain spectral purity. Interference filters are made by depositing thin semitransparent silver films on each side of a dielectric such as magnesium fluoride. A dielectric is an insulating material that does not allow electric current to flow. When light, perpendicular to the silvered surface, enters the interference filter, it will pass through the dielectric and be reflected from the second silver surface back through the dielectric to the first silver layer to be reflected again. Finally, the light is transmitted through the semitransparent silver film and into the photometer. Constructive and destructive interference will occur as the light is reflected between the transparent silver films. Constructive interference will occur only when the wavelength of the light is equal to the thickness of the magnesium fluoride layer or a multiple of that thickness. Interference filters will allow transmission of a range of wavelengths between 10 and 20 nm. wide and will allow 40 to 60 per cent of incident light to be transmitted. The band of wavelengths allowed to pass is called the band pass. The thickness of the magnesium fluoride can be carefully controlled and by varying this thickness filters that transmit different wavelengths are obtained. Multilayer interference filters can be prepared by depositing several layers of dielectrics on each other, each of which is a fraction of a wavelength thick. Multilayer interference filters such as this will have a band pass of 5 to 10 nm. and will allow 60 to 95 per cent of the incident light to be transmitted. Interference filters are inexpensive individually, but several sets are required to work at different wavelengths. The Technicon AutoAnalyzer colorimeter contains interference filters of this type with individual specifications on the filters. For example, the identification number 530 on a filter means that the peak transmittance occurs at 530 nm. A second number, e.g., 18, that appears below the first number refers to the band pass transmitted by that filter at one-half the height of maximum transmission.

GELATIN FILTERS

Less expensive photometers may use gelatin or colored glass filters in order to obtain some spectral resolution. The colored gelatin is layered between two plates of glass. The gelatin filter can be used to take out the color of light that is the complement of the color of the gelatin. Similarly, colored glass can be prepared to function the same way as the gelatin filters. Glass and gelatin filters permit only a small percentage of incident light to be transmitted and have relatively wide band passes of up to 50 nm.

ANALYTICAL CELL OR CUVET

The function of the cuvet is to hold the solution in the instrument while its absorption is being measured. Cells are made of soft or borosilicate glass, quartz, or plastic. The soft glass cells are used for solutions that are acidic and do not etch glass. Strongly alkaline solutions should be measured in borosilicate cells because of their higher resistance to alkali. As soon as measurements are completed, alkaline solutions must be rinsed from the cells. Glass cells are unsuitable for measurements in the short ultraviolet region of the spectrum. Some glass cells (Corex) can be used to make measurements at 340 nm. Only quartz or plastics that do not absorb ultraviolet radiation can be used for measurements at wavelength below 320 nm. One way to remember to use quartz for ultraviolet measurements is to recall that it is impossible to tan one's skin in sunlight that has passed through a glass window. Tanning of the skin is achieved by the ultraviolet radiation from the sun and ultraviolet radiation is absorbed by ordinary window glass. Recently, some plastic materials have been developed that show little or no absorption of radiant energy from 200 to 700 nm. Generally, these cells are inexpensive and in some cases disposable and will most certainly find increased usage in the near future.

Common errors in handling cuvets are failure to position the cell in the photometer properly and failure to match absorbance readings of the cells. When round cuvets are used, they should be marked near the top and always positioned in a predetermined manner. If inexpensive unmatched cells are used, blank readings should be taken to measure the tolerance of each cuvet at each wavelength used.

DETECTORS

The two most commonly used devices for measuring light intensity in the UV and visible regions of the spectrum are barrier layer cell and photomultiplier tubes. The barrier layer cells are rugged and are used in inexpensive instruments; photomultipliers are almost always used in the higher quality, more expensive spectrophotometers.

Barrier layer cells. These cells operate on the principle that when light falls on certain metals or semiconductors, electrons will flow in proportion to the intensity of the light. The barrier layer cell consists of a thin layer of silver on a layer of the semiconductor selenium. The silver and selenium metals are then mounted on an iron backing or support. The iron backing is deficient in electrons and is therefore the positive electrode. The silver mounted on top of the selenium where the light will be incident is the negatively charged electrode. When light falls on the thin semi-translucent silver metal, electron flow from the semi-conductor selenium into the iron backing occurs, but not in the reverse direction. The electron flow is measured and is directly proportional to the light intensity incident on the photocell.

The sensitivity to wavelength of the barrier layer cell is similar to the human eye. The maximum sensitivity of both occurs at 550 nm. Barrier layer cells are usually used at high levels of illumination and the output from these photocells is generally not amplified. Barrier layer cells are very stable, but are slow in responding to changes of light intensity. Because of their slow response time, they are not suitable as detectors in instruments that employ interrupted (chopped) light beams falling on the detectors. Another disadvantage of this photocell is that it tends to show fatigue. Fatigue occurs in a barrier layer cell when, at a constant extremely high level of intensity, the electrical output of the photocell decreases with time. Therefore, barrier layer cells should not be used at extremely high illumination.

The Coleman II spectrophotometer and the Technicon AutoAnalyzer interference filter colorimeter contain photocells of this type. The cells are very rugged, last for years, and perform well as inexpensive detectors. A potential problem with barrier layer cells is that their electrical output is very temperature dependent. If a heat producing, high intensity lamp or flame is used for the light source, instrument design must be such that thermal stability and rapid temperature equilibrium of the photocell is achieved. This is accomplished in the AutoAnalyzer colorimeter by having the light source far removed from the photocell. Use of heat shields or plastic materials that do not conduct heat readily are other ways of improving thermal stability of instruments.

Photomultiplier tubes. A photomultiplier is an electron tube that is capable of significantly amplifying a current. It is constructed by using a light sensitive metal as the cathode that is capable of absorbing light, and emitting electrons in proportion to the radiant energy that strikes the surface of the light sensitive material. The electrons produced by this first stage go to a secondary stage (surface) where each electron produces between four and six additional electrons. Each of these electrons from the second stage go on to another stage again producing four to six electrons. Each electron produced cascades through the photomultiplier stages, thus, the final current produced by such a tube may be one million times as much as the initial current. As many as 10 to 15 stages or dynodes are present in common photomultipliers.

When operating such a tube, voltage is applied between the photocathode and each successive stage. The normal increment of voltage increase of each photomultiplier stage is from 50 to 100 volts larger than that of the previous stage. A common photomultiplier tube will have in the neighborhood of 1500 volts applied to it.

Photomultiplier tubes have extremely rapid response times, are very sensitive, and do not show as much fatigue as other detectors. Because of their excellent sensitivity and rapid response, all stray light and daylight must be carefully shielded from the photomultiplier. A photomultiplier with the voltage applied should never be exposed to room light because it will burn out. Because of the fast response time of the photomultiplier, this detector is applicable to interrupted light beams, such as those produced by choppers and thus provides significant advantages when used as a UV-visible detector in spectrophotometers. The rapid response times are also needed when a spectrophotometer is being used to determine an absorption spectrum of a compound. The photomultiplier also has adequate sensitivity over a wider wavelength range than photocell detectors.

When voltage is applied to photomultipliers and all light has been blocked from them, some current will usually be produced. This current is called dark current. It is desirable to have the dark current of photomultipliers at their lowest level since it would also be amplified and would appear as background noise.

Double Beam Spectrophotometers

There are several optical and electrical configurations used in commercially available spectrophotometers. Each configuration has advantages for certain applications.

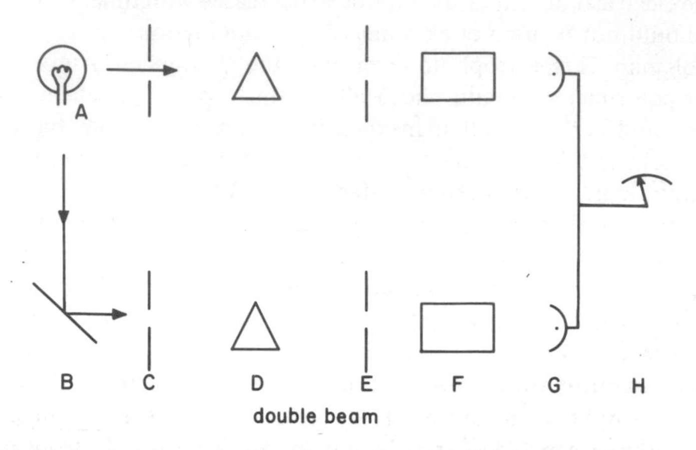

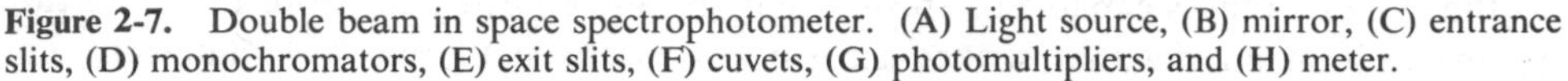

Figure 2-7. Double beam in space spectrophotometer. (A) Light source, (B) mirror, (C) entrance slits, (D) monochromators, (E) exit slits, (F) cuvets, (G) photomultipliers, and (H) meter.

Double beam instruments have been classified as double beam in space or double beam in time. Figure 2-7 shows a double beam system in space and Figure 2-14 shows a double beam in time instrument. Notice, in the instrument shown in Figure 2-7 that all components are duplicated except the light source. The two beams pass at the same time through different components separated in space. This arrangement would compensate for changes in intensity of the light source and also compensate for changes in absorptivity of the reagent blank as the wavelength is changed in a scanning operation.

A double beam instrument in time usually uses the same components as a single beam instrument. The two beams pass through the same components but not at the same time. Duplication of cuvet compartments is sometimes used. A light beam chopper (a rotating wheel with alternate silvered sections and cutout sections) is inserted after the exit slit. A system of mirrors would pass the reflected portion of the light off the chopper through a reference cuvet and then onto the common detector. Just as a single beam instrument is adjusted to 100%T with the blank before and between sample readings, the double beam system makes these adjustments automatically. The detector, as in Figure 2-14, is made to look alternately at the sample and then the reference beam of light. The difference or ratio of the timed signals is then amplified and is proportional to the substance of interest in the sample cuvet. The double beam in time approach using one detector compensates for light source variation as well as for sensitivity changes of the detector.

Various other combinations of components and parts of the two approaches just presented have been used in double beam spectrophotometers. Usually the design is conceived to solve a specific problem.

Although more expensive, double beam instruments provide increased quality measurements. Some double beam instruments use a recorder for the output. The recorder traces a plot of the absorbance or %T versus time or wavelength as the operator desires. The recording double beam spectrophotometer has its greatest advantage when scanning the spectrum. It automatically compensates for absorbance changes of the blank and for variations in the intensity of the light incident on the cells when scanning.

Selection of an Instrument for Photometry

The most important consideration in the selection of an instrument for spectrophotometric or photometric analysis is the use to which the instrument is to be put. If a high precision scanning instrument is required, a high quality recording double beam spectrophotometer is needed. If, on the other hand, it is necessary to measure changes in concentration at a limited number of wavelengths when it is not critical to have spectral purity and wavelength isolation, an instrument such as a Bausch & Lomb Spectronic 20, a Coleman II, or a Coleman Jr. Model 6C would work well. Such instruments are inexpensive and suffice when large changes in intensity per unit concentration are present.

If a UV spectrophotometer is to be used for measuring a barbiturate, it is desirable to have a double beam recording scanning instrument for such a determination. The advantage of a double beam instrument is that a continuous correction for optical errors or deficiencies can be made as the wavelength changes automatically. In the case of the barbiturate determinations, if a known negative serum sample is inserted in the reference beam and the sample to be analyzed is placed in the sample beam, the instrument automatically makes the correction for unwanted extracted substances and presents to the operator a corrected peak that is directly related to concentration without further calculations. This technique of using a serum sample in the reference beam of double beam UV-visible recording spectrophotometers also has the advantage of increasing the sensitivity for all measurements of barbiturates.

In the use of single beam colorimeters, such as the Bausch & Lomb Spectronic 20, the Coleman II and the Model 6A, the method of *ultimate precision* may be applied. This is accomplished by closely bracketing the unknown solution with two solutions of known concentration, then adjusting the darker colored solution to read zero transmittance and the lighter colored solution to read 100 per cent transmittance. The per cent transmittance of the unknown is read and the result obtained by interpolation. The maximum sensitivity and smallest error in measurement is achieved by using this method of ultimate precision.

Figure 2-8 is an example of an absorption spectrum, showing the plot of absorbance versus wavelength. If a single well defined absorption spectrum is obtained, the amount of absorption that occurs at the peak wavelength (B of Fig. 2-8) is to be preferred over an absorption measurement on the downslope (A of Fig. 2-8) of the peak. An increase in sensitivity and specificity results from a measurement at peak absorption; however, if interferences are present, it may be necessary, although less

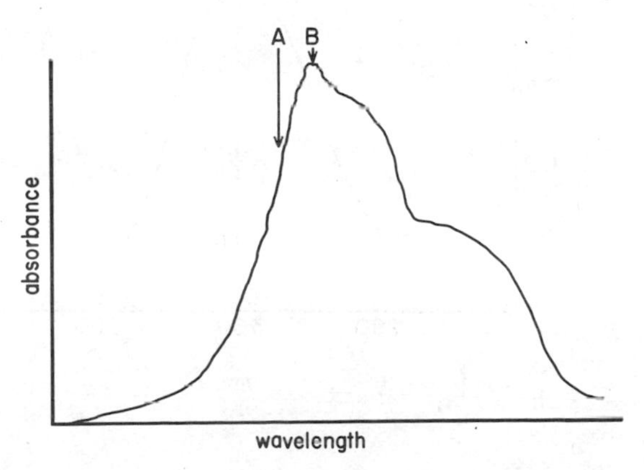

Figure 2-8. Example of an absorption spectrum (see text).

desirable, to make measurements of absorption off the peak wavelength and relate this measurement to concentration. If the absorption peak is sharp or if a shoulder wavelength is selected for absorbance readings, great care must be used in adjusting the proper wavelength each time the instrument is used. Shoulder operation can easily introduce large measurement errors and should be avoided.

Occasionally by not measuring at the peak absorption, the linear working range of a method can be expanded. This occurs because of the reduction in sensitivity and is frequently used in the cholesterol and glucose methods.

A method to correct for background interferences is to measure absorbance at the peak wavelength and at two other wavelengths, usually equidistant from the peak. Values for the latter are averaged to obtain a baseline under the peak, which is then subtracted from the peak reading. The value thus obtained is known as a "corrected" absorbance and can be related to the concentration provided that the background absorbance is linear with wavelength over the region in which readings are made. This technique of making corrections for interfering substances is called the *Allen correction* and is illustrated in Figure 2-9.

Salicylate extracted from acidified serum, for example, shows a peak absorbance at 300 nm. (see curve A of Fig. 2-9). An extract of salicylate-free serum also exhibits appreciable absorbance at this wavelength, but the absorbance is linear between 280 and 320 nm. (line B of Fig. 2-9). The corrected absorbance at 300 nm. is obtained from the Allen equation:

$$A_{\text{corr}} = A_{300} - \frac{(A_{280} + A_{320})}{2}$$

Similar corrections are applied in procedures for spectrophotometric determinations of porphyrins, steroids, and other compounds.

Before using the Allen correction, knowledge of the shape of the absorption curve for the substance of interest and of the interferences must be obtained. The linearity of the baseline shift should be verified by measuring the absorption spectrum of commonly encountered interferences. Care should be exercised in the use of the Allen correction because if it is not properly used, it may introduce larger errors than would be observed without correction. For example, such a situation may occur if the background reading is not linear over the region measured.

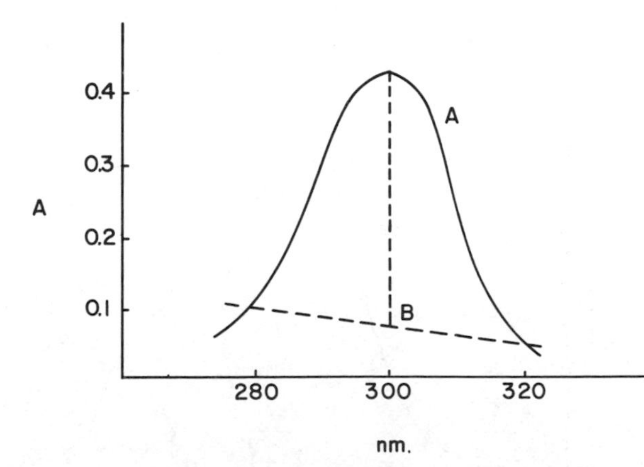

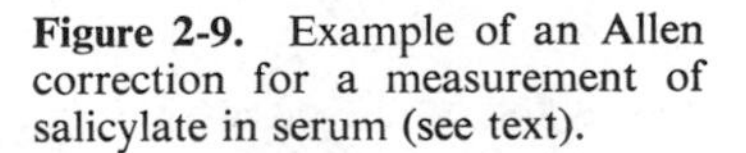
Figure 2-9. Example of an Allen correction for a measurement of salicylate in serum (see text).

FLAME PHOTOMETRY

The flame emission spectrophotometer is one of the earliest and most important instruments in clinical chemistry laboratories. It offers one of the most convenient, accurate, and precise measurements made in laboratories today. The basis of the measurement was illustrated in preliminary chemistry courses where qualitative analysis was performed using the flame spot test in which the color of a burning flame is different for certain cations.

The principle behind flame photometry involves the excitation of electrons in an atom by the heat energy of a flame. The electrons, being unstable in this excited stage, then give up their excess energy to the environment as they change from the higher energy state (excited) to a lower energy state. If the energy is dissipated as light, the light may consist of one or more than one energy level and therefore may possess different wavelengths. These different wavelengths or lines are the spectra of the atoms and are individually characteristic for each element. The wavelength to be used for the measurement of an element—as in spectrophotometry—depends upon the selection of a line of strong enough intensity to provide adequate sensitivity. It also depends upon freedom from other interfering lines at or near the selected wavelength.

Alkali metals are comparatively easy to excite in the flame of an ordinary laboratory burner. Lithium produces a red, sodium a yellow, potassium a violet, rubidium a red, and magnesium a blue color in a flame. These colors are characteristic of the metal atoms that are present as cations in solution.

Under constant and controlled conditions the light intensity of the characteristic wavelength produced by each of the atoms is directly proportional to the number of atoms that are emitting energy, which in turn is directly proportional to the concentration of the substance of interest in the sample. Thus, flame photometry lends itself well to direct concentration measurements of some metals.

Other cations, like calcium, are less easily excited in the ordinary flame. In these cases, the amount of light given off may not always provide adequate sensitivity for analysis by flame emission methods. The sensitivity can be improved slightly by using higher temperature flames. Of the more easily excited alkali metals like sodium, only 1 to 5 per cent of those present in solution become excited in a flame. Even with this small percentage of excited atoms, the method has adequate sensitivity for measurement of alkali metals for most bioanalytical measurements. Most metal ions are not as easily excited in a flame and flame emission methods are not as applicable for their measurement.

Essential Parts of the Flame Photometer

Figure 2-10 shows a schematic diagram of the basic parts of a flame photometer. A supply of gases, two-stage pressure regulators, and high pressure tubing must be used to lead the gases to the flame. An atomizer is needed to spray the sample as fine uniform droplets into the flame (see also Chapter 10). The monochromator entrance and exit slits and detectors are similar to those previously discussed in the spectrophotometer section.

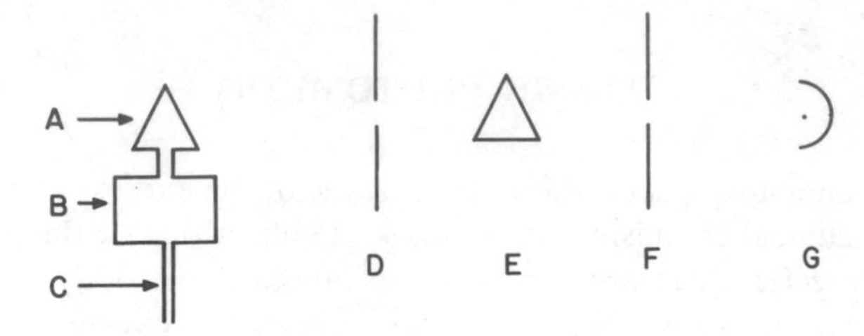

Figure 2-10. Essentials of a flame photometer. (A) Flame, (B) atomizer, (C) aspirator, (D) entrance slit, (E) monochromator, (F) exit slit, and (G) detector.

GASES FOR FLAME PHOTOMETRY

Various combinations of gases and oxidants have been proposed and are being used in flame photometers. A mixture of hydrogen and oxygen gas recommended for the Beckman DU flame photometer currently produces the hottest temperature flame commonly used in conventional flame photometers. In addition natural gas, acetylene, and propane with either air or oxygen are other combinations of gases frequently used. All these fuel gases and their various oxidants work well and the only real difference exists in the flame temperature and, therefore, the sensitivity that each combination provides. It is essential that the flame temperature be held constant, otherwise sensitivity changes will result. High quality gas regulation to maintain constant flame temperature is also essential for proper operation.

THE ATOMIZER

The atomizer and the flame are the two most critical components in a flame photometer. The function of the atomizer is to break up the solution into fine droplets so that the atoms will absorb heat energy from the flame and become excited. There are basically two types of atomizers commonly used in flame photometers. In one type of burner, the gases are passed at high velocity over the end of a capillary suspended in the solution, causing liquid to be drawn up the capillary into the flame. This type of burner is called the total consumption burner and details of it can be found in the section on atomic absorption. A second kind of burner involves the gravity feed of solution through a restricting capillary into an area of high velocity gas flow where small droplets are produced and passed into the flame. The large droplets in this type of burner are usually taken to waste and not all the sample is forced to go into the flame as in the capillary type burner.

THE FLAME

The purpose of the flame is to transfer energy to the unexcited atoms as just explained. The single most important variable of the flame is the temperature. Frequent standardization of flame photometers is essential because thermal changes do occur and affect the response of the flame photometer.

MONOCHROMATOR

The monochromators, including the entrance and exit slits, are similar to those previously described for spectrophotometers. Their function is to isolate the wavelength of interest from interfering light before it passes on to the detector. Ideally, monochromators in flame photometers should be of higher quality than those found in absorption spectrophotometers. When nonionic materials are burned, light of varying wavelength is given off. This is known as *continuous emission* and will be added to the *line emission* of the element being measured. For this reason, the narrowest band path that is achievable should be used to eliminate as much of the extraneous, continuous emission as is possible, but still permit a maximum of the line emission to pass through to the detector.

DETÈCTORS

The detectors used in flame photometers operate by the same principle and in the same way as those previously described in the spectrophotometry section. In designing a flame photometer, compensation or design features must be incorporated so that thermal equilibrium is achieved rapidly. Temperature change severely affects the output of photocell detectors and is an inherent source of error in flame photometry. Flame photometers that use photomultipliers for detectors have improved sensitivity and, because of improved design, seldom require long times to come to thermal equilibrium. Even this type of flame photometer, however, usually requires aspirating of water and standards for establishing flame thermal equilibrium before measurements are taken.

Operation of Flame Photometers

The major problems associated with flame photometers involve inadequate control over the flame and aspirator. Slight variations in gas pressure will change both the rate of aspiration of the sample and the temperature of the flame. A significant amount of design effort has been exerted to assure constant flame and aspirator conditions. The Perkin–Elmer Corporation (South Pasadena, Calif.) and Instrumentation Laboratories (Boston, Mass.) designed flame photometers that use an internal standard of lithium. In these instruments a single flame and multiple detectors to monitor the same flame are employed. The ratio of the sample and reference (lithium) detectors is proportional to the sample concentration. Therefore, any change in flame characteristics and aspirator conditions would simultaneously affect the signal to both the lithium reference detector and the sample detector. By using the ratio of the two detectors, errors due to flame fluctuations or changes in the aspiration rate are minimized.

Instrumentation Laboratories has produced a solid state, direct reading flame photometer for analysis of sodium and potassium that also uses an internal standard of lithium. In the IL flame photometer, the ratio between the lithium and sodium and the lithium and potassium channels is taken, amplified, and fed to a servomotor that produces a direct digital readout displayed on the front of the instrument. Lithium also acts as a *radiation buffer*. If potassium is measured, for example, the potassium signal is critically dependent upon the amount of sodium present unless a high concentration of another easily excited cation such as lithium is present. In the absence of a high concentration of lithium, energy will be transferred from an excited sodium atom to a potassium atom. This would produce a different percentage of excited potassium atoms, depending upon the sodium concentration and, therefore, analytical errors would result. A means of compensating for this error is to dilute the samples with an excessively high concentration of lithium so that the same percentage of potassium becomes excited regardless of the sodium concentration in the sample. The use of an internal standard and radiation buffer, the direct readout for sodium and potassium concentrations, and the simple 200-fold dilution of serum make the IL flame photometer highly suitable for use in the clinical chemistry laboratory.

ATOMIC ABSORPTION SPECTROPHOTOMETRY

The principle of atomic absorption had been known for a century before a useful analytical application evolved. Allan Walsh, an Australian physicist, suggested the idea as an analytical procedure in 1955. Soon after 1955, the production of commercial equipment resulted in a rapid expansion of applications. In February, 1965, less than 500 units were in use in the United States. The number in February, 1966 was over 1000 units and an estimate made in early 1968 indicated that over 5000 atomic absorption units are in operation in the United States.

TABLE 2-3. *Metal Ions Detectable Below 1 ppm by Flame Emission or Absorption Methods*

—																	—
Li	Be											—	—	—	—	—	—
Na	Mg											Al	Si	—	—	—	—
K	Ca	—	Ti	V	Cr	Mn	Fe	Co	Ni	Cu	Zn	Ga	Ge	As	Se	—	—
Rb	Sr	—	—	—	Mo	—	Ru	Rh	Pd	Ag	Cd	In	Sn	Sb	Te	—	—
Cs	Ba	—	—	—	—	—	—	—	Pt	Au	Hg	Tl	Pb	Bi	—	—	—
—	—	—															

The reason for this rapid growth in atomic absorption spectrophotometry is that instrumentally, the most advanced atomic absorption spectrophotometer is very simple compared to analytical tools such as mass spectrometry, neutron activation analysis, x-ray fluorescence, and arc or laser emission spectroscopy. Atomic absorption spectrophotometry has essentially no spectral interferences and only a few chemical interferences, frequently making preliminary separations unnecessary. Table 2-3 shows that a large number of elements have been detected in concentrations below 1 ppm by flame emission or flame absorption methods.

Atomic absorption spectrophotometry is basically the inverse of emission methods. In all emission methods, arc, spark, laser, flame, x-ray fluorescence, or neutron activation analysis, the sample is excited in order to measure the radiation energy of interest given off as the sample returns to its lower energy level. Extraneous radiation must be filtered out from the energy of interest if interference by these signals is to be avoided.

In atomic absorption spectrophotometry the element is not excited in the flame, but merely dissociated from its chemical bonds and placed in an unexcited, nonionized quantum mechanical ground state. This statement means that the atom is at a low energy state in which it is capable of absorbing radiation at the very narrow band pass width of from 0.01 to 0.1 Å. The source emitting such radiation is the hollow cathode lamp. The energy of the absorbed radiation is equal to that which is emitted if the element in question were excited. Although only a small percentage of atoms in the flame is excited (1 to 5 per cent), nearly all are converted to the dissociated form in which they are capable of absorbing light emitted by the hollow cathode lamp. As a general rule, the sensitivity of atomic absorption is approximately 100 times greater than that of flame emission methods.

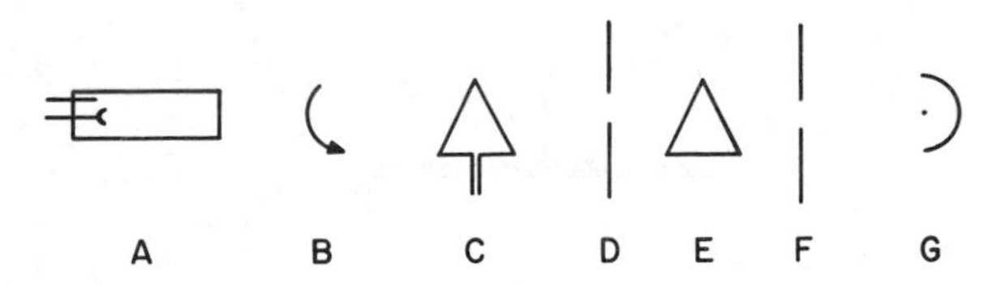

Figure 2-11. Essentials of an atomic absorption instrument. (A) Hollow cathode, (B) chopper, (C) flame, (D) entrance slit, (E) monochromator, (F) exit slit, and (G) detector.

Components of an Atomic Absorption Spectrophotometer

Figure 2-11 shows the basic components of an atomic absorption spectrophotometer. The hollow cathode lamp is the light source, the nebulizer (atomizer) sprays the sample into the flame, and the monochromator, the slits, and the detectors have their usual function as described previously in the spectrophotometer sections.

THE HOLLOW CATHODE LAMP

Figure 2-11 also shows the components of a hollow cathode lamp. The cathode is made of the metal of the substance to be analyzed and is different for each metal analysis. The hollow cathode lamp produces a wavelength of light specific for the kind of metal in the cathode. In some cases, an alloy is used to make the cathode; then a multielement lamp results.

Neon or argon gas at a few millimeters of mercury pressure is usually used as a filler gas. A neon filled lamp during operation produces a reddish orange glow, and the argon produces a blue to purple glow inside the hollow cathode lamp. Quartz or special glass that allows transmission of the proper wavelength is used as a window.

A current is applied between the two electrodes inside the hollow cathode lamp and metal is sputtered from the cathode into the gases inside the glass envelope. When the metal atoms collide with the gases, neon or argon, they lose energy and emit their characteristic radiation. Calcium has a sharp, intense, analytical emission line at 4227 Å, which is most frequently used for calcium analysis. In an interference-free system only calcium atoms will absorb the calcium light from the hollow cathode as it passes through the flame.

THE BURNER

Until now, only two kinds of burners have been used in most clinical applications. One is a *total consumption burner*, as illustrated by the Beckman burner and shown in Figure 2-12. With this burner, the gases, hydrogen and air, and the sample are *not* mixed before entering the flame. One disadvantage is that relatively large droplets are produced in the flame and cause signal noise by light scattering. Also, the amount of acoustical noise produced is very high, and may become uncomfortable after a few hours of operation. An advantage of this type of burner is that the flame is more

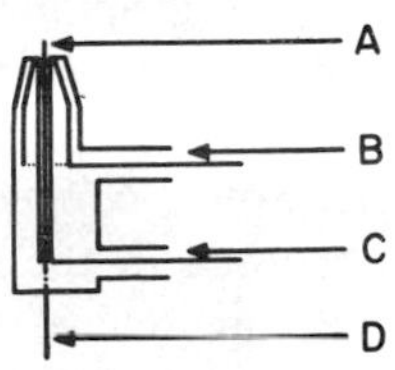

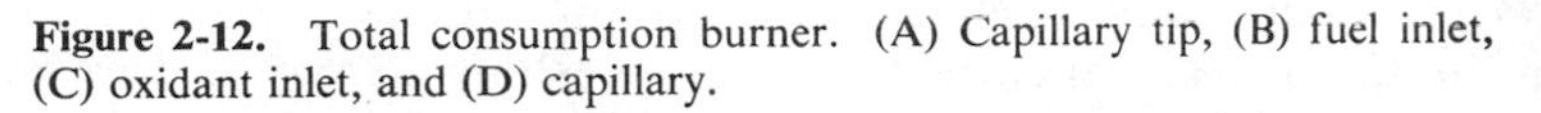
Figure 2-12. Total consumption burner. (A) Capillary tip, (B) fuel inlet, (C) oxidant inlet, and (D) capillary.

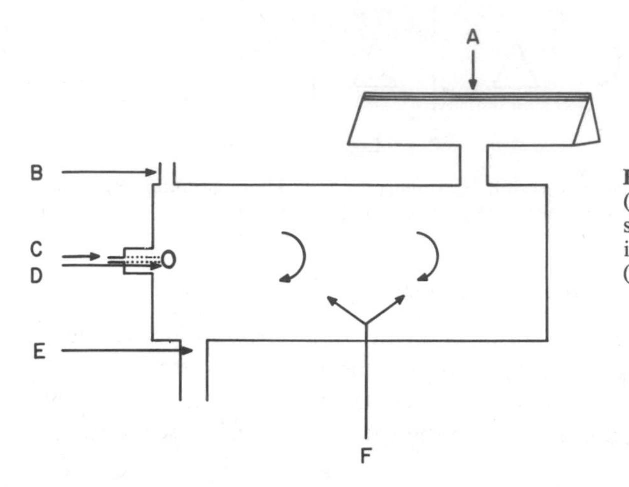

Figure 2-13. Laminar flow burner. (A) Boling head, (B) fuel inlet, (C) sample capillary and nebulizer (atomizer), (D) oxidant inlet, (E) drain, and (F) spoilers.

concentrated and it can be made hotter, causing molecular dissociations that may be desirable for some chemical systems.

Figure 2-13 shows a *premix burner* (laminar flow burner) and illustrates how the sample is aspirated, volatilized, and burned. Notice the gases are mixed and the sample atomized *before* being burned. An advantage of this system is that the large droplets go to waste and not into the flame, thus producing a less noisy signal. Another advantage is that the path length of the burner is longer than the total consumption burner. This produces greater absorption and increases the sensitivity of the measurement. A disadvantage of the premix burner is that the flame is usually not as hot as that of the total consumption burner, and it cannot sufficiently dissociate certain metal complexes in the flame (e.g., calcium phosphate complexes). Recently, nitrous oxide premix burners have been developed by Willis, an Australian chemist. These burners produce higher temperatures and will dissociate some calcium complexes. This makes the addition of competing cations (e.g., lanthanum or strontium) to the solutions unnecessary; however, an error is introduced into the calcium determination by the use of the high temperature nitrous oxide burner. Calcium becomes excited to a significant extent and emits in the flame. Thus, the use of this burner produces a new problem that must be solved and further evaluations must be conducted to determine the usefulness for the clinical chemistry laboratory.

Another type of burner, called the triflame burner, possesses the best characteristics of both the total consumption burner and the premix burner. Evaluation of this burner for clinical chemistry applications is incomplete.

MONOCHROMATOR AND DETECTOR

All atomic absorption systems use monochromators and photomultipliers for, respectively, isolating a spectrally pure light signal and measuring the intensity of that signal. The monochromator filters out extraneous light from the flame, and the photomultiplier converts that part of the light from the hollow cathode that was not absorbed in the flame to an electrical current and amplifies this current to drive a readout device or recorder. The monochromators and the photomultipliers have been discussed previously in the spectrophotmoetric section of this chapter.

Interferences in Atomic Absorption Spectrophotometry

There are three general types of interferences in atomic absorption spectrophotometry—chemical, ionization, and matrix effects.

Chemical interference refers to the situation when the flame cannot dissociate the sample into free atoms so that absorption can occur. An example of this is the phosphate interference in the determination of serum calcium caused by the formation of calcium phosphate complexes. These calcium phosphate complexes do not dissociate in the flame unless a special high temperature burner is used. The phosphate interference is overcome by adding a cation that will compete with calcium for the phosphate. Usually in determinations of calcium in serum, lanthanum or strontium is added to the serum and replaces and releases the calcium from the phosphate in solution. The free calcium is then capable of absorbing the calcium light from the hollow cathode. The freeing of calcium occurs because lanthanum and strontium form more stable complexes with phosphate than does calcium.

Ionization interference results when atoms in the flame become excited instead of only dissociated and then emit energy of the same wavelength that is being measured. Compensation for this condition can be achieved by adding an excess of a more easily ionized substance that will absorb most of the flame energy so that the substance of interest will not become excited. Another way to correct for ionization interference is to operate the flame at a lower temperature.

A third type of interference is the *matrix interference*. One example of a matrix effect is the enhancement of light absorption by organic solvents. An atom may absorb between two and five times more energy when dissolved in an organic instead of an aqueous solvent. A second kind of matrix effect is the light absorption caused by formation of solids from sample droplets as the solvent is evaporated in the flame. This will usually occur only in concentrated solutions of greater than 0.1 *M*. Refractory oxides of metals formed in the flame can also be classified as matrix interferences.

Commercial Atomic Absorption Spectrophotometers

A number of different instrument companies are now producing atomic absorption spectrophotometers. Perkin–Elmer markets models 403, 303, and the 290, and Beckman has developed the model 979. Instrumentation Laboratories and Jarrell-Ash (Waltham, Mass.) are two other manufacturers of atomic absorption spectrophotometers. The Perkin–Elmer Model 290 was designed principally for the clinical chemistry laboratory and similar applications. Several investigators have reported successful application of this instrument to calcium and magnesium determinations in serum and urine. The single beam instrument design characteristics of the Perkin–Elmer 290 provide one light path and, therefore, the instrument makes a direct measurement. Figure 2-14 shows the schematic diagram of the Perkin–Elmer Model 303. If one traces the light path in Figure 2-14 from the hollow cathode lamp, the chopper either reflects the light beam or allows the light beam to pass alternately between the reference beam in the rear and the sample beam in the front portion of the figure. The beams then pass alternately through the same monochromator to the one detector. When the beams arrive at the detector, they are out of phase with each other. In other words, the photomultiplier looks first at the reference beam and then uses this value to compare the reference and the sample beams. When more atoms are present in the flame, more light will be absorbed, and the greater the difference

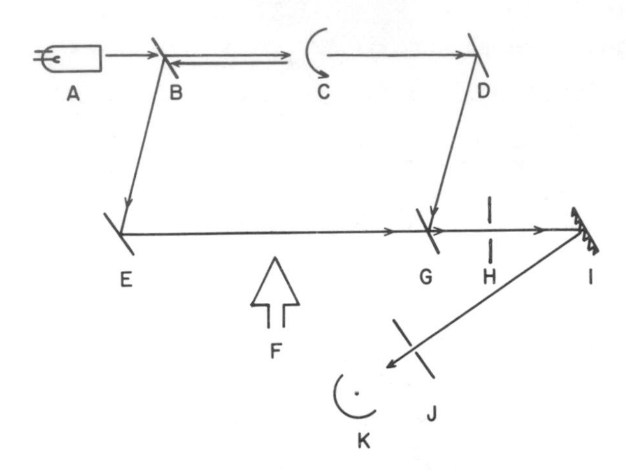

Figure 2-14. Schematic diagram of a double beam in time atomic absorption spectrophotometer. (A) Hollow cathode lamp, (B) half silvered mirror that reflects and transmits light, (C) chopper, (D) mirror, (E) mirror, (F) flame, (G) half silvered mirror, (H) slit, (I) grating, (J) slit, and (K) detector.

will be in light intensity between the sample and reference beams. This difference is then amplified and fed into a readout device. Note that in the double beam system, differences between the sample and reference beams are continuously being measured. This double beam arrangement compensates for changes in the output of the hollow cathode lamp and for changes in the detector system.

Summary

Atomic absorption spectrophotometry provides the analyst with instruments that are accurate and precise and that are obtainable at prices most laboratories can afford. To achieve comparable selectivity, sensitivity, and versatility with other instrumentation, the cost would be several times that of atomic absorption spectrophotometers.

Atomic absorption spectrophotometry is sensitive, accurate, precise and shows very high specificity. One of the reasons for these advantages is that the method does not require excitation of the sought-after substance and, thus, it is less affected by temperature variations in the flame and the transfer of energy from one atom to another. The high specificity results from the fact that the light used has an extremely narrow band pass (0.1 Å), which is selectively absorbed by atoms being measured.

The disadvantages of atomic absorption spectrophotometry are few. The most significant difficulty is the elimination of interferences. Although some suggestions have been presented to compensate partially for some of these factors, more work is needed to study and explain some of the interference problems that remain in atomic absorption spectrophotometry.

FLUOROMETRY

Fluorescence is energy emission that occurs when certain compounds absorb electromagnetic radiation, become excited, and then return to an energy level that is usually slightly lower than their original energy level. Since the energy given off is less than that absorbed, the wavelength of the light being given off will be longer than that absorbed for excitation. Only in rare situations will the emitted energy be equal to or higher than the absorbed energy. A delay time of between 10^{-8} and 10^{-4} seconds occurs between the absorption of the energy and the releasing of part of the energy in the form of light.

If the length of time is longer than 10^{-4} seconds from the time the chemical species absorbs the energy until the light is emitted, this process is called phosphorescence. The remainder of this discussion will be centered on fluorescence, since it is the more common process used in the laboratory.

Fluorometric Instrumentation

Figure 2-15 shows a schematic diagram of the components of a fluorometer. The energy source of a fluorometer is generally a mercury arc lamp or other suitable UV-visible light source that will produce enough energy so that, when absorption occurs, electron transitions to a higher energy level within the molecule will occur. In a fluorometer, the entrance and exit slits are similar to those described in spectrophotometers. Fluorometers that use a continuous source, such as a xenon source, have a monochromator so that isolation of the desired wavelengths is achieved before excitation of the substance (primary monochromator). All fluorometers have another (secondary) monochromator system that will selectively remove unwanted wavelengths of light before they fall upon the detector. Fluorometers are designed so that the secondary monochromator and the detector are at right angles to the incident light beam into the cuvet. This arrangement prevents the light from the high energy source of the mercury or xenon lamp from reaching the detector.

Advantages. The single most important advantage of fluorometry is its extreme sensitivity. Sometimes the sensitivity may be 1000 times that of colorimetric methods. Some molecules fluoresce directly, but a larger percentage must be complexed or chemically reacted to transform them into fluorescent compounds. The fluorescence of the new compound is then measured. By selection of different complexers, fluorescence of different wavelengths may be produced making it possible to work at a wavelength significantly removed from interferences.

Disadvantages. Fluorescent spectra are not as valuable for qualitative identification as are absorption spectra. Other disadvantages of fluorometry are quenching interferences, extreme sensitivity to pH change, temperature change, and interferences due to the presence of other foreign, undefined fluorescent materials. Frequently, energy is transferred from one molecule to another in solution and is dissipated in this manner rather than being given off directly as fluorescence energy. Quenching sometimes occurs when foreign materials form unwanted nonfluorescent complexes with the substance of interest.

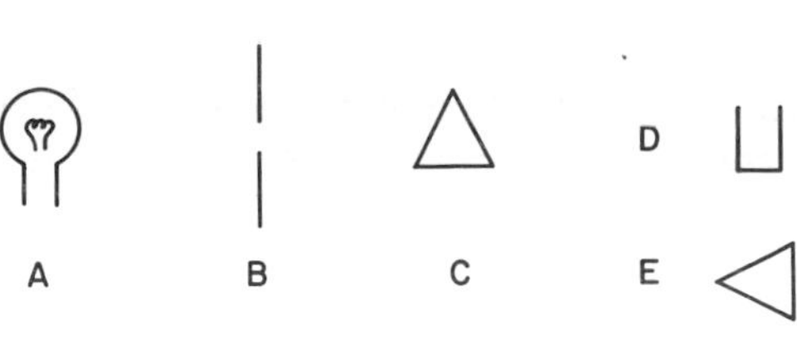

Figure 2-15. Essentials of a fluorometer. (A) Light source, (B) slit, (C) primary monochromator or filter, (D) cuvet, (E) secondary monochromator or filter, (F) slit, and (G) detector.

Examples of fluorometric methods used today are the determination of catecholamines, phenylalanine, porphyrins, and cortisol.

TURBIDIMETRY

The principle of the turbidity measurement is to determine the amount of light blocked by particulate matter as light passes through the cuvet. Several problems are inherent in making turbidimetric measurements and these problems are mostly associated with sample and reagent preparation rather than with the operation of the instrument. Turbidimetric measurements can be made with either a regular colorimeter or spectrophotometer.

The amount of light that is blocked by a suspension of particles in a cuvet depends not only upon the number of particles present, but also upon the cross-sectional area of each particle. If the particle size of the standards is not the same as the particle size in the samples being measured, errors in turbidimetric measurements result. Another problem with the turbidimetric measurement is the need to keep the length of time between sample preparation and measurements as constant as possible. Particles may settle out of solution while the measurements are being made, thus producing an error. Control of the rate of settling is usually accomplished by using gum arabic or gelatin, which provides a viscous medium that retards particle settling while measurements are being made.

Turbidimetric measurements are acceptable provided the number of particles and their size are in a reasonably narrow range. A high intensity light or a very low intensity light should not fall on the photodetector because errors in instrumentation would then augment other errors in the measurement.

Turbidimetric measurements are used in several liver function tests, the zinc and the thymol turbidity tests being two examples.

NEPHELOMETRY

Nephelometric measurements are similar to turbidimetric measurements. In nephelometry, the light that is scattered by the small particles is measured at right angles to the beam incident in the cuvet. The amount of scatter that occurs is related to the number and size of particles in the light beam. The particle size, their shape, and the wavelength of the incident light are important variables to control. The shorter the wavelength of the incident light, the greater the degree of dispersion. Nephelometry has an advantage over turbidimetric measurements in that nephelometric measurements are usually capable of somewhat greater precision. Turbidimetric and nephelometric measurements are not frequently used because they are inherently incapable of high precision. Specific chemical reactions are being sought to achieve higher accuracy and precision for the compounds on which these measurements are currently being used.

AutoAnalyzer

The introduction of electrically operated instruments and equipment into the clinical chemistry laboratory is very recent. The photoelectric colorimeter was introduced in the 1930's and the flame photometer in the 1940's. When it became apparent that the laboratory procedures were more precise and more readily available and when an increased knowledge of the clinical significance of laboratory results became known, more test requests were received by the laboratory.

In an attempt to solve the problem of the ever increasing workload in the laboratory, Dr. Leonard Skeggs and a friend began working in his basement and garage to develop an automated laboratory instrument that would continuously remove protein from the substance of interest. They made a dialyzer by machining mirror image grooves in two pieces of lucite and clamping the lucite blocks together with a piece of cellophane between them. A Coleman colorimeter was adapted for a flow-through cell that would measure the intensity of colored solutions as they passed through the system. Neither a sampler nor a recorder was used on this prototype. In the initial application, the serum sample was mixed with urease and was passed through a heating coil at 55°C. Ammonia produced by the urea and urease reaction diffused through the dialyzer membrane and was picked up by a water and air stream. The water and ammonia stream was then added to a solution of Nessler's reagent to produce a yellow color proportional to the amount of urea originally present in the sample. Dr. Skeggs first tried unsuccessfully to interest companies in producing and selling this instrument for clinical chemistry analysis. Finally, the Technicon Instrument Corporation (Tarrytown, N.Y.) accepted the challenge and the first commercial model appeared on the market in 1957. Technicon assigned the trade name of AutoAnalyzer to these instruments, which have found their way into most hospital laboratories. Approximately 8000 AutoAnalyzers were put into operation performing automated analysis in the first 10 years that the instrument was available.

The AutoAnalyzer is modular and components can be interchanged, which simplifies troubleshooting, servicing, and repair. The instrument is flexible and can be used for a large number of different determinations.

The AutoAnalyzer operations are patterned after the usual types of analytical procedures used in the chemistry laboratory. The function of AutoAnalyzers is to replace with automated devices the mundane tasks of pipetting, preparing protein-free filtrates, heating the color forming reagents in a water bath, and measuring the intensity of color. The AutoAnalyzer has made possible an increase in speed and an increase in the precision for many of the procedures that have been studied. Comparison between methods and precision data are usually available with new AutoAnalyzer methods, but recovery and accuracy data are frequently not complete.

Initially, there was some resistance to use of the AutoAnalyzer. Chemists as a group were very skeptical of the principles that the AutoAnalyzer employed. For example, the analytical measurement in most AutoAnalyzer methods is made on the small fraction of the sample that dialyzed across the membrane. For some of the analyses, the percentage of material dialyzed across the membrane may be as low as 5 to 20 per cent of the total sample. Another initial objection to the AutoAnalyzer was that the methods employed were not always based upon the best manual analytical chemistry procedures available; however, the AutoAnalyzer improved the precision of some of the poorer manual chemical procedures, thus making them acceptable for use in the laboratory. The principle that the standards and the samples are handled in precisely the same way and that any deficiency of the analyzer affects both the standards and the samples in the same manner is the basis of much of the success of the Technicon system.

Another reason for the resistance to the use of the AutoAnalyzer was the initial cost of the equipment. The Technicon Corporation was able to show that the cost per test was indeed decreased with the equipment compared with the manual methods.

The Technicon Corporation now manufactures and markets modules that permit the clinical chemistry laboratory to automate in excess of 90 per cent of the total number of laboratory tests that are performed. In addition to the modules, Technicon now has available several multichannel AutoAnalyzers, such as the SMA-12/30, SMA-12/60, SMA-6/60, SMA-4 and SMA-7. The letters used in the prefix, SMA, stand for *s*equential *m*ultiple *a*nalyzer, and the number 12, 6, 4, or 7, refers to the number of analytical procedures performed by that instrument on one sample.

Components of an AutoAnalyzer System

The AutoAnalyzer modules are electricomechanical devices that perform automatically the same functions that were previously performed manually. The function of each module and a description of how each operates will be covered in the following sections.

SAMPLER I AND SAMPLER II

The purpose of a sampler is to move a sample probe automatically into cups containing standard, sample, or wash solutions, respectively. On the Sampler I, the sample frequency is adjustable at rates of 20, 40, and 60 samples per hour. The sample probe is moved slowly in a vertical direction into a sample cup for a predetermined time (e.g., 40 seconds). The sample probe then moves slowly out of the sample cup and on into the next discrete sample, after the sample tray has moved to the next position. This slow movement of the sample probe was shown to introduce an error if the depth of liquid in the sample cup was not kept constant.[23]

The Sampler II by Technicon has incorporated into its design several desirable characteristics and it has replaced the Sampler I. The sample tray of the Sampler II is capable of handling all cup sizes currently available, the 1/2 ml., the 2 ml., the 3 ml., and the 8.5 ml. cups. Both the Sampler I and the Sampler II trays are capable of holding 40 discrete samples at one time.

The sample probe mechanism on the Sampler II operates vertically much more rapidly into and out of the sample than does the Sampler I probe. Therefore, the Sampler II is not as critically dependent upon the fluid level in the sample cup as is the Sampler I. In addition, the Sampler II has a wash reservoir containing water that is aspirated by the sample tube and introduced between each sample. Another advantage of the Sampler II over the Sampler I is that the range of sample frequency is greater and may be varied by simply inserting a drop-in cam. A characteristic of the cam is that it is possible to adjust the sampling to wash ratio for each of the sample frequencies. Sample frequencies are available in intervals of 10, from 20 to 120 samples per hour. The sampling time to wash time ratios at fixed sample frequency are available in a range of 6:1 to 1:6. Care should be taken to check the cams used so that timing errors are not introduced. Several of the commercially available cams do not achieve the required tolerances. Timing errors in the neighborhood of 3 to 5 per cent can be introduced in an analysis system by substandard cams.[25] Inaccurate timing produces an analytical error because in such a case the per cent of steady state achieved differs from cup to cup.

THE PERISTALTIC PUMP I AND PUMP II

The Pump I was the original design module sold by Technicon; Pump II is an improved model that is now available. The function of the pump is to aspirate and dispense portions of sample, reagents, and air into the AutoAnalyzer system in a reproducible manner. In both pump models, two parallel stainless steel chains similar to a bicycle chain are connected by the stainless steel rollers and are driven by a synchronous motor. The rollers run over a pressure sensitive tilting table on which a set of manifold tubes are placed. This tilting table is called a platen and is spring supported to allow compression of the polyvinyl (tygon) tubes by the stainless steel rollers. The Pump I has five rollers on a continuous chain, and the Pump II has eight rollers. As the rollers close the polyvinyl tubes by compression on the platen, solutions are forced along in the tubes, producing an aspiration on the downstream side and forcing the solution through the various modules on the upstream side of the pump. The Pump I has a capacity of 14 pump tubes (manifold tubes) and the Pump II is capable of handling 22 tubes at one time.

The pump is a very flexible module and flow rates and proportions of the sample volume to reagents volume can easily be modified by changing the diameter of the tubes in the pump. Flow rates between 0.001 ml./min. and 4 ml./min. are possible by selection of the proper size tube.

The Pump II has several advantages over the Pump I. Each of the rollers on the Pump II is larger in diameter and they are placed closer together. This permits the polyvinyl tubing to be more uniformly occluded, which in turn provides a more constant flow rate and longer tube life. The pumps on the SMA-12/60 have an additional device, which allows bubbles to be introduced into the flowing stream at the same frequency as that at which the pump rollers leave the platen. It has been experimentally determined that if this is done, constant volumes are maintained between roller segments of the tubes. Control of the frequency and size of the air bubbles that are introduced into the system have also been shown to reduce the amount of recorder pen noise and significantly improve the wash characteristics between samples.

THE DIALYZER

Until recently, the AutoAnalyzer was the only automated clinical chemistry system that was capable of separating large molecular weight molecules like protein automatically and continuously from the substances to be analyzed. This dialyzer, to a large degree, contributed to the success of the AutoAnalyzer. The principle of the dialyzer is shown in Figure 2-16. The dialyzer is constructed of a piece of lucite

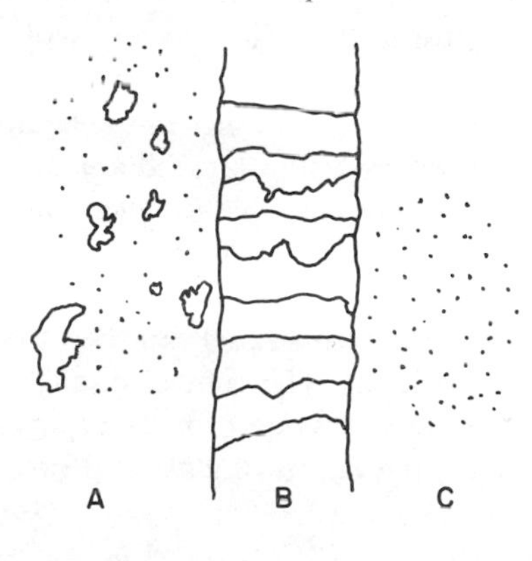

Figure 2-16. Schematic diagram of a dialyzer membrane. (A) Donor stream containing protein, (B) the cellophane membrane, and (C) recipient stream with protein removed.

plate, into which mirror image grooves are milled. The two mirror image grooves are then separated by a special cellophane membrane designed for the AutoAnalyzer. When a solution containing protein (donor or sample stream (A), as shown in Fig. 2-16) is passed on one side of the membrane, the small molecular weight molecules diffuse through the cellophane membrane into the recipient or reagent stream, and the large molecular weight molecules are held back and passed out to waste. The recipient stream may contain a reagent that will produce a color with the substance of interest or it may be any other reagent to which the color forming reagent is added at a later stage. Only the small molecular weight materials are allowed to diffuse through the cellophane.

If complete equilibrium conditions were to exist between the donor and the recipient streams, equal concentrations of the substance of interest would exist on each side of the membrane. Because equilibrium is not usually achieved in the constant-flow system, at least half of the substance to be measured (frequently much more) escapes dialysis and is lost with the protein as waste. This disadvantage has not been a severe limitation to the AutoAnalyzer system, since adequate sensitivity is available in the overall system. In the use of the AutoAnalyzer, it is critically important that the conditions between standardization and sample analysis remain constant. The percentage of material that diffuses across the membrane is temperature dependent and must remain constant for both standards and samples. Maintenance of a constant temperature becomes critical when the amount of material that diffuses across the membrane is only 5 to 20 per cent of the total amount of material being analyzed. The path length of the dialyzer, i.e., the distance that the solution is in contact with the cellophane membrane on the usual dialyzer systems now in the field, is 88 inches. The dialyzer path on the SMA-12/60 is only 6 or 12 inches long. This, of course, means that the percentage of material that passes across the dialyzer membrane on the SMA-12/60 is considerably less than that on the conventional dialyzer; however, the sensitivity remains similar since the sample does not undergo as high a dilution. Part of the instrument is thermostated at 37°C. and the solutions that pass through the dialyzer are preheated, thus eliminating the need for large water baths to maintain a constant temperature in the dialyzer.

THE HEATING BATH

The purpose of the heating bath is to continuously heat the recipient stream for a specified time to promote the color development or other chemical reactions (e.g., enzyme reactions). Commonly, two 40 ft. lengths of glass coil are submerged in a mineral oil or ethylene glycol solution and maintained at a fixed temperature between 37 and 95°C. The two coils of the heating bath may be used in parallel for two independent methods or in series if a longer time is needed for one chemical reaction. Longer reaction times in the heating bath can also be achieved by using a slower flow rate or by using large diameter glass coils. The three most common temperature regulators available on the heating bath are the 37°C., the 95°C., and the variable temperature controller. Most of the determinations in the laboratory today are run either at 37 or 95°C.

The SMA-12/60, on the other hand, has the tubing mounted in a heated aluminum block. The control of such a unit is not dependent upon a solution circulating around a glass coil. Further advantages of the 12/60 heating block are compactness and a reduction of glass coil breakage.

THE COLORIMETER

The function of the colorimeter is to measure continuously the intensity of the color that is present in the solutions flowing through the flow cell and to produce an electrical signal proportional to the increase or decrease in that color. The colorimeter is a single light source, split-beam instrument. One beam of light from the single light source passes through the sample cell cuvet to the sample photocell and another beam of the same light source passes to the reference photocell. The ratio of the two signals is fed to the recorder amplifier for strip chart recorder presentation. Before the liquid of the AutoAnalyzer system can be passed through the colorimeter, the bubbles must be removed. This is accomplished by pumping less solution through the flow cell than is being introduced into an inverted "*T*." The bubbles and extra solution are allowed to go to waste. If bubbles were allowed to pass through the flow cell, large amounts of light would fall intermittently on the sample photocell, causing very erratic pen movement on the strip chart recorder.

The adjustment of the strip chart recorder pen to 100%*T* can be made in two ways; either the 100%*T* potentiometer can be used, or the size of the aperture can be changed. If the energy passing through the sample beam is insufficient to enable the strip chart recorder pen to be adjusted to 100 per cent transmittance (100%*T*), it means that not enough light is passing through the sample cell or too much light is passing through the reference cell. This is easily compensated for by inserting an aperture of smaller diameter into the reference side of the colorimeter and making the proper adjustment with the 100%*T* potentiometer.

If, on the other hand, the 100%*T* potentiometer is unable to relieve the pen from a position that exceeds 100%*T*, it means that too much light is passing through the sample cell or too little is reaching the reference photocell. This is corrected by inserting a larger aperture into the reference beam of the colorimeter.

THE RECORDER

The recorder is a 10 mv. potentiometric null-balance recorder built for Technicon by the Bristol Company (Waterbury, Conn.). The recorder contains an amplifier for the signal from the colorimeter. It makes a continuous tracing corresponding to the change in color intensity of the liquid that flows through the colorimeter.

The multichannel analyzers, like the SMA-4, SMA-12/30, SMA-12/60, and SMA-6/60, all have direct readout of results in concentration units. The %*T* signal from the colorimeter undergoes a logarithmic conversion, and this signal in the proper units is put out on the strip chart recorder. This additional feature greatly reduces technologists' time in the laboratory by eliminating calculations and other clerical tasks.

Operation of AutoAnalyzer System

Without the introduction of bubbles of air into the sample and reagent streams of the AutoAnalyzer, the system would not function satisfactorily.

One or more of the manifold tubes pumps air bubbles into the flowing streams of liquid. The function of the bubble is to reduce laminar flow by segmenting the solution. The bubble wipes the inside of the tubing, which greatly reduces the interaction between samples in the system. Without the bubble, the wash and interaction characteristics would be so poor that the instrument would not be useful.

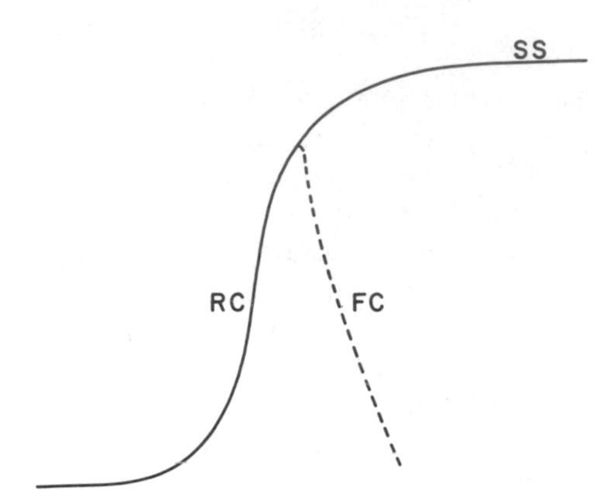

Figure 2-17. Autoanalyzer terminology. (RC) Rise curve, (FC) fall curve, and (SS) steady state.

It should be pointed out that sensors other than colorimeters are available for AutoAnalyzers. Flame photometers and fluorometers are examples.

For proper understanding of the operation of an AutoAnalyzer, a few terms should be defined so that conversation can take place between various operators of the systems. Figure 2-17 diagrammatically shows some of the terms that are commonly used in describing AutoAnalyzer operation. The highest point on the curve, labeled SS, is the sample *steady state.* It is the point at which the color intensity reaches its maximum and remains constant as long as the sample is being aspirated. The portion of the curve labeled RC indicates the *rise* portion of the curve, and the section labeled FC is the *fall* portion of the curve. The dotted line labeled FC indicates how a peak is formed on an AutoAnalyzer. Notice that the conventional manner in which AutoAnalyzers are operated does not always allow the instrument to come to a steady state reading. This is done to allow discrete, sharp peaks to form, which are more easily read and measured.

The time between samples (T_{bs}) can be defined by the following equation:

$$\frac{3600 \text{ sec./hr.}}{\text{no. samples/hr.}} = \text{sec./sample}$$

The time between samples (T_{bs}) is made up of two components, the length of time that the sample probe is in the sample cup (T_{in}) and the length of time that the sample probe is out of the sample cup (T_{out}).

In an AutoAnalyzer, a certain percentage of one sample remains in the system and is removed when the next sample passes through. A constant percentage of the concentration in one sample is then picked up by the sample that follows it. This percentage of the leading sample that is picked up by the sample behind it is called per cent interaction ($\%I$) of the system. The $\%I$ is independent of the amount of time that the sample probe is outside the sample (T_{out}), and is dependent only upon the sample frequency (T_{bs}) of the system under study. Figure 2-18*A* shows how a $\%I$ measurement is made, if sample frequency is fast. A low concentration is run initially (S_1), followed by the highest concentration standard available (S_m), followed by the same low concentration that was run before the high standard (S_2). The amount that the second low standard is elevated over the measured first standard value is then calculated. The following formula permits calculation of $\%I$:

$$\frac{S_2 - S_1}{S_m} \times 100 = \%I$$

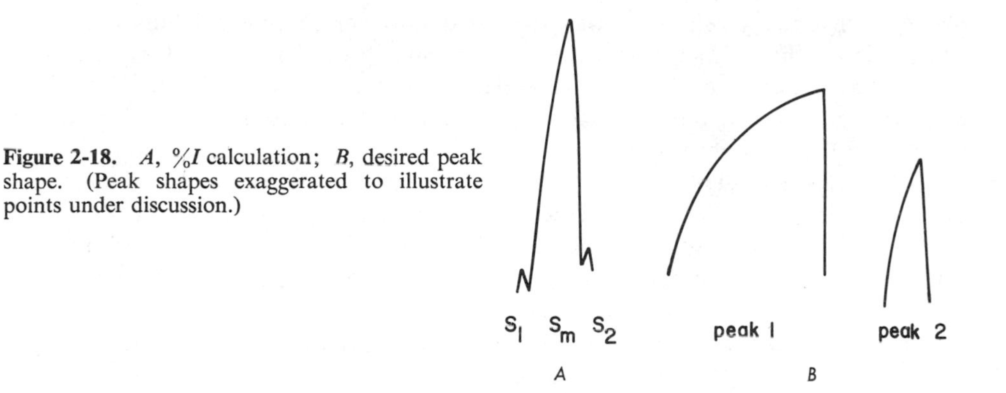

Figure 2-18. *A*, %*I* calculation; *B*, desired peak shape. (Peak shapes exaggerated to illustrate points under discussion.)

Experimentally it can be shown that if the AutoAnalyzer has a sample of high concentration followed by one of lower concentration, a certain percentage of the high peak should be subtracted from the lower peak value.

The %*SS* is defined as the relative peak height of the system running at the usual sample frequency at the conventional T_{in}-T_{out} ratio compared to the peak height that is obtained under steady state conditions. The Technicon Corporation recommends that AutoAnalyzers should operate in excess of 85 per cent of steady state for satisfactory precision. It can be shown experimentally that the closer one approaches steady state, the higher the precision of the analysis.

The shapes of peaks are very revealing of the operating characteristics of an AutoAnalyzer system. At regular or low sample frequency, low concentrations present low, flat peaks, but high concentrations exhibit sharp, pointed peaks. If a peak is shaped such as illustrated by peak 1, Figure 2-18*B*, we know that the analyzer is operating very close to steady state because of the lack of symmetry of the peak. If a peak shaped like peak 2 is obtained, however, we know that the analyzer is not operating as close to steady state as is desirable. Any change in peak shape during a run indicates a malfunction somewhere in the system, which should be investigated and corrected.

There are basically three kinds of noise that can be associated with an AutoAnalyzer system: *electronic noise*, *optical noise*, and *chemical noise*.

Electronic noise will usually be constant and independent of the pen position. Frequently, electronic noise will be harmonic in nature and will occur as a function of time. The magnitude of noise is usually independent of the pen position.

Optical noise can be caused, for example, by a precipitate in the system flowing through the flow cell or by small bubbles introduced through the flow cell. Optical noise is asynchronous and can occur without a pattern.

Chemical noise can usually be identified by a difference in the magnitude of noise at baseline compared with the noise at steady state. Chemical noise is usually caused by inadequate pump characteristics, surging of the solutions in the lines because of varying pressures, unclean glass coils and tubing, and poor bubble patterns. The bubbles introduced into an AutoAnalyzer system should be of consistent size and frequency for optimum performance.

Chromatography

The history of chromatography began when a Russian botanist named Tswett ground up some green leaves and passed them over a sorbent in a glass column. He observed that the green color was separated into various color bands that were adsorbed onto the solid support in the glass column. It was because of this experiment and of the separation of colors that the word chromatography was coined for this process. Today most applications of this technique do not involve separation of colors, but the term chromatography remains.

The purpose of chromatography in all cases involves the separation of a mixture on the basis of specific differences of their physical characteristics. The types of chromatography that will be discussed in this section include ion exchange chromatography, paper chromatography, gas-liquid chromatography, liquid-liquid chromatography, gel permeation chromatography, and thin-layer chromatography. Electrophoresis will also be discussed in this section since this separation technique in some respects is similar to chromatography.

In *paper chromatography* the physical characteristics that determine separations are the rate of diffusion, the solubility of the solute, and the nature of the solvent. In *liquid-liquid chromatography*, separation is based on differences in solubility between two liquid phases. One of the liquids is usually aqueous and the other is an organic solvent. Solubility can be modified by changes in ionic strength or pH of each of the liquid phases. In *gel permeation chromatography*, the molecular weight, and the size and charge of the ions or molecules are the characteristics that are responsible for separation. In *ion exchange chromatography*, the separation of substances depends principally upon the sign and ionic charge density. Ions with greatest charge density will be held most strongly on an ion exchange material. In an *electrophoretic* separation, the physical characteristic that differentiates between components is the mobility of the substance of interest in an electric field. In *gas-liquid chromatography*, the sample volatility, its rate of diffusion into the liquid layer of the column packing, and the solubility of the sample gas in the liquid layer (partition coefficient) determine the separation capabilities of this technique. *Thin-layer chromatography*, like paper chromatography, depends upon the rate of diffusion and solubility of the substance of interest in solvents as the components migrate through media such as silica gel.

ION EXCHANGE CHROMATOGRAPHY

Ion exchange chromatography is a very well established procedure that has been studied intensively. The purification of water as it percolates through soil is an example of this process.

Ion exchange chromatography has many uses in clinical chemistry. High concentrations of ascorbic acid, bilirubin, and uric acid, for example, inhibit the rate of the glucose oxidase reaction. The use of the proper strong ion exchange and gel permeation materials permits removal of these interferences and improves the accuracy of the glucose oxidase method for glucose. The DuPont (Wilmington, Del.) Automatic Clinical Analyzer (ACA) uses this separation for its glucose methodology. In a similar manner, creatinine and uric acid can be separated by ion exchange techniques before performing the color development step in the creatinine determination.

Figure 2-19 shows the functional groups that are attached to the styrene structure used in constructing a synthetic strong anion exchange resin. Nitrogen is bonded to the styrene and three methyl groups are bonded to the nitrogen. The chloride anion is electrostatically attracted to the positive charge that remains on the nitrogen atom. Any anion that passes through the solution that has greater affinity to the nitrogen

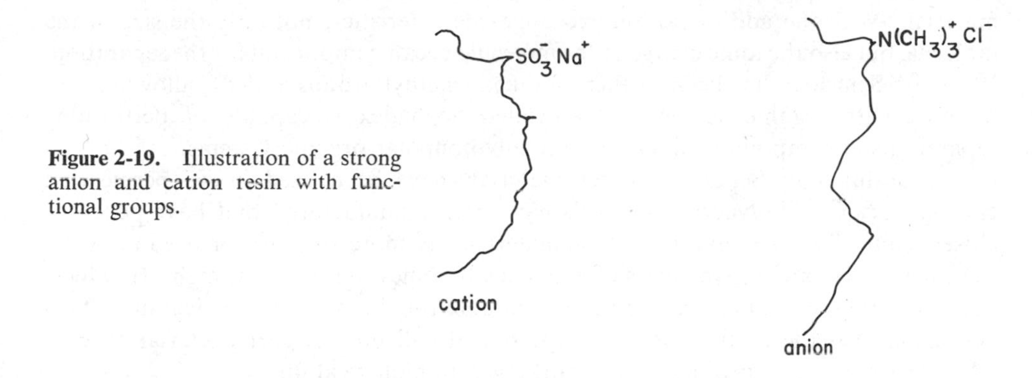

Figure 2-19. Illustration of a strong anion and cation resin with functional groups.

than the chloride ion causes a displacement of the chloride from the resin. The amount and the rate of exchange depend upon the relative affinities of the nitrogen atom for the chloride ion or the other anions in solution. This is a technique to remove unwanted ions from solutions.

Figure 2-19 also shows a typical strong cation exchange resin, again with the styrene structure, but with SO_3^- groups bound to the styrene. In this example, any cation that would be more strongly held to the SO_3^- than sodium would be preferentially removed from solution and the sodium ion would be discharged into the solution. As with the anion exchange material, a cation exchange process serves to separate unwanted substances from solution. In certain cases, very large volumes of dilute solutions may be passed through ion exchange materials to concentrate the solute of interest effectively. Ion exchange chromatography will find increasing uses in the clinical chemistry laboratory since it materially improves the specificity and accuracy of the method by removal of interfering substances.

GEL PERMEATION CHROMATOGRAPHY

Gel permeation chromatography became widely used in the early 1960's. The trade name of the material that appeared on the market is Sephadex (Pharmacia Fine Chemicals, Inc., Piscataway, N.J.). Sephadex is a dextran material that has been modified so that it contains pores of accurately controlled size. Various pore size materials are currently commercially available in Sephadex. When a mixture of small and large molecules is allowed to pass over small particles of the dextran in a glass column, the small molecules and ions diffuse into the gel. Larger molecules such as proteins are too large to diffuse into the interstitial cavities of the material and pass rapidly through the column. The smaller sized molecules and ions are then temporarily retained until they have time to diffuse back out of the dextran. Thus, the large molecular weight materials will appear in the effluent first and the smaller molecular weight materials will be delayed in the dextran packing of the column.

The use of Sephadex makes it possible to separate compounds by their molecular weight, providing that the pore size of the material is properly selected for the separation. Compounds that have a molecular weight difference of at least 1000 can be separated by this technique.

Gel permeation chromatography was further improved by introducing ion exchange groups on the dextran. The ion exchange characteristics of the dextran, in addition to the size consideration, greatly expanded the separation capability of this material. With the additional ion exchange characteristics, not only the size of the molecule but also the ionic charge of the molecule became important for the separation. Recently, Sephadex has been further modified (methyl groups added), allowing it to be used with organic solvents. Methylated Sephadex is capable of performing separations of compounds dissolved in highly nonpolar organic systems.

In addition to Sephadex, other materials have been used in gel permeation chromatography. Polyacrylamide gels have been manufactured that have pore size closely controlled like dextran. These materials are more suitable for use in a wider pH range. The rate of hydrolysis of Sephadex becomes significant at high pH values; therefore, polyacrylamide gels are more suitable for this type of application. Polyacrylamide has the additional advantage that it will not support bacterial growth, which is frequently a problem when working with biological fluids.

Gel permeation chromatography has been used principally for the separation of proteins from smaller molecular weight materials such as metal ions. In addition, gel permeation chromatography has been used quite extensively in the study of isoenzymes and enzyme chemistry.

GAS-LIQUID CHROMATOGRAPHY

Gas-liquid chromatography (GLC) was developed by a crash program instituted by one of the large industrial chemical companies during World War II. The first published work on GLC appeared in the early 1950's. Today GLC is one of the most versatile, powerful analytical tools available. This technique is capable of separating and measuring nanogram amounts of substances. It is used for the measurement and fractionation of steroids, lipids, barbiturates, drugs, and blood alcohol and the measurements of other toxicological substances. Gas-liquid chromatographic methods are rapid, sensitive, and accurate.

INSTRUMENTATION

The basic components of a gas-liquid chromatograph are shown in Figure 2-20. Although the basic principles are not difficult to understand, the technological requirements for columns, electronics, and temperature control were difficult to achieve until a few years ago.

The carrier gas is introduced near the top of the column, near the point where the sample is to be injected. The carrier gas is an inert gas, such as helium, nitrogen, or argon, that flows at a constant rate through the column. The column is usually glass or stainless steel and is packed with an inert solid support, such as diatomaceous earth,

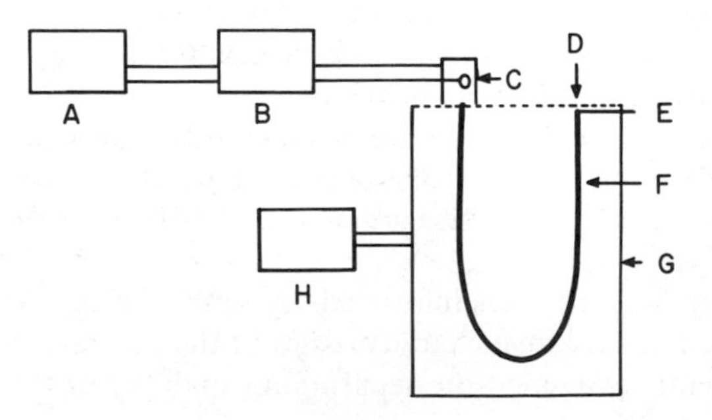

Figure 2-20. Essentials of a gas-liquid chromatograph. (A) Strip chart recorder, (B) electrometer to measure and amplify detector current, (C) hydrogen flame detector, (D) sample input, (E) carrier gas input, (F) column, (G) temperature controlled oven, and (H) programmer to control oven temperature.

that has been coated with a thin layer of a liquid phase. The liquid phase is usually some silicon material that is a nonvolatile liquid and does not react with the substance of interest.

The detectors of gas-liquid chromatographs generally are of three basic types: the hydrogen flame, the thermal conductivity, and the electron capture detectors. Since the hydrogen flame detector is most commonly used in the clinical chemistry laboratory, it is the only detector that will be discussed. It consists of a small platinum loop mounted approximately 1 inch above the exit port of the column. A metal gas jet is mounted over the exit port of the column, and a flammable gas such as hydrogen is passed through the gas jet. A small flame is ignited and burns at the tip of the jet where the column effluent and the hydrogen mix. A voltage is applied between the gas jet and the platinum ring mounted above it. The platinum ring is maintained at a positive voltage (up to 300 volts). When the hydrogen flame is burning, ions that are formed by burning the effluent in the flame will be collected by the platinum ring and produce a small current between the gas jet and the platinum loop. This current is detected by the electrometer, amplified, and the signal fed to the recorder.

The electrometer. Basically, the electrometer is an electronic device that is capable of measuring very small currents and amplifying them linearly. The amplified signal is then fed to the readout device, which is usually a strip chart recorder.

The programmer. The programmer is a combination of electronic controls whose basic function is to provide temperature control of the column oven. The oven is usually controlled to within 0.1°C. and is usually capable of being changed at a linear and constant rate ranging from 1°C. per minute to 50°C. per minute. This change in temperature during a determination helps produce separations of materials that have penetrated the liquid phase. As the temperature changes, the respective components of the sample are then selectively eluted from the column in accordance with their differences in volatility.

The recorder. The recorder is simply an electromechanical device that measures the voltage output of the electrometer and presents it on a strip chart for interpretation by the operator.

PRINCIPLES OF OPERATION

In a gas-liquid chromatography determination, the substance of interest must be volatile or a new compound (derivative) must be formed that is volatile. The derivative, in addition to being more volatile may protect the compound of interest from decomposing by binding with heat labile groups. A small quantity of the liquid material to be measured is injected onto the column and is immediately volatilized in the entrance port to the column. The carrier gas then transports the volatilized sample to the liquid phase on the solid support within the column. Separation occurs as a result of the different solubility and the different diffusion rates of the various components of the sample gas into and out of the liquid phase. Thus, the various fractions of the gaseous sample tend to move through the column at different rates and appear at different times at the detector. As these bands are eluted from the column by the carrier gas, the burning flame produces a large number of ions. The surge of ions causes an increase in detector current, which is amplified by the electrometer and sent to the recorder. Peaks will then appear on the strip chart recording. As each of the bands is eluted from the column, successive peaks will be presented on the strip chart recorder. The length of time for each of the peaks to appear on the strip chart recorder from the injection time is the *retention time* and is characteristic for the substance of interest. Two substances in a mixture generally do not have identical retention times. The retention time, therefore, qualitates the substance in the sample and the peak area quantitates the amount of each of the fractions present. If good resolution in the gas chromatograph is achieved and if the peaks formed on the strip chart recorder are very sharp, then the peak height may be found to be proportional to the concentration. Temperature changes, type or concentration of the liquid phase, volatility of different derivatives, or some other parameter can be introduced to effect an adequate fractionation of the sample.

Examples of the application of gas-liquid chromatography to clinical chemistry have been presented in Chapter 7 on lipids and in Chapter 9 on endocrinology.

In summary, because the gas chromatograph separates, detects, qualitates, and quantitates several fractions of a volatilized sample in a single step, its use has great potential in clinical chemistry.

LIQUID-LIQUID CHROMATOGRAPHY

When two liquid phases are mixed, as is commonly done in a separatory funnel, substances will distribute themselves between the two phases. The solubility principally determines the distribution of the species into the aqueous or the organic layer in a two phase system. Generally, the solubility will be determined largely by the relative polarity of the substance of interest and the polarity of the two liquid phases. A highly polar substance tends to be more soluble in a highly polar solvent such as water; the less polar substances tend to be more soluble in the less polar solvents such as organic solvents (see also Chapter 9).

The solubility of a species and, therefore, the control of what phase the species will be found in, is most easily controlled by changing the pH of the aqueous solution. For example, if an anion is found in the aqueous phase and if that anion is to be extracted into the organic phase, addition of hydrogen ion will usually produce the transfer. When H^+ is added to the aqueous layer and the anion becomes protonated, the polarity of that molecule is much less than what it was as an anion. Its solubility will be much greater in the organic phase because of the reduced polarity.

An example of how liquid-liquid chromatography is applied to clinical chemistry is the UV barbiturate method. The acid form (nonionized, nonpolar) of the barbiturate is first extracted from the aqueous phase (blood, urine) into an organic phase like chloroform. Next, the chloroform layer is contacted with a NaOH solution and the anion of the barbiturate goes into the alkaline aqueous phase, which is then analyzed for barbiturates.

The distribution ratio ($D_{o/w}$) defines the amount of material found in each phase at equilibrium.

$$D_{o/w} = \frac{[C_{\text{org}}]}{[C_{\text{water}}]}$$

$[C_{\text{org}}]$ = concentration in the organic phase

$[C_{\text{water}}]$ = concentration in the aqueous phase

A column liquid-liquid chromatography system can be established by impregnating a solid support with one liquid phase, then allowing another liquid to percolate through the column. A solid phase frequently is silica gel, the liquid phase is water, and an organic solvent is allowed to percolate through the column. Separations will occur with such an arrangement in a manner similar to other column chromatographic separations.

PAPER CHROMATOGRAPHY

In ascending paper chromatography, a strip of filter paper is usually hung vertically into a solvent. The solvent moves up through the paper by capillary action, with the paper serving as a wick. A spot of the substance to be fractionated is placed on the paper just above the solvent level and permitted to dry before the paper is inserted in the jar containing the solvent. As the solvent moves up through the paper, various fractions in the sample move at different rates. The relative solubilities of the components of the sample in the solvent mixture, the polarity of the solvent, and the polarities of the solutes of interest all affect the rate at which different components move. After the separation has taken place, the paper is removed, dried, and sprayed with a chemical for color development. The spots may then be quantitated by measuring their area or intensity, or both. Sometimes the spots may be visible in the ultraviolet region of the spectrum, in which case the paper is examined under such light.

Paper chromatography has been used in clinical chemistry in the past for fractionation of sugars, amino acids, and barbiturates. Today amino acid analyzers involving ion exchange chromatography are more frequently used for amino acid separations, and barbiturates are usually identified by UV spectrophotometry, thin-layer or gas chromatography. Sugars are usually separated by column chromatography with an ion exchange material as packing.

THIN-LAYER CHROMATOGRAPHY

The principles of thin-layer chromatography (TLC) are similar to those described for paper chromatography. The main difference between the two techniques is that glass or plastic plates to which is attached a thin layer of silica gel, alumina, polyacrylamide gel, or starch gel are used for the matrix instead of filter paper. The glass plates are then placed into a solvent solution and the solvent passes through the thinly layered material on the glass plate in the same manner as it passes up the paper, namely, by capillary action. Again, separation occurs because of differences in solubility, polarity of the solvent, polarity of the substance of interest, and rate of diffusion.

One advantage of thin-layer chromatography is that the spot may be scraped from the plate, easily redissolved in a solvent and then analyzed on an analytical instrument such as a gas chromatograph or a fluorometer. Another advantage of TLC is the rapid separation that reduces the time necessary to perform an analysis (30 to 90 minutes compared to 12 to 24 hours for paper chromatography). The main functions of thin-layer chromatography are the identification and separation of unknown substances in one step or preliminary purification of mixtures before performing the final analysis by another technique.

Electrophoresis

Electrophoresis is a valuable tool in the clinical chemistry laboratory for the separation and measurement of electrically charged substances such as protein fractions and isomers of enzymes. Such substances, in solution, will move in an electric field, depending upon their net charge either to the positively charged anode or the negatively charged cathode. The support medium usually consists of paper, starch gel, cellulose acetate, polyacrylamide gel, or some other porous material that is saturated with buffer.

Electrophoresis is most commonly used in clinical laboratories for the separation of proteins found in serum. A sample is applied to the buffer saturated support medium and a voltage is applied. The buffer solution carries a few milliamperes of current, which causes the molecules to separate, depending on their net charge density. After the current to the electrophoresis cell is turned off, the support medium is removed from the cell and put into a dye solution, which stains the various fractions on the matrix. Quantitation of the fractions is achieved by using a densitometer to measure either the intensity of the light reflected from the dyed fraction or the amount of light that is transmitted. The peak areas obtained from the densitometer for each of the bands found on the matrix support are proportional to the concentration of that fraction. An example and more details of an electrophoretic procedure are presented in Chapter 5, Proteins and Amino Acids.

INSTRUMENTATION

The instrumentation for electrophoresis basically consists of a power supply, an electrophoresis cell, electrodes submerged in a buffer, a matrix of paper, gel, or cellulose acetate, and an instrument (densitometer) for measuring the amount of each of the fractions after the separation and staining. The power supply commonly requires an output of about 120 to 250 volts with about 2 to 5 mAmp. of current being drawn per strip. The buffer is usually a carefully controlled ionic strength solution of borate, barbiturate, or acetate. A too low ionic strength buffer causes heat to be generated within the cell; a high ionic strength does not permit good separation of fractions. The electrodes that are immersed in the buffer are usually placed in separate compartments connected to the cell by salt bridges so that pH changes that occur in the electrode chambers do not alter the pH of the buffer saturating the matrix. Significant care must be taken by the operator to prevent contamination of the cell, the electrodes, or the matrix with fingerprints or any other foreign substance that will disturb the even current flow throughout the matrix.

Gasometry

This discussion will be limited to carbon dioxide determinations in clinical laboratories as measured by volumetric and manometric methods. Oxygen is most frequently measured electrochemically by the use of a P_{O_2} electrode, as discussed in the electrochemical section of this chapter. Spectrophotometric measurements are also used for the determination of oxygen in blood.

Two ideal gas laws were established by Boyle and Charles more than a century ago. Boyle's law states: At constant temperature, a fixed weight of gas occupies a volume inversely proportional to the pressure exerted upon it. Mathematically expressed, Boyle's law is described by the formula:

$$\frac{V_1}{V_2} = \frac{P_2}{P_1}$$

where V_1 = the initial volume
V_2 = the final volume
P_1 = the initial pressure
P_2 = the final pressure

This equation is the basis of the manometric blood gas measurement.

Charles' law states: At constant pressure, the volume occupied by a fixed weight of a gas is directly proportional to the temperature (on the Kelvin scale). The mathematical formula that describes Charles' law is:

$$\frac{V_1}{V_2} = \frac{T_1}{T_2}$$

where V_1 and T_1 indicate initial volume and temperature, respectively, and V_2 and T_2 indicate final volume and temperature, respectively. This equation is the basis for the volumetric method of blood gas analysis that evolved later.

D. D. Van Slyke and his co-workers were instrumental in making improvements on the direct measurement of blood gases in the early 1900's. The amount of gas can be measured directly as it is released chemically from the biological sample. In serum determinations of total CO_2 content, the CO_2 is released by the addition of lactic acid. The gas released can be absorbed and the differences in pressure produced between the released gas and the absorbed gas can be used to calculate the actual concentration in the sample ("manometric technique"). The second procedure is preferable in the clinical laboratory because gases other than the one of interest may be present and this technique corrects for the presence of such gases. In 1917 a technique was developed based on the measurement of the volume of a gas corrected to standard temperature and pressure. This measurement became known as the "*volumetric technique.*"

In the manometric technique the pressure of the gas is measured at a constant volume and constant temperature and the concentration of the gas (e.g., m*M*./L.) is then calculated. When the test is not performed at the specified temperature, a correction factor must be applied. Because of their greater accuracy, manometric techniques have largely replaced volumetric methods in clinical chemistry laboratories today. The Natelson method for determination of plasma CO_2 content is an example of a manometric technique and is discussed in the electrolyte chapter (Chapter 10).

Accuracy and Precision

The accuracy and precision of the volumetric and manometric techniques are within acceptable limits for clinical chemistry laboratories. Ideally, with stringent care, 2 per cent precision is claimed for the volumetric method and 0.3 per cent has been claimed for manometric measurements of gases in biological materials.

The method of choice in the laboratory depends upon the sample size available, the accuracy and precision desired, and the speed with which the analysis is to be performed. Manometric methods lend themselves well for the determination of carbon dioxide in biological samples (see Chapter 10).

Either the manometric or the volumetric method is probably satisfactory in today's laboratories. The most significant disadvantage to both methods is that they are slow and do not easily lend themselves to batch analysis or to automation. As a result, automated methods based on other analytical approaches have been developed and are replacing gasometric techniques. An example of an automated procedure for carbon dioxide is found in Chapter 10.

Osmometry

The *osmolality* of a solution is dependent only on the *number* of particles in solution. One osmol of a substance is equal to the gram-molecular weight divided by the number of particles or ions into which the substance dissociates in solution. Thus, since glucose molecules do not dissociate in aqueous solution, 1 osmol of glucose = 1 mole = 180 gm. For NaCl, which dissociates into 2 ions in aqueous solution, 1 osmol = 0.5 mole or 58.5/2 = 29 gm. For Na_2SO_4, 1 osmol = 0.33 mole, and so on, assuming 100 per cent dissociation for ions in solution.

A solution containing 1 osmol of solute per kilogram of solvent has a concentration of 1 osmolal. This concentration is independent of temperature since it is based on weight only. A solution containing 1 osmol of solute per liter of solution has a concentration of 1 osmolar. Osmolar concentrations vary with temperature since the volume varies with temperature. For aqueous solutions of low concentrations, such as body fluids, the difference between osmolality and osmolarity becomes negligible and it is customary to use milliosmols per liter (1 osmol = 1000 milliosmols).

The freezing point of water is depressed 1.86°C. when solute is added to make a 1 osmolal solution; hence, 1 osmol of any solute is that amount which will depress the freezing point of 1 kg. of water by 1.86°C. An osmolality measurement in clinical chemistry provides an estimate of the effective number of particles in solution, even though we do not know either the nature or concentration of the individual substances dissolved in the solution.

Instrumentation

Osmolality is determined by measuring the freezing point depression of a solution. The apparatus consists of a cooling bath to freeze the specimen and a thermistor. A thermistor is a device whose electrical resistance increases as the temperature decreases. With a constant current source and a balancing circuit, the resistance of a potentiometric balance bridge is adjusted to equal the resistance of the thermistor. This resistance of the thermistor is inversely related to the temperature being measured. Thus, a thermistor is an electrical thermometer.

The measurement is very accurate and precise and is easily calibrated. The precision of an osmometer is usually of the order of 0.3 per cent or less (see also Chapter 12 on renal function).

Principle of Operation

The amount that the freezing point is depressed is related to the concentration by the following equation:

$$\Delta T = \frac{RT_0^2 M_1 W_2}{H_f W_1 M_2}$$

where ΔT = the change in the freezing point
R = the gas constant, 1.987 cal./mole
T_0 = the freezing point of the solvent in °K.
M_1 = the molecular weight of the solvent
W_2 = the weight of the solute in grams
H_f = the heat of fusion of the solvent, cal./mole
W_1 = the weight of the solvent in grams
M_2 = the molecular weight of the solute

The weight of solute (W_2) divided by the molecular weight of the solute (M_2) is equal to the number of moles of solute present; hence, from the preceding equation the depression in freezing point is directly proportional to the number of moles of solute and is independent of the molecular weight.

The heat of fusion of water is 1436 calories per mole and the freezing point of water is 273.1°K. For example, a 1 molal solution of urea contains 60 gm. of urea dissolved in 1000 gm. of water. The change in freezing point, calculated from the preceding equation is:

$$\Delta T = \frac{(1.987)(273.1)^2(18.02)(60)}{(1436)(1000)(60)}$$

$$= 1.86°$$

For any aqueous solution, therefore, the molality (m) may be calculated from the simplified equation:

$$m = \frac{\Delta T}{1.86}$$

REFERENCES

Balances

1. Blaedel, W. J., and Meloche, V. W.: Elementary Quantitative Analysis: Theory and Practice. 2nd ed. New York, Harper & Row, Publishers, 1963.
2. Boutwell, J. H.: Clinical Chemistry. Laboratory Manual and Methods. Philadelphia, Lea & Febiger, 1961.
3. MacFate, R. P.: Introduction to the Clinical Laboratory. 2nd ed. Chicago, Year Book Medical Publishers Inc., 1966.
4. Sorum, C. H.: Fundamentals of General Chemistry. 2nd ed. Englewood Cliffs, N.J., Prentice-Hall, Inc., 1963.

Electroanalytical Chemistry

5. Delahay, P.: New Instrumental Methods in Electrochemistry. New York, Interscience Publishers, Inc., 1954.
6. Kolthoff, I. M.: Polarography. 2nd ed. New York, Interscience Publishers, Inc., 1952.
7. Lingane, J. J.: Electroanalytical Chemistry. 2nd ed. New York, Interscience Publishers, Inc., 1958.
8. Meites, L.: Polarographic Techniques. 2nd ed. New York Interscience Publishers, Inc., 1965.
9. Purdy, W. C.: Electroanalytical Methods in Biochemistry. New York, McGraw-Hill Book Co., Inc., 1965.

Spectrophotometric Analysis

10. Meloan, C. E.: Instrumental Analysis Using Spectroscopy. Philadelphia, Lea & Febiger, 1968, vol. 1.
11. Strobel, H. A.: Chemical Instrumentation. Reading, Mass. Addison-Wesley Publishing Co., Inc., 1960.
12. Willard, H. H., Merritt, L. L., Jr., and Dean, J. A.: Instrumental Methods of Analysis. 4th ed. Princeton, N.J., D. Van Nostrand Co., Inc., 1965.

Flame Photometry

13. Dean, J. A.: Flame Photometry. New York, McGraw-Hill Book Co., Inc., 1960.
14. Meloan, C. E.: Instrumental Analysis Using Physical Properties. Philadelphia, Lea & Febiger, 1968, vol. 2.

Atomic Absorption

15. Kahn, H. L.: J. Chem. Educ., *43*, 1966.
16. Robinson, J. W.: Atomic Absorption Spectroscopy. New York, Marcel Dekker, Inc., 1966.
17. Walsh, A.: Spectro. Chim. Acta., *7*, 108, 1955.
18. Willis, J. M.: Nature, *207*, 715, 1965.
19. Zettner, A., and Seligson, D.: Clin. Chem., *10*, 869, 1964.

Fluorometry

20. Hercules, D. M.: Fluorescence and Phosphorescence Analysis. New York, Interscience Publishers, Inc., 1966.
21. Udenfriend, S.: Fluorescence Assay in Biology and Medicine. New York, Academic Press, Inc., 1962.
22. Rubin, M.: Fluorometry in Clinical Chemistry. Progress in Clinical Chemical Methods. Basel/New York, Karger, A. G., S., 1968, p. 6.

Auto Analyzers

23. Thiers, R. E., Cole, R. R., and Kirsch, W. J.: Clin. Chem., *13*, 451, 1967.
24. Thiers, R. E., and Oglesby, K. M.: Clin. Chem., *10*, 246, 1964.
25. White, W. L., Erickson, M. M., and Stevens, S. C.: Practical Automation for the Clinical Laboratory. St. Louis, The C. V. Mosby Co., 1968.
26. Young, D. S., Montague, R. M., and Snider, R. R.: Clin. Chem., *14*, 993, 1968.

Chromatography

27. Giddings, J. C.: Dynamics of Chromatography. New York, Marcel Dekker, Inc., 1965, vol. 1.
28. Giddings, J. C., and Keller, R. A.: Advances in Chromatography. New York, Marcel Dekker, Inc., 1965.
29. Helfferich, F.: Ion Exchange. New York, McGraw-Hill Book Co., Inc., 1962.

Gas Chromatography

30. Dal Nogare, S., and Juvet, R. S., Jr.: Gas Liquid Chromatography. New York Interscience, Publishers, Inc., 1962.
31. Eik-Nes, K. B., and Horning, E. C.: Gas Chromatography of Steroids. New York, Springer-Verlag, Inc., 1968.
32. McNair, H. M., and Bonelli, E. J.: Basic Gas Chromatography. Walnut Creek, Calif., Varian Aerograph, 1968.
33. Purnell, H.: Gas Chromatography. New York, Wiley, 1962.
34. Wotiz, H. H., and Clark, S. J.: Gas Chromatography in the Analysis of Steroid Hormones. New York, Plenum Press, 1966.

Gasometry

35. Van Slyke, D. D., and Neill, J. M.: J. Biol. Chem., *61*, 523, 1924.
36. Van Slyke, D. D., and Stadie, W. C.: J. Biol. Chem., *49*, 1, 1921.
37. Woolmer, R. F., and Parkinson, J.: pH and Blood Gas Measurement. Boston, Little, Brown & Co., 1959.

Osmometry

38. Daniels, F., and Alberty, R. A.: Physical Chemistry. New York, John Wiley & Sons, Inc., 1955.
39. Haraway, A. W., and Becker, E. L.: JAMA, *205*:506, 1968.

Chapter 3

PRINCIPLES OF MICROTECHNIQUE

by Wendell T. Caraway, Ph.D.

In analytical chemistry, microchemical analysis usually involves the determination of the elemental composition of only a few milligrams of sample. *Ultra*microanalysis refers to the study of samples in the microgram range or the determination of a substance present in microgram quantities in a larger sample. In clinical chemistry these concepts are less applicable inasmuch as the limiting factor is most often the *volume* of sample that can be obtained.

Before 1912 the quantitative determination of glucose or uric acid required 25 to 50 ml. of blood for a single determination. Within the next decade a number of "microtechniques" were developed that required only 1 or 2 ml. of blood, a sample conveniently obtained from adults by venipuncture. A marked expansion in the number of tests for diagnostic purposes, plus the need for multiple determinations in infants and children, has required development of techniques in which the size of sample must be reduced to a few hundredths or even a few thousandths of a milliliter for a single determination. As a result, clinical microchemistry has evolved to the point where we can determine up to 10 constituents on a single sample of capillary blood.

An adequate capillary puncture can yield 0.4 to 1 ml. of blood. Since serum is required for nearly all chemical determinations, it follows that a total of 0.20 to 0.5 ml. of serum is available, or approximately 0.02 to 0.05 ml. of serum per determination if 10 determinations are to be done. One microliter (μl.) is equal to 0.001 ml.; consequently, we can refer to this level of operation (20 to 50 μl.) as microliter analysis. There is no real value in defining further such terms as micro-, semimicro-, submicro-, or ultramicromethods of analysis since such a division is relative and subject to change with further improvement in techniques.

It is of interest to note that in 1878 Gower described a method for determination of hemoglobin in 20 μl. of blood. In contrast, the determination of protein-bound iodine in serum is regarded clinically as a macromethod (1 ml. of serum), although the amount of iodine in the sample is only about 0.05 μg. Before we become complacent, however, we should consider that most current methods for determination of plasma estrogens require collection of 25 to 50 ml. of blood.

The value of clinical micromethods is well established. It may be either difficult or undesirable to perform venipuncture or to withdraw large volumes of blood for chemical analyses from infants, obese subjects, or patients in shock or with severe burns, especially when results may be obtained with equal reliability from capillary blood. Experimental laboratory work on small animals is also facilitated by the use of such techniques. The discussion that follows is intended to illustrate the general approach used in the development of micromethods, techniques for collection and handling of capillary blood, and specialized equipment currently available.

DEVELOPMENT OF MICROMETHODS

An approach frequently applied in the laboratory, when the sample size is limited, is to use one half the specified quantity of serum, make up the difference with water, analyze by the usual procedure, and multiply the final result by 2. Further scaling down of sample is undesirable for several reasons: (1) The sensitivity is decreased as a result of decreased scale deflections on spectrophotometers or divisions on burets. (2) Reagent blanks become more critical as their values approach that of the sample. (3) Any errors in technique will be multiplied in proportion to the dilution factor. (4) Decreased protein concentrations may lead to changes in pH, which invalidate the final readings, especially in enzyme studies in which the pH optimum may be relatively narrow.

Macroprocedures are usually designed to produce near optimal readings for specified reagents, wavelengths, cuvet diameters, and so on. Consequently, it is necessary to introduce modifications when working with smaller samples. These may be designated as *micromicro-* and *micromacro*modifications. The following considerations apply mainly to spectrophotometric procedures.

Micromicromethods

In this method the volumes of both sample and reagents are scaled down proportionately; i.e., both sample and reagent volumes are brought into the microliter range and micropipets are used for both. Since the final volume of solution is also reduced proportionately, it is desirable to use cuvets that maintain approximately the same light path as those used in the macroprocedure, but that permit readings on greatly reduced volumes. The Lowry-Bessey cell provides a 10 mm. light path, but requires only 0.1 ml. of solution for readings on a modified Beckman DU spectrophotometer (Beckman Instruments, Inc., Fullerton, Calif.).[3,7,8] Thus, a macroprocedure that requires 1 ml. of serum and produces a final volume of 10 ml. of solution could be scaled down, theoretically, to use 10 μl. of serum and provide a final volume of 0.1 ml. of solution with no sacrifice in absorbance. In practice, however, certain general principles must be observed when scaling down an existing method:

1. If possible, avoid precipitation of proteins so that the entire sample may be used rather than an aliquot of supernatant.

2. Avoid dilutions to volume. A small volume of diluent may be pipetted with greater accuracy than would be obtained by diluting serum to the mark in a small volumetric flask.

3. Centrifuge to remove precipitates. Avoid filtrations; appreciable volumes of filtrate are absorbed on the paper or lost by evaporation.

4. Maintain concentrations and relative volumes similar to those used in standardized macromethods. A change in reaction conditions requires more thorough evaluation of the modified procedure.

5. Use smaller glassware to minimize evaporation and surface effects.

6. Use special micropipets and microburets to maintain precision. As the volume delivered between two marks on conventional pipets or burets is decreased, the relative meniscus error is increased.

7. Combine reagents, when possible, to reduce the number of pipettings and associated errors.

8. Adapt procedures to commercially available equipment because specialized apparatus is expensive to construct and presents maintenance problems.

Micromacromethods

A micropipet is used only to sample serum in this approach. Subsequent volumes of reagents are added from conventional macropipets, and final volumes are kept to the minimum required by the available photometer or spectrophotometer. Consider, for example, in the determination of blood sugar, the classic Folin-Wu method in which 2 ml. of blood is brought to 20 ml. in the process of precipitating proteins. Next, 2 ml. of filtrate (equivalent to 0.2 ml. of blood) is used in a procedure that results in 25 ml. of solution, of which 5 ml. (equivalent to 0.04 ml. of blood) might be used for the final reading at 420 nm. It is readily apparent that most of the sample has been discarded along the way, and, in addition, a two to threefold increase in absorbance may be obtained by reading at 660 nm., rather than at 420 nm.

Equal sensitivity and an equivalent final volume can be achieved by starting with 0.05 ml. of sample. For example: Add 1 ml. of combined tungstic acid reagent to 0.05 ml. of serum, centrifuge, and remove 0.5 ml. of supernatant for analysis. Other reagent volumes are reduced to one fourth of those in the standard method; the final volume is brought to 5 ml. and the absorbance is measured at 660 nm.

The micromacromethod has been described for use with the Coleman Jr. spectrophotometer (Coleman Instruments, Maywood, Ill.).[2] It is readily integrated with existing macrotechniques since no special glassware or equipment is necessary other than the micropipets for initial sampling of serum. Adaptation of micromacromodifications to other instruments depends largely on the minimum volume of final solution required for reading. Modification of existing, well established macromethods is not always straightforward. Such modifications must be evaluated thoroughly against the original method and should include specimens with low, normal, and elevated values, as well as specimens showing hemolysis, lipemia, or jaundice.

Rather than modifying existing classic methods, we may employ a more sensitive chemical procedure that enables us to start with a few microliters of serum, but to end up with a conventional volume of final solution for spectrophotometric or fluorometric analysis. For example, the Clark-Collip method for serum calcium requires 2 ml. of serum, precipitation of calcium as the oxalate, and final titration of oxalic acid with permanganate. A recently described fluorometric method requires only 20 μl. of serum. Urea nitrogen determination by the urease-Nessler technique requires approximately 0.5 ml. of serum, but, by use of the very sensitive Berthelot reaction for ammonia, urea nitrogen may be determined on as little as 10 μl. of serum, eventually diluted to a final volume of 10 ml. The Berthelot reaction depends on the formation of intensely colored indophenol produced from phenol and alkaline hypochlorite in the presence of ammonia.

Such techniques are, of course, very sensitive to contamination. For example: calcium binds to glass easily, but may be removed by complexing agents used in the determination; distilled water or reagents may contain traces of ammonia that become critical at these minute levels of analysis. These methods, however, can be extremely sensitive, highly specific, and readily adaptable to the micromacroapproach.

Selection of Micromethods

The choice of approach to microchemical methods depends on the volume and variety of tests requested, the equipment and personnel available, and individual preference. In a children's hospital or in a large pediatric unit a separate microchemical laboratory with specialized personnel is practicable. When the volume of microwork is less and must be absorbed into the work load of regular personnel, it is preferable to adopt micromodifications of existing macromethods and as much as possible to use the same reagents and familiar equipment. This also provides more efficient coverage for emergency situations and lessens the demand for specially skilled personnel.

Many laboratories now perform a major portion of their clinical chemistry determinations by automation; however, since no automated equipment for microliter analysis is commercially available at this time, manual techniques are still required for microchemical analyses, as well as for certain emergency or nonautomated determinations. There is a trend at present to develop more sensitive procedures and to use smaller volumes of serum for both routine and microchemical analyses.

The sensitivity of a method depends upon the analytical approach and type of instrumentation. Some typical examples are shown in Table 3-1. The volume of sample shown is that normally taken for analysis; the absolute amount of the substance

TABLE 3-1. *Sensitivity of Chemical Procedures*

Determination	*Method*	*Initial Volume of Sample* (ml.)	*Final Amount of Substance Measured* (μg.)
Glucose	Folin-Wu	1.00	200
	Glucose oxidase	0.02	20
Urea nitrogen	Urease-Nessler	0.50	15
	Berthelot	0.02	3
Calcium	Clark-Collip	2.00	200
	EDTA titration	0.10	10
	Calcein-fluorometric	0.02	2
Chloride	Schales-Schales	0.50	1800
	Chloridometer	0.02	75
Cholesterol	Schoenheimer-Sperry	1.00	240
	Ferric chloride	0.20	80
	Fluorometric	0.02	8
	Gas chromatography	0.20	1
Phenylalanine	UV-spectrophotometric	1.00	5
	Fluorometric	0.025	0.25

given is that normally measured in the final step. In general, fluorometric methods are more sensitive than colorimetric methods. Gas chromatographic methods are capable of both high sensitivity and specificity.

COLLECTION AND HANDLING OF BLOOD SAMPLES

The major chemical difference between arterial and venous blood is the degree of oxygen saturation of the hemoglobin. Blood is oxygenated while passing through the lungs and is then pumped from the left ventricle of the heart into the arterial circulation. A slight hydrostatic pressure still exists in the arterial end of the capillaries; hence, blood collected by capillary puncture is essentially arterial blood. Oxygen is partially removed from the blood by tissues surrounding the capillary bed. The blood then returns through the venous circulation to the heart.

For most practical purposes there is little difference in the chemical composition of plasma obtained from arterial or from venous blood. Glucose is removed from arterial blood by the tissues and metabolized in part to lactate, which then appears in venous blood in higher concentration than in arterial blood. In the fasting state, the concentration of glucose in capillary (arterial) blood is approximately 5 mg./100 ml. higher than in venous blood. During a glucose tolerance test, however, this difference may approach 30 to 50 mg./100 ml. For this reason the source of the specimen should be consistent throughout the test.

Collecting Tubes

Capillary collecting tubes approximately 150 mm. in length with a 1.5 mm. bore and tapered ends are available commercially or can be made as follows: Clean, dry, borosilicate tubing (Pyrex or Kimax), 3 mm. O.D. (1.5 to 1.8 mm. I.D.), is cut into approximately 27 cm. lengths. The middle portion of the tubing is heated in a gas flame, drawn out to form a bore of about 0.8 mm., broken in two, and the ends fire polished. The tubes are cleaned thoroughly and dried in an oven at 110°C. Each tube has a capacity of approximately 0.25 ml. After filling with blood, the larger end may be sealed either with sealing wax, by heating in a small flame to close the larger end, provided care is taken to avoid overheating and hemolyzing the blood, or by covering with a special plastic cap (Critocap). The last technique is preferred. As a general rule anticoagulants are not used because fibrinogen interferes with electrophoretic separations and other protein fractionation techniques, and ammonium ions from ammonium heparinate interfere with the determination of urea by techniques that employ urease. Heparinized tubing, however, will provide a somewhat greater volume of plasma, and such specimens are easier to centrifuge and are preferred by some workers for certain determinations.

Similar collecting tubes, 75 to 100 mm. long, can be made from 4 mm. O.D. tubing with a bore of 2.5 to 2.7 mm. Blood may also be collected directly in 6 × 50 mm. glass test tubes or in small polyethylene centrifuge tubes (Beckman Instruments Inc., Spinco Div., Palo Alto, Calif.).

Collection of Blood

The site selected for puncture is cleaned with 70 per cent ethyl or isopropyl alcohol and allowed to dry. In adults, the fingertip is the usual site although some prefer to sample from the ear lobe; in infants, either the heel, the big toe, or the sole of the foot is satisfactory. Better flow of blood is obtained if the extremity is first placed in warm water or wrapped with a warm towel. The puncture is made with a Bard-Parker No. 11 scalpel blade or other type of lancet and should be sufficiently deep to permit a free flow of blood without squeezing. The first drop of blood is wiped off with dry gauze. When a large drop of blood has appeared, the tip of the collecting tube is touched to the drop while the tube is held slightly downward from a horizontal position. The blood enters the tube by capillarity. The tube is filled to within about 1 cm. of the larger end with care to avoid the introduction of air bubbles. Several tubes may be filled, if required, from a single puncture. Excessive squeezing of tissues must be avoided because this can cause hemolysis and contamination with tissue juices.

The tubes are capped and placed in a test tube for transporting and centrifuging. A free-swinging, horizontal type of centrifuge is recommended since the capillary tubes tend to break or leak in an angle head centrifuge. When analyses must be delayed, the tube is scratched with a file and broken above the juncture of clot and serum. The serum tube is then recapped and refrigerated.

To collect blood in small test tubes or centrifuge tubes, the surface of the puncture site is held in a nearly vertical position and the drops of blood are permitted to flow freely into the tube. The lip of the tube should not be scraped over the surface as this may cause hemolysis. Up to 1 ml. of blood may be collected from a satisfactory puncture. The tubes are placed in larger tubes for centrifuging and sampling or may be centrifuged directly in special microcentrifuges. As with macromethods, the serum should be removed from the clot and refrigerated when analysis must be delayed.

If desired, certain determinations, e.g., glucose and urea, may be performed on whole blood. In this event, blood may be sampled directly into a micropipet, then washed out into a measured volume of one component of a protein precipitating reagent that changes the pH sufficiently to prevent glycolysis.

MICROPIPETS

Various types of micropipets are illustrated in Figure 3-1. Volumes are often expressed in lambdas. One lambda (λ) = 1 μl. = 0.001 ml. Clinical laboratory personnel are familiar with the Sahli hemoglobin pipet calibrated to contain 0.02 ml. (20 μl.). Blood is drawn just above the mark, the outer surface of the pipet is wiped clean with tissue paper, and the blood is adjusted to the mark by touching the tip repeatedly to the paper. The contents are then delivered by rinsing several times into a diluent suitable for hemoglobin determination. The diluent should not be drawn above the calibration mark because this will add a film of sample in excess of the calibration. The same technique may be used to sample serum for microchemical analyses. Similar pipets calibrated to contain (TC) are available in various sizes. Larger sizes have a bulb to decrease the surface area to volume ratio and a small bore to facilitate accurate adjustment to the calibration mark.

To sample serum from a capillary tube, the tube is etched with a file above the clot, held horizontally, and broken in two. The tip of the pipet is held against the cut end of the tube of serum and the pipet is filled by gentle suction (Figure 3-2). The Sahli type of pipet is convenient because no new techniques need be introduced; however, the rather large blunt tip requires extra care in wiping all excess serum from the outer surface. In addition, the tip may be too large to sample serum from blood collected in 6 × 50 mm. test tubes. The Kirk pipet has the tip drawn out to a finer

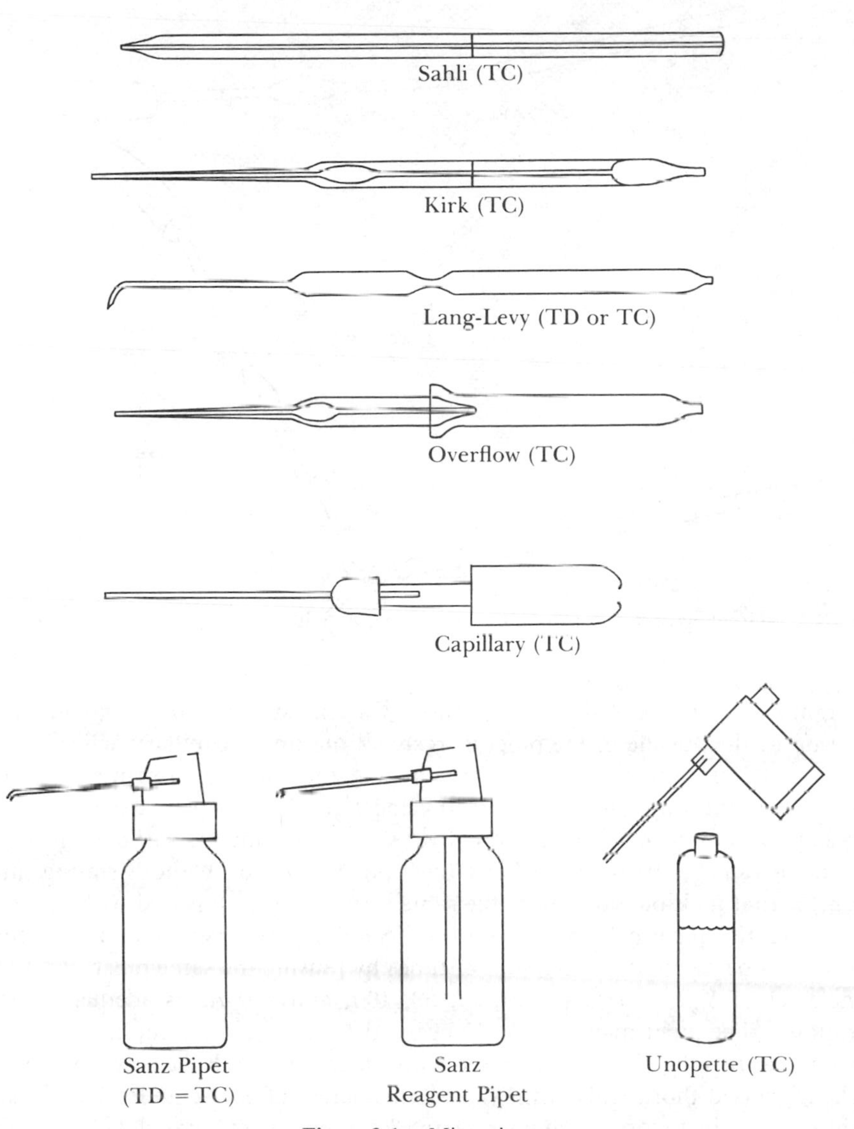

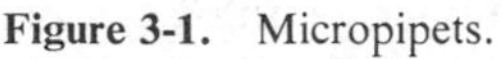

Figure 3-1. Micropipets.

point and thus overcomes these objections. Kirk pipets may be obtained in sizes as small as 1 μl (Microchemical Specialties Co., Berkeley, California); however, 10 μl. is a practical lower limit for microliter analysis of serum.

The Lang-Levy pipet is self-adjusting and may be calibrated either to contain (TC) or to deliver (TD). The upper portion has a constriction with a bore slightly less than the bore at the tip so that the liquid will be held in the pipet by surface tension. When used TD, a rubber aspirator tube is attached and the liquid is drawn into the pipet just above the constriction (Figure 3-2). Suction is released and, with the pipet held nearly vertical and with the tip of the pipet still below the surface of the liquid, the solution in the pipet stops automatically at the constriction. The pipet is withdrawn and the bent tip placed against the side and near the bottom of the receiving tube. Gentle pressure is now applied and, as the last of the liquid is forced out, the

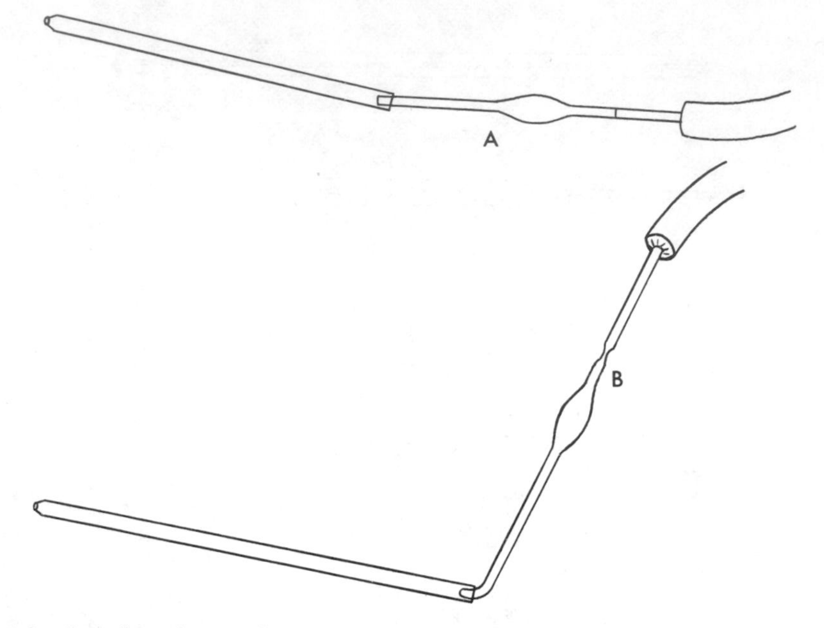

Figure 3-2. Sampling of serum from a capillary tube, A, with a Kirk pipet, B, with a Lang-Levy pipet.

tip is removed from the wall of the test tube. Care must be taken that no liquid clings in a drop to the outside of the pipet or reenters the tip by capillary action. For TC operation the procedure is the same except that the contents are rinsed out several times into a receiving solution. Polyethylene Lang-Levy pipets are also available. They have a nonwetting surface and deliver the entire contents without rinsing.

The advantage of using the Lang-Levy pipet TD (i.e., without rinsing into the diluent) is that multiple samples of the same serum may be pipetted without washing and drying the pipette between samples. Similarly, in micromicro-operations a reagent may be pipetted to a number of tubes by reusing the same pipet. In practice, a single set of pipets, sizes 10, 20, 25, 50, 100, and 200 μl., is adequate for most operations. The pipet may be rinsed once with serum before sampling for analysis. Between reagents the pipet is first rinsed with water, then with the succeeding reagent. Pipets are rinsed thoroughly with water immediately after use and stored in a horizontal position in a closed container. Lang-Levy pipets, calibrated TC, may also be used TD provided the standard and unknown are delivered from the same pipet.

Overflow type pipets are calibrated TC. Smaller sizes fill automatically by capillary action; gentle suction is applied to fill larger sizes. Contents are rinsed out in the manner that was just described. This type pipet is more difficult to clean when sample has overflowed into the holder around the upper tip. Precision-bore disposable capillary tubes calibrated TC are also available (e.g., Drummond Scientific Co., Broomall, Pa.). The capillary is inserted into a holder attached to a rubber bulb that has a small opening in the top. The bulb is squeezed gently, the opening closed with the fingertip, the tube filled with sample, and suction released by removing the fingertip. To dispense, the opening is again closed and contents expelled by gentle pressure. Rinsing is performed in the same manner.

An interesting variation of the overflow pipet has been described by Sanz.[12] A calibrated polyethylene pipet is fitted into a stopper inserted into the side of the cap of a small polyethylene bottle. The cap of the bottle is a transparent dome with an opening in the top. The operation is comparable to that described for the Drummond capillary tubes. By inserting a vertical tube into the bottle, the device is converted into a reagent pipet. To use, the bottle is squeezed until the liquid rises into the dome and fills the pipet. On release of pressure the remaining liquid flows back into the bottle. The opening at the top is then closed with a fingertip and gentle pressure is applied to expel the reagent. The polyethylene is nonwetting, when thoroughly cleaned, and contents are delivered completely. Thus, the TD and TC volumes are the same. The polyethylene pipetting devices are an integral part of the Spinco ultramicroanalytical system.

A disposable self-filling pipet attached to a polyethylene reagent reservoir has also been described.[13] This unit is sold under the trade name Unopette (Becton, Dickinson & Co., Rutherford, N.J.). A glass capillary pipet fitted in a plastic holder fills automatically with serum or whole blood. The reagent bottle is squeezed slightly while the pipet is inserted. On release of pressure the sample is aspirated into the diluent or precipitating reagent. Intermittent squeezing fills and empties the pipet to wash out the contents. The unit has been adapted for a number of chemical and hematological determinations.

Calibration of Micropipets

A pipet may be calibrated TC by filling with mercury, expelling the mercury into a tared container, and weighing it on an analytical balance. The weight of mercury is divided by its density to determine the volume of the pipet. The problem is to control the level of mercury while adjusting to the calibration mark. This is accomplished as follows. Grease the plunger of a 2 ml. syringe with stopcock grease to provide an airtight seal. Insert the hub of the barrel into a one-hole No. 0 rubber stopper; then mount the syringe in a clamp on a ring stand (Fig. 3-3). Insert the upper end of a

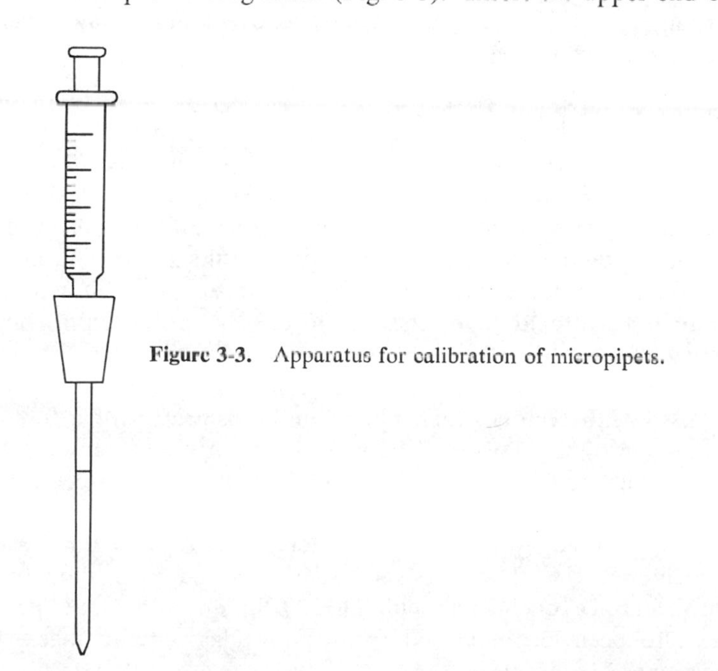

Figure 3-3. Apparatus for calibration of micropipets.

clean, dry micropipet into the other side of the stopper. Pour some clean dry mercury into a small beaker. Draw up the plunger slightly; then raise the beaker until the tip of the pipet is slightly below the surface of the mercury. Raise the plunger slowly until the mercury meniscus is adjusted exactly to the calibration mark; then withdraw the beaker downward in a clean sharp motion. Bring a small tared beaker under the tip and depress the plunger to expel all the mercury. Obtain the weight of mercury and divide by the density of mercury at its observed temperature to find the volume of the pipet. The density of mercury at 25 $\pm$ 5°C. is 13.5340 $\pm$ 0.0123 or $\pm$0.091 per cent; therefore, for temperatures between 20 and 30°C. it is sufficiently accurate to use the density at 25°C.

Micropipets may be calibrated TD (or TC) by various titrimetric or spectrophotometric procedures. Thus, a concentrated solution of sodium chloride may be sampled, diluted, and analyzed for either sodium or chloride and compared with a standard accurate macrodilution. Similarly, 1 volume of 4 per cent potassium ferricyanide solution may be delivered (for TD) or washed out (for TC) into 200 volumes of water to provide a 1:201 microdilution. Use 20 μl. of solution plus 4.00 ml. of water, 50 μl. of solution plus 10.00 ml. of water, and so on. A 1:201 macrodilution is prepared by adding 1.00 ml. of solution to a 200 ml. volumetric flask previously filled to the mark with water. The absorbance of each dilution is then measured in a spectrophotometer at 420 nm. against a water blank. An average of triplicate determinations should be used.

$$\frac{\text{Average } A\text{, micro}}{\text{Average } A\text{, macro}} \times \text{nominal capacity } (\mu\text{l.}) = \mu\text{l. of sample}$$

Similar micro- and macrodilutions may be made from radioiodinated serum albumin. Equal aliquots are pipetted and counted in a scintillation detector to the desired degree of precision. Calculations are the same as with the spectrophotometric method.

Pipets can be calibrated TC more accurately than TD. Of the various glass micropipets, only the Lang-Levy type is used TD and it may be calibrated to $\pm$1 per cent accuracy by the techniques just described.

Cleaning Glassware

Because of the small sample size in microwork the effect of possible contamination is much greater; thus, all glassware must be scrupulously cleaned. An entire sample of 20 μl. of serum contains only about 64 μg. of sodium. Detergents frequently contain trisodium phosphate or other sodium salts and preferably should be avoided. A satisfactory procedure is to immerse glassware in, or fill it with, a 1:100 dilution of ammonium hydroxide and let stand to dissolve protein films or precipitates. The glassware is then rinsed several times, in turn, with water, dilute hydrochloric acid (1:100), water, and distilled water before drying at 100°C. Pipets, after cleaning, may be rinsed with acetone, then placed on an aspirator for a few minutes and dried by a stream of air. Polyethylene pipets and other ware may be cleaned in a similar manner, but are never oven dried since softening and distortion occur above 80°C.

For more thorough cleaning, glassware is soaked in chromic acid (sulfuric acid) solution followed by rinsing with water, dilute ammonium hydroxide, and dilute hydrochloric acid. The glassware is then soaked for 2 days in distilled water to remove all traces of chromic acid. Dilute solutions of pepsin (1 per cent in 0.1 N HCl) have also been recommended for soaking glassware to digest protein films and precipitates.

ANALYTICAL METHODS

A detailed description of various analytical micromethods is beyond the scope of this chapter; however, there are several books on the subject.[2,6,7,9–11] The following discussion will be concerned with the application of various analytical techniques to microliter analysis and with special equipment used in micromethods.

Gravimetric Methods

Because of the limited sample size there has been little or no application of gravimetric methods to clinical microchemistry. Kirk[6] describes the quartz fiber microbalance, which permits weighings on the order of 1 μg. or less, but such measurements are rarely made in the routine clinical laboratory.

Titrimetric Methods

Commonly used procedures involve titration of chloride with mercuric nitrate and titration of calcium with EDTA solutions (see Chapter 10, Electrolytes). These methods have been modified for microchemical determinations. Ammonia and carbon dioxide may also be determined titrimetrically after separation by diffusion techniques.

Microburets are used in titrimetric procedures. They are constructed on the principle of a syringe. The plunger can be machined to precision diameter. The displacement by the plunger is indicated on a micrometer dial and is proportional to the volume delivered. A typical model is shown in Figure 3-4. Various sizes are available that permit accurate delivery of as little as 0.01 μl. per scale division. Several points should be noted.

1. The titrant is prepared at a relatively high concentration to reduce the volume necessary for a titration and to provide sharp end points. For example, if an end point can be determined to ± 1 scale division, the volume of titrant in an average titration should be equivalent to 100 scale divisions to provide a precision of ± 1 per cent. The concentration of titrant is a compromise between choosing a weaker concentration, which provides a larger volume and therefore a smaller reading error, and a stronger concentration, which minimizes end point error and avoids over-dilution of the sample.

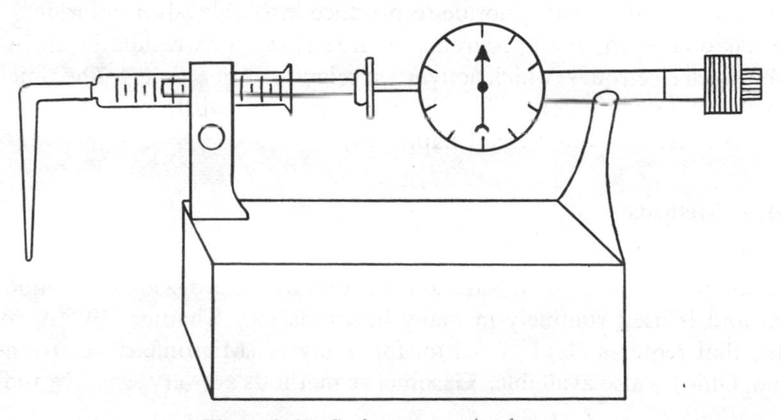

Figure 3-4. Syringe type microburet.

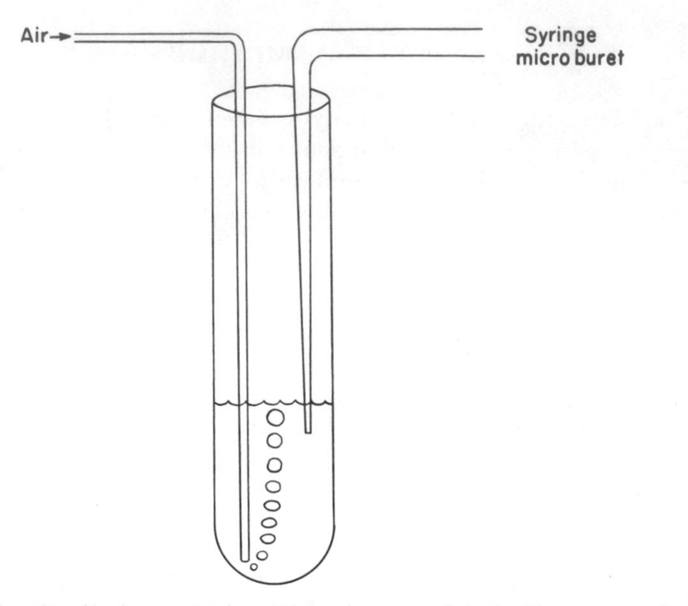

Figure 3-5. Microtitration in a test tube. Mixing is accomplished with a stream of air bubbles.

2. The tip of the pipet is placed beneath the surface of the solution to be titrated to permit even dispensing of reagent and to avoid discrete drop formation. The tip should be drawn to a fine point to prevent diffusion of liquid back into the pipet.

3. Titrations may be performed in white porcelain spot plates to permit easier detection of color changes. The solution is stirred with a small glass rod as the titration proceeds. For titrations in test tubes the solution may be mixed with a fine stream of air bubbles or an inert gas (Fig. 3-5). Other units employ vibrators or small cups that are rotated electrically to obtain mixing during the titration. Magnetic stirrers are also convenient. Very small stirring bars are available commercially.

4. Blanks and standards should be included routinely and titrated to the same end point as the unknown. Indicator error is more significant in microtitrations; it is therefore advisable to pipet a fixed volume of indicator rather than to add it dropwise. Duplicate determinations are recommended.

The Cotlove chloridometer is a coulometric instrument that generates silver ions in proportion to the time that a constant current is passed through the electrodes. The silver ions combine with chloride to produce insoluble silver chloride. When all chloride has combined, the appearance of free silver ions results in an increase in current between electrodes, which activates a relay to stop a timer. The time required is proportional to the chloride ion concentration. As little as 10 μl. of serum is sufficient for a determination (see Chapter 10).

Gasometric Methods

The major application of gasometric methods is in the determination of carbon dioxide content of serum. The Natelson manometric microgasometer requires 30 μl. of serum and is used routinely in many hospitals (see Chapter 10). A volumetric apparatus that requires 20 μl. of serum for analysis (Micrometric Instruments Co. Cleveland, Ohio) is also available. Gasometric methods are very sensitive and reliable.

They have been applied to the determination of various substances, but have limited use in routine work because only one determination can be performed at a time.

One problem encountered in the determination of carbon dioxide on capillary blood stems from exposure of the blood to air during collection. Some carbon dioxide may be lost during exposure; however, if collecting tubes are filled completely and promptly, there is little loss of carbon dioxide during the process. The tubes are capped or stoppered immediately and are kept stoppered during centrifugation and until ready for analysis. Results obtained in this manner are considered as reliable or more so than those obtained after transfer of serum and equilibration with alveolar air. Whole blood collected in this manner in heparinized tubes can also be used for pH determination provided the measurement is made without delay.

Flame Photometry

Most flame photometers are easily adapted to microtechniques. To conserve sample it is advisable, when possible, to perform both sodium and potassium determinations on the same dilution. It is more accurate to pipet the diluent rather than attempting to dilute the serum to the mark in a small volumetric flask. For example, if readings are normally made on a 1:100 dilution, the microdetermination would be made on a 1:101 dilution prepared by adding 50 μl. of serum to 5.00 ml. of diluent. Standards must also be prepared at 1:101 dilution. This is easily accomplished by filling a 100 ml. volumetric flask to the mark with diluent, and then adding 1.00 ml. of stock standard.

The diluent is 0.02 per cent Sterox for direct standard instruments such as the Coleman Jr. spectrophotometer. For internal standard instruments, such as the IL flame photometer, 25 μl. of serum or standard is added to 5.00 ml. of solution containing 15 mEq./L. of lithium. The final dilution is 1:201 rather than the usual 1:200. Similar principles can be applied to other flame photometers.

Excessive squeezing around the puncture site should be avoided when blood is collected, especially for electrolyte determinations because tissue juice contains a high concentration of potassium.

Spectrophotometric Methods

Despite the many recent developments in instrumentation, spectrophotometric methods are still the procedures of choice for most clinical chemical analyses. In general, methods based on color development are relatively specific, simple, rapid, and require less expensive or specialized equipment.

Criteria for microcuvets have been discussed in an earlier section. Various types of cuvets are illustrated in Figure 3-6. To obtain maximum absorbance on a limited amount of constituent, the effective light path is increased to the maximum and the final volume of solution is held to the minimum, consistent with ease of filling and handling the cuvets. To illustrate, assume that a given weight of colored substance is dissolved in the minimum volume of solution and absorbance readings then taken on various instruments. The data in Table 3-2 show that relative maximum absorbance for each solution is proportional to the light path of the cuvet and inversely proportional to the final volume required for measurement. Although horizontal cuvets provide maximum absorbance per unit volume it should be noted that these cuvets are

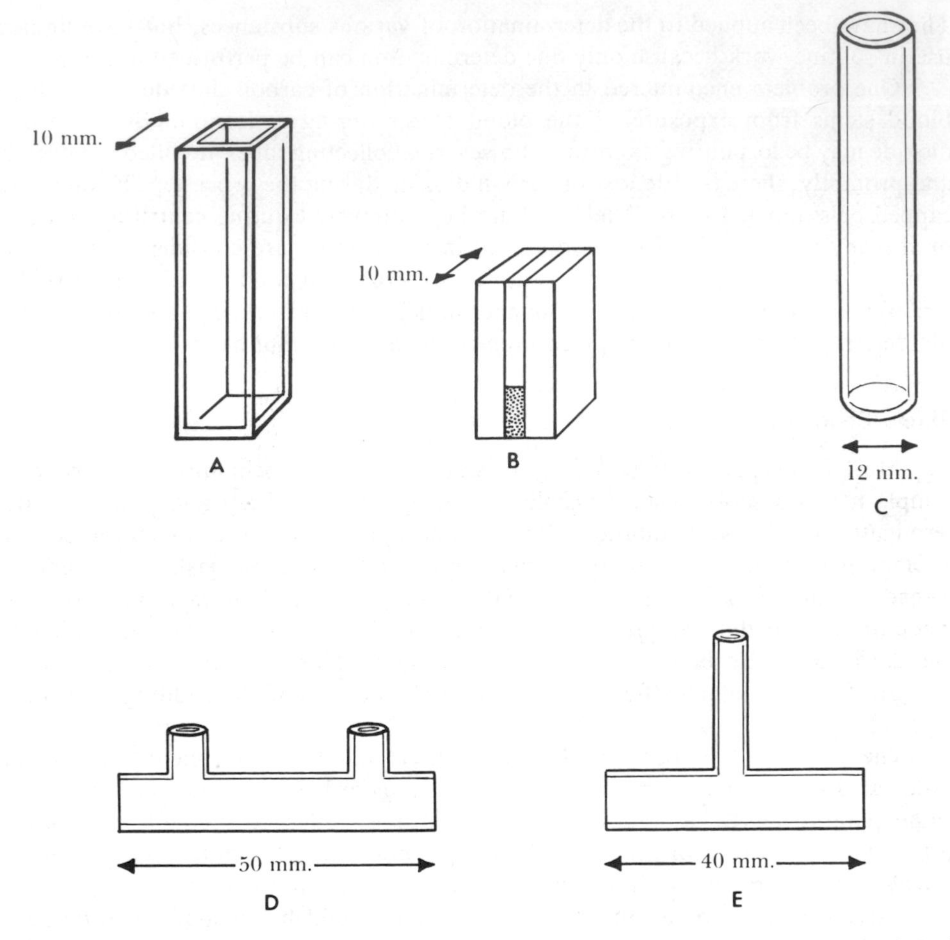

Figure 3-6. Cuvets. *A*, Beckman DU, 3 ml.; *B*, Lowry-Bessey, 0.1 ml.; *C*, Coleman Jr., 1.5 ml. *D*, Coleman-Universal, 2.5 ml.; *E*, Klett-Industrial, 2 ml.

TABLE 3-2. *Effective Light Path and Minimum Volume of Cuvets*

Instrument	*Cuvet*	*Light Path* (mm.)	*Minimum Vol.* (ml.)	*Efficiency* (mm./ml.)
Beckman DU	Regular	10	3.0	3.3
	Lowry-Bessey	10	0.1	100.0
Coleman Jr.	10 mm. O.D.	8	1.0	8.0
	12 mm. O.D.	10	1.5	6.7
	19 mm. O.D.	17	5.5	3.1
Coleman Universal	Horizontal	50	2.5	20.0
Klett, Clinical	Regular	12	5.0	2.4
	Micro	12	2.0	6.0
Klett, Industrial	Horizontal	40	2.0	20.0
Leitz, B & L	Regular	10	3.0	3.3
Spinco	Micro	6.4	0.1	64.0
Gilford	Micro	10	0.5	20.0

Note: For a given weight of constituent dissolved in the minimum volume shown, the absorbance is approximately proportional to the efficiency. The light path in the Coleman cuvets is assumed to be 2 mm. less than the outside diameter.

more difficult to fill, empty, and align in the light path than conventional upright cuvets. Certain spectrophotometers (Beckman Instruments Inc., Spinco Div., Palo Alto, Calif., and Gilford Instrument Labs Inc., Oberlin, Ohio) have been designed especially for microwork.

Some increase in absorbance as well as improved linearity with concentration are usually obtained with instruments that operate at narrower bandwidths of light. This is especially true for substances that exhibit a sharp peak of absorption. In Figure 3-7, curves are shown for two solutions, C and D, with markedly different spectral absorbance curves. It is apparent that solution C will show a much higher absorbance in a spectrophotometer that supplies monochromatic light with a 1 nm. bandwidth, as compared to an instrument with a 35 nm. bandwidth. On the other hand, solution D would have essentially the same absorbance with either instrument.

The wavelength selected is usually at the peak of maximum absorbance in order to achieve maximum sensitivity; however, it may be desirable to choose another wavelength to minimize interfering substances. For example, turbidity readings on a spectrophotometer are greater in the blue region than in the red region of the spectrum, but the latter region is chosen for turbidity measurements to avoid absorption of light by bilirubin (460 nm.) or hemoglobin (417 and 575 nm.). The color developed in the Liebermann-Burchard reaction for cholesterol in chloroform exhibits two peaks (420 and 660 nm.). The color developed in the alkaline picrate procedure for creatinine produces a relatively flat peak in the visible region at approximately 480 nm., but the reagent blank itself absorbs light strongly below 500 nm. A compromise is made by selecting a wavelength at 520 nm. to minimize the contribution of the blank.

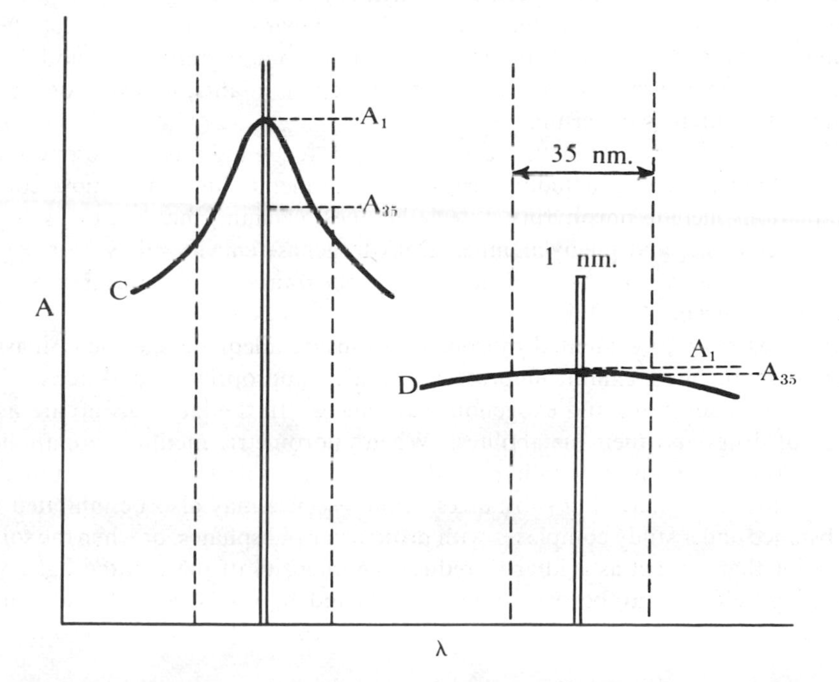

Figure 3-7. Dependence of absorbance on bandwidth of light and shape of absorption curve. Curve C exhibits a sharp peak of absorption; hence, the absorbance (A_1) with a 1 nm. band width is considerably greater than that obtained with a 35 nm. band width (A_{35}). Curve D exhibits a relatively flat absorption curve and the absorbance readings are essentially the same with either a 1 nm. or 35 nm. band width.

Blank readings should, of course, be kept as low as possible. A small difference between two large numbers is subject to greater uncertainty as shown in the following examples. The calculations indicate the extreme limits of error that could occur with a maximum relative error of ±3 per cent in reading the instrument. In practice, the random error should rarely reach these extremes.

	High Blank	*Low Blank*
Test	0.600 ± 0.018	0.233 ± 0.007
Blank	0.400 ± 0.012	0.033 ± 0.001
Test—blank	0.200 ± 0.030	0.200 ± 0.008
Maximum error	± 15%	± 4%

Fluorometric Methods

A substance is said to fluoresce when it absorbs light at one wavelength and emits light at a longer wavelength. In a photofluorometer, ultraviolet light (exciting source) is passed through a primary filter to screen out light above the desired wavelength. Part of the light energy is absorbed by the substance in solution. The emitted light, with less energy and longer wavelength, is passed through a secondary filter at right angles to the ultraviolet beam and is measured photometrically. A combination of secondary filters may be used to screen out undesired fluorescence from other compounds. More elaborate spectrofluorometers have facilities for dialing specific wavelengths for the exciting and secondary beams.

With appropriate methods and adequate instrumentation, fluorometric methods can be made very sensitive and specific. Phenylalanine, for example, can be measured readily on 25 μl. of serum by fluorometry; by comparison, an ultraviolet spectrophotometric method requires 1 ml. and a column chromatographic method 3 ml. of serum. Since the determination is usually requested on infants, it is very desirable to use microliter volumes of serum.

Fluorometric methods have been developed for hundreds of compounds of potential clinical interest, including drugs and drug metabolites. The more common determinations include porphyrins, catecholamines, calcium, magnesium, salicylate, quinidine, tyrosine, and phenylalanine. Dehydrogenase enzymes may be assayed in small amounts of serum by measuring the appearance or disappearance of the fluorescent coenzyme NADH.

Methods must be evaluated thoroughly to insure adequate specificity inasmuch as many compounds exhibit fluorescence under appropriate conditions. These include both endogenous and exogenous substances. In the latter group are a large number of drugs and their metabolites. When fluorometric methods are applied to measure products in an enzymatic procedure it is advisable to include serum blanks to correct for such interfering substances. Fluorescence may also be inhibited when the substance under study complexes with proteins or phosphates, or when the solution has a color that can act as a filter to reduce the intensity of the emitted light. Such "quenching" effects may be overcome or evaluated by use of an internal standard added to a second aliquot of sample and carried through the entire procedure.

Diffusion Methods

Relatively volatile substances can be removed from complex biological systems by isothermal distillation. Similarly, a reaction that results in formation of a gas, such

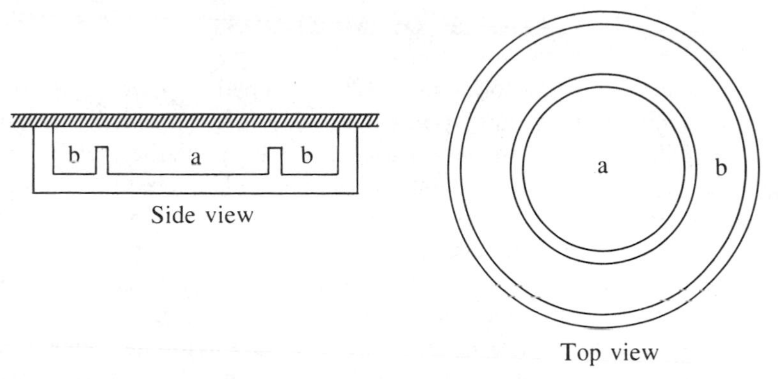

Figure 3-8. Conway diffusion cell; (a) inner well, (b) outer well.

as ammonia or carbon dioxide, can be followed by trapping the gas in a suitable absorbent for final analysis. The Conway diffusion cell has been designed to provide optimum rates of diffusion (Fig. 3-8).

In a typical analysis for urea nitrogen, a small volume of serum (20 μl.) and urease solution are pipetted into the outer ring and incubated to convert urea to ammonium carbonate. A volume of 50 per cent (w/v) potassium carbonate is pipetted into the opposite side of the outer ring in such a way that the solutions do not run together. Dilute boric acid, containing indicator, is pipetted into the inner well. The frosted surface of a glass plate is greased with stopcock grease and fitted on the dish to provide an airtight seal. The dish is then rotated at a slight angle to mix the solutions in the outer ring. The strong alkali releases ammonia, which diffuses out and is trapped in the boric acid solution. The rate of diffusion can be increased by placing the dish in a 37°C oven. When diffusion is complete the lid is removed and the ammonia in the inner well is determined by microtitration with dilute standard acid.

The solution in the inner well must have a positive trapping reaction for the gas in order to drive the diffusion to completion. The advantage of the diffusion methods is their relative specificity. The major disadvantage is that 30 to 60 minutes or longer may be required for diffusion. The cells also present problems in cleaning and storage to insure that no trace of residue remains from previous analyses. General theory of diffusion and description of methods are discussed by Conway.[4]

Electrophoresis and Chromatography

Electrophoretic separation of serum protein fractions is easily accomplished on 0.5 to 10 μl. of serum and is thus within the range of microliter analysis. Paper chromatography and thin-layer chromatography have been applied to the separation and estimation of amino acids in urine and to many other substances of clinical interest. Gas-liquid chromatography has also developed as a powerful tool to separate and measure microgram quantities of closely related materials. With this technique, some preliminary purification and preparation of derivatives is usually necessary before analyses can be performed, and the methods at present have not been adapted for routine application to microliter quantities of serum. Where appropriate, electrophoretic and chromatographic methods have been included elsewhere in the text.

CHOICE OF METHODS

Virtually all clinical laboratories should be equipped to perform determinations of glucose, urea nitrogen, sodium, potassium, chloride, carbon dioxide, pH, and bilirubin on capillary blood. In one large general hospital these eight procedures accounted for 75 per cent of all microtests performed. The next important group of tests includes total protein, albumin, calcium, phosphorus, and salicylate. Most enzyme determinations are easily adapted to the microliter scale by the selection of more sensitive techniques. Thus, NADH can be measured fluorometrically rather than spectrophotometrically at 340 nm. Simple scaling down of macroprocedures can be hazardous because of surface tension and colloid forces that tend to denature enzymes. Serum protein fractionation is readily performed by electrophoresis.

Results obtained by microtechniques should be comparable to, and reported in, the same units as macromethods to avoid confusion regarding normal values. It should be noted, however, that normal values for many constituents vary with age and may be considerably different than those for adults. The book by Behrendt is helpful in this respect.[1] Further information on normal values for children may be found in other texts[5,9,11] and in the table of normal values in the appendix of this book.

So many microchemical procedures have been developed that no two chemists are likely to agree as to which are the "best" methods. Once the decision has been made to introduce microtechniques, the analyst should survey the field and choose procedures or systems that appear most convenient for his laboratory. The methods chosen should have demonstrated accuracy and precision and, ideally, require not more than 50 μl. of serum for a single determination.

REFERENCES

1. Behrendt, H.: Diagnostic Tests in Infants and Children. 2nd ed. Philadelphia, Lea & Febiger, 1962.
2. Caraway, W. T.: Microchemical Methods for Blood Analysis. Springfield, Ill., Charles C Thomas, Publisher, 1960.
3. Caraway, W. T., and Fanger, H.: Ultramicro procedures in clinical chemistry. Am. J. Clin. Path., *25*:317, 1955.
4. Conway, E. J.: Microdiffusion and Volumetric Error. 5th ed. London, Crosby Lockwood & Son Ltd., 1962.
5. Henry, R. J.: Clinical Chemistry. Principles and Technics. New York, Harper & Row, Publishers, 1964.
6. Kirk, P. L.: Quantitative Ultramicroanalysis. New York, John Wiley & Sons, Inc., 1950.
7. Knights, E. M., Jr., MacDonald, R. P., and Ploompuu, J.: Ultramicro Methods for Clinical Laboratories. 2nd ed. New York, Grune & Stratton, Inc., 1962.
8. Lowry, O. H., and Bessey, O. A.: The adaptation of the Beckman spectrophotometer to measurements on minute quantities of biological materials. J. Biol. Chem., *163*:633, 1946.
9. Meites, S., and Faulkner, W. R.: Manual of Practical Micro and General Procedures in Clinical Chemistry. Springfield, Ill., Charles C Thomas, Publisher, 1962.
10. Natelson, S.: Microtechniques of Clinical Chemistry. 2nd ed. Springfield, Ill., Charles C Thomas, 1961.
11. O'Brien, D., Ibbot, F. A., and Rodgerson, D. O.: Laboratory Manual of Pediatric Micro- and Ultramicro- Biochemical Techniques. 4th ed. New York, Harper & Row, Publishers, 1968.
12. Sanz, M. C.: Ultramicro methods and standardization of equipment. Clin. Chem., *3*:406, 1957.
13. Walter, A. R., and Gerarde, H. W.: A rapid semiautomatic system of chemical analysis using true microspecimens. Clin. Chem., *10*:509, 1964.

Chapter 4

CARBOHYDRATES

by Wendell T. Caraway, Ph.D.

Carbohydrates account for the major food supply and source of energy for the people of the world. Depending on the economy, 50 to 90 per cent of the carbohydrates consumed come from grain, starchy vegetables, and legumes (rice, wheat, corn, potatoes, and so on). Other important sources are sucrose (cane and beet sugar, molasses), lactose (milk and milk products), glucose (fruits, honey, corn syrup), and fructose (fruits, honey). Meat products and sea food contain less than 1 per cent glycogen and do not contribute appreciably to the total carbohydrate intake. The glycogen content of fresh liver is approximately 6 per cent, but decreases rapidly to less than 1 per cent on standing at room temperature. Cellulose occurs in stalks and leaves of vegetables, but this carbohydrate is not digested by humans. Although a small portion may be split to glucose by bacterial action in the large bowel, this contribution is relatively insignificant.

Despite the major utilization of carbohydrate for energy, only a small amount is stored in the body. The average adult reserve is about 370 gm., stored chiefly as liver and muscle glycogen. Since 1 gm. of carbohydrate supplies 4 Calories, the total body store of carbohydrate would provide only 1480 Calories or approximately half the average daily caloric needs. When total caloric intake exceeds the daily expenditure the excess carbohydrate is readily converted to fat and stored as adipose tissue. This results in increased efficiency of food storage since 1 gm. of fat yields 9 Calories, i.e., more than twice that obtained from 1 gm. of carbohydrate.

The most common disease related to carbohydrate metabolism is diabetes, which is characterized by insufficient levels of active insulin. Deficiency of insulin results in impaired metabolism, an increase in blood glucose concentration, and secondary changes in fat metabolism, leading eventually to ketosis and possible diabetic coma (see Chapter 7, The Lipids). Other complications include hypercholesterolemia, atherosclerosis, and kidney disease. There are approximately two million recognized diabetics in the United States and probably two million more as yet undiagnosed cases. The overall incidence, therefore, is nearly 2 per cent of the entire population. The condition shows a strong familial tendency; the probability that an individual will develop diabetes is several times greater when there is a family history of the disease.

Early recognition of diabetes will permit earlier management and perhaps delay or minimize the complications of the disease. Overproduction or excess administration of insulin causes a decrease in blood glucose to levels below normal. In severe cases the resulting extreme hypoglycemia is followed by muscular spasm and loss of consciousness, known as "insulin shock." Measurements of blood glucose, therefore, assume considerable significance and the clinical laboratory must be prepared to furnish results rapidly and with a high degree of reliability.

The determination of blood glucose is the procedure most frequently performed in the clinical chemistry laboratory. In an average hospital this determination usually represents more than 20 per cent of the total number of chemical analyses requested. In addition, many glucose determinations are performed in clinics, independent laboratories, and physicians' offices as an aid in the diagnosis and treatment of diabetes. The determinations may be performed on patients in the fasting state, in the postprandial state, or in conjunction with glucose tolerance tests in accordance with the physician's request. As we shall see later, various factors other than insulin affect the blood glucose level and it is a mistake to assume that an elevated level is diagnostic for diabetes in the absence of other supporting information or confirmatory tests.

CHEMISTRY OF CARBOHYDRATES

The term carbohydrate refers to hydrates of carbon and is derived from the observation that the empirical formulas for these compounds contain approximately one molecule of water per carbon atom. Thus glucose, $C_6H_{12}O_6$, and lactose, $C_{12}H_{22}O_{11}$, can be written as $C_6(H_2O)_6$ and $C_{12}(H_2O)_{11}$, respectively. These compounds are not hydrates in the usual chemical sense, however, and noncarbohydrate compounds such as lactic acid, $CH_3CH(OH)COOH$ or $C_3(H_2O)_3$, can have similar empirical formulas. In more descriptive terminology the carbohydrates are defined as the aldehyde and ketone derivatives of polyhydric alcohols. The simplest carbohydrate is glycol aldehyde, the aldehyde derivative of ethylene glycol. The aldehyde and ketone derivatives of glycerol are, respectively, glycerose and dihydroxyacetone (Fig. 4-1).

Monosaccharides

Sugars containing three, four, five, and six or more carbon atoms are known, respectively, as trioses, tetroses, pentoses, hexoses, and so on, and are classified as monosaccharides. Aldehyde derivatives are called aldoses and ketone derivatives are called ketoses. Typical examples are the six-carbon sugars glucose (an aldose) and fructose (a ketose), as shown in Figure 4-2.

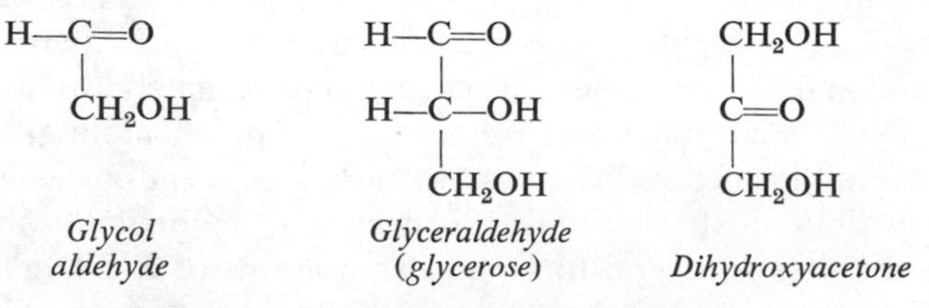

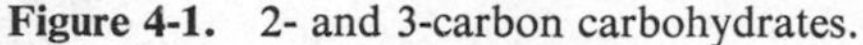

Figure 4-1. 2- and 3-carbon carbohydrates.

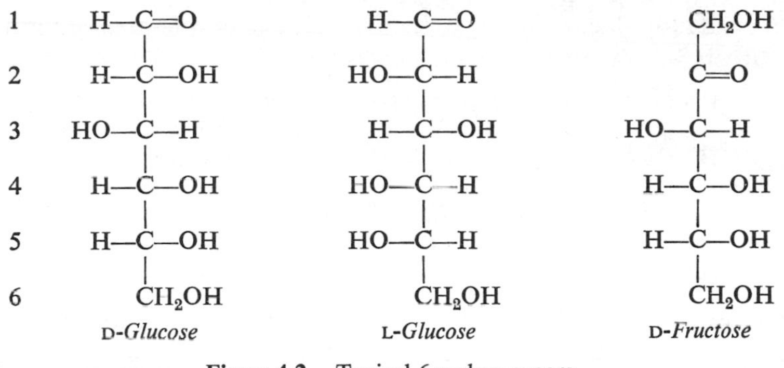

Figure 4-2. Typical 6-carbon sugars.

The carbon atoms in the chain are numbered 1 to 6 as shown by the numbers at the left of the formula for D-glucose. The designation D- or L- refers to the position of the hydroxyl group on the carbon atom next to the bottom —CH_2OH group. By convention, the D- sugars are written with the hydroxyl group on the right and the L- sugars with the hydroxyl group on the left. Compounds that are identical in composition and differ only in spatial configuration are called stereoisomers. The majority of the sugars occurring in the body are of the D- configuration. A number of different structures exist, depending on the relative positions of the hydroxyl groups on the carbon atoms. The D-hexose series is shown in Figure 4-3.

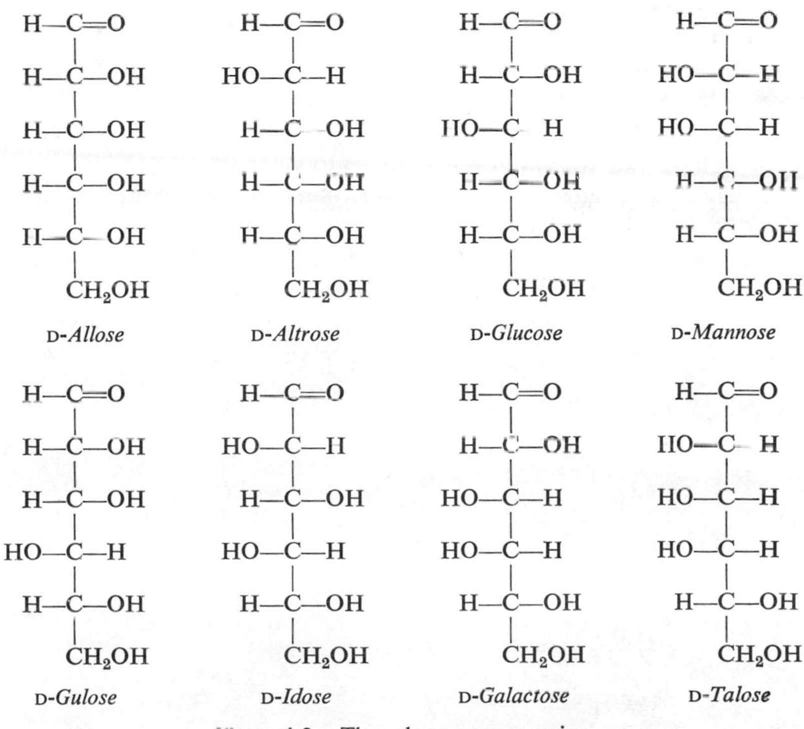

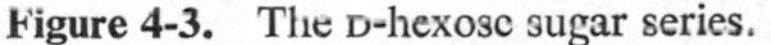

Figure 4-3. The D-hexose sugar series.

The formula for glucose can be written in either aldehyde or enol form. Shift to the latter structure is favored in alkaline solution:

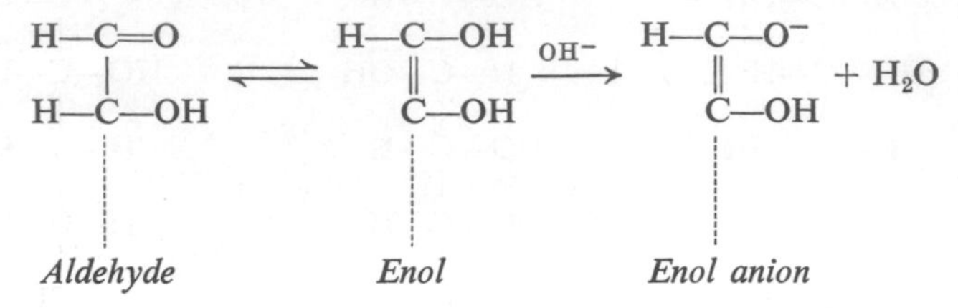

The presence of a double bond and a negative charge in the enol anion form make glucose an active reducing substance and provide a basis for its analytical determination. Thus, glucose in hot alkaline solution readily reduces metallic ions such as cupric to cuprous ion and the color change can be used as a presumptive indication for the presence of glucose. Sugars capable of reducing cupric ions in alkaline solution are commonly known as "reducing sugars."

Aldehyde and alcohol groups can react to form hemiacetals. In the case of glucose the aldehyde group reacts with the hydroxyl group on carbon 5 as shown in Figure 4-4. With this ring structure the hydroxyl group on the first carbon can be written to the right or to the left. By convention the form with the hydroxyl group on the right is called α-D-glucose and the form with the hydroxyl group on the left is called β-D-glucose. The common anhydrous crystalline glucose is in the α-D-form. The β-D-form is obtained by crystallization from acetic acid. The two forms differ with respect to optical rotation of polarized light. The specific rotation, $[\alpha]_D^{25}$, for the α-D-form is +113° and for the β-D-form +19.7°. Either form in aqueous solution gives rise to an equilibrium mixture that has a specific rotation of +52.5°. The equilibrium established at room temperature is such that about 36 per cent of the glucose exists in the α-form and 64 per cent in the β-form; only a trace remains in the free aldehyde form. The enzyme glucose oxidase reacts only with β-D-glucose. For this reason, standard solutions to be used in glucose oxidase methods for glucose determinations should be permitted to stand at least 2 hours in order to obtain equilibrium comparable to that in test samples to be analyzed.

From the ring structures shown in Figure 4-4 it is not apparent why the aldehyde group should react with the distant hydroxyl group on carbon 5. The spatial arrangement is better represented by a symmetrical ring structure, depicted by the Haworth formula, in which glucose is considered as having the same basic structure as pyran (Fig. 4-5). In this formula the plane of the ring is considered as perpendicular to the plane of the paper with the heavy lines pointing towards the reader. Hydroxyl groups

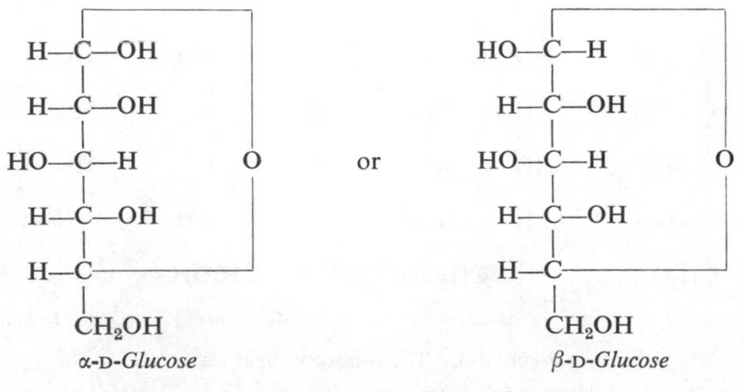

Figure 4-4. Optically active forms of D-glucose.

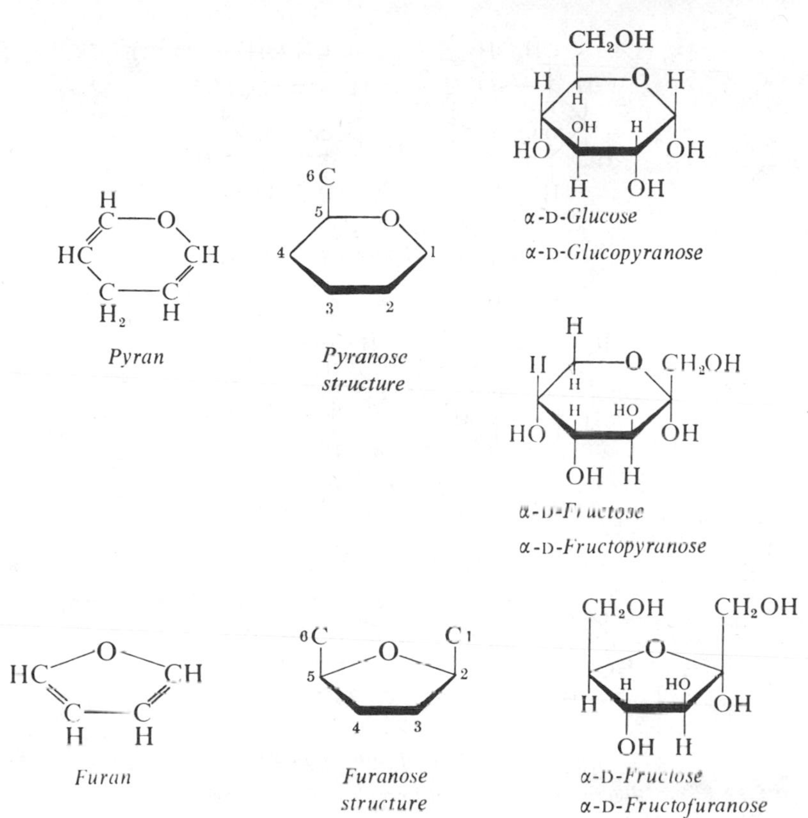

Figure 4-5. The Haworth formula for sugars.

in position 1 are then below the plane (α-configuration) or above the plane (β-configuration). A six-membered ring sugar, containing five carbons and one oxygen, is a derivative of pyran and is called a pyranose. When linkage occurs with formation of a five membered ring, containing four carbons and one oxygen, the sugar has the same basic structure as furan and is called a furanose. Representative formulas are shown in Figure 4-5. Fructose is shown with two cyclic forms. Fructopyranose is the configuration of the free sugar and fructofuranose occurs whenever fructose exists combined in disaccharides and polysaccharides as in sucrose and inulin.

Disaccharides

In addition to internal hemiacetal formation, we may have interaction of groups between two monosaccharides with loss of a molecule of water to form disaccharides. The chemical bond between the saccharides always involves the aldehyde or ketone group of one monosaccharide joined to the other monosaccharide, either by one of the latter's alcohol groups (e.g., maltose) or by the latter's aldehyde or ketone group (e.g., sucrose). The linkage of an aldehyde or ketone with an alcohol is called a glycosidic linkage. The most common disaccharides are:

Maltose (glucose + glucose)
Lactose (glucose + galactose)
Sucrose (glucose + fructose)

Structural formulas and chemical names are shown in Figure 4-6.

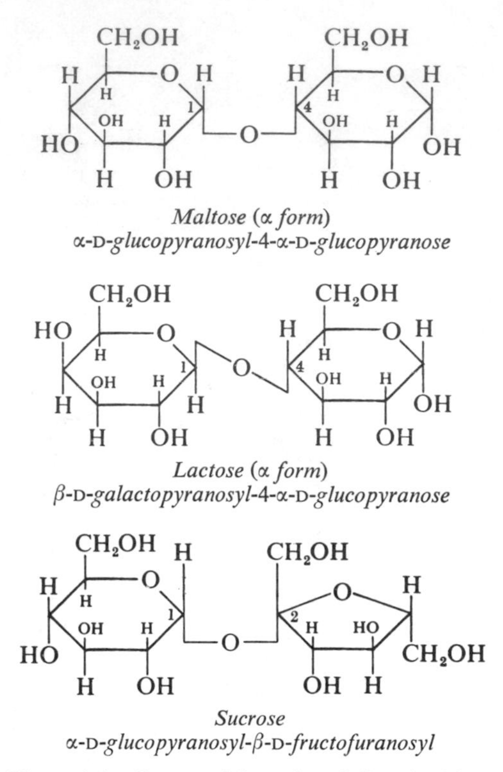

Figure 4-6. Structural formulas of disaccharides.

If the linkage between the two monosaccharides is between the aldehyde or ketone group of one molecule and the hydroxyl group of another molecule (as in maltose and lactose), there will remain one potentially free ketone or aldehyde group on the second monosaccharide. Consequently, the sugar will be a reducing sugar. The reducing power, however, is approximately only 40 per cent of the reducing power of the two single monosaccharides together since one of the reducing groups is not available. If the linkage between the two monosaccharides involves the aldehyde or ketone groups of both molecules (as in sucrose), a nonreducing sugar results since there is no remaining free aldehyde or ketone group.

Polysaccharides

The linkage of many monosaccharide units together results in the formation of polysaccharides. In starch and glycogen, the chief reserve carbohydrates of plants and animals, respectively, one molecule may contain from 25 to 2500 glucose units. The suffix *-an* attached to a name of a monosaccharide indicates the main type of sugar present in the polysaccharide. Starch and glycogen, for example, are glucosans since they are composed of a series of individual glucose molecules. Inulin, found in the tubers of the dahlia and the Jerusalem artichoke, is a polysaccharide consisting largely of fructose units and is known as a fructosan.

Nearly all starches are composed of a mixture of two kinds of glucosans called amyloses and amylopectins. The relative proportions of these two glucosans in a starch vary from approximately 20 per cent amylose and 80 per cent amylopectin in wheat, potato, and ordinary corn starch to nearly 100 per cent amylopectin in the starch of waxy corn. On the other hand, a few corn starches are known that contain

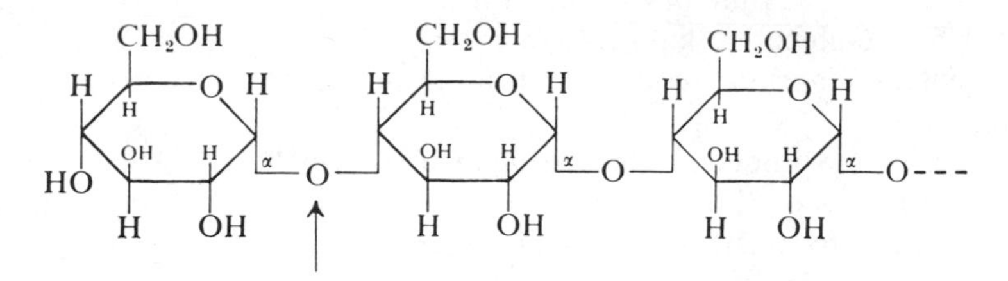

Chain of glucose molecules linked by 1,4-linkages as found in amylose

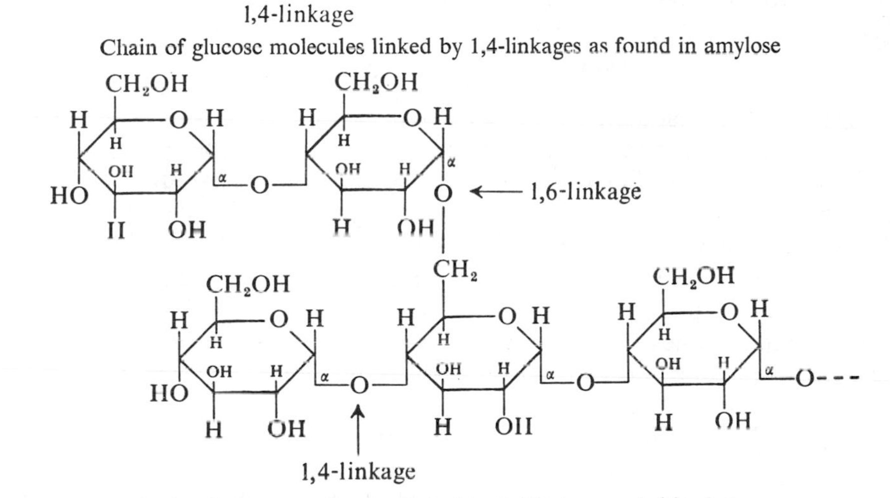

Chain of glucose molecules linked by 1,4-linkage and side chain linked by 1,6-linkage as found in amylopectin

Figure 4-7. Structures for amylose and amylopectin.

up to 75 per cent amylose. Iodine gives a typical deep blue color with amylose and a red to violet color with amylopectin. These characteristic colors disappear when the respective glucosans are hydrolyzed to smaller units such as dextrins and maltose. This disappearance of the starch-iodine color is utilized in some quantitative methods for amylase determination.

Although both amylose and amylopectin are made up of glucose molecules, there is one significant difference in their structure. Amylose, which has a molecular weight from 4000 to about 50,000, consists of one long unbranched chain of 25 to 300 units of glucose. These units are linked together by a 1,4-linkage with only the terminal aldehyde group free. In amylopectin the majority of the units are also connected with 1-4 α-links, but, in addition, there are 1,6-α-glycosidic bonds (amounting to about 4 per cent of the total) that form side chains. The structure is a branch-on-branch arrangement of 1000 or more D-glucopyranose units with a molecular weight for amylopectin ranging from 50,000 to about 1,000,000. Glycogens have structures similar to amylopectins except that branching is more extensive. The average length of a branch in a glycogen molecule is usually 12 or 18 D-glucopyranose units compared to about 25 units in amylopectin. Examples of polysaccharide linkages are shown in Figure 4-7.

The difference in structure between amylose and amylopectin becomes important in the proper selection of the starch substrate for amylase determination (see page 409). Any differences in the structure of starch will affect the rate of hydrolysis. The so-called α-amylase of pancreatic origin hydrolyzes the 1,4-glucoside linkage with special preference for the more central linkages. This results initially in the production of some maltose and a mixture of dextrins, which are subsequently also hydrolyzed to maltose. The 1,6-glucoside linkages are not attacked by α-amylase and relatively large molecules of so-called residual dextrins are left after the action of the enzyme on amylopectin.

Dextrins are the products of the partial hydrolysis of starch. They are a complex mixture of molecules of different sizes. Those formed from amylose are unbranched chains and those from amylopectins are branched chains of glucose molecules. Erythrodextrins are larger, branched dextrins that produce a reddish color with iodine.

METABOLISM OF CARBOHYDRATES

Starch and glycogen are partially digested by the action of salivary amylase to form intermediate dextrins and maltose. Amylase activity is inhibited at the acid pH of the stomach. In the small intestine the pH is increased by alkaline pancreatic juice and the amylase of the pancreas effects digestion of starch and glycogen to maltose. The latter, along with any ingested lactose and sucrose, is split by the disaccharidases in the intestinal mucosa (maltase, lactase, and sucrase) to form the monosaccharides glucose, galactose, and fructose.

Absorption of these monosaccharides is fairly complete and appears to occur by an active enzymatic transfer process. This is inferred because the rate of absorption for glucose and galactose is several times greater than that for xylose, which is thought to be absorbed by passive diffusion. Some conversion of fructose to glucose apparently occurs during the process of absorption, and the interconversion can be visualized in terms of the enediol form common to both as shown in Figure 4-8.

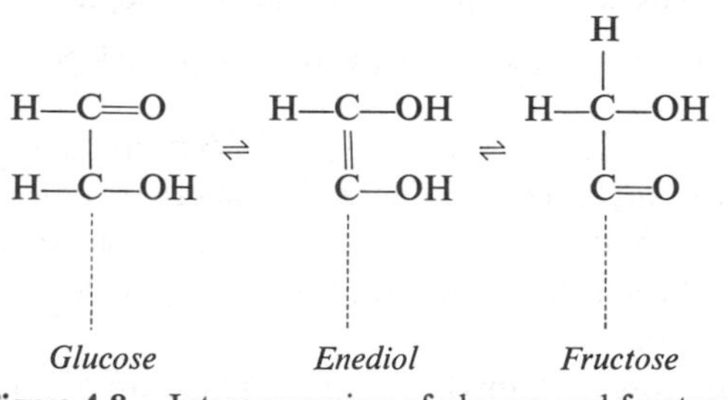

Figure 4-8. Interconversion of glucose and fructose.

Following absorption into the portal vein, the hexoses are transported to the liver. Depending on the needs of the body, the carbohydrates may be converted and stored as liver glycogen, metabolized completely to carbon dioxide and water to provide immediate energy, converted to keto acids, amino acids, and protein, or converted to fat and stored as adipose tissue. The complete picture of intermediary metabolism of carbohydrates is rather complex and interwoven with the metabolism of lipids and amino acids (see Chapter 7, Fig. 7-3). For a comprehensive treatise the reader should consult textbooks of biochemistry.

Factors involved in the regulation of blood glucose concentration are illustrated in the partial outline shown in Figure 4-9. Each step is catalyzed by a specific enzyme and, in some cases, different enzymes may be responsible for a given step, depending on which way the reaction proceeds For example, the initial phosphorylation of glucose is catalyzed by glucokinase, but the reverse reaction depends upon glucose-6-phosphatase. Fructose and galactose are phosphorylated and eventually enter the same metabolic pathway as glucose.

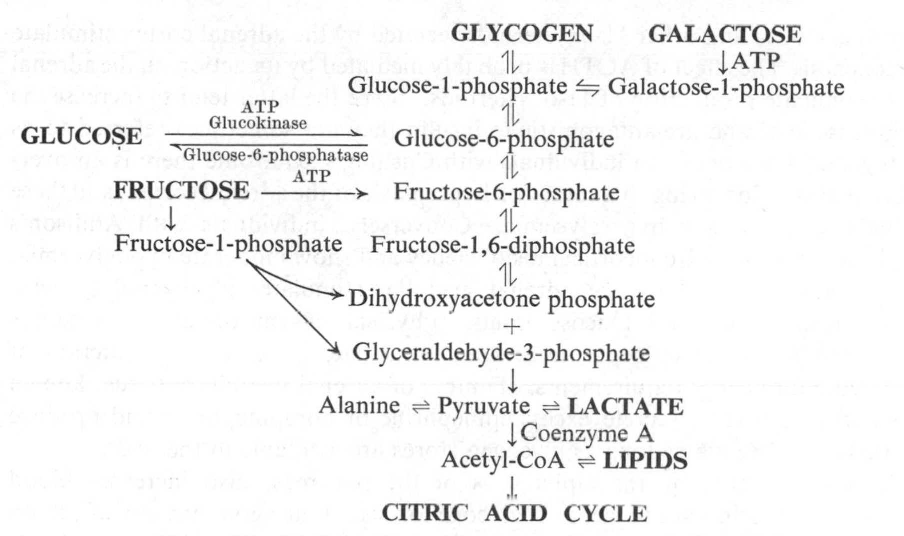

Figure 4-9. Partial outline of the intermediary metabolism of glucose.

Various terms are applied to describe general processes in carbohydrate metabolism. *Glycogenesis* refers to the conversion of glucose to glycogen and *glycogenolysis* is the breakdown of glycogen to form glucose and other intermediate products. The formation of glucose from noncarbohydrate sources, such as amino acids, glycerol, or fatty acids, is *gluconeogenesis*. The conversion of glucose or other hexoses into lactate or pyruvate is called *glycolysis*. The net result of factors affecting these various processes determines the level of glucose in the blood.

Regulation of Blood Glucose Concentration

In the fasting state the level of blood glucose is maintained by drawing upon the glycogen stores of the liver and a slight amount may also be derived from the kidney. Both of these organs contain the specific enzyme, glucose-6-phosphatase, necessary for the conversion of glucose-6-phosphate to glucose. Skeletal muscle, although it stores glycogen, is lacking in this enzyme and cannot directly contribute glucose to the blood. As blood glucose levels increase, usually by absorption of carbohydrates from the intestine, glycogenolysis is replaced by glycogenesis, whereby excess blood glucose is converted into liver and muscle glycogen.

A number of hormones are important in the regulation of blood glucose concentration.

Insulin, produced by the beta cells of the islets of Langerhans in the pancreas, promotes glycogenesis and lipogenesis (formation of fat from carbohydrate) with resultant decrease in blood glucose levels. Another important action of insulin is to increase the permeability of cells to glucose. With deficiency in effective insulin (diabetes), the fasting blood glucose level tends to increase (hyperglycemia) and the body shows less ability to metabolize carbohydrates. At the other extreme, an islet cell tumor can produce an excess of insulin, resulting in very low levels of blood glucose (hypoglycemia).

Growth hormone and *adrenocorticotrophic hormone* (ACTH) are secreted by the anterior pituitary. Both have an antagonistic action to insulin and tend to raise blood glucose levels.

Hydrocortisone and other 11-oxysteroids secreted by the adrenal cortex stimulate gluconeogenesis. The effect of ACTH is probably mediated by its action on the adrenal cortex to stimulate production of 11-oxysteroids. Since the latter tend to increase the blood glucose level and are antagonistic to insulin they are sometimes referred to as "diabetogenic" hormones. In individuals with Cushing's syndrome there is an overproduction of steroids owing to a tumor or hyperplasia of the adrenal cortex, and these individuals tend to show hyperglycemia. Conversely, individuals with Addison's disease have a primary adrenocortical insufficiency and show moderate hypoglycemia.

Epinephrine, secreted by the adrenal medulla, stimulates glycogenolysis with resultant increase in blood glucose levels. Physical or emotional stress causes increased production of epinephrine and an immediate increase in production of blood glucose for energy requirements. Tumors of adrenal medullary tissue, known as pheochromocytomas, secrete excess epinephrine or norepinephrine and produce moderate hyperglycemia as long as glycogen stores are available in the liver.

Glucagon, secreted by the alpha cells of the pancreas, also increases blood glucose levels by stimulating hepatic glycogenolysis. Glucagon has no effect on muscle glycogen as shown by the lack of elevation of blood lactate and pyruvate levels following its administration.

Thyroxine, secreted by the thyroid, also appears to stimulate glycogenolysis. Thyrotoxic individuals may show symptoms of mild diabetes and almost complete absence of liver glycogen. Thyroxine also increases the rate of absorption of glucose from the intestine.

DETERMINATION OF GLUCOSE IN BODY FLUIDS

NORMAL GLUCOSE LEVELS IN WHOLE BLOOD AND SERUM

The normal glucose concentration in whole blood, determined by highly specific methods, ranges from 65 to 95 mg./100 ml. Glucose concentration is uniform in the water phase of plasma and erythrocytes. Since plasma contains approximately 12 per cent more water than whole blood, it follows that the total glucose in 100 ml. of plasma is about 12 per cent greater than in 100 ml. of whole blood. The normal range for glucose in plasma (or serum) is usually given as 70 to 105 mg./100 ml. This unequal distribution can be illustrated by assuming a concentration of 1 mg. of glucose per ml. of the water phase in blood as follows:

Component Analyzed	*Per Cent Water*	*Mg. Glucose/100 ml. Component*
Erythrocytes	73	73
Whole blood	83	83
Plasma	93	93

From these values it is apparent that the "normal" range for glucose in whole blood will vary with the hematocrit. For this reason it is preferable to perform glucose determinations on plasma or serum. In addition, the erythrocytes contain sulfhydryl compounds, glutathione and ergothioneine, which also act as reducing substances and interfere in procedures for glucose based upon its reducing action in alkaline solution unless they are removed before analysis. These nonglucose reducing substances, called "saccharoids," increase the apparent glucose values by about 10 to 30 mg./100 ml. when whole blood is analyzed by the Folin-Wu copper reduction method. The resulting normal range, including saccharoids, becomes 80 to 120 mg./100 ml. and is reported as "blood sugar," although "total reducing substances" would be a more appropriate term.

Owing to the variation in normal values for true glucose in plasma compared to whole blood, and the relative lack of specificity of certain methods for glucose, confusion has arisen regarding the interpretation of values reported by different laboratories. It is not uncommon for a physician to perform blood sugars in his office by the Folin-Wu method on whole blood. He may then see patients at one hospital where the laboratory reports "true" whole blood glucose and patients at another hospital that reports serum glucose. Further difficulty arises when one attempts to interpret changes following a glucose tolerance test. If we assume that serum contains 12 per cent more glucose than whole blood and that saccharoids in normal whole blood average 20 mg./100 ml., the values are related approximately as follows:

True glucose (mg./100 ml.)		*Reducing "sugar" (mg/100 ml.)*
Whole Blood	*Serum*	*Whole Blood*
70	80	90
100	110	120
150	170	170
200	225	220
300	335	320

Thus, the values shown for serum were obtained by multiplying the values for whole blood true glucose by 1.12; values for reducing "sugar" in whole blood were obtained by adding 20 mg./100 ml. to the whole blood true glucose values. Fortunately, most methods applied to serum give results in fairly good agreement since the saccharoid content is low and relatively constant; hence, comparison of results from different laboratories would be facilitated by using serum or plasma uniformly for glucose determinations.

STABILITY OF GLUCOSE IN BODY FLUIDS

When blood is drawn, permitted to clot, and stands uncentrifuged at room temperature, the average decrease in serum glucose is approximately 7 per cent in 1 hour.[8] In separated unhemolyzed serum the glucose concentration is generally stable up to 8 hours at 25°C. and up to 72 hours at 4°C. Variable stability, related to bacterial contamination, is observed for longer storage periods. Plasma, removed from the cells after moderate centrifugation, contains leucocytes that also metabolize glucose. Cell-free plasma shows no glycolytic activity. From these data it follows that whole blood glucose determinations should be performed promptly after collection of the specimen. Plasma or serum from blood without preservative must be separated from the cells or clot within a half-hour after the blood is drawn if glucose values within 10 mg./100 ml. of the original value are to be obtained consistently.

Glycolysis can be prevented and glucose stabilized up to 24 hours at room temperature by adding sodium fluoride to blood. Fluoride ion also inhibits coagulation by precipitating calcium; however, clotting may occur after several hours and it is advisable to use a combined fluoride-oxalate mixture, such as 2 mg. of potassium oxalate and 2.5 mg. of sodium fluoride per ml. of blood. Either the whole blood or plasma may be analyzed. Fluoride ion inhibits urease activity; consequently, the specimens are unsuitable for determination of urea in those procedures that require urease.

Cerebrospinal fluids are frequently contaminated with bacteria or other cellular constituents and should be analyzed for glucose without delay. Glucose may be preserved in 24 hour collections of urine by adding 5 ml. of glacial acetic acid to the container before starting the collection. The final pH is usually between 4 and 5 and bacterial activity is inhibited at this level of acidity. The use of 5 gm. of sodium benzoate per 24 hour specimen is also effective in preserving the urine.

Alkaline Ferricyanide Methods

In hot alkaline solutions ferricyanide ion (yellow) is reduced by glucose to ferrocyanide ion (colorless).

$$Fe^{III}(CN)_6{}^{-3} \xrightarrow[\text{(glucose)}]{+e^-} Fe^{II}(CN)_6{}^{-4}$$

Ferricyanide *(Ferric iron)* — *Ferrocyanide* *(Ferrous iron)*

The decrease in color of ferricyanide ion, measured at 420 nm., is proportional to the glucose concentration (inverse colorimetry). The reagent blank, without glucose, has the greatest absorbance, and measurements in the low or normal range are inherently less accurate since we are basing our readings on a small difference between large absorbance values. More precision is obtained by automation and the alkaline ferricyanide method is currently in wide use where glucose determinations are performed on the AutoAnalyzer (see Chapter 2). Other reducing substances interfere; e.g., 1 mg. of creatinine reacts the same as 1 mg. of glucose and 1 mg. of uric acid reacts equivalent to 0.5 mg. of glucose. Thus, serum from a patient with uremia, in which both creatinine and uric acid are markedly elevated, would show a falsely elevated value for glucose. High concentrations of creatinine and uric acid render the alkaline ferricyanide method unsuitable for determination of glucose in urine.

Copper Reduction Methods

PRINCIPLE

In hot alkaline solution, glucose readily reduces cupric ion to cuprous ion with formation of (mainly) cuprous oxide (Cu_2O). The reaction is not stoichiometric, but depends on the alkalinity, time and temperature of heating, and concentration of reagents. Under carefully controlled conditions the reaction is reproducible and provides quantitative results when standards are analyzed in the same manner as protein-free filtrates. Reoxidation of cuprous ion by oxygen from the air is prevented by using a constricted tube to minimize surface area or by incorporating 18 per cent sodium sulfate in the reagent to decrease the solubility of oxygen. The next step is to add phosphomolybdic (or arsenomolybdic) acid, which is reduced by the cuprous ion to form lower oxides of molybdenum. These oxides produce a blue color suitable for photometric measurements.

In the Folin-Wu procedure, proteins are precipitated with tungstic acid and the water-clear, protein-free filtrate is used in the reaction. The method lacks specificity owing to the presence of glutathione and other nonglucose reducing substances that appear in the filtrate. The method is described in Chapter 8, Enzymes, where it is discussed in reference to the measurement of reducing sugars formed by the action of amylase on starch.

In the Somogyi-Nelson procedure, proteins are precipitated by the addition of barium hydroxide and zinc sulfate. Protein is removed as zinc proteinate, sulfhydryl compounds as zinc salts, and the remaining zinc and barium ions as zinc hydroxide and barium sulfate.

$$ZnSO_4 + Ba(OH)_2 \rightarrow Zn(OH)_2\downarrow + BaSO_4\downarrow$$

Uric acid and some creatinine are also precipitated and adsorbed on barium sulfate so that the resultant filtrate is virtually free of nonsugar reducing substances. When the reagents are properly balanced the filtrate has a pH of approximately 7.4. The relative specificity of the method has resulted in its adoption by many laboratories and clinics as a standard reference method for glucose in blood.

REAGENTS

1. Zinc sulfate solution, 5 per cent (w/v). Dissolve 50 gm. of $ZnSO_4 \cdot 7H_2O$ in water and dilute to 1 L.

2. Barium hydroxide solution, approximately 0.3 *N*. Dissolve 50 gm. of $Ba(OH)_2 \cdot 8H_2O$ in water and dilute to 1 L. Let stand for 2 days in a covered container; then decant or filter. Store in a polyethylene bottle and protect from carbon dioxide of the air. This solution must be balanced against the zinc sulfate solution as follows: Pipet 10.0 ml. of zinc sulfate solution into a flask and dilute with 25 ml. of water. Add phenolphthalein indicator and titrate slowly with the barium hydroxide solution, using vigorous mixing, to a definite permanent pink color. Dilute the stronger of the two reagents, if necessary, such that 10.0 ml. of zinc sulfate solution requires 10.0 $\pm$ 0.1 ml. of barium hydroxide solution for the end point.

3. Copper sulfate solution. Dissolve 29 gm. of anhydrous Na_2HPO_4 and 40 gm. of potassium sodium tartrate ($KNaC_4H_4O_6 \cdot 4H_2O$) in 700 ml. of water. Add 100 ml. of 1 *N* NaOH. Add, with stirring, 80 ml. of 10 per cent (w/v) $CuSO_4 \cdot 5H_2O$ solution. Add 180 gm. of anhydrous sodium sulfate, mix to dissolve, and dilute to 1 L. Let stand for at least 2 days; then filter to remove any copper salts that may precipitate. The reagent may be used indefinitely although, after prolonged standing, the solution may need to be refiltered.

4. Arsenomolybdate solution. Dissolve 50 gm. of ammonium molybdate, $(NH_4)_6Mo_7O_{24} \cdot 4H_2O$, in a mixture of 900 ml. of water and 42 ml. of concentrated sulfuric acid. Dissolve separately 6 gm. of sodium arsenate ($Na_2HAsO_4 \cdot 7H_2O$) in 50 ml. of water and add to the first solution. Mix and incubate at 37°C. for 48 hours. Store in a brown bottle at room temperature. The color should be yellow; if greenish-blue, discard.

5. Stock standard glucose, 1 per cent. Transfer 1.000 gm. of dry reagent grade glucose to a 100 ml. volumetric flask. Add 0.2 per cent benzoic acid solution; mix to dissolve; then dilute to the mark with benzoic acid solution. Mix thoroughly and store in a tightly stoppered bottle in the refrigerator. This solution is stable for many months.

6. Working standards. Warm a portion of the stock standard to room temperature.

Solution *a*. Pipet 10.0 ml. of stock standard to a 100 ml. volumetric flask and dilute to the mark with 0.2 per cent benzoic acid solution to provide a concentration of 100 mg./100 ml.

Solution *b*. Pipet 10.0 ml. of solution *a* to a 100 ml. volumetric flask and dilute to the mark with 0.2 per cent benzoic acid to provide a concentration of 10 mg./100 ml. Two milliliters of this working standard are treated in the same manner as a 1:20

protein-free filtrate and correspond to a blood glucose concentration of 200 mg./100 ml. Other working standards may be prepared by diluting proportionate volumes of solution *a* to 100 ml.

PROCEDURE

1. Pipet 0.5 ml. of blood, serum, plasma, or cerebrospinal fluid into a small Erlenmeyer flask. An Ostwald-Folin pipet is convenient for this measurement.
2. Add 7.5 ml. of water and 1.0 ml. of barium hydroxide solution. Mix and let stand at least 30 seconds.
3. Add 1.0 ml. of zinc sulfate solution, mix well, let stand 2 minutes or more, and filter or centrifuge. This provides a 1:20 protein-free filtrate.
4. Pipet 2.0 ml. of filtrate or supernatant into a Folin sugar tube calibrated at 25 ml. Include a blank (2 ml. of water) and standard (2 ml. of working standard *b*) with each series and treat the same as the filtrate.
5. Add 1 ml. of copper sulfate solution and mix.
6. Place tubes in a boiling water bath for 10 minutes; then cool thoroughly in cold water.
7. Add 1 ml. of arsenomolybdate solution and remix.
8. Allow to stand at least 2 minutes, dilute to 25 ml. with water, and mix by inversion at least three times.
9. Measure absorbance of test and standard at 490 nm. against the blank. The absorbance is stable up to 1 hour at this wavelength.

$$\frac{A_{\text{sample}}}{A_{\text{standard}}} \times 200 = \text{mg. glucose/100 ml. of sample}$$

When the glucose in the sample is greater than 300 mg./100 ml., dilute a portion of the "test" with an equal volume of water and measure the absorbance against a portion of the blank diluted in the same manner. Multiply the result by 2. If the glucose is greater than 600 mg./100 ml., the test should be repeated starting with 0.5 ml. of filtrate plus 1.5 ml. of water and multiplying the final result by 4. Glucose concentrations greater than 600 mg./100 ml. may exceed the oxidizing capacity of the copper reagent.

MICROMETHOD

Use a micropipet calibrated to contain 0.1 ml. Wash out the sample into a mixture of 3.5 ml. of water and 0.2 ml. of barium hydroxide solution in a heavy wall centrifuge tube. Mix and let stand until the solution turns brown. This may require several minutes on blood from infants owing to the high concentration of alkali-resistant fetal hemoglobin. Add 0.2 ml. of zinc sulfate solution, mix well, allow to stand 2 minutes, and centrifuge. Use 2.0 ml. of the clear supernatant and proceed as described in the macromethod. Since the dilution is 1:40 in the micromethod (rather than 1:20), the result must be multiplied by 2. Although the preceding description applies to whole blood, serum or plasma may be analyzed in a similar manner. The sensitivity may be increased further by measuring absorbance at 620 or 660 nm. but the readings tend to be less stable than at 490 nm. Sensitivity may also be doubled by making the final dilution to 12.5 ml. rather than 25 ml.

REFERENCES

Nelson, N.: J. Biol. Chem., *153*:375, 1944.
Somogyi, M.: J. Biol. Chem., *160*:61, 1945.
Somogyi, M.: J. Biol. Chem., *195*:19, 1952.

Glucose Oxidase Methods

PRINCIPLE

Glucose oxidase is an enzyme that catalyzes the oxidation of glucose to gluconic acid and hydrogen peroxide.

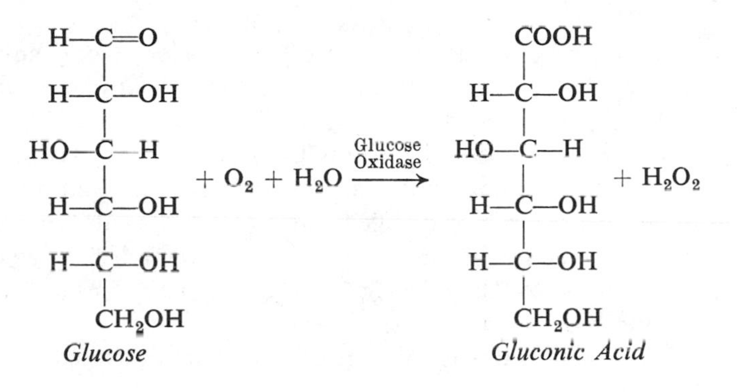

Introduction of the enzyme peroxidase and a chromogenic oxygen acceptor, such as *o*-dianisidine, provides color development.

$$o\text{-Dianisidine} + H_2O_2 \xrightarrow{\text{Peroxidase}} \text{Oxidized } o\text{-dianisidine} + H_2O$$
(*color*)

Glucose oxidase is highly specific for β-D-glucose. As noted earlier, glucose in solution exists as 36 per cent α- and 64 per cent β form. Complete oxidation of glucose, therefore, requires mutarotation of α to β form. Some commercial preparations of glucose oxidase contain an enzyme, mutarotase, that accelerates this reaction.

The second step, involving peroxidase, is less specific. Various substances, such as uric acid, ascorbic acid, bilirubin, and glutathione, inhibit the reaction, presumably by competing with the chromogen for hydrogen peroxide. For this reason, results obtained directly on serum tend to be lower than true glucose values. Most interfering substances are eliminated by use of a Somogyi zinc filtrate. Peroxides may be released in acid filtrates and cause positive errors. Some glucose oxidase preparations may contain catalase as a contaminant that competes with peroxidase for hydrogen peroxide and decreases the final color obtained. Standards and unknowns should be analyzed simultaneously under conditions such that the rate of oxidation is proportional to glucose concentration. The reaction is stopped and color developed after a standard incubation period.

In some methods the final mixture is acidified slightly to stop the reaction and the yellow color is measured at 395 nm. In stronger acid solution the color becomes pink with maximum absorbance at 540 nm. and both sensitivity and stability are improved. The method to be described provides results in close agreement with the Somogyi-Nelson method and can be used to determine glucose in the presence of other sugars.

The glucose oxidase methods are not directly applicable to urine specimens owing to the high concentration of enzyme inhibitors present. A method has been described in which the urine is first treated with an ion-exchange resin to remove interfering substances.[7]

REAGENTS

1. Zinc sulfate and barium hydroxide solutions are prepared as described in the Somogyi-Nelson method.

2. Glycerol-buffer solution, pH 7.0. Dissolve 3.48 gm. of anhydrous Na_2HPO_4 and 2.12 gm. of KH_2PO_4 in 600 ml. of water. Add 400 ml. of glycerol and mix thoroughly.

3. Glucose oxidase reagent. Preparation of this reagent from individual constituents has been described by Fales. A suitable packaged mixture known as Glucostat is available (Worthington Biochemical Corp., Freehold, N.J.). The smaller vial contains the chromogen (10 mg. of *o*-dianisidine), and the larger vial contains approximately 125 mg. of glucose oxidase and 5 mg. of peroxidase. Completely dissolve the soluble chromogen in 1.0 ml. of distilled water and drain into an amber bottle. Dissolve the contents of the larger vial in and dilute to 100 ml. with glycerol-buffer solution. Add this solution to the amber bottle and mix well. This solution is stable up to at least 1 month when stored in the refrigerator.

4. Sulfuric acid, 6 *N*. Add, with mixing, 200 ml of concentrated sulfuric acid to 1000 ml. of distilled water.

5. Standard glucose solution, 100 mg./100 ml. in 0.2 per cent benzoic acid solution as described for the Somogyi-Nelson method. Standard solutions prepared from dry glucose should stand at least 2 hours to insure that mutarotation has reached a state of equilibrium. Solutions to be administered for glucose tolerance tests should also be prepared at least 2 hours before use.

PROCEDURE

1. Prepare a 1:20 protein-free filtrate of whole blood, plasma, serum, or cerebrospinal fluid as described for the Somogyi-Nelson method. Heparin, oxalate, or EDTA are satisfactory anticoagulants. Sodium fluoride (2.5 mg./ml. of blood) may be used as preservative, but not thymol, which inhibits the reaction.

2. Prepare blank and standard filtrates in the identical manner as sample, using water and 100 mg./100 ml. of glucose standard, respectively, as starting materials.

3. The glucose oxidase reagent should be warmed to room temperature.

4. Pipet 0.20 ml. of each filtrate into respective test tubes.

5. Add 1.0 ml. of glucose oxidase reagent to the tubes, one at a time in timed sequence. Mix each tube immediately and place in a 37°C. water bath.

6. After exactly 30 minutes, remove the tubes one at a time, add 5.0 ml. of 6 *N* sulfuric acid, and mix. In this manner the incubation time for all tubes is kept constant.

7. After 5 minutes, but within an hour, measure the absorbance at 540 nm. against the blank.

$$\frac{A_{\text{sample}}}{A_{\text{standard}}} \times 100 = \text{mg. glucose/100 ml. of sample}$$

COMMENTS ON THE METHOD

The absorbance is usually linear with glucose concentrations up to 400 mg./100 ml.; however, this should be checked with each new lot of glucose oxidase reagent, using standards of 100, 200, 300, and 400 mg. of glucose/100 ml. When the sample glucose concentration is too high for accurate measurement, the filtrate is diluted twofold or fourfold and the analysis repeated.

Rubber tubing, used for dispensing deproteinizing solutions or distilled water, has been found to interfere with adequate color development; therefore, rubber tubing connections on dispensers or automatic sampling and diluting devices should be kept to a minimum and should be suspected if standards yield unexpectedly low color values.

Many of the common drugs likely to be encountered in blood have been tested and found to have no appreciable effect on the enzymatic method. Potential inhibitors or color producing compounds are apparently either removed with the protein precipitate or are diluted to the point that interference becomes negligible.

Commercially, there is a similar product, Galactostat (Worthington Biochemical Corp., Freehold, N.J.), suitable for the specific determination of galactose in the presence of other sugars (see Chapter 13, Liver Function Tests).

REFERENCES

Washko, M. E., and Rice, E. W.: Clin. Chem., 7:542, 1961.
Fales, F. W.: *In* Standard Methods of Clinical Chemistry. D. Seligson, Ed. New York, Academic Press, Inc., 1963, vol. 4, p. 101.

o-Toluidine Methods

PRINCIPLE

Various aromatic amines react with glucose in hot acetic acid solution to produce colored derivatives. Among those proposed are aniline, benzidine, 2-aminobiphenyl and *o*-toluidine. The latter condenses with the aldehyde group of glucose to form an equilibrium mixture of a glycosylamine and the corresponding Schiff base as illustrated in Figure 4-10. The green colored end product has an absorption maximum at 630 nm.

Dubowski applied this reaction to trichloroacetic acid filtrates of serum and obtained results in good agreement with those obtained by an automated ferricyanide procedure. The method appears to be highly specific since negligible values were obtained on serum, cerebrospinal fluid, and urine, following yeast fermentation to destroy glucose. The ratios of absorbance obtained with other sugars compared to glucose are as follows:

Glucose	1.00
Galactose	1.00
Mannose	1.00
Fructose	0.00
Lactose	0.33
Maltose	0.05
Arabinose	0.17
Xylose	0.19

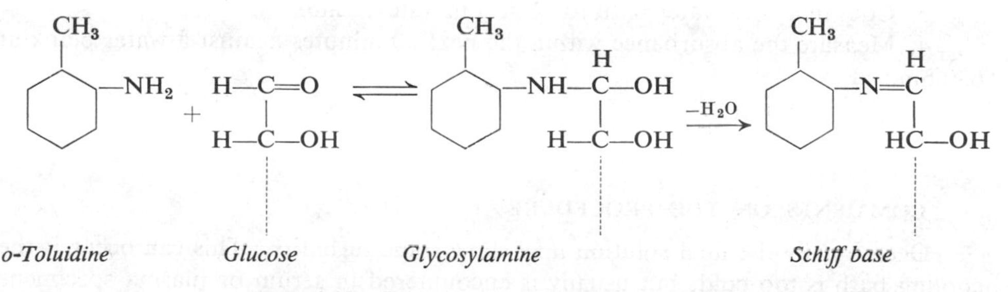

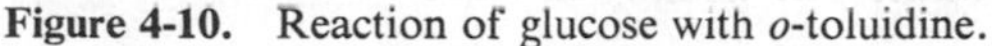

Figure 4-10. Reaction of glucose with *o*-toluidine.

Pentoses produce an orange color with maximum absorption near 480 nm. Since galactose produces color equivalent to glucose, the *o*-toluidine procedure may be used to evaluate galactose tolerance or lactose tolerance tests.

Other workers have noted that the reaction could be applied directly to serum without removal of protein. The following modification has been used in the author's laboratory for several years.

REAGENTS

1. *o*-Toluidine, 5 per cent solution. Transfer 3.0 gm. of thiourea to a 3 L. Erlenmeyer flask. Add 1900 ml. of glacial acetic acid and 100 ml. of *o*-toluidine. Mix until the thiourea is dissolved and store in a brown bottle at room temperature. Contact with the skin should be avoided. The reagent should be dispensed from an all glass automatic pipet such as a Repipet (Labindustries, Berkeley, Calif.).

This reagent may be used for months, although some variability occurs between batches and older reagent tends to produce increased absorbance values with glucose. For these reasons a standard should be included with each series of determinations. Thiourea decreases the color of the reagent blank to an absorbance of about 0.01. Blanks can usually be omitted and the standards and unknowns measured against a distilled water blank; however, each new lot of *o*-toluidine should be checked by substituting water for serum in the method. If the absorbance of the blank is appreciable, it will be necessary to include a blank with each series of determinations.

2. Standard glucose solution, 200 mg./100 ml. in 0.2 per cent benzoic acid solution. Prepare by dilution from stock 1 per cent standard described for the Somogyi-Nelson method.

PROCEDURE

An automatic dilutor, set to sample 0.10 ml. and dilute with 0.90 ml. of water, is convenient for this determination. If this is not available, the sample may be diluted 1:10 with water and 1.0 ml. of diluted sample used in the first step. For whole blood, prepare a 1:10 protein-free filtrate by mixing 0.20 ml. of specimen with 1.80 ml. of 3.0 per cent trichloroacetic acid. Allow to stand 5 to 10 minutes, centrifuge, and pipet 1.0 ml. for analysis. This procedure should also be followed for grossly hemolyzed or lipemic sera.

1. Sample 0.10 ml. of serum, plasma, cerebrospinal fluid, urine, or standard solution and dilute with 0.90 ml. of water in a test tube (or pipet 1.0 ml. of a 1:10 dilution or 1:10 protein-free filtrate). Large test tubes (19 × 150 mm.) are used for ease of mixing.

2. Add 7.0 ml. of *o*-toluidine reagent, mix, and place in a boiling water bath for 10 minutes.

3. Cool in a cold water bath for 2 or 3 minutes; remix.

4. Measure the absorbance within the next 30 minutes against a water blank at 630 nm.

$$\frac{A_{\text{sample}}}{A_{\text{standard}}} \times 200 = \text{mg. glucose/100 ml. of sample}$$

COMMENTS ON THE PROCEDURE

Occasionally the final solution may show some turbidity. This can occur if the cooling bath is too cold, but usually is encountered in serum or plasma specimens with a high lipid content. In this event add 2.0 ml. of isopropyl alcohol to the 8 ml. of final reaction mixture, mix, measure absorbance, and multiply by 10/8. If the mixture is still cloudy, repeat the test on a protein-free filtrate.

The color follows Beer's law and sufficient reagent is present to permit simple dilution of the final mixture with acetic acid for values up to 2000 mg./100 ml. Moderate hemolysis does not interfere. Bilirubin remains unchanged under these conditions and interference is negligible. In some modifications of this method undiluted serum is added directly to the reagent. More intense color is obtained with glucose under these relatively anhydrous conditions, but the interference from bilirubin becomes significant because of its conversion to the green pigment biliverdin.

REFERENCES

Dubowski, K. M.: Clin. Chem., *8*:215, 1962.
Feteris, W. A.: Am. J. Med. Tech., *31*:17, 1965.

GLUCOSE TOLERANCE TESTS

Patients with mild or diet-controlled diabetes may have fasting blood glucose levels within the normal range, but be unable to produce sufficient insulin for prompt metabolism of ingested carbohydrate. As a result, blood glucose rises to abnormally high levels and the return to normal is delayed. In other words, the patient has decreased tolerance for glucose. Therefore, glucose tolerance tests are most helpful in establishing a diagnosis of a mild case of diabetes.

When a standard dose of 100 gm. of glucose is given orally, absorption occurs rapidly and the blood glucose concentration increases. This stimulates the pancreas to produce more insulin with the result that after 30 to 60 minutes the blood glucose level begins to decrease. Since there now exists more insulin than necessary, the blood glucose tends to drop below the fasting level after 1.5 to 2 hours, and then returns to normal levels by approximately 3 hours. Response to glucose in various conditions is shown in Figure 4-11. Values refer to serum or plasma glucose concentrations. As noted earlier, these values are approximately 12 per cent greater than true glucose levels in whole blood.

In a normal response the fasting level of serum glucose is within normal limits; the peak concentration is reached by 30 or 60 minutes and does not exceed 170 mg./100 ml.; and the 2 hour level drops below 120 mg./100 ml. Corresponding values for true glucose in whole blood are 150 and 110 mg./100 ml.

The patient should be placed on a diet containing 1.75 gm. of carbohydrate per kilogram of body weight for 3 days before a glucose tolerance test. If carbohydrate intake has been too low preceding the test, a false diabetic type curve may be obtained.

Oral Glucose Tolerance Test

The test is usually performed in the early morning. The patient should not eat after the evening meal on the day before the test, although water may be taken. The patient should remain at rest during the test and also refrain from smoking or eating. A fasting blood sample and urine specimen are obtained. A solution containing 100 gm. of glucose is given to adults; for children, 0.5 gm. of glucose per pound of

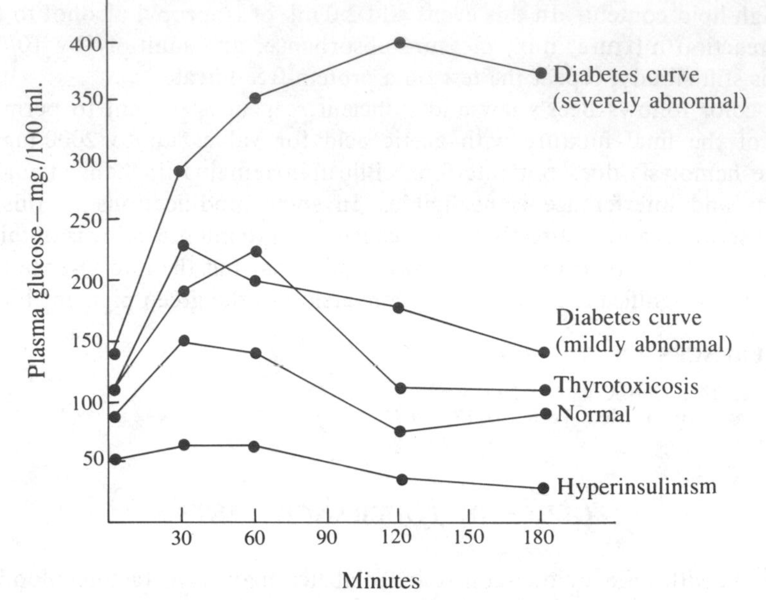

Figure 4-11. Response to oral glucose tolerance test.

body weight is satisfactory. Various commerical preparations are available; otherwise, 100 gm. of glucose is dissolved in about 200 ml. of water and flavored with lemon juice.

Blood specimens are collected at ½, 1, 2, and 3 hours after glucose ingestion. Urine specimens are collected at the same time and are analyzed semiquantitatively for glucose. Normally these should all show a negative reaction. The level of plasma glucose at which glucose appears in the urine is called the renal threshold and is approximately 180 mg./100 ml. Some individuals exhibit lower renal thresholds and excrete glucose in the urine even when the glucose tolerance curve is normal. Although this may be unrelated to any pathologic condition, about one third of these individuals eventually develop diabetes.

A 3 hour test is usually adequate for routine evaluation of diabetes. If hypoglycemia is suspected, additional specimens are taken at 4 and 5 hours. Patients with adrenal insufficiency or with islet cell tumors of the pancreas tend to have low fasting levels of blood glucose. Response to a glucose tolerance test may appear normal over the first 3 hours, but values continue to fall during the fourth and fifth hours. Some patients with latent diabetes tend to show hypoglycemia during this period also, probably associated with a delayed secretion of insulin.

Intravenous Glucose Tolerance Test

Poor absorption of orally administered glucose may result in a "flat" tolerance curve. In this event, glucose may be given intravenously, and preparation of the patient is exactly the same as for the oral test. A fasting blood sample and urine specimen are obtained. A physician then injects 50 per cent (w/v) glucose intravenously. The recommended dose is 0.333 gm. of glucose per kg. of body weight. A simple method to determine the proper amount of 50 per cent glucose to be injected is to divide the patients's weight in pounds by 3.3. For example, 165 lb./3.3 = 50 ml. No more than 50 ml. should be injected.

Blood and urine specimens are collected at ½, 1, 2, and 3 hours as for the oral test. The interpretation is similar except that the peak blood value occurs with the ½ hour specimen and has little numerical significance.

Two Hour Postprandial Glucose

Since the 2 hour specimen in a glucose tolerance test has the greatest significance in evaluating diabetes, the test may be shortened for screening purposes to a single determination. The patient consumes a breakfast, lunch, or glucose solution, containing 100 gm. of carbohydrate. Two hours after the meal, blood is drawn for a glucose determination. The patient should be instructed to consume the required amount of carbohydrate and to remain at rest during the 2 hour period following the meal. Many physicians now request 2 hour postprandial glucose determinations routinely in lieu of fasting glucose levels as guides to insulin requirements. Under usual hospital conditions it is often difficult, however, to control the 2 hour time interval very accurately since timing may start at the beginning, midway, or end of the test meal. To insure uniformity of carbohydrate intake and accurate timing, it is recommended that 100 gm. of glucose in solution be used routinely as a test meal for postprandial blood glucose determinations.

Interpretation, as with other glucose determinations, varies with the method and depends on whether whole blood or plasma is used for the analysis. In general, plasma or serum glucose at 2 hours postprandially should be less than 120 mg./100 ml; the true whole blood glucose should be less than 110 mg./100 ml.

Insulin Tolerance Test

This test is useful in evaluating patients with insulin resistance or certain endocrine disorders. The patient is placed on a diet containing at least 300 gm. of carbohydrate daily for 2 or 3 days before the test. With the patient in the fasting state, blood is taken for a baseline glucose level, after which regular insulin is injected intravenously by a physician in an amount corresponding to 0.1 unit per kg. of body weight. Blood specimens are then taken for glucose determinations at 20, 30, 45, 60, 90, and 120 minutes after the insulin was given. A syringe containing 50 ml. of 50 per cent glucose should be available for intravenous injection. The patient should be observed closely and a physician should be available to make the injection and terminate the test should a hypoglycemic reaction occur.

Normally the blood glucose decreases to about 50 per cent of the fasting level by 30 minutes, and then returns to normal fasting limits by 90 to 120 minutes. There are two types of abnormal response. The insulin resistant type shows only slight or delayed decrease in blood glucose and occurs with adrenal cortical hyperfunction (Cushing's syndrome), in acromegaly, and in some cases of diabetes. In the second type of response the blood glucose falls normally, but the subsequent rise is delayed or does not occur at all. This situation occurs with hypofunction of the anterior pituitary (Simmond's disease) or the adrenal cortex (Addison's disease), and in hyperinsulinism. In cases of suspected Simmond's disease it is recommended that half the usual dose of insulin be given and that the patient be watched carefully for signs of hypoglycemia. Glucose solutions or fruit juice should normally be given to patients at the end of insulin tolerance tests.

Other Tolerance Tests

Various other tests have been proposed that require serial determinations of blood glucose. Some of these will now be described briefly.

Tolbutamide (1-butyl-3-*p*-tolylsulfonylurea, Orinase) is a compound that stimulates the pancreas to produce insulin. Following intravenous injection, the normal response is similar to that observed with the insulin tolerance test: the blood glucose decreases to about 50 per cent of the fasting level by 30 minutes, then returns to normal. If the blood glucose level at 20 minutes is between 80 and 84 per cent of the fasting value the patient is said to have a 50 per cent probability of having diabetes.[5] In more severe cases the response will be even less inasmuch as the pancreas is unable to secrete adequate insulin. The test has also proved to be valuable in evaluating hypoglycemic states caused by insulomas.[2] In this condition injection of tolbutamide results in marked decrease in blood glucose to values in the range of 20 to 30 mg./100 ml. and persistent hypoglycemia up to 3 hours. As with the insulin tolerance test, patients must be watched carefully for hypoglycemic reactions and the test terminated, if necessary, by intravenous administration of glucose.

The *epinephrine tolerance* test is used to evaluate one form of glycogen storage disease (Type I, von Gierke's), a condition in which there is a deficiency or absence of the enzyme glucose-6-phosphatase in the liver. This enzyme is the catalyst for the final step in the formation of blood glucose from hepatic glycogen. Individuals with von Gierke's disease have low glucose levels in the blood, increased liver glycogen, but decreased *availability* of liver glycogen as shown by less than normal or no increase in blood glucose following administration of epinephrine. In a normal person, after intramuscular injection of 1 ml. of a 1:1000 solution of epinephrine hydrochloride, the blood glucose increases 35 to 45 mg./100 ml. in 40 to 60 minutes and returns to the fasting level by 2 hours. Blood specimens are taken at 30, 45, 60, 90, and 120 minutes after injection.

Deficiency of small bowel mucosal lactase has been found to be a rather common condition in healthy adults. Such deficiency may be associated with intolerance to lactose manifested by diarrhea and other symptoms following ingestion of milk. The diarrhea will usually disappear if lactose is eliminated from the patient's diet. A *lactose tolerance test* can be done to evaluate this condition.[1] A standard oral glucose tolerance test is performed first to provide a basis for comparison. On the following day the test is repeated except that 100 gm. of lactose is substituted for glucose. If lactase activity is present the lactose will be split to glucose and galactose and the resultant tolerance curve will be similar to that observed with glucose. With lactase deficiency the lactose tolerance curve will be flat with a rise not exceeding 20 mg./100 ml. over the fasting level. Either a copper reduction method or the *o*-toluidine method may be used to determine blood "sugar," since both galactose and glucose react in either method.

URINARY SUGARS

Occurrence of Sugars in Urine

Urine is examined routinely to detect or determine the presence or amount of glucose; this is done either as a screening procedure or as a guide to insulin therapy. Other sugars may also appear in the urine in certain conditions and interfere with the detection and determination of glucose. The sugars of clinical interest are all reducing sugars; that is, they readily reduce cupric ion in hot alkaline solution. Except for the very rare cases of galactosuria, glucose is the only sugar found in urine that is of pathological significance.

Galactose appears in the urine of infants with galactosemia, a condition characterized by inability to metabolize galactose. Such infants fail to thrive on milk since half of the milk sugar, lactose, is converted to galactose. Lactose is sometimes found in urine of women during lactation and occasionally toward the end of pregnancy. The laboratory may be required to differentiate this sugar from glucose. Fructose may appear in the urine after eating fruits, honey, and syrups, but has no significance. Fructosuria is a rare and harmless congenital defect that should not be confused with diabetes. Pentoses may occur in urine after eating such fruits as cherries,

TABLE 4-1. *Reducing Substances in Urine*

Fructose	Ketone bodies
Lactose	Sulfanilamide
Galactose	Oxalic acid
Maltose	Hippuric acid
Arabinose	Homogentisic acid
Xylose	Glucuronic acid
Ribose	Formaldehyde
Uric acid	Isoniazid
Ascorbic acid	Salicylates
Creatinine	Cinchophen
Cysteine	Salicyluric acid

plums, or prunes, or as a harmless congenital anomaly and, as with fructose, must be distinguished from glucose. Maltose has been reported to occur along with glucose in the urine of some patients with diabetes.

Many reducing substances other than sugars may also occur in urine. A partial list of the more important reducing substances is shown in Table 4-1.

Qualitative Methods for Total Reducing Substances

PRINCIPLE

Benedict's qualitative reagent contains cupric ion complexed with citrate in alkaline solution. Glucose, or other reducing substances, reduces cupric ion to cuprous ion with resultant formation of yellow cuprous hydroxide or red cuprous oxide.

REAGENT

Dissolve 17.3 gm. of $CuSO_4 \cdot 5H_2O$ in 100 ml. of hot water. Dissolve separately, with heating, 173 gm. of sodium citrate ($Na_3C_6H_5O_7 \cdot 2H_2O$) and 100 gm. of Na_2CO_3 in 800 ml. of water. Allow to cool, then add the citrate-carbonate solution, with mixing, to the copper sulfate solution. Dilute to 1 L. with water. This reagent is stable.

PROCEDURE

Add 8 drops (0.4 ml.) of urine to 5 ml. of reagent in a test tube. Mix and place in a boiling water bath for 3 minutes. Remove and examine immediately. Report as 0 to 4+ according to the following criteria:

Appearance	*Report*	*Approximate Glucose Concentration* (gm./100 ml.)
Blue to green, no precipitate	0	0–0.1
Green with yellow precipitate	1+	0.3
Olive green	2+	1.0
Brownish orange	3+	1.5
Brick red	4+	2.0 or more

A convenient adaptation of the preceding procedure is marketed in tablet form (Clinitest, Ames Co. Div., Miles Laboratories, Elkhart, Ind.). The tablets contain anhydrous cupric sulfate, sodium hydroxide, citric acid, and sodium bicarbonate. Five drops (0.25 ml.) of urine are mixed with 10 drops of water in a test tube. One tablet is added and the mixture is allowed to stand undisturbed for 15 seconds, remixed, and observed for color. A chart provided by the manufacturer is used to interpret the result. Heat is generated by contact of sodium hydroxide and water. The initial reaction between citric acid and sodium bicarbonate causes the release of carbon dioxide, which blankets the mixture and reduces contact with oxygen from the air.

Quantitative Methods for Total Reducing Substances

Although quantitative measurement of total reducing substances in urine provides information of limited diagnostic value, the test is still performed in a number of hospital laboratories. The Folin-Wu or Somogyi-Nelson methods may be used for this purpose. The urine usually needs diluting to below about 300 mg./100 ml. to bring the concentration of glucose within the range of the method. The dilution necessary can be estimated from the qualitative test. The preparation of protein-free filtrate is omitted; instead, the urine is further diluted with water and analyzed in the same manner as a protein-free filtrate. Results are corrected for the initial dilution. A small amount of reducing substances is found in urine specimens that are negative with the qualitative tests. Expressed as glucose, the concentration is less than 150 mg./100 ml. of urine.

SEPARATION AND IDENTIFICATION OF SUGARS

Techniques for separating and identifying sugars include fermentation, osazone formation with phenylhydrazine, specific chemical tests, and paper or thin-layer chromatography. The availability of glucose oxidase test strips has greatly simplified the differentiation of glucose from the many other reducing substances.

Glucose, fructose, maltose, and mannose are fermentable with yeast, but lactose, galactose, and pentoses do not ferment. Of the fermentable sugars, only glucose and, rarely, fructose are likely to occur in urine. The fermentation test can be used, therefore, to differentiate glucose from lactose or other nonfermenting sugars.

Fermentation Test

Bring a portion of the urine to boiling to destroy *E. coli*, which can ferment lactose. Cool to room temperature and reserve a portion for a qualitative test for reducing sugar. Add about 0.3 gm. of dry active baker's yeast to 10 ml. of the boiled specimen and mix with a stirring rod until the yeast is dispersed into a homogeneous mixture. Transfer to a large test tube and incubate unstoppered with occasional mixing for 1 hour in a 37°C water bath. Centrifuge the incubated specimen and perform a qualitative test for reducing sugar on the supernatant. Compare the results with a similar test performed on the unfermented specimen.

If the fermented specimen is negative, all the sugar in the urine is probably glucose. If the fermented sample is the same as the unfermented sample, some sugar or reducing substance other than glucose is present. When the fermented sample is positive, but lower than the untreated sample, the difference is considered to be glucose. It is good practice to include a control to check the activity of the yeast. Dissolve 0.1 gm. of glucose in 10 ml. of urine previously shown to be negative for

TABLE 4-2. *Results Obtained by Four Glucose Procedures on 20 Urine Specimens All Negative for Reducing Substances*

Method	Apparent gm. glucose/gm. creatinine *Range*	*Mean*
Ferricyanide	1.07–2.77	1.80
Folin-Wu	0.83–2.16	1.24
Somogyi-Nelson	0.26–1.26	0.53
o-Toluidine	0.10–0.37	0.20
Fermentation	0–0.18	0.06

reducing substances. This specimen, when carried through the fermentation test, should become negative for reducing sugar. A blank may also be included by mixing 0.3 gm. of yeast with 10 ml of water to rule out the possible presence of nonfermenting reducing substances in the yeast. If necessary, the yeast may be suspended in saline, centrifuged, and rewashed to remove reducing substances.

If desired, a quantitative determination of glucose in urine may be performed on the fermented and unfermented specimens. The difference between the two values is considered to be glucose. For the Somogyi-Nelson procedure, a filtrate is prepared in the same manner as described for blood. The *o*-toluidine method is performed as described for serum. The amount of glucose excreted normally, as determined by highly specific enzymatic methods, is less than 0.5 gm./24 hr. Random specimens show an upper limit of normal of approximately 30 mg./100 ml.

Generally, it is not necessary to resort to fermentation tests to determine total glucose excretion in a 24 hour urine specimen provided the glucose method has reasonable specificity. In one study, 20 morning urine specimens were selected that were all negative for reducing sugar by the copper reduction test. These were analyzed by an automated alkaline ferricyanide method, the Folin-Wu method, the Somogyi-Nelson method, and the *o*-toluidine method. Creatinine was also determined. After yeast fermentation, the specimens were again analyzed by the *o*-toluidine method and the decrease in value recorded as fermentable sugar. Yeast blanks were included. Results, shown in Table 4-2, are expressed as apparent gm. of glucose per gm. of creatinine. Of the four procedures, the *o*-toluidine method was found to be the most specific and the ferricyanide method the least specific for the estimation of glucose in urine.

Qualitative Tests for Individual Sugars

GLUCOSE

A convenient paper test strip is commerically available (Clinistix, Ames Co.). The filter paper is impregnated with glucose oxidase, peroxidase, and *o*-tolidine and provides a simple color test according to principles discussed in an earlier section. The test end is moistened with urine and examined after 10 seconds. A blue color develops if glucose is present. The sensitivity of the strip has been adjusted to take into account the presence of enzyme inhibitors normally occurring in urine. Thus, a positive test will be obtained with lower concentrations of glucose in water as compared to urine. For the same reason, a false positive test may be obtained with very dilute specimens.

In one study of 2000 urine specimens, 11 false negative enzyme paper tests were encountered. Among the inhibitors identified were ascorbic acid, dipyrone, and meralluride sodium (mercuhydrin). Several antibiotics contain ascorbic acid as a preservative. The acid is largely excreted unchanged and can cause false negative results. Contamination of urine with hydrogen peroxide or a strong oxidizing agent, such as hypochlorite, produces false positive results. For routine examinations, however, a negative stick test is considered negative for glucose. A positive stick test is further evaluated by one of the copper-reduction methods, such as a Clinitest tablet.

REFERENCE

Free, A. H., Adams, E. C., Kercher, M. L., Free, H. M., and Cook, M. H.: Clin. Chem., *3*:163, 1957.

SELIWANOFF'S TEST FOR FRUCTOSE

Hot hydrochloric acid converts fructose to hydroxymethyl furfural, which links with resorcinol to produce a red-colored compound. To make the reagent, dissolve 50 mg. of resorcinol in 33 ml. of concentrated hydrochloric acid and dilute to 100 ml. with water. Add 0.5 ml. of urine to 5 ml. of reagent in a test tube and bring to a boil. Fructose produces a red color within $\frac{1}{2}$ minute. The test is sensitive to 0.1 per cent fructose provided excess glucose is absent. A 2 per cent solution of glucose will produce about the same color as 0.1 per cent fructose after $\frac{1}{2}$ minute of boiling. A 0.5 per cent solution of fructose should be used as a control. With high concentrations of fructose, a red precipitate forms, which may be filtered and dissolved in ethanol to produce a bright red colored solution.

BIAL'S TEST FOR PENTOSE

By heating with hydrochloric acid, pentoses are converted to furfural, which reacts with orcinol to form green-colored compounds.

Dissolve 300 mg. of orcinol in 100 ml. of concentrated hydrochloric acid and add 0.25 ml. of 10 per cent ferric chloride solution. Glucose, if present in the urine, should be removed by fermentation. Add 0.5 ml. of urine to 5 ml. of reagent in a test tube and bring to a boil. Pentoses produce a green color. The test is sensitive to 0.1 per cent pentose. A 0.5 per cent solution of xylose should be used as a control. Glucuronates will produce a similar color if the boiling is prolonged. Fructose, as with Seliwanoff's reagent, produces a red color.

A combination of the preceding tests will usually differentiate the nature of the reducing sugar as summarized in Table 4-3. Lactose and galactose will not be differentiated from each other. More definitive separation can be achieved by use of paper chromatography.

TABLE 4-3. *Differentiation of Reducing Sugars*

Sugar	*Fermentation*	*Clinistix*	*Special Tests*
Glucose	+	+	
Fructose	+	−	Seliwanoff
Galactose	−	−	
Lactose	−	−	
Maltose	+	−	
Pentoses	−	−	Bial

Identification of Urinary Sugars by Paper Chromatography

PRINCIPLE

Sugars can be separated by ascending or descending chromatography on paper and located after color development with dinitrosalicylic acid. The variable rates of migration depend upon the solubility of the sugars in the particular solvent system. Presumptive identification is made by comparison of the R_f value of the unknown with those of authentic samples. The following procedure may be performed conveniently in a 6 × 18 inch Pyrex jar with a tightly fitting glass cover.

REAGENTS

1. Solvent. Perform the following procedure under a hood. Mix 60 ml. of *n*-butanol, 40 ml. of pyridine, and 30 ml. of water. The mixture is completely miscible. Pour into the bottom of the jar and allow to equilibrate at least 30 minutes before use.

2. Spray reagent. Dissolve 0.5 gm. of 3,5-dinitrosalicylic acid in 100 ml. of 4 per cent (w/v) sodium hydroxide.

3. Reference sugar solutions. Prepare 1.75 per cent (w/v) solutions of glucose, fructose, galactose, maltose, lactose, and xylose in 0.2 per cent benzoic acid solution. These solutions are stable for months when refrigerated.

PROCEDURE

1. Determine the concentration of sugar in the urine by means of one of the qualitative copper reduction tests. Dilute the specimen, if necessary, to a sugar concentration of approximately 1 per cent. If the concentration is only 0.5 per cent, use twice as much sample in the test.

2. Draw a pencil line 1 inch from and parallel to the 10 inch side of a 10 × 14 inch section of Whatman No. 1 filter paper. Place pencil marks at least 1 inch apart on the line to indicate starting positions for each reference sample and for the urine specimen.

3. Apply approximately 0.01 ml. of each solution to its respective point from a microhematocrit tube or a 10 μl. pipet. Half of this amount should be applied and permitted to dry before adding the remainder in order to keep the diameter of the spots as small as possible. Allow all samples to dry completely.

4. Staple the sheet into a 14 inch high cylinder so that the line of application is at the bottom. Insert the paper into the chromatography jar, tape on the cover, and allow to stand undisturbed for about 16 hours (overnight) at room temperature.

5. Remove the sheet and mark with a pencil along the solvent front. Allow to air dry under the hood for 4 hours.

6. Spray the sheet with dinitrosalicylic acid reagent from an atomizer and allow to air dry for 4 hours.

7. Heat the paper at 100°C. for 10 minutes in a drying oven. The reducing sugars appear as brown spots against a yellow background.

8. Measure the distance from the starting line to the center of each spot; similarly, measure the distance from the starting line to the edge of the solvent front. Calculate the ratio of fronts, R_f:

$$R_f = \frac{\text{distance traveled by solute spot}}{\text{distance traveled by solvent front}}$$

The R_f values vary slightly from run to run and, for this reason, known reference samples should be included each time. Average approximate values are as follows:

Sugar	R_f
Lactose	0.22
Maltose	0.28
Galactose	0.36
Glucose	0.41
Fructose	0.46
Xylose	0.52

A typical run is shown in Figure 4-12.

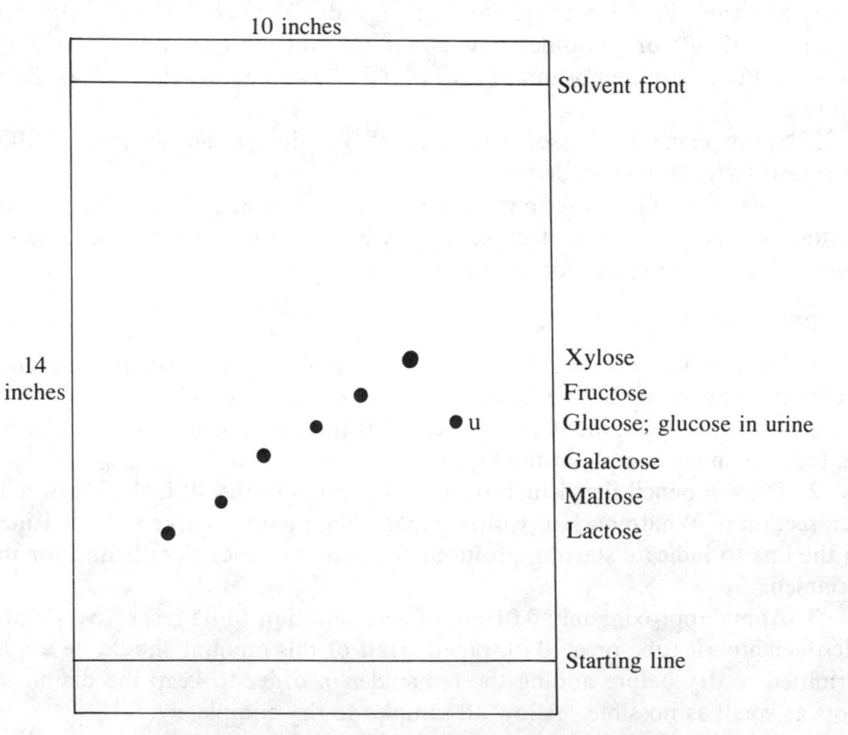

Figure 4-12. Paper chromatography of urine sugars.

The unknown sugar in the urine is presumed to be the same as a known reference standard when both migrate the same distance under the same test conditions. For confirmation, the urine specimen may be mixed with an equal volume of the known standard and rechromatographed. Only one sugar spot will appear on the paper if the two sugars are identical. Dinitrosalicylic acid is a highly specific reagent for reducing sugars.

REFERENCE

Sophian, L. H., and Connolly, V. J.: Am. J. Clin. Path., *22*:41, 1952.

ASCORBIC ACID

Ascorbic acid (vitamin C) has a chemical structure that justifies its classification as a carbohydrate. The presence of a double bond in the enediol grouping makes the compound an active reducing agent under mild reaction conditions. The compound is also a fairly strong acid ($pK_1 = 4.17$); a 0.5 per cent solution has a pH of approximately 3. Ascorbic acid is readily oxidized to dehydroascorbic acid, which, above pH 7, undergoes rapid base catalyzed hydrolysis to diketogulonic acid. Conversion to the latter also occurs in strong acid solution. Chemical structures are shown in Figure 4-13.

The oxidation of the monovalent ion of ascorbic acid by oxygen is markedly accelerated by trace amounts of cupric ion. Below pH 2 iron also acts as a catalyst; above pH 7, autoxidation occurs rapidly, even in the absence of metal catalysts, as a result of direct reaction between oxygen and the divalent ion of ascorbic acid. As expected, ascorbic acid in separated plasma or serum is unstable and samples must be analyzed very shortly after blood is drawn. The compound is somewhat more stable in whole blood, presumably because of the reducing action of glutathione in the erythrocytes; however, whole blood samples should also be analyzed without delay.

At 37°C., dehydroascorbic acid has been shown to have a half-life of 2 minutes at pH 7.4 and only 0.5 minute at pH 8.0. It follows, as has been demonstrated, that very little dehydroascorbic acid exists in blood.

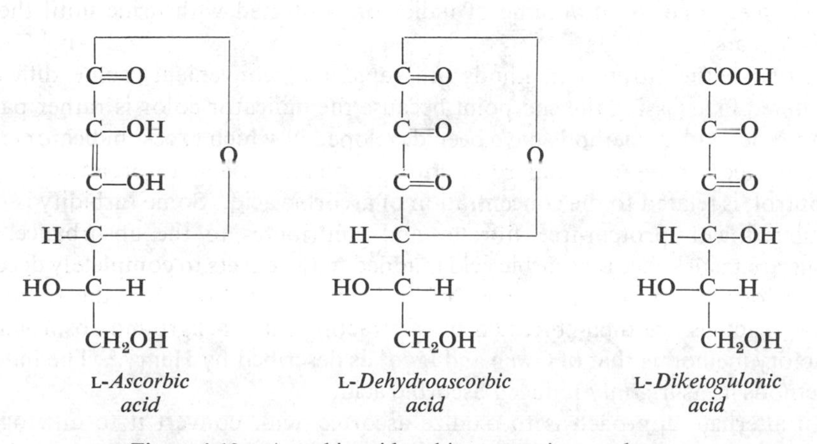

Figure 4-13. Ascorbic acid and its conversion products.

Clinical Significance

The fundamental role of ascorbic acid in metabolic processes is not known, although there is some evidence that it may be involved in the intermediary metabolism of tyrosine. Ascorbic acid is essential for the normal regulation of the colloidal conditions of intercellular substances, including the fibrils and collagen of connective tissues, and possibly the intercellular substance of the capillaries. Deficiency of vitamin C results in abnormal development and maintenance of tissue structures, capillary defects, and eventual development of scurvy.

Of many species investigated, only man, other primates, and guinea pigs are unable to synthesize ascorbic acid. Therefore, the entire human requirement must be supplied by the diet. There is no evidence that any particular organ or tissue serves as a storage reservoir; however, the body does maintain an overall reserve as shown by the fact that clinical manifestations of scurvy develop in man only after several months on an ascorbic acid–free diet.

Normal plasma contains about 0.5 to 1.5 mg. of ascorbic acid/100 ml. The red cells contain about twice this concentration and the white blood cells and platelets 20 to 40 times that in plasma. The plasma level gradually decreases as ascorbic acid deficiency develops. There is considerable evidence that the concentration in the leucocytes and platelets provides a much better index of body reserves and methods have been developed for measuring the ascorbic acid in the buffy coat formed at the erythrocyte-plasma interface after centrifugation.

Methods

Ascorbic acid may be determined by titration with an oxidation-reduction indicator such as 2,6-dichlorophenolindophenol in acid solution.[3] Ascorbic acid reduces the indicator to a colorless form. A protein-free filtrate is prepared from fresh blood or plasma by deproteinization with metaphosphoric acid or trichloroacetic acid. A portion of the filtrate is titrated directly with the indicator from a microburet. The indicator solution is blue, but becomes pink in acid solution. The titration is continued until a faint pink color persists. Similar titrations are carried out on fresh standard solutions and on a water blank, both treated in the same manner as the original blood specimen to provide an acid medium. For urine the titration is sometimes reversed; i.e., a constant volume of indicator is titrated with urine until the color just disappears.

Although the titration methods are rapid and convenient, some difficulty is encountered in assessing the end point because the indicator color is rather pale and tends to fade. Other methods have been developed in which excess indicator is added to protein-free filtrate and standard solutions. The decrease in absorbance, compared to a control, is related to the concentration of ascorbic acid. Some turbidity is usually encountered with protein-free filtrates and contributes to the absorbance. After readings are made, excess ascorbic acid is added to the cuvets to completely decolorize the indicator and permit measurement of the background absorbance due to turbidity. Original readings are then corrected by subtracting this background absorbance. A satisfactory method is that of Owen and Iggo[6] as described by Henry.[4] The indicator-dye methods measure only reduced ascorbic acid.

An alternate approach is to oxidize ascorbic acid, convert it to diketogulonic acid in strong acid solution, and form a diphenylhydrazone by reacting with 2,4-

dinitrophenylhydrazine. The hydrazone dissolves in strong sulfuric acid solution to produce a red color, which can be measured spectrophotometrically. This method measures both ascorbic acid and dehydroascorbic acid. As noted earlier, very little dehydroascorbic acid exists in blood so that either approach provides substantially the same result. The method to be described employs cupric ion as the oxidizing agent followed by hydrazone formation.

REAGENTS

1. Trichloroacetic acid, 10 per cent (w/v).
2. 2,4-Dinitrophenylhydrazine reagent. Dissolve 2.0 gm. of the crystalline compound in 100 ml. of 9 *N* sulfuric acid (75 ml. of water plus 25 ml. of concentrated sulfuric acid). Filter and store in a brown bottle in the refrigerator. Refilter a portion before use, or use the clear supernatant.
3. Thiourea solution. Dissolve 10 gm. of thiourea in 100 ml. of 50 per cent ethanol and refrigerate.
4. Cupric sulfate solution, 1.5 per cent. Dissolve 1.5 gm. of $CuSO_4 \cdot 5H_2O$ in water and dilute to 100 ml.
5. Combined color reagent. Prepare fresh on the day of use by mixing:
   5 ml. of 2,4-dinitrophenylhydrazine reagent
   0.1 ml. of cupric sulfate solution
   0.1 ml. of thiourea solution.
6. Sulfuric acid, 85 per cent. To 20 ml. of distilled water add 180 ml. of concentrated sulfuric acid. Mix, cool, and store in a glass-stoppered bottle in the refrigerator, or cool a portion in ice water before using.
7. Standard. Prepare a fresh 1.000 per cent (w/v) stock solution of ascorbic acid. Dilute this 1:500 with water just before use to provide a working standard of 2 mg./100 ml.

PROCEDURE

Blood must be fresh. Prepare the protein-free filtrate as soon as possible after blood is drawn (no later than 30 minutes). Ascorbic acid is stable for several hours in the filtrate.

1. Pipet 1.0 ml. of fresh whole blood or plasma into a small test tube.
2. Add 1.0 ml. of 10 per cent trichloroacetic acid and 0.5 ml. of chloroform.
3. Stopper, shake vigorously 10 to 15 seconds, and centrifuge.
4. Pipet 1.0 ml. of clear supernatant to a test tube.
5. Prepare a blank and standard by adding 0.5 ml. of 10 per cent trichloroacetic acid to 0.5 ml. of water and working standard, respectively.
6. To each tube add 0.4 ml. of freshly prepared combined color reagent and mix.
7. Stopper and place in a 56°C. water bath for 1 hour.
8. Cool in an ice bath for about 5 minutes.
9. Add slowly, and with mixing, 2.0 ml. of ice-cold 85 per cent sulfuric acid.
10. Let stand at room temperature for 30 minutes; then remix.
11. Transfer to cuvets with a minimum capacity not exceeding 3 ml. and measure absorbance against the blank at 500 nm.

$$\frac{A_{\text{sample}}}{A_{\text{standard}}} \times 2 = \text{mg. ascorbic acid/100 ml.}$$

The normal range for plasma or serum is 0.5 to 1.5 mg./100 ml. and for whole blood 0.7 to 2.0 mg./100 ml. Fresh urine specimens are diluted 1:5 and analyzed as described for plasma. The normal range is approximately 1 to 7 mg./100 ml., but this varies greatly with diet. Because of the instability of ascorbic acid, determinations on 24 hour urine collections are generally unreliable.

REFERENCES

Natelson, S.: Microtechniques of Clinical Chemistry. 2nd ed. Springfield, Ill., Charles C Thomas, 1961, p. 121.
Roe, J. H., and Kuether, C. A.: J. Biol. Chem., *147*:399, 1943.

REFERENCES

1. Basford, R. L., and Henry, J. B.: Postgrad. Med., *41*:A-70, 1967.
2. Fajans, S. S., Schneider, J. M., Schteingart. D. E., and Conn, J. W.: J. Clin. Endocrinol. & Metab., *21*:371, 1961.
3. Farmer, C. J., and Abt, A. F.: Proc. Soc. Exp. Biol. Med., *32*:1625, 1935; *34*:146, 1936.
4. Henry, R. J.: Clinical Chemistry: Principles and Technics. New York, Hoeber Medical Div., Harper & Row, Publishers, 1964, p. 715.
5. Kaplan, N. M.: Arch. Int. Med., *107*:212, 1961.
6. Owen, J. A., and Iggo, B.: Biochem. J. *62*:675, 1956.
7. Salamon, L. L., and Johnson, J. E.: Anal., Chem., *31*:453, 1959.
8. Weissman, M., and Klein, B.: Clin. Chem., *4*:420, 1958.

ADDITIONAL READINGS

Cantarow, A., and Schepartz, B.: Biochemistry. 4th ed. Philadelphia, W. B. Saunders Co., 1967.
Duncan, G. G. (Ed.): Diseases of Metabolism. 5th ed. Philadelphia, W. B. Saunders Co., 1964, chaps. 2, 13, 14.
Pigman, W. (Ed.): The Carbohydrates: Chemistry, Biochemistry, Physiology. New York, Academic Press, Inc., 1957.

Chapter 5 / # PROTEINS AND AMINO ACIDS

by John F. Kachmar, Ph.D.

Chemistry of Amino Acids and Proteins

AMINO ACIDS

Proteins constitute that class of biochemical compounds most characteristic of protoplasm and life. Many individual carbohydrates, lipids, and nucleotides are encountered in all animal and plant forms. Proteins, however, are specific; each species of organism is associated with a large number of proteins typical of itself and itself alone. Indeed, there occur proteins unique for a given organ or tissue, and even for an individual organism. All proteins contain carbon, hydrogen, oxygen, nitrogen, and sulfur; in addition, individual proteins may contain phosphorus, iodine, iron, copper, zinc, or other elements. The presence of *nitrogen* in all proteins sets them apart from carbohydrates and lipids: The average nitrogen content is approximately 16 per cent.

When proteins are broken down into individual elementary units by acid, alkali, or enzymatic hydrolysis, it is found that these basic units consist of *alpha-amino acids*. Some 40 different amino acids have been isolated from various proteins, but only about 20 are present in all proteins in varying amounts. These amino acids are linked together by *peptide bonds* (see equation 4) into long chains, which contain from 50 to many thousands of amino acids. The number of amino acids, the order in which they are joined together in the long chain, and the manner in which the chain is twisted, folded, and cross-linked make possible the many millions of proteins present in the multitude of living organisms.

Alpha-amino acids constitute that class of organic acids which contain an amino group located on the carbon atom adjacent to the carboxyl group, as illustrated in the general formula

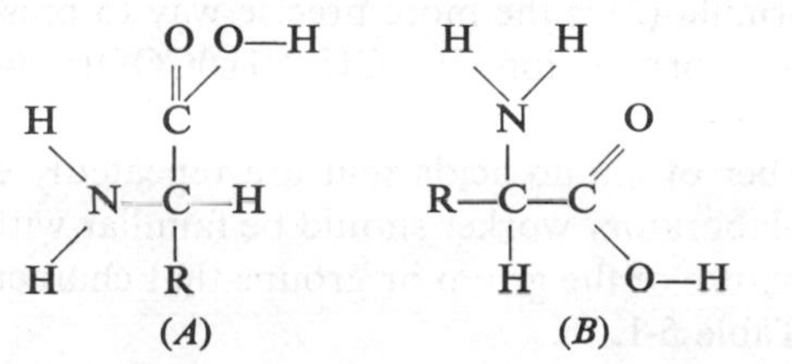

(1)

The R group represents the remainder of the molecule, and varies from H in *glycine* to the complex indole ring system in *tryptophane*. With the exception of the simplest amino acid, *glycine*, the central (α) carbon atom is asymmetric because all four groups linked to it are different. Thus, mirror image, *D*, and *L*, stereoisomeric forms are possible; however, in natural proteins all amino acids have the *L* configuration shown in formula (1), in which the structure (*A*), with the NH_2 group to the left of the α-carbon atom, is the conventional method of designating the *L* configuration, and the structure (*B*) is the more convenient form used in most texts.

Amino acids contain both the acidic *carboxyl* group (proton-donating) and the basic *amino* group (proton-accepting) within the molecule. Such compounds are referred to as *amphiprotic* or *amphoteric* compounds. They behave both as acids and bases. In the pH range encountered physiologically (bordering neutrality), the carboxyl group is dissociated, and the NH_2 group has bound the proton to give the following structure:

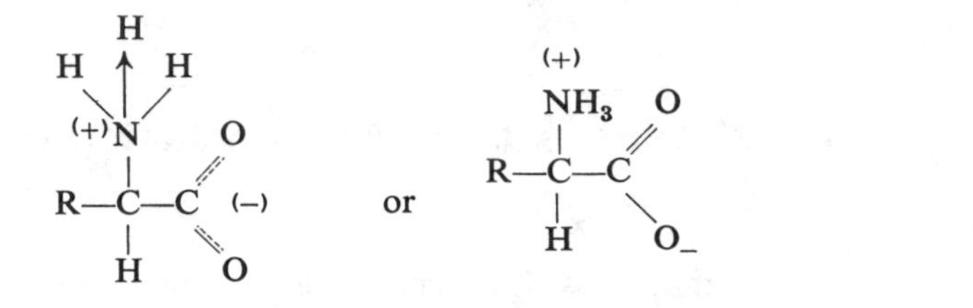

(2)

This type of ionized molecule with negative and positive charges is referred to as a "zwitterion." Proteins themselves are complex zwitterion type structures, containing many negatively and positively charged groups in each molecule. Because amino acids can react with either acids or bases, their solutions can be used as pH buffers. This is illustrated in the following equations:

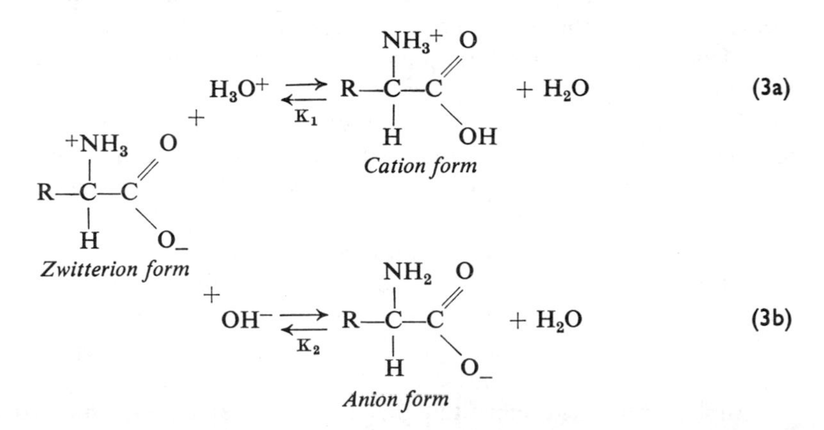

(3a)

(3b)

For glycine, the value of $pK_1 = 2.42$ and that of $pK_2 = 9.47$. Although the structure shown in formula (2) is the more precise way to present the structure of an amino acid, the simpler formulation, R—CH(NH_2)COOH, will be used most often in this chapter for purposes of convenience.

There are a number of amino acids that are repeatedly encountered in clinical chemistry. Thus, the laboratory worker should be familiar with the formulas of these compounds and the nature of the group or groups that characterize them. These are presented in part in Table 5-1.

THE PEPTIDE BOND

Amino acids can react with one another to form peptides by the linking of the amino group of one acid with the carboxyl group of another acid.

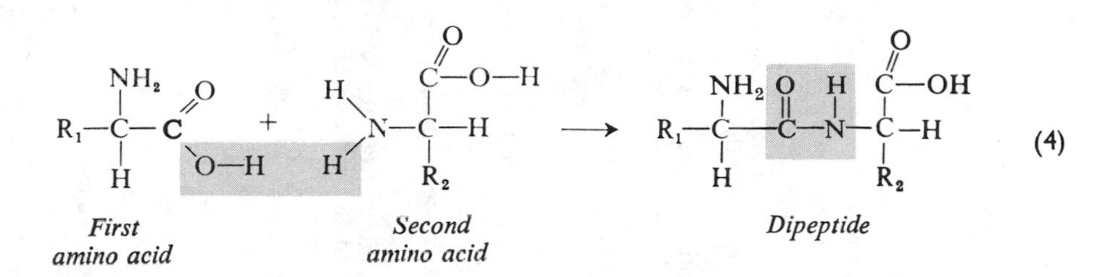

In this reaction the H from the NH_2 group of the one acid combines with the OH of the COOH group of the other to form H_2O, the two amino acid residues then being linked via the *peptide linkage*, $-\underset{\overset{\|}{O}}{C}-\underset{\overset{|}{H}}{N}-$ or $-CONH-$

Amino acids, in forming peptides, can react through either the amino or carboxyl ends. For example, glycine and alanine can react to form two different dipeptides, *glycyl-alanine* and *alanyl-glycine*.

$$\underset{\textit{Glycine}}{NH_2CH_2COOH} + \underset{\textit{Alanine}}{CH_3CHNH_2COOH} \rightarrow \underset{\textit{Glycyl-alanine}}{NH_2CH_2\overset{\overset{O}{\|}}{C}-NH-\overset{\overset{H}{|}}{\underset{\underset{CH_3}{|}}{C}}-COOH} \quad (5)$$

$$\underset{\textit{Alanine}}{CH_3CHNH_2COOH} + \underset{\textit{Glycine}}{NH_2CH_2COOH} \rightarrow \underset{\textit{Alanyl-glycine}}{CH_3CHNH_2\overset{\overset{O}{/\!/}}{C}-NH-CH_2COOH} \quad (6)$$

An important tripeptide is *glutathione*, γ-glutamyl-cysteinyl-glycine, in which the terminal COOH of glutamic acid is linked with the NH_2 of cysteine, and the COOH of the latter linked with the NH_2 of glycine.

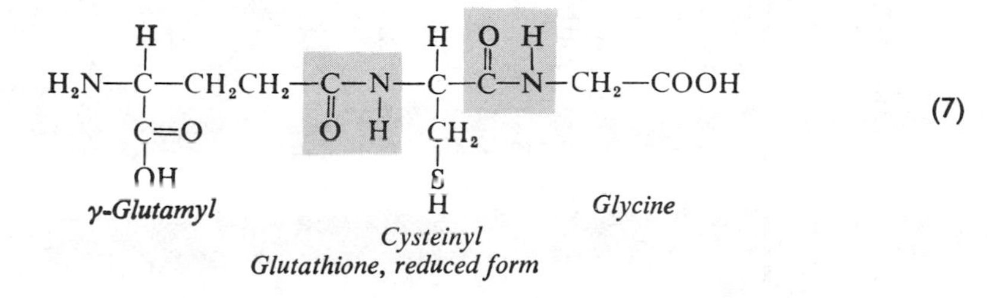

Glutathione, reduced form

Short chains of amino acids (6 to 30 residues) linked by peptide bonds are referred to as *polypeptides*. The hormone *oxytocin* is an octapeptide. The terms *proteose* and *peptone* refer to protein breakdown products containing large polypeptides, which differ from true proteins in not being coagulated by heat. When the number of amino acids linked together reaches 40 or more (molecular weight of 5000), the chain takes on the physical properties associated with proteins.

TABLE 5-1. *List of Some Amino Acids of Direct Interest in Clinical Chemistry*

Name	*Formula*	*Comments*
	Neutral Amino Acids: Contain one carboxyl and one amino group each	
Glycine	NH_2CH_2COOH	Simplest amino acid; used as buffer
Alanine	CH_3CHNH_2COOH	Substrate for GPT, one of the transaminases
Cysteine	$HS—CH_2CHNH_2COOH$	The (—SH) group referred to as sulfhydryl group; (—SH) necessary for activity of many enzymes; heavy metals inactivate proteins by combining with the SH
Cystine	$S—CH_2CHNH_2COOH$ | $S—CH_2CHNH_2COOH$	Oxidized form of cysteine; forms one type of kidney stone; in proteins, links two peptide chains together
Leucine	$CH_3—CH(CH_3)—CH_2CHNH_2COOH$	One of the branched chain amino acids; faulty metabolism in "maple-syrup" disease; R group is hydrophobic
Serine	$HOCH_2CHNH_2COOH$	In phosphoproteins, the PO_4 is linked through this acid; involved in active center of many enzymes
Phenylalanine	$C_6H_5—CH_2CHNH_2COOH$	A dietary essential amino acid; metabolism defective in phenylketonuria
Tyrosine	$HO—C_6H_4—CH_2CHNH_2COOH$	The phenol ring used to quantitate proteins; accumulates in tissues in tyrosinosis; step in synthesis of thyroxine and catecholamines
Tryptophane	(indole ring, N—H)—CH_2CHNH_2COOH	Essential; contains indole ring system; metabolites involved in carcinoid disease
Thyroxine	$HO—C_6H_2I_2—O—C_6H_2I_2—CH_2CHNH_2COOH$	Thyroid hormone (T_4); contains iodine

	Acidic Amino Acids: Contain two carboxyl groups and one amino group	
Aspartic acid	$HOOC—CH_2CHNH_2COOH$	Substrate for GOT, another transaminase;
Glutamic acid	$HOOC—CH_2CH_2CHNH_2COOH$	Substrate for both GOT and GPT and enzyme method for determining ammonia; Glutamine = amide ($CONH_2$ in place of second COOH)
	Basic Amino Acids: One carboxyl, one α-NH_2, and one other basic group or NH_2	
Arginine	$H_2N—C(=N—H)—NH—CH_2CH_2CH_2CHNH_2COOH$	Involved in urea synthesis; the basic group is guanidine
Lysine	$H_2N—CH_2CH_2CH_2CH_2CHNH_2COOH$	End NH_2 referred to as ϵ (epsilon) NH_2
	Imino Acids: (>N—H replaces NH_2)	
Hydroxyproline	H H HO—C——C—H H_2C CH—COOH N H	Important amino acid in collagen

Classification and Properties of Proteins

Proteins differ from one another in the number of the various amino acids linked together in the chain and in their arrangement or order in the chain. The molecular weight can vary from 5738 for each of the two equal subunits of *insulin* (51 amino acids) to as high as several million for certain structural proteins.[49] The larger protein moieties often consist of two or more similar or related chains linked together by weak chemical bonds. The five isoenzyme forms of *lactic dehydrogenase* are examples of such associations.

In solution, the protein chains coil and fold over and about each other to form complex globular, ellipsoidal, or needle shaped forms.[80] This folded chain constitutes the final *native protein*. The coils (helices) and folds are held together by hydrogen bonding, by weak covalent or electrostatic bonds between different R groups, and by the disulfide (—S—S—) bonds of cystine. Many physical and chemical properties of proteins derive from this "tertiary" (folded) structure of the protein. When these tertiary bonds are broken, the chain may be partly or completely unfolded, and the protein is said to be *denatured*, with a resulting change in properties. The most striking change is a decrease in solubility, and, in the cases of enzymes, partial or total loss of catalytic properties. Egg white is a native protein; it is easily soluble in water and gives a clear water-white solution. If the solution is vigorously stirred, however, such as with an egg beater, or heated to above 60 to 70°C., the mechanical or heat energy, respectively, will break the tertiary structure, causing the egg white to coagulate or set into a white insoluble form. *Protein solutions must be treated with extreme care to avoid the occurrence of denaturation.*

Laboratory Techniques used in Protein Studies[28,80]

Although the number of individual proteins is very large, proteins can be conveniently differentiated into a few classes by laboratory techniques based on their special chemical and physical properties. The methods and the protein properties that are utilized in these methods are as follows:

1. Salt or solvent fractionation: changes in solubility in the presence of charged ions or the presence of dehydrating solvents, or both.
2. Electrophoresis: differences in surface electrical charge density.
3. Ultracentrifugation: variations in molecular weight and molecular shape.
4. Chromatography: differences in size, shape, and charge as affecting rate of flow through various types of chromatographic media.

SALT FRACTIONATION[28,97]

Proteins are colloidal solutions. The individual molecules are charged because of the presence of ionized chemical groups on their surfaces. Included among such charged groups are carboxyl (—COO^-), and NH_3^+ and also, if the pH is favorable, the *guanidine*, *imidazole*, or ϵ-NH_3^+ groups of the amino acids *arginine*, *histidine*, and *lysine*. Some solvent water molecules are more or less bound to the surface of the protein molecules. These phenomena give proteins certain solubility properties that can be modified drastically by the presence of other charged ions. The earliest and still an important means of differentiating and separating proteins is based on such solubility relations.

Albumins. These are defined as those proteins that are soluble in water and soluble in dilute and concentrated salt solutions, but insoluble in highly concentrated salt solutions such as saturated ammonium sulfate, $(NH_4)_2SO_4$.

Globulins. These are proteins that are insoluble in water, soluble in weak neutral salt solutions (1 to 10 per cent), but insoluble in concentrated salt solutions (20 to 50 per cent). Ammonium sulfate is the salt most often used, but Na_2SO_4, Na_2SO_3, $MgSO_4$, and others are also used. The first three are still commonly used in the laboratory to separate protein compounds in blood serum.

Plasma and extracts of tissues contain many individual proteins that behave like globulins, i.e., in being insoluble in water, but soluble in weak (0.05 to 0.25 *M*) salt solutions. The individual globulins differ considerably in structure and size and as a result require different concentrations of divalent (Mg^{++}, $SO_4^{=}$) ions before they will again become insoluble. By varying the type of salt and, especially, salt concentration, an entire spectrum of globulins can be obtained. Salt fractionation is of little value for isolating individual pure proteins but it is a useful tool for separating a mixture into broad groups.

An analogous technique takes advantage of the fact that the solubility of globulins is decreased by adding certain organic solvents (alcohol, acetone) to protein solutions. To avoid denaturation, such solvent precipitation of proteins must be done at low temperatures (−5 to −30°C.). Often changes in pH and the addition of certain metal ions are used to improve specific separations. This technique is used commercially to prepare plasma protein fractions (γ globulin, albumin) for clinical purposes.

DIFFERENCES IN SURFACE ELECTRICAL CHARGE PROPERTIES[28,80]

Proteins are amphoteric compounds similar to the amino acids, having both positive and negative charges on the same molecule, the number depending upon the pH of the solution. The pH at which the number of positive and negative charges is equal is called the *isoelectric* point (pI). For most proteins this pH is on the acid side, varying from 4.6 for albumins to 5.1 to 6.2 for globulins. At a pH below the pI, the positive charges outnumber the negative charges and the protein behaves like a positive ion (cation); solutions will contain $(Pr)^{+n}$ ions, analogous to the Na^+ ion. At physiological pH values, however, the negative charges will be in the majority and the protein will behave like an anion $(Pr)^{-n}$, analogous to Cl^- and HCO_3^-. Different proteins will vary in the net positive or negative charge per unit area of surface. Thus, if a solution containing proteins is placed between a pair of electrodes connected to a voltage source, the proteins will migrate to either cathode or anode, depending on their net charge, at speeds depending on the amount of charge and the size and shape of the individual protein molecules. This technique is referred to as *electrophoresis*. At pH = 8.6, all serum proteins are negatively charged, but the albumins migrate most rapidly and are separated from the globulins. The latter can be split into three or more fractions, called alpha (α), beta (β) and gamma (γ) globulins, the γ being the slowest moving. Under proper conditions, each of these may be separated into subfractions.

DIFFERENCES IN MOLECULAR SIZE[80]

Because proteins are made up of different numbers and kinds of amino acids, individual proteins will differ in molecular weight. If a protein solution is placed in a centrifuge operating at a very high speed, the high centrifugal force will force the heavier molecules to sediment or settle out faster than the lighter molecules. The necessary high speeds can be attained only by use of an *ultracentrifuge*, usually not available in a routine clinical laboratory. This technique is particularly useful in separating various light lipoproteins from other (heavier) serum proteins.

CHROMATOGRAPHIC SEPARATIONS

Because of differences in size, shape, and chemical composition, different proteins will be adsorbed on and can be eluted from various adsorbents at different rates. By control of pH, buffer type, and concentration, and by the proper choice of adsorbents and eluting solutions, individual proteins can be separated from each other and obtained in a highly purified form. This technique is not yet important in routine work (except to separate thyroxine binding protein from contaminating iodine containing material), but it will probably become a very important tool in the clinical laboratory in the near future. Among promising adsorbents are ion exchange resins and molecular sieve materials such as Sephadex and Bio-Rad gels.

Each of the techniques just mentioned separates protein mixtures into individual fractions by use of a specific protein property. A fraction that may appear as a single homogeneous electrophoretic band may, on ultracentrifugation, give rise to several sedimentation fractions. Similarly, an albumin fraction obtained by salt precipitation will, invariably, when subjected to electrophoresis, form two or more individual bands in the albumin area of the electrophoretogram. Both large and small protein molecules may have the same net surface electrical charge density. Also, proteins of similar surface electrical properties will, because of differences in shape, size, and type of amino acid side chains on the surface, pass through an adsorbent column at considerably different rates. The various protein separation techniques thus complement each other.

Classification by Structure

Proteins can also be classified as *simple* and *conjugated* proteins. In the latter, some type of nonprotein moiety is linked with the folded amino acid chain. This added *prosthetic group* conveys new and characteristic properties to the complex formed. The colored *chromoproteins* contain an organic prosthetic group that is linked to some metal ion. In the *metalloproteins* some metal ion is attached directly to the protein. Examples of metalloproteins are *ferritin*, containing iron, and *ceruloplasmin*, containing copper. *Lipoproteins* contain bound cholesterol, phospholipids or glycerides, or both, whereas *glycoproteins*, such as *mucins* and *orosomucoid* contain complex carbohydrates in their structure. In the *nucleoproteins* (cellular chromatin material) the protein is associated with chains of deoxyribonucleic acids.

The *albuminoids* are a special group of proteins, characterized by being essentially insoluble in most common reagents. They comprise the various fibrous proteins, which have a supporting or protecting function in the organism. Among these are the *collagens* (skin, cartilage), *elastins* (tendons), and *keratins* (hair, feathers).

Plasma and Serum Proteins

CLINICAL SIGNIFICANCE[32,63,88,97]

Blood is made up of particulate cell forms suspended in a fluid medium called *plasma*, a very complicated mixture of inorganic and simple and complex organic materials dissolved in water. If the blood is collected without anticoagulant, in vessels with siliconized or nonpolar surfaces, the fluid separating is referred to as *native plasma*. This approximates very closely the plasma actually present in circulating blood. It is the practice, however, to use such anticoagulants as oxalate, citrate, EDTA, and heparin to prepare specimens of plasma for study or analysis. This differs from native plasma because it is modified by the presence of the added chemicals and by the partial loss of calcium bound by oxalate, if this is used. If blood is permitted to clot, the fluid separating is referred to as *serum*. It lacks the protein fibrinogen present in plasma, the fibrinogen having been transformed into insoluble fibrin in the clotting process. The fibrinogen constitutes only some 3 to 6 per cent of the total plasma proteins. It is satisfactory and much more convenient to use serum rather than plasma in clinical chemical studies.

Some 92 to 93 per cent of plasma or serum is solvent water; of the 7 to 8 per cent of total solutes present, the proteins occur in the greatest concentration, roughly 6.8 to 8.8 (6.5 to 8.5) gm./100 ml. of plasma (serum) *water*. Because of the large molecular weight of proteins, however, their molar concentration calculates to about 0.80 to 1.10 m*M*/L. In clinical work, the concentration of proteins is given in terms of grams per 100 ml. of *serum volume*. It would be more precise to express it in terms of *serum water*. The latter value is 1.07 times the serum volume figure.

Plasma proteins serve a number of different functions in the organism. They play a nutritive role, inasmuch as they constitute a portion of the amino acid pool of the body; thus, the proteins are a form of storage amino acids. If needed, these proteins can be broken down in the liver to produce amino acids for use in building other proteins. Alternatively, they can be deaminated to give keto acids that can be mobilized to provide caloric energy or be transformed into carbohydrates and lipids. Plasma proteins also act as transport agents: many vital metabolites, metal ions, hormones, and lipids are transported about the body, bound to and carried by certain specific proteins. Some proteins have important special functions of their own, such as enzymes, the immune antibodies among the globulins, and the various proteins associated with blood coagulation.

Another important function of proteins is a physicochemical one. The plasma proteins, being large, colloidal molecules, are nondiffusible; i.e., they cannot move through the thin capillary wall membranes as can most other blood solutes. They are thus entrapped in the vascular system and exert a *colloidal osmotic pressure*, which serves to maintain a normal blood volume, and a normal water content in the interstitial fluid and the tissues. The albumin fraction is most important in maintaining this normal colloidal osmotic or *oncotic pressure* in blood. If the albumin falls to low levels, water will leave the blood vessels and enter the extracellular fluid and the tissues, thus producing edema.

The maintenance of the acid-base balance in blood also involves the plasma proteins. As amphoteric compounds, they function as buffers to minimize sudden, gross changes in the pH of the blood.

Despite their presence together, the various proteins of plasma do not originate from the same source. The liver is the main organ for the synthesis of albumins and α and β globulins, and perhaps some nonimmune γ globulins; among these proteins are included the blood clotting and transport proteins. The cells of the reticuloendothelial system (spleen, bone marrow, lymph nodes) serve as the source of the antibody, immune γ globulins and perhaps some β globulins.

The level of total serum proteins found in healthy young and middle-aged adults is 6.0 to 8.2 gm./100 ml. of serum. In plasma, fibrinogen increases this value by an additional 0.2 to 0.4 gm./100 ml. A diurnal variation of 0.5 gm./100 ml. reflects small changes in the ratio of vascular to nonvascular fluid in the course of daily activity. In disease states both the total protein and the ratio of the individual protein fractions may change independently of one another. In states of dehydration, total protein may increase some 10 to 15 per cent, the rise being reflected in all protein fractions. Dehydration may result either from a decrease in water intake, as occurs in frank water deprivation (thirst), or from excessive water loss, as occurs in severe vomiting, diarrhea, Addison's disease, and diabetic acidosis. The absolute quantity of serum proteins is unaltered, but the concentration is increased because of the decreased volume of solvent water. In multiple myeloma, the total protein may increase to over 10 gm./100 ml., the increase being almost entirely due to the presence of markedly elevated levels of myeloma proteins (abnormal forms of γ globulins). The quantities of other proteins are essentially unaltered.

Hypoproteinemia, characterized by total protein levels below 6.0 gm./100 ml., is encountered in many unrelated disease states. In the nephrotic syndrome large masses of albumin may be lost in the urine as a result of leakage of the albumin molecules through the damaged kidney. In salt retention syndromes, water is held back to dilute out the retained salt, resulting in the dilution of all protein fractions. Large quantities of proteins are lost in patients with severe burns, extensive bleeding, or open wounds. Water is replaced by the body more rapidly than is protein, effecting a decreased total protein concentration. A long period of low intake or deficient absorption of protein may affect the level and composition of serum proteins, as in sprue and in other forms of intestinal malabsorption, as well as in acute protein starvation (kwashiorkor). In these conditions the liver has inadequate raw material to synthesize serum proteins to replace those lost in the normal turnover (wear and tear) of proteins and amino acids.

In general, changes in total proteins may occur in one, several, or all fractions. It is also possible for significant changes to occur in different directions in different fractions, without changes in the total protein concentration.

The physician may occasionally request only a value for total protein in serum; however, more commonly he will ask for total protein, the values for the albumin, and total globulin fractions, and for the *albumin-globulin (A/G) ratio*. In healthy young and middle-aged adults, the *albumin* may vary from 3.8 to 4.7 gm./100 ml., and the *total globulins* from 2.3 to 3.5 gm./100 ml. The range found for the albumin-globulin ratio is 1.1 to 1.8, averaging 1.5. The globulins are usually separated by a salt fractionation procedure, but may be estimated from the sum of electrophoretically separated fractions. Albumins can be estimated separately by virtue of their ability to bind certain dyes such as methyl orange and hydroxyazobenzoic acid. These techniques will be discussed later.

METHODS FOR THE DETERMINATION OF TOTAL PROTEINS IN SERUM[28,92,97]

Determination of Protein Nitrogen (Kjeldahl Technique)[1,31]

Inasmuch as all proteins contain nitrogen that is derived from their constituent amino acids, proteins can be determined by measuring the nitrogen present in the isolated protein. Until recently, it has been conventionally assumed that 16 per cent of the mass of serum proteins is nitrogen, and from this it follows that

$$\text{protein} = \frac{1.00}{0.16} \times N = 6.25 \times N \qquad (8)$$

The factor, 6.25, for converting N value to protein value is still used by most laboratories, but more recent, careful studies have indicated that this value is probably low and that the true value may be closer to 6.45. Until a more accurate value is accepted by a consensus of clinical chemists, however, it will be convenient to retain and use the value of 6.25 for the protein/N ratio.[86,97]

The protein nitrogen is usually measured by any one of many semimicro modifications of the classic *Kjeldahl nitrogen* technique. For precise work, trichloracetic acid or tungstic acid is employed to precipitate the proteins, the NPN (nonprotein nitrogen) being removed with the filtrate-supernate. The washed precipitate is transferred to Folin-Wu or Kjeldahl digestion tubes and the organic matter oxidized by hot refluxing H_2SO_4, with or without added catalysts such as $CuSO_4$, $HgSO_4$, selenium, or SeO_2 to speed up the oxidation reactions. For efficient oxidation, the temperature should be at about 340 to 360°C.; K_2SO_4 is added to raise the boiling point of the sulfuric acid. In this procedure, the carbon, hydrogen, and sulfur in the protein are oxidized to CO_2, CO, H_2O, and SO_2 and the nitrogen is converted to NH_4HSO_4.

$$\underset{[C,\,H,\,N,\,O,\,S,\,P]}{\text{Protein}} \xrightarrow[H_2SO_4 + \text{Catalyst}]{\text{Heat}} CO_2 + H_2O + SO_2 + P_2O_7^{-4} + CO + NH_4^+ + HSO^-_4 \qquad (9)$$

The NH_4HSO_4 formed is then measured by adding excess alkali and distilling the liberated NH_3 into excess standard acid and back titrating with standard NaOH (equations 10 and 11), or by distilling the NH_3 into boric acid, and measuring the NH_3 entrapped as $NH_4H_2BO_3$, by titrating with standard HCl (reactions 10, 12 and 13).

$$NH_4HSO_4 + 2OH^- \xrightarrow{\text{distill}} NH_3\uparrow + H_2O + SO_4^= \qquad (10)$$

$$NH_3 + H^+ \xrightarrow[HCl]{\text{excess}} NH_4^+ + H^+ \text{ (excess), measured with Std. NaOH} \qquad (11)$$

$$NH_3 + H_3BO_3 \xrightarrow[H_3BO_3]{\text{Excess}} NH_4H_2BO_3 \longrightarrow NH_4^+ + H_2BO_3^- \qquad (12)$$

$$H_2BO_3^- + H^+ \xrightarrow[HCl]{\text{Std.}} H_3BO_3 \qquad (13)$$

Alternatively, the NH_3 formed in the Kjeldahl digestion may be measured by direct nesslerization of the digestate or of the distilled NH_3. Neither of these alternatives is to be recommended; their precision is much less than that of the previous titration methods, and the Cu^{++} or Hg^{++} catalysts interfere with the Nessler reaction. In practice, the Kjeldahl reaction is performed directly on a serum aliquot and the total nitrogen value obtained is corrected for nonprotein nitrogen by a separate assay of the NPN.

The Kjeldahl procedure is too slow for routine analysis of a large number of specimens. It is used primarily as a standard method against which all other methods are calibrated because it is capable of high precision and accuracy. It has already been mentioned that there is some question about the correct average factor for converting Kjeldahl nitrogen into protein. The nitrogen content of the major individual protein fractions in serum varies considerably, ranging from about 15.1 to 16.8 per cent, with corresponding protein/N factors ranging from 6.00 to 6.65.[97] Until a better value is established, it will do little harm to accept the arbitrary factor of 6.25 (16 per cent N), as an adequate reflection of the average nitrogen composition of serum proteins in healthy individuals. For proteins in abnormal sera, this assumption is less likely to be valid. Because of this uncertainty, the Kjeldahl nitrogen value cannot be used as an absolute standard for quantitating serum proteins, if indeed any such standard is possible. It is conventional practice, however, to calibrate other protein methods against the Kjeldahl value as a standard.

Biuret Method for the Determination of Total Protein in Serum and Exudates

All proteins contain a large number of peptide bonds. When a solution of protein is treated with Cu^{++} ions in a moderately alkaline medium, a colored chelate complex of unknown composition is formed between the Cu^{++} ion and the carbonyl (—C=O), and the (=N—H) of the peptide bonds. An analogous reaction takes place between the cupric ion and the organic compound biuret, $NH_2—\overset{\overset{\displaystyle O}{\|}}{C}—NH—\overset{\overset{\displaystyle O}{\|}}{C}—NH_2$ and, therefore, the reaction is referred to as the *biuret reaction*. The reaction takes place between the cupric ion and any compound containing at least two NH_2CO—, NH_2CH_2—, NH_2CS—, and similar groups joined together directly, or through a carbon or nitrogen atom. Amino acids and dipeptides cannot give the reaction, but tri- and polypeptides and proteins react to give pink to reddish violet colored products. In the biuret reaction one copper ion is linked to between four and six nearby peptide linkages by coordinate bonds; the more protein present, the more peptide bonds available for reaction. The intensity of the color produced is proportional to the number of peptide bonds undergoing reaction. Thus, the biuret reaction can be used as a basis for a simple and rapid colorimetric method for determining protein.

Kingsley[47] introduced the first procedure simple enough to be practical in the clinical laboratory. Since then, the biuret method has become the method of choice for measuring proteins because of its simplicity and adequate precision and accuracy. The reaction is performed directly using either undiluted or diluted serum. Peptides are present in serum, but their concentration is so low that they contribute no increment to the biuret color. Ammonium (NH_4^+) ions interfere by buffering the pH at too low a level, but the NH_4^+ level in serum is too low to interfere. Most biuret methods can determine protein in the range of about 0.5 to 15 mg. of protein in the aliquot being measured. A large number of modifications of the procedure are available; all are alike in principle, but vary in details and in the composition of the biuret reagent. The method to be presented is based on the procedure of Reinhold,[71] using the Weichselbaum biuret formulation, and will serve as an example of the biuret technique.

REAGENTS

1. Biuret diluent, KI, 0.5 per cent (w/v) in 0.25 *N* (±0.02) NaOH. Prepare the NaOH in 8 L. quantities by dilution of standard 2.50 *N* (10.0 per cent w/v) NaOH with CO_2-free water. Water fresh from a still or a good ion exchange column is essentially CO_2 free, as is freshly boiled water. Store the reagent in stoppered polyethylene bottles.

2. Stock biuret reagent, Weichselbaum formula. Dissolve 15.0 gm. of finely pulverized $CuSO_4 \cdot 5H_2O$ in 70 to 80 ml. of water. Prepare a solution of 45.0 gm.

Rochelle salt (potassium sodium tartrate, tetrahydrate) in 600 to 700 ml. of biuret diluent and slowly add the $CuSO_4$ solution with stirring. Both solutions must be at room temperature when mixed to prevent reduction of Cu^{++} by the tartrate. Add biuret diluent to a volume of 1000 ml. Before use, filter the solution through a qualitative paper to remove any deposited cuprous oxide. Store in a polyethylene bottle, away from strong direct light.

3. Working biuret. Dilute the stock biuret 1:5 with biuret diluent. Store the same way as the stock biuret.

4. Alkaline tartrate, 0.9 per cent (w/v) Rochelle salt in biuret diluent. The solution is identical to working biuret solution except Cu^{++} is omitted. This is used to correct for serum pigment error as discussed under Procedure and Comments.

5. Saline, NaCl, 0.85 per cent (w/v) in water.

6. Standard protein. It is most convenient to use Armour Bovine Serum Albumin Standard (B.S.A.) solution, available in 3.0 ml. ampules containing pure bovine albumin. The analysis in terms of protein N is imprinted on the ampule and is in the range of 9.7 to 10.3 mg. of protein N/ml. This is equivalent to 60.6 to 64.4 mg. of protein/ml. (6.06 to 6.44 gm./100 ml.), if the customary protein/N ratio of 6.25 is used (see the discussion under the Kjeldahl procedure). Dilute the ampule solution 1:10 with saline by pipetting 2.50 ml. into a 25.0 ml. volumetric flask containing 20 ml. of saline; the albumin is delivered down the sides of the flask to avoid foaming. After diluting to the mark with saline, gently but thoroughly mix the flask contents.

7. Control serum. Any commercial control material may be used. If lyophilized, the reconstituted solution must be clear and free of turbidity. Alternatively, a laboratory composite serum may be used. Prepare this by pooling left over normal sera, which are free of hemolysis, significant icterus, turbidity, or chyle. Mix the pool, pass it through several layers of Pyrex wool to remove fine clots, transfer to 15 × 120 mm. test tubes (polycarbonate is best) in 1 ml. aliquots, and store frozen. Before assaying, remove a tube, thaw in a bath at 37°C., and carefully mix. The protein value may be measured by the Kjeldahl procedure or by assaying it against another serum of known value.

PROCEDURE

The reaction is carried out directly in 19 × 150 mm. Coleman cuvets. Set up a cuvet containing 2.0 ml. of saline for each unknown, and pipet 0.10 ml. (100 μl.) of specimen into the saline using 0.10 ml. Folin-Wu or 100 μl. micropipets. If TC micropipets are used, rinse the pipets three to four times with the saline. Carefully wipe the tips and ends of the pipets free of adherent serum before delivering the serum aliquot to the saline. An automatic dilutor is a convenience if a large number of specimens is to be analyzed. Also set up a blank cuvet containing only 2.10 ml. of saline.

To all cuvets add 8.0 ml. of working biuret. It is convenient to use an automatic pipet. Thoroughly mix the tubes and allow the color to develop for 30 minutes. Measure the absorbances of the solutions at 550 nm. using a spectrophotometer with the blank set at 100%T. Referring to the calibration curve, convert the absorbance data to gm. of protein/100 ml. of serum. Alternatively, if the calibration is linear, $C = F \times A$, where F is the calibration factor (see next section), A is the absorbance, and C is the protein concentration in gm./100 ml.

If a serum is highly pigmented (hemoglobin, bilirubin), milky (lipemic), or turbid in appearance, make a correction for the pigment or turbidity. Hemoglobin and, to a much lesser extent, bilirubin, absorb light at 550 nm. and this absorbance will be added to the true protein biuret color. If the 1:21 dilution of the serum in saline is visibly free of color, the correction may be omitted. It is best to attempt to correct for serum turbidity or opalescence whenever encountered. If correction is needed, set up an additional cuvet containing 0.10 ml. of serum and 2.0 ml. of saline. This cuvet receives 8.0 ml. of alkaline tartrate in place of working biuret. The corresponding blank contains 2.1 ml. of saline plus 8.0 ml. of alkaline tartrate. Read the absorbances of the sample containing cuvets at 550 nm. If any readings are 0.010 or higher, subtract them from the biuret color readings, and then use the corrected readings to calculate protein concentration.

CALIBRATION PROCEDURE

1. Transfer the following volumes, 0.50, 1.00, 1.50, and 2.00 ml., of the 1:10 diluted Armour B.S.A. standard to cuvets. Add saline to make the volume in each equal to 2.10 ml. Then add 8.0 ml. of working biuret to each, as in the serum procedure, and measure the absorbances of the standard solutions after 30 minutes of color development.

2. In the 10.1 ml. cuvet volume, 1.00 ml. of the 1:10 diluted standard will contain

$$S = 6.25 \times P \times 1/10, \quad \text{or}$$

$$S = (0.625) \times P \text{ mg. of protein} \tag{14}$$

where P represents the mg. N/ml. value marked on the ampule.

3. If the unknown serum contains T gm. of total protein/100 ml., the 0.10 ml. used for the determination will contain T mg. of protein, also dissolved in a cuvet volume of 10.1 ml. The absorbance of this serum must equal that of some standard, containing S mg. of protein. The T mg. $=$ S mg., or $S = T$. Thus, the numerical value of the protein in the cuvet in milligrams is identical with the serum protein concentration expressed in gm./100 ml. The standards represent 0.5, 1.0, 1.5, and $2.0 \times 0.625 \times P$ gm. protein/100 ml. Depending on the actual concentration in the Armour ampule, the standards represent approximately 3.1, 6.3, 9.5, and 12.6 gm. of protein/100 ml. The calibration factor F is equal to the slope of the calibration line and is given by C_s/A_s. The individual absorbances of the respective standards are used to calculate the slope by the following approximation formula:

$$F = \frac{\Sigma C}{\Sigma A} = \frac{\text{sum of concentrations represented by standards}}{\text{sum of absorbances of the standards}} \tag{15}$$

COMMENTS

1. An alternative procedure to correct for pigment or turbidity error, or both, is to proceed as follows: After the biuret color of the unknown has been read, add 0.25 ml. of a saturated solution of KCN to the tube and the biuret blank. The CN^- ion will remove the Cu^{++} from the red-violet copper protein complex by forming the nonionized $Cu(CN)_2$, thus clearing the tube of any biuret color. Read the absorbance of the cyanide treated solution and subtract this absorbance, representing pigment and turbidity, from the biuret color reading. In place of the KCN solution, about 50 to 70 mg. of dry KCN may be added from the end of a spatula. Utmost care must be the rule when working with the extremely poisonous KCN.

2. The precision (repeatability) expected for this procedure is about ±0.30 gm./100 ml., or about ±3.5 per cent (95 per cent confidence interval). The main source of error is in pipetting the initial sample aliquot of 0.10 ml. The biuret reagent should be filtered at least weekly. The presence of pigment or turbidity, even when the correction is applied, is a source of error. The calibration curve is reproducible for any given spectrophotometer, small differences in calibration factor F arising from small differences in individual biuret reagents.

3. The biuret technique can be used to assay proteins in the range of about 1.5 to 15 mg. in the sample aliquot. It is useful in the determination of serum and exudate proteins, but is not sufficiently sensitive for the assay of proteins in the majority of spinal fluid, urine, or transudate specimens. Specimen aliquots greater than 0.10 ml. may be used (up to 2.0 ml.), provided that saline is added to a final volume of 2.1 ml. This permits the use of the procedure with specimens of transudates and urines containing at least 200 mg./100 ml. of protein. The NH_4^+ ion in the urine will interfere unless it is removed by passing the urine through a cation ion exchange column. With urines, pigment correction is mandatory.

REFERENCE

Reinhold, J. G.: Total protein, albumin and globulin. *In* Standard Methods of Clinical Chemistry. D. Seligson, Ed. New York, Academic Press Inc., 1953, vol. 1, p. 88.

Total Protein Determination Using the Folin-Ciocalteu Reagent

Solutions of various types of phenols behave as weak reducing agents and can be oxidized to quinone-type compounds by many oxidizing agents. A complex phosphotungstomolybdic acid reagent, *phenol reagent*, devised by Folin and Ciocalteu,[13] oxidizes phenolic compounds under alkaline conditions; it is reduced from its initial golden yellow color to a deep blue. This reagent is used to measure phenolic compounds and to assay tyrosine, the amino acid that contains a phenol side chain. Inasmuch as all proteins contain some tyrosine, this reaction has been used to quantitate plasma, serum, and spinal fluid proteins. The indole and imidazole groups of tryptophane and histidine also react with the reagent, but the reaction is weaker than that given by the phenol structure.

Proteins vary in their tyrosine, tryptophane, and histidine content. Collagens and gelatin are quite low in tyrosine and tryptophane, whereas insulin and the γ globulins have a greater than average proportion of these amino acids. Thus, the Folin-Ciocalteu equivalent for various proteins will vary in proportion to their content of the reacting amino acids. Protein methods based on the use of the phenol reagent cannot be recommended for the assay of proteins in preparations containing mixtures of many proteins. Even though serum proteins show a much smaller spread in their tyrosine contents (5.5 per cent for albumin; 6.8 per cent for γ globulin), this spread is sufficient to give rise to a 20 to 30 per cent difference in Folin-Ciocalteu equivalents. The reagent may be used to assay serum proteins, but the accuracy obtained is less than that of the simpler biuret method, and the accuracy may be quite poor with serums that have considerably altered A/G ratios or that contain abnormal proteins. Methods employing the phenol reagent do have the advantage of being some five to ten times more sensitive than biuret procedures and can be used when only 0.30 to 1.50 mg. of protein are present in the available aliquot. They are quite suitable for the assay of samples containing single proteins of known, constant tyrosine content. A useful application is in the determination of plasma fibrinogen, after the protein has been isolated as a fibrin clot.

In a modified technique, introduced by Lowry and his associates,[54] the reagent is used to determine very small quantities of protein. Cupric ion in weakly alkaline tartrate is added to the protein solution. The biuret-type copper protein complex is then treated with dilute phenol reagent. Both the biuret complex and the aromatic side chains of the protein reduce the reagent, the combined reaction being about 100 times more sensitive than the biuret procedure. Protein quantities of the order of 30 to 100 μg. can easily be determined with a precision of ± 10 per cent.

Determination of Total Protein by Measurement of Refractive Index[74]

The refractive index of water at 20°C. is 1.3330. If a solute is added to the water, the refractive index, η, will be increased; the increment in η is directly proportional to the concentration of solute in dilute solutions and nearly proportional to concentration over a two- or threefold range in more concentrated solutions (5 to 20 per cent w/v). This proportionality still holds for mixtures of solutes if their refractive indices are similar in magnitude. In industry, the use of refractive index measurements for the assay of sugar, syrups, oils, and other materials is well established. The relation between refractive index and concentration is usually obtained by empirical calibration.

The increment in refractive index can be used to measure total solids present in urine. The solutes consist mainly of urea, Na^+, Cl^-, K^+, and $H_2PO_4^-$ ions, the combined concentration ranging from 2.5 to 7.0 gm./100 ml. Similarly, refractive index measurements can be used to obtain clinically useful, though not always accurate, measurements of serum protein concentrations. In this application, a reasonable assumption is made that the concentrations of individual inorganic electrolytes and simple organic metabolites do not vary appreciably from serum to serum, and that differences in refractive index are primarily a reflection of differences in protein concentration. Studies have shown that this is generally true, and refractive index measurements of clear, nonpigmented, nonturbid sera provide a very rapid and direct method for measuring serum protein when extreme accuracy is not essential. Samples from patients with azotemia, hyperlipemia or hyperbilirubinemia or hemolyzed samples may give cause for gross errors.

The TS Meter, model 10401 (American Optical Co. Scientific Instrument Div., Buffalo, N.Y.), is a simple and convenient instrument of the refractometer type, designed for measurement of serum protein. Only slightly more than 1 drop of protein solution is required, the sample being spread by capillary action as a thin film between the measuring prism and a cover plate. The refractive index of a solution has an appreciable negative temperature coefficient; therefore, there must be either precise measurement of temperature or some form of compensation for change of η with temperature. The TS Meter provides for such compensation so that the readings are valid for the entire temperature range of 15 to 37°C. After placing the sample drop on the prism, the cover plate is pressed down, and the instrument is pointed at a source of reasonably intense light. The light beam enters parallel to the prism, is refracted by the protein solution, and then projected against the eyepiece, which contains scales calibrated in both refractive index and grams of protein per 100 ml. The refracted rays light up a segment of the field viewed in the eyepiece, the field being separated by a sharp demarcation line into a light area and a dark area. The reading of the scale at the boundary line separating the two areas gives the protein concentration directly.

MISCELLANEOUS METHODS FOR PROTEIN DETERMINATION

Measurement of Specific Gravity

Aliquots of serum are dropped into a series of carefully prepared solutions of copper sulfate of exactly known specific gravity. The drops will rise in $CuSO_4$ solutions of density greater than that of serum and will drop rapidly to the bottom of cylinders containing solutions of density lighter than that of serum. Eventually a $CuSO_4$ solution will be located in which the drop remains suspended for nearly a minute before dropping. Nomograms and tables, prepared by D. D. VanSlyke,[67] relating the specific gravity of the $CuSO_4$ solutions and the concentration of protein are available in many laboratory manuals.[16,92] The $CuSO_4$ specific gravity standards are difficult to maintain. The results are rather crude and subject to large errors in the case of lipemic sera and sera from patients with abnormally high sugar and urea levels and with abnormal A/G ratios. The method can also be used to estimate total blood protein.

Measurement of Protein by Ultraviolet Light Absorption

Protein solutions show strong ultraviolet light absorption bands at 279 nm. and at 210 to 215 nm.[90,94] The absorption peak at 280 nm. is due to the aromatic rings of tyrosine and tryptophane, and that at 215 nm. to the peptide bond. A 1:100 dilution of serum is used for readings at 280 nm., and a dilution of 1:1000 for measurements at 215 nm. The absorbance of other solutes at these wavelengths is made negligible by the dilution of the specimen. If the procedure is standardized against normal serum, it is possible to derive accurate protein values on normal sera using specimens of 5 to 10 μl. Because of the varying concentrations of tyrosine in the different serum protein fractions, however, the method is less accurate when used with sera containing abnormal proteins, or with abnormal A/G ratios. A method employing absorbance readings at 215 nm. has been proposed for measuring spinal fluid protein.[65] It should be mentioned that precise measurement of absorbance in the low ultraviolet range (215 to 220 nm.) presents many problems of technique that are sufficient to discourage its use in a routine laboratory except when there is no other option.

Urine Protein

All urines contain some protein. The protein excreted by healthy individuals is of the order of 50 to 100 or perhaps 150 mg./24 hr. After considerable muscular exertion the value may be as high as 250 mg. Random specimens will contain under 10 mg./100 ml., although in overnight specimens the level may be as high as 15 to 20 mg./100 ml. *Proteinuria* is said to be present whenever the urinary protein output is greater than that reflected in these normal values. Not all proteinuria is clinically significant, but persistent abnormal levels of protein in the urine is an indicator of the presence of kidney and urinary tract disease. Thus, the detection of urinary protein and the quantitative assessment of the degree of proteinuria are very important laboratory procedures.

In general, the urinary proteins reflect those present in plasma, along with special proteins of renal tubular origin, and those exuded from the tissues that line the genitourinary tract. The glomerular filtrate contains small amounts of all plasma protein fractions, but most of the filtered protein is reabsorbed in transit through the kidney tubules. Some of the globulins present in urine are of smaller size than their electrophoretic counterparts in serum. The new techniques of immunoelectrophoresis and antibody-antigen diffusion in agar are rapidly changing our knowledge of urine proteins. The old concept that albumin was the only protein present in any significant amount, even in pathological urines, is no longer valid, and therefore the term *albuminuria* should no longer be used. Some 40 to 60 per cent of the proteins present in normal urines are mucins (glycoproteins) originating from the linings of the lower genitourinary tract, and are usually of no clinical significance.

Proteinuria is often a manifestation of primary renal disease,[32,42,88] although transient proteinuria may occur with fevers, thyroid disorders, and in heart disease, in the absence of renal disease. Proteinuria may be evident very early in the course of various renal disease states. With such conditions as pyelonephritis, reflecting bacterial infection in the kidney, and acute glomerulonephritis, often associated with recent streptococcal infections, the degree of proteinuria is slight, usually amounting to less than 2 gm. per day. In chronic glomerulonephritis and in the nephrotic syndrome, including lipoid nephrosis, and in some forms of hypertensive vascular disease (*nephrosclerosis*), protein loss may vary from a few grams to as much as 20 to 30 gm. per day. Proteinuria is encountered in certain other disease entities when and if they give rise to kidney lesions. Among these are lupus erythematosus, amyloidosis, toxemia of pregnancy, septicemia, and certain forms of drug and chemical poisoning. In multiple myeloma, not only does one find a modest increase in protein output, but in 40 to 60 per cent of the cases, the urine contains a group of abnormal proteins referred to as Bence-Jones proteins.

In healthy individuals, transitory elevations in urine protein output are encountered after intense exercise or work, and after exposure to cold. *Orthostatic proteinuria* is a benign condition in which protein excretion is normal when the patient is lying down (prone), but is mildly elevated when the patient walks or stands erect for any period of time. Persons subject to orthostatic proteinuria often are embarrassed during medical examinations for insurance because their urines are found to be positive, when tested for protein after the applicants had been on their feet and active for a good part of the day. On rechecking the urine in the morning after a night's rest, the urine is usually found to be negative for protein.

REFERENCES

Hoffman, W. S.: The Biochemistry of Clinical Medicine. 3rd ed. Chicago, Year Book Medical Publishers, Inc., 1964.
Kark, R. M. et al.: A Primer of Urinalysis. 2nd ed. New York, Hoeber Medical Div., Harper & Row, Publishers, 1963.
King, J. S. et al.: J. Clin. Invest., *37*:315, 1958; *37*:1658, 1958.
Thompson, R. H. S., and King, E. J.: Biochemical Disorders in Human Disease. 2nd ed., New York, Academic Press, Inc., 1964.

QUALITATIVE TESTS FOR URINE PROTEIN

In devising qualitative tests for urine proteins, it is desirable that the test be negative whenever protein is present in normal concentrations, but that the test be positive whenever protein concentration is greater than 20 to 25 mg./100 ml. The test should not be sensitive to mucins and other proteins of nonrenal origin and it should be capable of rough quantitation. A large number of such tests are available.

The older tests are based either on precipitating the proteins by heat or by reaction with anionic protein precipitants. A more recent test depends on a change in the color of an indicator in the presence of protein.

Qualitative tests are best done on fresh morning specimens of urine. Such specimens are usually fairly concentrated, thus making possible the detection of proteinuria early in disease. They will also be free of the orthostatic effect discussed previously. If urines must be held for examination at a more convenient time, and if timed specimens are to be examined, they should be refrigerated and examined within 48 hours. Layering with toluene may be used to minimize microbial growth. Bacterially contaminated specimens (pyelonephritis, lower genitourinary tract infections) should be examined only when fresh.

Heat Coagulation Test

The pH of the urine is checked with bromcresol green pH indicator paper, (pH range, 3.8 to 5.5) and if above pH 5.0, acetic acid, 30 per cent (w/v) is added dropwise until the pH is between 4.0 and 4.6. If any turbidity or insoluble matter is present or is formed (mucoproteins may precipitate on adding the acid), it is removed by centrifugation. About 7 to 10 ml. of clear urine are placed in a 15 × 120 mm. test tube and the upper one-third is gently heated over a Bunsen flame until boiling just begins. The heated part is compared in ordinary light with the lower unheated portion of the specimen. Any turbidity, cloudiness, or precipitate demonstrates the presence of protein. The degree of the reaction is customarily graded 0 to ++++. False positive results may occur and are discussed in the section on interpretation of results. The grading of the turbidity observed and its relation to the protein concentration is given in Table 5-2.

TABLE 5-2. *Interpretation of the Heat Coagulation Test for Urine Protein**

Appearance	*Reading*	*Approximate Protein Concentration* (mg./100 ml.)
No turbidity	negative	0–4
Slight, distinct turbidity	± (trace)	4–10
Definite turbidity; light print readable	+	15–30
Light cloud; heavy print readable	++	40–100
Moderate cloud with slight precipitate	+++	200–500
Heavy cloud with flocculent precipitation and coagulation	++++	over 800–1000

* After Kark, R. M. et al.: A Primer of Urinalysis. 2nd ed. New York, Hoeber Medical Division, Harper & Row, Publishers, 1963.

REFERENCE

Kark, R. M. et al.: A Primer of Urinalysis. 2nd ed. New York, Hoeber Medical Div., Harper & Row, Publishers, 1963.

Sulfosalicylic Acid Test

Into a 15 × 120 mm. test tube, place 5 ml. of clarified urine (pH 4.5 to 6.5). Then 0.50 to 0.80 ml. of 20 per cent (w/v) sulfosalicylic acid is added carefully down the sides of the tube to layer underneath the urine. After 1 minute the interface is examined for the presence or absence of turbidity or cloudiness. A barely evident turbidity (5 to 10 mg./100 ml.) is graded as ±, and increasing degrees of turbidity, cloudiness, or precipitation are graded + to ++++. False positives, due to non-specific precipitation, may be encountered and are discussed in the section on interpretation of results.

The reagent is prepared by dissolving 100 gm. of reagent grade sulfosalicylic acid (2-hydroxybenzoic, 5-sulfonic acid dihydrate) in about 350 ml. of water with the aid of heat. The warm solution is filtered through two layers of filter paper and the total filtrate diluted to 500 ml. when cool. The solution should be crystal clear and colorless. It is discarded when turbidity or darkening develops.

Albustix Test

Albustix[19,72] is the name of a commercial product from the Ames Company (Div. Miles Laboratories, Elkhart, Ind.). It consists of strips of special adsorbent paper, the ends of which have been treated with citrate buffer of pH = 3.0, containing bromphenol blue indicator (pH = 3.0, yellow; pH = 4.2, blue). The test is based on the so-called "protein error" of indicators. At the pH of the indicator strip, most of the indicator is in the yellow, nonionized acid form, HI, with only a small fraction in the blue ionized form, I^-. If protein is present, it will bind with the anion form, I^-, since the anion form has a greater affinity for the protein than for the hydrogen ion. The removal of I^- will cause more of the indicator acid to ionize and the concentration of HI will decrease, resulting in a change in the ratio of I^-/HI. The color presented by the paper strip will depend on the relative proportions (ratio) of the two forms. The quantity of indicator is fixed and, if sufficient protein is present, the larger fraction of the indicator will be in the colored form. The greater the quantity of protein, the deeper blue will be the hue of the indicator strip.

If the strip is dipped momentarily into a urine containing no or only traces of protein, it will show a yellow color. With increasing levels of protein, the strip will present colors ranging from yellow-green, to green, to increasingly deeper shades of blue. The strips come in a vial, to which is attached a chart relating color to protein concentration. A yellow-green color is graded as a trace and reflects 20 to 30 mg./100 ml. of protein in the urine. Various shades of blue are graded + to ++++ and reflect levels of 30 to 500 mg./100 ml. This test technique has been demonstrated to be a rapid and convenient laboratory aid, although it is not as sensitive as the heat and sulfosalicylic acid tests.

Interpretation of Results of Qualitative Tests

All three techniques are positive in the presence of Bence-Jones proteins—Albustix, the least so—as well as with albumins and globulins.[42,55] The heat and sulfosalicylic acid (SSA) tests will also be positive in the presence of certain proteins of unknown clinical significance. Very important, however, are the false positives obtained when certain organic x-ray contrast media are present in solution in the urine specimens. They interfere with the SSA test by producing a turbidity. Similar false positives may be obtained if patients are receiving tolbutamide, certain sulfa drugs (Gantricin), and medications such as benzoin. These materials may show up as an initial turbidity on acidifying the urine in the heat test, but such turbidity is usually dissolved on subsequent boiling. The Albustix test is not affected by the presence of such materials; however, the Albustix strips may give false positive reactions if urines are alkaline and well buffered. Dialysis of such urine specimens for several hours will often decrease interfering materials to negligible concentrations. A carefully performed heat test is the most sensitive of the three tests discussed; the other two are more convenient for the rapid testing of large numbers of urine specimens.

QUANTITATION OF URINARY PROTEIN

Urine Protein Determination by Turbidity with Naphthalenesulfonic Acid

REAGENTS

1. Sulfonic acid reagent, β-naphthalenesulfonic acid (Eastman Kodak Co.), 0.9 per cent (w/v), dissolved in acetic acid, 2.4 per cent (v/v). Dilute 6.0 ml. of glacial acetic acid to 250 ml. with H_2O and dissolve 2.25 gm. of the β-naphthalenesulfonic acid (BNSA) in the solution. Use a fresh lot of BNSA, with a minimum of discoloration. The solution should be clear, but may have a very slightly brown cast.

2. Saline, NaCl, 0.85 per cent (w/v).

3. Standard solutions. Dilute a composite of normal sera, clear, nonicteric, and free of hemolysis, 3:100 with saline in a volumetric flask. Then prepare working standards by diluting 1.0, 2.0, 4.0, 6.0, 8.0, and 10.0 ml. of the diluted serum to 10.0 ml. with saline. These are then treated exactly as are the unknowns. The protein content of the original serum is determined by the biuret method in quadruplicate and the average value is calculated. If this value is P gm./100 ml., the working standards will have concentrations of $3P$, $6P$, and so on, up to $30P$ mg./100 ml. or $0.03P$, $0.06P$ and so on, up to $0.30P$ gm./L. If a 6.85 gm./100 ml. of serum is used, the standards will represent concentrations of 0.21, 0.41, . . . , 2.06 gm./L.

4. Daily control standard. The same serum used to prepare the standards may be diluted 3:100 with 0.05 per cent sodium benzoate. This can be stored in the refrigerator. Daily, pipet about 2.5 ml. to a tube, warm to room temperature, and dilute 2.0 ml. to 10.0 ml. with saline. This is assayed along with the unknowns and should give a value within ± 5 per cent of the 0.06P gm./L. standard just discussed.

PROCEDURE

1. Preparation of urine: The urine should be either a 12 hour (night or day) or preferably, a 24 hour specimen. Gently but thoroughly mix the specimen and record the volume of the specimen to the nearest 5 ml. Filter a 70 to 100 ml. aliquot through a fluted filter paper in a 50 to 75 mm. funnel into a dry flask. Discard the first 10 ml. and set aide the remainder for assay. If the filtrate is turbid, centrifuge it at about 2500 rpm for 10 to 15 minutes to remove as much turbidity as possible. If available, a bacteriological membrane filter may be used to aid in clarification of the urine. If refrigerated, warm the specimen to room temperature (23 to 28°C.) before analysis.

2. The urine is checked by qualitative tests (Albustix are convenient) to estimate the level of protein present. If over 150 to 200 mg./100 ml., dilute the urine with saline to be in the range of 50 to 150 mg./100 ml.

3. The reaction is carried out directly in appropriate (e.g., 12 × 75 mm. Coleman) cuvets. One pair is needed for each unknown urine and an additional pair to correct for color (if any) in the reagent. Arrange the cuvets in parallel rows in a rack.

4. Pipet 0.50 ml. of urine or diluted urine into each pair of cuvets. The cuvets in the front row (F) receive 2.0 ml. of saline. Then, at $\frac{1}{2}$ minute intervals, add 2.0 ml. of BNSA reagent to the back row (B) tubes, one at a time, cover with parafilm, and mix immediately. After 5 to 6 minutes, mix the tubes again gently, to avoid air entrapment, and read their absorbances immediately in the spectrophotometer at 620 nm. against their F tube mates as blanks.

5. Measure the correction for any color in the sulfonic acid reagent by reading the absorbance of 2.0 ml. of BNSA reagent plus 0.50 ml. of saline against saline. This value may vary from zero to 0.030 and will increase very slowly and needs only to be verified every 2 weeks. Discard the reagent when the reagent absorbance is over 0.040.

CALCULATIONS

1. Subtract the reagent absorbance reading (if any) from the absorbance of the unknowns.

2. Convert the corrected absorbance reading to concentration by the use of a calibration graph, or less accurately, by the use of a calibration factor for the procedure. If the urine was diluted, multiply this value by the appropriate dilution factor. This gives the protein concentration in grams per liter. The 12 or 24 hour outputs of protein are obtained by multiplying the concentration value by the volume expressed in liters.

$$Q = (A_U - A_R) \times F \times D \times V \qquad (16)$$

where Q = gram protein per 12 or 24 hour period
A_U, A_R = absorbances of unknown and of reagent, respectively
F = calibration factor
D = dilution factor
V = 12 or 24 hour volume of urine, expressed in liters.

3. The results are reported in grams, if over 1 gram per volume, and in milligrams if under that value.

CALIBRATION

1. Calibration graph: Plot the corrected absorbances of the standards against their protein value in gm./L. The graph is slightly S-shaped, but the curve is essentially linear in the range of 0.20 to 2.0 gm./L.

2. The calibration factor F can be obtained by calculating the slope of the line from the measured absorbances of the standards.

$$F = \frac{\text{sum of protein concentrations of all standards}}{\text{sum of absorbances of all standards}} \qquad (17)$$

The factor should not be used with high absorbance values, where deviation from linearity is apparent.

REFERENCE

Waldman, R. K. et al.: J. Lab. Clin. Med., *42*: 489, 1952.

COMMENTS

1. The reproducibility of the method is about ± 5 per cent except with very low or very high absorbance readings, in which case it falls to ± 10 per cent. Overall accuracy is of the order of 10 to 15 per cent.

2. Significant figures in reporting results: Absorbances can be read to three significant figures, but the last figure is subject to considerable doubt. If $A = 0.165$, we are certain of the 1 and 6, but in replicate readings the 5 may be replaced by a 3, 6, 2, or even 9 ($A = 0.165 \pm 0.004$). Similarly, measurements of urine volumes in 1 or 2 L. cylinders, graduated at 10 or 20 ml. intervals, respectively, are only accurate to two or three significant figures. A reading of 1124 ml. could just as well be 1122 or 1130 ml. or worse, if the meniscus is read at the wrong angle. This assumes that no urine was lost during collection and in transfering of the urine contents from bottle to cylinder, and that, during the measurement of volume, sufficient time was allowed to permit complete drainage of bottle contents to the cylinder. The uncertainty in the measured volume is at least 10 ml. The number of significant figures in the product of two numbers can never be more than the least number in either of the two factors. The following is a sample calculation:

V = urine volume = 1125 ± 10 ml.; 3 significant figures; the per cent uncertainty is $\frac{10}{1125} \times 100 = \pm 0.9$ per cent.

A = absorbance = 0.124 ± 0.004; 2 significant figures; the uncertainty in the value = ± 3.2 per cent.

D = dilution factor = 5.0; 2 significant figures; uncertainty = 1/50 or ± 2 per cent.

F = calibration factor = 6.5; 2 significant figures; uncertainty = ± 5 per cent, the reproducibility of the procedure.

Q = quantity of protein in grams

$$Q = (1.12\mathit{5}) \times (0.12\mathit{4}) \times (5.\mathit{0}) \times (6.\mathit{5}) = 4.5\mathit{3375} \text{ gm. volume.}$$

In the preceding formula, the uncertain digits are printed in italics. At least two factors have uncertainties in the second figure; hence, the product of all factors can only be given to two significant digits.

$$Q = 4.\mathit{5} \text{ gm./volume*}$$

The procedure has a reproducibility of ± 5 per cent which is the minimum uncertainty expected in the final answer, and equals $\pm 0.2\mathit{3}$ gm. The total output then can be expressed as 4.5 ± 0.2 gm./volume.

Urine protein outputs should never be reported to more than three significant digits; in most cases only two will be warranted. Use of more will imply pretensions to a level of accuracy not possible with the methods now available. Obviously the preceding comments apply to the reporting of any laboratory results.

3. Sulfosalicylic acid (SSA)[14,61] at a concentration of 3 per cent may be used in place of the BNSA, and a number of procedures for urinary protein are based on the use of various concentrations of SSA. Although both SSA and BNSA precipitate urinary polypeptides and proteoses (dialyzable protein nitrogen), SSA precipitates a much larger fraction and tends to give appreciably higher apparent protein values, especially with urines from patients with both renal and liver disease. The BNSA results correlate somewhat better with those calculated from Kjeldahl nitrogen analyses of nondialyzable urine nitrogen. Both SSA and BNSA precipitate and measure Bence-Jones protein.

4. Trichloracetic acid (TCA)[29,61] at final concentrations of 2.5 to 3.0 per cent may also be used for the determination of protein in urine although some reports

* The per cent uncertainty of a product of several factors is given as the square root of the sum of the squares of the per cent uncertainties of each factor. In the sample calculation:

$$\text{per cent uncertainty in } Q = \sqrt{(0.9)^2 + (3.2)^2 + (2)^2 + (5)^2} = 6.3 \text{ per cent;}$$

6.3 per cent of $4.5\mathit{3} = 0.2\mathit{9}$. Thus $Q = 4.5\mathit{3} \pm 0.2\mathit{9}$, or $Q = 4.5 \pm 0.3$ gm./volume. (18)

suggest that TCA does not precipitate all nondialyzable amino acid N. One advantage of trichloracetic acid is that the reagent produces approximately equal turbidities with both albumins and globulins. Naphthalenesulfonic acid behaves similarly, but sulfosalicylic acid gives some 10 per cent more turbidity with globulins than with albumins. Within the limits of accuracy of turbidimetric procedures, this is a minor source of uncertainty.

BENCE-JONES PROTEIN IN URINE

Bence-Jones protein is a type of abnormal protein present in the urine of some 35 to 65 per cent of patients with multiple myeloma. This disease, *myelomatosis*, is characterized by a large increase in the number of plasma cells, which are probably of an abnormal type. Associated with the plasma cell proliferation is a moderate to large increase in the globulin fraction of serum, the total globulin level at times reaching values of 11 to 12 gm./100 ml. Generally, the increased globulin behaves electrophoretically as γ globulin. On occasions, a peak, the so-called "M" peak, is seen between the β and γ globulin fractions, and even more rarely, in the α globulin region. Recent studies have shown that γ globulins are made up of four peptide chains linked together as a unit. The inner two are the longer and are referred to as *H* (heavy) *chains*; the outer two are shorter and are called *L* (light) *chains*. In myelomatosis, there appears to exist a considerably increased and unbalanced production of the two forms, so that serum contains a high level of abnormal γ globulins. These may be either the normal type of γ globulin, designated as 7S or IgG, or the less common IgA type. The excess of short L forms is excreted and is detected in the urine in the form of those abnormal proteins first detected by Bence-Jones some 100 years ago.

Normal proteins, such as the common albumins and globulins, when heated, do not coagulate and precipitate out of solution until the temperature reaches 56° to 70°C. Bence-Jones proteins coagulate at a much lower temperature, 40 to 60°C.; furthermore, unlike normal proteins, the precipitated Bence-Jones proteins redissolve as the temperature of the solution is increased to between 85 and 100°C., and often reprecipitate as the temperature is again decreased to 45 to 50°C. Because of this peculiar behavior on heating, Bence-Jones proteins have been referred to as *pyroglobulins*. The Bence-Jones proteins have a molecular weight of the order of 45,000, roughly one-fourth that of complete γ globulin molecules. They are precipitated by sulfosalicylic and trichloracetic acids, and are thus included among proteins measured by the common methods used to quantitate urinary proteins.

The detection of the "M" band in serum and of Bence-Jones proteins in urine is valuable for confirmation of the clinical diagnosis of multiple myeloma. Although the newer techniques of immunoelectrophoresis[104] will be used more and more to demonstrate directly the presence of increased levels of IgG and IgA globulins in the serum and of L chains in the urine of patients with myeloma, the classic tests to be described next will still serve as useful screening procedures when attempting to establish a diagnosis.

REFERENCES

Cantarow, A., and Schepartz, B.: Biochemistry. 4th ed. Philadelphia, W. B. Saunders Co., 1967.

Hoffman, W. S.: The Biochemistry of Clinical Medicine. 3rd ed. Chicago, Year Book Medical Publishers, Inc., 1964.

Thompson, R. H. S., and King, E. J.: Biochemical Disorders in Human Disease. 2nd ed. New York, Academic Press, Inc., 1964.

TESTS FOR THE DETECTION OF BENCE-JONES PROTEIN

Bradshaw Test[92]

The urine specimen is treated with 1 or 2 drops of 33 per cent acetic acid to bring the pH down to 4.8 to 5.0, centrifuged clear, and diluted 1:1 with water. The diluted urine is layered carefully over concentrated HCl in a test tube. If Bence-Jones protein is present, it will be precipitated by the HCl and will form a fine or heavy ring at the interface of urine and HCl. The absence of a ring definitely rules out the presence of Bence-Jones protein. A positive test should be confirmed by the heating test because occasionally albumins may also precipitate, if present in large quantity. Protein, as evidenced by a positive sulfosalicylic acid test, must be present. If protein is negative, any HCl interface ring observed in the Bradshaw test is not due to Bence-Jones protein, but probably has formed from the reaction of the HCl with some unknown materials in the urine. The Bradshaw test serves as a useful screening test to determine which specimens should be subjected to the confirmatory heating test.

Heating Test (Jacobson and Milner)[37]

The urine specimen should be fresh or refrigerated under toluene. The pH is checked with pH paper, and adjusted to the pH range 4.8 to 5.0 with 33 per cent acetic acid. The urine is then centrifuged to remove any turbidity (filtration is avoided). If a previous qualitative test for protein was over 3+, the original urine specimen is diluted with normal urine (fresh, free of protein) until it tests 2+ or less. Five to seven ml. of the prepared urine are transferred to a test tube, which is immersed in a beaker of water, warming over a slow-heating hot plate or gas burner. A thermometer is immersed in the urine, and the beaker is heated slowly, with the temperature on the thermometer being watched closely. If Bence-Jones protein is present, turbidity will begin to form in the urine at about 42 to 47°C., indicating that coagulation (precipitation) is beginning. The volume of precipitate will increase until the temperature reaches 60°C. If no precipitate is evident at this point, no Bence-Jones protein is present. Any precipitate forming at temperatures above 57 to 60°C. is derived from urinary albumins and globulins.

After the Bence-Jones precipitate has formed, the beaker is rapidly heated to 100° C. As the temperature of the urine approaches 100° C, the precipitate should dissolve completely or show a diminution in volume. If the precipitate is still present at temperatures of 95 to 98°C., the tube contents should be filtered through a hot funnel. The funnel, filter paper, and receiving tube may be heated either in a drying oven or by passing boiling hot water through them several times. As the filtrate of dissolved Bence-Jones proteins cools to 55 to 60°C., the protein should begin to precipitate out again. If only Bence-Jones protein is present, the test is clear-cut. If albumins are also present, the results of the test may, at times, be equivocal. Low temperature coagulation of the protein is observed, but there is none or only little reprecipitation on cooling.

Alternatively, after the initial coagulation at 45 to 60°C., the precipitate may be centrifuged down, the supernatant poured off, the precipitate suspended in normal urine, and heated to near 100°C. Nearly complete re-solution of the precipitate should occur. Here, as with all qualitative tests, only unequivocal results are reported as positive; any doubtful or suspicious observations are best reported as negative, and repeat tests with new specimens attempted.

Cerebrospinal Fluid Protein

The cerebrospinal fluid (spinal fluid, CSF) is a clear, colorless fluid that fills the nontissue spaces of the brain and spinal cord. The fluid serves to maintain constancy of intracranial pressure and to provide a mechanical, water jacket type of protective coating for the delicate nerve tissue. It is formed as a secretion by cells in the cerebral ventricles. The total volume is about 150 ml. The fluid is a slightly modified ultrafiltrate of plasma. Its composition varies slightly, depending on where it is sampled. There is a gradual change in concentration of its components as the fluid flows down from the brain ventricles to the cisternal space and then to the lumbar area of the spinal cord, where the usual laboratory spinal fluid specimen is obtained by means of a spinal needle puncture.

Normally, a spinal puncture will supply a 6 to 10 ml. specimen. Because of the effort and skill demanded in obtaining good specimens and because of the trauma to the patient, repeat punctures are to be avoided. Spinal fluids are, therefore, very precious specimens; they are handled carefully to avoid loss and every effort is made to get as much information as possible from the volume of sample available. Any leftover portion is stored refrigerated or frozen for 7 to 10 days for possible repeat determinations or for other tests that might be dictated by the clinical status of the patient.

In routine chemical examinations of spinal fluids, the physician will rarely be interested in other than total protein, glucose and chloride. There will be occasional requests for sodium, calcium, bilirubin, and the enzyme, lactic dehydrogenase (LDH). In addition to these chemical data, the clinician will be interested in the CSF pressure, the color, turbidity, and cell count of the fluid, and the presence of old or new blood. The Lange colloidal gold test, serological tests for syphilis, and bacteriological studies may also be requested.

SPINAL FLUID PROTEIN VALUES IN HEALTH AND DISEASE

A normal spinal fluid specimen is clear, colorless, free of turbidity, contains a few lymphocytes, and has a protein content in the range of 15 to 45 mg./100 ml.[81,92] Ventricular fluids have the least protein (10 to 15 mg./100 ml.) and lumbar fluids the highest level of protein (20 to 45 mg./100 ml.).

It is important that every CSF specimen be centrifuged before it is subjected to chemical study. Any erythrocytes or leukocytes that are present must be removed in order not to include cellular protein in the determination of the true CSF protein level. If, because of a traumatic puncture, a considerable amount of fresh blood is present in the specimen, the results for protein (and LDH) will be less valid because they will reflect the proteins of blood more than those originally present in the spinal fluid. (Sugar and chloride assays, however, will still be meaningful.) Serum contains 200 to 400 times more soluble protein than does the CSF; therefore, as little as 0.20 ml. of blood can add to the CSF specimen four times the quantity of soluble protein present in 5 ml. of normal CSF. Gross blood, when present, should always be mentioned in the laboratory report. Specimens from patients with a recent brain hemorrhage may possess a light yellow color. Such yellow pigmentation is referred to as *xanthochromia*, and is due to bilirubin, present as a result of the breakdown of old hematoporphyrin. When xanthochromia is evident, this also should be noted in the laboratory report. Occasionally, surgical talc may contaminate the specimen; this must be removed, although often it is difficult to centrifuge down the light talc grains. Talc turbidity will invalidate any turbidimetric assay of protein.

All protein components present in serum are also present in the CSF. Because of the relatively low levels of protein present, it has not been convenient to estimate both albumin and globulin fractions using routine methods. As a rule, any increase in the total CSF protein is reflected in an increase in all protein components; this is certainly true of the increased protein associated with exudation across the inflamed central nervous system (CNS) linings encountered in various forms of meningitis. A possible elevation in globulin level is demonstrated by the use of the *Pandy*, *Nonne-Apelt*, and *Ross-Jones* tests. These are protein flocculating tests, relatively insensitive to increases of albumin level, but reflecting increased globulin concentrations. These crude globulin flocculation tests are still useful, but there is an increased demand for methods that measure CSF globulins quantitatively. Electrophoretic separations of spinal fluid proteins are being performed increasingly, but they can only be done after CSF specimens have been concentrated some 50- to 100-fold. Because the techniques used to concentrate fluids tend to alter some of the proteins present, results are not as satisfactory as desired. In multiple sclerosis, in some forms of encephalitis, and in nonpurulent meningitides, total CSF proteins may be normal or marginally elevated; however, they will contain elevated levels of γ globulins which may not be noted unless γ globulin concentrations as such are measured.

The normal range for total CSF globulin[36] is 6 to 16 (average = 10) mg./100 ml., of which about one-fourth is γ globulin. The albumin level is about 10 to 30 mg./100 ml. and reports for A/G ratios have varied from 1.6 to 2.2. CSF protein con-

TABLE 5-3. *Cerebral Spinal Fluid Protein in Various Disease States**

Clinical Condition	*Appearance*	*Total Protein (mg./100 ml.)*	*Globulin Reaction*
Normal	Clear, colorless	15-45	neg. or trace
Coccal meningitis	Purulent, cells, turbid, opalescent	100–500	1+ to 3+
Tuberculous meningitis	colorless, fibrin coagulum, cells	50–300 occ. up to 1000	1+ to 3+
Benign lymphocytic meningitis	clear, colorless, lymph cells	30–100	neg.
Encephalitis	clear, colorless, cells	15–100	neg.
Poliomyelitis	clear, colorless	10–300	neg.
Neurosyphilis	clear, colorless	50–150	3+ or 4+
Disseminated sclerosis	clear, colorless	25–50	neg. or 1+
Spinal cord tumor	clear, colorless, or xanthochromic	100–2000	1+ to 3+
Brain tumor	clear	15–200	neg. or 1+
Brain abscess	clear, or slightly turbid	30–300	neg. or 1+
Cerebral hemmorhage	colorless, or xanthochromic, or bloody	30–150	neg. or 1+

* After Varley, H.: Practical Clinical Chemistry. 4th ed. New York, Interscience Publishers, Inc., 1967, pp. 708–711; and Stewart, C. P., and Dunlop, D.: Clinical Chemistry in Practical Medicine. 6th ed. Edinburgh, Livingstone, Ltd., 1962, pp. 216–217.

centration is somewhat higher in infants than in adults. Protein concentrations found in a number of clinical conditions are listed in Table 5-3.

COLLOIDAL GOLD CURVE

The Lange colloidal gold curve flocculation test[16,81] uses the composition of the proteins in spinal fluid to aid in the diagnosis of certain types of CNS disease. Globulins act to cause the flocculation of metastable colloidal gold sols, whereas albumins tend to inhibit such flocculation. As long as both fractions are present in normal amounts, no or very little flocculation is observed when diluted spinal fluid is added to the gold sols. When the A/G ratio is decreased and abnormal albumins and globulins are present, however, the capacity of the albumin to inhibit agglutination of the gold sol particles is diminished and precipitation of the sol particles occurs. In the Lange test, a series of 10 tubes are prepared to contain a fixed volume of a buffered gold sol preparation. To these there is added a fixed volume of increasing serial dilutions of the CSF (1:1 to 1:512). With normal spinal fluids, no flocculation occurs and the solutions exhibit the normal, clear, pinkish violet color. With abnormal fluids, the sol particles are unstabilized and the flocculation is evidenced by the solutions undergoing changes to purple-blue and blue, and eventually to colorless, when precipitation is complete. The degree of color change for each of the 10 tubes is coded with a number from 0 to 5. For example, a normal CSF will commonly give a Lange reaction reading of 000 111 100 0, whereas a paretic fluid may read 555 544 210 0, and a meningitic fluid 001 344 311 0.

METHODS FOR THE DETERMINATION OF CEREBROSPINAL FLUID PROTEIN

The most popular and common procedures used to estimate CSF proteins employ measurement of turbidity when proteins are reacted with precipitating reagents such as trichloracetic and sulfosalicylic acids. The customary form of the biuret technique is insensitive to the very low levels of protein present in normal CSF. In a 100-fold more sensitive form proposed by Daughaday, a modified biuret reagent is used together with the Folin-Ciocalteu phenol reagent; however, amino acids and other nonprotein substances interfere by reacting with the reagents and this interference is taken into account by an empirical correction. Other approaches have been proposed, but none have the simplicity of the turbidimetric procedures.

The main difficulty with the turbidimetric procedures is the selection of the ideal precipitating reagent. Such a reagent should react similarly with both albumins and globulins. The same degree of turbidity should be formed with equal quantities of the various protein forms. The turbidity should be stable long enough for its convenient estimation, and it should be reproducible from day to day and from sample to sample, normal or abnormal. No such ideal reagent has been found to date; most reagents produce more turbidity either with albumin (sulfosalicylic acid) or with globulin (trichloracetic acid), and the amount of turbidity and the form of flocculated particles depend on the temperature and manner of mixing of the specimen and the reagent. A method using sulfosalicylic acid will be presented in detail and a TCA method will be described in outline only. It is important that the work be done carefully and the details of the technique rigidly adhered to if reproducible and correct results are to be obtained.

Determination of Total Protein by Sulfosalicylic Acid Turbidity

REAGENTS

1. Sulfosalicylic acid, (SSA), 3.0 per cent (w/v), A. R. After preparation, filter the reagent until clear and store away from direct light. The reagent is colorless and should be discarded when any color or turbidity develops.

2. Standard protein solutions. Prepare a composite of normal, hemolysis-free, nonicteric and nonturbid sera. Determine the total serum protein in quadruplicate by the biuret procedure. Carefully dilute the composite serum 3:100 with physiological saline; further dilute aliquots of this dilute serum (of approximately 180 to 200 mg./100 ml. concentration) with saline to form a series of standards of the order of 20, 40, 80, 120, 160, and 200 mg. protein/100 ml. Do all mixing gently to avoid protein denaturation. These standards are then treated in the same way as CSF specimens.

3. Physiological saline, NaCl, 0.85 per cent. (w/v).

PROCEDURE

1. Centrifuge the CSF specimen to sediment any cells or other particulate matter. Use small size centrifuge tubes or test tubes; 10 × 75 mm. tubes are convenient for specimens of 1 to 5 ml.

2. Use a pair of 12 mm. cuvets for each CSF specimen. Mix the reagents and specimen directly in the cuvets. Place the cuvets in parallel rows in a test tube rack. The front row (F) cuvets will serve as pigment controls for the back row (turbidity) cuvets.

3. Both cuvets receive 0.50 ml. of the CSF specimen. If a protein value over 150 mg./100 ml. is expected (turbid or purulent specimens), prepare a dilution of 1:3, 1:5, or 1:10 in saline, and use 0.50 ml. of the diluted specimen.

4. All front row (control) cuvets receive 2.0 ml. of saline. At one-half minute intervals, add 2.0 ml. of 3 per cent sulfosalicylic acid to each of the back row cuvets. After addition of the reagent, cover each cuvet with parafilm and mix by gentle inversion three to four times before treating the next specimen. Take care to avoid entrapment of air bubbles. Set the spectrophotometer at 620 nm.

5. After 6 to 7 minutes, again mix each turbidity (back row) cuvet by gentle inversion and read its absorbance at 620 nm. against its front row mate set at 100 per cent transmittance.

6. Check the turbidity of the reagent by diluting 2.0 ml. of reagent with 0.50 ml. of saline and measuring its absorbance against saline. If over 0.010, discard the reagent.

CALCULATIONS

The calibration curve is linear in the range of 0 to 150 to 200 mg./100 ml. of CSF protein. For the interval of linearity, a Beer's law calibration constant or factor may be calculated and used, or the data may be graphed and unknowns read directly from the calibration graph. The calibration constant F is obtained by calculating the average slope of the line relating absorbance to concentration, either by least squares statistical treatment or by the approximation formula presented in equations (15) and (17). Then

$$C_u = F \times (A_u) \times (D)$$

where C_u and A_u represent concentration of protein and absorbance of unknown, and D is the dilution factor, if the CSF was diluted before being assayed.

COMMENTS

1. Reproducibility is about ±5 per cent except at levels under 15 mg./100 ml. (± 10 per cent). Overall accuracy is about 7 to 10 per cent, being least accurate at the very low and very high extremes.

2. The control tube corrects for slight xanthochromia. The use of 620 nm. light further minimizes error due to yellow pigment being present.

3. Interference by contaminating hemoglobin (blood) pigment is not corrected for; the hemoglobin, being a protein, reacts with the SSA reagent to form a turbidity.

4. In turbidity measurements, true light absorbance is not measured. The particles in suspension, by diffracting and refracting light incident on them, decrease the light energy that reaches the photocell of the photometer. Although some light energy is also absorbed, it is primarily the loss of light energy that has been dispersed that is being measured. Any wavelength may be used, but the dispersion will increase as the wavelength is decreased. For uniformity of dispersion, the size and shape of the particles must be uniform. The latter is affected by temperature and by the rate and intensity of the mixing of the reagent and the specimen. Narrow cuvets are preferred because over a short light path, a lesser degree of variability in particle geometry is encountered. Temperature control is important; standardizations and determinations should be done at the same temperature, 25 ± 2°C. Particle sizes change with time and, therefore, time control is also important. It takes 5 minutes after reagent addition to obtain a reasonably uniform particle size at 23 to 28°C., and most investigators report that size dispersion is stable for up to about 10 minutes. With longer periods of time of reaction, agglomerates of particles begin to develop. At low levels of protein, turbidity develops more slowly; at high levels particles tend to sediment. Turbidity readings are most reproducible in the range of 25 to 90 mg./100 ml.

REFERENCES

Kachmar, J. F.: Unpublished procedure.
Cipriani, A., and Brophy, D.: J. Lab. Clin. Med., *12*:1269, 1943.

Determination of CSF Protein by Trichloracetic Acid Turbidity[61]

The reagent used is 3.0 per cent (w/v) trichloracetic acid (TCA) in water. The turbidity is developed over a 10 minute period, after mixing 2.0 ml. of the TCA with 0.50 ml. of specimen. Readings are made at 450 nm. against a blank containing saline and specimen. Precision and reproducibility are of the same order of magnitude as with the SSA procedure, but TCA appears to give more nearly equal turbidity increments for equal quantities of albumin and globulin than does the SSA reagent. The use of 450 nm. wavelength is advantageous because a given degree of turbidity will give a greater absorbance at 450 than at 620 nm. On the other hand, yellow xanthochromia pigment, when present, will also read much higher at 450 nm.

QUALITATIVE TESTS FOR GLOBULINS IN SPINAL FLUIDS

Pandy Test

The reagent consists of a saturated solution of pure, white-pink phenol in water. About 10 gm. of phenol (free of any brown-black coloration) is suspended in 100 ml.

of water and the mixture gently warmed with shaking at 50 to 60°C. The material is transferred to a wide-mouth brown bottle and allowed to settle at room temperature (25 ± 2°C.) for 48 to 72 hours before use. The reagent must be clear, colorless, and nonturbid. It is stored at room temperature in a brown bottle away from direct light and away from any cooling drafts.

PROCEDURE

To 1.0 ml. of Pandy reagent in a 10 × 75 mm. test tube, add 0.10 ml. of clear CSF fluid. Mix the tube by inversion and allow to stand for exactly 3 minutes, and then examine it for the presence of any opalescence or turbidity by viewing it against a black background. The turbidity formed indicates the presence of greater than normal amounts of globulins, primarily γ globulins. Grade the degree of turbidity from negative to ++++. The observed turbidity is scored similarly to the scheme described in Table 5-2 for the heat test for urine protein, but the protein equivalents are different. No consistent relationship between turbidity score or appearance and globulin concentration has been established. A Pandy reading of trace to 1+ will reflect a total protein concentration of 50 to 80 mg./100 ml., with a globulin level of about 20 to 40 mg./100 ml. For a ++ (2+) reading, the respective values are 100 to 160 and 35 to 70 mg./100 ml. In many cases an increase in albumin is accompanied by a proportionate increase in the globulins. In other cases (e.g. poliomyelitis) there is a disproportionate increase in the albumin fraction, giving cause to an increase in total protein without a corresponding Pandy reaction. In multiple sclerosis, in which only γ globulins are increased, Pandy readings of 1+ or 2+ are obtained even though the total protein is within the normal range.

Nonne-Apelt and Ross-Jones Tests

The reagent for both tests is a saturated solution of ammonium sulfate in water. In the Nonne-Apelt test, equal volumes (0.50 ml.) each of CSF and $(NH_4)_2SO_4$ are mixed in a small test tube, and after 3 minutes, the tube is examined for opalescence and turbidity, as in the Pandy test. In the Ross-Jones variant of the test, the CSF is carefully layered over the ammonium sulfate solution, and the interface between the two solutions is observed for the presence of a ring of fine precipitate. With all three tests, a negative or at most a trace reading is observed with normal spinal fluids. The Pandy test is somewhat more sensitive than the ammonium sulfate tests.

REFERENCES

Varley, H.: Practical Clinical Chemistry. 4th ed. New York, Interscience Publishers Inc., 1967, pp. 689–712.

Serum Protein Fractionation

Only two methods of separating serum protein components are sufficiently simple and convenient to be useful in routine work. *Salt fractionation* is easy to carry out, involves no special equipment, and, if done carefully, provides very useful clinical information. Commonly, only separation of albumin from the globulins as a group is carried out, but techniques are available for separating and quantitating the chief globulin components, especially γ globulin. When more precise values of the protein fractions are desired, *electrophoretic techniques* are used. These require more skill and the use of special equipment, and they are more time consuming, although with cellulose acetate membranes, separations may be made in less than an hour. *Ultracentrifugation*, requiring an expensive instrument, is not suitable for routine use, but is especially useful in the study of lipoproteins and macroglobulins.

Salt Fractionation; Albumin-Globulin Ratio

Albumins are soluble in water, whereas globulins are not. If an electrolyte solution of low concentration (0.1 *M*) is added to a globulin, the latter will go into solution by virtue of the phenomenon of *salting-in*. The cations and anions of the electrolyte bind to reactive groups on the protein molecules and break the bonds holding the protein micelles together, allowing the individual protein molecules to undergo solution. The salt concentration of plasma, approximately 0.15 *M*, is ideally suited to maintain globulins in solution. As the salt concentration is increased to high levels (of the order of 2 *M*), the various globulins will again become insoluble and will precipitate by *salting-out*. The level of electrolyte needed to precipitate the various globulins will vary with the individual protein and electrolyte and will also depend on pH and temperature.

Ammonium sulfate was the first salt to be used and it is still used in protein and enzyme investigations. Those proteins precipitating in one-third saturated ammonium sulfate solution were termed *euglobulins* (true globulins). As the salt concentration is increased, the pseudoglobulins begin to precipitate out, and at concentrations near 50 per cent saturation, all globulins become insoluble, the proteins still in solution constituting the albumins. The use of ammonium sulfate for the isolation and assay of γ globulins will be described later. The salt is not used in routine clinical work because it interferes with the biuret reaction and obviously cannot be used with the Kjeldahl nitrogen technique.

The only salts whose use has become established in routine clinical chemistry are sodium sulfate and sodium sulfite. The former (Na_2SO_4) was used as early as 1901, but its large scale use followed the work of Howe[33] in 1921, who recommended the use of 22 per cent (w/v) sodium sulfate to precipitate globulins. Total protein was assayed, and, after removing globulins by filtration, the albumin fraction was measured in the filtrate, globulin being calculated by difference. In 1940 Kingsley[46] introduced the use of ether as a means of separating the precipitated globulins from the soluble albumins, without resorting to the slow, tedious filtration. By carefully dispersing diethyl ether into the mixture of serum plus salt solution, the insoluble globulin micelles would entrap sufficient ether so that on centrifugation, instead of sedimenting as they would normally, the globulins would float above the albumin solution as a compact pellicle. With this type of procedure, serum albumin was found to range from 4.0 to 5.5 gm./100 ml., the globulin from 2.0 to 2.9 gm./100 ml., and the albumin/globulin ratio from 1.5 to 2.5 (average, 2.0).

With the development of Tiselius' moving boundary electrophoresis technique, it was logical to compare the electrophoretic albumin fraction with that separated by the use of Howe's 22 per cent sodium sulfate. It was immediately evident that the Howe technique yielded higher values because it included with the albumins, the α_1 and some of the α_2 globulins. Studies by Majoor and Milne showed that to free the albumin of all globulin fractions, the concentration of sodium sulfate must be of the order of 26 to 27 per cent (w/v)(1.9 *M*). Unfortunately, this concentration is beyond the solubility of sodium sulfate at room temperature (25°C.), and the salt solution must be kept at 35 to 37°C. to avoid crystallization of the salt on storage and during use. With the use of 26 per cent sodium sulfate, the albumin/globulin ratio is of the order of 1.1 to 1.8, averaging about 1.4; this agrees quite well with that obtained by electrophoretic methods. Other salts have also been employed. Campbell and Hanna used sodium sulfite, obtaining complete separation of globulins from albumin at 27 per cent (w/v) (2.2 *M*). This salt has the advantage of being more soluble than sodium sulfate, and, because it acts as a buffer, better pH control of the precipitation reaction is possible. Reinhold proposed the use of a mixture of sodium sulfate plus sodium sulfite and this procedure will be presented in detail.

By using varying concentrations of sodium sulfate (or sodium sulfite), it is possible to separate all of the main globulin fractions from one another. For example, Kibrick and Blonstein[44] used 15 per cent (w/v) sodium sulfate to precipitate the γ globulins and 26 per cent salt concentration to precipitate all globulins. The difference between these values gave the sum of α and β globulins. By further use of 19 per cent sodium sulfate, which precipitated the β- plus γ globulins, the amount of α globulins could also be determined. Such fractionations are tedious, however, and not too satisfactory, especially when abnormal sera are used, and they have never become popular. The electrophoretic separation is simpler and considerably more accurate.

CLINICAL APPLICATION AND NORMAL VALUES

Serum from a healthy adult contains about 3.8 to 4.7 gm. of albumin per 100 ml., averaging about 4.3 gm./100 ml. The total globulin values range from 2.3 to 3.5 gm., with an average of 3.0 gm./100 ml. There is no evidence of any differences between males and females. The slight diurnal variation in total protein that is observed (+0.5 gm./100 ml.) affects all fractions equally, with the result that the albumin/globulin ratio is quite constant for any one individual. In newborn infants, albumin and globulin average about 3.5 and 2.0 gm., respectively. These levels increase slowly until about the third year, when adult values for each are reached.

Although in dehydration the total serum proteins may be above normal and, in water-loading, below normal, the ratio of albumins to globulins may be essentially unchanged. Whenever *hypoalbuminemia* occurs, it usually reflects an impairment in the synthesis of albumin by the liver, or a sudden or chronic loss of the albumin through the urine. In most acute and chronic infections, the albumin is moderately depressed (3.2 to 4.2 gm./100 ml.), reflecting increased breakdown (catabolism) as a result of the infection and the accompanying fever. The globulins are usually slightly increased, especially the γ globulins, reflecting synthesis of antibodies to the infecting agent. In chronic infections, globulin values are higher, especially those of the γ variety. In such conditions as lupus erythematosus, sarcoidosis, and multiple myeloma large increases (up to 2 to 5 gm./100 ml.) in γ globulins and occasionally in β globulins are found very often. In *Laennec's cirrhosis*, with its diminution in functioning liver tissue, albumin may fall to levels of 2.4 gm./100 ml. or lower, and total globulins are often raised to 3.5 to 4.0 gm. or more. The A/G ratio falls from a normal average value of 1.4 to values from 0.7 to 1.0. Whenever globulins are present at higher concentrations than albumin, a reversal of the A/G ratio is said to have occurred. This ratio, though a time-honored parameter of disease, is not very meaningful. It is more important and significant to define the *absolute change* in albumin and globulin levels specifically, than to consider the ratio. The ratio may be reversed because the globulin level is high but the albumin normal, e.g., in sarcoidosis, or because the albumin is low but the globulin normal, e.g., in the nephrotic syndrome. It would enhance the analysis of disease if the concept of the A/G ratio were abandoned.

In renal disease with accompanying proteinuria, a fall in serum albumin occurs as a result of urinary losses. In the case of lipoid nephrosis and other types of the nephrotic syndrome, the albumin may fall to as low as 0.9 gm./100 ml., and is often in the range of 1.5 to 2.0 gm./100 ml., because of urinary losses of albumin of the order of 10 to 30 gm. per day. Characteristic of the nephrotic syndrome is an elevation in serum α_2 globulins, accompanying the hypoalbuminemia. A low value for serum albumin is one of the important causes of edema, which may be observed in the nephrotic syndrome, in starvation, in malabsorption, and, of course, in cirrhosis, in which a loss of albumin may be caused by the formation of ascitic fluid.

REFERENCES

Watson, D.: Albumin and "total globulin" fractions of blood. *In* Advances in Clinical Chemistry. H. Sobotka and C. P. Stewart, Eds. New York, Academic Press, Inc., 1965, vol. 8, p. 238.

Owen, J. A.: Effect of injury on plasma proteins. *In* Advances in Clinical Chemistry. H. Sobotka and C. P. Stewart, Eds. New York, Academic Press, Inc., 1967, vol. 9, p. 2.

Hoffman, W. S.: The Biochemistry of Clinical Medicine. 3rd ed. Chicago, Year Book Medical Publishers, Inc., 1964.

PROCEDURES FOR ALBUMIN DETERMINATION

The procedures currently being used for the estimation of serum albumin fall into three types: (1) assay of albumin by the biuret technique after removal of globulins by salt precipitation: (2) measurement of albumin directly, by virtue of the property of albumins to bind certain dyes, such as methyl orange, HABA, and bromcresol green; and (3) assay of albumin after separation by electrophoretic procedures. Detailed procedures for the first two will be presented in this section; electrophoretic separation will be discussed in the following section.

Determination of Total Protein and Albumin by Salt Fractionation Using the Reinhold-Kingsley Procedure

REAGENTS

1. Stock and working biuret reagents.
2. Protein standards.
3. Alkaline tartrate solution.
4. Potassium cyanide.

These reagents (1–4) are identical with those already described for the assay of total protein by the biuret method.

5. Sulfate-sulfite solution (globulin precipitating reagent), 20.8 per cent (w/v) Na_2SO_4 anhydrous, plus 7.0 per cent (w/v) Na_2SO_3 anhydrous. Acidify about 850 to 900 ml. of H_2O with 2 ml. of concentrated H_2SO_4 and rapidly pour with vigorous stirring into a 2 L. beaker containing freshly weighed salts (208 gm. anhydrous Na_2SO_4, plus 70 gm. anhydrous Na_2SO_3). Occasionally, the beaker must be heated to 40 to 45°C. to effect complete solution of the salts. Filter the solution through a double layer of rapid-flow filter paper into a 1 L. volumetric flask, allow to cool to about 27 to 30°C. and then make up to the mark with water. Check the pH to insure that it lies between 7.0 and 8.0. For pH measurement, dilute 1.0 ml. of the solution

to 25.0 ml. with water and check the pH of the diluted solution with a pH meter. If necessary, adjust the pH by the dropwise addition of 2.5 *N* NaOH or 2 *N* H_2SO_4. Store the solution at a temperature of 30 to 37°C. (water bath, 37°C. incubator).

6. Diethyl ether, A. R. If the ether is kept in the refrigerator, prewarm it to room temperature before use.

PROCEDURE

1. The analysis should be done in an area where the temperature is 25 to 30°C. Pipet 0.50 ml. of each serum into a 15 × 120 mm. test tube.

2. Treating *one serum at a time*, pipet 9.5 ml. of the Na_2SO_4–Na_2SO_3 solution and add slowly with continuous mixing to a tube containing the 0.50 ml. of serum. Immediately cover the tube with a stopper or Parafilm and mix gently by 3 to 4 inversions. Avoid vigorous shaking to minimize denaturation of the albumin. Immediately, using a volumetric pipet, remove 2.00 ml. of the mixture and transfer to a cuvet marked T for total protein. Then set aside the serum-salt tube until all sera have been similarly treated. It is important that the T aliquot be pipetted immediately after mixing the serum and salt solution, before any settling of the globulin flocs has occurred.

3. Pipet 2.0 ml. of the Na_2SO_4–Na_2SO_3 solution to another cuvet marked B (blank).

4. Treat the tubes containing the remaining 8 ml. of serum-salt solution mixture as follows: Add diethyl ether to the tubes to within 1.5 cm. of the top of the tube (approximately 3 ml.). Tightly stopper the tubes and then *gently* invert 40 times during a period of 20 seconds. Avoid vigorous mixing to prevent denaturing the albumin and the formation of an emulsion.

5. After allowing the tubes to settle for 2 to 3 minutes, during which time visible rising of the globulin floc should be evident, centrifuge the tubes, with stoppers removed, for 10 to 15 minutes at 2000 to 2500 rpm. A hard, thin pellicle of globulin should be present at the interface between the ether and albumin (water) layers. The bottom albumin solution should be clear or, at worst, show only a hint of opalescence. If turbidity is present, improper separation of globulins has occurred and the work should be repeated from the beginning.

6. Tilt the tubes, one at a time; then, gently tap each tube at the level of the globulin pellicle to loosen it from the sides of the tube. Insert a 2.00 ml. volumetric pipet through the ether into the albumin layer, by-passing the globulin pellicle, (avoid touching the pellicle or the sides of the tube), and withdraw 2.00 ml. of albumin solution. Tightly seal the pipet end with the fingertip until the pipet is beyond the ether layer, to avoid sampling ether as well as albumin. Wipe the pipet sides and tip clean of any possible contamination with globulin precipitate, and transfer the 2.0 ml. aliquot to a cuvet labeled A for albumin.

7. Add 8.0 ml. of working biuret reagent to all cuvets, B, T, and A, mix the contents of the cuvets, and permit color to develop for 30 minutes. Read the absorbances of the T and A tubes against the B tube, using a wavelength of 550 nm.

8. Referring to the calibration curve, convert the absorbance readings to protein values in terms of gm./100 ml.

CALCULATIONS

1. The T tube results give the value of total proteins in terms of gm./100 ml.
2. The A tube values are the measure of serum albumin in gm./100 ml.

3. Then (T–A) will give the concentration of total globulins in the serum.

4. The albumin-globulin ratio (A/G) is obtained by dividing the albumin value by the globulin value.

Serum-pigment correction: As in the case of total protein, the presence of excessive serum pigment (bilirubin, hemoglobin) or turbidity must be corrected for. Either of the techniques previously described may be used. If the technique using parallel readings with biuret and alkaline tartrate reagents is used, the precipitation of globulin is carried out in duplicate tubes—one tube being used to get T and A with biuret reagent, the other to get the serum color correction for the T and A tubes, respectively. The color corrections are subtracted from the crude T and A values to give the true total and albumin results. If KCN is used, the correction is obtained by bleaching out the copper color in all cuvets after the initial biuret readings have been made, and re-reading the KCN treated cuvets.

CALIBRATION

The calibration curve is made in exactly the same way as that described for total serum protein, except that the protein standard is diluted with the Na_2SO_4–Na_2SO_3 solution rather than with saline. This is necessary because the salts present depress the biuret color to a slight but measurable extent.

REFERENCE

Reinhold, J. G.: Total protein, albumin, and globulin. *In* Standard Methods of Clinical Chemistry. D. Seligson, Ed. New York, Academic Press, Inc., 1953, vol. 1, p. 88.

ALBUMIN DETERMINATION BY DYE BINDING

All proteins, and especially albumins, tend to react with many chemical species by means of electrostatic and tertiary van der Waal's forces and by virtue of hydrogen bonding. Bilirubin, fatty acids, most hormones, and many drugs are transported about the body bound to albumin. Many colored dyes in the anionic form also possess this protein-binding property. This property has been used in attempts to devise methods by which albumin could be measured directly, without previous precipitation of globulins.

In dye-binding methods, only those dyes or indicators can be used that bind very tightly to the albumin molecule, so that practically 100 per cent of the albumin present is bound to dye. The binding must be unaffected by reasonably small changes in ionic strength and pH. Further, the color of the protein-bound dye should be different from that of the free dye; i.e., there should be a substantial shift in the wavelength of light at which maximum absorption occurs in the two forms. The albumin-dye concentration can then be measured in the presence of excess dye. Finally, dye binding to other protein fractions (globulins) must be negligible if the dye binding is to be the basis for a valid assay of albumin. For practical reasons, the color characteristics of the dye should be such that it can be measured at wavelengths of light where bilirubin and hemoglobin will give negligible or minimal interference. The use of *methyl orange* was proposed by Bracken and Klotz[4] and Watson and Nankiville;[98] *HABA* (2-(4′-hydroxyazobenzene)-benzoic acid) by Rutstein, Ingenito, and Reynolds[75] and Martinek;[59] and *bromcresol green* by Rodkey.[73] The use of bromcresol green is most free of pigment interference. Bilirubin at levels over 5 mg./100 ml. interferes with HABA procedures and to a lesser degree with methyl orange methods.

The use of these dye reagents is now well established, and several manual and AutoAnalyzer procedures using methyl orange and HABA are available; however, there is considerable disagreement over the reliability of the dye-binding methods. Occasionally, albumin values obtained are unusually high as a result of the binding of the dye to proteins other than albumin. As an example of dye-binding technique, a methyl orange procedure will be presented in outline. It is a modification of the method proposed by Wrenn and Feichtmeir[103] and is designed for use with the Coleman spectrophotometer and 19 mm. cuvets.

Assay of Serum Albumin by Methyl Orange Binding

REAGENTS

1. Working dye reagent. Prepare by adding 5 to 10 ml. portions of 0.10 per cent (w/v) methyl orange to 1000 ml. of citrate buffer, 0.055 M, pH $= 3.50 \pm 0.03$; check the absorbance against water at 540 nm. after each addition. Continue addition of dye until the absorbance lies between 0.78 and 0.85. Use the reagent at 24 to 27°C., although it is stored in the refrigerator.

2. Standards. These are prepared from crystalline human serum albumin. Bovine albumin should not be used because there are significant differences in dye binding by albumins from different species. It is wise to assay the albumin standard against the biuret total protein method so that the total protein and albumin values will be consistent. Prepare the stock reagent, which has a concentration of 10.00 gm./100 ml., by dissolving the albumin in saline. Prepare the working standards by diluting the stock with saline to give levels of 1.0 through 6.0 gm./100 ml.

PROCEDURE

1. Pipet 10.0 ml. of working dye reagent into a series of cuvets. One cuvet will be needed for the albumin blank, one each for the six standards, and one for each of the unknowns. Transfer 0.20 ml. each, of saline, standards, and sera, to the respective cuvets. After mixing by gentle inversion, permit the cuvets to stand for 20 minutes.

2. Read the absorbances of the standard and serum specimens against the blank at 540 nm. If the specimens are slightly lipemic, icteric, or hemolyzed, pigment color corrections will have to be made. For this purpose, add 0.20 ml. of the test serum to 10.0 ml. of the citrate buffer and read the absorbance against water (or citrate buffer) at 540 nm. Subtract these readings from the previous readings.

3. Convert the absorbance readings to albumin concentration by use of the calibration curve, which follows Beer's law up to concentrations of about 4 to 5 gm./100 ml. Prepare new calibration curves daily.

COMMENTS

Normal globulins and most abnormal globulins have no effect on the albumin values. Occasionally paraproteins and macroglobulins will interfere by giving high results. Dye binding, being a dissociation-association equilibrium phenomenon, is influenced by changes in temperature; a rise in temperature increases the dye-albumin dissociation and thus gives lower results. Therefore, standards and sera should be run under identical conditions. Heparinized plasma should not be used because heparin, for reasons not understood, enhances dye binding by albumin. Serum albumin levels obtained by methyl orange dye binding techniques tend to be 0.1 to

0.2 gm./100 ml. higher than those obtained by salt-fractionation or electrophoresis, although some chemists have reported results as much as 0.5 gm./100 ml. higher.

REFERENCES

Wrenn, H. T., and Feichtmeir, T. V.: Am. J. Clin. Path., *26*:960, 1956.
Watson, D., and Nankiville, D.: Clin. Chim. Acta, *9*:359, 1964.

DETERMINATION OF γ GLOBULIN IN SERUM

The γ globulin fraction of serum proteins is most precisely determined by electrophoretic methods; however, the following chemical procedure by Wolfson is convenient, rapid, and sufficiently accurate for clinical work. Semiquantitative assays of γ globulin may be obtained by measuring zinc turbidity, as proposed by Kunkel,[52] or ammonium sulfate turbidity, as proposed by de la Huerga and Popper.[9]

REAGENTS

1. Wolfson's γ globulin reagent, 19.3 per cent (w/v) $(NH_4)_2SO_4$ in 4.0 per cent NaCl (w/v). Dissolve 193 gm. of ammonium sulfate and 40 gm. of sodium chloride in about 700 ml. of water. Filter the solution into a 1000 ml. volumetric flask and, after thoroughly rinsing the filter, dilute to a volume of 1000 ml. Store at room temperature.

2. Dilute working biuret reagent. Refer to the reagents in the method for total protein by the biuret procedure.

PROCEDURE

1. Pipet 9.60 ml. of Wolfson's reagent into a thick-wall 12.5 ml. conical centrifuge tube. Do each determination in duplicate. Any serum or commercial reference serum available with a known level of γ globulin can be used as a control.

2. Pipet 0.40 ml of serum and layer it on the top of the reagent in the centrifuge tube.

3. Cover with Parafilm and mix by inversion until no further increase in turbidity is observed.

4. Centrifuge the tube for 15 to 20 minutes at 2500 to 3000 rpm (or faster if a superspeed centrifuge is available). The supernate must be clear. If turbidity is still evident, cool the tube to 15°C. and recentrifuge. This assures that all the γ globulin is sedimented; at the salt concentration used (1.4 M $(NH_4)_2SO_4$), other proteins are not precipitated.

5. Carefully, decant most of the supernatant without disturbing the globulin pellet. Alternatively, the supernatant may be removed by suction.

6. Centrifuge again to repack the pellet.

7. Remove the remaining supernatant by careful decanting; then invert the tubes over a filter paper and allow the tubes to remain inverted for 5 minutes to promote complete drainage.

8. With a piece of rolled filter paper, wipe as much of the salt solution adhering to the sides of the tube as possible, taking care not to touch or disturb the pellet. This assures that all protein, except the precipitated γ globulin, has been decanted off and that the quantity of NH_4^+ remaining is too small to interfere significantly with the biuret reaction. None of the pellet may be lost because it contains the entire γ globulin fraction being isolated.

9. Add 2.0 ml. of water to each tube and shake vigorously to loosen and break up the pellet. Then add 8.00 ml. of working biuret reagent, mix vigorously, and let stand for 30 minutes to develop color.

10. Read the absorbance at 540 nm. against a blank containing 2.0 ml. of water and 8.0 ml. of working biuret reagent.

11. Using the biuret protein calibration curve, obtain the gm./100 ml. equivalent to the absorbance reading. This value is then divided by 4 because 0.40 ml. of serum was used, whereas the biuret curve is based on a serum aliquot of only 0.10 ml.

NORMAL VALUES

The normal value for serum γ globulin by this procedure is 1.05 gm./100 ml. (range: 0.55 to 1.30 gm./100 ml.). These values are more applicable to Caucasians; the level found in Negroes is somewhat higher, namely 0.88 to 1.60 gm./100 ml.

REFERENCES

Wolfson, W. Q., Cohn, C., Calvary, E., and Ichiba, F.: Am. J. Clin. Path., *18*:723, 1948.
Friedman, H. S.: Gamma globulin in serum. *In* Standard Methods of Clinical Chemistry. D. Seligson, Ed. New York, Academic Press, Inc., 1958, vol. 2, p. 40.

Zinc Turbidity and Ammonium Sulfate Turbidity

The zinc turbidity procedure[52] is discussed in detail in Chapter 13 on liver function tests. Serum is diluted 1/60 with a very weak solution of $ZnSO_4$ at pH 7.5. A turbidity is formed as a result of the precipitation of the γ globulin by zinc ions. The turbidity is measured and serves as a rough measure of the γ globulin present. In the analogous ammonium sulfate procedure of de la Huerga and Popper,[9] the precipitating reagent consists of 18.9 per cent (w/v) $(NH_4)_2SO_4$ in 2.93 per cent (w/v) NaCl solution.

Fibrinogen

PROPERTIES AND CLINICAL SIGNIFICANCE

Fibrinogen is one of the plasma protein factors involved in the process of blood coagulation. It is an elongated globulin with a molecular weight of about 350,000, and is synthesized only in the liver. The thrombin formed in the second stage of coagulation acts on fibrinogen to split off a peptide fragment, converting the soluble fibrinogen into insoluble fibrin, which constitutes the clot proper. The fibrino-peptide split-off from the fibrinogen molecule by the trypsin-like action of the thrombin contains some 3 per cent of the original protein nitrogen. The residual fibrinogen fibrils then aggregate and enmesh to form a three-dimensional, cross-linked fibrin gel. A transluscent colorless gel is produced from plasma, but in blood clotting the gel is red because of entrapped blood cells. On standing, the gel contracts (syneresis) to form a tougher mass, extruding serum in the process.

The plasma concentration of fibrinogen is about 200 to 400 mg./100 ml. (0.2 gm. to 0.40 gm./100 ml.); none is present in serum, since fibrinogen is removed as fibrin in the clotting process. Elevations of fibrinogen levels up to 700 mg./100 ml. are encountered in many inflammatory diseases, such as *rheumatic fever*, *pneumonia*, *septicemias*, and *tuberculosis*. An increase in the erythrocyte sedimentation rate (ESR) is intimately associated with increased plasma fibrinogen levels.

Decreased levels of plasma fibrinogen are rather uncommon. In *congenital hypofibrinogenemia*, very little fibrinogen is present because of a rare genetic defect that is reflected by the inability of the liver to synthesize fibrinogen. The condition of *acquired hypofibrinogenemia* is of most interest in clinical work. Low levels of fibrinogen may be observed as a consequence of severe liver disease, which is also associated with low levels of several other clotting factors, such as prothrombin, and Factor VII. However, low levels of fibrinogen are not clinically important in most patients with severe liver disease because liver disease that produces severe hypofibrinogenemia usually leads to rapid decline and death by other causes. The most serious cases of low fibrinogen levels are encountered in certain complications of pregnancy, and it is in connection with these conditions that the laboratory is most often asked to assay fibrinogen levels, frequently on an emergency basis. In premature separation of the placenta (antepartal hemorrhage), the high levels of thromboplastic agents present in the placenta are released into maternal blood, resulting in a rapid conversion of fibrinogen to fibrin in the blood, placenta, and other organs. This eventually depletes the fibrinogen stores and results in severe hemorrhage and often shock and death. A similar situation occurs as a result of fetal death in utero, although the process is slower. These conditions are emergencies, requiring prompt and appropriate treatment, which may include intravenous administration of fibrinogen in large doses. Plasma fibrinogen may fall to concentrations as low as 50 to 70 mg./100 ml. Severe fibrinogen depletion can also occur as a complication of lung and prostatic surgery.

METHODS FOR THE DETERMINATION OF FIBRINOGEN

One group of quantitative methods,[69,70] resorts to isolating the fibrinogen (as fibrin) from the other plasma proteins. Clotting is carried out by diluting the plasma with saline and adding either exogenous thrombin or excess Ca^{++} ion in a quantity sufficient to overcome the anticoagulant present. The isolated, washed fibrin clot is then assayed by one of the available protein methods: measurement of Kjeldahl nitrogen and conversion to protein by the factor 5.95 (16.9 per cent N), or re-solution of the clot in NaOH and measurement of protein with either the biuret, Folin-Ciocalteu, or Lowry copper-phenol methods, or measurement of ultra violet absorbance at 280 nm. There is considerable disagreement over the tyrosine equivalent of fibrinogen as measured by the Folin reagent; therefore, the biuret and copper-phenol methods are preferred.

The other group of quantitative methods[64,68] is based on adding sufficient salt solution to cause precipitation of the fibrinogen but not of any of the other plasma proteins, and measuring the turbidity of the suspension. The plasma is diluted 1:10 or 1:20 with the salt solution; 12.5 per cent Na_2SO_3, 10.5 per cent Na_2SO_4, and 13.4 per cent $(NH_4)_2SO_4$ are examples of precipitants used. These turbidimetric methods can be carried out rapidly and can provide clinically valuable results in a short time in cases of acute emergencies; however, the turbidity methods do not have the specificity of the methods based on fibrin isolation, and results obtained with them in the concentration range below 0.10 gm./100 ml. tend to be unreliable. As examples of methods for fibrinogen determination, a biuret procedure based on fibrin isolation and a turbidimetric procedure are presented.

Determination of Fibrinogen by Fibrin Isolation and Assay with Biuret Reagent

REAGENTS

1. Saline, NaCl, 0.90 per cent (w/v) in water.

2. Thrombin solution, 500 NIH units/ml. Add sterile saline with a sterile syringe and needle to a vial of 5000 units Thrombin-Topical. Mix gently to dissolve. The solution is stable for 3 to 4 weeks, if refrigerated. Withdraw aliquots using sterile technique.

3. NaOH, 3 per cent (w/v), 0.75 *N*.

4. Phosphate buffer, 0.2 *M*, pH = 6.3 to 6.4. Dissolve 1.85 gm. of KH_2PO_4 and 0.91 gm. of Na_2HPO_4 in water and dilute to 100 ml.

5. Ground pyrex glass. Obtain about 30 to 50 cm. of 3 to 4 mm. diameter Pyrex tubing and cut into 1 to 2 cm. pieces, place in a mortar, and grind with a pestle until pieces about 0.5 to 1.0 mm. in size are obtained. Suspend the ground glass in about 400 ml. of water, stir vigorously to suspend the particles, let settle for several seconds, and decant the very fine particles. Repeat the removal of the fine particles several times. Wash the glass particles with 0.05 *N* NaOH and 0.10 *N* HCl, and then rinse with water by decantation until free of acid. Collect on a filter paper, air dry, and transfer to a clean dry test tube.

6. Standard. Determine the protein content of a pooled serum specimen with an established method after diluting it 1:5 with saline. A commercial control serum of verified protein content may also be used.

PROCEDURE

1. The specimen consists of plasma, obtained by centrifuging blood containing oxalate or citrate as anticoagulant. Run unknown plasmas in duplicate.

2. Place about 0.50 gm. of ground glass into two 15 × 125 mm. test tubes, using the end of a spatula. Add 8.0 ml. of saline and 1.0 ml. of phosphate buffer to each of the tubes, followed by 0.50 ml. of the plasma to be analyzed. If a low level of fibrinogen is expected, use 1.0 ml. of plasma. Mix well by vigorous tapping.

3. Rapidly add 2 drops (0.10 ml.) of 500 unit thrombin solution to each tube, using a 1 ml. sterile syringe and needle. Mix immediately by tapping vigorously enough to suspend most of the ground glass several centimeters up into the solution in the tube. If more than one unknown is being analyzed, treat only two tubes at a time with thrombin.

4. Let the tubes set for 5 minutes; clotting is completed when a solid gel forms. Tap to loosen the gel from the sides of the tube and let set for another 5 to 10 minutes.

5. Centrifuge for 5 minutes at 2500 rpm. The glass particles enmeshed in the clot will force the clot to sediment to the bottom of the tube in the form of a clot-glass pellet. Decant off the supernatant. With care, practically all of the fluid can be poured off without disturbing the clot-glass particles. If some gel is still present, insert a thin glass rod and squeeze and press out any fluid in the gel and press the gel against and into the glass particles.

6. Wash the clot-glass particles by adding saline and resuspending the particles by vigorous shaking. Saline is added down the sides of the tubes while they are hand-rotated to wash the sides. Centrifuge, decant supernatant, and repeat the wash procedure a second time. Decant the saline as completely as possible.

7. To each tube add 2.0 ml. of 3 per cent NaOH. Place the tubes into a boiling water bath and heat for 15 minutes to dissolve the fibrin. Shake the tubes every few

minutes to aid solution. After heating, cool the tubes to room temperature and mix.

8. Working standards and blank. Pipet 0.20 ml. and 0.40 ml. each of the diluted standard to tubes and add 1.80 ml. and 1.60 ml., respectively, of the 3 per cent NaOH. The blank tube contains the powdered glass plus 2.0 ml. of the NaOH.

9. To all tubes add 4.5 ml. of working biuret reagent, mix, and allow the color to develop for 15 to 20 minutes. Read the absorbances of the samples, standards, and blank against an instrument blank containing 2.0 ml. of NaOH and 4.5 ml. of working biuret reagent at a wavelength of 550 nm. If any opalescence is present, centrifuge the tubes before reading their absorbances.

CALCULATIONS AND COMMENTS

If the diluted serum standard contains Z gm./100 ml. of protein, the two working standards will represent $Z/500$ and $Z/250$ gm. of protein in the cuvet volume of 6.5 ml. The unknown cuvet will contain P/100 gm. of fibrinogen if the unknown has P gm. fibrinogen/100 ml. and if 1.0 ml. of plasma is used for the assay. If the unknown is compared with the 0.20 ml. standard, then

$$\frac{P}{100} = \frac{A_u}{A_s} \times \frac{Z}{500} \quad \text{or} \quad P = \frac{A_u}{A_s} \times Z \times \tfrac{1}{5}$$

where P = fibrinogen concentration in gm./100 ml.,

A_u and A_s = absorbances of the unknown and standard, respectively

Similarly, if 0.50 ml. of unknown is used and the absorbance is compared with the 0.40 ml. standard, then

$$P = \frac{A_u}{A_s} \times Z \times \tfrac{4}{5}$$

The reproducibility of the procedure is fairly good; the coefficient of variation is about 8 per cent. Some other proteins are partly occluded in the clot despite the washing procedures. The error here has been estimated at anywhere from zero to 20 per cent (in extreme cases). On the other hand, fibrinolysins may destroy (solubilize) some fibrin, and there is some loss on heating the fibrin clot with the dilute alkali. Occasionally, losses occur because of incomplete clotting, especially when low levels are present. In such situations, when no real emergency exists, allowing a longer time in the clotting step will improve the fibrin recovery. Fibrinogen is stable at room temperature for several days, and for several weeks in the refrigerator.

REFERENCES

Ratnoff, O. D., and Menzie, C.: J. Lab. Clin. Med., *37*:316, 1951.

Henry, R. J.: Clinical Chemistry; Principles and Technics. New York, Hoeber Medical Div., Harper & Row, Publishers, 1964.

Reiner, M., and Cheung, H. C.: Fibrinogen, *In* Standard Methods of Clinical Chemistry, D. Seligson, Ed. New York, Academic Press, Inc., 1961, vol. 3, p. 114.

Turbidimetric Determination of Fibrinogen

The reagent used contains 13.3 per cent $(NH_4)_2SO_4$ (w/v) in 1.0 per cent NaCl (w/v), with the pH adjusted to 7.0. Pipet 6.0 ml. of reagent to 0.50 ml. of plasma and measure the turbidity at 515 nm. after 3 minutes against a blank containing saline plus reagent. Construct a standard curve using purified fibrinogen (95 per cent clottable protein) dissolved in fibrin-free serum. This approach encompasses all the weaknesses of turbidimetric methods.

The floc is rather coarse and the suspensions are not as stable as desirable. In cases of severe bleeding, however, the method does provide rapid information to the physician.

REFERENCES

Parfentjev, I. A., Johnson, M. L., and Clifton, E. E.: Arch. Biochem., *46*:470, 1953.
Powell, A. H.: Am. J. Clin. Path., *25*:340, 1955.

Separation of Proteins by Electrophoresis

When dissolved in water, proteins exist as charged molecules or as aggregates of charged molecules (micelles). The ionization of carboxyl and imidazole groups will give rise to negative charges and the binding of protons onto amino groups to positive charges. The extent to which both types of reactions will occur will depend on the

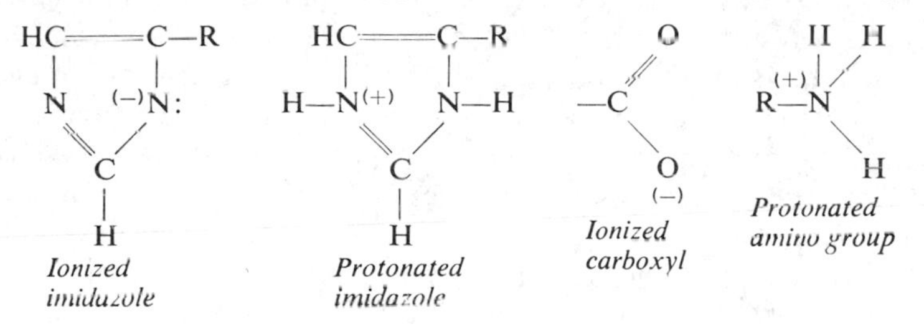

pH of the medium in which the protein is dissolved and the specific dissociation constants associated with the individual groups. In general both minus and plus charged groups will be found on any one molecule at the same time, but the given protein moiety will have either an overall plus or minus charge at any given pH. At the isoelectric point, both types of charges are present in equal quantity, and the protein shows minimum solubility at this pH. About the protein molecule there is found a layer of bound water, containing associated hydroxyl and hydrogen ions and other electrolyte ions, and these will also contribute to the net charge per unit area of the surface of the protein micelle. In an electric field the individual protein molecules will move in the direction of the electrode of charge opposite to that of the protein molecule itself; that is, a negatively charged molecule will move in the direction of the positive electrode or anode. This movement or migration of charged particles, relative to the solvent, under the influence of an electric charge is referred to as *electrophoresis*.

Electroendosmosis refers to the movement of the solvent in the opposite direction under the influence of an electric field, when the protein, paper, gel or other colloidal material is fixed and not free to move by itself. In paper electrophoresis both phenomena are observed.

The application of the principle of electrophoresis to clinical and biochemical problems became important with the work of Tiselius and Longsworth and their development of the moving boundary technique.[24,80] As a result of their work, electrophoresis became a precise laboratory tool and started the modern era of protein chemistry. In the moving boundary method the protein solution to be studied is dialyzed against an appropriate buffer. Extraneous ions are removed and the protein dissolved in the buffer of desired pH and known ionic strength. The solution is placed in a U-shaped, narrow, rectangular cell, and layered under a solution of the same buffer in such a way as to give a very sharp boundary between the buffer and protein solutions. The buffer solution is connected by side arms with the electrode chambers, which consist of an appropriate electrolyte and platinum or silver electrodes. The entire apparatus is cooled to 4°C. to minimize convection currents and diffusion, and the electrodes are then connected to a power supply. Under the influence of the electric field, the proteins, negatively charged at the working pH of 8.6, tend to migrate to the anode. The rate of migration will depend primarily on the charge density of the protein molecule—the greater the charge, the greater the velocity. The ionic strength of the buffer, the shape, and, to a smaller degree, the mass of the molecule will also have an effect on the speed of movement.

The *electrophoretic mobility* is defined as the velocity of migration in centimeters per second per unit field strength (volts/cm.). As a result of differences in mobility, after a period of time, some of the faster moving species will have left the original solution and entered the buffer phase, and the protein solution, initially homogeneous in composition, will now show bands of inhomogeneity. Such movement will take place in both arms of the U-tube in the same net direction, upward (ascending) in one arm and downward (descending) in the other. The changes in protein concentration along the tube as the result of the electrophoretic movement can be evaluated by measuring the refractive index of the solution, which will be a linear function of protein concentration. This is conveniently done by a special optical system, which presents the changes in protein concentration as a series of peaks and valleys, reflecting changes in the refractive index along the cell. The area under a peak will be proportional to the concentration of that component in the protein mixture. This is illustrated in Figure 5-1.

When a phosphate buffer of pH 7.4 is used, four peaks are obtained. The first, the fastest moving peak is the largest, and can be shown to contain protein having the properties of albumins. The other three peaks have the properties of globulins, and are labeled α, β, and γ globulins, in the order of diminishing mobility. It was

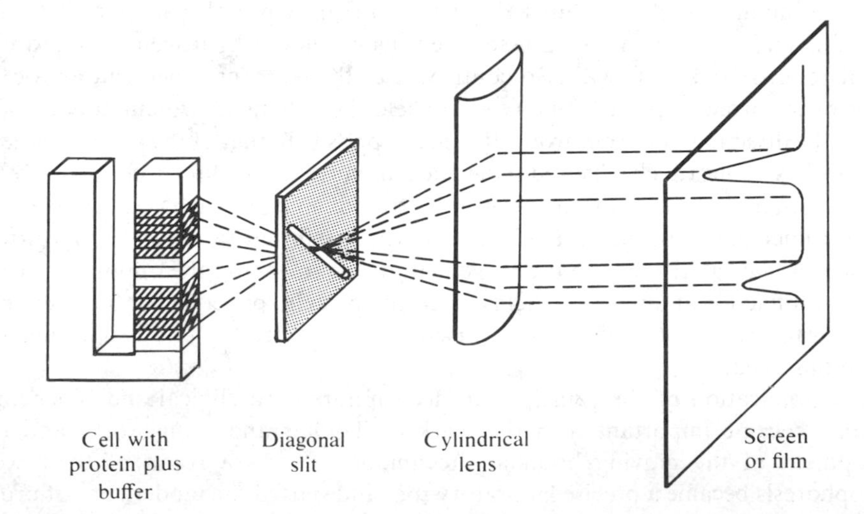

Figure 5-1. Simplified diagram showing the visualization of a pattern after an electrophoretic run using the cylindrical lens system. The migration of the protein components in the buffer in the cell gives rise to sections of changing refractive index (hatched areas), which are translated to two peaks (corresponding to the two areas of changing protein concentration) on the screen. (After Steiner, R. F.: The Chemical Foundations of Molecular Biology. Princeton, N.J. D. Van Nostrand, Inc., 1965; Gray, G. W.; Sci. Am., *185*:45, 1951.)

soon established that the size and mobility of the peaks and their number depended on pH, type of buffer, and ionic strength. When barbital (veronal) buffer at pH 8.6 was used, a fifth peak presented itself, between the albumin and the alpha peak; this was then referred to as the α_1 peak, and the original peak as the α_2 peak. Inasmuch as the most precise electrophoretic patterns of sera were obtained using pH 8.6 barbital buffer, the use of this buffer superseded the use of other buffers in clinical work. For special work different buffers and pH conditions are utilized to get separations otherwise not possible, for example, in the separation of some hemoglobin variants and in the isolation of glycoproteins.

Within a short time, several forms of the Tiselius apparatus were available from manufacturers; however the following factors mitigated against the technique becoming a routine clinical chemistry laboratory procedure: the requirement of a relatively large sample volume (2.0 to 5.0 ml.), the necessity for preliminary dialysis of the serum, the lengthy time interval required for the electrophoretic run, the inconvenience of only one specimen being examined at a time, and the high cost of the apparatus and specialized training required for its proper use. With the increasing use of paper as a chromatographic support medium, it was inevitable that efforts would be made to use paper as a supporting framework through which electrophoretic migration could be carried out. The initial difficulties were solved by Durrum[11] and other investigators[66] in the period from 1950 to 1953. Apparatus and techniques for successful electrophoretic separations of proteins and smaller ions became available, thus beginning the era of *zone electrophoresis*—electrophoretic separation through solid media. Within the next decade it was demonstrated that starch gel, agar gel, cellulose acetate, polyacrylamide, and cyanogum gel could also be used as supporting media. Separations were rapid, the technique and apparatus were relatively simple and inexpensive, and very small quantities of specimen were required. As a result, these techniques now make it possible for even a small laboratory to do electrophoretic studies of serum proteins and to use this laboratory procedure as a day to day routine clinical tool.

PAPER AND CELLULOSE ACETATE ELECTROPHORESIS

When paper is saturated with aqueous buffer solutions the fluid occupies the interstices between the cellulose fibers and flows or moves in and along these interlaced, zigzag channels. The rate of flow will depend on the size and shape of the channels and on the electrical charges present on the surface of the cellulose fibers. In practice the paper is soaked in appropriate buffer and then suspended between two supports.[66,84,93] The ends of the paper are in contact with paper wicks, which dip into a buffer compartment and provide electrical contact with the electrodes. The entire system is enclosed to permit saturation of the chamber with moisture to minimize evaporation from the surface of the paper strip. After a short flow of current to allow for equilibration, the current is turned off, and the protein specimen applied to the paper in the form of a linear streak across the paper. The current is then turned on and enough potential applied (100 to 400 volts) to allow for a voltage drop of 3 to 8 volts per centimeter of paper length (1.5 to 7.0 mamp./strip). The protein dissolves in the buffer on the paper and the various protein components begin to migrate in the buffer along the paper interstices toward the anode at rates proportional to their mobilities.

Endosmosis, however, complicates the migration; the water micelles about the negatively charged cellulose fibers carry a positive charge and, as the current flows, a fraction of the water tends to move toward the cathode (negative electrode), and in so doing carries the protein dissolved in it in the same direction. The net observed migration of the protein fractions is the resultant of both these effects. The motion is further distorted because of evaporative loss of water from the paper surface as a result of heat generated by the electrical current, this effect being greatest at about the middle of the paper strip. To minimize convection currents caused by heat generation, with resultant diffusion and smearing of patterns, the entire operation is best carried out at 4°C. (cold room or refrigerator), but room temperature is satisfactory.

After the electrophoretic run, the paper is removed rapidly (being kept taut or stretched out to avoid back diffusion of buffer), blotted dry, and then dried in an oven at 120 to 130°C. for a fixed period of time. This operation is critical because it denatures the proteins, and all fractions must be denatured completely. Unless the protein is colored (hemoglobin) the presence of the protein fractions cannot be seen on the paper. Visualization is accomplished by soaking the paper in a solution of some protein-binding dye. Among the dyes that have been used are *bromphenol blue*, *ponceau red*, *amido black 10B*, and *lissamine green*, the first two being used most frequently. After the staining procedure, excess stain or dye is washed off, and the paper is dipped into a fixative solution to fix the dyed protein onto the paper. The actual details for staining, washing, and fixing will vary with the dye used, but they must be adhered to very closely if reproducible patterns are to be obtained.

Quantitation is done in two different ways, by elution and by direct spectrophotometric scanning.[62,66,93] The first procedure is simplest and probably most accurate, although not as rapid or convenient. The various fractions are delineated, cut out with a pair of shears, and placed in tubes containing a fixed volume of dilute NaOH solution or some other appropriate solvent. After elution of the dye from the paper strips is complete, the paper is centrifuged off, and the amount of dye present is measured in a spectrophotometer. Using an appropriate calibration curve, the quantity of protein associated with the eluted dye can be ascertained. Some dye is bound by the paper itself, and a piece of the paper from an area containing no protein fraction is cut, eluted, and measured for dye content to serve as a correction for background.

In the scanning procedure, the paper is introduced directly into special types of spectrophotometers. The paper is made to move over a slit at a constant speed, and the absorbance of the dye on the paper along its entire length is determined. As the readings are made, the amplified signal from the detector governs the motion of a pen, which records a curve showing changes in absorbance with distance along the paper. This procedure can best be performed with a densitometer coupled with a built-in integrator, which will measure the area under the peaks of the scan. In the mechanism of the densitometer there may be included mechanical means (cams) to compensate for lack of linearity and for unequal dye binding by the various protein fractions. (See the discussion of accurate quantitation.) In the actual scan, the pattern formed is very similar to that produced with the moving boundary apparatus, a series of peaks and valleys, the areas under the peaks being proportional to the quantity of protein in that fraction.

Alternative, but less convenient, methods for quantitation include the following: (1) to actually count the squares (graph paper) under each peak, (2) to measure the area by use of a planimeter (a device for measuring area under curves), or (3) to actually cut out the patterns for the various protein fractions and to weight them on an analytical balance, the weights found being proportional to the protein content.

The problem of accurate quantitation is not as simple as just outlined. No dye binds equally, mol per mol, to each type of protein. Globulins bind less dye than do the albumins, and they differ in dye binding among themselves. The degree of binding will depend on the extent of denaturation of the proteins when the paper strip is oven dried. The quantity of dye associated with protein at any given area of the paper strip will also depend on how much protein is present at that point. If the protein layer is thick, the dye will have difficulty penetrating the top layers of protein and dye to reach the lower layers. The degree of binding will also depend on the time allowed for staining and on the properties of the given stain preparation.

Another difficulty encountered is that of *trailing*, especially in the case of albumin. As the fast moving albumin migrates, not all molecules are able to move at optimal speed. Some are denatured and precipitate out along the migration path. Others find "barriers," as a result of local properties in the paper. As a consequence, along with the main portion of albumin concentrated at its peak, some albumin is found over the entire length of the paper. This is evaluated as being part of any given globulin fraction, along with the true globulin located at that given spot. The extent of trailing encountered depends, in part, on the properties of the paper used, on temperature, and on the type and ionic strength of the buffer used. Trailing can be reduced by adding nonionic detergent to the buffer solution.

Further problems arise when strips are scanned. Dye does not combine with the protein uniformly over the protein area. There are light holes present that decrease effective measured absorbance over any given protein fraction area. Ideally, stained areas should have the shape of uniform regular rectangles. In practice, however, the areas have rough edges and are irregular ovals; thus the actual location of the scanning slit over the paper has an effect on the absorbance readings obtained. In addition, the relationship between absorbance and dye concentration is not linear, the degree of deviation varying with the dye used and the technique being followed. Surprisingly, however, the results obtained agree well with those obtained with the classic Tiselius moving boundary technique.

It is important to appreciate that the paper electrophoresis technique is to a very great extent an empirical methodology. The actual details of procedure will vary with the type of apparatus being used and the efforts that investigators using the apparatus have made in working out conditions leading to reasonably correct and reproducible results. It is important that all details of described procedures, as given by the instrument manufacturers or investigators, be followed precisely.

The sample size is usually in the order of 5 to 10 μl., containing about 0.3 to 0.8 mg. of protein. If the protein content of the specimen of serum is considerably above or below the normal range, the volume used should be adjusted accordingly. If electrophoretic patterns on urine or spinal fluid specimens are desired, these fluids will usually have to be concentrated by ultrafiltration either through special membranes under positive pressure (N_2 gas at 15 atmospheres) or into a vacuum (negative pressure). Much less satisfactory is osmotic concentration, in which the specimen is dialyzed against concentrated solutions of sucrose, Carbowax 2000 or PVP (polyvinylpyrrolidone).

Special applicators are available commercially, consisting of two fine parallel wires held in a small frame. The specimen is transferred to the wires, where it is held by surface tension as a thin uniform film. When this is applied to the paper, a much more uniform application is obtained than is possible with a pipet streak. A variety of papers are available, each manufacturer selling or recommending a particular brand or type that is best for his apparatus. All filter paper manufacturers offer brands specially processed for paper chromatographic work. Whatman No. 1 and No. 3MM are types commonly used. The strips used vary from 8 to 25 mm. in width. The paper is usually suspended horizontally in the cell, and is held as taut as possible. In the Beckman apparatus, following the example of Durrum, the paper is suspended in the form of an inverted "V."

Since, in addition to the advantages, there are disadvantages to both procedures, investigators kept searching for other materials for support vehicles. In 1955 Smithies[77] introduced *starch gel*, and Kohn[50] demonstrated the advantages of using *cellulose acetate* strips in 1957. Later agar, cyanogum, and polyacrylamide gels were also introduced. These materials have a more uniform structure than paper, and this provides much faster separations (30 minutes to 2 hours, as against 16 to 24 hours), producing sharper fractions and often separating serum proteins into subfractions. As a result, within the last several years, paper has been replaced by these other media, particularly cellulose acetate strips.

A group of cellulose acetate patterns obtained by Tietz[89] with the Beckman Microzone system are presented in Figure 5-2. The five protein bands are clearly shown in the pattern obtained with normal serum (Metrix control serum). The changes in the relative proportions of the various fractions, which are a result of several disease states, are evident in the other seven patterns.

The values for the various fractions on cellulose acetate are essentially the same as those obtained with paper. The elements of technique[5,25,78] just discussed, except for those relating to individual properties of the media themselves, are very similar to those described for paper.

Starch gel, polyacrylamide, and cyanogum separations can result in 8 to 20 different fractions as against the customary five, but this can be a source of inconvenience and confusion, insofar as it is difficult to assign the various subfractions to one of the common fractions, especially in the β and γ regions. The number of fractions obtained will vary with the buffer used, TRIS buffer giving especially good resolution. The various subfractions are real in the sense that serum contains proteins with a full spectrum of mobilities. At least three α globulins, two β globulins, and three γ globulin fractions can be obtained, as well as a band moving faster than the main albumin area, the so-called prealbumin band. In *immunoelectrophoresis*,[104] using agar gel, it is advantageous to have as complete a separation of individual

---

Figure 5-2. A set of cellulose acetate electrophoretic patterns obtained with the Beckman microzone system. A, albumin; B, α_1 globulin; C, α_2 globulin; D, β globulin; and E, γ globulin region.

Pattern 1, *Normal* serum (Reconstituted Metrix Control Serum). Pattern 2, *Chronic infection* (hepatitis): a relative decrease in albumin with a notable elevation in γ globulin. Pattern 3, *Destructive lesion* (acute reaction): pattern associated with a myocardial infarction with relative increased levels of α_2 and α_1 globulins. Pattern 4, *Hypogammaglobulinemia*: the considerable drop in γ globulin is readily observed. A slight increase in α_2 globulin is also seen on this pattern, but is not a common finding in hypogammaglobulinemia. Pattern 5, *Nephrotic syndrome*: the very low level of albumin is striking, as is the extremely elevated peak for α_2 globulin and the moderate rise in β globulin. Pattern 6, *Cirrhosis*: the elevation in both γ and β globulin, with a partial fusion of these bands, is present, along with a decrease in albumin. Pattern 7, *Multiple myeloma*: the clone of myeloma proteins is associated entirely with the β globulins. Pattern 8, *Multiple myeloma*: the myeloma protein is entirely in the γ globulin area. The fall in albumin is only apparent. It is interesting to compare this γ globulin region with those in patterns 2, 4, 6, and 7. Note the homogenous peak in patterns 7 and 8 and the heterogenous peak in pattern 2.

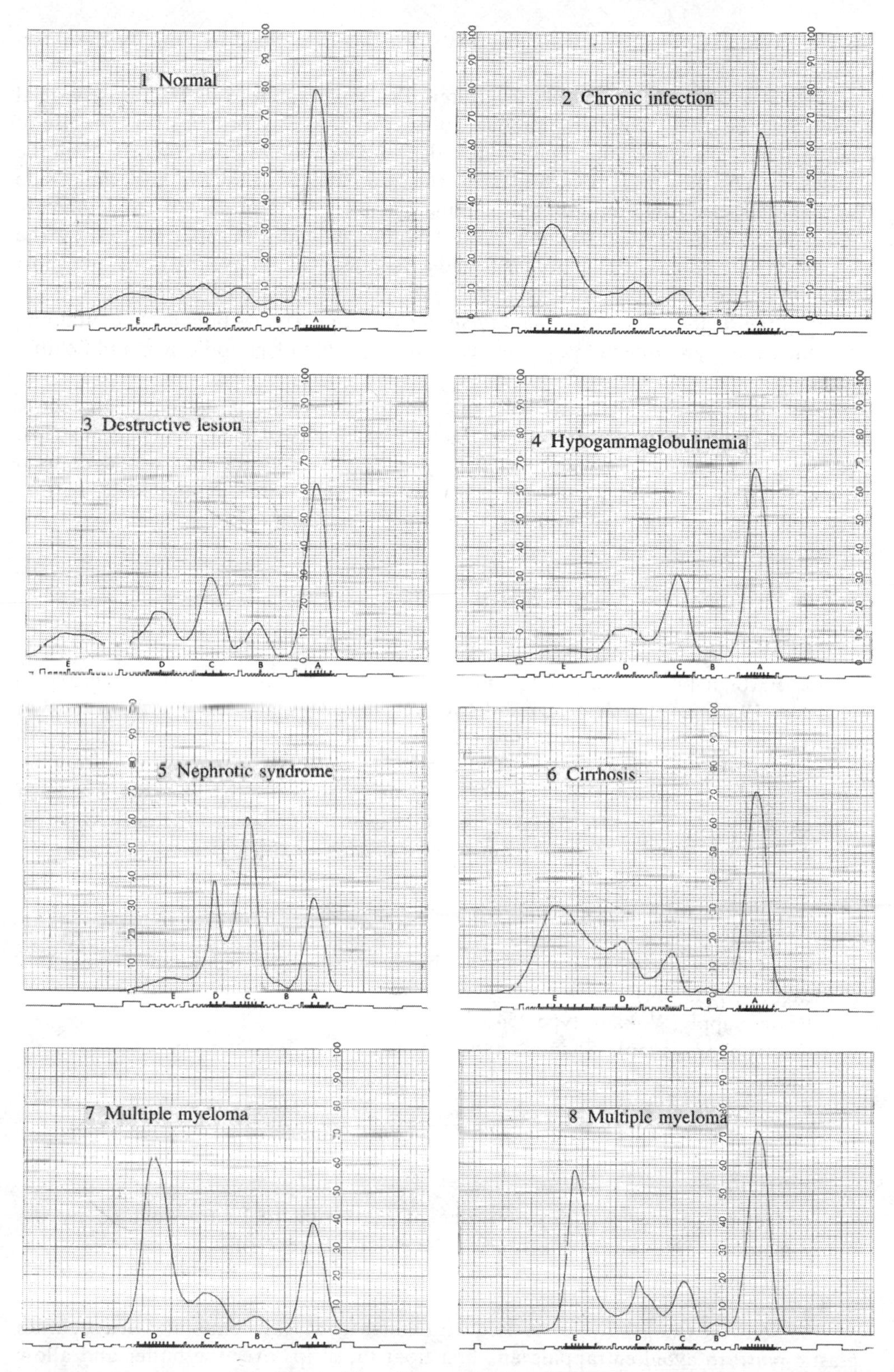

Figure 5-2. *See opposite page for legend.*

protein types as possible, but for routine work this is avoided to simplify the interpretation of the patterns.

Once a serum has been electrophoretically separated into its fractions, special stains may be used to show the presence of special classes of proteins. By the use of sudan Red III or some other lipid-soluble dye, the location of lipoproteins[22,82] on the pattern can be shown. It is customary to run two patterns of the same specimen side by side in the electrophoretic apparatus; one is stained with a protein dye to demarcate the location of the protein fractions, and the other stained with sudan III to show the lipid containing areas. Some lipid is found associated with all five fractions, but most is found in two areas, one between the α_1 and α_2 globulins, the α *lipoproteins*, and the other in the β globulin region, the β *lipoproteins*. Similarly, carbohydrate containing proteins, the *glycoproteins*,[43] can be localized by the use of periodic acid and Schiff's reagent.

As an illustration of the general method of zone electrophoresis[78] a procedure using cellulose acetate[5,51] is presented. The method uses Colab equipment and strips, but the overall procedure is not significantly different from one that would be used with other commerically available apparatus. The strips are quantitated by cutting and elution, but could be read using a scanning densitometer.

Serum Electrophoresis Using Cellulose Acetate

REAGENTS

1. Barbital buffer, pH = 8.6, ionic strength = 0.05 (per C. J. Brackenridge[5]). Dissolve 19.6 gm. of anhydrous sodium acetate, 50.0 gm. of the sodium salt of diethyl-barbituric acid, 34.2 ml. of 1.0 *N* HCl, and 3.84 gm. of calcium lactate in water and make the volume to 10 L. Add 50 ml. of 5 per cent (w/v) thymol in isopropanol as a preservative. Check the pH using a pH meter to verify that it is 8.6 $\pm$ 0.1; if not, carefully add concentrated HCl or NaOH to bring the pH to the desired value.

2. Staining solution, 0.2 per cent (w/v) ponceau S dye in 3 per cent (w/v) trichloroacetic acid.

APPARATUS

The Colab equipment listed is available from Consolidated Laboratories, Inc. (Chicago Heights, Ill.)

Power supply, Vokam, type 2541.
Electrophoresis tank, Shandon type.
Electrophoresis wicks (1-3/4 $\times$ 4-1/4 inch).
Plastic strip holders.
Oxoid cellulose acetate strips, 12.5 $\times$ 2.0 cm.

PROCEDURE

1. Determine the total protein value of the serum to be fractionated by the biuret method. This can be done during the electrophoretic run.

2. Fill the Shandon tank with 850 ml. of buffer. Set the bridge gap to a width of 8 cm.

3. Soak the Oxoid strips to be used with buffer for at least 15 minutes in a separate plastic pan; to avoid entrapping air, first float the strips over the buffer and allow them to fall into the buffer when saturated.

4. Blot the strips between sheets of filter paper to remove excess buffer.

5. Place single wicks along the bridge of the tank. Then place the Oxoid strips onto the wicks, across the bridges, making sure that they are parallel to each other and to the edges of the tank. Take care that the strips and wicks are in intimate contact with each other to insure good current flow. Between one and eight strips can be run at one time. If less than a full complement of strips is used, they are distributed uniformly over the width of the tank.

6. Place the curved holders against the outside edges of the bridges to insure tight contact between the strips and wicks and to hold the strips level and taut.

7. Apply about 10 μl. of serum to each strip approximately 3 cm. from the cathode end of the strip. Streak the serum in a micropipet across the strip with a steady but rapid motion, starting about 0.5 cm. from the side of the strip to within 0.5 cm. of the other side. With some practice it is possible to draw a smooth, uniform streak with no difficulty.

8. Place the slanted lid over the tank immediately after completing application of the serum specimens onto the strips. The slanting lid assures that any moisture that may have evaporated during the course of the run and condensed on the lid will flow to the edge of the tank, and not drop onto the strips.

9. Connect the electrical leads from the tank to the power supply, taking care to insure the correct polarity of the connections. It is wise to alter the direction of current flow on alternate runs to minimize chemical changes in the two buffer compartments. The point of application of the serum must be governed by the projected direction of current flow.

10. Adjust the current on the power supply to give 1.5 mamp./2 cm. width of Oxoid strips carrying the current. For a full complement of 8 strips a total current of 12 mamp. is used. This will require an initial voltage of about 200 volts, which will slowly drop to about 160 volts during the 90 minute run.

11. After the electrophoresis has proceeded for 90 minutes, turn off the power supply, rapidly remove the strips from the tank and place them in the ponceau S dye solution for about 5 to 10 minutes. During all processing handle the strips with plastic-tipped forceps.

12. Place the strips in 5 per cent (v/v) acetic acid to remove excess dye. The trichloroacetic acid in the dye will have fixed the stain onto the strips. Unlike paper, the cellulose acetate absorbs very little dye itself, and repeated washings to remove excess (nonprotein bound) dye are not necessary.

13. Place the strips on glass plates and dry in the oven at 90 to 100°C. until they appear brittle (about 40 minutes).

14. Examine the strips and mark off and cut out with a pair of scissors the zones containing the five protein fractions. Also cut out a piece of the strip about the same width to serve as a background blank. Insofar as possible, cut the fraction strips to about the same width, although the rather diffuse γ globulin zone may be considerably wider than the other fraction zones.

15. Place the cut, stained fraction zones into 12 mm. Coleman cuvets, to which then add exactly 3.0 ml. of 0.10 *N* NaOH. Tap the tubes to insure complete wetting of the Oxoid strip by the NaOH and then allow them to stand for 20 minutes to permit elution of the dye from the strips into the alkali.

16. Finally, add 0.3 ml. of 40 per cent (v/v) acetic acid to neutralize partially the NaOH. Mix the cuvets and read them in the Coleman 6D spectrophotometer against the blank tube at 520 nm. The eluted strips can be removed with a stirring rod, or

allowed to rest at the bottom of the cuvet, causing no interference in the absorbance readings.

CALCULATIONS

1. Add up the absorbance readings of all the protein fractions or zones to give the total absorbance, A_T, of all the protein on the strip.
2. Divide the absorbance, A_X, of each zone by the value for total absorbance and multiply by 100; this gives the per cent of the total protein found in that zone.
3. Divide the per cent value of each fraction by 100 and multiply by the value of the total protein; this gives the concentration of that fraction in gm./100 ml.

COMMENTS

It is helpful to run each serum specimen in duplicate. The patterns on the two strips should be identical on visual examination; if they are not, the patterns should be discarded and the procedure repeated. One of the two strips is cut for elution and the other preserved as such and submitted with the report to the physician or kept on file in the laboratory for examination by the physician.

A rather large sample (10 or 20 μl.) is used to permit elution of enough dye to give high enough absorbance readings when dissolved in 3 ml. of eluting solution. If the strips are to be read directly in a strip scanner and densitometer (the Photovolt Corporation instrument may be used), a much smaller application of specimen, of the order of 2 or 3 μl., is sufficient. The power supply can furnish current to two tanks at one time, and thus 16 patterns can be run in one operation, the entire procedure taking 2½ to 3 hours.

REFERENCES

Brackenridge, C. J.: Anal. Chem., *32*:1353, 1960; *32*:1359, 1960.
Kohn, J., and Feinberg, J. G.: Electrophoresis on Cellulose Acetate. Shandon Instruments Co., Applications Bull. 11, London, 1965.

NORMAL VALUES AND CLINICAL SIGNIFICANCE

A representative sample of results for the normal ranges of values for the various fractions is presented in Figure 5-2 and Table 5-4. These results were obtained over the last 15 years, by different investigators or laboratories, using different methods. The data are reported in per cent of the total protein found in any given electrophoretic fraction. Despite some evident differences, there is a general agreement among the various sets of data. The values for the globulin fractions vary more than do those for the albumin fraction.

There are few good studies of the precision attainable in doing replicates of the same serum by the same procedure. In one such study, Kaplan and Savory,[41] using cellulose acetate, found an average value of 61 per cent for the albumin fraction, based on 40 replicates, with a standard deviation, s, of 1.46 per cent, giving a coefficient of variation (C.V.) of 2.4 per cent. These data suggest a range of 58.0 to 64.0 per cent for the 95 per cent confidence limits for the albumin fraction. As can be seen by referring to the Table 5-4, this roughly corresponds to the range of values obtained for the serum albumin fraction in normal sera by the 7 laboratories for which ranges are given. Two conclusions are suggested by this data: that the degree of precision of the various methods is about the same, and that the observed range in

TABLE 5-4. *Values for Protein Fractions of Normal Serum Obtained by Various Types of Electrophoretic Techniques (Relative Concentration in Per Cent of Total Protein)*

Protein Fractions						
Albumin	*Globulins*					
	α_1	α_2	β	γ	*Method*	*Reference*
52.2–67.0	2.8–4.6	6.6–13.6	9.1–14.7	9.0–20.6	C.A.-scan	Kaplan and Savory[41]
55–64	3.6–7.2	7.6–10.1	11.6–15.2	10.1–17.2	M.B.	Ehrmantraut[12]
47–71	2.5–5.8	5.1–12.0	4.5–15.7	11.3–24.	Paper-scan	Ehrmantraut[12]
55–70	2.0–4.0	7.1–12.9	7.2–10.8	9.8–17.2	C.A.-scan	Grunbaum et al.[25]
55.7	7.8	9.9	11.1	15.3	M.B.	von Frijtag Drabbe and Reinhold[93]
58.9	4.3	9.0	10.3	17.3	Paper-scan	von Frijtag Drabbe and Reinhold[93]
58.6	3.9	8.0	12.5	17.6	Paper-scan	Klatskin et al.[48]
55–65	1.0–5.7	4.9–11.2	7–13	9.8–18.2	C.A.-elution	Kachmar[40]
54–70	2–5	7–11	8–14	10–20	C.A.	Sunderman[97]
53–65	2.5–5.0	7–13	8.0–14.0	12.0–22.0	C.A.-microzone	Tietz[89]
Absolute concentration in gm. protein/100 ml. serum. total globulin = 2.5–3.5.						
3.5–4.7	0.17–0.33	0.42–0.45	0.52–1.05	0.71–1.45		

Abbreviations:
C.A. = cellulose acetate
M.B. = moving boundary

albumin values may reflect the inherent degree of imprecision in the methods as much as it does the true physiological variation in the population.

For the α_1 globulin fraction, Kaplan and Savory[41] obtained a coefficient of variation (C.V.) of 14.2 per cent for an average value of 3.2 per cent of total protein. This corresponds to a 95 per cent confidence interval of 2.2 to 4.2 per cent protein. The C.V. values for the other three globulin fractions (α_2, β, and γ) were 6.0, 6.1, and 5.1 per cent, respectively. These correspond to 95 per cent confidence limits of 8.8 to 11.2 per cent for the α_2 globulins, and 9.7 to 12.4 and 13.1 to 16.2 per cent for the β and γ fractions. With abnormal sera the variation found was somewhat greater. In interpreting the results of electrophoretic fractionations,[12,38,78] very little clinical significance can be associated with small variations from the normal range of values. Serial patterns in intervals of several days may increase discrimination. They may show regular, progressive changes or relative constancy and thus be more meaningful than a single examination. Certainly, each laboratory should establish its normal range and the 95 per cent confidence limits of its data. If more than one technologist is entrusted with performing the separations, it should be established that all technologists can obtain essentially identical results, that is, approximately the same average value with the same standard deviation.

Reproducible values for the *lipoprotein* and *glycoprotein* fractions are even more difficult to obtain. Methods are far from standardized,[22] and the staining procedures are more difficult to perform. Lipophilic dyes will usually give a heavy stain in the β region, and a rather diffuse area of light staining in the α_1 and α_2 globulin areas and part of the albumin area. Straus and Wurm[82] found 60.2 per cent of the total lipid stain in the β fraction, 6.1 per cent in the α_2 fraction, and 19 per cent in the diffuse albumin-α_1 globulin portion of the pattern. The ratio of β lipid to α lipid usually varies from 2.5 to 3.0. All of the five protein fractions contain protein-bound carbohydrate. Kelsey[43] found the greatest quantity among the α_2 globulins, some 30 per cent of the total. This fraction includes the acute-reactive proteins,[63] which increase significantly (50 to 100 per cent) in acute infections and severe trauma. The α_1, β, and γ globulin fractions contain about 19 per cent each, and albumin the least of all, about 11 per cent.

Despite the initial hope that the ability to measure the changes in the individual serum protein fractions would facilitate clinical diagnosis, the actual experience over the last 15 years has been somewhat discouraging. Fractionation data, in certain instances, are certainly more useful than the rough albumin-globulin determinations, but often they convey no more useful information than the latter. Table 5-5 presents

TABLE 5-5. *Changes in Electrophoretic Patterns of Serum Proteins Associated with Various Disease States*

	Protein Fraction						
		Globulin					
Disease	*Albumin*	α_1	α_2	β	γ	*Glycoprotein*	*Lipoprotein*
Nephrosis	D, DD		E, EE		D	E	
Acute glomerulonephritis	±		E				
Acute infection	D, ±		E, ±		±		
Chronic infection	D				E	E, ±	
Rheumatoid arthritis	D, ±		E	±	E	E	
Hepatic cirrhosis	D		E±	±E	E, EE		
Viral hepatitis	D	D	D, ±	E	E		
Hypogammaglobulinemia					D, DD		
Obstructive jaundice	D, ±	±	E	E	±		E, ±
Myeloma				E±	E, EE	E	
Sarcoidosis	±		E	E	E, ±	E, ±	
Hodgkin's disease	DD	E±	E		E		E
Lupus erythematosus	D		E, ±		E	E	
Macroglobulinemia	D			E	E	E, ±	
Cryoglobulinemia					E		
Allergies	±		E		D, ±		
Diabetes	D, ±		E, ±	E±			
Leukemia, granulocytic	D				E, ±		

Symbols: D or E = decrease or elevation up to 25 per cent.
DD or EE = decrease or elevation over 25 per cent.
± = Marginal decrease or increase from normal.

a summary of the changes usually observed in various disease states. Numerical values are not presented because they vary within large limits. The symbol D represents a decrease in any given component up to 25 per cent below the normal range; DD represents decreases over 25 per cent. Similarly, the symbol E is used to represent an increase over normal, and EE increases over 25 per cent above normal. The symbol ± denotes marginal changes either above or below normal. Only in such disease states as nephrosis, multiple myeloma, hypo- or agammaglobulinemia, active postnecrotic cirrhosis, clinically active hepatitis, and sarcoidosis are the changes sharp enough to be helpful in suggesting a probable diagnosis from the values of the various fractions. Examples are presented in Figure 5-2. In these cases, simple visual examination of either the stained pattern on the strip or the scanned pattern may be sufficient. When other clinical information suggests that the patient's condition may be reflected in a significant change in one or more serum protein fractions, electrophoretic patterns should be obtained. Concurrent patterns for lipoproteins or glycoproteins, or both, may also be needed.

It must be kept in mind that in a number of specific diseases, the concentration of only one or a group of closely related specific proteins may deviate from normality.

Such a condition occurs in multiple myeloma, in which a clone of one cell type produces abnormal γ globulins in such large quantities that the increase is easily demonstrated. In most cases, however, the increase or decrease of any single protein species may not be large enough to affect the value of the entire fraction in which it is found electrophoretically, and the electrophoretic pattern may show no or at best marginal changes in the overall protein pattern. In the future, improved techniques for isolating and demonstrating changes in the concentration of individual, specific proteins may be expected to develop.

Electrophoretic patterns of *cerebrospinal fluid* have also become of interest, especially in the attempts to aid in the diagnosis of multiple sclerosis. In a normal CSF, the same five fractions are encountered as in serum,[36,65,76] in about the same proportions, with the exception that the percentage of total albumin is somewhat higher than in serum (62 $\pm$ 5 per cent of total protein). Significant increases occur primarily in the γ globulin fraction; in multiple sclerosis an appreciable increase will be found even when the total proteins are within normal limits. In bacterial and viral meningitis, a depression of albumin accompanies the γ globulin elevation.

Electrophoretic examinations of *urine* specimens have not been very useful; however, electrophoretic separation coupled with immunological identification of the individual protein moieties is proving to be very informative, especially in the identification of Bence-Jones protein.

The Ultracentrifuge and Its Application in Clinical Chemistry

If a suspension of sand in water is permitted to stand for a period of time, particles will settle to the bottom of the container, and a complete separation of the particles from the water will have occurred. The sedimenting force will have been the force of gravity acting on the relatively heavy sand particles and overcoming the resistive force of the viscosity of the water. If the particles are very fine and much lighter than sand, a considerably longer time will be required to effect a complete separation of the suspended particles from the suspending medium. If the particles are small enough, and, therefore, of very low mass, no visible separation may take place even after many days. Sedimentation may be speeded up by the use of the centrifuge, whereby the fixed gravitational force is replaced by a centrifugal force, which may be increased at will to a force equivalent to many times the gravitational force. Common laboratory centrifuges can produce centrifugal forces of 500 to 4000 gravities ($\times$ g), sufficient to effect separation of even very fine precipitates in the course of 10 to 15 minutes. The force obtained is governed by two factors: the radius of rotation, which is fixed for any given centrifuge head; and the speed of rotation, or angular velocity, which can be controlled and varied within the limits of the engineering capabilities of the instrument. If separation of particles of colloidal or molecular dimensions is desired, however, very high speeds of the order of 30,000 to 100,000 rpm are necessary, to produce the centrifugal forces of 50,000 to 200,000 ($\times$ g). Instruments capable of producing such high forces are referred to as ultracentrifuges,[23,80] but even with such powerful forces, runs of several hours or days may be required to obtain the separations or isolations desired.

With modern analytical ultracentrifuges, such as the Beckman Model E, it is possible to effect separation of soluble proteins because of the differences in the molecular weights of the protein micelles. Ultracentrifugation is used to demonstrate that a given protein solution contains at least two or more proteins of different molecular weights, and to calculate the molecular weights of dissolved proteins. In the *sedimentation velocity* technique, the solution of the protein in appropriate buffer is sedimented at very high velocities, and the progress of sedimentation observed through an optical glass slit in the rotor by means of a schlieren lens system. Initially, the solution has a uniform refractive index; however, as sedimentation proceeds and the solute is being forced toward the periphery of the sedimentation cell, the local packing of the protein results in a change in refractive index, which the schlieren lens will demonstrate as a peak, similar to that seen in moving boundary electrophoretic patterns. The area of the peak will be proportional to the quantity of protein present at that point. As sedimentation proceeds, the peak will move down the pattern. If two proteins are present, peaks reflecting both will be formed, the peak for the heavier protein first, and then the peak for the lighter protein. Different proteins are characterized by their *Svedberg*, or S, *numbers*. A material sedimenting at the rate of 1 Svedberg unit will sediment at a rate of 1×10^{-13} cm. per dyne per second per gram of material.

Both molecular weight and overall density of the protein will affect the S value obtained. The density of the protein will be governed in part by its quaternary structure and in part by the nature of the prosthetic groups bound to the protein. If the latter are lipids of a density less than that of water, the overall sedimentation rate may be less than that expected from the gross molecular weight of the molecule.

Serum albumins are classified as S_{4-5} proteins, inasmuch as the albumins, as a class, sediment at Svedberg values of about 4.3 to 5.4. The immune γ globulins (molecular weight = 165,000) that sediment with S values of about 6.8 to 7.5 are referred to as S_7 or 7S globulins. Other γ globulins (macroglobulin type), the so-called IgM globulins, are referred to as the 19S and 30S globulins. It should be emphasized that there is no correlation between the electrophoretic class to which a protein belongs and the sedimentation class into which it may fall.

The ultracentrifuge has been used clinically to demonstrate the presence of macroglobulins in patients with Waldenström's macroglobulinemia and in patients with myeloma. The resolution capable of being obtained with ultracentrifugation techniques is inferior to that possible with zone electrophoresis. This fact, in addition to the cost of the equipment, has limited the use of ultracentrifuges in the study of clinical problems. Outside of the use of sedimentation techniques in demonstrating macroglobulins, the only other extensive application has been in the study of plasma lipoproteins and their association with atherosclerotic disease. Indeed ultracentrifugal analysis of serum proteins remains a research rather than a clinical tool.

Serum proteins can be conveniently separated into three sedimentation groups, those with densities of less than 1.063, those with densities over 1.200, and an intermediate group with densities in the 1.063 to 1.200 range. Gofman[10,21] showed that the lighter fractions could be more conveniently separated by adding salt to the serum to increase its density to 1.063. The heavier fractions would sediment as before, but now the lighter fractions would rise and float over the main body of the serum plus salt solution. These fractions are characterized by their S_f or Svedberg flotation values, the S_f being the obverse of the usual Svedberg number. The serum protein fractions moving to the top of the solution (floating) are referred to as the light density fraction, LDF, and could be further separated into subclasses: $S_f = 0$ to 10, $S_f =$ 10 to 20, $S_f = 20$ to 400, and S_f over 400. These fractions, when separated and subjected to electrophoresis, migrate as β globulins and are referred to as β lipoproteins. They contain about 75 per cent of the protein-bound lipid. Although, the average molecular weight of the protein in this fraction is about 1,200,000, the high content of lipid (density = 0.80 to 0.85) counteracts the high molecular weight, with the result that the sedimentation properties are those of very light proteins.

The proteins with densities of 1.063 to 1.200 also contain lipid, and migrate electrophoretically with the α globulins. They are referred to as α_1 and α_2 lipoproteins and, particularly, as the high density fraction, HDF. Some of the properties of the LDF and the HDF are presented in Table 5-6.

The designation of some of the subclasses is rather arbitrary and has varied with investigators. Gofman[21] has emphasized that the $S_f = 12$ to 20 fraction is of the most interest, an increase in this class being most directly related to the degree of atherosclerotic disease and the incidence of myocardial infarctions. Another index that has been used is the ratio of β to α lipoprotein fractions (β/α ratio) in serum, which in young adults is about 2.5 to 3.0, and in persons with severe atherosclerosis can rise to 8.0 to 9.0.

The LDF and the HDF differ primarily in their content of triglycerides. The *chylomicrons* are lipid globules surrounded by a protein membrane; they can be rather large, up to 0.1 to 0.5 μ in diameter. When the blood specimen is drawn shortly after a heavy meal, the chylomicrons give serum a characteristic milky appearance (postprandial hyperlipemia). In health, none are present in a fasting blood specimen. After being transported by the blood and lymphatics to the liver, the triglyceride globules are broken down, and the individual fat molecules transferred to the α and β lipoproteins, with the resultant clearing of the plasma. The normal β proteins are those of the $S_f = 0$ to 12 class. As they bind triglycerides to transport them to

TABLE 5-6. *Properties of Serum Lipoprotein Fractions*

	Low Density Fraction (LDF)					*High Density Fraction (HDF)*		*Albumin*
	S_f designation							
	400–1000	100–400	20–100	12–20	0–12	α_1	α_2	
	Chylomicrons	β-lipoproteins				α-lipoproteins		
Concentration of fraction in serum (mg./100 ml.)								
	0–10		160	40	280	40	240	4000
Composition of fractions in per cent of total mass								
Protein	2–5	7		11	21	33	57	99
Triglycerides	87–99	52		26	10	11	6	0.4
Cholesterol	2	13		45		19		
Phospholipids	7	20		25		26		
Density	1.00	0.98–1.02		1.007		1.063–1.107	1.107–1.220	

the tissues or fat depots, they become LDF fractions of the 12 to 20, 20 to 100, and 100 to 400 classes, depending on the quantity of triglyceride being transported. After depositing the fats, they revert back to the $S_f = 0$ to 12 class. Cholesterol and its esters are similarly transported; the phospholipids aid in carrying other lipid classes. The role of the α lipoproteins is still not clear; their concentration in serum is relatively constant, whereas that of the β lipoproteins varies diurnally and with age.

The Gofman flotation technique as applied to serum proteins is a useful diagnostic tool in the study of patients with atherosclerosis, heart disease, diabetes, and diseases of lipid transport; however, the technique is too cumbersome for the routine clinical workup of patients.

REFERENCES

De Lalla, O. F., and Gofman, J. W.: Ultracentrifugal analysis of serum lipoproteins. *In* Methods of Biochemical Analysis. D. Glick, Ed. New York, Interscience Publishers, Inc., 1954, vol. 1, p. 459.

Hoffman, W. S.: The Biochemistry of Clinical Medicine. 3rd ed. Chicago, Year Book Medical Publishers, Inc., 1964.

Cantarow, A., and Schepartz, B.: Biochemistry. 4th ed. Philadelphia, W. B. Saunders Co., 1967.

Ultrafiltration

Ordinary filtration is used to separate small but visible particles suspended in some solvent or solution. The original suspension is poured into a funnel, which holds a filter paper or membrane containing pores or openings small enough to prevent passage of the suspended particles, but large enough to allow rapid flow-through of solvent or solution. The smaller the particles to be separated, the smaller must be the average size of the pores in the paper. In *ultrafiltration* the pore size is made so small that macromolecular solutes can be separated from the solvent, electrolytes, and other solutes. The size of the macromolecule that can be separated will depend on the pore size. In the most common laboratory application of ultrafiltration, it is desired either to separate a protein from an associated ion or to affect a concentration of a protein solution.

Ultrafilter membranes are usually prepared from collodion or viscose. Such membranes are permeable to all small solutes, but impermeable to proteins of a molecular weight of 20,000 to 40,000 or higher. Micropore filter membranes are now available from a number of manufacturers; a large variety of pore sizes are offered so that proteins of different molecular size can be separated from one another. Because of the small pore size, the flow of water or solution through the ultrafilter is very slow. To accelerate the process, filtration is usually carried out under positive or negative pressure. A common application of micropore filters is the removal of small particles, bacteria, and viruses from solutions and serum (sterilization).

The recent development of *molecular sieves* provides the clinical chemist with a novel, modified form of ultrafiltration. One such material is Sephadex (Pharmacia Fine Chemicals, Inc., Piscataway, N.J.), a polymeric form of dextran, prepared as fine globules, containing very small pores. If a solution is poured through a column of the material, the solvent and small molecules can enter the pores and be temporarily entrapped within the dextran particles, whereas the larger protein molecules, too large to enter the pores, flow through and can be collected free of the small ions and molecules. Addition of eluting solvent will then free the smaller solutes from inside the pores, and these can then be collected, separated from the protein originally present.

Glycoproteins and Mucoproteins

Conjugated proteins containing a complex carbohydrate moiety linked to the protein chain are referred to as *glyco-* and *mucoproteins*.[6,7,28] A large number are known with considerable individual variation in composition and properties. The carbohydrate chain linked to the protein is made up of the *hexoses*, galactose or mannose; the *hexosamines* (amino sugars), glucosamine or galactosamine; the *methylpentose*, fucose; and a *sialic acid*, such as N-acetylneuraminic acid. The sialic acid is located at one end of the carbohydrate chain; the other end of the chain is linked by a peptide bond to one of the peptide chains of the protein molecule.

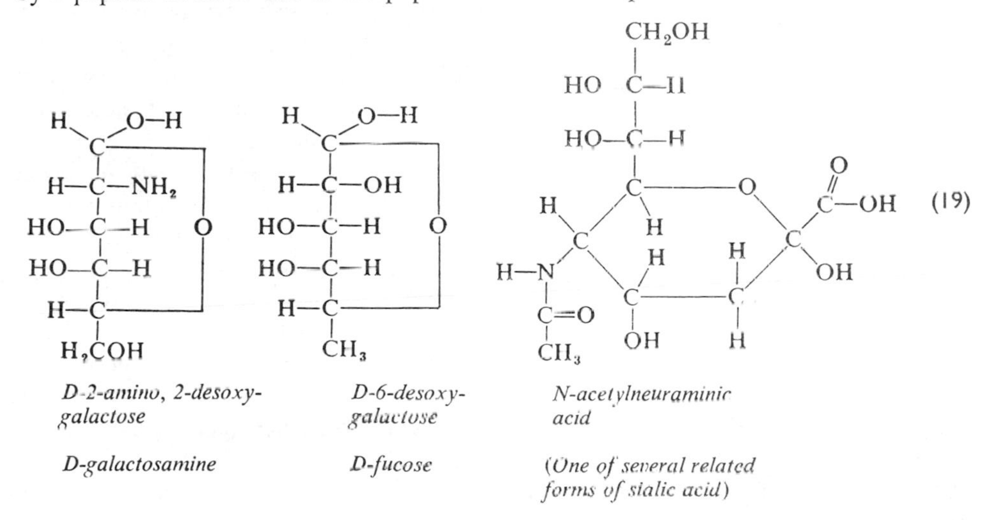

(19)

D-2-amino, 2-desoxy-galactose — *D-galactosamine*

D-6-desoxy-galactose — *D-fucose*

N-acetylneuraminic acid — *(One of several related forms of sialic acid)*

In a useful classification, those carbohydrate-proteins containing less than 4 per cent of hexosamines are termed *glycoproteins*. They contain from a trace up to about 15 per cent of total carbohydrate. Those with over 4 per cent of hexosamines and containing from 10 to 75 per cent of carbohydrate are termed *mucoproteins*. The expression *mucoids* is also used in characterizing those mucoproteins in which the bond linking the carbohydrate chain and the protein is split only with difficulty, in contrast to another group containing acidic mucopolysaccharides, in which the linkage is polar and easily split. The glyco- and mucoproteins differ from the albumins and globulins in that they are not precipitated by perchloric and sulfosalicylic acids.

Mucoproteins of clinical interest[6] are found in both serum and urine. Electrophoretically, most mucoproteins migrate with a mobility associated with α_1 globulins, although individual proteins may be found among all the usual electrophoretic groups. They are identified on electrophoretic strips or gels by their capacity to give a positive staining reaction with Schiff's reagent. The characteristic pink color is due to a reaction between the carbohydrate portion of the mucoprotein and the reagent. The mucoid present in serum in greatest amount is *orosomucoid;* it has a molecular weight of about 40,000 and is quite acidic, possessing an isoelectric point at pH 2.3. A number of other mucoproteins are also present in varying amounts.

Among the true glycoproteins are such important serum proteins as *transferrin*, *ceruloplasmin*, *haptoglobin*, and *prothrombin*. Although these proteins can best be identified by use of immunoelectrophoretic techniques,[104] they are customarily assayed by taking advantage of specific properties of the individual proteins, such as iron binding in the case of transferrin. *Mucins*, composed of mucoproteins and mucoids, are present along with mucopolysaccharides in mucus, a secretory material elaborated to lubricate the linings and cavities of many organs. A small amount is normally present in urine; increased amounts are excreted as a result of inflammatory conditions present in the genitourinary tract.

Mucoproteins constitute about 1 to 2 per cent of all total serum proteins. The range of concentration reported as present in normal serum depends on the method of measurement used. The most accepted range is 75 to 135 mg./100 ml. (average value = 120 mg./100 ml.). This includes orosomucoid and several less well defined components, plus a portion of the glycoproteins. Increases in serum levels are associated with the presence of inflammatory and proliferative diseases such as acute infections, rheumatoid arthritis, tuberculosis, carcinoma, and lymphosarcoma. This increase, however, presupposes normal liver and adrenal function, inasmuch as serum mucoprotein levels are depressed in acute liver disease, cirrhosis, and adrenal insufficiency. As a result, normal levels may be encountered even in patients with demonstrated acute reactive disease.[63] Because of the large spread in normal values, the inadequacy of available analytical techniques, and the nonspecific nature of the processes leading to elevated values, the determination of serum mucoproteins is of limited value in clinical diagnosis. Such determinations may provide useful corroborative data to the physician. In rheumatoid arthritis, values can range from 100 to 400 mg./100 ml. and in acute infections from 120 to 450 mg./100 ml.

Serum mucoproteins are isolated from other serum proteins by virtue of their solubility in perchloric acid, which precipitates other serum proteins. They are then precipitated with phosphotungstic acid. Following this, the mucoproteins can be assayed by measuring their total carbohydrate content by use of orcinol or diphenylamine, or by measurement of their protein content by the Folin, biuret or Kjeldahl methods, or by a turbidimetric technique. If carbohydrate is measured, the result may be reported in terms of carbohydrate content or converted to total mucoprotein by assuming 12 per cent carbohydrate in the protein. Methods are also available for assaying serum protein-bound hexose and serum hexosamines. The method for mucoproteins presented here is based on one devised by Winzler. Mucoproteins are separated and then measured by assaying their tyrosine content with the Folin-Ciocalteu reagent. It is assumed that the average mucoprotein contains 4.2 per cent tyrosine.

Determination of Serum Mucoproteins

REAGENTS

1. Perchloric acid, $HClO_4$, 1.80 *M*. Dilute 70 ml. of concentrated (72 per cent) $HClO_4$ to 500 ml. The reagent is stable.

2. Phosphotungstic acid, 5 per cent (w/v) in 2 *N* HCl. The reagent is stable.

3. Folin-Ciocalteu reagent. To a 1 L. glass, round bottom flask add 50.0 gm. of $Na_2WO_4 \cdot 2H_2O$, 12.5 gm. of $Na_2MoO_4 \cdot 2H_2O$, and about 350 ml. of water. Rotate the flask until the salts are dissolved. Then add 25 ml. of 85 per cent orthophosphoric acid, followed by 50 ml. of concentrated HCl. Attach a ground-glass reflux condenser (an all glass apparatus is used), insert the flask in a Glas-Col heating mantle, and let the contents of the flask be refluxed for 10 hours. Permit the flask to cool slightly, remove the condenser, and add 75 gm. of Li_2SO_4 along with 25 ml. of water

and 3 or 4 drops of Super-Oxol (30 per cent H_2O_2). Reheat the flask, allow it to boil for 10 to 15 minutes, and then filter. The reagent should be free of any greenish tint (if present, the addition of peroxide and boiling is repeated). Dilute one part of this stock reagent with two parts of water to prepare the working reagent. Store the reagent in a brown bottle.

4. Saline, NaCl, 0.85 per cent (w/v).
5. Sodium carbonate, Na_2CO_3, 20 per cent (w/v).
6. Tyrosine standard. Stock, 20 mg./100 ml. Dissolve 50 mg. of tyrosine in 250 ml. of 0.10 *N* HCl. The solution is stable if refrigerated. Working standard, 5 mg./100 ml. Dilute the stock standard 1:4 with 0.10 *N* HCl. Prepare the dilute standard fresh weekly.

PROCEDURE

1. To 0.50 ml. of serum add 4.50 ml. of saline, followed by 2.50 ml. of 1.8 *N* $HClO_4$, which is added slowly, dropwise with shaking of the tube. The conditions of precipitation must be maintained uniform and the temperature kept between 22 and 25°C.
2. After 10 to 12 minutes, centrifuge the specimen and filter it through a hardened filter paper (Whatman No. 50).
3. Pipet 5.0 ml. of the collected filtrate and transfer to another tube (13 × 100 mm.). Add 1.0 ml. of 5 per cent phosphotungstic acid (PTA), mix well, let stand for 10 minutes, and centrifuge in the clinical centrifuge at maximum speed (2500 to 4000 rpm).
4. Decant off the supernate. Wash the precipitate of mucoproteins by adding 2 ml. of PTA, breaking the precipitate, mixing it with a stirring rod, and then recentrifuging. Carefully pour off the supernate and invert the tube and permit it to drain dry.
5. Then add 1.0 ml. of 20 per cent Na_2CO_3 to the tube, shake it to aid solution of the precipitate, and add 3.50 ml. of water.
6. The blank consists of 3.50 ml. of H_2O plus 1.0 ml. of sodium carbonate.
7. Prepare the standards by pipetting 0.25 and 0.50 ml. of dilute (5 mg./100 ml.) tyrosine standard to two tubes, plus water, to make 3.50 ml., followed by 1.0 ml. of carbonate solution.
8. Add 0.50 ml. of Folin reagent to all tubes, mix, and warm in a water bath at 37°C. for 15 minutes.
9. Rapidly cool the tubes to room temperature and to each add 2.0 ml. of water, with mixing. Transfer the contents to cuvets, and read the absorbance level at a wavelength of 680 nm. Compare the absorbances of the unknowns with the closest matching standard.

CALCULATIONS

Since only 5.0 ml. of the 7.5 ml. of $HClO_4$ filtrate are used in the tyrosine assay, only 0.33 ml. of serum is used for the determination of the tyrosine value. The two standards contain 0.0125 and 0.025 mg. of tyrosine. The milligrams of tyrosine in the cuvet volume of the unknown are given by

$$\text{mg. tyrosine} = \frac{A_u}{A_s} \times C_s$$

and the concentration in the serum in mg./100 ml. is equal to

$$\frac{A_u}{A_s} \times \frac{100}{0.33} \times 0.0125$$

assuming that the weaker standard is used. This is simplified to

$$\text{mg. tyrosine/100 ml.} = \frac{A_u}{A_s} \times 3.75 \text{ (or } \times 7.5 \text{ respectively)}, \qquad (20)$$

depending on which standard is used. Accepting 4.2 per cent as the fraction of tyrosine in mucoprotein, the right hand side of equation (20) can be multiplied by 100/4.2 = 23.8 to convert the tyrosine values to protein values, giving

$$\text{mucoproteins in mg./100 ml. serum} = \frac{A_u}{A_s} \times 94 \text{ (or } \times 188 \text{ respectively).}$$

COMMENTS

Different investigators have reported somewhat differing normal values. Most normal sera will fall within the range of 75 to 135 mg./100 ml. The precision expected is about ±15 per cent. The results are probably low because a fraction of the mucoproteins is lost by coprecipitation with other proteins in the perchloric acid treatment. The separation of mucoproteins from the other proteins is at best a rough, empirical procedure, and reproducible results are obtained only by strict adherence to experimental protocol.

REFERENCE

Winzler, R. J.: Determination of serum glycoprotein. *In* Methods of Biochemical Analysis. D. Glick, Ed. New York, Interscience Publishers, Inc., 1955, vol. 2, p. 279.

Urinary Mucins—Detection and Measurement

Filter the urine specimen to be tested free of particulate matter and dilute with an equal volume of water. Use three tubes, putting 2 or 3 ml. of diluted urine in each. Put 5 to 6 drops of saline into the first tube, which serves as a control. Put 33 per cent (w/v) acetic acid dropwise into the second tube, until a precipitate or opalescence is observed, and then add additional acid dropwise, as long as additional precipitation occurs. Put this amount of acid plus 5 to 7 drops more into the third tube. Compare the quantity of precipitate in the third tube with that in the second. If the quantity of precipitate formed in the third tube appears not to have decreased with the addition of excess acid, mucin (mucus) is present. If the precipitate dissolves, it is probably a lipoid globulin. Urines containing normal amounts of mucin will give no precipitate with the acid. Urinary mucins can be quantitated using the procedure of Lockey, Anderson, and Maclagan.[57] These authors report excretions of the order of 100 to 150 mg./24 hr.

Serum Cryoglobulins

CLINICAL SIGNIFICANCE

The term *cryoglobulins*[56] refers to certain abnormal globulins, occasionally encountered in serum, that can precipitate or gel out when the serum is cooled to a low temperature, but that redissolve when the serum is rewarmed to 37°C. Cryoglobulins are not of hepatic origin, but are produced by some part of the reticuloendothelial system. Often the presence of cryoglobulins is associated with increased levels of plasmacytes or cells resembling plasma cells in bone marrow. Some cryoglobulins are indistinguishable from true immune γ globulins, whereas others present themselves as separate globulins migrating electrophoretically between the β and γ globulins. The molecular weight varies from 100,000 to 1,800,000, and the nitrogen content is of the order of 14 to 16 per cent.

The presence of cryoglobulins is associated with a number of different clinical conditions.[56,88,105] They were first found associated with multiple myeloma, but are also encountered in kala-azar, lupus, lymphosarcoma, rheumatoid arthritis, and other autoimmune diseases, and are often associated with vascular sclerosis and platelet defects in blood coagulation. In *essential cryoglobulinemia*, the cryoglobulins are not associated with any well defined disease states. Clinically, cryoglobulinemia presents the features of Raynaud's syndrome—intolerance to cold, purpura, gangrene of the extremities, and skin sores. The cryoglobulins tend to gel out in the blood when it circulates in the toes and fingers, impairing circulation in these areas. Death may result from blockage of key blood vessels in such vital organs as the kidneys, brain, and lungs. The specific symptoms depend on the amount present and the degree of anti-inflammatory response of the body.

Cryoglobulins are soluble at body temperature (37°C). Occasionally, they will precipitate out at room temperature, but as a rule, sera have to be cooled to 10°C. or lower, before precipitation occurs. The temperature at which gelling will occur depends on the concentration and type of cryoglobulins present. The cryoglobulins in serum may separate out in the form of a gel or as a flocculent precipitate. Rewarming the serum to 37°C. will redissolve the floc or gel, but heating to 57°C. will destroy or modify the cryoglobulins so that they will no longer gel on cooling.

Qualitative Test for Cryoglobulins

While processing blood specimens in which the presence of cryoglobulins is suspected, it is important to keep them at 37°C. during the procedure. After having been drawn from the patient, allow the specimen to clot at 37°C. in a water bath, clotting being permitted to proceed for at least an hour to insure removal of all fibrinogen. Centrifuge the tube containing the clot in cups containing water at 37 to 40°C. and store the separated serum at 37°C. until tested. To test for the presence of cryoglobulins,[56,105] cool the serum specimen overnight in a refrigerator at 4 to 10°C. If no precipitation or gelling is observed overnight at 4 to 10°C. keep the serum chilled for at least another 48 hours. Check it on the second and third days to be certain that no latent precipitation has occurred. Turbidity and lipemia present in the specimen make it difficult to detect low degrees of gelling. If such a serum must be tested, divide it into two portions; chill one and examine the other at room temperature as a control. To differentiate from *cryofibrinogen*, remix the gelled specimen and warm to 37°C. If the precipitate dissolves and the serum clarifies, the presence of cryoglobulins is confirmed.

Quantitative Assay of Cryoglobulins

The preferred procedure is to separate the gel as outlined under the qualitative test, and then to centrifuge the gelled serum in a refrigerated centrifuge. Pour off the supernatant serum and wash the precipitate with ice-cold saline and centrifuge again. After pouring off the cold wash saline, dissolve the precipitate in saline at 37°C. and then assay for nitrogen by the Kjeldahl technique or for protein directly by the biuret method. Alternatively, the protein can be analyzed by the biuret method in the serum before and after precipitating the cryoglobulins. It is probable that complete precipitation of the cryoglobulins does not take place, and some may be lost in the washing procedure. When present, the amount of cryoglobulins can range from 40 mg./100 ml. to about 4 gm./100 ml.[32]

Macroglobulins

The majority of plasma globulins have a molecular weight in the range of 150,000 to 200,000. A small proportion, less than 4 to 5 per cent of the total serum proteins, are very large size molecules, with molecular weights in the neighborhood of 1,000,000. These large proteins are referred to as *macroglobulins*.[30,88,95] If plasma is centrifuged at ultrahigh speeds, three classes of proteins can be differentiated as a result of differences in the rate of sedimentation (see the section on ultracentrifuge separations).

The macroglobulins constitute the heaviest, S_{19}, class, and they include 50 to 70 per cent of the α_2 globulins and 10 to 30 per cent of the γ globulins. The normal level present in plasma is about 0.20 gm./100 ml., ranging from 0.07 to 0.43 gm./100 ml. In a very small number of individuals, considerably higher levels may be found. In *primary macroglobulinemia*, first studied in detail by Waldenström, the level of S_{19} proteins is well over 15 per cent of the total serum proteins. Macroglobulinemia is a metabolic disease, characterized by weight loss, and susceptibility to infection, and involves the entire reticuloendothelial system. The serum electrophoretic pattern is similar to that found in multiple myeloma; Bence-Jones proteins are occasionally found, and some of the macroglobulins behave like cryoglobulins, with resultant impaired blood flow in the extremities. *Secondary macroglobulinemia* is more common, and is associated with a large number of different disease entities. The level of macroglobulins is always under 15 per cent. Secondary macroglobulinemia is found in patients with leukemia, lymphosarcoma, and rheumatoid arthritis. In myelomatosis, the concentration of S_{19} proteins is usually normal, or only slightly elevated.

Macroglobulinemia is best diagnosed by ultracentrifugation of a serum specimen. The peak demonstrating the presence of the high molecular weight IgM globulins shows up quite clearly. Isolation of the macroglobulins can also be achieved by gel filtration on Sephadex-200, using either column or thin-layer chromatography. Immunoelectrophoretic patterns can provide supportive evidence, the increased level of macroglobulins being demonstrated by the presence of heavy precipitant arcs with anti-IgM antisera.

The Sia[15,58] test is a simple qualitative test for macroglobulins based on the fact that the IgM globulins have decreased solubility in water that contains a low concentration of salts. In one procedure, 0.10 ml. of serum is added slowly down the sides of a test tube into a solution of 5.0 ml. of 0.01 *M* phosphate buffer, pH 7.1. If macroglobulins are present at levels over 0.7 gm./100 ml. a flaky, birefringent precipitate is formed, which dissolves on the addition of a spatula tip of NaCl. Normal, myeloma, and arthritis sera give no precipitation, and lipemic sera only a hazy opalescence. Malaria and kala-azar sera produce a positive result.

Amyloid

As a consequence of long term, chronic, suppurative disease and a number of other conditions there is deposited in the blood vessels and matrix of a number of organs a type of abnormal proteinaceous material called *amyloid.* Histologic specimens of such tissue stain brown with aqueous iodine, the color turning blue on acidification. Because of the starch-like reaction of this material, it was called *amyloid.* The material is not starch, but is believed to be a complex glycoprotein containing chondroitin-sulfuric acid. It is encountered in such organs as the liver, spleen, adrenals, and kidney, the amyloid degeneration occurring as an aftermath of such diseases as rheumatoid arthritis, syphilis, tuberculosis, and Hodgkin's disease. Amyloid material has the property of absorbing dyes such as Congo red or Evans blue. This property is used in the chemical detection of amyloid to confirm diagnoses made by histologic examination of biopsy specimens.

Two forms of amyloid disease are recognized: the more common, secondary amyloidosis, which has been discussed, and the rare primary amyloidosis, not associated with any previous disease.

The Congo Red Test

PROCEDURE

1. Withdraw a specimen (10 ml.) of fasting blood from the patient from one arm and note the time. This is the control specimen. Remove the syringe, with the needle in the vein, insert a syringe containing Congo red, and inject the dye.

2. Inject 10 ml. of a 1 per cent (w/v) solution of pure Congo red dye (a total of 100 mg.) into the patient. Alternatively, the patient receives 1.0 ml. of the dye (10 mg.) per 10 lb. of body weight, but not more than 10 ml. This is injected over a period of a minute.

3. Three to four minutes later, with a new needle, withdraw a specimen of blood from a vein in the opposite arm. This specimen will serve as the 100 per cent standard. It is assumed that the Congo red will have been thoroughly mixed with the blood in the 3 to 4 minute period, with minimal absorption by amyloid or reticuloendothelial tissue.

4. Exactly 60 minutes after the injection, withdraw a third specimen of blood. This is the test specimen. At the same time collect a specimen of urine.

5. Allow the three specimens of blood to clot, then centrifuge, and separate the clear serum. Pipet 4.0 ml. of acetone into each of three test tubes. Then to the acetone tubes add dropwise with shaking 1.00 ml. of each of the sera. Tightly stopper the tubes, mix vigorously several times over a 2 to 3 minute period, and then centrifuge for 4 to 5 minutes. The acetone removes all protein by precipitation, including any hemoglobin (hemolysis) that would interfere, and also solubilizes any lipid material present.

6. Rapidly pipet about 3 ml. of each supernate into appropriate cuvets (12 mm.

Coleman) and measure the absorbances of the 4 minute and 60 minute specimens against the control specimen using a wavelength of 515 to 520 nm.

7. *Calculation:*

$$\frac{\text{absorbance of 60 minute specimen}}{\text{absorbance of 4 minute specimen}} \times 100 = \text{per cent dye retained}$$

8. Examine the urine specimen for the presence of red pigment (dye); it may be acidified with a few drops of concentrated HCl and re-examined for the presence of a blue pigment. (Congo red is a pH indicator, the acid form being blue below pH 3.0, and the salt form red above pH 4.4). The presence of pigment in the urine may invalidate the test (see under interpretation).

9. *Interpretation:* Normally, some 65 to 70 per cent of the dye is still retained in the plasma after 1 hour; this remainder is then slowly cleared from the blood by the reticuloendothelial system. Customarily any retention of more than 60 per cent is considered normal. A 1 hour level (retention) of under 40 per cent is consistent with amyloid disease, and any retention of less than 15 to 20 per cent is quite suggestive of amyloidosis, provided no dye was lost in the urine.

Urinary loss of dye invalidates the significance of the 40 per cent value and makes interpretation of even the 15 to 20 per cent value equivocal. It is useful to obtain a fasting urine specimen as a control for checking for dye in the 1 hour specimen. Also, the fasting specimen should be checked for protein. The presence of 2+ proteinuria will strongly suggest that urinary loss of dye will occur because the dye is carried about in the blood bound to albumin. In this case the test should not be performed.

COMMENTS

1. The withdrawal of the fasting blood may be omitted, and a blank containing 4 parts of acetone and 1 part of water used instead.

2. Only parenteral Congo red should be used. Impurities present in untested dye lots may give rise to serious reactions in the patient. Also, to avoid possible allergic reactions, the test should not be performed on patients on whom previous tests had been done. Injections of dye should be given by a physician or trained nurse, never by a technologist.

REFERENCES

Taran, A.: J. Lab. Clin. Med., *22*:975, 1937.

Hoffman, W. S.: The Biochemistry of Clinical Medicine. 3rd ed. Chicago, Year Book Medical Publishers, Inc., 1964.

Thompson, R. H. S., and King, E. J.: Biochemical Disorders in Human Disease. 2nd ed. New York, Academic Press, Inc., 1964.

Amino Acids and Related Metabolites

Next to urea, amino acids constitute the second largest source of the nonprotein nitrogen (NPN) in serum. In general, serum urea-N and serum NPN parallel each other in serum; as the former rises, the latter increases in about the same proportion. The difference between the two includes nitrogen contained in a number of different metabolites; however, the largest portion (75 per cent) of nitrogen is accounted for as amino acid nitrogen (AAN). There is no convenient method for measuring total

AAN, although useful procedures for measuring α-amino-N are available. Determinations of AAN have not been as useful in clinical diagnosis as one would expect, in the light of the importance of amino acids in many aspects of human biochemistry. It is now appreciated that it is not the total of all amino acids that is important, but the changes in concentrations of certain individual amino acids or groups of related amino acids that is of significance in many disease states.

We now know of a large number of inborn errors of metabolism associated with abnormal amino acid metabolism, some of which result in mental retardation. In *oligophrenic phenylketonuria* the enzyme functioning to convert phenylalanine to tyrosine is deficient in quantity, resulting in the elevation of phenylalanine levels in the sera of affected infants from the average normal value of 2.0 mg. phenylalanine/100 ml. (0.16 mg./100 ml. AAN) to values as high as 40 mg./100 ml. (3.4 mg./100 ml. AAN). This large increase in one amino acid is accompanied by a small decrease in the tyrosine level and is not sufficient to increase the total amino acid nitrogen significantly above the normal level. Thus, a total AAN value would not be diagnostically informative.

Specific methods are now available for the routine assay of a number of individual amino acids in serum and urine, and more will be developed in the near future. In addition, chromatographic methods have been developed that are capable of measuring the actual levels of all or almost all amino acids in sera and urines. Most of these methods are too time-consuming and complex for use in the routine clinical laboratory, but simplified procedures of a screening type can be used in routine work to demonstrate qualitatively abnormal levels of single or groups of amino acids.

CLINICAL SIGNIFICANCE

The determination of total *plasma* or *serum amino acids* (as AAN) can at times provide clinically useful information. The normal value ranges from 2.5 to 4.5 mg. or 3.5 to 7.0 mg./100 ml., depending on the method of assay. Low values are rarely encountered; the deficiency in amino acid intake in starvation is countered by the breakdown of plasma and tissue proteins into their constituent amino acids. In the liver, amino acids are converted into protein or deaminated to keto acids. The amino group is excreted as urea, which is synthesized in the liver by means of the Krebs-Henseleit cycle. These processes tend to keep the amino acid nitrogen at a fairly constant level for any one individual. There is a small (1 to 2 mg./100 ml.) elevation 1 or 2 hours after a meal that is relatively heavy in proteins, with a fall to the fasting level thereafter. A significantly large increase in AAN is encountered only in very severe liver disease, particularly in *yellow atrophy* of the liver, the result of chemical poisoning by such agents as phosphorus, chloroform, and carbon tetrachloride. A slight increase is also found accompanying diabetes mellitus, and in patients with impaired renal function, along with elevations in all components of the NPN.

The urinary output of amino acids has been of more clinical interest than the plasma or blood levels of amino acids.[3] With the use of the more specific methods, the daily output of amino acid nitrogen is found to be in the order of 50 to 200 mg., some 1 to 2 per cent of the total daily urinary nitrogen excretion. With less specific techniques, values for urinary amino acid excretion of the order of 200 to 700 mg./24 hr. are obtained. The individual amino acids are freely filtered by the glomeruli, but the urinary concentration is quite low because the acids are actively reabsorbed by the renal tubules. Only a few amino acids are present in the urine in quantities of over 50 mg./day; among these are alanine, glycine, histidine, glutamine, and 1-methyl histidine. The remainder are present in small amounts, varying from traces to 5 to 20 mg./24 hr. Roughly one-third are present in the free form; the rest occur as peptides or in other bound or conjugated forms.

A general increase in urinary amino acid output accompanies severe liver disease, notably acute atrophy of the liver, and is seen in clinical situations reflecting protein breakdown, such as in starvation and various debilitating diseases. Increases in a single or a small group of amino acids are observed in conditions such as *cystinuria.* The de Toni-Fanconi syndrome defines a more generalized defect in tubular reabsorption. Impaired reabsorption of all amino acids is accompanied by a defect in glucose and phosphate reabsorption as well. In Wilson's disease (hepatolenticular degeneration) a generalized aminoaciduria accompanies deposition of large amounts of copper in the tissues. The nature of thc key defect is not understood but the disease is characterized by a very low level of copper-binding protein (ceruloplasmin) in the plasma.

Increases in certain specific amino acids may occur as in *alkaptonuria* in which tyrosine and homogentisic acid (HGA, 2,5-dihydroxyphenylacetic acid) are found to be increased as a result of a metabolic defect. The enzyme that normally oxidizes HGA by splitting the benzene ring is missing, with the result that HGA cannot be metabolized. Both HGA and its predecessor, tyrosine, accumulate and are excreted in the urine. Similarly, in *phenylketonuria*, increased amounts of both phenylalanine and its deamination product, phenylpyruvic acid, are excreted.

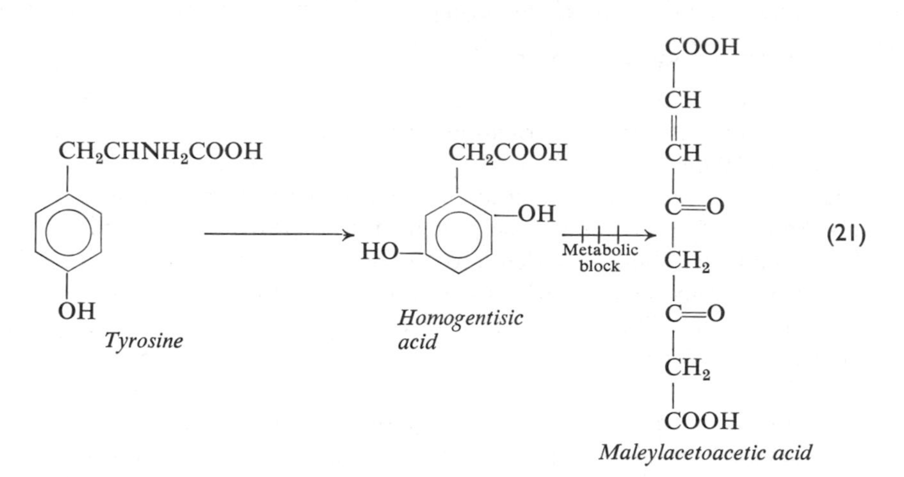

(21)

REFERENCES

Bigwood, E. J., Crokaert, R., Schram, E., Soupart, P., and Vis, H.: Amino aciduria. *In* Advances in Clinical Chemistry. H. Sobotka and C. P. Stewart, Eds. New York, Academic Press, Inc., 1959, vol. 2, p. 201.

Thompson, R. H. S., and King, E. J.: Biochemical Disorders in Human Disease. 2nd ed. New York, Academic Press, Inc., 1964.

ANALYTICAL PROCEDURES FOR AMINO ACIDS

The most precise method for assaying amino acids employs the gasometric ninhydrin reaction.[17] Ninhydrin (triketohydrindene hydrate) reacts with amino acids

as follows:

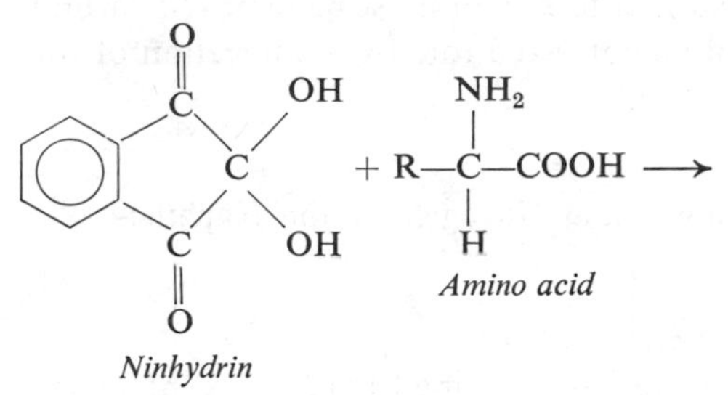

O
C
OH
C
H
C
O
+ R—C(=O)—H + CO_2 + NH_3 (22)

Hydrindantin *Aldehyde*

In this reaction the amino acid is oxidized to an aldehyde containing one less carbon atom, resulting in the release of ammonia and carbon dioxide. Exactly 1 mole of carbon dioxide is liberated per each mole of amino acid, with the exception of aspartic acid, which gives off 2 moles. All true amino acids undergo the ninhydrin reaction, as do peptides and proteins and any compounds containing free amino groups. Proline and hydroxyproline do not react. The carbon dioxide formed can be determined manometrically, and is a measure of the amino acids present in a given sample.

If the pH of the reaction environment is about 3 to 4, the ammonia, ninhydrin, and the reaction product, hydrindantin, react to form a bluish colored reaction product.

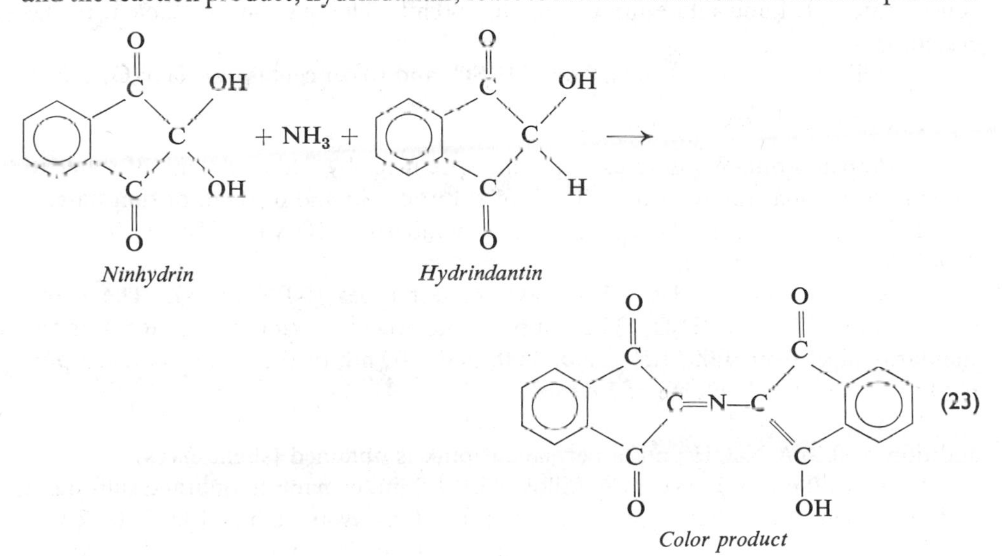

This color reaction can also be used for both the detection and quantitation of amino acids, especially in chromatographic procedures.

Sodium β-naphthoquinone-4-sulfonate will react with amino acids under weakly alkaline pH conditions to form a brownish orange colored product. Frame[18] developed this Folin reaction into a useful and practical method for amino acids in plasma, and, with care, in urine. Ammonia interferes and must be removed (urine). Uric acid gives a weak reaction, which can be corrected for. A modification of this procedure is now presented.

Determination of Plasma and Urinary Amino Acid Nitrogen by the Naphthoquinone Sulfonate Procedure

SPECIMENS

Plasma specimens are preferable. Serum values are some 10 to 20 per cent higher, partly because of the amino acids released by the clotting process. Urine should be collected in bottles containing 10 ml. of concentrated HCl to prevent microbial destruction.

REAGENTS

1. β-naphthoquinone-4-sulfonate, sodium salt, 0.5 per cent (w/v) solution in water, prepared fresh within 1 or 2 hours of use.
2. Borax, sodium tetraborate, 2 per cent (w/v) in water. The reagent is stable and is stored in a polyethylene bottle.
3. NaOH, 0.10 *N*.
4. Acid-formaldehyde reagent, 0.30 *N* HCl, containing 0.12 per cent HCHO (w/v). Add 3.0 ml. of 40 per cent formalin to 1 L. of 0.30 *N* HCl.
5. Sodium thiosulfate, $Na_2S_2O_3$, 0.05 *M*. Dissolve 3.10 gm. of $Na_2S_2O_3 \cdot 5H_2O$ in 250 ml. of water.
6. Amino acid stock standard solution (1.0 ml. = 0.20 mg. AAN). Add 134 mg. of glycine, 262 mg. of glutamic acid, and 0.5 gm. of sodium benzoate to about 300 ml. of H_2O and stir the mixture until the solids are all in solution. Add 2 ml. of concentrated HCl and dilute the volume to 500 ml. This standard is stable if refrigerated.
7. Folin-Wu filtrate reagents, 0.66 *N* H_2SO_4 and 10 per cent (w/v) $Na_2WO_4 \cdot 2H_2O$.

PROCEDURE FOR PLASMA AAN

1. Prepare Folin-Wu filtrates of plasma specimens by adding 8.0 ml. of water to 1.00 ml. of plasma, followed by 0.50 ml. of sulfuric acid and 0.50 ml. of tungstate.
2. Dilute standard. Dilute the stock standard 1:10 with H_2O (1.0 ml. = 20 μg. AAN). Prepare fresh.
3. Reaction set up. Use 25 × 200 mm. test tubes (NPN tubes). The blank tube contains 5.0 ml. of H_2O. The sample tubes receive 5.0 ml. of filtrate. For the standards pipet 0.50, 1.00, 1.50, 2.00, 3.00, and 4.00 ml. of the dilute standard plus H_2O to make a total volume of 5.0 ml.
4. Add 1 drop of phenolphthalein indicator to each tube; titrate by dropwise addition of 0.10 *N* NaOH until a permanent pink is obtained (slight excess).
5. Add 1.0 ml. of borax, mix, follow with 1.0 ml. of naphthoquinone sulfonate, and mix again. Immediately place all tubes in a Carbowax bath at 100°C. (±2°C.).
6. Heat the tubes for exactly 10 minutes. Cool in running water for exactly 5 minutes.
7. Add, in turn, followed by mixing, 1.0 ml. of acid formaldehyde, 1.0 ml. of thiosulfate, and 15.0 ml. of water. Let the color develop for a minimum of 10 minutes.
8. Pour solutions into cuvets and read in a spectrophotometer against the blank

within 15 minutes. The solutions should be clear; if not, add a drop of a clearing agent such as Duponol W.A. (4 per cent).

9. Plot a standard curve and obtain the mg./100 ml. value of the unknowns from the curve. The standard curve is not linear and will vary some with each reaction run; therefore, unknowns should be compared against standards run concurrently.

CALCULATIONS

If the plasma contains Z mg./100 ml. of AAN, the 5.0 ml. of filtrate used in the assay will contain $5 \times Z$ μg. of AAN. The working standards contain 10, 20, 30, 40, 60, and 80 μg. AAN. If the unknown color matches that of a standard containing S μg., then $5 \times Z = S$, or $Z = S/5$. The standards thus represent 2, 4, 6, 8, 12, and 16 mg./100 ml. of AAN.

PROCEDURE FOR URINARY AAN
REAGENTS

These are the same as those used for plasma AAN. In addition, Dowex 50 × 8 ion-exchange resin, 200 to 400 mesh, hydrogen form is needed.

PROCEDURE

1. Ion exchange column. Suspend the resin in double its volume of water. Stir the suspension, allow it to settle for a moment, and pour off the supernate. Repeat this operation twice. This washes the resin and removes some of the very fine particles. After the last wash, add water so that the total volume is about one half more than that of the resin alone. Pour the suspension into a brown glass bottle for storage. Place a 5 mm. plug of loose glass wool in the bottom of a 1 × 10 cm. chromatographic tube. Suspend the resin, pour into the column and allow it to settle by gravity to form a resin column about 5 to 6 cm. in height. Place a small glass wool plug on the top of the column. Wash the column twice with water, the water level never being permitted to drop below the level of the upper plug.

2. The urine specimen should have been collected in acid. If necessary, an aliquot (100 ml.) is further acidified to a pH between 1 and 2 and then filtered.

3. Pass 10 ml. of the prepared urine through the column, the level not being allowed to flow below that of the top glass plug.

4. Wash the column with 0.01 N HCl, twice with 1 ml. and once with 3 ml. Then wash out the acid by adding water through the column until the pH of the eluate is above 5. The pH can be conveniently checked with short range pH paper (Hydrion, range 4.0 to 5.7, the indicator being bromcresol green).

5. Elute the amino acids by the addition of 4 ml. of 1 N NaOH. Discard the first 2 to 3 ml. of eluate (water). Watch the resin, and when the top half of the column has taken on the brown color of the sodium form of the resin, begin the collection of the amino acid eluate. Add 7 ml. of additional NaOH and collect a total of 10.0 ml. of eluate in a test tube graduated at the 10 ml. mark.

6. Aerate the tubes for 20 to 30 minutes by drawing air through them with the help of a water suction pump. In this step, any free ammonia, which would interfere, is removed. On removal of the air inlet tube, rinse it with a small volume of water. Adjust the eluate in the graduated tube to a volume of 10 ml.

7. Transfer 1.00 ml. of eluate to a 50 ml. volumetric flask, and add about 30 to 35 ml. of H_2O, followed by 2 drops of phenolphthalein indicator. Add dropwise HCl (2 to 3 N) until the pink color has disappeared. Then add dropwise NaOH

(1 *N*) until the appearance of the first permanent pink color. Then make up the volume to 50 ml.

8. The remainder of the procedure is identical to that for plasma, except that 5 ml. of eluate replaces the 5 ml. of blood filtrate.

CALCULATIONS

If the urine contains Z mg. of AAN/ml., 1 ml. of the eluate will also contain Z mg. When this is diluted to 50 ml., 5.0 ml. of the diluted eluate will contain $0.10 \times Z$ mg. AAN. If the matching standard contains S mg. AAN, $0.10 \times Z = S$, or $Z = 10S$. The working standards contain 0.01, 0.02, . . . , mg. of N and thus represent 0.10, 0.20, 0.30, 0.40, 0.60, and 0.80 mg. of AAN/1.0 ml. urine. The 24 hour output of AAN is then calculated

$$\text{mg. AAN/24 hr.} = \left(\frac{A_u}{A_s}\right) \times C_{st} \times (\text{total urine volume in ml.})$$

COMMENTS

The normal value for serum AAN by this procedure is reported as 3 to 6 mg./100 ml. Uric acid reacts weakly, 1.0 mg./100 ml. giving an AAN equivalent of 0.10 mg./100 ml. One may correct for the uric acid present, although the error is of the order of the overall accuracy of the method. Free sulfonamides, if present, also interfere to a similar degree. The normal 24 hour urine output is 50 to 200 mg./24 hr. No correction for uric acid in the urine is required because it is removed by the ion exchange column. The overall precision of the method is ± 10 per cent.

REFERENCES

Frame, E. G., Russel, J. A., and Wilhelmi, A. E.: J. Biol. Chem., *149*:225, 1943.
Sobel, C., Henry, R. J., Chiamori, N., and Segalove, M.: Proc. Soc. Exp. Biol. Med., *95*:808, 1957.

CYSTINE AND CYSTEINE

On occasion a laboratory is asked to test urine specimens or kidney stones for the presence of the sulfur-containing amino acid, cystine. Both cysteine and cystine are normally present in small amounts in urine. The combined total 24 hour output of both is about 10 to 100 mg. Cysteine constitutes about 10 per cent of the total. The sulhydryl or thiol, —SH, group is easily oxidized to the disulfide (—S—S—), 2 moles of cysteine forming 1 mole of cystine. The reduction of cystine to cysteine is

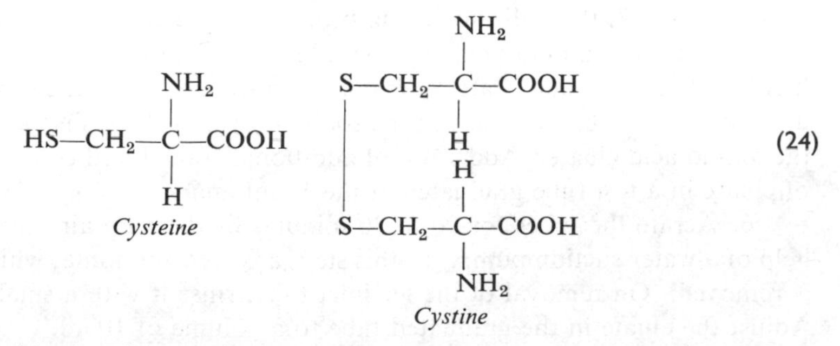

more difficult to achieve. The reactions by which this occurs physiologically are not well understood, although one known pathway includes an enzymatic reduction involving glutathione. The usual chemical procedure for cystine analysis is to add

NaCN to the cystine preparation, which results in the following reaction:

$$\begin{array}{ccc} \text{Cys—S} & & \text{Cys—S—CN} \\ | & + \text{NaCN} \longrightarrow & + \\ \text{Cys—S} & & \text{Cys—S—Na} \\ \textit{Cystine} & & \textit{Cysteine} \end{array} \qquad (25)$$

$$\text{Cys} = \text{HOOC—CHNH}_2\text{—CH}_2\text{—}$$

Since there is no convenient way to detect cystine as such, all routine procedures use this reaction to convert cystine to cysteine, the —SH group of which is reactive chemically. Cystine is relatively insoluble at near neutral pH, but is soluble in mineral acids and in alkalis, including NH_4OH. The limit of solubility in urine is about 30 mg./100 ml. In patients with cystinuria,[2,88] this limit is surpassed and cystine is precipitated out in the urine, and even in the renal pelvis, forming cystine stones (calculi). The cystine in urinary sediment can be identified by the characteristic appearance of the crystals—clear, flat hexagonal plates—soluble in HCl and NH_4OH, but insoluble in acetic acid.

Cystinuria is a disease characterized by the excretion of large quantities of the basic amino acids, lysine, arginine, ornithine, and cystine, and by a tendency to form and deposit cystine calculi (stones). The disease involves a genetic defect in the renal tubular reabsorption of lysine, arginine, and ornithine, which indirectly affects cysteine excretion and results in the excretion of increased levels of cystine. Several related forms of the disease exist, and some also include a defect in intestinal absorption of cysteine.

The disease *cystinosis* is unrelated to cystinuria, and its physiologic basis is even less well understood. A genetic defect in the metabolism or transport of cystine is present, resulting in the deposition of cystine in the body tissues, although the urinary output may be normal, or only slightly elevated.

Three different chemical tests are used to identify cystine in stones and urinary sediments. In the *nitroprusside test*, the unknown is treated with NaCN (see equation 25), followed by a few drops of a fresh solution of nitroprusside. If cystine is present, a deep reddish violet (magenta) color is formed.

In the *lead acetate test*, the specimen is heated with strong (40 per cent) NaOH to split the sulfide group from the amino acid. The free sulfide ion is then detected by the addition of lead acetate, the Pb^{++} ion reacting with the sulfide ion to form the black precipitate of PbS.

The *Sullivan and Hess test* uses naphthoquinone sulfonate. The reagent is added after treating the specimen with NaCN, followed by alkaline sulfite and dithionate (sodium hyposulfite). Although all amino acids will react with the reagent, on treatment with dithionate the cystine product is converted to a pink-red material, whereas the initial colored product formed with the other amino acids is bleached out.

Most quantitative methods are based on the measurement of the sulhydryl group as such or on the capacity of sulhydryl to reduce reagents such as Folin's phosphotungstic acid. These methods are nonspecific and tend to give high results; nevertheless, they are practical and there is no difficulty in differentiating normal from cystinuric urines.

Chromatographic methods are more specific, but are not practical for use in routine laboratories. To facilitate separation of cystine from other amino acids with this technique, it is oxidized with peroxide to cysteic acid ($HO_3SCH_2CHNH_2COOH$). Procedures for two qualitative tests will be presented, but no quantitative method is presented.

Tests for Cystine and Cysteine (Urine, Urine Sediments, and Renal Stones)

NITROPRUSSIDE TEST

Pipet about 3 ml. of fresh urine into a test tube or a small crucible. In the case of a renal stone, grind the stone to a fine powder, transfer a small amount to a crucible, and heat with 2 to 3 ml. of H_2O; allow the crucible to cool. Add 1.0 ml. of fresh 5 per cent (w/v) NaCN (handle with care) and allow the tube or crucible to stand for 10 minutes. It is wise to use a control consisting of 3 ml. of normal urine to which a few milligrams of cystine have been added. Then add dropwise a fresh 5 per cent (w/v) solution of sodium nitroprusside ($Na_2FeNO(CN)_5{\cdot}2H_2O$). If cystine or cysteine is present, a magenta color develops. Normal urines give a faint brown or reddish brown color.

LEAD ACETATE TEST

Boil about 5 to 7 ml. of urine with 1.0 to 1.5 ml. of 40 per cent NaOH. Cool the tube or crucible and add 3 to 5 drops of saturated lead acetate ($Pb(CH_3COO)_2{\cdot}2H_2O$) solution. The formation of a brownish black to black precipitate indicates organically bound sulhydryl. If the urine contains more than 2+ protein, the protein will interfere because the heating with alkali will split off sulfhydryl from the protein molecules. Protein is removed by heating the urine to 70 to 80°C., followed by filtering.

PHENYLKETONURIA

Oligophrenic phenylketonuria[88,102] is an inheritable disease characterized by a defect in the ability to metabolize the amino acid phenylalanine. The disease is caused by the absence or deficiency of the enzyme L-*phenylalanine hydroxylase*, which converts phenylalanine to the amino acid tyrosine. As a result, phenylalanine accumulates in the tissues of the body. A portion of this phenylalanine is excreted in the urine as such, but a larger fraction is deaminated to phenylpyruvic acid and excreted in that form.

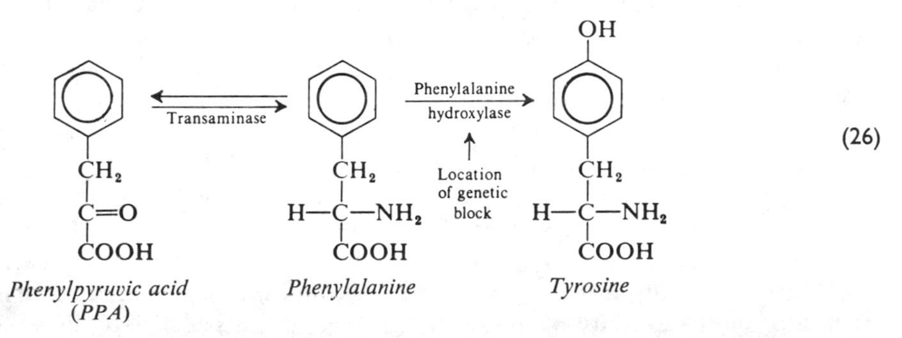

Phenylpyruvic acid (PPA) *Phenylalanine* *Tyrosine*

The phenylalanine accumulating in the tissues is apparently toxic, especially to developing brain tissue, with the result that brain damage and progressive mental retardation occurs. The injury to brain tissue begins within the second and third week of life and progresses with time, becoming maximal at about 8 to 9 months. Brain damage can be minimized if the newborn child is placed on a low-phenylalanine diet soon after birth. Even if diagnosis is made late, if phenylalanine is then removed from the diet, the rate of further mental deterioration can be decreased, provided that this is done before 4 to 6 months have passed. If no therapy is instituted, the child may develop as an idiot with an I.Q. of 20 to 30. Because the effects of the disease may be minimized by very early diagnosis and the use of phenylalanine-free diets, a number of states in this country now require, by statute, that all newborn infants be tested for the presence of this genetic defect.

All premature infants and also some full-term infants may show elevated levels of serum phenylalanine (up to 7 mg./100 ml.) as a result of a temporary immaturity of the liver at birth. These infants must be differentiated from the much smaller number of true phenylketonuric infants. This may be done by rechecking the blood within 2 to 3 weeks after birth and by determining the serum tyrosine level which is normal in the nonaffected group and low in the phenylketonuric group (see under tyrosine).

The defect is best detected by the assay of serum phenylalanine levels. Normal full-term infants will have levels ranging from about 1.2 to 3.4 mg./100 ml., averaging 2.1 mg./100 ml. With premature infants the level will often be higher, but will fall into the range for normal full-term infants within 7 to 20 days. Phenylketonuric infants will have levels of over 4.5 mg./100 ml. by the third or fourth day if fed a normal diet. This value will steadily increase as long as the infant is receiving a normal (phenylalanine containing) diet, and levels of over 20 to 30 mg./100 ml. will be found after 7 to 10 days. If a phenylalanine-free diet formula is instituted, the level will fall to near normal values within a short time, and remain there as long as phenylalanine intake is restricted.

Urinary Phenylpyruvic Acid

Until relatively simple and practical methods for the measurement of serum phenylalanine were developed, the clinician and technologist had to rely on the detection of increased levels of *phenylpyruvic acid* (PPA) present in the urine of affected babies. If ferric chloride is added dropwise to a urine containing PPA, a definite but transient blue-green color forms (Penrose-Quastel reaction). With high concentrations of PPA the color can be quite dark and may persist for 5 to 10 minutes, but with low or moderate levels the color may be hard to detect, especially in the presence of other urinary pigments. Detectable levels of PPA may not be excreted until after the third or fourth week of life, by which time the infant will have left the hospital. PPA usually cannot be detected in infant urines until the serum phenylalanine concentration has reached levels between 12 and 15 mg./100 ml. Infants affected with phenylketonuria will excrete up to 2 gm. of PPA/day, and urinary concentrations will be from 50 to 100 mg./100 ml.

A commercial impregnated test paper called *Phenistix* (Ames Co., Div. Miles Laboratories, Elkhart, Ind.) may also be used to test for PPA. In this procedure, also based on the ferric chloride reaction, the presence of PPA will change the color of the paper from a light gray to a gray-green or green-blue. Light, medium, and high readings correspond to PPA levels of 15, 40, and 100 mg./100 ml. It should be noted that the urine tests for PPA, though still used, have been replaced to a large degree by the more precise and reliable determinations of serum phenylalanine levels.

Qualitative Test for Urinary Phenylpyruvic Acid

REAGENTS

1. 10 per cent (w/v) $FeCl_3 \cdot 6H_2O$ in water.
2. Phosphate precipitating agent. This reagent contains 2.2 gm. of $MgCl_2 \cdot 6H_2O$, 1.4 gm. of NH_4Cl, and 2.0 ml. of concentrated NH_4OH dissolved in 100 ml. of solution.

PROCEDURE

Add with mixing 1.0 ml. of phosphate precipitant to 4.0 ml. of fresh urine. Then filter the specimen and acidify the filtrate with 2 to 3 drops of concentrated HCl. Add 2 to 3 drops of $FeCl_3$; observe the filtrate after each drop for any color formation. A green to blue-green color that persists for 2 to 4 minutes indicates a positive test. Very rapidly fading greens suggest homogentisic acid (HGA) or p-hydroxyphenylpyruvic acid (pHPPA), and should be read as negative. Imidazolepyruvic acid gives a positive test, identical with that for PPA, but it is encountered only rarely. Bilirubin will give a false positive reaction, but it normally would be absent. Other color hues are negative. The sensitivity is about 10 mg./100 ml. of PPA. By using serial dilutions of the urine filtrate, a rough quantitative measure of the PPA may be made.

Phenylalanine in Serum

Several chemical methods and a microbiologic procedure have been developed for the assay of phenylalanine in serum or blood. The most commonly used chemical method is that of McCaman and Robins.[60] This procedure is a fluorometric method in which phenylalanine is reacted with ninhydrin and the fluorescence formed is measured after enhancement by the addition of the dipeptide, leucylalanine. In the procedure of Udenfriend and Cooper[91] the phenylalanine is decarboxylated to phenylethylamine, which is measured by reaction with methyl orange. LaDu and Michael[53] used L-amino oxidase to convert phenylalanine to PPA, which is converted to a borate complex absorbing in the ultraviolet range (308 nm.).

The microbiologic procedure was introduced by Guthrie[26] and is based on phenylalanine counteracting the effect of a metabolic antagonist on the growth of a special strain of *B. subtilis.* The organism is streaked on an agar plate containing minimal growth nutrient plus β-2-thienyl alanine, a metabolic antagonist to phenylalanine. The similarity in the structures of the two compounds is evident from their formulas. In one form of the test, filter paper discs impregnated with the test serum

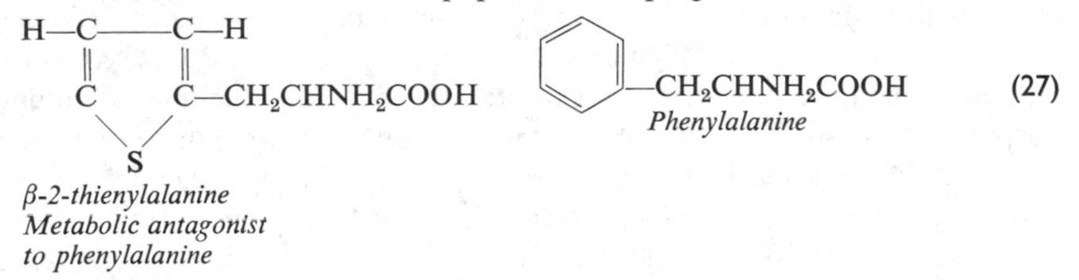

β-2-thienylalanine Metabolic antagonist to phenylalanine

Phenylalanine

are placed on the agar. If the serum has an elevated level of phenylalanine, enough will diffuse out into the medium to counteract the inhibitory effect of the thienylalanine, and a ring of bacterial growth will surround the disc. If the phenylalanine level is normal, the inhibition to growth will not be overcome, and no bacterial growth will occur. This is a useful screening procedure, and can be modified to a semiquantitative method. False positives can occur, and any positive test should be verified by a chemical determination of phenylalanine.

Fluorometric Determination of Serum Phenylalanine

REAGENTS

1. Succinate buffer, 0.6 *M*, pH 5.88 $\pm$ 0.03. Dissolve 70.9 gm. of succinic acid in about 600 ml. of H_2O; using a pH meter, carefully add 5 *N* NaOH (about 200 ml.)

with thorough stirring, until a pH of 5.88 is obtained. Dilute to 1 L. Store the reagent in a refrigerator; recheck its pH frequently.

2. Ninhydrin solution, 0.03 *M*. Dissolve 2.67 gm. of the best grade ninhydrin in water to make 500 ml. of solution. Keep this solution refrigerated.

3. L-Leucyl-L-alanine, 0.005 *M*. Dissolve 101 mg. in 100 ml. of water. This reagent has limited stability.

4. Buffered ninhydrin-peptide mixture. This is prepared fresh daily by mixing 5 volumes of succinate buffer, 2 volumes of ninhydrin soluion, and 1 volume of the peptide solution (L-leucyl-L-alanine).

5. Copper sulfate solution, 0.8 m*M*./L. Dissolve 50 mg. of $CuSO_4 \cdot 5H_2O$ in 250 ml. of water.

6. Alkaline tartrate solution. Dissolve 1.33 gm. of Na_2CO_3 and 57 mg. Rochelle salt, $KNaC_4H_4O_6 \cdot 4H_2O$, in 500 ml. of water. This reagent is stable; store in a polyethylene bottle.

7. Alkaline copper reagent. Mix 3 volumes of alkaline tartrate with 2 volumes of copper solution. Prepare this reagent fresh daily.

8. Trichloracetic acid, 0.6 *N*. Use fluorometric grade CCl_3COOH. The solution contains 98.0 gm. of TCA/L. and should be stored refrigerated.

9. Phenylalanine-in-serum standard. Prepare a stock standard of 200 mg./100 ml. water using only L-phenylalanine of the purest grade available. For the preparation of the working standards use a composite of nonhemolyzed, nonicteric, and nonchylous serum as shown below:

Standard No.	*Stock Standard* (ml.)	*Composite Serum* (ml.)	*Concentration of Phenylalanine* (mg./100 ml.)
1	none	5.00	A
2	0.10	4.90	0.98 A + 4.0
3	0.20	4.80	0.96 A + 8.0
4	0.30	4.70	0.94 A + 12.0
5	0.40	4.60	0.92 A + 16.0

The A stands for the concentration of phenylalanine in the composite serum. This is obtained by analyzing the composite serum against a previous set of standards or against a set of phenylalanine-in-water standards. If A = 2.1 mg./100 ml., the preceding five standards will contain 2.1, 6.1, 10.0, 14.0, and 17.9 mg./100 ml. These standards are treated exactly as are the unknown sera. The standards can be kept frozen for 4 to 6 weeks. The calibration curve is linear through 14 to 16 mg./100 ml.

Standards-in-water. Dilute 25 and 50 μl. of the 200 mg./100 ml. stock standard to 4.0 ml. with 0.3 *N* TCA. Treat in the same way as the sample filtrates; they represent 2.5 and 5.0 mg./100 ml. These standards are used only to assay the normal composite serum used to prepare the in-serum standards.

SPECIMEN

Draw specimens of infant blood into microcapillary tubes. If the assay is done immediately, whole blood may be used; however, if the assay is delayed, centrifuge the blood, separate the plasma or serum, and store frozen until analysis is undertaken. (50 μl. of sample, as required in the procedure, may be stored in plastic centrifuge tubes.) Whole blood values appear to increase on standing; the increase may be as much as 30 to 50 per cent in the course of 5 to 7 days, when in the normal range. Such an increase does not occur with frozen stored serum or plasma.

PROCEDURE

This is a microprocedure, and a set of microliter pipets, a microcentrifuge, as well as plastic microcentrifuge tubes (0.3 to 0.5 ml. capacity), should be available.

1. Into microcentrifuge tubes, pipet 50 μl. of sample and standards, respectively, and add 50 μl. of 0.6 *N* trichloroacetic acid. Mix the tubes by vigorous hand-tapping or with a mechanical mixer. Allow them to set for 10 minutes and then centrifuge in a microcentrifuge.

2. Pipet 50 μl. of each filtrate and transfer to a set of 13 $\times$ 100 mm. test tubes. Also set up a blank containing 50 μl. of 0.30 *N* trichloracetic acid.

3. Add 0.80 ml. of the buffered ninhydrin-dipeptide mixture to each tube and thoroughly mix the tubes. Cover the tubes with small marbles, or otherwise cap, transfer to a water bath at 60°C. $\pm$ 1°C., and heat for 2 hours.

4. While the tubes are incubating, turn on the fluorometer and warm it up. The Turner model 111 is a convenient instrument to use. The activating (primary) wavelength used is 365 nm. (Turner filter No. 7-60 (110–811)); the emission (secondary) wavelength is 515 nm. (Turner filter No. 58 (110–822)).

5. After the 2 hour incubation period, remove the tubes from the bath and place into a cooling bath at about 20°C.

6. Add 5.0 ml. of fresh alkaline copper reagent to each tube. Mix the contents of the tubes and transfer to fluorometer cuvets; make the fluorescence readings of the blank, standards, and samples. Zero the Turner fluorometer against air and choose the aperture or slit settings so that the 20 mg./100 ml. or highest standard will give a scale reading of 60 to 80. Correct all readings for the reading of the TCA blank.

7. Draw a calibration curve relating the corrected fluorometer scale reading to concentration of phenylalanine, and read the concentrations of the unknowns from the curve.

8. Run all standards in duplicate. Repeat any test with a value over 4 mg./100 ml. to verify the result. The pediatrician should be immediately informed and asked to provide a fresh, new specimen to confirm the result.

COMMENTS

Up to 40 tests can be conveniently carried out in a given run. Between runs, any specimens received should be spun down, serum or plasma separated, and the microtube portion containing the plasma or serum kept frozen until the assay is begun. The procedure can be scaled down, in which case filtrates are made with 20 μl. of serum and 20 μl. of trichloracetic acid, and 20 μl. of filtrate used for the reaction; 300 μl. of buffered ninhydrin-peptide and 2.0 ml. of alkaline copper reagent are used in place of the quantities previously mentioned.

The pH of the buffer used must be 5.88 $\pm$ 0.03. If the pH is lower, a loss of sensitivity occurs, and at higher pH values, other serum amino acids will give significant fluorescence. At pH 5.88, only tyrosine, leucine, and perhaps arginine give measurable readings, but these are less than 5 per cent of an equimolar level of phenylalanine. Obviously, if these amino acids are present at elevated levels, they will enhance the phenylalanine value obtained. The use of in-protein standards corrects for the presence of normal levels of other amino acids and assures the same pH in unknowns and standards.

NORMAL VALUES FOR SERUM PHENYLALANINE

Adults:	0.8 to 1.8 mg./100 ml.
Full-term, normal weight newborns:	1.2 to 3.4 mg./100 ml.
Prematures, low-weight newborns:	2.0 to 7.5 mg./100 ml.
Phenylketonuric newborns:	over 4.5 mg./100 ml. after 2 to 3 days, and, if not treated, rising to 15 to 30 mg./100 ml. in a period of 10 days.

REFERENCE

Wong, P. W. K., O'Flynn, M. E., and Inouye, T.: Clin. Chem., *10*:1098, 1964.

Test for the Heterozygosity of the Phenylketonuria Gene

Serum phenylalanine levels in persons who are heterozygotes (carriers) for the phenylketonuric gene are frequently within the range encountered in normal persons free of the genetic defect. At times, it is desirable to be able to establish which parent and which siblings of an affected child are carriers of the genetic defect. Since the serum level is noninformative in this respect, advantage can be taken of the fact that such carriers have a limited capacity to metabolize phenylalanine. The carrier to be tested is given an oral dose of phenylalanine, 100 mg./kg. of body weight. In a non-carrier, the phenylalanine concentration will rise from a fasting value of about 1.4 mg./100 ml. to a value of 9 mg./100 ml. at 1 and 2 hours, and then drop to 5 mg./ 100 ml. at 4 hours. In a person who is a heterozygote, the phenylalanine concentration will rise from normal to a value of about 19 mg./100 ml. at the first hour and fall much slower than in the normal person. A *phenylalanine tolerance index* is calculated by adding up the 1,3, and 4 hour values. The values of the index for normal individuals and carriers is sufficiently disparate that there is no difficulty in distinguishing between the two. In phenylketonurics the rise in serum level is higher and more prolonged, reaching 30 mg./100 ml. at 1 hour and 40 to 59 mg./100 ml. at 2 to 5 hours.

REFERENCES

Hsia, D.Y.-Y., Driscoll, K.W., Troll, W., and Knox, W.E.: Nature. *178*:1239, 1956.
Hsia, D.Y.-Y., and Inouye, T.: Inborn Errors of Metabolism. Chicago, Year Book Medical Publishers, Inc., 1966, pp. 69, 71, 222.

TYROSINE, TYROSINOSIS, AND ALKAPTONURIA

Tyrosine is derived partly from dietary proteins and partly from hydroxylation of phenylalanine to tyrosine, as discussed in previous sections. The further metabolism of tyrosine may follow a number of different pathways. In one such pathway, a minor though important one, tyrosine is converted to dihydroxyphenylalanine (DOPA) and then to the pressor hormones, norepinephrine and epinephrine. The following illustrates this scheme:

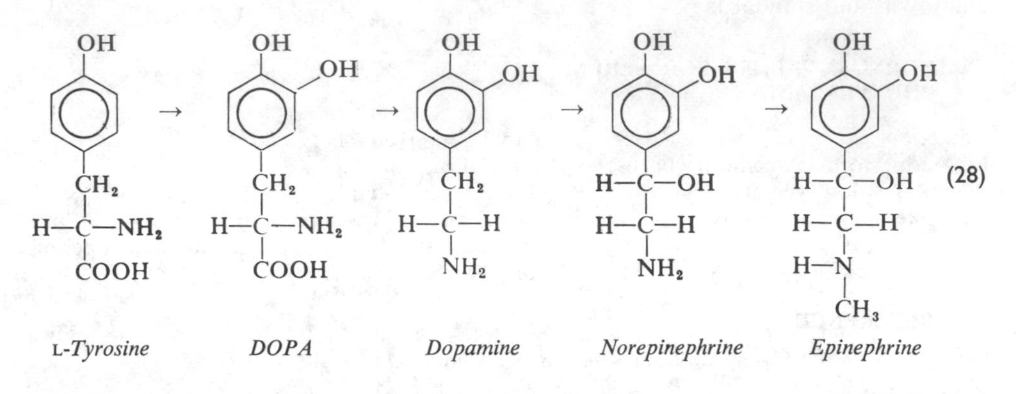

The major pathway leads, by deamination, to p-hydroxyphenylpyruvic acid (pHPPA) and then to homogentisic acid (HGA).[7,88,102] The aromatic ring in the latter is then split by homogentisic acid oxidase to form maleyl-acetoacetic acid, which then is hydrolyzed into fumarate and acetoacetate. These four carbon compounds are then metabolized via the tricarboxylic acid cycle. Deficiencies of pHPPA-oxidase and HGA-oxidase, of genetic origin, give rise to two rare but well recognized biochemical diseases, tyrosinosis and alkaptonuria. In the former, little or no HGA is formed and tyrosine and p-hydroxyphenylpyruvic acid as well DOPA and p-hydroxyphenyllactic acid accumulate and are found in significantly elevated amounts in the urine.

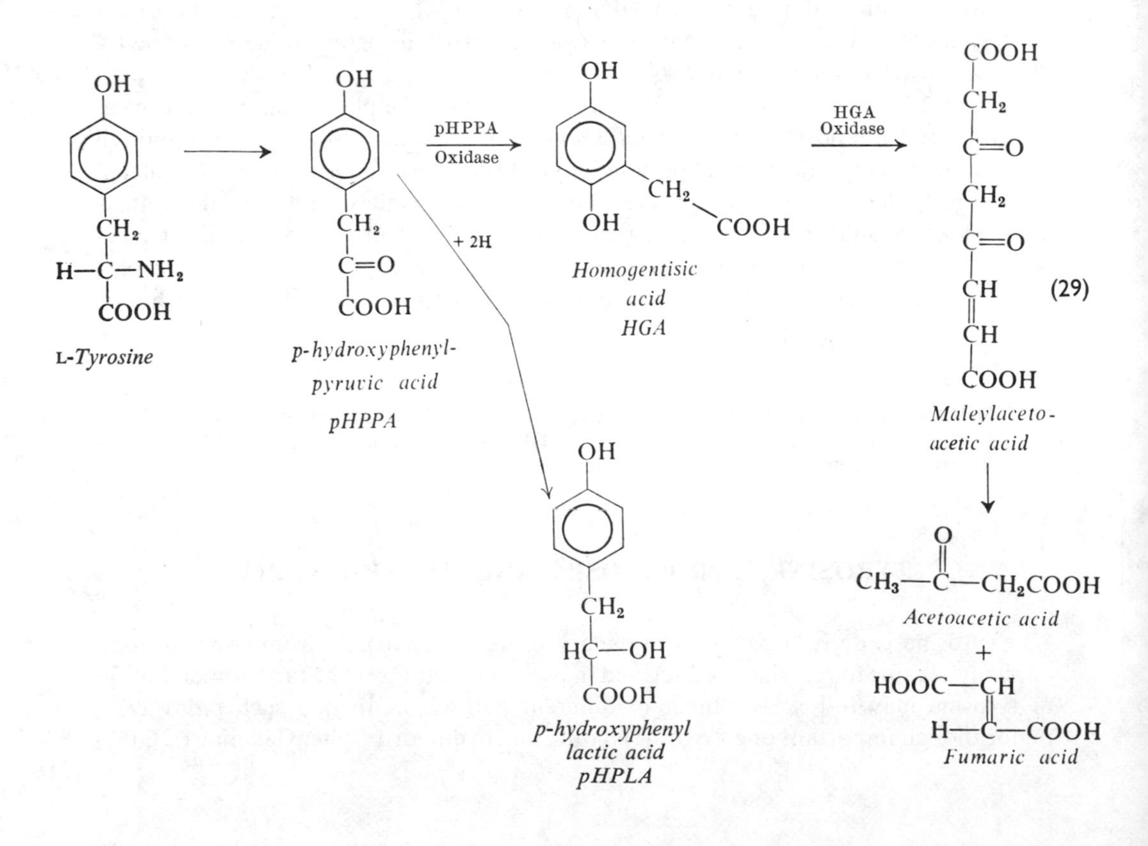

The elevated levels of homogentisic acid (HGA) in urine are demonstrated by the darkening of the urine on exposure to light and by the reducing properties of the hydroquinone ring in HGA. Occasionally, enough HGA may be retained in the tissues, especially in cartilage such as in the ear, to give the patient a characteristic stained appearance referred to as *oochronosis.* Both tyrosinosis and alkaptonuria produce no significant clinical symptoms and appear not to be detrimental to health.

SERUM TYROSINE LEVELS

The concentration of tyrosine in serum in normal adults is 0.8 to 1.3 mg./100 ml., the average being 1.1 mg./100 ml. In adults with phenylketonuria the tyrosine level is below normal, namely, 0.55 ± 0.2 mg./100 ml.

In full term newborns of normal weight, the normal serum levels for tyrosine range from 1.6 to 3.7 mg./100 ml; however, considerably higher concentrations, 7 to 24 mg./100 ml., are found in the sera of premature infants and in full-term infants of low birth weight. These (temporarily) elevated levels are due to the immaturity of the liver and its decreased ability to synthesize the appropriate enzymes. As the liver matures, the accumulated tyrosine is metabolized and serum levels will decrease to adult levels within 4 to 8 weeks. An immaturity of the liver may also be the cause of a phenylalanine hydroxylase deficiency and, therefore, a (temporary) increase in serum phenylalanine levels. Furthermore, the enzyme phenylalanine hydroxylase may also be inhibited as a result of the accumulation of tyrosine. Care must be taken in such cases to avoid an erroneous diagnosis of phenylketonuria.

Methods for the Determination of Serum Tyrosine

The most convenient method for assaying tyrosine in serum is the fluorometric procedure of Udenfriend and Cooper.[91] A trichloracetic acid filtrate of the serum is treated with a mixture of α-nitroso-β-naphthol (ANBN) and nitrite in the presence of nitric acid. Initially the ANBN and nitrite react to form a pinkish colored complex with the tyrosine. Treatment with HNO_3 converts this to a yellow fluorescing material. The yellow pigment is extracted into ethylene dichloride to separate it from the excess ANBN and its fluorescence is measured at 570 nm., after activation by light of 460 nm. wavelength. Hsia[35] has adapted the Udenfriend procedure to simple micro- and semi-micromethods. The method is not specific, and tyramine and other p-hydroxyphenyl compounds also react, but these are present in insufficient quantity to interfere in serum assays. Normal 24 hour urine output is 8 to 20 mg. for both adults and children. Interfering tyrosine derivatives are of more significance in urine assays.

Tests for Alkaptonuria

URINE DARKENING TEST

Urine from an individual with alkaptonuria, when freshly voided and exposed to strong light, will darken from the surface downward. The darkening process reflects the oxidation of the para dihydroxy aromatic ring of homogentisic acid by air to brown or black colored melanin-like pigments. The darkening is accelerated by alkalinization of the urine. Visible darkening may occur in several hours, but occasionally may require 12 to 24 hours. Urine from patients with melanomas may contain melanogens. Such urine will also darken on standing as the melanogens are oxidized to brownish black melanins, but the darkening is slower and is favored by acid rather than by alkali. Urine specimens containing phenols, gentisic acid (a salicylate metabolite), and indican may also exhibit the darkening phenomenon.

REDUCTION TESTS

HGA will reduce an ammoniacal solution of silver nitrate very rapidly. Add 5.0 ml. of 3 per cent (w/v) $AgNO_3$ to 0.50 ml. of urine, followed by a few drops of 10 per cent (v/v) NH_4OH. If HGA is present, a brown-black to black precipitate of reduced, elemental silver will be formed immediately, often even before the addition of the NH_4OH. With melanogens the reaction is very slow and excess NH_3 must be present. On heating the urine with Benedict's qualitative sugar reagent, the supernatant becomes dark and the usual precipitate of yellow cuprous oxide is observed. Melanogens do not reduce Benedict's reagent except if present in very large quantities. The supernatant is not darkened.

FERRIC CHLORIDE TEST

If 10 per cent (w/v) $FeCl_3 \cdot 6H_2O$ in 0.1 *N* HCl is added dropwise to several milliliters of urine, a transient blue color will form, as with phenylpyruvic acid, but PPA containing urines will not react positively with any of the reduction tests just described. Melanogens give a brownish black color with $FeCl_3$. Methods for the quantitative assay of HGA are available, but quantitation is seldom requested.

Test for Tyrosinosis

Tyrosinosis can be diagnosed most reliably by the isolation and identification of p-hydroxyphenylpyruvic acid (pHPPA) from urine by chromatographic means. Urine of affected patients may contain up to 1.6 gm. of pHPPA, some 25 times the quantity present in normal urine. More conveniently, the *Millon reaction* for tyrosine can be used as a means of detecting tyrosinosis urine. The test is given by tyrosine and other tyrosine derivatives, as well as by pHPPA, and is therefore not specific. Nevertheless, the test is useful because elevated levels of tyrosine and hydroxyphenyllactic acid accompany the increased urinary output of pHPPA.

The procedure in the test is to mix 2.5 ml. of urine and 2.0 ml. of 15 per cent (w/v) $HgSO_4$ in 5 *N* H_2SO_4, allow it to set for an hour, and then centrifuge. Transfer the supernatant to a 50 ml. beaker, add 10 ml. of 2 *N* H_2SO_4, and heat and gently boil the mixture for 10 to 15 minutes. Add a second 2 ml. of 2 *N* sulfuric acid and permit the beaker to cool for 30 minutes, after which add 1.0 ml. of fresh 2 per cent $NaNO_2$ (nitrite). The formation of an orange color indicates a positive test.

Melanuria; Urinary Melanogens

Melanin, an abnormal pigment occasionally observed in urine specimens, is also an end product of the metabolism of tyrosine. This complex pigment and its colorless precursor, melanogen, are formed as the result of the oxidation of DOPA (dihydroxyphenylalanine) via pathways that are still obscure. The pigment occurs normally in the skin and in most tissues to some degree. More is present in dark-skinned individuals than in light-skinned persons. Absence of the pigment gives rise to the condition of albinism. Melanomas are rare melanin-producing tumors, often developing in the retina of the eye. These tumors grow slowly and may or may not produce sufficient pigment to be detected in the urine. If metastasis to the liver has occurred, easily demonstrable quantities of the pigment or its colorless precursor are often present in the urine. Melanin-containing urine darkens when exposed to air and sunlight just as does homogentisic acid containing urine. A number of simple tests have been devised to aid in the diagnosis of melanuria and in differentiating it from alkaptonuria.

Tests for Melanogens

To perform the $FeCl_3$ test of Jaksch-Pollak, add dropwise a 10 per cent (w/v) solution of $FeCl_3$ to 5.0 ml. of urine. At first, a grayish precipitate of ferric phosphate forms; with additional $FeCl_3$, the precipitate darkens, and eventually dissolves, producing a brownish black supernatant fluid if melanogens are present. Harrison[92] has suggested the use of 10 per cent $FeCl_3 \cdot 6H_2O$ in 10 per cent (v/v) HCl. Because of the acid, no precipitate forms and, if melanogens are present, the urine rapidly changes to a dark brown.

For the *Thormählen test*, add 5 to 6 drops of fresh 5 per cent (w/v) sodium nitroferricyanide to 5.0 ml. of urine in a test tube, then add 0.5 ml. of 10 *N* NaOH, and mix vigorously. Rapidly cool the tube in cold tap water and to acidify the contents, add 33 per cent (v/v) acetic acid. With a normal urine, an olive or brownish green color is obtained; melanin-containing urine gives a color varying from a greenish blue to a bluish black, depending on the quantity of melanogen present. This is probably the most specific and sensitive test available. High levels of creatinine give a brown color.

To perform the *ammoniacal silver nitrate test*, add to 0.5 ml. of urine 5.0 ml. of 3 per cent (w/v) $AgNO_3$ and 2 per cent (v/v) NH_4OH until the AgCl precipitate is almost dissolved. The solution will darken as a result of the formation of both melanins and colloidal silver. The reaction develops slowly. In contrast, homogentisic acid darkens the silver solution rapidly, even before the addition of the ammonia.

In performing tests for the presence of melanins, doubtful results should be ignored. With few exceptions, positive urine will give unequivocal results with each of the tests. Strongly positive reacting urine suggests liver involvement by the tumor. Indican-containing urine may, on occasion, be mistaken for melanin urine, but a test for indican can easily be performed to resolve the issue.

Test for Indican

Indican is indoxyl sulfate, an intestinal decomposition product of tryptophane, which accumulates under conditions of intestinal stagnation. It is absorbed from the gastrointestinal tract and is excreted in the urine, the normal output being about 10 to 20 mg./24 hr. Indican-containing urine will have a brown appearance and for this reason may be mistaken for melanotic urine. For the *Obermayer* test mix 6 ml. each of urine and Obermayer's reagent (0.4 per cent $FeCl_3 \cdot 6H_2O$ (w/v) in concentrated HCl) in an 18 × 200 mm. test tube. Add some 3 to 4 ml. of chloroform, stopper the tube, and invert vigorously 10 to 12 times. If indican is present, the chloroform will be colored blue (indigo blue).

REFERENCES

Varley, H.: Practical Clinical Chemistry. 4th ed. New York, Interscience Publishers, Inc., 1967, pp. 724–730.

Henry, R. J.: Clinical Chemistry: Principles and Technics. New York, Hoeber Medical Div., Harper & Row, Publishers, 1964, pp. 340–344.
Woolf, L. I.: Inherited metabolic disorders. Errors of phenylalanine and tyrosine. *In* Advances in Clinical Chemistry. H. Sobotka and C. P. Stewart, Eds. New York, Academic Press, Inc., 1963, vol. 6, p. 98.
Thompson, R. H. S., and King, E. J.: Biochemical Disorders in Human Disease. 2nd ed. New York, Academic Press, Inc., 1964, pp. 743–808.

REFERENCES

1. Archibald, R. M.: Nitrogen by the Kjeldahl method. *In* Standard Methods of Clinical Chemistry. D. Seligson, Ed. New York, Academic Press, Inc., 1958, vol. 2, p. 91.
2. Berlow, S.: Abnormalities in the metabolism of sulfur containing amino acids. *In* Advances in Clinical Chemistry. H. Sobotka and C. P. Stewart, Eds. New York, Academic Press, Inc., 1967, vol. 9, p. 165.
3. Bigwood, E. J., Crokaert, R., Schram, E., Soupart, P., and Vis, H.: Amino aciduria. *In* Advances in Clinical Chemistry. H. Sobotka and C. P. Stewart, Eds. New York, Academic Press, Inc., 1959, vol. 2, p. 201.
4. Bracken, J. S., and Klotz, I. M.: Am. J. Clin. Path., *23*:1055, 1953.
5. Brackenridge, C. J.: Anal. Chem., *32*:1353, 1357, 1359, 1960.
6. Brimacombe, J. S., and Stacey, W.: Mucopolysaccharides in Disease. *In* Advances in Clinical Chemistry. H. Sobotka and C. P. Stewart, Eds. New York, Academic Press, Inc., 1964, vol. 7, p. 199.
7. Cantarow, A., and Schepartz, B.: Biochemistry. 4th ed. Philadelphia, W. B. Saunders Co., 1967.
8. Cipriani, A., and Brophy, D.: J. Lab. Clin. Med., *12*:1269, 1943.
9. de la Huerga, J., and Popper, H.: J. Lab. Clin. Med., *35*:459, 1950.
10. De Lalla, O. F., and Gofman, J. W.: Ultracentrifugal analysis of serum lipoproteins. *In* Methods of Biochemical Analysis. D. Glick, Ed., New York, Interscience Publishers, Inc., 1954, vol. 1, p. 459.
11. Durrum, E. L.: J. Am. Chem. Soc., *72*:2943, 1950.
12. Ehrmantraut, H. C.: Clinical Significance of Paper Electrophoresis. Beckman Instruments, Inc. Spinco Div., Palo Alto, Calif., 1958.
13. Folin, O., and Ciocalteu, V.: J. Biol. Chem., *73*:627, 1927.
14. Folin, O., and Denis, W.: J. Biol. Chem., *18*:273, 1914.
15. Franglen, G.: Clin. Chim. Acta, *14*:559, 1966.
16. Frankel, S., and Reitman, S.: Gradwohl's Clinical Laboratory Methods and Diagnosis. St. Louis, C. V. Mosby Co., 1963, vol. 2, p. 1114.
17. Frame, E. G.: Free amino acids in plasma and urine by the gasometric ninhydrin-carbon dioxide method. *In* Standard Methods of Clinical Chemistry. D. Seligson, Ed. New York, Academic Press, Inc., 1963, vol. 4, p. 1.
18. Frame E. G., Russel, J. A., and Wilhelmi, A. E.: J. Biol. Chem., *149*:255, 1943.
19. Free, A. H., Rupe, C. O., and Metzler, I.: Clin. Chem., *3*:716, 1957.
20. Friedman, H. S.: Gamma globulin in serum. *In* Standard Methods of Clinical Chemistry. D. Seligson, Ed. New York, Academic Press, Inc., 1958, vol. 2, p. 40.
21. Gofman, J. W.: Circulation, *2*:161, 1950.
22. Gottfried, S. P., Pope, R. H., Friedman, N. H., and De Mauro, S.: J. Lab. Clin. Med., *44*:651, 1954.
23. Gray, G. W.: Sci. Am., *184*:42, 1951.
24. Gray, G. W.: Sci. Am., *185*:45, 1951.
25. Grunbaum, B. W., Lyons, M. F., Carroll, N., and Zec, J.: Microchem. J., *7*:54, 1963.
26. Guthrie, R., and Susi, A.: Pediatrics, *32*:338, 1963.
27. Harper, H. A.: Review of Physiological Chemistry. 10th ed., Los Angeles, Lange Medical Publishers, 1965.
28. Henry, R. J.: Clinical Chemistry, Principles and Technics. New York, Hoeber Medical Div., Harper & Row, Publishers, 1964.
29. Henry, R. J., Sobel, C., and Segalove, M.: Proc. Soc. Exp. Biol. Med., *92*:748, 1956.
30. Heremans, J. F., Laurell, A. H., and Waldenstrom, J.: Acta Med. Scand., *170* (Suppl. 367) 1, 1961.
31. Hiller, A., Plazin, J., and Van Slyke, D. D.: J. Biol. Chem., *176*:1401, 1948.
32. Hoffman, W. S.: The Biochemistry of Clinical Medicine. 3rd ed. Chicago, Year Book Medical Publishers, Inc., 1964.
33. Howe, P. E.: J. Biol. Chem., *49*:93, 1921; *49*:109, 1921.

34. Hsia, D. Y.-Y., Driscoll, K. W., Troll, W., and Knox, W. E.: Nature, *178*:1239, 1956.
35. Hsia, D. Y.-Y., and Inouye, T.: Inborn Errors of Metabolism. Chicago, Yearbook Medical Publishers, Inc., 1966, pp. 69, 71, 222.
36. Igou, P. C.: Am. J. Med. Tech., *33*:354, 1967.
37. Jacobson, B. M., and Milner, L. R.: Am. J. Clin. Path., *14*:138, 1944.
38. Jencks, W. P., Smith, R. B., and Durrum, E. L.: Am. J. Med., *21*:387, 1956.
39. Kachmar, J. F.: Unpublished procedure.
40. Kachmar, J. F.: Unpublished data.
41. Kaplan, A., and Savory, J.: Clin. Chem., *11*:937, 1965.
42. Kark, R. M., Lawrence, J. R., Pollak, V. E., Pirani, C. L., Muercke, R. C., and Silva, H.: A Primer of Urinalysis. 2nd ed. New York, Hoeber Medical Div., Harper & Row, Publishers, 1963.
43. Kelsey, R. L., de Graffenried, T. P., and Donaldson, R. C.: Clin. Chem., *11*:1058, 1965.
44. Kibrick, A. C., and Blonstein, M.: J. Biol. Chem., *176*:983, 1948.
45. King, J. S., Boyce, W. H., Little, J. M., and Rartom, C.: J. Clin. Inv., *37*:315, 1958; *37*:1658, 1958.
46. Kingsley, G. R.: J. Biol. Chem., *133*:731, 1940.
47. Kingsley, G. R.: J. Lab. Clin. Med., *27*:840, 1942.
48. Klatskin, G. Reinmuth, O. M., and Barnes, W.: J. Lab. Clin. Med., *48*:476, 1964.
49. Klotz, I. M.: Science, *155*:697, 1967.
50. Kohn, J.: Clin. Chim. Acta, *2*:297, 1957; *3*:450, 1958.
51. Kohn, J., and Feinberg, J. G.: Electrophoresis on Cellulose Acetate. Shandon Instruments Co., Applications Bull. 11, London, 1965.
52. Kunkel, H. G.: Proc. Soc. Exp. Biol. Med., *66*:217, 1947.
53. La Du, B. N., and Michael, P. J.: J. Lab. Clin. Med., *55*:491, 1960.
54. Lowry, O. H., Rosebrough, N. J., Farr, A. L., and Randall, R. J.: J. Biol. Chem., *193*:265, 1951.
55. Leonards, J. R.: New England J. Med., *256*:230, 1957.
56. Lerner, A. B., and Watson, C. J.: Am. J. Med. Sci., *214*:410, 1947.
57. Lockey, E., Anderson, A. J., and Maclagan, N. F.: Brit. J. Cancer, *10*:209, 1956.
58. Martin, N. H.: Quart. J. Med., *29*:179, 1960.
59. Martinek, R. G.: Clin. Chem., *11*:441, 1965.
60. McCaman, M. W., and Robins, E.: J. Lab. Clin. Med., *59*:885, 1962.
61. Meulemans, O.: Clin. Chim. Acta, *5*:757, 1960.
62. Owen, J. A.: Paper electrophoresis and protein-bound substances in clinical investigations. *In* Advances in Clinical Chemistry. H. Sobotka and C. P. Stewart, Eds. New York, Academic Press Inc., 1958, vol. 1. p. 238.
63. Owen, J. A.: Effect of injury on plasma proteins. *In* Advances in Clinical Chemistry. H. Sobotka and C. P. Stewart, Eds. New York, Academic Press, Inc., 1967, vol. 9, p. 2.
64. Parfentiev, I. A., Johnson, M. L., and Clifton, E. E.: Arch. Biochem. Biophys., *46*:470, 1953.
65. Patrick, R. L., and Thiers, R. E.: Clin. Chem., *9*:283, 1963.
66. Peeters, H.: Paper electrophoresis. *In* Advances in Clinical Chemistry. H. Sobotka and C. P. Stewart, Eds. New York, Academic Press, Inc., 1959, vol. 2, p. 2.
67. Peters, J. P., and Van Slyke, D. D.: Quantitative Clinical Chemistry. Baltimore, The Williams & Wilkins Co., 1956, vol. 2, p. 941.
68. Powell, A. H.: Am. J. Clin. Path., *25*:340, 1955.
69. Ratnoff, O. D., and Menzie, C.: J. Lab. Clin. Med., *37*:316, 1951.
70. Reiner, M. and Cheung, H. C.: Fibrinogen. *In* Standard Methods of Clinical Chemistry. D. Seligson, Ed. New York, Academic Press, Inc., 1961, vol. 3, p. 114.
71. Reinhold, J. G.: Total protein, albumin and globulin. *In* Standard Methods of Clinical Chemistry. D. Seligson, Ed. New York, Academic Press, Inc., 1953, vol. 1, p. 88.
72. Rice, E. W., and Cicone, M. A.: Am. J. Clin. Path., *29*:90, 1958.
73. Rodkey, F. L.: Clin. Chim. Acta, *10*:643, 1964.
74. Rubin, M. E., and Wolf, A. V.: J. Biol. Chem., *225*:869, 1957.
75. Rutstein, D. D., Ingenito, E. F., and Reynolds, W. F.: J. Clin. Invest., *33*:211, 1954.
76. Saifer, A., and Gerstenfeld, S.: Clin. Chem., *8*:236, 1962.
77. Smithies, O. J.: Biochem. J., *61*:629, 1955.
78. Smithies, O. J.: Adv. Protein Chem., *14*:65, 1958.
79. Sobel, C., Henry, R. J., Chiamori, N., and Segalove, M.: Proc. Soc. Exp. Biol. Med., *95*:808, 1957.
80. Steiner, R. F.: The Chemical Foundations of Molecular Biology. Princeton, N.J., D. Van Nostrand Co., Inc., 1965.
81. Stewart, C. P., and Dunlop, D.: Clinical Chemistry in Practical Medicine. 6th ed. Edinburgh, Livingstone, Ltd., 1962.
82. Straus, R., and Wurm, M.: Am. J. Clin. Path., *28*:89, 1958.

83. Sunderman, F. W., Jr., and Sunderman, F. W.: Am. J. Clin. Path., *27*:125, 1957.
84. Sunderman, F. W. Jr., and Sunderman, F. W.: Am. J. Clin. Path., *33*:369, 1960.
85. Sunderman, F. W. Jr., and Sunderman, F. W.: Am. J. Clin. Path., *27*:125 1957.
86. Sunderman, F. W. Jr., Sunderman, F. W., Falvo, E. A., and Kallick, C. J.: Am. J. Clin. Path., *30*:112, 1958.
87. Taran, A.: J. Lab. Clin. Med., *22*:975, 1937.
88. Thompson, R. H. S., and King, E. J.: Biochemical Disorders in Human Disease. 2nd ed. New York, Academic Press, Inc., 1964.
89. Tietz, N. W.: Unpublished data.
90. Tombs, M. B., Souter, F., and Maclagan, N. F.: Biochem. J., *73*:167, 1959.
91. Udenfriend, S., and Cooper, J. R.: J. Biol. Chem., *203*:953, 1953.
92. Varley, H.: Practical Clinical Chemistry. 4th ed. New York, Interscience Publishers, 1967.
93. von Frijtag Drabbe, C. A. J., and Reinhold, J. G.: Anal. Chem., *27*:1092, 1955.
94. Waddell, J. W.: J. Lab. Clin. Med., *48*:311, 1956.
95. Waldenstrom, J.: Acta Med. Scand., *117*:216, 1944.
96. Waldman, R. K., Krause, L. A., and Borman, E. K.: J. Lab. Clin. Med., *42*:489, 1952.
97. Watson, D.: Albumin and "total globulin" fractions of blood. *In* Advances in Clinical Chemistry. H. Sobotka and C. P. Stewart, Eds. New York, Academic Press, Inc., 1965, vol. 8, p. 237.
98. Watson, D., and Nankiville, D.: Clin. Chim. Acta, *9*:359, 1964.
99. Winzler, R. J.: Determination of serum glycoprotein. *In* Methods of Biochemical Analysis. D. Glick, Ed. New York, Interscience, Publishers, 1955, vol. 2, p. 279.
100. Wolfson, W. Q., Cohn, C., Calvary, E., and Ichiba, F.: Am. J. Clin. Path., *18*:723, 1948.
101. Wong, P. W. K., O'Flynn, M. E., and Inouye, T.: Clin. Chem., *10*:1098, 1964.
102. Woolf, L. I.: Inherited metabolic disorders. Errors of phenylalanine and tyrosine. *In* Advances in Clinical Chemistry. H. Sobotka and C. P. Stewart, Eds. New York, Academic Press, Inc., 1963, vol. 6, p. 98.
103. Wrenn, H. T., and Feichtmeir, T. V.: Am. J. Clin. Path., *26*:960, 1956.
104. Wunderly, C.: Immunoelectrophoresis. Methods, interpretations, results. *In* Advances in Clinical Chemistry. H. Sobotka and C. P. Stewart, Eds. New York, Academic Press, Inc., 1961, vol. 4, p. 207.
105. Zinneman, H. H., Levi, D., and Seal, U. S.: J. Immunol., *100*:594, 1968.

Chapter 6

HEMOGLOBINS, PORPHYRINS, AND RELATED COMPOUNDS

by John F. Kachmar, Ph.D.

Hemoglobin and Its Derivatives

CHEMISTRY OF HEMOGLOBIN

Hemoglobin is the red pigmented protein that gives blood its characteristic red color. It is present almost entirely in the erythrocytes, and its main function in man and other vertebrates is to transport oxygen from the lungs or gills to the cells and tissues of the body. A similar protein, *myoglobin*, is present in muscle cells, functioning there to convey the oxygen from the blood and interstitial fluid to the individual cells.

Because of its physiological importance, hemoglobin has been extensively studied and it is one of the few proteins whose gross structure is reasonably well understood. Hemoglobin and myoglobin are examples of *conjugated proteins*, being constituted of a protein linked both to a complex organic prosthetic group and to a metal atom. The protein moiety is referred to as *globin;* the associated metal atom is a divalent *iron* and the organic moiety is a species of porphyrin, *protoporphyrin* IX (type III). (Porphyrins will be discussed in the second portion of this chapter.) The iron-porphyrin combination is referred to as *heme* or as the *heme group*.

Globin consists of four polypeptide chains, two α and two β *chains*, the former containing 142 and the latter 146 amino acid residues. The individual chains are twisted, like threads on a screw, to form a helical pattern (α helices) containing 3.6 amino acid residues per turn. The four chains are further folded and bent into three-dimensional structures. In the complete globin molecule, in which the four chains are associated together, each chain occupies one of the four corners of a roughly shaped tetrahedron, with the four porphyrin rings embedded in hollows in the folded chains. An iron atom is located in the center of each porphyrin ring and it is linked directly

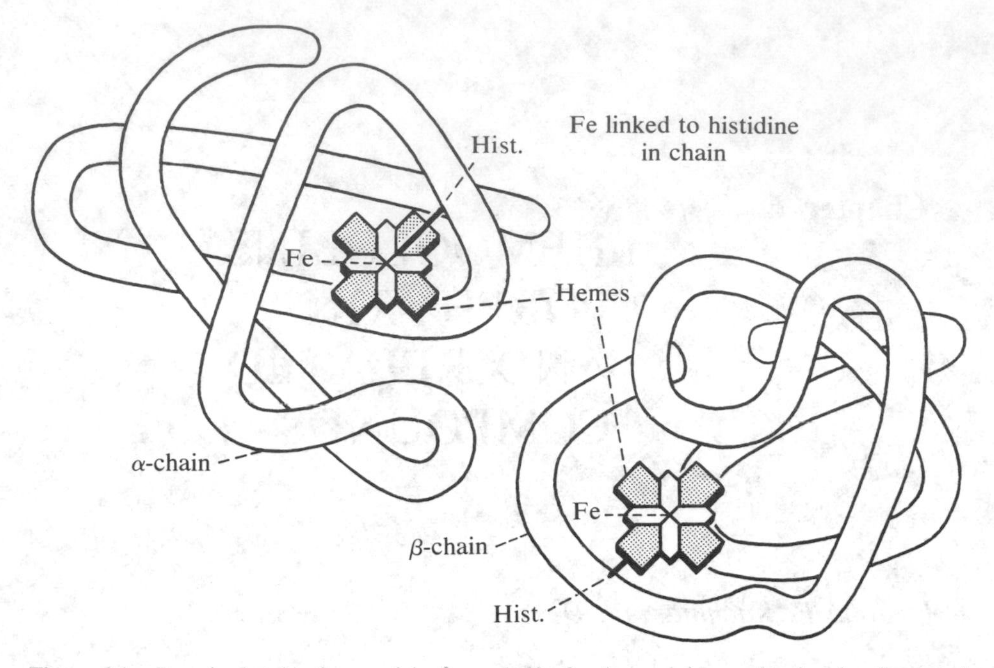

Figure 6-1. Rough sketch of a model of one half of a hemoglobin molecule, showing relative positions of the α and β chains and their associated heme groups. The other pair of chains are so positioned that the four form the corners of a crude tetrahedron. (After Steiner, R. F.: The Chemical Foundations of Molecular Biology. Princeton, N.J., D. Van Nostrand, 1965.)

to the polypeptide chain (via a histidine residue of the latter) and to the porphyrin ring.

The molecular weight of hemoglobin (Hb) is 64,456.[25] Each molecule contains four hemes (and therefore four ferrous ions) for each $\alpha_2\beta_2$ globin molecule. A diagram of a model of one half of a hemoglobin molecule, based on the x-ray diffraction studies of Perutz,[12,19] is presented in Figure 6-1. (The myoglobin molecule contains only one polypeptide chain and only one heme moiety, and has a molecular weight of about 16,120, one-fourth that of hemoglobin.)

OXYHEMOGLOBIN

Each heme in the hemoglobin molecule can react with and bind one molecule of oxygen (O_2) for a total of four oxygen molecules. The oxygen attaches directly to the ferrous iron by a coordinate bond involving d-orbital electrons, replacing a water molecule present in deoxygenated hemoglobin (Hb).[16] The resultant oxygenated hemoglobin is generally referred to as oxyferrohemoglobin or *oxyhemoglobin* (HbO_2). Since 1/4 mole of hemoglobin can bind with one mole of oxygen, one gram of hemoglobin can theoretically bind 22,390/16,120 = 1.38 ml. of oxygen, at 0°C. and 760 mm. barometric pressure. The actual value measured experimentally is 1.36 ml.[20] The average hemoglobin content ($Hb + HbO_2$) of the blood of a normal human adult male is 15.6 gm./100 ml.; thus, 100 ml. of such blood has an oxygen binding capacity of $15.6 \times 1.36 = 21.2$ ml. or 21.2 volumes per cent. This is equivalent to $212.0/22.4 = 9.4$ mmol. of either iron or oxygen per liter of blood. (The average hemoglobin content for an adult female is 14.0 gm./100 ml.; for newborn infants it may be as high as 22 gm./100 ml.*)

The oxygenation of hemoglobin is a dissociation reaction governed by an equilibrium constant, whose value varies slightly, depending on the number of oxygen molecules already bound by the hemoglobin.

$$(Hb)_4 + O_2 \overset{K_1}{\rightleftharpoons} (Hb)_3(HbO_2) \qquad \textit{First oxygen}$$

$$\text{or:} \quad K_1 = \frac{[Hb_4][O_2]}{[Hb_4O_2]}$$

$$(Hb)(HbO_2)_3 + O_2 \overset{K_4}{\rightleftharpoons} (HbO_2)_4 \qquad \textit{Last oxygen}$$

$$\text{or:} \quad K_4 = \frac{[Hb_4(O_2)_3][O_2]}{[(HbO_2)_4]}$$

The ratio of oxygenated hemoglobin to total hemoglobin, $[HbO_2]/[Hb + HbO_2]$, usually called the per cent oxygen saturation, is determined in part by the partial pressure of oxygen (Po_2) in the lungs, which is about 100 mm. Hg (14 per cent, v/v). As blood passes through the lung capillaries, hemoglobin associates with oxygen and will be saturated to about 95 to 98 per cent. In the capillaries of tissues, where the Po_2 may be as low as 10 to 20 mm. Hg, the majority of O_2 dissociates from the HbO_2 and molecular oxygen is released. This oxygen dissolves (physically) in the lymph fluid and then diffuses to the tissues where it is utilized.† As O_2 is released from HbO_2 (and CO_2 enters the blood), arterial blood changes to venous blood, which has an HbO_2 saturation of 60 to 80 per cent (see chapter 11).

HEMOGLOBIN DERIVATIVES

Hemoglobin can also bind with molecular species such as carbon monoxide, pyridine, and nitric oxide.[21] The compound formed with CO is referred to as carbonmonoxohemoglobin or *carboxyhemoglobin* (HbCO). Carbon monoxide is bound to reduced hemoglobin some 200 times more strongly than is oxygen, so that only 0.15 per cent carbon monoxide in air (PCO of 1 mm. Hg) in equilibrium with blood may keep the per cent oxygen saturation below 50 per cent. The subject of carbon monoxide poisoning is discussed in the chapter on toxicology.

The ferrous iron in heme can be readily oxidized to the ferric (trivalent) state by such oxidizing agents as potassium ferricyanide, quinone, and nitrite ions. Hemoglobin containing *ferric* (Fe^{+++}) iron is called ferrihemoglobin or *methemoglobin* (MetHb). It is also referred to as *hemiglobin*, the *i* in the second syllable denoting the presence of iron in the ferric state. Methemoglobin cannot bind oxygen and it is, therefore, ineffective as an oxygen transport agent. Methemoglobin can add on a cyanide ion (CN^-) to form ferrihemoglobin cyanide or *cyanomethemoglobin* (MetHbCN). This reaction, involving the conversion of hemoglobin to cyanomethemoglobin, is the basis of the most commonly used procedure for measuring blood hemoglobin levels.

---

* At adequate oxygen tension, the extent to which oxygen binds with hemoglobin is constant and, therefore, the amount of oxygen bound by a specimen of blood may be used as a method of measuring the hemoglobin content of blood or for standardizing a hemoglobin method (see also Chapter 11).

† Any reagents or conditions which decrease O_2 tension in a blood specimen *in vitro*, such as sodium dithionite ($Na_2S_2O_4$), ferrous ammonium citrate, or just displacement of oxygen with nitrogen, will serve to convert HbO_2 to Hb.

When hemoglobin is treated with acids at a pH below 3, the hemoglobin is denatured, the heme is cleaved from the denatured globin, and *acid hematin* is formed. Similarly, alkaline cleavage occurs at pH levels above 11 to form *alkaline hematin.* Both the acid and alkaline hematin reactions yield a brown colored product, and procedures based on these reactions have been used in measuring blood hemoglobin, but are now largely replaced by other methods.

All forms of hemoglobin and their derivatives are colored (reduced hemoglobin = purplish red, oxyhemoglobin = bright red, carboxyhemoglobin = cherry pink), and have characteristic absorption spectra in the visible range with characteristic maxima at significantly different wavelengths. These differences in spectral curves are utilized in the assay and identification of hemoglobin and its derivatives. Table 6-1 lists the *absorbance maxima* of these various hemoglobin derivatives. The

TABLE 6-1. *Absorbance Maxima of Hemoglobin and Hemoglobin Derivatives*

Hemoglobin Derivative	*Absorbance Maxima, Wavelength in nm.*				
Reduced hemoglobin	428–430		555		
Oxyhemoglobin	412–415		541	576–578	
Carboxyhemoglobin	417–418		537	568–572	
Methemoglobin (neutral pH)	404–407	500	540	578	630
Cyanomethemoglobin	413–418	480	541	580–590	
Sulfhemoglobin			540		618
Acid hematin	376				632 (HOAc) 665 (HCl)
Alkaline hematin	385			550–580 (NaOH)	610 (NH_4OH)
Heme			550	575	

absorbance bands in the far violet (405 to 425 nm.) are referred to as *Soret bands* and are about tenfold greater in magnitude than those in the 500 to 650 nm. interval. Complete spectral curves for oxygenated hemoglobin and carboxyhemoglobin are presented in the toxicology chapter.

NORMAL VALUES AND CLINICAL SIGNIFICANCE

The measurement of whole blood hemoglobin is one of the oldest and most important clinical procedures. Any serious deficiency in hemoglobin results in insufficient oxygen being available to the tissues to meet the needs of normal metabolism and energy production. The generally accepted normal value for young adult males is 13 to 18 gm./100 ml., and for young adult females from 11 to 16 gm. hemoglobin/100 ml. of blood.[20,21] Newborn infants have levels from 16 to 22 gm./100 ml. This concentration drops rapidly during the first months, and then somewhat more slowly until a level of about 12 gm. is reached at the second year of life. It slowly rises until puberty, at which period the level in females stabilizes at the approximate adult level, whereas that in males continues to increase to about the seventeenth year before reaching adult levels.

The only natural environmental factor affecting hemoglobin levels is high altitude; the reduced atmospheric oxygen is compensated for by an increase in hemoglobin concentration in the blood to levels ranging from about 16 to 23 gm./100 ml. In anemias, the decreased total hemoglobin levels primarily reflect a decrease in the number of erythrocytes, although a diminished hemoglobin content per cell may also occur. Since physiologically normal red cells contain their maximum content of hemoglobin, any increase in blood hemoglobin above normal values is generally the result of an abnormal increase in the number of cells, such as is encountered in polycythemia vera, unless the increase in concentration is caused by a decreased plasma volume.

Normally, about 0.5 to 1.5 per cent of the total hemoglobin is in the methemoglobin form. In heavy smokers as much as 6 to 10 per cent of the total hemoglobin may be in the carboxy form, and effective oxygen capacity will be less than that reflected by the total hemoglobin present.

Whenever intravascular hemolysis occurs, that is, red cells are being disrupted (lysed) while circulating in the vascular system, hemoglobin is liberated and is dissolved in the plasma. Normally, *plasma hemoglobin* is present only in very small amounts in the blood. In the past, values of up to 10 mg./100 ml. were accepted as normal; however, at present, with better techniques, and with extreme care being taken to avoid hemolysis during the actual sampling of the blood, it is felt that levels rarely exceed 1 mg./100 ml.[17] Since some intravascular hemolysis takes place at all times, a mechanism must exist to keep plasma levels low. Plasma contains a hemoglobin-binding protein, *haptoglobin*, which binds the free hemoglobin. The quantity present is such that it can bind some 70 to 140 mg. of hemoglobin/100 ml. of plasma. Any hemoglobin present above this quantity will exist in the free form and is excreted by the kidneys. Levels of haptoglobin and haptoglobin-bound Hb may occasionally be ordered and can be performed in laboratories equipped to do electrophoretic protein studies.

HEMOGLOBIN VARIANTS

In the various derivatives of hemoglobin, the protein (globin) portion of the molecule remains unaltered; changes encountered involve only the valence of the iron atom and the nature of the ligand bound to the iron. Another, more fundamental type of variation is associated with modifications in the protein moiety of the hemoglobin molecule. Both chains of either the α or β chain pair may be completely replaced by abnormal peptide chains, and small fragments, or even a single amino acid residue in the α or β chains, may be replaced by other amino acids. Such changes in the peptide chains may give rise to significant changes in the physical or chemical properties of the hemoglobin variants. These hemoglobin variants are inherited and are examples of "molecular diseases," in which the organism produces molecules that are different from the normal form with respect to chemical composition and physical properties. The presence of some of the variants may be physiologically innocuous, whereas the presence of other forms such as hemoglobin S and hemoglobin M is associated with disease states. The identification of the various forms by the techniques of hemoglobin protein electrophoresis has become a very important clinical laboratory procedure in chemical hematology and will be discussed later in this chapter.

PROCEDURES FOR THE DETERMINATION OF HEMOGLOBIN

Blood Hemoglobin

Several methods are available for the routine assay of hemoglobin in whole blood. In the *cyanomethemoglobin* technique, the blood specimen is diluted with a reagent containing ferricyanide and cyanide, which converts both reduced hemoglobin (Hb) and oxyhemoglobin (HbO_2) to the cyanomethemoglobin form. The absorbance of the latter at 540 nm. is then used for quantitation. Any carboxyhemoglobin or methemoglobin present is converted into cyanomethemoglobin and will be included in the value for total Hb. In 1958, the Panel on the Establishment of a Hemoglobin Standard of the Division of Medical Sciences of the National Academy of Science recommended that this procedure be adopted by all laboratories,[6] and the method is now widely used. Carefully standardized reagents and certified hemoglobin standards are now available from many commercial sources.

In the *oxyhemoglobin method*,[10] blood is diluted with a solution of the tetrasodium salt of EDTA and after vigorous mixing in the presence of air to convert all Hb to HbO_2, the absorbance of the HbO_2 at 540 nm. is measured. Although only part of any HbCO and none of the MetHb will be oxygenated, both of these hemoglobin derivatives absorb at 540 nm. (95 per cent and 80 per cent of that for HbO_2, respectively) and, therefore, will be included in the value for total Hb.

All hemoglobin forms are measured by the *iron method*, in which the hemoglobin is oxidized by a mixture of nitric, sulfuric, and perchloric acids to liberate the heme iron. The free iron, after reduction to the ferrous state, is measured colorimetrically as ferrous complex of α-dipyridyl or o-phenanthroline. From the accepted value for the percentage of iron in hemoglobin (0.338 per cent), the hemoglobin in the specimen can be calculated. The iron method is too slow and complicated for routine use, but it is used to standardize the other hemoglobin methods.[20,25]

Determination of Blood Hemoglobin by the Cyanomethemoglobin Procedure

REAGENTS

1. Hemoglobin Standard. This is now available commercially in the form of 5 ml. bottles containing 80 mg./100 ml. of hemoglobin, the concentration of which is certified to be correct within ± 2 per cent. The standard is reasonably stable if kept refrigerated. The hemoglobin is in the cyanomethemoglobin form.

2. Cyanomethemoglobin reagent. This contains 1.0 gm. of $NaHCO_3$, 50 mg. of KCN, and 200 mg. of potassium ferricyanide [$K_3Fe(CN)_6$] in 1 L. of solution. The solution is stable, but should never be frozen, since recent observations have shown that such treatment results in false low values. The reagent is also available commercially.

PROCEDURE

1. For instruments requiring cuvet volumes of 5.0 ml. or less, add 0.02 ml. (20 μl.) of blood to 5.0 ml. of reagent contained in the appropriate cuvet. If 19 mm. Coleman cuvets are used, add the 20 μl. of specimen to 6.0 ml. of reagent. (A Sahli pipet may be used, provided its calibration has been verified. Because it is a TC pipet, rinse it with the reagent at least three to four times.) The dilution of blood is 1:251 or 1:301 when diluent volumes of 5 and 6 ml., respectively, are used.

Alternatively, an automatic dilutor may be used. Precise delivery of 0.02 ml. of blood is not essential provided the specimen is diluted either exactly 1:251 or 1:301, respectively.

2. Mix the contents of the cuvet and allow to set for 20 minutes before reading.

3. Read the unknowns and standards against the blank (reagent), using a wavelength of 540 nm. (or equivalent filter). The hemoglobin of the blood specimen can be obtained from the calibration curve or by use of a calibration factor.

CALIBRATION

Dilute 1.0, 2.0, 3.0, 4.0, and 5.0 ml. of the 80 mg./100 ml. standard with cyanomethemoglobin reagent to 5.0 ml. or 6.0 ml., respectively. The five standards represent 4, 8, 12, 16, and 20 gm. of hemoglobin/100 ml. of blood, respectively. Read the standards against the blank as outlined in step 3. Construct a standard curve by plotting per cent transmittance or absorbance against the hemoglobin concentration.

Although the standard curve is reproducible with any given lot of reagents, it is good practice to run one working standard (highest) daily to verify the validity of the calibration.

NORMAL VALUES

The generally accepted range for normal values for blood hemoglobin of healthy males is 13 to 18 gm./100 ml., and for females 11 to 16 gm./100 ml.

COMMENTS

If the dilution of specimen is 1:250, the highest (undiluted) standard then represents 80 mg./100 ml. × 250 × 1/1000, or 20.0 gm. of hemoglobin/100 ml.

If 5.0 ml. of stock standard is diluted to 6.0 ml., the dilution will contain 80 × 5/6 mg. of Hb/100 ml. If a 1:300 dilution of specimen is compared against this standard, the latter will be equivalent to 80 × 5/6 × 300 × 1/1000 or 20 gm. of hemoglobin/100 ml.

Using a dilution of 1:251 in place of 1:250 involves an error of only 0.4 per cent.

REFERENCE

Hainline, A.: Hemoglobin. *In* Standard Methods of Clinical Chemistry. D. Seligson, Ed. New York, Academic Press, Inc., 1958, vol. 2.

Determination of Plasma Hemoglobin

The level of free hemoglobin in plasma is much too low to permit the use of the methods described for measuring Hb in whole blood. A different approach is necessary, and this is based on the chemical property of hemoglobin to catalyze the rapid oxidation of certain aromatic diamines by hydrogen peroxide. The most commonly used amine is benzidine(4,4'-diaminodiphenyl). The amine is oxidized to a

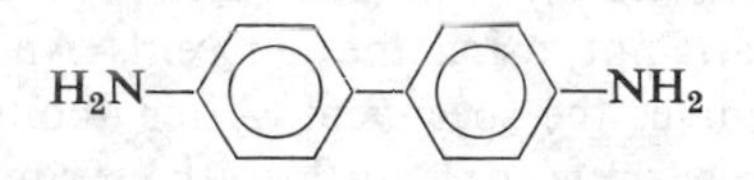

series of quinoid forms (see the section on occult blood), which initially are green to blue-black in color, but then in time are transformed to red colored compounds. The Hb functions much the same as the enzyme peroxidase, which also contains a heme group. Free or "nascent" oxygen is split from H_2O_2 to effect the oxidation of the amine. Inasmuch as only heme containing materials possess this catalytic property, it can be used as a test for the presence of hemoglobin, provided other heme proteins are absent or present only in comparatively low concentrations. The blue color initially formed in the reaction is not sufficiently stable to be used as a means of quantitating Hb. For this reason, measurement of the reddish purple color, which develops on standing, is preferred. This reaction is the basis of the most commonly used methods for assaying serum Hb because it is sensitive enough to measure as little as 2 μg. of Hb in a convenient serum aliquot. The procedure presented is that developed by Naumann.[17]

REAGENTS

1. Benzidine reagent, 10 per cent (w/v) in glacial acetic acid. Use benzidine base, not the hydrochloride salt. Benzidine base (Harleco 5447) is of sufficient purity. The material may be recrystallized from alcohol if necessary. The solution should be clear and nearly colorless; a little darkening occurs on standing. The reagent is stable for 7 to 10 days if refrigerated. Warm to room temperature before use.

2. Hydrogen peroxide, 3 per cent USP preparation or Superoxol (30 per cent H_2O_2, v/v) diluted 1:10. Renew once a month.

3. Hemoglobin standards. *Stock:* Obtain 10 ml. of citrated or heparinized blood, remove the plasma by centrifugation and wash the residual cells three to four times with saline. Lake (lyse) the cells by adding an equal volume of water and about 0.5 ml. of toluene, follow by vigorous shaking, and centrifuge at high speed (10,000 rpm). Pipet off the clear aqueous solution of Hb. Then analyze the Hb content by the cyanomethemoglobin method, and on the basis of the value found, adjust to a concentration of 10.0 gm./100 ml. Store the solution frozen in 1.0 ml. aliquots.

Dilute stock standard: Dilute the stock solution 1:100 (1.00 ml. = 1.00 mg.). This is stable refrigerated for a month.

Working standard: Dilute the dilute stock standard 1:100 with water, and then dilute 1 volume of this with 1 volume of benzidine reagent. Prepare this mixture fresh; it is stable for 3 to 4 hours (Concentration = 0.5 mg./100 ml.).

4. Benzidine blank. Dilute 1 volume of benzidine reagent with 1 volume of water. Stable for 1 week.

5. Acetic acid, 20 per cent (v/v). Dilute fresh glacial acetic acid, 1:5.

6. Thoroughly clean all glassware by soaking it in warm 10 percent (v/v) HNO_3. Avoid chromic acid. After thorough rinsing with distilled water, dry glassware in an oven, avoiding contact with iron rust.

SPECIMEN

Obtain the blood specimen with a heparinized syringe and a No. 18 gauge needle. The veni-puncture must be clean and effortless. Allow blood to fill the syringe by venous pressure; avoid aspiration of blood by pulling of the plunger. If any difficulty is experienced in drawing the blood, or the specimen has more than a few air bubbles, reject it and make a new attempt, using another vein. Any hemolysis arising during specimen collection invalidates the specimen. After about 5 ml. of blood has been collected, remove the needle, draw some air into the syringe and mix the blood with the heparin by gentle inversion of the syringe. Transfer the blood to a clean, dry tube, centrifuge it, and draw off the plasma with a Pasteur pipet to another tube, taking care not to draw off any of the buffy coat. Centrifuge the plasma, draw it off into a third tube, and recentrifuge. Only a minimum of sediment and no erythrocytes should be present, otherwise reject the specimen. Transfer the clear plasma to another dry tube for analysis or storage. Serum may replace plasma if care is taken to avoid hemolysis.

PROCEDURE

1. Dilute 1.0 ml. of the plasma with 2.0 ml. of water (see step 8).

2. Pipet 0.25 ml. of the diluted plasma into each of three 15 × 120 mm. test tubes, marked B, S, and T.

3. Add 0.25 ml. of 3 per cent peroxide to the B and S tubes. Mix these and set aside for 10 minutes.* Tube T is not treated.

4. Add 0.50 ml. of benzidine blank to the B and T tubes, and add 0.50 ml. of working standard (with benzidine) to tube S.

5. Now add 0.25 ml. of peroxide to the T tube. When H_2O_2 is added after the benzidine, the benzidine-blue reaction will take place.

6. Shake all tubes, and then set aside for 15 minutes.

7. Add 10.0 ml. of 20 per cent acetic acid to all tubes, cap the tubes with Parafilm, mix by inversion, and transfer the contents to cuvets.

8. Read the absorbances of the S and T cuvets against the B cuvet at 500 nm. If the absorbance of the T specimen is too high, quantitatively dilute its contents with up to 3 parts of acetic acid. If still too high, the work should be repeated beginning with a 1:10 or 1:20 dilution of the original plasma (step 1).

CALCULATION

In the procedure 0.25 ml. of a 1/D dilution of serum is compared with a 0.50 mg./100 ml. standard. Therefore

$$\text{mg. plasma Hb./100 ml.} = \frac{A_u}{A_s} \times \frac{0.50}{0.25} \times 0.50 \times \text{D}$$

or

$$\text{mg./100 ml.} = \frac{A_u}{A_s} \times \text{D}$$

COMMENTS

The normal range found using this procedure is 0.50 to 2.5 mg./100 ml. Some analysts, using different benzidine methods, report top normals nearer 1.0 mg./100 ml. Plasma proteins inhibit the benzidine red reaction; in the preceding procedure, protein is added to the standard so that the inhibition effect is identical in both the standard and unknown tubes.

Free hemoglobin in increased concentration appears to act as a toxic agent toward the renal tubules, and can cause severe kidney damage. Plasma hemoglobin levels are thus often requested for patients with hemolytic anemia or who experience blood transfusion reactions.

Elevations in plasma hemoglobin are found in patients with some types of hemolytic anemias (5 to 65 mg./100 ml.) and in patients with paroxysmal nocturnal hemoglobinuria (20 to 250 mg.). After transfusion reactions, levels may vary from 15 mg. to over several hundred mg./100 ml. Two or three week old blood bank blood may have as much as 50 mg./100 ml.

REFERENCE

Naumann, H. N.: The measurement of hemoglobin in Plasma. *In* Hemoglobin, its Precursors and Metabolites. F. W. Sunderman and F. W. Sunderman, Jr., Eds. Philadelphia, J. B. Lippincott Co., 1964.

---

* This step corrects for "plasma interference" by selectively inhibiting the reactivity of plasma hemoglobin in standard and blank reaction mixtures. Hydrogen peroxide will "deactivate" plasma hemoglobin without changing the effect of "plasma interferences." Subsequent addition of known amounts of hemoglobin and benzidine to the standard and of water and benzidine to the blank mixtures makes the ingredients of these reaction mixtures identical to those of the unknown.[17]

BLOOD METHEMOGLOBIN

The iron in the hemoglobin in erythrocytes is kept in the ferrous state by the action of an enzyme system by which the reduction of the ferric iron in methemoglobin is coupled to the oxidation of glucose. The key enzyme is *methemoglobin-cytochrome C reductase*. Some heme iron is always being spontaneously oxidized to the ferric state. As long as this proceeds at a normal rate, the previously mentioned enzyme system is able to maintain the bulk of the iron in the ferrous form, with the result that normally less than 1.5 per cent of the red cell Hb is in the methemoglobin, MetHb, form. The ingestion of certain types of chemical agents, however, can result in such a rapid oxidation of the iron that the enzyme reaction cannot keep pace, and *acquired* or *toxic methemoglobinemia* develops. Among such toxic agents are nitrite and nitrate (occasionally found in drinking water) and a large variety of organics and pharmaceuticals, such as aniline, acetanilid, phenacetin, and some of the sulfa drugs. In *congenital methemoglobinemia*[3] the enzyme is absent or present in deficient quantity and, as a result, 40 to 60 per cent of the Hb may be present in the MetHb form. MetHb itself is harmless, but it cannot bind oxygen, and individuals with elevated levels of MetHb have a cyanotic appearance and may develop a functional anemia. Cyanosis is characterized by bluish pigmentation, which replaces the normal pinkish red in the skin, and reflects a deficiency of oxyhemoglobin in the capillaries. The congenital disease is rare, and toxic methemoglobinemia is not common, but the laboratory may be asked to test for methemoglobinemia as one possible cause for cyanosis in a patient.

Methemoglobin is a brownish colored pigment; its solutions at neutral pH show an absorption band at 630 to 633 nm., characteristic of MetHb alone. If a solution of MetHb is treated with cyanide, the reddish cyanomethemoglobin formed has only a weak absorption band at 633 nm. This drop in absorbance at 633 nm. on conversion of MetHb to MetHbCN was used by Evelyn and Malloy (see Dubowski[5]) as a basis for a procedure to estimate the methemoglobin level in blood (see the following procedure).

Determination of Methemoglobin in Blood

REAGENTS

1. Phosphate buffer, *M*/15, pH 6.6. Dissolve 1.89 gm. of Na_2HPO_4 (anhyd.) and 2.85 gm. of KH_2PO_4 (anhyd.) in water to make 500 ml. of solution. Dilute this 1:4 with water to form fresh *M*/60 buffer. Store the reagent refrigerated and discard when microbial growth is evident.

2. Neutral sodium cyanide. This is prepared fresh as needed in 10 ml. lots from 10 per cent (w/v) NaCN and 12 per cent (v/v) acetic acid. Working in a well ventilated hood, add 5 ml. of the NaCN to a 10 ml. glass-stoppered cylinder, followed by 5 ml. of acetic acid. Stopper the cylinder and mix the contents by inversion. Use appropriate care in pipetting the poisonous NaCN. Carefully pour the unused reagent into the hood drain and thoroughly flush the drain.

3. Triton X-100 (Hartman Leddon Co.)

4. Potassium ferricyanide, 5 per cent (w/v), prepared fresh.

5. Two or three specimens of fresh, nonclotted blood which are to be used as normal controls.

6. Reagents for the determination of blood hemoglobin by the cyanomethemoglobin method (see the respective method).

SPECIMEN

Use whole blood, collected preferably in heparin or citrate, although oxalated blood is satisfactory. The specimen should be fresh, and should be analyzed within 2 hours. Lipemic bloods cannot be used directly; instead, use a saline suspension of washed cells. This is prepared by centrifuging the lipemic blood, discarding the plasma, washing the residual cells with saline, and then reconstituting to the original volume with saline.

PROCEDURE

1. Pipet 10.0 ml. of $M/60$ buffer into 15 × 120 mm. tubes and add 100 μl. each of the well mixed test and control specimens.

2. Add 1 drop of Triton X-100 to each tube, mix the tubes by gentle inversion six to eight times, and then allow to set for 5 minutes. (The Triton promotes lysis of the cells and minimizes turbidity.)

3. After 5 minutes, pour the solutions into cuvets and read the absorbances at 633 nm. in any convenient spectrophotometer. Record this reading as A_1.

4. Add 1 drop of neutral cyanide to the solutions in the cuvets, mix the cuvet contents, and re-read the absorbances after 5 minutes at 633 nm. Record these readings as A_2.

7. Then determine the total hemoglobin content of the blood specimens (tests and controls) using the cyanomethemoglobin procedure described in a previous section.

CALIBRATION

Lyse 100 μl. of each of two normal blood specimens in 10 ml. buffer (step 1). Then add 0.10 ml. of 5 per cent ferricyanide, mix the tubes, and allow to stand for 2 minutes. (All hemoglobin is converted to MetHb.) Read the absorbance ($= A_3$) at 633 nm. Add 1 drop of neutral cyanide (MetHb → MetHbCN), and read the absorbance ($= A_4$) after 2 minutes. Calculate the difference ($A_3 - A_4$), which represents the absorbance of all the hemoglobin, expressed as methemoglobin. The total Hb being known (step 7), calculate the calibration factor F_m for MetHb from

$$F_m = \frac{\text{gm. Hb/100 ml.}}{(A_3 - A_4)}$$

The F_m values of the two normal control bloods will usually check within 2 or 3 per cent. The average of the two is used in calculating the MetHb value of the unknown.

CALCULATIONS

$$\text{gm. MetHb/100 ml.} = (A_1 - A_2) \times F_m$$

$$\text{MetHb, as per cent of total Hb} = \frac{\text{gm. MetHb}}{\text{gm. Total Hb}} \times 100$$

INTERPRETATION

The quantity of MetHb found in presumably normal blood specimens is from 0.00 to 0.25 gm./100 ml. of blood, averaging about 0.10 gm./100 ml. This is equivalent to about 0.4 to 1.5 per cent of the total Hb.

In patients with *toxic methemoglobinemia*, values up to 20 or 30 per cent of total Hb are encountered.

REFERENCES

Dubowski, K. M.: Measurement of hemoglobin derivatives. *In* Hemoglobin, its Precursors and Metabolites. F. W. Sunderman and F. W. Sunderman, Jr., Eds. Philadelphia, J. B. Lippincott Co., 1964.

Van Kampen, E. J., and Zijlstra, W. G.: Determination of hemoglobin and its derivatives. *In* Advances in Clinical Chemistry. H. Sobotka and C. P. Stewart, Eds. New York, Academic Press, Inc., 1965, vol. 8.

IDENTIFICATION OF HEMOGLOBIN VARIANTS

Although a large number of variants of human hemoglobins have been encountered and are the subject of special investigations, only a few are of routine clinical interest. The most common, normal form of Hb is referred to as *hemoglobin* A_1 (HbA_1). Fetal blood contains HbA_1 as well as a sizable proportion of HbF, a species of hemoglobin characterized by its resistance to denaturation by alkali. The blood of patients with sickle cell disease contains HbS, in which valine replaces glutamic acid in position 6 of the β chain. In HbC, lysine replaces the No. 6 glutamic acid residue in the β chain. Thus, hemoglobins A_1, S, and C are identical except for the nature of the amino acid residue at position 6 in the β chain.

The structure of HbA_1 is denoted by the symbol $\alpha_2^{A_1}\beta_2^{A_1}$, and that of HbS and HbF by the symbols $\alpha_2^{A_1}\beta_2^{S}$ and $\alpha_2^{A_1}\gamma_2^{F}$. In HbF ($\alpha_2^{A_1}\gamma_2^{F}$), both β chains are replaced by a pair of different polypeptide chains referred to as γ chains, and in HbA_2 ($\alpha_2^{A_1}\delta_2^{A_2}$), which is present in normal individuals to the extent of 2.5 per cent, the β chains are replaced by δ chains. *Hemoglobin-Bart* is designated γ_4^{Bart} because it is built up of four γ chains, of the variety first seen in a patient with that name.

Paper electrophoresis of hemoglobin solutions, used until recently to separate and identify hemoglobin variants, has been replaced by cellulose acetate and starch gel electrophoresis, which provide faster and sharper separations. Hemoglobins S, C, A_2, and F all migrate at rates slower than the normal HbA_1, and can readily be distinguished and identified by electrophoresis. It is more convenient to quantitate HbF and HbS by taking advantage of some of their unique physicochemical properties. HbF is considerably more resistant to denaturation by strong alkali than the other hemoglobin forms and a procedure based on this property will be presented. The deoxygenated form of HbS is considerably less soluble than the reduced forms of the other hemoglobin species. It is this loss of solubility on the part of deoxygenated HbS in the red cells of patients with sickle cell anemia that gives the cells their singular (elongated, curved) shape. The reduced hemoglobin precipitates out inside the erythrocytes, and the pseudocrystals of reduced HbS distort the normal shape of the cells.

Hemoglobin Electrophoresis

Collect fresh whole blood, using any type of anticoagulant. Pack the cells by centrifugation, discard the plasma, and wash the cells three or four times with 0.85 per cent saline.

Suspend the washed residual cells in about twice their volume of water. Add about 0.5 ml. of toluene to aid lysis, and intermittently shake the suspension for about 10 minutes. After hemolysis is complete, centrifuge the material at 3000 to 4000 rpm for 10 minutes; with a pipet remove the top layer of toluene, containing most of the stroma (cell walls), and recentrifuge the solution of Hb to clear it from any remaining stroma. Assay the total Hb in the solution and adjust the concentration to about 10 gm./100 ml. Use this solution for the electrophoretic separation.

The buffer routinely used is barbital, pH 8.6, although, for separation of some of the rarer forms, phosphate buffers of pH 7.8 and 6.5 have been used. The amount of hemolysate applied to the cellulose acetate strip is about 2 to 5 μl. After the run, localize the separated fractions either by staining the Hb with benzidine-peroxide (see the section on occult blood), or by staining the protein with some suitable protein dye. At pH 8.6 all variants migrate to the anode; the order of migration, beginning with the fastest moving form, is: H = I, Bart, J = N, K, A_1 = M, F = P, S = D, A_2 = E, and C. Those forms separated by an equal sign (=) migrate at about identical rates.[10] There is, as a rule, no real difficulty in separating the four variants of most clinical interest (A_1, F, C, S); however, it must be appreciated that a fraction that appears to move as HbS may really be an HbD, or an A_2 may perhaps be an HbE. The former two may be differentiated by the solubility test, if the clinical picture is not compatible with the presence of HbS. Other tests that aid in the differentiation between the more common and the less common forms are also available. (The reader may consult Huisman,[11] Hutchinson, and the Sunderman-Sunderman Symposium on Hemoglobin.)

To identify the various forms of hemoglobin properly, it is advisable to run known specimens with and alongside the unknown. This is especially important when such rare forms as hemoglobin D, K, or Bart are suspected. The more common forms of hemoglobin are available from many laboratories that are engaged in the performance of hemoglobin electrophoresis but the rare forms can only be obtained from laboratories which possess banks of special cell types.

HbC will show an apparent migration to the cathode because of the considerable endosmotic flow occurring in Hb separations. At pH 6.6, all but the fastest moving form, H, are found on the cathodic side of the point of application.

Hemoglobin Variants and Associated Diseases

HEMOGLOBIN A_2

Normal blood specimens contain only the most common form, HbA_1, and a small fraction (1.8 to 3.5 per cent) of the slower moving HbA_2. Patients with Cooley's anemia (thalassemia) frequently have elevated values of A_2. The ratio of HbA_2 to A_1 hemoglobin can be determined after electrophoresis by cutting the strip or starch gel (the preferable medium) in order to isolate the HbA_1 and A_2 sections. Then the hemoglobins from each section are eluted with NaOH, treated with benzidine and H_2O_2, and measured spectrophotometrically. The ratio of A_2 to A_1 may then be calculated.

The nature of the Hb defect in *thalassemia* is still not well understood. There is no evidence for the existence of any specific defect or modification in any of the peptide chains. Thalassemia is relatively common among Mediterranean and Central and East European peoples and, therefore, to some degree is also found in the United States. Affected persons develop severe anemias and their hemoglobin patterns often show elevated C, F, and A_2 fractions. Apparently, a multiple gene defect is involved rather than a single gene abnormality.

HEMOGLOBIN S

About 8 to 10 per cent of American Negroes have blood containing HbS; among West African Negroes this percentage may be as high as 40 per cent. A homozygote for HbS has only this variant of hemoglobin and no HbA_1. Such a person will be affected with *sickle cell disease*, characterized by a severe hemolytic anemia, recurrent sickle cell crises, and a very short life expectancy. A heterozygous individual will have both HbA_1 and HbS, and is said to have "*sickle trait*" *disease*. The blood of many sickle cell anemia patients also contains some HbF and occasionally increased HbA_2 as well.

HEMOGLOBIN F

HbF is one of the embryonic forms of hemoglobin, facilitating efficient transport of oxygen to the tissues of the fetus from placental blood of limited oxygen content. During the last months of fetal development it is the predominant form of hemoglobin in fetal blood, and, at birth, newborns have some 80 per cent of their hemoglobin in the F form, with the rest in the A_1 form. Gradually, the HbF is replaced by the adult HbA_1, and by two years of age, the HbF concentration has dropped to the adult level of less than 1 per cent.

Procedures for performing the alkaline denaturation test for HbF, and the solubility test for HbS will be presented. Details for the electrophoretic procedures for identifying hemoglobins can be taken from the cited references.

Alkali-Resistant Hemoglobin (Hemoglobin F)

With the exception of HbF, all hemoglobin variants are rapidly denatured when exposed to an alkaline environment. The quantitative procedure for HbF takes advantage of this property.[8] Almost no HbF is denatured when exposed to a pH of 12.7 for 1 minute, whereas other hemoglobin varieties are completely denatured, the denaturation being characterized by loss of solubility. HbF is demonstrable in fetal blood beginning with the fifth week of gestation, and attains a level of 95 per cent of the total hemoglobin at the twentieth week. Thereafter there is a very slow decrease, so that by birth, the level has dropped to a range of 65 to 90 per cent of total Hb. Elevated concentrations of HbF may often be found, along with HbS and HbC, in both adults and children with sickle cell disease and thalassemia; occasionally HbF is encountered in hemolytic anemia patients, in whom it is apparently synthesized to compensate for the deficiency in ordinary HbA_1. In the very rare disease, *familial fetal hemoglobinemia*, it entirely replaces HbA_1. HbF can be detected and separated from all other hemoglobin variants except HbD by the use of starch gel electrophoresis, but it is simpler to establish an increased level of HbF by the alkali denaturation test.

Determination of Hemoglobin F by Alkali Denaturation

REAGENTS

1. KOH, 0.083 *N*. This can be prepared by careful dilution of 3.5 *N* KOH [20 per cent KOH (w/v)]. Check the concentration by titrating against standard 0.1 *N* HCl; it should be 0.083 ± 0.001 *N*. Store the solution in a tightly stoppered polyethylene bottle.

2. Ammonium sulfate (acidified, 50 per cent saturated). Prepare a saturated solution by adding with vigorous stirring 375 gm. of $(NH_4)_2SO_4$, A. R., to 500 ml. of H_2O. Dilute 250 ml. of supernatant with water to 500 ml. Then add 2.05 ml. of 6.0 *N* HCl. Check the acidity of the final reagent by adding 17.0 ml. of acid $(NH_4)_2SO_4$ to 8.0 ml. of 0.083 *N* KOH (6.8:3.2). The pH of the mixture should be between 6.5 and 7.8. Store this reagent in a polyethylene bottle.

3. EDTA-Na_4 solution, 0.3 per cent (w/v) in water.

PROCEDURE

1. Preparation of hemoglobin hemolysate: Any anticoagulant may be used in drawing the blood. After centrifugation, remove the supernatant plasma and wash the residual cell mass three or four times with 0.9 per cent (w/v) NaCl solution. Dilute the washed, packed cells with an equal volume of water and add 0.4 volume of toluene. Vigorously shake the mixture and set aside in a refrigerator overnight. Again shake the solution and then centrifuge for 10 minutes at 3000 to 4000 rpm. Carefully collect the hemoglobin solution under the toluene in a pipet, then recentrifuge, and use the clear solution in the test.

2. Working rapidly, pipet 0.20 ml. of hemolysate to 3.2 ml. of 0.083 *N* KOH in a 15 × 120 mm. test tube. Rinse the 0.20 ml. pipet three to four times with the KOH solution. The moment that the first half of the pipet contents have been transferred to the KOH, start a stopwatch, and rapidly but gently mix the tube contents.

3. At exactly 60 seconds, rapidly add 6.8 ml. of acid $(NH_4)_2SO_4$, mix the tube by inverting eight to ten times, and filter the contents through a Whatman No. 41 filter paper. Collect the filtrate.

4. While the material is filtering, set up a control by making a 1:250 dilution of the hemolysate in 0.3 per cent EDTA (0.10 ml. to 25.0 ml.).

5. Read the Absorbances of the control and filtrate at 540 nm., using any convenient spectrophotometer and appropriate cuvets.

CALCULATION

The control is the measure of the total hemoglobin. In determining the HbF, the hemolysate is diluted 0.2:10.2 = 1:51. HbF is customarily reported in terms of per cent of total hemoglobin as shown in the following formula.

$$\text{per cent Hb as HbF} = \frac{A_F}{A_C} \times \frac{50}{250} \times 100$$

$$\text{per cent HbF} = \frac{A_F}{A_C} \times 20$$

COMMENTS

By this procedure, the fraction of HbF in normal adult blood is under 2 per cent of total hemoglobin, of which probably only one half is true HbF. The precision of the result is of the order of 20 per cent at the normal level, and somewhat better at elevated HbF levels. Blood from patients with sickle cell anemia may show values for HbF ranging from normal to 25 per cent of total Hb, whereas that from sickle cell trait patients will usually give the same values as observed in normal adults.

REFERENCES

Goldberg, C. A. J.: Hemoglobins. In Standard Methods of Clinical Chemistry. D. Seligson, Ed. New York, Academic Press, Inc., 1961, vol. 3.

Huisman, T. H. J.: Normal and abnormal human hemoglobins. In Advances in Clinical Chemistry. H. Sobotka and C. P. Stewart, Eds. New York, Academic Press, Inc., 1963, vol. 6.

Hemoglobin S and the Sickling Phenomenon

Although HbS may be readily demonstrated by an electrophoretic study, it is more convenient to demonstrate the presence of HbS and to obtain an estimate of the quantity present by carrying out Itano's *ferrohemoglobin solubility test.*[8] (Ferrohemoglobin is another term or expression for deoxygenated or reduced hemoglobin.) Deoxygenated HbS is considerably less soluble in solutions of moderately concentrated salt solutions than are other hemoglobin variants. Although a number of different salts may be used, Itano developed a procedure using approximately 2.50 *M* phosphate as the precipitating agent. In the test, the hemoglobin solution is diluted with the phosphate solution, deoxygenated with $Na_2S_2O_4$, filtered to remove the precipitated HbS, and the remaining Hb measured.

Determination of Hemoglobin S by the Ferrohemoglobin Solubility Test

REAGENTS

1. Sodium dithionite, $Na_2S_2O_4$, Reagent grade, dry powder.
2. Phosphate buffer, 2.50 *M*, pH 7.0 ± 0.2. Dissolve 109 gm. of K_2HPO_4 and 85 gm. of KH_2PO_4 in water and dilute to 500 ml. Store the reagent at room temperature in a polyethylene bottle.

PROCEDURE

1. Dissolve approximately 20 mg. of $Na_2S_2O_4$ (sodium dithionite) in 1.8 ml. of the phosphate buffer in a 10 × 75 mm. test tube. The $Na_2S_2O_4$ may be added from the end of the spatula.
2. To this solution add 0.20 ml. of hemolysate with a TC pipet, rinse the pipet and mix the tube by gentle inversion. Allow the tube to set for 15 minutes.
3. Filter the contents of the tube through a Whatman No. 42 filter paper, in a 25 mm. funnel. Add 0.20 ml. of the filtrate to 4.8 ml. of phosphate buffer containing 20 mg. of $Na_2S_2O_4$.
4. Set up a control by diluting the hemolysate 1:250 as described in the procedure for alkali-resistant hemoglobin.
5. Read the absorbances of the diluted filtrate and control at 422 nm. against a water blank, using a narrow band-pass spectrophotometer, or at 420 nm. using a simpler instrument. If the absorbances are too high, dilute the solutions with an equal volume of buffer.

CALCULATION

The solubility test specimen has been diluted 1:10 × 1:25 = 1:250, identical to the dilution of the control.

$$\text{Per cent soluble Hb} = \frac{A_T}{A_C} \times 100$$

COMMENTS

The ranges for soluble hemoglobin found with various hemoglobin genotype patterns are as follows:[8]

Hemoglobin Genotype	*Per cent Total Hemoglobin that is Soluble*	*Clinical Condition*
A/A	90–105 per cent	normal
A/S	35–70 per cent	sickle cell trait
A/C	85–105 per cent	hemoglobin C trait
S/S	6–25 per cent	sickle cell disease
S/C	35–55 per cent	sickle cell C disease

TESTS FOR OCCULT BLOOD

The detection of sources of internal bleeding is often a challenging diagnostic problem. Recent and rather severe bleeding can usually be recognized visually on examination of a urine or stool specimen, or of a gastric or duodenal aspirate. The presence of small amounts of blood or of altered blood, however, cannot be so detected, and such blood is, therefore, referred to as hidden, "*occult*," blood. Normally, no blood should be present in a gastric or duodenal aspirate; testing of such material for occult blood will be discussed in the chapter on gastric analysis. Stool specimens and urines will always contain small amounts of blood or hemoglobin breakdown products. A normal 12 hour overnight urine may contain up to a half-million erythrocytes, but several times this amount may be necessary to insure detection by the usual microscopic examination or the available chemical tests. As much as 1 or 2 ml. of blood may normally enter the fecal matter being formed in the intestinal canal. A fair portion of this is degraded and is present in the stool as hematin, porphyrins, and other products. Any tests for occult blood should be able to differentiate between such normal, expected levels of blood and increased levels definitely suggestive of clinically significant bleeding, such as that from ulcers and malignant growths. Other possible sources of blood, however, such as that derived from the nose or mouth (tooth brush), from hemorrhoids, and from menstrual flow must be excluded. Clinically significant bleeding may be intermittent, and serial tests at regular intervals may be necessary to establish a diagnosis. Severe gastric or duodenal bleeding of the order of 60 to 90 ml. of blood per day will result in the formation of so-called "tarry stools," stools of shiny black color. Less severe bleeding will not affect the usual appearance of the stool, and the presence of blood must be confirmed chemically. The presence of obvious red streaks or spots will suggest lower colon or perianal bleeding. Ingested meat contains hemoglobin and myoglobin and the presence of these in the stool may give false positive tests for fecal occult blood. For best results, patients should be placed on a meat-free or low meat diet for several days before the evaluation of their stools for occult blood.

Blood in *urine* may arise from intravascular hemolysis, transfusion reactions, rupture of glomerular capillaries, and from bleeding points in the genitourinary system. *Hematuria* exists whenever *intact red cells* are present, as established by a careful microscopic examination of the urinary sediment. The acidity and osmotic strength of urine promote the rapid hemolysis of red cells and the release of the cellular hemoglobin into the urine (*hemoglobinuria*). Hematuria is seen only in genitourinary bleeding, but hemoglobinuria may also occur in conditions involving intravascular hemolysis, if the degree of hemolysis is severe enough to more than saturate the haptoglobin present in the plasma.

All tests for occult blood are based on the peroxidase action of hemoglobin, which has already been discussed. In addition to benzidine, chromogens that have been used include *o-tolidine* and *di-orthoanisidine*, analogs of benzidine, *guaiacol* (a component of the resin gum guaiac), and *leukophenolphthalein*, prepared by zinc reduction of the indicator dye phenolphthalein. With the exception of phenolphthalein, the color of the oxidized forms is blue, although the color observed may vary from green to a deep blue-black, depending on the quantity of Hb present and the proportions of chromogen and peroxide used. The formulas of the most commonly used reagents and that proposed for benzidine blue are given in Figure 6-2. All heme-containing materials give a positive reaction; among such materials are myoglobin

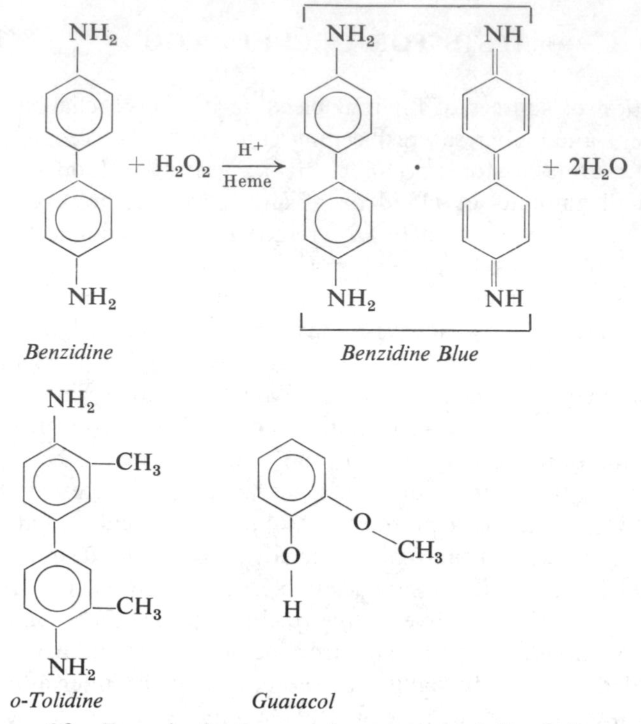

Figure 6-2. Formulas for the more important occult blood chromogens.

(meat) and the heme enzymes, catalase, and peroxidase, present in many plant foods. The presence of these enzymes in stool specimens is overcome by inactivating them by boiling suspensions of stool specimens.

Procedure for Detection of Occult Blood in Urine and Feces

REAGENTS

1. o-Tolidine stock solution, 4 per cent (w/v) in absolute ethyl alcohol. The reagent is stable for many months if refrigerated, but it should be discarded when any coloration appears. It is convenient to distribute the reagent into test tubes in 2 or 3 ml. quantities, and to remove tubes as needed to prepare fresh working reagent.

2. o-Tolidine working reagent, 1.33 per cent (w/v). Dilute the stock reagent with an equal volume of glacial acetic acid and an equal volume of water. The solution is stable for 30 days if refrigerated.

3. Hydrogen peroxide, 6 per cent. Prepare by diluting Superoxol (30 per cent H_2O_2) 1:5 with water. The reagent is usually stable for 1 month, but it should be tested for activity regularly.

TEST SPECIMENS

Stool. With the aid of an applicator stick, suspend a portion of stool, the size of a pea (5 mm. in diameter), in 5 ml. of water. Boil gently for a few seconds and cool before testing. Preferably, the original specimen should be obtained after the patient has been on a meat- and fish-free diet for 3 days.

Urine. Heat several milliliters of well mixed, uncentrifuged urine to boiling and allow to cool before testing. If the urine contains considerable amounts of ascorbic acid (over 25 to 30 mg./100 ml.), falsely weak or negative results may be obtained, since the ascorbic acid is preferentially oxidized by the peroxide. In such a case, acidify several milliliters of urine with concentrated acetic acid to a pH of about 3, and extract with an equal volume of diethyl ether. Collect the ether layer in a separate tube and evaporate to dryness by immersing the tube in a beaker of warm water. Dissolve the acid hematin residue in about 0.5 ml. water and then test it.

Boiling destroys any enzymes present in pus cells. Centrifugation of the urine will remove all intact red cells and thus only dissolved hemoglobin will be detected.

PROCEDURE

1. Place 4 drops or 0.2 ml. of working o-tolidine and 1 drop or 0.1 ml. of 6 per cent H_2O_2 in a 13 × 75 mm. test tube and let stand for 1 minute. If no color appears, the reagents and tube are not contaminated and the test is then completed. There should be no green-blue color in the mixed reagents, but a light yellow-green color will not invalidate the test.
2. Add 1 drop or 0.1 ml. of fecal suspension, or 3 to 4 drops (0.2 ml.) of urine and observe the color formed at 1 minute.

INTERPRETATION

If no or a light green color forms, the test is read as negative. A definite deep green, green-blue, or blue color is interpreted as positive for occult blood. The color will intensify with time. The rate and intensity of color formation increase with an increase in the quantity of blood in the specimen. The color seen may be graded 1+, 2+, and 3+; however, this is not recommended because the intensity of color obtained may be affected by the presence of inhibitors of an unknown nature.

COMMENTS

1. The benzidine reaction is not catalyzed by inorganic ions.
2. There has been a considerable difference of opinion as to the ideal concentration and the most favorable proportion of the two reagents in the test.
3. The order of the addition of reagents is not important, provided that sample and peroxide are not in contact with one another except in the presence of the chromogen.
4. Peroxide solutions may decompose and it is wise to test the reagents regularly using a 1:1000 dilution of blood in urine as test material.
5. Benzidine is now felt to be a possible cause of bladder cancer, and its frequent use has been discouraged. There is no advantage in replacing it with o-tolidine because all diaminodiphenyl derivatives are weakly carcinogenic materials.
6. Guaiacol is also used for testing feces, but is too insensitive for urine.

ALTERNATIVE PROCEDURES

Benzidine test. Dissolve 1 gm. of benzidine hydrochloride in 20 ml. of glacial acetic acid and dilute with 30 ml. of H_2O and 50 ml. of ethanol. This serves as the chromogen; use 3 per cent H_2O_2 as the oxidant. Mix 0.5 ml. each of benzidine and specimen, followed by 0.5 ml. of peroxide, and observe the color after 1 minute.

Occult test. For urine specimens, the commercially available *Occult Test* tablets (Ames Co., Div. Miles Laboratories, Elkhart, Inc.) may be used. The chromogen used is o-tolidine, and strontium peroxide is the source of peroxide. This material is too sensitive for use with feces.

REFERENCES

Kohn, J., and Kelly, T.: J. Clin. Path., *8*:249, 1955.
Varley, H.: Practical Clinical Chemistry. 4th ed. New York, Interscience Publishers, Inc., 1964, pp. 344–348.

DIFFERENTIATION BETWEEN MYOGLOBIN AND HEMOGLOBIN

Myoglobin (Mgb), the heme protein present in muscle cells, facilitates the transfer or transport of oxygen from the extracellular fluid into the muscle cells proper. Whenever there is severe injury to striated muscle tissue, considerable quantities of myoglobin are released into the extracellular fluid, eventually enter the blood, and are then excreted by the kidneys. *Myoglobinuria* may occur following crush injuries to the extremities, after high voltage electric shock, and occasionally after severe muscular exertion. Because both Hb and Mgb give a positive benzidine test, differentiation between the two materials can be an important laboratory problem.

Normally no Mgb is present in plasma. The small amount that is continuously released from muscle tissue is rapidly cleared by the kidneys into the urine. Unlike Hb, myoglobin is not bound to haptoglobin, which is a specific hemoglobin-binding protein. Mgb present in urine may be accompanied by Hb released from capillaries that were damaged by the same injury that affected the muscle. The clinician, especially in accident cases, will want to know whether the hemoprotein present in urine is one or the other of the two proteins or perhaps a mixture of both.

The two proteins can be distinguished by the properties that arise from their difference in molecular size. Myoglobin, having a molecular weight one-fourth that of hemoglobin, is soluble in 80 per cent saturated ammonium sulphate, whereas Hb is not. This is the basis of the simplest procedure for differentiating the two pigments. The two materials can also be separated by ultrafiltration with the proper choice of the filter membrane. The smaller Mgb molecule will pass through the pores, but Hb, the larger molecule, will be retained.

Within the last several years, more precise and delicate methods have become available. Electrophoresis permits separation of Hb and Mgb when both are present in concentrations as low as 10 mg. per 100 ml. The preferred technique uses cellulose acetate membranes and veronal buffer at pH 8.6.[2] The patient's urine is run in parallel with another aliquot to which some Hb from the patient's red cells is added. Both Hb and Mgb move to the cathode, but Mgb at a faster rate. After the run the strips are stained with di-o-anisidine and peroxide and compared. The position of the Hb is evident on the strip to which Hb had been added; a second, lighter band may suggest that Mgb is also present. The strip with the untreated urine may show one or two bands, and by comparing it with the control strip, it can be ascertained whether the patient's urine contained Hb, Mgb, or both.

Several immunochemical methods are now available. Antisera to human Mgb and human Hb can be prepared by injecting rabbits or other animals with the corresponding human heme protein. In the *Ouchterlony immunodiffusion technique*, the specimen and the antiserum are permitted to diffuse toward each other through agar. If the test specimen contains the hemoprotein that is the specific antigen to the antibody in the antiserum, a white arc of precipitate will form where the two diffusing fronts

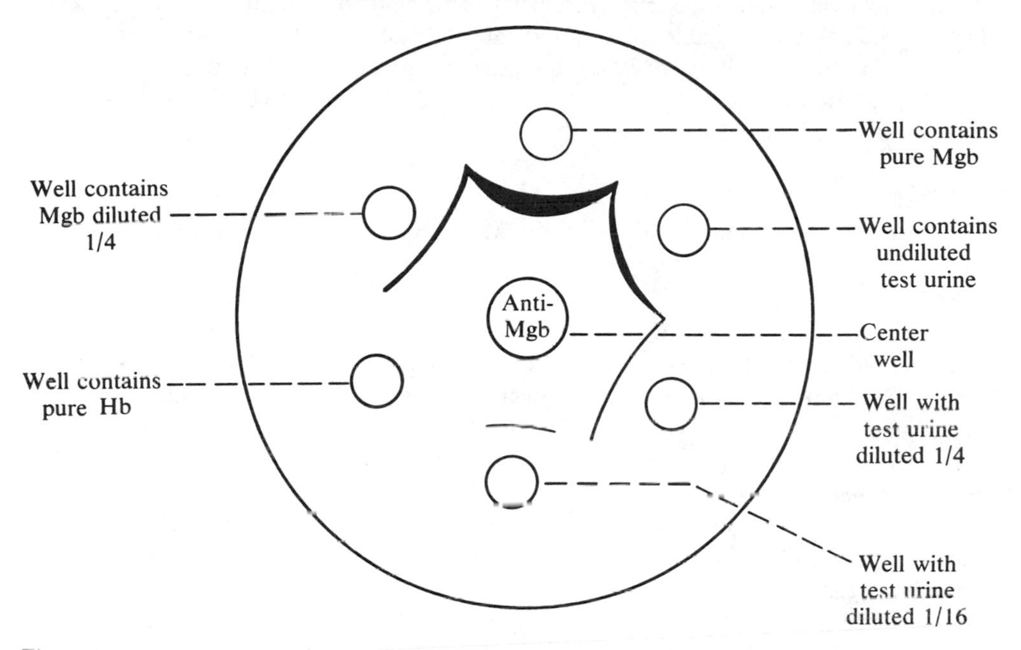

Figure 6-3. Drawing illustrating precipitation lines in an immunodiffusion plate. The urine specimen is assumed to contain myoglobin. Precipitation arcs are formed with material diffusing from the sample and myoglobin standard wells. Antiserum to myoglobin is contained in the center well. The position of the arc and its thickness will depend on the concentrations of antibody and antigen. No arc is formed between the anti-Mgb and hemoglobin.

meet. A drawing of such a plate is shown in Figure 6 3. Two plates are used, one with anti-Hb and the other with anti-Mgb in the center well. If Mgb is present, a precipitant line will form between the sample antigen well and the anti-Mgb center well. Appropriate controls are run along with the test specimen. If the urine contains both Mgb and Hb, precipitin lines will be seen on both plates. Rough quantitation is possible by using a series of dilutions of the unknown, and by using known amounts of Hb or Mgb, or both, in the control antigen wells. The reaction between Mgb in the unknown and its antiserum may also be carried out in a test tube or microscope slide, and as little as 0.08 mg./100 ml. of Mgb can be detected. This technique can also detect any Mgb released as a result of a severe myocardial infarction.

REFERENCE

Yakulis,V. J., and Heller, P.: Am. J. Clin. Path., *37*:253, 1962.

Although these immunological techniques will soon be routine in the bigger laboratories, smaller units will still have to rely on the simpler techniques, which, however, can be clinically very informative if carefully performed. The differential solubility procedure presented here was developed by Blondheim et al.[1]

Procedure for Differentiation Between Urinary Myoglobin and Hemoglobin

REAGENTS

1. o-Tolidine solution, 1.0 per cent (w/v) o-tolidine in absolute methanol. The material dissolves slowly. The solution is stable in the dark.

2. Acid peroxide. Mix 25 ml. of glacial acetic acid with 50 ml. of fresh 3 per cent H_2O_2. The solution is stable refrigerated for 30 days.

3. Ammonium sulfate, A.C.S. grade, dry chemical, fine crystals.

4. Sulfosalicylic acid, 3 per cent (w/v) in water. Dissolve A. R. grade chemical in warm water and filter if not clear.

SPECIMEN

The urine specimen should be fresh, with pH adjusted to between 6.5 and 7.5.

PROCEDURE

1. **Test for presence of hematuria.** The centrifuged sediment is examined microscopically for the presence of intact erythrocytes, ghosts of red cells, or both.

2. **Test for presence of soluble heme pigment.** A test for occult blood is performed. Add 2 drops of o-tolidine to 4 drops of sediment-free urine. After mixing, add 3 drops of acid-peroxide. The development of a green to blue-black color demonstrates the presence of heme.

3. **Test for protein.** To perform the *SSA* test for protein, add 3.0 ml. of sulfosalicylic acid to 1.0 ml. of sediment-free urine. A precipitate or turbidity establishes the presence of protein.

4. **Differentiation between Mgb and Hb.** If test steps 2 and 3 are negative, no Mgb can be present and test 4 need not be done.

a. Add 2.8 gm. of $(NH_4)_2SO_4$ to 5.0 ml. of clarified urine. Gently mix the urine and salt with a stirring rod or by gentle inversion (to avoid denaturation of proteins).

b. When the salt is all dissolved, filter the urine through a Whatman No. 42 filter paper and collect the filtrate. A reddish tinge to any precipitate on the filter is supporting evidence for the presence of Hb.

c. Test the filtrate with o-tolidine for heme pigment as noted before. The tolidine must be added first, before the peroxide. If the test is positive, the presence of Mgb is confirmed, since the urine contains a heme protein soluble in 80 per cent saturated ammonium sulfate solution.

COMMENTS

1. If the Mgb is denatured (old urine, extremes of pH, or vigorous mixing), it will be precipitated along with the Hb.

2. The test will detect Mgb if present at concentrations about 1 to 5 mg./100 ml. Hemastix (Ames Co.) may be used in place of the tolidine-peroxide.

REFERENCE

Blondheim, S. H., Margoliash, E., and Shafrir, E.: J.A.M.A., *167*:453, 1958.

Porphyrins and Related Compounds

CHEMISTRY OF THE PORPHYRINS

The porphyrins are complex ring compounds made up of four pyrrole rings linked together by methyne (—CH=) bridges. The formula for pyrrole is as follows:

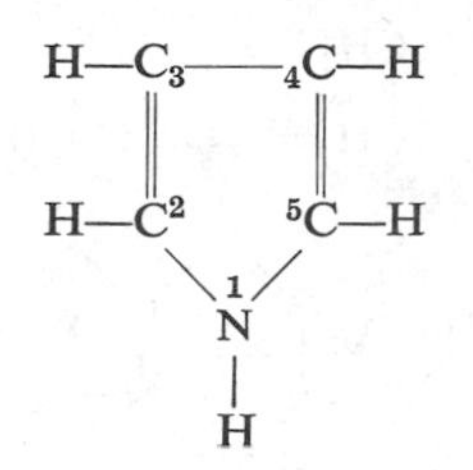

(3)

Carbon atoms 2 and 5 are involved in the methyne linking; the hydrogens on carbons 3 and 4 are replaced by a variety of other organic groups in the many individual porphyrins. In man and other mammals the most important porphyrin-containing compound is *heme*, which is present in hemoglobin, and in a number of important enzymes associated with respiration, such as those in the cytochrome system. The porphyrin in heme is termed protoporphyrin IX (Type III). The formula for heme, which contains ferrous iron linked to the protoporphyrin, is presented in Figure 6-4. The ferrous ion replaces two hydrogens on the pyrrole nitrogens, forming a structure in which the iron is linked to all four nitrogen atoms by covalent and coordinate bonds. In plants the photosynthetic pigment *chlorophyll* is also a porphyrin compound, although of even more complex structure, with magnesium ion rather than iron bound to the pyrrole rings. The porphyrins of clinical interest are formed in the chain of metabolic reactions directed toward the synthesis of heme, in contrast to bilirubin and the urobilinogens, which are metabolic breakdown products of heme.

The conventional system used in identifying the four pyrrole rings, the methyne groups, and the substituents on the pyrrole rings is indicated in the formula for heme (Fig. 6-4). In place of this detailed structural formula, it will be convenient to use the following shorthand structure:

Protoporphyrin-IX-(III)

Symbols identifying substituents on the pyrrole rings:

A, B, C, D = pyrrole rings

A	Acetic acid = carboxymethyl	$-CH_2COOH$
E	Ethyl	$-CH_2CH_3$
M	Methyl	$-CH_3$
P	Propionic acid = carboxyethyl	$-CH_2CH_2COOH$
V	Vinyl	$-CH{=}CH_2$

Porphyrins are synthesized in the liver and the marrow of the long bones by a series of enzymatic reactions, starting with glycine and acetate. The series of reactions is presented schematically in Figure 6-5. Acetate is converted to succinyl-coenzyme A, which condenses with glycine with the loss of a carbon dioxide to form *δ-aminolevulinic acid* (DALA). Two molecules of DALA are condensed to form *porphobilinogen* (PBG), an unstable pyrrole derivative containing acetic acid and propionic acid groups substituted on the 3 and 4 positions, respectively, of the pyrrole ring. Four molecules of PBG are then linked together into a ring to form a molecule of *uroporphyrinogen* (UPG). The presence of two different substituents on carbon atoms 3 and 4 in the PBG molecule permits the formation of four different isomers when the four PBG units join together to form the porphyrin ring. Using A and P to denote the acetic and propionic acid groups, the four possible isomers, differentiated by the order

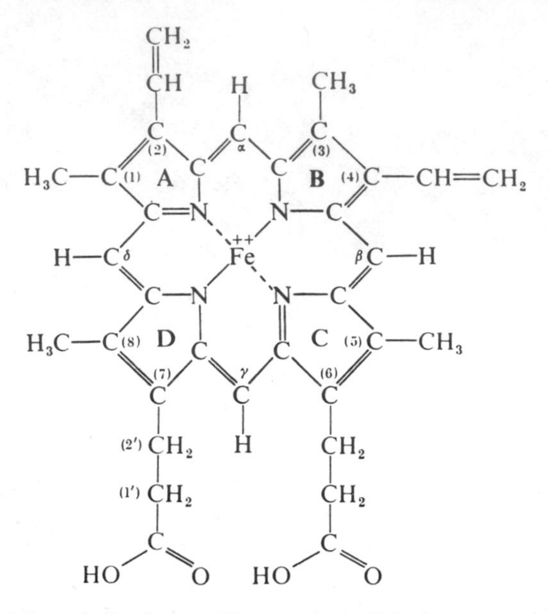

Figure 6-4. The structural formula for heme. The porphyrin linked to the ferrous iron is protoporphyrin IX, type III. Its formal chemical name is 1,3,5,8-tetramethyl,2,4-divinyl,6,7-bis (2′-carboxyethyl) porphin. The conventional labels for the pyrrole rings (**A, B, C, D**) and the methyne carbons (α, β, γ, δ) are indicated.

in which A and P occur around the ring, are designated as:

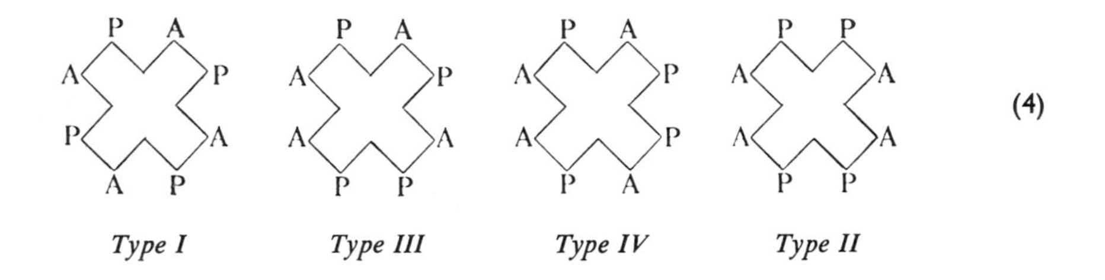

(4)

Only types I and III are formed physiologically; the normal pathway to heme synthesis leads only to type III. Although normally only trace amounts of the type I isomer are produced, in certain diseases this quantity may be greatly increased.

Uroporphyrinogen (UPG) is then converted to coproporphyrinogen (CPG), Type III, by decarboxylation of the four acetic acid groups, a CO_2 being removed from the —CH_2COOH to form a —CH_3 (methyl) group. The major portion of the UPG is converted to CPG, but a small fraction is oxidized to uroporphyrin (UP), the form usually found in the urine. In UPG methylene groups (—CH_2) link the pyrrole rings; these are easily oxidized to the methyne (—C═) links present in the porphyrins proper. UPG thus is a precursor of UP. Similarly, a part of the CPG is converted to coproporphyrin (CP), and excreted as such, both forms being found in urine. The main fraction of CPG undergoes further transformation to form protoporphyrin, PP, the propionic acid groups on positions 2 and 4 being decarboxylated to form ethyl

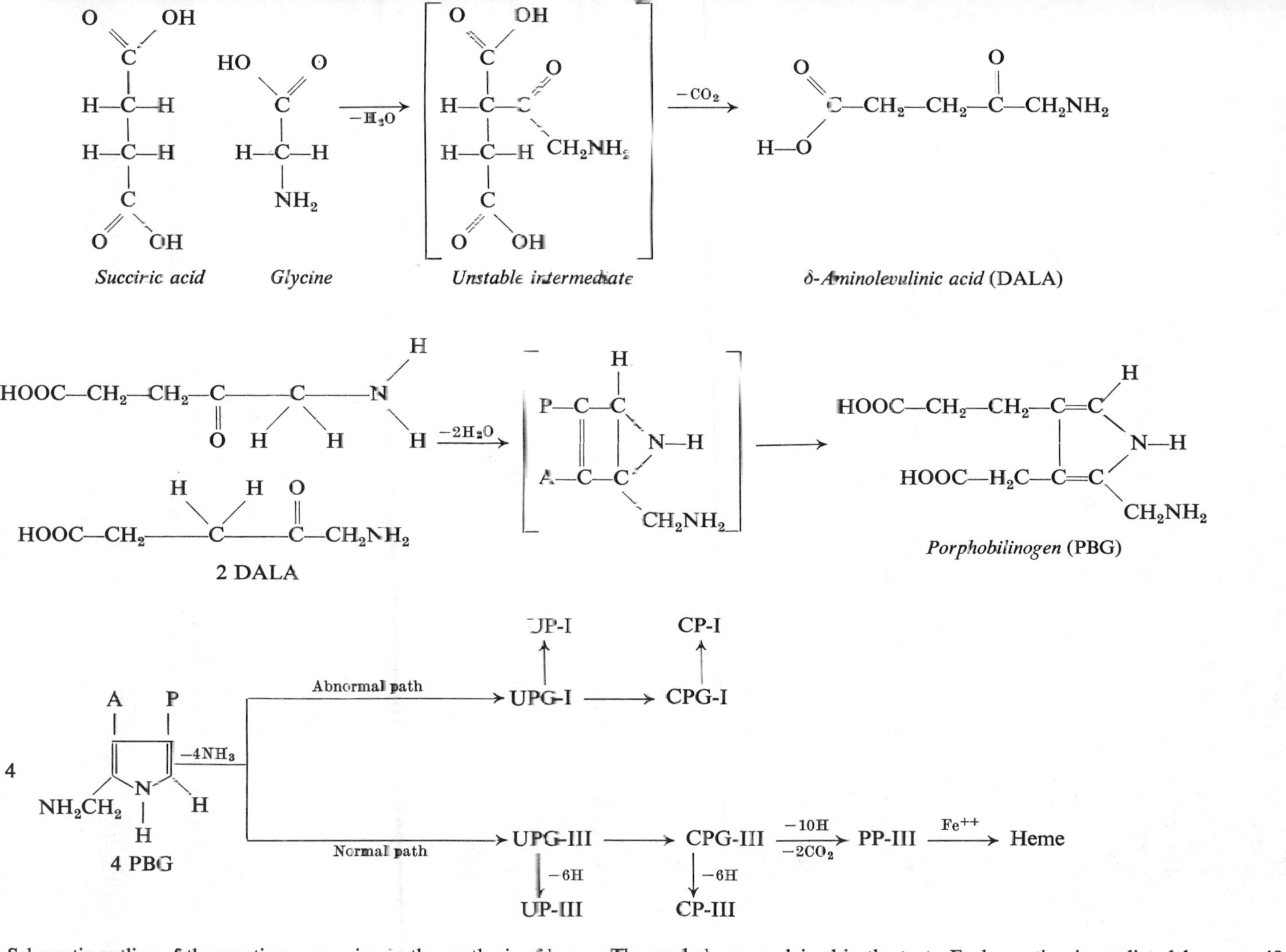

Figure 6-5. Schematic outline of the reactions occurring in the synthesis of heme. The symbols are explained in the text. Each reaction is mediated by a specific enzyme or group of enzymes.

groups, which are then dehydrogenated (loss of 2H) to form vinyl groups (—CH=CH_2). The PP then adds a ferrous ion to form heme.

CLINICAL SIGNIFICANCE

The heme group that ultimately becomes a part of the hemoglobin molecule is synthesized in the marrow of the long bones. Those heme groups that become the prosthetic groups of the heme enzymes, though synthesized in most cells of the body, are primarily of hepatic origin. Thus defects and abnormalities in the chain of reactions involved in porphyrin synthesis may occur as a result of defects either in the bone marrow or in the hepatic sites of synthesis. Such abnormalities are included in the group of diseases termed *porphyrias*, and are accompanied by increased urinary and fecal excretion of various *porphyrins*, or their precursors, or both. The *porphyrinurias* include those conditions in which an increased excretion of porphyrins, primarily coproporphyrin, is the result of some other primary disease state.

Erythropoietic porphyria is a very rare and fatal inherited defect, characterized by the excretion of very large quantities of both UP-I, and CP-I, the deposition of porphyrins in the teeth and bones, and an extreme sensitivity of the skin to light. *Protoporphyria* is a more common and relatively mild, recently recognized disease, characterized by skin sensitivity to light and increased fecal excretion of protoporphyrin. People affected by this disease develop symptoms only in summer, after exposure to sunlight on beaches. Both diseases are manifestations of a defect in the mechanism which controls the orderly production of the precursers of the heme ring in hemoglobin.

Several forms of *hepatic porphyrias* are recognized. In *acute intermittent porphyria* one finds increased excretion of DALA, PBG, and UP-III. The disease is also inherited, but usually manifests itself later in life (at age 30 or later). Acute attacks are often initiated by the intake of barbiturates, alcohol, and such drugs as Sulfanal and Sedormid. The chief clinical symptoms are abdominal colic and neurological disturbances. The disease is difficult to diagnose, and patients often undergo many medical and surgical manipulations unnecessarily, before the disease is recognized. In *porphyria cutanea tarda*, PBG execretion is normal, but the execretion of DALA, UP-III, and CP-III is moderately increased, and skin photosensitivity is present.

The most common condition leading to porphyrinuria is *lead poisoning*. Over 90 per cent of the urine specimens submitted to laboratories for porphyrin testing are from patients suspected of having lead poisoning. Adult patients are usually painters, using lead-base paints in areas without adequate ventilation and without masks, or workers in lead refineries and in lead battery plants. Most patients are children, however, many of whom, in the age range of 1 to 6 years, have a craving (referred to as *pica*) to chew on wood and plaster. If such material is coated with lead paint or otherwise contains lead salts, sufficient metal may be ingested to cause marginal or acute lead poisoning. If severe enough, such intoxication in young children may result in brain damage and death. Another source of lead poisoning is the inhalation of fumes from the careless burning of lead batteries. The chief symptoms are abdominal pain, porphyrinuria, and basophilic stipling of erythrocytes, along with neurological symptoms in very severe cases. The porphyrin present is coproporphyrin-III, but DALA execretion is also elevated. Porphyrinuria is also often, but not always, encountered in Hodgkin's disease, aplastic anemia, and hepatic disease.

Porphyrinuria urine specimens (from patients with lead poisoning) are usually tested only for the presence or absence of either CP or CP and UP. Recent experience,[4,9] however, has shown that quantitative measurement of DALA is a more sensitive diagnostic criterion. In investigating porphyrias, quantitative assays of 24 hour outputs of the various porphyrins or precursors are required before an adequate diagnosis may be made. Qualitative tests for PBG are useful in the diagnosis of intermittent porphyria, if equipment for quantitative measurement is not available. The color of the urine will often suggest the presence of porphyrins in abnormally large amounts, these urines having a characteristic "burgundy" color.

All porphyrins emit a characteristic orange-pink fluorescence when exposed to long wavelength ultraviolet light, such as that given off by common laboratory "black lights." This property is used for both the detection and assay of porphyrins. Porphyrins exhibit a strong Soret band absorbance in the 400 to 410 nm. interval. This property is used in the quantitative measurement of the individual porphyrins after they have been isolated from interfering compounds. In the following sections several representative methods for the detection and measurement of DALA, PBG, and porphyrins will be described.

PROCEDURES FOR THE DETERMINATION OF PORPHYRINS AND RELATED COMPOUNDS

Test for Porphyrinuria in Suspected Lead Poisoning

REAGENTS

1. Acetic acid, glacial.
2. Diethyl ether, reagent grade.
3. HCl, 5 per cent (w/v). Dilute 58 ml. of concentrated HCl to 500 ml. with water.

SPECIMEN

A fresh random urine specimen is best. If urine must be held overnight, add 0.5 gm. of Na_2CO_3/100 ml. of urine.

PROCEDURE

1. Transfer 10 ml. of urine to a 60 ml. Squibb type separatory funnel, preferably equipped with a Teflon stopcock and plastic stopper.
2. Acidify the urine with 4.0 ml. of acetic acid. (If the urine specimen contains Na_2CO_3, carefully swirl the funnel until all gas evolution has ceased.)
3. Add 20 ml. of ether, stopper the funnel, and shake vigorously for 4 to 5 minutes. In this step the coproporphyrin is extracted into the ether phase. After separation of phases, transfer the bottom (urine) layer to another funnel and re-extract with 20 ml. ether. Then discard the urine and combine the two ether extracts in the second funnel.
4. Add 2.0 ml. of 5 per cent HCl to the combined ether extract. Vigorously shake the funnel for 4 minutes. Under these more acid conditions, the CP enters the water (HCl) phase.
5. After separation of the phases, examine the HCl phase (in the lower end of the funnel) under a long wavelength ultraviolet lamp in a dark room or darkened corner for appearance of the characteristic pinkish orange fluorescence. The inexpensive

"Blak-Ray" (Wurtz light) lamp is a convenient UV source, which emits a band of light centering on 360 nm.

INTERPRETATION

Negative: No fluorescence observed; only the violet color of the lamp filter, or light blue or yellow-green fluorescence is seen.

Slightly positive (1+): Well marked fluorescence; if doubtful, the reading should be called negative.

Moderately positive (2+): Clear, strong fluorescence.

Strongly positive (3+): Very brilliant fluorescence.

Negative readings will be obtained if the CP level is under 15 μg./100 ml. The 1+, 2+, and 3+ readings will reflect CP concentrations of 15 to 25, 20 to 40, and over 50 μg./100 ml., respectively.

COMMENTS

Any uroporphyrins present will be retained in the original urine layer after the ether extraction of the CP. If present in large quantity, they can be detected by the fluorescence of the extracted urine. As a rule any UP fluorescence will usually be masked by the nonspecific fluorescence of other urinary pigments.

Procedure for the Isolation and Detection of Coproporphyrin and Uroporphyrin in Urine

This is an alternative method for detecting CP in urine, but it also tests for increased levels of UP, if present. Acidify 25 ml. of urine to a pH of 3.0 with glacial acetic acid and then extract with 30 to 35 ml. of a 1:1 mixture of ethyl acetate and n-butanol in a separatory funnel for 3 to 4 minutes. Both CP and UP enter the solvent layer. Separate and discard the urine phase. Wash the ethyl acetate–butanol extract twice with 40 to 50 ml. of water. Allow sufficient time for *clean separation* of the solvent and water phases. Next re-extract the washed solvent extract two or three times with 5 ml. portions of 10 per cent NaOH (w/v). Both the UP and CP enter the alkaline water layer, which is separated from the ethyl acetate–butanol. Acidify the NaOH extract with HCl to a pH of 2 to 3 (long range pH paper) and observe under UV light for pink-orange fluorescence. If this is seen, porphyrins are present. Then separate the CP from the UP by adding acetate buffer (4 part glacial acetic acid and 1 part saturated sodium acetate) to a pH of 4.8, and extract with two 50 ml. portions of ether. The uroporphyrins remain in the water (acetate-buffer) layer. Separate and re-extract the ether layer with 10 to 15 ml. of 5 per cent (v/v) HCl. (The CP enters the dilute HCl water phase.) Observe both the acetate buffer, containing the UP, and the HCl, containing the CP, for pinkish orange fluorescence.

REFERENCE

Sunderman, F. W., Jr., Hellman, E. S., and Tschudy, D. P.: Measurement of porphyrins and porphyrin precursors. *In* Hemoglobin, Its Precursors and Metabolites. F. W. Sunderman and F. W. Sunderman, Jr., Eds. Philadelphia, J. B. Lippincott Co., 1964.

Detection of Porphyrins in Feces

With 10 ml. of 95 per cent (v/v) ethyl alcohol, by vigorous stirring, emulsify a 5 gm. portion of a well mixed stool specimen. Then add 50 ml. 6 *N* HCl and, after thorough mixing, allow the preparation to stand for 6 hours or overnight. Dilute the acidified emulsion to 200 ml. with water and centrifuge and filter an aliquot. Test a 10 or 25 ml. aliquot for CP, or UP, or both, by either of the preceding urine methods. Any protoporphyrin present will separate with the CP fraction.

Quantitative Determination of Urinary Porphyrins

PRINCIPLE

Coproporphyrins (CP) can readily be separated from uroporphyrin type compounds by extraction from buffered urine into ether or ethyl acetate at pH 4.8. If fresh urines are used, this extract will predominantly contain preformed CP. If total CP (CP + CPG) is desired, and this is usually the case, allow the urine to age overnight. The precursors (CPG) are converted to CP on standing. Alternatively, the extract of the fresh urine can be oxidized with dilute iodine. The CP is then re-extracted into the water phase with dilute HCl and the CP content is estimated spectrophotometrically or fluorometrically.

The UP type components remain in the urine and can then be isolated by extraction into n-butanol at pH 3.2, or by adsorption on freshly prepared calcium phosphate or on an alumina chromatographic column. The determination of *total* UP is usually desired, and for this purpose the urine specimen is heated at 100°C. at pH 5.5 for 30 minutes to convert UPG to UP. After isolation, the UP is extracted into HCl solution and assayed by the same techniques used for CP.

Determination of Coproporphyrin in Urine

REAGENTS

1. Acetic acid, glacial.
2. Diethyl ether, peroxide-free (USP).
3. Iodine solutions. *Stock:* 1.0 per cent (w/v) in 95 per cent ethanol. *Dilute solution:* 0.005 per cent in H_2O. Make fresh as needed by diluting the stock solution.
4. HCl, 0.10 *N*. Dilute 0.85 ml. concentrated HCl to 100 ml. with water.
5. Acetate buffer, 4.5 *M*, pH = 4.8. Mix 1 part of glacial acetic acid with 4 parts of saturated sodium acetate solution and 3 parts of water.
6. Coproporphyrin standard. A 50 μg./100 ml. standard is available from Hartman-Leddon Co. (Harleco). Dilute this standard to 25 μg./100 ml. with 0.1 *N* HCl for spectrophotometric work and to 1, 2, 3, and 5 μg./100 ml. for fluorometric work. Make the dilute standards fresh, as needed.

SPECIMEN

Collect a 24 hour specimen of urine in a brown bottle containing 5 gm. of Na_2CO_3 to maintain the urine pH at 6.5 to 9.0. Keep the urine refrigerated while the total specimen is being collected. It is best to assay the specimen as soon as possible. Random specimens may be used, but they exhibit greater variation in concentration than do 24 hour samples.

PROCEDURE

1. Acidify a measured volume (50.0 ml.) of the urine with acetic acid to bring the pH in the range 4.0 to 4.5. Shake or stir the mixture to permit escape of CO_2 gas

formed from carbonate, which may be present. If more than a few drops of acid have been used, measure and note the volume of the acidified urine. Filter the urine and use for assay.

2. Pipet 15 ml. of the urine into a 125 ml. separatory funnel, followed by 5 ml. of pH 4.8 acetate buffer. Extract the buffered urine twice with 40 ml. portions of ether by shaking vigorously for at least 3 minutes. Pool the ether extracts and wash twice with 20 ml. portions of water. Combine the initial urine residue and the two water washes; this may be used for UP determination if this is requested, otherwise they are discarded.

3. Next wash the ether extract with 10 ml. of dilute iodine solution. This oxidizes all CPG to CP. After 5 min. completely draw off and discard the iodine wash water.

4. Extract the CP from the ether with 0.10 *N* HCl. Use 2.0 ml. of HCl the first time, and shake the ether with the acid for at least 2 minutes. Draw off the HCl into a 10 ml. graduated cylinder. Do the next extraction with 1.0 ml. of HCl and repeat this until the ether, when illuminated with a UV lamp, is free of all pinkish-orange fluorescence. Usually 4 to 7 ml. of HCl is needed, but occasionally as much as 15 to 20 ml. may be required before all the CP is extracted.

5. Measure the volume of the combined HCl extracts in the cylinder, and clear the extract by centrifugation in stoppered tubes.

SPECTROPHOTOMETRIC ASSAY

Transfer 3.0 ml. of the HCl extract to 1.00 cm. square cuvets and read the absorbance at 380, 401, and 430 nm. against 0.1 *N* HCl as a blank in a Beckman DU spectrophotometer. If a recording instrument is available, obtain the spectral scan from 350 to 500 nm. and read the absorbances from the recording. CP has a Soret peak at 401 nm.; if the wavelength calibration is in error, the peak may fall at some adjacent wavelength. The 401 nm. value should be that at the peak absorbance = A_{max}.

CALCULATIONS

The value of A_{max} obtained is not the true absorbance because it will include background absorbance due to other absorbing materials present in the extract. This is corrected for by use of the Allen "base-line correction" formula.[18,28] This formula assumes that the background absorbance is linear, which is usually, but not invariably, true. The value of A_{CP} is calculated:

$$A_{CP} = 2 \times A_{max} - (A_{380} + A_{430})$$

Then, if a and V are the volumes of the HCl extract and the 24 hour urine volume, the concentration of CP in the original urine specimen is given by

$$\mu g.\ CP/100\ ml. = \frac{A_{CP}}{1.835} \times \frac{1}{0.667} \times a \times \frac{100}{15}$$

or $$\mu g.\ CP/100\ ml. = 5.45 \times A_{CP} \times a = C_T$$

and $$\mu g.\ CP/24\ hr. = C_T \times \frac{V}{100}$$

The constant, K = 1.835, converts A_{CP} into the corrected absorbance of CP at A_{max} and also corrects for background absorption; 0.667 is the absorbance at 401 nm. of a solution containing 1 μg. of CP in 1 ml. of 0.1 *N* HCl.[18,30] If, in acidifying the initial urine, a significant dilution of the specimen results, correct for this by the use of the appropriate dilution factor.

REFERENCES

Haeger-Aronson, B.: Scand. J. Clin. Lab. Invest., *12*:Suppl. 47, 1960.
Sveinsson, S. L., Rimington, C., and Barnes, H. D.: Scand. J. Clin. & Lab. Invest., *1*:2, 1949.

FLUOROMETRIC ASSAY

Dilute the combined HCl extracts (step 5) with 0.1 *N* HCl to 25 ml. or some other convenient volume, if high levels are expected. Measure the fluorescence in a fluorometer, using a 405 nm. filter in the primary (activation) light path and a 615 nm. filter in the secondary (emission) path. Run a series of CP standards along with the unknowns.

$$\mu g.\ CP/100\ ml. = \frac{R_u}{R_s} \times S \times \frac{a}{15} \times D$$

where R_u and R_s = fluorometer readings of unknown and standard
S = concentration of standard, with reading nearest to that of unknown
a = volume of HCl extract (25 ml. or greater)
D = dilution factor to correct for dilution of urine on acidification, or for further dilution of HCl extract

REFERENCE

Talman, E. L.: Porphyrins in urine. *In* Standard Methods of Clinical Chemistry. D. Seligson, Ed. New York, Academic Press, Inc., 1958, vol. 2.

INTERPRETATION

Considerable disagreement exists between investigators as to what is the normal excretion of CP by healthy adults. This reflects differences in technique, the instability of CP, and the very great range found from person to person. In terms of concentration, a range of 3 to 12 μg./100 ml. may be taken as a fair compromise, although in random urine specimens, values as high as 20 μg./100 ml. may be found. In terms of normal 24 hour output, a range of 30 to 180 μg. may be taken as a consensus value. There are no good figures for CP output by children, but available data suggest it is somewhat lower than that found in adults.

In lead poisoning, individual specimens may show concentrations as high as 70 to 100 μg./100 ml., and 24 hour outputs may range from upper normal values to as high as 1.5 mg. per day.[9]

Determination of Uroporphyrins

The procedure has been adapted from that proposed by a group of Scandinavian investigators.[23,28] Only an outline of the procedure is presented. Precursors are converted to UP by heating the urine specimen, adjusted with acetate buffer to pH = 5.2 – 5.5, for 30 minutes in boiling water. The heating destroys CP and converts any PBG and UPG present to UP. Add to a 10 ml. aliquot of the treated urine 0.4 ml. each of 0.5 *M* Na_2HPO_4 and 30 per cent (w/v) $CaCl_2 \cdot 2H_2O$ and enough 10 *N* NaOH to raise the pH to between 10 and 11. Calcium phosphate is formed, which adsorbs all of the UP. The Ca^{++} is present in excess to insure absence of any free phosphate ion. Collect the $Ca_3(PO_4)_2$ precipitate by centrifugation, wash it in turn with 0.1 *N* NaOH and water and dissolve the washed precipitate in a minimum volume (5 to 8 ml.) of 1.5 *N* HCl to release the adsorbed UP. After removing any insoluble material, read the absorbances of the HCl extract at 380, 430, and 406 nm. in a 1.0 cm. cuvet, as described in the procedure for coproporphyrin. The Soret peak for UP is at 406 nm. The calculations are similar to those presented in the CP procedure, with K = 1.844, and 0.645 as the value of absorbance of 1.0 μg. of UP/ml. of 1.5 *N* HCl.

In the Talman-Schwartz procedure[24] the UP is adsorbed onto alumina, eluted with HCl, and then measured fluorometrically. In another approach,[7] the CP is first extracted from the urine, and then after readjusting the pH of the urine to between 3.0 and 3.2, the urine is extracted into n-butanol. The UP in the butanol extract is then extracted into aqueous HCl, where it is measured spectrophotometrically.

The normal 24 hour urinary output of uroporphyrins is quite small, ranging from none to about 35 μg. Small to moderate increases in output of UP-III are found in porphyria cutanea tarda and hepatic porphyria. Massive output, of the order of grams, primarily of type I, has been reported in congenital porphyria.

PORPHOBILINOGEN

Porphobilinogen (PBG) will react with p-dimethylaminobenzaldehyde (Ehrlich's reagent) in acid to form a red colored product. Many other materials present in urine will also react with the reagent. Among these are urea and indican (yellow color), bilirubin (green color), and tryptophane and indoxyl (orange to orange-brown color). The most troublesome metabolites are urobilinogen, indole, indoleacetic acid, and melanogen, all of which give a red color similar or identical with that derived from PBG. By controlling acidity and using chloroform, butanol, or both to remove a large fraction of the non-PBG color, the test can be made reasonably specific. Some materials such as sulfhydryl compounds may inhibit color formation. The best approach is to isolate the PBG chromatographically by adsorption on an appropriate ion exchange resin, as is done in the quantitative method to be described. If PBG is present in large quantities, simple 25- to 100-fold dilution of the urine will remove the effect of all types of interfering materials.

The same procedure may be used to detect urobilinogen (UBG) and PBG, both of which are found in most urines. Since the pigment formed with UBG is soluble in chloroform, and the PBG pigment is not, the two forms can be separated by extraction with chloroform. The reaction between Ehrlich's reagent and PBG is reported to be as follows:

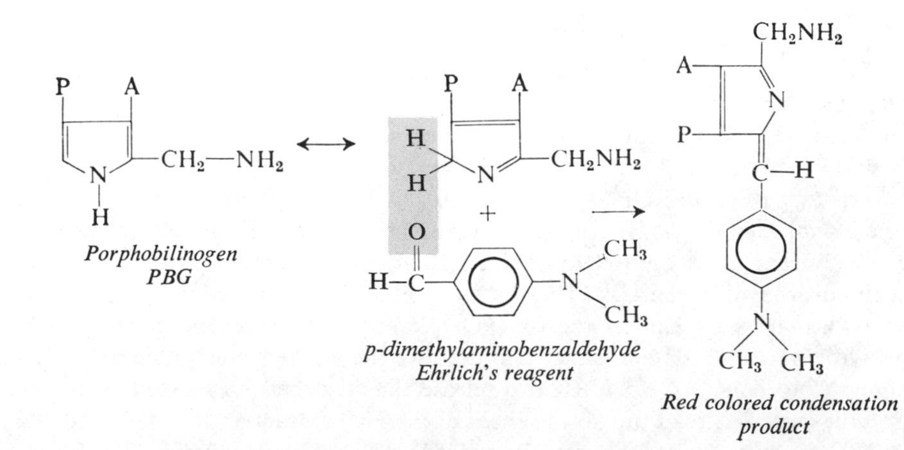

The red-colored product forms rapidly, but the color starts to fade slowly as the product reacts with additional PBG to form colorless dipyrrylphenylmethane and other products. Fresh urine should be used for the test, since the unstable PBG may, at times, disappear rapidly by conversion to other products. If testing of a urine specimen must be delayed, the specimen should be acidified to a pH of between 4 to 6, kept refrigerated, exposed to a minimum of light, and tested within 24 hours.

Procedure for Detection of Porphobilinogen

REAGENTS

1. Watson-Schwartz Ehrlich reagent. Dissolve 0.7 gm. of p-dimethylamino-benzaldehyde in 150 ml. of concentrated HCl and dilute the solution with 100 ml. of water. The chemical (crystals) must be colorless or have only a faint yellow color. Keep the reagent in a brown bottle.
2. Sodium acetate, saturated solution in water. Prepare and store at room temperature.
3. Chloroform, reagent grade.

PROCEDURE

1. Add 2.0 ml. of Ehrlich's reagent to 2.0 ml. of urine in a 18 × 200 mm. test tube. Mix, add 4.0 ml. of the saturated sodium acetate, and mix again. Check the pH with pH paper to confirm that it is in the range 4 to 5. The development of a red or pink color suggests either urobilinogen or PBG. With PBG the color will usually form immediately on the addition of Ehrlich's reagent, and will then intensify on the addition of acetate. If a color other than pink or red is obtained at this point, the test is reported as negative.
2. Add 5 to 6 ml. of chloroform, stopper the tube, and shake vigorously for 1 minute. Permit the phases to separate, using centrifugation, if necessary. If the water (upper) phase is colored red or pink, PBG is probably present. If the color is yellow-orange or a faint pink, the test result is considered negative. A red color in the lower (chloroform) phase suggests only that UBG is present. Transfer the water phase by careful pipetting to another test tube and add 2 to 3 ml. of n-butanol. After stoppering and shaking for a minute, allow the phases to clear and examine the lower (water) layer for a pink-red to purple color. If present, the test is reported as positive for PBG in the urine. The butanol soluble pigment is not significant.

REFERENCES

Watson, C. J., and Schwartz, S.: Proc. Soc. Exp. Biol. Med., *47*:393, 1949.
See also Henry, R. J.: Clinical Chemistry; Principles and Technics. New York, Hoeber Medical Div., Harper & Row Publishers, 1964.

Quantitative Determination of Porphobilinogen

The PBG in urine is separated from DALA, urea, and other interfering materials by adsorption onto an anionic ion exchange resin. It is then eluted with dilute acetic acid, treated with a modified Ehrlich reagent, and the absorbance of the pink color measured at 553 nm.

REAGENTS

1. Dowex-2 X-8 resin. This is an anion exchange resin that is used in the acetate form. The resin is available commercially as the chloride form. The conversion to the acetate form can be done either in cylinders on batch basis or, more conveniently, on a continuous flow basis, using a chromatographic tube (20 × 400 mm.). Make a suspension of resin in about four times its volume of water, allow it to settle, and remove the slow settling fines by decantation. Repeat this several times. Then pour the slurry into the chromatographic tube, the bottom end of which has been stoppered with a small plug of glass wool. Allow the resin suspension to settle and drain the water until its level reaches the top of the resin column. Add sodium acetate solution (3 *M*) and allow it to flow down the column. Check the eluate regularly for the presence of the chloride ion by testing with 3 per cent (w/v) $AgNO_3$ solution. Continue the passing of acetate through the column until the eluate is free of Cl^-, as evidenced by the absence of a precipitate of AgCl. In chromatographic work never permit the column to be free of the liquid phase; add more acetate whenever the level of fluid reaches the upper surface of the resin column. When Cl^- ion free, similarly wash the column with water until free of unbound acetate ion, as tested with 10 per cent $FeCl_3$ solution. (Ferric acetate has a brown color.) Then transfer the wet resin to a brown bottle and store in twice its volume of water. The suspension is stable for 3 to 4 months. In the batch process repetitive addition of acetate or water and decantation replace continuous flow through the column.

The resin in the acetate form can be represented as

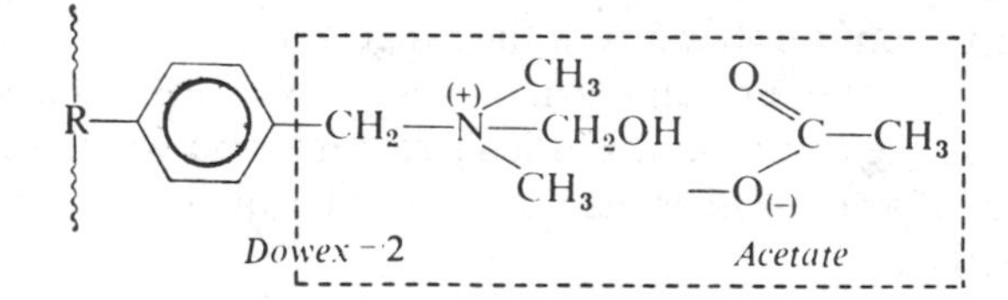

(5)

2. Ehrlich's reagent, Granick formulation. Dissolve 1.0 gm. of p-dimethylaminobenzaldehyde in 30 ml. of glacial acetic acid; to this add 8.0 ml. of 70 per cent $HClO_4$ and dilute the solution to 50 ml. with acetic acid. Prepare fresh and use within 3 hours.

3. Acetic acid, 1.0 *N* and 0.2 *N*. These solutions are prepared from glacial acetic acid (17 *N*).

PROCEDURE

1. Set up a 1 × 10 cm. chromatographic column, containing Dowex-2, acetate form, to give a resin column about 5 cm. long. Support the resin by a small plug of glass wool and a small circle of filter paper. Rinse the column with 5 ml. of water. Keep fluid in the column at all times.

2. Pipet 1.0 ml. of urine above the column. Allow the fluid to pass through at a rate of 6 drops per minute. When the urine level reaches the top of the resin, add 2 portions of 3.0 ml. of water to wash the column. Collect the urine eluting from the column and the wash water; these may be used for the assay of DALA, if this is projected, otherwise they are discarded. The PBG is retained on the column.

3. Elute the PBG from the resin in the column by adding 2.0 ml. of 1 *N* acetic acid, followed by several 2.0 ml. portions of 0.2 *N* acetic acid.

4. Collect the acid eluate into a 10 ml. cylinder and make the volume up to 10.0 ml. with 0.2 *N* acetic acid. Thoroughly mix the cylinder contents. If low levels

are expected, collect only 7 or 8 ml. of eluate. If the urine is not too deeply pigmented it is also permissible to put 2.0 ml. of urine specimen on the column (step 2).

5. Pipet 2.0 ml. of mixed eluate into a 13 × 100 mm. test tube. To another tube (blank) add 2.0 ml. of the 0.2 *N* acetic acid. Add 2.0 ml. each of fresh Granick-Ehrlich's reagent to both tubes, mix the tubes, and permit the color to develop for 15 minutes. Transfer the contents to 1.0 cm. square cuvets.

6. Using a Beckman DU spectrophotometer, measure the absorbances of the unknowns at 553 nm.

CALCULATIONS

Pure PBG is not available to serve as a standard. It has been established that ϵ (molar absorptivity) for PBG = 36,000.[9,15] This means that a solution of red condensation product equivalent to or containing 1.0 mole of PBG per liter (0.226 gm./ml.) will give an absorbance of 36,000 in a 1.0 cm. cell. From this it can be calculated that a solution containing 1.0 μg./ml. of PBG will have an absorbance of 0.160. The concentration of PBG in the initial urine specimen is then given by

$$\text{mg. of PBG/100 ml.} = \frac{A_T}{0.160} \times (1\ \mu\text{g./ml.}) \times \frac{4}{2} \times \frac{v}{a} \times \frac{100}{1000}$$

or

$$\text{mg. PBG/100 ml.} = 1.25 \times A_T \times \frac{v}{a}$$

where A_T = absorbance measured

v = volume of eluate

a = ml. urine put on column

If a = 1.0 ml. and v = 10.0 ml. then the mg. of PBG/100 ml. of urine = $12.5 \times A_T$.

COMMENTS

Normal urinary output of PBG ranges between 0.0 to 2.0 mg./24 hr., the usual values obtained ranging from 0.5 to 1.5 mg./24 hr. The reproducibility of the procedure is about ±20 per cent, but it is somewhat better if elevated levels of PBG are present. Significant increases are observed in hepatic porphyria.

REFERENCES

Mauzerall, D., and Granick, S.: J. Biol. Chem., *219*:435, 1956.

Haeger-Aronson, B.: Scand. J. Clin. Lab. Invest., *12*:Suppl. 47, 1960.

Sunderman, F. W., Jr., Hellman, E. S., and Tschudy, D. P.: Measurement of porphyrins and porphyrin precursors. *In* Hemoglobin, its Precursors and Metabolites. F. W. Sunderman and F. W. Sunderman, Jr., Eds. Philadelphia, J. B. Lippincott Co., 1964.

δ-AMINOLEVULINIC ACID (DALA)

Determination of δ-Aminolevulinic Acid in Urine

PRINCIPLE

The urine specimen is passed through an anionic ion exchange column to remove all PBG, which would otherwise interfere. The DALA is then isolated by adsorption on a cationic exchange resin, and after elution from the column, is heated with acetylacetone to form a condensation product, 2-methyl,3-acetyl,4-(2′-carboxyethyl)-pyrrole, chemically analogous to porphobilinogen. The acetylacetone condensation reaction proceeds as follows:

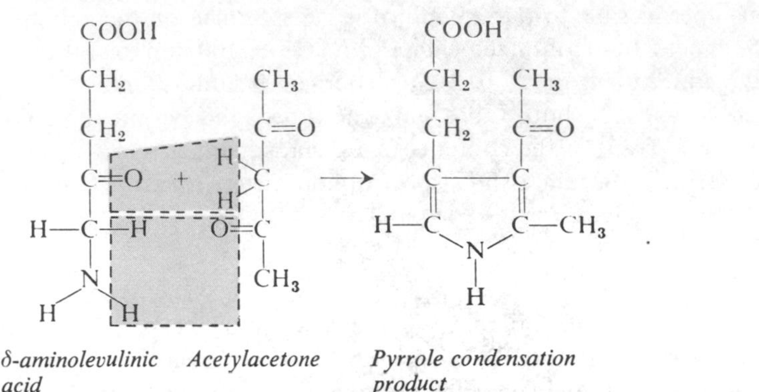

(6)

The condensation product is reacted with Ehrlich's reagent, and the pink-red colored product measured at 553 nm. The reaction with Ehrlich's reagent is identical to that with PBG.

REAGENTS

1. Dowex-2 X-8 Resin, acetate form. Refer to the previous procedure for PBG.

2. Dowex 50-W X-8, 200 to 400 mesh, hydrogen form. After removing fines, convert the resin into the sodium form by 20 hours contact with twice its volume of 2 *N* NaOH. Then wash the resin with water until the eluate is neutral. Convert it back to the hydrogen form by treating it, in turn, with 2 volumes of 4 *N* HCl and 6 volumes of 2 *N* HCl. Store the resin on the column or in a brown bottle in 1 *N* HCl. The prepared resin suspension will keep for 3 or 4 months.

DALA is adsorbed on the Dowex 50 column by the following exchange reaction:

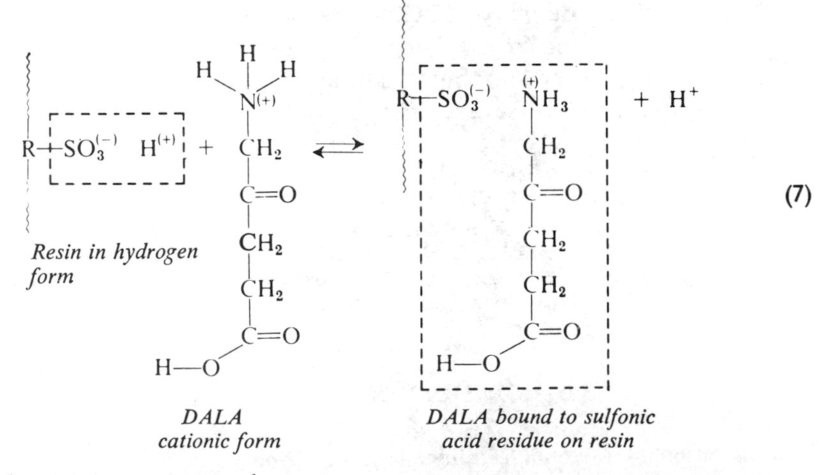

(7)

3. Acetylacetone, reagent grade.

4. Acetate buffer, pH 4.6, 0.50 *M*. Dilute 7.4 ml. of glacial acetic acid and 10.2 gm. of sodium acetate (anhydrous) to 500 ml. with water.

5. Sodium acetate, 0.50 *M*. Dissolve 20.5 gm. of anhydrous salt in water to make 500 ml.

6. Ehrlich's reagent (Granick). Refer to the PBG procedure in the previous section.

7. Standard DALA solution. The hydrochloride salt is available from the California Biochemicals Corporation (Los Angeles). Dissolve 100 mg. of DALA·HCl in 25.0 ml. of water to make a stock standard (1.0 ml. = 4.0 mg. DALA·HCl or 3.08 mg. DALA). This solution is stable for 6 months, if refrigerated. A 1:10 dilution in 0.25 *M* acetate buffer, pH 4.6, will keep for a week.

Prepare a working standard fresh as needed by diluting 0.81 ml. of the dilute stock to 25.0 ml. with acetate buffer (1.00 ml. = 10 μg. DALA).

SPECIMEN

Fresh urine should be used. Collect 24 hour specimens in bottles containing 10 ml. of glacial acetic acid. The pH of stored specimens should be between 4.0 and 6.5.

PROCEDURE

1. Pass a 1.00 ml. aliquot of urine through a Dowex 2 column as described in the PBG procedure. Collect the urine eluate and washings.
2. Introduce a 5 cm. long column of Dowex 50-W resin into a 1 × 10 cm. chromatographic tube. Wash the column with 25 to 30 ml. of H_2O to remove HCl.
3. Transfer the urine eluate and washings from the PBG isolation to the Dowex 50-W column. Rinse the column with 30 ml. of water. This washing will remove urea; continue washing until all urea has been removed, as confirmed by absence of yellow color when an aliquot of eluate is tested with the Watson Ehrlich reagent.
4. Add 3.0 ml. of 0.5 *M* sodium acetate and permit it to drain to the top of the resin column at an elution rate of 6 to 8 drops per minute. Discard the eluate (water).
5. Place a 10 ml. cylinder under the column and pass 7 ml. of 0.5 *M* sodium acetate down the column and collect the eluate containing the DALA until a total of 7 to 10 ml. is obtained.
6. Then add 0.20 ml. of acetylacetone to the collected eluate and dilute the volume to 10.0 ml. (or a smaller volume) with acetate buffer, pH 4.6. Thoroughly mix the cylinder contents.
7. Standards: Prepare a 10 μg./1.0 ml. working standard as described. Run 2 aliquots through the Dowex 50 column, 0.50 ml. and 2.0 ml. (5.0 and 20 μg. DALA). Treat the standard solutions exactly as the PBG urine eluates (steps 3 to 6).
8. Reagent blank: 10 ml. of acetate buffer, pH 4.6, plus 0.20 ml. of acetylacetone.
9. Transfer the 10 ml. portions of acetylacetone treated unknown, standards, and blank to 18 × 200 mm. Pyrex test tubes. Heat the tubes in a boiling water bath for 10 minutes, then quickly cool to room temperature.
10. Pipet 2.0 ml. aliquots of each to 13 × 100 mm. test tubes. Then add 2.0 ml. of Ehrlich's reagent and permit the color to develop for 15 minutes. Transfer the solutions to 1.0 cm. cuvets and read the absorbances of the unknowns and standards against the blank at 553 nm.

CALCULATIONS

$$\text{mg. DALA/100 ml.} = \frac{A_T}{A_S} \times C_S \times \frac{10}{v} \times \frac{100}{a} \times \frac{1}{1000}$$

where A_T and A_S = absorbances of unknowns and standard

C_S = concentration of standard, 5 or 20 μg./ml.

v = volume in ml., of acetate eluate

a = volume of urine used, in ml.

If v = 10 ml., a = 1.0 ml., and the 5 μg. standard is used, this simplifies to

$$\text{mg. DALA/100 ml. urine} = \frac{A_T}{A_S} \times \frac{1}{2}$$

The standard curve, if run, is linear. To minimize effort, run only two standards and compare the unknowns with the closest matching standard.

COMMENTS

The generally accepted normal range for adults is 0.1 to 0.6 mg./100 ml., with an average of about 0.24 mg./100 ml. The 24 hour output may vary from 1.5 to 7.5 mg. A fraction of this value is now known to represent aminoacetone normally present in all urine. In lead poisoning, in adults, the 24 hour excretion may be as high as 60 mg.

The Bio-Rad Laboratories (Richmond, Calif.) have placed on the market a test kit based on the described procedure, as adapted by Davis and Andelman.[4]

REFERENCES

Haeger-Aronson, B.: Scand. J. Clin. Invest., *12*:Suppl. 47, 1960.
Mauzerall, D., and Granick, S.: J. Biol. Chem., *219*:435, 1956.

REFERENCES

1. Blondheim, S. H., Margoliash, E., and Shafrir, E.: J.A.M.A., *167*:453, 1958.
2. Brodine, C. E., and Vertrees, K. M.: Differentiation of myoglobinuria from hemoglobinuria. *In* Hemoglobin, Its Precursors and Metabolites. F. W. Sunderman and F. W. Sunderman, Jr., Eds. Philadelphia, J. B. Lippincott Co., 1964.
3. Cawain, M. and Lappat, E. J.: Hereditary methemoglobinuria. *In* Hemoglobin, Its Precursors and Metabolites. F. W. Sunderman and F. W. Sunderman, Jr., Eds. Philadelphia, J. B. Lippincott Co., 1964.
4. Davis, J. R., and Andelman, S. L.: Arch. Environ. Health, *15*:53, 1967.
5. Dubowski, K. M.: Measurement of hemoglobin derivatives. *In* Hemoglobin, Its Precursors and Metabolites. F. W. Sunderman and F. W. Sunderman, Jr., Eds. Philadelphia, J. B. Lippincott Co., 1964.
6. Eilers, R. S.: Am. J. Clin. Path., *47*:212, 1966.
7. Fernandez, A. A., Henry, R. J., and Goldenberg, H.: Clin. Chem., *12*:463, 1966.
8. Goldberg, C. A. J.: Hemoglobins. *In* Standard Methods of Clinical Chemistry. D. Seligson, Ed. New York, Academic Press, Inc., 1961, vol. 3.
9. Haeger-Aronson, B.: Scand. J. Clin. Lab. Invest., *12*:Suppl. 47, 1960.
10. Hainline, A.: Hemoglobin. *In* Standard Methods of Clinical Chemistry. D. Seligson, Ed. New York, Academic Press, Inc., 1958, vol. 2; modified.
11. Huisman, T. H. J.: Normal and abnormal human hemoglobins. *In* Advances in Clinical Chemistry. H. Sobotka and C. P. Stewart, Eds. New York, Academic Press, Inc., 1963, vol. 6.
12. Kendrew, J. C.: Sci. Am., *205*:96, 1961.
13. Kohn, J., and Kelly, T.: J. Clin. Path., *8*:249, 1955.
14. Laurell, C. B., and Nyman, M.: Blood, *11*:493, 1957.
15. Mauzerall, D., and Granick, S.: J. Biol. Chem., *219*:435, 1956.
16. Murayama, M.: The chemical structure of hemoglobin. *In* Hemoglobin, Its Precursors and Metabolites. F. W. Sunderman and F. W. Sunderman, Jr., Eds. Philadelphia, J. B. Lippincott Co., 1964.
17. Naumann, H. N.: The measurement of hemoglobin in plasma. *In* Hemoglobin, Its Precursors and Metabolites. F. W. Sunderman and F. W. Sunderman, Jr., Eds. Philadelphia, J. B. Lippincott Co., 1964.
18. Rimington, C.: Biochem. J., *75*:620, 1961.
19. Steiner, R. F.: The Chemical Foundations of Molecular Biology, Princeton, N.J. D. Van Nostrand Co., Inc., 1965, pp. 197–208.
20. Sunderman, F. W.: Selected aspects of clinical hemoglobinometry. *In* Hemoglobin, Its Precursors and Metabolites. F. W. Sunderman and F. W. Sunderman, Jr., Eds. Philadelphia, J. B. Lippincott Co., 1964.

21. Sunderman, F. W., Copeland, B. E., MacFate, R. P., Martens, V. E., Naumann, H. N., and Stevenson, G. F.: Am. J. Clin. Path., *25*:489, 1955.
22. Sunderman, F. W., Jr., Hellman, E. S., and Tschudy, D. P.: Measurement of porphyrins and porphyrin precursors. *In* Hemoglobin, Its Precursors and Metabolites. F. W. Sunderman and F. W. Sunderman, Jr., Eds. Philadelphia, J. B. Lippincott Co., 1964.
23. Sveinsson, S. L., Rimington, C., and Barnes, H. D.: Scand. J. Clin. Lab. Invest., *1*:2, 1949.
24. Talman, E. L.: Porphyrins in urine. *In* Standard Methods of Clinical Chemistry. D. Seligson, Ed. New York, Academic Press, Inc., 1958, vol. 2.
25. Van Kampen, E. J., and Zijlstra, W. G.: Determination of hemoglobin and its derivatives. *In* Advances in Clinical Chemistry. H. Sobotka and C. P. Stewart, Eds. New York, Academic Press, Inc., 1965, vol. 8.
26. Varley, H.: Practical Clinical Biochemistry. 4th ed. New York, Interscience Publishers, Inc., 1964, pp. 344–348.
27. Watson, C. J., and Schwartz, S.: Proc. Soc. Exp. Biol. Med., *47*:393, 1949.
28. With, T. K.: Scand. J. Clin. Lab. Invest., *7*:193, 1955.
29. Yakulis, V. J., and Heller, P.: Am. J. Clin. Path., *37*:253, 1962.
30. Zenker, N.: Anal. Biochem., *2*:89, 1961.

ADDITIONAL READINGS

Cantarow, A., and Schepartz, B.: Biochemistry. 4th ed. Philadelphia, W. B. Saunders Co., 1967, pp. 120–141.
Gray, C. H.: Miscellaneous disorders of metabolism. III. Porphyrias. *In* Biochemical Disorders in Human Disease. R. H. S. Thompson and E. J. King, Eds. 2nd. ed. New York, Academic Press, Inc., 1964. pp. 848–864.
Henry, R. J.: Clinical Chemistry, Principles and Technics. New York, Hoeber Medical Div., Harper & Row Publishers, 1964, pp. 731–796, 797–831.
Hoffman, W. S.: The Biochemistry of Clinical Medicine. 3rd. ed., Chicago, Year Book Medical Publishers, Inc., 1964, pp. 463–487.
Hutchinson, H. E.: An Introduction to the Haemoglobinopathies. London, Edward Arnold, Publishers, 1967.
Levere, R. D., and Kappas, A.: Biochemical and clinical aspects of the porphyrias. *In* Advances in Clinical Chemistry. O. Bodansky and C. P. Stewart, Eds. New York, Academic Press, Inc., 1968, vol. 11, p. 134.
Sunderman, F. W., and Sunderman, F. W., Jr.: Hemoglobin, Its Precursors and Metabolites. Philadelphia, J. B. Lippincott Co., 1964.
Varley, H.: Practical Clinical Chemistry. 4th ed. New York, Interscience Publishers, Inc., 1967, pp. 571–608.
Watson, C. J.: Porphyrin metabolism. *In* Diseases of Metabolism. G. G. Duncan, Ed. 5th. ed. Philadelphia, W. B. Saunders Co., 1964.

Chapter 7 /

THE LIPIDS

by Robert L. Dryer, Ph.D.

The term *lipid* embraces a wide variety of compounds that are grouped together by virtue of similar solubility properties in chloroform, benzene, ether, carbon tetrachloride, and other agents termed "fat solvents." This broad and somewhat circular definition is made necessary by the heterogeneity of the substances involved, which include sterols, vitamins A, E, and K, bile pigments, waxes, carotenoid, and related dietary pigments, as well as the triglycerides, phosphatides, and free fatty acids. This definition is also necessary because of the difficulty of extracting any lipid from a natural source; it is a safe generalization that any extraction procedure competent enough to remove one sort of lipid will remove many or all types present in the sample.

For our purposes, we shall consider only those lipids that can be put into the following subgroups; nonesterified fatty acids, triglycerides, glycerophosphatides, sphingolipids, and sterols and sterol esters. Triglycerides, phosphatides, and sphingolipids can be degraded, by digestive processes or metabolic reactions, to several intermediate split products that can be detected by proper analytical procedures; these will also be mentioned.

FATTY ACIDS

The fatty acids of importance in human metabolism and nutrition are virtually all monocarboxylic straight chain compounds containing an even number of carbon atoms. The principal chain lengths range from 16 (C_{16}) to 24 (C_{24}) carbon atoms in length, and may be further characterized as saturated, monounsaturated, or polyunsaturated. Unsaturated acids contain double bonds connecting two adjacent carbon atoms. When more than one double bond exists, they are generally separated by at least two carbon atoms ($=CH—CH=$), and commonly by three ($=CH—CH_2—CH=$).

Although two carbon atoms may freely rotate about a single bond between them, a double bond more or less rigidly fixes the relative position of the two carbon atoms in space so that the remaining bonds of each carbon have specific spatial orientations. When a double bond occurs along the length of a carbon chain, the pair of hydrogen atoms on the participating carbon atoms may lie on the same side of the plane of the

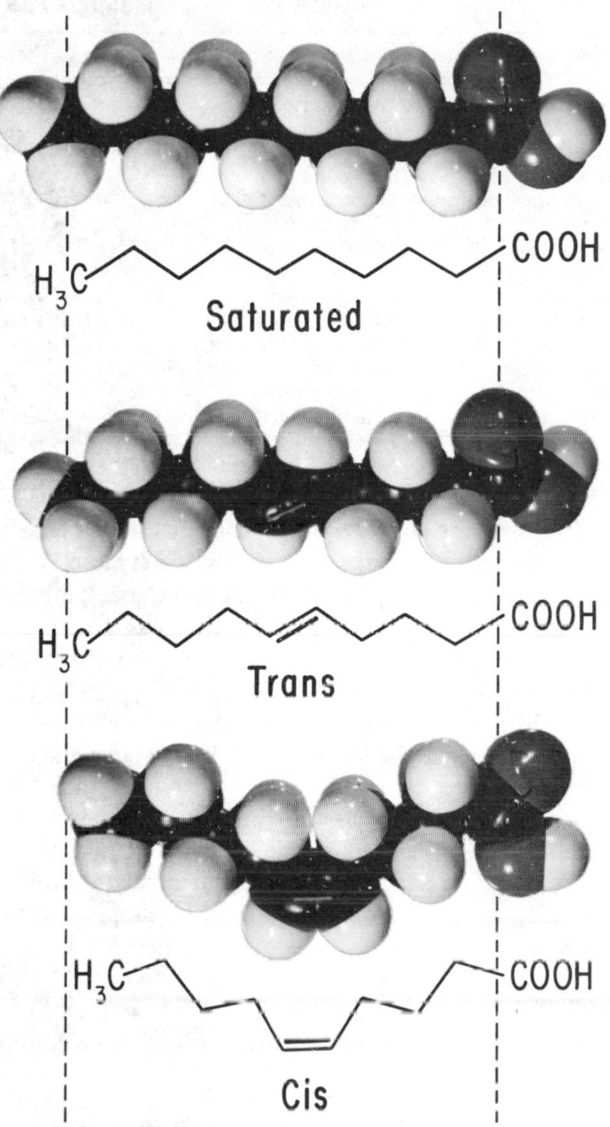

Figure 7-1. Representations of fatty acid structures by atomic models and by conventional line segments. The dotted vertical lines show that the saturated chain is the longest, that the *trans*-unsaturated chain is a little shorter, and that the *cis*-unsaturated chain is the shortest and the most bent.

double bond, in which case they are described as *cis*, or they may lie on opposite sides of the plane of the double bond, in which case they are described as *trans*. The double bonds of the naturally occurring fatty acids are of the *cis* configuration, as shown in Figure 7-1.

The *cis* double bond introduces a "kinking" into the molecule, and the more unsaturated the fatty acid the more "kinked" is the apparent structure. The presence of double bonds lowers the melting point compared to a saturated molecule of the same chain length, and acids with *cis* double bonds have lower melting points than the similar *trans* acids. This may be significant in keeping the melting points of fats consistent with animal body temperatures, and in the analysis of fatty acid mixtures by gas-liquid chromatography. The locus of double bonds along the chain length can be designated by two different systems. The first, and older, is the formal Geneva system whereby the double bond or bonds relate to the carboxyl end of the chain.

TABLE 7-1. *Representative Fatty Acids*

Common Name	*Geneva System Name*	*Formula*	*Sources*
Myristic	Tetradecanoic	$CH_3(CH_2)_{12}COOH$	Butter, oils
Palmitic	Hexadecanoic	$CH_3(CH_2)_{14}COOH$	Lard, etc.
Stearic	Octadecanoic	$CH_3(CH_2)_{16}COOH$	Tallow, etc.
Arachidic	Eicosanoic	$CH_3(CH_2)_{18}COOH$	Peanut oil
Lignoceric	Tetracosanoic	$CH_3(CH_2)_{22}COOH$	Brain, peanut
Myristoleic	9-Tetradecenoic	$C_{13}H_{25}COOH$	Human milk
Palmitoleic	9-Hexadecenoic	$C_{15}H_{29}COOH$	Milk, fish
Oleic	9-Octadecenoic	$C_{17}H_{33}COOH$	Many oils, fats
Selacholeic	15-Tetracosenoic	$C_{23}H_{45}COOH$	Brain
Linoleic	9,12-Octadecadienoic	$C_{17}H_{31}COOH$	Many oils
Linolenic	9,12,15-Octadecatrienoic	$C_{17}H_{29}COOH$	Many oils
Arachidonic	5,8,11,14-Eicosatetraenoic	$C_{19}H_{31}COOH$	Brain, etc.
Tuberculostearic	10-CH_3-Stearic	$C_{18}H_{37}COOH$	Tubercle sp.
Phthioic	3,13,19-Trimethyl-tricosanoic	$C_{25}H_{51}COOH$	Tubercle sp.
Cerebronic	2-OH-Tetracosanoic	$C_{23}H_{47}O_1COOH$	Brain
Hydroxynervonic	2-OH-9-Tetracosenoic	$C_{23}H_{45}O_1COOH$	Brain

According to this convention, linoleic acid can be described as $\Delta^{9,12}$-*cis*-octadecadienoic acid. One double bond, signified by the symbol Δ, lies between carbons 9 and 10, counting from the carboxyl end of the chain, and the second lies between carbons 12 and 13. A more recent convention has grown out of the interest in fatty acid metabolism, which has shown that as shorter chains are elongated, carbon atoms are introduced from the carboxyl end of the chain so that the relation of the double bonds to the hydrocarbon-like end of the chain remains unchanged. In this convention linoleic acid would be known as $\omega^{6,9}$-*cis*-octadecadienoic acid, in which the symbol ω represents the carbon atom farthest removed from the carboxyl group. By either of these conventions the saturated analog of linoleic acid, stearic acid, would be known as octadecanoic acid.

Man cannot readily synthesize the highly unsaturated fatty acids so that linoleic, linolenic, and arachidonic acids must be supplied in the diet. Fortunately, these are so abundant in natural foodstuffs that deficiencies are not common. A few of the more unusual fatty acids, especially those of the central nervous system, include a hydroxyl group, commonly in the position adjacent to the carboxyl end of the chain. Table 7-1 lists the names and structures of some common fatty acids.

TRIGLYCERIDES

Triglycerides are esters of fatty acids with the trihydroxyalcohol, glycerol. If the fatty acids are of such a length or degree of saturation that the triglyceride is a solid at room temperature, we speak of it as a fat; if the chain length or degree of unsaturation is such that at room temperature the product is a liquid, we speak of it as an oil. Triolein, for example, melts at −5°C. and tripalmitin melts at 65°C. Although oils such as triolein can be synthesized for laboratory use, most naturally occurring triglycerides contain more than one kind of fatty acid, and thus are described as mixed triglycerides. As a general rule, plant triglycerides have a fair degree of unsaturation and are oils, and terrestrial animal triglycerides are more nearly saturated and are fats. Triglycerides derived from marine animals are more highly unsaturated. Triglycerides contain only atoms of C, H, and O; therefore, triglycerides are sometimes called simple lipids.

Triglycerides can be readily hydrolyzed by strong alkalis or acids or by enzymes known as lipases, which are found in serum, pancreatic juice, and stools. Under drastic conditions the products include the ensemble of component fatty acids and free glycerol, but under milder conditions partial hydrolysis produces a mixture of one or more free fatty acids and mono- and diglycerides. Equations describing these reactions may be written as follows:

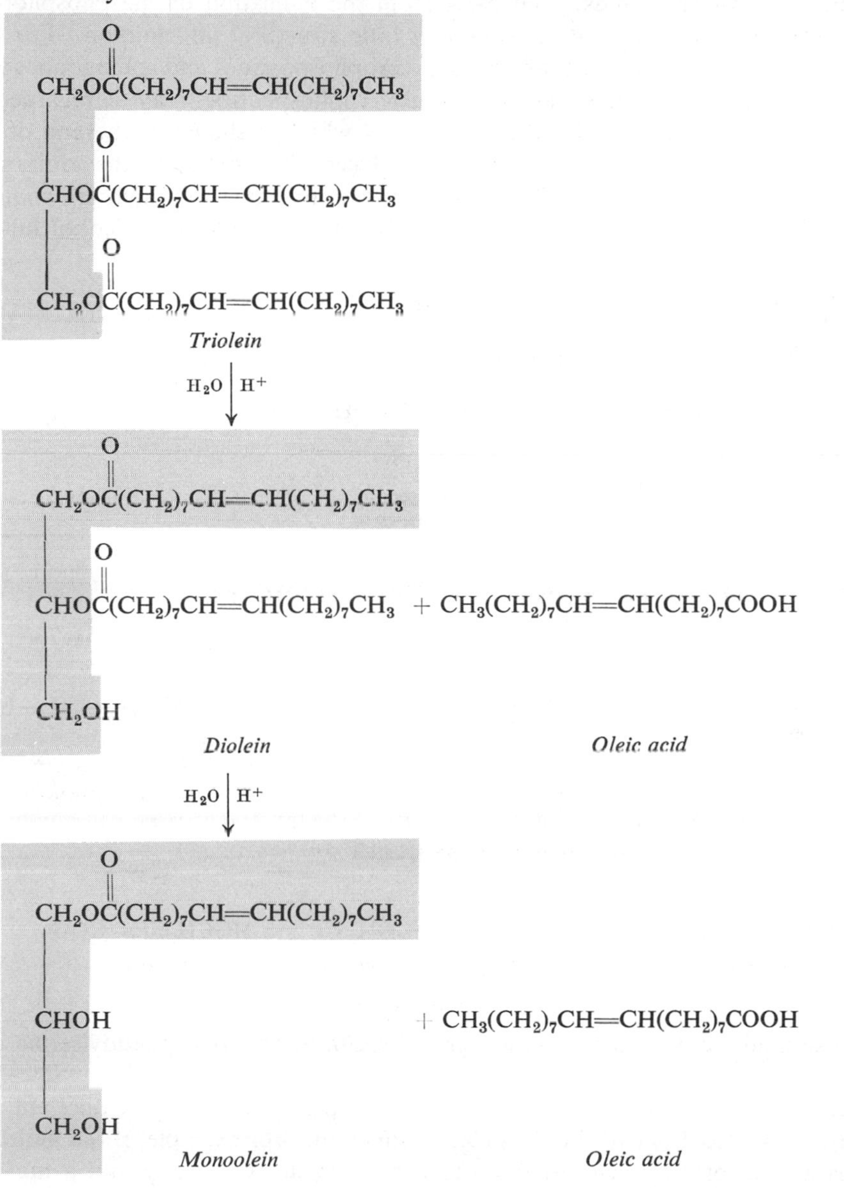

If H—OH is replaced by CH_3—OH and if boron trifluoride or anhydrous hydrochloric acid is used as the catalyst, the products are the methyl esters of the ensemble of fatty acids and free glycerol. This process of transesterification is termed methanolysis, and is frequently employed to prepare triglyceride fatty acids for study by thin-layer chromatography or gas-liquid chromatography. Details of this process will be described later (pp. 319, 320).

GLYCEROPHOSPHATIDES

These are the simplest members of the complex lipids, so-called because they contain atoms of P and N or S in addition to C, H, and O. The glycerophosphatides and the sphingophosphatides, which we shall consider next, were commonly described in the past as "phospholipids," but aside from the indication of the phosphorous content, the term phospholipid conveys very little structural information. For this reason we shall use the more precise terms glycerophosphatides and sphingophosphatides to describe the respective classes. The glycerophosphatides may be regarded as derivatives of phosphatidic acid, the structure of which is shown at the end of the paragraph. Phosphatidic acid is converted to a glycerophosphatide by the addition of one of a variety of amino alcohols, the three most important of which are choline, ethanolamine, and serine. The last of these is not only an amino alcohol but an amino acid as well.

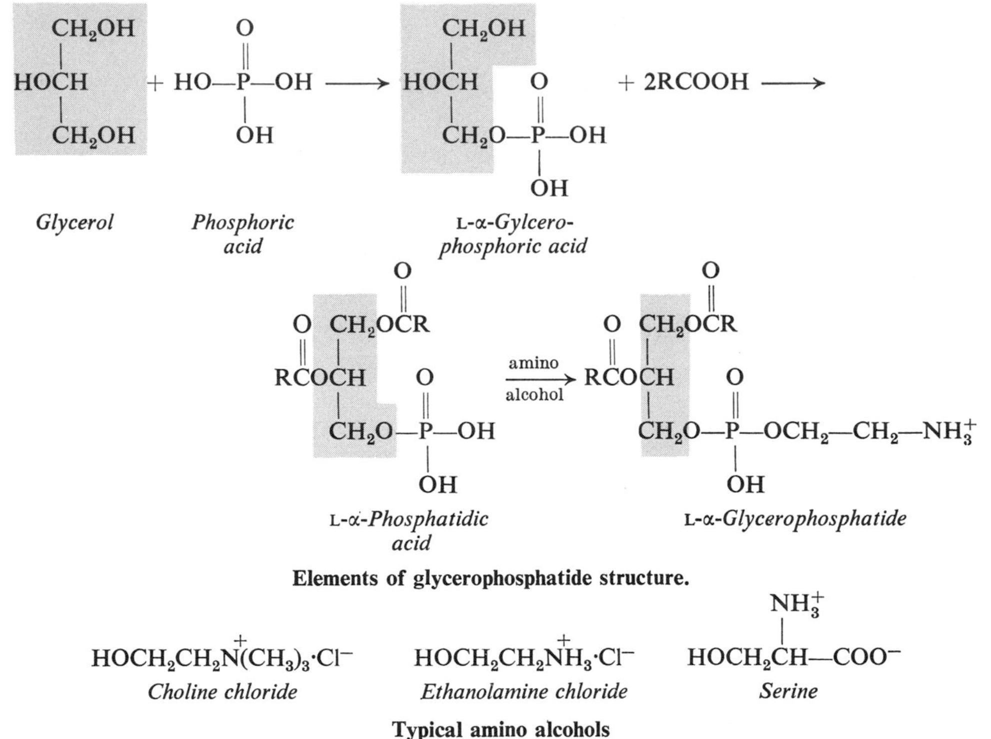

Elements of glycerophosphatide structure.

Typical amino alcohols

Phosphatidylcholine is frequently termed lecithin, and phosphatidylserine and phosphatidylethanolamine are frequently termed, collectively, cephalins. Since the fatty acids associated with these substances may be the same or may be different, the designation as a lecithin is still only a class distinction. For example, if the molecule contains a mole of oleic acid esterified to a primary alcohol group and a mole of arachidonic acid esterified to the secondary alcohol group, we could name the compound oleyl arachidonyl phosphatidyl choline.

A few further comments on the glycerophosphatide structures just shown are necessary. Glycerol has three alcoholic groups available for reaction. Two of these are primary alcohols, since the carbon to which they are attached also carries two hydrogen atoms. The primary groups obviously lie at either end of the glycerol molecule and, thus, are frequently described as being on an α-carbon. The remaining group is a secondary alcohol, since the carbon to which it is attached only carries one hydrogen atom. We can speak of it as attached to a β-carbon. Note that in the phosphatides and in phosphatide precursors we deal only with α-substituents. In other words, the phosphoric acid bearing the nitrogenous base is only found esterified to a primary alcoholic group of the parent glycerol.

Any carbon atom that bears four different substituent groups affects the plane of rotation of polarized light. The mechanism of this phenomenon is not important here, but it is significant that α-glycerophosphoric acid, phosphatidic acids, and the glycerophosphatides derived from them exist in two distinctly different forms. We indicate this by drawing the β-alcoholic group to the left of the carbon chain. If we were to draw this to the right we would be describing compounds with different chemical and physical properties, even though the atomic compositions of the two structures are precisely the same. Returning to the illustrative case just mentioned, we could precisely define the structure of the phosphatide as L-α-oleyl arachidonyl phosphatidyl choline, since all natural phosphatides are of the L configuration, have the phosphoric acid and the nitrogenous alcohol on an α-carbon, and have two fatty acid groups as specified.

A formally related phosphatide is worth brief mention here. It has been known for years that beef heart extracts contain a substance essential for the proper performance of serological tests for syphilis. In 1941, Pangborn isolated a material since known as cardiolipin, which fulfilled the requirement of binding the complement of beef heart extract with syphilitic serum. Note that cardiolipin is a phosphatide that contains no nitrogen.

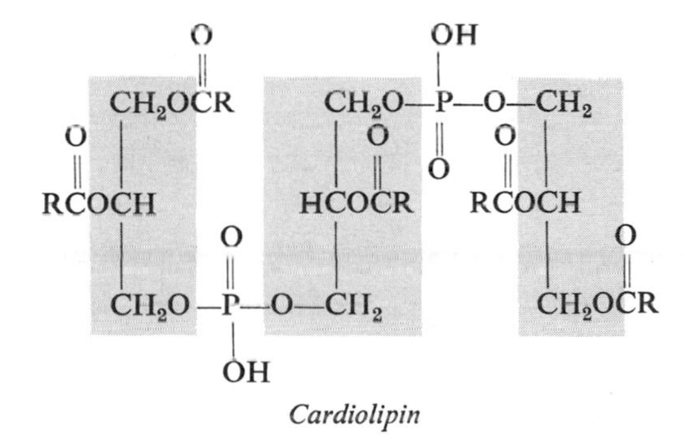

Cardiolipin

SPHINGOPHOSPHATIDES

Members of this group of phosphatides are also known as sphingomyelins. Instead of glycerol, these lipids contain the amino alcohol, sphingosine. Note that in

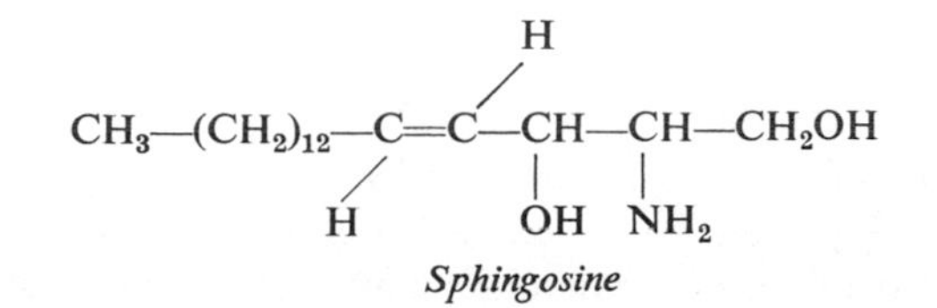

Sphingosine

this case the configuration around the double bond is *trans*. In sphingolipids only one mole of fatty acid is found per mole of lipid; it is attached as an amide to the nitrogen of sphingosine. The primary alcohol serves as a point of attachment for a mole of phosphoric acid and a mole of choline. The structure of sphingomyelin is therefore as follows:

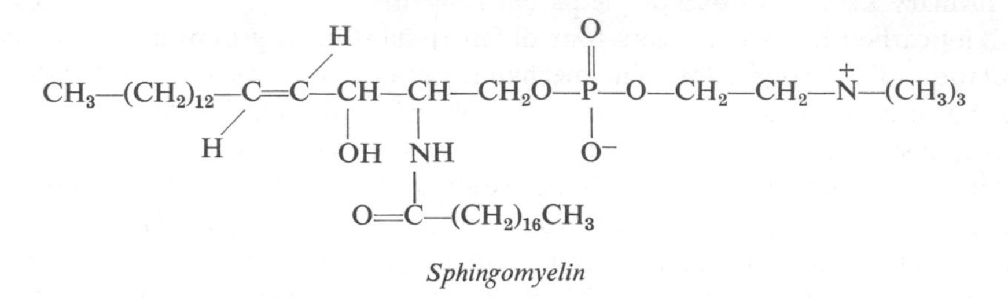

Sphingomyelin

Sphingosine can also be part of a group of lipids that contain no phosphorus or nitrogenous base. Instead, the primary alcohol of the sphingosine forms a glycosidic linkage with carbon atom 1 of a hexose, commonly galactose, and such structures are called cerebrosides or glycolipids. In Gaucher's disease part or all of the galactose may be replaced by glucose. The structure of a typical cerebroside is now given; note that cerebrosides contain no phosphorus. As a further modification, some

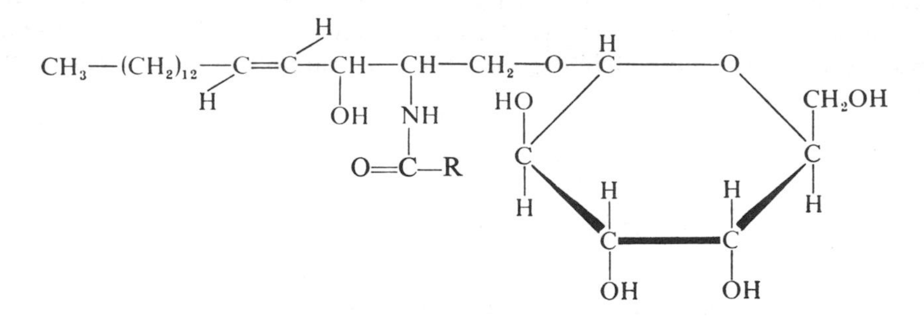

cerebrosides form sulfuric acid esters at C_6 of the galactose ring; these are known as sulfatides.

Most of the sphingolipids are found in the central nervous system and in cell membranes. The fatty acids involved are usually long chain (C_{20} to C_{24}), and commonly more saturated than the acids found in glycerophosphatides; they contain most of the hydroxy fatty acids found in human lipids.

BASIC CONCEPTS OF LIPID METABOLISM

The average American diet provides nearly 40 per cent of the total caloric intake as fat; in addition, the total daily caloric intake is often above the total expenditure, and the surplus is stored largely as fat. Our concern with lipid metabolism must, therefore, consider both the disposition of preformed lipid and the conversion and storage of excess energy taken in.

Digestive enzymes are normally capable of reducing ingested lipids to their fundamental moieties, glycerol, fatty acids, inorganic phosphates, and free nitrogenous bases such as choline, ethanolamine, and serine. Triglycerides are usually converted in part to monoglycerides, which are absorbed as such. A monoglyceride is an ester of glycerol with a single mole of fatty acid. This process of hydrolytic digestion, which takes place in the lumen of the small intestine, is aided considerably by the emulsifying properties of the bile salts, themselves derivatives of cholesterol. Therefore, in the absence of bile, for whatever reason, the digestion of fats is hindered and the stools passed contain significant amounts of unabsorbed fats. These are split only slightly by digestive enzymes, but are efficiently split by colonic microorganisms and produce foul-smelling and highly irritating stools. In addition, the fat that is not taken up by the small intestine contains much dissolved fat-soluble vitamins; this also represents a loss to the body.

Under the physiological conditions of the small intestine, the liberated fatty acids exist as soaps and as micellar complexes in the presence of bile salts. In this form they are taken up by the intestinal mucosa, along with the other digestive products, including glycerol and small amounts of mono- and diglycerides that escaped complete hydrolysis. Within the mucosal cells the fatty acids are reassembled to form triglycerides and glycerophosphatides. The same is probably true of the mono- and diglycerides. It is important to note, however, that the glycerol employed in resynthesis of lipid is not that freshly absorbed from the intestine; the glycerol is converted to glucose and other products. The reasons for this are too lengthy to review here, but the central fact is that the triglyceride-forming enzymes can transfer activated fatty acids only to glycerophosphoric acid, not to free glycerol. The re-synthesis of fat is an energy-requiring process. Careful studies have shown that the adult human can absorb as much as 150 gm. of fat per day, and that liquid fats are absorbed more quickly than solid fats.

Precisely what happens to the lipids formed within or taken up by the intestinal mucosa is still the subject of argument among the experts. Microscopic droplets of fat are expelled from the mucosal cells and transferred to the lymphatic vessels as chylomicrons, ranging in size from 35 nm. to 1 μm, which ultimately reach the blood via the thoracic lymph duct and hence reach the liver. It is claimed that the chylomicrons are simply fat droplets coated with protein, which prevents them from coalescing, and that, in general, chylomicrons do not contain shorter chain length fatty acids (C_4 to C_{10}). These shorter chains may reach the liver by direct transport in the portal blood, partly as unesterified fatty acids bound to albumin and partly as mono- or diglycerides. It is well established that the chylomicron count of blood and lymph rises sharply after fat ingestion. The presence of chylomicrons produces a turbidity of the serum (postprandial lipemia), and this phenomenon has been made the basis of a test for fat absorption efficiency.

A small amount of the fat reaching the liver is probably burned by that organ for energy, but it could not long support the burden of storing the total intake without degeneration (fatty liver). To avoid this, the liver attaches much of the lipid to α- and β-lipoproteins, which serve as transfer agents to the total distributed mass of adipose tissue. Lipoproteins differ in composition from the chylomicrons, which contain on the average 80 per cent triglyceride and 7 per cent phosphatide. The α-lipoproteins average merely 8 per cent triglyceride and about 20 per cent phosphatide, and the β-lipoproteins contain about 50 per cent triglyceride and 20 per cent phosphatide. Lipoproteins contain nearly three times as much cholesterol as chylomicrons. Although chylomicrons virtually disappear after absorption is complete, the lipoproteins are always present as components of the plasma or serum proteins. The fat delivered to the adipose tissue is stored as triglyceride, apparently without limit. In extreme obesity, more than 50 per cent of the total body weight can consist of fat.

Lipoproteins are generally classified according to their sedimentation constants in an *ultra*centrifuge after the density of the sample is increased by addition of certain salts. Under these conditions lipoproteins will float in the ultracentrifuge. The higher the flotation constant (S_f), the lower the density of the lipoprotein, and the larger the amount of lipid material it carries. The heaviest lipoproteins migrate with the α-globulins, and the lighter fractions are associated with the β-globulins.

Lipoproteins can also be detected by suitable staining procedures following electrophoretic separation, but the process is not as reliable, nor as quantitative, as the ultracentrifugal separation.

Lipoprotein transport is also available not only for carriage of preformed lipids, but for lipids made from excess carbohydrate or protein of the diet. The liver converts these energy forms to triglycerides and other lipids, and then handles them indistinguishably from those preformed in the food we eat. These facts are summarized in Figure 7-2.

Nonesterified Fatty Acids

The nonesterified fatty acids (NEFA) are sometimes called free fatty acids (FFA) or unesterified fatty acids (UFA). This small but important lipid moiety is generally carried by plasma albumin and not by the larger lipoproteins. The best evidence indicates that one mole of albumin can carry as much as 20 moles of fatty acid, all but about one third of which is readily removable under physiological conditions. The normal level of NEFA in human blood is only 0.8 to 1.0 mEq./L., amounting to about 20 to 25 mg./100 ml. of plasma, but the flux is very large and quite sensitive to exercise or other physical work, to the level of blood sugar, or to excitement or other emotional stress that liberates epinephrine. The NEFA are readily taken up by most tissues for satisfaction of energy requirements.

The oxidation of a mole of palmitic acid produces carbon dioxide and water plus a considerable quantity of energy, as shown in the following equation.

$$C_{15}H_{31}COOH + 23O_2 \rightarrow 16CO_2 + 16H_2O + 2{,}330{,}500 \text{ cal.}$$

The Calories produced by burning a mole of palmitic acid (16 carbon atoms) are approximately twice the Calories produced by burning an equivalent amount (2.5 moles, or 16 carbon atoms) of glucose. Of this large amount of energy, nearly 40 per cent can be trapped as useful chemical energy in our bodies. The chemical energy is contained in specialized molecules, which, by their structure, have the ability to hold unusually large quantities of energy. By means of suitable enzymes the chemical energy may be coupled to necessary metabolic processes. The most important of these specialized molecules is known as adenosine triphosphate, commonly indicated by the acronym ATP. Its structure and other information about this and related compounds may be found in any biochemistry text.[23] Keeping in mind that triglycerides contain three fatty acid residues, it is clear that fats make an excellent storage form for reserve energy. In addition to the high intrinsic energy content, fat molecules have a density less than 1.0. In a typical adult, stored fat is theoretically sufficient to provide for daily energy requirements for about 40 days, but it is abundantly evident from controlled starvation studies in normal subjects and from

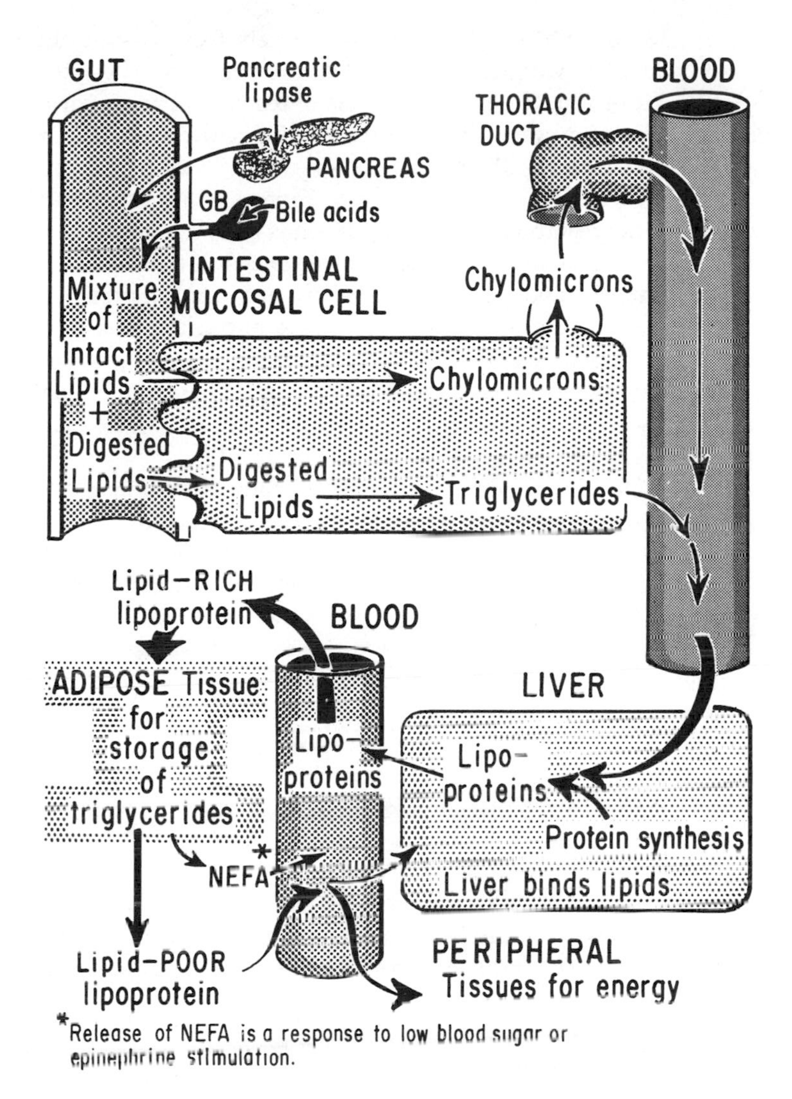

Figure 7-2. Factors involved in lipid transport in the body. The liver is central in the formation of lipid-rich lipoproteins, which carry triglycerides to the adipose tissue for storage. NEFA are released as needed to furnish available energy elsewhere. The lymphatic system cooperates in the absorption process.

studies of untreated diabetics that severe ketosis and other signs of disease are demonstrated in much shorter periods of dependence on fat metabolism. The reasons for this apparent discrepancy must be sought in the intermediary metabolism of the fatty acids.

Intermediary Metabolism of Fatty Acids

Long chain fatty acids are oxidized in the mitochondria of cells by a series of reactions that operate in a repetitive manner to shorten the chain by two carbon atoms at a time. As a result, each molecule is reduced to eight molecules of an intermediate known as acetyl-coenzyme A. These molecules do not normally accumulate, but are enzymatically combined with a material derived largely from carbohydrate metabolism, oxalacetic acid. The product of the condensation reaction is citric acid, which is a major component of still another cyclic series of reactions known variously as the Krebs cycle, the citric acid cycle, or the tricarboxylic acid cycle. The Krebs cycle serves as a common pathway for the final oxidation of nearly all food energy, whether derived from carbohydrate, fat, or protein. It is important to bear in mind that the smooth operation of the metabolic machinery depends on the availability of sufficient oxalacetic acid to serve as acceptor for acetyl-coenzyme A. In acute starvation or in deranged carbohydrate metabolism, represented by untreated diabetes mellitus, the supply of acetyl-coenzyme A is greater than the supply of oxalacetic acid. As a result, acetyl-coenzyme A must be handled by alternative pathways, which give rise to acetoacetic acid, β-hydroxybutyric acid, and acetone. These three substances are known collectively as ketonc bodies, and since the first two are acidic materials, they may cause a severe metabolic acidosis as they accumulate.

Ketosis can therefore be regarded as excessive production of acetyl-coenzyme A as the body attempts to obtain necessary energy from stored fat in the absence of adequate supply of carbohydrate metabolites. As fat flows from the depot stores, the long chain fatty acids are converted to coenzyme A derivatives in preparation for degradation, and the level of palmitoyl-coenzyme A and other long chain coenzyme A derivatives may also increase because of the failure of the acetyl-coenzyme A–oxalacetic acid condensation reaction. This compounds the problem, since the long chain fatty acid–coenzyme A derivatives inhibit the enzymes that might produce even minimal amounts of oxalacetic acid from glucose or glycogen. Thus, we see that the commingling of carbohydrate and lipid metabolism, which normally provides for efficient storage of excess energy as fat, carries with it built-in hazards when carbohydrate metabolism is disturbed and we are forced to depend only on stored fat. The body liberates large amounts of free fatty acids to override the glucose deficit. The liver can produce acetyl-coenzyme A faster and in greater quantities than the peripheral tissues can consume it. The excess acetyl-coenzyme A can then be coupled with itself (instead of with oxalacetic acid) in accordance with the following reaction.

$$\underset{\textit{Acetyl-Coenzyme A}}{2CH_3\text{—COO—Coenzyme A}} \rightarrow \underset{\textit{Acetoacetyl-Coenzyme A}}{CH_3\text{—CO}\cdot CH_2\text{—COO—Coenzyme A}} + \text{Coenzyme A}$$

Although this reaction normally has an equilibrium value far to the left, when excess acetyl-coenzyme A accumulates the equilibrium shifts to the right. Some of the acetoacetyl-coenzyme A can be reduced, as now shown, by reduced pyridine nucleotide. This equilibrium also usually lies to the left, but is reversed by the

$$\underset{\textit{Acetoacetyl-Coenzyme A}}{CH_3\text{—CO—}CH_2\text{—CO—Coenzyme A}} + NADH + H^+ \rightarrow \underset{\textit{β-Hydroxybutyryl-Coenzyme A}}{CH_3\text{—}\overset{OH}{\overset{|}{C}H}\text{—}CH_2\text{—CO—Coenzyme A}} + NAD^+$$

accumulation of acetoacetyl-coenzyme A.

The coenzyme A derivatives of fatty acid metabolism we have described are restricted entirely to an intracellular position; in order for the individual metabolites to leave the interior of cells, they must first be broken down to the free form, acetic acid, acetoacetic acid, and so on. An enzyme in liver and muscle facilitates the formation of the free acids, which can then diffuse from the cells into the blood as the levels of the compounds rise to appreciable values. Free acetoacetic acid is quite unstable, and some of it decomposes to form carbon dioxide and acetone. This decarboxylation reaction accounts for the formation of the third ketone body frequently observed in severe untreated diabetes mellitus. Muscle, heart, and brain can utilize ketone bodies carried to them via the blood because they can reform the coenzyme A derivatives of the acids by an enzyme that they contain, and ketone bodies can therefore maintain these vital organs, but only in an inefficient manner.

The entire process of ketosis can be reversed by restoring an adequate level of carbohydrate metabolism. In starvation this consists of adequate dietary intake; in diabetes mellitus this consists of insulin administration, which permits the high level of circulating blood glucose to be taken up by the cells, a process that apparently cannot proceed in the absence of the hormone. As the acceptor of acetyl-coenzyme A is again supplied, the normal operation of the Krebs cycle is restored, and the rush of fatty acids from adipose tissue slows and is finally reversed.

A more comprehensive idea of the metabolic reactions we have discussed may be gained from the scheme outlined in Figure 7-3, which relates the main features of

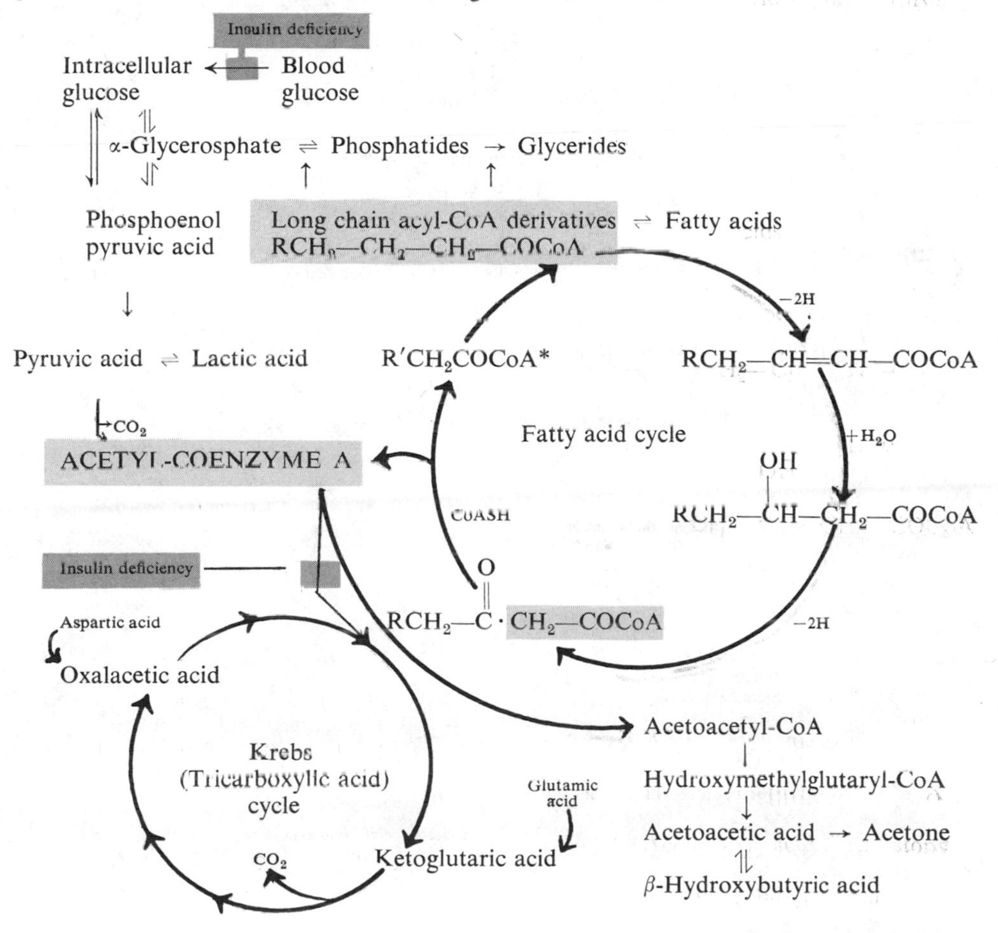

* R′ contains 2 carbons less than R, CoA represents Coenzyme A combined with the carboxyl end of the acyl chain.

Figure 7-3. Metabolic relations of fats, proteins, and carbohydrates. Note that both fat and carbohydrate metabolism produce acetyl-coenzyme A. Amino acids from proteins may feed by transamination into the Krebs cycle. Other amino acids can be converted to acetyl-coenzyme A also.

carbohydrate and lipid metabolism, and which further shows how even amino acids derived from proteins may be incorporated. The student may also note that oxalacetic acid can be obtained by transamination of the amino acid aspartic acid, and that transamination of glutamic acid can yield α-ketoglutaric acid, another intermediate in the Krebs cycle. Further details of the enzymes and mechanisms involved in the details of these processes may be found in any good biochemistry text.[23]

Fatty Liver

A second type of abnormality of lipid metabolism involves the accumulation of excessive fat in the liver. Fatty liver is often a correlate of alcoholism, malnutrition,

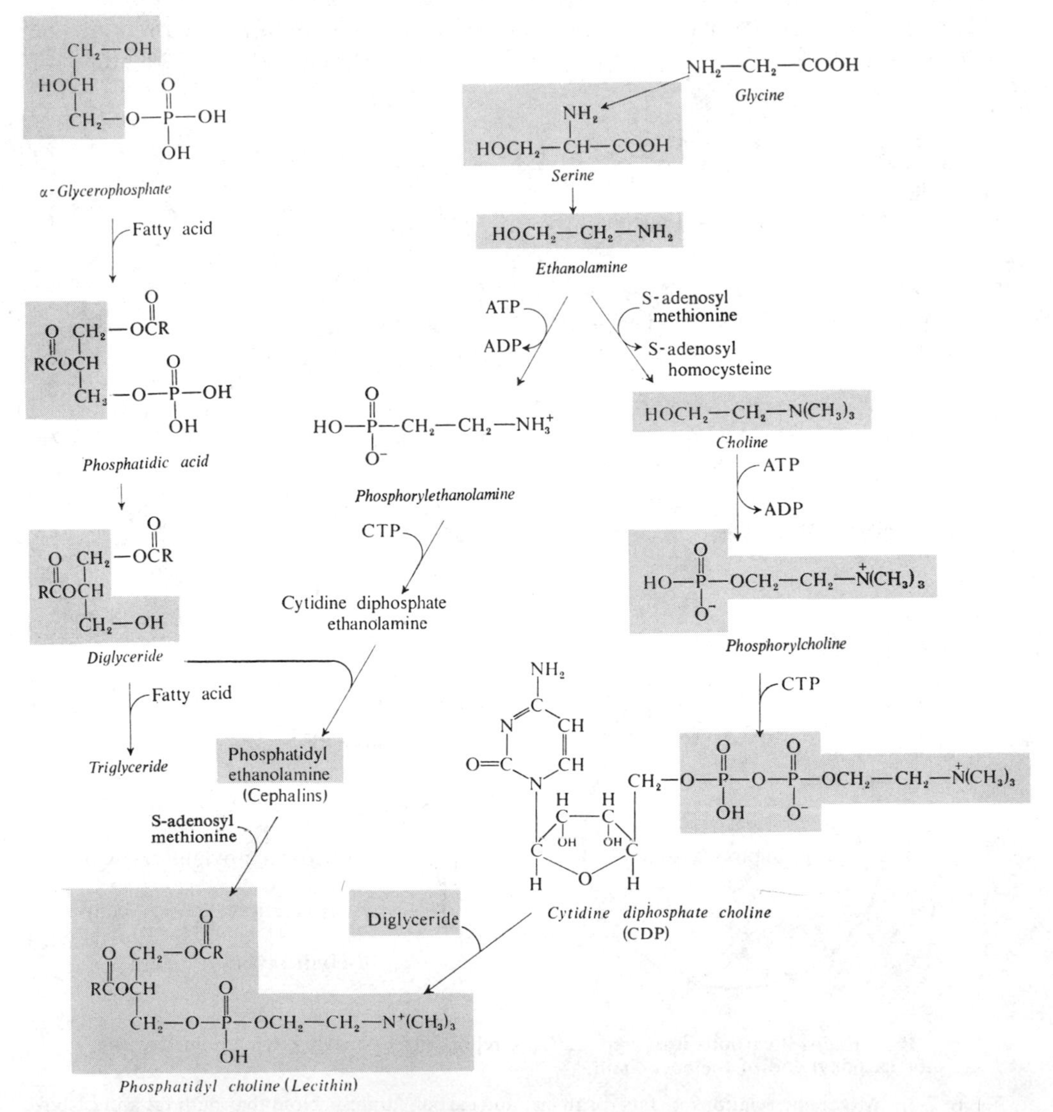

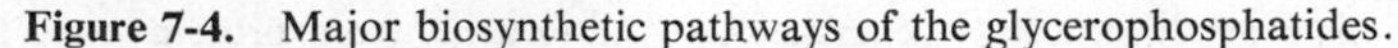

Figure 7-4. Major biosynthetic pathways of the glycerophosphatides.

obesity, and diabetes mellitus; however, it may also be caused by extremely high levels of fat in the diet or by such toxic agents as carbon tetrachloride and chloroform. Since these are frequently used under unhygienic conditions, intoxication by solvents may occur as an occupational hazard in some industries. The underlying mechanism in the production of fatty liver is a failure of adequate phosphatide synthesis, especially of phosphatidylcholine.

Fat ingested in the diet is absorbed and brought to the liver either directly through the portal blood or by the lymphatics as chylomicrons. The liver must then synthesize α- and β-lipoproteins, which must contain phosphatides, to transport the fat from the liver to the adipose tissue for proper storage. If the phosphatides are lacking in lecithin, or if the total production of phosphatides is insufficient, the triglycerides cannot be carried from the liver.

The main biosynthetic pathway for the phosphatides involves the enzymatic condensation of a diglyceride with a derivative of cytidine triphosphate, the details of which are set forth in Figure 7-4. As the figure shows, phosphatidylcholine may be formed directly by condensation of cytidine diphosphocholine with a diglyceride, or by condensation of cytidine diphosphoethanolamine with a diglyceride, after which methylation to the choline derivative can take place. For this reason, adequate synthesis of phosphatidylcholine can be insured by treatment with either choline or methionine, which serves as a donor of the methyl groups necessary to transform phosphatidylethanolamine to phosphatidylcholine. Choline and methionine are therefore known as lipotropic agents since they can prevent the fatty liver of nutritional deficiency, perhaps by enhancing the formation of the phosphatides required to transport triglyceride from the liver via the plasma lipoproteins.

PRINCIPLES AND PROBLEMS OF LIPID ANALYSIS

The handling and quantitative analysis of lipids require scrupulous attention to details and pose some problems not encountered in analysis of other types of materials. Since the lipids are usually water-insoluble, many volatile and flammable solvents are commonly employed, and these are a constant hazard. Even the nonflammable solvents create problems; volatile solvents evaporate so readily that the concentration of solutions may change appreciably unless considerable care is exercised. Volatile solvents have a high vapor pressure and a low surface tension so that special pipetting technique may be needed, especially if the room temperature is much above 20°C. Pipets should be carefully selected for fine tips; that is, not only is a tip with a small orifice to provide slow draining needed, but also a tip that does not have an excessive wall thickness to add to the surface at the tip, since the volatile solvents will "creep" up such a surface and evaporate, leaving a deposit of lipid material. Needless to say, chipped or imperfectly tipped pipets should not be used. In using even carefully selected pipets, the solution to be measured should be pulled into, and allowed to drain from, the pipet several times before a sample is taken; this will allow the free space above the solution to reach an approximate equilibrium vapor pressure with respect to the solvent. If this is not done, the content of the pipet may be forcibly ejected when the top is plugged with the forefinger. Since pipets are not calibrated for nonaqueous solutions, except on special order, it is a good idea to occasionally check the calibration of small pipets (0.1 ml. or less) by weighing. Errors of calibration of this sort may be very small, but for work of the highest precision they should not be neglected. For similar reasons, stoppered volumetric ware should be checked for tightness. Glass-stoppered flasks may be adequate for storage of aqueous solutions, but not for chloroform solutions. It is easy to check the tightness of fit by merely inverting a stoppered flask. If the solution seeps around the stopper, a better fit can be obtained by lapping the stopper into the flask neck with a little fine emery in water. Even better is the use of polyethylene or Teflon stoppers, which are soft enough to insure a good fit in most cases.

All solvents, including ether, chloroform, methanol, and hexane, should be at least analytical reagent grade (A.R.). For gas chromatography, we recommend the purest grade available (Spectroanalyzed). In general, it is neither economical nor safe to attempt recovery or initial purification of organic solvents. Ether of suitable grade (analytical reagent or for anesthesia) is generally purchased in metal cans, which are designed to inhibit peroxidation of the solvent. Other solvents are generally marketed in brown glass bottles. It is wise to procure these in containers of approximately 1 L. so that the contents can be consumed before they deteriorate. Since the purest chloroform is highly unstable, commercial material usually contains a trace of water or alcohol to inhibit decomposition, and this small amount of additive will not interfere. Hydrocarbon solvents are generally mixtures of pentanes and hexanes (low boiling petroleum ether, b.p. 30 to 60°C.) or of hexanes and heptanes (ligroin, b.p. 60 to 80°C.). Because these mixtures may be of variable composition and purity, depending on the source, they should be carefully examined for residue (by weighing after evaporation) and for acidity (by titration of the lower layer of a mixture of equal volumes of water and solvent shaken well in a separatory funnel) before use. Better still, they should be replaced by more reproducible solvents.

Lipid solutions should be stored at the lowest possible temperature. For most laboratories, this will be about −20°C., the temperature setting of "deep freeze" compartments of commercial refrigerators. For others, it may be −10°C., the common temperature of the freezing or "ice cube" section of household-type refrigerators. A considerable volume change occurs when lipid solutions are stored at these temperatures, and solutions should be allowed to come to room temperature before use. Storage vessels should be as nearly filled as is practical; when dealing with solutions of highly unsaturated lipids, the free space over the surface of the contents should be flushed with nitrogen or carbon dioxide to avoid air oxidation. Exposure of such solutions to light should also be minimized since the decomposition is photochemically accelerated. For shorter periods (up to several weeks) and for less unsaturated lipids (such as cholesterol), such precautions as flushing with inert gas may not be necessary, but there is still a real risk of decomposition that should never be overlooked.

When it is necessary to concentrate solutions of the complex lipids, solvents should be removed at the lowest possible temperature if the integrity of the recovered lipid is important. Under these conditions, temperatures higher than 40°C. should be avoided. For most accurate work, the evaporation should be done in a nitrogen or carbon dioxide atmosphere, preferably with a rotary flash evaporator, a variety of which are commercially available.

Removal of precipitates or separation of phases should be performed in the centrifuge rather than by filtration because the evaporation of solvents from the large exposed surface of filter papers may lead to considerable change in the concentration of a solution. Although this can be offset by repeated washing of the paper with fresh solvent, it is quicker and cheaper in the end to employ centrifugal separation. For precise work, the spun-down protein can be quickly washed by dispersion with fresh solvent, which is then added to the original supernatant fluid. Glass or Teflon stoppers should be fitted to the tubes, for reasons already mentioned. When separatory funnels are needed, they should be equipped with Teflon plugs since these are less likely to leak and do not require lubricating grease that might otherwise contaminate the sample.

A good review of these principles has been published by Sperry,[18] and his paper should be consulted for further details.

Standard Substances

A primary requisite for any analysis is a suitable standard sample of acceptable purity and stability; unfortunately, most lipid materials are extremely difficult to purify and many of them are not as stable as we might like. It is currently possible to purchase a considerable variety of materials of adequate purity from a limited number of sources. Those in the following list are recommended solely on the basis of the author's experience.

FREE FATTY ACID AND METHYL ESTERS

1. Hormel Foundation, Austin, Minn.
2. Applied Science Laboratories Inc., State College, Pa.
3. Sigma Chemical Co., St. Louis, Mo.

GLYCERIDES

A limited number of purified monoglycerides, diglycerides and triglycerides may be obtained from Distillation Products Industries, Rochester, N.Y., as well as from the suppliers just listed.

PHOSPHATIDES

Synthetic or highly purified natural phosphatides may be purchased from:

1. Analabs Inc., Hampden, Conn.
2. California Biochemical Corp., Los Angeles, Calif.
3. Sylvana Laboratories, Millburn, N.J.

STEROLS

At the present time, there is still no commercial source of cholesterol and related substances of purity sufficient to serve as a standard for analysis. Detailed instructions for the preparation of suitable material will therefore be presented as part of the analytical procedure for cholesterol.

Chromatographic Techniques

The most revolutionary advances in lipid analysis are dependent on the recent development of a variety of chromatographic separations. These physical methods have vastly extended our ability to quickly and reliably separate small samples of lipids into classes and even to determine individual compounds where previous chemical methods were inadequate or lacking. As a result, chromatographic separations are steadily replacing less advantageous chemical separations.

The older chromatographic separations were done in glass columns packed with either alumina, Florisil (a hydrated magnesium silicate), or silicic acid. Even more elegant methods have been provided by *thin-layer chromatography* (TLC), in which the same adsorbents are coated on flat glass or plastic plates. TLC is essentially the same as column chromatography, but since one dimension has been reduced extremely (to the thickness of the layer, or about 0.2 mm.), the separations are produced much more rapidly and are usually more sharply defined than in a column. Where column chromatographic separations might require 36 hours, TLC separations might require 2 hours. Space limitations preclude a lengthy theoretical discussion of these procedures, but they have become so important in lipid chemistry that some familiarity with them is essential to modern laboratory practice.

As applied to lipid chemistry, column chromatography and TLC most frequently operate on the principles of adsorption chromatography. The components of a lipid mixture are adsorbed more or less strongly to the surface of the silicic acid, alumina, Florisil, or similar adsorbent. The forces that hold the substances to the adsorbent are largely physical, and depend on slight degrees of polarity of the solute molecules, based on partial ionization. If the column or TLC plate is now developed with a solvent mixture properly chosen, those molecules that are adsorbed least tightly are eluted, or freed from attachment to the adsorbent surface, so that they then migrate with the developing solvent. The molecular species that are bound a little more tightly are then eluted, and so forth; thus, on completion of the development process the individual components have been moved to positions away from the starting point. In this fashion, a separation has been effected. When using columns, the separated molecular species are allowed to drip in turn from the lower end of the column, where they may be collected in suitable containers, but in TLC the components are merely moved to separate points on the surface of the plate, where they may then be detected by visualization under ultraviolet light or by special staining reagents. Once located, the individual spots may be scraped from the TLC plate and the separated lipid collected in a solvent. By use of suitable solvent mixtures, lipids may be separated into various classes; sterols, sterol esters, triglycerides, mono- and diglycerides, phosphatides, and free fatty acids can quickly be recovered from quantities as small as 5 mg. By different solvent systems, the phosphatides can be further separated into phosphatidylcholines, phosphatidylethanolamines, phosphatidylserines, and sphingomyelins. Nonesterified fatty acids can also be separated by TLC techniques, but for reasons to be described later they are generally first converted to their methyl esters. By using plates impregnated with silver nitrate and by other special means, it is also possible to separate fatty acids into classes according to degree of unsaturation.

TLC methods have the following distinct advantages:

1. Speed of operation
2. Ability to work with very small samples
3. Nondestruction of the sample
4. Relatively small cost of apparatus and materials
5. Precoated plates are now commercially available.

Another valuable and important technique is known as *gas-liquid chromatography* (GLC). This is a form of partition chromatography in which the molecules of a substance divide between a vapor state (a *gas*) and a dissolved state (a *liquid*), depending on the conditions established in the analysis. It is commonplace that every liquid has a tendency to evaporate, or lose molecules from its surface to the vapor phase. As a liquid is heated, the pressure of the vapor increases, until at the boiling point of the liquid the vapor pressure is equivalent to atmospheric pressure. When we speak of *low boiling* liquids, we are speaking of liquids with a *high vapor pressure*, and vice versa. Consider now a mixture of related compounds, such as fatty acids derived from plasma triglycerides. At a given temperature, each will have a finite vapor pressure, so that of the total amount of each molecular species present, some fraction will exist as a vapor and the rest will exist as a liquid, which may conveniently be dissolved in a very high boiling substance that has only a negligible vapor pressure itself. Given these conditions, a separation may be readily effected. In its simplest form, a gas-liquid chromatograph may be described as a long tube packed with an inert, particulate material, serving only as a mechanical support for a coating of a liquid having a very high boiling point. In a typical case, the inert support might consist of crushed fire brick, and the high boiling liquid might be polyethylene glycol succinate or a stopcock grease. The high boiling liquid is usually added at a level of 1 to 10 per cent by weight. The tube is uniformly heated to some temperature sufficient to insure an adequate vapor pressure under the partition conditions (in the case of fatty acid methyl esters this will be about 185°C.), and a slow stream of an inert gas such as nitrogen or argon is passed through the tube. A sample of mixed fatty acid methyl esters is now injected into one end of the column; a detector capable of sensing the presence of ester molecules in the vapor is connected to the other end of the tube so that positive signals produced by the detector can be recorded as a function of time. As the vapor molecules pass through the heated tube, under the influence of the inert carrier gas, each molecular species that has been separated by the partition process will enter the detector in an order dictated by its individual vapor pressure under the conditions of the experimental procedure. If the detector output is connected to a strip-chart recorder, the signals will produce a series of peaks, the location of which with respect to time is related to the structure of the vapor molecules. The area under each peak will be related to the quantity of each component in the mixture.

Several types of detectors have been developed, of which the most practical for general purposes is the flame detector. In this device a tiny burner, fueled by a mixture of hydrogen and air, is given a negative charge of about 300 volts. Just above the flame is a platinum ring that has a corresponding positive charge. As vapor phase molecules are fed into the base of the burner, they enter the flame, the energy of which produces ions from the vapor molecules. These are attracted to the ring electrode, and the current flow thus produced is then amplified and fed to a recorder. The flame detector is especially suited for biochemical studies since it is not sensitive to water, it is simple to maintain, it is difficult to damage, and it has a large linear range.

The student may have noticed that reference has been made in this brief discussion of GLC to methyl esters of the fatty acids rather than to the free acids themselves. This is because the common fatty acids in the free form are either solids or else have very high boiling points, but conversion of the free acids to the corresponding methyl esters is a convenient means of lowering the boiling point appreciably. Palmitic acid, for example, boils at about 350°C., but methyl palmitate boils at 196°C. It is clear, then, that whether we start from free fatty acids or from a triglyceride or phosphatide, some method of producing the methyl esters of the fatty acids is desirable. A useful reagent for this purpose is anhydrous boron trifluoride in methanol. Boron trifluoride acts as a very strong acid, and catalyzes a transesterification as follows:

$$\begin{array}{l} CH_2\text{—}OCOR \\ | \\ CH\text{—}OCOR'' + 3CH_3OH \\ | \\ CH_2\text{—}OCOR' \\ \textit{Triglyceride} \quad \textit{Methanol} \end{array} \xrightarrow{BF_3} \begin{array}{ll} CH_2\text{—}OH & RCOOCH_3 \\ | & \\ CH\text{—}OH & + R''COOCH_3 \\ | & \\ CH_2\text{—}OH & R'COOCH_3 \\ \textit{Glycerol} & \textit{Methyl Esters} \end{array}$$

or

$$\underset{\textit{Fatty Acid}}{RCOOH} + \underset{\textit{Methanol}}{CH_3OH} \xrightarrow{BF_3} H_2O + \underset{\textit{Methyl Ester}}{RCOOCH_3}$$

The esters can easily be washed free of the other products and the excess catalyst and recovered quantitatively for further study. The entire process takes less than 30 minutes.

GLC is a powerful analytical tool, not only for the analysis of lipids, but also for the analysis of blood and urinary alcohol, barbiturates, tranquilizers, and steroid hormones, to name a few other uses. As the literature of available methods has grown explosively in the past 10 years, the apparatus employed has also become more sophisticated. New liquid phases and new devices are constantly being developed and exploited. The current state of the art is such that samples of 3 μl., containing as many as 17 fatty acid methyl esters, can be qualitatively and quantitatively analyzed in approximately 25 minutes. The sample scale is sufficiently small so that individual spots eluted from TLC plates can be further characterized by GLC. The advantages of GLC are as follows:

1. Speed of both qualitative and quantitative analysis
2. Ability to handle very small samples
3. Adaptability to many types of analytical problems.

Against these advantages, however, must be weighed several disadvantages:

1. Relatively high initial cost of the apparatus, ranging from $3000 to $7000
2. A small, but definite, hazard of compressed gas cylinders, including hydrogen.

An exhaustive review of TLC has been edited by Stahl,[20] and a somewhat abbreviated version has been prepared by Randerath.[14] Similarly, a very thorough review of GLC has been written by Burchfield and Storrs,[5] and a brief review has been prepared by Littlewood.[12] These volumes include both theoretical discussions and detailed procedures.

Extraction of Lipids from Biological Materials; Sample Preparation

Except for the triglycerides of adipose tissue and some other cells, lipids exist in combination with protein, and in close association with other substances. Phosphatides, especially, are bound elements of tissues; thus, the first and most important requisite in lipid analysis is the preparation of an extract that contains as much as possible of the total lipid, and as little else as possible. If, for example, the phosphatides are to be determined by measuring the lipid phosphorus, it is essential that other phosphorus containing materials be eliminated from the extract. If the total lipid is to be measured by weight, no interfering substances can be tolerated. Older procedures described in the literature, and even some still used in various laboratory manuals, give extracts that are grossly contaminated with nonlipid material, such as urea, amino acids, and even sugar, carried into lipid extracts in chloroform by the emulsifying capacity of the phosphatides. Fortunately, recently developed procedures avoid these difficulties, and therefore are preferred.

Since much of the total lipid content of tissue is bound to protein, the first essential of an extraction procedure is to destroy such bonds. To this end, the protein must be dehydrated or denatured, which means in effect that it has to be altered sufficiently to free the lipid for solution by an extracting solvent. Even in the older literature this was recognized, and methanol, ethanol, or acetone have all been employed as protein denaturants. Acetone is a poor choice since it precipitates some phosphatide; ethanol is somewhat better since it does dissolve the lipids to some extent; however, methanol is, perhaps, the best denaturant. After denaturation, the lipids can be extracted by ether or chloroform. Ether is a fair solvent, but it also can produce a considerable degree of lipid peroxidation, besides suffering from the drawbacks of extreme flammability and a density less than 1.00. The latter property poses problems when the time comes to collect the lipid extract. The use of mixtures of alcohol and ether, as recommended by Bloor and others, suffers from these general characterizations, but perhaps the most severe criticism is still based on the observation that Bloor's solvent pair carries over considerable amounts of nonlipid contaminants, including urea, amino acids, and sugars. These problems can be minimized by the use of methanol and chloroform; it appears that the nonlipid impurities are less soluble in chloroform, and its high density makes it more convenient to recover the extracted lipid. Lipids are also less likely to be damaged in chloroform than in ether. Finally, the use of any solvent pair as a mixture is less efficient than applying the two solvents separately, since, in this method, each component of the mixture is maximally able to exert the effects of either protein denaturation or lipid extraction. Whatever mixture is finally employed, careful study has proven this point beyond doubt. In general, the final ratio of extracting medium to the sample of blood, or other tissue, should be at least 20:1. Two procedures of proven reliability will be presented.

METHOD OF FOLCH (MODIFIED)

This procedure, as we have modified it, is particularly useful for extraction of blood lipids, although it can also be used for solid tissues when they are homogenized.

1. Place 5 volumes of methanol in a centrifuge tube.

2. Add 1 volume of blood, plasma or serum, commonly 1 or 2 ml., to the methanol by expelling it into the solvent from a calibrated syringe fitted with a No. 25 needle. Pull some of the solvent up into the syringe; now flush it back into the tube. Mix well and allow to stand for 5 minutes.

3. Add a second 5 volume portion of methanol; then, mix well and, again, allow to stand for 5 minutes.

4. Add 10 volumes of chloroform, mix well, stopper, and allow to stand for 5 minutes, after which the solution is centifuged to collect the finely precipitated protein.

5. Transfer a measured aliquot of the clear supernatant solution to a clean centrifuge tube and add 0.2 volume (based on the aliquot taken, not on the original volume) of 0.1 *M* NaCl. Mix well; then centrifuge. The clear subnatant phase is taken for further analysis.

This procedure produces a lipid extract that is substantially free of nonlipid contaminants. As presented here, the method differs from the original by recommending the addition of the methanol in two separate aliquots. It has been our experience that this produces a protein precipitate that is finer, more granular and, therefore, more readily removed.

REFERENCE

Folch, J., Lees, M., and Sloane-Stanley, G. H.: J. Biol. Chem., *226*:497, 1957.

METHOD OF BLIGH AND DYER

This method uses the same solvents as the method of Folch, but the procedure differs in the purification step. Bligh and Dyer employ a solvent mixture that takes account of the water content of the sample, and that produces a monophasic mixture in the first step. This is then treated with sufficient water and chloroform to bring about phase separation. The subnatant chloroform solution contains about 99 per cent of the total lipid, substantially free of nonlipid contaminants. We have found it useful for muscle and stool analysis, but less useful than the method of Folch for blood analysis. It is best employed on samples that contain at least 80 per cent water. If the sample contains less water (dry weight determination), distilled water should be added to bring the level to that value; however, if the sample contains more water, a smaller aliquot is taken. Although the procedure is presented here, as in the original version, for large samples, it may be scaled up or down at will, depending on the sample requirements and on the equipment at hand. The procedure is as follows:

1. A 100 gm. sample is homogenized for 2 minutes in a Waring Blender, or equivalent homogenizer, with 100 ml. of chloroform and 200 ml. of methanol.
2. Add 100 ml. of chloroform and homogenize for 30 seconds longer.
3. Add 100 ml. of water; then homogenize for 30 seconds.
4. Filter the homogenate through Whatman No. 1 paper in a Buchner funnel with suction until the cake appears fairly dry.
5. Rehomogenize the filter cake, plus the paper, with 100 ml. of chloroform; then filter a second time through the same funnel. The second filtrate is added to the first.
6. Transfer the filtrate to a 500 ml. stoppered cylinder; then add 100 ml. of methanol. Mix well and allow the two phases to separate.
7. Aspirate and discard the upper aqueous methanolic layer, and collect the clear chloroform solution of the extracted lipid. Measured aliquots may be taken for further analysis.

REFERENCE

Bligh, E. G., and Dyer, W. J.: Canad. J. Biochem. Physiol., *37*:911, 1959.

Determination of Total Serum Lipids

The most direct and most reliable method for determination of total serum lipids is still the gravimetric procedure. Lipid extracts based on either of the total lipid extracts just cited are suitable for this purpose. A measured aliquot of the lipid extract is transferred to a tared disposable aluminum weighing dish (Arthur H. Thomas Co., Philadelphia, Pa., Catalog No. 4537, 70 mm.). The dishes containing the extracts are placed in a vacuum desiccator, then flushed several times with nitrogen or carbon dioxide. The solvent is removed by means of a water aspirator until the dishes again reach constant weight. The drying process may be speeded by applying heat to the outside of the closed desiccator with an infrared heat lamp. After weighing the dried residue, it may be dissolved in a little chloroform for other analytical purposes.

CLINICAL SIGNIFICANCE OF TOTAL SERUM LIPID ANALYSIS

The clinical significance of total serum lipid determinations is questionable. In the fasting state, values of 700 mg./100 ml., or more, may be considered elevated, but in the normal postprandial state, a range of 400 to 1000 mg./100 ml. is common. The exact level is dependent on the quantity and quality of the dietary intake. One school of thought holds that any rise in any component of the serum lipids is indicative of abnormality or disease, but this opinion is not universally accepted. It is true that changes in one component are usually mirrored by changes in the others, but the degree of increase, or decrease, is not consistent between the triglycerides, sterols, phosphatides, and so on. From the technical point of view, it is probably easier to measure the total serum lipids accurately than to measure cholesterol, but the latter does not show the marked variation in day to day levels, nor is it as sensitive to the time of sampling with respect to food intake. Because of these facts and because many clinicians are more familiar with the significance of cholesterol values, there is still good reason to depend on them.

Among the conditions that increase serum total lipids are diabetes mellitus, biliary cirrhosis, and late pregnancy. Nephrosis and hypothyroidism may also lead to very high levels of total blood lipids so that the serum may become opalescent as a result of dispersed lipid. In some nephrotics, total serum lipids (mainly cholesterol) may reach levels of 3200 mg./100 ml. Essential (idiopathic) hyperlipemia is a familial disease that may also cause an enormous increase in total lipid. We have seen levels of 6700 mg./100 ml. in such cases, and even higher values have been reported by others.

Few conditions cause a decrease in total serum lipids. Perhaps the most significant of these is a lowered caloric intake, especially if the fat or lipid intake is much reduced. It has also been demonstrated that total lipids decrease in hyperthyroidism, but there are much more precise measures of thyroid function so that this observation is of limited value.

Efforts are sometimes made to express total serum lipids in terms of equivalents of fatty acids, but little significance can be attached to such devices since the fatty acid components of triglycerides are mixed, and the molecular weights of the fatty acids present are not the same. Furthermore, the fatty acid content of triglycerides is at least three times the content in cholesterol esters, and as much as twice that of the phosphatides; any calculated value is, at best, an average of many factors. It therefore seems prudent to continue expression of these results in simple mass terms.

Total Fecal Lipids

Extraction of total lipids from fecal material is best done according to the method of Bligh and Dyer, just discussed, since the water content of individual stool evacuations is variable in quantity. Indeed, the overall quantity and quality of fecal samples is so variable that there is little reason for analyzing anything but a pooled sample collected over at least a 48 hour, or, preferably, a 72 hour period. As the individual stools are passed, they may be added to a tared glass container with a wide mouth, which should initially contain at least 100 ml. of methanol made 0.1 *N* in hydrochloric acid. There is a very real problem of collecting feces free of urine and other foreign matter, and the analyst should be aware of this problem, although he can rarely control it. Obvious foreign matter should be removed with forceps before any analysis is attempted. At the end of the collection period, and preferably even while additions are being made to the collection vessel, the fecal masses should be dispersed in the preservative methanol. When the collection has been accomplished, the tared weight of the container and the added methanol can be subtracted from the total final weight to obtain the weight of the collected stools.

Even while in the bowel, fecal lipids are subject to continued hydrolytic attack from intestinal and bacterial enzymes. Storage, even in the refrigerator, does not completely stop this reaction unless the pH is drastically changed, and this is the purpose of the acidic methanol preservative. In the presence of alcohol and acid, hydrolysis is reduced. At best, these precautions are only approximate; we must conclude that at present there is no way of keeping feces without change. Because of the problem of stool consistency and variable water content, it is best to describe the total lipid content in terms of dry weight, which means that the aliquot needs to be taken to dryness before analysis of the lipid content. Alternatively, the total 24 hour output weight may be used.

In the absence of any food intake, the fecal mass still contains a certain amount of fat and sterol. Part of this may be derived from the bacteria that inhabit the intestine and the bowel, but there is good evidence that the bowel, itself, serves as an excretory organ for lipids, especially for sterols. These may be derived from the desquamation of the mucosa, and from unresorbed bile.

It is generally agreed that total fecal fat levels of more than 25 per cent of the dry weight represent an abnormally high level, resulting from either diarrhea, Whipple's disease, pancreatic insufficiency, or some other form of malabsorption syndrome. Determination of fecal lipids is sometimes done as an index of pancreatic function since it is assumed that much of the splitting of fat depends on pancreatic secretions, particularly of lipase. If the split fats are less than 80 to 85 per cent of the total fat, the possibility of pancreatic insufficiency is raised, but other, more precise, tests are required to establish this diagnosis. In general, the reliability of fecal fat partition studies may be regarded as dubious, at least in any quantitative sense.

Normal values, based on the per cent relative to the dry weight of stool, average as follows:

Neutral fat	3.0
Split fat	
Mono- and diglycerides	4.3
Free fatty acids	5.6
Soaps	4.6
Total fat	17.5
Sterols	1.8
Total lipids	19.3

PROCEDURE FOR THE DETERMINATION OF TOTAL FECAL FAT

1. Prepare a wide mouth jar equipped with a tight lid by weighing it. Record the weight as J_0.

2. Add to the jar 100 ml. of methanol, which is 0.1 *N* in HCl; weigh the jar again and record this weight as J_m.

3. Collect the fecal sample by adding all evacuated material to the collection jar for at least 48 hours, taking note of the precautions just mentioned concerning foreign matter, urine, and so on. At the end of the collection period, weigh the container and contents again, and record this weight as J_s.

4. Calculate the weight of the collected feces by subtracting J_m from J_s. In other words, total sample weight $W_s = J_s - J_m$. Record this weight.

5. Thoroughly homogenize the contents of the jar with a Waring Blender, a paint mixing shaker, a high-speed stirrer, or a similar device. It is essential that the entire sample be reduced to a homogenous state.

6. Transfer a suitable aliquot (10 to 15 gm.) to a tared glass or porcelain weighing dish and evaporate to constant weight in a vacuum desiccator as previously outlined. Record the weight of the solid residue of the aliquot as W_a.

7. Prepare a total lipid extract of the residue by the method of Bligh and Dyer, and record the weight of the lipid contained in the extract as W_L.

8. Since the weight of the fecal homogenate aliquot taken included some mass resulting from the stool, and some from the methanol, we need to determine the actual fecal weight (W_f) in that weighed aliquot (W_a). It follows that

$$W_f = W_a \frac{J_s - J_m}{J_s - J_0}$$

9. The per cent of total lipid (expressed on a dry weight basis) is then determined by the expression

$$\text{per cent lipid} = \frac{W_L}{W_a \times \dfrac{J_s - J_m}{J_s - J_0}} \times 100$$

10. It is also possible to calculate the total lipid excreted for the entire collection period, and from that it is simple to calculate the total lipid excreted per 24 hours.

$$\begin{matrix}\text{total lipid} \\ \text{collection}\end{matrix} = W_L \frac{W_s}{W_f}$$

where the symbols have the same significance as used before.

11. If a measure of split fecal fat is desired, dissolve the residue from the total lipid extract obtained above in 2 ml. of hexane; transfer the solution to a stoppered separatory funnel.

12. Add to the solution 2 ml. of 0.2 N NaOH in 80 per cent ethanol and shake the stoppered tube thoroughly for 2 minutes. Allow the two phases to separate; then draw off and discard the lower ethanol phase, taking care to lose none of the supernatant hexane.

13. Transfer the extracted hexane solution to a tared weighing dish, and evaporate the contents until the dish has a constant weight.

14. The residue from the hexane solution contains the unsplit neutral fat plus sterols; thus, the difference between this weight and the initial total lipid weight is a measure of the split fat content of the sample.

Further details concerning alternative wet chemical methodology can be found in Henry.[10]

REFERENCE

Dryer, R. L., unpublished.

FECAL LIPID PARTITION BY TLC

Fecal lipid partition can also be done by TLC, using TLC plates coated with Silica Gel G (E. Merck Co., c/o Brinkmann Instruments, Westbury, N.Y.).[20] Either commercial, precoated plates or handcoated plates may be used for this purpose. Various suppliers now market precoated plates with the layers placed either on 8 inch square glass or plastic base. The plastic base can conveniently be cut with scissors to any desired smaller size, which obviously is not possible with glass plates. The plastic plates are designed for throw-away use, but the glass plates may be recovered for future use. Although the plastic plates are uniform and spare the analyst the tedium of plate preparation, the carrying capacity of the plastic plates is considerably lower. The glass plates can be obtained in a variety of layer thicknesses, and samples of the layer, bearing compounds of interest, can be readily removed for further analysis, which is not the case with the plastic plates.

If handcoated plates are desired, some form of spreading surface and some form of spreading device is also required. Several of these are commercially available from supply houses, and the details of their design and use may be found in the sources already cited.[14,20] Based on extensive experience, we shall describe the preparation of handcoated plates using the Desaga spreader and spreading board (C. Desaga Co., c/o Brinkmann Instruments, Westbury, N.Y.). Other devices may be equally satisfactory, but our experience with them has been minimal. It is possible, for example, to spread a thin layer by fastening two strips of masking tape to opposite edges of a glass plate, pouring a slurry over the surface, and then leveling the surface with a long glass rod supported on the strips of tape. This will give a layer as thick as the tape, and heavier layers may be applied by using more than one layer of tape. This crude process may give satisfactory layers, but it is not too reliable. A slightly more refined version uses glass plates with raised edges (Kontes Glass Co., Vineland, N.J.). The slurry is poured in the center of the plate, allowed to flow over the surface, and then smoothed with a glass bar that has been carefully ground to have a flat surface. This procedure is distinctly better than the tape process.

There is little doubt that the most successful spreaders, in terms of usable finished plates, are those that are designed to carry the slurry of silica gel in a chamber pulled over the surface of one or more clean dry glass plates in such a fashion that a layer of known thickness is produced on all of the plates. In the method that follows, we shall assume that such a spreading device is available, and that handcoated plates are to be employed. If the analyst wishes to use precoated plates, he may disregard the first steps in the procedure.

Plate preparation

1. Thoroughly wash five 8 inch square glass TLC plates, rinse in distilled water, and allow them to dry in the air. It is essential that the surface of the plates be free of grease. Clean plates should be handled only by the edges.

2. Align the clean dry plates against the edge guides of the spreading board. Each plate should closely touch the last.

3. Weigh out 20 gm. of dry Silica Gel G (or equivalent) and place it in a 125 ml. Erlenmeyer flask. Add 60 ml. of distilled water, stopper the flask, and shake the contents vigorously for at least 2 minutes.

4. Transfer the slurry of Silica Gel G in water to the chamber of the spreader, invert the chamber, and with a single, steady motion pull the spreader across the surface of the entire series of plates.

5. Allow the plates to dry in the air until the surface appears free of moisture; this will be accompanied by a lightening of the color as the layers dry.

6. Separate the plates from the spreading board, and with a spatula edge or similar tool carefully remove about 1/4 inch of the coated layer on all four sides of the plate. This will eliminate any possibility of layer thickness variability due to accumulation of slurry at the edges. The prepared plates may be stored indefinitely in a clean dry place.

Sample application

7. Coated plates must be activated just before use. The activation process involves heating to 110°C. for 1 hour to drive off moisture that decreases the adsorptive capacity of the Silica Gel G. Cool the activated plates in a desiccator.

8. By means of a fine-tipped micropipet, apply the sample to be separated as a series of drops along a line about 1/4 inch from one edge of the layer, which we will then regard as the bottom of the TLC plate. The pipet should be manipulated so that each drop does not spread to more than 1 to 2 mm., and each successive drop should be applied so that the material it deposits on the layer just touches the last application. In this way, a band of material is applied across the bottom of the plate, starting about 1/4 inch from the left edge, stopping about 1/4 inch from the right edge, and about 1/4 inch above the bottom of the layer. It is possible with just a little practice to apply 100 μl. of a solution containing as much as 50 to 100 μg. of total solute. Thicker layers have a higher capacity than thin ones.

Development of the chromatogram

9. Prepare a developing chamber by lining the inside with a sheet of filter paper or paper toweling so that it covers the inner surface of at least three sides of the chamber, from the bottom to within an inch of the top.[20]

10. Prepare a solvent system composed of hexane, ether, and glacial acetic acid (80:20:1, v/v). Pour enough of this solvent into the chamber to give a solvent level, when the TLC plate is inserted, just below the point of sample application. This will require a solvent depth of about 3/8 inch in the developing chamber. Make sure that the paper liner is wet with the solvent system, and cover the chamber for 15 minutes before using it.

11. Carefully insert the TLC plate, with the line of applied sample toward the bottom of the chamber; then close the chamber. Allow the solvent to rise up the surface of the plate until it has come to within 1 to 2 cm. of the upper edge. This will take about 1 hour.

12. Remove the plate and quickly make two small marks at the edges of the TLC layer to indicate the exact point of farthest solvent travel; then, place the plates in a hood to air dry. Finally, complete the removal of the solvent, especially the acetic acid, by briefly placing the plates in an oven at 110°C.

13. Using either an air brush, a perfume atomizer, or a commercial spray bottle (Kontes Glass Co., Vineland, N.J.), lightly spray the surface of the TLC layer with a solution of 2′,7′-dichlorofluorescein (0.05 per cent in ethanol) and allow the plate to dry briefly.

14. Observe the plate under ultraviolet light; with a pencil, mark the outlines of the spots that appear as a yellow or orange color against a much lighter background. Be sure to mark the line of application, or origin, as well.

15. Measure the distance from the origin, or point of sample application, and the highest rise of the solvent. Denote this value by R_s.

16. Measure the distance from the origin to the center of each visible spot. Denote these distances R_{u_1}, R_{u_2}, and so on.

17. Since the distance traveled is characteristic of the different components in the mixture, we can calculate a quantity known as the R_f for each component by the formula

$$R_{f_1} = R_{u_1}/R_s, \text{ and so forth.}$$

The quantity R_f, which may be read as relative front, can be used to identify the component found at a given spot, especially if standards are available for simultaneous comparison. For the system described here, the R_f values have been well established.

Interpretation of results

The preceding description is designed for qualitative separations, but with only very slight modification, involving only the thickness of the TLC layer and the size of the sample applied, it is possible to quantitatively separate samples of about 5 mg. The separated spots can be scraped off the TLC layer; the solute can then be eluted with chloroform, dried and weighed. If this is done, typical normal stool samples give data of which Table 7-2 is representative.

TABLE 7-2. *Composition of Fecal Lipids*

$R_f \times 100$	*Component*	*Weight Per Cent*
80–85	Sterol esters	8.5
65–75	Triglycerides	12.0
45–50	Free fatty acids	18.0
30–35	Uncharacterized pigment	27.0
25	Free sterols	10.0
10–20	Uncharacterized material	15.0
0–10	Phosphatides	9.0

The student should note that the extract positively contains a considerable amount of material that is lipid only from the standpoint of solubility, but that does not contribute to any of the recognized classes of lipids for which there is clinical concern. Since this is extracted along with the important lipids, and since it cannot be readily removed by simple means, there is always a real error involved in the gravimetric analysis. One of the more important reasons for the development of TLC procedures is that they offer a way around this problem.

REFERENCE

Stahl, E. (Ed.): Thin-Layer Chromatography. New York, Academic Press, 1965.

Determination of Fat in Urine by TLC

Normally, the urine is free of fat, but in extensive trauma, especially the type associated with crushing lesions or multiple fractures of the long bones, fat may find its way into the urine. Under these conditions, it may also find its way into the blood in the form of droplets large enough to produce emboli; thus, the diagnostic import of finding fat in the urine is considerable. Fat in the urine may also result from late nephrosis, but this is usually accompanied by high levels of protein from the lipoproteins that leak through the damaged glomeruli, and the consequences of this state are altogether different. Therefore, it is helpful to have a rapid determination of fat in urine, which TLC affords. In general, it is not too essential to have a quantitative value; the important result is basically the positive finding. A simple procedure follows:

1. Centrifuge a random specimen of urine at 2000 rpm for 5 minutes. The exact volume is not important; we generally take about 75 ml. in two tubes.
2. By means of a long capillary pipet connected to an aspirator, and inserted to the bottom of the tube, withdraw from below all but the last 5 ml. (top layer) from each tube.
3. Combine the residues, measure the volume, and prepare a Folch extract according to the method already outlined.
4. Evaporate the chloroform extract to approximately 0.2 ml.; then apply from 50 to 100 μl. to a TLC plate as described for the determination of fecal lipids.
5. Develop the plate, spray with dichlorofluorescein, and examine under ultraviolet light.
6. A spot with an R_f of 0.65 to 0.75 is indicative of fat in the urine.

REFERENCE

Dryer, R. L., unpublished.

Determination of Plasma or Serum Triglycerides

The determination of triglycerides, or neutral fat, in plasma or serum is of interest in following the course of diabetes mellitus, nephrosis, biliary obstruction, and various metabolic derangements due to endocrine disturbances. The level of triglycerides in plasma is a little higher than in serum, possibly because of entrapment of chylomicrons during the clotting process, and perhaps also because of lipolysis during clot formation. In either case, plasma is the preferred sample for analysis, and samples should be worked up promptly.

The level of triglycerides is quite variable, depending on the alimentary state, amount of exercise, and the general metabolic level. The quantity of triglyceride found probably is a better reflection of the dynamic state of metabolism than is the level of cholesterol, since the triglycerides can contribute freely to the energy pool of the body, but cholesterol cannot.

Normal values for plasma triglycerides also depend on the methods employed, some of which determine the triglycerides directly, some of which measure the glycerol derived from triglycerides, and some of which measure triglycerides as the difference between total lipids less sterols and phosphatides. Methods that depend on this difference tend to overestimate the triglyceride content. By the methods we will outline here, and on a normal mixed diet, we have found fasting triglyceride values ranging from 30 to 200 mg./100 ml.

Several problems arise in analysis of triglycerides. First, they must be separated from other lipids since the only possible chemical procedures depend either on the analysis of glycerol, which is also present in the phosphatides and in mono- and diglycerides, or they depend on the analysis of liberated fatty acids, which are also found in other lipids. The best procedures for separation of the triglyceride fraction depend on one form or another of chromatography. Second, if the method is to be based on analysis of the glycerol content of triglycerides, a procedure may be based on an enzymatic assay in which glycerol is a specific participant, or on a colorimetric assay of formaldehyde produced from the controlled oxidation of glycerol by chemical means. Finally, the estimate may be based on determination of the fatty acids liberated by hydrolysis of the triglycerides, either by direct titration with alkali, or by photometry of the colored fatty acid hydroxamate–ferric ion complex that may be chemically produced. Since the naturally occurring triglycerides are of the mixed acid type, and since these methods rarely separate the triglycerides from the split glycerides, they usually involve a series of approximations that might better be avoided. Henry[10] has reviewed a number of these methods.

The procedures presented here allow the separation of triglycerides by either TLC or column chromatography; since the samples handled are small, the analytical estimation requires a method of considerable sensitivity and specificity. We prefer the enzymatic assay of glycerol liberated from the triglycerides, but will also present a colorimetric procedure based on the periodate oxidation of glycerol.

The enzymatic determination of glycerol depends on the following reactions:[10]

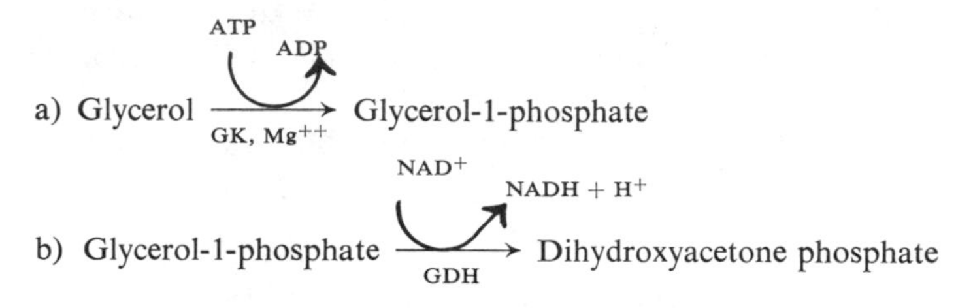

GK is the enzyme glycerol kinase, which, in the presence of magnesium ions, converts the substrate glycerol into glycerol-1-phosphate. The energy for this reaction comes from adenosine triphosphate (ATP), which is converted to adenosine diphosphate (ADP). Coupled to this reaction is a second system catalyzed by the enzyme glycerol-phosphate dehydrogenase (GDH), which produces dihydroxyacetone phosphate. This conversion is made possible by the addition of a hydrogen acceptor, nicotinamide dinucleotide (NAD^+), which is reduced ($NADH + H^+$). The reduction involves a loss of a double bond in the nucleotide structure, and this is accompanied by an increase in the absorption of light at 340 nm.* For each mole of glycerol converted to dihydroxyacetone phosphate, a mole of NAD^+ is converted to $NADH + H^+$. This is the basis of the enzymatic assay.

The chemical assay of glycerol depends on the cleavage of a carbon atom from glycerol by the periodate ion used as sodium periodate. In the cleavage process, the lost carbon atom is converted to formaldehyde. The reaction mechanism depends on the hydrated form of the periodate ion, as is set forth in the following equations.

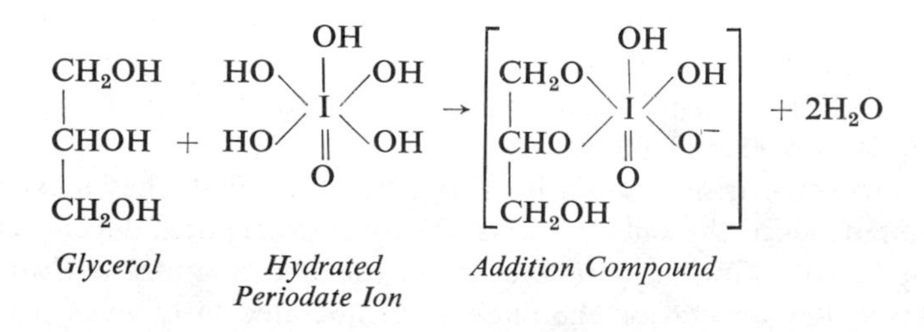

* The wavelength of 340 nm. is optimal, as it is the wavelength of maximal absorbance. Measurement may also be made at 366 nm., which in less expensive instruments is necessary if the light source is a mercury vapor lamp that has little radiant output at 340 nm. Reading at the longer wavelength involves some loss of sensitivity.

Two of the hydroxyl groups of the hydrated periodate ion condense with two adjacent hydroxyl groups of glycerol, with the simultaneous elimination of a molecule of water. The resulting cyclic structure is relatively unstable and decomposes by rupture of a C—C bond between two adjacent glycerol carbons. The remainder of the complex is thereby rendered still less stable, and it breaks down to iodate ion, glyoxylic aldehyde, and formaldehyde. When formaldehyde is treated with a sulfuric acid

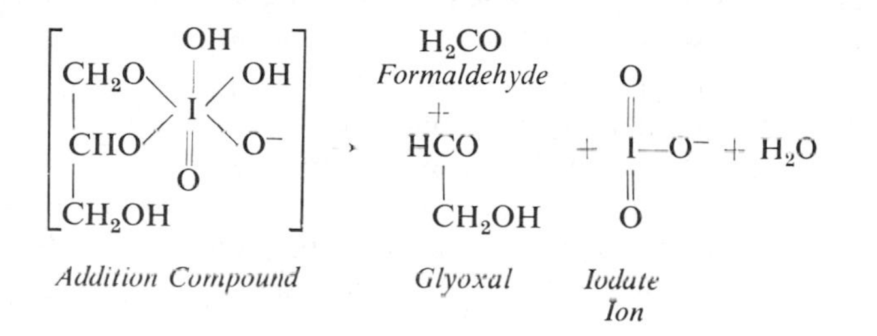

solution of chromotropic acid (4,5-dihydroxy-2,7-naphthalenedisulfonic acid), a pink color of unknown composition is formed. The absorbance of the colored solution is related to the amount of formaldehyde in accordance with Beer's law.

Both the enzymatic and the chemical methods require that the glyceride content of the sample first be hydrolyzed to free the glycerol, and also that the sample be free of glycerol derived from other sources. The enzymatic method is quite specific, but the chemical procedure is subject to interference from other materials that will also react with periodate. Interfering substances include glucose and some amino acids that might arise from phosphatides. It is, therefore, of considerable importance to separate carefully the triglycerides from other materials present in plasma or serum.

Enzymatic Method for Determination of Glycerol

PREPARATION OF THE TOTAL LIPID EXTRACT

1. Prepare a Folch extract from 1 ml. of plasma or serum, using 5 ml. of chloroform and 10 ml. of methanol, according to the procedure outlined previously.

2. With a pipet and a rubber bulb, transfer 8 ml. of the purified extract (equivalent to half the sample) to a clean, dry tube, carefully avoiding transfer of any of the supernatant phase.

3. Evaporate the solvent from the extract, then redissolve the lipid residue in 0.2 ml. of chloroform.

SEPARATION OF THE TRIGLYCERIDE FRACTION BY TLC

1. Using a fine-tipped micropipet, apply 0.1 ml. of the concentrated total lipid extract to a TLC plate coated with Silica Gel G (0.2 to 0.3 mm.). The pipet should be touched to the plate briefly so that a small spot (approximately 2 to 4 mm. in diameter) is formed. By moving the pipet slightly, a second spot is applied, just contiguous to the first. By repetition, a band of spots is formed that extends across the plate. This process is repeated until the entire volume of material has been applied.

2. Develop the TLC plate in a solvent system of hexane, ether, and acetic acid (90:10:1, v/v) until the solvent front is within 1 cm. of the plate top. The solvent system should be freshly made for each day's work.

3. Dry the plate by placing it *briefly* in an oven at 85°C.

4. Visualize the separated lipid zones by placing the dried plate in an empty developing tank, in the bottom of which a few iodine crystals have been placed. The iodine vapor will stain the lipid deposits a light brown or yellow. With the tip of a fine-pointed glass rod, mark the limits of each separated lipid zone. The triglycerides will have an R_f of approximately 0.24; however, depending on the concentration, this may extend from 0.18 to 0.30. The plate must be marked promptly after removal from the iodine vapor since the iodine stain quickly evaporates.

5. With a single-edged razor blade, carefully scrape off the silica gel containing the triglycerides. With a small funnel, transfer it to a stoppered centrifuge tube. Wash the last of the gel from the funnel with 1 to 2 ml. of chloroform. Stopper the tube and shake vigorously to elute the lipid; then centrifuge the solids so the solution can be decanted into a clean, dry 40 ml. stoppered centrifuge tube. (For this tube, we strongly recommend the tubes sold by Wilkens-Anderson Co., Chicago, Ill., Catalog No. 15845.)

6. The silica gel is washed with a second 1 ml. portion of chloroform, shaken well, and again centrifuged. The washing is added to the original eluate.

7. The combined extracts are evaporated to dryness, then promptly redissolved in 2 ml. of absolute methanol.

TRANSESTERIFICATION OF THE COLLECTED TRIGLYCERIDES

1. Add to each tube 0.05 ml. of 36 *N* sulfuric acid, with thorough mixing. (The sulfuric acid should be free of color, and should be fresh enough so that it has not taken up large amounts of water.)

2. Fit each tube with a 6 inch tube condenser equipped with a standard taper joint (14/35) at the lower end (Ace Glass Corp., Vineland, N.J., or other suppliers).

3. Place the tubes with condensers in a water bath set at 88°C. and allow the reaction to proceed for 1 hour. Instead of a water bath, we employ a special electrically operated heater known as a Dri-Bath, which is designed to accept the tubes specified (Thermolyne Corp., Dubuque, Iowa). A view of the assembled apparatus is shown in Figure 7-5.

4. Cool the tubes, and then add 0.2 ml. of 10 *N* NaOH; mix well.

5. Add 5 ml. of hexane, stopper the tubes, and shake thoroughly. Allow the two phases to separate; then collect the supernatant hexane phase as completely as possible, being careful not to lose any of the lower methanol phase. The hexane solution contains the fatty acid methyl esters, which may be examined by GLC if desired, or studied further by TLC.[20]

6. The lower methanolic phase contains the liberated glycerol, together with some inorganic salt. The methanol is evaporated to dryness, and the residue is redissolved in 0.1 ml. of water.

ENZYMATIC ASSAY OF GLYCEROL

The enzymatic assay of glycerol is precise and specific, and has been greatly facilitated by the commercial availability of suitably purified enzyme preparations. To employ this procedure it is essential to have a good spectrophotometer at least equivalent to a Beckman DB or DU. The requirement is based on the fact that absorbance readings must be made at a wavelength of 340 nm. A recording spectrophotometer is desirable, but not essential.

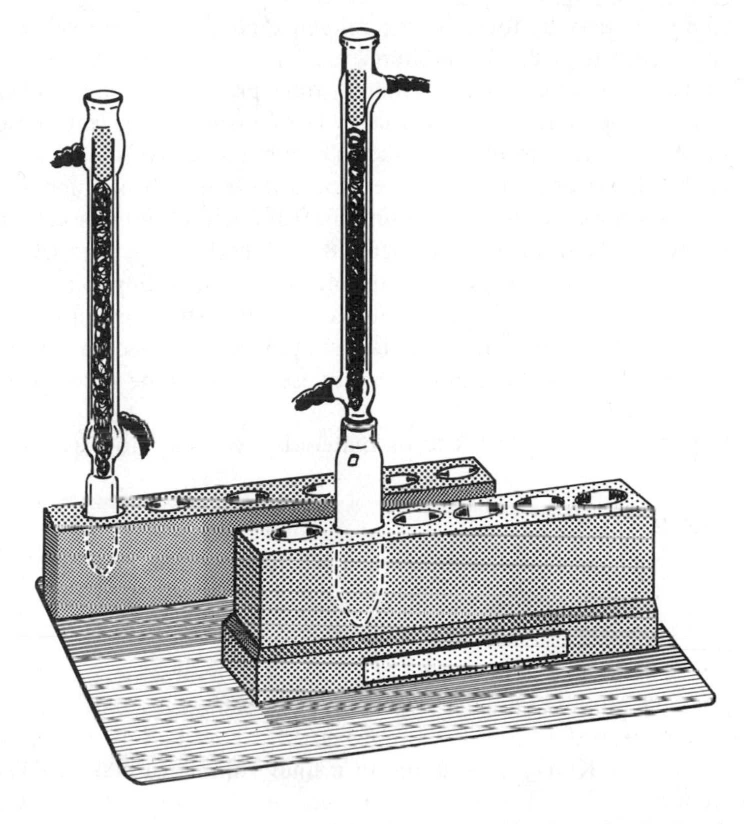

Figure 7-5. Illustration of a convenient apparatus for transesterification. For further details see text or *Clin. Chem.*, *13*:1014, 1967.

The determination is based on two coupled reactions:

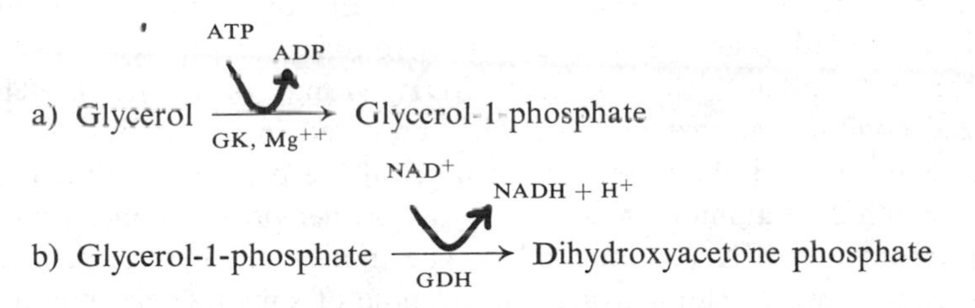

Glycerol is activated to form glycerol-1-phosphate in the presence of the enzyme glycerokinase (GK) and magnesium ions, with the high energy compound ATP as an energy source to drive the reaction.

The glycerophosphate is then reduced by the coenzyme NAD^+ to form dihydroxyacetone phosphate and the reduced pyridine nucleotide. This reaction is catalyzed by the enzyme GDH. The reaction is reversible, and the equilibrium usually lies to the left as written, but it is displaced to the right by working at fairly alkaline pH, and by chemically trapping the dihydroxyacetone produced with a hydrazine reagent.

The absorbance of NAD^+ at 340 nm. or 366 nm. is distinctly less than the absorbance of NADH, and this fact serves as a measure of the reaction system. For each mole of glycerol converted to dihydroxyacetone phosphate, a mole of NAD^+ is converted to NADH. Many other enzyme assays can be coupled to the reduction of

NAD^+, and details may be found in the reference cited, but here we shall concern ourselves only with the glycerol procedure.

Working with enzyme systems requires some precautions not always required in classic chemical methodology. The sensitivity of enzymes to traces of metal ions is well known. All solutions to be employed in enzymatic assays should be made in at least doubly-distilled water; all glassware should be free of heavy metals, especially mercury from Nessler's reagent and chromium from acid cleaning solution.

The enzymes and coenzymes, together with such high energy compounds as ATP, are unstable materials. They should be carefully stored according to the directions on each package if proper results are expected. Freezing and thawing may cause decomposition, and, therefore, the materials should be obtained in small packages. Solutions should be kept cold until immediate use; this can be accomplished with a beaker of cracked ice.

Many suppliers now offer kits of materials necessary for specific determinations.*

REFERENCE

Wieland, O.: Glycerol: *In* Methods of Enzymatic Analysis. H. U. Bergmeyer, Ed. New York, Academic Press, 1963, pp. 211–214.

PREPARATION OF REAGENTS

1. Buffered hydrazine solution (0.2 *M* glycine, 1 *M* hydrazine, 2×10^{-3} *M* Mg^{++}, pH 9.8). Mix 20.8 gm. of hydrazine hydrate (24 per cent, w/v) and 1.5 gm. of glycine with about 80 ml. of water. Add 0.2 ml. of 1 *M* $MgCl_2$ and adjust the pH to 9.8 with a few drops of 10 *N* KOH; then dilute to a final volume of 100 ml. The solution should be kept in a brown bottle in the refrigerator. It is stable for 4 weeks.

2. Adenosine triphosphate (0.05 *M* ATP). Dissolve 60.5 mg. of the disodium salt of ATP in about 1 ml. of water; neutralize very carefully with dilute NaOH to a pH of 7.0, and dilute to a final volume of 2 ml.

3. Nicotinamide dinucleotide (0.02 *M* NAD^+). Dissolve 34.6 mg. of NAD^+ in 2 ml. of water. This should not be used for more than 5 days since on standing it may be partially converted to an enzymatically inactive form.

4. Glycerophosphate dehydrogenase GDH (7 mg. protein/ml.). The twice-recrystallized commercial enzyme material is prepared in ammonium sulfate. The protein concentration will be stated on the label of the product. Dilute commercial preparations with 2 *M* ammonium sulfate to the proper protein concentration.

5. Glycerokinase, GK (60 units/ml.). The unit stated here is defined as that amount of enzyme which can activate 1 micromole of substrate per minute per mg. protein. In this instance, a bacterial enzyme is commonly employed as a source material, and the desired enzyme activity can be obtained at a protein concentration of about 1 mg. protein/ml. In preparing the enzyme, some procedures call for stabilization with glycerol; obviously, for the purposes intended here it is important to specify that the enzyme be glycerol free. Depending on the initial material, suspend the needed amount in 2*M* ammonium sulfate.

6. Glycerol stock standard. Glycerol is an extremely hygroscopic substance, and even analytical reagent grade is only 95 per cent pure, the remainder being water.

---

* Several major sources of enzymes and coenzymes in this country include: Sigma Chemical Co. St. Louis, Mo., California Biochemical Corp., Los Angeles, Calif., Worthington Biochemical Corp., Freehold, N.J., and Boehringer & Söhne, c/o Brinkmann Instrument Inc., Westbury, N.Y. From the last-named supplier, a complete kit for the enzymatic assay of glycerol is now available.

Therefore, weigh out 5.25 gm. of reagent grade glycerol in a 100 ml. volumetric flask and dilute to the mark with water. This can be labeled as equivalent to 50 mg./ml. It is stable in the refrigerator for 6 months.

7. Glycerol working standards. Because of the range of triglyceride levels that may be encountered, it is wise to set up several working standards. Dilute a volume of the stock standard 1:100 with water to provide an intermediate standard. Then dilute a portion of the intermediate standard 1:10 with water; this provides an aliquot of 2.5 μg./0.05 ml.

Dilute a second portion of the intermediate standard 1:25 with water; this provides an aliquot of 1.0 μg./0.05 ml.

The intermediate and the working standards should be made fresh before use.

ASSAY PROCEDURE

The procedure outlined here is designed for a Beckman spectrophotometer, either Model DB or DU. Suitable changes can easily be made to perform the procedure with other similar instruments.

1. Set the wavelength scale of the instrument at 340 nm.
2. Balance the instrument with water in cuvets which require not more than 3 ml. and with an optical path of 1 cm. The reference cuvet is left filled with water during the assay.
3. In each assay cuvet, pipet the following solutions in the indicated order:

2.0 ml. of buffer
0.10 ml. of GDH suspension
0.20 ml. of ATP solution
0.20 ml. of NAD^+ solution
0.05 ml. of sample (or standard)

Mix the contents by inversion, and measure the absorbance (A_1) against the water reference. Record the value as A_1.

4. Check the reading of the assay cuvet after a minute to be certain that the absorbance has reached a stable value.
5. Add to the assay cuvet 0.10 ml. of the glycerokinase suspension, mix thoroughly, and place the cuvet back in the spectrophotometer. The absorbance should show a considerable increase as the coupled reactions proceed. The absorbance will stabilize again at a higher reading in a few minutes; check the readings until a stable value is obtained. This should not require more than 5 to 10 minutes. Record the second value as A_2.

Repeat this process for each unknown and standard sample, recording the pair of values in each case. Any shift in the initial absorbance may be due to slight contamination with extraneous glycerol, by impure enzymes, or by excessive turbidity of the solutions.

CALCULATIONS

It is entirely possible to calculate the micromoles of glycerol present in each assay cuvet in terms of the known molar absorbance of NAD^+ and NADH; by this procedure it would theoretically not even be necessary to analyze standard samples. Nevertheless, in view of the possible contamination of enzymes and the other possible analytical errors just mentioned, we recommend the running of standards and the calculation of results in terms of the standard assay cuvet values.

The quantity of glycerol in plasma or serum can be expressed by the following equation:

$$Q_g = C_s \times \frac{(A_2 - A_1)_u}{(A_2 - A_1)_s} \times 100$$

where Q_g is the *quantity of glycerol* in mg./100 ml. of plasma or serum, C_s is the concentration of the glycerol standard in μg./ml., and $(A_2 - A_1)_u$ and $(A_2 - A_1)_s$ are the absorbance differences of the unknown and standard, respectively. The numerical factor converts the results from the 0.05 ml. aliquot analyzed to the basis of 100 ml.

The molecular weight of glycerol is 92, and the triglycerides of plasma or serum have an average molecular weight of 875; therefore, the quantity of triglycerides can be calculated by the following formula:

$$Q_{\text{triglyceride}} = 875/92\ Q_g = 9.5\ Q_g$$

REFERENCE

Wieland, O.: Glycerol: *In* Methods of Enzymatic Analysis. H. U. Bergmeyer, Ed. New York, Academic Press, Inc., 1963, pp. 211–214.

Chemical Method for Determination of Glycerol

REAGENTS

1. Glycerol stock standard. Weigh out 5.25 gm. of analytical reagent grade glycerol, 95 per cent, in a 100 ml. volumetric flask and dilute to the mark with water. Label this as containing 50 mg./ml. to correct for the water content of reagent grade glycerol. This stock solution can be kept in the refrigerator for 6 months.

2. Glycerol working standards. Because of the range of glycerol levels that may be encountered in serum or plasma samples, it is wise to set up several working standards. Dilute a volume of the glycerol stock standard 1:10 with water to provide a convenient intermediate standard. Then dilute a portion of the intermediate standard 1:10 with water; this provides a working standard with a concentration of 50 μg./0.1 ml. Dilute a second portion of the intermediate standard 1:25 with water; this provides a working standard that contains 20 μg./0.1 ml. The working standards do not keep and should be made fresh before use.

3. Alcoholic potassium hydroxide, 4 per cent. Weigh out 4 gm. of dry potassium hydroxide pellets; quickly dissolve them in 100 ml. of 95 per cent ethanol. Protect this solution from light and air. It does not keep well and should be replaced at weekly intervals, or sooner if color or turbidity appears.

4. Chromotropic acid solution. Dissolve 1 gm. of the solid chromotropic acid (4,5-dihydroxy-2,7-naphthalene disulfonic acid) (Distillation Products Industries, Rochester, N.Y.) in 100 ml. of water. Filter to remove any insoluble residue, then dilute to 500 ml. with 8 *M* sulfuric acid. Store the final solution in brown glass bottles, or otherwise protect from the light. Use only glass or polyethylene stoppers in the bottle. Different batches of chromotropic acid have different initial absorbancies, and the absorbance tends to increase with age of the solution. When fresh, the solution should have an absorbance that is not more than 0.045 when measured against distilled water at a wavelength of 570 nm. in a 1 cm. cell.

5. Sulfuric acid, 10 *N*. Dilute 140 ml. of concentrated sulfuric acid to 500 ml. with water. Store in a glass- or polyethylene-stoppered bottle.

6. Sodium periodate, 0.1 *M*. Dissolve 21.39 gm. of the solid sodium metaperiodate in water. Dilute to a final volume of 1 L. Store in the dark.

7. Sodium arsenite, 1 *M*. Dissolve 49.45 gm. of pure dry As_2O_3 in about 50 ml. of 10 *N* NaOH. Neutralize this solution with about 50 ml. of 10 *N* HCl; then dilute to 1 L. with water.

8. Florisil. Florisil is a proprietary designation for a magnesium silicate adsorbent (Floridin Co., Pittsburgh, Pa.). Obtain this material in a 60 to 100 mesh particle size, then dry about 100 gm. to constant weight in an oven at 105°C. Transfer exactly 100 gm. of the dried material to a 500 ml. Erlenmeyer flask; then add enough water in the form of a fine spray from a syringe fitted with a No. 25 needle to add 40 mg. of water per gram of adsorbent. Shake or rotate the flask during the period of water addition. When the addition is complete, store the final adsorbent in a tightly stoppered bottle. The exact water content is quite critical; if less than the indicated amount is added, the retention of phosphatides by the adsorbent will be incomplete.

9. Petroleum ether-diethyl ether. Mix petroleum ether (60 to 75°C.), or hexane, with peroxide-free diethyl ether (anesthesia grade) in the proportions of 95:5 (v/v). This should be kept cold or made fresh.

10. Chloroform acetone-water. Mix chloroform, acetone, and water in the proportions of 99:99:2 (v/v/v). Keep cold or mix just before use.

PREPARATION OF TOTAL LIPID EXTRACT

1. Take exactly 1 ml. of plasma or serum into a glass syringe fitted with a No. 25 needle at least 1 inch long; tuberculin syringes are recommended for this purpose. Forcefully deliver the contents into 24 ml. of a chloroform-methanol (2:1, v/v) mixture. Stopper the tubes, then shake mechanically for 10 minutes. Centrifuge the tubes for 5 minutes at 2000 rpm to pack the precipitated protein.

2. Collect a 5 ml. aliquot of the supernatant solution, and transfer it to a clean, dry test tube. Evaporate the contents to dryness in a stream of air, hastening the process by warming to about 55°C.

3. Dissolve the residue in 3 ml. of the petroleum ether-diethyl ether solvent. Close the tube with a marble or spare glass stopper to minimize solvent evaporation while the adsorbent columns are prepared.

PREPARATION OF CHROMATOGRAPHY COLUMNS

1. For each sample to be analyzed, weigh out aliquots of 750 mg. of the prepared Florisil. Weigh out three extra aliquots, one for each of the standards and one for the blank.

2. For each aliquot of Florisil, provide a funnel tube, 300 × 4 mm. I.D. (No. 46185, Kimble Products, Div. of Owens-Illinois, Inc., Toledo, Ohio). Clamp the tubes in an upright position, and under each place a 25 ml. volumetric flask. With a fine glass rod, push a small piece of glass wool into the lower end of each funnel tube to provide a support for the adsorbent powder, then add to each tube the weighed aliquot of Florisil. This will produce a column of powder about 125 mm. high. Tap the tube lightly with a pencil or similar object to settle the powder.

CHROMATOGRAPHY OF THE GLYCERIDES

1. Pour each of the total lipid extracts into successive columns of the series. Transfer the last of the extracts from the test tubes to the columns by rinsing with 2 ml. of the petroleum ether-diethyl ether solvent.

2. Wash the columns reserved for the blank and the standards with 5 ml. of the petroleum ether–diethyl ether solvent.

3. Add to all funnel tubes, 8 ml. of the chloroform-acetone-water solution.

4. Add to the standard volumetric flasks, 0.1 ml. of the two working standard glycerol solutions. These will now contain 50 μg. and 20 μg., respectively.

5. When all of the eluates have been collected, evaporate the solvents in an air stream at 55°C. It is important that all of the solvents be removed and that the residues be dry.

SAPONIFICATION OF THE TRIGLYCERIDES

1. Add to each of the volumetric flasks 0.25 ml. of alcoholic KOH and 2.5 ml. of 95 per cent ethanol. Swirl the flasks to mix the contents. Then place in a 55°C. water bath for 30 minutes to completely hydrolyze the triglycerides.

2. At the end of the saponification period, neutralize the excess alkali by adding to each flask 0.25 ml. of 10 *N* sulfuric acid.

OXIDATION OF GLYCEROL AND MEASUREMENT OF FORMALDEHYDE

1. To each flask add exactly 1.0 ml. of 0.1 *M* sodium periodate solution. Exactly 5 minutes later, add 1 ml. of 1 *M* sodium arsenite solution and swirl to mix. Timing is important here to minimize periodate attack of the glycerol beyond the desired stage. If the timing is ignored, more than 1 mole of formaldehyde per mole of glycerol may be formed, and such "overoxidation" may not be entirely reproducible.

When the arsenite is added, a yellow or brown color due to free iodine may appear momentarily; however, this will disappear on standing, or when the solutions are finally diluted with water to the mark.

2. Dilute all flasks to exactly 25 ml., stopper, and mix the contents well by inversion.

3. Transfer 2.0 ml. aliquots from each volumetric flask to clean test tubes, add 10 ml. of chromotropic acid reagent, and mix well. Place all the tubes in a 100°C. water bath in the dark (covered bath) for 30 minutes.

4. Cool the tubes to room temperature; then read the absorbance of the unknown and standard tubes against the reagent blank at 570 nm.

5. If the absorbance of any tube is too high or too low for proper reading against the standards, the photometry may be repeated on the oxidized glycerol preparations for as long as 36 hours later. As much as 0.3 ml. or as little as 0.1 ml. may be taken as an aliquot for photometry. Since the level of glycerol may vary widely, it is wise to save the oxidized extracts until after photometric measurements have been made.

CALCULATIONS

1. The average molar composition of blood triglycerides is such that the quantity of triglycerides may be estimated as 9.5 times the quantity of triglyceride glycerol. Therefore,

$$C_{\text{triglycerides}} = 9.5 \times 25 \times \frac{A_u}{A_s} = 237.5 \times \frac{A_u}{A_s}$$

where $C_{\text{triglycerides}}$ is expressed as mg./100 ml. and the high standard (50 μg./0.1 ml.) is employed. If the lower standard is used (20 μg./0.1 ml.), divide the numerical factor by 2.5 to obtain the proper results.

REFERENCE

Blankhorn, D. H., Rouser, G., and Weimer, T. J.: J. Lipid Res. *2*:281, 1961.

PLASMA OR SERUM NEFA

The determination of plasma or serum nonesterified fatty acids has already been mentioned. It is a measure of that portion of the total fatty acid pool that circulates in immediate readiness for metabolic needs. NEFA can be absorbed readily by muscle, heart, brain, and other organs as an energy source whenever insufficient quantities of glucose limit the usual carbohydrate energy source. In all probability, both glucose and NEFA are simultaneously taken up from the blood, even under normal conditions. By sampling the blood supply leading to and from a working muscle, it is evident that there is a considerable arteriovenous difference. When blood glucose levels are high, the NEFA level falls, and vice versa, emphasizing the reciprocal withdrawal of stored energy, under the simultaneous control of insulin and epinephrine. In acute starvation, the NEFA level may rise as much as threefold the normal value.

The *normal level* of NEFA ranges from 0.45 to 0.90 mEq./L., in adults; in children or hypermetabolic individuals the levels are somewhat higher. The values in paired plasma and serum samples indicate that the serum levels are substantially the same as plasma levels if sera are promptly prepared. On standing, plasma and serum levels tend to rise with time, probably by virtue of hydrolytic cleavage of both triglycerides and phosphatides.

Methods for the determination of NEFA depend on titration of the extracted fatty acids. The major difference in the various methods depends on how the end point is measured. A second difference is in the nature of the extraction procedure, which must be very precise if nonNEFA contamination by acidic material is to be avoided. Such contaminants include lactic acid, acetoacetic acid, and small amounts of the acidic phosphatides. Some recent studies have been made by esterifying the NEFA, then submitting them to study by GLC, but so far this technique has been limited to research rather than to routine clinical practice. We shall describe here an acceptable titrimetric procedure.

Determination of NEFA in Plasma or Serum

The NEFA are extracted from the plasma or serum by a mixture of heptane and isopropanol, which is then washed with dilute sulfuric acid to remove nonNEFA contaminants. The purified NEFA extract is titrated with dilute alkali, using thymolphthalein as an indicator. During the titration, the system must be protected from exposure to carbon dioxide from the air, which would also consume the dilute alkali. This is accomplished by bubbling a stream of alkali-washed nitrogen through the titration tube, which also stirs the system to insure thorough mixing of the added alkali. At least 85 per cent of the total NEFA are C_{16} and C_{18} acids so that very little error is introduced by using palmitic acid as a standard.

PREPARATION OF REAGENTS

1. Extraction mixture. Mix 40 volumes of isopropanol, 10 volumes of heptane, and 1 volume of 1 *N* H_2SO_4. The isopropanol and heptane should be spectroanalyzed grade, or else they should be freshly redistilled to eliminate traces of contaminating fatty acids.

2. Dilute sulfuric acid. Dilute 0.5 ml. of 36 *N* H_2SO_4 to a final volume of 1 L. with water.

3. Thymolphthalein indicator. Make a 0.01 per cent solution in 90 per cent ethanol.

4. Sodium hydroxide. Dilute 1 *N* NaOH with freshly boiled water to make a final concentration of 0.02 *N*. This solution should be stored out of contact with the air, and should be made fresh each day. The exact concentration is not important since it is standardized with known samples of palmitic acid.

5. Stock palmitic acid standard. Dissolve 32.0 mg. of pure palmitic acid (Hormel Foundation, Austin, Minn.) in 25 ml. of heptane. This stock solution, containing 5 μEq./ml., is stable indefinitely, but it must be tightly stoppered to prevent loss of volatile solvent.

6. Working palmitic acid standard. Dilute the stock standard solution 1:10 with heptane. This has a concentration of 0.5 μEq./ml. This is stable if protected against concentration change by loss of solvent.

EXTRACTION PROCEDURE

1. Label glass-stoppered centrifuge tubes for the blank, standard, and unknown. Add to each of them the indicated materials according to the following scheme.

Add	*Blank*	*Standard*	*Unknown*
NEFA	—	2 ml. working	2 ml. plasma or serum
Extraction mixture	10 ml.	10 ml.	10 ml.
Shake the unknown tube vigorously a few seconds, let stand 10 minutes, then proceed with the following steps.			
Heptane:	6 ml.	4 ml.	6 ml.
Water:	6 ml.	6 ml.	4 ml.
Invert all tubes 10 to 15 times gently, then let stand until the two phases separate.			

2. With a pipet and a rubber bulb, carefully transfer 4 to 5 ml. of the upper phase from each tube to a second set of clean stoppered centrifuge tubes, avoiding transfer of any of the lower phase.

3. Add 5 ml. of the 0.05 per cent H_2SO_4, shake the tubes vigorously, and separate the phases by centrifugation.

TITRATION PROCEDURE

The ultramicroburet used for titrating NEFA must be calibrated to at least 0.1 μL. or better if satisfactory accuracy is expected. (We have found the units made and sold by Micrometric Instruments Co., Cleveland, Ohio, or the unit sold by Beckman Instruments, Inc., Fullerton, Calif., to be equally satisfactory.) It is helpful to fit the outlet of the syringe with a fine hypodermic needle (26 gauge) over which thin-wall polyethylene tubing may be slipped. The tubing is cut of sufficient length to reach approximately to the bottom of the centrifuge tubes in which the NEFA samples are contained.

The nitrogen that stirs the solution and serves as an inert blanket must be washed with alkali to remove any traces of carbon dioxide. This is easily accomplished by a wash bottle of 10 per cent NaOH through which the gas is passed. The outlet of the wash bottle should be protected with a cotton or glass wool plug to entrap any chance droplets of alkali. The washed gas is then delivered into the NEFA solution via a fine polyethylene tube.

For the titrations, proceed as follows:

1. Place 1 ml. of indicator solution into a conical-tipped 15 ml. centrifuge tube, place the outputs of the nitrogen source and the microburet in the liquid, and start the flow of nitrogen through the solution.

2. Add a 3.0 ml. aliquot of the upper phase from extraction tubes, and proceed to add the alkali from the buret until the end point is reached. This is a greenish-yellow color, best observed with a fluorescent light and a white background. It is wisest to titrate the blank first, next the standard, and finally to match the unknown to the same color.

CALCULATIONS

The concentration of NEFA is obtained from the following formula:

$$\text{mEq. NEFA/L.} = 0.5 \times \frac{T_u - T_b}{T_s - T_b}$$

where T_b is the titration value of the blank, T_s is the titration value for the standard, and T_u is the value for the unknown plasma or serum sample.

REFERENCE

Trout, D. L., Estes, E. H., and Friedberg, S. J.: J. Lipid Res., *1*:199, 1960.

THE PLASMA GLYCEROPHOSPHATIDES (PHOSPHOLIPIDS)

The phosphatides are found in all tissues of the body. In most cells, they play an important structural role, forming an integral part of internal and external cell membranes. It is now believed that the matrix holding most of the oxidative enzymes of the mitochondria in their proper places is largely phosphatide in nature. Except in the brain and central nervous system, the fatty acids associated with the phosphatides tend to be more unsaturated than those found in the triglycerides.

In addition to the important structural function, the phosphatides are involved in the metabolism of other lipids and nonlipids. It was thought until recently that the phosphatides alone were essential in transporting triglyceride in the blood, but recent experiments have disproved that idea, as has already been mentioned. Neither is it currently accepted that phosphatides are essential in the absorption of fat from the gut. All tissues seem quite competent to form the amount and type of phosphatides that they require. Isotopic studies have revealed that the only important source of the plasma phosphatides is the liver, and the liver is also the only organ that takes up significant amounts from the blood. At the steady state, therefore, a normal level of plasma phosphatides reflects adequate liver function. It is, nevertheless, striking to note that the polar character of the phosphatides, and their tendency to form micellar aggregates in aqueous dispersion, lead to a surprisingly constant proportion between the level of cholesterol and the level of total phosphatides, provided the capacity of the liver to produce them is not severely damaged. It is argued by some that the phosphatides are essential to keep circulating cholesterol from depositing in the intimal lining of the circulatory system, as well as in other tissues, so that any condition that results in high levels of cholesterol will result in high levels of phosphatides. Conversely, if phosphatide levels are reduced, then cholesterol will have to be stored by the scattered cells of the reticuloendothelial system, at least until the capacity of those cells is overreached, after which the sterol will be deposited in areas where considerable damage may result. This argument is attractive, but has not yet been established beyond doubt. Paired measurements have determined that normally, whatever the explanation, there is an association between the level of cholesterol and phosphatide phosphorus, which can be expressed by the regression equation:

$$\text{mg. phosphatide-P/100 ml.} = 3.62 + (0.029 \times \text{mg. cholesterol/100 ml.})$$

The normal value of the ratio of total cholesterol per phosphatide ranges from 0.55 to 1.05. In atherosclerotic individuals, this ratio increases, and it is claimed that the measurement of the ratio has some predictive significance in preatherosclerotic individuals.

Serum contains many phosphatides; the distribution averages 60 to 70 per cent phosphatidylcholines (lecithins), 15 to 20 per cent sphingomyelins, 5 to 8 per cent phosphatidylethanolamines (cephalins) and phosphatidylserines (cephalins), and the remainder is made up of lysophosphatidylcholines (lysolecithins), plasmalogens, and other less well characterized materials. Because of the technical difficulties in fractionating the phosphatides, it is only recently, when the development of newer methods, especially those based on TLC, has simplified and shortened analytical procedures, that fractionation studies have been attempted.

The total phosphatide level of plasma increases with age from birth, as does the level of cholesterol, and this fact has been amply attested. Blood cells contain large amounts of phosphatides also, and it is most important that hemolysis be avoided in collecting samples for phosphatide determination. Pregnancy causes a distinct elevation of phosphatide levels. These facts are summarized in the following table of average normal values.

Age	*Total phosphatides* (mg./100 ml.)
Birth	61 ± 32
♂, up to 65 years	225 ± 50
♀, nonpregnant	195 ± 37
♀, early pregnancy	248 ± 43
♂, ♀, 65–88 years	281 ± 85

Very high fat diets tend to increase the levels, and low fat diets lead to lowered values.

Most of the methods for determination of the phosphatides involve the oxidative destruction of the phosphatides contained in a purified lipid extract followed by a measurement of the inorganic phosphorus liberated. It is not necessary to separate the phosphatides from the other lipids present. If the individual phosphatides are to be determined, some type of chromatography is employed, after which the separated fractions may also be degraded by oxidation and the phosphorus content of each determined. Specific methods for choline, ethanolamine, and serine are also available, but they are not generally employed in clinical practice. Instead the amount of phosphorus in each fraction is multiplied by an average factor, since the molecular weights of the various phosphatides are quite similar. For most purposes such a practical approach is quite justified.

Virtually all of the analytical methods depend on the formation of phosphomolybdate ion, which is then reduced by a suitable reagent to form the complex "heteropoly blue," a colloidal dispersion with a composition not precisely known. Most of the methods can, therefore, be described as variants of the Fiske and SubbaRow method for inorganic phosphorus. We shall present a chemical procedure for total phosphates and a TLC procedure for separation of total phosphatides into component classes.

Determination of Lipid Phosphorus

This method was originally devised using concentrated sulfuric acid and hydrogen peroxide as the oxidizing reagents to wet-ash the organic matter in a lipid extract; since then, we have modified the procedure to use concentrated perchloric and nitric acids because this combination is more efficient. Under the conditions described, perchloric acid is safe; however, it should always be kept in mind that wet-ashing procedures involving perchloric or sulfuric acid pose an occasional threat of explosive boiling, and the hot tubes should always be pointed away from the body and face.

The reductant employed is N-phenyl-p-phenylene diamine, or semidine, chosen because it produces a maximal color intensity in a short time, because the color is quite stable, and because it has a minimal tendency to reduce silicomolybdate formed by attack of the glass tubes. When the reductant is oxidized, it also forms a blue pigment, which increases the overall sensitivity of the method.

PREPARATION OF REAGENTS

1. Perchloric acid, A.R. Use 70 per cent as purchased.

2. Nitric acid, A. R. Use 70 per cent, as purchased.

3. Ammonium molybdate, $(NH_4)_6Mo_7O_{24} \cdot 4H_2O$, 0.025 *M*. Store in polyethylene bottles. Complete solution of the reagent may be aided by the addition of not more than 0.1 ml. of concentrated H_2SO_4/100 ml. of solution.

4. p-Semidine hydrochloride. Weigh out 50 mg. of the dry salt (Distillation Products Industries, Rochester, N.Y.); place it in a 100 ml. volumetric flask. Wet the powder with 2 to 5 drops of alcohol; then dissolve it in 100 ml. of 1 per cent (w/v) $NaHSO_3$. Filter off any small amount of insoluble residue. The solution should remain practically colorless for at least 10 days at room temperature in a clear glass bottle, after which a slight discoloration is evident. This does not impair the effectiveness of the reagent, but it does increase the blank absorbance.

5. Phosphorus standard, 1 mg./ml. Weigh out 438.1 mg. of dry KH_2PO_4 and dissolve it in sufficient 5 *N* H_2SO_4 to make a final volume of 100 ml. This stock standard has a concentration of 1 mg.P/ml. Use a 1:10 dilution of this as a working standard, which will contain 100 μg./0.1 ml. The working standard should be made freshly before use.

OXIDATION OF THE LIPIDS

1. Prepare a purified Folch lipid extract from 1 ml. of plasma or serum, according to the directions given earlier.

2. Transfer a volume of the purified extract equivalent to 0.2 ml. of the sample to an NPN tube calibrated at 12.5 ml.

3. Evaporate the solvents to dryness in a hood.

4. To two similar tubes, add 0.1 ml. and 0.2 ml., respectively, of the working standard phosphorus solution. A third tube is taken for a blank, to which nothing is yet added.

5. Place all the tubes in an electrically heated sand bath or Kjeldahl rack adjusted to 185 to 195°C. The tubes should be immersed at least $1\frac{1}{2}$ inches deep.

6. Add 0.4 ml. of concentrated perchloric acid, and heat until dense white fumes reflux to within an inch of the top of the tubes. The lipid residue will be charred and darkened in color at this point.

7. Cautiously add 0.2 ml. of concentrated nitric acid, and mix the contents of each tube well. Continue to heat the tubes until the dense white fumes rise close to the top again.

8. Add an additional 0.2 ml. of nitric acid; then allow the tubes to reflux for approximately 5 minutes longer. At this stage, the digests should be substantially free of color; if this is not the case, more nitric acid is added, as before, and the digestion continued.

9. When the lack of color indicates that digestion is complete, remove the tubes from the sand bath and allow them to cool to room temperature. Add 2 ml. of water and mix well; add the water cautiously!

10. Add 1 ml. of the molybdate reagent to all tubes and mix well.

11. Add 2 ml. of the semidine reagent, mix well, and then dilute to 12.5 ml. with water and mix again. Let the tubes stand for 10 minutes to allow complete color development. The color is then stable for at least 1 hour.

12. Using the blank tube as a reference, read the absorbance of the remaining tubes at a wavelength of 700 nm.

CALCULATION OF RESULTS

Since the range of phosphatide phosphorus is variable, it is wise to use two separate standards as outlined in the preceding procedure. The unknowns should be read against the standard that is closest in absorbance. If any of the unknowns show a density greater than 0.8, the analysis should be repeated with a smaller aliquot of the lipid extract. Depending on which standard is employed, the concentration of lipid phosphorus may be calculated by one of the following equations:

$$\text{mg. lipid-P/100 ml.} = A_u/A_s \times 50 \times 0.1 \quad (10\ \mu\text{g. standard})$$

or

$$\text{mg. lipid-P/100 ml.} = A_u/A_s \times 50 \times 0.2 \quad (20\ \mu\text{g. standard})$$

Since, on the average, phosphatides contain 4 per cent of phosphorus, the concentration of total phosphatide is obtained by multiplying total lipid-P by the factor 25.

REFERENCE

Dryer, R. L., Tammes, A. R., and Routh, J. I.: J. Biol. Chem., *225*:177, 1957.

Fractionation of Total Phosphatides by TLC

Separation of the total phosphatide content of plasma, serum, and other tissues has been greatly facilitated by the development of TLC procedures, and entirely new avenues of research and clinical practice are now possible. Although few specific needs for fractionation have yet been developed in the routine diagnostic laboratory, it seems safe to predict that the availability of the newer methodology will generate a greater interest in this problem. It has already been demonstrated, for example, that the effect of dietary fat intake has less effect on the fatty acid composition of the phosphatides than on the triglycerides.

It is not necessary to remove the triglycerides and sterols to perform a TLC separation of the phosphatide types. Instead, the solvent systems are designed to let the neutral lipids migrate with the solvent front, while the phosphatides migrate as distinct intermediate spots or bands. For the utmost precision, two-dimensional separations are employed,[15] but most purposes will be served by a one-dimensional separation. The separated spots may be visualized with spray reagents, or the spots may be eluted and wet-ashed for quantitative chemical determination according to the preceding procedure.

Earlier efforts to separate phosphatides on TLC layers of Silica Gel G were not completely successful because the calcium sulfate added to the silicic acid as a binder caused tailing of the spots and distortions of the separations; the success of newer methods depends largely on the use of Silica Gel H, which is free of added binders. The difference in polarity of the phosphatides is advantageously employed by including in the layers of Silica Gel H a small amount of alkali. Under these conditions, the more acidic phosphatides migrate faster than the more nearly neutral lecithins. The method we shall next describe has given good results, provided that the solvent system is fresh.

METHOD

1. In a tightly stoppered flask, disperse 20 gm. of dry Silica Gel H in 45 ml. of 0.001 *M* Na_2CO_3 by thorough shaking.
2. Coat the dispersed slurry on clean glass TLC plates at a thickness of 0.5 mm.
3. Allow the plates to air dry at room temperature for at least 30 minutes, then activate at 110°C. for 30 minutes. The plates may be stored in a desiccator after activation, but if they are not used within a day or two they should be reactivated at 110°C.
4. Prepare a Folch total lipid extract; with a micropipet, apply a volume equivalent to 0.2 ml. of serum as a line of small contiguous spots across one side of the activated TLC plate, about 1 cm. from an edge. The spots should form a band not more than a few millimeters high across the width of the dry plate. It may be necessary to go back and forth several times across the plate to apply the entire sample.
5. Develop the plate in a tank containing sufficient solvent to just cover the lower edge of the silica gel surface. The solvent is composed of $CHCl_3$:CH_3OH:CH_3COOH:H_2O in the proportions (v/v) of 50:25:7:3.
6. Remove and air dry the plate when the solvent has risen to within 1 inch of the top of the plate. This will require approximately 1 hour.
7. Spray the plate with 0.005 per cent 2′,7′-dichlorofluorescein, and observe it under an ultraviolet lamp. The phosphatide spots will stand out as slightly orange spots against a lighter yellow fluorescent background. The solvent front will also fluoresce because of the presence of the neutral lipids and sterols. These may be ignored.
8. Outline the phosphatide spots with the tip of a finely pointed glass rod.
9. Scrape off the outlined spots and transfer the coating thus recovered to individual digestion tubes.
10. Scrape an area of the coating approximating the size of a lipid-bearing spot into another tube for a coating blank, since the silica gel may sometimes contain small amounts of phosphorus.
11. Add 0.4 ml. of 70 per cent perchloric acid, and digest the lipids as described in the procedure for total lipid-P.

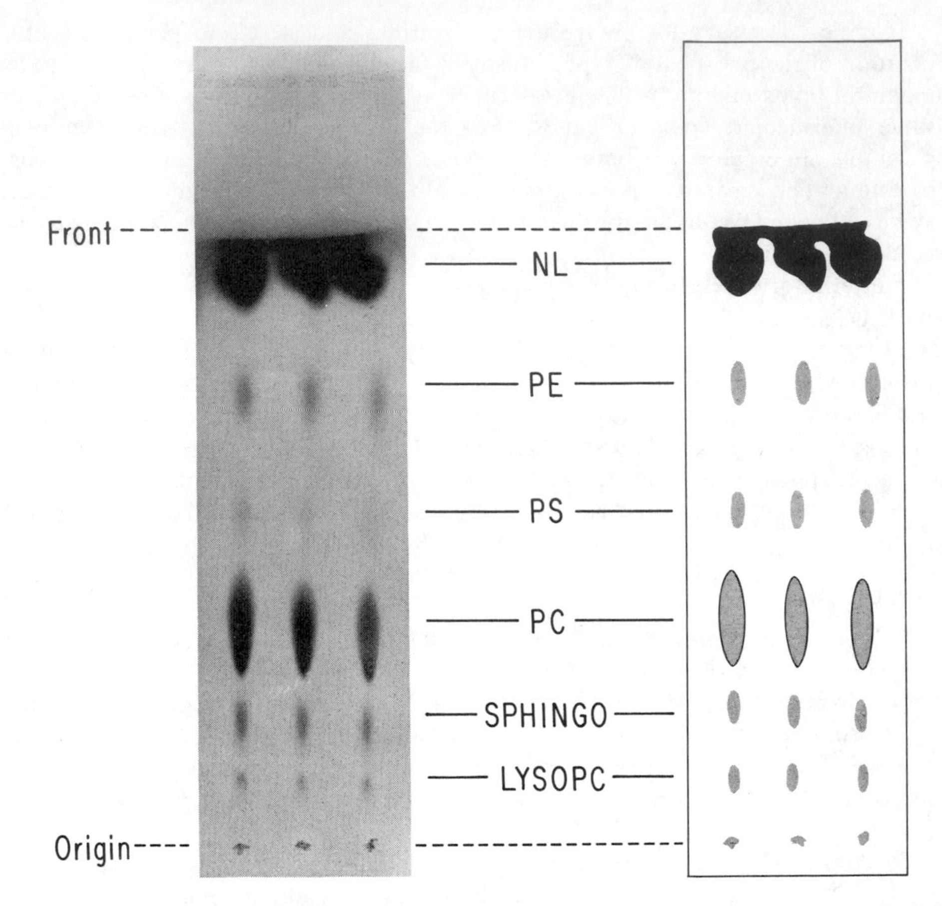

Figure 7-6. A typical thin-layer chromatographic separation of serum phosphatides. On the left is an actual photographic reproduction, on the right is a diagrammatic representation. LYSOPC is lysophosphatidylcholine, SPHINGO is sphingomyelin, PC is phosphatidylcholine, PE is phosphatidylethanolamine, PS is phosphatidylserine, and NL is neutral lipid. Dimensions of TLC plate are 5 by 20 cm.

14. Add 2 ml. of the semidine reagent, mix, and dilute to 5 ml. with water.

15. Centrifuge to spin down the suspended particles of silica gel; then decant the colored supernatant to cuvets for measurement of the absorbance.

16. Read the samples against standards, which contain 0.5, 1.0 and 2 μg. of phosphorus.

17. If only a qualitative fractionation is required, the plates may be sprayed with 10 per cent phosphomolybdic acid in 90 per cent ethanol, then heated to 100°C. for 3 minutes in an oven. The lipids will appear as blue-gray spots against a light background.

A photograph of the results obtainable by this procedure is shown in Figure 7-6. For illustrative purposes, the lipid spots were visualized with phosphomolybdic acid.

REFERENCE

Skipski, V., Peterson, R. F., Sanders, J., and Barclay, M.: J. Lipid Res., *4*:227, 1963.

CHEMISTRY OF THE STEROLS

The term sterol describes those alcoholic derivatives of the parent hydrocarbon, sterane, that are devoid of hormonal activity. The term steroid is reserved for the latter group of compounds, and they are considered elsewhere. As is shown, sterane can be further described as an alicyclic, saturated hydrocarbon consisting of four fused rings. Substitution may occur at various positions around the ring system, including the insertion of double bonds, the introduction of hydroxyl groups, angular methyl groups, and aliphatic side chains. Identification of the positions of substitution is by means of the following numbering system shown for the structure of cholesterol.

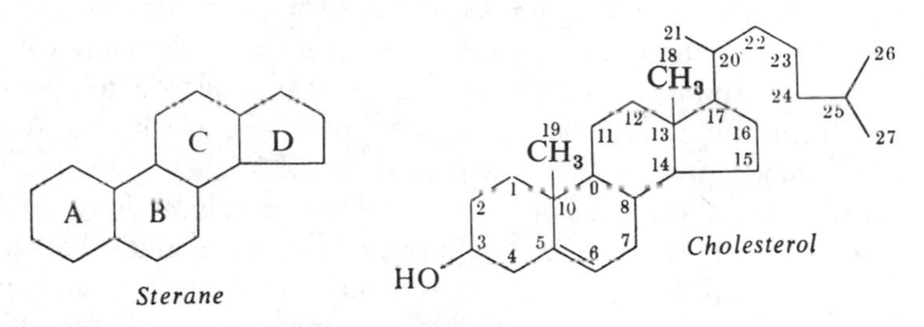

If the double bond of cholesterol is reduced, two different products can be formed, cholestanol and coprostanol, both of which may be found in human tissue or excreta. In the case of cholestanol, the atoms lie very nearly in a slightly pleated sheet; in coprostanol, the A ring is folded at nearly right angles to the remainder of the molecule. This alteration confers entirely different physical properties and chemical reactivities upon the two compounds. Coprostanol is formed by the reductive action of microorganisms of the gut, from which some of it may be reabsorbed.

The substituents around the ring may lie above or below the plane of the ring, as the molecular drawings show. This factor also affects the chemical reactivity of the molecules. The solid lines linking substituents to the ring in the drawings refer to bonds above the ring plane, and the dotted lines indicate substituents below the plane. In cholesterol, for example the —OH group is above the plane; under these conditions we speak of β-oriented groups, and if the group lies below the ring plane, it is termed α-oriented. This difference also affects chemical reactivity. For example, only β-oriented OH groups will form insoluble derivatives with digitonin, a most useful reagent for the analytical separation of serum cholesterol and other β-hydroxylated steroids.

The double bond in cholesterol has many of the general properties of double bonds. It can readily be brominated, and since the cholesterol dibromide is less soluble than free cholesterol or the impurities in commercial preparations, the dibromide is particularly useful in preparation of pure cholesterol for analytical standards. The isolated dibromide can be debrominated easily after separation. Pure cholesterol slowly oxidizes, even at refrigerator temperature, because of the presence of the double bond. Some workers prefer to prepare and store quantities of the pure dibromide, which can then be debrominated in small lots as needed.

In summary, the most descriptive name for cholesterol is Δ^5-cholestene-3β-ol. The symbol Δ^5 indicates the presence and position of a double bond between carbons 5 and 6, and the term 3β-ol indicates the presence and direction of orientation of the hydroxyl group. These are the most important features on which the analytical methods to be discussed later are based.

Metabolism of Cholesterol

Although the structure of cholesterol is complex, the biosynthesis of the total molecule can be accomplished from biochemically very simple precursors. Acetate radicals, chiefly in the form of acetyl-coenzyme A, are all that the body requires as starting material. Consequently, many amino acids, carbohydrates, and fatty acids, when supplied in excess of other metabolic needs, can contribute to the cholesterol pool. Details of the synthetic processes can be found in textbooks of biochemistry.[23] The liver is the main site of synthesis, but the skin, adrenals, gonads, intestine, and even the aorta can carry out the biosynthesis. It is estimated that the liver can produce 1.5 gm. per day, and the extrahepatic tissues can produce 0.5 gm. per day. The total produced from acetate or other sources is, therefore, about two to three times the amount consumed, preformed, in the typical American diet. Attempts to lower serum cholesterol by reducing the dietary cholesterol have not, in the long run, been successful unless the total caloric intake was reduced at the same time. This is presumably a result of increased rate of biosynthesis from an excessive energy intake. It is possible to lower the serum cholesterol by diets low in preformed cholesterol if the amount of fat and simple sugars (hexoses and disaccharides) is also curtailed. Oriental peoples, whose diets meet these standards, typically show cholesterol levels some 100 to 150 mg./100 ml. less than their American counterparts. If the Orientals switch to a typical Western diet, the level of cholesterol in their serum rises to typical American levels. The implication of cholesterol in atherosclerosis and heart disease has stimulated an enormous, growing literature, which we cannot review here. In the simplest terms, however, it would appear prudent to state that there is a statistically significant relation between high serum cholesterol levels and the incidence of coronary artery disease, which would make it desirable to maintain low, or at least low normal, levels of cholesterol in the serum. At one time, diets high in polyunsaturated fatty acids were very popular, since it was claimed that these substances tended to lower the serum cholesterol level. Recently, however, this idea has been attacked on the basis that the level was lowered by driving the cholesterol from the serum into the solid tissues, and since these include the vascular tissues, the consequences may not be entirely free from harmful results. It would probably be simpler to restrict the caloric intake or to burn the excess calories by more vigorous physical exercise or work.

There is good evidence that, other things being equal, the serum cholesterol level is affected by the output of several endocrine organs. There is an inverse relationship between the level of thyroxine and cholesterol in the blood. Hypothyroidism is associated with hypercholesterolemia to such a degree that measurement of serum cholesterol was used to monitor thyroid status routinely, until the development of the more specific PBI measurement some 15 years ago. The increased serum cholesterol is associated with an increase in serum β-lipoprotein and a decrease in α-lipoprotein. In hyperthyroidism the cholesterol levels may be lower than normal, and the corresponding shift in lipoproteins can be seen. Estrogenic hormones also seem to constitute an important control over serum cholesterol levels. Ovariectomized animals show a prompt and significant rise in circulating cholesterol. Estrogens appear to increase the biosynthesis of the sterol, but they increase the rate of excretion to an even greater extent, thus the net effect is a lower level in the blood. Postmenopausal women show higher cholesterol levels than premenopausal women, and then, parallel to men of the same age group, show a susceptibility to atheromatous lesions. The effect of estrogens is sufficiently impressive to have promoted their administration to men with known predispositions to coronary artery disease. As might be expected, estrogen treatment of men does lead to a lowering of the serum cholesterol and a simultaneous lowering of the β-lipoprotein in their serum. The side effects of estrogen therapy in men make this something less than an ideal treatment program.

Pregnancy may also be accompanied by a moderate increase in cholesterol level, perhaps due to altered endocrine function of several different organs. The change is usually regarded as entirely physiological, and the cholesterol values return to normal after parturition.

Diseases of the liver itself alter cholesterol levels. Early hepatitis produces an increase in serum cholesterol, but as the liver fails, or as the disease becomes increasingly severe, the level falls, probably due to decreased synthesis by the damaged or dying hepatic cells. In biliary disease, the bile canaliculi and ducts are blocked, and cholesterol excretion into the intestine via the bile is reduced. Biliary disease may also stimulate the biosynthesis of phosphatides, which will allow the serum to hold more cholesterol than normal, as a lipoprotein complex. Mechanical interference with the flow of cholesterol-containing bile is only part of the explanation for the rise of serum cholesterol.

Lipoid nephrosis induces a rise in total serum lipids, often of sufficient magnitude to give the serum a chylous or milky appearance. Much of the lipid increase is caused by accumulated cholesterol, but the milky appearance reflects increased levels of triglycerides.

Diabetes mellitus (effective insulin lack) is associated with hypercholesterolemia; the more uncontrolled the diabetes, the more the level of cholesterol is likely to be elevated. In these conditions there is a decreased capacity to synthesize cholesterol (and fatty acids), but there is an even greater decrease in the excretion or conversion of cholesterol to other products. This results in the elevation of serum cholesterol levels. Adequate treatment of the disease will return the cholesterol level to essentially normal levels.

DISPOSITION OF CHOLESTEROL

In view of the fact that the body can completely synthesize cholesterol from simple precursors, it is curious that so little can be done to degrade the molecule, or even to remove it from atheromatous plaques in the aorta and other vessels, for it is well established that once deposited in the intima of the blood vessels, there is no physiological capacity to remove it.

Circulating cholesterol can be eliminated through the liver in the form of bile acids or their salts. Figure 7-7 shows the structures of the major bile acids and their relations to cholesterol. The conversion of cholesterol to the individual bile acids involves introduction of one or two hydroxyl groups, all in the α-configuration, and the partial oxidation of the aliphatic side chain to introduce a carboxyl group. The simple bile acids can undergo further modification by condensation with glycine or with taurine (derived from the amino acid, cysteine) to form the so-called conjugated bile acids. For simplicity, the figure shows only the structures of glycocholic and taurodeoxycholic acids; the student should bear in mind that either simple bile acid can condense with either glycine or taurine. Chenodeoxycholic acid can undergo similar condensation reactions.

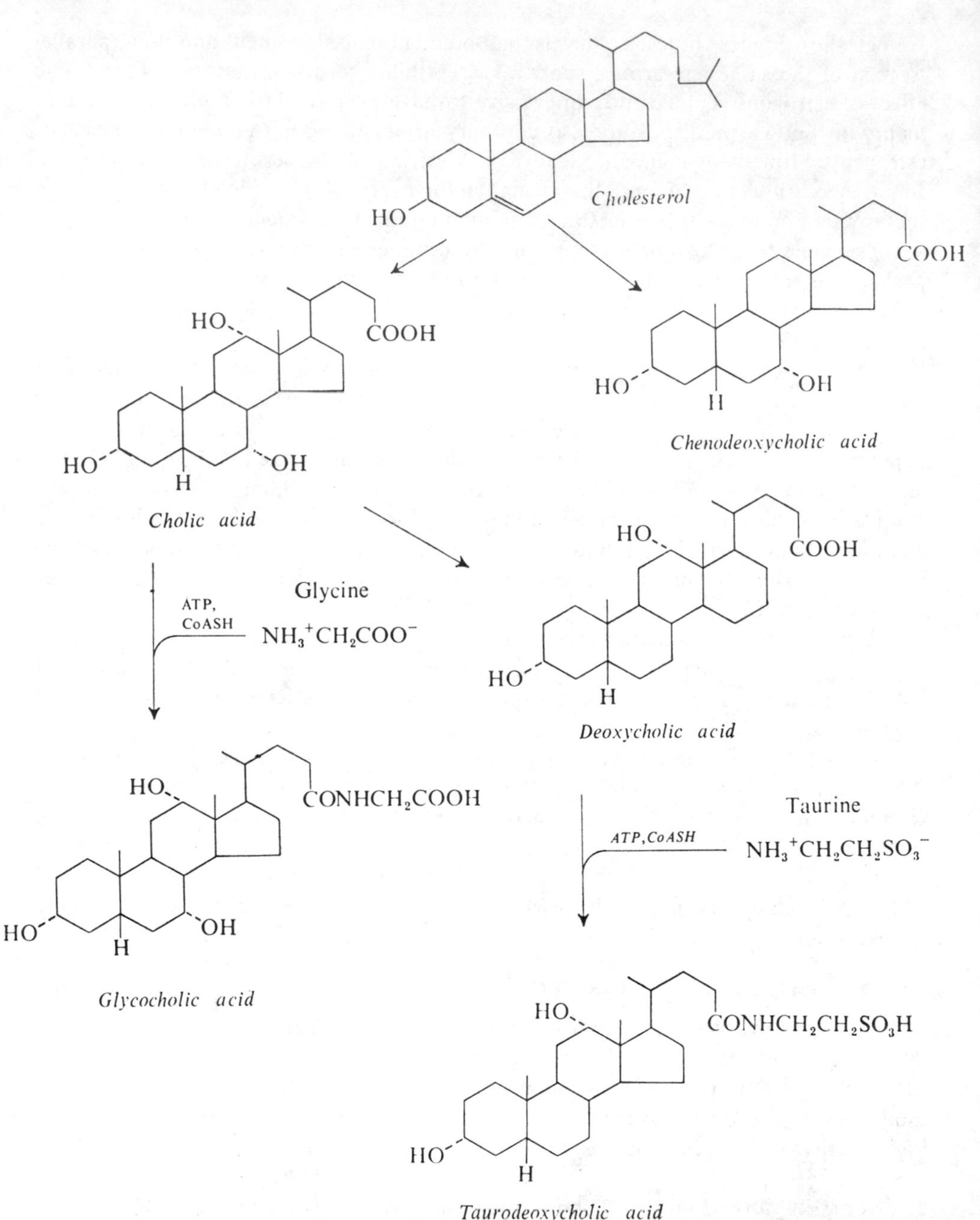

Figure 7-7. Structures of the major bile acids of man.

The bile acids not only require some cholesterol in their own formation, but they also can form molecular complexes with unchanged cholesterol; in so doing, they facilitate the excretion of still more cholesterol. If the molecular complex between the bile acids and cholesterol is broken down within the gallbladder, as sometimes happens during infectious processes, the cholesterol may deposit about some microscopic nidus to form gallstones, which may grow to the size of large marbles, and which frequently contain 60 to 80 per cent, by weight, of cholesterol.

In the intestinal lumen, the bile salts serve to emulsify ingested fats and thereby promote digestion. During the absorptive phase of digestion, some of the bile salts are reabsorbed with the fatty acids, which result from lipid digestion. The bile salts form molecular complexes with fatty acids as well as with sterols. Thus the bile acids, produced by the liver, pass to the intestine and are then partly returned to the liver. This process is termed the enterohepatic circulation, and, as we have mentioned, it is important in facilitating the excretion of sterols and the absorption of fatty acids. Failure of the enterohepatic circulation of the bile results in malabsorption of lipids and the fat-soluble vitamins.

FREE AND ESTERIFIED CHOLESTEROL

As an alcohol, cholesterol can form esters with many long chain fatty acids, and normally, about two thirds of the total circulating cholesterol exists in such form. The liver seems to be the major site of esterification, although cholesterol esterase is also found in the pancreas and in the intestinal mucosa. The esterase reaction appears to be reversible, but the optimum pH for the hydrolysis of esters is about 6.6, and the optimum pH for ester formation is lower, from 4.7 to 6.1. Bile salts are essential for the reaction to proceed in either direction.

A second enzyme exists in the plasma that can form cholesterol esters from the free alcohol in the presence of phosphatidylcholine. This enzyme is best described as a transferase, since it seems to take unsaturated fatty acids from the β-position of phosphatidylcholine and transfers them to the alcohol function of the sterol. Both of these enzymes are of analytical significance since they continue to operate after blood has been drawn. This implies that the ratio of free and esterified cholesterol in collected blood samples will change with time, even at refrigerator temperatures. For analyses in which correct values for free and esterified sterol are important, samples should be handled promptly after collection. In the past, great store was set on the partition of the total sterol between the two forms, but in recent years this enthusiasm has diminished since the ratio is substantially unchanged except in severe liver disease. Partition studies may be of some modest value in differentiating intra- and posthepatic disease.

PURIFICATION OF CHOLESTEROL

The first important problem in the analysis of cholesterol is the provision of material suitably purified for use as a standard. Commercial cholesterol preparations are not, at the moment, adequate. Even highly purified material is not stable, but undergoes oxidation by air to a complex mixture of products. Sterols notoriously form mixed crystals, and thus even recrystallization from various solvents is not sufficient to separate the impurities. A simple and rapid purification procedure has been developed by Fieser,[8] and it is recommended that this be used to prepare cholesterol for standards. The procedure is based on the fact that cholesterol dibromide is distinctly less soluble than free cholesterol, or the impurities associated with it. The impure sample can be brominated, the dibromide collected, and then debrominated. When the double bond is blocked by the addition of bromine, the molecule is protected from oxidation. For this reason, some workers actually store cholesterol as the purified dibromide. This is not employed as a standard, but as a safe and convenient storage form; it can be debrominated in small amounts as needed when fresh standard material is required in the laboratory. Details of the procedure follow.

1. Dissolve 5 gm. of commercial cholesterol in 35 ml. of dry, peroxide-free ether (anesthesia grade) by gentle warming; then cool the solution to room temperature. Filter off any insoluble residue.

2. Dissolve 0.166 gm. of anhydrous sodium acetate and 2.27 gm. (0.73 ml.) of bromine in 20 ml. of glacial acetic acid; then slowly add this solution to the cholesterol solution while stirring. Cholesterol dibromide will precipitate from the reaction mixture.

3. Collect the precipitate and wash on a Buchner funnel with small fresh portions of glacial acetic acid until the washings are colorless; save the washings. A second crop of crystals may be collected by adding 27 ml. of water to the combined washings. This may again be washed with small volumes of glacial acetic acid and added to the first crop of crystals. The combined crystals are then sucked dry at the aspirator and may be stored in a brown bottle. Alternatively, the material may be debrominated as indicated next. If less than the entire crop of dibromide is to be debrominated at a time, reduce the indicated volumes of reagents accordingly. The directions are for an entire batch of dibromide.

4. Suspend the cholesterol dibromide in 40 ml. of peroxide-free ether in a 250 ml. Erlenmeyer flask. Cool the flask and contents in an ice bath since the debromination reaction is exothermic. A mechanical or magnetic stirrer is desirable.

5. Weigh out 1.3 gm. of zinc dust, and slowly add this to the suspension of the dibromide in ether over a 5 minute period, stirring constantly. A vigorous reaction takes place, and care should be taken to avoid spattering.

6. When the reaction subsides, stir for an additional 15 minutes; then add 10 ml. of water to dissolve the zinc salts. Decant the ether layer into a separating funnel.

7. Wash the ether solution with 15 ml. of 2.5 per cent HCl, then wash three times with 10 ml. portions of water, and finally wash with 10 ml. of 10 per cent NaOH.

8. Evaporate the ether solution to approximately 20 ml. and add 20 ml. of methanol. Evaporate again until crystals just begin to appear. Allow the solution to cool and remain undisturbed until crystallization is complete. Place the mixture into a refrigerator for an hour to allow maximal crystallization; the larger the crystals, the smaller the surface exposed to the air, and the longer the shelf life that may be expected from the product.

9. Collect the solid purified cholesterol and dry it. The melting point of suitably purified material should be 149.5 to 150.0°C. The usual yield of product is about 70 per cent of the starting material. It should be stored in tightly stoppered bottles in the refrigerator. If solutions are to be stored, they should preferably be kept in the freezing compartment of a refrigerator.

REFERENCE

Fieser, L. F.: J. Am. Chem. Soc., *75*:5421, 1953.

FUNDAMENTAL CHEMISTRY OF THE CHOLESTEROL ANALYSIS

The number of procedures available for the determination of cholesterol and its esters is overwhelming, and yet the fundamental chemistry on which they depend is basically identical. It has been known for a long time that the two reactive centers in the molecule are represented by the double bond and the hydroxyl group; of these, the double bond is the most important.

Cholesterol reacts with strongly acid reagents to produce colored substances, chiefly cholestadiene sulfonic acids. In virtually all procedures, acetic acid and acetic anhydride are used as solvents and dehydrating reagents, and sulfuric acid or sulfuric

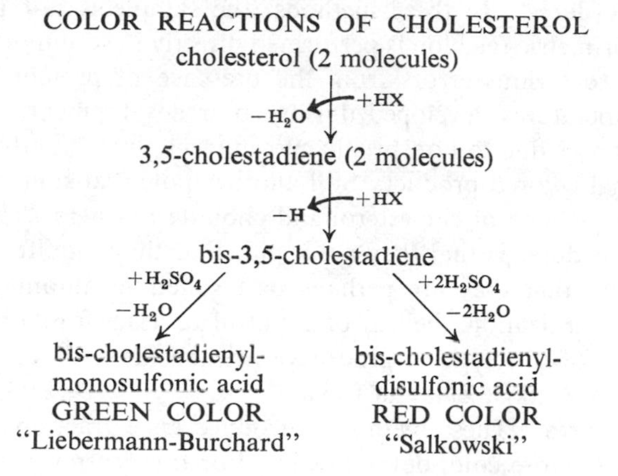

Figure 7-8. Proposed reaction mechanisms for production of color in the determination of cholesterol.

acid and p-toluenesulfonic acids are used as dehydrating and oxidizing reagents. In some procedures, the reaction of these agents is further accelerated by the addition of various metal ions, including among others iron, aluminum, or cobalt. Vanzetti has written an extensive review of the many methods.[22] He points out that by 1935 over 150 different methods had been proposed, and since then, still more have appeared. The general reaction mechanism of the color tests for cholesterol is presented in Figure 7-8. According to this scheme, cholesterol is first attacked by strongly acid reagents, generalized as HX, where X might stand for the sulfate ion or the acetyl radical. Such reagents first remove a molecule of water, then oxidize the intermediate to produce 3,5-cholestadiene (two double bonds). The oxidizing agent is usually sulfuric acid, which is converted to sulfur dioxide. The cholestadiene is attacked still further to form the dimer, *bis*-cholesta-3,5-diene, and is finally converted by the excess sulfuric acid to a mono- or a disulfonic acid, which is a highly colored molecule. Depending on the relative concentration of the oxidant, sulfuric acid, one gets either a green color (Liebermann-Burchard) due to a monosulfonic acid, or a red color (Salkowski) due to the formation of a disulfonic acid. The addition of iron or other metal ions favors the formation of the red color due to cholestadiene disulfonic acid. As Vanzetti points out, the reactions are not entirely specific, nor can they be controlled with sufficient precision to yield always the exact products shown, but with proper care, the reactions can be the basis of an accurate and reproducible analytical system. Insofar as the chemistry is concerned, there is little reason to prefer a "red" color reaction over a "green" color reaction, except that the "red" reactions generally produce a compound with a higher molar absorbance than the "green." Choice of a specific procedure is, therefore, based on other issues. These may include the interference by such substances as bilirubin, by considerations of speed and facilities available, and factors such as the manipulations required in the sample preparation. We shall next attempt to classify several representative procedures, and present them in detail. The choice of methods is entirely arbitrary. The range of available procedures is so great, and advancing so rapidly, that the student can only be urged to continue to consult the current literature to keep abreast of developments. The student should also develop a healthy scepticism, however, and not accept a procedure until it has been carefully evaluated against a recognized reference method.

Single-step procedures. In these methods, the sample is not purified to any degree, and the colorimetric reaction is performed directly in serum or plasma. These methods are liable to serious errors from the presence of protein, which may be charred by the temperatures developed during color development. They may also suffer from interferences due to nonspecific chromogens, including bilirubin, and by instability of the final colored product. Still another potential source of error is the difference in chromogenicity of cholesterol and cholesterol esters. The best that can be said of these procedures is that they are rapid, that they require the least degree of manipulation, and that they are perhaps best suited to automation; however, without internal standardization, they are of doubtful accuracy for more than screening purposes. Examples of this type of procedure are the methods of Pearson, Stern, and McGavick,[13] and Zlatkis, Zak and Boyle.[26]

Two-step procedures. These methods introduce an extraction step, which extracts the cholesterol before color development. For this reason, some of the interfering substances, especially protein, are removed, but the methods still are subject to error caused by nonspecific chromogens, such as bilirubin, and by the unequal chromogenicity of the free and esterified forms of the sterol. Of several methods in this class, that of Carr and Drekter[6] is perhaps the best. Except for highly icteric samples (high bilirubin levels), it gives results that are in close correspondence to those obtained by the most precise reference procedures, and it is simple enough to involve little manipulative error. It is, therefore, suited for routine work on a large scale basis.

Three-step procedures. These involve an extraction of the cholesterol, followed by a saponification of the esters before color development. Consequently, they do not suffer from serious error because of protein or of differences in chromogenicity of the free and ester forms; furthermore, the saponification step tends to destroy the nonspecific chromogens. The resultant increase in accuracy more than makes up for the extra manipulation of saponification. This class of procedures is best exemplified by the method of Abell et al.,[1] which is acceptable as a standard method in many laboratories. The manipulations are direct, and the method is suitable for large scale application. Unfortunately, the final colored product is of limited stability, and we have found it useful to employ the slightly modified Liebermann-Burchard reagent, stabilized by addition of sodium sulfate, recommended by Huang et al.[11]

Four-step procedures. These are the most complicated, but the most reliable procedures. The cholesterol is extracted, the esters are saponified, and the total sterol is then further purified by collection as the digitonide. The digitonide is decomposed by saponification, which again frees the cholesterol, and the product of this step is subjected to color development.

By introduction of the digitonin step, the effect of nonspecific chromogens is considerably reduced or eliminated, but then the cholesterol digitonide must be decomposed and the digitonin removed; otherwise, it too will give a positive reaction with the usual reagents since the digitonin structure includes a sterol ring. Cholesterol and digitonin, however, do not have the same chromogenicity, and it is therefore advisable to remove the digitonin before performing the chromogenic reaction. Four-step procedures involve many separate steps, each of which requires careful control; nevertheless, methods of this type are of high accuracy, and either the method of Schoenheimer and Sperry[16] or of Sperry and Webb,[19] a perfected version, are the most widely accepted reference methods for the determination of cholesterol.

Many other methods in each category could be cited, most of which represent slight modifications of fundamental procedures that will be presented in detail, but it has been our experience that many of the proposed modifications do not always perform as expected; therefore, before accepting modifications of the more standardized procedures, the modification should be carefully checked against a method of accepted performance. The review of Vanzetti contains a wealth of critical detail, and Henry's text[10] may also be consulted.

An interesting fluorescent method, based on the Liebermann-Burchard reaction, has been published by Albers and Lowry.[2] It is a true micromethod, capable of measuring as little as 0.1 μg. of cholesterol with high accuracy and precision. Free and esterified cholesterol produce nearly the same molar fluorescence, and the fluorescence readings are relatively stable for some time. The original paper may be consulted for further details.

On the assumption that the circumstances of use may dictate the choice of methodology, we shall next present the details of some typical methods with which we have had considerable experience, illustrative of the classification just outlined. Although the original specifications do not in all cases call for purified cholesterol, we shall assume that the standard material has been purified by the method of Fieser already described.

DETERMINATION OF TOTAL SERUM CHOLESTEROL

Method of Pearson, Stern, and McGavack

REAGENTS

1. Acetic anhydride, A.R.
2. Glacial acetic acid, A. R.
3. p-Toluenesulfonic acid (PTS). Use special grade for cholesterol determination (Fisher Scientific Co., Pittsburgh, Pa.). Dissolve 12 gm. of the solid sulfonic acid in 100 ml. of glacial acetic acid. Keep well stoppered.
4. Sulfuric acid, A.R. Use the commercially available 36 *N* acid. It should also be stoppered tightly to prevent uptake of water.
5. Standard cholesterol. Dissolve 200 mg. of purified cholesterol in 100 ml. of glacial acetic acid.

PROCEDURE

For each serum or plasma sample to be analyzed, set up two tubes, one of which will serve as a blank. Each run of analyses should also include a tube for a standard sample and one for a reagent blank. In the proper order, add to the indicated tube the volume (ml.) of the following materials.

	Reagent blank	*Standard*	*Unknown serum*	*Serum blank*
1. Standard solution	—	0.1	—	—
2. Serum	—	—	0.1	0.1
3. Water	0.1	0.1	—	—
4. Acetic acid	0.1	—	0.1	—
5. PTS	0.5	0.5	0.5	2.3
6. Acetic anhydride	1.5	1.5	1.5	—

7. Allow the contents of all tubes to stand without shaking until they have cooled to room temperature.

8. To all tubes except the serum blank, add 0.2 ml. of concentrated sulfuric acid; mix thoroughly, immediately after addition of the sulfuric acid, until the precipitate is completely dissolved.

9. Allow the tubes to stand at room temperature for 20 minutes before measuring the absorbance at 550 nm.

10. Set the zero point of the spectrophotometer with the reagent blank, then read the standard, the unknown samples, and the corresponding sample blanks against the reagent blank.

CALCULATIONS

$$\text{mg. cholesterol/100 ml. serum} = \frac{A_{\text{unkn}} - A_{\text{serum blk}}}{A_{\text{std}}} \times 200$$

PRECAUTIONS

Although the absorbance at 650 nm. is higher than at 550 nm., readings at the latter wavelength are recommended because of a greater color stability, and because the readings are linear over a longer portion of the standard curve.

There has been one report of an explosion with the PTS reagent, but in many analyses with this method we have had no adverse experiences.

REFERENCE

Pearson, J., Stern, S., and McGavack, T. H.: Anal. Chem., *25*:813, 1953.

Method of Carr and Drekter

REAGENTS

1. Glacial acetic acid, A.R.
2. Acetic anhydride, A.R. Keep tightly stoppered.
3. Sulfuric acid, A.R., 36 *N*. Keep tightly stoppered.
4. Cholesterol standard. Dissolve 200 mg. of purified cholesterol in approximately 50 ml. glacial acetic acid; then dilute to 100 ml. exactly.
5. Sulfuric acid-acetic acid reagent, 1:1 (v/v). Slowly, and carefully, pour 100 ml. concentrated sulfuric acid, preferably precooled in a refrigerator, into a 500 ml. Pyrex flask containing 100 ml. of glacial acetic acid. Mix the contents by gentle rotation until the addition is complete; then allow the mixture to cool to room temperature before using. If kept tightly stoppered, this reagent is stable at room temperature for at least 6 months.
6. Dehydrating reagent. Mix equal volumes of reagent 1 (acetic acid) and reagent 5 (sulfuric-acetic acids). Make fresh daily.

PROCEDURE

For each serum or plasma sample to be analyzed, set up two tubes, one of which will serve as a blank. For reasons to be discussed, the tubes should be lipless 16 × 125 mm. round bottom tubes (Kimax No. 45048 or Corning No. 9820 are recommended). Each run of analyses should also include a tube for a standard sample and for a reagent blank. Add in the proper order the indicated volume (ml.) of the following materials. Protect all tubes from exposure to intense light.

	Reagent blank	*Standard*	*Unknown serum*	*Serum blank*
1. Standard solution	—	0.2	—	—
2. Serum	—	—	0.2	0.2
3. Water	0.2	0.2	—	—
4. Acetic acid	0.8	0.6	0.8	0.8

5. Allow all the tubes to stand at room temperature 1 to 2 minutes. Make the addition of the acetic acid so that it falls freely into the liquid, and avoid letting the acid dribble down the walls of the tubes. Next add 4.0 ml. of acetic anhydride (No. 2).

6. Centrifuge all of the serum-containing tubes for 5 minutes at 2000 rpm, then decant the supernatant solution as completely as possible into clean, dry tubes properly labeled. Discard protein precipitate.

7. Acetic acid	—	—	—	0.1
8. Dehydrating reagent (No. 6)	0.1	0.1	0.1	—

9. Mix the tubes by rotation. The contents should become hot in 1 to 2 minutes; if not, add a second drop of the dehydrating reagent, then place all the tubes in a water bath at 25°C. Allow them to remain there for 5 to 10 minutes.

10. At 1 minute intervals, add 1 ml. of the dehydrating reagent (No. 6) to all tubes *except* the serum or plasma blanks which receive 1 ml. of acetic acid (No. 1). Begin with the reagent blank. After addition, mix the contents well; then return them to the water bath for 20 minutes before measuring the absorbance. Readings should start with the reagent blank, then be made in serial order at 1 minute intervals. Set the photometer at 620 nm. and adjust to zero with the reagent blank solution.

CALCULATIONS

$$\text{mg. cholesterol/100 ml. serum} = \frac{A_{\text{unkn}} - A_{\text{serum blk}}}{A_{\text{std}}} \times 200 \times 1.01^{*}$$

NOTES AND PRECAUTIONS

The 25°C. water bath is conveniently made of an enameled pot or pan of 6 quart capacity, half-filled with tap water. This volume of water is sufficient to maintain the required temperature within suitable limits (±1°).

Particular care should be taken to avoid slow addition of the various reagents by drainage down the sides of the tubes; for proper results, it is important to have the reagents mix quickly by direct addition to the bulk of the liquid.

The factor of 1.01, indicated in the formula for calculation of the result, is based on the estimated loss in drainage of the supernatant acid solution from the precipitated protein. According to the authors, use of this factor to offset the slight error incurred by the drainage procedure amounts to less than 1.0 per cent. If tubes of dimensions different from those just recommended are used, it will be necessary to establish a different factor to take into account the liquid held up on the walls during the decantation step.

---

* See under Notes and Precautions for an explanation of the factor of 1.01.

Although the authors do not present detailed instructions, they indicate that the method described can be modified for determination of free cholesterol after precipitation of cholesterol as the digitonide. We recommend this be done according to the following method of Abell et al.

REFERENCE

Carr, J. J., and Drekter, I. J.: Clin. Chem., *2*:353, 1956.

Method of Abell, Levy, Brodie, and Kendall

REAGENTS

1. Absolute ethanol. Material from Commercial Solvents Co., Terre Haute, Indiana, U.S. Industrial Chemicals Co., or similar sources may be used as received. We have substituted methanol for ethanol.

2. Hexane. Spectroanalyzed grade may be used as received, but less refined grades may require purification by washing with concentrated sulfuric acid followed by redistillation.

3. Acetic acid, A.R.

4. Sulfuric acid, A.R. Keep tightly stoppered.

5. Potassium hydroxide, A.R. Prepare a 33 per cent solution by dissolving 10 gm. of the solid pellets in 20 ml. of water.

6. Acetic anhydride, A.R.

7. Alcoholic KOH solution. Make this immediately before use by adding 6 ml. of the 33 per cent KOH solution (step 5) to 94 ml. of absolute ethanol.

8. Standard cholesterol. Dissolve 200 mg. of purified cholesterol in some warm absolute ethanol; then dilute to exactly 100 ml. with ethanol.

9. Modified Liebermann-Burchard reagent. Cool 30 volumes of acetic anhydride to 5°C. in a glass-stoppered cylinder; to this add 1 volume of cooled concentrated sulfuric acid, slowly and with mixing. When the temperature has again returned to 5°C., add 3 volumes of glacial acetic acid. To the final mixture add sufficient anhydrous sodium sulfate to give a concentration of 2 per cent (w/v). This reagent is stable for at least 2 weeks at room temperature.

PROCEDURE

1. Each sample of serum or plasma to be analyzed will require a 40 ml. glass-stoppered centrifuge tube (the tubes already mentioned in the section on transesterification procedures will suffice). One additional tube will be required for a reagent blank, and two more will be required for standards, one of which should be included at the beginning and one at the end of each run of analyses. Add, in the order indicated, the prescribed volumes of the reagents listed below.

	Reagent blank	*Standard*	*Unknown serum*
1. Standard solution	—	1.0	—
2. Serum	—	—	0.5
3. Ethanol	5.0	4.0	—
4. Alcoholic KOH	0.5	0.5	5.0

5. Stopper the tubes, shake well, and incubate all tubes in a water bath at 37 to 40°C. for 55 minutes. Cool the tubes to room temperature.

6. Add 10 ml. of hexane to all tubes, stopper, and shake the tubes vigorously for at least 1 full minute and add 5 ml. water. Centrifuge the tubes for 5 minutes at about 2000 rpm, or until the emulsions break and the two liquid layers are clearly separated.

7. Transfer 4 ml. aliquots from each hexane layer (upper phase) to clean, dry test tubes. Evaporate the solvent in a gentle stream of air or nitrogen.

8. Place all the tubes in a water bath at 25°C. for 5 minutes.

9. Add 6.0 ml. of the modified Liebermann-Burchard reagent to the reagent blank tube; then mix well. Use this tube to zero the photometer.

10. At timed intervals of 1 minute, add 6.0 ml. of the reagent to each of the remaining tubes, mixing after addition.

11. Allow all of the tubes to remain in the 25°C. water bath for exactly 30 minutes; then read the absorbance at 620 nm.

CALCULATIONS

$$\text{mg cholesterol/100 ml. serum} = \frac{A_{unkn}}{A_{std}} \times 2 \times \frac{100}{0.5}$$

NOTES AND PRECAUTIONS

This method, as just presented, is slightly modified from the original version; in our experience, the modifications minimize some of the sources of error intrinsic in the chromogenic reagent. The authors point out that the color reaction does not always strictly follow Beer's law, and recommend that standards spread over the expected range of serum concentrations be prepared.

Although this method is somewhat tedious, it does give results for total cholesterol of a high order of accuracy, provided the timing of the chromogenic reaction is carefully controlled.

REFERENCE

Abell, L. L., Levy, B. B., Brodie, B. B., and Kendall, F. E.: J. Biol. Chem., *195*:357, 1952.

DETERMINATION OF FREE AND TOTAL SERUM CHOLESTEROL

Method of Zak, Dickenman, White, Burnett, and Cherney

REAGENTS

1. Digitonin solution. Dissolve 1.0 gm. of digitonin in 50 ml. ethanol. Dilute to exactly 100 ml. with water.

2. Cholesterol standard solution. Dissolve 100 mg. of purified cholesterol in a little glacial acetic acid, then dilute to exactly 100 ml. in a volumetric flask with glacial acetic acid.

3. Iron stock solution. Dissolve 2.5 gm. of $FeCl_3 \cdot 6H_2O$, A.R., in 25.0 ml. of glacial acetic acid. Store in the freezing compartment of a refrigerator until needed. No precipitate forms in the frozen solution, which, therefore, keeps well.

4. Color reagent. Pipet 1.0 ml. of the iron stock solution into a 100 ml. volumetric flask and dilute to the mark with concentrated sulfuric acid, A.R., while swirling the contents of the flask. Discard any solution in which a precipitate has formed.

5. Extracting solvent. Mix equal volumes of 95 per cent ethanol and acetone, A.R., and stopper tightly to prevent change in concentration by evaporation.

PROCEDURE

1. Prepare a series of working standards by pipetting 0.1, 0.2, and 0.4 ml. of the cholesterol standard solution into individual 30 ml. test tubes. Dilute the contents of each tube to 3.0 ml. with glacial acetic acid. Set aside until the unknowns are ready for the addition of the color reagent.

2. Prepare a blank by placing 3.0 ml. of glacial acetic acid in a 30 ml. test tube; set this aside with the standards.

3. Pour about 10 ml. of extracting solvent into each of as many 25 ml. volumetric flasks as there are sera to be analyzed. Pipet, in a dropwise manner, 1.0 ml. of serum into each flask, while vigorously swirling the contents; then bring to the boiling point in a hot water bath. Mix the flasks constantly to avoid explosive bumping. Cool the flasks and dilute exactly to the mark with more of the extracting solvent.

4. The precipitated protein should be finely divided so that the contents of each flask can easily be filtered into clean, dry 30 ml. test tubes. Use Whatman No. 41-H filter paper, and cover the funnel tops with watch glasses to minimize evaporation.*

5. Pipet 2.5 ml. aliquots of the clear filtrate into a 15 ml. conical tipped centrifuge tube (for free cholesterol), and into a 30 ml. test tube (for total cholesterol).

6. Evaporate the contents of the (15 ml.) test tubes to dryness, then add 3.0 ml. of glacial acetic acid to each tube to dissolve the lipid residue.

7. Evaporate the contents of the (30 ml.) centrifuge tubes to about 0.5 to 1.0 ml. The evaporation may be assisted by a slow stream of air or nitrogen.

8. To the contents of each centrifuge tube, add 1.0 ml. of digitonin solution, mix well, and let stand for at least 30 minutes.

9. Centrifuge the tubes at about 3500 rpm for 10 minutes, decant and discard the supernatant, and allow the inverted tubes to drain on clean paper toweling for several minutes.

10. With a blowout pipet, forcefully deliver 4.0 ml. of acetone into each collected precipitate, then disperse the precipitate with a vibrating mixer (such as a Magna-Whirl) to wash the dispersed precipitate thoroughly in the acetone.

11. Centrifuge the washed precipitate for 10 minutes at 3500 rpm, decant the supernatant, and again allow the inverted tubes to drain on clean paper toweling for a few minutes.

12. Pipet into each tube 3.0 ml. of glacial acetic acid and warm briefly to assist in the solution of the precipitate.

13. To all of the tubes add 2.0 ml. of the color reagent, mix, and let stand for 20 minutes before photometry.

14. Set the zero point of the photometer with the blank, then read the absorbance of the standard and the unknown tubes at 560 nm.

CALCULATIONS

$$\text{mg. free (or total) cholesterol/100 ml. serum} = \frac{A_{\text{unkn}}}{A_{\text{std}}} \times \frac{25}{2.5} \times 100 \times \text{A}\dagger$$

* This is specified in the original paper; however, we recommend centrifugation instead of filtration for reasons already set forth.

† A is a factor depending on the aliquot of the standard solution in a given tube. For an aliquot of 0.1 ml., A = 0.1; for an aliquot of 0.2 ml., A = 0.2, and so on. The authors recommend that the absorbance of the various standards be plotted on graph paper, and that the concentrations of the unknown be read from this.

NOTES AND PRECAUTIONS

This procedure is considerably more rapid and less laborious than the classic procedure of Sperry and Webb, or Schoenhemier and Sperry, and yet it gives results that are in good agreement with them.

REFERENCE

Zak, B., Dickenman, R. C., White, E. G., Burnett, H., and Cherney, P. J.: Am. J. Clin. Path., *24*:1307, 1954.

REFERENCES

1. Abell, L. L., Levy, B. B., Brodie, B. B., and Kendall, F. E.: J. Biol. Chem., *195*:357, 1952.
2. Albers, W. and Lowry, O. H.: Anal. Chem., *27*:1829, 1955.
3. Blankhorn, D. H., Rouser, G., and Weimer, T. J.: J. Lipid Res., *2*:281, 1961.
4. Bligh, E. G. and Dyer, W. J.: Canad. J. Biochem. Physiol., *37*:911, 1959.
5. Burchfield, H. P. and Storrs, E. E.: Biochemical Application of Gas Chromatography. New York, Academic Press, 1962.
6. Carr, J. J. and Drekter, I. J.: Clin. Chem., *2*:353, 1956.
7. Dryer, R. L., Tammes, A. R., and Routh, J. I.: J. Biol. Chem., *225*:177, 1957.
8. Fieser, L. F.: J. Am. Chem. Soc., *75*:5421, 1953.
9. Folch, J., Lees, M., and Sloane-Stanley, G. H.: J. Biol. Chem., *226*:497, 1957.
10. Henry, R. J.: Clinical Chemistry, Principles and Technics. New York, Hoeber Division, Harper and Row, Publishers, 1964.
11. Huang, T. C., Chen, C. P., Wefler, V., and Raftery, A.: Anal. Chem., *33*:1405, 1961.
12. Littlewood, A. B.: Gas Chromatography. New York, Academic Press, 1962.
13. Pearson, J., Stern, S., and McGavack, T. H.: Anal. Chem., *25*:813, 1953.
14. Randerath, K.: Thin-Layer Chromatography. New York, Academic Press, Inc., 1964.
15. Rouser, G., Siakatos, A. N., and Fleischer, S.: Lipids, *1*:85, 1966.
16. Schoenheimer, R., and Sperry, W. M.: J. Biol. Chem., *106*:745, 1934.
17. Skipski, V., Peterson, R. F., Sanders, J., and Barclay, M.: J. Lipid Res., *4*:227, 1963.
18. Sperry, W. M.: Lipid analysis: *In* Methods of Biochemical Analysis. D. Glick, Ed. New York, Interscience Publishers, Inc., 1955, Vol. 2, pp. 83–111.
19. Sperry, W. M. and Webb, M.: J. Biol. Chem., *187*:97, 1950.
20. Stahl, E. (Ed.): Thin-Layer Chromatography. 2nd ed. New York, Academic Press, Inc., 1969.
21. Trout, D. L., Estes, E. H., and Friedberg, S. J.: J. Lipid Res., *1*:199, 1960.
22. Vanzetti, G.: Clin. Chim. Acta, *10*:389, 1964.
23. West, E. S., Todd, W. R., Mason, H. S., and Van Bruggen, J. T.: Textbook of Biochemistry. New York, The Macmillan Co., 1966, pp. 985–1035.
24. Wieland, O.: Glycerol: *In* Methods of Enzymatic Analysis. H. U. Bergmeyer, Ed. New York, Academic Press, Inc., 1963, pp. 211–214.
25. Zak, B., Dickenman, R. C., White, E. G., Burnett, H., and Cherney, P. J.: Am. J. Clin. Path., *24*:1307, 1954.
26. Zlatkis, A., Zak, B., and Boyle, A. J.: J. Lab. Clin. Med., *41*:486, 1953.

Chapter 8 / ENZYMES

by John F. Kachmar, Ph.D.

Clinical enzymology is undoubtedly the most rapidly developing field in contemporary clinical chemistry. Although the science of enzymes is at least 70 years old and the key roles that enzymes play in many biochemical reactions were well understood 30 years ago, the application of enzymology to clinical diagnosis has developed rather slowly and haphazardly. It was as recently as 1954 that Karmen demonstrated the association between the recent occurrence of an acute disease (myocardial infarction) and a sudden increase in the serum concentration of transaminase. Up to that time, only some half a dozen enzymes had been used clinically, and none of these were enzymes intimately involved in intracellular biochemical reactions. Lipases and amylases, for example, had been known for many years, and the relation between an elevation in the serum levels of these enzymes and acute pancreatic disease had been established by the early 1930's. Bodansky, Gutman, King, and other investigators had studied the phosphatases present in serum, and had demonstrated the importance of assaying these enzymes in the clinical study of liver function, bone dyscrasias, and prostatic carcinoma. Duodenal fluid levels of pepsin, trypsin, amylase, and lipase were useful in evaluating pancreatic function, and uropepsin levels in assessing gastric function. These applications nevertheless constituted only a small fraction of the total clinical chemical work, rarely amounting to more than 3 to 5 per cent of the total number of tests performed in the chemistry laboratory.

Today, in contrast, in the larger hospital laboratories enzyme assays may account for as much as 20 to 25 per cent of the total work load. Whereas only four or five enzymes were assayed in the past, now as many as 15 to 18 different enzyme determinations may be performed routinely.

Several factors are responsible for the sudden blossoming of clinical enzymology. An entire generation of biochemists had to isolate, identify, and characterize the enzymes present in tissue cells, and also had to clarify the complex chains of biochemical reactions which these enzymes catalyze. Methods and instrumentation had to be developed to make possible accurate measurement of very small quantities of enzymes, and laws governing the precise measurement of enzymes had to be established. Technology had to develop to the point at which many complex and rare metabolites and coenzymes, utilized in enzyme assays, could be made available commercially and at reasonable cost. It took time for basic knowledge and techniques to emerge from the research laboratories and make their way into the clinical laboratories, and to move from medical school classrooms into the hospitals and clinics.

These developments came to a head in about 1955. French investigators were already studying aldolase and showing that muscle cell disease was reflected in changes in serum aldolase levels. Shortly after Karmen's report, Wroblewski and others began studying the changes in serum lactic acid dehydrogenase levels in association with liver disease and coronary infarction. Since that time, practically every enzyme known to be important in cellular metabolism has been studied for possible relation between changes in its serum concentration and disease of one or more tissues or organs in the body.

In this discussion of clinical enzymology we will begin by delving sufficiently into the basic chemistry of enzymes to be able to understand the prerequisites for precise enzyme assays. These assays will be illustrated by a number of clinically useful methods and procedures. A discussion of the problems encountered in relating serum levels of various enzymes and diseases associated with specific organs and tissues will also be included in the presentation.

ENZYMES AS CATALYSTS

A chemical catalyst may be defined as a substance which increases the rate of a particular chemical reaction without itself being consumed or permanently altered. In other words, at the end of a catalytic reaction the catalyst appears unchanged in form and quantity, whereas the main reaction materials have undergone transformation into new products. Enzymes are special chemical catalysts of biological origin.

A catalyst changes only the *rate* at which equilibrium is established between reactants and products; it does not alter the equilibrium constant of the reaction. In a reaction in which only one set of products is chemically possible, the catalyst cannot effect any change in the nature of the products, but in a reaction in which several different possible pathways exist (as in many organic reactions), the catalyst may favor one pathway over other possible pathways, resulting in different yields of the various reaction products.

Only a small quantity of catalyst is needed for the catalyst to exercise its effect on a reaction, and this small quantity can carry on its catalytic role over and over again. Thus, the mass of reactants consumed and products formed is many times greater than the mass of the catalyst present.

Catalysts are used widely in a variety of industrial chemical processes, for example, in the manufacture of sulfuric acid (catalysis of $SO_2 + O_2 \rightarrow SO_3$ by platinum), and in the synthesis of gasoline from high boiling oils. It is believed that in the laboratory determination of calcium the titration of oxalate with permanganate is catalyzed by a trace of manganous ion (Mn^{++}), without which the reaction will not proceed. The catalytic action of iodide in the oxidation of arsenite by ceric ion is the basis for the determination of protein-bound iodine (PBI).

Enzymes are synthesized by all living organisms, and they accelerate the multitude of metabolic reactions upon which life depends. Without exception, all enzymes are proteins—often very fragile, labile proteins. Their catalytic activity depends on the presence of a precise conformational structure in the folded polypeptide chains; even minor alterations in this structure may result in the loss of activity. Denaturation may be detected by the loss of enzymatic activity long before other physical or chemical evidence of denaturation is demonstrated. In living organisms, enzymes are rapidly

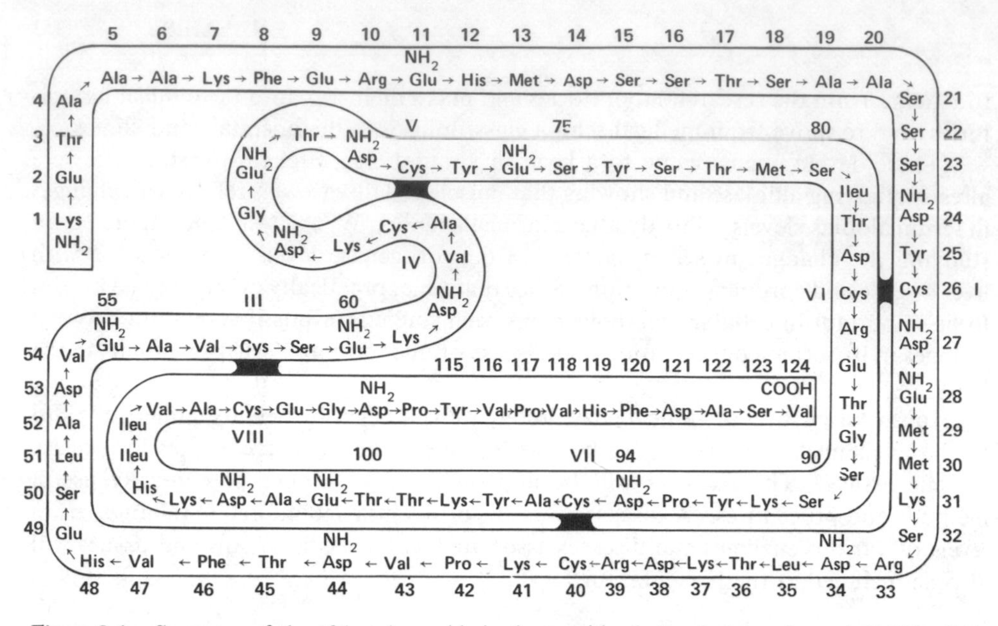

Figure 8-1. Sequence of the 124 amino acids in the peptide chain of ribonuclease A (I.U.B. Code No. 2.7.7.16) as formulated by Moore and his associates.[82] Note the cystine bridges cross-linking different segments of the amino acid chain. The numbers along the sides of the chain refer to the numerical order of the amino acids in the chain. (Reproduced with permission, Smyth et al.: J. Biol. Chem., *238*: 227, 1963.)

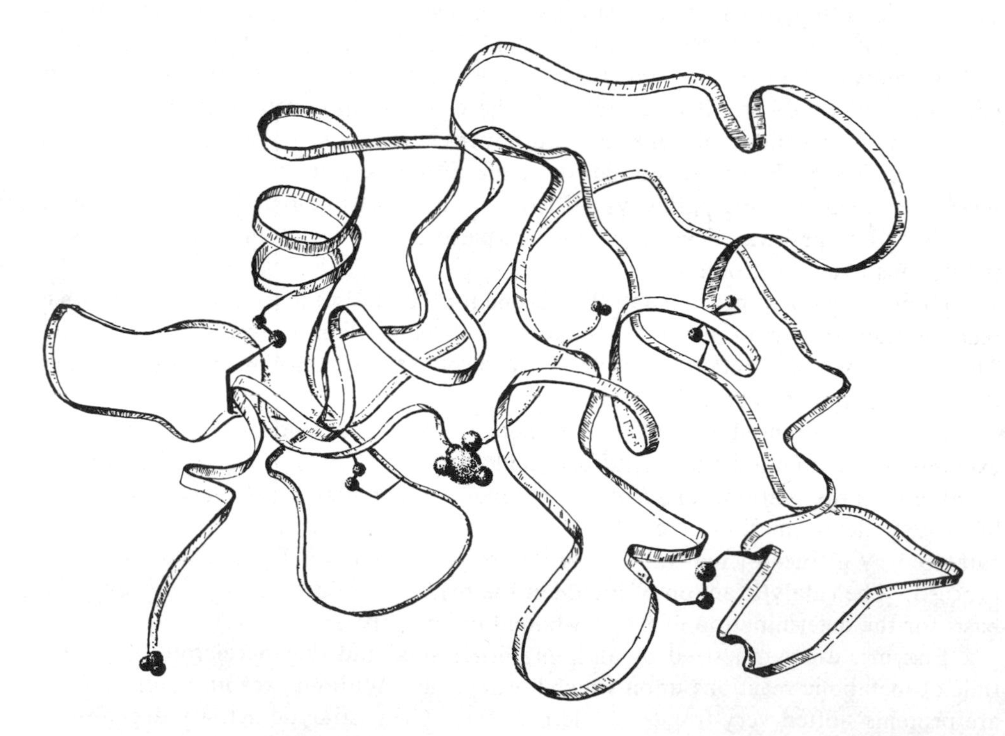

Figure 8-2. Plane projection of the three dimensional structure of ribonuclease A, as proposed by Harker from x-ray studies. Note the cystine bridges also shown in Figure 8-1. The phosphate group indicated in the figure lies in the "active center" of the enzyme, situated in a cleft in the three dimensional structure. (Reproduced with permission, from Kartha: Accounts of Chemical Research, *1*: 374, 1968.)

degraded, and their supply is replenished by new synthesis. Because they are generally unstable, enzyme preparations must be handled with special care. They are usually stored at low temperatures.

Over 700 enzymes have been isolated, many in a pure or crystalline form. Each of these is capable of catalyzing a specific type of organic or inorganic reaction. Some enzymes are relatively small molecules, with molecular weights in the order of 10,000, whereas others are very large molecules, with molecular weights ranging from 150,000 to over a million. Some enzymes are albumins, others have the properties of globulins. Because of their prime importance in biochemical phenomena, enzymes have been the subjects of intensive study. Not only have the amino acid sequences of a number of enzymes been elucidated, but even the conformational structures of a few such as ribonuclease and carboxypeptidase have been well established.

The sequence of amino acids in the chain of ribonuclease A, the enzyme which catalyzes the depolymerization of ribonucleic acid (RNA), as worked out by Smyth, Stein, and Moore,[82] is shown in Figure 8-1. This relatively small enzyme consists of 124 amino acids and has a molecular weight of 12,700. Figure 8-2 depicts the three-dimensional structure (conformation) of the amino acid chain, as elucidated by Harker from x-ray studies.[37]

ENZYME TERMINOLOGY

An enzyme-catalyzed reaction can be depicted in a simplified or schematic way as follows:

$$\underset{S}{\text{Substrate}} \xrightleftharpoons{\text{Enzyme }(E)} \underset{P}{\text{Product}} \tag{1}$$

Here it is assumed that only one substance participates in the reaction. This starting material, referred to as the *substrate*, S, undergoes chemical change to form P, the *product* of the reaction. In the absence of the enzyme, very little or no P will be formed, but in the presence of the catalyzing enzyme, the reaction will proceed and the product will be formed at a rate which will depend on the concentration of the enzyme, E, and other factors, such as temperature and pH. The chemical reaction is reversible, and the reaction may proceed in either the forward or reverse direction, depending on the value of the equilibrium constant, $K_{eq} = [P]/[S]$, and the conditions prevailing. If the reverse reaction is being studied, the labels "substrate" and "product" are interchanged.

Most enzyme-catalyzed reactions involve two substrates, and usually two products are formed. The more general reaction, therefore, may be written as:

$$\underset{S_1}{\text{Substrate 1}} + \underset{S_2}{\text{Substrate 2}} \xrightleftharpoons{\text{Enzyme }(E)} \underset{P_1}{\text{Product 1}} + \underset{P_2}{\text{Product 2}} \tag{2}$$

Very often, water, the solvent, serves as one of the substrates. Many enzymes require the presence of one or more activators. These, as well as inhibitors, will be discussed later in this section.

For the sake of convenience we will refer to enzymes as substances *acting on substrates*. In one sense no enzyme acts on a substrate; it only catalyzes the reaction involving the substrate as an initial reactant. On the other hand, the enzyme combines with a specific substrate to form an enzyme-substrate (*ES*) complex, which is an intermediate in the transformation of the substrate (*S*) into the product (*P*).

When an enzyme acts on a substrate, as a rule only a single chemical bond or chemical group in the substrate is involved in the reaction. Proteinases, for example, split the peptide bond in proteins to form products with free amino and carboxyl groups, whereas the chemical bonds in the remainder of the protein molecule remain unchanged.

ENZYME NOMENCLATURE

By the turn of the century it had become customary to identify individual enzymes by using the name of the substrate or group on which the enzyme acts, and then adding the suffix *-ase*. Thus the enzyme hydrolyzing urea was called ure*ase*, that acting on starch, amyl*ase* (from amylum, starch), and that acting on phosphate esters, phosphat*ase*. In a few instances, for purposes of clarity, the type of reaction involved was also identified, as in carbonic *anhydrase*, D-amino *oxidase*, and succinic *dehydrogenase*. Enzymes known prior to this attempt at systemization had already been given empirical names. Trivial names of this type, such as trypsin, steapsin, ptyalin, pepsin, and emulsin have been retained and remain in use.

This combination of a few trivial, common names and the larger number of semisystematic names was found serviceable until recently. With the rapid discovery of a multitude of enzymes, many acting on the same substrate and catalyzing different or related reactions, and with the characterization of others with a specific requirement for a single substrate or pair of substrates, this simple system became inadequate. The need for a definitive and standardized system of identifying enzymes was acute. In 1955 the International Union of Biochemistry appointed a commission to study the problem of enzyme nomenclature. The proposals were reported in 1961, and after further discussion they were published in 1964.[27,28a,71] These proposals have been accepted by all workers in the field. They provide a rational and practical basis for identifying all enzymes now known and for those which will be discovered in the future.

Two names are provided for each enzyme: (1) a *systematic* name, which clearly describes the nature of the reaction catalyzed, and has associated with it a unique *numerical code designation;* and (2) a working or *practical name*, which may be identical with the systematic name or a modification thereof which is more suitable for routine use. The unique numerical designation for each enzyme consists of four numbers, separated by periods, as for example: 2.2.8.11. The first number defines the class to which the enzyme belongs. All enzymes are assigned to one of six classes, characterized by the type of reaction they catalyze: (1) oxidoreductases, (2) transferases, (3) hydrolases, (4) lyases, (5) isomerases, and (6) ligases. The next two numbers indicate the subclass and sub-subclass to which the enzyme is assigned. These may differentiate the amino-transferring subclass from the phosphate-transferring group, and one type of acceptor sub-subclass from another. The last number is the specific serial number given each enzyme in its sub-subclass.

The name of each enzyme consists of two parts: the first gives the name of the substrate or substrates acted upon, and the second, a word ending in *-ase*, indicates the type of reaction catalyzed by all enzymes in the group. If two substrates are involved, both names are used and are separated by a colon; e.g., L-lactate: NAD

TABLE 8-1. *Code Designations, Systematic Names, and Trivial or Practical Names of Some Enzymes of Clinical Interest*

Trivial or Common Name	*Practical Name*	*I.U.B. Code Designation*	*Systematic Name*
Aldolase	Aldolase	4.1.2.7	Ketose-1-phosphate aldehyde lyase
Alkaline phosphatase	Alkaline phosphatase	3.1.3.1	orthophosphoric monester phosphohydrolase
Amylase, Diastase, Ptyalin	Amylase	3.2.1.1	α-1,4-glucan 4-glucano-hydrolase
Creatine phosphokinase (CPK)	Creatine kinase	2.7.3.2	Adenosine triphosphate: creatine phosphotransferase
Glutamic Oxalacetic Transaminase (GOT)	Aspartate transaminase	2.6.1.1	L-aspartate: 2-oxoglutarate aminotransferase
Isocitrate Dehydrogenase (ICD)	Isocitrate Dehydrogenase	1.1.1.42	L_s-isocitrate: NADP oxidore-ductase (decarboxylating)
Lactate Dehydrogenase (LDH)	Lactate Dehydrogenase	1.1.1.27	L-lactate: NAD oxidore-ductase
Lipase (Steapsin)	Lipase	3.1.1.3	Glycerol ester hydrolase
Pseudocholinesterase (Nonspecific, Type II)	Cholinesterase	3.1.1.7	Acylcholine acyl-hydrolase
Trypsin	Trypsin	3.4.4.4	None given. Sub-class: peptide hydrolases; Sub-subclass: peptide peptidohydrolases

oxidoreductase. Occasionally an expression in parenthesis, such as (decarboxylating), may be inserted to further identify the reaction. Because of the precise rules concerning terminology, any enzyme can be identified by both its code number and its systematic name. In Table 8-1 are listed a number of enzymes of clinical interest, identified by trivial and systematic names and by code numbers.

It has been a common and convenient practice to use capital letter abbreviations for certain enzymes, such as GPT for Glutamic pyruvic transminase (2.6.1.2); other examples are GOT, LDH, and CPK, as illustrated in Table 8-1. This practice is not permitted by the Commission rules, but it is so well established, and the convenience is so real, that it will undoubtedly persist. Such capital letter abbreviations are used in this chapter, but only after their meaning has been made unmistakably clear.

ELEMENTARY ASPECTS OF ENZYME CATALYSIS

A chemical reaction involving the transformation of a chemical material, S, into a product, P, can proceed spontaneously only if there is a decrease in *free energy* or *chemical potential* in the course of the reaction. Stated simply, chemical reactions proceed downhill, from the energy viewpoint. The symbol for Gibbs' free energy is G, and ΔG represents the difference between the chemical potential of the products, P, minus that for the initial reactants, S. A chemical reaction will tend to proceed spontaneously to completion or equilibrium only if there is a decrease in chemical potential in the course of the reaction, i.e., if ΔG has a negative value. This concept is illustrated in Figure 8-3. If ΔG is zero, no reaction will occur—the system is in equilibrium, and P and S have the same free energy content. If ΔG is positive, P will tend to react to form S; the uphill reaction will take place only if energy is provided

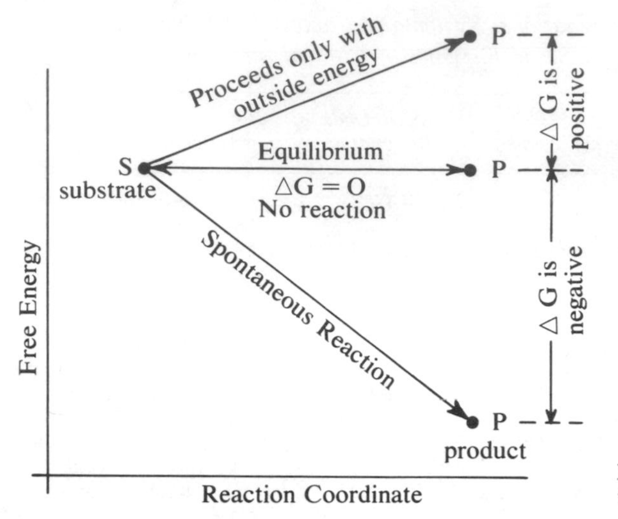

Figure 8-3. Free energy change and the course of chemical reactions.

from the outside to push the reaction uphill. In the living organism this outside energy is often provided by adenosine triphosphate, ATP.

Even though a chemical reaction is thermodynamically possible (i.e., ΔG is negative), the reaction may not proceed spontaneously, since only those molecules which are "excited" or "activated" will undergo reaction. The "active" molecules are those which have absorbed extra energy in some way, with the result that the bonds linking some or all of the atoms in the molecule are weakened or labilized. Most chemical materials are stable at ordinary conditions because only a very small fraction of the molecules are in such an "activated" state, that is, have enough extra energy to cross the so-called "activation barrier." In the spontaneous reaction

$$S \rightleftharpoons S^* \rightleftharpoons P^* \rightleftharpoons P \tag{3}$$

the molecules of S must first be activated by absorbing ΔA units of activation energy to form the "activated" molecules, S^*, as shown schematically in Figure 8-4 and equation 3. The activated, S^* molecules then undergo reaction to P^* and P. Many reactions require only a relatively low activation energy, and the energy of the thermal motion of the solute and solvent molecules is sufficient to activate enough molecules to initiate the reaction. The energy released in the reaction then makes possible the activation of all the reactants and allows the reaction to proceed to completion. Many reactions with very high activation barriers can be initiated by heating the reactants, as is often done in the chemical laboratory.

If the activation energy needed for a given reaction pathway is too high to permit rapid reaction, the reactant may still be able to undergo chemical change if some other reaction pathway having a lower activation energy requirement is available or provided. Catalysts provide such alternate reaction paths. In enzyme reactions, the substrate combines with the enzyme to form the enzyme-substrate complex, ES, already alluded to.

$$E + S \rightleftharpoons ES \rightarrow P \tag{4}$$

The activation energy required for both E and S to form ES is lower than that required to activate S alone, and the reaction will proceed via the alternate enzyme-substrate complex pathway. Such an alternate pathway is indicated in Figure 8-4. The ES complex is a form of an activated state with a low activation energy requirement.

Glucose-6-phosphate is an important intermediate in the metabolism of glucose. In water solution it can be hydrolyzed to form glucose and phosphate ion.

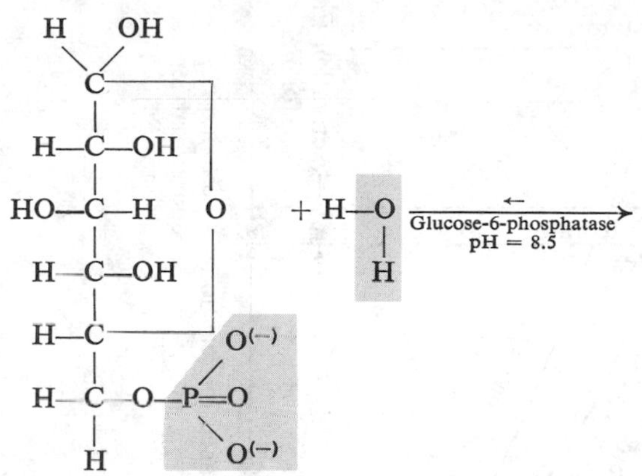

α-D-*glucose-6-phosphate*

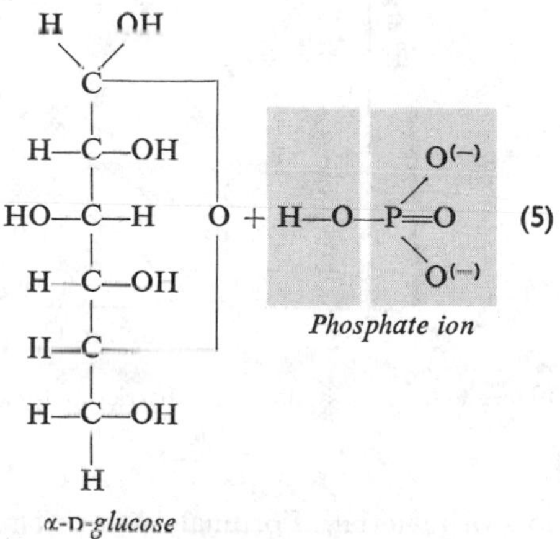

α-D-*glucose*

Phosphate ion

The free energy decrease for the hydrolysis of this ester at 37°C. and pH 8.5 is approximately −3000 calories,[58] which is large enough to permit rapid and spontaneous hydrolysis of the ester to glucose and HPO_4^{-} ion. Yet a sterile solution of glucose-6-phosphate can be kept at room temperature for many weeks, with no or very little hydrolysis taking place. However, if a few milligrams of a phosphatase are added to the solution, the ester is split very rapidly.

This rather simple hydrolytic reaction requires the severing of a (H—O—) bond in a water molecule, and a (P—O) bond in the ester. The rupture of these chemical bonds demands a rather extensive degree of activation of the water and glucose-6-phosphate molecules. The enzyme-catalyzed reaction proceeds by a different pathway. The enzyme couples with the glucose-6-phosphate, liberating the glucosyl residue and forming an enzyme-phosphate intermediate, which rapidly transfers the phosphate residue to a hydroxyl ion, freeing the enzyme to repeat the process over and over again. The complicated, enzyme-mediated pathway has a significantly lower activation energy requirement and therefore proceeds easily.

Enzyme Kinetics

Some of the techniques used in the analysis of inorganic and organic materials are not applicable to measuring enzymes, since the enzyme protein is present in a large mass of other proteins and compounds. Partial or total isolation is impractical, because suitable procedures for such steps are time-consuming and entail considerable

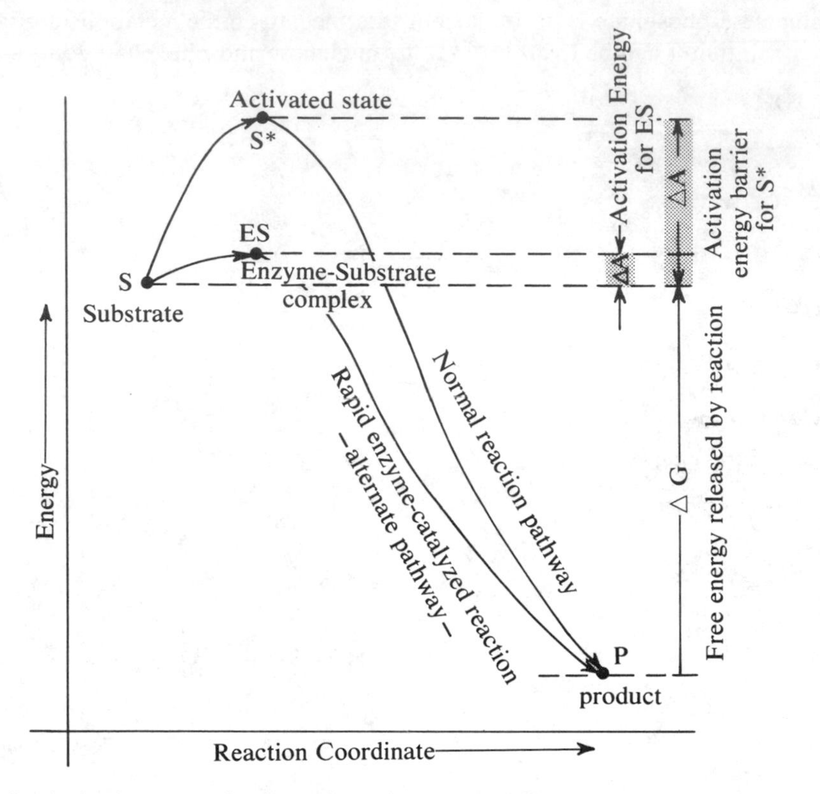

Figure 8-4. Activation energy barrier and reaction course, with and without enzyme catalysis.

loss of material. Fortunately, each enzyme provides us with a tool or "tag" for its measurement, namely its unique biochemical property of catalyzing a specific chemical reaction. We can thus quantitate a specific enzyme, even in complex mixtures containing other enzymes, by measuring what the enzyme can do, rather than by measuring the enzyme as a chemical entity in terms of mass. Even a small quantity of enzyme can, over a period of time, catalyze the transformation of a large quantity of substrate. If the substrate changes are measured by sensitive procedures, it is possible to quantitate minute amounts of a given enzyme.

The intensity of a catalytic activity can be utilized as a precise measure of enzyme concentration only if such activity is measured under well-defined experimental conditions. Thus, the main challenge in working with enzymes is to define the conditions applicable to each specific enzyme reaction.

The acid phosphatase in serum can hydrolyze phenyl phosphate (substrate) to phenol and phosphate ion (products). As the reaction proceeds, the substrate,

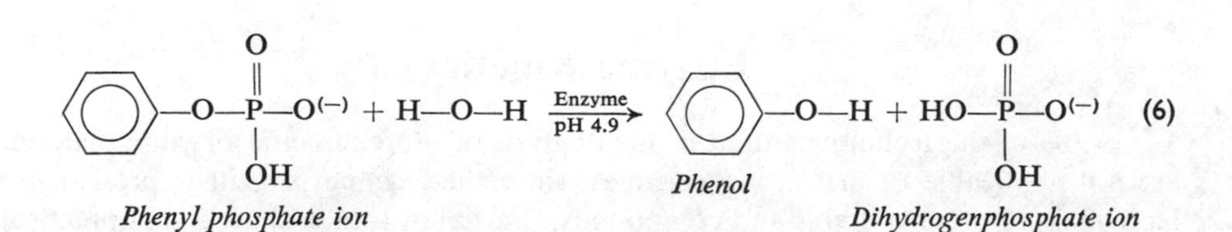

phenyl phosphate, is consumed, and its concentration in the reaction tube will decrease. At the same time the concentration of the products, phenol and inorganic phosphate, increase over their initial value. The more enzyme present, the greater the rate of catalysis and, therefore, the greater the rate of change in the concentrations of the reaction components. Inasmuch as one mole of each product is formed from one mole of substrate, under ideal conditions the rate of increase in product concentration should be identical to the rate of decrease in substrate concentration. Therefore, the reaction may be followed by measuring either the drop in substrate concentration or the increase in concentration of either product. (In our example shown in equation 6, there is no convenient method for determining phenylphosphate, whereas simple and convenient methods are available for measuring inorganic phosphate or phenol.) In general, whether one follows the decrease in substrate or the increase in product will depend on analytical convenience.

Both the concentration of enzyme present and the time interval during which the reaction proceeds will affect the quantity of substrate consumed or product formed. If the concentration of substrate is large compared to the amount of enzyme present, the reaction will be independent of substrate concentration, and doubling the quantity of enzyme will double the rate of reaction and thus double the quantity of product formed in a fixed time interval. Similarly, if the enzyme concentration is kept constant, the amount of product formed during the course of the reaction will be doubled, if the reaction period is doubled. The reaction rate is thus proportional to both time (t) and enzyme concentration (E):

$$Q = k_1 \times E \times t \tag{7}$$

where Q is the quantity of product formed (or increase in concentration of product), E is the concentration of enzyme present, and t is the interval of time during which reaction has been permitted to proceed. If the time interval (t) is fixed, then $Q = (k_1t) \times E$; if on the other hand, the enzyme concentration is kept constant, then $Q = (k_1E) \times t$. Thus, the quantity of product formed (or substrate consumed) will be a straight line function of either enzyme concentration or time, if the other is kept fixed. Figures 8-5 and 8-6 are graphs illustrating such straight line functions. In the Shinowara–Jones–Reinhart procedure for alkaline phosphatase, the quantity of enzyme present in the reaction tube is evaluated in terms of the phosphate-phosphorus produced by hydrolysis of β-glycerophosphate, whereas in a commonly used method for lactic dehydrogenase, the enzyme is assayed by measuring the decrease in absorbance as the reduced coenzyme I (NADH) is oxidized during the course of the reaction.

If the reaction time is fixed, the quantity of product formed is proportional to the concentration of enzyme present in the reaction tube. From equation 7 it follows that Q/t = rate of product formed = $k_1 \times E$. Thus, the enzyme concentration can also be determined by measuring the *rate* of product formation or substrate consumption, over any convenient time period. For any fixed enzyme concentration the rate of reaction will be constant with time.

In the preceding discussion we assumed that the substrate was present in large excess. With this assumption prevailing, the rate of reaction is constant with time and is dependent only on the concentration of enzyme in the system. Reactions obeying these conditions are referred to as *zero order reactions.* Whenever possible, enzymes are assayed under experimental conditions in which zero order kinetics prevail.

In the Wohlgemuth procedure for amylase, the enzyme is measured by the time

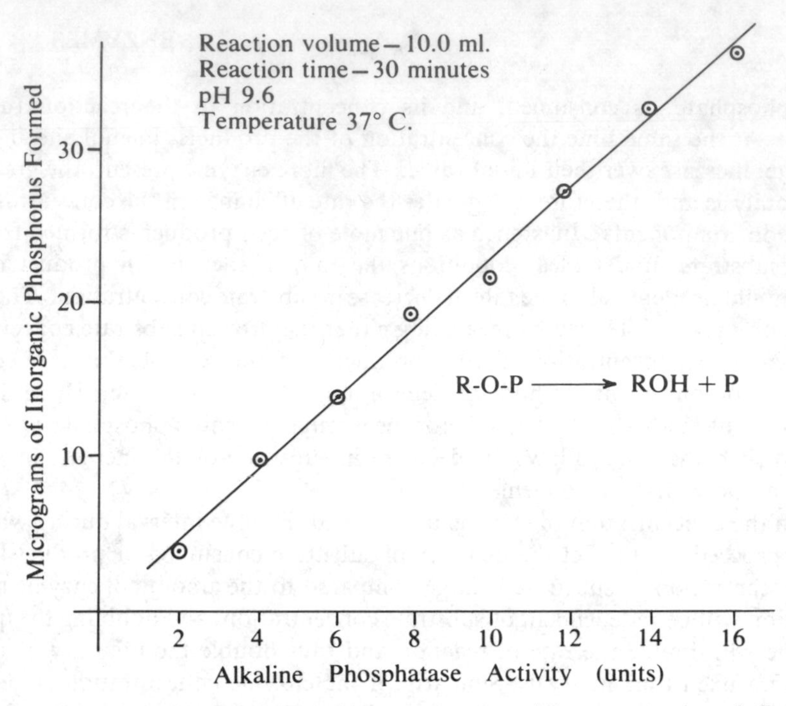

Figure 8-5. Formation of inorganic phosphate as a function of the concentration of alkaline phosphatase. Substrate: β-glycerophosphate.

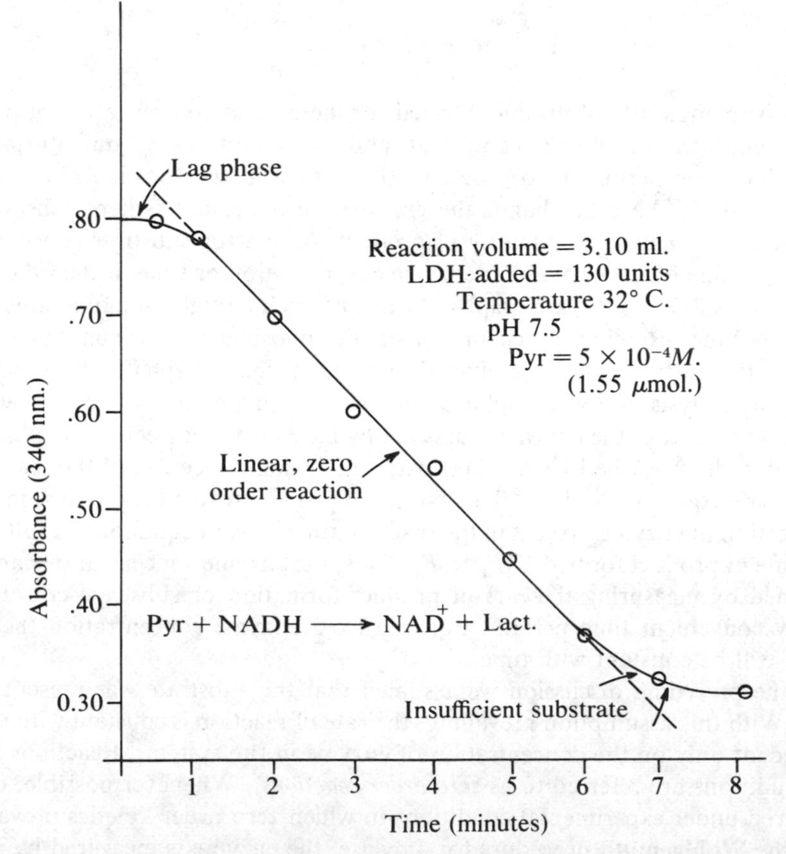

Figure 8-6. Rate of oxidation of NADH as a function of time for LDH acting on pyruvate. The concentration of NADH is expressed as its absorbance at 340 nm.

required to *consume all of a fixed quantity of substrate.* In this case equation 7 becomes

$$E = \left(\frac{Q}{k_1}\right) \times \frac{1}{t} \tag{8}$$

Thus, there is an inverse relationship between enzyme concentration and the time required for the reaction to go to completion, the curve relating E and t having the shape of a hyperbola.

Rates of all chemical reactions are governed by the concentrations rather than the absolute quantities of the reactants participating in the reaction. Therefore, in the previous discussion it would have been more precise to have referred only to concentrations of substrates and enzymes, and not to the quantities of these materials. In practice, for any given assay, the volume of solution in which the enzyme, substrates, and products are contained is kept fixed at some convenient value, so that the quantities of the components present are proportional to their concentrations. Under these conditions, concentration and quantity can be used interchangeably. In precise studies of enzyme reactions, it is important to employ and measure only true concentrations of each of the components involved in the reaction.

FACTORS GOVERNING THE RATE OF ENZYME REACTIONS

Hydrogen Ion Concentration (pH)

If an enzyme-catalyzed reaction is carried out at a series of different pH values with the concentrations of enzyme and substrate kept constant, no reaction occurs at certain pH ranges, whereas varying rates of reaction are observed at other pH values. When *enzyme activity* (rate) is plotted against pH, curves of the types presented in Figures 8-7 and 8-8 are obtained. In case of amylase (Fig. 8-7), no activity is measurable at values of pH below 4.0 or above 9.5, and maximum activity occurs at pH 7.0. Similar curves relating reaction rate and pH can be obtained for all enzymes, although the pH at which maximum activity occurs, and the shape of the curve, vary from one enzyme to another. The curve for monoamine oxidase[36] is a curve which extends over a wide range of pH, with a sharp drop-off in activity on the alkaline side, reflecting enzyme denaturation. The curves for urease, investigated by Howell and Sumner[48] (the latter is one of the early pioneers in enzyme science), show sharp optima, and also demonstrate that the type of ion (buffer) environment in which the enzyme is functioning has an effect on the reaction rate and the optimal pH for the enzyme (see Fig. 8-8).

The form of the pH-dependence curve is a result of a number of separate effects. The enzyme may catalyze the reaction only if the substrate is in either the undissociated or dissociated form, and the extent of either is dependent on the pH of the reaction system and the pK of the dissociating acid or base group. The activity of the enzyme will also be affected by the extent of dissociation of certain key amino acids in the protein chain, both at the "active center" (to be discussed later) and elsewhere on the molecule. Both pH and ionic environment will have an effect on the three-dimensional conformation (structure) of the protein and therefore on enzyme activity. At extreme values of pH, enzymes may even denature as shown in the example of monoamine oxidase. The extent of denaturation depends on the period of exposure to the extreme pH. Some enzymes, however, such as pepsin, are not only stable at

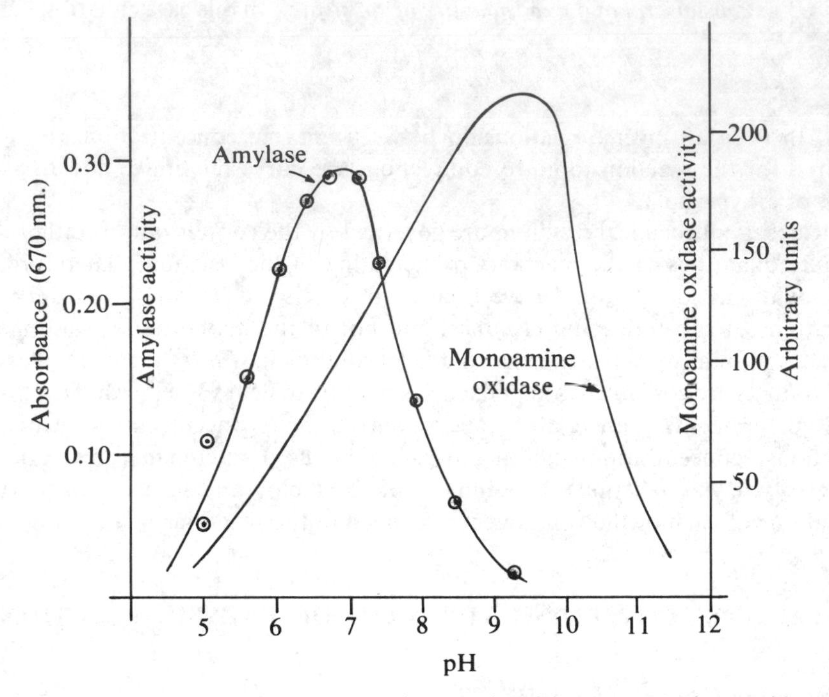

Figure 8-7. pH activity curves. Amylase data from student laboratory experiment. Monoamine oxidase curve from Hare.[36]

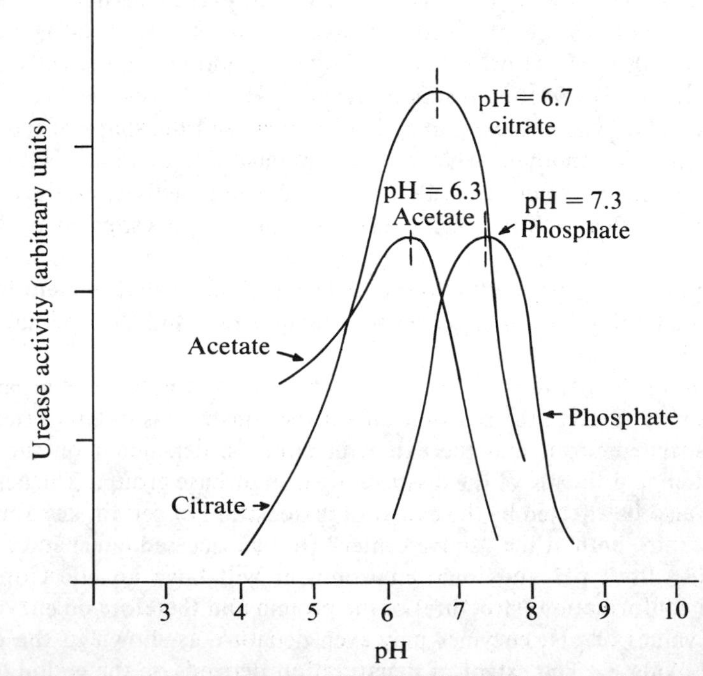

Figure 8-8. pH activity curves for urease, showing effect of buffer species on pH optimum. Adapted from Howell and Sumner.[48]

extreme values of pH (e.g., strong acid solution) but exert maximum activity at hydrogen ion concentrations at which many enzymes would be entirely inactivated.

A large majority of the enzymes in sera show maximum activity somewhere in the range of pH 7 to 8. It should also be noted that the optimal pH for a reaction may be different from the optimal pH found for the reverse reaction (see Lactic dehydrogenase, p. 434). Some enzymes are known to be associations of two or four individual peptide chains; at extreme values of pH these associations are disrupted, with an ensuing loss of catalytic activity.

Enzyme assays should be carried out at the pH of optimal activity because (1) the sensitivity of the measurement is maximal at this pH, and (2) the pH-activity curve usually has a minimum slope near this pH so that a small variation in pH will cause a minimal change in enzyme activity.

In some reactions acids or bases are formed or released during the reaction (e.g., formation of fatty acids by the action of lipase, or formation of ammonia from urea by urease). In these cases it is important to use buffers of sufficiently great buffer capacity to efficiently counteract the changes in pH resulting from such acid or base formation. The pK of the buffer system should be within one pH unit or less of the optimal pH of the enzyme system. Unfortunately, these rules have often been ignored in devising procedures for assaying clinically important enzymes.

Temperature

For each enzyme there is a temperature optimum for any given set of experimental conditions. The rates of all chemical reactions increase as the temperature at which the reaction is taking place increases. For most chemical and enzymatic reactions an increase in temperature of 10°C. will approximately double the rate of reaction. The actual Q_{10} value (the relative increase in rate per each 10°C. temperature rise) for enzymatic reactions varies from 1.7 to 2.5. At temperatures above 40 to 45°C. the enzyme protein undergoes increasingly rapid *heat* denaturation. Thus, the increasing rate of substrate reaction is counteracted by the even greater rate of loss of active enzyme. The observed rate of enzyme activity shows a maximum value in the vicinity of 35 to 45°C. (usually 37 to 39°C.), as shown in Figure 8-9.

Heat denaturation is a continuous process, and the rate of denaturation increases with temperature. Some denaturation of enzymes occurs even at room temperature, and the rate of denaturation increases at the temperature usually used in enzyme reactions (37°C.). To prevent loss of activity, enzyme preparations or samples are stored at refrigerator temperatures (2 to 5°C.) for short periods of time, and below freezing temperatures (−10 to −20°C.) for long periods of time. The presence of other proteins and other impurities tends to inhibit enzyme denaturation, and therefore highly purified preparations are often very unstable. There is considerable variation in the stability of individual enzymes. As a general rule enzyme specimens should be assayed immediately, or before significant loss of activity occurs. Permissible storage conditions and times will vary with the enzyme. Alkaline phosphatase, for example, is stable at room temperature for 24 hours, whereas acid phosphatase is exceedingly unstable, even refrigerated, unless kept at a pH below 6.0. On the other hand, a few enzymes are inactivated at refrigerator temperatures. Clinically important enzymes of this type are the liver-type isoenzymes of lactic dehydrogenase (LDH-4 and LDH-5). As a result, sera for LDH determinations *should not be refrigerated.*

The optimal temperature for any given enzyme is also dependent on the duration

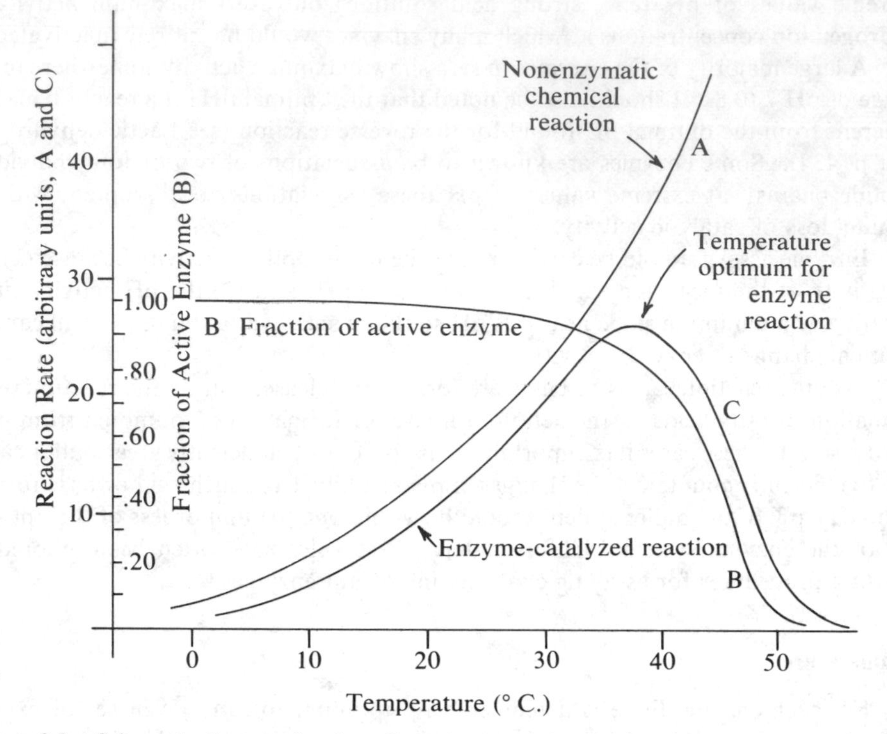

Figure 8-9. Schematic diagram showing effect of temperature on rate of nonenzyme-catalyzed and enzyme-catalyzed reactions.

of the exposure of the enzyme to the given temperature; the shorter the period of exposure, the higher the optimal temperature found. An enzyme may demonstrate an optimal temperature of 50 to 52°C. for a period of 4 to 5 minutes, but a temperature optimum of only 35°C. for a 3 to 4 hour run. The large majority of cellular and plasma enzymes are reasonably stable at 37°C., and since this is the temperature at which they function in the body, it has been customary to carry out enzyme assays at 37°C. However, this is not done universally, and many procedures were designed with 25°C. as the reaction temperature. The International Commission on Enzymes initially proposed the use of 25°C., where practical.[71] This proposal was not generally accepted in the United States, because it is much easier to maintain a higher water bath temperature, which requires only a source of controlled heat; a temperature of 25°C. requires both controlled cooling and heating.

An increasing number of enzyme assays are carried out in cuvets which remain in the cuvet compartment of spectrophotometers, thus permitting continuous mechanical recording of the change in absorbance with time, and even computer calculation of reaction rates. The continuous illumination heats up the cuvet compartments, which reach a steady state temperature of 30 to 32°C.[44] Many laboratories, therefore, use 30 or 32°C. to assay enzymes by methods involving continuous monitoring of change in absorbance with time. Today, satisfactory accessories for temperature control are available that can maintain the desired temperature without difficulty.

Recently the International Commission amended its proposal and recommended the use of 30°C., where practical.[28a] Thus, new procedures should provide for a 30°C. temperature, even though it can be expected that many laboratories will still carry out the older, well-established procedures for certain individual enzymes at the originally prescribed temperatures.

For precise work, water bath temperatures or cuvet compartment temperatures should be maintained constant to within ±0.3°C., and care must be taken that the solutions in the reaction tubes are truly equilibrated with the bath temperature before the enzyme run is initiated. This will insure that errors due to variation in temperature will be less than ±5 per cent.

Substrate Concentration

If one gradually increases the concentration of substrate in an enzyme reaction system, keeping all other factors constant, the rate of reaction will increase with the increase in substrate concentration until a maximum value is reached. Any further increase in substrate concentration, no matter how great, will elicit no further increase in reaction rate. This phenomenon is illustrated in Figure 8-10. At low substrate concentrations, the rate of the reaction increases linearly with an increase in concentration of substrate; then, as moderate levels of substrate concentration are approached, the rate of increase falls off until the maximum rate of reaction is obtained.

The significance of substrate-rate curves was first emphasized by Michaelis and Menten, and such curves are referred to as Michaelis-Menten plots. The shape of the curve is explained by the assumption that the substrate and enzyme first associate to form an enzyme-substrate complex, which then undergoes reaction. The rate of reaction depends on the concentration of the complex, rather than on the concentration of the enzyme itself. At low concentrations of substrate, only a fraction of the

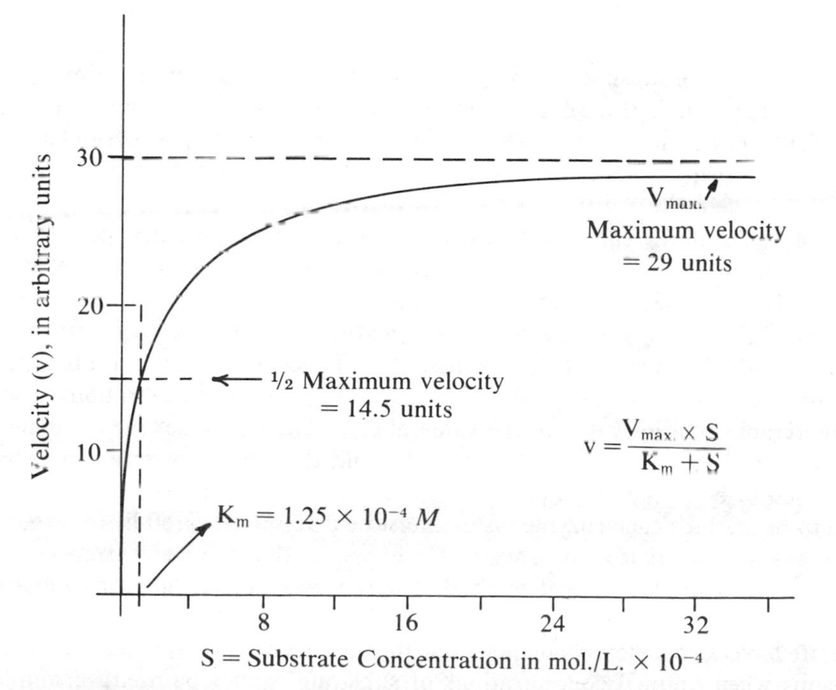

Figure 8-10. Michaelis-Menten curve relating enzyme reaction velocity (rate) to substrate concentration. The value of K_m is given by the substrate concentration at which one-half of the maximum velocity is obtained.

enzyme is associated with substrates and the rate observed reflects the low concentration of enzyme-substrate complex. At very high substrate concentrations, all the enzyme is bound to substrate, and a much higher rate of reaction is obtained. Moreover, inasmuch as all the enzyme is in the form of the complex, no further increase in complex concentration, and thus no further increment in reaction rate, is possible; the maximum velocity of the reaction has been reached.

The enzyme reaction is formulated as follows:

$$\begin{gathered} E_f + S \underset{k_{-1}}{\overset{k_1}{\rightleftarrows}} ES \\ ES \xrightarrow{k_2} E_f + P \end{gathered} \tag{9}$$

In these equations E_f refers to free enzyme concentration, and k_1 and k_{-1} to the rate constants for the association and dissociation of the complex. The complex breaks down into P and free enzyme at a rate governed by k_2. The rate or velocity (v) of the enzyme reaction can then be formulated as:

$$v = \frac{k_2 \times [E_t] \times [S]}{K_m + [S]} \tag{10}$$

where $K_m\left(= \frac{k_{-1} + k_2}{k_1}\right)$ is the Michaelis-Menten constant, and E_t is the total enzyme concentration (free plus substrate bound enzyme). The Michaelis constant, while not a true dissociation constant, is a measure of the affinity between enzyme and substrate; the smaller the value of K_m, the stronger the bond between enzyme and substrate. When S is much larger than K_m (over a hundredfold greater), equation 10 simplifies to

$$v = \frac{k_2 \times E_t \times S}{S} = k_2 \times E_t = V_{\max} \tag{11}$$

Thus when the substrate is present in great excess (all E is in the form of ES), the rate of the reaction depends only on enzyme concentration, and is not only independent of substrate concentration, but also has the greatest possible value under the given conditions. When the concentration of S is equal to K_m, the value of v becomes equal to $\frac{1}{2} \times V_{\max}$ (See Fig. 8-10).

The value of K_m can be found by determining the concentration of substrate at which one-half the maximum velocity is obtained. In practice, the Michaelis equation is rearranged in the form of $(1/v)$ as a linear function of $(1/S)$, or (S/v) as a function of S. Details can be found in any textbook on enzyme chemistry.

Zero order kinetics will be maintained if the concentrations of all substrates (and of all obligate activators) are present in large excess, i.e., at concentrations at least 50 and preferably 100 times that of the value of K_m. The K_m values for the majority of enzymes are of the order of 10^{-5} to 10^{-3} M, and therefore substrate concentrations are chosen to be in the range of 0.001 to 0.10 M. The actual concentration of substrate to be used for each enzyme system must be determined carefully by experiment. Henry and his associates have shown, for example, that the concentration of some substrates used in the original methods for the assay of serum transaminase was appreciably below the optimal level.[44] If the assay for GPT is run at the optimal substrate level, a considerably higher value for the enzyme level is obtained. There are occasions when optimal concentrations of substrate cannot be used; examples are when the substrate has limited solubility or when the concentration of a given substrate will inhibit the activity of another enzyme needed in a coupled reaction system. One must then compromise and use the non-ideal system; but to obtain reproducible results, it is necessary to adhere precisely to all details of the procedure.

In the course of the reaction, the substrate is consumed to form products of the reactions. If a reaction has proceeded for a considerable period of time, enough substrate may have been consumed (even though it was present initially in excess) so that the rate of reaction will no longer be independent of substrate concentration, and the rate will fall off continuously. If enough product has formed, the rate of the reverse reaction may become great enough to counteract the forward reaction, and the measured rate of the latter will decrease. The presence of a considerable amount of product may in itself inhibit enzyme activity. For these reasons, the rate of the enzyme reaction is maximal only during the first short period after the reaction has been initiated, and for precise enzyme assays, only the initial rate (at zero time) may be used as a measure of enzyme action. In actual practice accurate values are generally obtained if not more than 20 per cent of the substrate has reacted in the course of the enzyme assay. If more than this fraction of the substrate has been consumed, it is necessary to repeat the assay, using a smaller quantity of enzyme.

ENZYME ACTIVATORS AND INHIBITORS

If an active preparation of a transaminase is subjected to prolonged dialysis, the activity of the enzyme decreases until all activity is lost. If some of the dialyzate, suitably concentrated, is added to the enzyme in the dialysis sac, enzyme activity is restored almost completely. The activity of many enzymes is associated with the presence of some water-soluble, dialyzable chemical factor along with the protein moiety itself. The pure inactive protein is referred to as the *apoenzyme*, the dialyzable material as a *cofactor*, and the active material containing both as the *holoenzyme*. If the cofactor is an organic compound, it is termed a *coenzyme*, especially if the holoenzyme catalyzes group transfer reactions.

Activators

Although some enzymes (trypsin and fumarase) apparently function without any associated cofactor, others require the presence of certain metal ions. If the metal ion is weakly bound to the enzyme protein, and can be readily removed by dialysis, it is referred to as an *activator*. All phosphate transfer enzymes require the presence of Mg^{++} ions; without the magnesium, the enzymes are inactive. Other common activating cations are Mn^{++}, Fe^{++}, Ca^{++}, Zn^{++}, and K^{+}. Anions may also act as activators. Amylase will function at its maximal rate only if Cl^{-} (0.01 *M*) or other anions such as Br^{-} or NO_3^{-} are present; citrate and phosphate increase the activity of fumarase. Some enzymes require the obligate presence of two activating ions: both K^{+} and Mg^{++} are essential for the activity of pyruvic kinase. The velocity of the reaction depends on activator ion concentration in a fashion similar to its dependence on substrate concentration. A Michaelis-Menten constant can be determined from data relating enzyme activity with increasing metal ion concentration in the presence of excess substrate, and the value of K_A, the measure of the degree of affinity between enzyme and metal ion, can be calculated. It is important that activator-dependent enzyme reactions be run with both excess activator and excess substrate present. However, in some cases the addition of activator beyond a certain optimal concentration may result in a decrease in reaction rate (inhibition by excess activator); in such cases the optimal concentration of activator must be used.

The mechanism by which cations and anions activate enzymes varies from case to case. The ion may, for example, alter the spatial configuration of the protein necessary for proper binding of the substrate to the enzyme. This is the most likely explanation of the activity of monovalent ions like K^+ and Cl^-. Divalent ions such as Mg^{++} may link substrate or coenzyme to the enzyme protein. Other metal ions undergo oxidation and reduction, as in the case of iron in the cytochrome enzymes.

In a more general sense the term activation is applied to any process whereby an inactive enzyme is made catalytically active. The proteolytic enzymes are synthesized in the body in the form of inactive precursors, termed proenzymes or *zymogens*, which are then transformed by chemical agents into active enzymes. The precursor form of trypsin is called *trypsinogen*. When a small hexapeptide fragment is split from the end of the protein chain, by the action of enterokinase or active trypsin (see trypsin procedure, p. 461) the active proteolytic enzyme is formed. Similarly, H^+ ions in the stomach convert pepsinogen to the active gastric enzyme, pepsin.

Apparent activation of an enzyme will be observed whenever some material is added which can counteract the presence of some inhibiting agent.

Coenzymes

Coenzymes are relatively stable water-soluble organic compounds which are associated with the activity of certain types of enzymes. Unlike most ion activators, coenzymes actually participate in the reaction catalyzed by the enzyme; i.e., they are co-substrates in the reaction. As is true with any substrates, they must be present in excess concentration for the reaction to proceed at a maximum rate, and they are consumed, or their structure is altered, as the reaction proceeds.

The majority of the water-soluble vitamins are components of important coenzymes. Vitamin B_1 in the form of the pyrophosphate ester, thiamine pyrophosphate, is the coenzyme associated with decarboxylases, enzymes which remove carbon dioxide from oxocarboxylic acids. Niacinamide is a fragment of the molecule of the coenzyme associated with hydrogen transfer reactions, and pyridoxine, as pyridoxal- and pyridoxamine phosphate, participates in amino transfer reactions. Not all coenzymes are associated with vitamin activity; among those that are not are adenosine triphosphate (ATP) and related compounds which are involved in phosphate transfer reactions, and glucose-1,6-diphosphate, the coenzyme of hexosephosphate isomerase.

Dehydrogenases are a class of enzymes which catalyze hydrogen transfer reactions, such as the oxidation of an α-hydroxy acid to the corresponding oxoacid. A clinically important enzyme of this type is lactic dehydrogenase. Two related coenzymes are involved in these hydrogen transfer reactions: one is nicotinamide adenine dinucleotide (NAD), previously referred to as coenzyme I or as DPN (*d*iphospho*p*yridine *n*ucleotide), and the other is NADP (coenzyme-II, TPN) a phosphorylated derivative of NAD.

A *nucleotide* is an organic compound containing a heterocyclic nitrogen base, linked to a ribose unit, which is in turn joined to a phosphate residue at the C-5 position. In NAD one nucleotide contains adenine and the other contains niacinamide; both bases are linked to the sugar at the C-1 position of the ribose unit. The two nucleotides are joined through the phosphate residues to form a diphosphate (pyrophosphate) group. The formula for NAD is presented in Figure 8-11. The star (*) indicates the position of the third phosphate group in

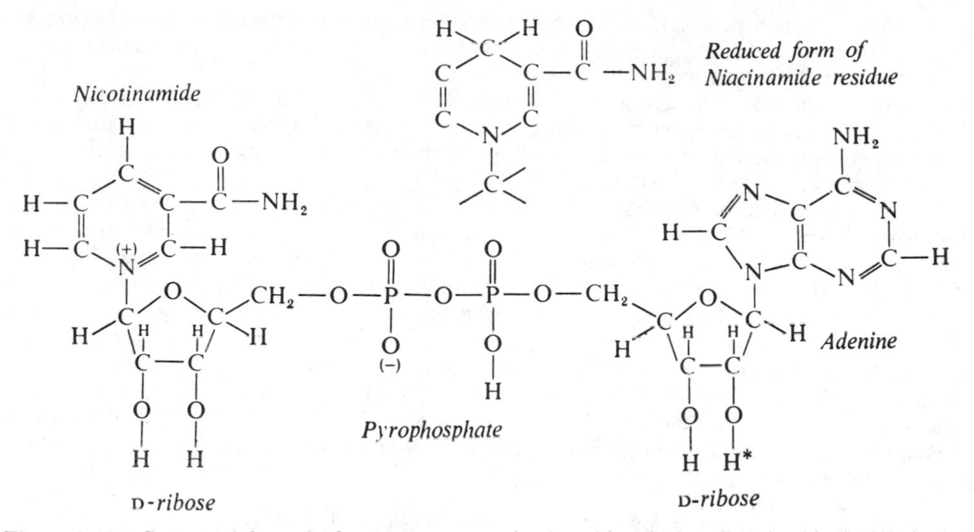

Figure 8-11. Structural formula for coenzyme I, nicotinamide adenine dinucleotide (NAD^+). In coenzyme II, $NADP^+$, a phosphoryl residue, $—P(=O)(OH)_2$, replaces the (H) at the position indicated by the star (*). The structure of the niacinamide portion in the reduced form of the coenzyme is also shown.

*n*icotinamide *a*denine *d*inucleotide *p*hosphate (NADP). Only the niacinamide (pyridine) moiety is involved in the hydrogen transfer reaction. When the ring is reduced, two electrons are transferred to the ring by the two hydrogens removed from the alcohol (α-hydroxy acid). One electron and one hydrogen go to the C-4 position of the pyridine ring; the other electron neutralizes the charge on the nitrogen and the hydrogen remains as a proton. Reduced coenzyme is formed, the customary symbol for which is NADH. The reaction

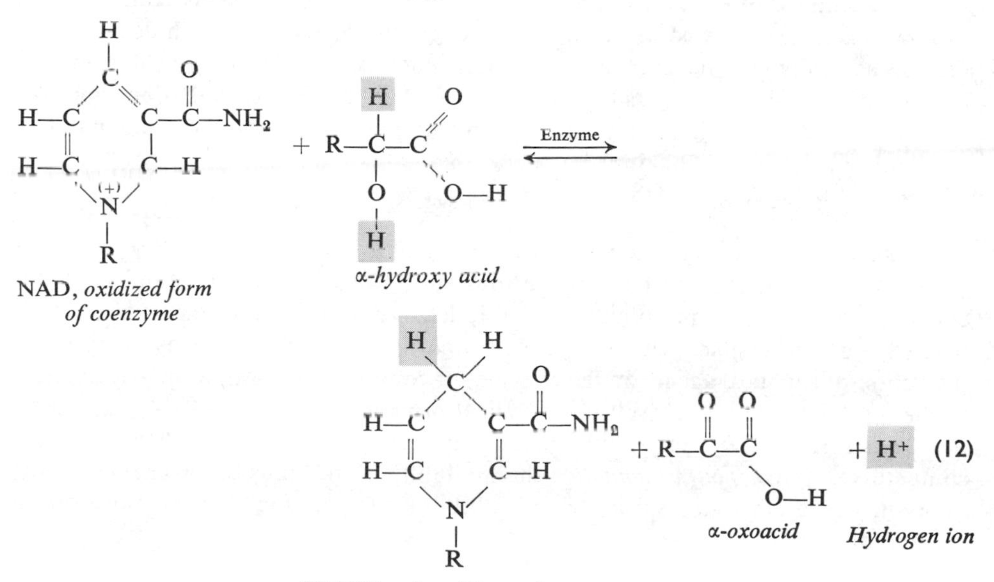

is reversible; NADH will reduce an oxoacid to an alcohol, and NAD will oxidize the alcohol to the oxoacid. The reaction proceeds to equilibrium, which is determined by the relative oxidation-reduction potentials of the NAD-NADH system and the oxoacid alcohol system. Glucose-6-phosphate dehydrogenase is an example of an enzyme requiring coenzyme II, NADP, as its hydrogen transfer coenzyme.

Other coenzymes function in an analogous way. It is to be emphasized that coenzymes are co-substrates in the transfer reactions. If a coenzyme can be regenerated continuously by another enzyme system, then a very small amount can serve to promote the reaction of a large quantity of specific substrate.

The first step in the metabolism of glucose is the phosphorylation of glucose to glucose-6-phosphate by hexokinase. Adenosine triphosphate (ATP) is the source of transferred phosphate, with adenosine diphosphate (ADP) being formed, one mole per mole of glucose. The reaction would cease the moment that the small quantity of ATP present was consumed, if it were not for the fact that further along in the series of reactions by which glucose is converted to pyruvate, 1-3 diphosphoglycerate and phosphopyruvate are formed. The phosphate transfer potential of these phosphorylated compounds is higher than that for the ATP-ADP reaction. They will transfer phosphate to ADP to regenerate ATP for re-use in the glucose-hexokinase reaction, thus permitting a very large quantity of glucose to be phosphorylated. This is illustrated in equation 13 which is also an example of the method devised by Baldwin[7] to portray cyclic reactions.

D-Gluc ⟶ D-Gluc-6-P (Mg^{++} | HK) with ADP-P ⟶ ADP; ADP ⟶ ADP-P (PK | K^+, Mg^{++}) with Pyr-P ⟶ Pyr (13)

D-Gluc =	D-glucose	ADP =	Adenosine diphosphate
D-Gluc-6-P =	Glucose-6-phosphate	ADP-P =	ATP = Adenosine triphosphate
HK =	Hexokinase	Pyr =	Pyruvate
PK =	Pyruvic kinase	Pyr-P =	Phospho(enol)pyruvate

Enzyme Denaturation and Inhibition

DENATURATION

Denaturation of the enzyme protein by the action of heat, mechanical agitation, or irradiation will be reflected in partial or complete loss of activity. Such denaturation may occasionally be reversed, but if the denaturing process is severe or prolonged, the loss of activity will be irreversible. Hence, it is emphasized again that enzyme preparations must be handled with care (minimal exposure to heat and light, minimal mechanical agitation or shaking).

ENZYME INHIBITORS

Whenever an enzyme reaction is proceeding at a rate less than that expected for the existing conditions of pH, temperature, substrate concentration, and activator concentration, the enzyme is said to be inhibited. Inhibition may be partial or total, reversible or irreversible. The term refers primarily to loss of activity as the result of the action of chemical agents on the enzyme, in contrast to denaturation, which refers to the loss of activity as a result of a physical process.

Two forms of enzyme inhibition may be encountered, competitive and non-competitive. In *competitive inhibition* the inhibiting agent binds to the enzyme at the same site as the substrate; substrate and inhibitor compete for the same position on the enzyme surface:

$$\begin{aligned} E + S &\underset{}{\overset{K_m}{\rightleftharpoons}} ES \longrightarrow E + P \\ E + I &\underset{}{\overset{K_I}{\rightleftharpoons}} EI\ (\textit{Inactive}) \end{aligned} \qquad (14)$$

The more of the enzyme present in the *EI* form, the less will be available to bind with substrate. If enough inhibitor is present it may saturate the enzyme so that no active enzyme-substrate complex can be formed, and no activity will be observed. If more substrate is added, it will displace the inhibitor from the enzyme and activity will be increased. The rate of activity observed will depend on the ratio of inhibitor concentration to substrate concentration, and on the relative degree of binding of each to the enzyme. For a fixed inhibitor concentration the velocity of the enzyme reaction increases with an increase in the concentration of substrate, but the maximum velocity obtained will be less than that observed in the absence of an inhibitor. Competitive inhibitors have chemical structures very similar or analogous to that of the substrate. Hexokinase, which acts on D-glucose, is inhibited by D-xylose and 6-deoxy-D-glucose. Enzymes requiring coenzymes can be inhibited by chemical analogs of the coenzymes. Thus, transaminases which require pyridoxine (as pyridoxal phosphate) will be competitively inhibited by deoxypyridoxine.

If an enzyme acts on two similar substrates, each will act as a competitive inhibitor of the other. The reduction of pyruvate to lactate by lactic dehydrogenase is inhibited by α-ketobutyrate, a substrate containing an additional $—CH_2$ group. Competitive inhibition by metal ions will be discussed in a separate paragraph.

In *noncompetitive inhibition* the inhibitor binds to the enzyme at a point other than that at which the substrate is bound. In some cases the enzyme remains active but the bound inhibitor has so modified its structure that its rate of activity is diminished.

An important example of noncompetitive enzyme inhibition is the inactivation of enolase by fluoride (F^-) ion. In one step in the metabolism of glucose, enolase catalyzes the conversion of 2-phosphoglycerate to phosphopyruvate. By inhibiting this reaction, F^- ion stops the glycolytic breakdown of glucose at this point, preventing formation of pyruvate. This is the biochemical rationale for the use of F^- as a preservative in the glucose determination.

A unique form of enzyme inhibition is manifested by *antienzymes*, as represented by a variety of trypsin inhibitors. These are proteins which bind to trypsin irreversibly, nullifying its proteolytic activity. One such inhibitor is present in the α_1-globulin fraction of serum proteins; others are found in soy beans and lima beans. Similar proteolytic inhibitors prevent the accumulation of excess thrombin and other coagulation enzymes, thus maintaining the coagulation process under control.

ENZYME INHIBITION BY METAL IONS

In many enzymes the presence of free cysteine sulfhydryl groups (—SH) is essential for activity. Oxidation of these groups to disulfide groups (—S—S—) may alter the spatial structure of the enzyme, resulting in loss of activity. Papain and bromelin, for example, are activated by treatment with cysteine, which reduces the disulfide links in the protein to the free —SH form. Very often the —SH groups on the enzyme will react with any heavy metal ions present as contaminants in the enzyme system, resulting in loss of enzyme activity. Such heavy metal–poisoned enzyme preparations can frequently (but not always) be reactivated by treatment with or dialysis against a solution of EDTA or BAL, or some inert chelating agent having a greater affinity to the metal than the enzyme. Competitive inhibition by metal ions can arise when two metal ions compete for the same binding side on the enzyme. Thus, Ca^{++} is an inhibitor for some enzymes which depend on Mg^{++} activation. Sodium and lithium are potent inhibitors of pyruvic kinase, for which potassium is an obligatory activator. Metal ions can be either competitive inhibitors (if they compete with an activator ion for the same site on the enzyme surface), or noncompetitive inhibitors. Heavy metal–poisoning of enzymes is usually of the latter type.

In order to avoid either type of inhibition, glassware used in the laboratory for the assay of enzymes must be washed and rinsed very thoroughly. Most detergents are enzyme denaturants; repeated rinsing is necessary to insure the removal of all detergent. Chromic acid is to be avoided in cleaning glassware, for it is difficult to remove the last traces of chromic ion from glass. Warm 10 per cent nitric acid soaks will efficiently remove heavy metal ions such as copper, mercury, lead, iron, and arsenic. Because reagents containing these metal ions are used daily in the clinical laboratory, it is prudent to segregate equipment used in enzyme work from that used elsewhere in the laboratory. This applies even to spectrophotometer cuvets. Obviously, only deionized water should be used both for rinsing glassware and preparing all reagents.

SUBSTRATE SPECIFICITY OF ENZYMES

Early in the study of enzymes it was appreciated that each enzyme catalyzes only one or a limited variety of chemical reactions. In fact, enzymes show a rather high degree of specificity for the substrates and types of reactions they catalyze, but the degree of specificity varies from one enzyme to another. Some enzymes show *absolute specificity*—they catalyze a single unique reaction, and no others. Pyruvic kinase, for example, mediates the transfer of a phosphate group between phosphopyruvate and ADP, and can function in no other reaction.

A somewhat lesser degree of substrate specificity is found in *hexokinase*. This enzyme transfers a phosphate group from ATP to D-glucose but it will also phosphorylate D-fructose, D-mannose, and D-2-deoxyglucose at almost equivalent rates. It will not act, however, on D-galactose, or on a variety of other hexoses or pentoses, although some of these are bound to the enzyme and can (competitively) inhibit enzyme action. Disaccharides, methylated sugars, and sugar alcohols are inactive.

The *phosphatases* are examples of enzymes with *group* specificity. These enzymes split phosphate from any of a large variety of organic phosphate esters, although at somewhat different rates. Materials as varied as glucose-6-phosphate, phenyl phosphate and β-glyceryl phosphate may serve as substrates.

The *esterases* and *proteinases* are groups of enzymes with even less specificity. The former hydrolyze esters to alcohols and carboxylic acids. A considerable range of chain length in both the alkyl (alcohol) and acyl (acid) portions of the esters is permitted (see lipase, p. 416). The proteinases attack peptide bonds, and will hydrolyze a variety of large or small proteins, and polypeptides (and even simple compounds containing only one peptide bond) to form free amino and carboxyl groups. (See trypsin, p. 461, uropepsin, p. 459.)

Stereo specificity is characteristic of many enzymes. The enzymes involved in glycolysis act only on the D-stereoisomers of glucose derivatives, never on the true L-forms. The transaminases convert oxoacids only to the L-isomers of the amino acids, and fumarase hydrates fumaric acid to the L-form of malate, rather than the D-glucose related mirror image (D-) form. Amylase hydrolyzes only starches, in which the residues are linked by α-1,4-linkages, and is inactive to cellulose, in which the sugar residues are connected by beta linkages.

UNITS FOR MEASURING ENZYME ACTIVITY

Enzyme concentrations are measured in terms of *activity* units present in a convenient volume or mass of specimen. The unit of activity is the measure of the rate at which the reaction proceeds, e.g., the quantity of substrate consumed or product formed in an arbitrary or convenient unit of time. The quantity of reacted substrate may be given in any convenient unit—milligrams, micromoles, change in absorbance, change in viscosity, or microliters of gas formed; time may be expressed in seconds, minutes, or hours. Since the rate of the reaction will depend on experimental parameters such as pH, type of buffer, temperature, nature of substrate, ionic strength, concentration of activators, and other variables, these must be specified in the definition of the unit.

In the course of many decades a multiplicity of units of enzyme activity have been introduced. Even for the same or similar enzymes, each investigator defined his unit in terms of quantities analytically or otherwise convenient for him at the time. A classic example is encountered in the types of units used in measuring phosphatase activity. Bodansky[13] defined his unit as the amount of enzyme that will split one milligram of phosphate-phosphorus from β-glycerophosphate at pH 8.6 in 60 minutes at 37°C. In their alkaline phosphatase method King and Armstrong[56] employed phenyl-phosphate as a substrate, and defined their unit as the quantity of enzyme that would liberate one milligram of phenol at pH 9.6 in 30 minutes at 37°C. Their unit for acid phosphatase (pH 4.9) was, however, defined in terms of an hour's reaction time. The Bessey-Lowry-Brock[11] unit of alkaline or acid phosphatase activity is expressed in terms of one millimole of substrate (*p*-nitrophenyl phosphate) hydrolyzed in 60 minutes. Thus, if a given phosphatase preparation hydrolyzes one millimole of each of the above mentioned substrates per minute, the following numerical values in terms of the individual units defined above would be obtained:

Bodansky	pH 8.6	31 mg. phosphorus/mmol. × 60 minutes	1860 units
King-Armstrong	pH 9.6	94 mg. phenol/mmol. × 30 minutes	2820 units
King-Armstrong	pH 4.9	94 mg. phenol/mmol. × 60 minutes	5460 units
Bessey-Lowry-Brock	pH 10.2	1 mmol. × 60 minutes	60 units

The concentration of enzymes is often expressed in terms of activity units per volume or mass of specimen, i.e., per mg. of specimen (wet liver tissue), per mg. of protein, or mg. of fat-free protein-nitrogen in the specimen. In clinical work, the concentration is generally reported in terms of some convenient unit of volume, such as activity per 100 ml. of serum or 1.0 ml. of packed erythrocytes.

Just as a multiplicity of units for reporting activity developed, a similar multiplicity of units of volume evolved. The Commission on Enzymes,[27,71] proposed that the *unit of enzyme activity* be defined as that quantity of enzyme that will catalyze the reaction of one micromole (μmol.) of substrate per minute, and that this unit be called the International Unit, U. Concentration is to be expressed in terms of U./ml. or mU./ml. (= U./L.), which ever gives the more convenient numerical value. In this chapter we use the symbol I.U., rather than U., to clearly differentiate the International Unit from other (conventional) units of activity.

In those instances in which there is some uncertainty as to the precise nature of the substrate (as with starch, proteins, complex lipids), the unit is to be expressed in terms of the chemical groups or residues measured in following the reaction (e.g., reducing sugars or amino acid units formed). The temperature recommendation of the Commission (30°C.) has already been discussed.

Although the proposals of the International Commission have been accepted by many scientists working with enzymes, it is not likely that laboratories using well-established enzyme procedures will cease to report enzyme values in terms of the older, empirical units to which they have become accustomed. But as new methods are devised, it is anticipated that the activity units will be established in accordance with the recommendations of the Commission.

In the presentation of specific enzyme procedures later in this chapter, the enzyme units used will be those introduced by the originators of the methods, but for each procedure the units and ranges of normal values will be converted into international units. Examples showing the conversion from empirical units to international units follow.

1. *Shinowara-Jones-Rinehart* (*SJR*)[78] unit for alkaline phosphatase activity: The unit of activity is defined as the quantity of enzyme that will split 1.0 mg. of phosphate phosphorus from β-glycerophosphate in 60 minutes at 37°C. and pH 9.6, under specified conditions of barbital buffer and substrate concentration. The enzyme level is customarily reported in terms of units per 100 ml. of serum. The 1.0 mg. of phosphorus is equal to 1/30.98 m-mols or 32.3 μmols of *P*. One mole of substrate will give off one mole of phosphorus. Thus 32.3 μmol. of substrate are reacted in 60 minutes, which calculates to 0.538 μmol. per minute. 1.0 SJR unit is then equivalent to 0.54 I.U., and 1.0 SJR unit per 100 ml. equates to 10 SJR units per liter, or 5.4 I.U./L. (= 5.4 mI.U./ml.).

2. *Sibley-Lehninger* (*SL*) unit[79] for aldolase activity: This is defined as the quantity of enzyme that will split 1.0 μl. (microliter) of fructose-1,6-diphosphate into the two trioses, dihydroxyacetone phosphate and 3-phosphoglyceraldehyde in 60 minutes under the conditions of the test. One μl. of substrate is then equal to 1/22.4 or 0.0446 μmol. and 1.0 SL unit equates to 0.0446/60 or 0.743×10^{-3} I.U. Finally, 1.0 SL unit/ml. is identical to 0.74 I.U./L. = 0.74 mI.U./ml.

3. *Karmen-LaDue* (spectrophotometric) unit[53] for transaminase activity: This unit is defined as a change (decrease) of 0.001 in the absorbance of NADH per minute, when the NADH is present in a total volume of 3.0 ml. and is measured at 340 nm. across a 1.0 cm. light path.

The molar absorbance* (mol/L.) of NADH is 6.22×10^3. A solution containing 1.0 μmol./ml. will have an absorbance of 6.22, and a drop of 0.001 in the absorbance will reflect a concentration change of $(1/6.22) \times 10^{-3}$ μmol./ml. In the total volume of 3.0 ml. this will represent a decrease in the quantity of NADH (substrate consumed) of three times this value, or 4.82×10^{-4} μmol. The Karmen unit is defined in terms of absorbance change per minute, so that 1.0 Karmen unit per ml. is equal to 0.48 mI.U./ml.

USE OF ENZYMES AS REAGENTS

The relatively recent availability of pure enzyme preparations at reasonable cost has made it practical to use enzymes as laboratory reagents for analytical work.

* There appears to be some inconsistency among authorities as to the definition of molar absorbance. Chemists define a molar property as that pertaining to a solution containing one mole of solute/liter, and with this meaning the molar absorbance of NADH is as given here. Bergmeyer and some others define molar absorbance as the absorbance of a solution containing one mole per ml. With this meaning the molar absorbance of NADH at 340 nm. is 6.22×10^6.

Their substrate specificity gives enzymes unique value in the determination of certain biological compounds. An important application is the use of *glucose oxidase* for the assay of blood glucose.[29,32] The common chemical methods, based on the reducing property of glucose, give more or less erroneously high results, because other sugars and other materials present in blood will also reduce the customary sugar reagents. The enzyme glucose oxidase, which catalyzes the oxidation of glucose by molecular oxygen, is not only inactive toward the usual interfering agents, but is also inactive toward all other reducing sugars, except 2-deoxy-glucose, which is not present in blood in measurable amounts. This then permits assay of "true" blood glucose, as discussed in the chapter on carbohydrates. Similarly, the use of galactose oxidase permits assay of galactose in the presence of glucose and other reducing sugars. (See Chapter 13 on liver function.)

The chemical assay of serum uric acid, based on the reduction of phosphotungstic acid, is also subject to interferences. The use of uricase, which is almost substrate-specific for uric acid (oxidation to allantoin), makes it possible to measure "true" uric acid in serum. (The enzyme also slowly attacks 6-thiouric acid, a metabolite of 6-mercaptopurine, which may be present in sera from patients receiving this anticancer drug.) The enzyme is not as useful for urine uric acid assays, because many urines contain materials which interfere with or inhibit the enzyme. Similar inhibiting effects are encountered in the use of other enzymes, and the possibility of such interference must be kept in mind when using enzymes with materials as complex as urine.

Other important serum constituents that can be rapidly and accurately analyzed by use of enzymes are lactic and pyruvic acids (lactic dehydrogenase), ethyl alcohol (alcohol dehydrogenase), glycerol (glycerol dehydrogenase), creatine (creatine kinase), and ammonia (glutamine synthetase).

In using enzymes for the assay of metabolites, the enzyme reaction system is so set up that all components are present in excess with the exception of the material to be determined. Sufficient enzyme is added to insure that the reaction will proceed to completion in a reasonably short period of time. If the equilibrium of the reaction is not favorable, other reagents (or another enzyme system) may be added to force the reaction in the desired direction. In measuring lactate, hydrazine or semicarbazide are added to "trap" the pyruvate formed, thus forcing the reaction to complete oxidation of the lactic acid (see Chapter 10).

Because of the cost of reagent enzymes, methods for determining compounds of clinical interest should be developed and used only in those instances, when convenient or accurate chemical methods are not available.

Blood and Serum Enzymes

Enzymes are the essential elements which enable the many biochemical reactions that constitute life to proceed in the cells of the body. Changes in enzyme concentrations in tissue cells should, therefore, reflect states of health and disease. Unfortunately, it has not been practical to assay enzymes in tissues on a routine basis. At best, changes in enzyme concentrations in the blood can be followed, with the assumption that these may parallel changes that have taken place in specific organs or tissues. Serum and all the cellular elements in blood have been studied for possible value in diagnostic medicine. Erythrocytes contain a large number of enzymes, but alterations in the levels of only a few, such as glucose-6-phosphate dehydrogenase and glutathione reductase, have as yet been correlated with the presence of specific disease states. The concentration of alkaline phosphatase in leukocytes has been of some value in the differentiation of certain types of leukemias. However, most of the effort has been expended in studying plasma and serum enzymes.

Hess[45] differentiates plasma enzymes into two classes: the *plasma specific* enzymes and the *non-plasma specific enzymes.* The first group includes those enzymes which have a very definite and specific function in plasma. Plasma is their normal site of action, and they are present in it at higher levels than in most tissue cells. Among these are the enzymes involved in blood coagulation, and such enzymes as ceruloplasmin, pseudocholinesterase, and lipoprotein lipase. These enzymes are synthesized in the liver and are constantly liberated into the plasma to maintain an optimal and effective steady state concentration. They are of clinical interest when present in plasma at levels that are below normal as a result of impaired liver function, or when entirely absent as a result of an inborn genetic defect. Hepatolenticular degeneration (Wilson's disease), for example, is associated with deficiency of ceruloplasmin, a copper-containing oxidase, and sensitivity to the anesthetic succemethonium results from deficiency or absence of pseudocholinesterase.

The second group, the *non-plasma specific enzymes*, have no known physiological function in plasma. They are present in plasma at concentrations much lower than their concentrations in certain tissues. In many cases plasma is deficient in activators or coenzymes necessary for the activity of these enzymes. This group can also be divided into two subclasses—the *enzymes of secretion* and the *enzymes associated with cellular metabolism.* Among the enzymes of secretion are such enzymes as amylase, lipase, and the phosphatases. Though secreted at high rates, they are rapidly disposed of through the usual excretory channels such as the urine, the bile, and the intestinal tract, and normal plasma levels are relatively low and constant. If, however, any of the usual pathways of excretion are blocked, or the rate of liberation into the extracellular fluid is suddenly accelerated, or the rate of production increased, a significant increase in plasma levels of these enzymes will occur.

The enzymes of cellular metabolism are located within the tissue cells, and are present there at very high concentrations. Some exist free in the cellular fluid, and others are contained in cellular structures such as the mitochondria and lysosomes. As long as the cell membrane is intact, such enzymes are contained within the cell walls, and the level of these enzymes in the extracellular fluid and the plasma is extremely low or absent. The cell membranes are impermeable as long as the cells are metabolizing normally. If cell activity is impaired or destroyed as a result of deficiencies in oxygen or glucose, or if the cell is damaged in some way (bacterial or viral infection), the membrane becomes permeable or ruptures. The cell contents, including their enzyme complement, are released into the extracellular fluid, and eventually reach the plasma. If a large volume of cells are so affected, the plasma level of the enzymes may increase very suddenly, to levels many times greater than normal.

If some given enzyme, such as alcohol dehydrogenase or sorbitol dehydrogenase, is present to any appreciable concentration in only one organ (the liver in the case of these two enzymes), an increase in the plasma level of the enzyme would pinpoint the source of the enzyme and thus identify the diseased or affected organ. Such enzymes are referred to as *tissue-* or *organ-specific enzymes.* Other enzymes of this type are the acetylcholinesterase of erythrocytes, the acid phosphatase present in the prostate gland, and the osteoblastic alkaline phosphatase of bone cells. The enzymes associated with glycolysis, the pentose cycle, and the citric acid cycle are probably present in all body cells. They cannot be thought of as being organ-specific, although concentrations in different tissues may vary considerably.

Aspartate transaminase, aldolase, lactic dehydrogenase, and transketolase are among the many such enzymes. An increase in plasma concentration of these enzymes can result from release of the enzyme from a variety of possible tissue sources. Although the level of any one enzyme may not be informative about its tissue of origin, a comparison of the levels of several different enzymes may be useful, since the several enzymes are present in different tissues in different ratios. If they are derived from a single cell type, their ratio in the plasma will approximate that in the tissue of origin. An examination of the serum isoenzyme pattern (see the next section) for the specific enzyme may also aid in pinpointing the source of the released enzyme.

It is understandable that it is the dream of the diagnostic clinician to be able to associate one specific enzyme with each tissue or disease state. A large number of enzymes and the variations in their plasma levels in health and disease have been studied in the search for such unique enzymes. Some have been found to be poor qualitative or quantitative indicators and have not been studied further. But about 12 enzymes have been established as being very useful clinical parameters; these will be discussed further in the methods section of this chapter.

Isoenzymes

Many enzymes which catalyze a specific chemical reaction are widely distributed throughout the cells of different species, and even throughout different kingdoms of living matter, animal, plant, and bacteria. Although they possess identical or very similar activity, these enzymes from different sources are *not identical proteins* since they exhibit readily demonstrable differences in physical, biochemical, and especially immunological properties. It is now generally believed that the structure of the "active center" is identical (or at least very similar) for all enzymes of a given specificity, regardless of origin, but that the amino acid sequences of many sections of the peptide chains may differ considerably.

Between 1952 and 1959, Pfleiderer, Wieland, Wieme, Markert,[55] and others established that even a given enzyme obtained from the same individual organism could exist in multiple forms, referred to as isoenzymes or heteroenzymes. If specimens of sera or extracts of different tissues are subjected to electrophoresis, using starch or agar blocks, specific enzymatic activity can be demonstrated at several areas or bands along the electrophoretogram. Examples of such electrophoretic patterns are shown in Figure 8-12. Wroblewski[105] proposed that different proteins present in the same individual, with similar or like enzymatic activity, be referred to as *isoenzymes*. Although the isoenzymes of lactic dehydrogenase have been most thoroughly investigated, other enzymes of clinical interest have also been shown to exist in multiple forms. Among these are alkaline phosphatase, amylase, creatine kinase, ceruloplasmin, glucose-6-phosphate dehydrogenase, and aspartate transaminase. It is conceivable that all enzymes occur in multiple, isoenzyme forms.

Isoenzymes[59] differ not only in their rates of electrophoretic migration but also in such properties as stability to heat denaturation, resistance to various chemical inhibiting agents, and affinity for substrates and coenzymes. Of considerable interest

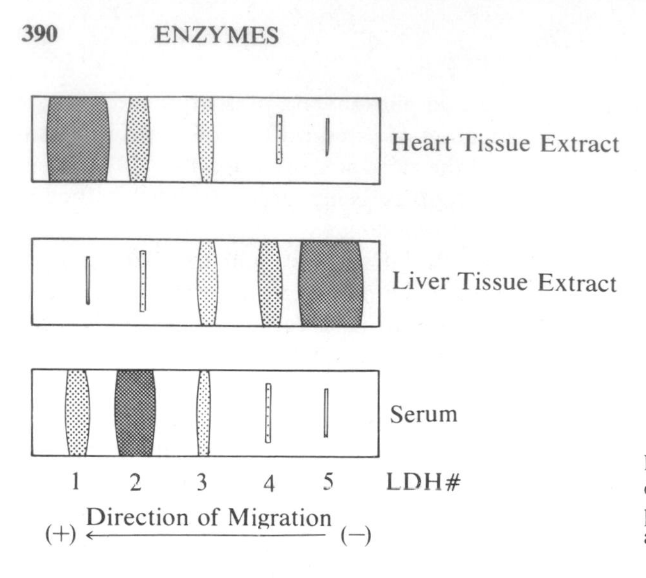

Figure 8-12. Examples of lactic dehydrogenase (LDH) isoenzyme patterns as separated on cellulose acetate.

is the fact that the proportions of the various isoenzymes present will vary from one tissue to another. Figure 8-12 shows that the LDH pattern obtained from extracts of human heart tissue is characteristically different from that obtained from human liver. The type of pattern observed serves to identify the tissue source of the enzyme. In general, the pattern obtained with normal serum is different from that associated with the various tissues. When a specific organ or tissue is injured, the intracellular enzyme released from the tissue cells diffuses into the plasma and imposes its pattern over that normally present. The type of serum bands obtained will then identify the tissues from which the increased level of the particular enzyme originated. In certain instances several organs may have the same or very similar isoenzyme profiles. If two or more organs are exuding their enzymes into the plasma, the pattern will be composite.

It is clear that isoenzyme patterns can be of significant value in differential diagnosis. Unfortunately, not all enzymes can be resolved into their individual isoenzymes as easily as lactic dehydrogenase. Because of its importance, the development of convenient isoenzyme fractionation techniques is a field of intense investigative activity.

It has now been established that lactic dehydrogenase is a protein composed of four subunits. If the band from heart tissue (which moves fastest during electrophoresis) is treated with urea to dissociate it into its monomeric units, it is found that the units are all identical. Similarly, those units obtained from the slowest moving band (from liver tissue) also are identical, but they differ from those obtained from heart tissues. The other isoenzyme bands dissociate into mixtures of heart H units and liver M units, the proportion of H units increasing with an increase in the electrophoretic mobility of the isoenzyme. The five isoenzymes of LDH thus differ in the number of H and M units in their active tetrameric forms. The fastest moving form, designated LDH-1, is composed of four H units, and can also be designated HHHH or H_4. The next fastest band is LDH-2, or HHHM or H_3M. The other forms are similarly indicated: LDH-3 (HHMM or H_2M_2), LDH-4 (HM_3), and LDH-5 (M_4), the slowest band. The designation M for the liver unit derives from the fact that it is also the predominant unit found in skeletal muscle. Formation of both the H and M forms is under genetic control, and genetic variants of the M form have been encountered. The proportion of the five LDH isoenzymes will not only vary from tissue to tissue, but it has been found to be different in fetal tissue and adult tissue.

Not all enzymes can be fractionated into five forms. Malic dehydrogenase, isocitric dehydrogenase, and creatine kinase appear to exist in only three electrophoretic forms. This may only mean that the exact experimental conditions necessary for separation of all forms have not yet been discovered. As many as seven forms have been reported for alkaline phosphatase.[18] If an enzyme exists as a dimer of two subunits, only three forms would be expected: AA, AB, and BB. If genetic variants of a tetrameric enzyme are present, more than five bands would be expected. The number of electrophoretic forms observed will vary not only with the technique used in separating them but also with the presence or absence of different ions.

Enzymes of Clinical Interest

In this chapter, in which the student is being introduced to the science of enzymology, it would be inappropriate to discuss each and every enzyme that has been of interest in clinical medicine. Table 8-2 contains a list of the most important enzymes that are used or have been studied for possible use in clinical investigations. For an enzyme procedure to become established in medical practice, it is necessary that one or more of the following conditions be met: (1) The assay method should be technically simple, adequately precise and reproducible, and not too time-consuming, so that many analyses can be performed in a reasonable length of time. (2) Data obtained should be diagnostically significant; i.e., the normal range must be reasonably narrow and well-defined, with minimal overlap between normal and abnormal values, and abnormal values should be consistently found in specific pathology. (3) The enzyme must be present in blood, urine, or some readily available tissue fluid. Tissue biopsy specimens may be more informative, but their routine procurement is not practical in the present state of medical art. (4) The assay procedure should not require complex instrumentation. However, if the test is sufficiently useful and in great demand, the acquisition of advanced and expensive instruments may be justified.

Of the enzymes listed in Table 8-2, only the first 15 to 18 have established themselves as valuable diagnostic tools. Some of the remainder are being used only to a limited extent, and are still being evaluated, perhaps to replace some of those in the first group. Others in the list will remain of interest only to research or special purpose laboratories, because assay procedures for them are difficult and their clinical value has yet to be established. In some instances the enzymes are associated with disease states which occur rarely or are not well understood. The list of enzymes of clinical interest will undoubtedly grow, because investigators are ever looking for enzymes that are specific for a single organ.

In the next section we will present methods for only a selected group of enzymes. The selection of enzymes chosen for presentation is guided by the following considerations: (1) methods will be presented for those enzymes of greatest proven clinical value, and thus most frequently requested; (2) it is intended to present methods based on varied experimental techniques, in order to acquaint the student with a diversity of approaches to enzyme assays; (3) the procedures will be discussed critically, so as to enable the reader to evaluate new procedures himself, as these procedures appear in the literature. For details of procedures for other enzymes, and,

TABLE 8-2. *List of Enzymes of Demonstrated or Assumed Usefulness in Clinical Diagnosis*

Enzyme	*Organ or Disease of Interest*	*Specimen**
Acid phosphatase	Prostate (Carcinoma)	S
Alkaline phosphatase	Liver, bone	SUW
Amylase	Pancreas	SUF
Lipase	Pancreas	SUF
Aspartate transaminase	Liver, Heart	SFC
Alanine transaminase	Liver, Heart	SFC
Lactic dehydrogenase	Liver, Heart, red blood cells	SUFC
Creatine kinase	Heart, muscle, brain	S
Isocitric dehydrogenase	Liver	S
Ceruloplasmin	Copper transport protein (Wilson's disease)	S
Aldolase	Muscle, heart	S
Trypsin	Pancreas, Intestine	FS
Glucose-6-phosphate dehydrogenase	Anemia (genetic defect)	R
Guanase	Liver	S
Sorbitol dehydrogenase	Liver	S
Ornithine carbamyl transferase	Liver	S
Pseudo-cholinesterase	Liver (insecticide poisoning)	S
Leucine aminopeptidase	Liver, pancreas	S
5′ Nucleotidase	Liver	S
Pepsin	Stomach	UF
Hexose isomerase	Liver	S
Pyruvic kinase	Liver	S
Hexose-1-phosphate uridyl transferase	Galactosemia	R
Glucose-6-phosphatase	Liver	S
Malic dehydrogenase	Liver	S
Glutathione reductase	Anemia, cyanosis	R
Arginase	Liver	S
β-Glucuronidase	Bladder tumors	U
Fructose-1-phosphate aldolase	Liver	S
Lipo-protein lipase	Liver	
Elastase	Collagen diseases	
Acetylcholinesterase	Insecticide poisoning	
Esterase	Liver	
Plasmin	Blood coagulation	

* Symbols: S—Serum, P—Plasma, U—Urine, F—Body fluids, C—Cerebrospinal Fluid, R—Erythrocytes, W—Leukocytes

indeed, for much useful information, the reader is referred to a number of excellent textbooks. King,[55] Varley,[96] Henry,[40] Henry and Bowers,[42] and Bergmeyer[10] give methods for a large number of enzymes, and references to others can be found in Wilkinson[101] and Abderhalden.[1] Material published and distributed by the instrument and reagent manufacturers is often informative and useful.[26]

The Phosphatases

I.U.B. No. 3.1.3.1 and No. 3.1.3.2

The clinically important phosphatases are a group of enzymes of low specificity which are characterized by their ability to hydrolyze a large variety of organic phosphate esters with the formation of an alcohol and phosphate ion. This group is composed of those enzymes which attack only monoesters of orthophosphoric acid.

$$R{-}O{-}\underset{\underset{O^{(-)}}{|}}{\overset{\overset{O}{\|}}{P}}{-}O^{(-)} + H{-}O{-}H \underset{\text{Phosphohydrolase}}{\xrightarrow{\text{Monoester}}} R{-}O{-}H + H{-}O{-}\underset{\underset{O^{(-)}}{|}}{\overset{\overset{O}{\|}}{P}}{-}O^{(-)} \quad (15)$$

The diesterases (I.U.B. No. 3.1.4.x, orthophosphoric diester hydrolases) constitute a separate group which splits compounds in which H_3PO_4 is esterified at two positions as shown in equation 15. Esters of pyrophosphoric and metaphosphoric acids are not attacked by phosphatases.

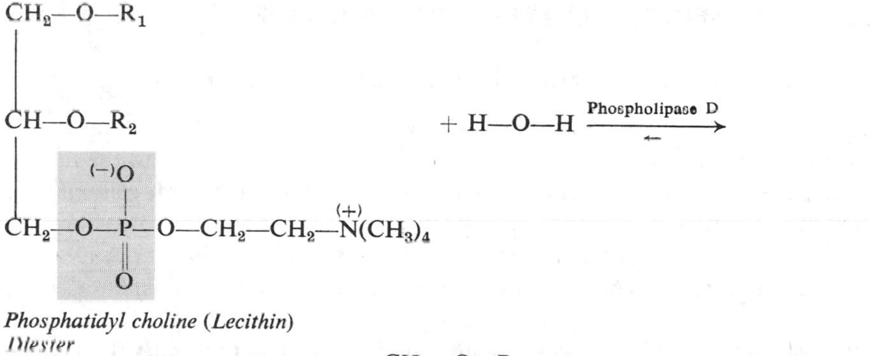

Phosphatidyl choline (Lecithin)
Diester

$$\begin{array}{l} CH_2{-}O{-}R_1 \\ | \\ CH{-}O{-}R_2 \\ | \quad\quad ^{(-)}O \\ | \quad\quad\quad | \\ CH_2{-}O{-}P{-}OH \\ \quad\quad\quad\quad \| \\ \quad\quad\quad\quad O \end{array} \quad + (CH_3)_4\overset{(+)}{N}{-}CH_2{-}CH_2{-}OH \quad (16)$$

Choline

Phosphatidic acid
Monoester

The alcohol esterified to the orthophosphoric acid, $(HO)_3P{=}O$, may be a simple aliphatic alcohol, a polyhydric alcohol such as sugar, or any one of a variety of aromatic hydroxy compounds. The phosphatases are not one enzyme but a group of related enzymes. Clinically, it has been practical to recognize three types: *alkaline phosphatase* (of serum, liver, bone, and intestines), with optimal activity at about pH 9.8; *acid phosphatase* (of serum, prostate and liver), with optimal activity at pH 4.9–5.0; and red cell phosphatase, with optimal activity at pH 5.5 to 6.0. The International Commission on Enzyme Nomenclature concluded, however, that only the first two were sufficiently well characterized to warrant their listing as unique enzyme species.

The probable function of the phosphatases is the transfer of the phosphate group from a donor substrate to an acceptor compound containing an (—OH) group. If the acceptor is water, (HOH), the net effect observed is hydrolysis. Schematically, the reactions are:

$$R_1\text{—O—P} + R_2\text{—O—H} \xrightarrow[\leftarrow]{\text{Phosphate transfer}} R_1\text{—O—H} + R_2\text{—O—P}$$

$$R_1\text{—O—P} + \text{H—O—H} \xrightarrow[\leftarrow]{\text{Hydrolysis}} R_1\text{—O—H} + \text{H—O—P} \qquad (17)$$

Donor ester phosphate *Acceptor alcohol* *Donor alcohol* *Acceptor phosphate*

If simple hydrolysis occurs, one mole of substrate, R_1—O—P, should produce one mole of each of the two products, alcohol and phosphate ion. If an acceptor alcohol molecule is present in the system, less than the stoichiometric quantity of inorganic phosphate is formed, the difference being accounted for by the phosphate transferred to the organic acceptor.

SERUM ALKALINE PHOSPHATASE

I.U.B. No. 3.1.3.1; Orthophosphoric acid monoester phosphohydrolase

This group of enzymes has optimal activity when the pH is in the neighborhood of 9.8, but the optimum observed varies with the substrate acted upon and the nature of the buffer present. At least three and as many as seven isoenzyme forms may be present in serum.[18] These isoenzymes have properties of those specific forms present in intestinal, liver, bone, placental, and renal tissue, but there is still uncertainty as to the source of the dominant form present in normal serum. Alkaline phosphatases act on a large variety of physiologic and nonphysiologic substrates, but the natural substrates on which they act in the body are not known. The fact that individuals born with an apparent genetic absence of the enzyme excrete large quantities of ethanolamine phosphate suggests that this (or perhaps phosphatidyl ethanolamine) may be one of the true physiologic substrates of the enzyme.

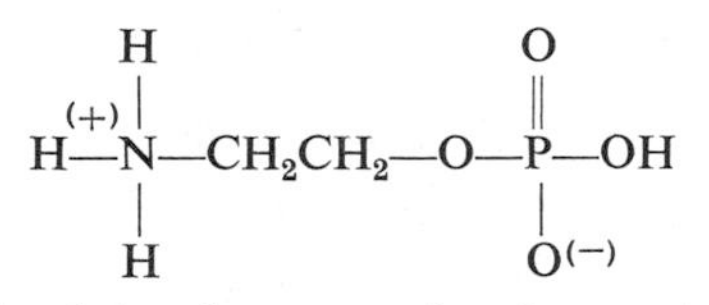

Phosphorylethanolamine or ethanolamine phosphate

Both Mg^{++} and Mn^{++} are activators for the enzyme, the optimal activity being obtained with approximately 0.001 *M* Mg^{++}. The phosphate ion is an inhibitor of the enzyme, as are Zn^{++}, Be^{++}, AsO_4^{-4}, oxalate, and CN^- ions. Variations in Mg^{++} and in substrate concentrations cause variations in the pH optimum. The type of buffer present (except at low concentrations) affects the rate of enzyme activity; buffers used include barbital, glycine, piperazine, 2-methyl-2-aminopropanol-1 (MAP), trishydroxymethyl aminomethane (TRIS), and diethanolamine.[39] Glycine depresses activity, perhaps by binding Mg^{++}, whereas the reagents containing alcoholic hydroxyl groups enhance activity, perhaps by acting as phosphate acceptors. Sulfhydryl compounds inhibit the enzyme.

The enzyme is present in practically all tissues of the body, and occurs at particularly high levels in the intestinal epithelium, kidney tubules, bone (osteoblasts), leukocytes, liver, and placenta. The serum enzyme is now believed to originate in the liver, although much evidence supports possible bone origin. The enzyme found in urine is derived from renal tissue and does not represent serum enzyme cleared by the kidney. Serum alkaline phosphatase is rapidly denatured at 57°C. but is relatively stable at lower temperatures (the placental forms are most stable). At room temperatures activity is maintained for 24 hours, and occasionally a 10 per cent increase in activity is observed. On refrigeration, activity falls very slowly, and in the frozen state full activity is retained for long periods of time.

Assay of Alkaline Phosphatase

Whenever the natural substrate of an enzyme is known, the assay of such an enzyme is usually carried out with this substrate, even though the enzyme may show a low level of specificity to this substrate. When the true substrate is unknown, the analyst is justified in using any substrate that is analytically convenient and gives a reasonably rapid rate of reaction. In 1930 Kay introduced the use of β-glycerophosphate as a substrate, and Bodansky used this as the basis for his classic procedure in 1932.[13] The reaction was followed by measuring the rate of liberation of inorganic phosphate ion as the enzyme hydrolyzed the glycerophosphate to glycerol. Bodansky used a weak barbital buffer which gave a resultant pH for the reaction mixture of approximately 8.8 at 37°C. He defined his unit of phosphatase activity as that quantity of enzyme which liberates 1.0 mg. of phosphate-phosphorus in 1.0 hour at 37°C. under the conditions of his procedure. The concentration was reported in terms of units per 100 ml. of serum. Since serum contains phosphate, the assay entails two measurements of phosphate-phosphorus, namely, before and after the one hour incubation period. Although this procedure had proven clinically valuable, it was soon appreciated that the pH of the assay was below the optimal value. Shinowara, Jones, and Reinhart[78] modified the procedure by using a more alkaline barbital buffer, which provides a pH of 9.8 for the enzyme reaction.

The concentration of buffer, however, is too weak to insure an identical pH with all sera. The pH of serum increases rapidly on standing, from about 7.4 to as much as 8.5, as a result of loss of carbon dioxide and liberation of ammonia from the hydrolysis of metabolites. The buffer value of serum is appreciable, and a buffer used in enzyme assays must be of a high enough concentration to give a fixed pH, regardless of the initial pH of the serum specimen to be assayed. On the other hand, too strong a buffer may inhibit the reaction.

In 1934 King and Armstrong[56] proposed the use of phenylphosphate as substrate. The rate of reaction is followed by measuring the phenol formed. Since there are

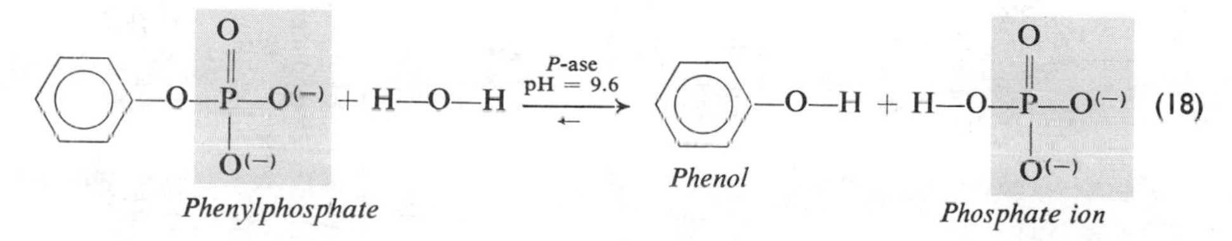

only traces of phenol-like compounds in serum, only one analysis is needed, rather than the two phosphate analyses required in the Bodansky method. The phenol which is formed can be assayed using the Folin-Ciocalteu reagent (King-Armstrong[56]) or 4-aminoantipyrine (Kind and King[54]), or by reaction with a diazo reagent. The method employing the Folin-Ciocalteu reagent requires a 30 minute incubation period followed by a deproteinization of the incubation mixture. Using antipyrine as a

TABLE 8-3. *Comparison of Methods for Assay of Serum Alkaline Phosphatase Activity**

	Bodansky	*Shinowara-Jones-Reinhart*	*Kind-King*	*Bessey-Lowry-Brock*	*Babson*	*Bowers-McComb*
Type	Manual	Manual	Manual	Manual	Manual	Automated
Buffer	Barbital	Barbital	Carbonate	Glycine	MAP	MAP
pH of reaction	8.5–8.7	9.4–9.8	9.5–9.7	9.8–10.1	10.1	10.1
Assay unit	1 mg. P produced in 60 minutes	1 mg. P produced in 60 minutes	1 mg. phenol produced in 15 minutes	1 mmol. PNP formed in 60 minutes	1 μmol. phenolphthalein produced in 1 minute	1 μmol. PNP split in 1 minute
Substrate	β-Glycerophosphate	β-Glycerophosphate	Phenylphosphate	p-Nitrophenyl phosphate	Phenolphthalein monophosphate	p-Nitrophenyl phosphate
Normal range	1.5–4.0 u/100 ml.	2.2–6.5 u/100 ml.	3.5–13 u/100 ml.	0.7–2.7 u/Liter	9–35 mIU/1 ml.	6–110 mIU/1 ml.
Normal range, converted to IU, μmol./min/ml.	8–22 mIU/ml.	15–35 mIU/ml.	25–92 mIU/ml.	13–38 mIU/ml.	9–35 mIU/ml.	6–110 mIU/ml.

Abbreviations: MAP = 2-methyl,2-aminopropanol-1
PNP = p-nitrophenol
u = unit
mIU = milli-International unit

* References:

King, J. Practical Clinical Enzymology. Princeton, D. Van Nostrand Co., 1965.
Wilkinson, J. H.: An Introduction to Diagnostic Enzymology. Baltimore, Williams and Wilkins Co., 1962.
Bowers, G. N., and McComb, R. B.: Clin. Chem., *12*:70, 193, 1966.

chromogenic reagent, deproteinization is not needed, and a 15 minute reaction period is sufficient.

In 1947 Bessey, Lowry, and Brock[11] proposed the use of a chromogenic, substituted phenylphosphate (p-nitrophenylphosphate) as the substrate. This ester is colorless, but the reaction product is colored at the pH of the reaction; thus, the enzyme reaction can be followed by observing the rate of formation of the yellow color of the p-nitrophenylate ion. No time-consuming deproteinization or further chemical reaction is required, and the reaction is linear with time.

Similarly, Huggins and Talalay[50] introduced the use of phenolphthalein diphosphate; the indicator phenolphthalein, released in the reaction, is measured by its red color. Unfortunately the substrate (a diphosphate ester) undergoes two separate hydrolytic reactions, which proceed at different rates. This results in complex kinetics and a nonlinear enzyme reaction. This material has been available as the commercial product Phosphatabs (Warner-Chilcott Laboratories).[57] Recently Babson and his associates[4] have prepared phenolphthalein monophosphate, and they and others[102] have demonstrated that this substrate gives linear reaction kinetics and is a useful and sensitive substrate for assaying phosphatase. The Shinowara, King-Armstrong, and Bessey procedures are those most widely used, and all have been adapted for automated analysis.

Phosphatase methodology is now undergoing a thorough reexamination[16,17,39] under the impetus of the need for automation, and as a result of efforts to reduce incubation times. The classic methods referred to above are "two point" methods, depending on the measurement of the reaction product at "zero" time, and after an enzyme incubation period of 30 to 60 minutes. It is difficult to maintain a linear reaction rate for such a long period (enzyme denaturation, depletion of excess substrate), and therefore short period "multiple point" assays with actual graphic presentation of the linearity of the reaction are preferred. The recent availability of automatic positioning spectrophotometers, such as the Gilford 2000 and analogous instruments by other manufacturers, has made such assays practical.

Concentration units for alkaline phosphatase. A multiplicity of units confounds the reporting of the concentration of this enzyme, more so than for any other enzyme of clinical interest. Each early investigator reported his results in the most convenient analytical units. Table 8-3 lists the units associated with the most commonly used procedures as well as the normal values for adults obtained with these methods. The diversity of the normal ranges is evident, even when expressed in international units. The Bodansky and Shinowara methods report enzyme activity in terms of milligrams of phosphate-*phosphorus* split from glycerophosphate in the course of an hour's reaction, whereas the King-Armstrong and the Kind-King units are based on the milligrams of *phenol* released from phenylphosphate in 30 and 15 minutes respectively.

Since one mole of phenol (C_6H_5OH; M.W. = 94) has thrice the formula weight of phosphorus (M.W. = 31), the phenol (Kind-King; KK) unit should be three times greater than the phosphorus unit. However, the Bodansky-Shinowara (B-S) unit is based on a reaction time that is four times as great. Thus, the KK unit should be $(94/31) \times (1/4) = 0.75$ the value of the BS unit, and a given enzyme quantity expressed in KK units should be 1.33 times greater than that expressed in BS units. The table shows that the normal range in KK units is about 2.5 to 3.0 times greater than in Shinowara units and 3 to 4 times greater than in Bodansky units. The former discrepancy is due to the greater rate of hydrolysis of phenylphosphate in carbonate buffer as against glycerolphosphate in barbital buffer. The difference between the

Bodansky and Shinowara normal values reflects the more favorable pH in the latter procedure.

The differences in the ranges of normal values by the six different methods, even when expressed in uniform units, reflect the effects that the differences in the nature and concentration of substrate, the choice and stability of pH, and the nature of the buffer have on the rate of the phosphatase reaction.

In the United States, the movement toward the use of the international unit has been very slow, and alkaline phosphatase values are still, to a large extent, expressed in Bodansky, King-Armstrong, or Bessey-Lowry units. The units used have been a source of inconvenience and confusion to physicians.

It is not acceptable to convert from one type of unitage to another by use of some average conversion factor, although this is a rather common practice. Thus it has been said that a Bessey-Lowry value may be converted to a Shinowara value by multiplying by 2.5. However, this factor, which is applicable to normal sera with an error of less than 8 per cent, may give grossly erroneous results when applied to abnormal sera with highly elevated values for the enzyme.

The Determination of Alkaline Phosphatase by the Procedure of Bessey, Lowry, and Brock

PRINCIPLE

Paranitrophenyl phosphate (PNPP) is colorless. The enzyme splits off the phosphate group to form free p-nitrophenol (PNP), which in the acid form in dilute solution is also colorless. Under alkaline conditions, this is converted to the nitrophenolate ion, which assumes a quinoid structure with a very intense yellow color. At the pH of the reaction most of the free nitrophenol is in the yellow-colored quinoid form. Thus, the course of the reaction can be followed by observing the increase in

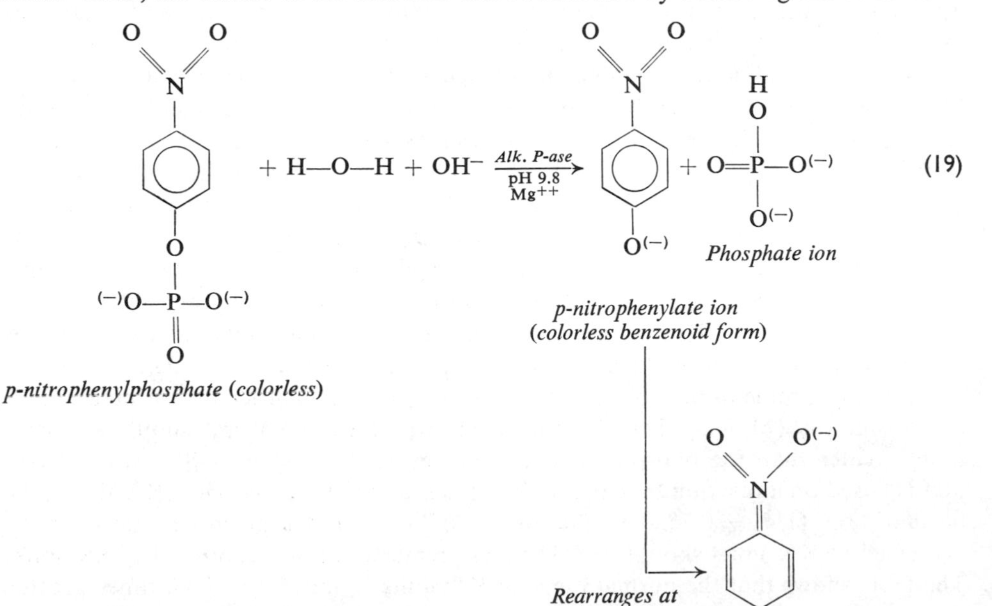

yellow color. The reaction is permitted to proceed for exactly 30 minutes, and is then stopped by adding sodium hydroxide, which inactivates the enzyme, and at the same time dilutes the nitrophenolate color, which is measured by its absorbance at 410 nm. The quantity of nitrophenol formed in the 30 minute reaction is calculated from a standard curve. The unit for enzyme activity is defined as the number of millimoles of p-nitrophenol formed in 60 minutes per liter of serum.

SPECIMENS FOR ENZYME ASSAY

Enzyme assays, unless otherwise specified, should be carried out on serum rather than on plasma samples. The common anticoagulants inhibit the activity of many enzymes, although the degree and type of inhibition encountered may vary with the specific enzyme under investigation and with the assay method used. Since the concentration of respective enzymes in red cells is frequently different from serum, most enzyme determinations require specimens free of hemolysis.

In the case of alkaline phosphatase the specimen required is serum, free of hemolysis.

REAGENTS

1. Glycine buffer, 0.10 *M*, containing 0.001 *M* Mg^{++}, pH of 10.3 to 10.4 at 37°C. Dissolve 7.50 gm. of highest purity glycine and 0.203 gm. of $MgCl_2 \cdot 6H_2O$ in about 750 ml. of water. To this solution add 85 ml. of 1.0 *N* NaOH. Check the pH with a pH meter, and adjust the pH to 10.5 to 10.7 at 25°C. (or 10.3 to 10.4 if measured at 37°C.) by adding more NaOH or 1.0 *N* HCl. (Glycine buffer has a large temperature coefficient.) Dilute the solution to 1 liter. A few drops of chloroform may be added as a preservative. If refrigerated, the reagent will be stable for 4 to 6 months.

2. Stock substrate: a freshly prepared solution of the disodium salt of paranitrophenyl phosphate, concentration = 4.0 mg./ml. = 15.2 μmol./ml. Prepare just enough to provide a two weeks' supply of the buffered substrate (reagent 3). The chemical is unstable, but high purity, stabilized preparations are available from a number of commercial sources in gram lots and in the form of 100 mg. capsules. The solid should be kept dry and stored in the dark at below 4°C.

3. Buffered substrate. Prepare the buffered substrate by mixing equal portions of glycine buffer and stock substrate. Check the pH as before, and adjust the pH if necessary. Pipet 1.0 ml. portions into a series of 15 × 125 mm. test tubes. Stopper the test tubes, store them in a freezer and use as needed. Alternately, prepare enough for the day's work and use immediately. Buffer concentration = 0.050 *M*; $Mg^{++} = 5 \times 10^{-4}$ *M*; p-nitrophenyl phosphate = 7.6 μmol./ml.

4. Sodium hydroxide, 0.02 *N*. Prepare by diluting any standardized stock NaOH solution.

5. Stock standard. Use 0.010 *M* p-nitrophenol in water. Dissolve 0.3479 gm. of pure crystals in 250 ml. of water. Specially purified crystals are available from several commercial sources. The standard is stable for six months if kept frozen and in the dark.

6. Hydrochloric acid, concentrated (12 *N*).

PROCEDURE

1. Remove a sufficient number of test tubes containing 1.0 ml. of frozen substrate from the freezer and place in a water bath at 37°C. Alternately pipet 1.0 ml. of

freshly prepared material into test tubes, and equilibrate these at 37°C. One tube is needed for each serum to be assayed, and two tubes are used as blanks.

2. At timed intervals, pipet exactly 0.10 ml. of the sera to be assayed into the tubes containing the buffered substrate; mix the contents of the tubes rapidly and thoroughly. If the sera have been refrigerated, first warm the specimens to between 25 and 30°C. Place 0.10 ml. of water in the blanks instead of serum.

3. Permit the tubes to incubate at 37°C. for exactly 30 minutes. At the end of the 30 minute interval, add 10.0 ml. of 0.02 *N* NaOH to each tube to stop the enzyme reaction. In the course of the incubation, examine the tubes periodically, and if one of the tests demonstrates a strong yellow color, add NaOH at some convenient time interval less than 30 minutes (10, 15, or 20 minutes) to stop the enzyme reaction. Mix the contents of the tubes and transfer to cuvets.

4. Read the two blanks against water as a reference blank at a wavelength of 410 nm. Both should give the same absorbance, a measure of free p-nitrophenol (PNP) present as the result of nonenzymatic hydrolysis of the substrate. Breakdown may have occurred in the tubes to a different degree. If the readings differ by more than 0.020 the quality of the substrate is suspect. Choose that blank which has the lowest absorbance as a reference for the unknowns.

5. Read the cuvets against the blank at 410 nm., and record the absorbances as R_1. Retain solutions in the cuvets.

6. Add one drop of HCl to all cuvets, and mix by tapping or inverting the tubes. The acid converts the quinoid form of the nitrophenol to the colorless acid form. Any remaining yellow color is derived from serum pigment.

7. Read the absorbances of the solutions again at the same wavelength and record these readings as R_2.

8. If the absorbance is higher than 0.600 (high activity present), repeat the run with a shorter incubation time, or with 0.10 ml. of a 1/5 or 1/10 diluted specimen. Recent experience has shown that water or saline should not be used as a diluent, since these will result in a change in the pH of the reaction mixture; rather, the unknown serum should be diluted with another serum of low phosphatase activity and normal inorganic phosphorus content.

9. Using the calibration curve, convert the R_1 and R_2 values to enzyme units, E_1 and E_2, respectively. The difference between the two, $(E_1 - E_2) = \Delta E$ gives the enzyme activity in Bessey-Lowry-Brock (BLB) units. Multiply this value by appropriate factors to correct for shortened reaction time or for dilution of specimen. The time factor is $(30/t)$, where (t) is the actual reaction time in minutes. The dilution factor is D, if the serum is diluted 1/D with water. If the diluent is another serum with an activity of (Z) units, and the dilution is 1/D, then the corrected activity is $[(D) \times (A) - (D - 1) \times Z]$, where (A) is the activity found with the diluted specimen.

CALIBRATION

Prepare a dilute standard solution of PNP by diluting the 0.010 *M* stock PNP 1:200 with water to make the PNP concentration $= 5.0 \times 10^{-2}$ mmol/L. Use the solution within 4 to 6 hours. Prepare working standards by diluting 1.0, 2.0, 4.0, 6.0, 8.0, and 10.0 ml. of the stock solution with water to a volume of 10.0 ml. The instrument blank consists of 10.0 ml. of water. Alkalinize all the tubes with exactly 1.1 ml. of 0.20 *N* NaOH, mix the solutions, and read their absorbances at 410 nm. The standards represent 1.0, 2.0, etc., BLB units of alkaline phosphatase activity per liter of sample.

This is demonstrated as follows: If the quantity of PNP split off by 0.10 ml. of enzyme in 30 minutes is Z mmol., then 1000 ml. of enzyme preparation will hydrolyze

$$Z \times \left(\frac{1000}{0.1}\right) \times \left(\frac{60}{30}\right) = 20{,}000 \times Z \text{ mmol}$$

in 1.0 hour's time. The working standard containing 1.0 ml. of dilute stock standard contains $0.001 \times 5.0 \times 10^{-2}$ mmol. of PNP, or 5.0×10^{-5} mmol. in the 11.1 ml. solution of the standard. This value multiplied by 20,000 gives $(2.0 \times 10^{4}) \times (5.0 \times 10^{-5})$ or 1.0 mmol. Thus the 1.0 ml. standard represents 1.0 mmol. PNP hydrolyzed from PNPP by 1 liter of sample in one hour. The calibration curve deviates slightly from linearity, except with narrow band pass instruments.

COMMENTS

1. The quantity of substrate in the enzyme reaction system is 7.6 μmol. $\times$ $(\frac{1.0}{1.1}) = 6.90$ μmol., with buffer concentration at 0.045 M glycine and Mg^{++} at 0.00045 M. For precise enzyme assays, less than 10 to 15 per cent of the substrate should be permitted to react if the rate is to remain constant with time. Thus, no more than 0.69 to 1.04 μmol. of substrate should be hydrolyzed. The highest standard contains 0.50 μmol. of PNPP, and unknowns giving absorbances as great as the highest standard (10 BLB units/L.) should still be following zero order kinetics. However, some of the PNPP may have decomposed, and it is best not to accept readings over 8.0 units.

2. The other product of the reaction, the PO_4^{-3} ion, inhibits the reaction; the more PO_4^{-3} present, the greater the degree of inhibition. This is another reason for repeating the analysis on specimens with high activities.

3. By reading the unknown specimens against the blank, the presence of non-enzymatically split PNP is corrected for. The difference in readings, before and after the addition of one drop of HCl, corrects for serum pigments. Bilirubin has an appreciable absorbance at 410 nm., and when present at levels above 8 to 10 mg./100 ml., it is better corrected for by diluting the appropriate volume of serum with 11.0 ml. of 0.02 N NaOH, and using this for reading R_2. Small amounts of hemolysis are adequately corrected for, but corrections are uncertain for severely hemolyzed samples, and also for chylous sera.

4. Conversion of BLB units to international units. One mmol. PNP released in 60 minutes equals 1000/60 or 16.6 μmol/min. Thus 1 BLB unit = 16.6 I.U./L. or 16.6 mI.U./ml.

NORMAL VALUES

The range encountered in healthy adults is 0.7 to 2.8 units/ml. The range for children is 2.2 to 6.7 units through the first year, followed by a fall to 1.5 to 4.5 units. During puberty, values again increase to 1.7 to 6.5 units, dropping to adult levels at the age of 16 to 18 years.

REFERENCES

Bessey, O. A., Lowry, O. H., and Brock, M. J.: J. Biol. Chem., *164*:321, 1947.

Tech. Bull. No. 104, Sigma Chemical Co., St. Louis, Mo., 1960.

Berger, L., and Rudolph, G. N.: Alkaline and acid phosphatase. In Standard Methods of Clinical Chemistry. S. Meites, Ed. New York, Academic Press, 1963, Vol. 5.

Determination of Serum Alkaline Phosphatase by the Procedure of Shinowara, Jones, and Reinhart, Modified

REAGENTS

1. Stock buffered substrate. Dissolve 10.0 gm. of sodium β-glycerophosphate and 18.0 gm. of sodium diethyl barbiturate in about 800 ml. of water and transfer to a 1 liter volumetric flask. Add 10 ml. of petroleum ether as a preservative, bring the volume (aqueous level) up to the 1000 ml. mark, and mix the solution. The solution is stable for 2 to 3 months if refrigerated. Double strength buffer is used as recommended by Henry.

2. Buffered working substrate, pH 10.8 $\pm$ 0.1. To a 200 ml. volumetric flask add 100 ml. of stock substrate, followed by 5.8 ml. of 0.10 N NaOH (added slowly while mixing) and 5 ml. of petroleum ether. After diluting to volume (aqueous level) and mixing, the pH of the solution is checked with a pH meter and if necessary, adjusted with 0.10 N NaOH or 0.10 N HCl. This solution is stable for four weeks if kept refrigerated. The concentration of β-glycerophosphate is 0.023 M, and that of the barbital buffer 0.044 M.

3. Trichloroacetic acid, 50 per cent TCA (w/v) in water. If refrigerated, the acid is stable for about three months.

4. Gomori inorganic phosphorus reagents.

   a. Acid molybdate, 2.5 per cent (w/v) $Na_2MoO_4 \cdot 2\,H_2O$ in 2.5 N H_2SO_4.

   b. Reducing agent, 1.0 per cent (w/v) of Metol in 3 per cent (w/v) solution of $NaHSO_3$. Metol, also known as Elon, is p-methylaminophenol hydrogen sulfate, a photographic developer.

   c. Dilute stock phosphorus standard. (1.00 ml. contains 100 μg. phosphorus in 8 per cent (w/v) TCA solution.) This can be prepared from any available stock phosphorus standard solution.

   d. Working phosphorus standard (1.0 ml. = 8 μg. phosphorus). This is prepared by a 2:25 dilution of the dilute stock in water. It is unstable, and is made fresh daily.

   e. Eight per cent (w/v) TCA solution.

PRINCIPLE AND OUTLINE OF PROCEDURE

Mix 9.0 ml. of the buffered substrate (pH 10.8) with 1.0 ml. of serum; the resulting pH will be 9.6 to 9.8. The activator, Mg^{++}, is provided by the serum. Set up two tubes for each unknown: into the first (test = T_1) put both serum and substrate, and in the second (serum control = T_2), only substrate. The phosphatase reaction proceeds in the first tube while the tubes are incubated at 37°C. for exactly 60 minutes.

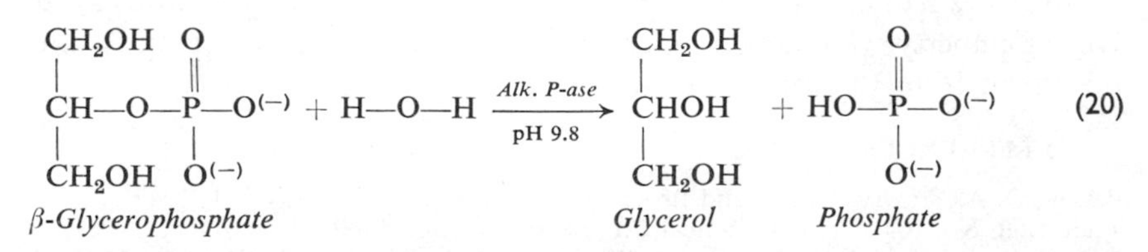

Add 2.0 ml. of 50 per cent TCA (which stops the reaction in tube 1) to both tubes. Then add 1.0 ml. of the serum to the second tube. Mix the tubes and collect the protein-free

filtrates after centrifugation. Also run a third tube, the substrate control (T_3) containing 9.0 ml. of buffered substrate and 1.0 ml. of H_2O (and after 60 minutes incubation, 2.0 ml. of TCA), with each set of unknowns. Analyze all filtrates for inorganic phosphorus.

Phosphorus analysis. Prepare "standard" cuvets by adding 2.0 ml. of 8 per cent TCA, to 0.0, 1.0, 2.0, etc. up to 6.0 ml., of the working phosphorus standard, respectively, and add water to a volume of 8.5 ml. Pipet 2 ml. of the filtrate from T_3 and 2 ml. of the filtrate from each pair of unknowns (T_1 and T_2) into cuvets, and add 6.5 ml. of water. Treat all cuvets with 1.0 ml. of molybdate and 0.5 ml. of Metol, mixing after the addition of each reagent. After a 30 minute color development, read the absorbances at 680 nm.

The standards represent 4.8, 9.6, etc., to 24.0 mg. phosphorus per 100 ml. The phosphorus values for each of the T_1, T_2 pairs and for the T_3 tube are calculated as follows:

If we denote enzymatically split-P by E-P, nonenzymatically split-P by NE-P, serum inorganic-P by S-P, and free inorganic-P in the substrate by F-P, the three tubes measure the following components:

$$T_1 = \text{E-P} + \text{NE-P} + \text{S-P} + \text{F-P}$$
$$T_2 = \text{NE-P} + \text{S-P} + \text{F-P}$$
$$T_3 = \text{NE-P} + \text{F-P}$$

The difference $T_1 - T_2$ is the measure of the enzymatic activity; $T_2 - T_3$ is a measure of serum phosphorus. β-glycerophosphate hydrolyzes slowly on standing, and solutions, if old, may contain some free phosphate. Fresh buffered substrates, prepared carefully with fine quality reagents, contain only traces of phosphorus. Whenever a T_3 absorbance of over 0.02 is obtained, the substrate is discarded.

In place of the Gomori procedure, any suitable phosphorus method may be used, with appropriate changes in the calculations, if needed. The occasional opalescent or milky TCA filtrates can usually be cleared by extraction with 1 ml. of cholorform.

For sera with activities higher than 20 units, the assay should be repeated with 1.0 ml. of a diluted specimen of the serum.

NORMAL VALUES

The normal value for adults is 2.5 to 6.5 Bodansky (Shinowara) units per 100 ml. The range for children is about 4 to 14 units; the variation with age is similar to that given for the Bessey-Lowry-Brock procedure.

International unit values are obtained by multiplying the Bodansky (Shinowara) value by 5.4.

REFERENCES

Shinowara, G. Y., Jones, L. M., and Reinhart, H. L.: J. Biol. Chem., *142*:921, 1942.
Henry, R. J.: Clinical Chemistry, Principles and Technics. New York, Hoeber Med. Div., Harper and Row, 1964.

PRECISION AND ACCURACY OF THE PHOSPHATASE PROCEDURES

Within the knowledge of the author, no careful studies of precision of the methods have been made. In the author's experience with the Bessey-Lowry-Brock method, the precision for specimens in the normal range is 10 per cent. It is somewhat better in the range of 4 to 10 units, but with sera with very high levels, requiring dilution or short incubation times, the precision is somewhat less. In the case of the Shinowara-Jones-Reinhart method, the variability encountered is of the order of 5 to 8 per cent in samples in the upper normal and moderately elevated range; there is greater variability in the low and very high ranges of activity. So many factors affect the accuracy of enzyme assays that no objective measure of accuracy is available. Some confidence in the accuracy of results follows from running reference control sera daily, using both a control in the normal range and one in the moderately elevated range. Either commercial controls or pooled laboratory sera may be used. The actual results need not check with those assigned by the reference serum manufacturer if assay methods or conditions are different, but daily assays in one laboratory, for any given lot, should replicate themselves within ±10 per cent.

CLINICAL SIGNIFICANCE

Since alkaline phosphatase is excreted through the biliary system, a moderate to high elevation (10 to 40 BLB units, 25 to 100 SJR units, or 40 to 150 KA units) in the serum enzyme level is encountered in *obstructive jaundice* associated with blockage of the bile ducts as a result of stones or tumors in the ducts or adjacent tissues. The rise in enzyme values usually parallels the degree of jaundice (bilirubin). After surgical removal of the cause of the obstruction, the enzyme levels slowly drop to normal. In *parenchymal liver disease*, only modest or moderate increases are observed (3 to 10 BLB units; 10 to 25 SJR units, or 15 to 40 KA units). Variation from patient to patient is considerable, depending on the degree of liver involvement. Higher values are observed in toxic hepatitis, and especially in chloropromazine hepatitis (15 to 50 BLB units or 35 to 120 SJR units). Elevated values are also demonstrated whenever the liver is infiltrated with foreign tissue (amyloid, sarcoid, carcinoma, tuberculosis). The liver is often the first organ to be infected by metastasizing cancer. Thus, an elevation in alkaline phosphatase, in the absence of primary liver disease, is suggestive of such spreading cancer. Alkaline phosphatase determinations are relied on routinely to aid in differentiating obstructive and parenchymal jaundice, but they are most informative when coupled with bilirubin level determinations and other tests of liver function.

The highest levels of serum alkaline phosphatase are encountered in *Paget's disease* (osteitis deformans), as the result of the action of the osteoblastic cells as they try to rebuild bone that is being dissolved by the uncontrolled activity of the osteoclasts. Values as high as 60 to 100 BLB units or 150 to 250 SJR units are not unusual. Only moderate rises are observed in *osteomalacia*, the levels dropping slowly in response to vitamin D therapy. Levels are normal in osteoporosis. In rickets, levels that are two to four times the normal are obtained and these fall slowly to normal on treatment with vitamin D. Moderate elevations are seen in primary *hyperparathyroidism* (3.5 to 8.5 BLB or 9 to 25 SJR units), and often, but less regularly in secondary hyperparathyroidism and in the *Fanconi syndrome*. Very high enzyme levels in serum are present in patients with carcinoma of the bone. Elevations are often but not always found with healing bone fractures. A rise to 2 to 3 times normal is observed in women in the third trimester of *pregnancy*; this additional enzyme is presumed to be of placental origin.

It is occasionally of clinical interest to establish whether the observed increased plasma enzyme level is of liver or bone origin. Isoenzyme separations have not been too useful, because with starch gel electrophoresis as many as six isoenzyme bands can be isolated.[18] In nonpregnancy sera only two or three bands can be located, and all fall in the alpha-beta globulin region. Bone, liver, and perhaps intestinal enzymes can be demonstrated, but there is considerable overlapping of bands.

There has been interest in demonstrating the predominant presence of one or another isoenzyme form by chemical and physical means.[30] Intestinal enzyme is specifically inhibited by L-phenylalanine, whereas the liver enzyme is denatured more rapidly than the bone enzyme in 20 per cent ethanol.

The various forms also differ in the rate of denaturation on being heated for 10 minutes at 56°C. Values of more than 35 per cent of original phosphatase activity suggest hepatic disease, whereas values of less than 25 per cent suggest skeletal disease.

Low levels of the enzyme are only rarely encountered. The enzyme concentration is low in persons suffering with kwashiorkor (generalized nutritional protein deficiency), and in cases of dwarfism. In congenital hypophosphatasemia, enzyme levels are below 0.5 BLB or 2 Bodansky (SJR) units, and are associated with high serum calcium and normal phosphorus levels. Both the serum and the urine of these infants contain high concentrations of phosphatidylethanolamine.

There has been some clinical interest in *urinary alkaline phosphatase* determinations, but values appear to be insufficiently discriminatory and still not well established. A tenfold increase has been found in patients with invasive cancer of the bladder or the kidney, and occasionally a rise is seen in patients with renal disease or renal infarcts. Urinary assays are complicated by the relatively high phosphate levels in urine which inhibit enzyme activity. Thus, urine must first be dialyzed to remove phosphates and perhaps other inhibitors.

ACID PHOSPHATASE

I.U.B. No. 3.1.3.2; Orthophosphoric monoester phosphohydrolase

Under the name of acid phosphatase are included all phosphatases with optimal activity at a pH below 7.0. Thus, the name refers to a group of similar or related enzymes rather than to one particular enzyme species. The greatest concentrations are present in the liver, spleen, milk, erythrocytes, platelets, and the prostate gland. The latter is the richest source, and it contributes about a third to a half of the enzyme present in serum of normal males. The source of the remainder of the serum acid phosphatase in healthy males and in women is not known, but there is some evidence that it derives from disintegrated platelets, red cells, and the liver. The optimal pH for acid phosphatases varies, depending on the tissue from which they are obtained. The prostatic enzyme, which is of the greatest clinical interest, has a well-defined optimal pH at 4.8 to 5.1. The observed pH optimum also varies with the substrate on which the enzyme acts; the more acidic the substrate, the lower the pH at which maximum activity is obtained. The enzymes can hydrolyze a variety of phosphate esters, and indeed, every substrate utilized in measuring serum alkaline phosphatase has also been used in evaluating acid phosphatase activity.

The acid phosphatases are unstable, especially at temperatures above 37°C. and at pH levels above 7.0. Some of the serum enzyme forms (especially the prostatic enzyme) are particularly labile, and over 50 per cent of the acid phosphatase activity may be lost in an hour's time at room temperature. Acidification of the serum specimen to a pH below 6.5 aids in stabilizing the enzyme. This may be conveniently accomplished by adding 10 mg. of disodium hydrogen citrate for each ml. of serum. Small 10 mg. pellets or tablets are available from several commercial sources. Alternately, 2 or 3 drops of 30 per cent (v/v) acetic acid per 5 ml. of serum will also lower the pH to the level at which the enzyme is stable. Under these conditions activity will be maintained at room temperature for several hours, and for up to a week if the serum is refrigerated. Both fluoride and oxalate ions inhibit enzyme activity and should not be used as anticoagulants.

Because of the clinical importance of elevated serum acid phosphatase levels in the diagnosis of prostatic cancer, it is desirable to be able to differentiate between an increase in the concentration of the specific prostatic enzyme and an increase in the activity of the nonspecific forms. Abul-Fadl and King[2] and others have demonstrated that addition of certain chemical agents can inhibit the activity of one or the other types of acid phosphatase. The prostatic enzyme is strongly inhibited by tartrate ions, whereas the red cell enzyme is inhibited by formaldehyde or the presence of cupric ions. The most common procedure for differentiating between prostatic and nonspecific activity is to perform the assay in the presence and absence of tartrate ion. Attempts have been made to discover differential substrates i.e., substrates which are hydrolyzed rapidly by the prostatic enzyme, but at a significantly slower rate by the other forms of the enzyme. Babson[5] has presented evidence that β-glycerophosphate and α-naphthyl phosphate are more sensitive to the action of the prostatic enzyme than are such substrates as phenyl phosphate and p-nitrophenyl phosphate. The former substrates are also less sensitive to the action of the red cell enzymes which may be present in serum.

Assay Methods

In 1938 Gutman and Gutman used phenyl phosphate as the substrate for serum acid phosphatase, and first demonstrated the extreme value of this determination in the diagnosis and therapy of prostatic cancer. Their procedure was a modification of the King-Armstrong method for alkaline phosphatase determination in which the acid phosphatase activity was measured at a pH of 4.9. Shinowara adapted his alkaline phosphatase procedure for acid phosphatase. Similarly, methods measuring the hydrolysis of p-nitrophenyl phosphate at pH 4.9 were devised by Andersch and Szczypinski and others.[55] Fishman and Lerner recommended the use of phenyl phosphate but converted the liberated phenol into a diazo dye for measurement. Babson urged the use of α-naphthyl phosphate as a prostatic sensitive substrate and also measured the hydrolyzed α-naphthol as a diazo color. Several of these approaches have been adapted to automated analysis. The method to be presented uses PNPP as substrate and is essentially that described in the Sigma Chemical Company Bulletin No. 104.[90]

Determination of Total and Prostatic Acid Phosphatase using p-Nitrophenyl Phosphate (PNPP)

PRINCIPLE

The enzyme is permitted to act on PNPP in citrate buffer at pH 4.9, in both the presence and the absence of 0.04 *M* tartrate ion, using a 30 minute reaction time. The liberated PNP is measured spectrophotometrically after conversion to the quinoid form by NaOH. The assay tube without tartrate gives a measure of all forms of the enzyme (platelet, red cell, prostatic, etc.). The tartrate inhibits the activity of the prostatic form, and therefore the tube containing tartrate measures the activity of the nonprostatic forms. The level of the prostatic enzyme is obtained by difference. The calibration curve prepared for the alkaline phosphatase procedure is used also for this enzyme, with a change in the enzyme unit ordinate as a result of differences in assay conditions. Results are reported in "total" and "prostatic" BLB units per liter, i.e., mmol./hr./L.

REAGENTS

1. Citrate buffer, 0.09 *M*, pH 4.85 ± 0.05. To a solution of 18.91 gm. of citric acid monohydrate in 500 ml. of water, add 180 ml. of 1.0 *N* NaOH and 100 ml. of 0.01 *N* HCl. Check the pH, adjust to 4.85, and dilute the solution to 1 liter. Add a few drops of chloroform as a preservative. The reagent will be stable for about six months if refrigerated.

2. Tartrate-citrate buffer, 0.09 *M* citrate, 0.04 *M* tartrate, pH 4.85. Dissolve 1.50 gm. of L(+) tartaric acid in 250 ml. of citrate buffer. The reagent will be stable for 6 months, if refrigerated.

3. Stock substrate: p-nitrophenyl phosphate, disodium salt (— 4.0 mg./1.0 ml. or 15.2 μmol./ml.). Refer to the same item under Alkaline phosphatase, page 399.

4. Buffered substrates. Mix one part of each of the buffers described previously with one part of stock substrate, and adjust the pH of each to 4.9. Pipet 1.0 ml. portions of each into 15 × 125 mm. test tubes, stopper the test tubes and store frozen. Identify the citrate-substrate tubes and the tartrate-citrate substrate tubes appropriately. These substrates are less stable than those at an alkaline pH; thus, the volumes of the buffered substrates prepared should not exceed one week's needs.

5. Sodium hydroxide, 0.10 *N* prepared by dilution from any available stock standard NaOH.

6. Standard solutions of p-nitrophenol. Refer to the same item under Alkaline phosphatase, page 400.

7. Enzyme stabilizer tablets. These contain 10 mg. of disodium citrate monohydrate. They may be prepared in the laboratory or purchased.

SPECIMENS

Hemolyzed serum specimens should be rejected, since they are contaminated with considerable amounts of red cell acid phosphatase. Serum from all specimens should be immediately separated from red cells and stabilized by the addition of disodium citrate monohydrate (10 mg./ml.). Chylous sera should not be used.

Any prostatic manipulation will cause a temporary but sizable elevation of serum acid phosphatase levels. Blood specimens from such patients should be obtained at least 24, and preferably 48, hours after such manipulation.

PROCEDURE

Use two citrate-substrate tubes and one tartrate-substrate tube for each unknown serum to be assayed. Mark one of the citrate tubes "B" (for blank), and the other "T" (total). In addition, obtain one pair of each type of substrate tube to serve as substrate controls and instrument blanks. Bring *all* tubes to 37°C., and also the sera to be tested. At timed intervals, pipet 0.20 ml. of each unknown serum to the "T" tube (total) and to one tartrate tube. Mix contents by swirling the tubes rapidly, and permit them to incubate for exactly 30 minutes. Incubate the "B" tubes and the substrate control tubes without adding serum. At the end of the 30 minute reaction period, add 4.0 ml. of 0.10 *N* NaOH to each tube to stop the enzyme reaction. Remove all tubes from the water bath, and then add 0.20 ml. of each appropriate serum to the "B" tubes, and 0.20 ml. of water to the substrate control tubes. Mix the contents of all tubes, and transfer the contents to cuvets.

Read the pair of citrate tubes used as substrate controls against each other at 410 nm., and select the cuvet with the lowest absorbance reading to serve as an instrument blank for the "T" and "B" tubes. (If the two control tubes differ in absorbance by more than 0.040, the quality of the substrate is suspect.) Read the absorbances of the "T" and "B" tubes against the selected instrument blank, and convert the readings to enzyme units, using the calibration curve. Record these as T_u and B_u. Similarly, select the tartrate instrument blank, read the tartrate tubes, and convert their values to enzyme units (N_u).

Then $(T_u - B_u)$ = total acid phosphatase activity in BLB units, and $(T_u - N_u)$ = prostatic acid phosphatase activity in the same units. The "B" tubes correct for any serum pigments absorbing at 410 nm.

CALIBRATION CURVE

Use the same curve obtained for the alkaline phosphatase procedure. The absorbance representing 1.0 alkaline phosphatase unit will represent 0.24 acid phosphatase unit, etc.

In the reaction, 0.20 ml. of serum is used, and the final volume after color development is 5.2 ml. The 1.0 ml. standard contains 0.05 μmol. of PNP in a final volume of 11.1 ml. If the enzyme present in the 0.20 ml. of serum splits Z μmol. of PNP in the 30 minute reaction, then

$$(Z) \times 60/30 \times 1000/0.20 \times 5.2/11.1 \times 1/1000 \times 0.05$$

or $Z \times 0.24$ represents the mmol. of PNP split by 1000 ml. of serum in 60 minutes.

COMMENTS

If the absorbance obtained is over 0.600, the assay should be repeated using either a shorter incubation time (15 or 10 minutes), or a smaller aliquot of serum (0.20 ml. of a 1/5 or 1/10 dilution of the serum). The identical volume of the same diluted serum must be used in all three tubes (citrate, tartrate, and blank). The correction factor is $(30/t) \times (0.20) \times D$, where t is the (shorter) incubation time, and D is the dilution factor.

NORMAL VALUES

Total acid phosphatase	Males:	0.15–0.70 BLB units
	Females:	0.02–0.55 BLB units
Prostatic acid phosphatase	Males:	0.01–0.30 BLB units
	Females:	0.00–0.05 BLB units

The Shinowara procedure with β-glycerophosphate as substrate gives a normal range for total phosphatase of 0 to 1.5 SJR-Bodansky units, and the Gutman-Gutman procedure with phenylphosphate as substrate gives 0.6 to 3.5 King-Armstrong units.

REFERENCES

Tech. Bull. No. 104, Sigma Chemical Co., St. Louis, Mo., 1960.
Author's laboratory.

CLINICAL SIGNIFICANCE

Large elevations of the prostatic acid phosphatase (and thus also total acid phosphatase) are found in the sera of males with *prostatic cancer* with metastases.

Values range from 1.1 to as high as 40 BLB (5-200 KA) units. With localized carcinoma, values may range from 0.7 to about 1.5 BLB units and differentiation from *benign hypertrophy*, in which levels up to 1.0 unit may be found, is at times difficult. After surgery or estrogen therapy, the levels slowly approach normal, with a subsequent rise, if the treatment is unsuccessful. Even though the prostatic enzyme may increase some ten to thirty fold in metastatic cancer, the level of the nonprostatic enzymes may increase only two or three fold.

High elevations in total acid phosphatase are encountered in women with metastatic breast cancer, the metastases being the source of the enzyme. Elevated levels of acid phosphatase have also been observed in patients with Gaucher's and Niemann-Pick disease. Elevations of the nonprostatic enzymes are seen in Paget's disease (although such elevations are more pronounced if phenylphosphate is used as substrate), in a number of prepubertal conditions, and in myelocytic leukemia.

Acid phosphatase is present in very high concentrations in semen, a fact utilized in forensic medicine in investigations of rape and similar offenses.

SUPPLEMENTARY READING (Alkaline and Acid Phosphatase)

Gutman, A. B.: Am. J. Med., *27*:875, 1959.
Woodward, H. Q.: Am. J. Med., *27*:902, 1959.

AMYLASE

I.U.B. No. 3.2.1.1; α-1,4-glucan, 4-glucanohydrolase

Amylase belongs to the class of enzymes referred to as *hydrolases*. These enzymes are characterized by their ability to catalyze the hydrolytic splitting of a variety of chemical bonds, such as —C—O (esters, ethers, hemiacetals), —C—N (peptide bonds), and —P—O (phosphate esters). The phosphatases have already been discussed. The hydrolytic reactions are reversible, although the equilibria are weighted in the direction of hydrolysis.

Amylases are a group of hydrolases which split polysaccharides such as starch and glycogen. These are constituted of α-D-glucose units linked through carbon atoms 1 and 4 located on adjacent glucose residues. Both straight chain (linear) polyglucans, such as amylose, and branched starches, such as amylopectin and glycogen, are hydrolyzed, but at different rates. The enzyme splits the chains at alternate α-1,4 links, forming maltose and some residual glucose in the case of amylose, and both these sugars plus a residue of limit dextrin in the case of the branched chain polyglucans. The α-1,6 linkage at the branch points in the latter is not attacked by the enzyme. (See Figure 4-7, p. 151.)

Two types of amylases are recognized. The *beta amylases* (bacterial exoamylases) can only act at the terminal reducing end of a polyglucan chain, splitting off a section of two glucose units (maltose) at a time. Animal amylases, including human serum amylase, are *alpha amylases*. They are also referred to as *endoamylases*, because they can attack α-1,4 linkages in a random manner anywhere on the polyglucan chain. Large polysaccharide molecules are thus rapidly broken down into smaller units—maltose residues and some glucose units. Since both maltose and glucose are reducing sugars, the course of the hydrolytic reaction is paralleled by an increase in soluble reducing materials.

Linear starch chains (in helical form) react with molecular iodine to form the well known, deep blue, starch iodine complex. The hydrolysis of starch thus is also paralleled by a gradual loss of ability to bind iodine, and the hue of the glucan-iodine complex changes to a light blue, then to violet, and finally to red. No iodine color is formed when the chain size is six glucose units or less.

Human serum amylase has a moderately sharp pH optimum at 6.9 to 7.0. The enzyme is active even at temperatures as high as 50°C., but it is customarily assayed at 37 or 40°C. Full activity is displayed only in the presence of a variety of univalent anions, such as chloride, bromide, nitrate, chlorate, or monohydrogen phosphate. Chloride and bromide are the most effective activators. Optimal activity is obtained at a Cl^- level above 0.01 *M*. Human serum amylase has a molecular weight of 45,000; the enzyme is thus small enough to pass through the glomeruli, and some amylase is normally found in the urine. The several amylase isoenzymes have been reported to migrate electrophoretically as γ-globulins.

The enzyme is quite stable; negligible activity is lost at room temperature in the course of a week, or at refrigerator temperatures over a two month period. With the exception of heparin, all common anticoagulants inhibit amylase activity; citrate and oxalate by as much as 15 per cent. As a consequence, amylase assays should be performed only on serum.

Amylase was originally referred to as *diastase*. This term is still used commercially, and occasionally even in clinical work, especially in reference to the urinary enzyme.

In the body, amylase is present in a number of organs and tissues. The greatest concentration is present in the pancreas where the enzyme is synthesized by the acinar cells and then secreted into the intestinal tract for digestion of starches. The salivary glands secrete a potent amylase to initiate hydrolysis of starches while the food is still in the mouth and esophagus. The action of the salivary enzyme, once referred to as *ptyalin*, is terminated by the acid in the stomach. In the intestinal tract, effective action of pancreatic and intestinal amylase is favored by the mildly alkaline conditions in the duodenum. Intestinal maltase then further hydrolyzes maltose into glucose. Most of the pancreatic amylase is destroyed by trypsin activity in the lower portions of the intestinal tract, although some amylase activity is present in stool. Significant amylase activity is present in liver, kidney, and muscle tissue, and some in cerebrospinal fluid, urine, and milk. The amylase found in normal serum may have its origin, at least in part, in the liver, because its concentration is not affected by pancreatectomy. The urinary enzyme is derived from plasma.

Assay Methods

Development of precise assay methods for measuring amylolytic activity is confounded by the difficulties encountered in preparing a substrate of known and reproducible composition and concentration. Starches vary considerably in their proportion of amylose and amylopectin, and the average chain length of the starch molecule depends on the method by which the substrate is prepared. Starch does not disperse in water to form a true molecular solution, but forms instead a colloidal sol containing hydrated starch micelles of varying size. The degree of dispersion varies with temperature; at lower temperatures amylose chains aggregate (retrograde) into large micelles. It is essential that starch substrates be prepared and treated in exactly the same manner if data from different assays are to be comparable.

Potato, corn, and Lintner's "soluble" starch are most commonly used, although pure amylose, amylopectin, and glycogen are preferred by some analysts. Starch sols deteriorate rather rapidly as the result of mold contamination; benzoic acid or p-hydroxy propylbenzoate may be added as preservatives. Sterile (autoclaved) sols keep well for several months.

In "*saccharogenic*" assays[43,83,84] the course of the enzyme reaction is followed by measuring the quantity of reducing sugars formed. Assays based on this approach were the first to be used, and they are still the most accurate. Any sensitive reducing sugar method, such as the Folin-Wu or the Somogyi-Nelson method, may be used. Recently, methods based on the reduction of picrate, ferricyanide, and 3,5-dinitrosalicylic acid,[77] and on the anthrone-sugar reaction have been advocated. The results are reported in terms of milligrams of "apparent glucose" formed, although the chief reducing sugar present is maltose, which has about 40 per cent of the reducing capacity of glucose, weight for weight. Reducing sugars are determined on protein-free filtrates of the reaction mixtures. At times, with some starch preparations, it is difficult to obtain clear, nonopalescent filtrates. This does not affect the assay, provided that the turbidity is due to starch and not due to incompletely precipitated proteins. (Turbidity due to starch clears up at a later stage of the procedure.)

Enzyme activity can be evaluated by following the decrease in substrate (starch) concentration, rather than by measuring the product. Methods based on this approach are referred to as "*amyloclastic*" *methods*. In the "*chronometric*" procedures (Somogyi[84]), the time required for amylase to completely hydrolyze all the starch present in a reaction mixture is measured. The endpoint is reached when there is absence of any material capable of forming the blue starch-iodine color. In the Wohlgemuth method serial dilutions of the enzyme preparation are added to a fixed quantity of starch, and that dilution is found which is just able to hydrolyze all the starch present in a fixed-time period. The "*amylometric*" procedures of Van Loon[94] and others,[81,85] measure the amount of starch hydrolyzed in a fixed period of time, using the blue starch-iodine color as the means for quantitating starch.

About 30 per cent of the requests for serum amylase determinations are ordered on a "stat" basis to confirm or rule out acute pancreatitis as a possible cause of acute upper abdominal pain. Thus, the method of Peralta and Reinhold,[97,98] which measures the change in turbidity of a starch sol over a short reaction period, offers certain advantages. This procedure can be performed in 15 to 20 minutes, as compared to 60 minutes for the saccharogenic technique.

In most assay methods the pH is maintained at about 6.9 to 7.2 using 0.02 *M* phosphate buffer; in some methods, Cl^- ion is added, but in others, the Cl^- present in serum or urine is assumed to be sufficient to provide maximum activation.

UNITAGE IN AMYLASE DETERMINATION

Somogyi[83] defined his unit in the saccharogenic assay as the quantity of enzyme that is able to liberate sugars with a reducing value equivalent to 1 mg. of glucose in the course of a 30 minute reaction at 40°C. and at pH of 6.9 to 7.0. The concentration is expressed as the number of units per 100 ml. of specimen. All methods based on measuring the formation of reducing sugars report results in terms of Somogyi units. There is no such uniformity however with the amyloclastic methods.

The Wohlgemuth *diastatic index* is defined as the number of ml. of 0.1 per cent (w/v) starch (i.e., mg. of starch) digested by the enzyme present in 1.0 ml. of specimen at 37°C. in 30 minutes. The Huggins and Russell[49] unit is the quantity of enzyme that will hydrolyze

1.0 mg. of starch to the dextrin stage in 60 minutes at 37°C. whereas one Street and Close[85] unit is that amount which dextrinifies 20 mg. of starch in 15 minutes. There is no direct relationship between any of these units and the Somogyi unit. The latter has been so generally accepted in this country that values obtained by other types of procedures are often multiplied by an appropriate factor to convert them into equivalent Somogyi units. In the chronometric methods, amylase units are calculated from the formula, units $= C/t$, where t is the time required to decolorize the starch preparation, and C is a factor so chosen that the values obtained are again comparable with those obtained with saccharogenic methods. Inasmuch as neither the substrate nor the products of the amylolytic reaction are single, well-defined chemical entities, there is no advantage gained in converting common amylase units into international units.

NORMAL VALUES

It is generally accepted that the amylase level in the serum of healthy persons is 50 to 150 Somogyi units per 100 ml., with the lower limit set by some workers at 30 or 40 and the upper limit at values varying from 120 to 160. The normal range in terms of Street and Close units is 6 to 33 units, and that for the Russell-Huggins method is 9 to 35 units. Assays for urinary amylase are best done on 1, 2, or 24 hour timed specimens rather than on random samples, and results are reported in terms of units per hour, per 2 hours, or per 24 hrs respectively. The lower range of normal is between 1000 and 1250 Somogyi units per 24 hours, and the upper range between 5000 to 6500 units per 24 hrs. The average daytime hourly output may be taken as between 45 to 275 units.[40] Amylase levels in duodenal fluid specimens may vary in various conditions from nearly zero to concentrations of the order of 50,000 to 80,000 Somogyi units per 100 ml.

Two methods will be presented as examples of procedures being used. The Henry-Chiamori modification of Somogyi's procedure is an example of a saccharogenic procedure, and the Roe and Smith procedure is an example of a simple amyloclastic method.

Determination of Amylase by an Amyloclastic Procedure

REAGENTS

1. Stock starch solution, 1.2 gm./100 ml. in H_2O. With stirring, suspend 3.0 gm. of Lintner's soluble starch in 20 to 30 ml. of water to form a homogeneous suspension, free of any clumps. To this, slowly add 150 to 200 ml. of boiling water with vigorous stirring. As the temperature is increased to 80 to 85°C., the starch particles will dissolve to form a slightly opalescent, homogeneous starch sol. Heat the solution, with stirring, to boiling, and then transfer it to a 250 ml. volumetric flask. Permit the sol to cool to 30 to 35°C., and increase the volume to 250 ml. with water. This solution is stable for three to four weeks when refrigerated. The solution should be discarded when any sign of microbial growth is evident.

2. Phosphate buffer pH 7.0, 0.20 M. This solution contains 10.2 gm. of KH_2PO_4 and 17.3 gm. of Na_2HPO_4 in 1000 ml. of solution. Keep the solution refrigerated, and discard on evidence of mold growth. Check the pH with a pH meter and adjust with 1 N H_3PO_4 or 1 N NaOH, if necessary.

3. Sodium chloride, 0.50 M. This solution contains 14.6 gm. NaCl per 500 ml.

4. Buffered substrate. Prepare a sufficient volume to meet the day's needs. Warm 50.0 ml. of stock starch to 85 to 90°C. and to this add 30.0 ml. of buffer and 10.0 ml. of the NaCl solution. Concentrations of the components of the reagent are as follows: starch, 6.67 mg./ml.; buffer, 0.067 M; NaCl, 0.056 M.

5. Iodine solutions. Stock iodine contains 3.0 gm. of I_2 and 30 gm. of KI per liter. Grind the iodine and KI together in a mortar, add water, and transfer the dissolved iodine to a one liter volumetric flask. Continue until all the iodine is dissolved. Then dilute the solution to the mark. The working iodine solution is a 1:10 dilution of this stock solution (concentration 0.0024 *N*).

6. Hydrochloric acid, 1.0 *N*. Dilute 41.5 ml. of concentrated HCl to 500 ml.

PROCEDURE

1. A pair of 18 × 200 mm. test tubes, labeled C and T, are needed for each serum to be assayed. After adding 4.5 ml. of buffered substrate at 85 to 90°C., place the tubes into a 37°C. water bath for 10 minutes to equilibrate to temperature.

2. Pipet 0.50 ml. of serum to the T tube. Note the time and incubate both tubes for exactly 30 minutes. The reaction tube contains 30 mg. of starch (6.0 mg./ml.), buffer at 0.06 *M*, and Cl at approximately 0.053 *M*.

3. At the end of the reaction period (30 minutes), stop the reaction by adding 1.0 ml. of *N* HCl, with mixing, to each tube; then add 0.50 ml. of serum to the C tube.

4. For each T and C tube prepare a 50 ml. volumetric flask (appropriately labeled) to contain about 35 ml. water and 1 ml. *N* HCl. Measure 0.5 ml. from each T and C tube into its correspondingly labeled flask and mix. Finally, add 1.0 ml. working iodine solution and make up to volume with water After mixing, pour into appropriate cuvets and read the absorbance at 620 nm. against water set at 0 absorbance. If the T has a green or yellow color, the starch has been nearly or totally exhausted, and the assay should be repeated with a 1:4 dilution of serum with saline.

CALCULATIONS

If A_T and A_C are the absorbance readings of the T and C cuvets, then the amylase level in units per 100 ml. is

$$\frac{A_C - A_T}{A_C} \times 600$$

Smith and Roe defined their amylase unit empirically as the quantity of enzyme that would hydrolyze 10 mg. of starch in 30 minutes, under the conditions of the assay. This unit was chosen because it gave values comparable with Somogyi units. The value of A_C represents the initial 30 mg. of starch, and $A_C - A_T = \Delta A$, the mg. of starch hydrolyzed by the enzyme in 0.50 ml. of serum specimens. Then

$$\frac{\Delta A}{A_C} \times 30 \times \frac{1}{10} \times \frac{100}{0.5} = 600 \times \frac{\Delta A}{A_C} \text{ units/100 ml.}$$

The factor 600 will be valid only if A_C represents 30 mg. of starch.

COMMENTS

The range of values found for apparently healthy persons is 50 to 140 units/100 ml. This range is comparable with that expressed in terms of the Somogyi units obtained with the saccharogenic method. Reproducibility is of the order of ±10 per cent. The procedure can be applied directly to urines and duodenal fluids, except that the specimens may require dilution. Even with sera, dilution is required for levels over 500 units per 100 ml. If elevated values are expected, the specimen should be run undiluted, diluted 1:4, and perhaps also diluted 1:16.

Some sera and urines contain some material that apparently inhibits formation of the starch-iodine color. The nature of the material is not known (except that it is dialyzable). With specimens containing this interfering agent, the value of A_C is found to be lower than that for other sera run that day, and for these specimens, the measured blue color is not a true estimate of the starch actually present. Because of this, some workers have suggested that the amylase procedures based on measuring starch by the starch-iodine color should no longer be used. However, the procedure just described is rapid and simple, and its reliability is not appreciably inferior to the saccharogenic methods.

REFERENCES

Smith, B. W., and Roe, J. H.: J. Biol. Chem., *179*:53, 1949.
Van Loon, E. J., Likens, M. R., and Seger, A. J.: Am. J. Clin. Path., *22*:1134, 1952.

Determination of Serum Amylase by a Saccharogenic Method

This procedure is an adaptation and modification of Somogyi's procedure by Henry and Chiamori.

REAGENTS

1. *Phosphate buffer*, 0.10 *M*, pH 7.0. Dissolve 4.55 gm. of KH_2PO_4 and 9.35 gm. Na_2HPO_4 in distilled water and dilute to one liter.

2. Buffered substrate. Dissolve 2.15 gm. of Lintner's soluble starch in the buffer and dilute to 200 ml. (Starch concentration = 10.7 mg./ml.) Prepare the starch solution as described in the previous procedure, but boil the solution for 3 minutes, before cooling and diluting to the mark.

3. Glucose standards. Prepare two standards containing 200 and 400 mg./100 ml. of glucose in saturated benzoic acid.

4. Folin-Wu protein-free filtrate reagents. 0.67 *N* H_2SO_4 and 10 per cent (w/v) $Na_2WO_4 \cdot 2H_2O$.

5. Folin-Wu sugar reagents.

a. Alkaline copper reagent. Dissolve 40.0 gm. of Na_2CO_3, anhydrous, in about 400 to 500 ml. of water. To this add 7.5 gm. of tartaric acid, with stirring, followed by a solution of 4.5 gm. of $CuSO_4 \cdot 5H_2O$ in approximately 100 ml. of water. Then dilute the solution to 1000 ml. The reagent is stable.

b. Phosphomolybdic acid reagent. Add 400 ml. of 10 per cent (w/v) NaOH, with stirring, to 70 gm. of molybdic acid (MoO_3) and 10 gm. $Na_2WO_4 \cdot 2H_2O$ in a 2000 ml. beaker, followed by 400 ml. of water. Boil the solution for 30 to 40 minutes, until it is free of ammonia; then cool, and dilute with water to about 700 ml. Add 250 ml. of concentrated H_3PO_4 and dilute the solution to 1000 ml. The reagent is stable.

PROCEDURE

Use two 15 × 120 mm. tubes, marked T and C, for each serum specimen. To each add 3.50 ml. of buffered substrate and equilibrate to 37°C. in a water bath. To the T tube add 0.50 ml. of serum. After mixing, incubate both tubes for exactly 30 minutes. During this period, prepare the blank and standards to contain 3.50 ml. of buffered starch, 0.50 ml. of H_2O or standards, respectively, 0.75 ml. of 0.67 *N* H_2SO_4 and 0.25 ml. of tungstate. At the end of the 30 minute reaction time, add 0.75 ml. and

0.25 ml. respectively of 0.67 N H_2SO_4 and 10 per cent $Na_2WO_4 \cdot 2H_2O$ to the T and C tubes to stop the reaction. Then add 0.50 ml. of serum to the C tube. After mixing, centrifuge the tubes and filter the supernatants. Transfer 1.00 ml. each of the blank, standards, and each filtrate to Folin-Wu sugar tubes graduated at the 12.5 ml. mark. After adding 1.0 ml of copper reagent, heat the tubes in a boiling water bath or Carbowax bath at 100°C. for 8 minutes, and cool to room temperature. Add 1.0 ml. of phosphomolybdic acid to each tube, with mixing, to allow the escape of all effervescent gas. Dilute the tubes to 12.5 ml., mix thoroughly, and transfer the contents to 10 or 12 mm. cuvets. Read the absorbances of the solutions against the blank at 680 nm. In the calculations that standard is used which more nearly matches the absorbance of the T tube color product. Then

$$\frac{(A_T - A_C)}{(A_S - A_B)} \times C_S = \text{amylase units/100 ml.,}$$

where C_S represents the glucose concentration of the standard in mg./100 ml. If the amylase value is over 550 units, the assay should be repeated using 0.50 ml. of a dilution of the serum in saline. The procedure is directly applicable to urine or other body fluid specimens. The pH of the specimen should be checked, and adjusted to 6.9 to 7.5, if necessary.

The precision of the procedure is of the order of ±10 per cent in the range of 80 to 200 units, somewhat better in the 200 to 450 unit range and a little worse at low or very high levels.

REFERENCES

Somogyi, M. L.: Clin. Chem., *6*:23, 1960.
Henry, R. J., and Chiamori, N.: Clin. Chem., *6*:434, 1960.

CLINICAL SIGNIFICANCE

In patients with *acute pancreatitis* serum amylase rises generally to values over 550 Somogyi units, and sometimes to levels as high as 2000 to 4000 units. The rise is transient, increasing during the first 24 to 30 hours and falling back to within the normal range within the next 24 to 48 hours. Hyperamylasemia is also observed in patients with perforated *gastric or duodenal ulcers*, intestinal obstruction or strangulation, and obstruction of the pancreatic or common bile duct. In general, any acute disease process in the areas adjacent to the pancreas may result in elevations of serum amylase. Similarly, surgical manipulations or procedures carried out near the pancreas will cause a transient rise in amylase. These values may reach levels as high as 2000 units, but they are usually under 600 units.

In *chronic pancreatitis* values range from normal levels to about 250 to 400 units. *Pancreatic carcinoma* is only occasionally accompanied by an amylase elevation. The injection of morphine causes a temporary rise in amylase by contracting the sphincter of Oddi. *Mumps* and bacterial parotitis, which block the secretion of salivary amylase, are associated with mild elevations of serum amylase (250 to 600 units). Levels in the low range of normal and even below normal may be encountered in patients with hepatitis and obstructive jaundice, and also those with liver tumors and abscesses.

In pancreatitis, the normal *urinary amylase* excretion of 50 to 300 units per hour may increase two to three fold, and the hyperamylasuria may persist for three to five days, long after the serum amylase has reached normal levels. Impaired excretion as a result of renal disease may be reflected in a rise in serum levels.

LIPASE

I.U.B. No. 3.1.1.3; Glycerol ester hydrolase

Lipases are defined as that group of enzymes which hydrolyze the glycerol esters of long-chain fatty acids. It is now believed that the products of the reaction are two moles of fatty acids and one mole of β-monoglyceride per mole of substrate. A small amount of the ester is hydrolyzed completely to glycerol. The enzyme attacks preferentially the end ester bond of the glycerol ester.

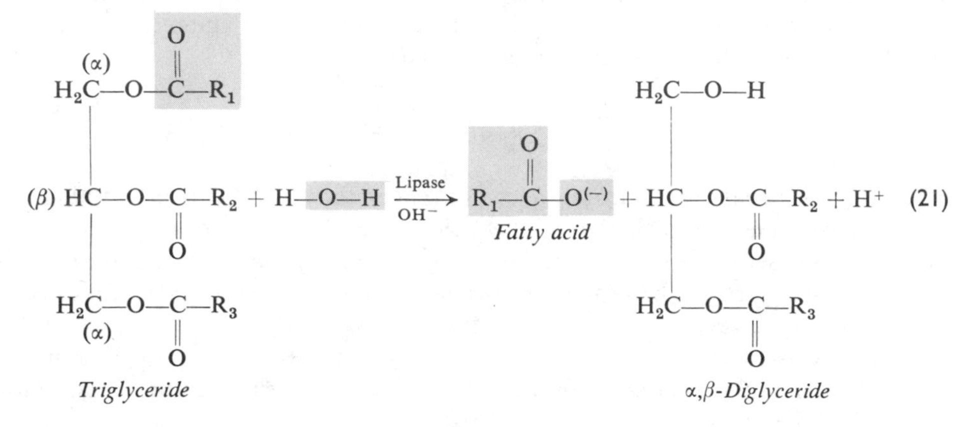

Desnuelle[25] has demonstrated that the enzyme acts only at the interface of water and substrate, and thus acts only on substrates which are present in an emulsified form. The same author has also demonstrated that the rate of enzyme activity is not related to the absolute concentration of substrate present, but rather to the surface area of the dispersed substrate. The preparation of a reproducible, stable emulsion of uniform particle size is extremely difficult; this is one reason for the considerable disagreement among investigators about the properties of the enzyme and its utility in clinical diagnosis. The optimal pH is about 7.8 to 8.0, although reported values range from 7 to 9 and depend on the nature of the substrate and on the presence or absence of certain reaction accelerators. Lipases are sulfhydryl enzymes and are activated by sulfhydryl compounds, such as cysteine and thioglycolic acid. Apparent activation is effected by bile salts under alkaline conditions, by albumin, and by calcium ions. The bile salts, being excellent emulsifying agents, probably promote the formation of a stable and finely dispersed emulsion of the fat in water, and the Ca^{++} ions function to remove fatty acids liberated in the reaction by forming insoluble calcium soaps. Fatty acids are inhibitors of the reaction, and their removal promotes the forward reaction. Among the inhibitors of the enzyme can be listed heavy metal ions, quinine, many aldehydes, eserine, and diisopropylfluorophosphate (see cholinesterase, p. 421). The enzyme is characteristically not affected by fluoride, triorthocresylphosphate, and atoxyl (sodium arsanilate), which are inhibitors of the aliesterases (to be discussed shortly). Some investigators have reported an anti-lipase in serum and a lipase inhibitor in urine.

In the human organism the most important source of lipase is the pancreas, although some lipase is secreted by the gastric mucosa and intestinal cells. Lipase activity is found in plasma, red cells, leukocytes, normal urine, and spinal fluid.

True lipase activity is differentiated from the activity of two or perhaps three other related enzymes. These are *carboxylic ester hydrolase* (I.U.B. No. 3.1.1.1), commonly referred to as true esterase or aliesterase; *aryl ester hydrolase* (I.U.B. No.

3.1.1.2), and *lipoprotein lipase*, not included in the International Commission list, perhaps because it is not sufficiently characterized. The aliesterases hydrolyze glycerol esters of short chain fatty acids (e.g., tributyrin) as well as esters of monohydric alcohols (e.g., ethyl acetate) and esters of dibasic acids (diethyl adipate). The esterases are inhibited by atoxyl, fluoride, and triorthocresylphosphate, and are not activated by bile salts. The aryl esterases hydrolyze such esters as phenyl acetate. Lipoprotein lipase[22] (clearing factor) is activated by heparin and hydrolyzes protein-bound triglycerides to free acids and monogylcerides, which are then transferred to an acceptor protein such as serum albumin. All three, along with lipase, are present in normal plasma. By the use of various combinations of activators and inhibitors, the activity of one or several of these related enzymes may be potentiated or suppressed in part or almost completely.

Assay Methods

Cherry and Crandall,[23] who first appreciated the clinical value of plasma lipase determinations, used a 50 per cent emulsion of olive oil in 5 per cent (w/v) gum acacia and a phosphate buffer of pH 7.0 as substrate, and measured the amount of carboxylic acid formed after a 24 hour reaction period at 37°C. The acid was titrated with 0.05 *N* NaOH to a phenolphthalein endpoint, and a unit of lipase activity was defined as the quantity of enzyme which liberated acid equivalent to 1.0 ml. of 0.05 *N* NaOH (50 μEq.). The enzyme concentration was reported in units per 1.0 ml. of serum. This technique is still the only practical method available, although details have been modified in the course of 30 years experience. It has been established that over 50 per cent of hydrolytic activity occurs in the course of the first four hours, and various investigators have reduced the incubation time to six, four, and three hours.[93] Tris (trishydroxymethylaminomethane) and veronal buffers (pH 7.4 to 8.0) have been used in place of phosphate buffer, and thymolphthalein has been recommended as a superior indicator, since the endpoint (pH 10.5) is a better stoichiometric measure of the acid produced. Tietz and Fiereck[93] demonstrated the increased precision possible if the titration is performed potentiometrically.

Most methods proposed the use of olive or corn oil as substrate. The use of tributyrin, as proposed by Goldstein, Finstein, and Roe, was shown to be inappropriate, since the substrate was responsive to aliesterase activity rather than to lipase activity.[40]

Colorimetric approaches have been attempted by some investigators. Seligman and Nachlass used β-naphthyl laurate, along with appropriate activators and inhibitors (to counteract esterase action), assaying the liberated β-naphthol by diazotizing with diorthoanisidine. Similarly, Gomori proposed the use of α-naphthyl laurate, and Saifer and Perle the use of phenyl laurate. These compounds, however, are not true lipase substrates, and it is likely that they measure arylesterase rather than lipase activity. Kim and Tietz have recently shown that methods employing these substrates are less discriminatory, and therefore less useful clinically, than procedures using emulsions of the olive oil type.

Determination of Serum Lipase

SPECIMEN

Serum specimens used in the determination of lipase should be free of significant hemolysis, since hemoglobin inhibits lipase activity. Lipase in serum is stable at room temperature for a week and may be kept for long periods in the refrigerator.

PRINCIPLE

An aliquot of serum is incubated with an olive oil emulsion, buffered at pH 8.0, for three hours at 37°C. The liberated fatty acids are titrated with 0.050 *N* NaOH to a light blue color with thymolphthalein as indicator, or electrometrically to a pH of 10.5.

REAGENTS

1. Purified olive oil. Add 300 ml. of the best quality olive oil to 60 gm. of chromatographic grade alumina (Al_2O_3) (Merck No. 71207) with stirring. Stir the suspension at 10 minute intervals over the course of an hour. Permit the alumina to settle out, and filter the oil through a qualitative filter paper (Whatman No. 1 or equivalent).

2. Oil emulsion. Add 7.0 gm. of gum acacia (emulsifier) and 0.2 gm. of sodium benzoate (preservative) to 100 ml. of water in a high speed blender, and dissolve with gentle blender action. Then add 100 ml. of purified olive oil, and emulsify the mixture by operating the blender at top speed for 10 minutes. Store the emulsion at 6 to 10°C. It should never be permitted to freeze. On standing over a period of time, some creaming may occur but the emulsion can be fully reconstituted by shaking it vigorously 10 times. If complete separation of the oil and water phases occurs, discard the emulsion.

3. Buffer base, stock solution, 0.80 *M* Trishydroxymethylaminomethane (TRIS). Dissolve 48.554 gm. TRIS in 500 ml. of water. Keep the reagent refrigerated.

4. TRIS-hydrochloride buffer, 0.20 *M*, pH 8.0 at 27°C. (pH 7.85 at 37°C.). To 50 ml. of buffer base in a beaker add 21 ml. of 1.0 *N* HCl and water to a volume of about 150 to 160 ml. Permit the solution to cool to 25°C. check the pH with a pH meter, and then adjust to pH 8.0 by careful addition of more acid. Adjust the volume to 200 ml. Note that TRIS buffers have a large temperature coefficient, $\Delta pH/\Delta T = \frac{0.15}{10} = 0.015$, which is about 15 times greater than that of a phosphate buffer at the same pH.

5. Standard sodium hydroxide, 0.050 *N*. Prepare by diluting any laboratory stock NaOH to exactly 0.050 *N*.

6. Thymolphthalein indicator, 1 per cent (w/v) in 95 per cent (v/v) ethyl alcohol.

7. Ethanol, 95 per cent (v/v).

PROCEDURE

1. Use a pair of 25 × 200 mm. test tubes, labeled T (test) and B (blank), for each unknown. Into each tube place 2.50 ml. of water, 10.0 ml. of olive oil emulsion, and exactly 1.00 ml. of TRIS buffer. The water and emulsion may be added by automatic pipets or with serological pipets, but the buffer should be measured out precisely.

2. Place the tubes in a water bath at 37°C. and permit to equilibrate to temperature for 10 minutes.

3. Then add 1.0 ml. of unknown serum or other specimen to the T tube, cover the tube, mix vigorously, and place into the water bath. Permit the reaction to proceed for three hours.

4. At the end of the incubation period, remove the tubes from the water bath and pour the contents of each into 50 ml. Erlenmeyer flasks. Rinse the tubes with 3.0 ml. of 95 per cent (v/v) ethanol, and add the rinse ethanol to the flasks. Then add 1.0 ml. of unknown serum to the B tubes, and mix both flasks by vigorous rotation.

5. Add 5 drops of indicator to each flask, and titrate the contents with 0.050 *N* NaOH. Feed the NaOH rapidly, dropwise, with stirring, until the first hint of blue is seen, then more slowly, until a definite blue color (greenish-blue with icteric sera) is obtained. The B flask titration measures the NaOH needed to bring the pH of the buffer and any acids in the serum to 10.5. The T flask titration measures this value plus the acid liberated from the triglycerides in the lipase reaction. The alcohol serves to stop the enzyme reaction and to transfer the fatty acids from the oil phase to the water phase.

6. Calculate the difference (T − B) = ΔV. This difference gives the value of the lipase activity in the customary Cherry-Crandall (Tietz-Fiereck) units. The unit is defined as the quantity of enzyme in 1.0 ml. of serum which will produce fatty acids equivalent to 1.00 ml. of 0.05 *N* NaOH, under the conditions of the test. Units may be converted to International Units, as follows: One conventional unit represents 1.0 ml. × 0.050 *N* (or 50 μEq. = 50 μmols.) of fatty acid split off in the three hour reaction period, or 50/180 = 0.277 μmol./minute. Thus 0.277 I.U. or 277 mI.U. are present in 1.0 ml., and 1.0 Tietz-Fiereck unit = 277 mI.U./ml.

NORMAL RANGE

Tietz and Fiereck report a normal range of 0.05 to 1.00 conventional units, with an average value of 0.42 units/ml. The equivalent values in international units are 14 to 280 (mean = 116) mI.U./ml.

COMMENTS

1. The reported reproducibility in the normal range is 10 per cent; this approaches ±5 per cent with sera containing increased values for lipase activity.

2. The 10 ml. of olive oil emulsion are sufficient to maintain maximum activity within the conditions of the procedure. The actual pH of the enzyme reaction mixture after mixing emulsion, buffer, and serum is about 7.65 at 37°C. The reaction is not linear with time over the three hour period, although the deviation is not pronounced and is significantly less than that obtained in methods using 6 or 24 hour reaction runs or 3 ml. of emulsion.

REFERENCE

Tietz, N. W., and Fiereck, E. A.: Clin. Chim. Acta, *13*:352, 1966.

CLINICAL SIGNIFICANCE

In general, lipase values parallel amylase values. Elevations are observed in *acute pancreatitis* (up to 10 units/ml. or more), and also in obstruction of the pancreatic duct. Elevated values usually persist longer than amylase elevations (up to 7 to 10 days). Occasionally, increased lipase values are found in patients with ileus, duodenal ulcers, and intestinal obstruction. Elevated levels are present in 50 to 60 per cent of patients with *carcinoma of the pancreas* and in some with chronic pancreatitis. Administration of opiates or morphine may cause a rise in lipase levels because of spasms of the duodenal musculature and the sphincter of Oddi. Normal values are usually found in patients with *mumps* (without pancreatitis) or liver disease, although reported results are conflicting (possibly because of the use of unspecific methods). Lipase is cleared by the kidneys, and normal urine may contain up to 0.7 units/ml. Appreciable increases in urine lipase may be found in cases of acute pancreatitis or obstruction of the pancreatic duct. If renal clearance is impaired. serum lipase levels may increase significantly (up to 4.0 units).

LIPASE IN URINE AND DUODENAL FLUID

Urine has been reported to contain a dialyzable inhibitor. Thus, methods for the determination of lipase in urine should include a dialysis of the urine sample. Procedures have been reported, however, in which urine samples were solely adjusted to a pH of 7.0 to 7.5 prior to analysis.

Duodenal fluid samples for lipase analysis should be diluted 1:5 or 1:10 with saline and analyzed without delay to minimize destruction of lipase by trypsin. The pH of the specimen should be adjusted to approximately 8.0, especially if duodenal content is contaminated with gastric HCl.

In view of the unsatisfactory nature of lipase methods as applied to urine and duodenal fluid samples, it is questionable whether such determinations are clinically useful.

SUPPLEMENTARY READING

Henry, R. J.: Pancreatic lipase. In Standard Methods of Clinical Chemistry. D. Seligson, Ed. New York, Academic Press, 1958, Vol. 2, pp. 86–93.

SERUM CHOLINESTERASE

I.U.B. No. 3.1.1.8; Acylcholine acylhydrolase

Serum cholinesterase is a plasma-specific enzyme which hydrolyzes acetylcholine and other acyl esters of choline, but it can also hydrolyze alkyl esters of short chain fatty acids. The hydrolysis of acetylcholine may be formulated as follows:

$$(H_3C)_2(CH_3)N^{(+)}-CH_2CH_2-O-\overset{\overset{O}{\|}}{C}-CH_3\ Br^{(-)} + H-O-H \xrightarrow[\text{Cholinesterase}]{}$$

Acetylcholine bromide

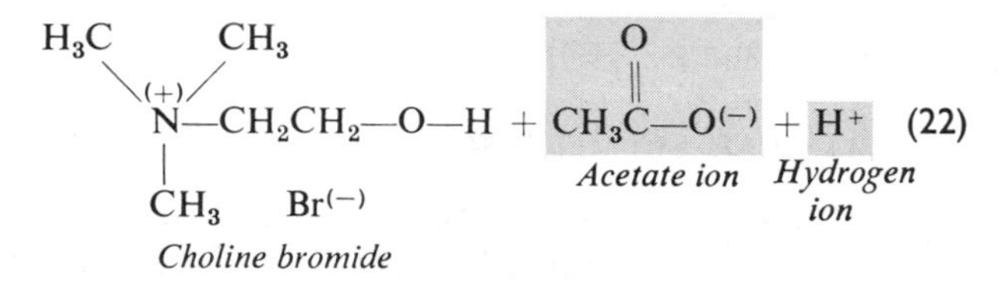

Choline bromide

Acetylcholine is very important physiologically. It is synthesized at nerve endings and acts to transmit impulses from nerve to muscle fiber. Cholinesterase destroys the acetylcholine after the impulse transmission has been mediated, so that additional impulses may be transmitted, if needed. Otherwise, the nerve would remain electrically charged and further conduction would not be possible.

Two cholinesterases of somewhat different specificity and origin are present in man. One of these enzymes occurs in erythrocytes, in nerve endings, and in the lungs, spleen, and gray matter of the brain. It is referred to as the *red cell*, "*true*," or "*Type I*" *cholinesterase*, and as acetylcholinesterase. The International Commission lists it as a separate enzyme, I.U.B. No. 3.1.1.7, acetylcholine acetylhydrolase. The second, serum enzyme, has been referred to as *pseudocholinesterase*, or "*Type II*" *cholinesterase*. It is present in the liver, the pancreas, the heart, and the white matter of the brain, as well as in serum.

These two enzymes differ in specificity towards some substrates but behave similarly toward others. The "serum enzyme" acts on benzoylcholine, but cannot hydrolyze acetyl-β-methylcholine; the red cell enzyme acts on the latter but not on the former.

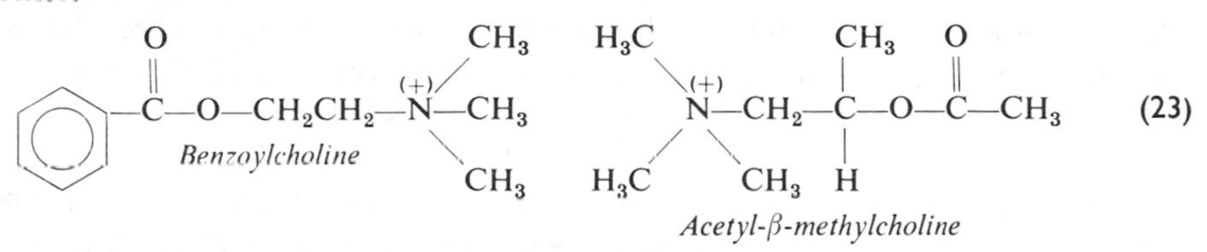

The red cell enzyme is inhibited by its substrate, acetylcholine, if present at concentrations above 10^{-2} M; the serum enzyme is not substrate inhibited. Both enzymes possess nonspecific (aliesterase) activity against simple esters, but acetyl cholinesterase attacks acetyl esters (ethyl acetate) more rapidly than butyryl esters (ethyl butyrate), whereas the serum enzyme is more reactive toward the latter.

Both enzymes are inhibited by the alkaloids prostigmine and physostigmine, both of which contain quaternary nitrogen (also present in choline) in their structures. These are typical competitive inhibitors, competing with the choline residue of acetylcholine for its binding site on the enzyme surface. Both enzymes are irreversibly

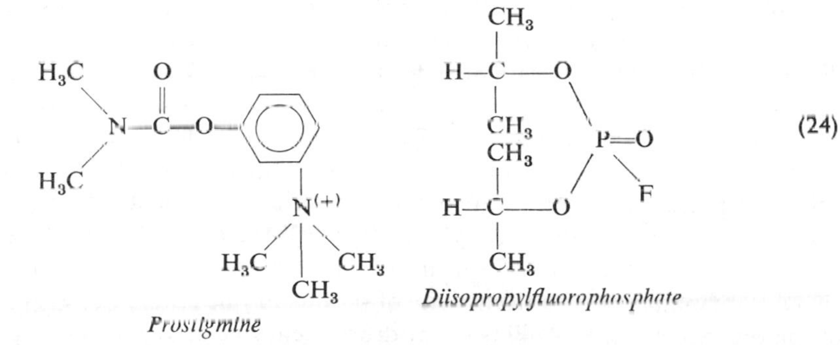

inhibited by some organic phosphorus compounds, such as diisopropylfluorophosphate. The phosphoryl group links very tightly to the enzyme at the site at which binding of the acyl group normally occurs, thus preventing attachment of the acetylcholine. The enzymes are also inhibited by a large variety of other compounds, among which are morphine, quinine, tertiary amines, phenothiazines, pyrophosphate, bile salts, citrate, and fluoride.

In electrophoretic patterns, two to seven bands of enzyme activity can be located, the number depending on the technique employed. Of more interest are the atypical (genetically variant) forms of the serum enzyme, characterized by weak activity (about one-sixth of normal), which are present in the sera of a small fraction (one in 5000) of apparently healthy persons. These forms also possess an increased resistance to inhibition by fluoride ion and dibucaine (see below).

Both the red cell and serum enzymes are quite stable, activity being unchanged at room temperature for 24 hours, and at refrigerator temperatures for 60 days. Frozen preparations maintain their activity for many months. For red cell enzyme assays oxalated blood is used, whereas pseudocholinesterase is best measured in serum.

CLINICAL APPLICATION

Serum cholinesterase levels may be requested as tests of liver function, as indicators of possible insecticide poisoning, or for the detection of patients with atypical forms of the enzyme. The spread of values encountered in apparently healthy people is rather wide, ranging between 130 and 310 de la Huerga units[24] (2.2 to 5.2 I.U./ml.), but the level in any given person is fairly constant. Levels in women are about 10 per cent lower than in men. Birth levels are low (one-fourth that of adults), but these levels increase rapidly, reaching adult levels by the second month of life. No enzyme is found in urine. In normal spinal fluid the enzyme level is very low and is of the red cell type; the serum enzyme, however, predominates if the protein level is over 100 mg./100 ml.

Serum cholinesterase concentrations serve as sensitive measures of liver function, if a patient's normal (base-line) level is known, which, unfortunately, is rarely the case. A 30 to 50 per cent decrease in level is observed in acute hepatitis, and decreases of 50 to 70 per cent occur in advanced cirrhosis and carcinoma with metastases to the liver. Essentially normal levels are seen in chronic hepatitis, mild cirrhosis, and obstructive jaundice.

Decreased levels of serum enzyme are also found in patients with acute infections, pulmonary embolisms, and muscular dystrophy, and after surgical procedures. After myocardial infarctions, the enzyme level decreases until the fifth day and then begins a slow rise to normal.

Moderately elevated levels of enzyme are observed in nephrosis. Synthesis of albumin to replace that lost in the urine is accompanied by synthesis of additional cholinesterase. Marginal increases in enzyme are also seen in thyrotoxicosis, in obese diabetics, and in patients with anxiety states.

Among the organic phosphorus compounds which inhibit cholinesterases are many organic insecticides, such as Parathion, Sarin, and tetraethyl pyrophosphate. Workers engaged in agriculture, and those working in organic chemical industries, are subject to poisoning by inhalation of these materials or contact with them. Obviously, if enough material is absorbed to inactivate all the acetylcholinesterase, death will result. It is the red cell (nerve) enzyme which is important here, although both types of enzyme are inhibited, with the activity of serum enzyme falling more rapidly than that of the red cell enzyme. A 40 per cent drop in enzyme activity occurs before the first symptoms are felt, and a drop of 80 per cent is required before serious neuromuscular effects become apparent. Near zero or zero levels of enzyme require emergency treatment of the patient with such enzyme reactivators as pyridine-2-aldoxime.

Succinyldicholine (Suxemethonium) is a drug used in surgery as a muscle relaxant. Because it is very similar to acetylcholine, it is also hydrolyzed by cholinesterase, and its physiological effect persists only long enough to meet the needs of the surgical procedure. In patients with low levels of enzyme activity, or in those with the atypical, weakly active form, this destruction of the drug will not occur rapidly enough, and the patient may enter a period of prolonged apnea. Thus, physicians may order cholinesterase levels on patients for whom Suxemethonium anesthesia is planned, to be certain that use of the drug is safe. The presence of abnormal enzyme may be confirmed by determining either the dibucaine (Nupercaine) or fluoride numbers. These parameters indicate the per cent inhibition of enzyme activity (toward benzoylcholine) in the presence of a standard concentration of these reagents. The average values of the dibucaine numbers for normals, heterozygotes, and homozygotes are 78, 60, and 16 per cent respectively.[52,101]

ASSAY PROCEDURES

The procedure of de la Huerga, Yesnick, and Popper[24] measures the quantity of acetylcholine remaining unreacted after incubation for 60 minutes with the sample at 37°C. The determination is performed at pH 8.6, the optimal pH for the enzyme reaction. The substrate mixture contains barbital buffer (0.075 *M*), 0.05 *M* acetylcholine bromide or iodide, 0.0045 *M* Mg^{++}, and 2.7×10^{-4} *M* K^{+} ions. After incubation, the reaction mixture is treated with hydroxylamine, which converts esters (in this case acetylcholine) to hydroxamic acid derivatives, which can then be measured colorimetrically at 540 nm. by virtue of the orange-brown complex formed with ferric ion in acid solution.

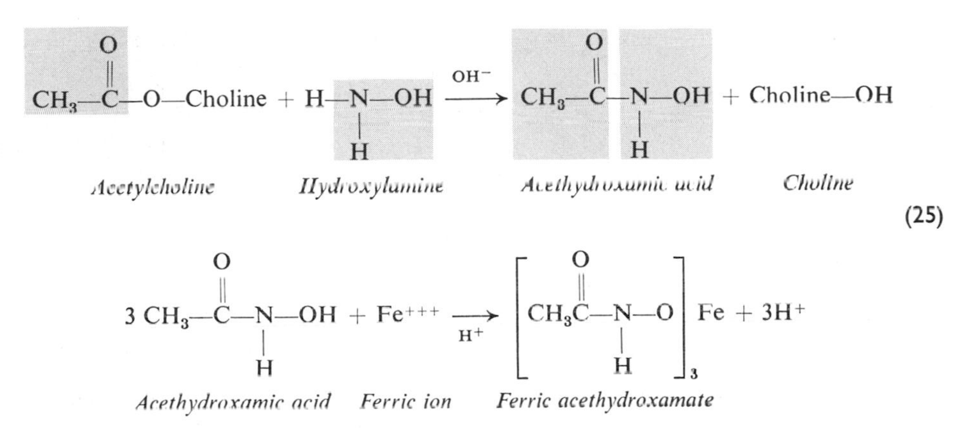

The de la Huerga unit of activity is defined as that quantity of enzyme present in 1.0 ml. of serum which will hydrolyze 1.0 μmol. of acetylcholine in 60 minutes, under the conditions of the assay. Thus, 1.0 I.U./ml. = 60 Huerga units/ml. and 1.0 Huerga unit/ml. = 16.7 mI.U./ml. The precision of the procedure is about 6 to 8 per cent in the normal range, and about ±10 per cent with low enzyme levels.

Other assay methods measure the amount of hydrogen ion formed in the reaction illustrated in equation 22. In manometric methods the hydrogen ion reacts with bicarbonate buffer to release carbon dioxide, which is measured. The potentiometric methods, such as that of Michel,[63] measure the drop in pH in the course of a fixed reaction period. A weak buffer is used to permit a limited degree of pH change; as a result, the reaction is proceeding at a varying pH, and constancy of rate is not maintained. If pH indicators, which change color in the pH range of 6.0 to 8.5, are used, the change in color, measured spectrophotometrically, can serve as measure of enzyme activity. In the titrimetric procedures, standard alkali is added to keep the pH constant as the reaction proceeds, and enzyme activity is measured by the equivalents of alkali required in a fixed reaction period.

Substrates other than acetylcholine have been used. Kalow and Genest[52] used benzoylcholine and measured the decrease in absorbance (substrate) at 240 nm. With acetylthiocholine, the liberated thiocholine is measured iodometrically. Phenylbutyrate may be used, in which case the split phenol is measured with the Folin-Ciocalteu reagent.

SUPPLEMENTARY READING

Wetstone, H. J., and Bowers, G. K.: Serum cholinesterase. In Standard Methods of Clinical Chemistry. D. Seligson, Ed. New York, Academic Press, 1963, Vol. 4, pp. 47–56.

ALDOLASE

I.U.B. No. 4.1.2.7; Ketose-1-phosphate aldehyde lyase

The enzyme aldolase belongs to a class of enzymes called *lyases.** The important reaction catalyzed by aldolase is the splitting of fructose-1,6-diphosphate (FDP) to glyceraldehyde-3-phosphate and dihydroxyacetone phosphate. The equilibrium of the reaction favors the formation of the fructose diphosphate.

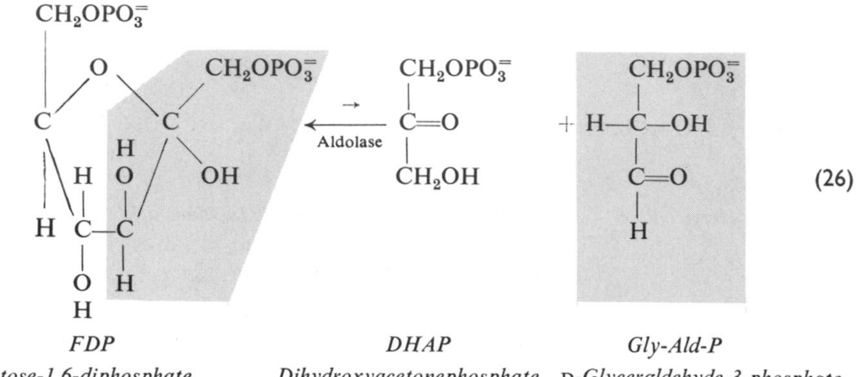

FDP *DHAP* *Gly-Ald-P*

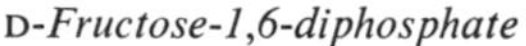

D-*Fructose-1,6-diphosphate* *Dihydroxyacetonephosphate* D-*Glyceraldehyde-3-phosphate*

This is one of the reactions in the glycolytic breakdown of glucose to lactic acid. The enzyme shows absolute specificity only for dihydroxyacetone phosphate (DHAP). The glyceraldehyde-3-phosphate (Gly-Ald-P) can be substituted by glyceraldehyde or formaldehyde, to form fructose-1-phosphate (F-1-P) or erythrose-1-phosphate, respectively.

The optimal pH for the enzyme is rather broad, ranging from 7.0 to about 9.6. It depends on the buffer used and on the nature of other reagents which are added to the enzyme system to push the reaction to completion. Heavy metal ions (Cu, Ag, Fe) inhibit activity, but chelating agents such as EDTA do not counteract the inhibition. Phosphate and borate buffers inhibit activity and should not be used. Optimal activity is observed at 46°C., but assays are generally performed at temperatures between 25 and 37°C.

The enzyme is present in all cells in the body, but differences are observed in the rates at which the enzymes of various organs act on the two substrates, fructose-1,6-diphosphate (FDP) and fructose-1-phosphate (F-1-P). The muscle enzyme, though very active against the diphosphate, shows no activity against F-1-P, whereas those of liver, kidney, and leukocytes show the same activity toward both substrates. The enzyme present in serum, heart, brain, lung, and erythrocytes shows substrate specificity similar or identical to that of the enzyme found in muscle.

The *serum enzyme* is quite stable. Activity is unchanged for 48 hours at room temperature, and remains unchanged for at least three to four weeks if the enzyme is refrigerated. Recent studies by Thompson[91] indicate that the normal range is 1.7 to 3.2 mI.U./ml. for assay values at 25°C. Assuming a temperature conversion factor of

---

* The lyases as a class include a miscellany of enzymes which reversibly cleave substrates into two parts, *without hydrolysis*, with the formation of a double-bonded carbon (—C=C=, or —C=O—) in one or both products.

12 per cent per degree, the corresponding values for 37°C. are 1.5 to 7.0 mI.U./ml. These latter figures are equivalent to 2.0 to 9.3 Sibley-Lehninger[79] (SL) units/ml. Recent experience suggests, however, that the normal range for inactive persons and patients at bedrest is only 50 to 70 per cent of the values observed in active normals.

NORMAL VALUES AND CLINICAL APPLICATION

The serum aldolase level at birth is about 15 to 22 SL units (11 to 17 mI.U./ml.). Values fall slowly with increase in age until adult levels are reached at the age of 18 to 20 years.

The level in red cells is about 150 times higher than that in serum; thus, hemolyzed serum should not be used for analysis. The enzyme is also found in spinal fluid and in serous effusions, but not in urine unless associated with proteinuria.

Serum aldolase determinations are of greatest value in muscle diseases, in which increases of 5 to 10 times the normal level may be seen. The highest levels are found in patients with progressive muscular dystrophy (Duchenne type, Erb's paralysis). The greatest serum elevations occur early in the course of the disease, but as the capacity of muscle cells to synthesize enzyme decreases, serum levels also decrease. Normal aldolase values are observed in muscular dystrophy arising from neurogenic causes, as well as in limb girdle dystrophy, poliomyelitis, polyneuritis, multiple sclerosis, and myasthenia gravis.

Increases in serum aldolase are also observed in myocardial infarction, the pattern of rise and fall paralleling that of GOT (aspartate transaminase) with a peak at 24 to 50 hours and a fall to normal levels at four to six days. Increases of 5 to 20 times the normal level are associated with acute hepatitis, but only marginal increases are seen with chronic hepatitis, cirrhosis, and obstructive jaundice.

In general, serum aldolase determinations do not provide more information than more commonly assayed enzymes such as GOT and LDH. Even in muscular dystrophy, measurement of CPK is more useful because elevations are greater and more easily differentiated from those due to other diseases.

Injections of deoxycorticosterone, cortisone, and ACTH will raise serum aldolase to levels between 12 and 20 SL units. This physiological response to hormone therapy must be kept in mind when interpreting elevated aldolase levels.

Methods for the Determination of Aldolase Activity

All assay methods are based on the forward reaction, as written in equation 26. The discoverers of aldolase, Meyerhof and Lohmann, assayed the enzyme by measuring the quantity of *alkali-labile phosphate* formed from FDP. The phosphate group present in the two triose-phosphate products of the reaction, glyceraldehyde-3-phosphate (Gly-Ald-P) and dihydroxyacetone phosphate (DHAP), can be split off by 1 *N* NaOH in 15 minutes, whereas the phosphate groups in FDP are not affected. Sibley and Lehninger[79] carried out the reaction in TRIS buffer at pH 8.6, reacted the dephosphorylated glyceraldehyde and dihydroxyacetone with 2,4- dinitrophenylhydrazine, and measured the colored nitrophenylhydrazones at 540 nm. The procedure was standardized against glyceraldehyde. A Sigma Chemical Company kit[80] is based on this procedure.

Bruns[20] suggested a method in which the aldolase reaction is coupled with an "indicator enzyme reaction," namely that of glyceryl-phosphate dehydrogenase

(1-Gly-PDH). The latter, with the mediation of NADH, catalyzes the reduction of DHAP to α-glyceryl phosphate (equation 27b). Triosephosphate isomerase (TPI) is added to the system as part of the 1-Gly-PDH reagent to insure that all triosephosphate formed in equation 26 is reduced to glycerylphosphate. Since serum contains

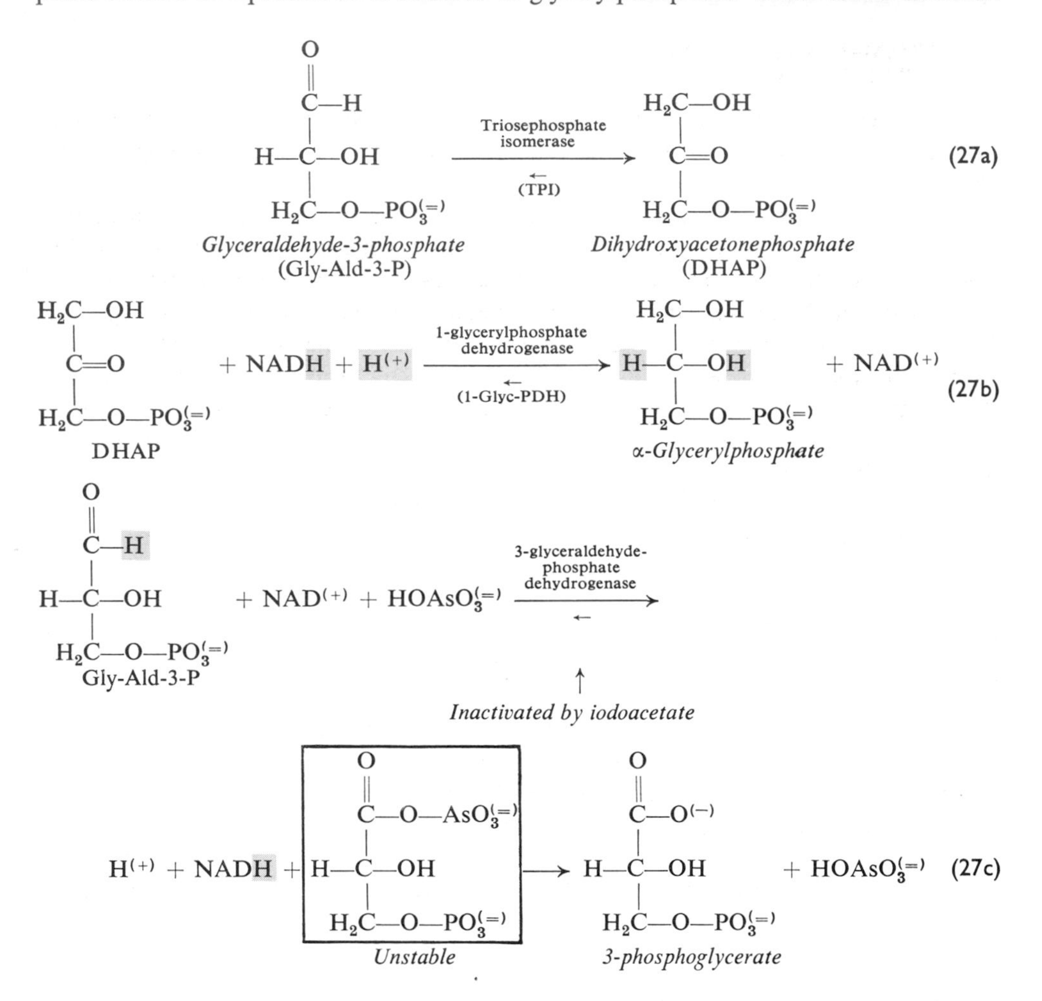

the enzyme glyceraldehyde-phosphate dehydrogenase, some of the Gly-Ald-3-P may follow a different reaction path and be oxidized to 3-phosphoglycerate as shown in equation 27c. Iodoacetate inhibits this reaction and is therefore added to the system to prevent this side reaction.

The enzyme reaction is followed by measuring the decrease in absorbance at 340 nm. as the concentration of NADH decreases, as shown in equation 27b. The Boehringer procedure[12] is based on this principle.

Determination of Aldolase by the Dinitrophenylhydrazone Reaction

REAGENTS

1. Collidine buffer, 0.10 *M*, pH 7.4. Dissolve 3.03 gm. (3.30 ml.) of collidine in about 200 ml. of water, and adjust the pH to 7.4 by adding 2 *N* HCl. Dilute the

volume to 250 ml. The reagent is stable refrigerated, but because collidine is volatile, the pH should be checked weekly.

2. Hydrazine, 0.56 *M*, pH 7.4. Dissolve 7.29 gm. of $NH_2NH_2 \cdot H_2SO_4$ in about 50 ml. of warm water, raise the pH to 7.4 by adding 10 per cent (w/v) NaOH, and make up the volume to 100 ml. The hydrazine "traps" the triosephosphates as they are formed and prevents any change in the ratio of the two forms by virtue of the TPI reaction (equation 27a).

3. Iodoacetate, 2×10^{-3} *M*, pH 7.4. Dissolve 42 mg. of sodium iodoacetate in 100 ml. of water, and adjust the pH to 7.4 with dilute NaOH. This reagent serves to inhibit the oxidation of Gly-Ald-3-P by NAD, if any 3-phosphoglyceraldehyde dehydrogenase activity is present in the serum or assay system.

4. Frustose diphosphate substrate, approximately 0.06 *M*, pH 7.4. The solution contains 32 mg. of the tetrasodium salt of fructose-1,6-diphosphate/ml. Check the pH and adjust to 7.4, if necessary. Prepare only a week's supply, and store the reagent in the freezer.

5. Trichloroacetic acid, TCA, 10 per cent (w/v). Stable.

6. Sodium hydroxide, 0.75 per cent (w/v), or 0.20 *N*. Stable.

7. Color reagent, 2,4-dinitrophenylhydrazine, 0.1 per cent (w/v) in 2 *N* HCl. Dissolve 200 mg. of reagent grade chemical in 200 ml. of warm acid (33 ml. concentrated HCl per 200 ml. acid solution). Store the solution in a brown bottle at room temperature.

8. Calibration reagents. See reagents for the determination of inorganic phosphate.

Note: Unless otherwise indicated, store all reagents at 2 to 4°C.

PROCEDURE

1. Into each of two 15 × 120 mm. test tubes, marked T (test) and B (blank), pipet 1.0 ml. of buffer plus 0.25 ml. each of water, hydrazine, and iodoacetate solutions. If a large number of specimens are assayed, it is convenient to premix these four reagents in the proportion 4:1:1:1, and to pipet 1.75 ml. to each tube. Then place the tubes in a 37°C. bath to come to temperature.

2. Pipet 1.0 ml. of serum into each tube, and then add 0.25 ml. of FDP substrate to tube T. Replace all tubes in the water bath at 37°C., and permit the reaction to proceed for 60 minutes.

3. Remove the tubes from the bath, and to each add 3.0 ml. of the 10 per cent TCA to stop the reaction. Immediately add 0.25 ml. of substrate to tube B, and mix each tube thoroughly, allow to set for 10 minutes, and then centrifuge.

4. Transfer 1.0 ml. of each supernatant to another 15 × 120 mm. test tube. Add 1.0 ml. of 0.75 per cent NaOH to each tube and set the tubes aside for 15 minutes to permit phosphate hydrolysis to occur. Then add 1.0 ml. of the color reagent (2,4-dinitrophenylhydrazine), warm the tubes at 37°C. for 10 minutes, and dilute the contents with 8.0 ml. of 0.75 per cent NaOH.

5. Read the absorbances of the T tubes against the B tubes at 540 nm., within 10 minutes. If the absorbance readings are over 0.300, repeat the analysis with a dilution of the serum specimen.

CALCULATION AND CALIBRATION

The Sibley and Lehninger unit of aldolase activity is defined as that quantity of enzyme in 1 ml. of serum which, under the conditions of the enzyme assay, hydrolyze

1 μl. of FDP. (In accordance with the suggestion of Warburg, all substrates—in this case FDP—are assumed to be gases.) One mole of a gas under standard conditions occupies 22.4 liters, and 1 μl. is equivalent to 1/22.4 = 0.0445 μmol. Also, since one mole FDP forms two equivalents of alkali-labile phosphate, we may redefine the Sibley and Lehninger unit as that amount of aldolase per ml. serum which will form (from FDP) 0.045 × 2 = 0.089 μmol. of alkali-labile phosphate.

Standardize the procedure by analyzing a number of serum samples with low, intermediate, and high enzyme levels as outlined in the procedure, steps 1 to 5. Treat a separate 1.0 ml. aliquot of the TCA filtrate obtained in step 3 of the procedure with 1.0 ml. of *N* NaOH, and allow the mixture to stand at room temperature for 20 minutes.* Then neutralize the filtrate with TCA and assay for inorganic phosphorus by a convenient phosphorus procedure. The difference in the inorganic phosphorus content of the T and B tubes represents the triose phosphate phosphorus.

Since 1 mole of FDP forms two equivalents of alkali-labile phosphate, 1 μg. of alkali-labile phosphorus is equivalent to 1/(31 × 2) = 0.0161 μmol. or 16.1 nmol. FDP. Thus, if 1.0 ml. of filtrate contains 1.0 μg. of phosphorus, the 6.0 ml. of filtrate contain the phosphorus from 16.1 × 6 = 96.6 n mol FDP. This, in turn, indicates that 99.6/44.5 or 2.17 μl. of FDP have been hydrolyzed during the 60 minute incubation (= 2.17 SL units). If 96.6 nmol. FDP are hydrolyzed in 60 minutes, then:

$$\text{nmol./min./ml. (= mI.U./ml.)} = \frac{96.6}{60} = 1.61$$

Thus, 1.0 SL unit = 1.61/2.17 = 0.745 mI.U./ml.

If the absorbance readings (procedure step 5) are carried out in a 1.0 cm. cuvet in an accurately calibrated, narrow band pass spectrophotometer, they may be converted into SL units of aldolase activity directly by multiplying by the factor F = 37 (Bruns[19]).

REFERENCES

Bruns, F. H.: Biochem. Z.: *325*:156, 429, 1954.

Bergmeyer, H.-U. (Ed.): Methods of Enzymatic Analysis. D. H. Williamson (Trans.) New York, Academic Press, 1963.

SUPPLEMENTARY READING

Thomson, W. H. S.: The clinical chemistry of muscular dystrophies. In Advances in Clinical Chemistry. H. Sobotka and C. P. Stewart, Eds. New York, Academic Press, 1964, Vol. 7, p. 138.

Fleisher, G. S.: Aldolase. In Standard Methods of Clinical Chemistry. D. Seligson, Ed. New York, Academic Press, 1961, Vol. 3, pp. 14–22.

CERULOPLASMIN

The enzyme ceruloplasmin (no I.U.B. number or name) is a blue colored α_2 globulin with a molecular weight of about 151,000, containing 0.32 per cent of copper. There are eight atoms of copper per mole, four of which are irreversibly bound to the protein. The other four are only loosely associated and are readily removable by mild proteolytic digestion. Of the total plasma copper, about 94 to 95% is bound

* The concentration of the NaOH is higher and the duration of the incubation period longer than in step 4 of the procedure to insure complete hydrolysis of *all* triosephosphates into inorganic phosphate.

to ceruloplasmin; the remainder is bound to albumin. Only a trace is present as free Cu^{++}.

Although the protein ceruloplasmin was initially studied as one of the "acute phase reaction" proteins,[67,74] its oxidase properties were soon recognized. Oxidases are enzymes which catalyze the oxidation of a substrate by molecular oxygen in a reaction in which only oxygen can serve as an acceptor for the hydrogen, and in which water is the product of the reaction.

$$2R{-}H_2 + O_2 \xrightarrow{\text{Oxidase}} 2R + 2H_2O \qquad (28)$$

$$2\ \text{L-ascorbic acid} + O_2 \xrightarrow{\text{Oxidase}} 2\ \text{dehydroascorbic acid} + 2H_2O$$

In addition to ascorbic acid, the following materials can serve as substrates: benzidine, p-phenylenediamine (PPD), N,N-dimethyl-p-phenylene diamine (DPD), epinephrine, serotonin, dihydroxyphenylalanine (DOPA), and guaiacol.

Cyanide (CN^-) and azide (N_3^-) complex and remove the loosely bound copper from the protein molecule and thus inhibit the enzyme. A number of anions (Cl^-, Br^-, CNS^-) and the cations Fe^{++}, Co^{++}, and Ni^{++} act as weak activators of the enzyme at low concentrations (0.01 *M*) but are all inhibitory at higher concentrations (0.5 *M*).

This enzyme is not included among the list compiled by the International Union of Biochemistry, possibly because the Commission felt that the enzyme was not as yet sufficiently characterized to warrant inclusion in the list.

CLINICAL SIGNIFICANCE AND NORMAL VALUES

The true function of ceruloplasmin in the body is as yet unknown. It has been postulated that it acts as a copper transport protein, and that it regulates the absorption of ingested copper from the gastrointestinal tract. (See also the discussion of copper in Chapter 10.) Most investigators doubt that ceruloplasmin acts as a plasma oxidase, although a number of important metabolites, such as DOPA, epinephrine, and serotonin, normally present in the blood, can serve as substrates for the enzyme.

The average serum level of ceruloplasmin is about 34 mg. per 100 ml., and it is believed that this enzyme is derived from the liver. The level of ceruloplasmin at birth is about one-fourth that found in adults. Enzyme levels then increase rapidly, reaching a maximum concentration at the second year of life (25 per cent above the adult level), followed by a slow drop through adolescence. Table 8-4 lists serum ceruloplasmin and copper levels at various ages, and also demonstrates how copper and ceruloplasmin concentrations change in parallel fashion.

The enzyme is stable at room temperature for two to three days, and for at least two weeks when refrigerated.

Since ceruloplasmin is one of the α_2-globulin "acute reaction" proteins,[67] its serum concentration is moderately increased (50 to 100 per cent) in conditions of physiological stress.[74] *Elevations* are encountered in chronic infections (tuberculosis and pneumonia), lupus erythematosus, rheumatic arthritis, and rheumatic fever. Increased levels are also seen after myocardial infarctions and after the stress of surgery. Both the physiological stress and the effect of increased estrogen levels account for the significantly increased levels of ceruloplasmin encountered in pregnancy.

Diminished levels of the enzyme are encountered in three conditions: kwashiorkor

TABLE 8-4. *Ceruloplasmin and Copper Concentrations in Serum in a Number of Clinical Conditions*

Conditions	*Ceruloplasmin (mg./100 ml.)* Mean Value and Range	*Copper (μg./100 ml.)* Mean Value and Range	*Reference*
Healthy newborns	7 (2–13)	16 (12–26)	Sass-Kortsak[74]
Two-year-olds	43 (31–54)	140 (95–186)	
Ten-year-olds	34 (22–45)	117 (72–162)	
Young Adults	31 (21–41)	109 (69–150)	
Adults	33 (25–43)	114 (89–147)	
Pregnancy at term	55 (39–89)	216 (118–302)	
Pregnancy	84 (62–107)		Markowitz[62]
Infectious disease	68 (60–83)		
Nephrosis	30 (12–54)		
Wilson's Disease	9 (2–19)	Usually very low in homozygotes; rarely low in heterozygotes	Sass-Kortsak[74]

(nutritional protein deficiency), nephrosis, and Wilson's disease. In nephrosis some 50 to 75 mg. of the enzyme may be lost daily in the urine, along with albumin, but blood levels, although usually decreased, may be normal.

Ceruloplasmin is of most interest clinically because of its relation to Wilson's disease, also known as hepatolenticular degeneration. This disease was first described by Wilson in 1912 and is characterized by low levels of serum copper and ceruloplasmin, the deposition of copper in many tissues of the body, a marked increased excretion of copper and amino acids, and muscular rigidity and lack of co-ordination as a result of injury to the cerebral basal ganglia. The copper found in the liver, brain, and kidney may be some 20 times that normally present. Copper is also deposited in the corneas of the eyes, giving rise to so-called Kayser-Fleischer rings. The neurological symptoms and the aminoaciduria are the result of the toxic effects of the copper deposited in the brain and kidneys. The nature of the biochemical defect which gives rise to the symptoms of the disease is not known. Some abnormality in copper metabolism is obviously involved, and it was first postulated that the defect was associated with the inability of the body to synthesize ceruloplasmin and regulate absorption of copper from the intestinal tract. It is now hypothesized that the metabolic defect may be in the mechanism by which copper is made available for synthesis into ceruloplasmin and for transport into and out of liver cells for excretion into the bile. The disease is treated by giving the patient EDTA, Dimercaprol (BAL), or penicillamine, agents which complex copper and facilitate its removal from the tissues in which it has been deposited. Although the disease is congenital, the clinical features usually do not develop until adolescence or early adulthood.

Methods for the Determination of Ceruloplasmin

Early methods for the determination of ceruloplasmin were based on the measurement of $\Delta A_{605\ nm.}$, before and after bleaching the blue color of the ceruloplasmin protein-copper complex with ascorbic acid or cyanide. Assays presently in use are based on measurement of the catalytic activity of ceruloplasmin. The substrate most frequently employed is p-phenylenediamine, (PPD), which, in the presence of oxygen and ceruloplasmin, is converted into a colored product, postulated by Rice[72] to be

Bandrowski's base (equation 29), and by Henry[42] to be the free radical semi-quinone, Wuerster's red (equation 30).

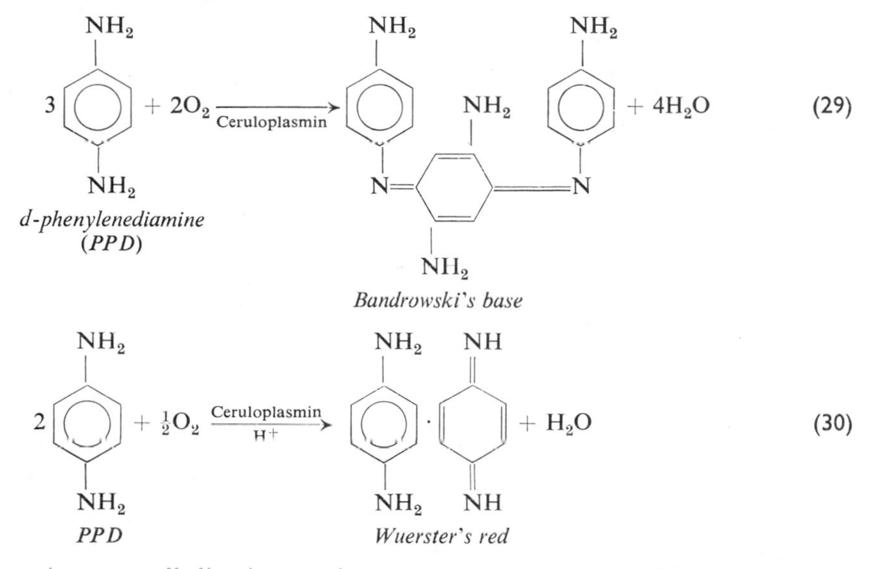

During the assay all diamines undergo some *nonenzymatic* oxidation, at a rate which depends on the purity of the substrate and the presence of catalyzing ions such as Cu^{++} and Fe^{++}. The reaction, therefore, is frequently carried out with and without addition of azide (NaN_3). The assay carried out with azide (enzyme poison) measures the nonenzymatic oxidation; the difference between the two assays is a measure of the rate of the enzyme catalyzed oxidation. Biological samples generally contain ascorbic acid and perhaps other oxygen acceptors whose oxidation precedes that of PPD. Thus, a "lag phase" of variable length has been observed in the ceruloplasmin assay, resulting in low results if so-called "one-point" procedures are used.

Henry,[40] therefore, recommended a two-point procedure, summarized as follows:

A cuvet containing 2.0 ml. of 0.10 *M* acetate buffer (pH 6.0) and 1 ml. of 0.25 per cent (w/v) solution of p-phenylenediamine dihydrochloride in acetate buffer (substrate) is labeled T, and a second test tube containing 1 ml. of buffer, 1 ml. of 0.1 per cent (w/v) of NaN_3 in buffer, and 1.0 ml. of PPD is marked C (control). Place both cuvets in a 37°C. water bath (in the dark) and add a serum aliquot to both tubes. After 10 and 40 minutes of incubation read the absorbance of the cuvet marked T against the cuvet marked C. Multiply the difference in results obtained between the 40 minute and 10 minute reading by 1000 to give the ceruloplasmin level, expressed in empirical units. With this procedure normal sera give values between 250 and 570 units. King[55] suggests that these units can be converted to mg. ceruloplasmin/100 ml. by using an empirical conversion factor of 0.060. The precision is of the order of ± 8 per cent. Glassware should be acid-washed thoroughly to remove all trace metals.

The enzyme has also been assayed by quantitative determination of the immunochemical precipitate formed with specific antihuman ceruloplasmin antiserum.[62]

SUPPLEMENTARY READING

King, J.: Practical Clinical Enzymology. Princeton, D. Van Nostrand Co., 1965.

Wilkinson, J. H.: An Introduction to Diagnostic Enzymology. Baltimore, Williams and Wilkins Co., 1962.

Sass-Kortsak, A.: Copper metabolism. In Advances in Clinical Chemistry. H. Sobotka and C. P Stewart, Eds. New York, Academic Press, 1965, Vol. 8, pp. 1–68.

LEUCINE AMINOPEPTIDASE

I.U.B. No. 3.4.1.1; No systematic name

Leucine aminopeptidase (LAP) is a peptidolytic or proteolytic enzyme which catalyzes the hydrolysis of N-terminal residues from certain peptides and amides containing a free amino group.

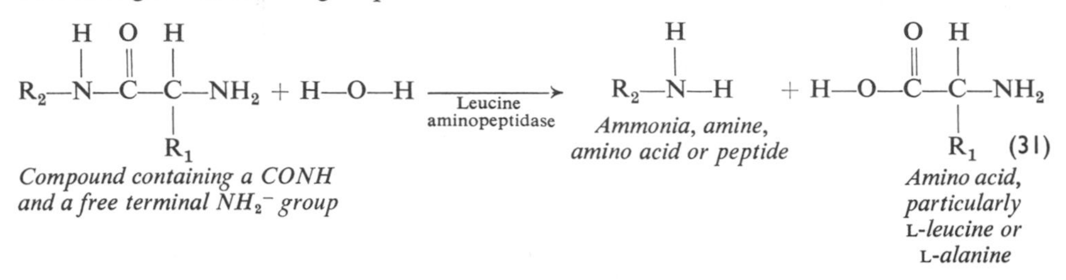

Activity is especially favored when the N-terminal residue is leucine. The enzyme is widely distributed in human tissues and is found, in order of increasing activity, in kidney, small intestine, brain, large intestine, spleen, liver, gastric mucosa, and pancreas. Significant activity has also been demonstrated in plasma, urine and bile.

It is probable that LAP activity is associated with a group of closely related enzymes, rather than with just one single enzyme. Suitable substrates are L-leucyl amide, L-leucylglycine, L-leucylglycylglycine, and both leucyl- and alanyl-naphthylamide. Either magnesium or manganese ions can serve as activators. EDTA inhibits the enzyme, but citrate, fluoride, and oxalate do not. The optimal pH is 7.2 to 7.5.

Numerous substrates have been used to determine aminopeptidase activity. One of the most commonly used compounds is L-leucyl-β-naphthylamide. The β-naphthylamine split off can be measured by the Bratton-Marshall procedure for sulfonamide assay, i.e., by diazotization and coupling with N-(1-naphthyl)ethylenediamine dihydrochloride.[34] The peptidolytic reaction is presented in equation 32, and the Bratton-Marshall diazotization reaction in equation 33.

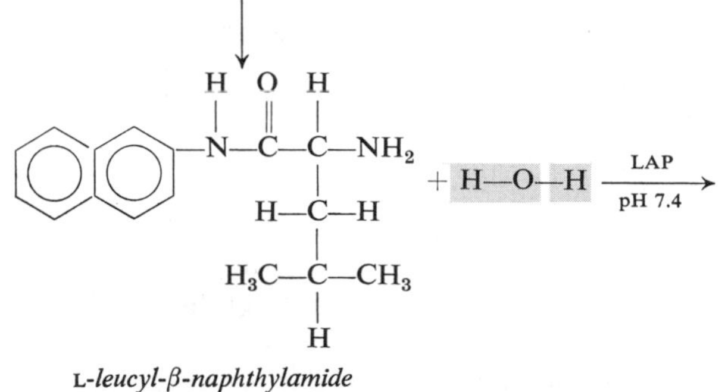

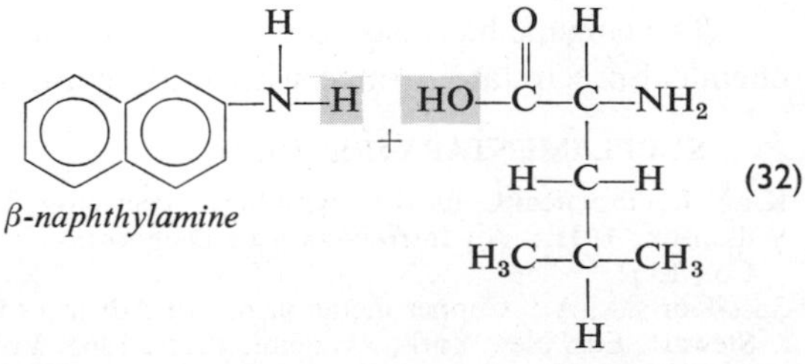

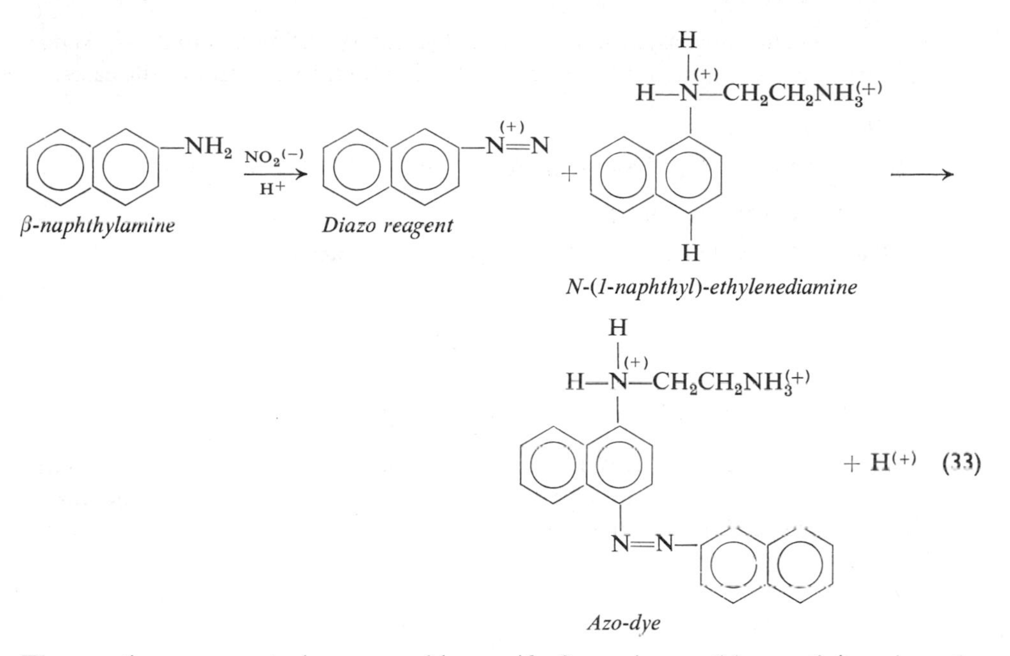

The reaction appears to be reasonably specific for aminopeptidase activity, since the substrate is not hydrolyzed by trypsin, chymotrypsin, pepsin, or carboxypeptidase.

CLINICAL APPLICATION

The determination of leucine aminopeptidase in serum has limited diagnostic significance. Early reports suggested that such assays were of value in the diagnosis of cancer of the pancreas and in the differential diagnosis of jaundice. In one survey, serum LAP was found to be elevated in all of 14 patients with carcinoma of the pancreas. Other investigators, however, have reported normal serum levels in patients with pancreatic cancer without metastases or accompanying jaundice.

The present consensus appears to be that serum LAP activity is not a reliable guide to the diagnosis of pancreatic carcinoma; nor can the test be used to differentiate hepatocellular jaundice from obstructive jaundice. In general, elevated values have been found in a number of conditions involving the liver, pancreas, and biliary tract, including acute pancreatitis, carcinoma of the pancreas, common duct stones, and viral hepatitis.

Serum LAP activity is also elevated during pregnancy; during the third trimester values range from two to four times the upper limit of normal. Moderate increases have also been observed between two and five months of pregnancy, but patients with hydatidiform mole had values in the normal range. Further observations are needed to establish the value of serum LAP determinations in distinguishing these two conditions.

NORMAL VALUES

Normal values depend upon the method used. Goldbarg and Rutenburg originally defined their units (GR) in terms of Klett readings but later converted these values to equivalent micrograms of standard β-naphthylamine. The GR unit is defined as the quantity of enzyme in 0.02 ml. of serum that will split off 1/12 μg. of naphthylamine in two hours. It follows that one GR unit = 0.24 mI.U./ml.

The normal range for serum LAP activity is 84 to 200 GR units for males and 76 to 184 GR units for females. The enzyme is also found in urine, but because of the presence of other chromogens, urines must be subjected to preliminary dialysis. When serum LAP is elevated, urine LAP is also elevated in about 98 per cent of the cases.

REFERENCE

Goldbarg, J. A., Pineda, E. P., and Rutenberg, A. M.: Am. J. Clin. Path., *32*:571, 1959.

SUPPLEMENTARY READING

King, J.: Practical Clinical Enzymology. Princeton, D. Van Nostrand Co., 1965.

LACTATE DEHYDROGENASE

I.U.B. No. 1.1.1.27; L-Lactate: NAD oxidoreductase

Lactate dehydrogenase (LDH) is a hydrogen transfer enzyme which catalyzes the oxidation of L-lactate to pyruvate with the mediation of NAD as hydrogen acceptor. The reaction is reversible and the reaction equilibrium strongly favors the reverse reaction, namely the reduction of pyruvate to lactate.

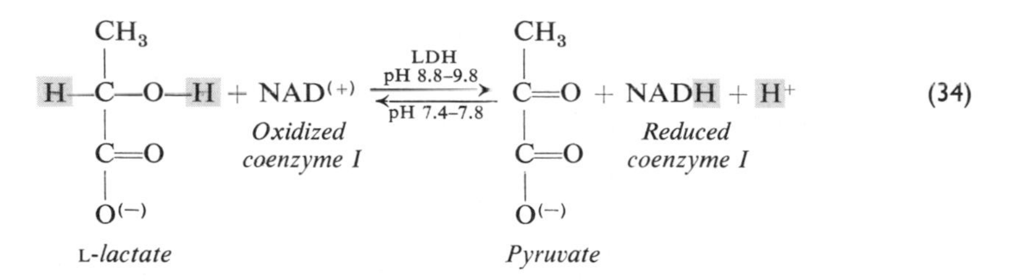

The pH optimal for the forward reaction is 8.8 to 9.8; for the reverse reaction it is 7.4 to 7.9. The optimal pH varies with the source of enzyme and depends on the temperature as well as on substrate and buffer concentration. The specificity of the enzyme extends from L-lactate to a variety of related α-hydroxy and α-hydroxy-γ oxoacids, although only α-hydroxy butyric acid (AHBA), the next higher homolog, reacts at a rate approximating that for lactic acid. This catalytic oxidation of α-hydroxy butyric acid to α-oxo-butyric acid is referred to as α-hydroxy butyric acid dehydrogenase (HBD) activity.[101] The enzyme does not act on D-lactic acid, and only NAD will serve as coenzyme.

The enzyme has a molecular weight of 140,000 and is composed of four peptide chains of two types, *M* and *H*, each under separate genetic control. (Refer to the section on Isoenzymes, p. 389.) Recent studies have demonstrated the existence of another LDH form (LDH-C), present only in postpubertal testicular tissue. It is located on electrophorctic patterns between LDH-3 and LDH-4, and is apparently controlled by a separate genetic locus.

Lactate dehydrogenase is inhibited by sulfhydryl reagents such as mercuric ions and p-mercurichlorobenzoate, the inhibition being reversed by the addition of cysteine or glutathione. Borate and oxalate inhibit by competing with lactate for its binding site on the enzyme; similarly, oxamate competes with pyruvate for its binding site. Both pyruvate and lactate in excess inhibit enzyme activity, although the effect of pyruvate is greater. Substrate inhibition decreases with increase in pH. EDTA inhibits, perhaps by binding Zn^{++}, but the postulated activator role for zinc ions is not fully established.

STABILITY OF LACTATE DEHYDROGENASE

The purified enzyme is very unstable when dissolved in water, but it is quite stable in concentrated ammonium sulfate and in the presence of other proteins. The different isoenzymes differ in their *sensitivity to cold*, LDH-4 and LDH-5 being especially labile. All activity of these two forms in tissue extracts is lost, if the extracts are stored at −20°C. overnight.[109] Loss of activity may be prevented by addition of NAD or glutathione. Both types of monomers bind a mole of NAD, but the binding of NAD to the *M* form is weaker and some dissociation occurs, with concomitant exposure of sulfhydryl groups to oxidation. In serum, the sulfhydryl in albumin and other proteins retards inactivation of the *M* rich isoenzymes, LDH-4 and LDH-5. Serum specimens may be stored at room temperature for two to three days, without significant loss of activity. If specimens of sera must be stored for longer periods, they should be kept at near freezing temperatures, with NAD (10 mg./ml.) or glutathione added to decrease the rate of inactivation of LDH-4 and LDH-5.

Significant loss of activity as a result of *heat denaturation* occurs even at 45°C., and at temperatures above this, activity drops off rapidly, particularly that of the LDH-4 and LDH-5 forms.[105] All LDH-5 activity in serum is lost if the specimen is incubated at 57°C. for 30 minutes. In contrast, the heart form, LDH-1, is stable at 65°C. for 30 minutes.

DISTRIBUTION AND CLINICAL SIGNIFICANCE

LDH content in various tissues and serum. In the literature one finds a multiplicity of values for the range of LDH activity in the sera of healthy persons. Normal values will be discussed in the section on methods, but for the present purpose, the range of 200 to 430 Wroblewski-LaDue (WL) units/ml., measured at 32°C., and established in the author's laboratory over a period of several years, will be used as a basis for discussing changes in LDH associated with various disease states.

Enzyme levels present in various tissues (units/gm.) are very high as compared to serum: liver, 260,000 units/gm.; heart, 160,000 to 240,000 units; kidney, 250,000 to 300,000 units; skeletal muscle, 133,000 units; and whole blood, 16,000 to 62,000 units.[55] Thus, tissue levels are about 1000-fold higher than those normally found in serum, and leakage of the enzyme from even a small mass of damaged tissue can increase the observed serum level. Each tissue has its own pattern of isoenzymes, and diffusion from a given tissue may impress its pattern onto the pattern found in the serum. Isoenzyme patterns found by various investigators for some selected tissues are presented in Table 8-5. Values for the distribution of the isoenzymes in serum, culled from the literature, are also included. The variations in the values obtained reflect both real variations in sera and differences in methods used to quantitate the isoenzyme fractions.

LDH serum levels in different disease states. *Myocardial infarcts* are associated with elevations up to as much as 2500 units (10 times normal) but are usually only 5 to 6 times normal. The rise in serum level begins at 48 to 72 hours after the onset of pain, and the level remains elevated for 10 to 14 days. This contrasts with GOT values, which begin their rise at 6 to 12 hours but are back within normal range within five days after reaching elevations of 10 to 20 times normal. LDH values above 1800 to 2000 units indicate a poor prognosis. Values are moderately elevated in cardiac

TABLE 8-5. *Isoenzyme Pattern in Some Selected Tissues in Per Cent of Total LDH Activity*

Organ or Tissue	*Lactate Dehydrogenase Isoenzyme, LDH No.* 1	2	3	4	5	*Reference No.*
Heart	35–70	28–45	2–16	0–6	0–5	69
Kidney	28	34	21	11	6	69
Liver	0–8	2–10	3–33	6–27	30–85	69
Skeletal muscle	1–10	4–18	8–38	9–36	40–97	69
Brain	21–25	21–26	36–54	15–20	2–8	69
Erythrocytes	39–46	36–56	11–15	4–5	2	69
Lung	10	20	30	25	15	69
Spleen	6–10	11–15	35–40	20–28	5–20	69
Normal serum	25–31	38–45	17–22	5–8	3–6	68
Normal serum	31–54	37–54	3–15	0–5	0–3	95
Normal serum	26–40	40–54	10–18	3–9	0	
Normal serum	31–32	40–42	21–22	5–6	0	60

failure and in pericarditis with hepatic congestion. Levels are moderately to markedly elevated in patients with severe shock and anoxia.

Elevations are also observed in *liver disease*, but these are again of a smaller degree than the elevations of transaminases. Elevations up to 4000 units are found in toxic jaundice and infectious mononucleosis; in viral hepatitis, the values are slightly lower. In cirrhosis and obstructive jaundice, levels may be normal or reach values up to 800 units.

Elevations in serum LDH are seen in about one-third of patients with *renal disease*, but they are not well correlated with other parameters of renal disease such as proteinuria. Elevations in *urine LDH* levels are found in the majority of cases of active glomerulonephritis and acute tubular necrosis, and in most patients with renal or bladder cancers. Wacker and Dorfman report normal urine LDH levels of between 475 and 2000 units per eight hour urine output; these may rise in pathological conditions to levels between 4000 and 6500 units.

Elevations in serum LDH are seen in about 50 per cent of patients with various types of *carcinoma* and are especially high in cases of abdominal and lung cancers. However, the elevations associated with cancers are too erratic to be of use in clinical diagnosis. More useful, to a degree, are assays of LDH in effusions obtained from cancer sites. LDH levels in these fluids are often higher than serum, whereas LDH levels in fluids bathing healthy tissue are lower than serum levels. Elevations, up to two to three times normal, are also encountered in leukemias, especially those of the myelocytic type.

Significant elevations of serum LDH are found in patients with untreated *pernicious anemia*, with levels reaching as high as 20,000 units, although values usually range from 2500 to 15,000 units. With treatment the LDH levels approach normal values. In other types of hemolytic anemias, levels range from normal to about 2000 units.

LDH Isoenzymes

Areas of necrosis in any tissue will release LDH from that tissue with a resultant rise in the serum level of the enzyme. Where only a single tissue is known to be involved, the change in LDH level will reflect the severity of the insult to that tissue or organ. However, frequently it is not possible to determine clinically which organ is contributing to the rise in LDH, particularly in the early stage of a disease process, when the LDH may be only marginally elevated. In such cases, a study of the isoenzyme pattern is often helpful. In *myocardial infarctions*, such patterns will usually show only bands 1, 2 and 3, with the proportion of the total LDH in the first two bands higher than that present in normal serum. The relatively high proportions of LDH-1 and LDH-2 in heart tissue will augment that normally present in these bands in serum. In the case of hepatic disease, the high proportion of LDH-5 in liver will add to the normally low serum level of this isoenzyme and give a pattern in which the first two bands are relatively lower than normal; bands 4 and 5, normally just barely visible, become distinct and clearly elevated. Quantification of the fractions will show the actual changes in the proportions of the various isoenzymes.

A rough but often clinically useful evaluation of a serum LDH pattern may be obtained by determining the proportion of LDH activity destroyed by heating serum at 57°C. for 30 minutes, and that fraction which retains its activity after 30 minutes incubation at 65°C.[105] If total activity (at 32°C.) is denoted by T, 57°C. activity by L, and 65°C. activity by H, ($T - L$) will represent the heat labile (LDH 4 and LDH 5) fractions, with a value of 10 to 25 per cent in normal sera, increasing to 33 to 80 per cent in patients with liver disease. The stable fraction is given by H, with a normal range of 20 to 40 per cent, rising to 45 to 65 per cent in patients with myocardial infarctions.[61]

Rosalki and Wilkinson[101] advocated measurement of α-hydroxy butyric acid dehydrogenase (HBDH) activity and calculation of the LDH/HBDH ratio. LDH-1 appears to be more active with α-ketobutyrate than with pyruvate as substrate. With normal serum, the ratio varies from 1.2 to 1.6; in liver disease the ratio is increased and ranges from 1.6 to 2.5, and in myocardial infarction it is decreased to between 0.8 and 1.2.

Lesions in other tissues will also alter the serum isoenzyme pattern, but the changes observed are more difficult to correlate with the specific tissue concerned. The pattern for *pernicious anemia* is much like that for normal serum, except for an increase in LDH-2 and a slight decrease in LDH-1. The kidney pattern is very much like that for normal serum. In *pulmonary infarcts* the LDH-3 band is elevated, and is almost equal to LDH-2, whereas the level of LDH-1 is clearly decreased. But in general, except for myocardial infarcts and hepatic disease, isoenzyme patterns are of little value in pinpointing the tissue of origin responsible for the increased serum LDH level.

The normal value for *spinal fluid* LDH is 15 to 60 units.[44] Elevations are associated with subarachnoid hemorrhage and with cerebrovascular thrombosis and hemorrhage. CSF-LDH is usually normal in patients with brain or meningeal tumors, but may be elevated in cases of invasive cancers originating from primary sources elsewhere.

Assay Methods for Lactate Dehydrogenase

A multiplicity of individual procedures have been introduced over the last fifteen years, using both the forward and reverse reactions. Wroblewski and Ladue[106] adapted the classic assay of Kubowitz and Ott to the determination of LDH in serum specimens. This method is based on the reverse (pyruvate-lactate) reaction and uses a pH of 7.4 and a temperature of 25°C. The reaction is followed by measuring the decrease in absorbance at 340 nm. as NADH is oxidized to NAD. The unit of activity is defined as that quantity of enzyme which causes an absorbance change of 0.001 per minute if present in a total volume of 3 ml. and measured in a cuvet with a 1.0 cm. light path. This method was subsequently modified and improved by Henry et al.[44] Procedures based on the measurement of changes in NADH or NADPH concentration are generally referred to as "spectrophotometric" or "kinetic" methods.

Valee and his associates[3] recommended the use of the spectrophotometric method with the forward "lactate to pyruvate" reaction at a pH of 9.0 to 9.5. These authors argued that both lactate and NAD are more stable than pyruvate and NADH, and that furthermore, the inhibitory effect of pyruvate (substrate in the reverse reaction) would prevent maximum reaction rates. Nevertheless, procedures based on the forward reaction have not become popular.

Colorimetric methods available are of two types. In the first group, pyruvate is reacted with 2,4-dinitrophenylhydrazine (2,4-DNPH) to form the corresponding phenylhydrazone, which has a golden-brown color at alkaline pH. The colorimetric procedure of Cabaud and Wroblewski[21] is based on the reverse reaction and is

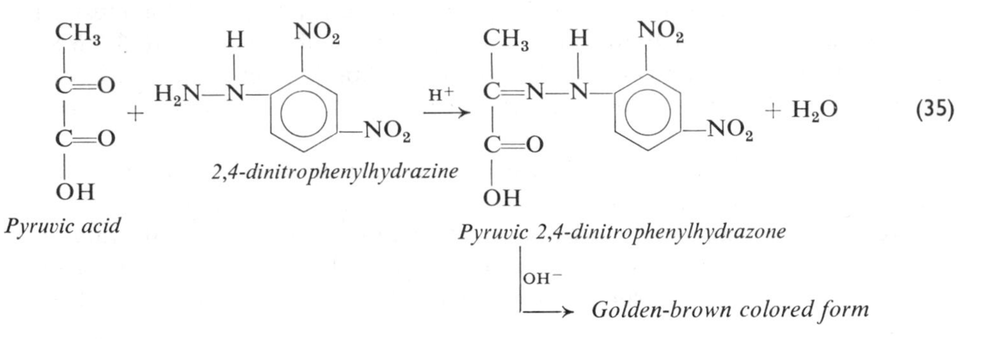

carried out at a pH of 7.8 to 8.0 and at 37°C. The color of the pyruvic 2,4 dinitrophenylhydrazone is measured at 440 or 525 nm., but the units are expressed in terms of the WL "spectrophotometric" unit. The normal range for this method is 200 to 600 W-L units/ml. Several commercial kits are based on this method. In the colorimetric procedures of King[55] and of Zimmerman and Weinstein,[107] 2,4-DNPH is used to measure the pyruvate formed in the forward reaction.

The second group of colorimetric procedures is based on the reduction of such dyes as 2,6-dichlorophenol-indophenol or 2-p-iodophenyl-3-p-nitrophenyl tetrazolium chloride (INT) by the NADH formed in the forward reaction. Phenazine methosulfate (PMS) serves as an intermediate electron carrier between the NADH and the dyes. The colorimetric methods using tetrazolium salts are frequently used to visualize the various LDH isoenzyme fractions after electrophoretic separation.

Determination of Serum Lactate Dehydrogenase by Measurement of NADH Consumed in the Reverse Reaction

REAGENTS

1. Phosphate buffer, 0.10 *M*, pH 7.4. Dissolve 13.95 gm. of anhydrous K_2HPO_4 and 2.70 gm. of anhydrous KH_2PO_4 in water and dilute to 1000 ml. Store the buffer in the refrigerator.

2. Sodium pyruvate, 1.0 mg./1.0 ml. of phosphate buffer. The solution is stable for 1 to 2 weeks when refrigerated, and for at least two months when frozen. Tubes

containing a day's needs (e.g., 5 to 10 ml.) may be prepared, frozen, and thawed out as needed. Concentration = 9.1 μmol./ml.

3. β-NADH, sodium salt (90 to 95 per cent grade), 2.5 mg./1.0 ml. phosphate buffer (= 3.3 μmol./ml.). This reagent is best prepared fresh daily but will keep for a week if frozen. The concentration of the solution may be verified by measuring the absorbance at 340 nm.

4. Dichromate blanks. Stock, 150 mg. $K_2Cr_2O_7$ per 500 ml. of water containing 3 to 4 drops of concentrated H_2SO_4. This is diluted 1:4, 1:7, etc., as needed. (See step 2 of procedure.)

PROCEDURE

1. Use a narrow band pass spectrophotometer (Beckman DU or equivalent), preferably one equipped with a circulating constant temperature water bath. Henry et al. recommend for the assay a temperature of 32°C., which is approximately the equilibrium temperature of the cuvet compartment of the Beckman DU spectrophotometer. If no temperature control is available, measure the temperature of the test mixture at the beginning and the end of the reaction and calculate the mean temperature. Then correct the measured activity to 32°C. The International Union for Biochemistry has recommended the use of 30°C.

2. Pipet 2.5 ml. of buffer, 0.20 ml. of NADH solution, and 0.10 ml. of serum (use a micropipet) into a 1.0 cm. square cuvet. After mixing the contents, place the cuvet in a 32°C. incubator or water bath for 20 to 30 minutes. This incubation permits a reduction by the NADH of any pyruvate and other keto acids present in the serum. During this period, measure the absorbance (at 340 nm.) of the reaction mixture and select a dichromate blank that will give an absorbance reading for the test cuvet of between 0.45 and 0.60. (See comment 2.)

3. At the end of the incubation period, add 0.20 ml. of pyruvate solution, which has also been prewarmed to 32°C., mix rapidly, and then take absorbance readings at one minute intervals for 5 to 8 minutes. The fall in absorbance per minute should be constant over the period. If the ΔA per minute begins to fall off either gradually or sharply before more than four readings have been made, exhaustion of NADH has occurred and the run should be repeated with 0.050 ml. of serum or a dilution (1:5 or 1:10) of the serum with buffer.

CALCULATION

The LDH activity in Wroblewski units at 32°C. is given by

$$\text{units/ml.} = \frac{\Delta A}{\text{min.}} \times \frac{1.0}{v} \times (1000) \times F_T$$

where v = ml. of serum used in the assay, F_T = temperature correction factor, the values of which are listed in Table 8-6; and ΔA/min. is the absorbance change

TABLE 8-6. *Temperature Correction Factors for LDH Based on a Reference Temperature of* 32°C.*

Temperature, °C.	25	27	29	30	32	35	37	40
Temperature factor, F_T	1.67	1.44	1.24	1.16	1.00	0.81	0.70	0.57

* After Henry, R. J. et al.: Am. J. Clin. Path., **34:** 381, 1960.

(decrease) per minute measured at temperature T. The value of ΔA/min. can be calculated by using (total ΔA in t minutes)/(t minutes), or by plotting the data on graph paper (absorbance against time), drawing the best fitting line and then calculating ΔA/min.

Table 8-6 lists temperature correction factors given by Henry and associates for some selected temperatures. These are based on their measured Q_{10} of 2.1 for the 30 to 40°C interval. Values of the factors for the temperatures not listed can be obtained by extrapolation.

NORMAL VALUES

Values for the normal range of serum LDH differ considerably, depending on the type of reaction (forward or reverse) and the type of units employed. Normal values are further affected by the varying assay conditions such as pH, temperature, and substrate concentration, as employed in the various methods. Whichever procedure is employed, each laboratory should establish that the assay reaction used proceeds at optimal conditions, and it should be verified that the published range for normal values is consistent with the experience of the laboratory.

Henry and his associates[44] reported the following values for serum LDH in healthy adults. The values, given in WL units/ml. at 32°C., are in the 95 per cent confidence range and are based on a sample of 40 sera: Serum LDH: males, 235 to 440; females, 210 to 425, with geometric mean values of 320 and 300 respectively. The data fitted a logarithmic normal distribution. Spinal fluid LDH: 15 to 70 units/ml.

Thompson,[91] in his study of a sample of 50 sera with assays performed at 25°C. and expressed in international units, reported values which, when converted to WL units at 32°C., are: males, 190 to 450 (mean = 320); females, 245 to 495 (mean = 370).

In the author's laboratory, the normal range obtained is 200 to 430 WL units/ml. with assays run at 32°C. The median value is about 320. These are based on a sample of over 10,000 sera assayed between 1967 and 1969. Thus the experience of the three laboratories is very similar. Sex differences have been reported but findings are contradictory.

If the reaction is run at 30°C., as per the recent recommendations of the International Commission, the normal range becomes 190–400 WL units/ml.

COMMENTS AND DISCUSSION

1. In the reaction mixture the concentrations of pyruvate and NADH are 6.0×10^{-4} M and 2.2×10^{-4} M respectively. These concentrations permit activity at optimal rates and provide sufficient excess so that linearity of rate with time will persist for 6 to 7 minutes. The comparable values given by Bergmeyer[10] are 3.0 and 1.3×10^{-4} M, respectively.

2. Reaction Blank. The blank is used to compensate for the absorbance of serum pigments. Some analysts use water, or even omit the blank cuvet. Others have used a blank containing serum and buffer only. However, the dichromate blanks are convenient. With spectrophotometers with an absorbance offset mechanism, blanks are not necessary.

3. Some analysts omit the preincubation step, claiming that the endogenous reactions do not significantly alter the value of ΔA/min.; others use a preincubation period of 10 minutes.

4. Hemolyzed serum should not be used since red cells contain 150 times more LDH than serum. For the same reason, it is best to separate serum from the clot within two hours after the blood specimen has been drawn. Heparinized plasma is satisfactory, but plasma containing other anticoagulants, especially oxalate, should not be used.

REFERENCE

Henry, R. J., Chiamori, N., Golub, O., and Berkman, S.: Am. J. Clin. Path., *34*:381, 1960.
Wroblewski, T., and LaDue, J. E.: Proc. Soc. Exp. Biol. Med., *90*:210, 1955.

SUPPLEMENTARY READINGS

Wilkinson, J. H.: An Introduction to Diagnostic Enzymology. Baltimore, Williams and Wilkins Co., 1962.
Bergmeyer, H.-U. (Ed.): Methods of Enzymatic Analysis. D. H. Williamson (Trans.) New York, Academic Press, 1963.
Bowers, G. N.: Lactic dehydrogenase. In Standard Methods of Clinical Chemistry. D. Seligson, Ed. New York, Academic Press, 1963, Vol. 4, pp. 163–172.
Latner, A. L.: Isoenzymes. In Advances in Clinical Chemistry. H. Sobotka and C. P. Stewart, Eds. New York, Academic Press, 1967, Vol. 9, pp. 57–71.
Diagnostic Enzymology, Dade Reagents, Inc., Miami, Fla., 1966.
Manual of Clin. Chem. Procedures. Dade Reagents, Inc., Miami, Fla., 1965.

The Transaminases

GLUTAMIC OXALACETIC TRANSAMINASE (GOT)

I.U.B. No. 2.6.1.1; L-aspartate: 2-oxoglutarate aminotransferase

GLUTAMIC PYRUVIC TRANSAMINASE (GPT)

I.U.B. No. 2.6.1.2; L-alanine: 2-oxoglutarate aminotransferase

The transaminases constitute a group of enzymes which catalyze the interconversion of amino acids and α-ketoacids by transfer of amino groups. L-glutamic acid acts as the amino group donor in most transamination reactions. Phosphopyridoxal and its amino analog, phosphopyridoxamine, function as coenzymes in amino transfer reactions. The transaminases have also been referred to as aminotransferases and aminopherases. The nomenclature for specific transaminases varies, and may be based on the two amino acids involved, on one of the amino acids and the oxoacid with which it reacts, or on only one amino acid, it being understood that the other amino acid is glutamic acid. Thus, aspartate transaminase, commonly known as glutamic oxalacetic transaminase (GOT), catalyzes the reaction shown in equation 36. Alanine transaminase, more familiar as glutamic pyruvic transaminase (GPT), catalyzes the

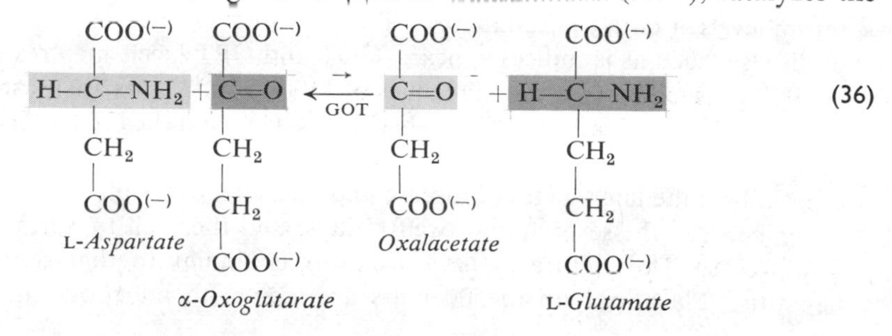

analogous reaction presented in equation 37. The reactions are reversible but the equilibrium of both the GOT and GPT reactions favors formation of aspartate and alanine respectively.

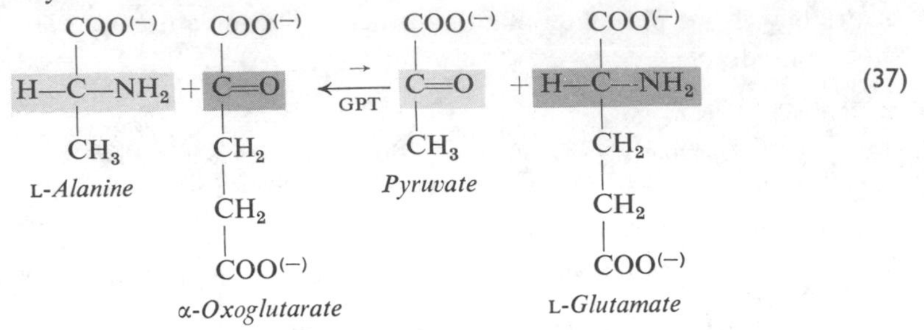

Transaminases are widely distributed in animal tissues. Both GOT and GPT are normally present in human plasma, bile, cerebrospinal fluid, and saliva, but not in urine unless a kidney lesion is present. Activities in various tissues, relative to serum, are shown in Table 8-7.

TABLE 8-7. *Transaminase Activities of Human Tissues Related to Serum as Unity*

	GOT	*GPT*
Heart	7800	450
Liver	7100	2850
Skeletal Muscle	5000	300
Kidney	4500	1200
Pancreas	1400	130
Spleen	700	80
Lung	500	45
Erythrocytes	15	7
Serum	1	1

CLINICAL SIGNIFICANCE

Following *myocardial infarction* GOT appears in serum in increased activity, as might be expected from its relatively high concentration in heart muscle (see Table 8-7). Serum levels, however, do not begin to rise until 6 to 8 hours after the onset of pain. Peak values are reached after 48 to 60 hours, and the level falls to within the normal range by the fourth or fifth day, provided that no new infarct has occurred. Levels above 400 to 500 Karmen units (15 to 25 times normal) are usually associated with fatal infarcts. The peak values are roughly proportional to the degree of cardiac tissue damage. GPT levels are within normal limits, or only marginally elevated.

In *hepatitis* and other forms of liver disease with associated hepatic necrosis, serum levels of both transaminases will be elevated, even before the clinical symptoms of disease (such as jaundice) appear. GOT and GPT levels may reach values as high as 4000 Karmen units, although values of 1500 to 2000 units are most frequently seen. In the majority of cases GPT is higher than GOT, and the GPT/GOT ratio (less than unity normally and in cases of myocardial infarction) becomes greater than 1.0, especially in the later stages of disease and in the recovery phase. Peak values usually occur between the seventh and twelfth days, and then fall to normal by the third to fifth weeks. The picture in *toxic hepatitis* is similar to that seen with infectious hepatitis. Elevations up to 500 units are seen with infectious mononucleosis. In *cirrhosis* the levels observed will vary with the activity of the cirrhotic process, ranging from upper normal to 200 units, with GOT greater than GPT. Similar values are found in obstructive jaundice, but with GPT greater than GOT.

GOT (and less frequently, GPT) levels are also increased in *muscular dystrophy* and dermatomyositis, reaching levels up to 150 to 200 units, but they are usually normal in other muscle diseases, especially those of neurogenic origin. *Pulmonary emboli* can raise GOT levels to two to three times normal, and slight to moderate elevations are seen in acute pancreatitis, crushing muscle injuries, and liver metastases.

TECHNIQUES FOR THE ASSAY OF TRANSAMINASES

As is evident from equations 36 and 37, the assay system for measuring transaminase activity will contain two amino acids and two keto acids. This presents certain problems in evaluating enzyme activity. There is no convenient method available for assaying either of the amino acids in the reaction system. Recourse must therefore be to measuring the keto acids, either the acid consumed or the acid formed in the reaction. The keto acids can be assayed colorimetrically by coupling with 2,4-dinitrophenylhydrazine, as discussed in connection with the colorimetric procedures for LDH (equation 35). There are several disadvantages however to this colorimetric approach. Two keto acids are present, both of which can form the colored phenylhydrazone product. In the case of the GOT reaction, these are α-ketoglutarate and oxalacetate. As the reaction proceeds, one keto acid is increasing in concentration while the other is decreasing. The reaction begins with a keto acid present (one substrate), and hence reagent blanks have a high absorbance. Substrate concentration is kept low, to minimize the high blank readings, and this may become a factor limiting the reaction rate. In the GOT reaction, one of the products, oxalacetate, is relatively unstable. Keto acids such as pyruvate, normally present in serum, produce phenylhydrazone color and contribute to the high blanks. Despite these limitations, the colorimetric approach is still feasible, since the hydrazones of the products of both the GPT and GOT reactions (pyruvate and oxalacetate respectively) are considerably more chromogenic than is the α-ketoglutarate. The colorimetric methods are relatively simple, and though they are less accurate than the "spectrophotometric" or "reaction rate methods" to be described next, they are widely used and are available in many proprietary "kit" forms. Calibration of the colorimetric methods is done empirically so that results obtained are comparable to those obtained by the "reaction rate" methods.[70]

A very useful and highly specific "rate of reaction" technique for measuring keto acids is to reduce them to the corresponding hydroxy acids by use of reduced coenzyme I (NADH) in the presence of a specific dehydrogenase. Oxaloacetate, formed in the GOT reaction, is reduced to malate in the presence of malic dehydrogenase (MDH) (equation 38). Pyruvate, formed in the GPT reaction, is reduced to lactate by lactate dehydrogenase (LDH) (see equation 34). Substrate, MDH, LDH, and NADH are

$$\underset{\textit{Oxalacetate}}{\begin{matrix} COO^{(-)} \\ | \\ C{=}O \\ | \\ CH_2 \\ | \\ COO^{(-)} \end{matrix}} + NADH + H^{(+)} \xrightarrow[\text{(MDH)}]{\text{Malic dehydrogenase}} \underset{\textit{L-Malate}}{\begin{matrix} COO^{(-)} \\ | \\ H{-}C{-}OH \\ | \\ CH_2 \\ | \\ COO^{(-)} \end{matrix}} + NAD^{+} \qquad (38)$$

present in large excess, so that the reaction rate is limited only by the amount of GOT and GPT respectively. As the reactions proceed, NADH is oxidized to NAD^+. The disappearance of NADH per unit of time may be followed by measuring the decrease in absorbance for several minutes at 340 nm. The change in absorbance per minute (ΔA/min.) may be related directly to micromoles of NADH oxidized and, in turn, to micromoles of substrate transformed per minute (international units). The advantages that this technique offers are that it is possible to use the initial, linear phases of the reaction to determine reaction rates; multi-point assays are obtained; substrate concentrations can be set high enough so as not to be rate-limiting; sufficient coenzyme and dehydrogenase may be added to drive the coupled reaction to completion and minimize reverse reactions and flexibility in timing or recording the reaction is available. The disadvantages are that the substrate mixtures are more complex, expensive instruments are required, and the measurements may be rather time consuming, especially if done manually.

Reaction Rate Methods for Assay of Transaminases

A spectrophotometer with good resolution at 340 nm. is required. Temperature of the reaction mixture in the cuvet must be controlled at a constant known level. A thermostatically controlled cuvet compartment is recommended. (Refer also to the discussion under Lactate dehydrogenase methods, p. 439.) A preliminary incubation period should be included to destroy endogenous keto acids in the serum (side reaction) prior to adding the keto acid involved in the reaction. Following this, there is a lag phase of 2 or 3 minutes as with most coupled reactions. Consequently, several readings (5 to 7, at 1 minute intervals), are taken to establish the lineaı portion of the curve. A recorder which produces a curve directly related to absorbance is very useful. Because the reaction mixture has a rather high absorbance initially, it is customary to use a blank reference solution, such as potassium dichromate, so as to produce an initial absorbance for the test of about 0.50, as discussed in the LDH procedure.

Determination of GOT

REAGENTS

1. Phosphate buffer, 1.0 *M*, pH 7.4. Dissolve 136 gm. KH_2PO_4 and 33 gm. NaOH in water and dilute to 1000 ml.

2. Phosphate buffer, 0.10 *M*, pH 7.4. Dilute the 1.0 *M* buffer ten-fold with water. Store the buffers refrigerated.

3. NADH solution, 2.5 mg. β-NADH, disodium salt (90 to 95 per cent grade)/ml. of 0.1 *M* phosphate buffer. It is best to prepare enough for a day's needs. Aliquots may be stored frozen up to one week.

4. α-ketoglutarate solution, 0.1 *M*, in 0.1 *M* phosphate buffer. To approximately 35 ml. of distilled water in a beaker add 5 ml. of 1 *M* phosphate buffer and 0.73 gm. of α-ketoglutaric acid. Adjust to pH 7.4 ± 0.1 with 1 *N* NaOH (approximately 8.1 ml. are required). Dilute to 50 ml. with water. The solution is stable in the refrigerator.

5. MDH solution. Dilute a stock solution of malic dehydrogenase (free of transaminase activity) with 0.1 *M* phosphate buffer to yield a preparation containing 10,000 units/ml. This dilute solution should be prepared within four hours before use.

6. L-aspartate, 0.375 *M*, in 0.1 *M* phosphate buffer. Add 5.0 gm. L-aspartic acid to a 250 ml. beaker containing 50 ml. of water and 35 ml. of 1 *N* NaOH. Mix and warm on a steam bath until crystals are dissolved, then cool to room temperature and add 10 ml. of 1 *M* phosphate buffer. Adjust to pH 7.4 ± 0.1 with 1 *N* NaOH, and dilute to 100 ml. with water. Store the reagent in the refrigerator.

7. Dichromate blanks. Prepare a stock 0.001 *M* solution by dissolving 30 mg. of $K_2Cr_2O_7$ in 100 ml. of water and adding a few drops of concentrated sulfuric acid. This solution is further diluted with water as required in the procedure.

PROCEDURE

1. Add the following to a cuvet with a 1 cm. light path:

1.3 ml. 0.1 *M* phosphate buffer
1.0 ml. aspartate solution
0.2 ml. NADH solution
0.1 ml. MDH solution
0.20 ml. serum

If a large number of determinations are to be done, it is convenient to mix a batch of the first three reagents, sufficient for one days' work, and to pipet 2.5 ml. of the composite reagent to the cuvet, followed by the MDH and serum. The components listed may be mixed in a small test tube and preincubated (step 2). After addition of the ketoglutarate, transfer the contents of the tube to a cuvet and maintain at reaction temperature.

2. The reaction is run preferably at 32°C. or 30°C. Refer to the LDH procedure for comments pertaining to temperature control.

Mix and incubate for approximately 20 to 30 minutes in the spectrophotometer or in a water bath at the same temperature. The "endogenous" reactions are usually completed in this time period. This can be verified, if necessary, by measuring the absorbance at 340 nm. (A_{340}) against a reference dichromate solution and then measuring again a few minutes later. The two readings should be the same if the side reactions have ended. If the second reading is less than the first, continue the readings until no further change in A_{340} is observed. The reference dichromate is chosen to provide an A_{340} of approximately 0.5.

When the absorbance has stabilized, add 0.2 ml. of 0.1 *M* α-ketoglutarate solution, which has been prewarmed to the temperature of the reaction, and mix. Transfer to a cuvet, if this has not yet been done, and take readings every minute for at least 7 to 10 minutes (or use a suitable recorder). Discard the early readings in the lag phase and use the linear period to calculate the average ΔA per minute, either by graphing, or averaging the ΔA/min. values for each reading.

CALCULATION

$$\text{GOT in Karmen units/ml.} = \frac{\text{average } \Delta A/\text{minute}}{\text{ml. of serum used}} \times 1000$$

Correct the results to 32°C. if necessary, by using the factors given in Table 8-8 ($Q_{10} = 1.92$).

The "spectrophotometric" (Karmen) unit[53] is defined as that quantity of enzyme in 1.0 ml., that will cause a change in absorbance (of NADH) of 0.001 unit per minute. This unit may be converted to international units by multiplying by 0.484.

TABLE 8-8. *Temperature Correction Factors for GOT and GPT*

Temperature,* °C.		25	27	30	32	33	35	37	40
Temperature factor, T_F	GOT	1.59	1.39	1.14	1.00	0.94	0.82	0.73	0.60
	GPT	1.57	1.38	1.13	1.00	0.94	0.83	0.73	0.61

* Factors for intermediate temperatures are obtained by extrapolation.

Determination of GPT

REAGENTS

1. Phosphate buffer, 1 *M* and 0.1 *M*. These are the same as those used in the GOT determination.

2. NADH solution. Refer to GOT determinations.

3. DL-alanine, 1 *M*, or L-alanine, 0.5 *M*, in 0.1 *M* phosphate buffer. To approximately 75 ml. of distilled water in a beaker add 10 ml. of 1 *M* phosphate buffer and 8.9 gm. DL-alanine (or 4.45 gm. L-alanine). Adjust the pH to 7.4 ± 0.1 with 1 *N* NaOH (approximately 0.8 ml. is required). Dilute to 100 ml. with water. This solution is stable in the refrigerator.

4. LDH solution. Dilute a stock solution of lactic dehydrogenase with 0.1 *M* phosphate buffer to yield a preparation containing 10,000 units/ml. This dilution can be used for approximately one week if refrigerated.

5. α-ketoglutarate solution, 0.1 *M*, in 0.1 *M* phosphate buffer, as described for the GOT determination.

PROCEDURE

1. Add the following to a cuvet with a 1 cm. light path:

1.3 ml. 0.1 *M* phosphate buffer
1.0 ml. alanine solution
0.2 ml. NADH solution
0.1 ml. LDH solution
0.20 ml. serum

Prepare a composite containing the first three or four reagents, as suggested in the GOT procedure.

2. The procedure and calculations from this point on are identical with those described for GOT (page 445).

COMMENTS PERTAINING TO THE SPECTROPHOTOMETRIC PROCEDURE

1. Occasionally the side reactions in the preincubation stage may be quite extensive, thus consuming a large portion of the available NADH. If the ΔA per minute in the test decreases with time, suboptimal NADH concentration is indicated. In this event, the test may be rerun using a higher initial concentration of NADH.

2. Very turbid or icteric serums may be diluted in order to obtain the appropriate initial absorbance.

3. Hemolyzed serum should not be used. As shown in Table 8-7, GOT and GPT activities in erythrocytes are, respectively, 15 and 7 times greater than those of serum.

4. Serum GOT is stable for at least two or three weeks, and GPT up to one week, when refrigerated.

NORMAL VALUES

The normal ranges for serum levels of GOT and GPT in this procedure, based on the reports of Henry and of Thompson, and on experience in the author's laboratory, are as follows (32°C. activity):

	Males	*Females*
Serum GOT	15–38	12–30
Serum GPT	6–35	6–25

The normal range for GOT in cerebrospinal fluid is 7 to 49 units; the range for GPT is approximately 0 to 5 units. The range in international units is approximately half (0.48 ×) the values shown.

REFERENCES

Henry, R. J., Chiamori., N., Golub, O., and Berkman, S.: Am. J. Clin. Path., *34*:381, 1960.
Karmen, A., Wroblewski, F., and LaDue, J. E.: J. Clin. Inv., *34*:126, 133, 1955.

Colorimetric Method for Assay of Transaminase

REAGENTS

1. Phosphate buffer, 0.1 *M*, pH 7.4. Dissolve 23.86 gm. of Na_2HPO_4 and 4.36 gm. of KH_2PO_4 in water and dilute to 2000 ml.

2. Pyruvate, 2.0 mmol./L. (for standard curve). Dissolve 0.110 gm. of sodium pyruvate in 500 ml. of phosphate buffer. Make fresh as needed.

3. GOT substrate. Transfer 0.146 gm. of α-ketoglutaric acid and 13.30 gm. of DL-aspartic acid to a beaker. Add 1 *N* sodium hydroxide until solution is complete. Adjust to pH 7.4 with sodium hydroxide, transfer quantitatively with phosphate buffer to a 500 ml. volumetric flask and dilute to the mark with buffer. This solution is stable when refrigerated.

4. GPT substrate. Transfer 0.146 gm. of α-ketoglutaric acid and 8.90 gm. of DL-alanine to a beaker. Add 1 *N* sodium hydroxide until the solution is complete. Adjust to pH 7.4 with sodium hydroxide, transfer quantitatively with phosphate buffer to a 500 ml. volumetric flask and dilute to the mark with buffer. This solution is stable when refrigerated.

5. 2,4-dinitrophenylhydrazine, 1 mmol./L. Dissolve 0.198 gm. of 2,4-dinitrophenylhydrazine in 1000 ml. of 1 *N* hydrochloric acid. Store in a dark bottle in the refrigerator.

6. Sodium hydroxide solution, 0.4 *N*. Dissolve 16 gm. of sodium hydroxide in water and dilute to 1000 ml.

PROCEDURE

1. Pipet 1.0 ml. of the GOT or GPT substrate into a test tube and place in a constant temperature water bath at 37°C. for 10 minutes.
2. Add 0.2 ml. of serum and mix.
3. Incubate exactly 60 minutes for GOT (30 minutes for GPT).
4. At the end of the incubation period, add 1.0 ml. of 2,4-dinitrophenylhydrazine reagent, remove tubes from bath, and mix.
5. Let stand at room temperature for 20 minutes.
6. Add 10 ml. of 0.4 *N* NaOH solution, mix by inversion, and let stand 10 minutes.

7. Measure the absorbance, using water as a blank, with a spectrophotometer at 505 nm. Any wavelength or filter in the range of 490 to 530 nm. may be used.

8. Obtain units of activity from respective standard curves. Sera with values which exceed the limits of the standard curves should be diluted with water and the analysis repeated.

CALIBRATION (STANDARD CURVE)

1. Add pyruvate standard, GOT substrate, and water to 5 tubes as follows:

No.	*ml. pyruvate*	*ml. GOT substrate*	*ml. water*	*GOT units*	*GPT units*
1	0	1.00	0.2	0	0
2	0.10	0.90	0.2	24	28
3	0.20	0.80	0.2	61	57
4	0.30	0.70	0.2	114	97
5	0.40	0.60	0.2	190	150

2. Add 1.0 ml. dinitrophenylhydrazine to each tube, mix, and let stand 20 minutes. Then add 10 ml. of 0.4 *N* NaOH, mix by inversion, and let stand 10 minutes.

3. Set the spectrophotometer to zero absorbance with a water blank at 505 nm. and record the absorbance for each tube.

4. Plot the readings against the corresponding units for GOT and GPT. Connect the points by a smooth curve.

COMMENTS PERTAINING TO THE COLORIMETRIC PROCEDURE

1. Reading the samples against a reagent blank prepared by substituting water for serum in the procedure is often not practical, because the blank may have such a high absorbance that the spectrophotometer cannot be set to zero absorbance (100 per cent T). Readings are customarily made against a water blank, as are the readings in the calibration procedure. A reagent blank should be included with each run as a check on the quality and stability of the reagents. It should read approximately the same in each run.

2. The concentrations of the substrate solutions and the dinitrophenylhydrazine solution are very critical. These solutions are available commercially (Sigma Chemical Company, St. Louis, Missouri, and Dade Reagents, Miami, Florida).

3. The reaction also measures any endogenous serum keto acids. This source of error is usually not significant.

4. Grossly lipemic or icteric sera may contribute substantially to the final absorbance and require use of a serum blank. This is prepared by mixing 1.0 ml. of substrate, 0.2 ml. of serum, and 1 ml. of 2,4-dinitrophenylhydrazine in a test tube. (Addition of the last reagent inactivates the enzyme.) After 20 minutes add 10 ml. of 0.4 *N* NaOH, and measure the absorbance after 10 minutes against a water blank. This absorbance of the serum blank must be subtracted from the absorbance of the test *before* obtaining units from the calibration curve, since the curve does not follow Beer's law.

5. It is convenient to start the GOT determination 30 minutes before the GPT determination so as to synchronize subsequent color development.

NORMAL VALUES

Values established for the method described are 8 to 40 units for serum GOT and 5 to 35 units for GPT.

REFERENCE

Reitman, S., and Frankel, S.: Am. J. Clin. Path., *28*:56, 1957.

Alternate Procedure for Assay of GOT

Babson and coworkers[6] have proposed a colorimetric reagent which is highly specific for oxaloacetate. This compound, 6-benzamido-4-methoxy-*m*-toluidine diazonium chloride (Azoene Fast Violet B), is sufficiently sensitive so that the assay for GOT requires only 20 minutes incubation time. Reagent blanks are minimal compared to the Reitman-Frankel procedure, and a wider range of assay is possible before specimens must be diluted. Reagents are available from the Warner-Chilcott Laboratories (Morris Plains, New Jersey) and are referred to as TransAc method reagents. The method has the disadvantage that it must be standardized against pre-assayed samples.

SUPPLEMENTARY READING

Wilkinson, J. H.: An Introduction to Diagnostic Enzymology. Baltimore, Williams and Wilkins, 1962.
Friedman, M. M., and Taylor, T. H.: Transaminases. In Standard Methods of Clinical Chemistry. D. Seligson, Ed. New York, Academic Press, 1961, Vol. 3, pp. 207–216.
Bergmeyer, H.-U. (Ed.). Methods of Enzymatic Analysis. D. H. Williamson (Trans.) New York, Academic Press, 1963.
King, J.: Practical Clinical Enzymology. Princeton, D. Van Nostrand Co., 1965.

ISOCITRATE DEHYDROGENASE

I.U.B. No. 1.1.1.42; L_S-isocitrate: NADP oxidoreductase (decarboxylating)

Serum isocitrate dehydrogenase (ICD) catalyzes the oxidative decarboxylation of isocitric acid to α-oxoglutaric acid. The hydrogen transfer coenzyme in this reaction is NADP (TPN). It is not clear whether the oxidation and decarboxylation reactions

$$\underset{\textit{L-Isocitrate}}{\begin{array}{c} COO^{(-)} \\ | \\ HO{-}C{-}H \\ | \\ H{-}C{-}COO^{(-)} \\ | \\ CH_2 \\ | \\ COOH \end{array}} + \underset{\substack{\textit{Coenzyme II} \\ (\textit{TPN})}}{NADP^{(+)}} \xrightarrow[\overleftarrow{\;Mn^{++}?\;}]{\text{Isocitric dehydrogenase}} \underset{\substack{\textit{Reduced coenzyme II} \\ (\textit{TPNH})}}{NADPH} + H^{+} + \underset{\textit{Oxalosuccinate}}{\begin{array}{c} COO^{(-)} \\ | \\ C{=}O \\ | \\ H{-}C{-}COO^{(-)} \\ | \\ CH_2 \\ | \\ COOH \end{array}} \quad (39)$$

Hydrogen Transfer Reaction

$$\underset{\textit{Oxalosuccinate}}{\begin{array}{c} COO^{(-)} \\ | \\ C{=}O \\ | \\ H{-}C{-}COO^{(-)} \\ | \\ CH_2 \\ | \\ COOH \end{array}} + H^{+} \xrightarrow[\overleftarrow{\;Mn^{++}\;}]{\text{Isocitric dehydrogenase}} \underset{\alpha\textit{-Oxoglutarate}}{\begin{array}{c} COO^{(-)} \\ | \\ C{=}O \\ | \\ H{-}C{-}H \\ | \\ CH_2 \\ | \\ COOH \end{array}} + \underset{\textit{Carbon dioxide}}{CO_2} \quad (40)$$

Decarboxylation Reaction

occur simultaneously or in separate steps. The serum enzyme is primarily of liver origin, although the enzyme is found in all cells in soluble form. Another similar enzyme (1.1.1.41) is also present in cells but is bound to particulate matter in the mitochondria, and requires NAD (DPN) as the hydrogen acceptor. The two enzymes also differ in that the serum enzyme can decarboxylate added oxalosuccinate (equation 40), whereas the mitochondrial enzyme cannot, i.e., it can only act on isocitrate.

The serum type enzyme is found in high concentrations not only in liver, but also in heart, skeletal muscle, kidney, adrenals, platelets, and red cells. The enzyme has a molecular weight of 64,000. Manganous ion (Mn^{++}) is the preferred activator, but it can be replaced by Mg^{++} or Co^{++}. The optimal concentration of Mn^{++} for activation is 0.001 to 0.003 *M*; only 80 and 40 per cent of optimal activity is obtained with Mg^{++} and Co^{++}, respectively, and this only at levels of 0.006 to 0.008 *M*. The Mn^{++} appears to be required for the decarboxylation reaction. Sulfhydryl binding reagents such as Cu^{++}, Hg^{++}, and parachloromercuribenzoate ($COOH\text{-}C_6H_4\text{-}HgCl$) are potent inhibitors, but the inhibition can be reversed by sulfhydryl compounds such as glutathione. Fifty per cent inhibition of activity occurs in the presence of 0.01 *N* CN^- and azide. The enzyme has a broad range of optimal pH of 7.0 to 7.8.

Reports of isoenzyme forms of ICD have been published. Apparently four forms exist, although only two can be demonstrated in liver and heart tissue. The fast moving form is the most abundant form in liver tissue, whereas the predominant type in heart extracts has a slow mobility and is very heat-labile.

STABILITY

The serum enzyme will maintain activity unchanged for 24 hours at ambient temperature, or for three weeks at refrigerator temperatures. Freezing may result in the loss of some enzyme activity.

CLINICAL APPLICATION AND NORMAL VALUES

Only relatively low levels of isocitric dehydrogenase are found in sera of healthy adults. In the author's laboratory, with the enzyme assayed by the reaction-rate procedure at 32°C., normal values have been found to range from 50 to 250 Wolfson-Williams-Ashman (WWA) units,[103] with the large majority falling between 50 and 150 units. This is equivalent to 0.8 to 4.0 mI.U./ml.

There is considerable discrepancy in the published reports as to the value of the temperature coefficient for the reaction. The Q_{10} values reported vary from 1.7 to 3.2. Assuming a fair average Q_{10} of 2.2 (= 8 per cent increase in rate per degree), the 32°C. values become 35 to 175 at 25°C., and 80 to 390 at 37°C., respectively. These are equivalent to 0.6 to 3.0, and 1.2 to 6.5 mI.U./ml. at these temperatures. Other values reported in the literature are as follows: 0.8 to 4.4 mI.U. at 25°C. and 2.0 to 8.7 mI.U. at 37°C., and 55 to 225 and 30 to 192 WWA units at 25°C. Despite this apparent discordance in the normal values as measured at different temperatures, there should be no difficulty in differentiating ICD serum levels in health and disease. Levels in serum of cord blood are 2 to 5 times that found in the serum of adults.

In general, *elevations* in serum ICD will be encountered only in diseases involving the liver parenchymal tissue. The highest levels are found with infectious hepatitis (5 to 25 times the normal level). The rise in enzyme level can be detected even in the incubation phase of the disease. Except in chronic hepatitis, in which increased enzyme values persist for 3 or 4 months, the level drops down to within the normal range in 18 to 21 days. Almost equally high elevations are seen with hepatitis associated with infectious mononucleosis, although levels of only 300 to 1000 WWA units are observed with toxic and serum hepatitis. Very high serum levels are also common in neonatal biliary duct atresia.

In cirrhosis, serum enzyme concentrations will vary from normal to about 800 WWA units. Similar results are found with obstructive jaundice associated with cancer or pancreatitis; somewhat higher levels occur with liver neoplasms. Normal levels are encountered with early obstructive jaundice, and in a variety of other diseases involving the heart, lungs, kidney, skeletal muscle, and other tissues. A marginal increase has been reported in protein deficiency, and appreciable increases in cases of kwashiorkor.

It is remarkable that despite a very high level of the enzyme in cardiac tissue, no elevations are observed in myocardial infarctions, unless the infarctions are accompanied by congestive failure and the resultant diminished blood flow, with secondary effects on the liver. Presumably the heart enzyme is either very labile or very easily inactivated. However, increases in enzyme level are observed in the majority of cases of moderate to severe pulmonary infarctions.

Red cells contain some hundred-fold more enzyme than does serum. Therefore, sera showing hemolysis should not be used for assay. However, normal values are found with all forms of anemias, except pernicious anemia, in which values can range from normal to about 1000 units.

Very low levels have been reported in cerebrospinal fluid (2 to 12 WWA units). These levels are in fact too low to measure with the usual procedures used for serum. Moderate elevations of ICD in cerebrospinal fluid have been reported in cases involving cerebral tumors, vascular cerebral lesions, and purulent meningitis.

ASSAY METHODS

Wolfson and Williams-Ashman measured the enzyme with a kinetic procedure (at 25°C.) by determining the increase in NADPH.[103] The WWA unit is defined as the quantity of enzyme in 1.0 ml. of serum that will produce one nanomol of NADPH in one hour under the conditions of the test. NADPH is measured by its absorption at 340 nm. Colorimetric procedures based on measuring the oxoglutarate formed in the reaction as its 2,4-dinitrophenylhydrazone were devised by Bell and Baron[8] as well as Taylor and Friedman.[89] The colored oxoglutarate phenylhydrazone is measured at 390 nm. These colorimetric methods are carried out at 37°C.

Determination of Serum Isocitrate Dehydrogenase by the Method of Wolfson and Williams-Ashman (at 30°C.)

REAGENTS

1. TRIS buffer, 0.10 *M*, pH 7.5. Add 12.10 gm. of TRIS (trishydroxymethylaminomethane) to a beaker containing about 75 ml. of 1.0 *N* HCl in about 500 ml. of water. Warm the solution to 30°C. and add additional 1 *N* HCl until the pH reaches 7.5. A pH meter standardized at 30°C. is used for the pH adjustments. The solution is then diluted to 1000 ml. (If the reaction is to be run at another temperature, such as 25 or 37°C. the buffer must be adjusted to pH 7.5 at that temperature.)

2. Substrate, 0.10 M. Dissolve 265 mg. of the trisodium salt of D,L-isocitric acid in 10 ml. of 0.85 per cent (w/v) NaCl. Store the reagent in the refrigerator at 2 to 4°C.

3. Saline, 0.147 M = 0.85 per cent NaCl (w/v) in water. Keep refrigerated.

4. Activator solution, 0.015 M Mn^{++}. Dissolve 300 mg. of $MnCl_2 \cdot 4H_2O$ in 100 ml. of the saline. Store refrigerated.

5. NADP solution, 0.001 M. Dissolve 0.83 mg. of NADP in 1.0 ml. of saline. Prepare only enough for one day's needs. Any solution remaining may be stored for 2 to 3 days, if kept frozen.

PROCEDURE

1. Mix the following together in a 13 × 100 mm. test tube, and then place the tube in a water bath to equilibrate to 30°C. (5 to 8 minutes).

Buffer	1.50 ml.
Activator	0.50 ml.
NADP	0.50 ml.
Serum	0.50 ml.

Transfer the contents of the tube to a 1.00 cm. square cuvet.

2. Add 0.10 ml. of the substrate to the solution, mix the contents of the cuvet with a thin glass rod, and place the cuvet into the cuvet chamber of the spectrophotometer.

3. Read the absorbance of the solution at 340 nm. against a water blank at regular one or two minute intervals for a period of 10 to 15 minutes. An initial lag phase lasting from 1 to 4 minutes may be encountered, but thereafter the absorbance should increase at a constant rate. Take enough readings (6 to 10) to insure that the enzyme reaction is proceeding at a linear rate.

4. Calculate the rate of increase in absorbance per minute from the recorded readings. This is best done by drawing the best straight line relating absorbance and time, and calculating the change in A (ΔA) over a 6 or 10 minute interval. If a recording spectrophotometer is available (Unicam, Gilford, Kintrac, etc.) the ΔA/min. can be calculated directly from the spectrophotometer trace.

5. If the trace is not linear, or if the rate of change of A is greater than 0.05/min., repeat the run with a 1:5 or 1:10 dilution of the serum in saline.

6. The assays should be done preferably in an instrument in which the cuvet chamber is temperature-controlled. If such control is not available, measure the temperature of the cuvet at the end of the run, and calculate the average temperature during the run, using $\frac{T + 30}{2}$. The appropriate temperature factor (F_T) is then used in the calculations. (30°C. = initial reaction temperature.)

CALCULATIONS

ICD activity in WWA units = nanomols NADPH formed/ml. serum in 60 minutes at 30°C. =

$$\frac{\Delta A/\text{min.}}{6.22} \times \frac{3.1}{0.5} \times 1000 \times 60 \times \frac{1}{F_T} \times D$$

where 6.22 = Absorbance of NADPH at a concentration of 1 μmol/ml.;
3.1 = volume in the cuvet, in ml.;
1000 converts μmol into nanomol;
0.50 = ml. of serum used;
D = serum dilution;
and F_T = temperature correction factor

TABLE 8-9. *Temperature Correction Table for Isocitrate Dehydrogenase**

Temperature, °C.	25	26	28	30	32	33	35	37
Temperature factor, F_T	0.68	0.73	0.85	1.00	1.17	1.26	1.47	1.70

* After King, J.: Practical Clinical Enzymology. Princeton, D. Van Nostrand Co., 1965.

This simplifies to:

$$\text{WWA units at } 30°\text{C.} = 60{,}000 \times D \times \frac{1}{F_T} \times \Delta A/\text{min.}$$

Activity, in mI.U./ml. = (1/60) × WWA units/ml.

The temperature correction table (8-9) is based on a Q_{10} of 2.2.

COMMENTS

1. A temperature of 30°C. was chosen in accordance with the 1964 recommendation of the International Union of Biochemistry to measure and report enzyme activity at this temperature.

2. Instead of a water blank, a solution of dichromate, of such concentration that its absorbance at 340 nm. is just less than that in a "test" cuvet containing normal serum, may be used. For more precise work the blank should be identical in composition with that of the test, except that saline replaces the substrate aliquot.

REFERENCES

Wolfson, S. K., and Williams-Ashman, H. G.: Proc. Soc. Expt. Biol. Med., 96:231, 1957.
Bowers, G. N.: Clin. Chem., 5:509, 1959.

SUPPLEMENTARY READING

King, J.: Practical Clinical Enzymology. Princeton, D. Van Nostrand Co., 1965.

GLUCOSE-6-PHOSPHATE DEHYDROGENASE

(I.U.B. No. 1.1.1.49; D-glucose-6-phosphate: NADP oxidoreductase)

Glucose-6-phosphate dehydrogenase (G-6-PDH) is a hydrogen transfer enzyme which mediates the reversible transfer of hydrogen from glucose-6-phosphate to coenzyme II (NADP). The products of the reaction are 6-phosphogluconolactone and the reduced form of the coenzyme, NADPH. G-6-PDH has also been identified with the names zwischenferment and hexose phosphate dehydrogenase. The following reaction is involved:

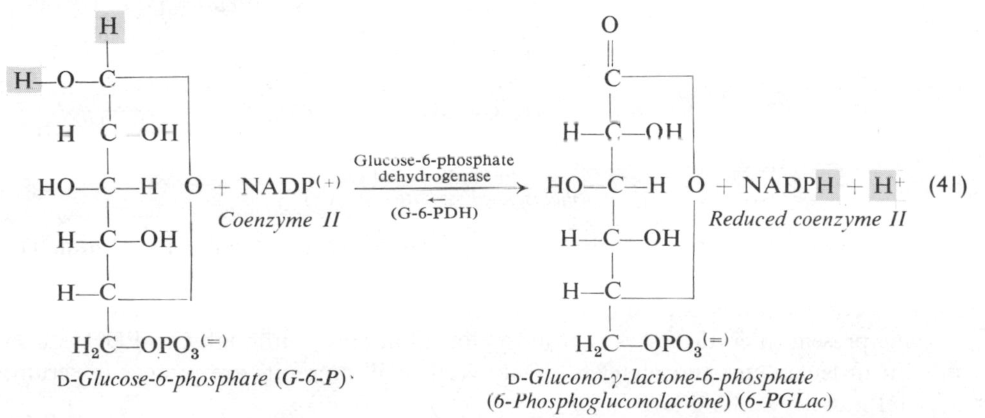

If pure G-6-PDH is present the reaction stops at the lactone stage. In tissues, however, a lactonase is present which catalyzes the rapid hydrolysis of the lactone to the 6-phosphogluconic acid (6-PGA) which is the substrate of another enzyme, 6-phosphogluconic acid dehydrogenase (6-PGDH). This latter enzyme is also a NADP-mediated hydrogen transferase which oxidizes 6-PGA with decarboxylation to ribulose-5-phosphate and carbon dioxide.[20] These two reactions are presented in equations 42 and 43.

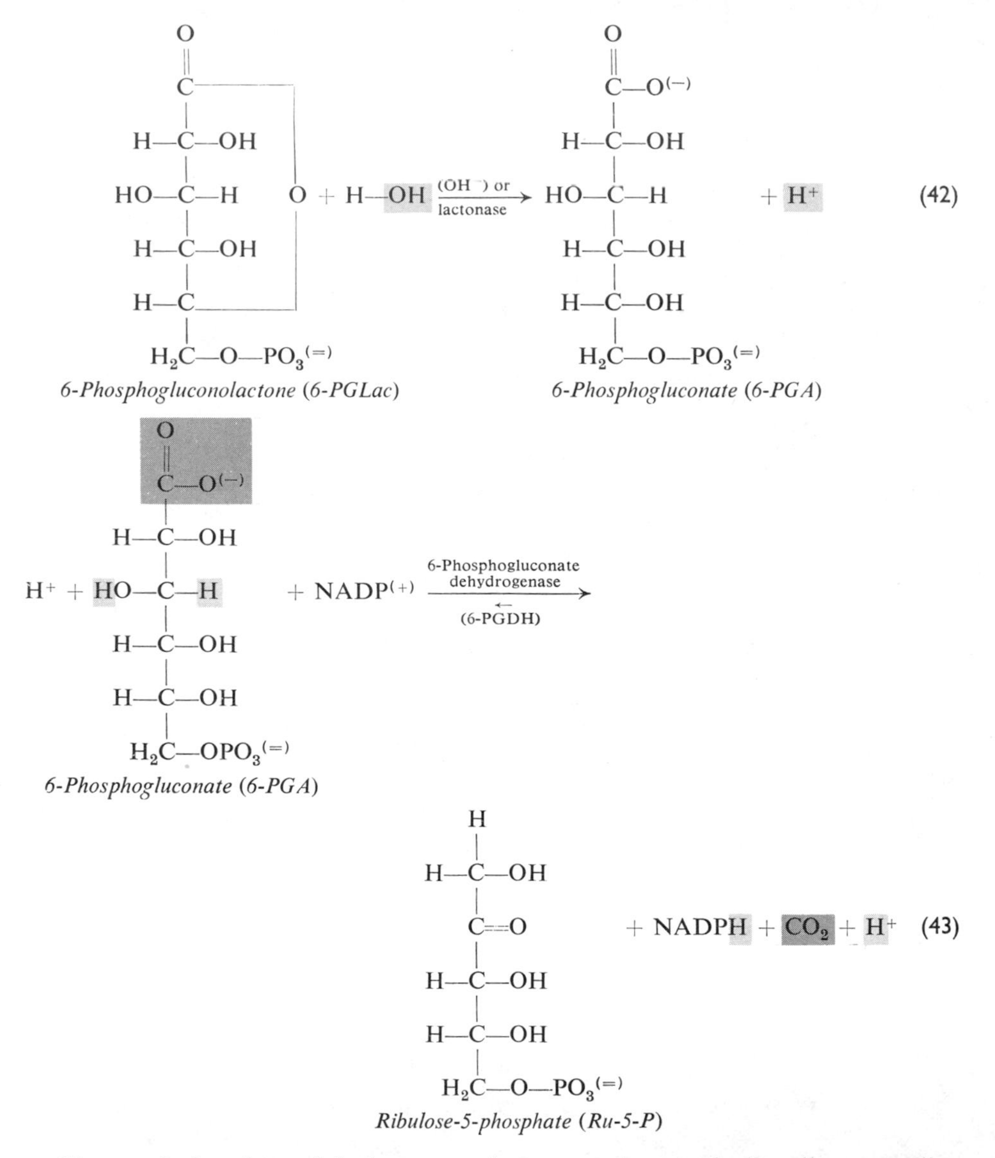

Glucose-6-phosphate dehydrogenase is present in practically all mammalian cells. Normally, the highest levels are present in the adrenal glands, although even higher concentrations are found in the mammary glands during lactation. The enzyme present in erythrocytes is of most clinical interest, although G-6-PDH is also present in leukocytes and platelets. Only very small amounts are present in serum, skeletal muscle, and kidney.

The red cell enzyme has a molecular weight of 190,000. The enzyme acts preferentially on glucose-6-phosphate and coenzyme II (NADP), although galactose-6-phosphate can be oxidized at about 14 per cent of the rate for G-6-P, and NAD can serve as a weak hydrogen acceptor (3 per cent of activity of NADP). The red cell enzyme requires Mg^{++} as an obligate activator; Ca^{++} and Mn^{++} can replace Mg^{++} in enzyme preparations from other sources. All heavy metals except copper and zinc are powerful inhibitors of the enzyme. The inhibition can be counteracted by EDTA, but not by sulfhydryl agents such as cysteine. The enzyme is also inhibited by sulfonamides, tolbutamide, phenothiazines, primaquine (and other 8-amino-quinoline antimalarials), vitamin K analogs, and by 17- and 20-oxosteroids. Cyanide, iodoacetate and fluoride do not inhibit this enzyme.

The oxidation of glucose-6-phosphate by G-6-PDH is the first step in the so-called pentose-phosphate shunt, which accounts for about 10 per cent of glucose utilization. The main portion of glucose is metabolized by the Meyerhof pathway to pyruvate and lactate, and is eventually oxidized via the citric acid cycle. It appears that the function of the pentosephosphate shunt is to produce NADPH and pentose phosphates; the former is needed for vital synthetic reactions, and the latter as starting materials for nucleotide and nucleic acid synthesis. More specifically, high levels of NADPH are required by red cells to maintain sufficient glutathione in the reduced state, without which cell integrity cannot be preserved. By referring to equations 41, 42, and 43, it can be seen that the successive G-6-PDH and 6-PGDH reactions produce 2 moles of NADPH for each glucose molecule oxidized to pentose phosphate.

CLINICAL APPLICATION

Normal values for serum and erythrocyte concentration of G-6-PDH are not well established. The enzyme is somewhat unstable. The level varies with the age of red cells and is also affected by hormonal influences. The normal range for serum concentration has been reported as 0.5 to 6.0 mI.U./ml., whereas for red cells the normal range has been variously reported as 1200 to 2000 mI.U./ml. packed cells;[20] 120 to 240 mI.U./1.0 billion (10^9) cells; and 150 to 280 μmol./min./100 ml. red cells[108] (which equals 150 to 280 mI.U./10^9 cells, assuming a hematocrit of 45 per cent and a normal red count of 4.5×10^6 cells/μl.).

Serum elevations are encountered only in pulmonary and myocardial infarctions. Alterations in *erythrocyte* concentration of the enzyme are of most clinical interest. The red cell level in newborns is about double that in adults, and it is even higher in premature infants. Young red cells have more enzyme than older, mature cells. Elevations are observed in hyperthyroidism and in various types of anemias.

A small proportion of individuals are born with no or very low levels of G-6-PDH in their red cells.[20,87] If this group is treated with such antimalarials as pamaquine or primaquine, they suffer a severe hemolytic episode, males being much more severely affected than females. The hereditary defect is transmitted by a partially dominant gene which is sex-linked, with transmission affected through the female. It is more common among Negroes, people in the Mediterranean area, non-Ashkenazic Jews, Malayans, and people in some sections of China. Among Sardinians and Italians, the hemolytic reaction (favism) is set off by the ingestion of fava beans which contain some sensitizing ingredient. The genetic defect is present in about 4 per cent of persons in the United States (7.2 per cent of Negroes, 1.4 per cent Caucasians). The disease is self-limiting; after an acute attack, essentially full recovery occurs until a new attack is precipitated by an initiating agent. Besides the antimalarials already mentioned, Aspirin, nitrofurantoin, sulfa drugs and a variety of organic compounds, including tolbutamide (an oral antidiabetic), can precipitate the hemolytic episodes. These chemicals are all inhibitors of G-6-PDH, with the result that no or insufficient enzyme activity is available to produce sufficient NADPH to maintain the necessary levels of glutathione in the affected cells. It has been speculated that the affected cells are deficient in an activator of the enzyme, since stroma from normal cells added to hemolyzates of affected cells results in almost normal enzyme activity. Some cases of hemolytic jaundice observed in newborn infants without evidence of Rh or ABO incompatibility may be associated with the presence of cells deficient in G-6-PDH activity.

Assay Methods for Glucose-6-Phosphate Dehydrogenase

G-6-PDH activity is most precisely measured by the Warburg technique. The NADPH formed in the reaction reduces molecular oxygen with the aid of an added cytochrome-c oxidation system, and the consumption of oxygen is measured with the Warburg apparatus. Although capable of high sensitivity and accuracy, this procedure is too slow for routine laboratory use.

A number of procedures are based on the measurement of the increase in absorbance at 340 nm., reflecting the formation of NADPH in the G-6-PDH reaction. Clinical specimens, whether sera or hemolysates or red cells, will contain some 6-phosphogluconic acid dehydrogenase (6-PGDH) activity. This reaction, as shown in equations 42 and 43, uses as substrate the product of the G-6-PDH reaction, and also produces NADPH. Thus, the measurement of NADPH reflects the action of both enzymes. In precise work, the enzyme determination is set up in duplicate: one cuvet receives 6-PGA as substrate, and the other G-6-P as substrate. The former measures the NADPH from the 6-PGDH reaction, and the latter, that from both reactions; the difference is then a measure of G-6-PDH activity alone. However, by using a pH unfavorable for the 6-PGDH reaction, and by running the reaction for only a short period of time, the effect of any 6-PGDH activity in the sample can be minimized. As a consequence, most published procedures and commercial kits omit the 6-PGA assay tube. For specimens with normal levels of G-6-PDH, the error is insignificant, but with specimens with low G-6-PDH activity, the amount by which the measured activity exceeds that of true G-6-PDH (because the former includes some 6-PGDH activity) may be appreciable. However, differentiation between normal specimens and those specimens deficient in G-6-PDH is still unmistakable. An example of such a procedure will be presented.

In the third group of procedures the NADPH formed in the G-6-PDH reaction reduces certain dyes to their colorless form and the color decrease is measured colorimetrically. In the screening procedure of Motulsky the time required to decolorize a given quantity of brilliant cresyl blue is measured. In the Ellis and Kirkman[28] approach, an electron transfer intermediate, phenazine methosulfate (PMS) is added to mediate the reduction of the purple redox dye 2, 6-dichloroindophenol to its colorless form by the NADPH. The change in absorbance is measured at 620 nm.

Determination of Erythrocyte Glucose-6-Phosphate Dehydrogenase Activity

REAGENTS

1. Tris buffer, 0.05 *M*, pH 7.6, containing 0.005 *M* EDTA. Dissolve 1.212 gm. of TRIS (trishydroxymethylaminomethane) in 75 ml. of 0.10 *N* HCl contained in a

400 ml. beaker. Adjust the pH to 7.6 at 30°C. (or 32°C.) by careful addition of HCl. Then add 380 mg. of EDTA (disodium salt, dihydrate) and dilute the solution to 200 ml. Keep the solution refrigerated, prewarming to 30°C. before use.

2. NADP solution, 0.012 *M*. Dissolve 5 mg. of NADP in 5.0 ml. of water. The reagent is stable for two or three days if refrigerated, but is best kept frozen.

3. Substrate, 0.042 *M*. Dissolve 12.5 mg. of the anhydrous disodium salt of glucose-6-phosphate in 1.0 ml. of water. Prepare only enough for a week's requirement, and store frozen.

4. Digitonin, saturated solution in water, approximately 20 mg./100 ml.

5. Saline, 0.85 per cent (w/v).

6. Activator, 0.4 *M* $MgCl_2 \cdot 6H_2O$ in water. This contains 4.1 gm. of the salt in 50 ml. of water.

PROCEDURE

1. Preparation of red cell hemolysate. Centrifuge about 1.5 to 2.0 ml. of heparinized or oxalated blood to pack the cells and draw off the plasma, buffy coat, and top layer of red cells (reticulocytes and young cells). Wash the red cells in the tube two or three times with saline solution, and repack with a final centrifugation. Aspirate off the last saline wash as completely as possible, and resuspend 0.50 ml. of the packed cells in 2.50 ml. of digitonin solution. Allow hemolysis to proceed for 20 to 30 minutes at 4°C. and then remove the stroma by centrifugation at 3000 r.p.m. for 15 minutes. The stroma-free hemolysate is best used immediately, but it may be stored under refrigeration for one to two hours. (Alternately, 0.20 ml. of fresh blood may be drawn, washed in 2 to 3 ml. of saline several times, using a conical centrifuge tube, and the cells repacked by centrifugation. After draining off the saline completely, add 0.50 ml. of digitonin to the tube to hemolyze the cells, and prepare the hemolysate as just described.)

2. Pipet the following volumes of reaction components directly into a 1.00 cm. square cuvet:

	Test	*Blank*
Buffer—EDTA	2.85 ml.	3.00 ml.
NADP	0.10	—
Hemolysate	0.05	0.05

Prewarm the cuvets in a water bath at 32°C. (or 30°C.). Then add 0.05 ml. of substrate to each "test" cuvet, mix the contents with a fine stirring rod, and read the absorbance of the test cuvette against the blank at 340 nm. in a Beckman DU or comparable spectrophotometer, every minute over a period of 10 to 15 minutes. After an initial lag period of 2 to 4 minutes, the absorbance should increase linearly with time. Plot the data (or use a recorder tracing), and calculate the average ΔA/min. over the linear portion.

CALCULATIONS

Enzyme concentration is usually expressed per ml. of packed cells, but some investigators use other volume units, such as 1.0 billion (10^9) cells; 4.5×10^9 cells; 1.0 ml. whole blood; 100 ml. packed red cells; and 100 ml. whole blood. How these various volume units are interrelated is presented below. If a patient's hematocrit or red cell count varies more than 10 per cent from normal, the actual magnitudes should be taken into account in the calculations.

The nanomols of NADH formed per minute (= mI.U.) are obtained in the following manner:

$$\text{mI.U.} = (\Delta A/\text{min.}) \times \frac{3.1}{6.2} \times 1000 = (\Delta A/\text{min.}) \times 500.$$

To prepare the hemolysate, 0.50 ml. of packed cells were diluted with 2.50 ml. of digitonin for a dilution of approximately 1/6. Thus, the 0.05 ml. of hemolysate used in the assay represents (1/6) × (0.05), or 1/120 ml. of packed cells. The activity of the enzyme in the red cells is then given by:

$$\text{mIU/1.0 ml. } \textit{packed cells} = (\Delta A/\text{min.}) \times 500 \times 120 = (\Delta A/\text{min.}) \times 60{,}000.$$

Assuming an hematocrit of 0.45 and a whole blood red cell count of 4.5×10^6 RBC/μl. (= 4.5×10^9 RBC/ml.), 1.0 ml. of packed cells will contain 10×10^9 RBC, or 1.0×10^{10} RBC. Thus

$$\text{mI.U./}\textit{billion cells} = \text{mI.U./}1.0 \times 10^9 \text{ cells} = (\Delta A/\text{min.}) \times 6000.$$

and,

$$\text{mI.U./ml. } \textit{whole blood} = \text{mI.U./}4.5 \times 10^9 \text{ cells} = (\Delta A/\text{min.}) \times 27{,}000.$$

Also, 1.0 mI.U./ml. packed cells = 0.10 I.U./100 ml. packed cells, = (0.05) × (ΔA/min.) × 60,000, and therefore

$$\text{I.U./100 ml. } \textit{packed cells} = (\Delta A/\text{min.}) \times 6000.$$

COMMENTS

1. Glycyl-glycine and triethanolamine may be used as buffers in place of TRIS. A pH of 7.6 is used to minimize the activity of 6-PGDH present in the hemolysate.

2. The enzyme is very unstable in the hemolysate. If the assay must be delayed, it is best to store the entire blood specimen as such, because the enzyme in intact cells bathed in plasma will retain activity unchanged for several days, if refrigerated.

TABLE 8-10. *Temperature Correction Factors for Glucose-6-Phosphate Dehydrogenase*

Temperature, °C.	25	28	30	32	35	37
Temperature factor, T_F	1.36	1.12	1.00	0.90	0.78	0.70

3. To assay serum activity, use the procedure just described with the following modifications: use 1.0 ml. of serum in place of the hemolysate, reduce the buffer volume to 1.85 ml., and use 0.10 ml. of activator solution.

$$\text{mI.U./ml. serum} = (\Delta A/\text{min.}) \times 500$$

4. In Table 8-10 are listed temperature correction factors for use in converting activity measured at a given temperature to equivalent activity at 30°C. (or some other temperature).

REFERENCES

Zinkham, W. H., and Lenhard, R. E.: J. Pediatrics, *55*:319, 1959.

Bergmeyer, H.-U. (Ed.) Methods of Enzymatic Analysis. D. H. Williamson (Trans.) New York, Academic Press, 1963.

SUPPLEMENTARY READING

Bruns, F. H., and Werners, P. J.: Dehydrogenases. In Advances in Clinical Chemistry, H. Sobotka and C. P. Stewart, Eds. New York, Academic Press, 1962, Vol. 5, pp. 242–271.

Tarlov, M. L. et al.: Arch. Int. Med., *109*:209, 1962.

UROPEPSIN

I.U.B. No. 3.4.4.1 (Pepsin); no systematic name

That human urine contains an enzyme capable of splitting proteins under acid conditions was known as early as 1861. Bendersky (1890) was the first to use the name *uropepsin* in referring to the enzyme. The use of the term pepsin to differentiate those proteinases which hydrolyze proteins under the acid conditions present in the stomach from other proteolytic enzymes had already become well established. Several such proteolytic enzymes are present in urine, with the optimal pH varying from 1.5 or 2.0 to 4.5 or 4.8. The first is a true pepsin, characterized by its ability to split proteins to proteoses and peptones in the pH range 1.5 to 5.0. Pepsins preferentially split peptide bonds connecting two aromatic amino acids (tyrosine, phenylalanine) or connecting a glutamic acid and an aromatic amino acid.[27] This is illustrated in Figure 8-13. Simple peptides containing these amino acid residues are hydrolyzed, although large protein molecules are not normally broken down completely to individual amino acids.

The chief cells in the stomach (refer to the chapter on Gastric Analysis) are the site of synthesis of pepsin, which is secreted in the form of an inactive precursor (zymogen, or proenzyme) called pepsinogen (or in the new terminology, propepsin). At acid pH pepsinogen (molecular weight 42,500) is broken down into fragments

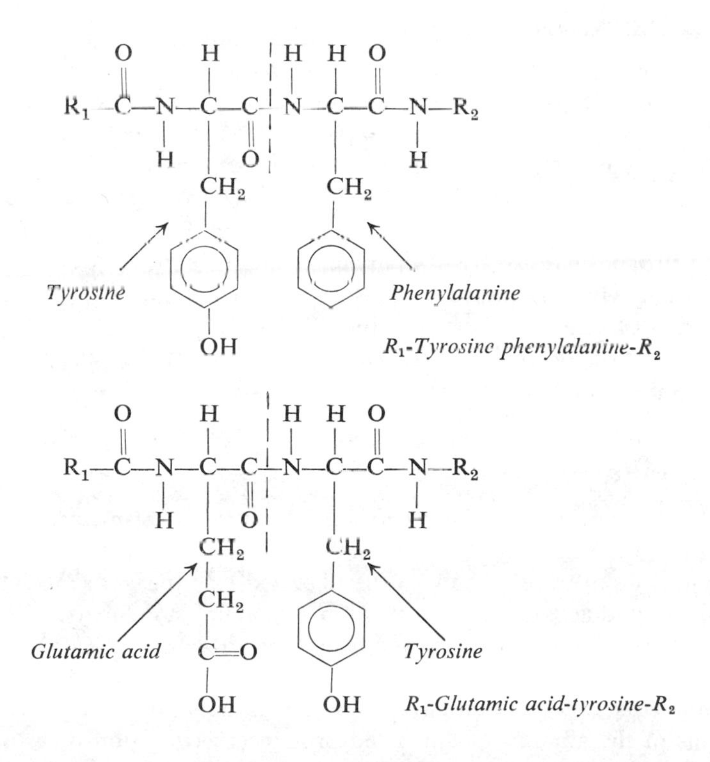

Figure 8-13. Two segments of peptide chains illustrating peptide bonds preferentially hydrolyzed by pepsin. The dotted line indicates the position of attack.

consisting of six small polypeptides and the active enzyme, pepsin, with a molecular weight of 34,500. Once formed, the pepsin itself will convert more pepsinogen to pepsin; i.e., the conversion of the proenzyme to pepsin is autocatalytic. Pepsinogen and pepsin differ markedly in their stability in alkaline solution: free pepsin is rapidly and permanently destroyed at pH 12 even at room temperature, whereas pepsinogen at this pH is stable even at 50°C. and can readily be converted to active pepsin on subsequent acidification.

The optimal pH for pepsin is 1.5 to 2.0, although the actual value will vary with the nature of the protein substrate. It is generally accepted that gastric juice contains at least two forms of pepsin, each with a somewhat different optimal pH. Presumably both may also be found in the urine.

Of the pepsinogen released from the gastric mucosa, about 99 per cent is secreted into the stomach to become part of the gastric fluid. The remaining one per cent diffuses into the intracellular fluid about the chief cells, eventually reaches the blood, and is ultimately excreted by the kidneys into the urine. Any active pepsin that may get into the blood is rapidly inactivated by the pH conditions prevalent there. Under the mildly acid conditions current in urine most of the time, part or all of the uropepsinogen may be converted to uropepsin. Daily excretion is nearly constant for any one individual, although there is a diurnal variation in output which parallels the diurnal variation in adrenal activity. The quantity excreted is independent of urine pH, volume, and specific gravity. If the pH is kept at between 5.0 and 6.5, the urine enzyme is stable at room temperature for 2 or 3 days, and in the refrigerator for two weeks. No inhibitor of the enzyme is present in urine.

CLINICAL SIGNIFICANCE

Uropepsin determinations are used mainly as an aid in the diagnosis of diseases of the stomach. However, decreased levels are also found in patients with myxedema, Addison's disease, and hypopituitarism, and elevated levels are encountered in patients with hyperthyroidism, Cushing's syndrome, and in physiological stress situations.[88]

The rates of excretion of free HCl by the gastric mucosa, and of uropepsin in the urine, roughly parallel each other, both in health and in disease. Occasionally, however, uropepsin is excreted by patients with histamine resistant achlorhydria (but not in true achylia). In patients with gastric resections, uropepsin output is related to the quantity of gastric tissue remaining.

Injection of cortisone or corticotropin stimulates uropepsin output to levels between 100 and 200 per cent above normal.

Assay Methods

Two approaches have been used to assay uropepsin activity. Mirsky[64] and Gray measured the amount of tyrosine split as result of the action of the enzyme on hemoglobin as substrate. The tyrosine liberated was measured with the Folin-Ciocalteu reagent, and activity was expressed in Anson units, defined as the quantity of enzyme which splits 0.04 mg. of tyrosine from hemoglobin at 37°C. and pH 1.5 in a reaction period of 30 minutes. In analogous procedures, serum proteins, edestin, and radioiodinated serum albumin have been used as substrates, and activity was measured in terms of the amount of unreacted protein (precipitation of proteins with trichloroacetic acid and assays with Biuret reagent), or by determining the increase in absorbance at 280 nm. as tyrosine is liberated.

In the second approach, the enzyme is quantitated by measuring the rate at which it clots milk casein under standard conditions (0.75 *M* acetate buffer, pH 4.9; homogenized milk; 37°C.). Pepsin possesses some rennin-like activity; i.e., it attacks soluble casein and modifies it to a form which is insoluble and which thus clots or precipitates out. This procedure was introduced by West and his associates,[99,40] and though it is not as precise as the Anson-Mirsky method, it is adequate for most clinical purposes.

SUPPLEMENTARY READING

King, J.: Practical Clinical Enzymology. Princeton, D. Van Nostrand Co., 1965.
Abderhalden, R.: Clinical Enzymology. P. Oesper (trans.) Princeton, D. Van Nostrand Co., 1967.

TRYPSIN

I.U.B. No. 3.4.4.4; no systematic name

Trypsin and pepsin, already discussed, are termed proteinases, or endopeptidases, because they hydrolyze proteins by splitting certain peptide bonds placed anywhere along the aminoacid chain of the protein. Trypsin preferentially splits those bonds which involve the carboxyl group of either lysine or arginine.[27] This specificity is illustrated in Figure 8-14, which presents the structures of two synthetic substrates benzoyl-lysine-amide (BLA), and p-toluenesulfonyl-arginine methyl ester (TAME), which are rapidly hydrolyzed by the enzyme. Although the prime specificity of trypsin is directed toward peptide bonds, the enzyme does possess some esterase and amidase activity.

Trypsin is synthesized in the acinar cells of the pancreas in the form of the inactive proenzyme, trypsinogen. The latter is stored in the zymogen granules, and is secreted into the duodenum under the stimulus of either the vagus nerve, or the intestinal hormone, pancreozymin. In the intestinal tract the trypsinogen is converted to the active enzyme by the intestinal enzyme enterokinase, or by preformed trypsin. For more information the reader is referred to Chapter 15.

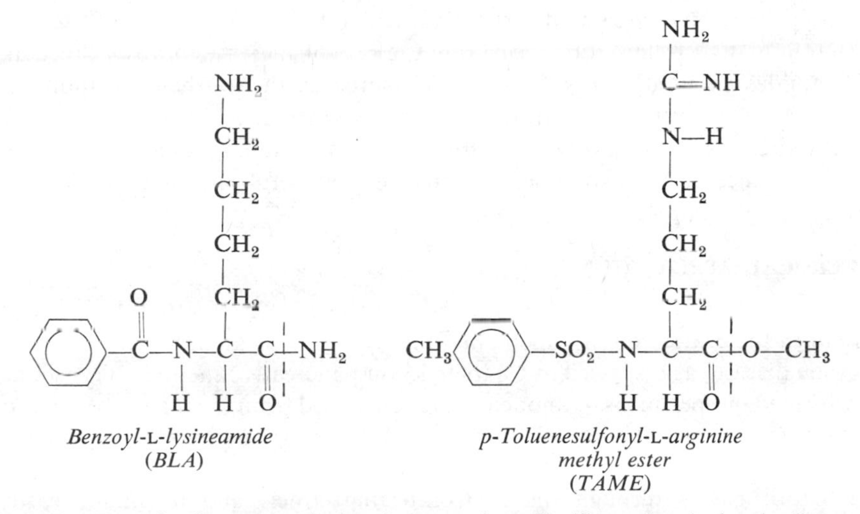

Figure 8-14. Formulas for two synthetic substrates of trypsin, illustrating the bond splitting specificity of the enzyme, which attacks the bonds involving the carboxyl of lysine or arginine. The dashed line indicates the bond split. BLA hydrolysis is an example of amidase activity; the TAME hydrolysis is an example of esterase activity.

The molecular weight of trypsin is 23,800, and the enzyme contains a predominance of basic amino acids (isoelectric point = pH 10). The pH for optimal tryptic activity toward protein substrates is in the range of 8.0 to 9.0, but with synthetic substrates such as TAME and BLA, referred to above, the optimal pH is 7.8. Activity is stimulated by calcium and magnesium ions, and to a lesser extent by cobalt, manganese ions, and aliphatic alcohols. Enzyme activity is inhibited by cyanide, sulfide, citrate, fluoride, and heavy metals, and by organic phosphorus compounds. Egg white, soy beans, lima beans, and human serum contain trypsin inhibitors, small-sized proteins which combine irreversibly with trypsin, and inactivate it by blocking out the "active center." The serum inhibitor (an α_1-globulin) presumably protects the serum proteins against attack by trypsin if, for some reason, any appreciable quantity of the enzyme enters the vascular system. The enzyme is stable in weak acid but is rapidly destroyed at an alkaline pH. The enzyme digests itself, and when functioning at its optimal pH of 9, its activity falls off rapidly.

The enzyme attacks preferentially unfolded protein chains; therefore, denatured proteins are used as substrates for this enzyme. In clinical work casein, gelatin, hemoglobin, denatured serum proteins and I^{131}-labeled serum albumin have been used. Willstätter and Lagerlöf used casein and measured activity in terms of the quantity of alkali needed to neutralize the carboxyl acids liberated. Gowenlock used denatured serum proteins and expressed activity in terms of tyrosine units, measuring the tyrosine with the Folin-Ciocalteu reagent. Hemoglobin was used by Mirsky et al.[64] who also measured tyrosine (see Uropepsin p. 460), whereas Gordon preferred casein and used micro Kjeldahl determinations to measure the trichloroacetic acid–soluble NPN–compounds (amino acids and peptides).

The units used to express activity are all empirical and as a rule cannot be converted to international units; nor can they be related to one another. The synthetic substrates such as *TAME*, already referred to, *BAEE* (benzoyl-L-arginine ethyl ester), and *BAPNA* (benzoyl-arginine-p-nitroanilide) are used in research work to quantitate trypsin. Where possible, the change in absorbance at some fixed wavelength is followed during the course of the reaction. For example, since benzoyl-arginine, a reaction product, has a greater absorbance at 253 nm. than does the substrate, BAEE, the reaction may be followed by measuring the rate of increase in absorbance at 253 nm. If amides are used, the NH_3 can be measured by the Berthelot method. These procedures using synthetic substrates are rapid and simple, but their application to clinical studies has not been very successful because of contaminating materials present in duodenal juice or stool specimens. Thus, only a useful screening method will be described in this chapter.

CLINICAL APPLICATION

Although they are of limited clinical value, trypsin determinations have been used as an aid in the evaluation of pancreatic function[47] and as an aid in the diagnosis of fibrocystic disease, as discussed in Chapter 15 on pancreatic function. Trypsin assays are performed on specimens of duodenal juice aspirated from the intestinal tract, or on fresh stool specimens. All investigators agree that determinations of trypsin in stool specimens from adults are of little value. Much pancreatic trypsin may be destroyed as the stool passes through the gastrointestinal tract, and it cannot easily be differentiated from trypsin and other proteinases associated with intestinal bacteria.

In hemorrhagic and other forms of acute pancreatitis, some trypsin may leak out into the interstitial fluid and eventually enter the plasma and be detected there.

Attempts have been made to assay trypsin in serum, but the results have not been promising. All or part of the trypsin may be inactivated by the natural serum trypsin inhibitor, and furthermore, thrombin and plasmin, present to some extent in all plasmas, possess trypsin-like activity, and there is no simple method of differentiating this activity from true trypsin activity.

Fibrocystic disease (cystic fibrosis, mucoviscidosis) in children[76] is accompanied by deficient secretion of trypsin by the pancreas. Although the disease is diagnosed clinically and by measurement of sweat electrolytes, the demonstration of no or little tryptic activity in stool specimens (or in duodenal juice samples) is useful supporting information. This can be done conveniently with the Harrison x-ray film test described next.

Semi-Quantitative Determination of Trypsin by the X-Ray Film Test

PROCEDURE

1. The test sample may be either a duodenal fluid, or a fresh stool specimen. Because of ready availability, the latter is more frequently used. Homogenize the stool specimen with four volumes of Na_2CO_3 solution, 1 per cent (w/v), to give a stool dilution of 1:5. Mix the duodenal fluid 1:1 with glycerol. Since trypsin is unstable at alkaline pH, the test must be continued without delay.

2. Place a set of 10 test tubes (10 × 75 mm.) into a rack, and to each tube add 0.30 ml. of a buffer solution which contains 0.2 per cent Na_2CO_3 (w/v) and 0.8 per cent $NaHCO_3$ (w/v).

3. Into the first tube place 0.30 ml. diluted duodenal specimen or stool homogenate; then mix the specimen and the Na_2CO_3-$NaHCO_3$ diluent thoroughly by drawing in and out of a 1 ml. serological pipet. Transfer exactly 0.30 ml. to the next tube, mixing in a similar manner. Continue the dilution and transfer procedure through the fourth tube.

4. The fifth tube is a new 1:64 dilution of the duodenal specimen (0.10 ml. of a 1:1 dilution plus 3.1 ml. diluent) or a 1:160 dilution of stool (0.1 ml. of 1:5 dilution plus 3.1 ml. diluent). Continue serial dilutions from these tubes through the tenth tube.

The tubes containing dilutions of duodenal fluid range from 1:4 through 1:2048. For a stool specimen, the dilutions range from 1:10 through 1:5120.

5. Obtain a piece of unexposed x-ray film and cut out a piece about 5 cm. × 12 cm. With a pencil, write the numbers 1 through 12 lightly over the area of the film, roughly dividing it into 12 equal areas. The ends of the film strip may be clipped to a thin piece of cardboard to prevent curling.

6. Place a drop each of saline and diluent on the film near the numbers 11 and 12 to serve as controls. Then, starting with the highest dilution, place a drop of each dilution onto the film, starting with position 10. All drops should cover about the same area.

7. Place the x-ray film with the diluted specimens into a Petri dish which contains a circle of moistened filter paper. Cover the dish and place in an incubator at 37°C. for 30 minutes. The wet paper serves to keep the dish moist and to prevent caking of the sample.

8. After the incubation, place the dish for 10 minutes into the freezer compartment of a refrigerator to harden the softened gelatin. Remove the film and place under

a light stream of cold tap water. The washing removes the drop of specimen and any loose or solubilized gelatin.

9. Examine the strip for the presence of digested (clear) areas, which indicate hydrolysis of the gelatin by the action of the enzyme. Note the highest dilution of the specimen which gives rise to a clear area.

INTERPRETATION OF RESULTS

The stool of normal infants under one year of age will show tryptic activity through the dilution of 1:80 or higher. With older children, activity may be evident only through a dilution of 1:40. Infants with fibrocystic disease will rarely give tests positive beyond the 1:10 dilution (stool specimens, tube 1; duodenal specimens, tubes 2 or 3).

REFERENCES

Schwachman, H., and Kulczycki, L. L.: Am. J. Dis. Child., *96*:6, 1958.
Harrison, G. S.: Chemical Methods in Clinical Medicine. 3rd Ed. London, Churchill Press, 1948.

SUPPLEMENTARY READING

King, J.: Practical Clinical Enzymology. Princeton, D. Van Nostrand Co., 1965.
Bergmeyer, H.-U. (Ed.): Methods of Enzymatic Analysis. D. H. Williamson (Trans.) New York, Academic Press, 1963.

CREATINE PHOSPHOKINASE (CREATINE KINASE)

I.U.B. No. 2.7.3.2; Adenosine triphosphate: creatine phosphotransferase

Creatine phosphokinase (CPK) or creatine kinase catalyzes the reversible phosphorylation of creatine by adenosine triphosphate (ATP). (See Figure 8-15.) The equilibrium position for the reaction is dependent upon the hydrogen ion concentration. The optimal pH values for the forward and reverse reactions, as written, are 9.0 and 7.2 respectively. At the optimal pH the reverse reaction proceeds six times faster than the forward reaction.

Magnesium ions are required for maximum activity, although manganese, and to a lesser extent, calcium, can also act as activators. Excess of any activator will inhibit the reaction. Zinc, copper, silver and mercuric ions are inhibitory, as are L-thyroxine, iodoacetate, fluoride, citrate, and EDTA. Although the enzyme is relatively unstable in serum, full activity can often be restored by the addition of a suitable sulfhydryl compound; the most commonly used are cysteine and Cleland's reagent (dithiothreitol).

CLINICAL SIGNIFICANCE

CPK activity is greatest in striated muscle tissue, brain, and heart tissue. Determination of serum CPK has proved more sensitive than any other enzyme procedure in the investigation of skeletal muscle disease, and is also useful in the diagnosis of myocardial infarction and cerebrovascular accidents.

Serum CPK activity is elevated in all types of *muscular dystrophy*.[92] The highest values are seen in the Duchenne type, in which levels up to 50 times the upper limit of normal may be encountered. Levels are highest in infancy and childhood and may become elevated long before the condition is apparent clinically. CPK levels characteristically fall as the patient gets older, because the mass of functioning muscle is

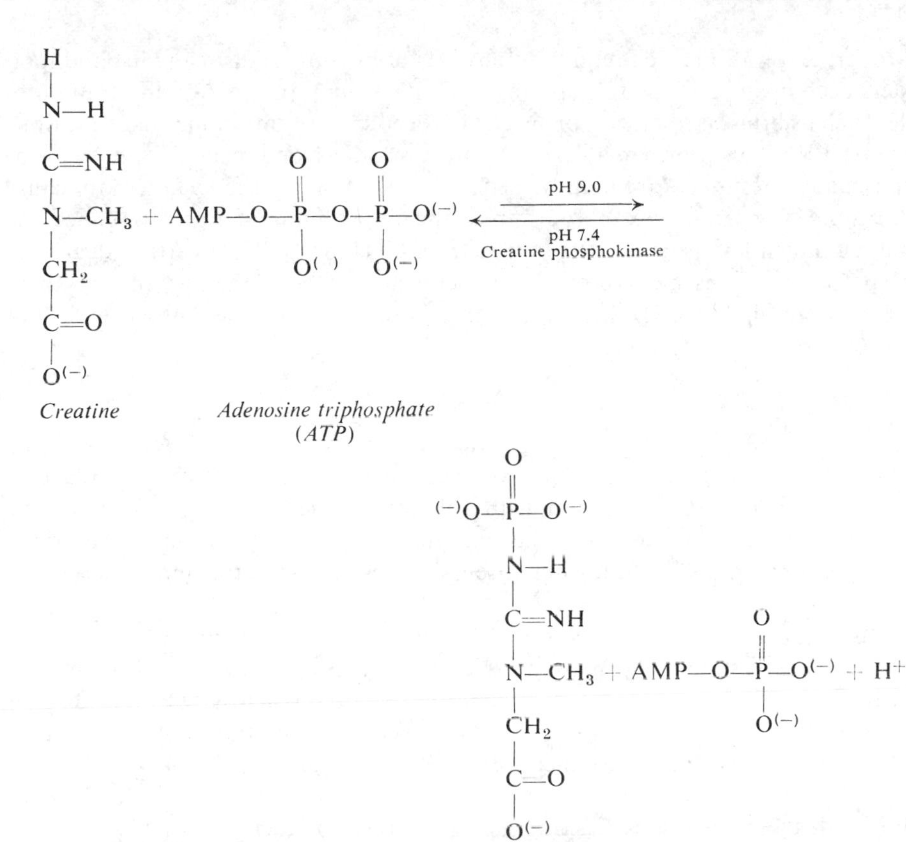

Figure 8-15. Phosphate transfer between creatine and ATP. The reverse reaction is favored because the phosphocreatine has a higher phosphate transfer potential than ATP; the N—P bond is higher in energy than the O—P bond in ATP. AMP denotes adenosine monophosphate.

diminished as the disease process progresses. About 80 per cent of symptomless female carriers of Duchenne muscular dystrophy show moderate elevation of serum CPK values.

Following a *myocardial infarction*, serum CPK becomes elevated within 4 to 6 hours, reaches a peak between 24 and 36 hours, and then returns to normal as early as the third day. The average maximum elevation is 10 to 12 times the upper limit of normal, and the test is one of the earliest and most sensitive indicators of myocardial infarction. However, if determinations are made two or more days after onset of pain, the peak level may be missed.

Liver has negligible CPK activity. Hepatic congestion and liver disorders which may accompany cardiac disease do not result in elevated CPK values, in contrast to increased values observed here for GOT and LDH. Normal levels are also observed following pulmonary infarction, a condition which frequently simulates myocardial infarction and which is occasionally accompanied by elevation of other enzymes indistinguishable from those of cardiac origin.

Serum CPK activity may be increased in patients with acute cerebrovascular disease. Values reach a maximum about three days after the acute episode and return to normal by the end of two weeks.

Moderate elevations are found in some cases of hypothyroidism and in childhood disorders accompanied by muscle spasms or convulsions. In general, any trauma to muscle, such as bruises, fractures, or surgical procedures, results in marked elevations in serum CPK levels which may persist for seven days or longer. Even frequent intramuscular injections (antibiotics, sedatives, etc.) have been shown to cause elevations ranging from two to four times the upper limit of normal. Since the enzyme has much less general distribution than GOT or LDH, a certain degree of organ specificity is present. Another significant advantage is the fact that erythrocytes exhibit negligible CPK activity; hence, the determination in serum is not affected by moderate hemolysis.

CPK Assay Methods

Numerous procedures have been developed for the assay of CPK. The reaction may be initiated in either direction with subsequent measurement of creatine or creatine phosphate. Colorimetric, fluorometric, and coupled enzymatic reactions have been proposed. The following discussion summarizes the more important approaches.

1. In the first type of procedure, ATP and creatine are incubated with the sample, and after a suitable incubation period the reaction is stopped by bringing the acidity to 1 *N*. Creatine phosphate, formed in the reaction, is acid-labile and is hydrolyzed under these conditions, whereas ATP and ADP are stable. The inorganic phosphate, thus released, is determined colorimetrically and serves as an index of CPK activity. Serum blanks are run to correct for preexisting phosphate. The forward reaction with these substrates is slower than the reverse reaction; in addition, the production of ADP gradually inhibits the reaction, and the rate tends to become nonlinear. On the other hand, creatine and ATP are much less expensive than creatine phosphate and ADP.

2. The color reaction between creatine, diacetyl, and α-napthol[51] has been used in colorimetric assays, which usually start with creatine phosphate and ADP as substrates. Incorporation of cysteine in the reaction system stabilizes the enzyme and gives higher activities. In order to prevent spontaneous hydrolysis of creatine phosphate (which would occur in acid solution) proteins in this procedure are precipitated with barium hydroxide and zinc sulfate, which provide a nearly neutral pH. Cysteine interferes in the color reaction, but this is overcome by the addition of p-chloromercuribenzoate, a potent inhibitor of sulfhydryl compounds, which also inhibits further CPK activity. Serum blanks are included to correct for endogenous creatine.

3. Sax and Moore[75] proposed a fluorometric method based on a reaction between creatine and ninhydrin in strongly alkaline solution. Serum is mixed with creatine phosphate substrate, and the reaction is initiated by addition of ADP. After the reaction run, a Somogyi filtrate is prepared and mixed with ninhydrin solution and potassium hydroxide. Maximum fluorescence is reached between 5 and 15 minutes. Creatine serves as a standard. The normal range by this method is approximately 13 to 55 mI.U./ml. Serum blanks may be included by substituting water for ADP but these blanks average only 3 units, and Sax and Moore suggest that they could be omitted in routine determinations. Since a number of drugs or medications, including vitamins and antibiotics, may also produce fluorescence, it would appear advisable to run serum blanks, especially with all specimens that show abnormally high values for CPK activity.

4. The procedure of Tanzer and Gilvarg,[86] developed to assay creatine, has been modified into a method for CPK. The forward reaction (Fig. 8-15) is coupled to two other enzyme reactions, leading eventually to the oxidation of NADH, which is followed spectrophotometrically at 340 nm.

$$\text{creatine} + \text{ATP} \xrightleftharpoons{\text{CPK}} \text{creatine phosphate} + \text{ADP} \quad (44a)$$

$$\text{phosphoenolpyruvate} + \text{ADP} \xrightleftharpoons{\text{PK}} \text{pyruvate} + \text{ATP} \quad (44b)$$

$$H^+ + \text{pyruvate} + \text{NADH} \xrightleftharpoons{\text{LDH}} \text{lactate} + \text{NAD} \quad (44c)$$

The main (CPK) reaction is run in the unfavorable direction, but it is "pushed" by the pyruvic kinase (PK) used in the second, "auxiliary" reaction which regenerates ATP and forms pyruvate. Pyruvate is measured with the LDH indicator reaction. The pH used is optimal for the CPK reaction but not for the other reactions. However, reactions 44b and 44c proceed in the favorable direction, and are not rate-limiting. Because ATP is regenerated, it is possible to maintain a constant Mg^{++}/ATP ratio, which, for optimal rates, must be 1:1. This assay system illustrates the difficulties that may be encountered in using coupled enzyme systems: it is not always possible to have conditions optimal for all the enzyme systems involved. In any case, the secondary reactions must not be rate-limiting.

In a modification of this procedure, Nuttal and Wedin omitted the step shown in equation 44c and measured the accumulated pyruvate by forming the 2,4-dinitrophenylhydrazone which can be estimated colorimetrically. (See GPT procedure.)

5. Another coupled enzyme system is based on the use of the reverse reaction, with ADP and creatine phosphate as substrates. The ATP formed is used to phosphorylate glucose in the presence of hexokinase (HK). The glucose-6-phosphate product is then oxidized by NADP (TPN) in the presence of glucose-6-phosphate dehydrogenase (G-6-PDH) to form 6-phosphogluconic acid and NADPH. The reaction is measured by following the increase in absorbance at 340 nm., as NADPH is generated.

$$\text{Creatine phosphate} + \text{ADP} \xrightarrow{\text{CPK}} \text{creatine} + \text{ATP} \quad (45a)$$

$$\text{ATP} + \text{glucose} \xrightarrow{\text{HK}} \text{glucose-6-phosphate} + \text{ADP} \quad (45b)$$

$$\text{glucose-6-phosphate} + \text{NADP} \xrightarrow{\text{G-6-PDH}} \text{6-phosphogluconate} + \text{NADPH} + H^+ \quad (45c)$$

This is a completely "down-hill" system, i.e., all reactions proceed in the favorable direction. The optimal pH for the entire system is 6.8 to 6.9.

This assay system, first used by Nielsen and Ludvigsen,[65] and by Oliver,[66] was studied and modified by Rosalki.[73] To simplify the work, Rosalki lyophilized a solution of all the reaction components and distributed the dry reagent into gelatin capsules, each capsule containing, when reconstituted with water, sufficient material for one assay.* Hess and his coworkers[46] have published an evaluation of this product and, subsequently, have described their own detailed modification of the Oliver procedure by which CPK assays can be performed on as little as 10 μl. of serum. The method which follows is that given by Rosalki.[73]

* These capsules were prepared by Calbiochem, Los Angeles, California. They were available commercially, but have since been replaced by a kit of two vials, one containing NADP, and the other the remaining components. The contents of one vial contain sufficient material for five assays.

Determination of Creatine Phosphokinase

REAGENT

Prepare a bulk substrate which contains the stated concentrations of the following ingredients dissolved in 0.05 *M* tris buffer, and adjust to pH 6.8 at the temperature at which the reaction is to be carried out.

Component	*Concentration*	*Quantity present in 100 ml.*
ADP	0.001 *M*	50 mg.
Creatine phosphate	0.01 *M*	280 mg.
Glucose	0.02 *M*	360 mg.
$MgCl_2 \cdot 6H_2O$	0.03 *M*	405 mg.
AMP	0.01 *M*	395 mg.
Cysteine·HCl	0.005 *M*	80 mg.
Hexokinase	0.6 I.U./ml.	60 I.U.
Glucose-6-phosphate dehydrogenase	0.3 I.U./ml.	30 I.U.
NADP	0.0008 *M*	65 mg.

These concentrations are the actual values in the composite. The composite reagent may be prepared from stock solutions of the individual reagents. The AMP, ADP, creatine-phosphate, and NADP are available as sodium salts in grades of purity of 95 to 99 per cent. Stock TRIS buffer (0.20 *M*) may be prepared by dissolving 12.12 gm. of trishydroxymethylaminomethane in 300 ml. of water and adding 6 *N* HCl to a pH of 6.8 and making up the volume to 500 ml. The HK and G-6-PDH are available in grades containing 100 to 300 units per ml. These are diluted appropriately. The presence of AMP is explained in Comment No. 2.

Enough of the mixture may be prepared for a day's or several days' use; any excess may be stored at 2 to 4°C. for 2 to 3 days.

PROCEDURE

Perform the assay preferably at 30°C. (or 32°C.) as discussed in the LDH procedure. Correct the results obtained at other temperatures, using the factors listed in Table 8-11.

As with other coupled reactions, there is an initial lag phase before intermediate compounds accumulate to produce a steady state. A recording device to monitor changes in absorbance is highly recommended.

Pipet 3.0 ml. of the prepared reaction mixture into a 1.0 cm. square cuvet, and place this into a 30°C. water bath to equilibrate to temperature. (If commercial capsules or kit vials are used, reconstitute these in accordance with the manufacturer's instructions.) With a delivery pipet, add 0.1 ml. of sample and mix the contents of the cuvet by tapping. Avoid shaking. Return the cuvet to the water bath and hold the mixture at constant temperature for six minutes. Then measure the increase in absorbance per minute at 340 nm. against the distilled water blank. Obtain readings at 1 minute intervals for 8 to 15 minutes until the linear reaction rate is well defined.

TABLE 8-11. *Temperature Correction Factors for Creatine Phosphokinase**

Temperature, °C	25	27	29	30	31	32	33	35	37
Temperature factor, F_T	1.47	1.27	1.08	1.00	0.93	0.85	0.79	0.68	0.58

* After Rosalki, S. B.: J. Lab. Clin. Med., **69:** 696, 1967.

CALCULATIONS

The activity of CPK is expressed in International Units. Since for each mole of phosphocreatine consumed, one mole of NADP is reduced, the rate of NADPH formation may be used to calculate the activity. With readings made at 340 nm., a reaction volume of 3.1 ml., and a serum specimen of 0.10 ml.

$$\text{mI.U./ml.} = \Delta A/\text{min.} \times \frac{3.1}{6.2} \times 1000 \times \frac{1.0}{0.1} \times F_T = \Delta A/\text{min.} \times 5000 \times F_T,$$

where F_T is the temperature correction factor listed in Table 8-11.

COMMENTS

1. No serum blanks need be used. Highly icteric sera may have a high initial absorbance, but such sera are rarely encountered in conditions in which the test is performed. If indicated, dichromate blanks may be used (see LDH or GOT procedure).

2. One should be concerned with possible side reactions which may consume reaction intermediates and prevent maximal reactions. The AMP is added to inhibit myokinase interference (AMP + ATP → ADP); the excess AMP forces the myokinase reaction to the right, preventing ADP from forming ATP, except by accepting phosphate from creatine phosphate. Interference from phosphatases, which may break down the glucose-6-phosphate, and ATP-ase, which hydrolyzes ATP, is negligible.

3. CPK loses activity in some but not all sera, presumably as a result of reversible inactivation because of the oxidation of essential sulfhydryl groups. This inactivation may be reversed in total or in part by reacting the enzyme with sulfhydryl-containing reagents such as cysteine, β-mercaptoethanol, and dithiothreitol. Some workers recommend adding the agent to the serum, and indeed, this (at concentrations of 0.005 *M*) will prevent loss of activity on storage. Others, including Rosalki, prefer to incorporate the reagent in the enzyme assay mixture. There is some evidence that the isoenzyme predominant in brain tissue is stable, but that the enzyme of heart tissue is unstable.

4. Serum is the preferred specimen, alhough EDTA plasma has also been used. However, since enzyme inhibition has been noted with EDTA and other anticoagulants, it appears safer to use only serum. Very hemolyzed serum should not be used owing to the possible liberation of intermediate substrates such as ATP and glucose-6-phosphate from the red cells.

5. Serum CPK activity is increased up to 48 hours following severe or prolonged exercise. Moderate elevations may occur under normal everyday physical stress. This should be considered, especially when dealing with outpatients.

NORMAL VALUES

The correct range of normal values has still to be established. Early reports stated that normal serum contains no CPK activity; Nielsen and Ludvigsen reported 0 to 17 mI.U./ml. Assay methods were inadequate, and no provision was made for restoring enzyme activity lost through —SH oxidation.

With the procedure just presented and with measurements taken at 37°C., Rosalki reported a normal range of 12 to 99 mI.U./ml. for males and 10 to 60 mI.U./ml. for females. Converted to activity at 30°C. these values become 7 to 54, and 6 to 40 respectively. Thompson,[91] using an almost identical method, but with assays run at 25°C, found for males a range of 5 to 50 (mean = 27), and for females 9 to 31 mI.U./ml. (mean = 20). These values, converted to activity at 30°C., become respectively 7 to 73 and 13 to 45 mI.U./ml.

The normal values discussed apply to individuals engaged in average activities. There are indications that values for individuals at bedrest are lower than these, but no well-designed studies have been carried out to verify this point.

REFERENCES

Rosalki, S. B.: J. Lab. Clin. Med., *69*:696, 1967.
Hess, J. W., et al: Clin. Chem., *13*:994, 1967.
Oliver, I. T.: Biochem. J., *61*:116, 1962.

SUPPLEMENTARY READING

King, J.: Practical Clinical Enzymology. Princeton, D. Van Nostrand Co., 1965.
Thomson, W. H. S.: Clin. Chim. Acta, *21*:469, 1968.
Thomson, W. H. S.: The clinical chemistry of muscular dystrophies. In Advances in Clinical Chemistry. H. Sobotka and C. P. Stewart, Eds. New York, Academic Press, 1964, Vol. 7, p. 138.

Concluding Remarks

Enzyme assay techniques are changing rapidly. The slow, cumbersome, and inherently inaccurate two-point procedures are being replaced by rapid, automated or semi-automated multiple-point methods, which will soon also include computer calculation of reaction rates, and direct print-out of results. The potential for increased productivity and accuracy will only be realized if the machines are used by informed technologists and chemists.

No analysis is better than the quality of the specimen to be analyzed and the quality of the reagents used. Hence, the emphasis in this chapter has been placed on the type of specimen to be used, its stability, the conditions governing storage and preservation, and on precise preparation of reagents and their stability. The use of certain high quality commercial "kits" and commercial reagents will increase, and is to be encouraged, but appropriate controls on reagent quality and the accuracy of the procedures must always be instituted.

The enzymes discussed and the procedures presented were selected to acquaint the student with examples of all classes of enzymes, and with the large variety of enzyme assay methods available. The references include excellent books and monographs, as well as journal papers, many of which should be available in the laboratory library for repeated consultations.

The use of enzymology in diagnostic medicine is still in its early phases. Much has still to be learned. Practical techniques for evaluating isoenzyme patterns for most enzymes have still to be developed. The natural history relating to the rise and fall of serum enzyme levels is to a large degree unexplored. Accurate ranges for normal values under various conditions and differentiated by sex, age, race and culture need to be established or reevaluated, using precise methods. The standardization proposals of the International Commission will be an invaluable aid in this task.

REFERENCES

1. Abderhalden, R.: Clinical Enzymology. P. Oesper (trans.), Princeton, D. Van Nostrand Co., 1961.
2. Abul-Fadl, M. A. M., and King, E. J.: Biochem. J., *42*:28, 1948; *45*:51, 1949.

3. Amador, E., Dorfman, C. E., and Wacker, W. E. C.: Clin. Chem., *9*:301, 1963.
4. Babson, A. L., Greeley, S. J., Coleman, C. M., and Phillips, G. E.: Clin. Chem., *12*:482, 1966.
5. Babson, A. L., Reed, P. A., and Phillips, G. E.: Am. J. Clin. Path., *32*:83, 1959.
6. Babson, A. L., and Shapiro, P. O.: Clin. Chim. Acta, *8*:326, 1963.
7. Baldwin, E.: Dynamic Aspects of Biochemistry. 4th ed. London, Cambridge U. Press, 1963.
8. Bell, J. L., and Baron, D. N.: Clin. Chim. Acta, *5*:740, 1960.
9. Berger, L., and Rudolph, G. N.: Alkaline and Acid Phosphatase. *In* Standard Methods of Clinical Chemistry, Vol. 5. S. Meites, Ed., New York, Academic Press, 1965, p. 211.
10. Bergmeyer, H.-U. (Ed): Methods of Enzymatic Analysis. D. H. Williamson (trans.), New York, Academic Press, 1963.
11. Bessey, O. A., Lowry, O. H., and Brock, M. J.: J. Biol. Chem., *164*:321, 1947.
12. Biochemica Test Combination, Aldolase, No. 15974, Boehringer Mannheim Corp., New York.
13. Bodansky, A.: J. Biol. Chem., *99*:107, 1932; *101*:93, 1933.
14. Bowers, G. N.: Lactic dehydrogenase. *In* Standard Methods of Clinical Chemistry, Vol. 4, D. Seligson, Ed., New York, Academic Press, 1963, pp. 163–172.
15. Bowers, G. N.: Clin. Chem., *5*:509, 1959.
16. Bowers, G. N., Kelley, M., and McComb, R. B.: Clin. Chem., *14*:595, 608, 1967.
17. Bowers, G. N., and McComb, R. B.: Clin. Chem., *12*:70, 193, 1966.
18. Boyer, S. H.: Science, *134*:1002, 1961.
19. Bruns, F. H.: Biochem. Z., *325*:156, 429, 1954.
20. Bruns, F. H., and Werners, P. H.: Dehydrogenases. *In* Advances in Clinical Chemistry, Vol. 5, H. Sobotka and C. P. Stewart, Eds., New York, Academic Press, 1962, pp. 242–271.
21. Cabaud, P. G., and Wroblewski, F.: Am. J. Clin. Path., *30*:234, 1958.
22. Cantarow, A., and Schepartz, B.: Biochemistry, 4th ed., Philadelphia, W. B. Saunders, 1967.
23. Cherry, I. S., and Crandall, L. A.: Am. J. Physiol., *100*:266, 1932.
24. De la Huerga, J., Yesnick, C., and Popper, H.: Am. J. Clin. Path., *12*:1126, 1952.
25. Desnuelle, P.: Pancreatic lipase. Adv. Enzymology, *23*:129, 1961.
26. Diagnostic Enzymology, Dade Reagents, Inc., Miami, Fla., 1966.
27. Dixon, M., and Webb, E. C.: Enzymes. 2nd ed., New York, Academic Press, 1964.
28. Ellis, H. A., and Kirkman, H. N.: Proc. Soc. Expt. Biol. Med., *106*:607, 1961.
28a. Enzyme Nomenclature. Recommendations 1964 of Internat. Union Biochemistry, New York, Elsevier Publ. Co., 1965.
29. Fales, F. W.: Glucose (Enzymatic). *In* Standard Methods of Clinical Chemistry, Vol. 4, D. Seligson, Ed., New York, Academic Press, 1963, p. 101.
30. Fishman, W. H., and Ghosh, N. K.: Isoenzymes of alkaline phosphatase. *In* Advances in Clinical Chemistry, Vol. 10, O. Bodansky and C. P. Stewart, Eds. New York, Academic Press, 1968.
31. Fleisher, G. A.: Aldolase. *In* Standard Methods of Clinical Chemistry, Vol. 3, D. Seligson, Ed., New York, Academic Press, 1961, pp. 14–22.
32. Free, A. H.: Enzymatic determination of glucose. *In* Advances in Clinical Chemistry, Vol. 6, H. Sobotka and C. P. Stewart, Eds. New York, Academic Press, 1963, pp. 67–97.
33. Friedman, M. M., and Taylor, T. H.: Transaminases. *In* Standard Methods of Clinical Chemistry, Vol. 3. D. Seligson, Ed. New York, Academic Press, 1961, pp. 207–216.
34. Goldbarg, J. A., Pineda, E. P., and Rutenberg, A. M.: Am. J. Clin. Path., *32*:571, 1959.
35. Gutman, A. B.: Am. J. Med., *27*:875, 1959.
36. Hare, M. L. C.: Biochem. J., *22*:968, 1928.
37. Harker, cited from Chem. Eng. News, Feb. 6, 1967, pg. 60.
38. Harrison, G. S.: Chemical Methods in Clinical Medicine, 3rd ed., London, Churchill, 1948.
39. Hausman, T.-U., Hekger, R., Rick, W., and Gross, W.: Clin. Chim. Acta, *15*:241, 1967.
40. Henry, R. J.: Clinical Chemistry, Principles and Technics. New York, Hoeber Med. Div., Harper and Row, 1964.
41. Henry, R. J.: Pancreatic lipase. *In* Standard Methods of Clinical Chemistry, Vol. 2, D. Seligson, Ed., New York, Academic Press, 1958, pp. 86–93.
42. Henry, J. B., and Bowers, G. N., Jr.: An Introduction to Enzymology and Serum Enzymes. Chicago, American Society Clinical Pathologists, 1960.
43. Henry, R. and Chiamori, N.: Clin. Chem., *6*:434, 1960.
44. Henry, R. J., Chiamori, N., Golub, O., and Berkman, S.: Am. J. Clin. Path., *34*:381, 1960.
45. Hess, B.: Enzymes in Blood Plasma. K. S. Henley (trans.), New York, Academic Press, 1963.
46. Hess, J. W., MacDonald, R. P., Natho, J. W., and Murdock, K. J.: Clin. Chem., *13*:994, 1967.
47. Howat, H. T.: Gastroenterology, *42*:72, 1962.
48. Howell, S. F., and Sumner, J. B.: J. Biol. Chem., *104*:619, 1934.
49. Huggins, C., and Russell, P. S.: Ann. Surg., *128*:668, 1948.
50. Huggins, C., and Talalay, P.: J. Biol. Chem., *159*:399, 1945.
51. Hughes, B. P.: Clin. Chim. Acta, *7*:597, 1962.
52. Kalow, W., and Genest, K.: Canad. J. Biochem., *35*:339, 1957.

53. Karmen, A., Wroblewski, F., and LaDue, J. E.: J. Clin. Inv., *34*:126, 133, 1955.
54. Kind, P. R., and King, E. J.: J. Clin. Path., *7*:322, 1954.
55. King, J.: Practical Clinical Enzymology. Princeton, D. Van Nostrand Co., 1965.
56. King, J., and Armstrong, A. R.: Canad. Med. Assn. J.: *31*:376, 1934.
57. Klein, B., Read, P. A., and Babson, A. L.: Clin. Chem., *6*:269, 1960.
58. Klotz, I. M.: Energy Changes in Biochemical Reactions, New York, Academic Press, 1967.
59. Latner, A. L.: Isoenzymes. *In* Advances in Clinical Chemistry. Vol. 9, Eds., H. Sobotka and C. P. Stewart, New York, Academic Press, 1967, pp. 57–71.
60. Latner, A. L., and Turner, D. M.: Clin. Chim. Acta, *15*:97, 1967.
61. Manual of Clin. Chem. Procedures. Dade Reagents, Inc., Miami, Fla., 1965.
62. Markowitz, H. C., Gubler, C. J., Maloney, J. P., Cartwright, G. E., and Wintrobe, M. M.: J. Clin. Inv., *34*:1498, 1955.
63. Michel, H. O.: Cholinesterase in Human Red Blood Cells and Plasma. *In* Standard Methods of Clinical Chemistry, Vol. 3, D. Seligson, Ed., New York, Academic Press, 1961, pp. 93–98.
64. Mirsky, I. A., Block, S., Osher, J., and Broh-Kahn, R. H.: J. Clin. Inv., *27*:818, 1948.
65. Nielsen, L., and Ludvigsen, B.: J. Lab. Clin. Med., *62*:159, 1963.
66. Oliver, I. T.: Biochem. J., *61*:116, 1962.
67. Owen, J. A.: Effect of Injury on Plasma Proteins. *In* Advances in Clinical Chemistry, Vol. 9, H. Sobotka and C. P. Stewart, Eds., New York, Academic Press, 1967, p. 2.
68. Papadopoulos, N. M., and Kintzios, J. A.: Am. J. Clin. Path., *47*:96, 1967.
69. Pfleiderer, G.: Recent developments concerning the nature of isoenzymes. *In* West European Symp. Clin. Chem., Vol. 4, R. Ruyssen, Ed., New York, Elsevier, 1965, pp. 105–110.
70. Reitman, S., and Frankel, S.: Am. J. Clin. Path., *28*:56, 1957.
71. Report of the Commission on Enzymes of the International Union of Biochemistry, New York, Pergamon Press, 1961.
72. Rice, E. W.: Ceruloplasmin assay in serum. *In* Standard Methods of Clinical Chemistry, Vol. 4, D. Seligson, Ed., New York, Academic Press, 1963, pp. 39–46.
73. Rosalki, S. B.: J. Lab. Clin. Med., *69*:696, 1967.
74. Sass-Kortsak, A.: Copper metabolism. *In* Advances in Clinical Chemistry, Vol. 8, H. Sobotka and C. P. Stewart, Eds., New York, Academic Press, 1965, pp. 1–68.
75. Sax, S. M., and Moore, J. J.: Clin. Chem., *11*:951, 1965.
76. Schwachman, H., and Kulczycki, L. L.: Am. J. Dis. Child., *96*:6, 1958.
77. Searcy, R. L., Hayashi, S., and Berk, J. E.: Am. J. Clin. Path., *46*:582, 1966.
78. Shinowara, G. Y., Jones, L. M., and Reinhart, H. L.: J. Biol. Chem., *142*:921, 1942.
79. Sibley, J. A., and Lehninger, A. L.: J. Biol. Chem., *177*:859, 1949.
80. Sigma Tech, Bul. No. 70, Aldolase, Sigma Chem. Co., St. Louis, Mo., Aug., 1961.
81. Smith, B. W., and Roe, J. H.: J. Biol. Chem., *179*:53, 1949.
82. Smyth, D. G., Stein, W., and Moore, G.: J. Biol. Chem., *235*:648, 1960; *237*:1845, 1962.
83. Somogyi, M.: Clin. Chem., *6*:23, 1960.
84. Somogyi, M.: J. Biol. Chem., *125*:399, 1938.
85. Street, H. V., and Close, J. R.: Clin. Chim. Acta, *3*:301, 1956.
86. Tanzer, M. L., and Gilvarg, C.: J. Biol. Chem., *234*:3201, 1959.
87. Tarlov, M. L., Brewer, G. J., Carsons, P. E., and Alving, A. S.: Arch. Int. Med., *109*:209, 1962.
88. Taylor, W. H.: Physiol. Rev., *42*:519, 1962.
89. Taylor, T. H., and Friedman, M. E.: Clin. Chem., *6*:208, 1960
90. Tech. Bull. No. 104, Sigma Chem. Co., St. Louis, Mo., 1960.
91. Thomson, W. H. S.: Clin. Chim. Acta, *21*:469, 1968.
92. Thomson, W. H. S.: The clinical chemistry of muscular dystrophies. *In* Advances in Clinical Chemistry, Vol. 7, H. Sobotka and C. P. Stewart, Eds., New York, Academic Press, 1964, p. 138.
93. Tietz, N. W., and Fiereck, E. A.: Clin. Chim. Acta, *13*:352, 1966.
94. Van Loon, E. J., Likens, M. R., and Seger, A. J.: Am. J. Clin. Path., *22*:1134, 1952.
95. Van der Helme, H. J.: Clin. Chim. Acta, *7*:124, 1962.
96. Varley, H.: Practical Clinical Chemistry, 4th ed., New York, Interscience, 1967.
97. Ware, A. G., Walberg, C. B., and Sterling, R. E.: Turbidimetric Measurement of Amylase. *In*: Standard Methods of Clinical Chemistry, Vol. 4, D. Seligson, Ed., New York, Academic Press, 1963, pp. 39–46.
98. Webster, P. D., and Zieve, L. N.: New Eng. J. Med., *267*:604, 654, 1962.
99. West, P. M., Ellis, F. W., and Scott, B. L.: J. Lab. Clin. Med., *39*:159, 1952.
100. Wetstone, H. J., and Bowers, G. K.: Serum cholinesterase. *In* Standard Methods of Clinical Chemistry, Vol. 4, D. Seligson, Ed., New York, Academic Press, 1963, pp. 47–56.
101. Wilkinson, J. H.: An Introduction to Diagnostic Enzymology. Baltimore, Williams and Wilkins, 1962.
102. Wilkinson, J. H., and Vodden, A. V.: Clin. Chem., *12*:701, 1966.
103. Wolfson, S. K., and Williams-Ashman, H. G.: Proc. Soc. Expt. Biol. Med., *96*:231, 1957.

104. Woodward, H. Q.: Am. J. Med., *27*:902, 1959.
105. Wroblewski, F., and Gregory, F. F.: Ann. N.Y. Acad. Science, *94*:921, 1961.
106. Wroblewski, F. and La Due, J. E.: Proc. Soc. Expt. Biol. Med., *90*:210, 1955.
107. Zimmerman, H. J., and Weinstein, H. G.: J. Lab. Clin. Med., *48*:607, 1961.
108. Zinkham, W. H., and Lenhard, R. E.: J. Pediatric, *55*:319, 1959.
109. Zondag, H. A.: Science, *142*:965, 1963.

Chapter 9 / ENDOCRINOLOGY

Part I

by Sati C. Chattoraj, Ph.D.

Endocrinology is a science which deals with the products of a group of glands and their action in maintaining the chemical integrity of cell environment. These glands are specialized, structurally and functionally. Unlike other multicellular glands, they are devoid of ducts. Secretions are released directly into the blood stream; hence the designations *ductless* or *endocrine* (internally secreting). The glands secrete one or more specific types of active principles which are called *hormones* (Greek: *hormon*—exciting, setting in motion). For efficient transport of the secreted material, these glands are highly vascularized.

According to Bayliss,[7] a hormone is defined as any substance normally produced by specialized cells in some part of the body and carried by the blood stream to another part from which it affects the body as a whole. For example, adrenocorticotrophin (ACTH) is secreted by the pituitary, but it affects the functional activities of the adrenal cortex. Similarly, blood-borne hormones of the adrenal cortex regulate carbohydrate, fat, protein, and mineral metabolism of the body. Figure 9-1 shows the approximate location of the endocrine glands in the body.

NATURE AND ACTIONS OF HORMONES

The hormones vary widely in chemical composition, ranging from amines (thyroxine, epinephrine, etc.) through complex steroid ring structures (corticosteroids, androgens, estrogens, etc.) to proteins (adrenocorticotrophic hormone, chorionic gonadotrophin, thyrocalcitonin, etc.). Hormones possess a high degree of structural specificity. Any alteration in the molecular composition of a hormone brings dramatic changes in its physiological activity. For example, although the structural difference between the female sex hormones estradiol and estriol is only an additional α-hydroxyl group (- -OH) at the C-16 position, estradiol is the most potent estrogen whereas

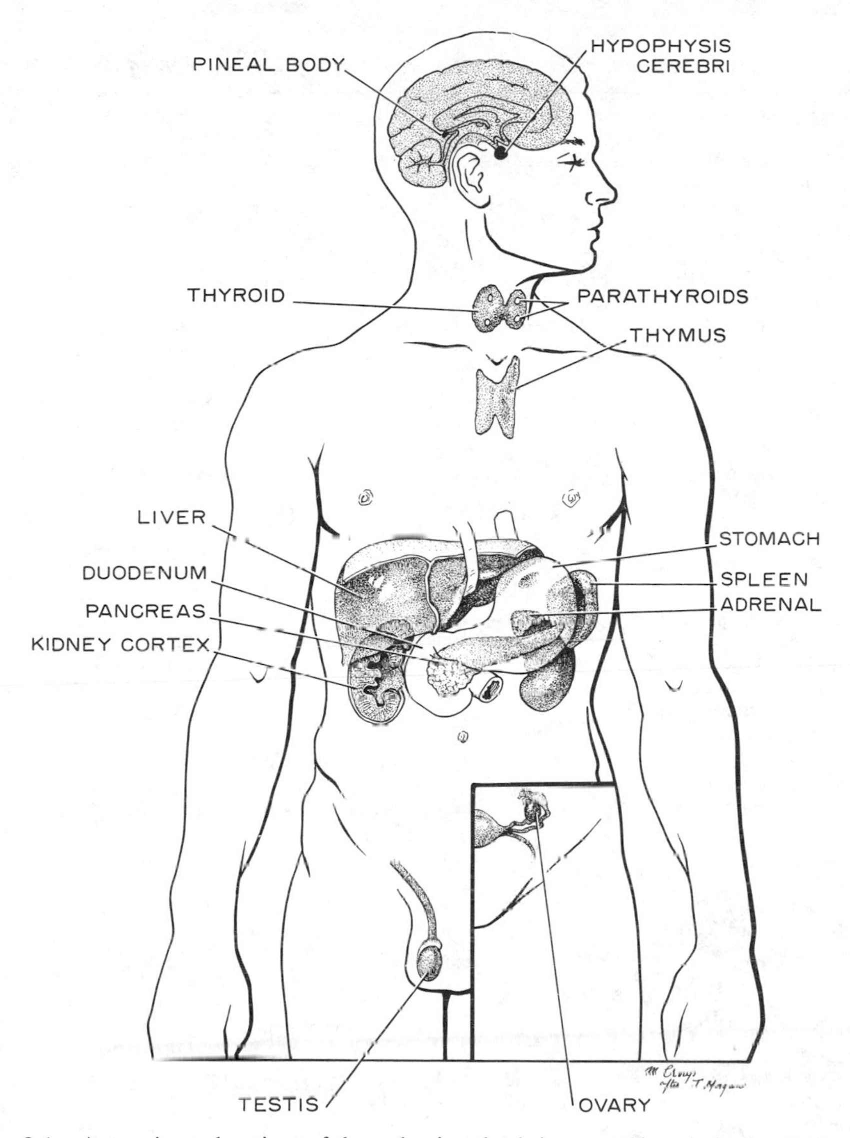

Figure 9-1. Approximate locations of the endocrine glands in man. Though the liver, kidneys, and spleen add important materials to the blood, they are not definitely known to be organs of internal secretion. (From Turner, C. D.: General Endocrinology, 4th edition, W. B. Saunders Co. Philadelphia, 1966.)

estriol is almost inert so far as its effect on accessory sex organs is concerned. Similarly, when norepinephrine, a hormone of the adrenal medulla, is methylated on its amine-N to produce epinephrine, this minor structural change not only diminishes its potency but also alters the nature of its biological activity.

Actions of the hormones are complex and diverse. (The source, chemical nature, and possible site of the actions of individual hormones are summarized in Table 9-1.) They may, however, be broadly divided according to three general aspects.

1. Regulatory function. One of the major functions of the endocrine system is to maintain constancy of chemical composition (homeostasis) of plasma and interstitial and intracellular fluids for proper and efficient function and growth of the

TABLE 9-1. *The Hormones, Their Source and a Brief Description of Their Action*

Endocrine Gland and Hormone	*Nature of Hormone*	*Site of Action*	*Principal Actions*
Hypothalamus			
Various releasing factors	Polypeptides	Anterior pituitary	Release of trophic hormones
Anterior pituitary			
Somatotrophin, growth hormone (STH, GH)	Protein	Body as a whole	Growth of bone and muscle
Adrenocorticotrophin (ACTH)	Polypeptide	Adrenal cortex	Stimulates formation and secretions of adreno-cortical steroids
Thyrotrophin (TSH)	Glycoprotein	Thyroid	Stimulates formation and secretion of thyroid hormone
Follicle-stimulating hormone (FSH)	Glycoprotein	Ovary	Growth of follicles, with LH secretion of estrogens and ovulation
		Testis	Development of seminiferous tubules; spermatogenesis
Luteinizing or interstitial cell-stimulating hormone (LH or ICSH)	Glycoprotein	Ovary	Formation of corpora lutea, secretion of progesterone
		Testis	Stimulation of interstitial tissue—secretion of androgen
Prolactin (lactogenic hormone, luteotrophin)	Protein	Mammary gland	Proliferation of mammary gland and initiation of milk secretion.
		Corpus luteum	Maintenance of corpus luteum
Posterior pituitary			
Vasopressin (ADH, antidiuretic hormone)	Octapeptide	Arterioles	Elevates blood pressure
		Renal tubules	Water reabsorption
Oxytocin	Octapeptide	Smooth muscle (uterus, mammary gland)	Contraction, action in parturition and in sperm transport; ejection of milk
Intermediate Lobe			
Melanophore-stimulating hormone (MSH, intermedin)	Polypeptide	Skin	Dispersion of pigment granules; darkening of skin
Thyroid			
Thyroxine and triiodothyronine	Iodoamino acids	General body tissue	Stimulates oxygen consumption and metabolic rate of tissues
Parathyroid			
Parathyroid hormone (PTH, Parathormone)	Polypeptide	Skeleton, kidney, gastrointestinal tract	Regulates calcium and phosphorous metabolism
Adrenal Cortex			
Adrenal cortical steroids—cortisol, aldosterone	Steroids	General body tissue	Carbohydrate, protein and fat metabolism; salt and water balance, inflammation, resistance to infection; hypersensitivity

TABLE 9-1. *(continued)*

Endocrine Gland and Hormone	Nature of Hormone	Site of Action	Principal Actions
Adrenal Medulla			
Norepinephrine and epinephrine	Aromatic amines	Sympathetic receptor	Mimic sympathetic nervous system
		Liver and muscle	Glycogenolysis
		Adipose tissue	Release of lipid
Thyrocalcitonin			
(Calcitonin)	Polypeptide	Skeleton	Inhibits calcium resorption; lowers plasma calcium and phosphate
Ovary			
Estrogens	Phenolic steroids	Female accessory sex organs	Development of secondary sex characteristics
Progesterone	Steroids	Female accessory reproductive structures	Preparation for ovum implantation; maintenance of pregnancy
Relaxin	Polypeptide	Symphysis pubis, uterus	Relaxation, aids in parturition
Testis			
Testosterone	Steroid	Male accessory sex organs	Development of secondary sex characteristics, maturation and normal function
Pancreas			
Insulin	Polypeptide	Most cells	Regulation of carbohydrate metabolism; lipogenesis
Glucagon	Polypeptide	Liver	Glycogenolysis
Placenta			
Estrogens, Progesterone, gonadotrophins (HCG)	Same as above	Same as above	Same as above
Growth hormone-prolactin, relaxin			
Gastrointestinal Tract			
Secretin and pancreozymin	Protein	Pancreas	Secretion of alkaline fluid and digestive enzymes
Cholecystokinin	Protein	Gallbladder	Contraction and emptying
Enterogastrone	Protein	Stomach	Inhibition of motility and secretion
Gastrin	Protein	Stomach	Secretion of acid

organism. This homeostatic mechanism is maintained through the sensitively regulated metabolism of salt, water, carbohydrate, fat, and protein by secretion of appropriate hormones. When there is derangement in the salt and water balance, hormones such as vasopressin and aldosterone come into play, whereas, if there is an increased concentration of blood glucose (hyperglycemia), as for example, after a carbohydrate-rich meal, insulin is promptly secreted from the pancreas so that the glucose will be utilized at a faster rate until the concentration decreases to its normal level.

2. Morphogenesis. Some hormones play an important part in controlling the type and rate of growth of an organism. The development of the male and female sex characteristics under the influence of the respective sex hormones (testosterone and estradiol-17β) is perhaps the best example.

3. Integrative action. This aspect of hormonal function is the most complex and the least understood. Broadly speaking, each hormone has a specific function. For example, the biological action of corticotrophin (ACTH), secreted by the pituitary gland, controls the functional status of the adrenal cortex; estrogens and progesterone, produced in the ovaries and called female sex hormones, regulate the development of secondary sex characteristics; the adrenal hormone, aldosterone, controls salt and water balance; the pancreatic hormone, insulin, regulates carbohydrate metabolism; and so on. However, even though a particular hormone dramatically influences a single biochemical event, or dramatically changes the morphology and rate of metabolism of a single organ, the presence of other hormones produced by different endocrine glands is also important for efficient functioning. Insulin alone would not be adequate to maintain the balance of carbohydrate metabolism. The concerted action of glucagon (from the pancreas) as well as the action of other hormones from glands such as the pituitary (corticotrophin, somatotrophin) adrenals (glucocorticoids, epinephrine), thyroid (thyroxine), and even gonads (estrogens) are important. This interrelation is not limited to the endocrine glands but extends to the nervous system as well. While it is true that the mineralocorticoids (deoxycorticosterone, aldosterone) have a profound influence on the maintenance of salt and water balance, this control mechanism would fail without the simultaneous adjustment of the rate of blood flow, blood pressure, and vasoconstriction by the autonomic nervous system. Therefore, under normal circumstances, there exists an integrative functioning of the endocrine and nervous systems which is reflected by the maintenance of a constant body environment. Derangements of such interdependence give rise to disease states.

CONTROL OF HORMONE SECRETION

There are several mechanisms for maintaining the delicate balance between the products of hormones and need of the organism for hormones. Detailed discussions of these mechanisms are beyond the scope of this book and students are referred to other texts.[81,93] The following is a brief outline, covering only the salient features of these mechanisms.

The anterior pituitary, often called the "master gland," occupies a central position in the control of hormone secretion. It secretes several trophic* hormones, which stimulate and maintain certain other endocrine glands. In the absence of these trophic hormones, the target glands† are unable to maintain a normal rate of secretion. The main target organs of the trophic hormones are the thyroid gland, adrenal cortex, testis, and ovary. The trophic hormones for the thyroid and the adrenal cortex are known as *thyrotrophin* (TSH—thyroid stimulating hormone) and *corticotrophin* (ACTH—adrenocorticotrophic hormone), respectively; those involved in the functional activity of either the ovary or the testis are follicle stimulating hormone (FSH), luteinizing hormone (LH), and luteotrophin (LTH—luteotrophic

---

* There is some controversy about the spelling of the anterior pituitary hormones. Many workers prefer to use the ending "tropic" meaning "a turning," whereas others prefer the ending "trophic" meaning "to nourish." In the present writing the latter ending will be used.

† The term "target" is often used to refer to the site of action of any hormone, whether "trophic" in nature or not. For example, the thyroid gland and the uterus are the target organs for TSH and estrogens, respectively.

hormone), also called prolactin or lactogenic hormone. These last three hormones are grouped together under the generic term *gonadotrophic* hormones. In addition to these five pituitary hormones, there is *somatotrophin* (STH—somatotrophic hormone, also called growth hormone) which does not have a specific target organ but exerts a wide variety of metabolic effects and promotes general body growth.

To understand the interplay of hormone secretion control, whether directed by the pituitary or not, a description of certain basic concepts is warranted. Most important is the *feedback* loop or *servo* mechanism. (The concept is originally derived from the operation of an electrical network, specifically, a relay device consisting of input and output signals to actuate the automatic control of a complex machine, instrument, or operation.) In simplified terms, the feedback system may be described as one in which the degree of function of two variables, A and B, are interdependent. When A is directly proportional to B (i.e., when B increases, A also increases), the relationship is described as *positive feedback*, whereas if A is inversely proportional to B (i.e., when B increases, A decreases), then a *negative feedback* exists. It is important to note in this connection that feedback regulation, especially negative feedback, seems to be a cardinal feature not only in the control of hormone secretion but also in the growth and metabolism of unicellular bacteria to highly complex multicellular organisms. The feedback relationship may be represented diagrammatically as follows:

1. Negative feedback:

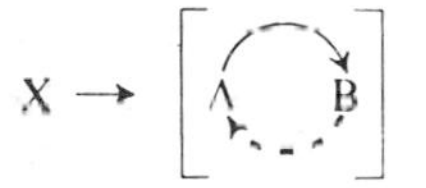

"X" increases A which in turn increases (solid line) the concentration of B to a level at which B decreases (dashed line) the concentration of A.

2. Positive feedback:

X → [A B]

"X" increases A, leading to the increase of B which in turn increases A.

The functional relationship between the pituitary and the target glands (adrenal cortex, gonads, thyroid) is based on feedback regulation, primarily negative feedback. As stated earlier, the effect of negative feedback control is characteristically opposite to the initial stimulus. Thus, a diminished level of cortisol in the blood (initial stimulus) triggers the pituitary to secrete more ACTH which in turn enhances the activity of the adrenal cortex to secrete more cortisol (final response). Conversely, an elevated level of blood cortisol causes the pituitary to secrete less of the trophic hormone, resulting in diminished production of the hormone by the adrenal cortex. The importance of such feedback control lies in the delicate maintenance of an optimal concentration of hormones in the blood, which in effect maintains the constancy of other blood constituents. Recent studies have shown, however, that the release of trophic hormone is controlled by the feedback loop existing between the hormones of the target glands and the hypothalamus. This portion of the brain manufactures specific polypeptides known as *releasing factors* for specific trophic hormones (CRF—corticotrophin releasing factor; LRF—LH releasing factor; TRF—thyroprotrophin releasing factor, etc.). Under appropriate stimuli the releasing factors reach the pituitary through the hypothalamo-hypophyseal vascular system and cause the release of the respective trophic hormones.

Steroid Hormones

Steroids are compounds containing the cyclopentanoperhydrophenanthrene ring system (Figure 9-2). The three six-sided rings (A, B, C) constitute the phenanthrene nucleus to which is attached a five-sided ring (D), cyclopentane. The prefix "perhydro" refers to the fact that all the necessary hydrogen atoms have been added to the compound to make it fully saturated. This class of compounds includes such natural products as sterols (e.g., cholesterol), bile acids (e.g., cholanic acid), sex hormones (e.g., estrogens, androgens), corticosteroids, cardiac glycosides (e.g., digitoxigenin), sapogenins (e.g., tigogenin) and some alkaloids (e.g., solasodine). The steroid hormones we are concerned with, contain up to 21 carbon atoms (C_{21} steroids) numbered as shown in Figure 9-2. Each carbon atom of a ring bears two hydrogen atoms, except when it is common to two rings, in which case it bears only one hydrogen atom (i.e., at C-5, C-8, C-9, and C-14). Carbon-17 bears one hydrogen atom, while C-10 and C-13 are bound only to other carbons (i.e., C-19 and C-18 respectively). The carbon atoms composing the rings and the hydrogen atoms attached to them are not usually written into the structure unless it is required to draw special attention to configuration. Furthermore, in all naturally occurring steroid hormones, the projected solid line from the carbon atom at position 10 or 13 usually designates the presence of an angular methyl ($—CH_3$) group, unless otherwise indicated.

Steroids consisting of tetracyclic rings are three-dimensional. Thus, the constituent carbon atoms and the hydrogen atoms or their substituents lie in different planes, giving rise to *isomers*. The direction of the hydrogen atoms, the substituents, and the side chain play a much more important role in the distinction of various isomers of the steroid compounds than does the relative position of the carbon atoms in the rings. Thus the isomerisms resulting from fusion of two rings are decided on the basis of the spatial relationship between the hydrogen atoms or the substituents at common carbon atoms. When rings A and B are fused, two isomers are possible, depending upon whether the hydrogen atom at C-5 and the methyl group at C-10 are on the same or opposite side of the plane of the rings. If the hydrogen atom points in the same direction as the angular methyl group at C-10, the compound is said to be the

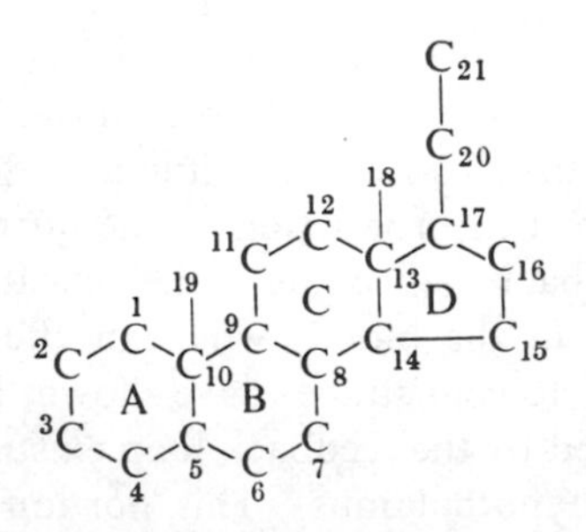

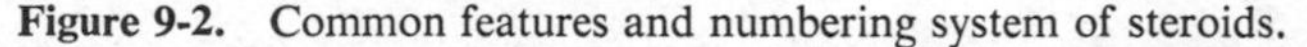
Figure 9-2. Common features and numbering system of steroids.

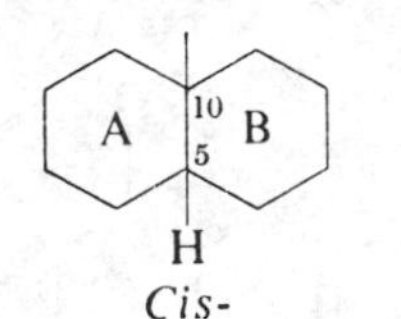

Cis-

Hydrogen atom and methyl group are on the same side (above the plane of the paper-5β-isomer).

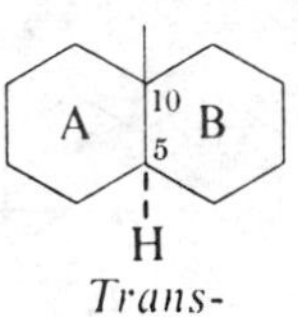

Trans-

Hydrogen atom is on the opposite side (below the plane of the paper-5α-isomer).

Figure 9-3. Fusion of rings A and B in naturally occurring steroids.

cis or *normal* form. If, however, they are on the opposite sides, the compound is said to be the *trans* or *allo* form.

While the rings A and B may be either *cis* or *trans*, the rings B/C and C/D have *trans* configuration in all naturally occurring steroid hormones.

The two methyl groups attached to C-10 and C-13 lie above the plane of the molecule and are customarily the points of reference for describing the spatial orientation of other substituents in the steroid nucleus. Substituents on the same side as these two methyl groups are said to possess a β-configuration which is indicated by a solid line (—) joining them to the appropriate carbon atoms in the nucleus. Substituents on the opposite side are attached by a broken line (--) to denote an α-configuration. Thus in the structures shown in Figure 9-3, when the hydrogen substituent at C-5 is *cis*, the isomer is the 5β-isomer, and when it is *trans* the isomer is accordingly the 5α-isomer. Similarly, the substituents at C-3, C-11, C-17, or any other carbon atoms are indicated as either α or β configuration depending on their spatial orientation relative to these methyl groups (C-10 and C-13).

The innumerable steroids containing the cyclopentanoperhydrophenanthrene nucleus differ from one another by the introduction of double bonds between certain pairs of carbon atoms, by the introduction of substituents for the hydrogen atoms, or by the addition of a specific type of side chain. On the basis of such structural characteristics, the steroidal compounds are classified as derivatives of certain parent hydrocarbons, namely *estrane* (for estrogens), *androstane* (androgens), and *pregnane* (for corticosteroids and progestins).

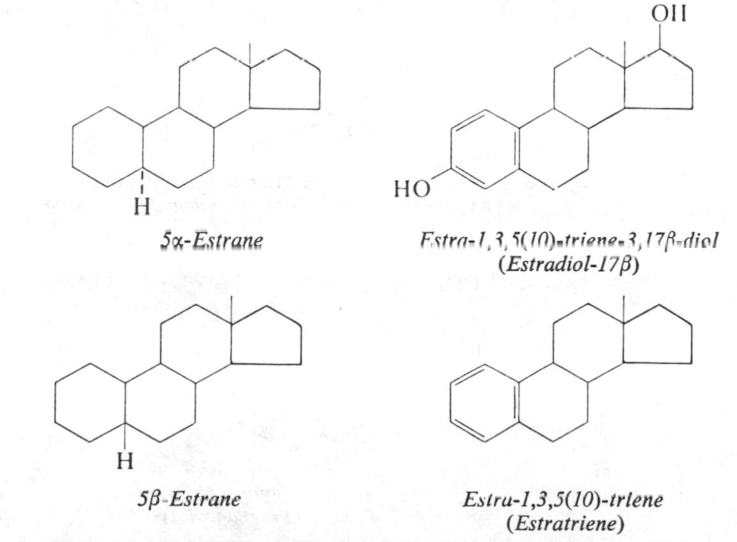

Figure 9-4. Parent hydrocarbons of estrogens.

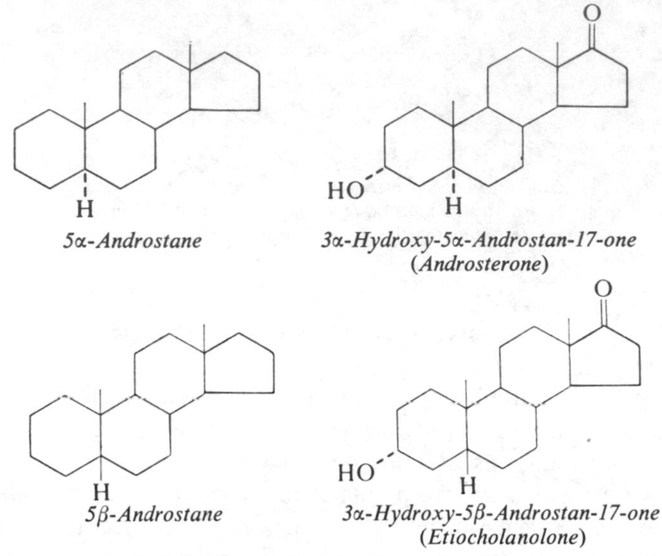

Figure 9-5. Parent hydrocarbons of androgens.

It should be noted that the parent substance *estrane* lacks one methyl group at C-10, and hence it is a C-18 compound. Furthermore, estrogens are actually derivatives of the compound estratriene, since the benzenoid ring structure is a common feature of all the naturally occurring estrogens.

The parent substance *androstane* is a C-19 compound and possesses 5α or 5β configuration. Naturally occurring *androsterone* and *etiocholanolone* are the examples of the respective derivatives of these isomers.

The special feature of the hydrocarbon *pregnane* is an ethyl side chain (—CH_2—CH_3) attached to C-17, making it a C-21 compound. The side chain is in *cis* relationship to the methyl groups at C-10 and C-13 and is, therefore, β-oriented. Like

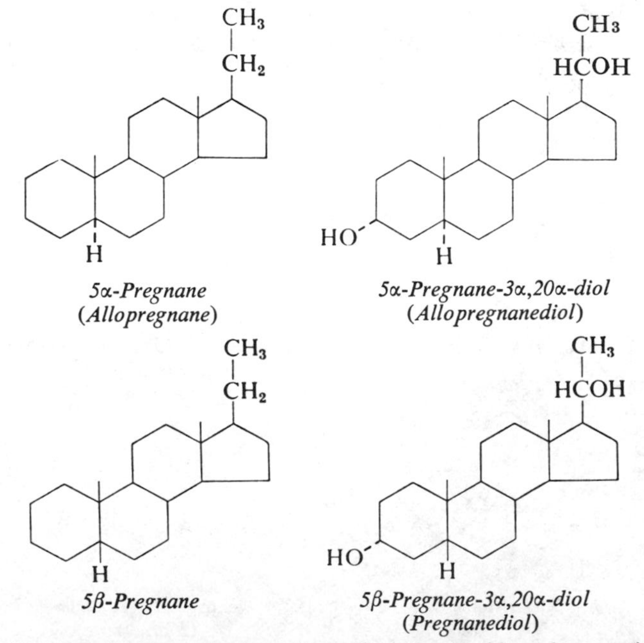

Figure 9-6. Parent hydrocarbons of corticosteroids and progestins.

androstane, this parent substance has also two isomers, 5α- and 5β-pregnane. *Allopregnanediol* and *pregnanediol* are the respective derivatives of these two isomers.

It should be noted that the prefix *allo* refers only to stereoisomerism of the hydrogen atom at C-5. When the configuration differs at any other carbon atom, the prefix *epi* is used; for example, androsterone possesses a 3α-hydroxyl group, and epiandrosterone a 3β-hydroxyl group; testosterone possesses a 17β-hydroxyl group, and epitestosterone a 17α-hydroxyl group.

To describe a compound with chemical nomenclature, a variety of other *suffixes* and *prefixes* are used. The suffix *-ane* indicates a fully saturated compound (e.g., pregn*ane*), *ene-* the presence of one double bond (e.g. pregn*ene*), *-diene*, two double bonds, *-triene*, three double bonds; the terminal 'e' is omitted before a vowel, e.g., -4-en-3β-ol. The position of the double bond is indicated by the number of the carbon atom from which it originates, and it is understood to terminate at the next higher carbon atom (i.e., 4-ene means that a double bond lies between C-4 and C-5). However, when an alternative is possible (for example, a double bond originating at C-5 in the *androstane* nucleus can terminate at C-6 or C-10), the number of the carbon atom at which the bond ends is written in parenthesis. Thus a double bond at C-5 terminating at C-10 is designated as 5 (10) -ene. A formerly used prefix for a double bond is the symbol Δ with a superscript indicating the position of the double bond (e.g., Δ^5). An alcohol (—OH substituent of the nucleus) is indicated by the suffix *-ol,* (two alcohol groups as *-diol*, three as *-triol*, etc.), or by the prefix *hydroxy* or *oxy* (dihydroxy for two, trihydroxy for three, etc.) Ketones (C═O) are identified by the suffixes *-one* for one keto group, *-dione* for two keto groups, etc., or by the prefix *oxo-* (see Table 9-2).

In naming a compound containing double bonds, hydroxyl groups, and ketones, priorities are given to the use of suffixes and prefixes. Thus, hydroxyl groups are indicated by the prefix followed by the suffixes for other substituents. Accordingly, the systematic name of dehydroepiandrosterone (Figure 9-7) is written as 3β-hydroxyandrost-5-en-17-one. Note that to denote the bond of unsaturation the first part of the parent hydrocarbon is followed by the position of the bond and the suffix (i.e., androst-5-en). If the prefix Δ for unsaturation is chosen, then suffixes for both the

TABLE 9-2. *Common Suffixes and Prefixes for Steroids*

Suffix or Prefix	*Definition*
Suffix	
-ane	Saturated hydrocarbon
-ene	Unsaturated hydrocarbon
-ol	Hydroxyl group
-one	Ketone group
Prefix	
hydroxy-(-oxy-)	Hydroxyl group
keto-(oxo-)	Ketone
deoxy-(desoxy-)	Replacement of hydroxyl group by hydrogen
dehydro-	Loss of two hydrogen atoms from adjacent carbon atoms
dihydro-	Addition of two hydrogen atoms
cis-	Spatial arrangement of two substituents on the same side of the molecule
trans-	Spatial arrangement of two substituents on opposite sides of the molecule
α-	Substituent which is *trans* to the methyl group at C-10
β-	Substituent which is *cis* to the methyl group at C-10
epi-	Isomeric in configuration at any carbon atoms except at the junction of two rings
Δ^n-	Position of unsaturated bond

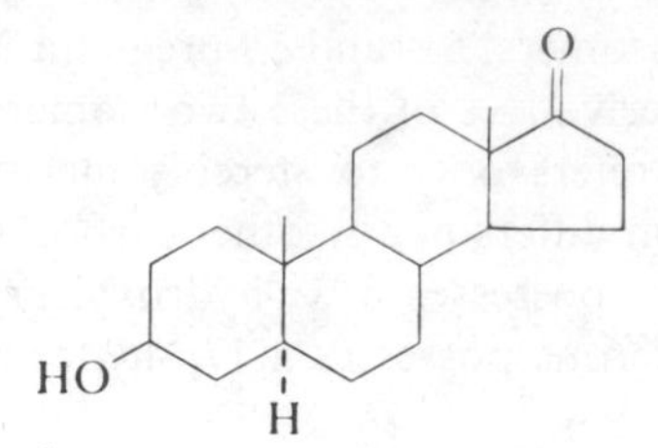

3β-Hydroxy-5α-androstan-17-one
(Epiandrosterone)

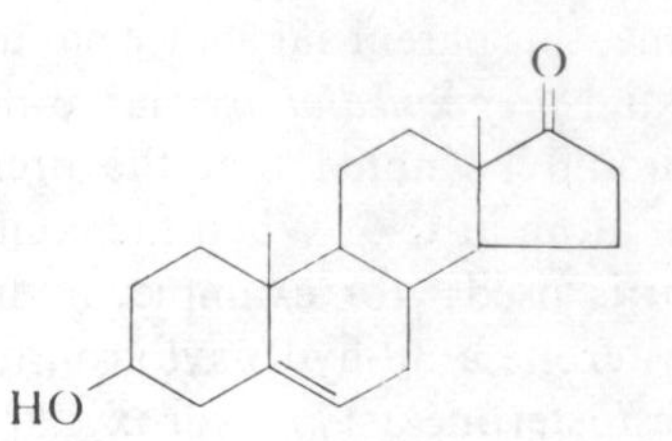

3β-Hydroxyandrost-5-en-17-one
(Dehydroepiandrosterone)

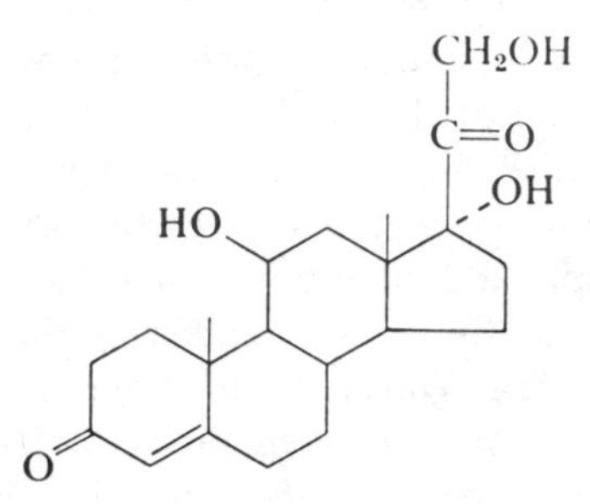

11β,17α,21-Trihydroxypregn-4-ene-3,20-dione
(Cortisol)

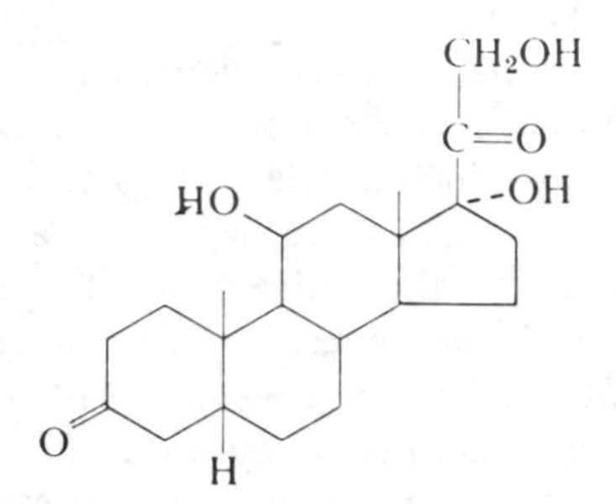

11β,17α,21-Trihydroxy-5β-pregnane-3,20-dione
(Dihydrocortisol)

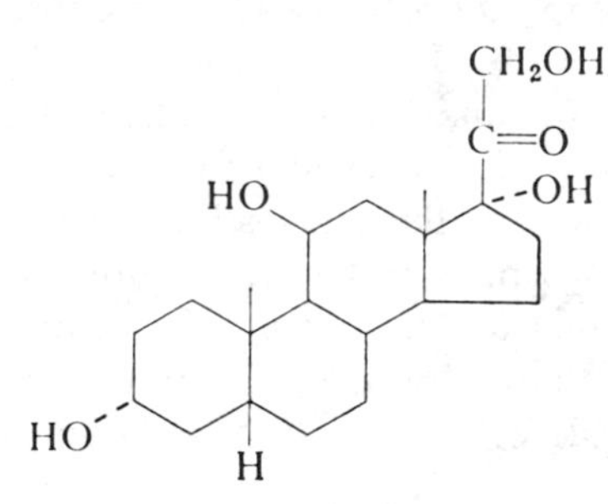

3α,11β,17α,21-Tetrahydroxy-5β-pregnan-20-one
(Tetrahydrocortisol, Urocortisol)

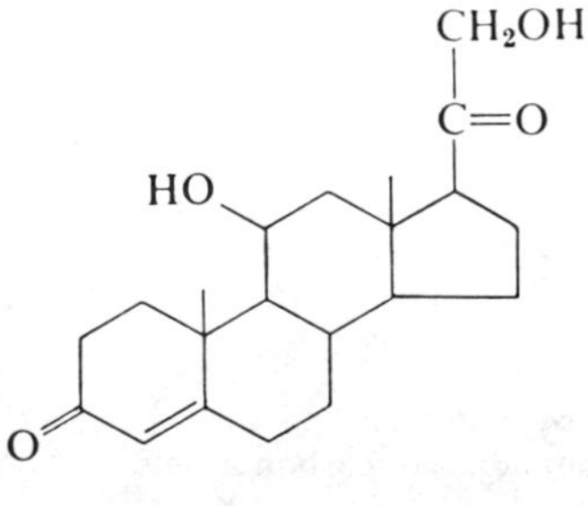

11β,21-Dihydroxypregn-4-ene-3,20-dione
(Corticosterone)

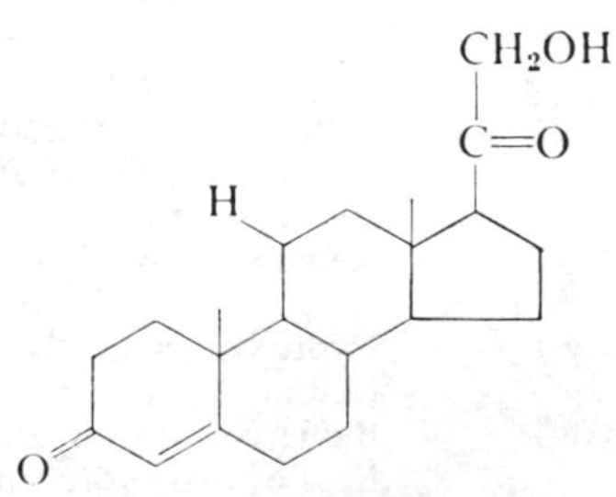

21-Hydroxypregn-4-ene-3,20-dione
(11-Deoxycorticosterone)

Figure 9-7. Illustration of semitrivial names.

TABLE 9-3. *Trivial and Systematic Names of Some Important Steroid Hormones*

Trivial Name	*Systematic Name*
Estradiol-17β	Estra-1,3,5(10)-triene-3,17β-diol
Estrone	3-Hydroxyestra-1,3,5(10)-trien-17-one
Estriol	Estra-1,3,5(10)-triene-3,16α-17β-triol
Testosterone	17β-Hydroxyandrost-4-en-3-one
Androsterone	3α-Hydroxy-5α-androstan-17-one
Etiocholanolone	3α-Hydroxy-5β-androstan-17-one
Dehydroepiandrosterone	3β-Hydroxyandrost-5-en-17-one
Adrenosterone	Androst-4-ene-3,11,17-trione
Progesterone	Pregn-4-ene-3,20-dione
Pregnanediol	5β-Pregnane-3α,20α-diol
Cortisol	11β,17α, 21-Trihydroxypregn-4-ene-3,20-dione
Urocortisol (tetrahydro F)	3α,11β,17α, 21-Tetrahydroxy-5β-pregnan-20-one
Aldosterone	11β,21-Dihydroxy-3,20-dioxopregn-4-en-18-al

hydroxyl and the ketone groups are used, e.g., Δ^5-androsten-3-βol-17-one. When there is only one kind of substituent, the use of a suffix is customary, e.g., pregnanediol (5β-pregnane-3α,20α-diol), androstenedione (androst-4-ene-3,17-dione).

The usefulness of the systematic name of a compound lies in the fact that it gives information about the parent substance, the position of unsaturation, and the nature, position, and orientation of substituents. The trivial name, as the term suggests, conveys little or no information about the chemical origin and characteristics of a compound, e..g, cortisol, progesterone, testosterone, etc. The trivial and systematic names of some of the important steroid hormones are given in Table 9-3.

In addition to the usual suffixes and prefixes just discussed, there are some special prefixes which are generally used for the semitrivial names of the compounds. Thus, the prefix *dehydro* is used to indicate the loss of two hydrogen atoms from adjacent carbon atoms with the formation of a double bond, e.g., dehydroepiandrosterone (Figure 9-7). The prefix *dihydro-* or *tetrahydro-* indicates the addition of two or four hydrogen atoms to the molecule respectively, as in dihydrocortisol and tetrahydrocortisol (Figure 9-7). The replacement of a hydroxyl group by hydrogen (C—OH → CH) is prefixed by *deoxy-* (or *desoxy-*), for example, 11-deoxycorticosterone.

GENERAL ASPECTS OF BIOSYNTHESIS AND METABOLISM OF STEROID HORMONES

The advent of radio-labeled compounds has played a very important role in the elucidation of biogenesis and metabolism of steroid hormones. The use of radioactive acetic acid and cholesterol for the study of steroidogenesis *in vivo* and *in vitro* has produced radioactive steroid hormones, lending support to the concept that both compounds are precursors of steroid hormones. Similarly, the administration of radioactive steroid hormones followed by the separation and identification of radioactive metabolites in the urine has helped to delineate the metabolic pathways. Even though acetate and cholesterol are both precursors of steroid hormones, they do not, in all probability, constitute a separate pathway of biosynthesis but follow the general sequence, acetate → cholesterol → steroid hormones. It has been amply documented that acetate is the sole precursor of cholesterol; among the 27 carbon atoms constituting cholesterol, 12 originate from the carboxyl carbon (C) and 15 from the methyl

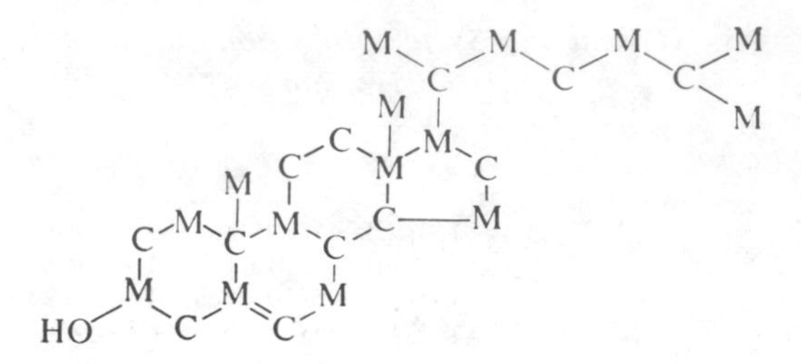

Figure 9-8. Carbon atoms of cholesterol derived from carboxyl carbon (C) and methyl carbon (M) of acetic acid.

carbon (M) of acetic acid (CH_3COOH) as shown in Figure 9-8. Stepwise degradations of radioactive steroid hormones (e.g., cortisol, corticosterone) synthesized from ^{14}C-labeled acetate indicate that the individual carbon atoms of steroid hormones correspond to those of cholesterol originating from carboxyl and methyl carbon atoms of acetic acid. There are, however, 30 separate biochemical reactions involved in the biosynthesis of cholesterol from acetate, and many more enzymatic reactions come into play to convert cholesterol to a variety of steroid hormones. For a detailed discussion of these topics students are referred to an excellent authoritative monograph by Dorfman and Ungar.[24]

In normal men and women, steroid hormones are produced in the adrenals, ovaries, and testes. All these glands utilize the same precursors, such as acetate and cholesterol, but the quality and the quantity of steroid hormone produced by each gland are different. The difference is inherent in the degree of activity and the presence or absence of certain enzymatic systems. For example, the enzymes 11β-hydroxylase and 21-hydroxylase are uniquely present in the adrenals to synthesize the characteristic hormones of the glands, the corticosteroids. Similarly, the enzymatic distinction between the ovaries and the testes lies in the fact that the ovaries contain an active aromatizing enzyme system (necessary to convert male sex hormones, e.g., testosterone, to female sex hormones, e.g., estradiol), in addition to the enzymes found in the testes.

The different enzymes participating in the biogenesis of steroid hormones may be broadly classified into the following functional groups.

1. Hydroxylases. These enzymes catalyze the substitution of the hydroxyl group (—OH) for hydrogen (—H). For example, 21-hydroxylase introduces an hydroxyl group at C-21. Similarly, 11β -hydroxylase substitutes a hydroxyl group at the β position of C-11. There are numerous examples of other important hydroxylases, such as 20α-hydroxylase, 19-hydroxylase, 17α-hydroxylase, etc. The cofactors are NADPH and molecular oxygen. The reaction is irreversible.

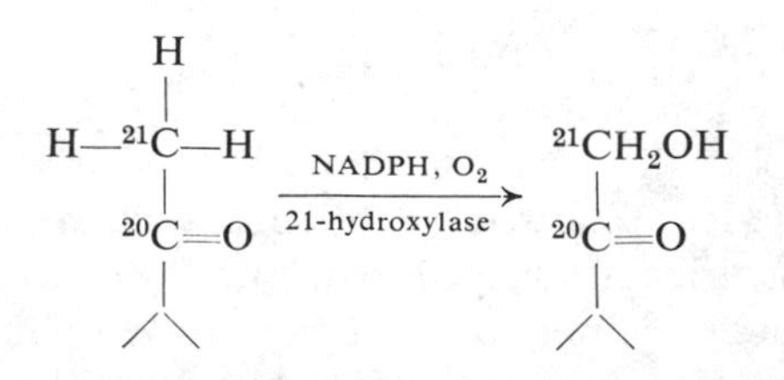

2. Desmolases. These enzymes are required for splitting off the side chain. There are two desmolases, 20,22-desmolase and 17,20-desmolase, which are very important in steroidogenesis. The former participates in the conversion of the C-27

carbon compound, cholesterol, to a C-21 compound, pregnenolone, whereas the latter transforms C-21 steroid hormones to C-19 steroid hormones. The required cofactors are thought to be NADPH and molecular oxygen.

$$\underset{\textit{Cholesterol}}{R{-}\overset{\overset{^{21}CH_3}{|}}{^{20}C}{-}^{22}C{-}(CH_2)_2CH(CH_3)_2} \xrightarrow[\text{20, 22-desmolase}]{} \underset{\textit{Pregnenolone}}{R{-}\overset{\overset{^{21}CH_3}{|}}{^{20}C}{=}O} + \underset{\textit{Isocaproic aldehyde}}{H{-}\overset{\overset{O}{\|}}{C}{-}(CH_2)_2CH(CH_3)_2}$$

3. Dehydrogenases. This group of enzymes catalyzes the transfer of hydrogen (oxidation and reduction). The reaction is generally reversible. The cofactor is either NAD or NADP (oxidized or reduced form, depending on the direction of the reaction). Some examples are 3β-hydroxysteroid dehydrogenases, 11β-hydroxysteroid dehydrogenase, 17β-hydroxysteroid dehydrogenase, Δ^5-3β-hydroxysteroid dehydrogenase, 3α-hydroxysteroid dehydrogenase, etc.

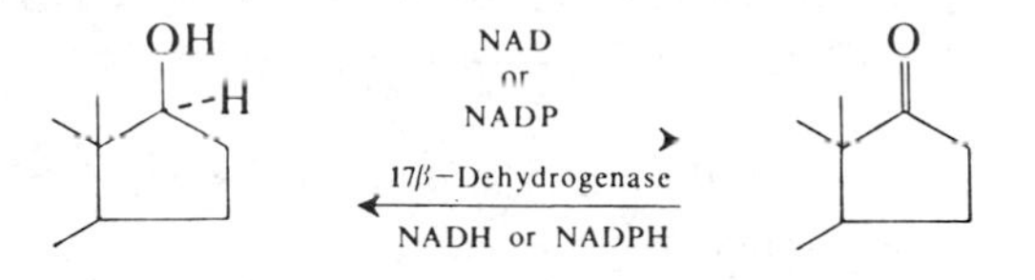

4. Isomerases. These enzymes catalyze the migration of a double bond. The most important enzyme of this group that is involved in steroidogenesis is Δ^5-ketosteroid isomerase. The concerted action of Δ^5-3β-dehydrogenase and Δ^5-ketosteroid isomerase on pregnenolone produces progesterone through the oxidation of the 3β-hydroxyl group and the migration of a double bond ($\Delta^{5-6} \rightarrow \Delta^{4-5}$).

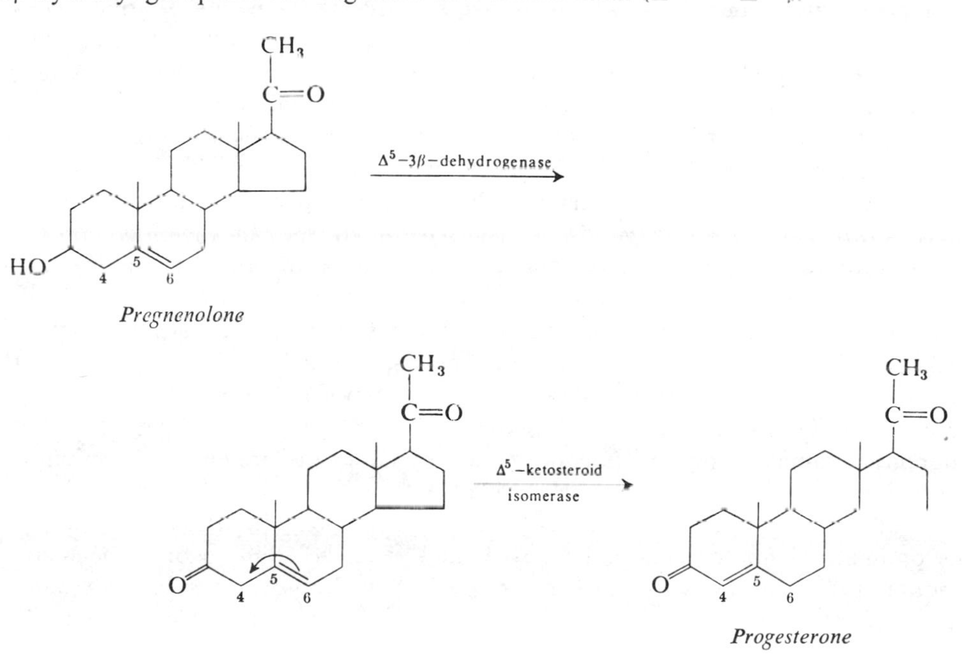

Pregnenolone

Progesterone

The liver is the major site of steroid metabolism. There is some evidence, however, that the kidney and the gastrointestinal tract may also carry out some of the metabolic transformations of steroids. Physiologically active steroid hormones have very high structural specificity. Any alterations in the number or nature of the substituents, or in the steroid nucleus, are liable to make the hormones inactive, or they may change their specific activity. Introduction of a new hydroxyl group (e.g.) estradiol → estriol), dehydrogenation (e.g., testosterone → androstenedione), reduction of a double bond (e.g., cortisol → dihydrocortisol) or conjugation of an essential hydroxyl group(s) with a chemical moiety such as glucuronic acid (glucosiduronic acid) (e.g., testosterone → testosterone glucuronide) are not only important biochemical steps for neutralizing the effectiveness of hormones but also for their rapid elimination from the systemic circulation. The conjugation of these hormones and their metabolites with sulfuric or glucuronic acid is, by far, the most efficient single metabolic process for their excretion in the urine, by virtue of the high water solubility of these conjugates. Almost all the steroid hormones and their metabolites are excreted as either glucuronides or sulfates.

GENERAL COMMENTS ON THE METHODS OF STEROID DETERMINATION

The clinical significance of the determination of steroid hormones and their metabolites in urine or in plasma is the assessment of the secretory activity of the glands producing these hormones. Unfortunately, very few available methods for plasma determination are suitable for use in a routine clinical laboratory. As a result, for clinical purposes only, urinary measurements have been a widely accepted practice. In most instances the urinary excretion of a hormone, its metabolites, or both does not account for the total amount of hormone secreted by the gland, but it does represent an approximate proportion of the amount secreted during the period of urine collection. Thus, urinary assays indirectly reflect the secretory activity of the endocrine glands. However, factors such as incompleteness of collection, altered renal function, and contribution of more than one gland to the same hormone(s) or metabolites may warrant special precaution in the interpretation of urinary values.

Although steroid hormones differ greatly in their physiological activities in the body, the assay procedures have many similarities and require the following general steps: *hydrolysis*, *extraction*, *purification and separation*, and *final quantitation.*

Hydrolysis. The majority of the steroid hormones and their metabolites are excreted in the urine as the water-soluble conjugates of glucosiduronic acid (glucuronic acid) and sulfuric acid. In the absence of any direct method for the estimation of such conjugated steroids in the urine, the splitting (hydrolysis) of such ether and ester linkages is an obligatory first step. Two general types of procedures are available, namely, *acid hydrolysis* and *enzyme hydrolysis.* In acid hydrolysis, an aliquot of a 24 hour urine sample is boiled, generally under reflux, in the presence of a specified concentration of mineral acid for a specified length of time (10 to 60 minutes). For enzymatic hydrolysis, a portion of a 24 hour urine specimen is adjusted with buffer to the optimal pH for the enzyme employed, and after the addition of an adequate amount of the respective enzyme (β-glucuronidase to hydrolyze glucosiduronates, and sulfatase to hydrolyze sulfate conjugates), the test sample is incubated for 18 to 76 hours at a specified temperature (e.g., 37°C.).

From the technical point of view, acid hydrolysis is always preferable (except for acid-labile hormones) because of its simplicity, speed, and completeness of reaction irrespective of the nature of conjugates. Enzyme hydrolysis, on the other hand, requires special attention regarding the optimal concentration and type of enzyme, the pH, the temperature, and the duration of incubation. In addition, the possible presence of enzyme inhibitors, varying in amount and nature with different urine samples, will always throw some doubt on the degree of completeness of hydrolysis. In spite of such drawbacks, enzymatic hydrolysis is employed in many procedures, particularly when the steroids are labile in strong acid solution (e.g., pregnanetriol, corticosteroids).

Extraction. Following hydrolysis, the free steroids become sparingly soluble in aqueous solution. Thus, when an immiscible organic solvent in which steroids are highly soluble is added to the hydrolyzed urine and is shaken, the vast majority of steroids are extracted into the organic layer. Repeating the process of extraction with a fresh volume of the organic solvent ensures the maximum recovery of steroids from the urine. The selection of the organic solvent is based on the polarity of the steroid hormones in question. One should bear in mind that even though the nonpolar component, the 4-ring system, is common to all steroids, the polarity increases as the number of oxygen groups (i.e., ketone and hydroxyl groups) and double bonds increases. Steroids with one or two oxygens (e.g., androgens, estrogens) are of low polarity, and the best choice of solvent for their extraction would be one of a nonpolar nature, such as ether or benzene. Similarly, steroids with three or more oxygens (e.g., corticosteroids and their metabolites) are quite polar, and for their extraction polar organic solvents such as chloroform, dichloromethane, or ethyl acetate would be most suitable.

The property of relative solubility of substances in two immiscible solvents has not only been exploited for the extraction process but has also been the basis for the separation and purification of substances (partition chromatography). The ratio of the concentration of a substance in a nonpolar phase to the concentration of the same compound in the polar phases is known as the "partition coefficient." Substances with high K values will mostly be in the nonpolar phase, whereas those with low K values will preferably move into the polar phase. In the extraction process, the solvent system is composed of an organic solvent (a relatively nonpolar phase) and the hydrolyzed urine (polar phase). Steroids of high partition coefficients will consequently be extracted into the organic layer.

Purification and separation. Although a proper choice of solvent improves the selectivity of extraction, nonetheless a large number of urinary pigments, chromogenic substances, and other nonspecific materials are invariably extracted along with the steroids. Removal of such contaminants, especially those which will interfere in the final estimation, is very important.

The *solvent partition* method is the method most simple and suitable for a clinical laboratory. Thus, it is widely used for preliminary purification and separation of compounds of interest. The basic principle is the same as that of the aforementioned extraction. The organic solvent containing steroids and other urinary impurities is treated with weakly *basic solution* (sodium bicarbonate, sodium carbonate). By virtue of their greater solubility, strongly acidic components migrate into the aqueous layer to be discarded. The separation of neutral and phenolic steroids can be achieved in a similar way. Because of the acidic nature of the phenolic steroids (estrogens) they are readily extractable with an aqueous solution of sodium hydroxide. After lowering the pH of the alkaline solution, estrogens are re-extracted with a suitable solvent (diethyl ether) and are processed further for final estimation. Most often the organic extract is washed to neutrality with water to ensure complete removal of alkali which, if allowed to remain, might interfere in subsequent work-up of the materials.

The degree of purification and separation needed prior to quantitative measurement will depend, of course, on the method used for final quantitation. For example, if the final mode of estimation is a color reaction which is very specific for an individual steroid or for a group of steroids, further purification of the steroid extract may be omitted. In fact, many colorimetric assays of steroid hormones are performed on crude urinary extracts. Although such determinations yield fairly adequate information for most clinical purposes, occasionally, specific measurement of an individual steroid or group of steroids necessitates further purification and separation of the extracts. There are various techniques available for this purpose. For detailed description of theory and application of these methods to steroid analysis the reader is referred elsewhere.[22] It suffices to mention here that these methods are based either on some selective chemical reaction (e.g., Girard derivative formation for the separation of ketonic and nonketonic steroids, digitonin formation to separate 3β-hydroxy and 3α-hydroxy steroids) or on physical techniques such as counter current distribution (CCD), paper and column partition chromatography, adsorption chromatography (column and thin layer), or gas chromatography.

Estimation. Methods for the quantitative estimation of steroids may be divided into three categories—*colorimetric*, *fluorimetric*, and *gas chromatographic procedures*.

COLORIMETRIC PROCEDURES. Colorimetric methods are by far the most commonly used clinical methods for the quantitation of steroid hormones. In these methods a certain functional group of a steroid is made to react with a particular chemical reagent to form a specific colored product. For example, steroids containing a keto

$$\left(\begin{array}{c} O \\ \| \\ C \\ / \quad \backslash \end{array} \right)$$

group in position 17 (17-ketosteroids) react with *meta*-dinitrobenzene in alcoholic alkali to produce a reddish-purple compound. The intensity of the color is proportional to the concentration of the steroid and is measured with a spectrophotometer or a colorimeter, using the wavelength of maximum absorption. The principal source of error in this procedure is the interference from nonspecific chromogens derived from other steroidal and nonsteroidal components of urinary extracts. Various measures have been suggested to correct for such interference.

In later sections some specific methods for steroidal hormone measurement are discussed. It is important to note that in some instances the nonspecific chromogens may account for as much as 90 per cent of the actual value.

FLUORIMETRIC ESTIMATION OF STEROIDS. Many steroids are known to produce a characteristic fluorescence if present in a suitable medium such as sulfuric acid or phosphoric acid. The activation and emission wavelengths, under specified experimental conditions, are relatively specific for a given substance. Although this technique offers certain advantages (sensitivity and, at times, increased specificity), from the practical point of view it is afflicted with many drawbacks, such as nonspecific fluorescence, and quenching effects caused by solvent residues, urinary contaminants, and improperly purified reagents. The need for delicate reaction conditions and the instability of fluorescence have further limited the usefulness of this technique in routine clinical laboratories. Nevertheless, the fluorimetric assays of certain hormones appearing in microgram quantities in biological fluids (e.g., cortisol in plasma, estrogens in the urine of men and nonpregnant women) have been proved to be quite useful. In some instances they have replaced previous colorimetric procedures.

GAS CHROMATOGRAPHIC ESTIMATION. This technique is one of the latest and most promising additions to steroid methodology. Speed, sensitivity, accuracy, precision, and specificity are demanding criteria for selection of procedures to measure compounds of clinical interest. Gas chromatographic techniques appear to fulfill these requirements. As a result, within a short period of time this technique has become an important tool for clinical laboratories and has replaced many of the time-consuming and occasionally nonspecific chemical methods. In recent years, gas chromatographic methods for the analysis of most of the steroid hormones have been reported. Many of them have been found to be quite suitable for routine clinical use (e.g., pregnanediol, pregnanetriol, 17-ketosteroids, estriol), and these methods are described in greater detail in the appropriate sections of this chapter.

The theory and the instrumentation involved in gas chromatography have been described in Chapter 2. The following paragraphs describe briefly the features pertinent to steroid analysis.

A glass or stainless steel column of any size and shape may be used, limited only by the ease of packing and operation. The analytical columns for steroid work vary from 1/8 to 1/4 inch in diameter and 4 to 12 feet in length. The *solid supports* are deactivated (acid, base washed, silanized) diatomaceous earths of small and uniform particle size with great surface area. The mesh size of these supports is in the range of 80 to 100 and 100 to 120, and gives adequately efficient columns for steroid analysis. Commercially available supporting materials such as Diatoport S (F and M Scientific Company, Avondale, Pennsylvania), Gaschrom P, Gaschrom Z, and Gaschrom Q (Applied Science Labs, Inc., State College, Pennsylvania) are widely used. Both selective and nonselective liquid (stationary) phases are employed, depending on the nature of the steroids being analyzed. The separation with a *nonselective* phase depends primarily on the molecular size and shape of the compounds. Most of the useful nonselective phases are methyl-substituted silicone polymers which are marketed under various trade names, such as SE-30, OV-1, F-60, JXR, etc. Among them, SE-30 and OV-1 appear to be the most suitable because of their excellent thermal stability and chemical inertness towards the steroids. The separation of compounds which vary in the nature, number, or stereochemical arrangement of the functional group(s) is best achieved with *selective phases*. The selective phases that have been recommended for steroid separations include XE-60 (Cyanoethyl methyl silicone polymer), QF-1 (Fluoroalkyl silicone polymer), OV-17 (methyl phenyl silicone polymer), and the polyester NGS (neopentyl glycol succinate). The nonselective phases are more stable thermally, the operation temperature limit being in the vicinity of 320°C. in comparison to 235°C. for most selective phases. The silicone polymer OV-17, however, may be used up to about 300°C. For steroid analysis the most suitable concentration of the stationary (liquid) phase is generally found to be between 1 and 4 per cent (w/w) of the solid support. Different techniques of coating the stationary phase onto the solid support, and packing and conditioning of the column will be described in a later section (17-ketosteroids).

The most suitable and most versatile *detection system* for routine steroid analysis is the flame ionization detector (for principle of operation see Chapters 2 and 7). Since the rate of ion production is proportional to the concentration of combustible substances, the detector response depicted as a peak for an individual compound on the chart paper is related to the amount present in the sample. The linearity of response in the expected concentration range of analysis is preestablished with pure reference compounds. The quantitation is carried out by comparing the peak size of the unknown to that obtained for a known concentration of its reference standard.

Optimal performance of a flame-ionization detector is dependent on the ratio of hydrogen to carrier gas flow rate. The efficiency of a specific column also varies with the operating temperature and the flow rate of carrier gas. For this reason, whenever a new column is used, the optimal conditions of these parameters should be determined by trial using a set of reference standards.

There are two methods of *introduction of samples* into a gas chromatograph. The first is the so-called *solution method.* Depending on the anticipated concentration, the materials to be chromatographed are dissolved in a known volume of an organic solvent (benzene, iso-octane, acetone, ethyl acetate, carbon tetrachloride, carbon disulfide, etc.). Usually 1 or 2 μl. of the final solution is drawn into a Hamilton microliter syringe and is injected into the gas chromatograph.

The other method for sample introduction is the newly developed *solid injection technique.* Generally, the sample is dissolved in a suitable solvent and a desired amount of this solution is applied onto an inert stainless steel or platinum gauze, which is located in a dimpled Teflon plate. The solvent evaporates, leaving a solid residue on the gauze which can then be introduced either manually or by automatic solid injection techniques.[69] The technique has the advantage of greater sensitivity, since the sample size can be increased without having an undesirable large solvent peak.

Irrespective of the methods of sample introduction, the use of an *internal standard* is predicated for several reasons. With the solution method it is difficult to inject exact aliquots of dissolved substances on each occasion. The variation may arise from either the mode of injection or the continual changes of sample concentration caused by evaporation of the solvent. With the solid injection technique, variation may also be caused by the loss of an unknown quantity of the materials left on the Teflon plate. The presence of an internal standard of known concentration compensates for such variations. It also corrects for the effect of variation in the detector response. Furthermore, the retention time of compounds relative to the internal standard may be used as a means of their identification in an unknown sample. The selection of a compound is generally based on the following considerations: it should approximate the physico-chemical properties of the component or components to be measured; it must yield a completely resolved peak; and it must not be present in the original sample. The method of computation of results using internal standards is described elsewhere. (See methods for 17-ketosteroids, estriol, pregnanediol.)

The *derivative formation* of steroids for the gas chromatographic methods has been found to play an important role in quantitative analysis. The purposes of such procedure are numerous and may be summarized as follows: (1) decrease of polarity minimizing tailing and adsorption (e.g., trimethyl silyl ether or acetate derivative of estriol), (2) stabilization of thermally reactive structural arrangements preventing thermal breakdown (e.g., O-methyloxime trimethyl silyl ether derivative of 17-hydroxycorticosteroids and their metabolites), (3) increase in volatility so that the compounds will be eluted faster under allowable experimental conditions (e.g., trimethyl silyl ether derivatives of steroids), and (4) alteration of physico-chemical property to improve gas chromatographic separation of closely related compounds (e.g., trimethyl silyl ether derivatives of 17-ketosteroids, estrogens, and progesterone metabolites as well as acetate derivatives of estrogens and progesterone metabolites).

RELIABILITY CRITERIA FOR A METHOD OF STEROID HORMONE ASSAY

Over the past three decades, a very large number of methods for the quantitative determination of steroids in biological fluids have appeared in the literature. It is, therefore, important to establish, for a method, certain reliability criteria such as accuracy, precision, specificity, and sensitivity, as suggested by Borth.[10] Whenever a method is to be introduced for routine analysis it should first be examined along these guidelines to determine whether or not it is suitable for its intended use. It is also necessary to repeat these evaluations at frequent intervals for the maintenance of good quality control.

ACCURACY

This term refers to the degree to which measurements approach the true value of the quantity being measured. Evaluation of this criterion is generally performed by determining the percentage of recovery of added steroids. Ideally, the compound to be recovered should be added in the form in which it occurs in the sample. Because of lack of complete knowledge of the nature of the conjugates, and because of their unavailability, free steroids are usually added to the sample for such evaluations. Although experiments of this type are not entirely satisfactory, they still give some information about the losses incurred during processing of the sample.

PRECISION

This term refers to the magnitude of the random errors and thus demonstrates the reproducibility of the measurements. The precision is frequently evaluated by performing replicate determinations on the same specimen or by carrying out multiple recovery experiments, each time adding the same concentration of the desired compound. Usually, for a particular method there is a range of measurement at optimal concentration, judged by the maximum reproducibility (minimum variation between replicate analyses). According to Marrian,[49] a standard deviation of ± 10 percent at optimal concentration is considered to be quite acceptable for a method of steroid assay.

SPECIFICITY

This refers to the exclusive measurement of a compound or compounds for which the method has been designed. In other words, when a method is designed to measure 17-ketosteroids, it should measure only those steroids and nothing else. The clinical usefulness of the determination of hormones, their metabolites or both, in blood or urine, lies in the proper assessment of their production in the body. A method without specificity defeats this purpose. Generally, the specificity of steroid assay hinges on some selective color reactions (e.g., Zimmermann color reaction for 17-ketosteroids) or on the chromatographic behavior of the compounds (e.g., gas chromatographic measurement of 17-ketosteroids, pregnanediol, estrogens). It should be mentioned here that the selectivity of the color reaction or the chromatographic property of the compounds does not necessarily impart specificity. The interference from many drugs and other materials in the biological extract may yield spurious results. For this reason, a method to be followed is given careful consideration in regard to its sources of error, and in regard to the different tests applied in order to prove the validity of measurement.

SENSITIVITY

This term is defined as the minimum amount of a substance in biological medium that can be determined with accuracy, precision, and specificity by a particular method. This is largely limited by the degree of sensitivity of the final method of quantitation. Generally, the methods involving fluorescence are more sensitive than those based on color reactions. However, in many instances, the highest sensitivity seems to be obtained by the flame ionization and electron capture detectors used in gas chromatographic methods. For each method involving colorimetry, fluorimetry, or a gas chromatographic detection system, there is an intrinsic range of sensitivity. The determination of a substance above or below that level may become unreliable. Thus, the choice of an assay procedure having an adequate range of sensitivity for the purpose for which it is to be used is very important.

ADRENOCORTICAL STEROIDS

More than 40 different steroids have been isolated from the adrenals. Among these are the corticosteroids, which are produced exclusively by the adrenals, as well as the androgens, progesterone, and the estrogens, which are also secreted by the gonads.

Adrenal activity is regulated by an anterior pituitary hormone, adrenocorticotrophic hormone (corticotrophin, ACTH). Of the various hormones released as a result of corticotrophin stimulation, only cortisol has a feedback inhibitory effect. The mechanism and principle of feedback control have been described in the introductory section of this chapter. The control of aldosterone secretion is not entirely dependent on corticotrophin; agents derived from liver (angiotensinogen) and kidney (renin) also come into play here. A detailed discussion of the control mechanism of aldosterone secretion may be found elsewhere.[93]

Corticosteroids

The corticosteroids are, from the physiological as well as the quantitative points of view, the most important groups of adrenal steroids. The structural formulas of some of the biologically most active corticosteroids are shown in Figure 9-9, along with their trivial and systematic names. These compounds all possess a Δ^4-3-keto group (unsaturation between carbon atoms 4 and 5 and a keto group at carbon atom 3); a side chain ($H_2\text{—}\overset{\overset{HO}{|}}{C}\text{—}\overset{\overset{O}{\|}}{C}\text{—}$) substituted at C-17 in the β position (above the plane of the paper); and, with the exception of compound S and deoxycorticosterone, an oxygen function (keto or β-hydroxyl) at C-11. Cortisone and hydrocortisone (cortisol) also have a 17α-hydroxyl group (below the plane of the paper).

The corticosteroids show maximum structural specificity. Structural alterations, especially the reduction of Δ^4-3-keto group, make them biologically inactive.

The major corticosteroids—namely, cortisol, corticosterone, and aldosterone—are secreted by the adrenals at the rate of 25 mg., 2 mg. and 200 μg./day, respectively. There is diurnal variation in the secretion of cortisol and corticotrophin. Soon after midnight the blood level of cortisol starts to rise, reaching a maximum at about early morning, after which there is a gradual decline to the lowest level between early evening and midnight. In certain adrenal diseases, diurnal rhythmicity of blood levels of cortisol is absent, e.g., in Cushing's syndrome. Functionally the adrenal cortical

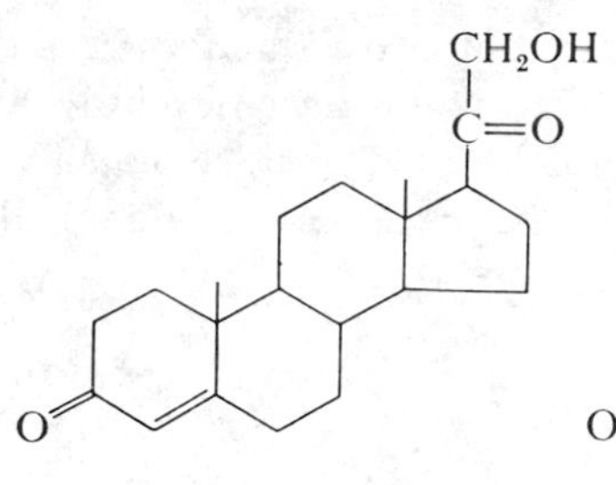

Desoxycorticosterone
("DOC," "Cortexone," "Q")
(21-Hydroxy-pregn-4-ene-3,20-dione)

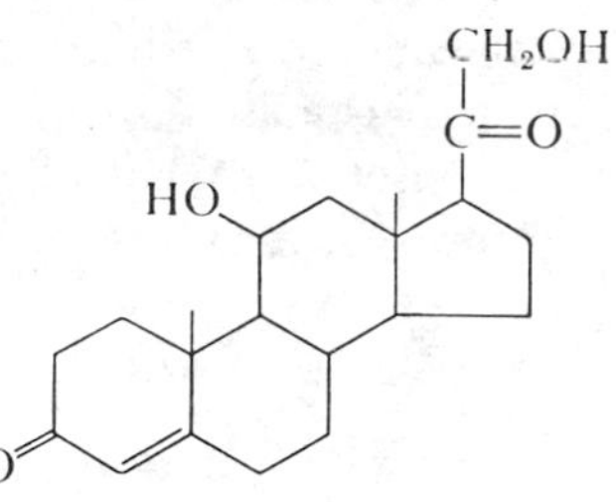

Corticosterone
("B")
(11β,21-Dihydroxy-pregn-4-ene-3,20-dione)

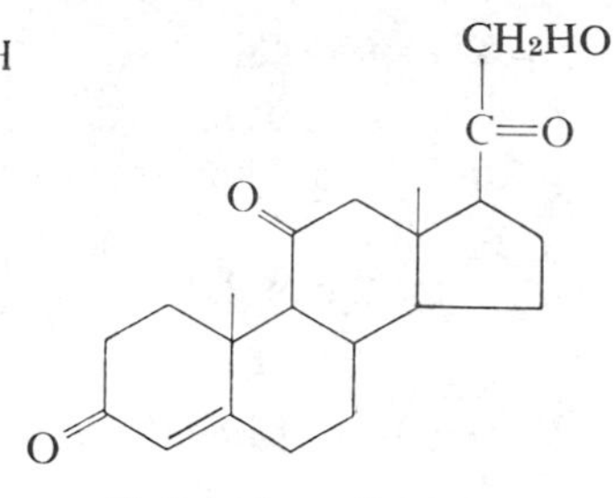

11-Dehydrocorticosterone
("A")
(21-Hydroxy-pregn-4-ene-3,11,20-trione)

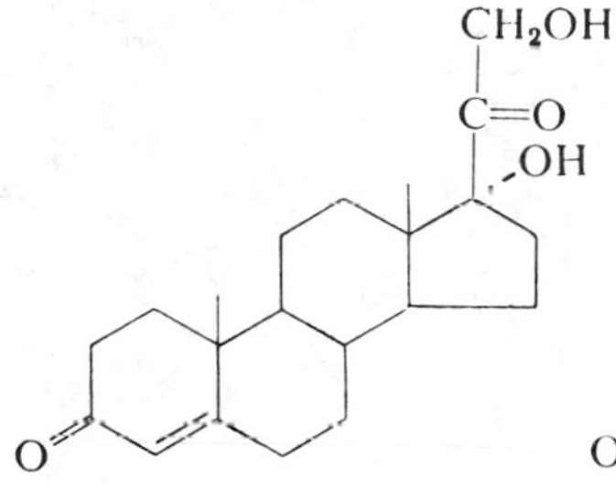

11-Desoxycortisol
("Cortexolone," "S")
(17α,21-Dihydroxy-pregn-4-ene-3,20-dione)

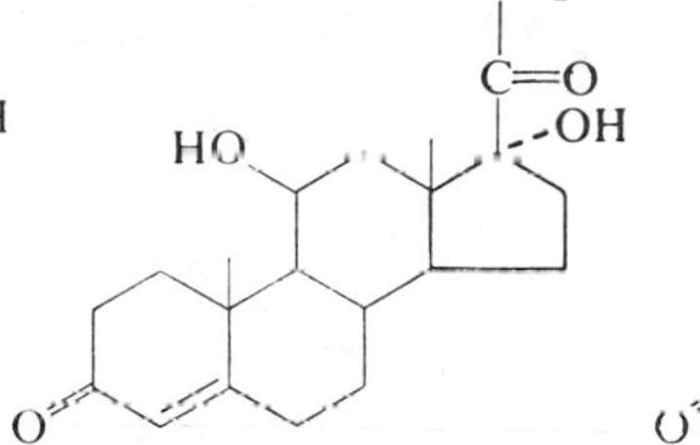

17α-Hydroxycorticosterone
("Cortisol," "F")
(11β,17α,21-Trihydroxy-pregn-4-ene-3,20-dione)

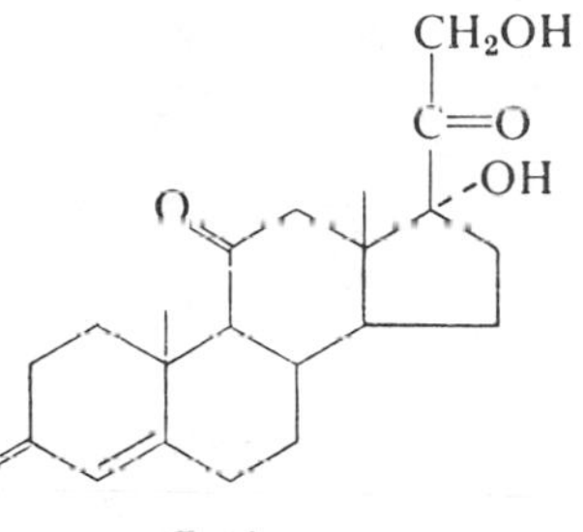

Cortisone
("E")
(17α,21-Dihydroxy-pregn-4-ene-3,11,20-trione)

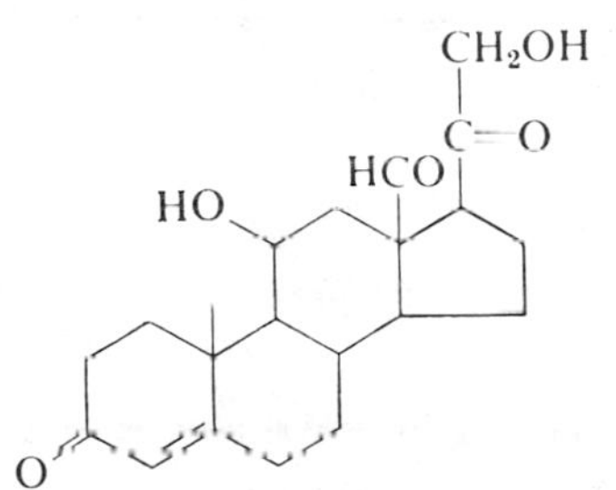

Aldosterone
(11β,21-Dihydroxy-3,20-dioxo-pregn-4-en-18-al)

Figure 9-9. Structural formulas of biologically active corticosteroids.

steroids may be subdivided into *glucocorticoids* and *mineralocorticoids*. Glucocorticoids participate in controlling carbohydrate metabolism and include the compounds cortisol, cortisone, corticosterone, and 11-dehydrocorticosterone. Mineralocorticoids regulate salt and water metabolism and include the compounds 11-deoxycorticosterone and aldosterone. It should be noted that aldosterone, in addition to having the general structural characteristic of corticosteroids, uniquely possesses an aldehyde substituent at C-18 (See Figure 9-9).

Other Adrenal Hormones

Besides corticosteroids, the adrenals also secrete androgens, progesterone, and estrogens, all of which are known to be produced by the gonads as well. The major androgens are androsterone, etiocholanolone, androstenedione, testosterone, dehydroepiandrosterone (DHA or DHEA), and 11β-hydroxyandrostenedione (Figs. 9-10, 9-11).

DHA and 11β-hydroxyandrostenedione are, from the quantitative standpoint, the most important adrenal 17-ketosteroids and are believed to be produced exclusively by this gland. The rate of secretion of DHA has been calculated to be as much as 25 mg./day. The estrogens (estrone, 18-hydroxyestrone) and progesterone of adrenal origin are quantitatively very insignificant.

Biogenesis of Adrenal Corticosteroids

Investigations with radio-labeled compounds have shown that acetate, cholesterol, pregnenolone and progesterone are all precursors of corticosteroids. In all probability, these precursors constitute a single biosynthetic pathway as shown in Figure 9-10. The

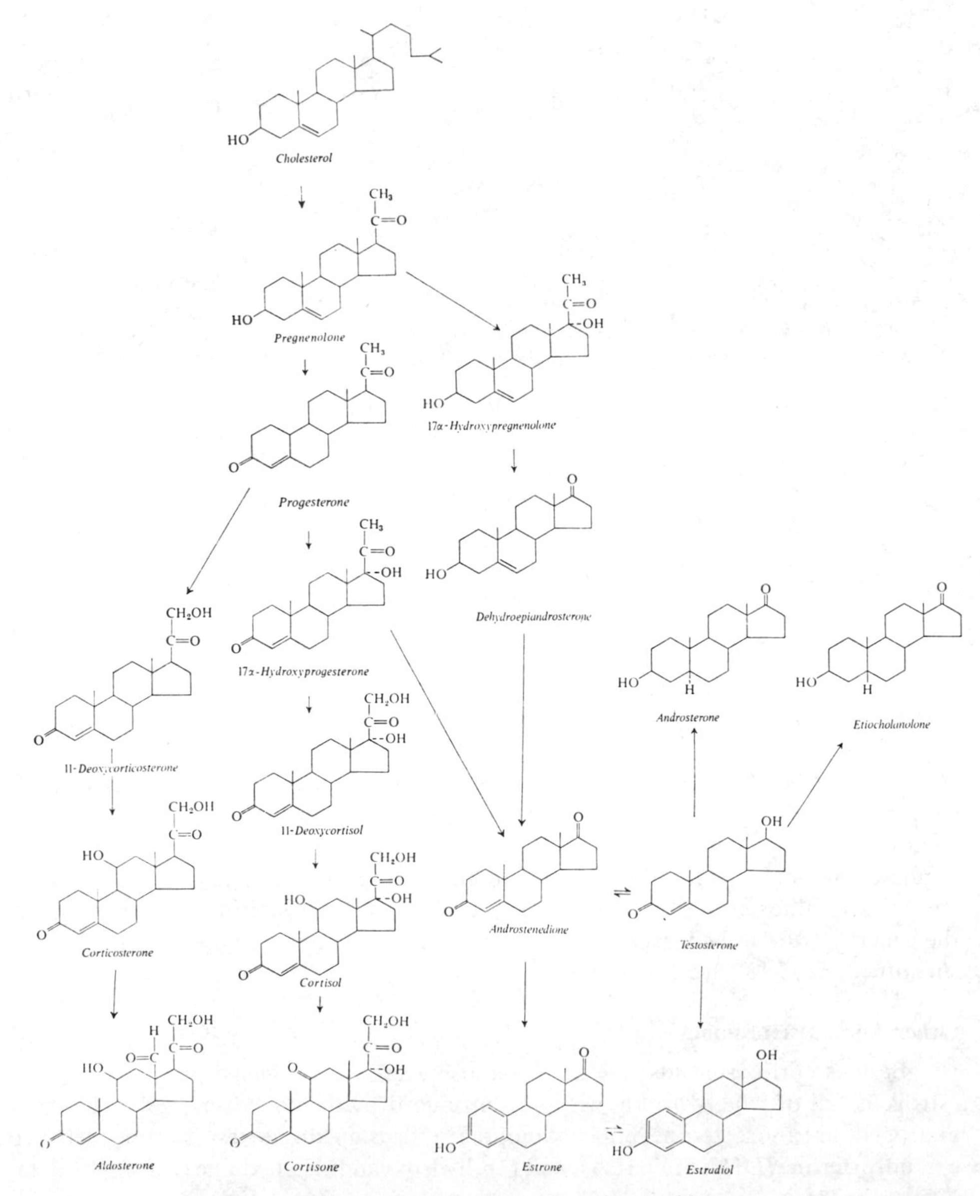

Figure 9-10. Biogenesis of corticosteroids.

important and characteristic biochemical events in the formation of adrenal steroids are the introduction of hydroxyl groups (—OH) at C-21, C-17, and C-11, catalyzed by the specific enzyme systems known as hydroxylases. The biosynthesis of aldosterone has not yet been completely elucidated, but it is generally believed that corticosterone is hydroxylated at C-18, followed by dehydrogenation to produce aldosterone. Androstenedione and testosterone are formed from 17α-hydroxyprogesterone catalyzed by the enzyme 17,20-desmolase. Another route of their synthesis is from dehydroepiandrosterone, which is synthesized directly from 17α-hydroyxpregnenolone following removal of the side chain. The subsequent hydroxylation of androstenedione at C-11 forms 11β-hydroxyandrostenedione. The aromatization of androstenedione or testosterone produces estrogens.

Metabolism of Adrenal Cortical Steroids

Common aspects of steroid metabolism have been discussed before (page 485). The metabolism of cortisol is shown in Figure 9-11. The reduction of the double bond between carbon atoms 4 and 5 by the enzyme Δ^4-5β or Δ^4-5α-reductase produces dihydro or allodihydrocortisol (Dihydro F). The direction of reduction is predominantly towards 5β in the human. Little or no dihydro compounds are found in the urine. Further preferential reduction of the ketone group at C-3 by 3α-hydroxysteroid dehydrogenase forms tetrahydro compounds. While the latter products are the major excretory metabolites, the hydrogenation of ketone groups at C-20 produces some hexahydro compounds (cortols). A small percentage of tetrahydro and hexahydro cortisol is converted to 11-oxygenated (11β-hydroxy or 11-keto) 17-ketosteroids by the removal of the side chain (see Figure 9-11). The latter compounds are also formed from 11β-hydroxyandrostenedione following the reduction of ring A. The metabolism of other androgens has been considered elsewhere (see Androgen metabolism).

Corticosterone, aldosterone, and other minor C-21 steroids such as 11-deoxycorticosterone and 11-deoxycortisol (compound S) follow the same sequence of reductive catabolism as cortisol. However, the compounds devoid of 17α-hydroxyl groups (corticosterone, aldosterone, 11-deoxycorticosterone) are not metabolized to C-19 17-ketosteroids. The transformation of cortisol to cortisone, corticosterone to 11-dehydrocorticosterone, and 11β-hydroxyandrostenedione to 11-keto-androstenedione is reversible. Consequently, the reduced metabolites with corresponding keto groups at C-11 are also excreted in the urine.

In the *adrenogenital syndrome*, characterized by the absence or deficiency of biosynthetic enzymes leading to the formation of cortisol, the metabolites that are normally present in minute quantities are produced in excess. For example, if the deficiency lies in the function of 21-hydroxylase, then there is a build-up of 17α-hydroxyprogesterone and 21-deoxycortisol, which give rise to the excessive production of urinary metabolites such as pregnanetriol, 11-keto and 11β-hydroxypregnanetriol and 17α-hydroxypregnanolone (Fig. 9-12). Similarly, when the enzyme 11β-hydroxylase is blocked, there is overproduction of 11-deoxycortisol (compound S) and its metabolites. It should be noted in this connection that the negative feedback control exists mainly between the plasma levels of free cortisol and corticotrophin (ACTH). So, in the absence of cortisol, there is an unrestrained secretion of corticotrophin, causing stimulation of the adrenal gland. This constant stimulation of the gland leads to hyperplasia accompanied by excessive production of androgens and C-21 intermediate products.

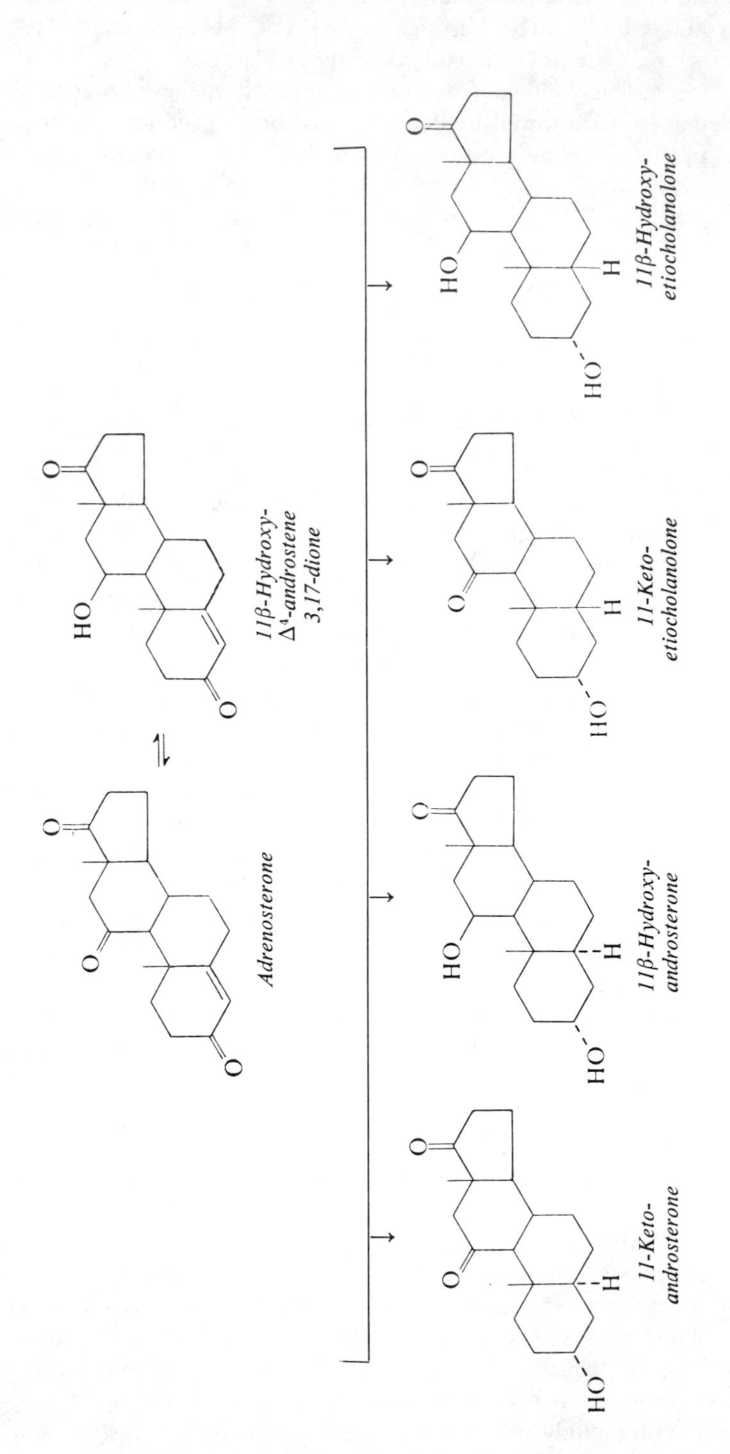
Adrenosterone
11β-Hydroxy-Δ4-androstene 3,17-dione
11-Keto-androsterone
11β-Hydroxy-androsterone
11-Keto-etiocholanolone
11β-Hydroxy-etiocholanolone

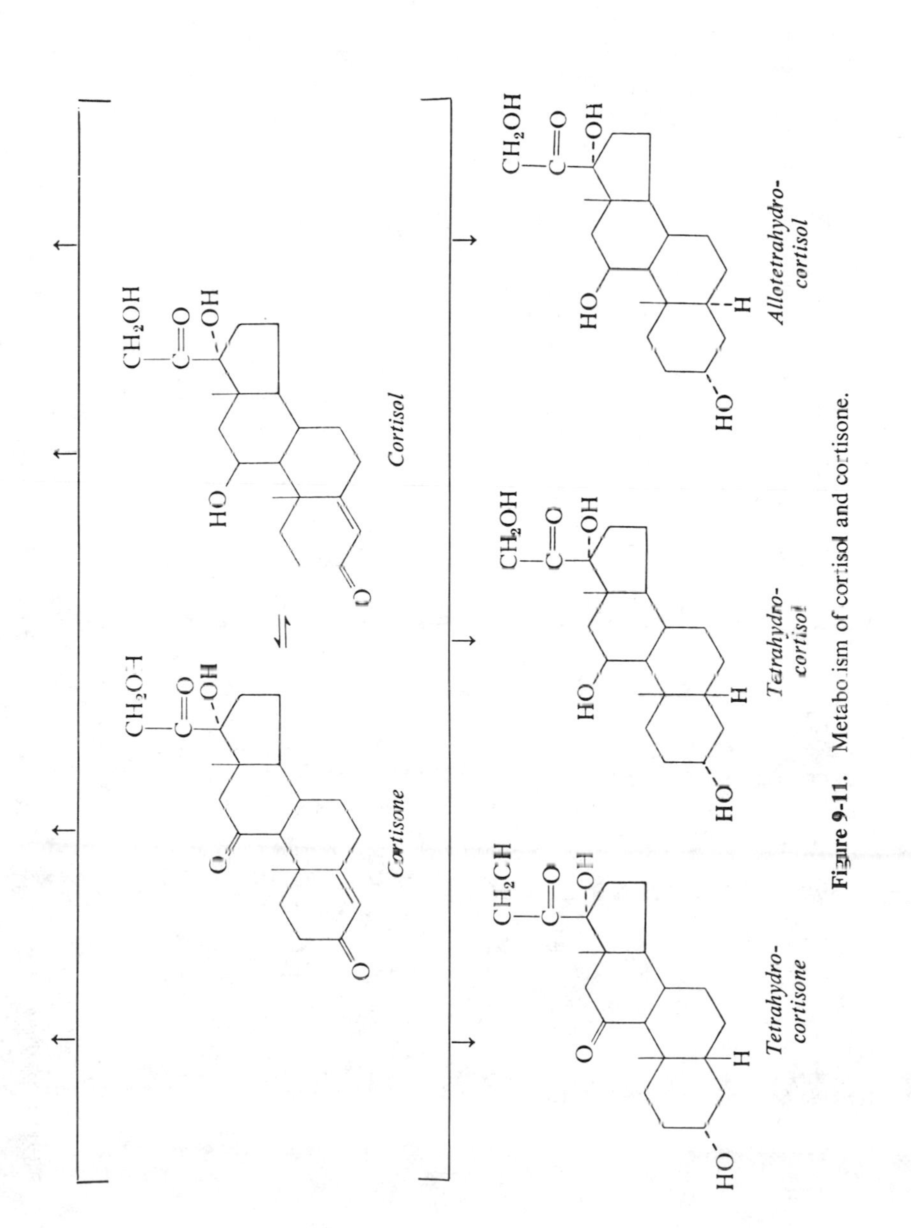

Figure 9-11. Metabolism of cortisol and cortisone.

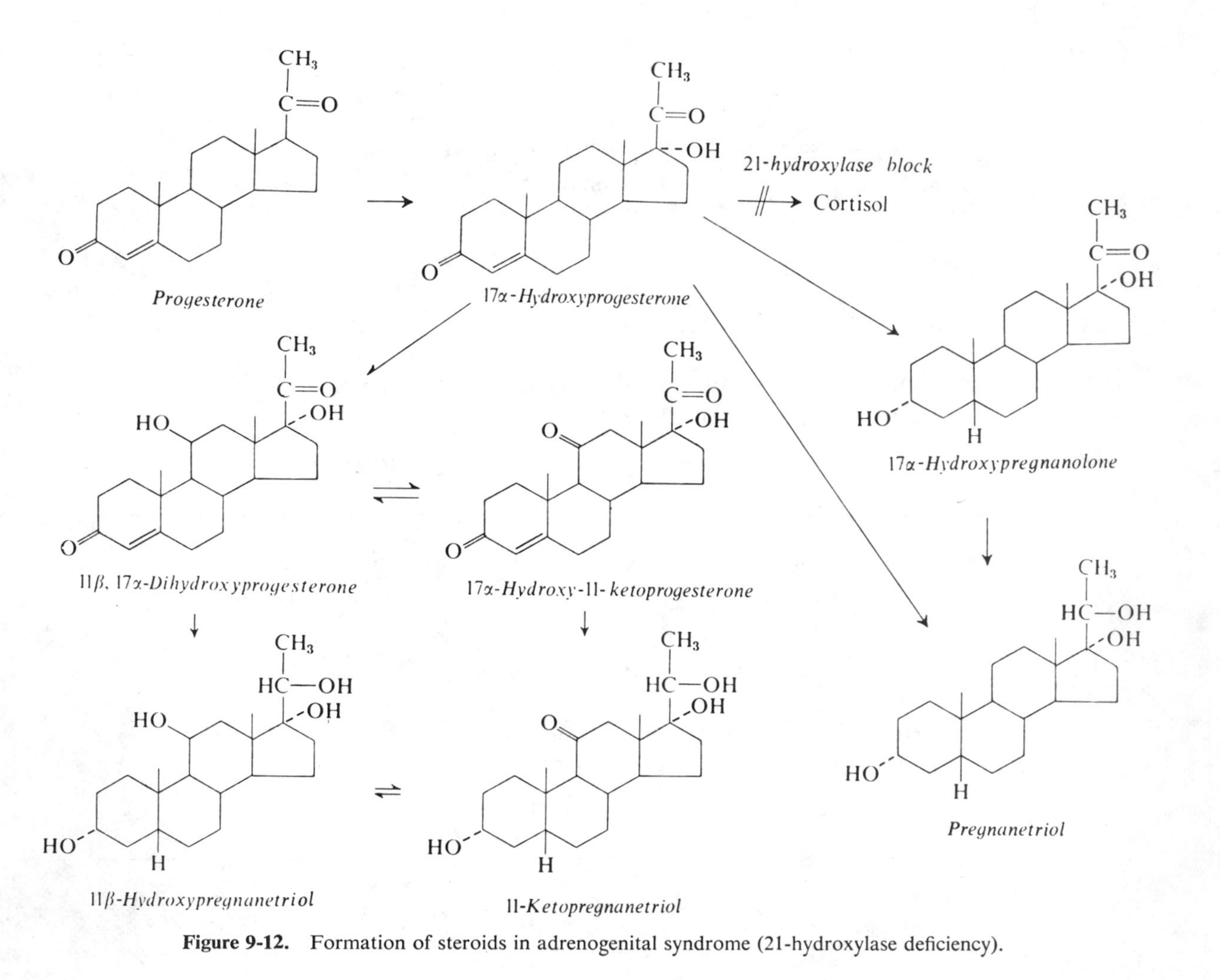

Figure 9-12. Formation of steroids in adrenogenital syndrome (21-hydroxylase deficiency).

The Estimation of Corticosteroids in Urine

Since cortisol is the principal corticosteroid secreted by the adrenal cortex, the excretion values of its metabolites in the urine are used as an index of the functional status of this gland. The urinary metabolites derived from cortisol may be grouped as follows:

1. Tetrahydro metabolites: tetrahydrocortisol (THF), tetrahydrocortisone (THE), and allotetrahydrocortisol.
2. Hexahydro metabolites: α and β cortols, and cortolones.
3. 11-oxygenated 17-ketosteroids: 11β-hydroxyetiocholanolone, 11-ketoetiocholanolone, 11β-hydroxyandrosterone, and 11-ketoandrosterone.

In addition to these cortisol metabolites, the urinary products of clinical importance also include tetrahydro 11-deoxycortisol (THS), pregnanetriol, 11-keto- and 11β-hydroxypregnanetriol, and 17α-hydroxypregnanolone.

The methods for estimation of 11-oxygenated 17-ketosteroids and pregnanetriol are considered in the respective sections. In the present section, the methods related to the determination of corticosteroids and their C_{21} metabolites are described.

The methods based on the Porter-Silber[72] reaction given by the steroids containing the dihydroxy acetone side chain are the methods most widely used in a routine clinical laboratory.[22] It should be noted, however, that this color reaction does not include all the C-21 metabolites of cortisol (e.g., α and β cortols, α and β cortolones). The assay procedures based on the chemical oxidation to 17-ketosteroids (see 17-ketogenic steroids) measure all the major C-21 metabolites and have been shown to render more clinically useful information. Both these methods are described in detail. There are, however, many additional chemical methods for the quantitative determination of corticosteroids but because of their limited application to the clinical field they are not discussed here. The routine analysis by gas chromatographic methods[26] appears to be very promising but cannot be considered presently as a routine assay procedure. The determination of aldosterone and its metabolites by chemical methods is too involved and complex for a common routine laboratory.

Determination of Urinary Corticosteroids by a Modified Porter-Silber Method[70]

PRINCIPLE

In 1950 Porter and Silber[60] described a color reaction based upon the formation of a yellow pigment (absorption maximum at 410 nm.) when certain corticosteroids react with phenylhydrazine in the presence of alcohol and sulfuric acid. They demonstrated that this color reaction is given primarily with corticosteroids that possess a dihydroxy acetone side chain as illustrated below:

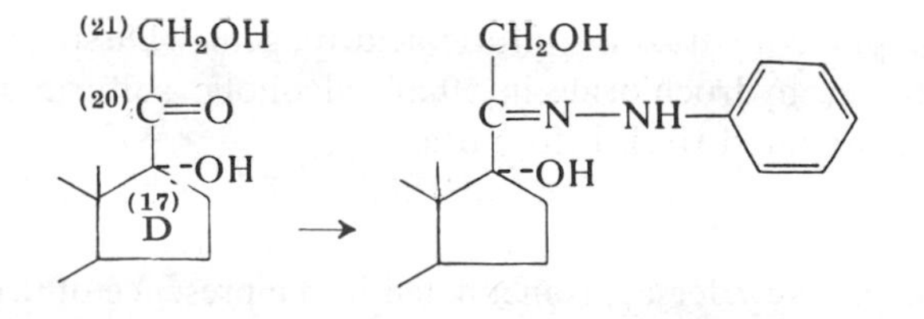

17,21-Dihydroxy-20-ketone *Yellow pigment*

Corticosteroids with this configuration include cortisol (compound F), cortisone (compound E), 11-deoxycortisol (compound S), and their tetrahydro derivatives (see Fig. 9-11). In urine, major corticosteroids reacting with the Porter-Silber reagent are tetrahydro E and F; in certain adrenogenital syndromes (e.g., 11β-hydroxylase deficiency) and during the metopirone test, which blocks 11β-hydroxylase activity, tetrahydro S comprises the bulk of the Porter-Silber chromogens.

The basic steps of the procedure consist of hydrolysis of conjugates by β-glucuronidase, extraction with chloroform, washing the chloroform extract with dilute alkali to remove estrogens, bile acids, and interfering chromogens, and color reaction with alcoholic phenylhydrazine-sulfuric acid reagent.

REAGENTS

1. Chloroform, A. R. The solvent should be freshly distilled from anhydrous potassium carbonate (K_2CO_3). Store freshly distilled chloroform in an amber bottle. To prevent formation of phosgene, 1 per cent of ethanol should be added.

2. Sodium hydroxide, 0.1 *N*. Dissolve 4 gm. of sodium hydroxide pellets in 1 L distilled water.

3. Ethanol, purified. Absolute ethanol is purified as follows: to 1 L. of absolute ethanol add 2 gm. 2,4-dinitrophenylhydrazine hydrochloride and 0.5 ml. concentrated hydrochloric acid (HCl). Let stand for approximately 48 hours. Distill through a 10 inch Vigreaux column, discarding the first and last 100 ml. Redistill through the same column, again discarding the first and last 100 ml. The purity is determined by mixing this ethanol with the phenylhydrazine-sulfuric acid reagent. No color should develop on standing overnight at room temperature.

4. Sulfuric acid, 64 per cent (v/v). To 360 ml. distilled water slowly add 640 ml. of concentrated sulfuric acid, A. R., with constant swirling. Prepare only in a pyrex container (2 L. Erlenmeyer flask) immersed in an ice water bath; the solution becomes extremely hot.

5. Alcoholic–sulfuric acid reagent (blank reagent). Mix 100 ml. 64 per cent sulfuric acid with 50 ml. absolute ethanol. The reagent is stable indefinitely.

6. Phenylhydrazine hydrochloride, recrystallized. A commercially available, chemically pure grade of phenylhydrazine hydrochloride is purified further as follows: add 100 gm. of phenylhydrazine hydrochloride to 500 ml. of warm water at 70°C. Add 1 gm. activated charcoal. Heat 1 L. ethanol to boiling and add to the dissolved phenylhydrazine in the water. Quickly filter while hot through Whatman No. 2 filter paper. Cool the filtrate in the refrigerator and collect the crystals in a sintered glass filter with medium porosity. Repeat the procedure of recrystallization, dissolving the crystals in proportionally less water. Wash the last collection of crystals with cold ethanol and dry thoroughly. Store in a tightly stoppered brown bottle in a desiccator over anhydrous calcium chloride. The purified material should have a melting point of 240°C. to 243°C.

7. Alcoholic phenylhydrazine–sulfuric acid reagent. Dissolve 50 mg. recrystallized phenylhydrazine hydrochloride in 50 ml. alcoholic–sulfuric acid reagent. The reagent should be prepared fresh before use.

8. β-Glucuronidase. The optimal pH and buffer to be used will vary with the source of the enzyme. Beef liver β-glucuronidase (Ketodase, Warner-Chilcott Laboratories, Morris Plains, New Jersey) is incubated in the presence of 0.1 *M* acetate buffer at pH 5. Bacterial β-glucuronidase (Sigma Chemical Company, St. Louis, Missouri) is incubated in 0.1 *M* phosphate buffer at pH 6.8. Prepare the enzyme solution in the concentration of 1000 units/ml. This should be prepared fresh before use.

9. Buffer solution. Phosphate buffer, pH 6.8, 0.5 *M*. To 500 ml. 1.0 *M* solution of KH_2PO_4 (68.0 gm. dissolved in 500 ml.) add 1 *N* NaOH to bring the pH to 6.8. Adjust the solution to a final volume of 1 L.

Acetate buffer, 1.0 *M*, pH 5. Dissolve 95 gm. of sodium acetate · $3H_2O$ and 17.2 ml. of glacial acetic acid in water, and dilute to a volume of one liter.

10. Steroid standard, stock solution. Transfer 25 mg. cortisol (compound F) or tetrahydro cortisone to a 250 ml. volumetric flask, and dilute to the mark with absolute ethanol. This stock standard solution contains 100 μg./ml.

11. Working standard solution. Transfer 5 ml. of stock standard solution to a 100 ml. volumetric flask, and dilute to the mark with distilled water. This working standard solution contains 5 μg./ml.

COLLECTION OF SPECIMEN

Collection of a complete 24-hour urine specimen is very important. Creatinine determination is believed to be a fair check for completeness of the specimen. The urine may be stored without any preservative in the refrigerator for a few days. Alternatively, the addition of 1 gm. boric acid per liter of urine will preserve the specimen at room temperature without any bacterial decomposition of the steroids.

PROCEDURE

Hydrolysis, Extraction, and Washing

1. Transfer 10 ml. urine to a 250 ml. glass-stoppered cylinder. Adjust the pH of the urine to 6.8 using indicator paper. Add 1 ml. β-glucuronidase (bacterial) solution (1000 units), 2 ml. 0.5 *M* phosphate buffer, and 0.1 ml. chloroform.

2. In a similar manner, prepare the water blank and standard using 10 ml. of distilled water and 10 ml. of working standard solution, respectively, instead of urine.

3. Mix the samples well and incubate at 37°C. for 18 to 24 hours.

4. Add 100 ml. chloroform to each glass-stoppered cylinder and mix the contents by repeated inversion for 30 seconds. Let the cylinders stand for 5 minutes in order to separate the aqueous and the organic phases.

5. Remove the aqueous supernatants by aspiration.

6. Add 10 ml. of 0.1 *N* NaOH to each cylinder and shake for 30 seconds. Allow to stand for 5 minutes. Aspirate off the alkali layer.

7. In a similar manner wash the chloroform extracts twice with 10 ml. of distilled water.

Porter-Silber Reaction

1. Transfer 40 ml. aliquots of each chloroform extract to properly labeled 50 ml. glass-stoppered centrifuge tubes as follows:

Blank-Blank	*Phenyl-Blank*	*Standard-Blank*	*Standard-Phenyl*	*Test-Blank*	*Test-Phenyl*
40 ml. blank extract + 5 ml. blank reagent (alcoh. H_2SO_4)	40 ml. blank extract + 5 ml. phenyl hydrazine reagent	40 ml. standard extract + 5 ml. blank reagent	40 ml. standard extract + 5 ml. phenyl hydrazine reagent	40 ml. urine extract + 5 ml. blank reagent	40 ml. urine extract + 5 ml. phenyl hydrazine reagent

2. Tightly stopper all tubes, shake vigorously for 30 seconds, and allow to stand for 15 to 20 minutes. Alternatively, centrifuge the tubes at 2000 r.p.m. for 10 minutes.

3. Transfer approximately 2.5 ml. of the supernatant phase from each tube into correspondingly labeled 10 × 75 mm. Coleman cuvets.

4. Incubate the tubes in a water bath at 60°C. for 30 minutes, or overnight in the dark at room temperature.

5. Measure the absorbance (A) with the Coleman spectrophotometer at wave length 410 nm. as follows:

Adjust the photometer to zero optical density using the blank-blank, and read the standard and test-blanks. Similarly, set the phenyl blank at zero absorbance and read the standard and test phenyl tubes.

CALCULATION

The standard sample contains 0.05 mg. of cortisol. Incorporating this value and the appropriate dilution factor (10) to calculate the concentration of corticosteroids/100 ml. of urine, the following equation is derived:

$$\text{Corticosteroids (mg./100 ml.)} = \frac{A \text{ test} - A \text{ test blank}}{A \text{ standard} - A \text{ standard blank}} \times 10 \times 0.05$$

$$= \frac{A \text{ test} - A \text{ test blank}}{A \text{ standard} - A \text{ standard blank}} \times 0.5$$

The excretion of corticosteroids per 24 hrs. is calculated as follows:

$$\begin{matrix}\text{Corticosteroids} \\ \text{(mg./24 hrs.)}\end{matrix} = \frac{\text{conc. (mg./100 ml.)} \times \text{urine volume (ml./24 hrs.)}}{100}$$

COMMENTS

Acid hydrolysis is unsuitable because the free corticosteroids are labile in a strongly acidic medium. The metabolites of cortisol contain numerous hydroxyl and keto groups, making them relatively hydrophilic. The use of a polar organic solvent such as chloroform ensures quantitative extraction of these steroids from hydrolyzed urine. To remove acidic components and phenols including estrogens, the solvent extract is washed with *dilute* alkali. The use of a strong alkali (stronger than 0.1 N) destroys the corticosteroids. The alkali-washed extract, termed the *neutral fraction*, contains metabolites of cortisol and of all other steroids excreted as glucuronides, as well as any other neutral lipid-soluble materials of urine. The selectivity of the color reaction toward the steroids with dihydroxyacetone side chains obviates the need for further purification. The impurities present in the extract form nonspecific brown chromogens in the presence of sulfuric acid. The use of a "urine blank" corrects for such background interference.

Various nonsteroidal substances, including acetone, fructose, and dehydroascorbic acid, also form a colored complex with the Porter-Silber reagent. In addition, the following drugs and their metabolites have been reported to cause interference with the colorimetric estimation: iodides, paraldehyde, chloral hydrate, Furadantin, bilirubin, colchicine, coffee, most sulfa drugs, chlorophenothiazines, spirolactones, quinine, and Darvon. Administration of these drugs should be withheld for several days prior to determination of corticosteroids.

NORMAL VALUES

Adult male	3–10 mg./24 hrs.
Adult female	2–8 mg./24 hrs.

REFERENCE

Sunderman, F. W., Jr.: *In* Lipids and the Steroid Hormones in Clinical Medicine. F. W. Sunderman and F. W. Sunderman, Jr., Eds. Philadelphia, J. B. Lippincott Co., 1960, p. 162.

Determination of Total 17-Hydroxycorticosteroids (Total 17-Ketogenic Steroids)[75]

PRINCIPLE

In 1952 Norymberski[55] reported that sodium bismuthate oxidizes several groups of 17-hydroxy corticosteroids to 17-ketosteroids, which can then be measured by the Zimmermann reaction (see determination of total 17-ketosteroids). He termed these steroids "17-ketogenic steroids." The characteristic side chains which are oxidized by sodium bismuthate are shown below:

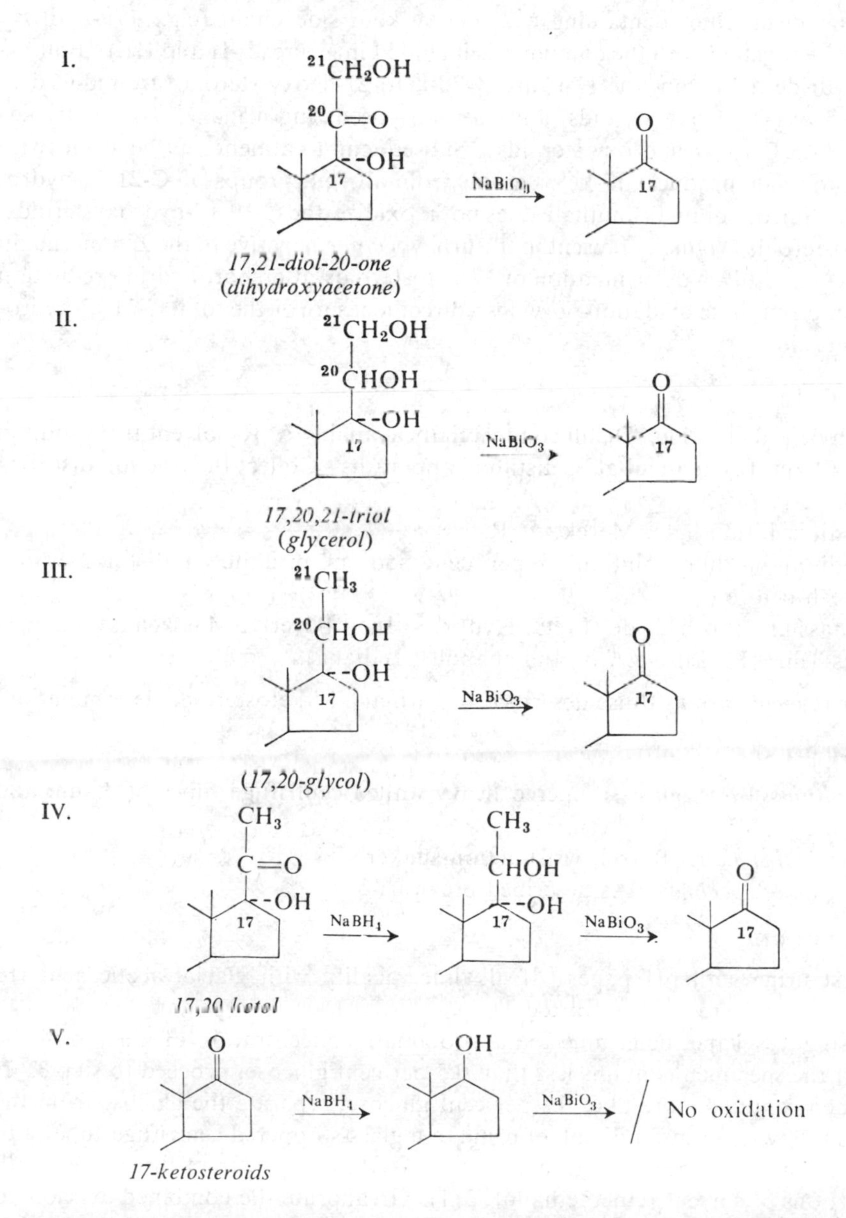

Group I includes cortisol, cortisone, their tetrahydro derivatives, 11-deoxycortisol (compound S) and tetrahydro S; Group II includes cortols and cortolones; Group III constitutes pregnanetriol and its 11-oxygenated derivatives. Group IV includes 17α-hydroxyprogesterone and 17α-hydroxypregnanolone.

It should be noted that the first two groups consist of active corticosteroids (cortisol, cortisone) and their metabolites, whereas group III and IV comprise mainly the metabolites of the precursors of active corticosteroids (e.g., 17α-hydroxyprogesterone). The excretion of the latter is quantitatively very significant in certain forms of the adrenogenital syndrome. Sodium bismuthate does not oxidize the 17-hydroxy compounds containing a ketone at C-20 and a methyl group at C-21 as shown in IV. In later modifications Norymberski and his coworkers[4] introduced a reduction step, using sodium borohydride prior to bismuthate oxidation. This made it possible to measure the metabolites containing a 21-deoxy keto side chain (e.g., 17α-hydroxy pregnanolone together with the compounds included in groups I, II and III.). Following borohydride reduction, the 17-hydroxy-20-keto-21-deoxy steroids are reduced to 17,20-dihydroxy-21-deoxy steroids, and naturally occurring urinary 17-keto steroids are reduced to C-19 17-hydroxy steroids. Subsequent treatment of the urine with sodium bismuthate produces 17-ketosteroids from all four groups of C-21 17-hydroxysteroids. Since sodium bismuthate does not reoxidize the C-19 17-hydroxysteroids, the 17-ketosteroids originally present in the urine become negative to the Zimmermann reaction. As a result, a determination of 17-ketosteroids after borohydride reduction and sodium bismuthate oxidation provides a direct measure of the total C-21 17-hydroxycorticosteroids.

REAGENTS

1. Ethylene dichloride. Distill commercially available A. R. solvent from sodium carbonate (2 gm./L.) in an all-glass distilling apparatus. Collect the fraction distilling between 83° and 84°C.
2. Sodium bismuthate, Merck, A. R.
3. Sodium bisulfite solution, 5 per cent sodium bisulfite in distilled water. Prepare fresh before use.
4. Potassium borohydride (Metal Hydrides, Inc., Beverly, Massachusetts).
5. Tes-Tape (Eli Lilly & Co., Indianapolis, Indiana).

Other reagents are as those described for urinary 17-ketosteroids determination.

APPARATUS

Special glassware: glass-stoppered heavy-walled centrifuge tubes of 35 ml. and 50 ml. capacity.

Mechanical shaker: Burrel, wrist action shaker.

Collection of specimen: As described previously.

PROCEDURE

1. Test urine with pH paper. If alkaline, acidify with glacial acetic acid (to dissolve phosphate precipitate if present).
2. Using Tes-Tape, determine the approximate concentration of glucose in the sample. If the specimen contains less than 0.5 per cent glucose, proceed to step 3. If the specimen contains more than 0.5 per cent glucose, separate the glucose from the steroids as follows: Transfer 20 ml. of urine to a glass-stoppered centrifuge tube, add 10 gm. of ammonium sulfate, and mix to dissolve the salt. Extract three times with 20 ml. portions of solvent (ether:ethanol, 3:1). Evaporate the combined extracts to dryness under nitrogen in a water bath at 50°C. Add 10 ml. of ethanol to the residue and warm the solution in hot water to dissolve the steroids. (Ignore the insoluble material). Cool, centrifuge, and transfer two 4 ml. aliquots (equivalent to 6 ml. urine) of the supernatant fluid (for duplicate analysis) to 50 ml. centrifuge tubes. Evaporate

the ethanol to dryness. Redissolve the residue in 0.5 ml. methanol and dilute to 8 ml. with water. Proceed to step 3, beginning with addition of potassium borohydride.

Reduction, oxidation, hydrolysis, and extraction

3. Place 8 ml. of urine in a 125 ml. Erlenmeyer flask. Add 100 mg. potassium borohydride. Check the pH. If the pH is not over 8, add an additional 25 mg. of borohydride. Let it stand for 2 hours or overnight at room temperature. (Preferably, instead of adding solid borohydride, 0.8 ml. 10 per cent freshly prepared solution of potassium borohydride may be used.)

4. Add 8 ml. of glacial acetic acid and allow to stand for 15 minutes. (The acid decomposes the excess borohydride.)

5. Transfer to a 50 ml. centrifuge tube. Add 2 gm. sodium bismuthate. Stopper and shake mechanically for 30 minutes away from direct sunlight. (The samples may be covered with a heavy black cloth during the treatment with bismuthate.) Add 2 gm. fresh sodium bismuthate and shake for an additional 15 minutes. Leave the samples overnight at room temperature. The following morning shake the tubes for 15 minutes.

6. Centrifuge for 10 minutes at 2000 r.p.m., and transfer 6.0 ml. of the supernatant fluid to 35 ml. glass-stoppered centrifuge tubes containing 1.5 ml. of freshly prepared sodium bisulfite solution. Mix the solution and allow to stand for 5 minutes.

7. Add 5 ml. of distilled water and 3.6 ml. of concentrated hydrochloric acid. Let stand for 15 minutes.

8. Place in a boiling water bath for ten minutes. Remove and cool the samples in a cold water bath.

9. Add 12 ml. of ethylene dichloride, and shake mechanically for 15 minutes. Centrifuge for 2 minutes at 2000 r.p.m.

10. Aspirate off the upper phase as completely as possible without losing any organic solvent.

11. Add to the organic extract 25 to 30 pellets of sodium hydroxide. Place in shaking machine for 15 minutes, centrifuge, and filter through 7 cm. Whatman No. 1 filter paper into a test tube.

12. Transfer 4 ml. of filtrate (= 1 ml. of urine) to a test tube and evaporate to dryness under nitrogen in a water bath at 50 to 55°C. (In the case of a 24-hour collection of large volume, use 8 ml. of filtrate.)

Color reaction

13. Perform the Zimmermann color reaction and measure the absorbance as described in the method for total 17-ketosteroid determination.

CALCULATIONS:

Total 17-ketogenic steroids (mg./24 hr.) =

$$\frac{\text{corrected } A \text{ of sample}}{\text{corrected } A \text{ of standard}} \times 0.05 \times \text{total volume of urine (in ml.)}$$

COMMENTS

Since the 17-ketosteroids formed from the 17-hydroxycorticosteroids are fairly stable in a hot acid medium, the hydrolysis of steroid conjugates can now be performed with acid as opposed to the enzymatic hydrolysis used in the direct method based on the Porter-Silber reaction. The presence of glucose in urine interferes with the bismuthate oxidation. All urine specimens should, therefore, be routinely tested with Tes-Tape, and the glucose removed before the determination is begun.

Most suitable means to rid the sample of the glucose appears to be the procedure outlined in step 2. Errors due to the presence of glucose may, however, also be avoided by increasing the amount of sodium bismuthate (1 gm. for each per cent of glucose above 0.5 per cent). The presence in urine of varying amounts of reducible substances other than glucose makes the use of a large excess of borohydride necessary. Addition of sufficient borohydride is indicated by effervescence on the addition of acetic acid (step 4). The absence of effervescence is suggestive of an insufficient amount of borohydride, which may yield misleading results because of the incomplete reduction of different ketone groups. Instead of sodium bismuthate, the oxidizing agent sodium meta periodate (ten volumes per cent of 10 per cent solution in 0.1 *N* NaOH) may also be used.[29,52] The advantage of using this reagent lies in the fact that in addition to oxidizing 17-hydroxycorticosteroids to 17-ketosteroids, it oxidizes glucuronides to the free steroids or to their formates, which are easily hydrolyzed in alkaline solution; thus, the need for acid hydrolysis is eliminated. Necessary precautions and drug interference in the color reaction have been discussed elsewhere (see determination of 17-ketosteroids).

Since a greater number of metabolites of cortisol (cortol and cortolone) are estimated by the ketogenic method, the normal urinary excretion values are generally higher than those obtained by the Porter-Silber method. This method also yields high values in the adrenogenital syndrome because of the presence of excessive amounts of urinary metabolites of the cortisol precursors, which go undetermined by the latter procedure.

NORMAL VALUES

Adult male	5–23 mg./24hrs.
Adult female	3–15 mg./24hrs.

REFERENCE

Sobel, C. S., Golub, O. J., Henry, R. J., Jacobs, S. L., and Basu, G. K.: J. Clin. Endocrinol., *18*:208, 1958.

The Estimation of Corticosteroids in Plasma

The main purpose of the estimation of corticosteroids in blood or in urine is to evaluate the rate of secretion of cortisol by the adrenal cortex as well as the actual level of the hormone to which the tissues are exposed. While the estimation of the urinary excretion renders indirectly information regarding the overall activity of the gland (i.e., secretion rate), to ascertain whether the tissues are exposed to proper amounts of cortisol, the blood estimation appears to be more useful. It should be noted in this connection that the urinary excretion may be elevated by the increased rate of production and metabolism of the hormones without the physiological level in the blood being enhanced. For example, in obesity and hyperthyroidism, the urinary excretion of 17-hydroxycorticosteroids is elevated even though the plasma level of cortisol is within the normal range. The measurement of plasma cortisol is also of value in the study of the existence of normal diurnal variation and in obtaining quick information in the performance of functional tests employing stimulation and suppression of the adrenals.

Cortisol represents almost 80 per cent of the total 17-hydroxycorticosteroids in the blood, and the majority circulates in its original form along with small amounts of unconjugated reduced derivatives. The biologically active unconjugated cortisol in the plasma is bound to some extent by albumin and to an α-globulin derived mainly from the liver. This latter protein is called *transcortin*, or corticosterone-binding globulin (CBG). While the precise function of such protein binding is still obscure, it is generally suggested that this mechanism assures a ready source of available circulating hormone and protects it from inactivation and conjugation in the liver. The concentration of CBG in the plasma rises during pregnancy and during estrogen therapy, with a concomitant increase of total 17-hydroxycorticosteroids in plasma. However, since the amount of free hormone, unassociated with protein, remains at physiological levels there are no untoward effects. It should be noted here that the unbound portion is biologically most significant, because this is the amount which is available for immediate physiological action.

There are two general methods for the estimation of blood corticosteroids; one is based on the Porter-Silber color reaction, and the other is based on the measurement of the sulfuric acid–induced fluorescence (see Dixon[21]). The latter method appears to possess greater merit and is widely used in routine clinical laboratories. Some of the relevant features are as follows: The fluorimetric method is more sensitive and thus requires only small quantities of plasma. Several steroids, such as prednisone, prednisolone and dexamethasone, if administered to the patient, interfere in the colorimetric method but not in the fluorimetric method. The presence of ketones, hexoses, and other commonly used drugs (see comment, page 504) which are potent sources of error in colorimetric estimation does not affect the fluorimetric measurement of plasma cortisol.

The major drawback of the fluorimetric technique is an overestimation of cortisol by about 2.5 μg. per 100 ml. plasma because of the presence of nonspecific fluorogens.[51] The nature of the nonspecific fluorogens is still unknown. It has been observed, however, that they are closely correlated to the true plasma cortisol level and are dependent upon the functional state of the adrenal glands.[54] The existence of nonspecific plasma fluorescence does not limit the clinical usefulness of the results obtained with the method.

Following the first report by Sweat[79] on the application of sulfuric acid–induced fluorescence to the quantitative determination of corticosteroids, a large number of methods with various modifications and applications has been published (see Dixon[21]). The fluorimetric method described here is based on the one reported by Mattingly.[51] The procedure appears to be simple, accurate, rapid, and suitable for routine clinical use.

Estimation of 11-Hydroxycorticosteroids in Plasma[51]*

PRINCIPLE

Free and protein-bound cortisol and corticosterone are extracted from the plasma with dichloromethane. The organic extract is shaken with a sulfuric acid-ethanol reagent. After removing the supernatant dichloromethane, a resulting fluorescence of

* It should be borne in mind that the use of this fluorimetric technique is not suitable in patients undergoing the Metopirone test, because 11-deoxycortisol (compound S), which is increased following the drug administration, does not produce fluorescence. Under these circumstances the plasma method based on the Porter-Silber reaction for 17-hydroxycorticosteroids (Peterson[58]) or the urinary determination by the methods described here will have to be employed.

the acid solution is read in a fluorimeter at a specified time, and is compared with that of a known concentration of cortisol treated in the same manner. Maximum fluorescence of corticosteroids is produced by excitation at 470 nm. (436 nm. when a mercury lamp is used as the exciting source). Maximum emission of fluorescence occurs at 530 nm.

REAGENTS

Use all reagents of A.R. quality.

1. Dichloromethane purified. Purify commerically available dichloromethane as follows: Let the solvent stand for several days over concentrated sulfuric acid, shaking occasionally. Wash further by shaking with concentrated sulfuric acid (100 ml./L.) followed by 1 *N* NaOH (100 ml./L.), and twice with distilled water (200 ml./L.). Dry over anhydrous sodium sulfate for 24 hours. Distill in an all-glass apparatus and collect the fraction boiling between 39 and 40°C. Store in an amber bottle.
2. Sulfuric acid (concentrated), A.R.
3. Sodium hydroxide (pellets), A.R.
4. Sodium sulfate (anhydrous), A.R.
5. Ethyl alcohol, purified, as described for Porter-Silber reagent (page 502).
6. Fluorescence reagent. Slowly add 7 volumes of concentrated sulfuric acid to 3 volumes of ethyl alcohol in a flask which is kept cold in iced water. The solution obtained should be colorless. If the ethyl alcohol is not purified enough, a brown color may develop. This reagent remains stable for a month at room temperature.
7. Cortisol standards. Dissolve 50 mg. cortisol in 50 ml. of purified ethyl alcohol. Take 1 ml. of this solution and dilute to 100 ml. with distilled water (10 μg./ml.). These solutions remain stable for months at 4°C.

For the working standard solution dilute 1 ml. of the 10 μg./ml. standard to 10 ml. with distilled water. This solution contains 1 μg./ml.

APPARATUS

All glassware should be cleaned with chromic acid, followed by thorough washing with tap water and finally distilled water. If a rubber Propipette is used to pipet solvent and reagents, plug all pipets with cotton wool to prevent rubber particles from contaminating the solutions.

COLLECTION OF BLOOD

Collect 10 ml. of blood in a test tube containing heparin as an anticoagulant. Separate the plasma as soon as possible, because the uptake of corticosteroids by red cells increases on standing. If necessary, plasma can be stored for 72 hrs. at 0 to 4°C. Do not freeze plasma since this may produce a precipitate which may trap or absorb steroids, causing low results.

PROCEDURE

Extraction of free steroids from plasma

1. Pipet 2 ml. of plasma into a 25 ml. glass-stoppered centrifuge tube. Into two separate tubes place two ml. of distilled water (water blank) and 1 ml. of working standard solution diluted with 1 ml. of distilled water (standard), respectively.
2. Add 15 ml. dichloromethane to each tube. Stopper and shake very gently by hand, or place in a slow moving mechanical shaker for 10 minutes (vigorous shaking causes emulsion formation).
3. Centrifuge for 2 minutes at 2000 r.p.m. Remove the supernatant plasma by suction.

4. (Fluorimetry should be performed in batches of not more than six plasma extracts, a blank, and a cortisol standard. Careful timing is necessary to keep non-specific fluorescence as low and as uniform as possible.) At one minute intervals, beginning with the blank, add 10 ml. of the extract to 5 ml. of the fluorescence reagent in a suitable glass-stoppered tube. Shake vigorously for 20 seconds.

5. Carefully suck off the supernatant dichloromethane from each tube in the same order as before, starting with the blank.

6. Thirteen minutes after mixing with the fluorescence reagent, read the fluorescence at 530 nm. (emitted wavelength) following excitation by 470 nm. Set the water blank to read zero on sensitivity range 1 of the instrument. One minute later set the standard to read 100 on sensitivity range 2. Then read the samples on the same sensitivity range at one minute intervals in the order in which the fluorescence reagent was added.

CALCULATION

$$\mu g.\text{ of cortisol/100 ml. plasma} = \frac{\text{reading of sample}}{\text{reading of standard}} \times 1.0 \times \frac{100}{2}$$

COMMENTS

Scrupulous cleaning of glassware and rigorous purification of dichloromethane are very important, because extraneous materials from the solvent and glassware may quench the fluorescence. The nonspecific fluorogens in plasma increase linearly with time relative to the fluorescence of the cortisol standard. For this reason, careful timing is necessary to insure the constancy of this nonspecific fluorescence from estimation to estimation. The principal corticosteroids measurable by the method are cortisol, corticosterone, and 20-dihydrocortisol. However, the last two steroids normally are not quantitatively as significant as cortisol in blood. Estrogens, especially estradiol, produce a considerable amount of fluorescence, but again their presence—even in the plasma of pregnant women—is negligible in comparison to the amount of cortisol. Therapy with triparanol or spironolactone may lead to falsely high results, and therapy with sulfadimine to falsely low values. Spuriously high results may also be obtained in patients taking contraceptive drugs, which increase CBG.

NORMAL VALUES

The values range from 6.5 to 26.3 μg. (average 14.2 μg.) per 100 ml. plasma between 9 and 10 a.m., and 2 to 18 μg. (average 8 μg.) per 100 ml. plasma at 4 p.m.

There is no significant difference between age and sex in adults.

REFERENCE

Mattingly, D.: J. Clin. Path., *15*:374, 1962.

CLINICAL SIGNIFICANCE

The usefulness of the estimation of these steroids in blood or in urine has been discussed in the appropriate sections. It may suffice to say here that the measurements are used primarily for the evaluation of adrenal or pituitary dysfunction.

Decreased values are found in adrenal insufficiency, as exemplified by Addison's disease. Similar values can also be noted in panhypopituitary states, such as Sheehan's syndrome.

Increased values are found in numerous conditions which clinically can present

Cushing's syndrome. The latter entity was originally considered to be the result of a pituitary basophil adenoma but it is now known to be associated with ACTH-like secreting tumors (bronchial adenoma, carcinoma, islet cell adenoma of the pancreas, Zollinger-Ellison syndrome, multiple endocrine adenomas). In addition, markedly elevated values are found in cases of adrenal carcinoma.

In the adrenogenital syndrome, total urinary 17-ketogenic steroids are elevated, whereas the Porter-Silber chromogens are low to normal when 21-hydroxylation is blocked and high when 11β-hydroxylation is blocked.

It should be noted here that high or low values are not diagnostic as to primary or secondary dysfunction of the adrenal cortex. More sophisticated tests based on the stimulation and suppression of adrenal function using various agents such as ACTH, dexamethasone and Metopirone are needed in differentiating between the primary and secondary sites of the adrenal dysfunction.

Stimulation and suppression tests. Stimulation tests using ACTH (corticotrophin) are most useful in differentiating between Addison's disease and hypopituitarism. The ACTH is administered either intramuscularly or intravenously. Most frequently, 25 units of aqueous ACTH in 500 ml. normal saline is administered intravenously over an exact 8 or 6 hour interval on two successive days. In normal patients, a two- to five-fold increase in 17-hydroxycorticosteroid and a two-fold increase in 17-ketosteroid excretion levels are noted. In Addison's disease (primary adrenal insufficiency), ACTH stimulation causes little or no rise in 17-OH-corticosteroid, 17-ketosteroid, or 17-ketogenic steroid excretion. In patients with adrenal insufficiency secondary to pituitary hypofunction, a gradual rise ("*stair case*" response) in steroid excretion is seen on successive days of ACTH stimulation. Such a "stair case" response signifies that the adrenals are responsive to exogenous ACTH and that the cause of adrenal insufficiency is due to pituitary or hypothalamic dysfunction.

The stimulation tests have also some diagnostic importance in hyperadrenalism. Thus, when Cushing's syndrome is present because of adrenocortical hyperplasia, the excretion of corticosteroids is increased three- to five-fold over control values following ACTH stimulation. In Cushing's syndrome secondary to adrenal carcinoma, on the other hand, no response to stimulation is noted because of the inherent functional autonomy of the tumor. However, the use of a *suppression test* to differentiate between adrenocortical hyperplasia and adrenal carcinoma is the most common and widely accepted clinical procedure.

The principle of this test is based on the feedback control mechanism normally existing between the release of pituitary ACTH and the circulating blood level of unbound cortisol. When the cortisol levels in blood are increased in the normal individual, the pituitary release of ACTH diminishes, with a consequent decrease of steroid output by the adrenal glands. In a diseased state, such as adrenal carcinoma, this feedback mechanism is completely deranged, and consequently the lack of suppression of adrenal activity, even after administration of large doses of cortisol or its potent synthetic analogues, is a good indication of the autonomous functioning of the adrenal tumor.

In clinical practice, a potent cortisol analogue such as dexamethasone (9α-fluoro-16α-methylprednisolone) is utilized in order that the administered compound may be given in such small amounts that it will not contribute significantly to the analysis of urinary steroids. A standard method of testing is to administer 0.5 mg. of dexamethasone every six hours for 48 hours. If no suppression is obtained, the test is repeated with consecutive doses of 2 mg. and 8 mg. In normal patients the administration of the 0.5 mg. dose reduces by 50 per cent the basal excretion level of 17α-hydroxycorticosteroids and total 17-ketogenic steroids. The increased steroid excretion which occurs in patients with adrenocortical hyperplasia is generally suppressed by 2 mg. doses, whereas the elevated steroid excretion of patients with self-sustaining adrenal carcinoma persists even with higher doses (8 mg.).

The *metyrapone test* is employed to assess the pituitary reserve for ACTH and thus it is valuable for the differential diagnosis of adrenocortical insufficiency. Metyrapone (SU 4885, Metopirone) is a drug that selectively inhibits the enzyme action of 11β-hydroxylase. Consequently, when this drug is administered, the conversion of 11-deoxycortisol (compound S) to cortisol is inhibited (see biogenesis of cortisol). The secretion of ACTH is controlled solely by cortisol through the feedback mechanism. The inhibition of cortisol synthesis caused by the drug allows unrestricted release of ACTH, which in turn stimulates the adrenal cortex to produce increased amounts of 11-deoxy corticosteroids. Since the test demands an endogenous supply of ACTH, the degree of adrenocortical stimulation demonstrated by the increased urinary excretion of 11-deoxy C_{21}(metabolities of compound S and 17α-hydroxyprogesterone) and C_{19} steroids (17-ketosteroids) reflects the reserve capacity of the pituitary to release ACTH.

In common practice, 750 mg. of metyrapone is administered orally every 4 hours over a 48-hour period. The urinary steroid excretion in normal subjects shows an increase of twice the amount of the basal level. In patients with hypoadrenocorticalism due to pituitary deficiency, no elevation in urinary excretion is noted. Since the net effect of this test is to increase the endogenous level of ACTH, the results in Cushing's syndrome are similar to the stimulation test as previously described. In adrenocortical hyperplasia the urinary excretion of steroids increases as much as two-fold, whereas with adrenal carcinoma no such change occurs. It is noteworthy that in Cushing's syndrome, caused by the ACTH-secreting nonendocrine tumors, the suppression test using dexamethasone and the metyrapone test do not elicit any response, because the site of excessive production of ACTH is beyond the pituitary-hypothalamic axis.

Generally, levels of urinary excretion of corticosteroids and 17-ketogenic steroids are measured in these tests. However, blood cortisol levels may be used instead, since they correlate well with urinary findings and provide a rapid diagnostic test.

ANDROGENS

Androgens are a group of C_{19} steroids which exert profound influence on the male genital tract and are concerned with the development and maintenance of secondary sex characteristics of the male; hence they are called "male sex hormones." The structural formulas of the four most important biologically active androgens—testosterone, androstenedione, dehydroepiandrosterone, and androsterone—are shown in Figure 9-13. These hormones are secreted by the testes, the adrenals, and the ovaries. The principal and biologically most active naturally occurring androgen is testosterone, which is derived mostly from the testes. The structural characteristic of this steroid is an unsaturated bond between carbon atoms 4 and 5 and a ketone group at C-3 (Δ^4-3-keto) and a hydroxyl group in the β position at carbon 17. The hydroxyl

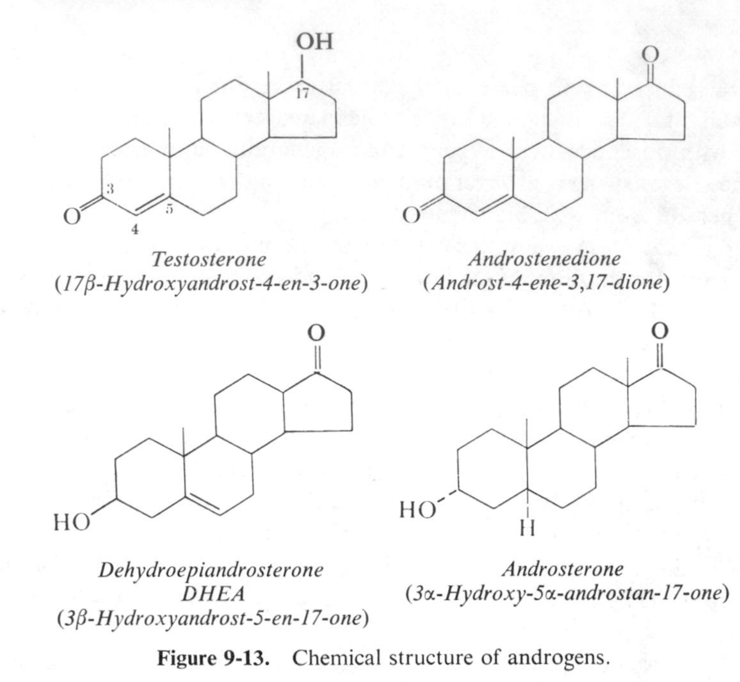

Figure 9-13. Chemical structure of androgens.

group in the β position (above the plane of the paper) is essential for biological activity, since epitestosterone (17α-hydroxyl) and androstenedione (17-keto) have very little or no activity.

The relative potency of the four androgens may be expressed in per cent as follows: testosterone = 100, dehydroepiandrosterone = 16, androstenedione = 12, and androsterone = 10 per cent. From the quantitative point of view, androstenedione is the major ovarian androgen. Dehydroepiandrosterone, on the other hand, is derived exclusively from the adrenals. Androsterone is a metabolic product of testosterone and androstenedione. The 11-oxygenated C-19 steroids derived from the adrenals are biologically inactive.

Biosynthesis

Cholesterol and acetate serve as precursors of androgens, and the postulated pathway is shown in Figure 9-14. Note that it involves the same steps, starting from acetate to cholesterol and cholesterol to 17α-hydroxyprogesterone, as in the biosynthesis of the adrenal cortical hormones. Oxidative removal of the side chain in 17α-hydroxyprogesterone, catalyzed by the enzyme desmolase (see general discussion), produces androstenedione, which is reduced by the 17β-hydroxysteroid dehydrogenase to testosterone. The transformation of androstenedione into testosterone is a reversible reaction. In the ovary, the equilibrium lies far toward the formation of androstenedione, so that very little testosterone is produced. In the testes, on the other hand, the forward reaction, favoring the production of testosterone, is more active. In the adrenals androstenedione and testosterone are formed through the same biochemical sequences, in addition to the formation of dehydroepiandrosterone from 17α-hydroxypregnenolone following the removal of the side chain. The biosynthesis of 11-oxygenated adrenal androgens can occur through 11β-hydroxylation of

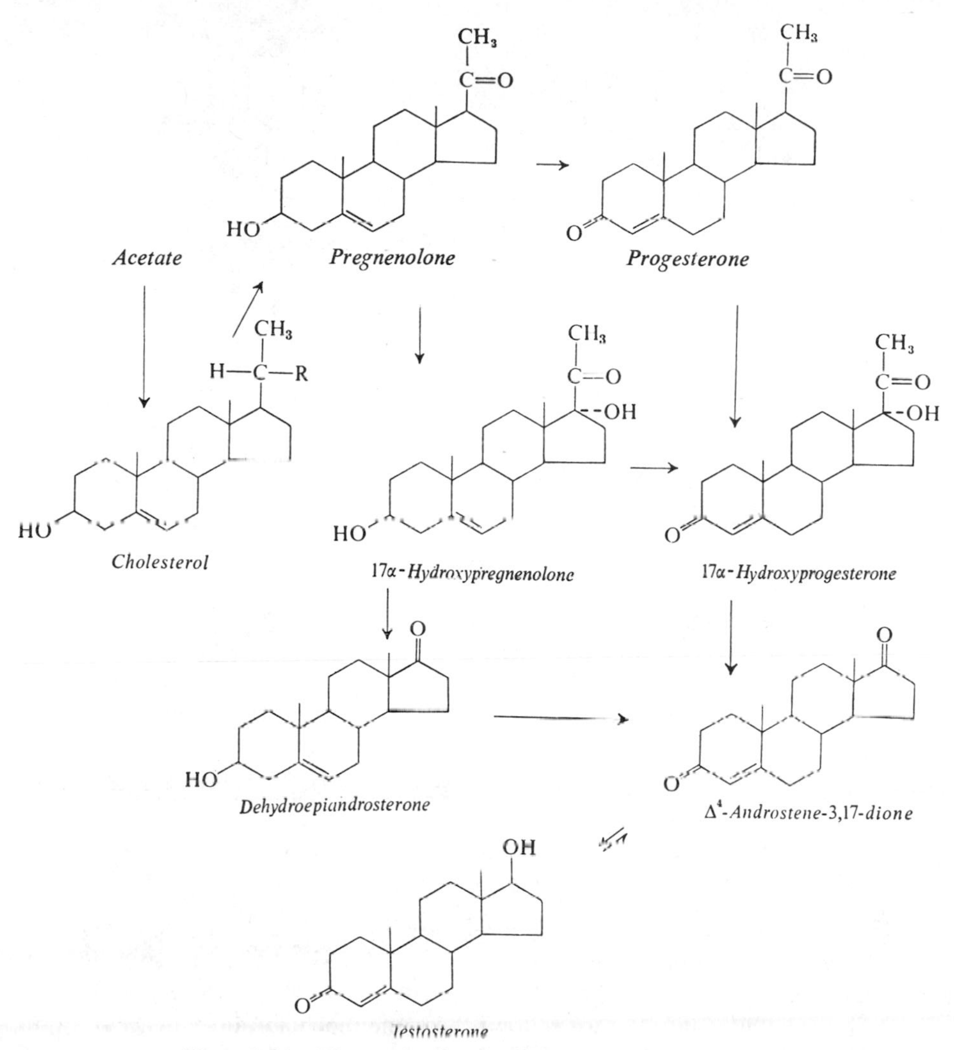

Figure 9-14. Biogenesis of androgens (ovary, adrenal, testis).

androstenedione or from 17α-hydroxyprogesterone followed by 11β-hydroxylation and removal of the side chain.

Metabolism

The main metabolic products of androstenedione, testosterone, and dehydroepiandrosterone are shown in Figure 9-15. Similar to corticosteroid metabolism, the reduction of ring A, and subsequent oxidation and conjugation are the major metabolic steps. The reduction of the double bond gives rise to two isomers, differing in the spatial configuration of H at C-5. When the H is on the same side of the methyl group at C-10 (cis) then it is a 5β isomer (etiocholanolone), and when it is on the opposite side (trans), the isomer is a 5α isomer (androsterone). Similarly, the hydrogenation of the ketone group at C-3 produces 3α-hydroxy (androsterone, etiocholanolone) or 3β-hydroxy (epiandrosterone) compounds, the reaction being catalyzed by the corresponding enzymes, such as 3α-hydroxy or 3β-hydroxysteroid dehydrogenase.

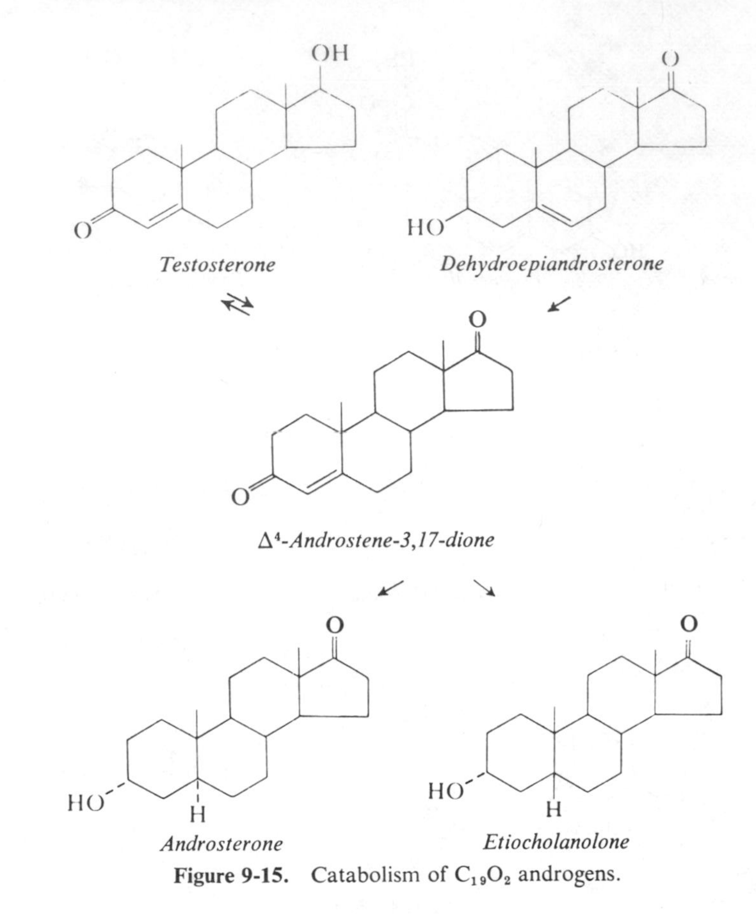

Figure 9-15. Catabolism of $C_{19}O_2$ androgens.

Quantitatively, the 3α-hydroxysteroids are predominant in urine. Dehydroepiandrosterone, containing a β-hydroxyl group at C-3 and an unsaturated bond between carbon atoms 5 and 6 (3β-hydroxyandrost-5-ene), is first irreversibly converted to androstenedione, which in turn follows the same biochemical sequences described previously. Dehydroepiandrosterone is also excreted unchanged, as the sulfate conjugate. The metabolism of 11-oxygenated androgens have been described in a previous section (see cortisol metabolism). All these catabolites constitute the group of steroids known as the 17-ketosteroids, by virtue of their ketone group at carbon atom 17. While the conjugation of all these steroids may occur with sulfuric acid and glucuronic acid, the glucuronide is predominant in androsterone, etiocholanolone, and 11-oxygenated 17-ketosteroids; dehydroepiandrosterone is present exclusively as the sulfate conjugate.

Estimations of Urinary 17-Ketosteroids

The 17-ketosteroids are metabolites of androgens secreted by the adrenals, by the testes, and possibly to some extent by the ovaries. In men, approximately one-third of the total urinary 17-ketosteroids represent the metabolites of testosterone secreted by the testes, whereas most of the remaining two-thirds are derived from the steroids produced by the adrenals. In women, who usually excrete smaller quantities than men, the total 17-ketosteroids are derived almost exclusively from the adrenals.

The bulk of the urinary 17-ketosteroids consists of androsterone, epiandrosterone, etiocholanolone, dehydroepiandrosterone, 11-keto and 11β-hydroxyandrosterone, and 11-keto and 11β-hydroxyetiocholanolone. It should be recalled that dehydroepiandrosterone and 11-oxygenated 17-ketosteroids are the products of adrenals only, while the others also arise from the precursors elaborated by the gonads. Thus, the main purpose of the quantitation of these steroid metabolites is to assess gonadal and adrenal function.

There are a number of chemical methods available for the estimation of total 17-ketosteroids.[17,22] The final quantitation in most of these is based on the color reaction originally described by Zimmermann.[100] The method described by Drekter et al.[25] with modifications by Sobel et al.[75] has been shown to be most adequate for routine clinical use and is given below.

PRINCIPLE

The 17-ketosteroids are excreted as water-soluble conjugates of glucuronic acid and sulfuric acid. Cleavage of these conjugates with acid is followed by extraction, washing with alkali, and finally the color reaction. The reaction is based on the treatment of 17-ketosteroids with meta-dinitrobenzene in alcoholic alkali to produce a reddish-purple color with maximum absorption at 520 nm. Marlow[48] has demonstrated that the development of color depends on the presence of an active methylene group adjacent to a carbonyl group, most likely giving the following product:

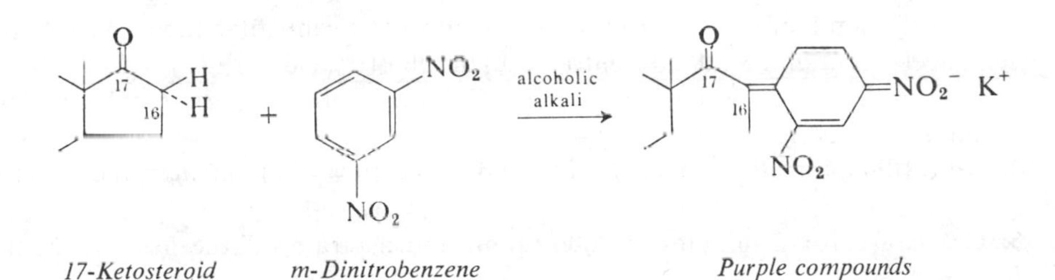

When the ketone group is situated at other positions (e.g., Δ^4 3 keto in testosterone, progesterone, cortisol) the color development is less intense and the absorption maxima differ. The quantitation is carried out by the comparison of the color density of the sample with that of a known amount of pure standard, such as dehydroepiandrosterone.

REAGENTS

1. Ethanol, purified. (See 17-hydroxycorticosteroid determination.)
2. Ethanol, 70 per cent (v/v). Dilute 700 ml. of purified ethanol to 1 liter with distilled water.
3. Ethylene dichloride, redistilled.
4. Potassium hydroxide, A. R. grade. Prepare a saturated aqueous solution.
5. Potassium hydroxide–ethanol solution. Add 1 volume of saturated solution of potassium hydroxide to 4 volumes of purified ethanol. Centrifuge and use the supernatant. Prepare this reagent just before use.
6. m-Dinitrobenzene, 1.16 per cent (w/v) in purified ethanol. Purify the commercially available material as follows: Dissolve 30 gm. of the substance in a minimal

volume of ethanol by warming in a steam bath. Cool, add 30 ml. of 20 per cent sodium hydroxide solution and allow to stand for 30 minutes. Add 3 volumes of distilled water with mixing and let it stand 15 minutes. Filter off the crystalline precipitate on a Buchner funnel. Wash the crystals on the funnel with distilled water and suck dry. Redissolve the crystals in a minimal volume of ethanol and add 3 volumes of distilled water as before. Wait for 15 minutes and filter on a Buchner funnel. Wash the crystals with distilled water and suck as dry as possible. Transfer crystals to a Petri dish and dehydrate in a desiccator over anhydrous calcium chloride. Store the final product in an amber bottle.

7. Dehydroepiandrosterone (DHEA) standard. Dissolve 10 mg. of DHEA in 100 ml. of purified ethanol. This solution contains 100 μg./ml.

COLLECTION OF SPECIMEN

(See method for corticosteroids in urine.)

PROCEDURE

1. Test urine with pH paper. If alkaline, acidify with glacial acetic acid to dissolve phosphate precipitate if any.

Hydrolysis, extraction, and washing

2. Transfer 8 ml. of urine to a 35 ml. glass-stoppered centrifuge tube. Add 2 ml. of glacial acetic acid and 3 ml. concentrated hydrochloric acid.

3. Stopper the tube and place it in a 100°C. bath for 10 minutes. Cool it under cold running tap water.

4. Add 10 ml. of ethylene dichloride and place in a shaking machine for 15 minutes.

5. Centrifuge for 2 minutes at 2000 r.p.m. and aspirate off the urine as completely as possible.

6. Add to the solvent 25 to 30 pellets of sodium hydroxide and place in shaking machine for 15 minutes. Alternately, the extract may be washed with 10 per cent NaOH solution followed by two water washes. Centrifuge as before and filter the solvent through Whatman No. 1 filter paper.

7. Transfer 2.5 ml. of the filtrate (equivalent to 2 ml. urine) to a test tube and evaporate to dryness under nitrogen in a water bath at 50 to 55°C. (If very low concentrations are expected, use 5 ml. of the filtrate.)

Color reaction and spectrophotometric reading

8. Perform the Zimmermann reaction as follows: (a) To a blank tube, the sample tube, and a standard tube containing 50 μg. DHEA, add 0.2 ml. of m-dinitrobenzene solution. (b) Add 0.2 ml. of freshly prepared alcoholic potassium hydroxide solution and mix. (c) Place the tubes in a water bath at 25°C. in the dark for 30 minutes. (d) To each tube add 5 ml. of 70 per cent ethanol and mix.

9. Measure the absorbance of the standard and the sample in a spectrophotometer or a colorimeter at 480, 520, and 560 nm., setting the instrument at 100 per cent transmission with the blank solution.

CALCULATION

Calculate the corrected optical density of the standard and sample using the following formula:

$$\text{corrected absorbance} = (A_{520}) - \frac{(A_{480} + A_{560})}{2}$$

Calculate 24 hr. excretion of 17-ketosteroids as follows:

mg. of 17-ketosteriods/24 hrs.

$$= \frac{\text{corrected } A \text{ of sample}}{\text{corrected } A \text{ of standard}} \times 0.05 \times \frac{\text{24 hr. urine volume (ml.)}}{2}$$

COMMENTS

Most of the urinary 17-ketosteroids are excreted as sulfate and glucuronide conjugates which are hydrolyzed by strong acid and heat. The duration of hydrolysis is very critical. Less than 10 minutes will cause incomplete hydrolysis and more than 10 minutes will lead to gradual destruction of the steroids and formation of an increased amount of nonsteroidal chromogens. Addition of glacial acetic acid helps to minimize the formation of nonspecific chromogens during the hydrolytic procedure, particularly in the case of an alkaline urine specimen. Although solvents such as benzene, carbon tetrachloride, and ether are suitable for extraction, ethylene dichloride is aptly suitable because it can extract the steroid hormones from hydrolyzed urine more quantitatively at a relatively low ratio of the solvent to urine. This is technically advantageous for a routine laboratory method since it avoids the need for handling and evaporating large quantities of solvent.

According to Drekter et al.[25] the treatment of the extract with pellets of sodium hydroxide is superior to the customary treatment with aqueous sodium hydroxide. In contrast to the latter practice, solid NaOH removes phenols and other urinary pigments more completely. It should be noted that estrone, which is a 17-ketosteroid, is removed by alkali treatment because of its phenolic nature (see p. 539) and thus is eliminated prior to the colorimetric reaction of the "neutral" 17-ketosteroids.

For the Zimmermann reaction two different alkaline reagents have been in common use, namely, aqueous and alcoholic KOH. The former yields colors of much less intensity than the alcoholic reagent and the latter has the disadvantage of being unstable. As suggested by Sobel et al.[73] the saturated solution of KOH is stable and yields very low blank absorbance. According to the same authors, the time, temperature and the dilution with 70 per cent ethanol give maximum color development and stability. To avoid undue effects in the color reaction, the ethanol must be of highest quality and purified following the procedure given in the text. However, in spite of meticulous care in the preparation of reagents and in the color development, there is always the formation of background nonspecific chromogens arising from other ketonic steroids and nonsteroidal ketones. The reading of the absorbance at three wavelengths and the use of the correction formula (Allen correction) serve to eliminate the effect of such background interference in the estimation.[3] The correction is based on the assumption that the absorbances of nonspecific materials at the three chosen wavelengths lie on a straight line. In other methods the preparation of a urine blank and subtraction of its reading from the sample is supposed to serve the same purpose.*

---

* The following drugs and their metabolites in urine are known to yield fictitious results causing either under- or over-estimation of 17-ketosteroids: ascorbic acid, Doriden, morphine, meprobamate, and penicillin G.

NORMAL VALUES

Children: excretion values are the same for both sexes through childhood.

Up to 1 year	less than 1 mg./24 hrs.
1–4 years	less than 2 mg./24 hrs.
5–8 years	less than 3 mg./24 hrs.
8–12 years	3 to 10 mg./24 hrs.
13–16 years	5 to 12 mg./24 hrs.
Young adult male	9 to 22 mg./24 hrs.
Adult male	8 to 20 mg./24 hrs.
Adult female	6 to 15 mg./24 hrs.

In both sexes, the rate of excretion declines progressively after about age 60.

REFERENCES

Sobel, C. S., Golub, O. J., Henry, R. J., Jacobs, S. L., and Basu, G. K.: J. Clin. Endocrinol., *18*:208, 1958.

Drekter, I. J., Heisler, A., Scism, G. R., Stern, S., Pearson, S., and McGavack, T. H.: J. Clin. Endocrinol., *12*:55, 1952.

CLINICAL SIGNIFICANCE

Decreased values are obtained in primary hypogonadism (Klinefelter's syndrome, castration), secondary hypogonadism (panhypopituitarism), and primary hypoadrenalism (Addison's disease, especially in women).

Increased values are obtained in testicular tumors (interstitial cell tumor, chorioepithelioma), adrenal hyperplasia, and adrenal carcinoma.

Fractionation of 17-Ketosteroids

CLINICAL IMPORTANCE

The estimation of total urinary neutral 17-ketosteroids serves as a screening test for the diagnosis of an adrenal or gonadal disease. But to derive meaningful information, the determination of *individual* components of this group of steroids is very important.

When the 17-ketosteroids are measured as a group, no distinction is made between the metabolites derived mainly from the testes and those derived primarily from the adrenals. It is known that androsterone and etiocholanolone are primary metabolic products of testosterone. The increased excretion of these two compounds in a male, without proportionate changes of DHEA and 11-oxygenated 17-ketosteroids, will yield a positive indication of testicular dysfunction. When only total 17-ketosteroids are estimated, such specific changes will go unobserved. Similarly, in both adrenocortical hyperplasia and carcinoma the excretion value of 17-ketosteroids is quite high. The fractionated estimation shows that whereas in hyperplasia all components of 17-ketosteroids are elevated, in carcinoma the increased excretion value is mainly the result of the presence of excessive DHEA. Thus, to differentiate adrenocortical carcinoma and hyperplasia, individual estimation of dehydroepiandrosterone, rather than the determination of the 17-ketosteroids as a group, is of considerable diagnostic value. Furthermore, the fractionation provides also the opportunity to determine the ratio of $C_{19}O_2$ (androsterone, etiocholanolone, DHEA) to $C_{19}O_3$ (11-oxygenated 17-ketosteroids). The importance of such study lies in differentiating the type of adrenogenital syndrome. For example, an increased ratio of urinary $C_{19}O_2$ to $C_{19}O_3$ will signify a block in 11β-hydroxylation. The efficacy of the administration of metyrapone (metyrapone inhibits 11β-hydroxylase) may also be properly evaluated on the basis of changes in the ratio.

METHODS FOR FRACTIONATION

Before the modern techniques involving chromatography became available, the use of chemical reactions that were selective toward a characteristic group in the steroid molecule was the primary means of separation. Two such widely used procedures are the separation by Girard's Reagent T and separation by digitonin. The former reagent separates the ketonic neutral compounds containing the 17-ketosteroids from the nonketonic neutral compounds. The principle is based on the fact that Girard's Reagent T forms water-soluble derivatives with the 17-ketosteroids so that the ketones can be separated from the nonketonic material by distribution between water and ether. The treatment of the aqueous solution containing the ketone-derivatives with mineral acid (e.g., HCl) followed by re-extraction with ether, will yield free ketonic compounds.

The separation of the 3α-hydroxy- and 3β-hydroxy-17-ketosteroids in urine is performed by the use of digitonin. This substance is capable of forming insoluble complexes with steroids containing a hydroxyl group with the β-configuration at C-3 (dehydroepiandrosterone), but it does not react with 3α-hydroxysteroids (androsterone, etiocholanolone). In patients with adrenocortical tumors, very high values for the 3β-hydroxy 17-ketosteroids, particularly dehydroepiandrosterone, are almost always obtained. Thus, in the investigation of patients with adrenal hyperfunction the digitonin separation aids in the differential diagnosis. However, from the technical point of view, the quantitativeness of this approach is occasionally doubtful, and at present it is very seldom used. The separation of the ketonic and nonketonic fractions by Girard's Reagent is similarly of little practical value. During the past decade, the fractionation of individual 17-ketosteroids, employing adsorption and partition chromatography (see Dorfman[22]) followed by color reaction, has provided valuable clinical information. In spite of improved reproducibility and resolution, the time-consuming and sophisticated nature of such procedures make them unsuitable for the common, routine laboratory. In recent years, the advent of gas chromatography as an analytical tool appears to have overcome these methodological deficiencies. Speed, sensitivity, accuracy, and simplicity can be achieved with this technique.

The practical usefulness of gas chromatographic analysis has perhaps been realized to the fullest extent in the fractionation and quantitation of the 17-ketosteroids. Following the single injection of an urinary neutral extract, all the 17-ketosteroids can be analyzed individually. Other compounds of clinical interest, such as pregnanediol and pregnanetriol, can be analyzed as well.

Several procedures have been reported on the gas chromatographic analyses of 17-ketosteroids.[26,98]

The use of trimethylsilyl ether (TMSi) derivatives and a gas chromatographic column containing a selective phase (e.g., NGS, XE-60, etc.) has been found to be the most suitable for the adequate separation and reliable quantitation of these steroids. The following is a gas chromatographic procedure as described in the manual of a workshop on gas chromatography of steroids (Tietz). The method is claimed to be suitable for routine clinical use. A description of the preparation of a gas chromatographic column obtained from the same source is also given. The importance of the characteristics of the column in the gas chromatographic analyses of steroids has been discussed in the general comments on steroid determination.

GAS CHROMATOGRAPHIC SEPARATION AND DETERMINATION OF 17-KETOSTEROIDS, PREGNANEDIOL, AND PREGNANETRIOL

PRINCIPLE

The 17-ketosteroids are excreted in the urine as a mixture of glucuronides and sulfates. The hydrolysis is generally performed either by sulfuric acid in the presence of overlying benzene, to minimize the destruction particularly of dehydroepiandrosterone, or by the enzyme Glusulase, containing both β-Glucuronidase and sulfatase. The organic extract (benzene) is washed with sodium hydroxide solution to remove strongly acidic components and phenolic steroids (estrogens). The benzene extract is further washed with water to remove excess alkali, is dehydrated and evaporated to dryness. The dried residue containing neutral steroids is dissolved in tetrahydrofuran (THF) and transferred to a centrifuge tube. Following evaporation of THF, the trimethylsilyl ether derivatives of 17-ketosteroids, pregnanediol, and pregnanetriol are formed by reacting the steroids in THF with hexamethyldisilazane and trimethylchlorosilane. The TMSi derivatives are then injected into a gas chromatographic column containing a selective stationary phase such as XE-60. The quantitation of individual steroids is carried out by comparing the ratio of peak area of individual steroids and the internal standard in the sample to that obtained from the known concentrations of the corresponding standard and internal standard.

I. Preparation of the Column

A. SILANIZING THE COLUMN

(Perform all manipulations with organic solvents and silanizing reagents under a hood.)

Reagents

1. Acetone, A. R.
2. One per cent (w/v) KOH.
3. Methanol, A. R.
4. Toluene, A. R.
5. Silanizing reagent: 5 per cent (v/v) solution of dimethyldichlorosilane (G.E. No. SC-3002) in toluene.

CAUTION: Prepare this solution under a hood; use a rubber bulb to pipet; avoid introduction of moisture, as poisonous fumes may be liberated.

Reagent check: All these reagents should be of highest purity and all organic solvents should be redistilled unless there is evidence that their purity is sufficient. The purity may be checked by evaporating an appropriate amount of solvent (e.g., 50 ml.) to approximately 0.5 ml. and injecting this solvent onto the column. The presence of peaks would indicate impurities in the reagent.

Procedure

1. With the aid of a syringe or a small funnel, fill and rinse the column with acetone. Remove traces of acetone with a vacuum.
2. Fill the column with 1 per cent KOH and let it stand approximately 5 minutes. Drain the KOH solution from the column.

3. Rinse the column 3 times with methanol and then once with toluene.
4. Place the silanizing reagent into the column and let it stand for 15 minutes.
5. Drain the column and then rinse it twice with toluene.
6. Rinse the column quickly with methanol using a vacuum.
7. Evaporate traces of methanol by placing the column into the heated column oven (at approximately 100°C.) of the gas chromatograph.
8. Prior to filling the column with support, dry it thoroughly with a stream of nitrogen. This is necessary in case any moisture has condensed in the column during cooling.

B. SIEVING THE SUPPORT

Screen the support to decrease the variation in the sizes of individual particles by placing the solid support (e.g., Gas Chrom-P) into a set of standard sieves and shaking it on a mechanical shaker (U.S. Standard Sieve Series, W. S. Tyler Co., Cleveland, Ohio).

To obtain a support of 100–120 mesh, use only the particles which passed through the 100 but not the 120 mesh size sieve.

C. ACID-WASHING THE SUPPORT*

(The support prepared by the following procedure is sufficient to fill two 6 foot × 4 mm. columns.)

1. Place 20 gm. of the support (100–120 mesh) into a 1 liter beaker and add approximately 400 ml. of concentrated HCl, A. R.
2. Let stand for 12 hours; stir occasionally.
3. Remove the acid with a filter stick (Corning No. 39533).
4. Wash the support in this manner three more times. Allow 1 hour for each of these washings.
5. After the fourth acid wash, pour 750 ml. of water into the beaker with the support, stir, and let it stand for several minutes. Then decant the water.
6. Wash the support again with water by repeating step 5.
7. Place the support into a Buchner funnel and wash it with water (pH of the wash should be the same as that of the water).
8. Dry the support by suspending it in methanol; decant the methanol along with the fragmented finer particles.
9. Preliminary drying is done at room temperature by spreading the support over the entire surface of a large watch glass, Petri dish, or porcelain dish at least 150 mm. in diameter.
10. After the last traces of methanol vapors are gone, complete the drying in an oven at 80°C. Store the support in a desiccator.

D. SILANIZING THE SUPPORT

Reagents

1. Five per cent (v/v) dimethyldichlorosilane in toluene (General Electric No. SC-3002).
2. Toluene, A. R.
3. Methanol, A. R.

---

* This step, as well as step D, can be omitted, if the employed support has already been treated in this manner by the manufacturer.

Procedure

1. Place all the support from procedure C into a 500 ml. filtering flask.

2. Add a 5 per cent solution of dimethyldichlorosilane to the support. The amount of solution added should be such that there is an adequate volume of liquid ($\frac{1}{2}$ to 1 inch) above the support.

3. Use a vacuum to reduce the pressure in the flask for several minutes, and swirl the flask to free any air bubbles trapped on the support.

4. After air bubbles cease to form, shut off the vacuum and remove the support from the flask by filtration through a Buchner funnel.

5. Wash the support well with toluene and dry it with methanol.

6. Air dry the support and then place it in an oven at 80°C. as outlined under procedure "C".

E. COATING THE SUPPORT

1. Place the acid-washed and silanized support into a filtering flask.

2. Add the solution of the liquid phase in the appropriate solvent, in this case 4-5 gm. XE-60 (Applied Science Lab.) in 100 ml. acetone, A. R. There should be an adequate level of the solution of the liquid phase above the support. *Note:* Many other liquid phases (NGS, Hi-Eff 8B, OV-17, etc.) have been used successfully for the separation of 17-ketosteroids. In our experience, XE-60 has given columns with the highest efficiency.

3. Reduce the pressure with a vacuum and swirl the flask to remove air bubbles from the surface of the support; disconnect the vacuum, and allow the suspension to stand for 5 minutes.

4. Swirl the flask again and pour the entire contents on a Buchner funnel. Reduce the vacuum and let the solution drain through the funnel until excess moisture is removed.

5. Carry out the preliminary drying at room temperature, and then dry in an oven at 80 to 100°C. Do not place the support into the oven until it is air-dried.

F. PACKING THE COLUMN

1. Pour a small amount (4 to 6 inches) of the coated support into a 6 ft., 3 to 4 mm. diameter, U-shaped column, using a small funnel.

2. Pack the support firmly by tapping the walls of the column with a rubber coated, heavy glass rod.

3. Pour 4 to 6 additional inches of support into the column, and pack the column by tapping until it is filled to within 1 inch of both the end and the side arm.

4. Place a glass wool plug at each end of the column. The glass wool should first be silanized by a procedure identical to that outlined under "Silanizing the Column."

Packing the coil columns. Place a silanized glass wool plug into the lower opening of the column. Attach this end of the column to a vacuum. Then follow the procedure just described (F, 1–4).

G. CONDITIONING THE COLUMN

1. Place the column into the column oven and connect it to the carrier gas but not to the detector.

2. Heat the column overnight at a temperature of 250°C. (for XE-60).

3. The carrier gas flow should be set at 40 ml. per minute.

4. At the end of the conditioning period, connect the column to the detector and adjust the temperatures to the operating conditions (column 205-225°C., detector 260°C., injection port 260°C).

II. Acid Hydrolysis and Extraction

(This procedure may be used for the determination of 17-ketosteroids, and pregnanediol. Pregnanetriol cannot be determined by this procedure.)

Reagents

1. Benzene, A. R., redistilled.
2. Tetrahydrofuran (THF), A. R. Store over anhydrous Na_2SO_4.
3. Forty per cent (v/v) H_2SO_4, A. R.
4. Concentrated H_2SO_4, A. R.
5. Ten per cent (w/v) NaOH.
6. Na_2SO_4, A. R., anhydrous.
7. Internal standard solution (20 mg./100 ml.). Place 40 mg. epicoprostanol into a 200 ml. volumetric flask and make up the volume to the mark with tetrahydrofuran.

Procedure

1. Mix, measure, and record the volume of the 24 hour urine collection.
2. Pipet a 50 ml. aliquot into a 100 ml. beaker and adjust the pH to 0.8 with 40 per cent H_2SO_4. (It is recommended that the procedure be carried out in duplicate.)
3. Transfer the specimen into a 500 ml. round bottom flask. Add 50 ml. of redistilled benzene and a few boiling chips.
4. Attach to a reflux condenser (with a glass fitting), and heat the mixture to boiling using a heating mantle. The heat should be just high enough to allow the urine to boil gently; too intense a temperature will char the organic layer.
5. After 15 minutes, remove the flask from the mantle and cool the mixture by immersion in an ice bath.
6. Pour the contents into a separatory funnel and drain the aqueous urine layer (bottom layer) into a round bottom flask. Save the organic layer.
7. Using the urine layer, repeat this procedure (steps 4 to 6) twice; first refluxing at a pH of 0.8 for 1/2 hour, then refluxing at a pH of 0.2 for 1 hour. Pool the benzene extracts from steps 6 and 7.
8. After the third refluxing, add 5 ml. of concentrated H_2SO_4 to the urine and reflux without benzene for 20 minutes.
9. Cool, add 50 ml. benzene, and shake in a separatory funnel. Discard the urine layer. Add the benzene extract to the previous three extracts (steps 6 and 7).
10. Wash the combined benzene extracts with three 15 ml. portions of 10% NaOH.
11. Wash the benzene extract with water in a similar manner until the wash water is neutral.
12. Dry the benzene extract by pouring it quantitatively into a round bottom flask through a glass funnel which has been fitted with a glass wool plug and 2 gm. anhydrous Na_2SO_4. Wash the Na_2SO_4 with 10 ml. benzene.
13. Add 1 ml. of the internal standard solution. (In case of a very concentrated urine, or when high values are expected, use 2 ml. of the internal standard.)
14. Evaporate the benzene under vacuum on a rotary evaporator at approximately 60°C.

15. Transfer the residue with two approximately 5 ml. portions of THF to a conical, glass-stoppered centrifuge tube.

16. Evaporate the solvent to dryness under a stream of pure dry nitrogen by placing the tube into a water bath at 60°C.

17. Place the tube in a desiccator until the ether derivatives are prepared.

III. Enzymatic Hydrolysis and Extraction

(May be used for 17-ketosteroid, pregnanediol, and pregnanetriol analysis.)

Reagents

1. Glusulase (150,000 units/ml. of β-glucuronidase and 75,000 units/ml. of sulfatase). Endo Labs., Inc., Garden City, N.Y.
2. Benzene, A. R., redistilled.
3. Tetrahydrofuran, (THF), A. R.
4. Glacial acetic acid, A. R.
5. Five per cent (w/v) $NaHCO_3$, A. R.
6. Na_2SO_4, A. R., anhydrous.

Procedure

1. Mix, measure, and record the volume of the 24 hour urine collection.
2. Pipet a 50 ml. aliquot (duplicates) into a 100 ml. beaker. Adjust the pH to 5 by adding 2 gm. sodium acetate and 0.5 ml. glacial acetic acid. Check the pH with a pH meter, and add more glacial acetic acid if needed.
3. Transfer the sample to a 100 ml. glass-stoppered flask, add 0.5 ml. of Glusulase, and stopper.
4. Incubate at 37°C. for at least 36 hours.
5. Place the urine into a separatory funnel and extract it three times with 60 ml. portions of benzene.
6. Wash the combined extracts twice with 15 ml. of 5% $NaHCO_3$. Discard the aqueous layer.
7. Wash the extract with 15 ml. portions of distilled water until the pH of the wash is the same as the pH of the distilled water. Discard the washes.
8. Proceed as in acid hydrolysis and extraction (II, steps 12 to 17).

IV. Preparation of Trimethylsilyl Ether Derivatives

NOTE: Precautions should be taken to exclude moisture from all equipment and reagents.

Reagents

1. Tetrahydrofuran (THF), A. R.
2. Prepare a standard mixture by placing into a 200 ml. volumetric flask 10 mg. each of androsterone, etiocholanolone, dehydroepiandrosterone, pregnanediol, pregnanetriol, 11-keto-androsterone, Δ^5-pregnenetriol, 11-ketoetiocholanolone, 11β-hydroxyandrosterone, 11β-hydroxyetiocholanolone, and 11-ketopregnanetriol. Dilute to volume with THF.
3. Hexamethyldisilazane.
4. Trimethylchlorosilane.

Procedure

1. Remove the centrifuge tube with the urine extract (II, step 17 of extraction procedure) from the desiccator, and dissolve the residue in 2 ml. of THF.

2. Set up a standard (in duplicate) by pipetting 1 ml. of the standard mixture solution and 1 ml. of internal standard solution (II, Reagent No. 7) into a glass-stoppered centrifuge tube.

3. Add 0.3 ml. of hexamethyldisilazane and 0.1 ml. of trimethylchlorosilane to all tubes. Stopper, mix, and place in the desiccator overnight at room temperature. Alternatively, incubate the sample for 1/2 hour at 60°C.

4. After incubation, mix and centrifuge.

5. Transfer the supernatant into another centrifuge tube. (This transfer does not have to be quantitative. Any loss of extract will be compensated for by an equal loss of internal standard.) The supernatant may be cloudy at this point.

6. Place the tubes into a water bath at 60°C., and evaporate the solution under nitrogen.

7. Add 1 ml. THF, mix, and centrifuge again.

8. Transfer the supernatant to another centrifuge tube. It should be clear (but may be colored) at this point. Repeat step 6.

9. Add 0.2 ml. THF to the redissolved residue. If a high concentration of 17-ketosteroids or pregnanediol and pregnantriol is expected, use larger amounts of THF (e.g., 0.4 ml.).

V. Operation of the Gas Chromatograph

Adjust the temperature of the injection port, column, and detector as follows: Injection port 260°C., column 205-225°C., detector 260°C. Adjust the flow of carrier gas (e.g., helium or nitrogen) in the Barber-Colman Model 5000, to 65 ml./min., in the Beckman GC 5 to 45 ml./min. Adjust the hydrogen and air flow in accordance with the procedures provided by the manufacturer. For further details see the manufacturer's instruction manual.

VI. Injection Technique

Many techniques for injection have been recommended, but for this application the following technique appears to provide the most reliable results.

1. Draw approximately 1 microliter of THF into a 10 μl. Hamilton syringe. Pull the needle out of the solution and withdraw the plunger slightly so as to introduce a very small air bubble.

2. Draw 1 to $1\frac{1}{2}$ μl. of sample into the syringe (depending on the expected concentration of the components).

3. Hold the plunger securely in position and place the needle through the septum of the injection port.

4. When the needle is in the proper position, inject the sample with one fast movement, and hold the plunger in position until the solvent front appears on the recorder.

5. Remove the syringe and wash it immediately with THF.

VII. Calculation

Determine the areas (peak height $\times$ peak width at 1/2 peak height) under the peaks corresponding in retention time to authentic steroids and the internal standard, and use the formula as described under the gas chromatographic analysis of pregnanediol.

COMMENTS

The requisites for gas chromatographic analysis of steroids have been discussed before (see general comments on the Methods of Steroid Determination). It should be emphasized here that the preparation of the column has much to do with the efficient separation and quantitation of steroids. The deactivation of active sites on the column tubing and the solid supports through acid-base washing and silanization, the use of a uniform mesh size, and the coating of the stationary phase onto the solid support are essential for the preparation of an efficient column. Solid supports are rather brittle. The support material must be handled as gently as possible at all times in order to minimize fragmentation. The coating of the support with the stationary phase as described in the text is sufficiently gentle. The only disadvantage of this particular method[83] lies in the uncertainty of the final concentration of the stationary phase. Frequently this is of little significance, unless, as occurs with an occasional column, the phase concentration is too low. Another method[95] of coating the solid support consists of evaporating the requisite amount of dissolved stationary phase onto the solid support while constantly stirring the mixture. This generally leads to significant fragmentation. It has been suggested that good results may be obtained by suspending the dry, coated support in absolute ethanol and gently floating the fragments off the top of the liquid.

The treatment of dehydroepiandrosterone sulfate with hot mineral acid leads to artefact formation. For colorimetric estimation involving the Zimmermann reaction, such structural rearrangements may not influence the results drastically so long as the reactive group in ring D is intact (see total 17-ketosteroid estimation). The gas chromatographic separation, on the other hand, requires the structural integrity of the compound, and hence proper precaution is needed to keep the structure intact while the sample is being processed. The mild acid hydrolysis with overlying benzene minimizes the losses caused by artefact formation. The enzyme hydrolysis is very specific and does not cause any undue structural rearrangements but is time consuming. In addition, the presence of varying quantities of β-glucuronidase inhibitors in the urine may lead to incomplete hydrolysis. Several non-enzymic procedures for the apparently complete and safe hydrolysis of these steroid conjugates suitable for gas chromatographic methods may be found elsewhere.[26,98]

The fractionation and determination of individual 17-ketosteroids, pregnanediol, and pregnanetriol in crude urinary extracts most often yield clinically useful results. It should be noted, however, that the detection system in a gas chromatograph is nonspecific, and to achieve precise and specific analysis, adequate and efficient prepurification of the urinary extracts is very important. It is interesting to note in this connection that, even after considerable purification of the urinary extract, the gas chromatographic method yields lower values for etiocholanolone and dehydroepiandrosterone and a higher value for androsterone in comparison to the results obtained by the well-standardized gradient elution chromatographic methods followed by the Zimmermann reaction. The overestimation of androsterone by gas chromatography is attributed to the presence of epiandrosterone.[80]

At present, no satisfactory explanations are known for the low yield of etiocholanolone and dehydroepiandrosterone. The precautions suggested in the text for the preparation of trimethylsilyl ether derivatives should be carefully followed.

The use of tetrahydrofuran as the solvent may cause occasional problems if it is not freshly distilled. Other solvents such as chloroform, hexane, pyridine, may be used instead. Various commercially available silylation reagent mixtures (Tri-sil, BSA—Pierce Chemical Co., Rockford, Illinois) seem to be conveniently suitable in a routine laboratory.

NORMAL VALUES DETERMINED BY GAS CHROMATOGRAPHY (in mg./24 hr.)

	Males	*Females*
1. Pregnanediol	0.2–1.2	0.2–6.0
2. Androsterone	2.0–5.0	0.5–3.0
3. Etiocholanolone	1.4–5.0	0.8–4.0
4. Dehydroepiandrosterone	0.2–2.0	0.2–1.8
5. Epicoprostanol (Internal Standard)		
6. Pregnanetriol	0.5–2.0	0.5–2.0
7. Δ^5-Pregnenetriol		0.2–0.9(?)
8. 11-Ketoandrosterone	0.2–1.0	0.2–0.8
9. 11-Ketoetiocholanolone	0.2–1.0	0.2–0.8
10. 11β-Hydroxyandrosterone	0.1–0.8	0–0.5
11. 11β-Hydroxyetiocholanolone	0.2–0.6	0.1–1.1
12. 11-Ketopregnanetriol		0–0.3(?)
13. Total 17-Keto Steroids	5.0–12.0	3.0–10.0

Note: Compounds are listed in the order in which they come off the XE-60 column. Open spaces indicate that normal values have not yet been established.

REFERENCES

Vestergaard, P., and Clausen, B.: Acta Endocr., Supplement 64, 1962.

Creech, B. G.: J. Gas Chromat. *2*:195, 1964.

Wotiz, H. H., and Clark, S. J.: Gas Chromatography in the Analysis of Steroid Hormones. New York, Plenum Press, 1966.

Eik-Nes, K. B., and Horning, E. C. (Eds.): Gas Phase Chromatography of Steroids. New York, Springer-Verlag, 1968.

Tietz, N. W.: Gas Chromatographic Separation and Determination of 17-Ketosteroids, Pregnanediol, and Pregnanetriol. Workshop Manual: "Gas Chromatography in Clinical Chemistry". The Chicago Medical School and Mount Sinai Hospital Medical Center, Chicago, Illinois. November 13 to 16, 1967.

PROGESTERONE AND ITS METABOLITES

Progesterone is a female sex hormone. This compound, in conjunction with estrogens, regulates the accessory organs during the menstrual cycle. The importance of this hormone also lies in preparing the uterus for the implantation of blastocysts and in maintaining pregnancy. In nonpregnant women, progesterone is secreted mainly by the corpus luteum (a yellow glandular mass in the ovary, formed by an ovarian follicle following the discharge of its ovum), whereas during pregnancy the placenta becomes the major source. It should be recalled that progesterone is also an obligatory precursor for the synthesis of corticosteroids and androgens. Consequently, the adrenals and the testes are also considered minor sources of this steroid hormone.

The structural formula of progesterone, a C_{21}-compound, is shown in Figure 9-16. Similar to corticosteroids and testosterone, progesterone (pregn-4-ene-3,20-dione) contains a keto group (at C-3) and a double bond at C-4-5(Δ^4) in ring A, and this structural characteristic is considered to be essential for progestational activity. The two-carbon side chain ($CH_3—\overset{\overset{O}{\|}}{C}—$) on C-17 does not seem to be very important for its

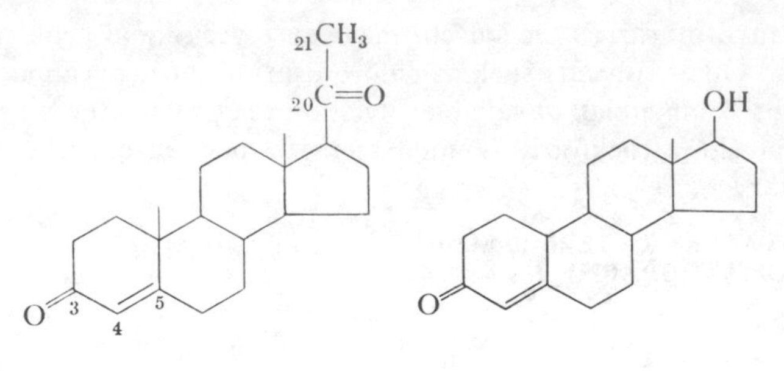

Progesterone
(Pregn-4-ene-3,20-dione)

19-Nortestosterone
(17β-Hydroxy-19-norandrost-4-en-3-one)

Figure 9-16. Structural formula of progesterone and 19-nortestosterone.

physiological action. Indeed, the synthetic compound, 19-nortestosterone (absence of a methyl group at C-10) and its derivatives (Fig. 9-16) are (orally effective) more potent progestational agents than progesterone itself.

Biogenesis and Metabolism

The biosynthetic pathway of progesterone is believed to involve the same enzymatic sequences from acetate through cholesterol and pregnenolone to progesterone as described in earlier sections (see corticosteroids and general aspects of steroidogenesis). The important metabolic events leading to the inactivation of progesterone are reduction and conjugation. An examination of the chemical structure shows that there are three different sites in the molecule of progesterone which are liable to hydrogenation (reduction). These are the double bonds between the carbon atoms 4 and 5 and the keto groups at C-3 and at C-20. It should be further noted that the reduction of each site may give rise to two isomers differing in the spatial orientation of the hydrogen at C-5 or the hydroxyl groups at C-3 and at C-20. The formation of an α or β isomer is catalyzed by the specific enzymes (e.g., Δ^4-5α- or Δ^4-5β-reductase, 3α or 3β hydroxysteroid dehydrogenase, 20α- or 20β-hydroxysteroid dehydrogenase). The main metabolic pathway is outlined in Figure 9-17.

The metabolites of progesterone may be classified into three groups based on the degree of reduction. (1) Pregnanediones. The double bond is reduced, producing two compounds: pregnanedione (H at C-5 is in β orientation—above the plane of the paper) and allopregnanedione (H at C-5 is in α orientation—below the plane of the paper). (2) Pregnanolones. The keto group at C-3 is reduced to a hydroxyl group. The hydroxy group may again be either α or β in orientation. However, the urinary metabolites are preponderantly in α configuration. (3) Pregnanediols. Further reduction of the keto group at C-20 gives rise to these metabolites. As in the case of the hydroxyl group at C-3, the metabolites containing the 20α-hydroxyl group are quantitatively most important. In fact, urinary measurement of pregnanediol (5β-pregnane-3α-20α-diol) is used as an index of endogenous production of progesterone because it has been shown that this metabolite is quantitatively most significant and correlates fairly well with a majority of clinical conditions. The reduced metabolites are finally conjugated with glucuronic acid and excreted as water-soluble glucuronides.

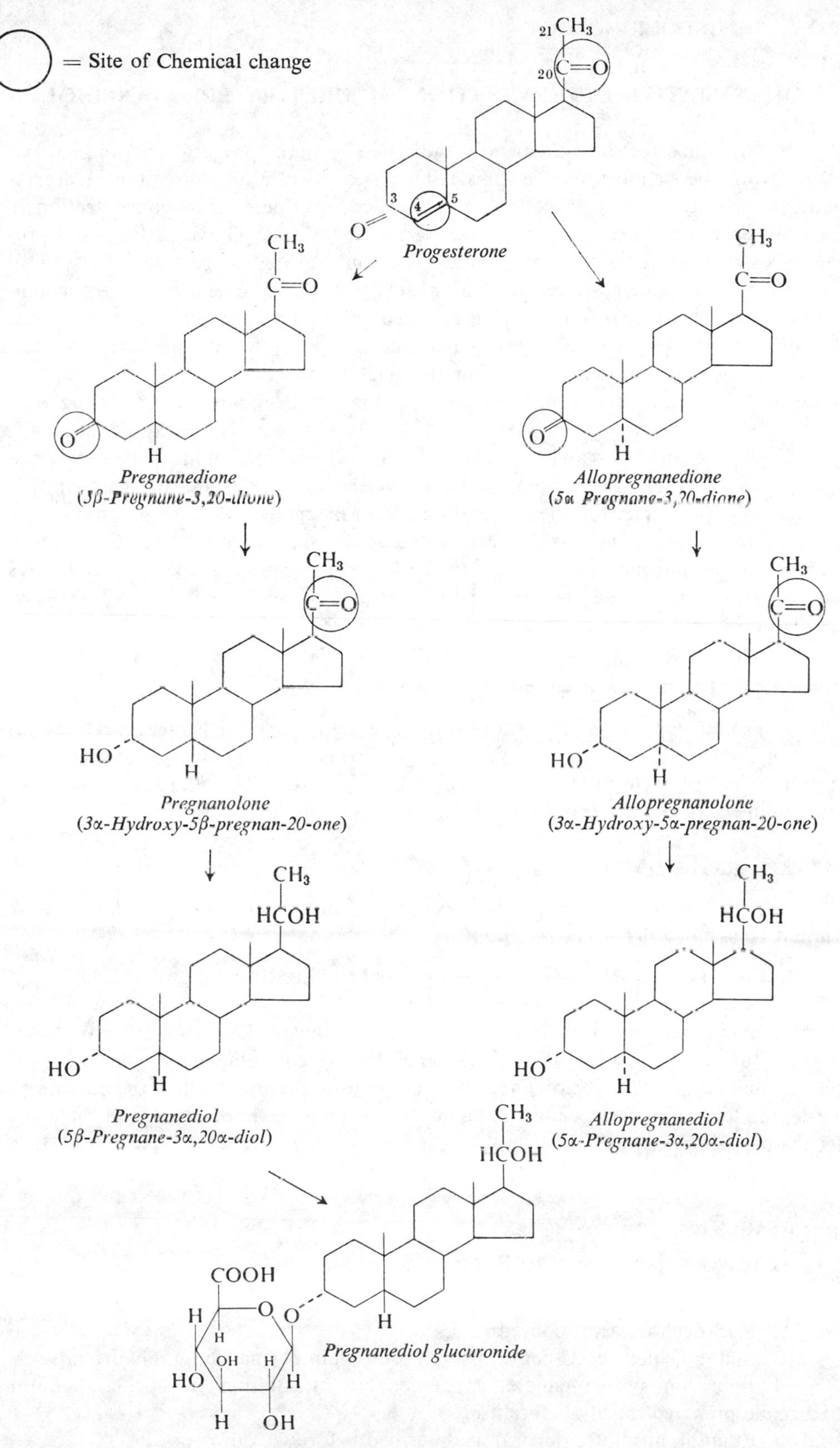

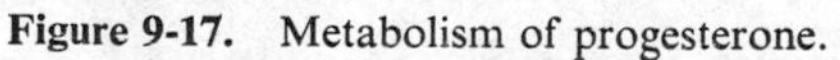
Figure 9-17. Metabolism of progesterone.

QUANTITATIVE DETERMINATION OF URINARY PREGNANEDIOL

Ideally, the direct measurement of the active hormone progesterone, in blood or urine, would be the most reliable approach to assess its rate of production. Unfortunately, very little or no progesterone is excreted in urine, and its concentration in blood even during pregnancy is in microgram quantities. The sensitive assay procedures available at the present time are too sophisticated to be used in a clinical laboratory (see Vander Molen[84]). Consequently, the measurement of urinary metabolites, especially pregnanediol which is excreted in milligram amounts, has been used to evaluate the rate of production of the original hormone. Studies on the recovery of urinary pregnanediol after the injection of progesterone have shown that this metabolite does indeed bear a proportionality to the total amount of circulating progesterone.[38]

Since pregnanediol is an inactive metabolite, its measurement has to be performed by chemical methods rather than by bioassay techniques. During the past decade, numerous improvements in the purification and in the mode of final estimation of this compound have been achieved. Detailed discussions on this topic may be found in review articles and monographs.[39,46] In the following section, a chemical and a gas chromatographic method are described in detail. Both of these methods have been shown to be suitable for routine clinical use.

Chemical Determination of Urinary Pregnanediol[40]

The main steps in the procedure consist of acid hydrolysis, toluene extraction, permanganate oxidation of the extract, first chromatography on alumina column, acetylation, second chromatography on alumina, color reaction with sulfuric acid, and measurement of the chromogen at 430 nm. in a spectrophotometer.

APPARATUS

All glassware should be cleaned thoroughly. Contamination with cork or rubber should be avoided.

1. *Chromatographic column:* glass tubes 12 cm. long, 10 mm. in diameter with a 75 ml. reservoir and a sealed-in sintered disc of porosity no. 3 to support the column of alumina. An alumina column is prepared as follows: Partly fill the chromatographic tube with specified solvent (benzene or petroleum ether) and pour 3 gm. of deactivated alumina (see below) in a thin stream into the tube. Allow the column to settle, tap to level its surface, and add a protective 5 mm. layer of silver sand. Do not let the column dry at any time.
2. Spectrophotometer

REAGENTS

All reagents should be of A. R. grade

1. Toluene, redistilled
2. Hydrochloric acid, concentrated
3. Sodium chloride, 25 per cent (w/v) solution in normal sodium hydroxide.
4. Potassium permanganate, 4 per cent (w/v) solution in normal sodium hydroxide prepared freshly before use.
5. Ethanol, absolute, purified as described before (p. 502).

6. Deactivated alumina (100/150 mesh): deactivate by exposing alumina for ten days in layers with occasional mixing to an atmosphere saturated with water vapor at room temperature. The convenient way of doing this is to place the alumina in a petri dish in a desiccator containing water at the bottom. Store the deactivated alumina in an airtight container. Before use, each batch should be standardized with authentic standard solution using the solvent systems as described in the text.
7. Benzene redistilled, water-saturated.
8. Silver sand, (40/100 mesh) purified by boiling with 30 per cent hydrochloric acid and then washing thoroughly with tap water and then distilled water.
9. Light petroleum ether (b.p. 40 to 60°C.) redistilled, water-saturated.
10. Acetyl chloride.
11. Sodium bicarbonate, 8 per cent (w/v) solution in distilled water.
12. Sodium sulfite, solid.
13. Sulfuric acid, concentrated.
14. Standard pregnanediol diacetate solution (20 mg./100 ml.) in ethanol.

PROCEDURE

Collect a 24-hour urine specimen, without using a preservative, and store at 4°C.

Acid hydrolysis, extraction, and permanganate oxidation. Measure one twentieth of the total volume into a 500 ml. round bottom flask and dilute to 150 ml. with distilled water. (This aliquot of urine is adequate for specimens from non-pregnant women. Smaller aliquots, such as 1 or 2 per cent of the total volume, should be used when urine from pregnant women is used.) Add 50 ml. of toluene and a few glass beads to prevent bumping. Heat to boiling under a reflux condenser. Add 15 ml. concentrated hydrochloric acid through the condenser and continue boiling for ten minutes. Cool the flask rapidly under running tap water and transfer its contents to a 500 ml. separating funnel. Remove the urine layer and re-extract with an additional 50 ml. of toluene. Combine the toluene extracts and discard any urine which has settled out from the emulsion.

Shake the toluene extract with 25 ml. of a 25 per cent solution of sodium chloride in normal sodium hydroxide. Discard the aqueous layer including the curdy precipitate at the liquid interface. Add 50 ml. of freshly prepared 4 per cent potassium permanganate in normal sodium hydroxide, and shake for 10 minutes. (This step may be carried out conveniently in a wrist-action mechanical shaker. The toluene extract is transferred to a 250 ml. Erlenmeyer flask. To minimize losses, the separating funnel and the flask should be adequately rinsed with fresh toluene.) Discard the permanganate layer and wash the toluene layer 4 to 5 times with successive 50 ml. quantities of distilled water to remove all permanganate color. Filter the toluene through a Whatman No. 1 filter paper into a 250 ml. round bottom flask, pouring from the separating funnel so as to leave behind a few drops of emulsion. Evaporate the toluene under reduced pressure to a volume of approximately 10 ml., and cool to room temperature.

First alumina chromatography. Prepare a column of deactivated standardized alumina using benzene as the solvent, as described previously. Apply the toluene extract to the column. Rinse the flask with 5 ml. of benzene and add this to the column. When all the solvent has percolated through, add to the column 25 ml. of 0.8 per cent ethyl alcohol in benzene. Discard the eluate, which contains some pigmented material and relatively non-polar steroids. Elute with 12 ml. of 3 per cent ethanol in benzene. Collect the eluate (which contains all the pregnanediol) into a 6×1 inch test tube. Evaporate the solvent to dryness under nitrogen.

Acetylation and second chromatography. Dissolve the residue in the tube in 2 ml. of benzene, and add 2 ml. of acetyl chloride. Loosely stopper the tube, swirl briefly, and let stand at room temperature for approximately 1 hour. Add 25 ml. of light petroleum ether and transfer to a separating funnel. Wash the solvent mixture once with 50 ml. of distilled water, once with 25 ml. of 8 per cent sodium bicarbonate solution, and twice with 25 ml. distilled water. Carefully run off the last drop of water and then pour the light petroleum ether onto an alumina column prepared as above, but using petroleum ether instead of benzene. When the solution has percolated through, elute the pregnanediol diacetate with 15 ml. of benzene into a 6 × 1 inch test tube. Evaporate the solvent to dryness as before, and place the tube for at least an hour in a desiccator over anhydrous calcium chloride.

COLORIMETRY

Add approximately 10 mg. of sodium sulfite to the residue in the test tube and then 10 ml. of concentrated sulfuric acid. Stopper the tube, shake well, and place in a water bath at 25°C. for 17 hours. Prepare a standard tube containing 100 μg. of pregnanediol diacetate (0.5 ml. of a 20 mg./100 ml. standard solution in ethanol). Evaporate the ethanol to dryness, and add to the residue 15 ml. of benzene, as before. Desiccate the tube and develop the color as above. Read the color density of the standard as well as the unknown against the sulfuric acid blank at 430 nm.

CALCULATION

If one-twentieth of the 24-hour urine specimen was taken, mg. of pregnanediol/24 hr. $= \frac{\text{Reading of unknown}}{\text{Reading of standard}} \times 20$ (aliquot of urine) $\times\ 0.1$ (concentration of standard in mg.) $\times\ 0.8$ (ratio of Mol. Wt. of free and acetylated pregnanediol).

The color density of the unknown may also be converted to mg. of pregnanediol diacetate by means of a calibration curve constructed by employing standard concentrations which cover the ranges expected for the analyses of the samples. This will also verify the proportionality of color density to the concentration.

COMMENTS

Pregnanediol is excreted in urine almost exclusively as pregnanediol glucuronide. Acid hydrolysis liberates pregnanediol, which can be extracted easily with an organic solvent such as toluene. The exposure of free pregnanediol in hot acid solution increases its destruction. The overlaying toluene facilitates the removal of pregnanediol from acid solution as quickly as it is liberated. Exact timing of hydrolysis is very important if destruction is to be minimized. A second extraction of the hydrolyzed urine will ensure quantitative extraction.

Emulsions are likely to form during extraction with toluene. Washing the extract with sodium chloride in sodium hydroxide solution not only breaks the emulsion but also removes the acidic and phenolic components. The potassium permanganate oxidation step removes some of the steroidal artefacts formed during acid hydrolysis, which otherwise would be eluted from the column along with pregnanediol and produce falsely high sulfuric acid chromogens. In fact, the color reaction with sulfuric acid is very nonspecific. Almost all steroids, and for that matter many organic compounds, are liable to form a yellow color in the presence of concentrated sulfuric acid. Thus, to achieve specificity, the rigorous purification of pregnanediol by two alumina chromatography steps is very important.

The presence of oxidizing agents in commercially available sulfuric acid and impurities from the solvents interfere with the color development. Addition of a reducing agent such as sodium sulfite helps to overcome the untoward effects of oxidizing agents. The effects of solvent impurities are minimized by adding the same amounts of solvents to the standard tubes that are used in the final stage of the method (15 ml. of benzene).

REFERENCE

Klopper, A., Michie, E. A., and Brown, J. B.: J. Endocr. *12*:209, 1955.

Gas Chromatographic Determination of Urinary Pregnanediol[19,69]

A general discussion of the gas chromatographic analysis of steroids has already been presented (page 491). The basic steps involved in this technique are as follows: acid hydrolysis, extraction with toluene, washing with alkali and water, acetylation, and injection into a gas chromatograph. The quantitation is carried out by comparing the ratio of peak height or peak area obtained from the known concentrations of authentic standard and internal standard with that of unknown and internal standard in the sample.

REAGENTS

All reagents are of A.R. grade.

1. Hydrochloric acid, concentrated.
2. Toluene, redistilled.
3. Pyridine, redistilled under anhydrous conditions.
4. Acetic anhydride, redistilled under anhydrous conditions.
5. Sodium chloride, 25 per cent (w/v) in normal sodium hydroxide.
6. Sodium sulfate, anhydrous.
7. Acetone, spectro grade.
8. Pregnanediol (gas chromatographically pure) standard solution, 20 mg./100 ml. ethanol.
9. Cholesterol propionate (gas chromatographically pure) solution, 20 mg./100 ml. ethanol. Used as internal standard.

PROCEDURE

Collect a 24-hour specimen without using a preservative and store at 4°C. With the urine obtained from men and nonpregnant women use 5 per cent of the total volume; with women in the first trimester of pregnancy use 2.5 per cent of the total volume, in the second and third trimesters use 1 per cent of the total volume.

Hydrolysis and extraction. Transfer the appropriate aliquot of urine into a 500 ml. round-bottom flask and dilute to 150 ml. with distilled water. Add 50 ml. of toluene and a few glass beads. Place under a reflux condenser and heat to boiling. Add 15 ml. of concentrated hydrochloric acid through the condenser, and continue boiling for 10 minutes. Cool the flask rapidly under running tap water and transfer the contents to a separating funnel. Remove the urine layer and re-extract with 50 ml. fresh toluene. Combine the toluene extracts, run off and discard any urine which has settled out from the emulsion. Shake the extract with 25 ml. of a 25 per cent solution of sodium chloride in normal sodium hydroxide. Discard the aqueous layer, including the curdy precipitate at the interface. Wash twice with 25 ml. of distilled water. Add

20 μg. of cholesterol propionate (0.1 ml. of 20 mg./100 ml. ethanol) to the toluene extract in the separating funnel. Dehydrate the toluene extract in a 250 ml. Erlenmeyer flask containing approximately 10 gm. anhydrous sodium sulfate. Then filter the toluene through a Whatman No. 1 filter paper into a 250 ml. round-bottom flask. Evaporate the toluene to dryness under reduced pressure, and dissolve the residue in 5 ml. of acetone. Transfer the solution with a Pasteur pipet into a 15 ml. centrifuge tube. Rinse the flask thoroughly two more times with 2 ml. of fresh acetone and transfer as before to the centrifuge tube. Evaporate the acetone to dryness at 50°C. under a stream of nitrogen. At this point prepare a standard tube by adding 20 μg. each of pregnanediol and cholesterol propionate (0.1 ml.) to a centrifuge tube. Evaporate the ethanol to dryness.

Acetylation. Add to the dried residue 0.2 ml. of pyridine and 0.2 ml. acetic anhydride. Shake to dissolve the contents in the tube. Stopper. Incubate for one hour at 68°C. or let it stand overnight at room temperature. Treat the tube containing standard and internal standard in a similar manner. Evaporate the acetylating reagent mixture to dryness at 60°C. under a stream of nitrogen.

Gas chromatography. The operating condition will vary with the column characteristics. The following is a typical description.

Column: 6 foot × 4 mm. (internal diameter) glass tube packed with 3% SE-30 or OV-1 on 80–100 mesh diatomaceous earth (acid washed, silanized).
Temperature: 250°C.
Carrier gas (nitrogen): 40 ml./min.
Flash heater: 270°C.
Detector (flame ionization): 280°C.

Dissolve the contents of the standard tube in 100 μl. of acetone. Draw 2 μl. of solution into a 10 μl. syringe (Hamilton) and inject into the gas chromatograph. Dissolve the acetylated sample residue in 100 μl. of acetone and inject 2 μl. as before. Repeat the injection of standard at the end of the sample run.

CALCULATION

Measure the peak heights of the pregnanediol diacetate and the internal standard in the standard and sample gas chromatograms. Calculate the amount of pregnanediol in the unknown sample according to the following formula:

$$\text{mg. of pregnanediol per 24 hours} = \frac{R_{St} \times R_u \times I \times \dfrac{\text{Total volume of urine}}{\text{Volume of urine used}}}{1000}$$

$$\text{where } R_{St} = \frac{\text{Peak height of 20 } \mu\text{g. cholesterol propionate}}{\text{Peak height of 20 } \mu\text{g. acetylated pregnanediol}}$$

$$R_u = \frac{\text{Peak height of acetylated pregnanediol in sample}}{\text{Peak height of internal standard in sample}}$$

$$I = \mu\text{g. of internal standard added to sample } (= 20\ \mu\text{g.})$$

COMMENTS

The remarks regarding the acid hydrolysis, extraction, and alkali wash as outlined for the chemical determination of pregnanediol are equally pertinent to the present method. The rigorous purification of the urinary extract prior to gas chromatography is not required. A gas chromatographic column of relatively high efficiency separates pregnanediol diacetate adequately from other peaks arising from urinary contaminants, with the exception of allo-pregnanediol diacetate. However, the amount of the latter compound in the urine is minute, and thus it does not affect the quantitative yield of pregnanediol. For detailed discussions of accuracy, precision, specificity, sensitivity, and the applicability of this technique to clinical conditions, the reader is referred to the original publications.[19,69,70,96] The purpose of adding internal standard to the sample prior to GLC has already been discussed (see page 491). The addition of internal standard directly to the toluene extract in the separatory funnel also corrects for the procedural losses incurred during transfer. Detection sensitivity and column efficiency in relation to flow of carrier gas, hydrogen, air, as well as column temperature, must be optimized with authentic standard whenever a new column is introduced. Furthermore, constancy of peak height ratio of standard and internal standard over a concentration range of analysis should be established for individual columns. The amount of sample to be injected depends largely on the concentration of pregnanediol expected. It has been our experience that injection of more than 5 μl. of solution of crude urinary extract causes overlapping of peaks and often obscures minor peaks. It also prevents the detector response from approaching the baseline between peaks, thus preventing readings at low attenuation. It is generally more desirable to inject small volumes and decrease the attenuation. The quantitation on the basis of peak height ratio has been shown to be quite accurate and reproducible in the laboratory. However, since the influence of temperature on the peak height is quite marked, the measurement of the peak area (peak height times peak width at half the peak height) is suggested in cases in which the temperature in the oven is not stable and uniform. Since the calculation is based on the comparison of known mass (20 μg.) of free pregnanediol, the relative change of molecular weight following acetylation is compensated and thus the need for correction for molecular weight change is unnecessary.

REFERENCES

Chattoraj, S. C., and Wotiz, H. H.: Fertil. Steril., *18*:342, 1967.
Scommegna, A., and Chattoraj, S. C.: Obstet. Gynec., *32*:277, 1968.

NORMAL VALUES AND CLINICAL SIGNIFICANCE

Pregnanediol is a major metabolite of progesterone, the hormone of the corpus luteum and the placenta. Thus, the utility of pregnanediol determinations lies mainly in gynecological disorders and in cases of abnormal pregnancies. In children, little or no pregnanediol can be found in the urine, and in men it is derived mainly from progesterone and 11-deoxycorticosterone from the adrenals, the amount never exceeding more than 1 mg./24 hours.

Pregnanediol excretion during normal menstrual cycles. The normal menstrual cycle of 28 days duration may be divided broadly into two phases, the follicular phase and the luteal phase (progestational or secretory phase). The follicular phase lasts from the onset of bleeding to the rupture of the follicle and discharge of the ovum; the luteal phase begins with ovulation and formation of the corpus luteum, and continues until bleeding starts again. During the follicular phase the level of urinary pregnanediol excretion remains near 1 mg./24 hours or slightly higher. During the luteal phase the excretion of pregnanediol increases gradually to its maximum (luteal peak) between the twenty-first and twenty-fourth days, depending on the interval of the menstrual cycle. The value declines three to four days before menstruation. The values of urinary pregnanediol during the normal menstrual cycle are as follows:

Proliferative (follicular) phase	0.10–1.26 mg./24 hr.
Luteal phase	1.17–9.50 mg./24 hr.

In gynecological disorders the estimation of urinary pregnanediol provides a means of investigating ovarian function during the cycle. If ovulation does not occur, there is no cyclical rise of pregnanediol excretion.

Pregnanediol excretion during pregnancy. During the first three months of pregnancy, the amount of pregnanediol excreted is comparable to that found during the luteal phase of the menstrual cycle, although the values may be a little higher. When the placenta increases the secretion of progesterone, there is a steady rise of pregnanediol excretion until about the thirty-second week of gestation when the excretion curves level off. Within 24 hours of delivery, excretion values start to drop, reaching nonpregnancy levels in four or five days.

In pregnant women there are wide individual variations in the excretion of pregnanediol. The following is the normal range of excretion at different weeks of gestation:

Weeks	*Range (mg./24 hours)*
10–12	5–15
12–18	5–25
18–24	13–33
24–28	20–42
28–32	27–47

Generally, in threatened abortions, toxemias of pregnancy, and intra-uterine fetal deaths involving the placenta, values that are lower than normal at corresponding weeks of gestation are obtained. For an authoritative discussion and information on the excretion of pregnanediol in abnormal pregnancies and other gynecological disorders the reader is referred to the monograph by Loraine and Bell.[46]

ESTROGENS

Estrogens, like progesterone, are female sex hormones. They are responsible for the development and maintenance of the female sex organs and secondary sex characteristics. They also participate in the regulation of the menstrual cycle and in the maintenance of pregnancy. In women, estrogen is secreted mainly by the ovarian follicles and, during pregnancy, by the placenta. Adrenals and testes are also believed to secrete estrogens, but only minute quantities. Until recently, only three estrogens, estrone, estradiol-17β, and estriol were known to be produced in the body. Estrone and estradiol-17β, which are interconvertible, were considered to be the secretory products of the ovary, and estriol was considered to be the ultimate end product of both (see Fig. 9-18). While estrone and estradiol-17β are still considered to be the only estrogens secreted by the ovary, their metabolism appears to follow a more complicated biochemical sequence, giving rise to more than a dozen metabolites. In Figure 9-18 the names and structural formulas of some of the newly discovered estrogens are shown. It should be noted, however, that the three classical estrogens, i.e., estrone, estradiol-17β, and estriol, are quantitatively most important, and the discussion will mainly be confined to these estrogens.

Among the three major estrogens, estradiol-17β is by far the most potent and is considered to be the true ovarian hormone in women. Structurally, estrogens are derivatives of the parent hydrocarbon estrane. They consist of 18 carbon atoms and possess the following characteristic features: (1) the aromatic nature of ring A, (2) the presence of a ketone (e.g., estrone) or hydroxyl group (e.g., estradiol) at C-17 and

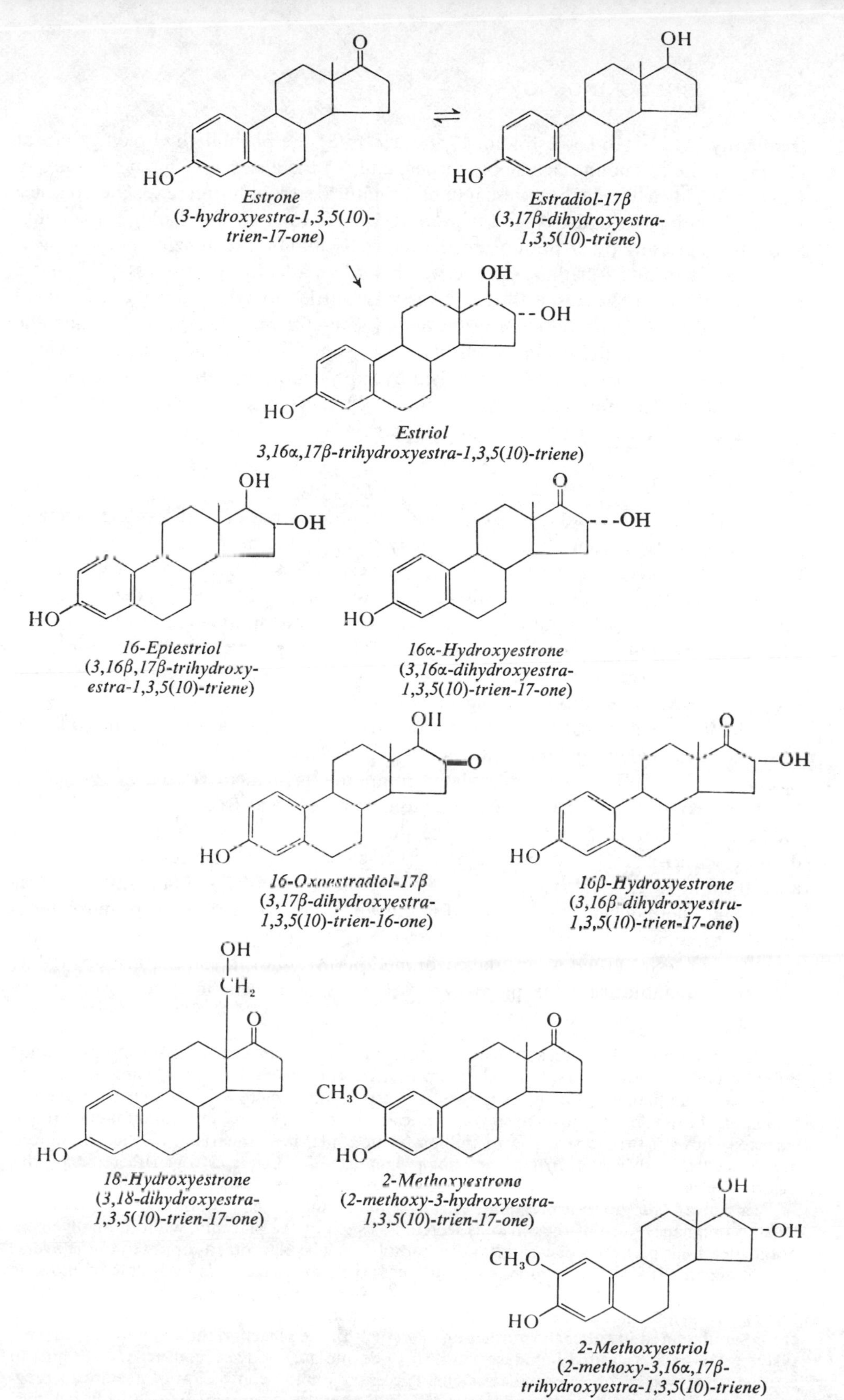

Figure 9-18. Structural formula of the three classical estrogens (estrone, estradiol-17β, and estriol) and some of the newly discovered estrogens.

frequently at C-16 (16-ketoestradiol-17β, estriol), (3) the phenolic hydroxyl group at C-3 giving the compounds acidic properties, and (4) the absence of the methyl group (carbon 19) at C-10, which is present in other natural steroid hormones. The presence of the phenolic ring and oxygen function at C-17 is essential for biological activity. Substituents at any other position in the molecule diminish feminizing potency. For example, estriol and 2-methoxyestrone which contain a hydroxyl group at C-16 and a methoxy group at C-2 respectively possess very little biological activity. Chemical names of estrogens are derived by means of standard nomenclature, denoting the position of unsaturation and substituents based on the estrane structure. Thus, the chemical nomenclature of estrone is 3-hydroxy-estra-1,3,5(10)-trien-17-one. The suffixes and prefixes for various substituents and the systems for chemical names have been described in the first part of this chapter.

Biogenesis

In vivo and *in vitro* studies with radioactive compounds have shown that acetate, cholesterol, progesterone, and, interestingly enough, the male sex hormone, testosterone, can all serve as precursors for the synthesis of estrogens. It has been demonstrated, furthermore, that these precursors do not operate independently in the biosynthesis but constitute a single pathway leading to the formation of estrogens. The biosynthesis of progesterone and testosterone have been described in the appropriate sections. Here, only the biochemical transformation of testosterone to estrogen (i.e., the formation of aromatic ring A) will be discussed.

The first biochemical event in the aromatization of testosterone is hydroxylation of the C-19 methyl group by the enzyme 19-hydroxylase to produce 19-hydroxytestosterone (see Fig. 9-19). The hydroxylated compound is further oxidized by 19-oxidase to 19-oxotestosterone. The C-19 carbon atom and the C-1 hydrogen atom of this intermediate are eliminated as formaldehyde through the action of the enzyme 10, 19-desmolase leaving a C-1(10) double bond (i.e., a double bond between carbon atoms 1 and 10). The resulting 3-oxo-androst-1(10),4-diene aromatizes spontaneously to estradiol-17β. The biochemical sequence for the formation of estrone from androstenedione is the same as that described for estradiol-17β. It should be noted here that the reactions of testosterone to androstenedione and estradiol to estrone are reversible. However, quantitatively the pathway involving the conversion of testosterone to estradiol is more significant in the ovary.

It has been shown in recent years that the biosynthesis of estrogens during pregnancy differs qualitatively and quantitatively from that in women who are not pregnant. During pregnancy, the major source of estrogens is the placenta, whereas in women who are not pregnant, the ovaries are the main site of synthesis. During pregnancy the amount of estrogen excreted shows a progressive rise to milligram amounts, in comparison to the microgram quantities excreted in women who are not pregnant. The major estrogen secreted by the ovary is estradiol, whereas the major product secreted by the placenta is estriol.

In women who are not pregnant, estriol is the metabolic end product of estradiol-17β. During pregnancy most of the estriol is secreted as such by the placenta, without involvement of the metabolic pathways of estrone and estradiol-17β. The placenta, as opposed to the ovary, cannot accomplish *de novo* synthesis of estrogens; i.e., the placenta is incapable of utilizing simple precursors such as acetate, cholesterol, or progesterone. Consequently, it must depend for its estrogen formation on the immediate precursors (e.g., testosterone, androstenedione, etc.) manufactured in either the mother or the fetus. In the placenta, the aromatizing enzyme systems (19-hydroxylase, 19-oxidase, and 10,19-desmolase) actively convert C_{19} steroids to estrogens. However, other enzyme systems necessary for the conversion of acetate to pregnenolone and progesterone to androgens (C_{19} compounds) are absent or relatively inactive.

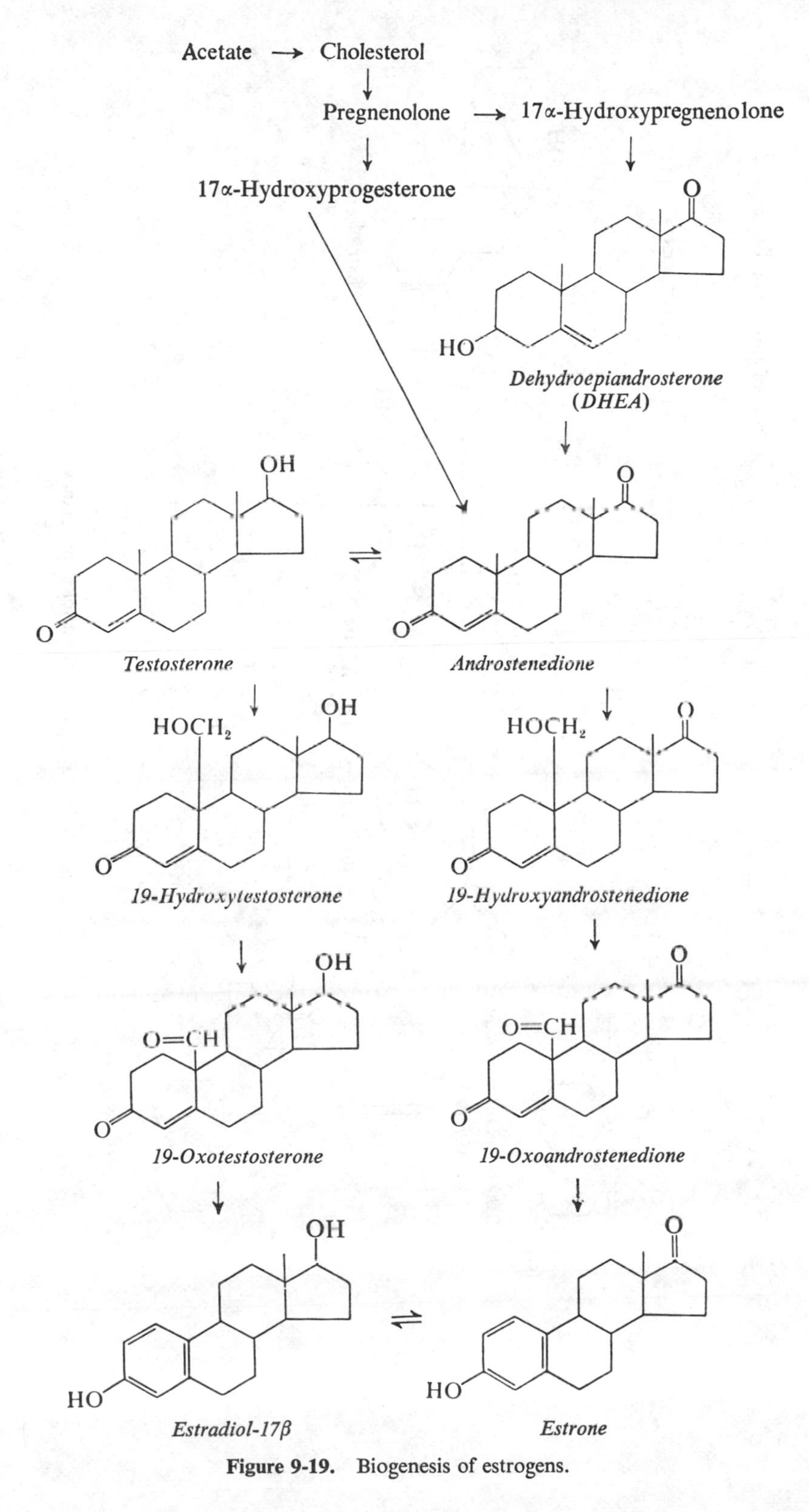

Figure 9-19. Biogenesis of estrogens.

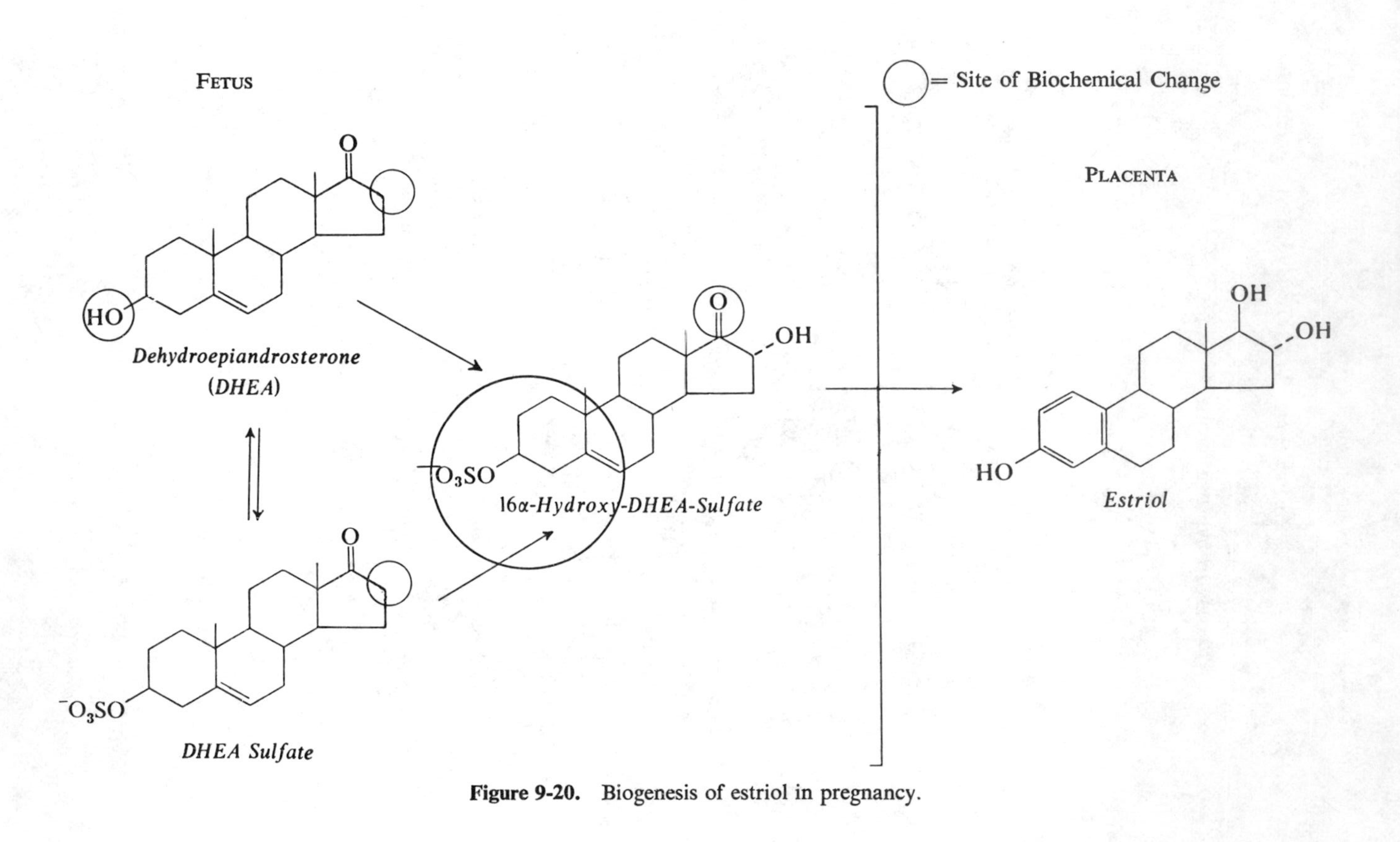

Figure 9-20. Biogenesis of estriol in pregnancy.

According to experimental and clinical evidence, the C_{19} steroid dehydroepiandrosterone, which is secreted by the fetal adrenals as the sulfate conjugate, serves quantitatively as the most important precursor for placental estriol synthesis (Fig. 9-20). In addition to providing the supply of the precursor, the fetus also performs an important enzymatic reaction, the α-hydroxylation at C-16, which is a necessary step for estriol formation. Whereas the enzyme 16-α hydroxylase is very active in fetal tissues (liver, adrenals) the placenta is practically devoid of this enzyme. Therefore, an examination of estriol biosynthesis in pregnancy reveals that both the fetus and the placenta join in a concerted action. The fetus supplies the precursors and carries out the 16α-hydroxylation, and the placenta aromatizes the C_{19} neutral steroids. A defect in either will be reflected in a decreased production of estriol. For this reason, estriol excretion in pregnancy is used as an indicator of feto-placental status.

Metabolism

With the discovery of new estrogens, the metabolic pathway has become far more complex than the hitherto known simple reaction sequence: estradiol ⇄ estrone → estriol. An extensive discussion of the new metabolic pathway is beyond the scope of this book. The reader is referred to the excellent review by Breuer[13] and the monograph by Dorfman and Ungar.[24] It should be noted in this connection that the significance of all the newly discovered metabolites in health and disease has not yet been clarified. Chemical assays of the three classical estrogens, estrone, estradiol-17β, and estriol, is adequate in most situations.

Figure 9-21 summarizes the present state of knowledge concerning the metabolism of estrone and estradiol-17β in the human body. As for other steroid metabolism, the liver is the primary site for the inactivation of estrogens. The main biological chemical reactions are hydroxylation, dehydrogenation and hydrogenation, methylation, and conjugation. Hydroxylation is one of the most important reactions in the metabolism of estrogens. In fact, the formation of estriol from estradiol-17β and estrone is actually initiated by the enzyme, 16α-hydroxylase (Fig. 9-21). *In vivo* and *in vitro* experiments have indicated the presence of

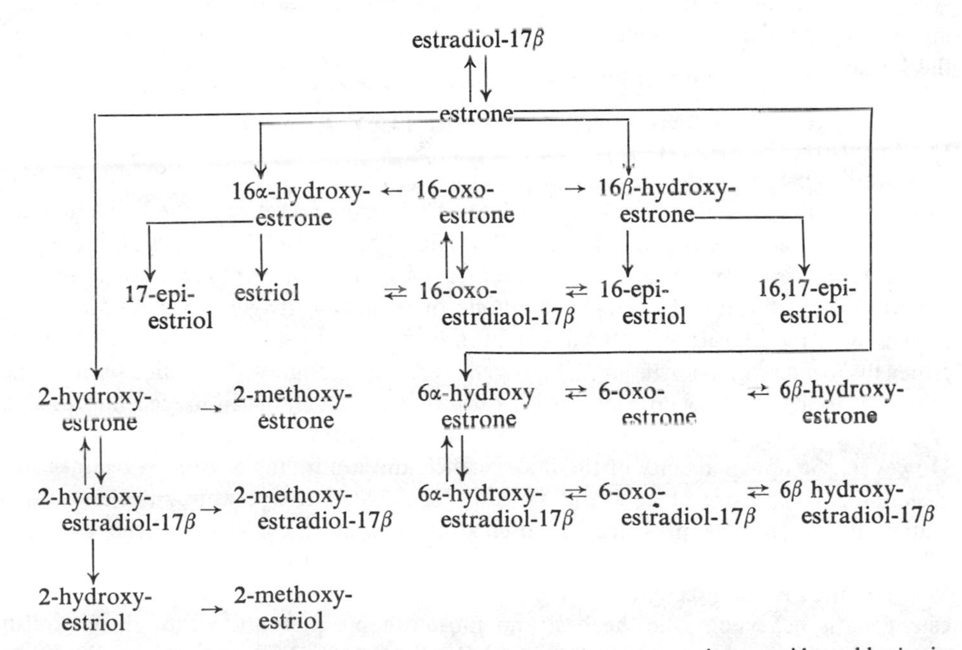

Figure 9-21. Metabolism of estradiol-17β and estrone in the human organism as evidenced by *in vivo* and *in vitro* experiments.

several enzymes necessary for hydroxylation: 16β-hydroxylase for the formation of 16-epiestriol and 16,17-epiestriol, 6α-hydroxylase for 6α-hydroxyestrone, and 2-hydroxylase for 2-hydroxyestrone. The interconversion of estrone and estradiol is the most important and best example of the *dehydrogenation* (estradiol 17β → estrone) and *hydrogenation* (estrone → estradiol 17β) reaction catalyzed by the enzyme 17β-dehydrogenase. In addition, the enzymes 16α- and 16β-dehydrogenases and 6α- and 6β-dehydrogenases, which are needed to carry out the oxidation and reduction of the oxo-(keto) group at the respective carbon atoms, have been shown to be present in the liver. *Methylation* is considered to be a unique reaction of estrogen metabolism. The formation of 2 methoxy-estrogens is accomplished in two separate stages, namely, hydroxylation followed by methylation (Fig. 9-21). Methylation is catalyzed by the enzyme O-methyl transferase while S-adenosylmethionine (SAME) acts as a methyl donor.

The following is a brief outline of the salient features of estrogen metabolism. Estradiol-17β is rapidly oxidized to estrone, which, in turn, is hydroxylated to form 16α-hydroxyestrone. The subsequent reduction of this compound to estriol (see Fig. 9-21) appears to be the most important reaction quantitatively. It is to be noted further, that the formation of estriol is achieved exclusively via this pathway; direct hydroxylation of estradiol does not appear to contribute to its formation significantly. It should be recalled, however, that in pregnancy estriol can be synthesized directly from 16α-hydroxylated dehydroepiandrosterone arising from the fetus followed by subsequent aromatization in the placenta (see Fig. 9-20). The next important sequence of reactions is the oxidation of estriol to 16-oxo(-keto) estradiol-17β as well as the formation of 16β-hydroxyestrone with the subsequent reduction to 16-epiestriol. There is evidence of the formation of 17-epiestriol (17α-hydroxy group) and 16,17 epiestriol (16β-hydroxy and 17α-hydroxy groups) from their respective precursors (16α-hydroxyestrone and 16β-hydroxyestrone). From a quantitative point of view these two epimeric estriols are not considered to be significant metabolites. *Conjugation* is the final step in the metabolic processes. Estrogens are conjugated with glucuronic and/or sulfuric acid. It should be recalled that the conjugation imparts more water solubility to these lipids, allowing them to be eliminated rapidly though the kidney.

DETERMINATION OF URINARY ESTROGENS

Studies based on the administration of tracer doses of radioactive estrogens have shown that the urine is the principal route of estrogen excretion; the urinary determination of three estrogens—namely, estrone, estradiol-17β, and estriol—renders adequate information about the endogenous production of estrogenic hormones. In recent years a variety of chemical methods of urinary estrogen determination have appeared in the literature.[46] In most methods, the final estimation is carried out by either fluorimetry or colorimetry. Estrogens form an orange-yellow color with intense greenish fluorescence when heated with sulfuric acid. Such acid-induced fluorescence is extremely sensitive and permits detection of as little as 0.005 μg. of estrogens. However, the nonspecificity of the fluorescence and a number of other variables such as length of exposure of acid and steroids to elevated temperature, the amount of water added, and the presence or absence of solvents have always been causes of difficulty in utilizing this kind of reaction for routine analysis.

The method of Preedy and Aitken[61] involving fluorimetric estimation of urinary estrogens is believed to be the best and most reliable method, although its routine clinical application has been very limited. Of all the colorimetric reactions, the Kober reaction is the best known and most accepted for the quantitative determination of estrogens. The original method proposed by Kober[41] consisted of three stages: (1) heating with concentrated sulfuric acid, (2) reheating after dilution with water, and (3) addition of phenol to decrease the fluorescence and increase the formation of the characteristic red color of estrogens. These reaction conditions, however, did not give stable and reproducible colors, and thus Brown[15] suggested modifications which result in reproducible and maximum color formation.

The specific groups in the steroid molecule required for the development of the pink color in the Kober reaction are the phenolic or phenolic ether groups at C-3 of ring A and an intact ring D oxygenated at C-17. Thus the reaction is specific for estrogens. If the specimen, however, contains urinary contaminants, a nonspecific yellow-brown color, produced during the reaction sequence, is superimposed on the pink color caused by estrogens. For this reason, extensive purification of the urinary extract prior to color reaction, in addition to the background color correction, becomes a necessity for reliable measurement of estrogens.

The modification of the Kober reaction by Brown[15] is presently the most widely applied method. This method has the distinction of being the first published procedure that is sensitive and specific enough to measure individually estrone, estradiol-17β, and estriol in the urine of nonpregnant women, in whom the excretion values lie in the microgram range. The main steps in this method are: acid hydrolysis, extraction, separation into neutral and phenolic fractions, formation of 3-methyl ethers of the three estrogens, purification and separation by alumina chromatography, and final quantitation by the Kober color reaction using the Allen correction. Although Brown's method has been used successfully in many laboratories to obtain valuable information regarding the urinary excretion of estrogens in normal menstrual cycles and in menstrual disorders,[46] a method which is simple, rapid, and adequately sensitive to become suitable for routine clinical use is still to be achieved.

Recently, Ittrich[35] has introduced an ingenious method for avoiding the nonspecific portions of chromogens. The products of the second stage in the Kober reaction are diluted with water to an acid concentration of 20 to 30 per cent and are shaken in the cold with chloroform containing 2 per cent p-nitrophenol and 1 per cent ethanol. The color resulting from the estrogens is extracted into the chloroform layer, whereas the yellow-brown nonspecific color remains in the acid layer. The novelty of this modification lies in the fact that the color complex in chloroform can be measured either colorimetrically or fluorimetrically, since it emits an intense yellowish-green fluorescence when excited with visible light. When it is measured colorimetrically, an amount equivalent to 0.2 μg. of estrogen can be determined. Fluorimetrically, as little as 0.005 μg. of estrogens can be detected. This procedure has been incorporated in various methods for the estimation of estrogens in urine and blood.[46] Indeed, following the "Ittrich extraction," the fluorimetric measurement of total estrogens in crude urinary extracts without chromatographic purification has been shown to be highly suitable and fairly specific for routine analysis.[34,16,68] In the following paragraphs a method based on such a principle for urinary determination of total estrogens in women who are not pregnant is described.

Determination of Urinary Total Estrogens[34,16,68]

PRINCIPLE

The following basic steps are involved: acid hydrolysis, extraction with diethyl ether, washing the ether extract with carbonate buffer, separation into phenolic and neutral steroids by partition between sodium hydroxide solution and the organic extract, re-extraction of phenolic steroids with diethyl ether from aqueous solution, development of the Kober color followed by extraction of the color complex with chloroform containing 2 per cent p-nitrophenol, and measurement of the fluorescence of the organic phase in a fluorimeter using 530 nm. as the wave length for excitation and 550 nm. as that for emission. The amount of estrogens is calculated by comparing the intensity of fluorescence of the sample with that obtained from standard mixtures (estrone, estradiol-17β, and estriol) of known concentration.

REAGENTS

All reagents should be of A. R. grade.

1. Hydrochloric acid, concentrated.
2. Diethyl ether, freshly distilled in all glass apparatus.
3. Sodium carbonate buffer, pH 10.5, prepared by mixing 150 ml. of 20 per cent (w/v) sodium hydroxide solution with 1 liter of 8 per cent (w/v) sodium bicarbonate.
4. Light petroleum (B.P. 40 to 60°C.), redistilled in all glass apparatus.
5. Sodium bicarbonate, 8 per cent (w/v) solution.
6. Sodium sulfate, anhydrous.
7. Ethanol, absolute, purified (see Corticosteroid estimation).
8. Hydroquinone, recrystallized from ethanol, 4 per cent (w/v) solution in ethanol.
9. p-Nitrophenol, recrystallized from benzene; dissolve 2 gm. of p-nitrophenol in 98 ml. of chloroform containing 1 per cent (v/v) ethanol.
10. Sulfuric acid, concentrated.
11. Chloroform, freshly distilled in all glass apparatus.
12. Standard solution: Weigh 8 mg. each of estrone and estriol and 4 mg. of estradiol-17β and dissolve in 100 ml. absolute ethanol. Dilute 0.1 ml. of this stock solution with absolute ethanol to 100 ml. in a volumetric flask. (This working standard contains 0.008 μg. of estrone and estriol and 0.004 μg. estradiol-17β/0.1 ml.) The stock solution should be stored at 4°C.

PROCEDURE

Collect the specimen of urine, and test the urine for glucose as described for the total 17-ketogenic steroids (see page 506). If the glucose concentration is more than 0.5 per cent, follow the procedure as described there, using 1 per cent of total volume of urine. Dilute the glucose-free residue to 25 ml. with distilled water and proceed as follows.

Hydrolysis and extraction. Transfer 1 per cent of total volume of urine into a 250 ml. round-bottom flask, and dilute to 25 ml. with distilled water. Add several glass beads (to prevent bumping) and heat to boiling under a reflux condenser. Add 5 ml. of concentrated hydrochloric acid through the condenser and continue boiling for 30 minutes. Cool the flask rapidly under running tap water and transfer the contents to a separatory funnel. Extract the hydrolyzed urine once with 25 ml. of ether and twice with half the volume (12.5 ml.) of ether. Then shake the combined ether layers with 10 ml. of sodium carbonate buffer (pH 10.5) solution. Discard the aqueous layer.

Separation of phenolic steroids. Add 50 ml. of petroleum ether to the ether extract in the separatory funnel. Extract the organic solvent mixture two times with 25 ml. of 1 *N* NaOH, collecting the alkali layer in an Erlenmeyer flask. Partly neutralize the alkaline solution by adding solid $NaHCO_3$ in portions until the pH is 10. Transfer the aqueous solution into a separatory funnel. Extract the solution three times with ether, once with an equal volume (50 ml.) and twice with half the volume (25 ml.). Then shake the combined ether layers with 20 ml. of 8 per cent sodium bicarbonate. Discard the aqueous layer. Wash the ether extract with 10 ml. of distilled water and drain off the water as completely as possible. Transfer the ether extract to an Erlenmeyer flask containing 5 gm. of anhydrous sodium sulfate. Rinse the separatory funnel with a few ml. of fresh ether and add to the flask. Filter the ether extract through a Whatman No. 1 filter paper into a 250 ml. round bottom flask. Evaporate the ether to dryness at 30°C. under reduced pressure.

Fluorimetry. Dissolve the residue in the flask in 5 ml. of absolute ethanol. Transfer 2 ml. aliquots to two glass-stoppered tubes. Prepare standard tubes (in duplicate) containing 0.1 ml. (= 0.02 μg. total steroids; estrone = 0.008 μg., estradiol = 0.004 μg., and estriol = 0.008 μg.) of working standard solution and a blank tube containing pure ethanol. Add 0.5 ml. of 4 per cent hydroquinone solution to each tube, and evaporate to dryness under nitrogen in a water bath at 50°C. Place the tubes in an ice-water bath. Add 0.4 ml. of distilled water and 0.75 ml. of concentrated sulfuric acid. Stopper the tubes and heat in a boiling water bath for 40 minutes with occasional shaking. Place the tubes in an ice bath. Add to the tubes 1.5 ml. of distilled water and mix thoroughly. Allow the tubes to stand in ice not less than 5 minutes and not more than 25 minutes. Add 2.5 ml. of 2 per cent p-nitrophenol in chloroform. Stopper the tubes and shake vigorously for 30 seconds. Centrifuge the tubes for 3 minutes at 2500 r.p.m. Aspirate off the upper layer. Transfer the organic phase into the cuvets. Read the fluorescence at 550 nm. following excitation at 530 nm. Set the instrument to zero absorbance with the blank, and read the standards and the samples.

CALCULATION

If one per cent of the total volume of urine is used, then

$$\mu g.\text{ of total estrogens per 24 hrs.} = \frac{\text{Reading of sample}}{\text{Reading of standard}} \times 0.02 \times 5/2 \times 100$$

COMMENTS

Since the bulk of the estrogens are excreted as water-soluble conjugates of glucuronic and sulfuric acids, hydrolysis is necessary to allow extraction of the steroids with an organic solvent such as ether. As a matter of expediency, acid hydrolysis is generally used. However, the presence of glucose in the urine causes destruction of estrogens during acid hydrolysis, which in effect will yield low values. Thus, the test for glucose must be performed on each urine specimen.

Washing the specimen with sodium carbonate buffer (pH 10.5) removes strongly acidic components. As described before, estrogens are slightly acidic in nature because of the presence of a phenolic hydroxyl group at C-3. This acidic property has been utilized for the separation of phenolic estrogens and neutral steroids. Shaking the organic solvent with sodium hydroxide solution moves the estrogens into the alkali layer while the neutral steroids (17-ketosteroids, pregnanediol, corticosteroids, etc.) stay in the organic phase. Petroleum ether is added to achieve better recovery of some of the estrogens (estrone, estradiol-17β) which would otherwise be left in some quantity in the ether layer when partitioned with sodium hydroxide solution. Since estrogens are quite soluble at a pH above 11, the adjustment of the pH to below 10.5 is very important for quantitative re-extraction of estrogens with ether.

The addition of hydroquinone protects the estrogens from oxidation and facilitates the formation of the Kober-color complex. The directions for addition of exact aliquots of different reagents for color development, the duration of heating, and extraction of the color complex with p-nitrophenol solution after the specified length of time should be followed as closely as possible. The tubes should be kept in ice at all times. The intensity of fluorescence decreases with time; thus, the fluorimetric reading should be completed within half an hour after extraction of the color complex. Since the relative intensities of fluorescence of the three estrogens, estrone, estradiol-17β, and estriol, are different, determination of the total estrogen content without a separation of the individual estrogens may result in some error. The use of a standard solution of three estrogens mixed in a ratio generally excreted in the urine of a normal nonpregnant woman (estrone, estradiol-17β, estriol, 2:1:2) reduces the possibility of such error.

NORMAL VALUES AND CLINICAL SIGNIFICANCE

The excretion of estrogens in children is very low, being generally less than 1 μg. per 24 hours. In men, a constant amount of estrogens excreted derive from the adrenals and probably also from the testes. The average value is approximately 11 μg. with a range of from 5 to 18 μg. per 24 hours.

Estrogen excretion during the normal menstrual cycle. Although there is wide individual variation, the excretion of estrogens in normally menstruating women has a definite pattern. During the first seven to ten days of the cycle the excretion is generally low. The estrogen level then rises steadily to a well-defined peak on or about the thirteenth day. This peak is called the *ovulatory peak* and is assumed to coincide with the rupture of the follicle. Following the peak excretion of estrogens, there is a rapid fall succeeded by another rise around the twenty-first day of the cycle (*luteal peak*). The excretion value then declines gradually to the lowest level at the onset of the next menstrual cycle. Brown[15] has studied extensively the excretion of estrone, estradiol-17β, and estriol during normal menstrual cycles. The following are the excretion values of total estrogens, computed from his data:

Onset of menstruation:	4–25 μg. (mean 13 μg.)/24 hours
Ovulation peak:	28–99 μg. (mean 56 μg.)/24 hours
Luteal peak:	22–105 μg. (mean 43 μg.)/24 hours
Menopausal women:	1.4–19.6 μg. (mean 6.4 μg.)/24 hours

Decreased excretion of estrogens is found in women with amenorrhea due to agenesis of the ovaries, primary ovarian malfunction, dysfunction of the pituitary, and other metabolic disorders.

Increased excretion of estrogens is generally associated with ovarian tumors (solid or cystic), hyperovarianism in females, and testicular tumors in males.

REFERENCES

Ittrich, G.: Acta Endocr. *35*:34, 1960.
Scholler, R., Leymarie, P., Heron, M., and Jayle, M. F.: Acta Endocr. *52* (Suppl. 107), 1966.
Brown, J. B., Macnaughton, C., Smith, M. A., and Smyth, B.: J. Endocr. *40*:175, 1968.

Determination of Urinary Estriol in Pregnancy

Methodological difficulties have always been a stumbling block in the clinical utilization of any investigational procedure. This fact has also been true for estriol assays, and only recently have chemical procedures been simplified for practical routine use. However, some degree of purification is still necessary, and any short cut in methodology is by necessity accompanied by the loss of one or more reliable criteria of a method. The introduction of gas chromatography for the analysis of estriol, because of its speed, precision, and sensitivity, has opened new vistas for clinical application. It should be emphasized here that the speed of analysis is of the utmost importance in the clinical management of obstetrical patients. A variety of methods for gas chromatographic analysis of estriol have been reported.[98] The method which in the experience of the author has been found very suitable for clinical use is described below.

Gas Chromatographic Analysis of Estriol[69,97]

PRINCIPLE

The basic steps in the procedure include acid hydrolysis, extraction, carbonate buffer wash, separation of phenolic and neutral steroids by partition between sodium hydroxide solution and ether extract, re-extraction of estrogens from alkaline solution, acetylation, and injection into a gas chromatograph. As in pregnanediol analysis, quantitation is carried out by comparing the peak height ratio of known concentration of standard and internal standard with that obtained from the sample.

REAGENTS

All reagents should be of A. R. grade. Reagents 1 to 7 are the same as described for the estimation of total estrogens (p. 546).

8. Pyridine, redistilled in all glass apparatus under anhydrous conditions.
9. Acetic anhydride, redistilled as above.
10. Estriol standard, 20 mg. estriol (gas chromatographic purity) dissolved in 100 ml. absolute ethanol (purified).
11. Internal standard, 20 mg. cholesterol (gas chromatographic purity) dissolved in 100 ml. absolute ethanol.

PROCEDURE

Collection of urine specimen, test for glucose, and removal of glucose from the urine are followed as described in the procedure for determination of 17-ketogenic steroids (page 506). Use 2.5 per cent of the total volume of urine obtained in the first and second trimesters of pregnancy, and 1 per cent of the total volume obtained in the last trimester.

Hydrolysis and extraction. The procedure is the same as described for the determination of total estrogens.

If the urine aliquot to be used is more than 25 ml., dilute to 50 ml. with distilled water and use double the amount of all reagents described before.

Separation of phenolic steroids. Follow the procedure as described for determination of total estrogens, including the step of sodium bicarbonate and water wash. Add 20 μg. of cholesterol (0.1 ml. of 20 mg./100 ml. ethanol) to the ether extract. Transfer the ether extract to an Erlenmeyer flask containing 5 g. of anhydrous sodium sulfate. Filter the ether extract through a Whatman No. 1 filter paper into a 250 ml. round bottom flask. Evaporate the ether to dryness at 30°C. under reduced pressure. Dissolve the residue in 5 ml. of acetone. Transfer the solution with a

Pasteur pipet into a 15 ml. centrifuge tube. Rinse the flask two more times with 2 ml. of fresh acetone and add to the centrifuge tube. Evaporate the acetone to dryness at 50°C. under a stream of nitrogen. At this point prepare a standard tube by adding 20 μg. each of estriol and cholesterol (0.1 ml. of 20 mg./100 ml. standard solution) to a centrifuge tube. Evaporate the ethanol to dryness.

Acetylation. To the dried residue add 0.1 ml. of redistilled anhydrous pyridine and 0.5 ml. of distilled acetic anhydride. Shake gently to dissolve the contents of the tube. Stopper. Incubate for one hour at 68°C., or let it stand overnight at room temperature. Treat the tube containing the standard and the internal standard in a similar manner. Evaporate the acetylating reagent mixture to dryness at 60°C. under a stream of nitrogen.

Gas chromatography and calculation. Follow the same experimental conditions and the procedure as described for the gas chromatographic analysis of pregnanediol.

COMMENTS

For the remarks regarding the acid hydrolysis, extraction, separation of phenolic steroids, see the method for determination of total estrogens. For relevant discussions on the gas chromatographic procedure, refer to the sections *Comments on gas chromatographic analysis of pregnanediol* (p. 536) and *Gas chromatography of steroids* (p. 491).

TABLE 9-4. *Normal Values for Estriol Excretion during Pregnancy**

Weeks (of gestation)	*Average (mg./24 hrs.)*	*Range (mg./24 hrs.)*
19	3.45	1.90–5.00
20	5.00	2.80–7.25
21	5.40	3.00–8.00
22	6.00	3.30–8.60
23	6.40	3.60–9.50
24	7.00	3.80–10.50
25	7.50	4.20–11.25
26	8.20	4.55–12.25
27	9.00	5.00–13.50
28	9.85	5.40–14.50
29	10.50	5.80–16.00
30	11.50	6.30–17.50
31	12.75	6.90–19.00
32	13.50	7.50–20.80
33	14.80	8.15–22.50
34	16.50	8.80–24.60
35	17.50	9.70–27.00
36	19.00	10.50–29.00
37	21.00	11.50–32.00
38	22.80	12.40–35.00
39	24.60	13.50–38.00
40	27.00	14.50–41.20
41	29.00	16.00–44.00
42	32.00	17.25–49.00

*Data from Scommegna, A., and Chattoraj, S. C.: Am. J. Obstet. and Gynec., *99*:1087, 1967.

NORMAL VALUES AND CLINICAL SIGNIFICANCE

In normal pregnancy, the values of urinary estriol excretion increase with each week of gestation. The range and the average excretion values (mg./24 hours) from 19 to 42 weeks are given in Table 9-4. Because of individual variation, the range of normal values is very wide. It should be noted in this connection that a clinical decision should not be formulated on the basis of a single, isolated determination of estriol excretion. Serial assays in a pregnancy with complications are much more reliable and meaningful, since they allow study of the trend of estriol production by the fetoplacental unit, reflecting its functional status. In *pathologic* pregnancies, such as those complicated by toxemia, hypertension, diabetes mellitus, and postmaturity, where the fetus may be in jeopardy, estriol determination can be used by obstetricians as an aid in deciding whether to allow a pregnancy to continue or to terminate it to remove the fetus from a hostile intrauterine environment. It should be recalled that the biogenesis of estriol in pregnancy involves both the fetus and the placenta (see Figure 9-20), and any anomaly in the function of either the fetus or the placenta is reflected by the decreasing maternal excretion of estriol.

REFERENCES

Wotiz, H. H., and Chattoraj, S. C.: *In* Gas Chromatography of Steroids in Biological Fluids. Lipsett, M. B. (Ed.), New York, Plenum Press, 1965, p. 195.

Scommegna, A., and Chattoraj, S. C.: Obstet. Gynec. *32*:277, 1968.

Protein Hormones

The hormones secreted by the pituitary, the pancreas, and the parathyroids are all proteins in nature. The number and the type of hormones secreted by each gland are enumerated in Table 9-1. These hormones differ from each other not only in their physiological action but also in their chemical structure, ranging from a simple octapeptide (e.g., oxytocin) to complex protein molecules (e.g., FSH, LH, TSH). Their assay involves associated problems. The hormones of low molecular weight, such as steroids and catecholamines, possess distinctive chemical groups toward which a specific color reaction may be directed as a means of assay (see chemical methods for the determination of 17-ketosteroids, catecholamines, etc.). In the case of protein hormones, no such singularity in chemical composition exists. Consequently, the direct chemical determination of these hormones in biological fluids such as blood and urine cannot be made. In the absence of any chemical assay procedures, determinations of protein hormones are limited to biological and immunological techniques. The basic principles and the prerequisites of these two techniques are discussed briefly in the following paragraphs.

BIOASSAY TECHNIQUES

In this type of procedure a concentrate of the hormone to be measured is injected into suitably prepared animals to elicit a specific physiological response, which is then compared to the response produced by a reference standard preparation. However, until very recently there was no recognized standard preparation available for comparative assay of the protein hormones. The results were therefore expressed in terms of arbitrary animal units. A unit is defined as the amount of a specific hormone required to produce a specific effect. Thus a "mouse uterine unit" is the dose of total gonadotrophins (FSH and LH) necessary to cause a 100 per cent increase in uterine weight of immature mice. Further examples of arbitrary animal units are "rat ovarian unit" for human pituitary gonadotrophin (HPG), "rat ovarian hyperemia unit" for human chorionic gonadotrophin (HCG), and "Junkmann-Schoeller unit" for thyroid stimulating hormone (TSH).

Because of the nonreproducibility of quantitative bioassay results in different laboratories, it soon became apparent that the degree of response to a minimum dose of hormones differs not only with the animal colony but also between individual animals in the same colony maintained under rigidly controlled environmental conditions. It was, therefore, necessary to establish some kind of international standard or international reference preparation (IRP) with specified biological activity which could be utilized throughout the world for the purpose of comparing bioassay results. In recent years international standards have become available for the following protein hormones: insulin, TSH, ACTH, prolactin, HCG, posterior pituitary hormones, HPG, and human growth hormone (HGH). The use of such international standards for the bioassay technique has greatly mitigated ambiguity and confusion of bioassay results.

Since the bioassay technique is based on the physiological response specific for the hormone being measured, this is the most direct means of analysis. To obtain reliable and reproducible results by this procedure, proper precautions and careful design of experiments are very important. The factors causing variation of biological response in experimental animals are numerous. Among them, the strain and the number of animals used for each experiment (four-point assay using two groups of animals for standard and two groups of animals for unknown is generally recommended), the number of injections (the same amount of hormone injected in two doses usually gives a larger response), the time of the day the assay is made (the diurnal variation in response, as for example of the adrenals to ACTH), and the medium used to administer the hormone are perhaps the most noteworthy.

Other problems exist, as well. It has been shown that in biological fluids there are substances other than the test material which may augment, decrease, or abolish the specific response. In addition, the probability of chemical change of the hormone before or after injection is always a cause of uncertainty. It is believed, however, that the establishment of a dose-response curve for the standard preparation and the test material is very important to validate the experiment. A significant deviation from parallelism between the slopes of the dose-response curves of the standard and the unknown preparations makes the results invalid. In fact, the computation of bioassay results in animal units without reference to a recognized standard has doubtful quantitative significance, even though, for clinical purposes, such practice is still in vogue (see urinary gonadotrophins). The sources of errors and statistical treatment of the bioassay data have been discussed in a recent monograph.[28]

IMMUNOASSAY TECHNIQUES

The principle of immunoassay of protein hormones depends on the fact that these hormones are antigenic, i.e. when introduced into an animal they cause the production of specific proteins known as antibodies (antihormones). Antigen and antibody combine in a specific manner to form an antigen-antibody complex, which can be detected and measured by various serological methods, such as agglutination, precipitation, complement fixation, lysis, etc. The hemagglutination inhibition, complement fixation, and radio-immunoassay are by far the most commonly used serological techniques for the qualitative and quantitative determination of the protein hormones in biological fluids. Technical details and authoritative discussion of these assay methods may be found elsewhere.[94] Following is a brief discussion of their principles and prerequisities.

Hemagglutination Inhibition Test

If a suspension of red cells treated with tannic acid is exposed to a dilute solution of protein hormone, e.g., HCG, (human chorionic gonadotrophin) the cells become coated with a layer of the hormone. If these coated cells are mixed with an antihormone serum, anti-HCG, they will be agglutinated and will settle as a mat of cells to the bottom of the tube. This phenomenon is known as *hemagglutination*. In order to use such a serological reaction for the detection and quantitation of hormones, the following procedure is employed. Blood or urine containing the hormones to be measured is first mixed with an optimum amount of specific antiserum (antihormone). The hormone in the unknown sample combines with and neutralizes the antibodies, even though no visible reaction occurs. When the neutralized antihormone serum is then mixed with hormone-coated cells, there will be no antibody left to cause hemagglutination. As a result, the cells will settle to the bottom of the tube in a "doughnut" pattern, as do normal, nonagglutinated cells (positive-test). The hemagglutination is being inhibited by the presence of the hormone in the unknown sample. The process is commonly known as *hemagglutination inhibition reaction*. An illustration of the method for the detection of HCG in urine is shown in Figure 9-22.

For the quantitative hormone assay, the sample of serum or urine is serially diluted to reach an endpoint, at which further dilution will not cause inhibition of hemagglutination. A standard of known concentration is tested in the same manner at the same time. Since the dilution of a sample that just prevents agglutination

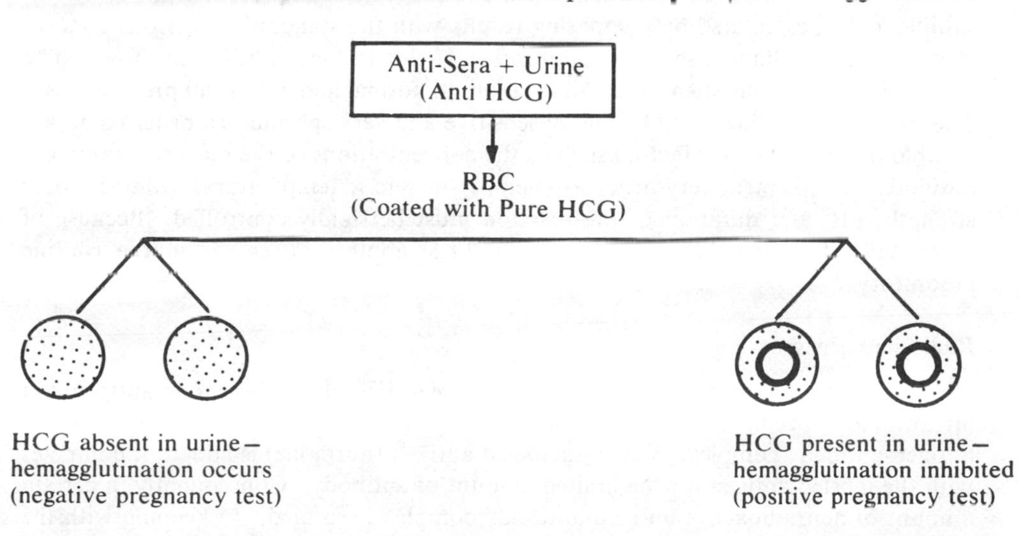

Figure 9-22

(endpoint) is expected to contain the same quantity of hormone as that of the standard preparation producing the same effect, a direct comparison gives the amount of hormone present in the unknown.

In addition to the qualitative and quantitative determinations of HCG in urine, this type of procedure has been successfully applied to the assay of levels of other hormones such as insulin, growth hormone, prolactin, and thyrotrophin in plasma. The procedure is highly sensitive, rapid, simple, inexpensive, and well suited for a routine clinical laboratory. Limitations of the method include lack of reproducibility, occasional nonspecificity, and possibility of hemolysis of the cells. However, the use of *latex particles* instead of tanned red cells as a carrier of hormone antigen overcomes the problem of hemolysis.

Complement Fixation Test

The complement present in fresh serum refers to a number of substances that participate in the hemolytic reaction. The conventional source of complement is fresh guinea pig serum. The test is based on the finding that an antigen-antibody complex *fixes* or binds complement. In other words, if complement is added to a solution containing HCG antigen and anti-HCG (test system), then it will be bound by the antigen-antibody complex and will not be available for the hemolytic reaction (*positive test*). A hemolytic *indicator system* is employed to study whether or not complement has been fixed. The indicator system consists of an appropriate mixture of sheep red blood cells and rabbit anti-sheep hemolysin (antibody against sheep erythrocyte stroma produced in rabbit). In the presence of complement and hemolysin, sheep red blood cells hemolyze. If the test system does not contain HCG antigen (for example, nonpregnancy plasma), anti-HCG alone cannot fix complement, and thus it remains available to cause hemolysis upon addition of the indicator system. This is then a *negative test*—indicating the absence of HCG antigen.

For a qualitative test, the presence or absence of hemolysis using a single undiluted sample is confirmatory. However, the quantitative analysis involves serial dilutions of the unknown sample to an endpoint showing little or no hemolysis. The degree of hemolysis is determined by the spectrophotometric measurement of the hemoglobin content of the solution. The amount of hormone present in the unknown sample is then calculated by comparing results with the standard at various concentrations run simultaneously. The application of this technique has been shown to be valuable in the measurement of HCG in serum in normal and abnormal pregnancies.[14] The method is considered to be highly sensitive and very specific. In order to obtain reliable results, however, factors such as the concentrations of the reactants (antigen, antibody, complement, erythrocytes, hemolysin, etc.), temperature, volume, ionic strength, pH, and duration of the reaction must be rigidly controlled. Because of these difficulties, the technique has found little application so far in the routine laboratory.

Radio-immunoassay

This method depends on the reaction between I^{131}-labeled hormone antigen and an antibody specific to the hormone resulting in the formation of a radioactive antigen-antibody complex. When unlabeled antigen (hormone) is added, it competes with the labeled antigen for the limited amount of antibody. Consequently, a certain amount of nonradioactive antigen-antibody complex is formed. In keeping with the principle of mass action, the ratio of radioactive antigen bound to antibody [B] to the free radio-labeled antigen [F], i.e., B:F, decreases progressively in proportion to the increasing concentration of the nonradioactive antigen [h]. The reaction based on Berson and Yalow[9] can be shown schematically as follows:

$$\begin{matrix} Ag^* \\ [F] \end{matrix} \searrow \qquad \qquad \qquad$$
$$+ \text{Ab} \rightleftharpoons \begin{matrix} \text{Ag*Ab} \\ [B] \end{matrix} + \text{AgAb}$$
$$\begin{matrix} Ag \\ [h] \end{matrix} \nearrow \qquad \qquad \qquad$$

Where Ag* = I^{131}-labeled hormone (antigen)
Ag = unlabeled hormone (antigen)
Ab = antibody specific to the hormone
[F] = concentration of free labeled hormone
[B] = concentration of labeled hormone bound to antibody
[h] = concentration of unlabeled hormone

$\frac{[B]}{[F]}$ is inversely proportional to [h]. $\frac{[B]}{[F]}$ is maximum when [h] = 0.

For carrying out a hormone assay, a standard calibration curve is first obtained as follows:

1. The requisite amounts of I^{131}-labeled protein hormone (e.g., insulin) and its antibody (e.g., anti-insulin) are determined so as to give 60 to 70 per cent binding of the hormone to the antibody.
2. Varying concentrations of the unlabeled standard preparation are then added to the system and allowed to stand for one or more days.
3. Following equilibration, the bound and free hormones are separated by a physicochemical method such as electrophoresis, chromatography, gel diffusion, solvent precipitation, etc., and the radioactive isotope (I^{131}) content of each is determined.
4. The ratio of radioactivity of bound and free fractions is then plotted against the concentrations of standard as shown in Figure 9-23.

For quantitative analysis of a hormone in plasma or urine, the procedure is similar, except that the various dilutions of the sample are incubated with predetermined amounts of I^{131}-labeled hormone and its antibody in order to bring the concentrations of the unknown within the range of the calibration curve.

The radio-immunoassay technique has been used with considerable success in assays of specific protein hormones when the biological assays presented numerous problems related to sensitivity (e.g., growth hormone, glucagon, insulin), specificity (e.g., insulin), or practicality (e.g., thyrotrophin). Detailed discussions of the immunoassay of hormones will be found in the review and original articles referred to in this section. It may be safely predicted that in clinical laboratories the bioassays which are alleged to be poorly sensitive, nonspecific, imprecise, inaccurate, or time-consuming will rapidly be replaced in the near future by immunoassays.

At the present time, the number of studies comparing the results obtained by immunological and biological methods is small, and no general conclusion can be drawn. Thus, the results obtained by immunological assays for HCG were found to be considerably higher than those obtained by bioassays.[11] The plasma levels of insulin determined by radio-immunoassay were, on the other hand, lower than those obtained by biological methods.[9]

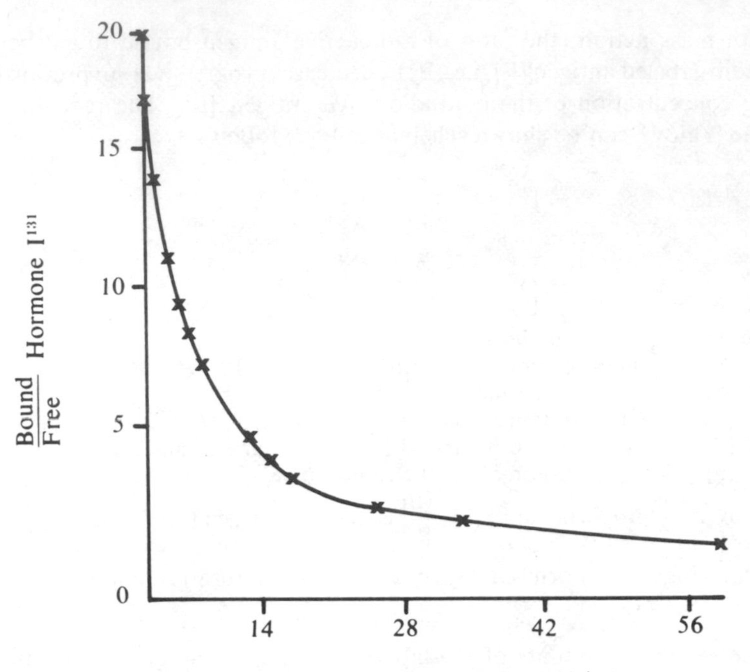

Figure 9-23. Illustrative standard curve for radio-immunoassay. Ratio of bound to free I^{131} labeled protein hormone plotted against the concentration of unlabeled standard preparation.

PITUITARY HORMONES

The pituitary gland (or hypophysis) secretes at least nine hormones, all of which are protein or peptide in nature. Anatomically the gland is divided into three zones: the anterior pituitary (pars distalis), the intermediate lobe (pars intermedia), and the posterior pituitary (neurohypophysis). The first two zones together are also called the adenohypophysis. Of the nine hormones, six are produced in the anterior pituitary (growth hormone, TSH, FSH, LH, prolactin and ACTH), one in the intermediate lobe (melanophore-stimulating hormone, MSH), and two in the posterior pituitary (vasopressin—also known as the antidiuretic hormone—ADH, and oxytocin). For a full account of each of these hormones the reader is referred to textbooks and other monographs.[23,93]

Growth Hormone

Somatotrophin, somatotrophic hormone (STH). The hormone is secreted by the anterior pituitary and promotes the growth of bone and all other tissues in the body. It stimulates protein synthesis (anabolic effect) and affects fat and carbohydrate metabolism (diabetogenic effect). In recent years, highly purified human growth hormone (HGH) has been prepared. It is an acidic protein having a molecular weight of 21,500. There are some controversies as to the separate existence of the hormone, *prolactin*. The physical, chemical, and immunological properties suggest that the biological activities of both prolactin and growth hormone are contained in the same molecule. The physiological action of prolactin is involved primarily in the proliferation of the mammary gland, initiation of milk secretion, and maintenance of the corpus luteum in the ovary. For further discussion of the hormone and the method of assay, the reader is referred elsewhere.[46] It may be mentioned, however, that the significance of prolactin in health and disease is not yet adequately known.

The determination of growth hormone in blood and urine has been reviewed by Loraine.[46] It suffices to say that available biological assay methods are not adequately sensitive to detect the presence of this hormone in human blood or urine. A sensitive radioimmunological procedure reported by Hunter and Greenwood[33] appears to be the method of choice. The basic steps of the technique involve the labeling of the pure growth hormone (antigen) with I^{131}, incubation of the optimal amount of labeled antigen with the test sample and with a predetermined amount of anti-HGH (antibody) for seven days, followed by the separation of the free antigen and the antigen-antibody complex by electrophoresis and the determination of radioactivity of the two fractions. The amount of unlabeled antigen (HGH) in the test sample is then calculated by comparison with a standard curve. According to Hunter and Greenwood,[33] the plasma concentration shows a wide fluctuation throughout the day and also varies with such factors as the nutritional condition of the subject, exercise, and sleep. The plasma HGH level is generally higher in children than in adults. A *decreased* level is found in dwarfism and an *increased* level in gigantism and acromegaly. The absolute amount of growth hormone excreted in the urine is not yet known.

Thyroid Stimulating Hormone

TSH, thyrotrophin, thyrotrophic hormone. This is an anterior pituitary hormone which controls the functional activity of the thyroid gland. In recent years it has become evident that TSH is not the only thyroid stimulator. Substances such as long-acting thyroid stimulator (LATS), detected in the serum of thyrotoxic patients, is produced outside the pituitary and mimics the biological action of TSH. The properties of this material have been discussed in detail by Kirkham.[36] Although highly purified preparations of TSH have been obtained, the hormone has not been isolated in pure form. According to the present evidence, it is a glycoprotein with a molecular weight of approximately 25,000, 8 per cent of which is represented by carbohydrates.

A great number of bioassay methods of varying sensitivity, reliability and ease have been developed for the measurement of TSH. Unfortunately none of them are suitable for routine clinical application.[36] The methods that appear to be most sensitive are radio-immunoassays.[50] The principle of such an assay technique has been described before. Because of the methodological deficiency, it has not yet been established whether TSH is excreted in urine. Reliable quantitative plasma values in normal men and women are also not available in the literature. According to the present evidence, plasma concentrations are *elevated* in all patients with primary hypothyroidism, whereas the concentrations *decrease* to nondetectable levels in patients with secondary hypothyroidism (pituitary hypothyroidism). In hyperthyroidism, the results reported in the literature are so contradictory that it is difficult at present to come to a conclusion as to its exact plasma level.

Adrenocorticotrophic Hormone

ACTH, corticotrophin, adrenocorticotrophin. This is a trophic hormone which regulates the functional activity of the adrenal cortex. It has also some extra-adrenal metabolic effects in the body. The administration of adrenocorticotrophin to adrenalectomized animals has shown that it mobilizes fat depots, increases the rate of fatty acid oxidation, and enhances muscle-glycogen synthesis, resulting in hypoglycemia. The isolation of ACTH in pure form from the pituitary of several species, including man, has been accomplished during the past decade. The primary structure of the hormone, including the amino acid sequence, has also been established. ACTH of all species examined so far consists of a single, unbranched polypeptide chain with 39 amino acid residues, having a molecular weight of approximately 4,500. Among species the difference in the sequence of amino acids lies mainly between residues 25 to 33; the biologically active portion of the molecule comprising the amino acid residues from 1 to 25 is identical in all species. Recently the synthesis of biologically active ACTH has been achieved. For further information the review article by Li[42] is suggested.

The so-called melanophore-stimulating hormone (MSH, intermedin), produced by the intermediate lobe of the pituitary, has a close structural relationship to ACTH. Two general types of MSH (α and β) are generally found. Beta-MSH is comparatively less active than α-MSH. It has been shown that the first 13 amino acids in α-MSH and amino acids 11 to 17 in β-MSH are identical to those found in the same numbered sequence in ACTH. It should be noted that the considerable pigmentation associated with Addison's disease of the adrenals is most likely caused by one or both of these portions of the ACTH molecule. Apart from the possible darkening effect of the skin, the biologic effect of MSH in man has not been completely investigated.

Even though ACTH is available in pure form, chemical procedures for its measurement are not yet available. Accordingly, all assays are confined to biological or immunological techniques. Bioassays for this hormone suffer, however, from the lack of sensitivity necessary to detect its presence in blood and urine. As a result, very little information is available about the concentration of ACTH in body fluids in health and disease. It is hoped that the application of immunological techniques will yield quantitative information in the near future.

Among the bioassays employed so far, the *Sayers test*, depending on the depletion of adrenal ascorbic acid, is one of the most sensitive, specific, and widely used methods.[66] The test is conducted in hypophysectomized rats. The procedure is as follows: After approximately 24 hours following hypophysectomy of the rats (the purpose of hypophysectomy is to remove the endogenous source of ACTH), the left adrenal is removed, and immediately an intravenous injection of standard ACTH or the test material is given. One hour later the right adrenal is removed. On the following day the concentration of total ascorbic acid (dehydro and reduced) of both adrenals is determined. The response is evaluated on the basis of the difference between the ascorbic acid content of the right (test) and left (control) adrenal glands. With this technique, as little as 0.2 mI.U. (milli-International Unit) can be detected. One *international unit* is defined as the activity contained in one mg. of the first International Standard for ACTH.

The radio-immunoassay as reported by Yalow et al.[99] appears to be very promising. The procedure and the principle have been described previously. It should be noted, however, that the amount of ACTH excreted in urine is still uncertain, and the concentration in blood is reported to be very small. Extraction of the body fluids to obtain the hormone in concentrated form appears to be a general preliminary step for the majority of assay methods. The different methods available for ACTH determination have been recently reviewed and discussed by Vernikos-Danellis.[85] The reports on the normal levels of ACTH in blood are quite variable and contradictory.

Though variable quantitative values have been reported, it may be stated that in pathological conditions, such as Addison's disease, in bilateral adrenalectomy, and in "stress" conditions, the plasma levels are *high*, whereas in adrenal carcinoma and in panhypopituitarism the levels are *low*.

GONADOTROPHINS

The pituitary gonadotrophins consist of three hormones: follicle stimulating hormone (FSH), luteinizing or interstitial cell stimulating hormone (LH or ICSH), and prolactin (otherwise known as luteotrophic hormone or luteotrophin—LTH, lactogenic hormone, mammotrophin). The latter hormone is elaborated by acidophilic cells, which are also believed to produce growth hormone; the former two hormones are elaborated by the basophilic cells of the pituitary. The function of prolactin and its structural relationship with growth hormone have already been briefly outlined (see growth hormone) and will not be discussed further. Here, the discussion will mainly be confined to FSH and LH, henceforth to be denoted together as HPG (human pituitary gonadotrophin), and the human chorionic gonadotrophin (HCG) of placental origin.

In the female, follicle stimulating hormone (FSH) causes growth of the ovarian follicles, and in the presence of small amounts of LH it promotes the secretion of estrogens by the maturing follicles. In the male, the same (pituitary) gonadotrophin stimulates spermatogenesis. Luteinizing hormone (LH) in the female causes ovulation of the follicle, which has previously been ripened by FSH. An optimum ratio of FSH and LH is believed to be important in producing ovulation, even though the effective concentration of each in the human has not yet been established. Following ovulation, the transformation of the ruptured follicle into the corpus luteum and its secretion of progesterone are also believed to be under the influence of LH. In the male, LH is called the interstitial cell stimulating hormone (ICSH) and is responsible for the production of testosterone by the interstitial cells of the testes. The secretion of FSH by the pituitary is inversely related to the estrogen level (see Control of hormone secretion). In the absence of ovarian production of estrogen, for example, in postmenopausal women, the release of FSH is high and relatively large amounts are found in the blood and urine.

The hormonal regulation of LH secretion is not yet fully known. In animal experiments it has been shown, however, that large doses of progesterone inhibit the release of LH, resulting in the failure of ovulation. In women, such a reciprocal relationship may exist between LH and progesterone secretion. In the male, testosterone regulates secretion of LH by a feedback mechanism. It should be noted, however, that in normally menstruating women there is a rhythmicity in the secretion of gonadotrophins and ovarian hormones, which is regulated by hormonal and hypothalamic mechanisms. In the male no cyclic production of testicular hormone occurs; it is a continuous process, much like the production of adrenal corticosteroids with a continuously operating feedback mechanism. Because of the intricate interplay of gonadotrophins and gonadal hormones throughout the menstrual cycle, the evaluation of these parameters in gynecological disorders are important diagnostic measures.

Although FSH and LH have not yet been isolated in absolutely pure form, the preparations so far obtained from animal and human pituitaries and from the urine of postmenopausal women indicate that these are glycoproteins with molecular weights of approximately 30,000.

Methods for the Assay of Human Pituitary Gonadotrophin (HPG) in Urine

At the present time bioassay methods appear to be most suitable for the determination of gonadotrophins in urine. The immunological assays,[53] which are more sensitive, have not yet found wide clinical application. It is expected, however, that the quantitative determination of FSH and LH by bioassay will slowly be replaced by radio-immunoassay. It should be noted that bioassays are not sensitive enough to measure these hormones in blood, and because of this, all clinical investigations have been conducted on urine, which can be extracted and concentrated by a variety of procedures prior to injection into animals. The determination of urinary total gonadotrophic activity, rather than the measurement of FSH and LH separately, appears to render adequate clinical information, and a procedure which is currently in wide use will be described in detail. It should be noted, however, that there are several assay methods available which are claimed to be specific for the individual determination of FSH and LH. Among them, the *augmentation test* for FSH by Steelman and Pohley[77] and the ovarian ascorbic acid depletion test (OAAD) for LH by Parlow[57] are those most widely used. Further information on these topics may be derived from the monographs by Albert[1] and Bell and Loraine.[8]

Determination of Total Gonadotrophic Activity in Urine[12,37]

PRINCIPLE

The method depends on the increase in uterine weight of immature mice injected subcutaneously with a urinary concentrate of gonadotrophin twice a day for two and one-half days. The result is expressed in mouse uterine (M.U.) units. One M.U. unit is defined as the amount of gonadotrophins required to cause a 100 per cent increase in uterine weight. Since the amount of gonadotrophins excreted in the urine is very low, an extraction and concentration of the sample is required. During this procedure the unwanted toxic and interfering substances are eliminated. The main steps are as follows: Adsorption of gonadotrophins onto Kaolin at pH 4.5, followed by elution with 2 *M* ammonium hydroxide (NH_4OH). The eluate is adjusted to pH 5.5 and the HPG is precipitated with 2 volumes of cold acetone. The precipitated material is further washed with ether and ethanol. The dry material is dissolved in normal saline in varying dilutions and injected into immature mice. The mice which serve as controls receive concurrently the same aliquot of saline.

REAGENTS

1. Glacial acetic acid, A. R.
2. Kaolin, Tamms Industries, Chicago, Oxford-English brand.
3. Ammonium hydroxide, 2 *M*.
4. Acetone, A. R.
5. Ether, A. R.
6. Ethanol, purified.
7. Saline, 0.85 per cent (w/v).

PROCEDURE

A complete 24 hour collection of urine is needed. During collection and prior to analysis, the urine should be kept refrigerated but not longer than 24 to 48 hours. Prolonged standing even at 5°C. may destroy HPG. It is best to start processing the urine as soon as it is received, at least through the step of Kaolin treatment, after which the preparation may be kept overnight in the refrigerator.

Extraction. Adjust the pH of the 24 hour urine specimen to 4.5 with glacial acetic acid. Add 20 gm. of Kaolin per liter of urine. Mix well with a stirring rod. Pour the mixture onto a Buchner funnel and filter under mild suction. (Alternatively, refrigerate overnight, aspirate off the urine layer on the following morning, add acidified water, mix, transfer onto the Buchner funnel and proceed as follows.) Discard the filtrate. Wash the Kaolin cake in the Buchner funnel with 2 liters of water acidified to pH 4.5 with glacial acetic acid.

Discard the wash. Elute the gonadotrophins from the Kaolin cake with 100 ml. of 2 *M* ammonium hydroxide. Again pour the eluate over the Kaolin cake to completely elute the HPG. Transfer the eluate to a 500 ml. centrifuge bottle. Wash the Kaolin cake with 50 ml. of distilled water and add the water wash to the ammonium hydroxide eluate.

Precipitation. Adjust the pH of the eluate to 5.5 with glacial acetic acid. Add 2 volumes (300 ml.) of cold acetone, and refrigerate for one hour. Centrifuge for 15 minutes at 2000 r.p.m. Pour off the supernatant. Add 20 ml. of ethanol, stir to suspend the precipitate, and centrifuge as above. Discard the supernatant. Repeat the process of washing with 20 ml. of ether. Let the precipitate dry completely under a stream of air, taking care not to lose any flakes of the precipitate. (The precipitate may be stored overnight or longer under desiccation in a freezer.)

Preparation of the solution for injection. Pulverize the precipitate with a glass rod and dissolve in 10 ml. of normal saline. Centrifuge for 10 minutes at 2000 r.p.m. Aspirate the supernatant with a 10 ml. syringe and transfer to a clean glass vial. Prepare two different dilutions, namely, 1 to 5 and 1 to 10, as follows: transfer 1 ml. aliquots of the supernatant to two separate glass vials containing 4 ml. and 9 ml. of normal saline respectively. Refrigerate all three solutions.

Injection. The immature mice to be used should be 21 to 23 days old and weigh 8 to 10 gm. Using three animals for each solution (undiluted, diluted 1:5, and diluted 1:10), inject each animal with 0.2 ml. twice a day (A.M. and P.M.) for two days, and once the last day. Inject simultaneously three more mice with normal saline to serve as controls. Sacrifice the animals on the fourth day. Dissect out the uterus from each animal, avoiding the adjoining fatty tissues, and weigh on a torsion balance.

Result. The response is considered to be positive when the average uterine weight of the test animals is at least twice the average uterine weight of the control animals (i.e., one M.U. unit). For example, if 1 ml. (5×0.2 ml.) of the extract dissolved in 10 ml. saline (undiluted supernatant) gives a positive response, the excretion of gonadotrophins per 24 hours is 10 M.U. units. Similarly, the positive response to the injection of 1 ml. of the solution diluted to 1:5 or 1:10 indicates the excretion of 50 or 100 M.U. units of gonadotrophins per 24 hours respectively.

COMMENTS

Mice (or rats) are chosen as the experimental animals for gonadotrophin assays because they are more sensitive to stimulation by these hormones than are other animals, such as rabbits, guinea pigs, and amphibia. The increase in the weight of the uterus depends on the stimulation of the ovaries by gonadotrophins to secrete estrogens, which in turn act on the uteri. Previously, it was erroneously believed that the mouse uterus test was specific for FSH activity, but recent evidence suggests that this test measures both FSH and LH.[67] However, in case of elevations of LH, the ovaries of the injected mice will be increased out of proportion. Thus, the result is designated "total gonadotrophic activity" in the urine.

It should be pointed out that the quantitation of gonadotrophic activity without reference to the international standard is not very accurate. A suitable standard for the mouse uterine test (e.g., NIH-HPG-UE) has recently become available from the Endocrinology Study Section, National Institutes of Health, Bethesda, Maryland. It is highly important to establish frequently, and preferably routinely, the parallelism of the dose-response curve for a suitable range of concentration between the reference standard and the test material prepared in a specific laboratory. Since gonadotrophins are glycoproteins, the experimental precautions concerning pH, temperature, ion concentration, heavy metal contamination, etc., are very important. Otherwise, denaturation of these hormones may occur, resulting in spurious results. The presence of toxic substances in the extract affects biological response and occasionally causes death of the experimental animals. Among the various measures employed to reduce toxicity, additional purification of the crude Kaolin extract, using ammonium acetate and ethanol[1] or tricalcium phosphate,[47] appears to be quite helpful.

NORMAL VALUES

Adults: 10–50 Mouse Uterine units/24 hrs.
Menopausal: 50–190 Mouse Uterine units/24 hrs.
Children (before puberty): Less than 6 Mouse Uterine units/24 hrs.

REFERENCES

Borth, R., Lunenfeld, B., and Manzi, A.: *In* Human Pituitary Gonadotrophins. Albert, H. (Ed.) Springfield, Illinois, Charles C. Thomas, 1961, p. 13.
Klinefelter, H. F., Jr., Albright, F., and Griswold, G. C.: J. Clin. Endocrin., *3*:529, 1943.

CLINICAL SIGNIFICANCE

The excretion of gonadotrophins in children is very low. Detectable amounts of HPG are present in the urine of normal men and normally menstruating women, and higher levels are found in postmenopausal women. Wide individual variation is generally observed, even though a mid-cycle peak in women with a normal menstrual cycle is a consistent pattern. Because of the nonfunctioning ovaries in postmenopausal women, relatively large amounts of gonadotrophins, consisting mainly of FSH, are excreted in the urine. The clinical usefulness of the measurement of gonadotrophins lies mainly in the differential diagnosis of primary gonadal failure and gonadal failure secondary to lesions of the anterior pituitary in both men and women.

Increased excretion is generally associated with ovarian agenesis, Klinefelter's syndrome (prepubertal or pubertal seminiferous tubule failure), and male climacteric. *Decreased* excretion may be found in patients with hypogonadotrophic eunuchoidism, panhypopituitarism, and anorexia nervosa.

HUMAN CHORIONIC GONADOTROPHINS (HCG)

Chorionic gonadotrophin, previously called "pregnancy prolan" or "anterior pituitary-like" (APL) substance, is secreted by the placenta and is found in urine, blood, amniotic fluid, colostrum, milk, and fetal tissues. This hormone appears within a few days after conception, and because of this, early confirmation of pregnancy is possible through its detection in urine and blood (pregnancy test). The level of HCG is highest during the first trimester and stabilizes at a lower level as the pregnancy progresses, becoming almost undetectable following parturition. The physiological activity of this hormone mimics that of the pituitary gonadotrophin LH, and it is assumed that during early weeks of pregnancy it maintains an active corpus luteum to supply progesterone for the maintenance of the uterus in pregnancy. Chemically, it is also a glycoprotein with an approximate molecular weight of 30,000.

Methods of Assay of HCG

There are a number of assay procedures available involving both biological and immunological techniques. Review articles on biological methods,[45] and immunoassays[9] are suggested for detailed information on these topics. Here, only the important tests (qualitative and quantitative) based on the biological and immunological methods are described.

Biologic Pregnancy Tests

Since HCG closely resembles LH of the pituitary in its physiological activities, the underlying principle of its bioassay is the same as that of LH. Thus injection of HCG produces corpora lutea and hemorrhagica in the ovaries of mice and rabbits, hyperemia in the ovaries of rats, and secretions of sperm and ova in toads and frogs. Biological methods based on each of these responses are available.[46] Until recently, methods depending on the ovarian hyperemia in immature rats[101] and expulsion of spermatozoa in amphibia[30] were most commonly used in clinical laboratories. The rat hyperemia test and the frog test as described by Albert et al.[2] and Henry et al.[31] respectively, are given below.

COLLECTION AND PREPARATION OF URINE SPECIMEN

For qualitative pregnancy tests, preferably the first morning void of urine is collected in a clean container. For quantitative determination, a 12 or 24 hour collection is required. The patient should be instructed to restrict intake of fluids from 8:00 p.m. until the morning collection. A large urine volume may yield a false negative reaction. Drugs such as salicylates (aspirin), barbiturates, and other sedatives should be withdrawn for at least 48 hours prior to collection of urine. These drugs may increase toxicity of the urine, causing false negative results or death of the animals. The urine specimens should be clear; presence of turbidity or urine sediments requires filtration or centrifugation. The pH of the urine should be adjusted to approximately 6.8; too acidic or too alkaline urine is detrimental to biologic responses.

Rat Hyperemia Pregnancy Test (Qualitative)

Use five Sprague-Dawley immature female rats, 21 to 28 days old and 40 to 65 gms. in weight. Inject 2 ml. of urine subcutaneously into each of the rats.

Four hours later sacrifice the animals. Remove the ovaries and place them on wet, clean filter paper. Place the ovary ventral side up so that it faces the observer, with the fallopian tube and fat attached. Moisten with saline.

If four or more of the ten ovaries are pink to red (hyperemic), then the test is positive. When most of the ovaries are very small and pale, the result is negative.

Frog Test[31]

Injection of human chorionic gonadotrophin into male frogs or toads causes discharge of sperms. The presence of spermatozoa in the cloacal fluid or urine of the male toad or frog generally occurs 1 to 4 hours after injection of urine or plasma and is a positive test for pregnancy.

PROCEDURE

Use the urine specimen as described above. Use two male toads (Bufo americanus) for the test. Wash the animals under the tap to remove adherent dirt. Place each animal in a separate jar. Remove a drop of cloacal fluid from each toad with separate medicine droppers and place on a clean slide. Examine under a microscope to ensure the absence of spermatozoa (the presence of spermatozoa will make them unsuitable for the test). Inject 1 ml. of urine per 10 gm. of body weight subcutaneously. Return the toads to the jars. Examine the voided urine for spermatozoa at hourly intervals. (If urine does not become available, a capillary pipet may be inserted into the cloaca to obtain fluid.)

If spermatozoa are found in the urine of both the animals, the test is positive. The presence of spermatozoa in one of them renders the results inconclusive, in which case the experiment should be repeated with more animals.

COMMENTS

Amphibia are relatively insensitive to HCG, and for this reason concentration of the urine specimen is occasionally necessary. Moreover, there is a seasonal variation in response, the maximum being in late winter and early spring and the minimum in autumn. To obtain reliable results, the number of animals for each experiment should be as great as possible.

REFERENCE

Henry, J. B., Krieg, A. F., and Davidsohn, I.: *In* Clinical Diagnosis by Laboratory Methods, 14th Edition. Davidsohn, I., and Henry, J. B. (Eds.) Philadelphia, W. B. Saunders Company, 1969, p. 1181.

Quantitative HCG Test Based on Rat Hyperemia Test[2]

PRINCIPLE

The quantitative analysis of HCG is generally suggested for clinical conditions, other than pregnancy, in which large amounts of this hormone are secreted (e.g., chorionepithelioma, hydatidiform mole, seminoma, and other extragenital teratoma). The procedure is the same as that described for the pregnancy test except that the serially diluted urine is injected into groups of immature rats until the test shows a negative response. The percentage of the hyperemic ovaries for each group is compared with that obtained with the International Standard preparation of HCG, and the total amount of HCG, expressed in terms of international units (I.U.), is calculated. One international unit (I.U.) for HCG is defined as the activity contained in 0.1 mg. of the First International Standard Preparation.

PROCEDURE

Obtain a 24-hour collection of urine specimen as described above. Measure the total volume of urine. Prepare different dilutions as given in the following table:

		Urine (ml.)	+	*Saline* (ml.)	*Dilution Factor*
Set 1	1	1		0	1
	2	3		6	3
	3	1		9	10
Set 2	4	1		9	10
	5	1		29	30
	6	1		99	100
Set 3	7	1		99	100
	8	1		299	300
	9	1		999	1000

Each set of experiments comprises three groups of rats. Start with Set 1. Inject five animals in each group with 1 ml. of the dilutions shown in the table. After four hours, sacrifice the animals and remove the ovaries. Determine the number of ovaries that are hyperemic. If the number of hyperemic ovaries in Set 1 is more than 80 per cent, proceed with Set 2 and, if necessary, with Set 3, until the number of hyperemic ovaries in successive groups falls above 50 per cent in one dilution (e.g., 1:100) and below 50 per cent in the next dilution (e.g., 1:300).

CALCULATION

Let us assume that the results obtained are as follows:

	Dilution	*Hyperemic Ovaries (per cent)*
Set 1	1:0	100
	1:3	90
	1:10	50
Set 2	1:10	50
	1:30	20
	1:100	0 (negative)

The useful dilutions in this case are 1:3, 1:10, and 1:30 with the respective percentages of hyperemic ovaries, 90, 50, and 20. Determine the unitage for each of them from the reference table below. (The data of the dose-response relationship of the international standard shown in the table should not be construed as a conversion table for other HCG determinations. Such data with the International Standard preparation should be established in the specific laboratory. The strain of animals, mode of injection, environment of the animal colony and many other factors, too numerous to mention, elicit variable responses.) The results in I.U. are as follows:

90 per cent = 3.60 I.U./ml.
50 per cent = 0.97 I.U./ml.
20 per cent = 0.37 I.U./ml.

Multiplying by the appropriate dilution factor, we get 10.8 (3.60 × 3), 9.7 (0.97 × 10), and 11.1 (0.37 × 30) I.U. per ml., respectively. Ideally, one would expect an equal number of I.U./ml. The variation is caused by experimental error and the inherent quality of the assay procedure, although the precision of the ovarian hyperemia test is claimed to be very high. The average of these three figures is 10.5 I.U./ml. If the volume of the 24-hour urine specimen is 1200 ml., then the excretion of HCG = 10.5 × 1200 = 12,600 I.U./24 hour.

TABLE 9-5. *Dose-response Relationship for an International HCG Standard in the Rat Hyperemia Test**

Hyperemic Ovaries (Per cent)	*I.U./ml. of Urine*
10	0.27
20	0.37
30	0.51
40	0.71
50	0.97
60	1.34
70	1.86
80	2.60
90	3.60
100	5.00 or more

* From Albert, A., et al.: *In* Textbook of Endocrinology. Williams, R. H. (Ed.), Philadelphia, W. B. Saunders Co., 1968, p. 1187.

REFERENCE

Albert, A., Mattox, V.R., and Mason, H.L.: *In* Textbook of Endocrinology, Williams, R. H. (Ed.) Philadelphia, W. B. Saunders Co., 1968, p. 1181.

CLINICAL SIGNIFICANCE

In an abnormal pregnancy, such as one complicated by preeclamptic toxemia, the urinary excretion of HCG is significantly *higher* than that of a normal pregnancy. In threatened and ectopic pregnancy the excretion is generally *below* the expected value.

Most of the usefulness of quantitative HCG assays is in diagnosis and treatment of trophoblastic diseases. Patients with hydatidiform mole or chorionepithelioma excrete large amounts of HCG. Serial determinations of HCG excretion provide an excellent indicator of response to therapy. Patients with ovarian and testicular teratomas and seminomas have also been reported to excrete HCG in large quantities.

Normal values, counting from the first day after the last menstrual period (LMP) in normal pregnancy are as follows:

20 to 40 days	less than 10,000 I.U. per 24 hours
41 to 100 days	10,000 to 100,000 I.U. per 24 hours
101 to 280 days	less than 40,000 I.U. per 24 hours.

Immunological Methods

Tests depending on hemagglutination inhibition[90] and complement fixation[14] are used widely for qualitative (pregnancy test) and quantitative determination of HCG. The principle and procedures for the techniques have already been discussed. In recent years a number of commercial preparations for the HCG test based on hemagglutination inhibition have become available ("UCG Test," Wampole Laboratories, Stamford, Conn.; "Gravindex," Ortho Pharmaceutical Co., Raritan, N.J.; "Pregnosticon", Organon, Inc., West Orange, N.J.) Because of wide use of commercially available test kits in clinical laboratories, the procedure for the pregnancy test by immunological methods is not described here. The specific protocols for these test kits are supplied by the manufacturers.

COMMENT

The test based on the agglutination inhibition of latex particles is more rapid (results obtained in 2 minutes) but less sensitive and less reliable than the 2-hour hemagglutination inhibition of HCG-coated red cells. A new *direct* agglutination test (DAP Test from Wampole, Stamford, Conn.) is also available (2 minute slide test). Latex particles are coated with antiserum (*not HCG*). Specimens with high HCG values cause agglutination. This is an excellent slide test but it is not as reliable as the UCG test. The drugs and toxic substances present in urine are less liable to interfere than in bioassays However, the precautions and procedures for the preparation of urine specimens as outlined for biological methods should be maintained, since denaturation of HCG, as well as dilute urine (sp. gr. less than 1.010), may yield false negative results. HCG antisera do not possess absolute specificity. Consequently, interfering substances, especially in proteinuria, may neutralize the antisera and cause a false positive test. This may also occur if the sensitivity of the test exceeds 0.7 I.U. per ml., because of the cross reactions with pituitary gonadotrophins. Denaturation of HCG antiserum enhances sensitivity, with a concomitant increase of false positive reactions; with loss of activity, relatively less HCG is required to neutralize the antibody. Under the circumstances, the use of known positive and negative controls and occasional restandardization of the reagents are extremely important if reliable results are to be obtained.

INSULIN

This hormone, elaborated by the beta-cells of the islets of Langerhans in the pancreas, is responsible for the control of carbohydrate metabolism and maintenance of the blood-glucose level (see the chapter on carbohydrates and regulation of blood glucose). Insulin is the protein hormone whose structure was first elucidated. It is composed of two long polypeptide chains, A and B, with specific amino acid sequences bound together by two disulfide linkages. There is also an intrachain disulfide bridge linking the amino acids at positions 6 and 11 in chain A. Although no definite part of the insulin molecule can be designated as the active center, it has been shown that the intact disulfide bridge is essential for full biological activity. The excellent review article by Sanger[65] provides further information.

Methods of Assay

The determination of plasma levels of insulin is a necessary practice in clinical fields. A discussion of the methods for determining levels of insulin in blood may be found in the article by Randle and Taylor.[62]

As with other protein hormones, the assay of this hormone is carried out by biological and immunological methods.

The *biological methods* are of two types, *in vivo* and *in vitro*. The *in vivo* methods depend on the ability of the hormone to produce hypoglycemic convulsions in mice and to cause hypoglycemia in rabbits. Although previously widely used, the quantitative measurement of plasma insulin with both of these procedures is unsatisfactory, and their use in the clinical laboratory has been very limited.

Among the *in vitro* methods, the rat epididymal fat pad test[50] and the rat diaphragm test are the most noteworthy. The *rat diaphragm technique* consists of incubating isolated diaphragms in a suitable glucose-containing medium and measuring the resultant glucose uptake during a given period of time. Although the technique is capable of detecting approximately 100 μI.U. of insulin per ml. of plasma, the lack of specificity is its greatest drawback. It has been shown that substances other than insulin present in the plasma fraction interfere with the measurement. In addition, the degradation of insulin by proteolytic enzymes in the diaphragm and adsorption of the hormone onto glassware are other important sources of error.

The principle of the *epididymal fat pad test* depends on the ability of insulin to affect the glucose metabolism in the fatty tissue. The procedure consists of incubating the fat pad of a rat with C^{14}-labeled glucose and determining either the rate of glucose uptake by the tissue or the rate of production of $^{14}CO_2$. The measurement of either of the end points correlates with the concentration of insulin in the medium. This method has greater sensitivity (10 μI.U./ml.) than the rat diaphragm test. As with other methods, it also suffers from lack of specificity, and thus measures "*insulin-like*" activity (ILA) of nonpancreatic origin.

IMMUNOASSAY

Because of severe limitations of biological methods in regard to sensitivity and specificity, in recent years the radio-immunoassay has become the technique of choice for measurement of insulin in blood. The subject has been reviewed by Berson and Yalow.[9] The principles of radio-immunoassays have been described previously. Briefly, the main steps of the insulin method are as follows: pork insulin is labeled with I^{131} and an antibody to pork insulin is produced in guinea pigs (a guinea pig anti-pork insulin serum reacts identically with pork and human insulin). A serial dilution of the standard human insulin and the plasma are prepared and incubated with a predetermined amount of radio-labeled insulin and anti-insulin for three to five days at 4°C. At the end of the incubation period, the aliquots are removed and subjected to electrophoresis in order to separate free insulin and antibody-bound insulin. The radioactivity of each of these two fractions is determined, and the ratio of the radioactive count (I^{131} content) of bound and free insulin is plotted against the concentrations of the standard, as shown in Figure 9-23 (page 556). The amount of insulin in the plasma is determined by comparing the results with the standard curve.

The normal values for plasma levels of insulin obtained by bioassays are higher than those obtained by immunoassays. Using the isolated rat diaphragm test, the average level of insulin in fasting plasma of normal subjects is 37 μI.U. per ml. (micro international unit) with a range from 11 to 240 μI.U. per ml., whereas that obtained by radio-immunoassay is 21 μI.U. per ml. with a range from 0 to 60 μI.U. per ml.[9] In diabetes mellitus the plasma level is generally *low*, and the pattern of insulin secretion following administration of glucose is different from that found in normal subjects. In the former the highest insulin levels are reached at two hours after glucose intake, whereas in normal subjects the corresponding time is 30 to 60 minutes. In patients with *islet cell tumor of the pancreas*, the values are *high*, which, in effect, causes hypoglycemia. Acromegaly and gigantism are also associated with a high level of plasma insulin.

CATECHOLAMINES

Catecholamines are amines which are distinguished by the presence of two hydroxyl groups on a benzene ring (catechol). The most important endogenously produced compounds of this group are epinephrine (adrenaline), norepinephrine (noradrenaline), and dopamine [β(3,4-dihydroxyphenyl) ethyl amine]. The structural formulas and the numbering system of these compounds are shown in Figure 9-24.

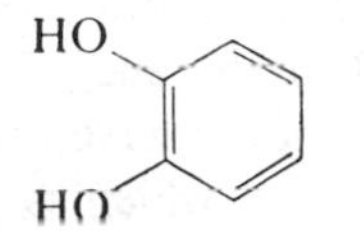

Catechol (Dihydroxybenzene)

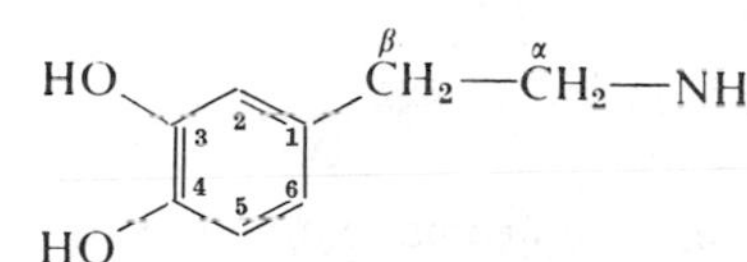

Dopamine[β(3,4-dihydroxyphenyl)ethyl amine]

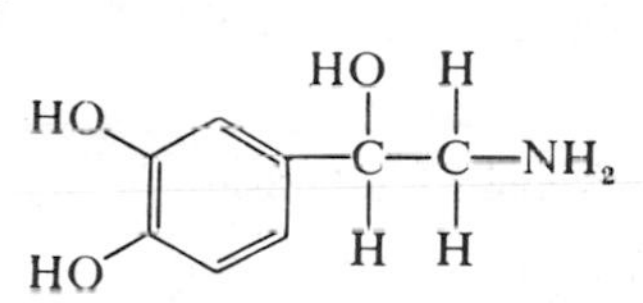

Norepinephrine (Noradrenaline)

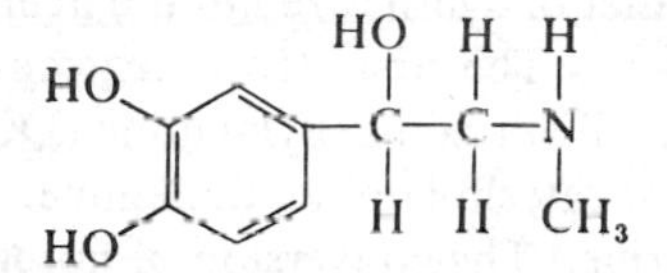

Epinephrine (Adrenaline)

Figure 9-24. Structural formulae and the numbering system of catecholamines.

Norepinephrine differs from dopamine in that it possesses a hydroxyl group on the β-carbon atom; epinephrine is distinguished from norepinephrine by the presence of a methyl group on the nitrogen of the terminal amino group. Dopamine and norepinephrine are primary amines; epinephrine, on the other hand, is a secondary amine because of the substitution of a hydrogen atom in the amino group by a methyl group. Epinephrine and norepinephrine are dihydroxylated phenyl (catechol) β-ethanolamine, and therefore exhibit the chemical properties of phenols, alcohols, and amines. Chemically, epinephrine is rapidly oxidized in neutral and alkaline solutions. Norepinephrine is much more resistant to oxidation. These differences in properties of the two compounds have been utilized for their individual estimations in biological fluids. For further information on the chemical properties of catecholamines the review article by Euler[87] is suggested.

The main sources of catecholamines in the body are the chromaffin cells in the adrenal medulla, the heart, lung, liver, intestines, and prostate glands, the chromaffin bodies of the fetus, the brain, and the adrenergic nerves. The chromaffin cells are so designated for their distinctive brown or black color on staining with chromic acid.

Epinephrine is quantitatively the most important substance produced by the adrenal medulla. Norepinephrine is the major substance liberated by the postganglionic sympathetic nerves. In addition, norepinephrine is the predominant catecholamine produced in fetal life of mammals. Dopamine is considered to be produced significantly in the brain and other viscera such as the lungs, liver, and intestines. The hormones are produced in mitochondria and are stored in special granules of the chromaffin cells. Administration of reserpine or other pharmacological agents, insulin-induced hypoglycemia, and stressful situations have all been shown to cause depletion of the catecholamine content of storage granules. For authoritative discussions on the property of chromaffin granules and the storage and release of catecholamine, the article by Weiner[89] should be consulted. Each of the three catecholamines, dopamine, norepinephrine, and epinephrine, has characteristic physiological functions and pharmacologic actions. Detailed consideration of these topics may be found in the textbooks of endocrinology listed at the end of this chapter.[81,93] It may suffice to mention here that the first two hormones have, in general, a marked influence on the vascular system, whereas epinephrine, which is considered to be the true adrenal medullary hormone, exerts a profound influence on the metabolic processes, especially carbohydrate metabolism.

Biosynthesis and Metabolism

In vivo and *in vitro* studies have shown that the aromatic amino acid L-tyrosine plays the most important role as precursor for the biogenesis of catecholamines (Fig. 9-25). This amino acid is found in abundance in plasma and tissues. Normally, tyrosine is derived from the diet or synthesized in the liver following hydroxylation of the amino acid, phenylalanine. The first step in catechol synthesis is the hydroxylation of tyrosine to produce dihydroxyphenylalanine (DOPA). The decarboxylation of DOPA gives rise to the first catecholamine, dopamine. Dopamine is then hydroxylated to produce norepinephrine. The conversion of norepinephrine to epinephrine is mediated by phenylethanolamine-N-methyl transferase, which is present almost exclusively in the adrenal medulla. Because of such limited distribution of methyl transferase, a pheochromocytoma (tumor of chromaffin tissue) secreting a large amount of epinephrine is generally associated with the adrenal medulla.

There are two important metabolic events for the ultimate disposition of catecholamines in the body: catechol-O-methylation and oxidative deamination (Fig. 9-26). A certain percentage of the methylated amines are excreted either in the free form or conjugated with sulfuric or glucuronic acid. The majority of 3-methoxy derivatives, however, undergo deamination and are then oxidized to 3-methoxy-4-hydroxymandelic acid (vanilmandelic acid, VMA), which is excreted in the free form. When oxidative deamination occurs first, epinephrine and norepinephrine are converted to a common metabolite, 3,4-dihydroxymandelic acid (DHMA), which is subsequently O-methylated to VMA. There are some controversies about the sequence of the enzymatic inactivation of catecholamines. According to present evidence, O-methylation is considered to be quantitatively the most significant step and is followed by oxidative deamination. It should be noted, however, that irrespective of the nature of initial enzymatic attack, the ultimate end product of both epinephrine and norepinephrine is VMA. For this reason, the urinary measurement of the latter reflects the total rather than the differential production of these two catecholamines in the

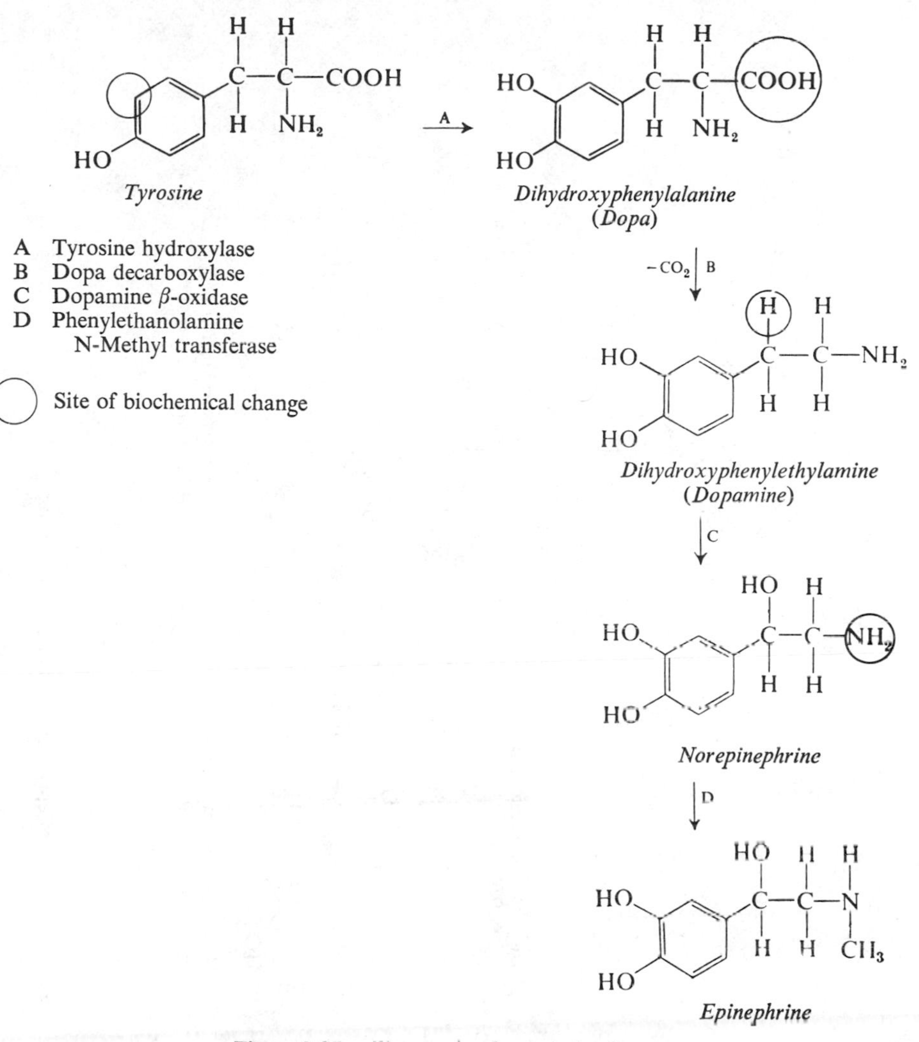

Figure 9-25. Biogenesis of catecholamines.

body. The final metabolite of dopamine following O-methylation and oxidative deamination is homovanillic acid, HVA (see Fig. 9-27).

Urinary Determination of Catecholamines

Because of the minute concentration of catecholamines in blood, the urinary assay of these hormones and their metabolites, metanephrine, normetanephrine, VMA, and HVA, is common practice in the clinical field. Bioassay techniques, which are limited to the measurement of physiologically active catecholamines such as epinephrine and norepinephrine, have been replaced in recent years by simpler, less costly, more precise, more sensitive, and more specific chemical methods. A large number of chemical methods are available in the literature. The subject has been reviewed by many investigators.[27,87,88]

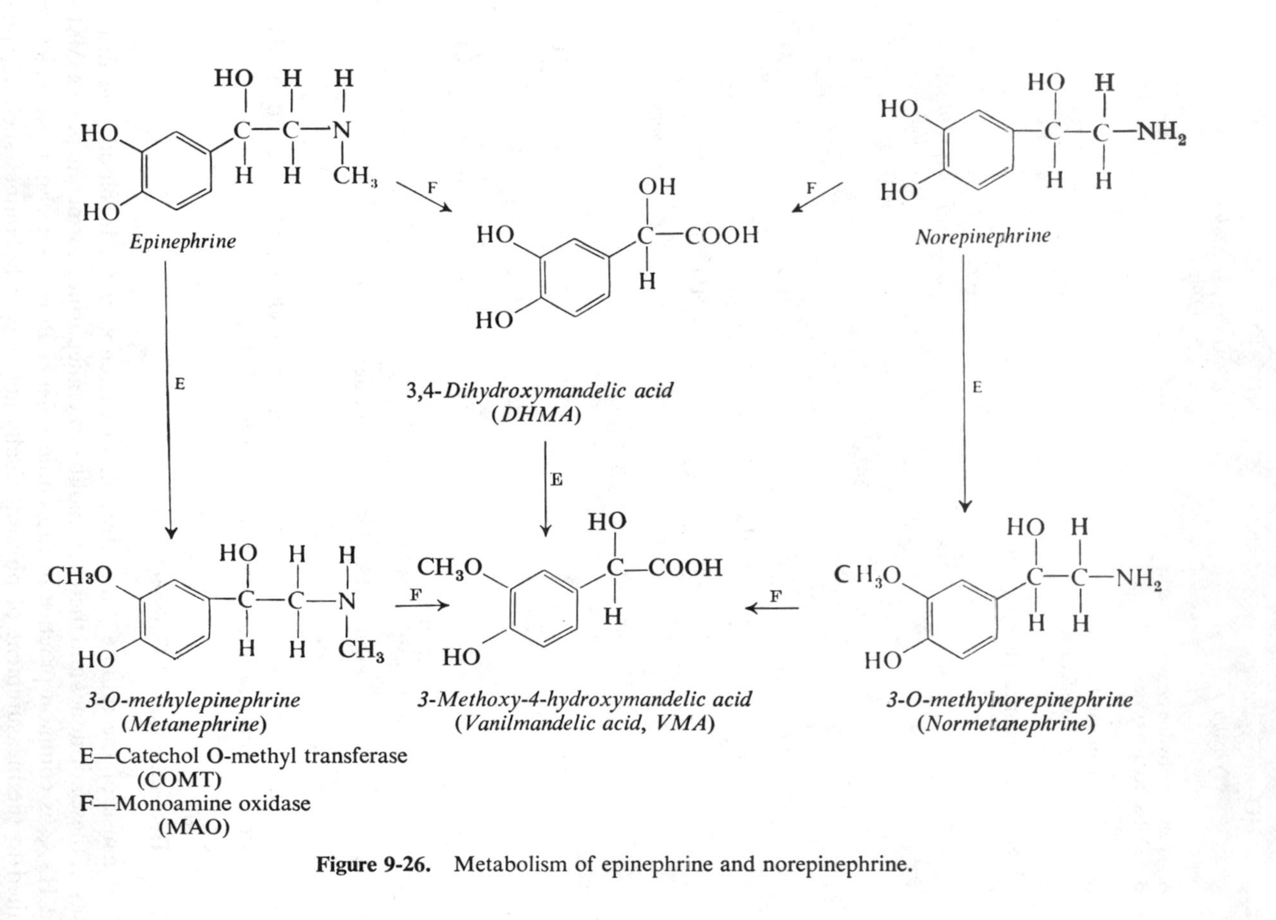

Figure 9-26. Metabolism of epinephrine and norepinephrine.

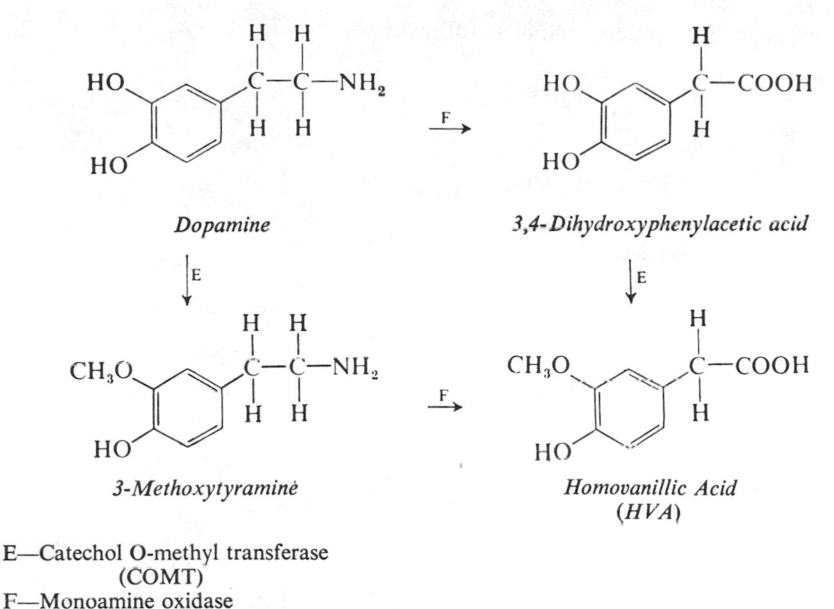

Figure 9-27. Metabolism of dopamine.

Individual determinations of urinary catecholamines, vanilmandelic acid, 3-methoxy metabolites (metanephrine and normetanephrine), and homovanillic acid (HVA) are generally indicated for clinical investigations. Methodological complications and deficiencies however do not allow all of them to become routine clinical tests, in spite of their usefulness for the diagnosis of catecholamine-secreting tumors (see Weil-Malherbe[88]). The application of gas chromatography appears to have the promise of circumventing many technical difficulties.[20,91,92] At the present time, however, measurements of total catecholamines and VMA are by far the most common routine clinical procedures.

Estimation of Total Catecholamines[76,88]

PRINCIPLE

Catecholamines are excreted partly in free form and partly in conjugated form, conjugated predominantly as ethereal sulfates which are easily hydrolyzed by heating with acid. Because of the high solubility of free catecholamines in aqueous media, customary extraction with organic solvents cannot be employed. They may, however, be adsorbed selectively on alumina (or appropriate resins), either by passing the hydrolyzed urine adjusted to pH 8.5 through a column of the adsorbent, or by mixing it with a batch of suspended alumina. The adsorbed catecholamines are then eluted with acid solution. The final method of quantitation by fluorimetry is based on the oxidation of catecholamines (epinephrine and norepinephrine) to stable fluorescent derivatives. Epinephrine and norepinephrine are oxidized to corresponding adrenochromes, which rearrange in alkaline medium to the fluorescent compounds adrenolutin and noradrenolutin respectively. These lutins are very unstable unless they are protected from oxidation by a suitable reducing agent such as ascorbic acid. The reaction sequence for epinephrine is as follows:

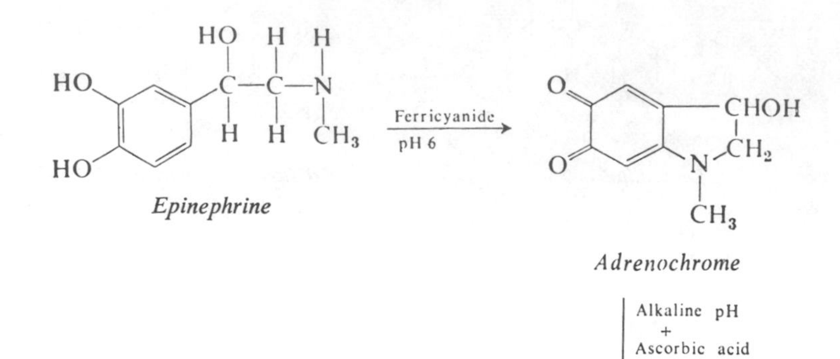

Alkaline pH + Ascorbic acid

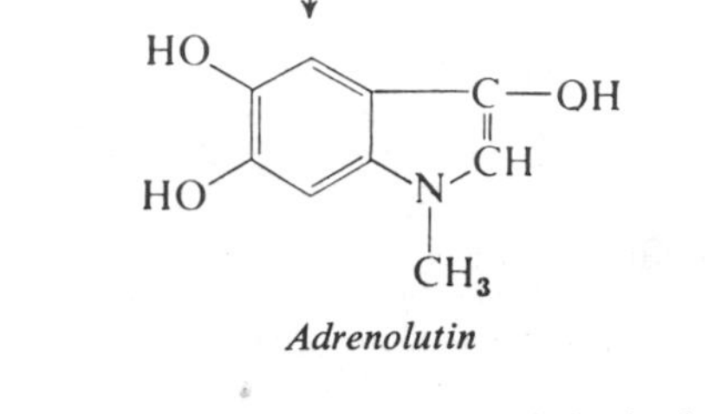

The procedure to be described consists of the following steps: (1) hydrolysis of urine at pH 1.5; (2) adsorption of catecholamines on a column of alumina at pH 8.5 and elution with 0.2 *M* acetic acid; (3) oxidation of catecholamines to lutine derivatives with potassium ferricyanide; and (4) measurement of fluorescence at 505 nm. following excitation at 400 nm.

REAGENTS

1. H_2SO_4, $6N$·
2. Aluminum oxide. (Alcoa activated alumina, grade F-20, 60–200 mesh.) Store tightly stoppered. If the activity has declined because of water adsorption, heat in a muffle furnace for 4 hours at 400°C.
3. Stock standard (noradrenaline). Dissolve 20.1 mg. levarterenol bitartarate, U.S.P., in 100 ml. of 0.1 *N* hydrochloric acid. This solution contains 100 μg./ml. This remains stable at least 6 months when refrigerated. (The reaction condition described here is optimal for the formation of a fluorescent derivative from noradrenaline. For this reason, noradrenaline rather than adrenaline is suitable for a reference standard. Also, noradrenaline is normally present in urine in larger quantity.)
4. Working standard. Dilute the stock standard 1:100 with distilled water. Prepare fresh before each use. (1 ml. $\equiv$ 1 μg. norepinephrine)
5. Sodium carbonate, 14 per cent.
6. Sodium acetate, 0.2 *M*. Dissolve 27.2 gm. $CH_3COONa{\cdot}3H_2O$ or 16.41 gm. CH_3COONa (anhydrous) in water and dilute to 1 liter with water. Adjust the pH to 8.5 with 0.5 *N* Na_2CO_3.
7. Acetic acid, 0.2 *M*. Dilute 6 ml. glacial acetic acid to 500 ml. water.
8. Sodium hydroxide, 10 *N*. Dissolve 40 gm. NaOH pellets and make volume up to 100 ml. with water.
9. Sodium acetate, saturated solution. Add 500 gm. anhydrous salt to 900 ml. water.
10. Acetate buffer solution. To 100 ml. saturated sodium acetate add 6.0 ml. 10 *N* NaOH. Do not keep this solution more than one month.

11. Potassium ferricyanide solution, 0.25 per cent. Prepare only small quantities and store in a dark bottle. Do not store for more than one month.

12. Sodium hydroxide, 20 per cent (w/v).

13. Ascorbic acid. Weigh out 20 mg. quantities and place into test tubes. Cover with parafilm until required for use.

14. Ascorbic acid–NaOH. Dissolve the ascorbic acid above in 1 ml. water, and add 9.0 ml. 20 per cent NaOH. Prepare fresh immediately prior to use.

PROCEDURE

Hydrolysis and alumina chromatography

1. Adjust two 10 ml. aliquots of urine to pH 2 with 6 N H_2SO_4.
2. Reflux the acidified urine for 20 minutes.
3. Cool and adjust the pH to 8.5 with 14 per cent Na_2CO_3. Centrifuge at 2000 r.p.m. for 5 minutes.
4. Prepare two alumina columns by adding an aqueous suspension of alumina to two chromatography tubes (8 × 130 mm.) until a height of approx. 4 cm is reached. Wash the columns with 5 ml. 0.2 M sodium acetate.
5. Allow all the urine to run through the columns, and discard the eluate. Never allow the column to run dry.
6. Wash the column with 20 ml. of 0.2 M sodium acetate followed by 5.0 ml. of water.
7. Elute the catecholamines with 50 ml. of 0.2 M acetic acid, and collect the initial 20 ml. of eluate.

Oxidation with ferricyanide and fluorimetry

8. Transfer 4 ml. aliquots of urine eluate and the reagents to each of three test tubes as shown in the following table.

Reagent added	*Standard + sample*	*Sample*	*Blank*
Eluate from column	4.0 ml.	4.0 ml.	4.0 ml.
Standard (1 μg./ml.)	0.5 ml	—	—
Distilled water	—	0.5 ml.	0.5 ml.
Acetate buffer	1.00 ml.	1.00 ml.	1.00 ml.
Potassium ferricyanide	0.1 ml.	0.1 ml.	0.1 ml.

Mix and allow to stand for two minutes.

9. To all tubes *except the blank* add 1.0 ml. freshly prepared ascorbic acid-NaOH. Mix and centrifuge for two minutes. To the blank add 1.0 ml. 20 per cent NaOH, and allow to stand for 20 minutes; centrifuge.

10. Read the fluorescence at 505 nm. following excitation at 400 nm.

CALCULATION

$$\mu\text{g of catecholamines per 24 hours} = \frac{A_t - A_b}{A_{st} - A_t} \times 0.5 \times \frac{20}{4} \times \frac{\text{T.V. Urine}}{10}$$

Where A_t = reading of the test
A_b = reading of the faded blank
A_{st} = reading of the internal standard (Standard + test)

COMMENTS

Fluorimetric methods are highly sensitive, but their susceptibility to extraneous factors demands scrupulous care in the cleaning of glassware and the preparation of reagents. Two common sources of contamination causing nonspecific fluorescence are introduced by the use of detergents and lubricants. Whenever these are used, measures should be taken to remove them completely. Other deleterious effects arise from the presence of heavy metals. Water to be used for the preparation of reagents must be de-ionized and distilled twice over alkaline potassium permanganate in an all-glass apparatus. Thorough washing of alumina with hydrochloric acid and water is also very important in order to remove alkaline, fluorescent, heavy metal impurities. The use of an internal standard is suggested because of the fact that pure noradrenaline standard produces greater fluorescence than the same standard added to urine eluates from alumina. This has been alleged to be caused by the presence of urinary "quenching" substances.

Although the present method has been shown to correlate fairly well with the clinical picture, certain inherent drawbacks of this procedure need to be mentioned. The single alumina column treatment of hydrolyzed urine does not achieve the necessary purification. As a result, nonspecific fluorescence in the so-called *faded-blank* of these preparations is high. The quenching of the fluorescence of the alumina eluates is also liable to cause serious error. Furthermore, the experimental conditions employed for the formation of the fluorescent compounds do not give maximum fluorescence. In addition, the reaction does not include dopamine. Weil-Malherbe[88] advocates the use of a second purification step, consisting of adsorption of amines on a cation exchange resin to eliminate interfering impurities. According to the same author, the addition of cupric ions catalyzes the formation of lutines and increases the intensity of fluorescence to a considerable extent. The method described appears to improve the intensity of fluorescence, the stability of the blank, and the discrimination between epinephrine and norepinephrine. Since the rates of oxidation of epinephrine and norepinephrine are different at different pH, a differential oxidation at pH 3 and pH 6 will enable one to measure epinephrine and norepinephrine individually, even if they are present in a mixture. The determination of norepinephrine in the presence of epinephrine may also be carried out by napthoquinone condensation.[6]

Since the method described here is not absolutely specific, falsely elevated results may be obtained because of the presence of urinary metabolites of medications such as adrenaline, adrenaline-like drugs, tetracyclines, quinidine and Aldomet (methyldopa). High values may also occur in patients with progressive muscular dystrophy, myasthenia gravis, and widespread burns. Individuals undergoing vigorous exercise may also show elevated urinary catecholamines. Malnutrition, transection of the cervical spinal cord, and familial dysautonomia may cause decreased excretion of catecholamines.

NORMAL VALUES

Normal values vary with the method of analysis. The excretion of total catecholamines in random urine samples ranges from 0 to 14 μg./100 ml. and up to 100 μg./24 hours.

REFERENCES

Sobel, C., and Henry, R. J.: Am. J. Clin. Path., *27*:240, 1957.
Weil-Malherbe, H.: *In* Methods of Biochemical Analysis. D. Glick (Ed.), N.Y., Interscience Publishers, 1968, vol. *16*, p. 293.

Urinary Estimation of VMA
(Vanilmandelic Acid; 3-Methoxy-4-hydroxymandelic Acid[59])

Since urinary VMA is quantitatively the most important metabolite of catecholamines, it has served as a useful index of endogenous production of these amines. Various methods depending on chromatography,[5] isotope dilution,[74] and colorimetry[64] have been used for its quantitative measurement. Among them, colorimetric estimations based on the oxidation of VMA to vanillin have found widest application in routine clinical laboratories. Since the first introduction of such a procedure,[64] a great number of methods with various modifications have appeared in the literature.[88] The method of Pisano et al,[59] which is considered to be the most simple and best suitable for routine clinical use, is described.

PRINCIPLE

VMA, along with other phenolic acids, is extracted from acidified urine with ethyl acetate. It is then extracted from the organic solvent with aqueous potassium carbonate solution. The potassium carbonate extract is treated with sodium metaperiodate to oxidize VMA to vanillin. To separate it from contaminating urinary phenolic acids, vanillin is selectively extracted into toluene and is determined spectrophotometrically at a wavelength of 360 nm.

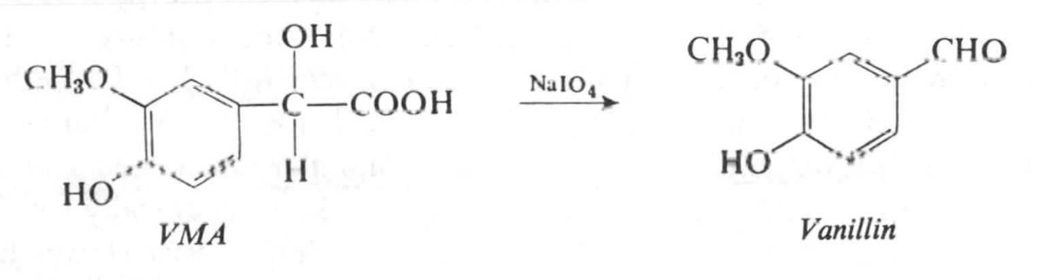

REAGENTS

All reagents should be of A.R. quality.

1. Hydrochloric acid, 6 *N*. Slowly add 500 ml. concentrated HCl to a 1 liter volumetric flask containing approximately 300 ml. water, and dilute to mark with water. Water to be used for the preparation of all reagents should be distilled twice in an all-glass apparatus.
2. Sodium chloride.
3. Ethyl acetate.
4. Potassium carbonate, 1 *M*. Dissolve 138 gm. of potassium carbonate in 1 liter of distilled water.
5. Sodium metaperiodate, 2 per cent (w/v) in distilled water. Make fresh weekly and store in an amber bottle.
6. Sodium metabisulfite, 10 per cent (w/v) in distilled water.
7. Acetic acid, 5 *N*. Dilute 286 ml. glacial acetic acid with distilled water to 1 liter.
8. Phosphate buffer, 1 *M*, pH 7.5. *Solution A:* Dissolve 178 gm. of disodium phosphate ($Na_2HPO_4 \cdot 2H_2O$) in distilled water and dilute to 1 liter. Store in a refrigerator. *Solution B:* Dissolve 27.22 gm. of potassium dihydrogen phosphate (KH_2PO_4) in 200 ml. of distilled water. Mix 168.2 ml. of Solution A with 31.8 ml. of Solution B. Check pH on pH meter, and make any necessary adjustment to obtain a pH of 7.5.
9. Hydrochloric acid, 0.01 *N* Dilute 0.83 ml. of concentrated HCl to 1 liter with distilled water.

10. Standard solution. Stock solution (1 mg./ml.): Accurately weigh 100 mg. of VMA and dissolve in 100 ml. of 0.01 N HCl in a volumetric flask. The solution is stable approximately three months under refrigeration.

Working solution (10 μg./ml.): Dilute 1 ml. of the stock solution to 100 ml. with 0.01 N HCl. Prepare fresh before use.

COLLECTION OF SPECIMEN

To preclude false elevations of urinary VMA, the intake of chocolate, coffee, bananas, foods containing vanilla, citrus fruits, and drugs such as aspirin and antihypertensive agents (e.g., Aldomet) must be restricted three days prior to and during collection of the urine specimen. The pH of the urine should be kept at approximately 2 during the collection by placing 10 ml. of 6 N HCl into a suitable container (dark brown bottle). After measurement of the total volume, 100 ml. aliquots may be stored at 4°C. for subsequent analysis. The specimen so preserved is stable for several weeks.

PROCEDURE

Pipet 0.2 per cent of the 24-hour volume into 50 ml. glass-stoppered (or screw-cap) centrifuge tubes marked previously as "tests," "internal standards" and "unoxidized blanks" in duplicate. To the internal standard tubes add 1 ml. of the working standard. Dilute the contents of all these tubes to 5.5 ml. with distilled water, and further acidify with 0.5 ml. of 6 N hydrochloric acid. Add a saturating amount of sodium chloride (approximately 3 gm.), mix, and extract with 30 ml. of ethyl acetate by shaking on a mechanical shaker for 30 min. Centrifuge for 5 minutes. Transfer 25 ml. of the organic extract (upper layer) to a second glass-stoppered centrifuge tube containing 1.5 ml. of 1 M potassium carbonate. Shake mechanically for 3 minutes and centrifuge for 5 minutes. Pipet 1 ml. of the carbonate phase (lower layer) to a third glass-stoppered centrifuge tube. To the test and standard tubes, add 0.1 ml. of 2 per cent sodium metaperiodate, mix, and stopper loosely; place all tubes including the tubes marked "unoxidized blank" (metaperiodate solution is omitted at this stage) into a water bath of 50°C. for 30 minutes. At the end of the incubation period, remove the tubes and cool to room temperature. To the unoxidized blanks, add 0.1 ml. of 2 per cent potassium metaperiodate and mix. Without delay add to all tubes 0.1 ml. of 10 per cent metabisulfite solution to reduce residual periodate. Neutralize with 0.3 ml. of 5 N acetic acid, and add 0.6 ml. of 1 M phosphate buffer at pH 7.5. (The pH can be checked at this point by adding one drop of 0.04 per cent aqueous cresol red. The solution should be yellow, indicating a pH of less than 8.8.) Shake mechanically for 3 min. with 20 ml. of toluene to extract vanillin, the oxidized product of VMA. Centrifuge for 5 minutes, and transfer 15 ml. of the toluene extract into a fourth glass-stoppered centrifuge tube containing 4.0 ml. of 1 M potassium carbonate. Shake mechanically for 3 minutes and centrifuge for 5 minutes. Transfer the carbonate layer containing vanillin into a microcuvet, and determine the absorbance at 360 nm. against a water blank.

CALCULATION

$$VMA \text{ excreted mg/24 hours} = \frac{A_t - A_b}{A_{st} - A_t} \times \frac{10}{1000} \times \frac{100}{0.2} = \frac{A_t - A_b}{A_{st} - A_t} \times 5$$

Where A_b = absorbance of "unoxidized" urine blank
A_t = absorbance of test
A_{st} = absorbance of internal standard (Standard plus test)

COMMENTS

Necessary care in the collection and preservation of the urine specimen as outlined in the text is very important. Diets and drugs contributing to the excretion of vanillin and other closely related phenoxy acids will yield falsely elevated results. However, as a precautionary measure, it is advisable to prepare an unoxidized blank for every sample, to correct for the presence of vanillin in urine even when the dietary restrictions prior to and during collection of the specimen have been followed. The absorbance may be measured against the unoxidized blank instead of the water blank, and in that case the need for subtraction of the absorbance of the urine blank from the absorbance of the test samples is obviated. The internal standard (addition of a known amount of VMA to the test specimen) compensates for procedural losses and for decomposition of vanillin and the relative inhibition of its formation because of the presence of unknown urinary factors. Indeed, at room temperature the oxidation of VMA by periodate proceeds smoothly in pure solutions, whereas an elevated temperature (50°C.) is required for the oxidation of VMA in urinary extracts. In occasional urinary samples, the oxidation may be strongly inhibited even at 50°C.[60]

The oxidation of VMA to vanillin is also sensitive to hydrogen ion concentration. In neutral and acidic solutions, oxidation results in the formation of a yellow pigment; strongly alkaline solutions, on the other hand, delay the formation of vanillin and cause its decomposition. Optimal conditions are obtained in 1 to 15 per cent (w/v) sodium or potassium carbonate solution at an approximate pH of 11 (Pisano et al.[60]). The maximum absorption of vanillin occurs at 348 nm. However, at this wavelength considerable absorbance of the oxidation product (p-hydroxybenzaldehyde) of p-hydroxymandelic acid, a normal constituent of urine, necessitates measurement at 360 nm., where the absorbance of vanillin is 80 per cent of its peak value and interference is minimal. It should be pointed out that the absorbance of vanillin drops sharply from 350 to 380 nm., and it is important that the wavelength setting remains exactly at 360 nm.[59]

NORMAL VALUES

Normal values range from 1.8 to 7.1 mg. of VMA per 24 hours, or 1.5 to 7.0 μg./mg. creatinine.

REFERENCE

Pisano, J. J., Crout, R. J., and Abraham, D.: Clin. Chim. Acta, 7:285, 1962.

CLINICAL SIGNIFICANCE

The adrenal medulla is a major source of catecholamines (especially epinephrine). The clinical importance of quantitation of these amines and their metabolites most often lies in the evaluation of the functional status of this gland. It should be noted, though, that the adrenal medulla is not essential for the maintenance of life, as is the adrenal cortex, and consequently, the consideration of deficient activity (hypofunction) of the gland has very little clinical significance; medullary hyperfunction, on the other hand, has great diagnostic importance particularly in patients with hypertension. Pheochromocytoma, a tumor of the chromaffin tissue in the adrenal medulla, is associated with the presence of greatly *increased* excretion of total catecholamines and VMA.

Elevated values have also been found in tumors of neural origin, such as neuroblastomas and ganglioneuromas. However, the findings in these pathological conditions are not consistent, and for proper diagnosis, the differential determinations of catecholamines and their metabolites, including homovanillic acid (HVA), are considered to be most helpful.

SEROTONIN AND ITS METABOLITE: 5-HYDROXYINDOLEACETIC ACID (5-HIAA)

Serotonin (5-hydroxytryptamine, 5-HT), a powerful smooth muscle stimulant and vasoconstrictor, is a derivative of the amino acid, tryptophan (see Fig. 9-28). This

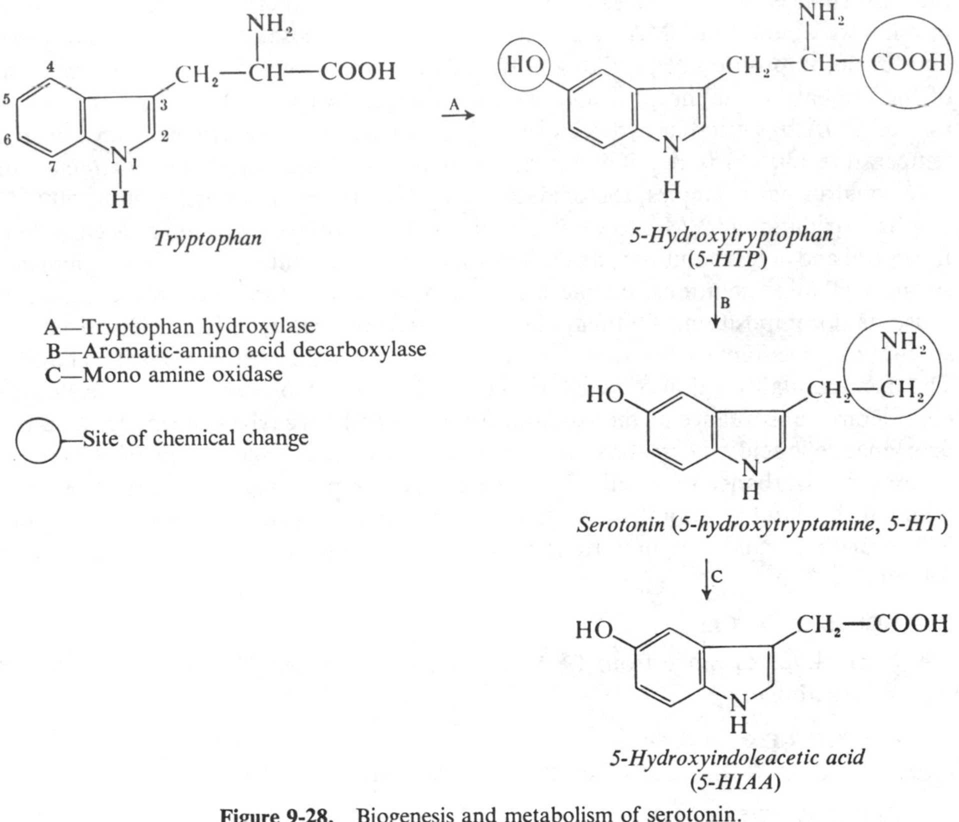

Figure 9-28. Biogenesis and metabolism of serotonin.

compound is formed predominantly in the enterochromaffin cells (otherwise known as argentaffin cells, because of their affinity for silver salts) of the gastrointestinal tract. It is transported in the blood by the platelets and is present in the brain and other tissues. In recent years, interest in this substance and other related hydroxy indoles has grown considerably because of the discovery that they are excreted in large amounts by patients with metastatic carcinoid syndrome (argentaffinoma).

The formation and breakdown of serotonin is depicted in Figure 9-28. The essential amino acid tryptophan is hydroxylated to form 5-hydroxytryptophan (5-HTP). Approximately 1 to 3 per cent of dietary tryptophan is normally metabolized by this pathway, but as much as 60 per cent of this amino acid is converted to 5-HTP in carcinoid tumors. The 5-hydroxytryptophan is decarboxylated to serotonin (5-hydroxytryptamine). While the enzymatic decarboxylation is most active in carcinoid tumors, it may also take place in the liver, kidney, lung, and brain.

Serotonin is pharmacologically the most active indole amine; however its biological activity is apparently lost when it is bound to tissues or platelets. It may rapidly undergo oxidative deamination in a tumor or in the blood after release from a tumor. The oxidative deamination of serotonin by the enzyme monoamine oxidase (MAO) leads to the formation of 5-hydroxyindole acetic acid (5-HIAA), which is quantitatively the most significant metabolite of the 5-hydroxyindole pathway. The majority of the 5-HIAA is excreted in the free form, although a small amount may be conjugated as the O-sulfate ester before excretion.

Urinary Determination of 5-Hydroxyindole Acetic Acid (5-HIAA)

Estimation of the parent hormone, serotonin, in blood and urine has been severely limited in the clinical laboratory because of its very low concentration and because of methodological complications. As a result, the urinary determinations of 5-HIAA continue to be the most useful means for the diagnosis of carcinoid tumors. In such cases, this metabolite of serotonin is excreted in very large amounts, often exceeding 350 mg. per 24 hours, and a positive result is obtained on simple qualitative (screening) tests. However, for early diagnosis, when tumors are small and have not metastasized, and in some carcinoid tumors where the excretion values barely exceed 8 mg., the more sensitive and specific quantitative test is required. Here, both qualitative and quantitative procedures based on the methods reported by Udenfriend and his associates[82] are described.

Screening Test[73]

The test is based on the development of a purple color, specific for 5-hydroxyindoles, on the addition of 1-nitroso-2-naphthol and nitrous acid. Other interfering chromogens are extracted in ethylene dichloride.

COLLECTION OF SPECIMEN

A random specimen is usually suitable for the screening test. For quantitative analysis, a complete collection over 24 hours should be used. The specimen should be collected in a bottle containing 25 ml. of glacial acetic acid. Acidification is important to prevent decomposition of 5-hydroxyindole acetic acid. Falsely negative results may occur in patients taking phenothiazine drugs. The ingestion of bananas, pineapples, walnuts, other foods containing serotonin, or cough medication containing glycerol guaiacolate may produce falsely positive results. Therefore, these drugs and diets should be restricted 3 to 4 days prior to the collection.

REAGENTS

1. 1-nitroso-2-naphthol, 0.1 per cent (w/v) in 95 per cent ethanol.
2. Sulfuric acid, 2 *N*.
3. Sodium nitrite, 2.5 per cent solution in water. Prepare freshly at frequent intervals. Refrigerate.
4. Nitrous acid reagent. Prepare fresh before use by adding 0.2 ml. of 2.5 per cent sodium nitrite to 5 ml. of 2 *N* sulfuric acid.
5. Ethylene dichloride, redistilled.

PROCEDURE

Pipet into a test tube 0.2 ml. of urine, 0.8 ml. of distilled water, and 0.5 ml. of 1-nitroso-2-naphthol. Mix. Similarly prepare another tube with *normal* urine to serve as a negative control. Add 0.5 ml. of freshly prepared nitrous acid reagent to both tubes and mix again. Let the tubes stand at room temperature for 10 minutes. Add 5 ml. of ethylene dichloride and shake. If turbidity results, centrifuge. A *positive test* shows a purple color in the top aqueous layer. The negative control with normal urine produces a slight yellow color.

COMMENTS

Dietary and drug restrictions as outlined are important if falsely negative or positive results are to be avoided. The substance p-hydroxyacetanilide derived from acetanilide or related drugs reacts similarly and adds to the color. Color formation may be inhibited in clinical conditions accompanied by excretion of large amounts of keto acids.

A purple color (positive test) will be seen at levels of 5-HIAA excretion as low as 40 mg. per 24 hours. At higher levels, the color is more intense and is almost black at levels above 300 mg. per 24 hours. A positive result should be verified with a quantitative method.

Quantitative Test[82]

PRINCIPLE

This method is considered to be specific for 5-hydroxyindoleacetic acid (5-HIAA). The procedure involves preliminary treatment of the urine with dinitrophenylhydrazine to react with any keto acids which may interfere later. The urine is then extracted with chloroform to remove indoleacetic acid. After saturation of the aqueous phase with sodium chloride, the 5-hydroxyindole acetic acid is extracted into ether and is then returned to a buffer of pH 7.0 for the color reaction, as described previously. The absorbance of the reaction product is measured at 540 nm. in a spectrophotometer.

COLLECTION OF SPECIMEN

Follow the procedure described for the screening test.

REAGENTS

Prepare reagents 1 to 4 as described for the screening test.

5. 2,4-dinitrophenylhydrazine, 0.5 per cent (w/v) in 2 *N* HCl.
6. Chloroform, reagent grade, freshly redistilled.
7. Sodium chloride.
8. Diethyl ether, reagent grade. Wash with a saturated solution of acidified ferrous sulfate to destroy peroxides, then with water (twice). Distill in an all-glass apparatus.
9. Ethyl acetate, redistilled.
10. Phosphate buffer, 0.5 *M*, pH 7.0. Mix 61.1 ml. of a solution of $Na_2HPO_4 \cdot 2H_2O$ (89.07 gm. per liter) and 38.9 ml. of a solution of KH_2PO_4 (68.085 gm. per liter). Check the pH on the pH meter, and make any necessary adjustments to obtain pH 7.0.
11. Standard solutions. *Stock solution:* Weigh accurately 5 mg. of pure 5-hydroxyindole acetic acid and dissolve in 20 ml. of glacial acetic acid. *Working solution:* Transfer 1 ml. of stock solution and dilute to 25 ml. with distilled water. This solution contains 10 μg./ml.

PROCEDURE

Transfer 6 ml. of urine into a 50 ml. glass-stoppered centrifuge tube. Prepare 'blank' and 'standard' tubes using 6 ml. of distilled water and 6 ml. (60 μg.) of working standard solution respectively. Add 6 ml. of 2,4-dinitrophenylhydrazine reagent (if a large precipitate occurs following addition of the reagent, centrifuge and use the supernatant for the subsequent steps). Allow to stand for 30 minutes. Add 25 ml. of chloroform, shake for a few minutes and centrifuge. Remove the organic layer, add another 25 ml. of chloroform, and repeat the extraction. After centrifuging, transfer a 10 ml. aliquot of the aqueous layer to a 40 ml. glass-stoppered centrifuge tube. Add approximately 4 gm. of solid sodium chloride and 25 ml. of ether. Shake for 5 min. and centrifuge. Transfer a 20 ml. aliquot of the ether to another 40 ml. glass-stoppered centrifuge tube. Add 1.5 ml. of phosphate buffer at pH 7.0. Shake for 5 minutes, centrifuge, and aspirate off the ether layer. Transfer 1 ml. of the aqueous phase to a 15 ml. glass-stoppered centrifuge tube containing 0.5 ml. of nitroso-naphthol reagent. Add 0.5 ml. of nitrous acid reagent, mix well, and warm at 37°C. for 5 minutes. Add 5 ml. of ethyl acetate, shake, allow the layers to separate, and remove the organic layer. Add another 5 ml. of ethyl acetate and repeat the above process. Transfer the final aqueous layer to a microcuvet. Read the absorbance of the standard and test samples against the blank preparation at 540 nm.

CALCULATION

$$\text{mg. 5-hydroxy indole acetic acid per liter of urine} = \frac{A \text{ of unknown}}{A \text{ of standard}} \times \frac{60}{1000} \times \frac{1000}{6}$$

$$= \frac{A \text{ of unknown}}{A \text{ of standard}} \times 10$$

COMMENTS

The important points relevant to the urinary determination of 5-hydroxyindole acetic acid have already been discussed. The salient feature of the quantitative method is the removal of keto acids through the formation of phenylhydrazones. The presence of excessive amounts of keto acids interferes with color formation. Reaction of 5-hydroxyindole acetic acid with nitroso-naphthol to form a violet chromophore is claimed to be very specific. Serotonin and 7-hydroxyindoles do not respond to this reaction.[83] The use of organic solvents such as chloroform and ethyl acetate at different steps of the procedure ensures the removal of other indoleacetic acids and nonspecific substances. The distribution of 5-HIAA between ether and water is very low; the addition of a saturating amount of sodium chloride aids in the quantitative transfer of 5-HIAA into the ether phase. The pH of the buffer solution should be exactly 7, since at higher pH values the compound becomes progressively more unstable.

NORMAL VALUES

The normal values for urinary excretion of 5-HIAA range from 2 to 8 mg. per 24 hours.

REFERENCES

Sjoerdsma, A., Weissbach, H., and Udenfriend, S.: J. Am. Med. Assoc., *159*:397, 1955.
Udenfriend, S., Titus, E., and Weissbach, H.: J. Biol. Chem., *216*:499, 1955.

CLINICAL SIGNIFICANCE

Metastatic carcinoid tumors (argentaffinoma) arising from the argentaffin cells produce excessive amounts of serotonin. Since the bulk of urinary 5-hydroxyindole acetic acid (5-HIAA) is derived from serotonin (5-hydroxytryptamine), the amount of this acid excreted reflects the secretion of the amine. Urinary levels of 5-HIAA ranging from 50 to 100 mg. per day have been reported in patients with carcinoid tumors. While marked *increases* are always associated with carcinoidosis, a slight elevation may be observed in some patients with nontropical sprue, or transiently following the administration of reserpine. In renal insufficiency and in some instances of phenylketonuria, *decreased* excretion of 5-HIAA has been noted.

REFERENCES

1. Albert, A. (ed.): Human Pituitary Gonadotrophins, Springfield, Illinois, Charles C Thomas, 1961.
2. Albert, A., Mattox, V. R., and Mason, H. L.: *In* Textbook of Endocrinology. Williams, R. H. (ed.), Philadelphia, W. B. Saunders, 1968, p. 1181.
3. Allen, W. M.: J. Clin. Endocrinol., *10*:71, 1950.
4. Appleby, J. I., Gibson, G., Norymberski, J. K., and Stubbs, R. D.: Biochem. J., *60*:453, 1955.
5. Armstrong, M. D., McMillan, A., and Shaw, K. N. F.: Biochem. Biophys. Acta, *25*:422, 1957.
6. Auerbach, M. E., and Angell, E.: Science, *109*:537, 1949.
7. Bayliss, W. M., and Starling, E. H.: J. Physiol., *28*:325, 1902.
8. Bell, E. T., and Loraine, J. A. (eds.): Recent Research on Gonadotrophic Hormones, Edinburgh and London, Livingstone, Ltd., 1967.
9. Berson, S. A., and Yalow, R. J.: *In* The Hormones. Pincus, G., Thimann, K. V., and Astwood, E. B. (eds.), New York, Academic Press, 1964, vol. *4*, p. 557.
10. Borth, R.: *In* Ciba Foundation Colloquia on Endocrinology, Vol. *2*. London, Churchill, 1952, p. 45.
11. Borth, R., Ferin, M., and Menzi, A.: Acta Endocrinol., *50*:335, 1965.
12. Borth, R., Lunenfeld, B., and Menzi, A.: *In* Human Pituitary Gonadotrophins. Albert, A. (ed.), Springfield, Charles C Thomas, 1961, p. 13.
13. Breuer, H.: Vitam. and Horm., *20*:285, 1962.
14. Brody, S., and Carlstrom, G.: J. Clin. Endocrinol., *22*:564, 1962.
15. Brown, J. B.: Biochem. J., *60*:185, 1955.
16. Brown, J. B., Macnaughton, C., Smith, M. A., and Smyth, B.: J. Endocrinol., *40*:175, 1968.
17. Callow, R. K.: *In* Hormone Assay. Emmens, C. W. (ed.) New York, Academic Press, 1950, p. 363.
18. Creech, B. G.: J. Gas Chromat., *2*:195, 1964.
19. Chattoraj, S. C., and Wotiz, H. H.: Fertil. Steril., *18*:342, 1967.
20. Clarke, D. D., Wilk, S., Gitlow, S. E., and Franklin, M. J.: J. Gas Chromat., *5*:307, 1967.
21. Dixon, P. F., Booth, M., and Butler, J.: *In* Hormones in Blood, Vol. *2*. Gray, C. H., and Bacharach, A. L. (eds.), New York, Academic Press, 1967, p. 305.
22. Dorfman, R. I. (ed.): Methods in Hormone Research, Vol. I, New York, Academic Press, 1962.
23. Dorfman, R. I. (ed.): Methods in Hormone Research, Vol. II. New York, Academic Press, 1962.
24. Dorfman, R. I., and Ungar, F.: Metabolism of Steroid Hormones. New York, Academic Press, 1965.
25. Drekter, I. J., Heisler, A., Scism, G. R., Stern, S., Pearson, S., and McGavack, T. H.: J. Clin. Endocrinol., *12*:55, 1952.
26. Eik-Nes, K. B., and Horning, E. C. (eds.): Gas Phase Chromatography of Steroids. New York, Springer-Verlag, 1968.
27. Elmadjian, F.: *In* Methods in Hormone Research, Vol. I. Dorfman, R. I. (ed.): New York, Academic Press, 1962, p. 337.
28. Emmens, C. W.: *In* Methods in Hormone Research, Vol. II, Dorfman, R. I. (ed.), New York, Academic Press, 1962, p. 3.
29. Few, J. D.: J. Endocrinol., *22*:31, 1960.
30. Galli-Mainini, C.: J. Clin. Endocrinol., *7*:653, 1947.
31. Henry, J. B., Krieg, A. F., Davidsohn, I.: *In* Clinical Diagnosis by Laboratory Methods. Davidsohn, I., and Henry, J. B. (eds.) Philadelphia, W. B. Saunders Co., 14th edition. p. 1181.
32. Horning, E. C., Vanden Heuvel, W. J. A., and Creech, B. G.: *In* Methods of Biochemical Analysis. Glick, D. (ed.), New York, Interscience Publishers, 1963, Vol. 11.
33. Hunter, W. M., and Greenwood, F. C.: Biochem. J., *91*:43, 1964.
34. Ittrich, G.: Acta Endocrinol., *35*:34, 1960.
35. Ittrich, G.: Z. Physiol. Chem., *312*:1, 1958.

36. Kirkham, K. E.: Vitamins and Hormones, *23*, 1965.
37. Klinefelter, H. F., Jr., Albright, F., and Griswold, G. C.: J. Clin. Endocrinol., *3*:529, 1943.
38. Klopper, A., and Michie, E. A.: J. Endocrinol., *13*:360, 1956.
39. Klopper, A. L.: *In* Methods in Hormone Research, Vol. I. R. I. Dorfman (ed.), New York, Academic Press, 1962. p. 139.
40. Klopper, A., Michie, E. A., and Brown, J. B.: J. Endocrinol., *12*:209, 1955.
41. Kober, S.: Biochem. Z., *239*:209, 1931.
42. Li, C. H.: Recent Progr. Hormone Res., *18*:1, 1962.
43. Lipsett, M. B. (ed.): Gas Chromatography of Steroids in Biological Fluids, New York, Plenum Press, 1965.
44. Loraine, J. A.: *In* The Pituitary Gland, Vol. 2. Harris, G. W., and Donovan, B. T. (eds.), London, Butterworth, 1965.
45. Loraine, J. A.: Vitamins and Hormones, *14*:305, 1956.
46. Loraine, J. A., and Bell, E. T.: Hormone Assays and Their Clinical Application, Second Edition, The Williams and Wilkins Co., Baltimore, 1966.
47. Loraine, J. A., and Brown, J. B.: J. Endocrinol., *18*:77, 1959.
48. Marlow, H. W.: J. Biol. Chem., *183*:167, 1950.
49. Marrian, G. F.: *In* Proceedings of Third Internat. Congress in Biochem., Brussels, 1955, p. 205.
50. Martin, D. B., Raynold, A. E., and Dagenais, Y. M.: Lancet, *2*:76, 1958.
51. Mattingly, D.: J. Clin. Pathol., *15*:374, 1962.
52. Metcalf, M. G., J. Endocrinol., *26*:415, 1963.
53. Midgley, A. R., Jr.: Endocrinol., *79*.10, 1966.
54. Nielsen, E., and Asfeldt, V. H.: Scan. J. Clin. Lab. Invest., *20*:185, 1967.
55. Norymberski, J. K.: Nature, *170*:1074, 1952.
56. Odell, W. D., Wilbur, J. F., and Paul, W. E.: Metabolism, *14*:465, 1965.
57. Parlow, A. F.: Fed. Proc., *17*:402, 1958.
58. Peterson, R. E.: *In* Lipids and Steroid Hormones in Clinical Medicine. Sunderman, F. W., and Sunderman, F. W., Jr. (eds.), Philadelphia, J. B. Lippincott Co., 1960, p. 164.
59. Pisano, J. J., Crout, R. J., and Abraham, D.: Clin. Chim. Acta, *7*:285, 1962.
60. Porter, C. C., and Silber, R. H.: J. Biol. Chem., *185*:201, 1950.
61. Preedy, J. R. K., and Aitken, E. H.: J. Biol. Chem., *236*:1297, 1961.
62. Randle, P. J., and Taylor, K. W.: *In* Hormones in Blood. Gray, C. H., and Bacharach, A. L. (eds.), New York, Academic Press, 1961, p. 11.
63. Randle, P. J.: Brit. Med. J., *1*:1237, 1954.
64. Sandler, M., and Ruthven, C. R. J.: Lancet, *2*:114, 1959
65. Sanger, F.: Science, *129*:1340, 1959.
66. Sayers, M. A., Sayers, G., and Woodbury, L. A.: Endocrinol. *42*:379, 1948.
67. Schmidt-Elmendorff, H., Loraine, J. A., and Bell, E. T.: J. Endocrinol., *24*:349, 1962.
68. Scholler, R., Leymarie, P., Heron, M., and Jayle, M. F.: Acta Endocrinol., *52*, Supplementum 107, 1966.
69. Scommegna A., and Chattoraj, S. C.: Obstet. Gynec., *32*:277, 1968.
70. Scommegna, A., Chattoraj, S. C., Wotiz, H. H.: Fertil. Steril., *18*:342, 1967.
71. Scommegna, A., and Chattoraj, S. C.: Am. J. Obstet. and Gynec., *99*:1087, 1967.
72. Silber, R. H., and Porter, C. C.: *In* Methods of Biochemical Analysis. Glick, D. (ed.), New York, Interscience, 1957, Vol. 4, p.139.
73. Sjoerdsma, A., Weissbach, H., and Udenfriend, S.: J. Am. Med. Assoc., *159*:397, 1955.
74. Smith, A. A., Schweitzer, J. W., and Wortis, S. B.: Fed. Proc., *18*:145, 1959.
75. Sobel, C. S., Golub, O. J., Henry, R. J., Jacobs, S. L., and Basu, G. K.: J. Clin. Endocrinol., *18*:208, 1958.
76. Sobel, C., and Henry, R. J.: Am. J. Clin. Path., *27*:240, 1957.
77. Steelman, S. L., and Pohley, F. M.: Endocrinol., *53*:604, 1953.
78. Sunderman, F. W., Jr.: *In* Lipids and the Steroid Hormones in Clinical Medicine. Sunderman, F. W., and Sunderman, F. W., Jr. (eds.), Philadelphia. J. B. Lippincott Co., 1960, p. 162.
79. Sweat, M. L.: Anal. Chem., *26*:773, 1954.
80. Thomas, B. S.: *In* Gas Chromatography of Steroids in Biological Fluids. Lipsett, M. B. (ed.), New York, Plenum Press, 1965, p. 1.
81. Turner, C. D.: General Endocrinology, 4th Edition. Philadelphia, W. B. Saunders Co., 1966.
82. Udenfriend, S., Titus, E., and Weissbach, H.: J. Biol. Chem., *216*:499, 1955.
83. Vanden Heuvel, W. J. A., Sweeley, C. C., and Horning, E. C.: J. Am. Chem. Soc., *82*:3481, 1960.
84. Vander Molen, H. J.: *In* Gas Phase Chromatography of Steroids, Eik-Nes, K. B., and Horning, E. C. (eds.), New York, Springer-Verlag, 1968, p. 150.
85. Vernikos-Dannelis, J.: Vitamins and Hormones, *23*:97, 1965.
86. Vestergaard, P., and Clausen, B.: Acta Endocr., Supplement 64, 1962.
87. von Euler, U.S.: *In* Hormones in Blood. Gray, C. H., and Bacharach, A. L. (eds.), New York, Academic Press, 1961, p. 515.

88. Weil-Malherbe, H.: *In* Methods of Biochemical Analysis, Vol. *16*, Glick, D., (ed.), New York, Interscience Publishers, 1968, p. 293.
89. Weiner, N.: *In* The Hormones. Pincus, G., Thimann, K. V., and Astwood, E. B. (eds.), New York, Academic Press, 1964, Vol. *4*, p. 403.
90. Wide, L., and Gemzell, C. A.: Acta Endocrinol, *35*:261, 1960.
91. Williams, C. M., and Greer, M.: Clin. Chim. Acta, *11*:495, 1965.
92. Williams, C. M., and Greer, M.: Clin. Chim. Acta, *7*:880, 1962.
93. Williams, R. H., (ed.): Textbook of Endocrinology. Philadelphia, W. B. Saunders Co., 1968, 4th Edition.
94. Wolstenholme, G. E. W., and Cameron, M. P.: *In* Ciba Foundation Colloquia on Endocrinology, Vol. *14*, London, Churchill, 1962.
95. Wotiz, H. H.: *In* Gas Chromatography of Steroids in Biological Fluids. Lipsett, M. B. (ed.), New York, Plenum Press, 1965, p. 301.
96. Wotiz, H. H., and Chattoraj, S. C.: *In* Gas Chromatographic Determination of Hormonal Steroids. Polvani, F. (ed.), New York, Academic Press, 1968, p. 261.
97. Wotiz, H. H., and Chattoraj, S. C.: *In* Gas Chromatography of Steroids in Biological Fluids. Lipsett, M. B. (ed.), New York, Plenum Press, 1965, p. 195.
98. Wotiz, H. H., and Clark, S. J.: *In* Gas Chromatography in the Analysis of Steroid Hormones. New York, Plenum Press, 1966.
99. Yalow, R. S., Glick, S. M., and Berson, S. A.: J. Clin. Endocrinol., *24*:1219, 1964.
100. Zimmermann, W.: Ztschr. Physiol. Chem., *233*:257, 1935.
101. Zondek, B., Sulman, F., Black, R.: J. Amer. Med. Assn., *128*:939, 1945.

Part II: Thyroid Function Tests

by Sheldon Berger, M.D.

The fully developed human thyroid gland, as viewed from the front, has been likened to a butterfly. The "wings of the butterfly" are the two lobes, and their connecting piece the isthmus. The gland is wrapped tightly around the anterior and lateral surfaces of the trachea and larynx, the isthmus crossing the trachea just below the cricoid cartilage. Each of the lobes measures approximately 4.0 cm. in height, 1.5 to 2.0 cm. in width, and 2.0 to 4.0 cm. in thickness; the isthmus is about 2.0 cm. in both length and width and 0.2–0.6 cm. in thickness. The average gland weighs 25 to 30 gm.

The *secretory unit* of the gland is the follicle. Each follicle is a sphere having a diameter of approximately 300 μ., and its walls consist of an epithelial cell monolayer. These cells manufacture and secrete the two thyroid hormones, L-thyroxine (3,5,3′,5′-L-tetraiodothyronine) and L-triiodothyronine (3,5,3′-L-triiodothyronine). These hormones, commonly designated T_4 and T_3 respectively, are stored within the lumina of the follicles.

Hormone synthesis and release is controlled by a polypeptide which originates in the anterior lobe of the pituitary, the thyroid-stimulating hormone, or TSH. Pituitary secretion of TSH is, in turn, regulated by two factors: a humoral signal released by the hypothalamus into the hypophyseal-portal venous system, and the concentration of free T_4 (and probably of free T_3 as well) in the interstitial fluid which bathes both the pituitary and the hypothalamus. More will be said about "free T_4" later. Since an *increased* concentration of hormone *suppresses* secretion of TSH, and a *decreased* concentration *augments* secretion of TSH, this regulatory arrangement has been designated a negative feedback system.

Thyroid hormone biosynthesis involves five distinct steps (Fig. 9-29):

1. Thyroidal trapping of serum iodide, catalyzed by a trapping enzyme. This step, thought to be oxygen-dependent, is inhibited by respiratory inhibitors such as the cyanide and azide ions.

2. Enzymatic oxidation of iodide to some reactive intermediate ("active I"). The enzyme responsible for iodide oxidation is a peroxidase, and the reaction has been written as follows:

$$H_2O_2 + 2\,I^- + 2\,H^+ \xrightarrow[\text{peroxidase}]{\text{iodine}} 2\text{ Active I} + 2\,H_2O$$

Active I may be the iodinium ion (I^+).

3. Iodination of tyrosyl residues present in follicular thyroglobulin by active I to produce mono- and diiodotyrosine (MIT and DIT). Steps 2 and 3 are closely coupled. The product of the peroxidase reaction, active I, is quickly incorporated into the 3-, or the 3- and 5- positions of the tyrosyl residues by a tyrosine iodinase.

4. Oxidation and condensation of MIT and DIT to form the iodothyronines, T_3 and T_4. This step is presumed to represent an enzymatic coupling of two molecules of

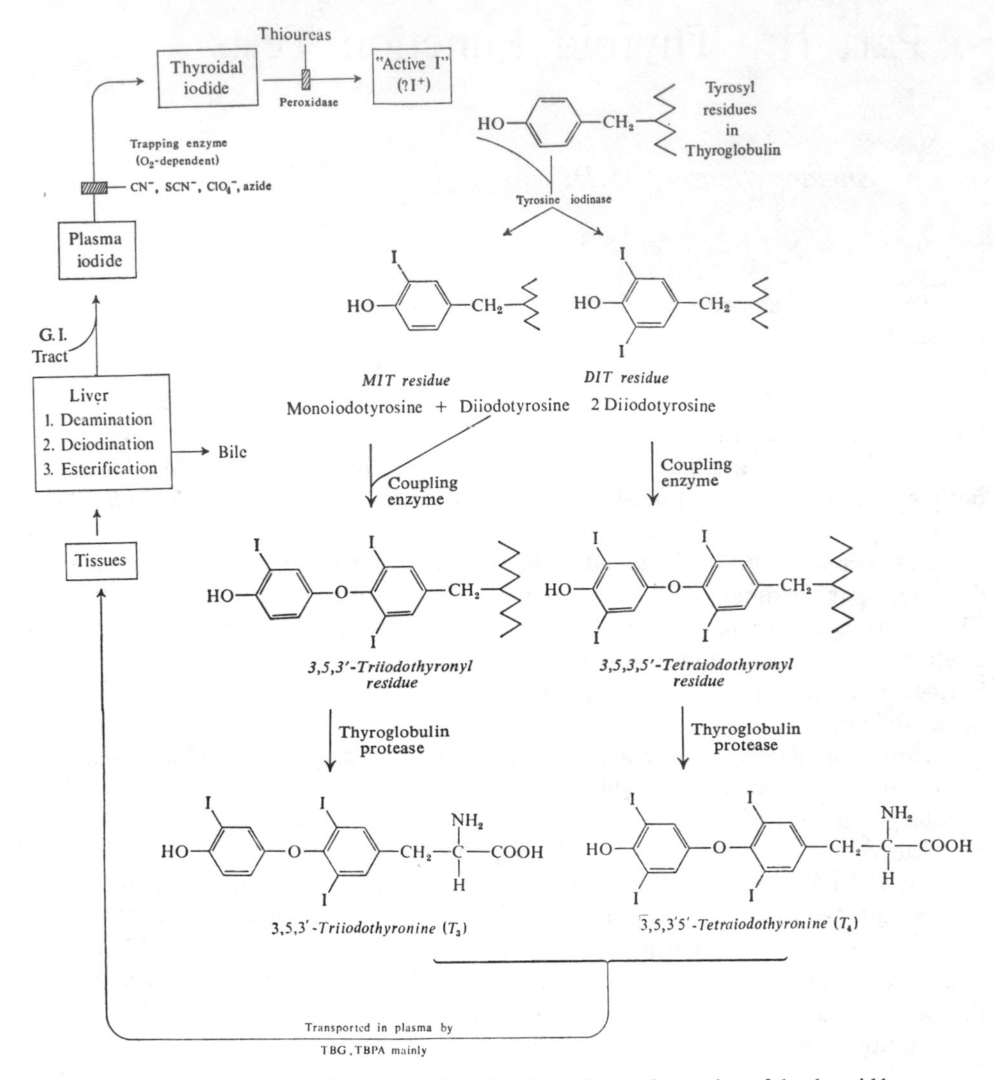

Figure 9-29. The metabolism of iodine, emphasizing formation and secretion of the thyroid hormones.

iodotyrosine with extrusion of an alanine side chain. It is important to appreciate that this coupling occurs while the iodotyrosines are bound in peptide linkages to other amino acids, and not in the free form. Thus, the T_3 and T_4 formed as a result of iodotyrosyl coupling are also stored as peptide-linked iodothyronyl residues in thyroglobulin.

5. Proteolytic cleavage of follicular thyroglobulin to "free" the MIT, DIT, T_3, and T_4 incorporated, by peptide bonds, into the parent protein. The MIT and DIT are promptly deiodinated, and the iodide recycled.

Proteolytic cleavage is achieved by thyroglobulin protease. MIT and DIT are deiodinated by a specific deiodinase (free iodotyrosine deiodinase), which is an NADP-enzyme found in the microsomes of thyroid epithelial cells. This deiodination is considered a conservation step since it prevents loss of iodine from the gland in the form of biologically inactive material (i.e., MIT and DIT).

After T_3 and T_4 diffuse into the blood they are bound to three proteins: an interalpha globulin (thyroid-binding globulin, TBG), a prealbumin (thyroid-binding prealbumin, TBPA), and an albumin (thyroid-binding albumin, TBA). More will be said about the carrier proteins, particularly TBG.

The metabolism of T_3 and T_4, like that of all amino acids, involves their deamination, either oxidatively or by transamination. The resulting thyropyruvates are then decarboxylated to form thyroacetates, which still exhibit some hormonal activity. The hormones may also be conjugated as glucuronides in the liver, in which form they enter the bile. Proper iodine balance in the organism requires that iodine be removed from the various iodine-containing catabolites by specific deiodinases and returned to the thyroid as iodide.

The biologic effects of the thyroid hormones are profound. They influence the rate of oxygen consumption and heat production in virtually all tissues. The mechanism of this effect may involve an uncoupling of oxidation and phosphorylation. The hormones are also indispensable for the growth, development, and sexual maturation of growing organisms. The capacity of the thyroid hormones to trigger the metamorphosis of amphibian tadpoles is, in fact, one of the classical bioassay systems in experimental endocrinology.

Evaluating Thyroid Function

Characterization of thyroid function has become, in recent years, an elegant laboratory exercise. The development of ingenious biochemical procedures for estimating the concentration and biological effectiveness of the thyroid hormones in plasma, and the general availability of radioactive iodine for clinical use, have broadened the understanding of thyroid physiology to a degree which few would have imagined 20 years ago. Still, the newer tests which have been developed, though increasingly specific in their capacity to measure thyrometabolic status, are not without important limitations. These limitations will therefore be considered in some detail. None of the tests can be interpreted casually, since correct interpretation often depends on a subtle point in iodine metabolism. The modern tests now require an order of awareness of basic matters which was quite unnecessary in the (not too remote) days when measurement of the basal metabolic rate (BMR) was the "court of final appeal." But progress is not without its associated problems and responsibilities, and one simply must learn what must be known!

Unlike the BMR, none of the tests which have supplanted it directly assesses the "metabolic impact" of the thyroid hormones on peripheral tissues, and, after all, it is this impact which ultimately determines the clinical thyroid status. A word about the BMR is perhaps in order then, at the outset, to explain why it has been largely superseded.

Following the classic studies of Magnus-Levy, Benedict, the DuBois, and Boothby, the BMR became *the* test for assessing the calorigenic effect of the thyroid hormones, and in general—when performed with due care—it served reasonably well. The great problems connected with the test involved the need for careful preparation of the patient, meticulous performance of the test, and the frequency of nonspecific results caused by interference by a variety of extra-thyroidal systemic disorders. Extensive experience with the procedure permits one to compile an impressive

TABLE 9-6. *Extrathyroidal Factors which Affect the BMR**

Factors which Increase the BMR

1. Faulty preparation for the test
   a. anxiety
   b. inadequate rest
   c. recent food ingestion
   d. calorigenic drugs (xanthines, sympathomimetics, thyroid preparations).
2. Errors in test performance
   a. increased environmental noise
   b. extremes in room temperature
   c. uncomfortable table or bed
   d. insufficient elevation of patient (patients with cardiac failure, obesity)
   e. tight nose piece, girdle or collar
   f. oxygen leaks (test system, perforated ear drum)
   g. improper breathing pattern
   h. beginning test with overexpansion of chest
3. Systemic disorders which increase oxygen consumption
   a. involuntary motor disorders
   b. cardiac failure
   c. pulmonary disease
   d. leukemia
   e. skin disease
   f. fever of any cause

Factors which Decrease the BMR

1. Errors in test performance
   a. saturated soda lime
   b. beginning test with underexpansion of chest
2. Systemic disorders which decrease oxygen consumption
   a. malnutrition
   b. Addison's disease
   c. nephrotic syndrome
   d. shock

* After Ingbar, S. H., and Woeber, K. A.: *In* Textbook of Endocrinology. Williams, R. H. (Ed.) Philadelphia, W. B. Saunders Co., 1968.

catalogue of the many factors which influence, and therefore obscure, interpretation of the BMR. These are shown in Table 9-6.

A search for laboratory alternatives was clearly in order. This chapter will be limited to those alternatives which most laboratories either rely upon now or are likely to institute soon. Each of the tests rests on important principles of thyroid physiology and iodine metabolism, as these are now understood. In an age when no principle is sacrosanct, there is no reason, of course, to consider the following material uniquely immune to challenge.

Serum Protein-Bound Iodine (PBI)

Iodine circulates in plasma in two forms: first, as a constituent of the two thyroid hormones, T_3 and T_4; and second, as free iodide. The average concentration of the former, i.e., hormonal iodine, in sera from normal subjects is 5.4 μg./100 ml.; free iodide concentration is generally 1.0 μg./100 ml. or less.

Unlike free iodide, both T_3 and T_4 are almost completely bound in dissociable linkage to various carrier proteins. Approximately 99.95 per cent of total serum T_4 is, in fact, now known to be protein-bound. If, therefore, one measures the iodine content of a protein precipitate of serum, one will have measured the concentration of hormonal iodine in the sample. Further, one will have measured mainly T_4 iodine, since T_3 is cleared from serum so rapidly that it does not contribute significantly to the protein-bound iodine, or PBI.

Barker and his co-workers,[2] in 1951, were the first to report a practical technique for measuring serum PBI. Their method, and all subsequent modifications of it, involves four basic steps: (1) precipitation of serum proteins, (2) repeated washing of the precipitate to remove trapped iodide, (3) oxidation of the serum protein-thyroid hormone complexes to liberate free iodine, and (4) measurement of the iodine so liberated on the basis of its ability to catalyze the reduction of ceric sulfate by the arsenite ion.

Serum PBI values range between 3.5 or 4.0 and 8.0 μg./100 ml. (mean = 5.4 μg./100 ml.) in normal subjects. There is no significant difference in PBI values between adult men and women; both sexes, however, tend to show a slight fall beyond the age of 50. Duplicate samples should agree to within 0.6 μg./100 ml., and different sera from the same individual should agree to within 1.0 μg./100 ml. The manual method of Barker, Humphrey, and Soley is very satisfactory, and its modification by Henry will be described in detail.[1]

Specimens should be collected carefully in thoroughly acid-cleaned test tubes. Since red cells are essentially devoid of thyroxine, hemolysis lowers the PBI level slightly because of dilution of serum by hemolysate. Sera, once separated from red cells, are extremely stable, even at room temperature.

PRINCIPLE

Serum proteins are precipitated with zinc hydroxide. The washed precipitate is then alkalinized by the addition of sodium carbonate to minimize iodine loss during incineration, dried, and incinerated at 620°C. Iodide present in the alkaline ash is measured by the ceric-arsenite reaction.

A variety of *wet digestion* methods for the determination of protein bound iodine have also been proposed. Only two, however, have found application in the routine laboratory. In the method by Zak,[24] serum proteins are precipitated with trichloroacetic acid, and are then digested with a chloric-perchloric acid digestion mixture in the presence of a small amount of chromate as catalyst. After evaporation of the perchloric acid excess, the iodine in the sample is determined by the classic ceric-arsenite reaction. Although this procedure offers advantages with respect to speed, it has not been widely used because of two problems: inaccuracy (approximately ±15 per cent error), and the need for extremely close control of temperature in *all* tubes processed (tests and standards).

The digestion of organic material by a mixture of sulfuric, nitric, and perchloric acids has been widely employed in the Technicon procedure for automated determination of serum PBI. Complete digestion requires only three minutes at 280°C., and the conditions of the automated system are extremely well-controlled. These factors confer acceptable accuracy to the procedure.

The iodine-catalyzed reaction between ceric ions (Ce^{+4}) and arsenious acid (As^{+3}) is a two-step reaction:*

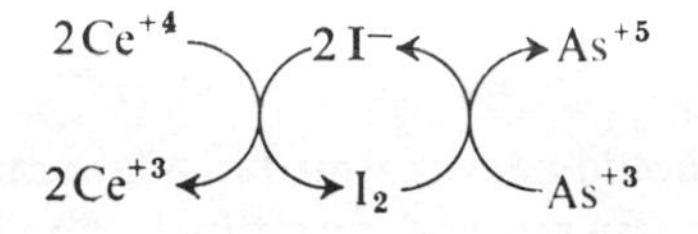

* After Pileggi and Kessler[19]

Ce^{+4} reacts with I^- to form Ce^{+3} and elemental iodine. Elemental iodine then reacts with As^{+3} to form $As^{+5} + I^-$. The I^- thus generated can react again with Ce^{+4}, etc. The reduction of the yellow-colored ceric ion (Ce^{+4}) to the colorless cerous ion (Ce^{+3}) is used to measure the rate of reaction, which is dependent, obviously, on the concentration of iodide present.

Chloride ions are known to enhance the catalytic action of iodide in the ceric-arsenite reaction. Thus, to eliminate interference by exogenous chlorides, an excess of chloride ions is added to the arsenious acid reagent.

The yellow color of the ceric ion exhibits maximum absorption in the ultraviolet range at about 317 nm. but is measured, in most procedures, at 415 or 420 nm. Use of the longer wavelength has several advantages: it obviates the need for an ultraviolet light source and quartz cuvets, and it extends the working range for the procedure. The ceric reaction follows Beer's law only if spectrophotometers with a narrow band width are used.

REAGENTS

1. Deionized water is used for the preparation of all reagents.

2. Zinc sulfate, 10 per cent (w/v). Slowly add 100 gm. $ZnSO_4 \cdot 7H_2O$ to hot deionized water. After the solution is cooled, dilute to 1 liter with water.

3. Sodium hydroxide, 0.5 *N* (20 gm. NaOH per liter). Adjust this solution so that 10.8 to 11.2 ml. are required to titrate 10 ml. of the zinc sulfate solution to a faint pink phenolphthalein endpoint.

4. Sodium carbonate, 4 *N*. Slowly add 424 gm. of anhydrous Na_2CO_3 to hot deionized water. After the solution has cooled, dilute to 2000 ml. with deionized water and mix well.

5. Sodium carbonate, 4 *N*, potassium chlorate, 2 per cent. Dissolve 20 gm. of $KClO_3$ in 1000 ml. of the 4 *N* Na_2CO_3. $KClO_3$ acts as an oxygen donor in the oxidation of organic material.

6. Arsenious acid reagent. Dissolve 3.944 gm. of As_2O_3 in 40 ml. 0.5 *N* NaOH with the aid of heat. Add this solution to approximately 3000 ml. of deionized water. Add 80 ml. of concentrated HCl and 158 ml. of concentrated H_2SO_4. Cool the solution, dilute to 4 L. with deionized water, and mix well.

7. Ceric ammonium sulfate, 0.0316 *N*. Add 48.6 ml. concentrated H_2SO_4 to approximately 600 ml. deionized H_2O. Add 20 gm. of ceric ammonium sulfate to the hot acid solution. Cool the solution, dilute to 1 L. with deionized water, and mix well.

8. Stock iodide standard (100 μg. iodide/ml.). Dissolve 130.8 mg. of desiccator-dried KI in deionized water and dilute to 1 L.

9. Working stock iodide standard (4 μg. iodide/ml.). Dilute 1 ml. of the stock iodide standard to 25 ml. with deionized water, and mix well.

10. Working iodide standards. Add 0.0, 1.0, 2.0, 3.0, 4.0, and 5.0 ml. of the working stock iodide standard (4 μg. iodide/ml.) to 100 ml. volumetric flasks containing 50 ml. 4 *N* Na_2CO_3. (The sodium carbonate must be prepared from the same lot used to prepare the 4 *N* Na_2CO_3-2 per cent $KClO_3$ reagent). Dilute the solutions to 100 ml. with deionized water, and mix well. These solutions contain 0.00, 0.04, 0.08, 0.12, 0.16, and 0.20 μg. iodide/ml. respectively.

PROCEDURE

Optimally, PBI tests should be performed in a laboratory reserved exclusively for this determination. All glassware and test tubes should be washed with approximately 6 *N* HNO_3 and rinsed well with distilled and deionized water.

1. Run each sample in duplicate. Pipet 1.0 ml. serum into a 15 × 125 mm. Pyrex test tube.

2. Add 7.0 ml. water, 1.0 ml. $ZnSO_4$, and 1.0 ml. 0.5 *N* NaOH, and mix thoroughly with a glass stirring rod. Rinse off the stirring rod with a small amount of water, and centrifuge the mixture for approximately 10 minutes.

3. Decant the supernatant solution. (A screening test for iodide contamination may be performed on the supernate if desired.)

4. Wash the protein precipitate by resuspending it in 5.0 ml. water with the aid of a glass stirring rod. Wash off the rod and centrifuge. Decant the wash and repeat this two more times.

Note: The water washes may be omitted if the serum sample is pretreated with a resin such as Amberlite IRA-401 (Mallinckrodt) as follows: Add to the serum approximately 1/10 of the volume of resin, mix, and allow this mixture to stand for 5 minutes with occasional stirring. Then centrifuge and use the supernatant serum for step 1 in the PBI procedure. It is also possible to pass the serum sample through a column prepared with the same resin.

5. To the washed protein precipitate add, without mixing, 0.5 ml. 4 *N* sodium carbonate-2 per cent potassium chlorate solution. Place the tubes in a dry bath or oven at 110°C. Since complete drying is essential, the tubes are best left in the drying oven overnight.

6. Incineration is carried out in the following manner: Place the tubes containing the dry precipitate in a metal (stainless steel) rack able to withstand a temperature of 620°C. Place the tubes in the incinerator for 2 hours at a temperature of 620°C.

7. Remove the tubes from the incinerator and allow to cool.

8. Add 10 ml. arsenious acid reagent to dissolve the ash. Allow the tubes to leach for 15 minutes, and then shake vigorously and centrifuge for 20 minutes.

9. Add 10 ml. of arsenious acid reagent to tubes containing the dried working iodide standards. (Prepare these in duplicate by adding 1 ml. of each of the working iodide standards: 0.00, 0.04, 0.08, 0.12, 0.16, and 0.20 μg. iodide/ml. to two tubes. Place the tubes in the 110°C. drying oven and dry.) Allow the tubes to leach, then shake vigorously and centrifuge for 20 minutes.

10. Transfer 5.0 ml. of the dissolved ash-arsenious acid solution and each standard-arsenious acid solution to clean tubes. Place the tubes in a 37 ± 0.1°C. water bath.

11. After the tubes have reached 37°C., add 1.0 ml. ceric ammonium sulfate reagent to each tube at precisely 30 second intervals, mix the contents of the tubes well, and return to the water bath.

12. Exactly 20 minutes after the addition of ceric ammonium sulfate to the first tube, read the color at 420 nm. against a distilled water blank. Chemicals and glassware should be sufficiently free of iodine to maintain a blank of 0.5 μg./ml. or less.

CALCULATION

A standard curve is prepared by plotting the absorbance readings, or Klett readings, against the iodide concentrations of 0, 4, 8, 12, 16 and 20 μg. iodide/100 ml., which, respectively, represent the iodide standards of 0.00, 0.04, 0.08, 0.12, 0.16, and 0.20 μg. iodide/ml. A smooth curve is drawn through the points and the PBI content of the unknown is read from this curve. The plot should be prepared on semilog paper or on regular graph paper if Klett readings are used. Normal range: (3.5) 4.0 to 8.0 μg./100 ml.

COMMENTS

1. Several pooled sera of known PBI value or commercial sera are also included with each group of unknowns to serve as control sera. Since the standards are neither precipitated nor incinerated, these phases of the analysis are not controlled by the standards. Control sera, on the other hand, are treated exactly as the unknowns, and either iodide loss or iodide contamination during analysis can be detected if values obtained on the control sera deviate from their established value.

2. Time and temperature of incineration should be controlled carefully, since iodine loss and severe bending of the Pyrex tubes can occur if the temperature exceeds 650°C. Temperatures below 620°C., on the other hand, result in either incomplete digestion of proteins or the production of carbon particles which exert an accelerating effect on the reduction of ceric ion, and therefore produce a positive error.

3. The reaction rate of the ceric-arsenite reaction is temperature-dependent. Thus, the temperature during this step should be 37 $\pm$ 0.1°C.

4. The standard curve is not precisely reproducible from day to day. This is perhaps because of variation in the amounts of iodine and/or reaction-inhibitors in the air.

5. Hemolysis results in a slight lowering of the PBI by a dilution effect, since the T_4 content of red cells is negligible.

REFERENCE

Barker, S. B., Humphrey, M. J., and Soley, M. H.: The clinical determination of protein-bound iodine. J. Clin. Invest. *30*:55, 1951, as modified by Henry, R. J.: *In* Clinical Chemistry, Principles and Technics. New York, Hoeber Med. Div., Harper and Row, 1964.

EXTRA-THYROIDAL FACTORS AFFECTING THE SERUM PBI LEVEL

The excitement which followed the development of the PBI was both considerable and justified, since serum PBI was observed to be rather consistently elevated among patients with hyperthyroidism, and depressed among hypothyroid subjects. The eclipse of the BMR had begun. Increased experience with the PBI, however, led to a more restrained enthusiasm, and the reasons for this transition must be understood very clearly in order to appreciate the subsequent history of thyroid methodology.

Under a variety of conditions a discrepancy between the serum PBI and the patient's clinical thyrometabolic state was observed. Sometimes the PBI was disproportionately elevated, and sometimes it was disproportionately depressed. The causes of such discrepancies, as presently understood, are classified in Table 9-7.

The circumstances which produce elevation of the PBI without a proportional increase in metabolic rate will be detailed below. Rationalization of the mechanisms of disproportionate depression of the PBI will then require only inverse reasoning.

Both euthyroid and hypothyroid subjects commonly receive a variety of thyroid preparations for a variety of reasons. The extent of, and direction of, their effect on the PBI depends upon their relative content of T_4 and T_3. This is so for the following reasons. The normal thyroid secretes approximately 100 μg. of T_4 and 25 μg. of T_3 each day. T_3 is, however, far more potent metabolically on the basis of weight comparison; the potency ratio T_3:T_4 is approximately 4:1. Despite the difference in amounts secreted, the relative contributions of T_3 and T_4 to thyrometabolic status in the normal individual are therefore equal. Should a physician elect to restore a

TABLE 9-7. *Causes of Discrepancy Between Serum PBI and Thyrometabolic Status*

- I. Disproportionate Elevation of the PBI
  - A. Thyroid replacement therapy (L-thyroxine)
  - B. Circulating iodoproteins
  - C. Circulating iodotyrosines
  - D. Administration of iodine-containing drugs or radiographic media
  - E. Increased number of available thyroxine-binding sites* on serum TBG
    - 1. Idiopathic or familial
    - 2. Estrogen-induced
- II. Disproportionate Depression of the PBI
  - A. T_3 the predominant, or only, circulating thyroid hormone
  - B. Thyroid replacement therapy (desiccated thyroid, purified thyroglobulin, 1-triiodothyronine)
  - C. Decreased number of available thyroxine-binding sites on serum TBG
    - 1. Idiopathic or familial
    - 2. Nephrotic syndrome
    - 3. Drug-induced
      - a. Androgens
      - b. Salicylates (large doses)
      - c. Diphenylhydantoin
      - d. O-p′-DDD

* Present TBG assays measure number of active binding sites rather than protein concentration. It is not possible, therefore, to distinguish increased TBG binding due to an increased concentration of this carrier protein in serum from increased binding activity of the individual TBG molecules.

hypothyroid individual to a euthyroid state using T_4 alone, he must replace 25 μg. of T_3 with its T_4 equivalent, i.e. 100 μg., in addition to replacing the 100 μg. of T_4 normally secreted. He prescribes, therefore, at least 200 μg. of T_4. (In fact he prescribes 300 μg. of T_4, since approximately one-third of an orally administered dose of T_4 is not absorbed.) By substituting T_4 for T_3 he is prescribing four molecules where one would do! Further, T_4 contains 33 per cent more iodine per molecule than does T_3 and, in addition, T_4 molecules are cleared from the serum far more slowly. For these reasons T_4 replacement produces a disproportionate elevation of the PBI. Replacement by pure T_3 would obviously produce disproportionate lowering of the PBI. The effect of commonly used thyroid hormone preparations on the PBI in euthyroid or hypothyroid subjects receiving normal replacement dosage is shown in Table 9-8. If one modifies the normal range in relation to both preparation and dosage, the PBI will still provide useful information.

The remaining four causes of spurious elevation are not so easily dismissed. Three involve matters of non-specificity and the fourth reflects increased T_4 binding by TBG.

THE PBI: PROBLEMS OF NONSPECIFICITY

T_4 in serum is, for the most part, loosely adsorbed to various carrier proteins. The principal carrier protein is an alpha-globulin called the thyroxine-binding globulin, or TBG. A very small fraction of total serum T_4 (about 0.05 per cent) circulates in the free state. More will be said about both TBG and free T_4 later in this discussion, but, for the moment, it will suffice to indicate that the principal iodine-containing components in sera from normal subjects are T_4-serum protein complexes (Fig. 9-30).

TABLE 9-8. *Effect of Various Thyroid Hormone Preparations on the PBI**

Hormone Preparation	*Daily Dosage (mg.)*	*Expected Serum PBI (µg./100 ml.)*
Desiccated Thyroid (Armour)	120–180	Normal to Low (2.9–7.2)
Desiccated Thyroid (Warner-Chilcott Spec. Prep.)	120–180	Low (3.3–3.7)
Purified Thyroglobulin (Proloid, Warner-Chilcott)	120–180	Low to Low-Normal (1.6–4.8)
L-Thyroxine (Synthroid, Flint)	0.2–0.3	Normal to High (5.6–11.2)
L-Triiodothyronine (Cytomel, SKF)	0.05–0.1	Low (0.4–1.4)

* After Sisson, J. C.: Principles of, and pitfalls in, thyroid function tests. J. Nucl. Med., *6*:853, 1965.

Under certain circumstances, however, other iodine-containing components may appear in serum and falsely elevate the PBI.

Intrafollicular T_4, i.e. T_4 stored within the thyroid follicles, is protein-bound as is T_4 in serum, but the nature of the intrafollicular T_4-protein bond is quite different from the serum T_4-protein complex (Fig. 9-29). T_4 is an amino acid—an iodoamino acid—and it is but one of many amino acids within the thyroid follicles which associate to constitute the several follicular proteins. (The principal follicular protein is called thyroglobulin.) T_4 is, in fact, but one of four iodoamino acids which are incorporated into the follicular proteins; the others are T_3, monoiodotyrosine (MIT), and diiodotyrosine (DIT). The amino acid residues which form these proteins are linked to one another by means of peptide bonds (—CONH—), and one property of such bonds is important to the thyroidologist. Peptide bonds are considerably stronger than adsorption bonds, and the various thyroxine-responsive tissues cannot cleave them. The tissues cannot respond to T_4 when it is supplied to them in this form, i.e., as

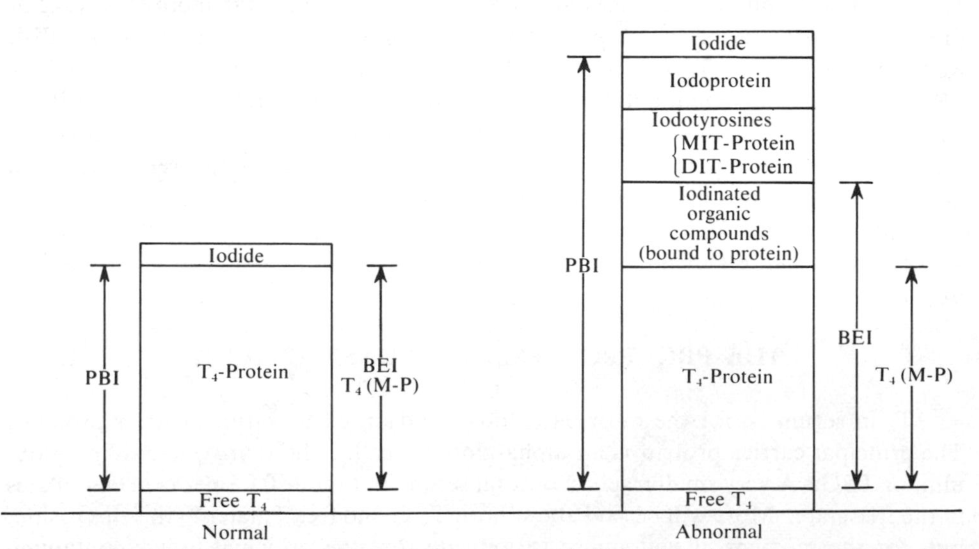

Figure 9-30. Principal forms of serum iodine. Arrows indicate those iodine-containing moieties which are measured by the respective laboratory procedure.

iodoprotein. In several thyroidal disorders, Hashimoto's thyroiditis in particular, the follicles may leak and release iodoprotein into the circulation. Since iodoprotein will co-precipitate with the plasma proteins to which T_4 is normally adsorbed, the four iodoamino acids represented in the follicular iodoproteins will be measured as PBI although they are hormonally inert. The PBI thus suffers what has been designated an iodoprotein problem.

Under other pathological circumstances, thyroid follicles may release the hormonally inactive T_3 and T_4 precursors, MIT and DIT. This situation is probably less common than iodoprotein escape. Since both MIT and DIT in serum are adsorbed to various proteins they too contribute to the PBI, and this unwelcome contribution has been called the iodotyrosine problem.

TABLE 9-9. *Classification of Iodinated Radiographic Contrast Media Based on Rate of Biological Removal (Duration of PBI Elevation)*

Short-lived (less than 6 weeks)
1. Iodopyracet (Diodrast, Hypaque)
2. Bunamiodyl (Orabilex)

Intermediate-lived (6–12 weeks)
1. Iodopanoic acid (Telepaque)
2. Acetrizoate sodium (Urokon)

Long lived (greater than 12 weeks)
1. Iodized poppy seed oil (Lipiodol)
2. Ethyl iodophenylundecylate (Pantopaque)
3. Iophenoxic acid (Teridax)
4. Iodoalphionic acid (Priodax)
5. Iodopamide methylaglucamine (Cholografin)

The most common cause of spurious elevation of the PBI, however, is exposure to a variety of iodine-containing drugs and radiographic contrast media. Most contrast media in common use are iodine-containing organic compounds which exhibit two annoying characteristics so far as the thyroid diagnostic laboratory is concerned. First, they adsorb to the serum proteins and thus raise the PBI. Second, their rate of clearance from serum is slow and they interfere, therefore, for long periods of time. These compounds have been classified, in fact, according to the time required for their biological removal from serum (Table 9-9).

Several drugs in common clinical use also contain iodine. These include diiodohydroxyquin (Diodoquin), isopropamide iodide (Darbid), the topical preparation providone-iodine (Betadine), and inorganic iodine itself. Iodine-containing organic compounds raise the PBI because they adsorb to serum proteins and release iodide in the course of their metabolism.

Increased concentrations of inorganic iodide in serum are thought to raise the PBI by two mechanisms, "iodide trapping" and nonspecific iodination of serum proteins. Routine washes of precipitated serum proteins in the various PBI methods remove about 97 per cent of iodide trapped in the protein precipitate. If one assumes then that 3 per cent remains trapped, at serum iodide concentrations of approximately 1.0 μg./100 ml. a 3 per cent washing inefficiency will elevate the PBI by only 1.0×0.03, or 0.03 μg./100 ml. At iodide concentrations of 100 μg./100 ml., on the other hand, the situation is serious since $100 \times 0.03 = 3.0$ μg./100 ml. Secondly, increased levels of serum iodide, by a mechanism which is not well understood, produce some iodination of all the serum proteins. In other words, hyperiodidemia produces iodoproteinemia! This consequence of iodide ingestion may be more important, in fact, than iodide trapping. In any case, iodide trapping may be obviated by resin pretreatment of the serum sample, as described in the PBI procedure.

In doses of 125 mg. per day or less, for periods of up to seven weeks, iodides do not raise the PBI by either mechanism. Daily doses of 200 to 600 mg. interfere for two to ten weeks, and massive doses (i.e., daily doses exceeding 3000 mg.) raise the PBI for up to four months after they are discontinued.

Three developments in recent years have contributed greatly to the solution of the foregoing problems of nonspecificity.

IMPROVED SPECIFICITY OF THE PBI

1. Serum Butanol-Extractable Iodine (BEI)

The first major innovation was advanced by Man et al.,[14] in 1951. These investigators appreciated that *n*-butyl alcohol, or *n*-butanol, has two properties on which an improved serum T_4 method might rest. Firstly, it is an excellent solvent for the iodothyronines T_3 and T_4. Secondly, although it can cleave an adsorption bond, it cannot cleave a peptide bond. If, then, instead of measuring the iodide content of a protein precipitate of serum, one were to measure the iodide content of a butanol extract of a protein precipitate of serum, one would have excluded iodoprotein from such an extract and thus solved the iodoprotein problem. Since, however, *n*-butanol extracts of serum also contain both iodides and iodotyrosines (MIT and DIT) there was work yet to be done.

Solution of both the iodide and iodotyrosine problems, based on the differential solubility of iodotyrosines and iodothyronines in alkali, was at hand, Whereas iodotyrosines are quite soluble in alkali, iodothyronines are not, and aqueous alkali washes would also remove iodide.

Iodoprotein, iodotyrosine, and inorganic iodide interference could all be circumvented by measuring the iodide content of alkali-washed butanol extracts of protein precipitates of serum! Butanol extracts of acidified serum were washed, therefore, with 3.8 *N* sodium hydroxide containing 5 per cent sodium carbonate. The resulting extracts, "purified" of both iodides and iodotyrosines, were expected to contain only T_4 and T_3. (The extracts were evaporated to dryness and analyzed for iodide content by procedures identical to those listed under PBI.) These expectations were confirmed, and the butanol-extractable iodine, or BEI, in normal subjects was found to range between 3.2 and 6.4 μg./100 ml.

When the *same* serum sample is examined, the difference between the PBI and BEI should be negligible, theoretically, in the absence of contaminating iodoprotein, iodotyrosine, and/or iodide. In fact, however, the BEI is, on the average, about 20 per cent lower. If the observed difference exceeds 20 per cent one must suspect the presence of abnormal iodo-compounds.

The BEI has never achieved great popularity despite its theoretical appeal, because it is cumbersome and subject to considerable experimental error. The recent development of other serum T_4 methods will, undoubtedly, further reduce the popularity of this test. For these reasons only the principles of the analysis will be presented.

Principles of the BEI Procedure. In the original procedure of Man et al.,[14] the diluted serum sample was extracted with *n*-butanol. Virtually all T_4, T_3, MIT, DIT, and organic and inorganic iodides move into the butanol phase. Washing the butanol extract with an alkaline solution (e.g., 4 *N* NaOH-5 per cent Na_2CO_3) removes inorganic iodine, MIT, and DIT; but all iodine-containing organic compounds remain with T_4 in the butanol extract and thus interfere with the test.

Benotti and Pino[4] simplified the procedure by precipitating the proteins and extracting the well-drained protein precipitate with *n*-butanol, obviating the need for alkaline washes, since inorganic iodine is removed with the supernatant obtained after protein precipitation. Masen[15] further simplified the procedure by replacing the *n*-butanol with a mixture of isoamyl alcohol (3-methyl-1-butanol) and trimethyl pentane. This solvent extracts neither iodides, iodotyrosines, nor certain iodine-containing organic compounds. Thus again, alkaline washing is unnecessary. This author further recommended pretreatment of the extract with bromine before its iodine content is measured by the ceric-arsenite reaction. Such treatment eliminates the need for destruction of the organic compounds by wet or dry digestion, as will be explained later in connection with the determination of thyroxine (T_4) by column.

Despite these developments, iodinated organic compounds—contrast media and drugs—remained the major contamination problem. In 1961, Pileggi[20] and his co-workers described another serum T_4 method which appeared to exhibit the theoretical advantages of the BEI, but unfortunately some of its limitations as well. This technique was called "thyroxine-by-column" (T_4-by-column) and it involved determination of the iodide content of Dowex resin column eluates.

2. Determination of Thyroxine (T_4)-by-Column

The usefulness of ion-exchange resins for the separation of serum iodoamino acids and inorganic iodide has been known for many years, but the method of Pileggi et al.[20] was the first T_4 method suitable for the routine clinical hospital laboratory. The method found wide acceptance and came to be considered by many the method of choice for the determination of serum T_4. Several modifications which have been reported confer a slight increase in accuracy but a considerable increase in both speed and convenience.[5-8,17-19] Until greater experience accumulates with respect to each of the methods now in use, it is difficult to decide which of these is most suitable for the routine laboratory. Thus, the principles common to all methods will be presented, but no specific method will be detailed. Such information can be obtained either from the original reports or from any of the procedures now supplied commercially in kit form.[5,8,17]

PRINCIPLES

1. Serum is diluted with an alkaline solution, generally 0.1 *N* sodium hydroxide, to raise the pH above 11.0 (earlier methods[20] recommended acetate buffer diluent which provided a final pH of 4.0). *Rationale:* At a pH above 11.0 T_4 dissociates completely from its carrier proteins. Ionization of the carboxyl group (and perhaps the hydroxyl group, too) of the T_4 molecule is thought to facilitate its release from proteins and enhance its attraction to the resin.[7]

2. The diluted specimen is then poured onto a strongly basic anion exchange resin (generally of the Dowex AG-1, X-2 type). Since preparation of the resin, as well as selection of a suitable and well-controlled mesh size, appears to be extremely

critical, the use of commercially prepared, disposable columns is recommended for those operators inexperienced in preparation of resin columns. Most methods currently employ resin columns in either the acetate or hydroxyl form. *Rationale:* As diluted serum passes through the column, T_4 and related compounds, as well as proteins and iodide, are adsorbed by the ion exchange resin. The affinity of the column for iodide, and the iodine moiety of the T_4 molecule, exceeds its affinity for acetate.

3. The column is washed with an acetate-alcohol solution. Either acetate-isopropyl alcohol (pH 8.0) or acetate-methyl alcohol (pH 5.5) is generally employed. *Rationale:* Washing at pH 8.0 completely removes all serum proteins from the column, and also removes both carbonate and bicarbonate, thus preventing CO_2 gas from disrupting the column when acetic acid is added in the next step. T_4, T_3, and inorganic iodide are retained by the column. Methods which incorporate the alkaline wash step also retain thyronines and most iodinated organic compounds on the resin and thus require an additional wash with glacial acetic acid at a pH of approximately 5.5 in order to remove these compounds. The wash with acetate-methyl alcohol, recommended by some, removes proteins, iodotyrosines, and many iodinated organic compounds with a single elution, but does not remove bicarbonate and carbonate prior to acidification.

4. The columns are primed with a small and accurately measured amount of glacial acetic acid. *Rationale:* T_4 and T_3 are held in the upper portion of the resin bed. Addition of glacial acetic acid moves these down the column. It is important to add a precise amount of glacial acetic acid, since an excess would cause a loss of T_4.

Note: In steps 1 through 4 all effluents are discarded.

5. The column is permitted to drain completely and a precisely measured amount of 50 per cent acetic acid is added to the column (in most cases 3.0 ml.) to provide a pH of about 1.4. *Rationale:* Under these conditions, 80 to 95 per cent of column T_4 (and T_3) is eluted. The exact fraction is rather constant for a given method and depends on both the type of resin used and the nature of the wash solutions. The column effluent is saved in an appropriate container. Inorganic iodide, in concentrations up to 1000 μg./100 ml., is retained by the column and will not interfere with the procedure.

6. An identical amount of 50 per cent acetic acid is added to the column a second time, and once more the effluent is saved in a separate cup. *Rationale:* The second acetic acid wash removes the remainder of the T_4 and T_3 from the column. Separate collection and analysis of two eluates provides a pattern of elution which is useful in the detection of contaminated samples. If, for example, eluate No. 2 contains more than the standard percentage of T_4 and T_3, contamination by exogenous iodide is suspected and the result disregarded.

Note: The procedure outlined provides effluents which are sufficiently concentrated to permit direct analysis for iodide; the tedious evaporation of acetic acid eluates which was required in earlier procedures is now unnecessary.

7. The T_3 and T_4 content of the two eluates is determined directly by any of the standard procedures employing the ceric-arsenite reaction without preliminary wet or dry digestion.[5,7,8,17–19] The eluates are treated with a solution of $KBrO_3$ and KBr, followed by a precisely timed addition of arsenious acid and then ceric reagent. *Rationale:* Although pure T_4 has the ability to catalyze the ceric-arsenious acid reaction (though to a lesser degree than iodide) T_4 eluted from a column has little or no such catalytic activity. This lack of activity may be due to the presence of inhibiting substances in the eluate which either bind to or are inactivated by Br_2.

The catalytic activity of T_4 in the ceric-arsenite reaction is thought by some to depend on the release of iodine from the T_4 molecule by the action of the ceric ion. Thus, pretreatment of the eluate with another halogen, such as Br_2, could result in the binding of those compounds which would otherwise have removed iodide from the ceric-arsenite catalytic cycle (see principle of the PBI method). Some authors feel that treatment with Br_2 releases iodine from thyroxine and thus makes it available for catalytic action. No direct evidence has been adduced, however, in support of this hypothesis.

Pretreatment of the eluate with Br_2 can be done using bromine water, but most procedures recommend a combination of $KBrO_3$ and KBr which, in an acid medium, generates Br_2 according to the following equation:

$$5\,Br^- + BrO_3^- + 6\,H^+ \rightarrow 3\,Br_2 + 3\,H_2O$$

Formation of elemental bromine can be recognized by the appearance of a yellow color in the reaction mixture. After addition of arsenious acid, Br_2 is immediately reduced to Br^-, which results in decolorization of the solution. Br^- has little or no catalytic activity in the ceric-arsenite reaction and thus causes no interference.

Note: Procedures which do not require digestion were originally designed as manual methods but recent modifications permit analysis of the eluate by the Auto Analyzer.[5,7,17,18]

STANDARDIZATION

Some authors recommend the use of aqueous KI standards. Others, considering these unsatisfactory, employ instead commercial control sera for preparation of the standard curve. Both approaches now appear to be undesirable, and the use of thyroxin in protein solution for standards is recommended.

CONCLUSIONS

Procedures which employ the principles just outlined are now sufficiently simple, rapid, and well-standardized so as to be suitable for the routine clinical chemistry laboratory. Since the T_4 by column method offers greater specificity than the classic PBI method, it may soon become the method of choice in most laboratories. The increased specificity is due to the elimination of many troublesome iodine-containing organic compounds in the first wash (step 3) and the retention of inorganic iodide by the column. Thus, only a limited number of iodine-containing compounds appear to interfere with the test. However, since many of the experiments on which the matter of iodide interference rests are based on *in vitro* rather than *in vivo* studies,[17,19,20] further clinical experience will be necessary.

NORMAL VALUES

The normal values for the various T_4-by-column methods range between 3.2 and 6.4 μg./100 ml. Nonincineration techniques are, on the average, 0.13 μg./100 ml. higher.[19]

3. Determination of Serum T_4 (Murphy-Pattee)[16]

In 1964 Murphy and Pattee reported an ingenious method for measuring serum T_4 which, for the first time, did not depend on iodide analysis.[16] Their method was based, rather, on a property of the T_4 molecule, specifically its capacity to displace radioactive T_4 from a standard radio T_4-protein complex, i.e. T_4-TBG.

In the Murphy-Pattee method, T_4 in a serum unknown is dissociated from its carrier proteins and is extracted from the serum with ethanol. The ethanol extract is dried, and a standard radio T_4-serum protein-barbital buffer mixture is added to the dried residue. The amount of radio T_4 incorporated into this mixture is calculated to saturate all T_4 binding sites on TBG. (Thyroxine-binding pre-albumin, or TBPA, the second major carrier protein, is inactivated by barbital buffer and is hence removed from competition with TBG.) As unlabeled T_4 in the reaction mixture is increased, fractional binding of radio T_4 to TBG *decreases*, since both labeled and unlabeled T_4 compete equally for available binding sites. Per cent radio T_4 bound to TBG is determined by use of an anion exchange resin and a standard well counter. Serum T_4 levels, expressed as T_4-iodine, range from 2.9 to 6.4 μg./100 ml. Thus, the results are virtually identical to those provided by the T_4-by-column method.

Neither iodoprotein, iodotyrosines, inorganic iodide, iodinated drugs, nor iodinated contrast media apparently interfere with this procedure. The anticonvulsant Dilantin lowers the result by reducing the binding affinity of the subject's TBG for T_4. On the other hand the d-isomer of T_4, d-T_4 (Choloxin), now commonly used clinically to lower serum cholesterol concentration, displaces radio T_4 from TBG in the reaction mixture and thus produces a falsely elevated result. Despite these instances of drug interference, the serum T_4 method of Murphy and Pattee appears to be the most specific serum T_4 method. It is yet too early to say, however, whether this method will satisfy the demanding requirements for a routine procedure. In any case this method does share, with the PBI, BEI, and T_4-by-column methods, the final cause of non-correspondence between the PBI and clinical thyrometabolic status, discussed below.

HYPERTHYROXINEMIA WITHOUT HYPERTHYROIDISM (AND HYPOTHYROXINEMIA WITHOUT HYPOTHYROIDISM)

In 1959 Beierwaltes and Robbins described a clinically euthyroid 48 year old man who, during a routine periodic health examination, was found to have PBI levels which ranged between 11.8 and 16.0 μg./100 ml.[3] There was no history of exposure to either iodine-containing drugs or radiographic contrast media. Serum BEI determinations were 12.3 and 13.0 μg./100 ml. The BMR (−20 per cent), serum cholesterol (212 mg./100 ml.), and thyroidal uptake of radioiodine (22 per cent at 24 hrs.) were all considered essentially normal. The man had three children; one, a 15 year old daughter, also exhibited elevated PBI and BEI levels. These appeared to be two euthyroid people with increased concentrations of T_4 in their serum.

Ingbar, in 1961, observed the reverse situation. He reported a euthyroid man with serum PBI concentrations of 2.0 μg./100 ml.[11] This man had extra-thyroidal disease, but none known to depress the PBI. Again the BMR (+3 per cent), serum cholesterol (210 mg./100 ml.), and radioiodine uptake (28 per cent at 24 hrs.) were normal. This man was an apparent example of a person who was euthyroid despite subnormal concentrations of T_4 in his serum.

These reports suggested that clinical thyrometabolic status evidently depends on factors other than total serum T_4 concentration, perhaps, rather, on that fraction of total serum T_4 which circulates in the free state and is, therefore, more accessible to the peripheral T_4-responsive tissues. Although it was not possible then to measure free T_4 in serum, it was possible to measure accessibility of serum T_4 to tissues (since this could be inferred from the T_4 degradation rate), and both authors proceeded to do just that.

Though measurement of T_4 degradation rate by the peripheral tissues was not difficult it did require a clear understanding of the kinetics of T_4 metabolism and the terminology generally applied to these kinetics. These terms are:

TDS = Thyroxine distribution space. The volume of body fluids which would be required to contain exchangeable thyroxine were it present throughout at the same concentration at which it exists in the plasma.

ETT = Extra-thyroidal thyroxine. The quantity of exchangeable thyroxine in terms of its content of iodine; believed to coincide closely with the quantity of extra-thyroidal hormone.

k = Fractional rate of turnover of thyroxine. The fraction of hormone within the TDS which is degraded and replaced per unit time, calculated as $0.693/t_{1/2}$ where $t_{1/2}$ = the thyroxine half-time, or time required for half the exchangeable thyroxine to be degraded and replaced.

TABLE 9-10. *Studies of Thyroxine Kinetics in Normals and Two Subjects with "PBI-Thyrometabolic Dissociation"*

Parameter	*Normals*	*Case of Beierwaltes and Robbins*[3]	*Case of Ingbar*[11]
BMR	15 to +15	−20	+3
PBI (μg./100 ml.)	3.5–8.0	11.8–16.0	2.0
Chol (mg./100 ml.)	140–250	212	210
RaI uptake (% dose, 24 hrs.)	15–45	22	28
BEI (μg./100 ml.)	3.4–6.8	12.3–13.0	——
TDS (L)	9.4	7.5	13.2
k (%/day)	10.6	5.7	23.1
C (L./day)	1.0	0.43	3.05
ETT (μg. I)	508	1099	264
D (μg.I/day)	54	63	61

C = Thyroxine clearance rate. The volume of the TDS which contains a quantity of thyroxine equal to that being degraded per unit time.

HI = The concentration of hormonal iodine in the plasma. The PBI is most commonly employed as an approximation, and thyroxine considered to be its major component.

D = Thyroxine degradation rate. The quantity of thyroxine undergoing degradation per unit time, in terms of its content of iodine.

The equations which express the relationships among these terms are:

$$\text{ETT} = \text{HI} \times \text{TDS} \qquad \text{(Equation 1)}$$

$$\text{C} = \text{TDS} \times \text{k} \qquad \text{(Equation 2)}$$

$$\text{D} = \text{HI} \times \text{TDS} \times \text{k} \qquad \text{(Equation 3)}$$

$$\text{D} = \text{ETT} \times \text{k} \qquad \text{(Equation 4)}$$

$$\text{D} = \text{HI} \times \text{C} \qquad \text{(Equation 5)}$$

Thus, by determining the PBI, TDS, and $t_{1/2}$ one may calculate D (Equation 3). Consideration of the techniques normally employed to determine TDS and $t_{1/2}$ is beyond the scope of this chapter. Such measurements were made, however, both in normal subjects and in the patients reported above and the results are shown in Table 9-10.

Note the similarity of the D values in the three columns. One patient (B/R) had a pool size approximately twice normal which was turning over half as fast, the other had a contracted pool, about half normal, with a turnover rate approximately twice normal. Both subjects had similar T_4 degradation rates and both were, therefore, euthyroid. The cause of the expanded ETT pool in the first instance was thought to be a genetically determined increase in T_4-binding sites on the thyroxine-binding alpha-globulin, TBG. The second case displayed a contracted ETT pool because of a decrease (idiopathic) in such binding sites.

The several messages to be gleaned from these patients are very important ones: (1) Serum T_4 concentration (whether measured as PBI, BEI, T_4-by-column, or T_4

Murphy-Pattee) is determined mainly by the number of binding sites on TBG which are occupied by T_4 molecules. (2) Thyrometabolic status, on the other hand, is evidently determined by the concentration of free T_4 in serum, and this very small fraction (about 0.05 per cent) of the *total* serum T_4 also regulates the hypothalamic and pituitary "sensors" of serum thyroid hormone concentration.

$$[T_4]\ [TBG] = k\ [T_4 \cdot TBG]$$

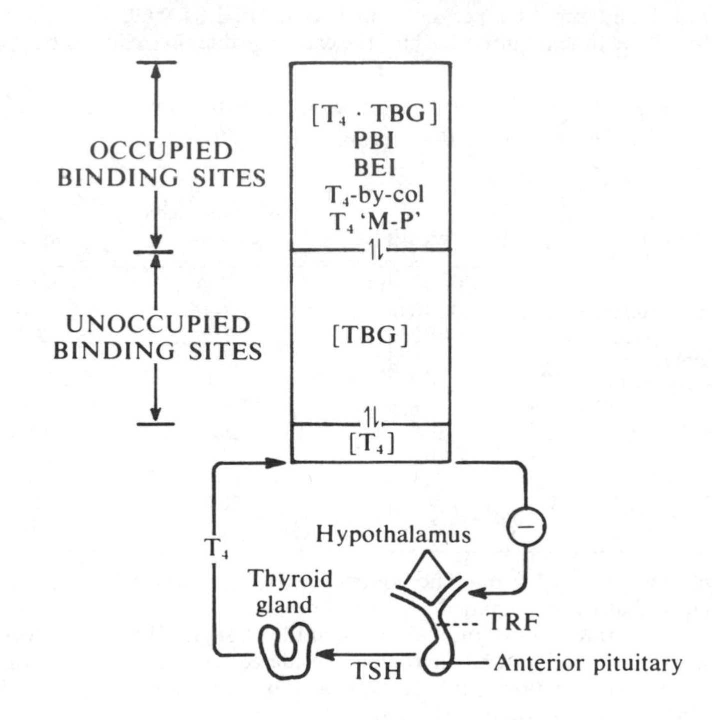

TRF = Thyrotropin – releasing factor
TSH = Thyroid – stimulating hormone

Figure 9-31. Schema illustrating the relationship between serum T_4 and its principal carrier protein, TGB. Note that an increase in *either* $[T_4]$ or [TBG] will shift the equilibrium towards $[T_4 \cdot TBG]$ and hence will raise any of the serum T_4 tests.

The interaction between serum free T_4 and the unoccupied or available binding sites on its principal carrier protein TBG conforms to the law of mass action and can be described by the following equation (see Figure 9-31):

$[T_4] \times [TBG] = k \times [T_4 \cdot TBG]$, where
$[T_4]$ = concentration of free T_4 in serum:
$[TBG]$ = concentration of unoccupied binding sites on serum TBG;
k = an association constant; and
$[T_4 \cdot TBG]$ = concentration of T_4-occupied binding sites on serum TBG

The relationship among these moieties has important diagnostic applications. A primary increase in either $[T_4]$ or [TBG] would drive this reaction to the right, increase serum $[T_4 \cdot TBG]$, and hence raise the PBI. Hyperthyroidism produces the former circumstance (primary increase in $[T_4]$), and estrogens and idiopathic or genetic influences produce the latter (primary increase in [TBG]). In the former case the patient is ill and requires treatment; in the latter case the patient is euthyroid and requires nothing but an understanding of the equation. It is obviously necessary to be able to distinguish these two situations.

The differential diagnosis may be achieved in four ways:

1. Measurement of the concentration of unoccupied binding sites on TBG ([TBG]) in serum.
2. Measurement of [T_4] in serum.
3. Measurement of thyroid gland activity (in terms of its avidity for radioiodine).
4. Measurement of metabolic impact of the thyroid hormones on the peripheral tissues (i.e., BMR).

The BMR has been discussed. The other diagnostic options will now be considered.

Measurement of [TBG]: The T_3 Test

If an elevated serum [T_4·TBG] were due to a primary increase in [T_4], the increased [T_4·TBG] would occur at the expense of [TBG] and, since serum [TBG] is not regulated by a feedback loop such as the hypothalamic-pituitary system which

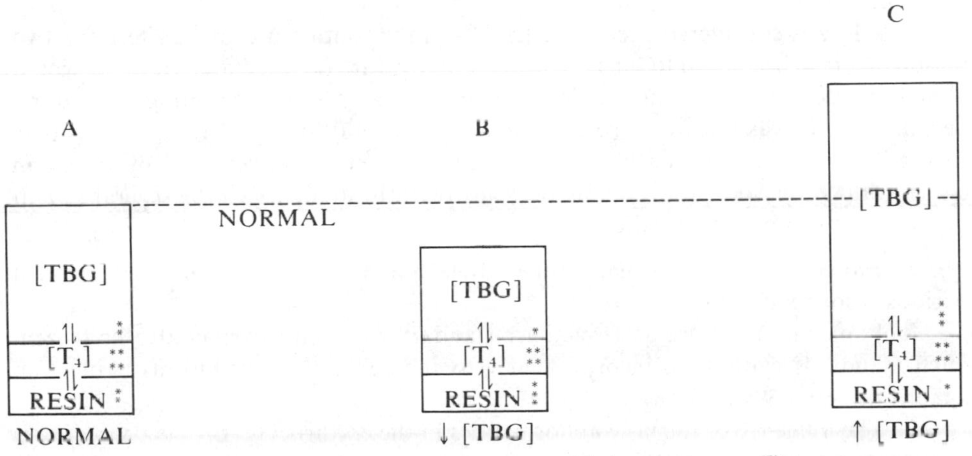

Figure 9-32. Estimation of [TBG] by use of a "competing" inert T_4 (or T_3) receptor.

protects [T_4], [TBG] will decrease. On the other hand, a primary increase in [TBG] will generate an increase in [T_4·TBG] at the expense of [T_4]. The fall in [T_4] is, however, only transient since this moiety is "protected" (Fig. 9-31). Further, the increased [TBG] is not fully saturated by combination with T_4, and the serum will exhibit an increased [TBG]. So, [TBG] is reduced in the first instance and elevated in the second.

[TBG] may be estimated indirectly by an interesting group of tests which have been designated collectively as T_3 tests. All are based on an ingeniously conceived *in vitro* competition for thyroid hormone between serum TBG and an added inert receptor (Fig. 9-32).

Hamolsky and his co-workers,[9] in 1957, developed the first of the T_3 tests, the *in vitro* red blood cell uptake of I^{131}-T_3, on the basis of the following considerations. The red blood cell was known to be capable of binding thyroid hormones on the basis of experiments in which red blood cells were incubated in saline-radio T_3 and saline-radio T_4 mixtures. Red blood cells were therefore considered possible models for all the thyroid hormone-responsive tissues. A sample of whole blood was considered, then, to house two binding systems: the various carrier proteins in serum, and the red blood cells. These compete in a sense for secreted hormone, and each system was thought to possess a characteristic hormone-binding affinity. The binding affinity of the carrier proteins was thought to be variable, depending upon the number of unoccupied binding sites on TBG, i.e. [TBG]. For example, the greater the number of unoccupied sites, or the higher the [TBG], the *poorer* the competitive position of the red cells for added hormone. The *smaller* the number of available sites, or the lower the [TBG], the *better* the competitive position of the red cells. The partition of *both endogenous and added* thyroid hormone in blood between carrier proteins and red cells was thought to depend upon these considerations and since tracer amounts of labeled hormone added to blood were known to mix thoroughly with the endogenous hormone pool, the partition of added labeled hormone was expected to mirror the partition of endogenous hormone between carrier proteins and red cells. Thus, the fractional uptake of added labeled hormone by red cells was expected to increase in hyperthyroidism, where [TBG] is reduced, and decrease in hypothyroidism, where [TBG] is increased. A large clinical experience with the procedure soon validated these predictions.

I^{131}–T_3 was considered preferable to I^{131}–T_4 for routine use in this test for two reasons. First, because TBG has a lesser affinity for T_3, a greater fraction would attach to red cells and thus confer greater accuracy to the counting procedure. Second, I^{131}–T_3 was known to be a more stable preparation, i.e., less photosensitive. In the T_3 procedure the red blood cells, however, were soon replaced by resins, in either granular or sponge form. Sisson[22] has recently summarized the several advantages of resins over red blood cells:

1. Either serum or plasma, unlike whole blood, may be frozen for extended periods prior to analysis.
2. Pooled sera for use as a reference standard for each run may also be frozen. Such standards cannot be incorporated into the red cell procedure as protection against day-to-day variability in the test.
3. The influence of intrinsic abnormalities in the erythrocytes on the procedure is avoided.
4. No hematocrit correction is necessary.
5. Changes in arterial P_{CO_2}, which influence the red cell procedure, are evidently without effect on resin T_3 tests.

Combination of the PBI, or any of the tests which measure [T_4·TBG], with a T_3 test would appear to represent an ideal laboratory tandem, since such combination would permit elegant physiologic interpretations. A patient who exhibits an elevation in both the PBI and T_3 tests evidently has an increase in [T_4·TBG] but a decreased [TBG]. This circumstance suggests the primary problem of increased T_4 secretion, i.e., hyperthyroidism. An increase in PBI associated with a decreased T_3 test indicates an increase in both [T_4· TBG] and [TBG] in serum. This suggests a primary increase in [TBG], and such patients are euthyroid. Such a circumstance is seen during pregnancy, with exposure to exogenous estrogen (commercial estrogen preparations, ovulation-suppressants), and in cases of genetic or idiopathic increases in [TBG]. A reduction in both the PBI and T_3 tests indicates a decrease in [T_4·TBG] in the face of an increased [TBG]. Such a situation suggests a primary decrease in T_4, i.e., hypothyroidism. Finally, a low PBI but elevated T_3 test suggests a reduction in both

[T_4·TBG] and [TBG]. This circumstance suggests a primary decrease in [TBG] caused by androgen excess, idiopathic reduction of binding sites, or successful competition for T_4 binding sites by certain drugs (large doses of salicylates, diphenylhydantoin, and the adrenolytic agent o-p′-DDD).

Thus, concordant variance of the PBI and T_3 test suggests altered thyroid function; discordant variance suggests altered [TBG] and a euthyroid state. These are very useful rules in thyroid diagnosis.

The binding capacity of TBG for T_3 is influenced by a variety of extra-thyroidal factors, and for this reason the T_3 tests are often said to provide too many false positives and false negatives. One may take exception to use of the adjective "false"

TABLE 9-11. *Extra-Thyroidal Factors Which May Affect the T_3 Test*

Factors which Elevate the Result

1. Loss of TBG in urine (nephrotic syndrome)
2. Reduced TBG synthesis
   a. hereditary-familial (idiopathic)
   b. hormone-mediated (androgens, anabolic steroids after 7–21 days of therapy; effect persists for 7–21 days after cessation of treatment)
3. Competition by drugs for T_4 binding sites
   a. on TBG—diphenylhydantoin (Dilantin sodium)
   b. on TBPA
      (1) salicylates (large doses)
      (2) dinitrophenol
4. Decrease in TBPA binding (causing endogenous T_4 to "shift" to TBG and thus displace T_3): acute illness (non-specific)
5. Mechanism uncertain
   a. anticoagulants (both heparin, coumarins)
   b. cardiac arrhythmias, supraventricular
   c. elevated arterial P_{CO_2}

Factors which Lower the Result

1. Increased TBG synthesis
   a. hereditary-familial (idiopathic)
   b. estrogen mediated
      (1) endogenous: pregnancy (by 2–3 wks. gestation; returns to normal 1–2 wks. post-partum)
      (2) exogenous (after 7–21 days of treatment)
         (a) all estrogenic preparations (incl. stilbesterols)
         (b) ovulatory suppressants, oral contraceptives
2. Mechanism uncertain
   a. drug-induced: perphenazine, prolonged (Trilafon)
   b. hepatic disease, acute and chronic

in this connection, but in any case the old problem of nonspecificity definitely haunts the T_3 test. The principal extra-thyroidal causes of abnormal T_3 tests are classified in Table 9-11.

Additional comment is necessary concerning the effects of treatment with the various commonly used thyroid hormone preparations on the T_3 test. When the normal or hypothyroid individual is given full replacement or suppressive doses of either desiccated thyroid (3 grains, or 180 mg.) or L-thyroxine (0.3 mg.), the T_3 test is generally normal. Administration of comparable doses of T_3, on the other hand, produces low results in about 50 per cent of cases so treated. The influence of these preparations on the PBI was shown in Table 9-8.

The T_3 test is a useful complementary parameter to the PBI, BEI, or serum T_4 if it is interpreted with an awareness of the many factors which influence it. Neither inorganic nor organic iodides interfere, with the apparent exception of the oral cholecystographic contrast medium, sodium ipodate, or Oragrafin sodium, and this property renders the T_3 test particularly useful when both the PBI and thyroidal radioiodine uptake studies are invalidated by iodine interference.

Normal values for this procedure have been expressed in many ways, but "per cent resin uptake" is perhaps the most common parameter and normal sera generally range from 25.5 to 37.5 per cent.

Serum Free Thyroxine (FT_4)

Free thyroxine in serum has been characterized as the moiety which probably determines clinical thyrometabolic status, and it is now possible to directly measure its concentration. This development, truly an important one, is somewhat disappointing as well, since the principal methods, those of Ingbar et al.,[12] Sterling and Brenner,[23] and Lee et al.,[13] are, in fact, hardly routine procedures.

TABLE 9-12. *Per Cent Free Thyroxine* (% FT_4), *Protein-Bound Iodine* (*PBI*), *and Absolute Concentration of Free Thyroxine* (FT_4) *in Sera of Normal Subjects and Subjects with Various Abnormal States**

Diagnosis	% FT_4	*PBI* (μg./100 *ml.*)	FT_4 (*as* T_4) *ng.*/100 *ml.*
Normal	0.050 ± 0.009	5.4 ± 0.8	4.03 ± 1.08
Myxedema	0.037 ± 0.010	1.6 ± 0.5	0.88 ± 0.52
Pregnancy	0.026 ± 0.006	8.0 ± 0.9	3.21 ± 0.56
Thyrotoxicosis	0.110 ± 0.072	12.9 ± 2.9	20.56 ± 13.07
General Extra-thyroidal Illness	0.078 ± 0.033	4.6 ± 1.4	5.12 ± 2.99

* $\bar{x} \pm$ S.D.; After Ingbar, S. H., et al.: J. Clin. Invest., *44*:1679, 1965.

The method of Ingbar and his co-workers is a two-stage "double dialysis" procedure. The first stage involves equilibrium dialysis of a test serum—radiothyroxine mixture against standard phosphate buffer, producing a dialysate which contains both free radiothyroxine and a radioiodide contaminant. A second dialysis, to remove this contaminant, is therefore necessary. Accordingly, a mixture of first stage dialysate and indifferent plasma is dialyzed against phosphate buffer containing Amberlite IRA-400 anion exchange resin. This resin adsorbs only the radioiodine contaminant and thereby maintains an effective diffusion gradient for iodide until all the contaminant is removed. Because of the strong binding affinity of the proteins in the indifferent plasma for T_4, little if any radiothyroxine is available to the resin. All radioactivity which remains within the dialysis bag is assumed, therefore, to represent free radiothyroxine.

This technique measures per cent FT_4. Absolute [FT_4] is determined from the following equation:

$$[FT_4], \text{ng./100 ml.} = \frac{\text{PBI} \times \text{per cent } FT_4}{0.65}$$

Typical FT_4 studies in normals and subjects having a variety of abnormalities are shown in Table 9-12.

Sterling and Brenner[23] simplified the FT_4 method significantly by replacing the second dialysis with a much simpler precipitation procedure. Their method involves

(1) dialysis of a serum—radiothyroxine mixture against phosphate buffer (as above), and

(2) precipitation of the radiothyroxine in the dialysate with $MgCl_2$ after addition of carrier T_4. This method is often called the magnesium precipitation method (after its second stage), and that of Ingbar, the resin dialysis technique.

The method of Lee et al.[13] is perhaps the simplest of the three. In this procedure radiothyroxine is added to serum, and the free and protein-bound fractions are separated on microcolumns of Sephadex G–25.

FT_4 correlates better than the PBI with clinical thyroid status, since FT_4 is unaffected by states of altered TBG capacity. However, in cases in which the PBI is invalidated by iodine contamination, FT_4 cannot be estimated with confidence since serum PBI is required in the calculation. Substitution of serum T_4 (Murphy-Pattee) for the PBI may, of course, obviate this problem.

FT_4 methods are relatively new and crucial questions remain unanswered. Their ultimate place in the thyroid laboratory is, therefore, uncertain.

Thyroidal Uptake of Radioiodine

The iodide pool in man contains approximately 280 μg. of iodide. From this pool, the normal thyroid gland extracts about 70 μg. each day for purposes of hormone synthesis. The normal thyroid, therefore, incorporates 70/280, or 25 per cent of the iodide pool each day.

Fractional uptake of the iodide pool by the thyroid is measured routinely today by uniformly labeling this pool with a small tracer dose of radioiodine (RaI) administered orally. I^{131}, in either liquid or capsule form, is the most commonly used isotope. At the desired interval following administration of the RaI (generally 24 hrs.), thyroidal radioactivity is measured by an external scintillation detector. Under identical conditions of geometry, collimation, and time, a standard (i.e., a tracer dose identical to that given the patient) is also counted. After background radiation is subtracted from both measurements, thyroidal radioactivity is expressed as a percentage of the standard. The accepted normal range in most laboratories is 15 to 45 per cent.[21]

The relationship between thyroidal uptake of iodine (I uptake, μg./24 hr.), iodide pool size (P_I, μg.), and fractional uptake of RaI (RaI uptake, per cent) may be expressed as follows:

$$\text{I Uptake, } \mu g = P_I \times \text{RaI Uptake}$$

Under normal circumstances a gland which accumulates 70 μg. of iodine each day for hormonogenesis, in the presence of a P_I of 280 μg., will exhibit an RaI uptake of 25 per cent, as we have said. Both hyper- and hypothyroidism naturally alter the iodine requirements of the gland, and if the P_I is normal, the RaI uptake will reflect these changed requirements. Thus, a hyperthyroid gland which doubles its rate of hormonogenesis will have an RaI uptake of 140/280, or 50 per cent, and a hypothyroid gland with half the normal rate of hormone synthesis will have an uptake of 35/280, or 12.5 per cent.

These very simple calculations highlight the first of the two major limitations of the RaI uptake. Since P_I is *not* measured routinely (and it is not always normal) there is no necessary connection between RaI uptake and rate of hormone synthesis. For example, a patient with an iodide pool expanded to 700 μg., but having normal thyroid function, will exhibit an RaI uptake of 70/700, or 10 per cent, and a subject with a contracted pool (100 μg) but normal thyroid function will have an uptake of 70/100, or 70 per cent! The importance of these considerations varies directly, of course, with the frequency of abnormal iodide pools in the population and, sad to say, abnormal pools are all too frequent. The many contemporary sources of iodine contamination have been considered in connection with previously described methods. These compounds all expand the P_I and thus depress RaI uptake.

Contraction of the iodide pool, because of deficient iodine intake, is a far less common circumstance in the United States, since the average diet includes at least 100–300 μg. of iodine in food and water. Although iodine intake may fluctuate moderately from day to day, these changes exert little effect on the RaI uptake. Iodized salt, to which most of us subscribe, is "iodized" in this country in a KI:NaCl weight ratio of 1:10,000; that is, each 10 gm. of iodized salt contains 1 mg. of KI, or 760 μg. of iodide, ten times the normal daily requirement. Thus, iodine prophylaxis very effectively eliminates "contracted poolers." Most of the high mountainous districts of the world (e.g., the Alps, the Himalayas, the Andes) and some nonmountainous areas as well (the Uele region of the Congo, Holland, and interior of Brazil) still harbor numerous iodine-deficient inhabitants.

Just as it is potentially hazardous to assume a necessary connection between RaI uptake and rate of iodide accumulation, it is equally dangerous to link rate of iodine accumulation with rate of hormonogenesis. Iodide trapping is but the first step in hormone synthesis; there are five others. Since the posttrapping steps may be interfered with on either a congenital or acquired basis (thyroiditis, foods, drugs), trapped iodide may be forced to leave the gland in an incompletely metabolized state. It is possible, therefore, to observe a normal or even elevated RaI uptake in hypothyroid subjects. Further, hyperthyroidism may result from excessive ingestion of thyroid hormone preparations (thyrotoxicosis factitia) or hormone production by a functioning tumor of the ovary (struma ovarii). In such instances pituitary secretion of TSH would be suppressed by the increased serum FT_4, thyroid function would compensatorily diminish, and the RaI uptake would be correspondingly depressed. Thus, one may observe hyperthyroidism in association with a reduced RaI uptake.

SUMMARY

Those tests of thyroid function which seem most relevant to the comtemporary clinical laboratory have been reviewed. Rationales and pitfalls have been emphasized. It is evident that despite the many remarkable advances which have led to the development of these procedures, there is none as yet which is sufficiently specific to permit facile interpretation. Care and caution must characterize both the performance and interpretation of these studies. The next era in thyroid diagnosis may witness the application of gas chromatography to presently unresolved problems.

REFERENCES

1. Barker, S. B., as modified by Henry, R. J.: *In* Clinical Chemistry, New York, Hoeber Medical Division, Harper and Row, 1964, p. 937.
2. Barker, S. B., Humphrey, M. J., and Soley, M. H.: The clinical determination of protein-bound iodine. J. Clin. Invest,. *30*:55, 1951.
3. Beierwaltes, W. H., and Robbins, J.: Familial increase in the thyroxine-binding sites in serum alpha globulin. J. Clin. Invest., *38*:1683, 1959.
4. Benotti, J., and Pino, S.: Simplified method for butanol-extractable iodine and butanol-insoluble iodine. Clin. Chem., *12*:491, 1966.

5. Bio-Rad T-4 by Column TM Test, Bio-Rad Laboratories, Richmond, California, 1969.
6. Bittner, D. L., and Grechman, R. J.: Semi-automated column analysis of serum thyroxine (T-4). *In* Automation in Analytical Chemistry, Technicon Symposia, 1966, Vol. 1, New York, Mediad, Inc., 1967.
7. Bittner, D. L., Young, D. P., Maffe, M. R., and Grechman, R. J.: An automated method for the bromination and quantitation of thyroxine in resin column eluates. Thyroxine Round Table, 1968 Annual Meeting, Am. Soc . Clin. Path., 1968.
8. C:T4, Procedure for thyroxine by column, Curtis Nuclear Corp., Los Angeles, California.
9. Hamolsky, M. W., Stein, M., and Freedberg, A. S.: The thyroid hormone-plasma protein complex in man, II. A new in vitro method for study of "uptake" of labelled hormonal components by human erythrocytes. J. Clin. Endocrinol., *17*:33, 1957.
10. Ingbar, S. H., and Woeber, K. A.: The thyroid gland. *In* Williams, R. H.: Textbook of Endocrinology. Philadelphia, W. B. Saunders Co., 1968, 4th Ed.
11. Ingbar, S. H.: Clinical and physiological observations in a patient with an idiopathic decrease in the thyroxine-binding globulin of plasma. J. Clin. Invest., *40*:2053, 1961.
12. Ingbar, S. H., Braverman, L. E., Dawber, N. A., and Lee, G. Y.: A new method for measuring the free thyroid hormone in human serum and an analysis of the factors that influence its concentration. J. Clin. Invest., *44*:1679, 1965.
13. Lee, N. D., Henry, R. J., and Golub, O. J.: Determination of the free thyroxine content of serum. J. Clin. Endocrinol. *24*:486, 1964.
14. Man, E. B., Kydd, D. M., and Peters, J. P.: Butanol-extractable iodine of serum. J. Clin. Invest., *30*:531, 1951.
15. Masen, J. M.: A simplified procedure for serum butanol extractable iodine. Am. J. Clin. Path., *48*:561, 1967.
16. Murphy, B. E. P., and Pattee, C. J.: Determination of thyroxine utilizing the property of protein-binding. J. Clin. Endocrinol., *24*:187, 1964.
17. Oxford T-4 Manual, Oxford Laboratories, San Mateo, California, 1969.
18. Passen, S., and von Saleski, R.: A semiautomated method which does not require digestion for the determination of serum thyroxine iodine (T_4). Am. J. Clin. Path., *51*:166, 1969.
19. Pileggi, V. J., and Kessler, G.: Determination of organic iodine compounds in serum. IV. A new nonincineration technic for serum thyroxine. Clin. Chem., *14*:339, 1968.
20. Pileggi, V. J., Lee, N. D., Golub, O. J., and Henry, R. J.: Determination of iodine compounds in serum. I. Serum thyroxine in the presence of some iodine contaminants. J. Clin. Endocrinol., *21*:1272, 1961.
21. Silver, S.: Radioactive isotopes in medicine and biology. Philadelphia, Lea and Febiger, 1963, 2nd Edition.
22. Sisson, J. C.: Principles of, and pitfalls in, thyroid function tests. J. Nucl. Med., *6*:853, 1965.
23. Sterling, K., and Brenner, M. A.: Free thyroxine in human serum: simplified measurement with the aid of magnesium precipitation. J. Clin. Invest., *45*:153, 1966.
24. Zak, B., Willard, H. H., Myers, G. B., and Boyle, A. J.: Chloric acid method for determination of protein-bound iodine. Anal. Chem., *24*:1345, 1952.

Chapter 10 /

ELECTROLYTES

by Norbert W. Tietz, Ph.D.

Electrolytes are classified as either anions or cations, depending upon whether they move in an electric field toward the anode or the cathode, that is, whether they have a negative or positive charge. They are essential components of all living matter and include the major electrolytes Na^+, K^+, Cl^-, HCO_3^-, $HPO_4^=$, Ca^{++}, and Mg^{++}, as well as the trace elements $Fe^{2+,3+}$, $Cu^{+,++}$, $Mn^{2+,4+}$, Co^{2+}, $Cr^{3+,6+}$, Cd^{++}, Zn^{++}, Br^-, and I^-. Although amino acids and proteins in solution also carry an electrical charge, in clinical chemistry they are usually classified separately from electrolytes. The major electrolytes occur primarily as free ions, while the trace elements occur primarily in some special combination with proteins and thus are also frequently classified separately.

The dietary requirement for electrolytes varies widely; most need to be consumed only in small amounts or at rare intervals, and are retained when in short supply. Some, like calcium and potassium, are continuously excreted and must be consumed regularly in order to prevent deficiency. Excessive consumption leads to corresponding increased excretion, mainly in the urine. Abnormal loss of electrolytes, which occurs through excessive perspiration, vomiting, or diarrhea, is readily assessed and can be corrected by administration of salts.

The role of electrolytes in the human body is manifold. There are almost no metabolic processes that are not dependent on or affected by electrolytes. Among other functions of the electrolytes are maintenance of osmotic pressure and hydration of the various body fluid compartments, maintenance of the proper body pH, regulation of the proper function of the heart and other muscles, involvement in oxidation-reduction (electron transfer) reactions, and participation as an essential part or cofactor of enzymes. Thus, it becomes quite apparent that abnormal levels of these electrolytes and trace elements may either be the cause or the consequence of a variety of disorders.

Determination of electrolytes is one of the most important functions of the clinical laboratory. Progress in this field, and especially in the field of trace elements, was hampered by the lack of suitable methods for their determinaion. In recent years, however, a number of instruments have been made available to facilitate analytical determination, and it is to be expected that much more work will be done and much knowledge will be gained in the next few years. Among the tools that we use to determine these elements are a variety of micro- and macromethods based on spectrophotometry, emission spectrography, flame spectrophotometry, neutron activation analysis, atomic absorption spectroscopy, and coulometric analysis. More specific information about the role of individual electrolytes, trace elements, and their determination is given in the respective paragraphs that follow.

Collection of Blood for the Determination of pH, CO_2 Content, and P_{CO_2}*

The P_{CO_2} of air (about 0.2 mm. Hg) is much less than that of blood (about 38 mm. Hg). Thus, when blood is exposed to air, the CO_2 content and the P_{CO_2} decreases and the pH increases correspondingly. For the determination of pH, CO_2 content, or P_{CO_2}, it is, therefore, necessary to collect, transfer, and manipulate blood (serum or plasma) under conditions in which exposure to air is avoided or kept at a minimum. Before performing these techniques, one should be aware of certain other pertinent properties of blood.

EFFECT OF TEMPERATURE AND STORAGE ON BLOOD pH

The pH of freshly drawn blood decreases on standing at a rate of 0.01 unit/10 minutes at 38°C. and about half this rate at 27°C. There is no measurable change in pH in 3 hours if the blood is kept at 4°C. in an ice water bath (all measurements, however, are made at 38°C.). Simultaneous with such decrease in pH is a corresponding decrease in glucose and an equivalent increase in lactate; the primary cause of these changes is thought to be glycolysis.

Respiration in freshly drawn blood (protected from air) causes a decrease of 0.1 m*M*./L./hour in total O_2 content at room temperature (double this value at 38°C.) and a nearly equivalent increase in total CO_2. Because of compensation between the alkalizing effect of oxygen decrease (conversion of $HHbO_2$ to the weaker acid HHb) and the simultaneous acidifying effect of the increase in CO_2 (carbonic acid), the effect on pH of spontaneous respiration of blood in vitro is negligible in the first hour, even at 38°C. (less than 0.01 pH).

Both glycolysis and respiration in freshly drawn blood depend on the metabolism of cells, and the effect can be avoided either by appropriate cooling of the blood in ice water at 4°C. (not over 3 hours), or by prompt centrifugation and anaerobic transfer of the supernatant serum or plasma to a separate vessel. The pH, CO_2 content, and P_{CO_2} of *separated* serum, not exposed to air, remain constant for several hours at 4°C.

The most effective method of avoiding the adverse effects of glycolysis and respiration on pH, CO_2 content, and P_{CO_2} of blood (plasma, serum) is to make all measurements within 30 minutes after the blood is drawn. Another method is to place the vessel containing the freshly drawn blood (with appropriate anticoagulant) immediately in ice water at 0 to 4°C. and carry out the analysis within the next 3 hours.† Immediate centrifugation and transfer of the plasma (serum) to a separate container, as just outlined, is also effective, but is rarely done.

---

* This material is taken in part from the manual: Electrolyte Institute and Workshop. Chicago, Ill. Sept. 16–18, 1965. N. W. Tietz, Ed. St. Louis, Catholic Hospital Association.

† If blood is stored in ice, it must be warmed to 38°C. before centrifugation (separation of red cells from plasma) or false high pH values are obtained. In practice, blood is warmed to room temperature only, since its temperature will further increase during centrifugation. Thus, only a small error is introduced.

CHOICE BETWEEN VENOUS, ARTERIAL, AND CAPILLARY BLOOD SAMPLES

The CO_2 content of *venous* blood (usually collected from the radial vein of the arm) is approximately 2 m*M*./L.* more and the O_2 content about 2 m*M*./L. less than that of arterial blood simultaneously drawn. This arterial-venous difference varies with the metabolic activity of the organ or tissue from which the venous blood is obtained. For this reason, arterial blood is of more uniform composition than venous blood and is, therefore, preferred for some studies. The arterial-venous pH difference, however, is extremely small, 0.01 to 0.03 pH, and is minimized by the compensatory effect of *increased* acidity due to increased CO_2 (carbonic acid levels) and *decreased* acidity due to decreased O_2 content as arterial blood is changed to venous blood. Under the extreme conditions of an elevated respiratory quotient of 0.95 (in case of high carbohydrate and low fat diet) and an elevated blood pH of 7.6, the arterial-venous pH difference reaches a maximum of 0.03. In conditions of low respiratory quotient, e.g., 0.70 (low carbohydrate diet or decreased carbohydrate utilization), and a pH of 7.2 (e.g., in diabetic acidosis), venous blood may be more alkaline than arterial blood by 0.01 pH unit. The important fact is that the arterial-venous pH difference in the majority of patients does not exceed 0.01 pH unit; therefore, for clinical purposes, the pH of venous blood has the same clinical significance as that of arterial blood.

Arterialized blood in respect to O_2 and CO_2 content, as well as pH, approximates arterial blood. It may be obtained by collecting the blood specimen from a limb that has been warmed to 45°C. for several minutes immediately before the blood is drawn. The entire forearm, including the hand and fingers, is immersed in a water bath at 45°C., or a towel wetted with water of 45°C. is wrapped around the forearm. The heat dilates the capillaries and, by accelerating the rate of blood flow, decreases the changes in blood composition caused by tissue respiration. Procurement of arterialized blood is rarely resorted to because of the inconvenience.

Capillary blood obtained by pricking the fingertip, earlobe, heel, or toe with a small sharp blade (Bard-Parker blade No. 11 is suitable) resembles arterial blood in its composition more closely than it resembles venous blood. Capillary blood may also be arterialized by warming the skin in the area of the puncture, but this arterialization procedure is rarely used. Speedy manipulation is required to avoid clotting and to prevent significant loss of carbon dioxide and gain of oxygen in capillary blood samples.

COLLECTION OF ARTERIAL BLOOD

Arterial blood should be collected by a skilled and experienced physician, since puncture of an artery is not only difficult but, in fact, may be dangerous. The blood should be collected in a Luer-Lok syringe, which has previously been coated with a solution of heparin (0.20 mg./ml. or 20 U.S.P. units/ml.). The heparin will serve not only as an anticoagulant but will also fill the air spaces in the syringe and the needle. After collection of the blood, the tip of the syringe is sealed either with a metal or other suitable cap, or with a mercury button. A slight excess of mercury may be introduced into the syringe to facilitate the mixing of the specimen after collection and before analysis. This is accomplished by swirling the syringe rapidly and, thus, moving the mercury drop within the syringe. If whole blood is used for analysis, it is important that the specimen is mixed adequately before analysis.

---

* Regarding the use of m*M*./L. for millimoles/Liter see comment in the Preface.

COLLECTION OF VENOUS BLOOD

Most technologists apply a tourniquet to the patient's arm before performing the venous puncture. If this is the case, the patient should *not flex his fingers*; the blood for pH and gas studies should be drawn immediately after applying the tourniquet, and blood required for other tests should be drawn thereafter. This will prevent any changes in the composition of the blood caused by stasis and accompanying accumulation of metabolites and decrease in oxygen in the blood. If, for any reason, the tourniquet is left on for more than a few seconds, it should be removed after successful venous puncture and the blood permitted to flow freely for at least 1 minute before withdrawing the sample. Aspiration of the sample immediately after removing the tourniquet introduces substantial errors since compounds accumulated during stasis will flush into the blood and rapidly change its composition. These adverse changes may also be circumvented by performing the venous puncture without applying a tourniquet.

There are three containers in which the venous specimen can be collected.

1. The blood may be collected in a heparinized Luer-Lok syringe as outlined under the collection of arterial blood.

2. The specimen may be drawn into a heparinized Vacutainer, a technique which has become very popular. In this case, however, the tube must be filled completely to avoid pH changes due to loss of CO_2 into the vacuum. If the pH is done on plasma, the tube must be centrifuged with the stopper in place and determinations must be made immediately after removing the stopper. Once the stopper has been removed for more than a few seconds, the specimen is unsuitable for any repeat analysis of pH, P_{CO_2}, or CO_2.

3. The blood may also be collected in a tube containing "light mineral oil," which is used to provide a barrier between blood and air to retard the exchange of O_2 and CO_2. Many investigators have expressed objections against the use of such oil because CO_2 is slightly soluble in this medium. The error introduced is negligible, however, if a thin oil layer is used, if the blood is not agitated excessively, if it is centrifuged for only 3 minutes, and if it is analyzed within 30 minutes after collection. Contamination of electrodes by mineral oil can be prevented by withdrawing the sample from under the oil with the aid of a 1 ml. Tuberculin syringe, fitted with a 26 gauge needle. After removal of the needle, the sample may be aspirated into the electrode directly from the syringe.

COLLECTION OF CAPILLARY BLOOD

Thoroughly cleanse the skin area where the puncture is to be performed and remove any excess moisture to prevent dilution of the sample and hemolysis. The puncture should be made deep enough to provide a good blood flow. (Sterile Bard-Parker blades No. 11 are satisfactory for this purpose.) The first drop of blood should be wiped off because it frequently contains some tissue juice that tends to hemolyze the red cells and alter the pH and other blood constituents. The blood is then collected with the aid of a heparinized capillary tube, by placing the end of the capillary into the freshly formed drop of blood. Spreading of the blood over a wide area should be avoided, since this will expose a large surface of blood to air, causing a loss of CO_2 and a change in pH. It should always be remembered that it is better not to do a determination at all than to do a determination on an unsatisfactory specimen.

After collection of the specimen, a small metal bar is placed into the capillary and the latter is sealed with a suitable sealing substance. Mixing of blood is provided by moving a magnet along the capillary. Thus, the metal bar will move up and down the capillary, mixing the blood with heparin while keeping the red cells suspended. The mixing should be repeated shortly before the actual pH determination.

If a pH meter with an ultramicroelectrode is available, the blood may be drawn into the electrode directly from the capillary or from the site of the puncture. In the latter case the electrode must be cleaned immediately after the measurement is taken to prevent clotting of the blood inside the electrode.

Collection of Specimens for the Determination of Anions and Cations

Specimens for the determination of various electrolytes in serum, such as Na^+, K^+, Cl^-, Fe^{+++}, Cu^{++}, Ca^{++}, and Mg^{++}, should be collected in tubes that are free of even trace contamination. If necessary, tubes should be washed with diluted nitric acid (1:3), rinsed in deionized water, and well dried. Hemolysis must be avoided since this will interfere to some degree with all tests measuring electrolyte concentration. In some cases, for example for the K^+ analysis, a hemolyzed specimen is unsuitable.

Except in emergencies, all specimens, including those for pH and CO_2 analysis, should be collected when the patient is in a fasting state.

REFERENCES

Gambino, S. R., Astrup, P., Bates, R. G., Campbell, E. J. M., Chinard, F. P., Nahas, G. G., Siggaard-Andersen, O., and Winters, R.: Am. J. Clin. Path., *46*:376, 1966.

Gambino, S. R.: Workshop Manual on Blood pH, P_{CO_2}, Oxygen Saturation and P_{O_2}. American Society of Clinical Pathologists, 1963.

SODIUM

Sodium is the major cation of extracellular fluid. It occupies a central role in the maintenance of normal hydration and osmotic pressure. The normal daily diet contains approximately 8 to 15 gm. (130 to 250 mEq.) of sodium chloride, which is nearly completely absorbed from the gastrointestinal tract. Except for a series of factors listed in the following paragraphs, the excess is excreted by the kidneys, which, therefore, are the ultimate regulators of the sodium content of the body. Sodium is a threshold substance with a normal renal threshold* of 110 to 130 mEq./L. Sodium is initially filtered by the glomeruli and is then reabsorbed to 80 to 85 per cent in the proximal portion and to some extent also in the distal portion of the tubules. The reabsorption is greatly affected by adrenal cortical hormones, mainly aldosterone, which enhances the tubular reabsorption of sodium and also of Cl^- (but decreases the tubular reabsorption of K^+). The exchange of Na^+ for H^+ in the renal tubules is an important mechanism of the acidification of urine and is discussed in detail in Chapter 11, Electrolyte (Cation-Anion) Balance.

Hyponatremia (low serum sodium level) is found in a variety of conditions including the following: (1) extreme urine loss, as seen in diabetes insipidus; (2) in metabolic acidosis, e.g., diabetic acidosis, in which cations are excreted along with the anions (see Chapter 11); (3) in Addison's disease, in which decreased secretion of corticosteroids, mainly aldosterone, causes decreased reabsorption of sodium by

* Threshold is the level in serum below which any sodium present in the glomerular filtrate is completely reabsorbed by the tubules (see also Chapter 12, Renal Function Tests).

the tubules and, thus, loss of serum sodium; (4) in diarrhea, in which an excessive amount of sodium is lost through the stool as a result of insufficient absorption of dietary sodium and sodium of pancreatic juice; and (5) in renal tubular disease, in which there may be a defect in either the reabsorption of sodium or the Na^+–H^+ exchange.

Hypernatremia (increased serum sodium level) is found in the following conditions: (1) hyperadrenalism (Cushing's syndrome), in which there is an increased production of corticosteroids and, thus, an increased absorption of sodium by the renal tubules; (2) severe dehydration due to primary water loss, in which there is a relative increase of sodium in serum; (3) certain types of brain injury; (4) diabetic coma after therapy with insulin, in which it is believed that the removal of excess glucose from the serum causes a retransfer of cellular sodium into the extracellular fluid in order to maintain equal osmotic pressure in both compartments; (5) excess treatment with sodium salts.

NORMAL VALUES

The range of normal values for serum sodium is 135 to 148 mEq./L. There is less agreement on the level of normal urinary sodium excretion. In part this is due to the great effect of dietary intake of sodium on urine levels. Values given in the literature for normal individuals on an average diet vary from 40 to 90 mEq./24 hr. to 43 to 217 mEq./24 hr. The rate of sodium excretion during the night is only one fifth of the peak rate during the day, indicating a large diurnal variation.

The sodium concentration in cerebrospinal fluid is almost identical to that found in serum (138 to 150 mEq./L.). The concentration of sodium in sweat will be discussed in connection with the sweat test (see Chapter 15).

Although a number of gravimetric and titrimetric methods for the determination of sodium in body fluids are available,[11] almost all laboratories today employ either emission flame photometry or atomic absorption spectrophotometry for this test because of the speed and simplicity of these methods. Since the principle of both techniques has been discussed in detail in Chapter 2, and since most manufacturers supply detailed procedures with their equipment, no attempt will be made here to outline a method. It may be stated, however, that sodium determination, even in the routine laboratory, may be performed with an accuracy of ± 2 per cent if an appropriately designed instrument is used. As a general rule, a flame photometer, employing an internal standard, is slightly more accurate than a direct reading instrument. Serum specimens for sodium analysis are generally diluted 1:100 or 1:200; thus, the dilution error may be as great or greater than the instrumental error, especially if very small volumes of sample are used (e.g., 0.025 ml. diluted to 5.0 ml. with a mechanical dilutor).

COMMENTS

An approximate check on the accuracy of sodium, chloride, and carbon dioxide determinations can be made by calculating the difference between the sodium value and the combined chloride and carbon dioxide values (in mEq./L.). This difference is normally 12 ± 2 mEq./L. Table 11-1 shows that in serum there are in addition to Na^+, 12 mEq./L. of other cations of which Ca^{++}, Mg^{++}, and "others" are not normally determined as part of an "electrolyte" study (unmeasured ions). There are also 24 mEq./L. of "undetermined anions" (HPO_4^{--}, SO_4^{--}, organic acids$^-$, proteins$^-$) present in serum.

Therefore:

Na^+	142	Cl^-	103
+ unmeasured		HCO_3^-	27
cations, incl. K^+	12	unmeasured	
	154	anions	24
			154

or:

$$
\begin{array}{lr}
Na^+ & 142 \\
-\ (Cl^- + HCO_3^-) & 130 \\
\hline
 & 12
\end{array}
$$

An alternative calculation is:

$$(Na^+ + K^+) - (Cl^- + CO_2) = 16 \pm 2$$

These calculations are based on the assumption that all unmeasured ions are close to the average normal value. This is true in many cases, but may not be true in conditions such as diabetes and renal impairment, in which the organic acid fraction is increased as a result of increased production of acetoacetic acid or retention of metabolites, respectively (see also under organic acids). Inspection of test results on blood (serum) glucose, urea nitrogen, or acetone performed on the same specimen may help to detect the situations in which the preceding formula does not hold true.

POTASSIUM

Potassium is the major intracellular cation, having an average cellular concentration for tissue cells of 150 mEq./L. and a concentration in red cells of 105 mEq./L. This is approximately 23 times higher than the concentration of potassium in extracellular fluid. The high intracellular concentration, in the presence of the low extracellular concentration, is believed to be maintained through an active transport mechanism that utilizes oxidative energy of the cells. In addition, the permeability of cell membranes for potassium is extremely slow, and rapid shifts of potassium in or out of cells have not been observed. There is, however, some movement of potassium ions across the cells. This is important to the practicing clinical chemist since shifts of potassium from the cells to the serum may invalidate serum potassium levels, if the serum is not separated from the cells shortly after collection of blood.

The requirements of the body for potassium are satisfied by a normal diet and to some extent by potassium arising from intracellular or interstitial fluid. Potassium, once absorbed by the intestinal tract, is partially removed from the plasma by glomerular filtration and is then nearly completely reabsorbed by the tubules. Unlike sodium and chloride, however, it is then effectively reexcreted by the distal tubules. There is no threshold level for potassium.

Any potassium reabsorbed by the intestinal tract causes only a slight and temporary increase in serum potassium levels. Only a fraction of this potassium moves into the red cells (and tissue cells), and the remainder is rapidly excreted by the kidneys as just explained. It is believed that this mechanism protects the body against high serum potassium levels, which could cause serious (inhibitory) disturbances in muscle irritability, respiration, and myocardial function as well as characteristic electrocardiographic changes. Such symptoms may appear with potassium levels above 7.5 mEq./L. Levels of 10 mEq./L. may be fatal, although patients with slightly higher levels have recovered.

Low potassium levels cause excitatory changes in muscle irritability and myocardial function, which are also accompanied by characteristic electrocardiographic changes. For these reasons the serum potassium determination has become a most important diagnostic tool in situations in which extremely high or low potassium levels are suspected. Prompt performance of the test by the laboratory is mandatory.

The mechanism of potassium excretion and the absence of a threshold level have the disadvantage, however, that the body has no effective mechanism to protect itself from excessive loss of potassium. Even in potassium deficiency, the kidney continues to excrete potassium. For maintenance of normal potassium levels, a regular daily intake of potassium is essential.

Hypokalemia or *hypopotassemia* (low serum potassium levels) may be seen in cases of potassium deficiency, although the large intracellular potassium reserves may maintain normal serum potassium levels despite an actual potassium deficiency. Aside from inadequate intake of potassium, deficiencies in this ion may occur as a result of excessive loss of potassium through the feces (in prolonged diarrhea) or through vomitus (after prolonged periods of vomiting). Increased secretion of mineral corticosteroids, especially aldosterone, results in a decreased reabsorption of potassium (the reverse of the behavior of sodium, which under such conditions is reabsorbed at an increased rate) and, thus, results in decreased serum potassium levels. Patients on a regular diet who exhibit serum potassium levels below 3 mEq./L. for several days are likely to have hyperaldosteronism. Urine potassium levels in these cases are increased. In alkalosis there is a K^+ movement into the cell that is matched by a movement of H^+ from the cell into the extracellular fluid. Thus, all other things being equal, alkalosis in itself results in decreased serum potassium levels; the opposite is the case in acidosis. Intracellular potassium deficiency may lead to alkalosis, a process which is explained in more detail in Chapter 11.

Hyperkalemia or *hyperpotassemia* (increased serum potassium level) is generally observed in cases of oliguria, anuria, or urinary obstruction. Renal failure due to shock results in decreased removal of potassium from serum; renal tubular acidosis interferes with the Na^+–H^+ exchange and, thus, also results in a retention of potassium in serum. In renal failure, one of the important purposes of renal dialysis is the removal of accumulated potassium from plasma.

SPECIMENS

Specimens for serum potassium analysis should be free from hemolysis, since the high concentration of potassium released from red cells significantly increases the serum levels and thus invalidates the test results. Blood specimens should also be separated from the red cells shortly after collection to prevent any leakage of potassium from the intracellular into the extracellular fluid. The shift of potassium is greater at refrigerator temperatures.

NORMAL VALUES

Normal serum potassium levels range from 3.5 to 5.3 mEq./L. In the newborn, levels are somewhat higher than those of adults, and most authors give a range of 4.0 to 5.9 mEq./L. Higher normal values for this group have been reported by some authors, but errors in specimen collection or handling may have been involved in these cases.

Urinary excretion of potassium varies greatly with the potassium intake, but commonly observed levels of persons on an average diet are 30 to 90 mEq./24 hr.

Determination of Potassium in Body Fluids

A wide range of methods has been proposed for the determination of potassium in body fluids.[12] These include gravimetric, turbidimetric, emission flame spectrophotometric, and atomic absorption spectrophotometric methods. Almost all modern laboratories use either of the last two procedures because they are fast, convenient, and reliable, as explained previously under the determination of sodium in serum. A modern, well designed flame photometer in a routine laboratory can deliver an accuracy of ± 0.2 mEq./L. for serum samples in the physiological range (approximately 4 per cent error) or better. The principles of flame photometry and atomic absorption spectrophotometry have been explained in Chapter 2 and detailed methodology is supplied with the respective instruments.

Practical Considerations Related to the Determination of Sodium and Potassium by Emission Flame Photometry

1. Sodium and potassium are the two most frequently determined metallic ions in a clinical laboratory that are analyzed by emission flame photometry. Lithium is generally used as an internal standard, although analysis of lithium (e.g., in the blood of psychiatric patients under treatment with lithium compounds) may also be performed by this technique. Metallic ions other than these are more practically determined by atomic absorption spectrophotometry or other techniques (see respective procedures). Each of these three ions (Na, K, Li) emits light of a variety of distinctive wavelengths, but the spectral lines most frequently used in clinical determinations are 590, 768, and 672 nm., respectively. The respective light is isolated (separated from interfering light) either by filters, gratings, or prisms.

2. When analyzing biological material it becomes necessary to dilute the samples before analysis. The extent of dilution depends on the type of instrument, the type of specimen, and the concentration of the ions to be analyzed. Such dilution is necessary to adjust the concentration of the measured ion so that the light intensity is within the range of optimum sensitivity and accuracy of the photometer. By decreasing the protein concentration and viscosity, dilution promotes a uniform flow of sample and decreases plugging of the atomizer. Dilution also decreases or eliminates interference by other sample constituents, which may cause extraneous light emission.

3. Reagents and standards, as well as dilutions of the sample, must be made with high purity water, preferably deionized water with an electrolyte content less than 0.5 ppm (resistance of at least 600,000 ohms), but preferably 0.1 ppm (one million ohms).

4. Sodium and potassium standards are generally prepared from their chloride salts since chloride ions do not affect the analysis. Lithium solutions are generally prepared from nitrates or sulfates. Most lots of lithium nitrate have an assay of less than 100 per cent. This must be considered when weighing out the salt for the preparation of the standards. Lithium standards for lithium analysis are generally prepared from Li_2CO_3.

Most manufacturers recommend that sodium be added to the potassium standard in a concentration comparable to that in normal serum (e.g., 140 mEq./L.). Addition of sodium compensates for the slight enhancement by sodium in the measurement of potassium.

Standard solutions are best stored in polyethylene squeeze bottles. The bottles should be kept relatively full, since this type of bottle "breathes," which means that water vapor escapes, causing a slow increase in the concentration of the standard.

5. Accuracy of data depends on following the manufacturer's instructions explicitly in order to secure uniformity with respect to flame size, rate of introduction of the sample, and patency of the analyzer. Variations in these three conditions cause corresponding variation in temperature or in the size of atomized droplets, which in turn alters the sensitivity of the instrument and may lead to erroneous results.

6. Both burner and atomizer must be cleaned frequently with a thin wire, and both parts must be rinsed well with deionized water, between samples and after use. Such a procedure is necessary to prevent or remove protein and salt deposits. Failure to do so will result in an uneven flow of the sample into the flame and thus in erroneous results. Addition of nonionic surface active agents such as Acationox to reagents and standards helps to provide an even flow into the burner and the formation of small, uniform droplets of the sample solution during atomization.

7. The flow of fuel and oxidant into the flame must be strictly controlled, especially in direct reading instruments. Any irregularity in the flow rate of either one or both these components results in a different flame temperature and flame size and therefore affects the sensitivity of the test procedure.

Internal standard instruments are less affected by these changes because of the compensatory effect of the internal standard, but even under these circumstances high quality flow regulators should be used to prevent any gross fluctuations in the gas flow. Difficulties have been reported with the use of city gas as a result of the varying concentrations of the different components of this gas or gross fluctuations in the gas pressure. To avoid this, many laboratories use gas tanks with adjustable pressure valves, if this is in accordance with the local fire laws.

8. Most manufacturers recommend that the electronic part of the flame photometer be left on at all times to eliminate warm-up periods and, in some instances, to prolong the life of the equipment. The sensitivity of photosensing devices will change with temperature; thus, insufficient warm up periods affect the accuracy of the results.

9. Some manufacturers request that, at a beginning of the test run, a standard or a lithium solution be aspirated for a certain time period (e.g., 2 to 5 minutes) in order to bring the burner, or the chamber (which houses the atomizer), or both, to thermal equilibrium, which is necessary for obtaining reproducible results. Evaporation of water in the atomizer chamber and in the flame causes a drop in temperature.

10. If an internal standard such as lithium is used, the quantitation of the unknown ion is based on the ratio between the signal obtained from the internal standard and that obtained from the unknown. Thus, it is important that the lithium concentration in the standards and the unknowns is identical. If a new batch of lithium reagent is prepared, old standards must be discarded and a new set prepared with the new lithium solution.

CHLORIDE

Chloride is the major extracellular anion, with 103 mEq./L. of the total anion concentration of 154 mEq./L.; thus, it is significantly involved in maintaining proper hydration, osmotic pressure, and normal anion-cation balance in this compartment. The content of chloride in red cell fluid is 49 to 54 mEq./L. and that of whole blood is 77 to 87 mEq./L. Tissue cells contain approximately 1 mEq./L.

Chloride ions ingested with food are almost completely absorbed by the intestinal tract and are removed from the blood by glomerular filtration. From the glomerular filtrate, chloride ions are passively reabsorbed by the tubules.

Chloride, to some extent, may be lost through excessive sweating during hot weather periods; however, in these situations aldosterone is secreted. This causes the sweat glands to secrete sweat of lower sodium and chloride concentration than that excreted during normal temperatures. Thus, the loss of Cl^- under these conditions is minimized.

Low serum chloride values are observed in salt losing nephritis as associated with chronic pyelonephritis. This loss is probably due to a lack of tubular reabsorption despite a body deficit of chloride. In Addison's disease chloride values are generally maintained close to normal except in Addisonian crisis, in which chloride (and Na^+) levels may drop significantly. Low serum chloride values may also be observed in those types of metabolic acidosis (e.g., diabetic acidosis and renal failure) that are caused by excessive production or diminished excretion of acids. In these cases chloride ions are partially replaced by the accumulated anions, for example, aceto-acetate and phosphate. Prolonged vomiting, from any cause, may result in a significant loss of Cl^- (through gastric HCl and Cl^-) and therefore a decrease in serum and body chloride.

High serum chloride values are observed in dehydration and in conditions causing decreased renal blood flow, such as congestive heart failure. Hyperchloremic acidosis may be a sign of severe renal tubular pathology. Excessive treatment with or intake of Cl^- obviously also results in high serum levels.

Methods for the Determination of Chloride in Body Fluids

MERCURIMETRIC TITRATION (SCHALES AND SCHALES, MODIFIED)

PRINCIPLE

A Folin-Wu protein-free filtrate of the specimen is titrated with mercuric nitrate solution in the presence of diphenylcarbazone as the indicator. The mercuric ions combine with chloride ions to form soluble but essentially nonionized mercuric chloride.

$$2Cl^- + Hg^{++} \rightarrow HgCl_2$$

After all chloride ions have reacted with mercuric ions, any excess Hg^{++} combines with the indicator diphenylcarbazone to form a blue-violet colored complex. The first appearance of this blue-violet color is considered the titration end point.

NORMAL VALUES

Normal values for chloride in serum or plasma range from 98 to 108 mEq./L. The serum chloride values vary little throughout the day although there is a slight decrease in chloride ions after meals because of the chloride required for the formation of gastric HCl. Values for spinal fluid are 118 to 132 mEq./L. Urine values vary greatly with Cl^- intake, but generally range between 110 and 250 mEq./24 hrs. Sweat chloride values will be discussed in connection with the sweat test (Chapter 15).

REAGENTS

1. Sulfuric acid, 0.72 *N*. Add 20 ml. of concentrated sulfuric acid, A. R., to approximately 700 ml. of distilled water in a 1000 ml. volumetric flask. Cool and dilute to volume.

2. Sodium tungstate, 10 per cent (w/v). Place 700 ml. of deionized water into a 1000 ml. volumetric flask and add 100 gm. of sodium tungstate ($Na_2WO_4 \cdot 2H_2O$). After all the sodium tungstate is dissolved, dilute to volume with deionized water. Allow to stand for several days and decant or filter if any precipitate has formed.

3. Mercuric nitrate solution, 0.01 *N*. Place 1.0833 gm. of HgO (red) into a small beaker or flask; add 3 ml. of concentrated HNO_3 and 20 ml. of deionized water. Stir until dissolved and transfer to a 1000 ml. volumetric flask. Fill up to volume with deionized water. Alternatively, 1.6681 gm. of $Hg(NO_3)_2 \cdot \frac{1}{2}H_2O$ may be used.

4. Diphenylcarbazone solution. Dissolve 250 mg. of s-diphenyl-carbazone in 100 ml. of methanol or ethanol. Store in a dark bottle in the refrigerator. The solution should have an orange-red color. If the color changes, e.g., to dark cherry red or to yellow, discard the solution. The reagent is generally stable for 2 months.

5. Chloride standard, 100 mEq./L. Place 5.845 gm. of NaCl (dried at 110°C.) into a 1000 ml. volumetric flask; add 3 ml. of concentrated HNO_3 and approximately 100 ml. of deionized water. After all the salt is dissolved, dilute up to volume with deionized water.

METHOD

1. Place 0.5 ml. of serum or standard, respectively, into an Erlenmeyer flask or test tube.* Add, in succession, 3.5 ml. of water, 0.5 ml. 0.72 *N* sulfuric acid, and 0.5 ml. of 10 per cent sodium tungstate. Mix well, allow to stand for 5 minutes, and centrifuge. (This procedure gives a modified Folin-Wu protein-free filtrate.)

2. Transfer 2 ml. of clear supernatant fluid into a suitable titration vessel and add 0.1 ml. (2 drops) of diphenylcarbazone solution. Titrate with mercuric nitrate solution from a microburet, calibrated in intervals of 0.05 ml. and capable of delivering drops equal to not more than 0.02 ml.† Burets with a fine glass tip are satisfactory, but hypodermic needles should not be used as tips since the metal will react with the mercuric nitrate solution.

3. When approaching the end point (first appearance of a faint blue-violet color), add the mercuric nitrate solutions in amounts not greater than 0.02 ml. at a time.

CALCULATION

Chloride concentration in mEq./L. =

$$\frac{\text{titration of unknown}}{\text{titration of standard}} \times 0.02 \times \frac{1000}{0.2}$$

or

$$\frac{\text{titration of unknown (in ml.)}}{\text{titration of standard (in ml.)}} \times 100$$

COMMENTS ON THE METHOD

1. The titration end point is sensitive to pH. There is some disagreement as to the best pH to be used, but most authors use a pH between 3 and 4.5, which is

* If sweat is to be analyzed in connection with the sweat test outlined in Chapter 15, the following modification is applied: Titrate 2.5 ml. of the extract from the gauze directly without making a protein-free filtrate. Also titrate 0.2 ml. of standard directly, and enter both results into the formula provided with the procedure for the sweat test.

† If spinal fluid or urine with low protein content is to be analyzed, 0.2 ml. of unknown and standard may be titrated directly without making a protein-free filtrate. One drop of approximately 0.07 *N* HNO_3 should be added before titration in order to provide an acid pH. (At alkaline pH, the indicator has a color similar to that observed at the titration end point.)

obtained if the protein-free filtrate is prepared as just described. Therefore, it is important that all tests and standards be titrated at this pH range.

2. Many laboratories titrate specimens directly with $Hg(NO_3)_2$ without preparing a protein-free filtrate. Such an approach yields rapid results, but has an inherent positive error of approximately 2 per cent due to the reaction of Hg^{++} with —SH groups of proteins and the masking of the end point by pigments in the sample. The end point is especially difficult to detect in highly icteric sera. Some authors have reported positive errors of as much as 15 mEq./L.

3. Other halogens, such as Br^-, I^-, and CN^-, CNS^-, as well as —SH groups will also react with Hg^{++} and thus give a positive error. In bromide poisoning, however, the bromide replaces some chloride and the titration may give an apparent normal level since it measures the sum of the two halogens.

REFERENCE

Schales, O., and Schales, S. S.: J. Biol. Chem., *140*:879, 1941.

Colorimetric Measurement of Chloride with Mercuric Thiocyanate

A procedure by Zall, Fisher, and Garner,[45] which was adapted to the AutoAnalyzer by Skeggs,[33] is based on the following principle: The sample is mixed with a solution of $Hg(SCN)_2$. As a result of the high affinity of Cl^- to Hg^{++}, there is a formation of undissociated $HgCl_2$, resulting in the release of free $(SCN)^-$. The $(SCN)^-$ reacts with Fe^{+++} of the ferric nitrate reagent to form a highly colored, reddish color complex of $Fe(SCN)_3$ with an absorption peak at 480 nm.

$$Hg(SCN)_2 + 2Cl^- \rightarrow HgCl_2 + 2(SCN)^-$$
$$3(SCN)^- + Fe^{+++} \rightarrow Fe(SCN)_3$$

PROCEDURE NOTES

The accuracy of this procedure, if carried out on the AutoAnalyzer as recommended by the manufacturer, is relatively poor. The use of secondary standards for calibration (serum samples previously analyzed for chloride content) has increased the accuracy of the method significantly and yields satisfactory results. It is not clearly understood why standards prepared in water are less satisfactory than standards containing protein. One possible explanation may be the significant difference in osmolality of the two standard solutions, which may have an effect on the water movement across the dialyzing membrane.

This procedure has the same interference from other halogens as has been described for the mercurimetric titration method.

Coulometric-Amperometric Titration of Chloride

In the coulometric-amperometric determination of serum chlorides, the sample is diluted in an acid solution (HNO_3—CH_3COOH mixture) containing a small amount (e.g., 25 mg./100 ml.) of gelatin. The nitric acid provides for good electrolytic conductivity; the acetic acid renders the solution less polar, thus, reducing the solubility of silver chloride and providing a sharper end point; the gelatin provides for a smoother and more reproducible titration curve by being adsorbed preferentially to high spots of the electrode and thus equalizing the reaction rate over the entire electrode surface. In addition to this, the acid solution aids in preventing reduction of precipitated silver chloride at the indicator cathode. Excess of protein, however, may introduce some error due to reaction of silver ions with sulfhydryl groups of protein.

If the titration is carried out in the Chloridometer (American Instrument Co., Inc., Silver Spring, Md. and Buchler Instruments Inc., Fort Lee, N.J.), the volume of the solution has to be at least 2.5 ml. so that the electrodes are fully covered with the solution. When the titration is started, a silver generator electrode is fed by a constant current that oxidizes silver to Ag^+ at a constant rate proportional to Q (coulombs). The silver ions thus produced combine with Cl^- to form a precipitate of silver chloride. After the equivalence point is reached (sufficient Ag^+ has been generated to react with all chloride present), any additional generation of Ag^+ will result in an increase in electroactivity of the titrant, which is measured amperometrically by a set of silver indicator electrodes. The increase in current activates a relay, which in turn will stop an automatic timer and also stop the generation of additional Ag^+. Since the current feeding the silver generator electrode is constant, the rate of generation of Ag^+ is also constant; thus, the time necessary to reach the titration end point can be taken as a measure of the chloride concentration. The titration time of standard solutions analyzed in the same manner can be used to calculate the chloride concentration of the unknown according to the following formula:

$$\frac{\text{titration time of unknown} - \text{titration time of blank}}{\text{titration time of standard} - \text{titration time of blank}} \times \text{concentration of standard}$$
$$= \text{chloride concentration}$$

COMMENTS ON THE PROCEDURE

The procedure just described is probably the most accurate method for the determination of chlorides employed in routine clinical chemistry laboratories, as demonstrated by the very close correlation with isotope dilution techniques (results of the titration technique were 99.7 per cent of those found with the isotope dilution technique). Other halogens, as well as CN^-, CNS^-, and —SH, interfere with the determination.

Greatest accuracy is obtained if the titration time is held between 70 and 160 seconds. If solutions with extremely low chloride content are to be analyzed (e.g., in microchloride determinations), the titration speed is reduced by reducing the current serving the silver generator electrodes. Standard solutions used in the determination of specimens with low chloride content should be diluted to a range that corresponds to the approximate concentration of the unknown, or, as in microanalysis, the amount of standard used should be reduced (e.g., 20 μl., if the serum sample used is also 20 μl.).

REFERENCES

Cotlove, E., Trantham, H. V., and Bowman, R. L.: J. Lab. Clin. Med., *51*:461, 1958.
Cotlove, E.: Determination of chloride in biological materials. *In* Methods of Biochemical Analysis. D. Glick, Ed. New York, Interscience Publishers Inc., 1964, vol. 12.
Cotlove, E.: Chloride. *In* Standard Methods of Clinical Chemistry. D. Seligson, Ed. New York, Academic Press, Inc., 1961, vol. 3, pp. 81–92.

Bicarbonate, Carbonic Acid, P_{CO_2}, Total CO_2, and pH

Definition of Terms

Bicarbonate. The bicarbonate (HCO_3^-) is the second largest fraction of the anions in plasma. It is customary to include in this fraction the ionized bicarbonate

(HCO_3^-) and the carbonate CO_3^{--}, as well as the carbamino compounds. At the pH of blood, the concentration of carbonate is only 1/1000 that of bicarbonate and the carbamino compounds are also present in such small amounts (less than 0.5 mEq./L.) that they are generally not mentioned specifically.

Carbonic acid. This fraction of blood, plasma, or serum includes the undissociated carbonic acid ($HHCO_3$),* and the physically dissolved (anhydrous) CO_2.

P_{CO_2}. The pressure of a mixed gas, such as air, is the sum of the partial pressures of the individual gases (see Chapter 11, equation 15). That part of the pressure which is contributed by CO_2 is called the partial pressure of CO_2 (P_{CO_2}). It is usually expressed in mm. of Hg. The only place in the body where the blood is in contact with a gas phase is in the lung alveoli (see Chapter 11). The P_{CO_2} of the blood not in contact with a gas phase (e.g., arterial, capillary, and venous blood) refers to the P_{CO_2} in a hypothetical gas phase with which the blood *would be* in equilibrium.

Total CO_2. The total CO_2 content of blood, plasma, or serum consists of an ionized fraction that contains HCO_3^- (and CO_3^{--}, carbamino compounds) and an un-ionized fraction that contains $HHCO_3$ and physically dissolved (anhydrous) CO_2.

$$[\text{Total } CO_2] = [HCO_3^-] + [HHCO_3]\dagger$$

CO_2 combining power. The value of the CO_2 combining power is an index of the amount of CO_2 that can be bound by serum, plasma, or whole blood as HCO_3^- at a P_{CO_2} of 40 mm. Hg at 25°C.

pH. The pH is the negative logarithm of the hydrogen ion concentration ($pH = -\log [H^+]$). Thus, the average pH of blood (7.40) corresponds to a hydrogen ion concentration of 0.000 000 04 (4×10^{-8} moles/L.). It can readily be seen that the pH value is a more convenient figure than the hydrogen ion concentration and this was one reason for introducing the concept of pH.

The Interrelation of Total CO_2, Bicarbonate, Carbonic Acid, P_{CO_2}, and pH in the Henderson-Hasselbalch Equation

$$K = \frac{[H^+][A^-]}{[HA]}$$

or

$$[H^+] = K\frac{[HA]}{[A^-]}$$

or

$$-\log [H^+] = pH = pK + \log\frac{[A^-]}{[HA]}$$

or

$$pH = pK + \log\frac{[\text{conjugate base}]}{[\text{nonionized acid}]}$$

This equation shows that the pH of a solution is determined by the pK (or pK′) of the acid and by the ratio of the conjugate base to the nonionized acid.

---

* It is customary to use the symbol $HHCO_3$ for carbonic acid to indicate that the first hydrogen atom is not ionized.

† A compound shown in square brackets indicates that reference is made to the concentration of the compound. In this case the concentration of all listed compounds is expressed in m*M*./L.

The concentrations of total CO_2, bicarbonate, carbonic acid, Pco_2, and pH are related to each other in the important and well-known Henderson-Hasselbalch equation*, which for blood serum at 38°C. is

$$pH = 6.10 + \log \frac{[HCO_3^-]}{[HHCO_3]}$$

For pure water in the absence of electrolytes, the pK of carbonic acid is 6.33 ($K = 10^{-6.33}$), but in serum, Hastings, Sendroy, and Van Slyke[11] found an *average* value for the pK of carbonic acid at 38°C. of 6.10. This figure is generally referred to as the pK′. The lower pK′ value for serum (in comparison to the pK for carbonic acid in pure water) is chiefly due to the greater ionic strength (μ) of serum ($pK' = pK - 0.5\sqrt{\mu}$), but it is also influenced by the effect of proteins and lipids on the solubility of CO_2.

For plasma, the value of pK′ varies with the nature of the anticoagulant; with a progressive increase in ionic strength, for example due to oxalate or citrate, the pK′ will fall progressively below the value of 6.10. The application of the Henderson-Hasselbalch equation to whole blood, a biphasic system of cells and plasma, is very complicated and less accurate; thus, it is not advisable to apply the Henderson-Hasselbalch equation to whole blood. Because of this mathematical relationship, any one of the values (pH, $[HCO_3^-]$, or $[HHCO_3]$) can be calculated if two of the three values are known.

In practice, there are convenient methods for measuring total CO_2 (CO_2 content) and pH and these procedures are described in detail in a later portion of this section. The concentration of $[HHCO_3]$ is not measured directly, but is readily determined for serum from the measurement of Pco_2 according to the equation:

$$[HHCO_3] = 0.0301 \times Pco_2$$

where $[HHCO_3]$ = the carbonic acid concentration expressed in m*M*./L.,

Pco_2 = the partial pressure of CO_2 gas expressed in mm. Hg, which is in equilibrium with the serum at 38°C. (see also under Measurement of Pco_2).

Thus, for normal serum at 38°C. at Pco_2 of 40 mm. Hg,

$$[HHCO_3] = 0.0301 \times 40 = 1.204 \text{ m}M\text{./L.}$$

The concentration of bicarbonate ($[HCO_3^-]$) is usually not measured directly, but is calculated as the difference between total CO_2 and $HHCO_3$.

$$[HCO_3^-] = [\text{total } CO_2] - [HHCO_3]$$

Thus, the Henderson-Hasselbalch equation for serum at 38°C. may be rewritten by substituting [total CO_2] − $[HHCO_3]$ for $[HCO_3^-]$ and by substituting $0.0301 \times Pco_2$ for $[HHCO_3]$.

$$\text{pH of serum at 38°C.} = 6.1 + \log \frac{[\text{total } CO_2] - (0.0301 \times Pco_2)}{(0.0301 \times Pco_2)}$$

This equation expresses the relationship between three experimentally determined values, namely, pH, [total CO_2], and Pco_2 for serum at 38°C. The mathematical application of this relationship is illustrated in the following examples:

Example 1: A freshly drawn blood sample has an experimentally determined serum pH of 7.44 and a CO_2 content of 26.7. Calculate the Pco_2, HCO_3^-, and $HHCO_3$.

* The Henderson-Hasselbalch equation is a logarithmic expression of the ionization constant equation of a weak acid.

Answer:

$$\text{pH} = \text{pK}' + \log \frac{\text{total } CO_2 - (0.0301 \times P_{CO_2})}{(0.0301 \times P_{CO_2})}$$

Therefore:

$$7.44 = 6.1 + \log \frac{26.7 - (0.0301 \times P_{CO_2})}{(0.0301 \times P_{CO_2})}$$

or

$$7.44 - 6.1 = \log \frac{26.7 - (0.0301 \times P_{CO_2})}{(0.0301 \times P_{CO_2})}$$

$$\text{antilog of } 1.34 = \frac{26.7 - (0.0301 \times P_{CO_2})}{(0.0301 \times P_{CO_2})}$$

$$21.88 = \frac{26.7 - (0.0301 \times P_{CO_2})}{(0.0301 \times P_{CO_2})}$$

$$(0.0301 \times P_{CO_2}) \times 21.88 = 26.7 - (0.0301 \times P_{CO_2})$$

$$0.658\ P_{CO_2} = 26.7 - (0.0301 \times P_{CO_2})$$

$$0.658\ P_{CO_2} + 0.0301\ P_{CO_2} = 26.7$$

$$0.688\ P_{CO_2} = 26.7$$

$$\mathbf{P_{CO_2}} = \frac{26.7}{0.688} = \mathbf{39\ mm.\ Hg}$$

The [$HHCO_3$] and the P_{CO_2} at 38°C. are related as follows:

$$[HHCO_3] = 0.0301 \times P_{CO_2}$$

Therefore:

$$[HHCO_3] = 0.0301 \times 39$$

$$\mathbf{[HHCO_3] = 1.2\ m\mathit{M}./L.}$$

The HCO_3^- is the difference between the total CO_2 and the [$HHCO_3$]; consequently,

$$\mathbf{[HCO_3^-]} = 26.7 - 1.2 = \mathbf{25.5\ m\mathit{M}./L.}$$

Example 2: Experimentally determined data on a blood serum are [total CO_2] = 25.2 m*M*./L.; P_{CO_2} = 37.5 mm. Hg. Calculate the pH, the [$HHCO_3$], and the [HCO_3^-] of this serum.

Answer:

$$\mathbf{[HHCO_3]} = 0.0301 \times 37.5 = \mathbf{1.13\ m\mathit{M}./L.}$$

$$\mathbf{[HCO_3^-]} = 25.2 - 1.03 = \mathbf{24.07\ m\mathit{M}./L.}$$

$$\text{pH} = 6.1 + \log \frac{24.07}{1.13}$$

$$\text{pH} = 6.1 + \log 21.30$$

$$\mathbf{pH} = 6.1 + 1.33 = \mathbf{7.43}$$

The preceding calculations can be applied to any type of combination of values and the results thus obtained can be used for the construction of a nomogram as shown in Figure 10-1, or tables can be prepared that permit easy and fast calculation of the missing parameter. Weisberg has constructed a slide rule for the same purpose and it is available from the Baxter Laboratories, Inc., Morton Grove, Illinois.

COMMENTS

Results obtained by calculation will, in general, agree with experimentally determined figures; however, there are some limitations in this method that should be clearly understood.

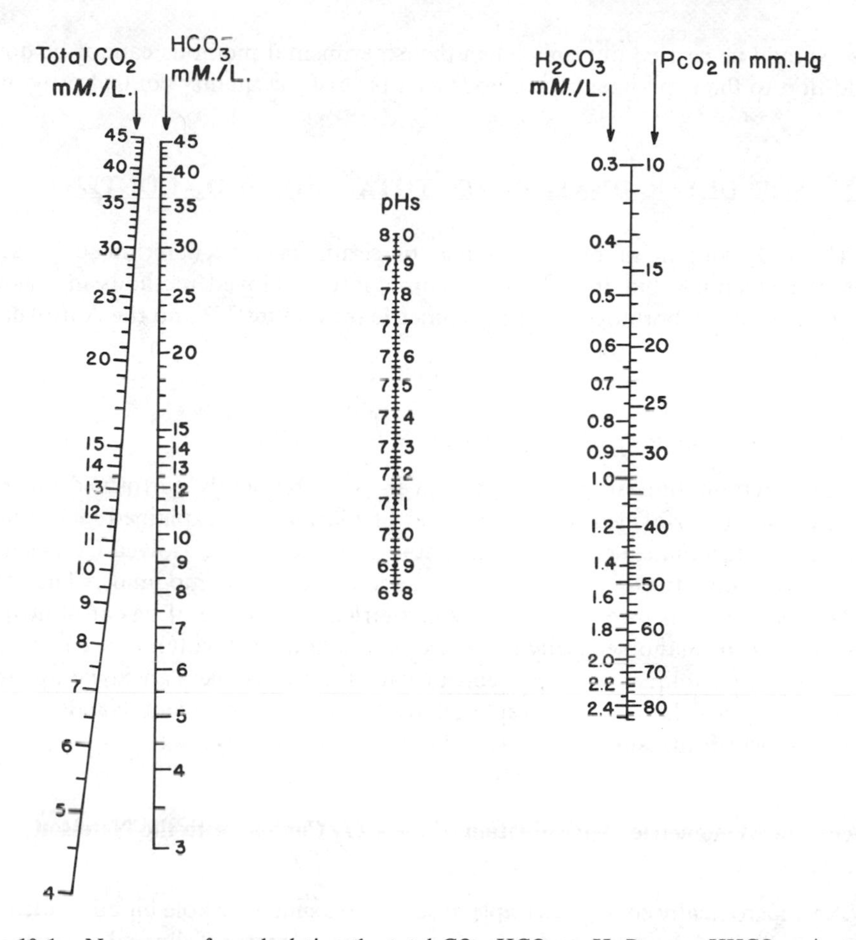

Figure 10-1. Nomogram for calculating the total CO_2, HCO_3^-, pH, P_{CO_2} or $HHCO_3$. A straight line is drawn between the two known points and the desired value is read from the other two scales. Example: the total CO_2 and the pH are known and their values are 13 m*M*./L. and 7.2, respectively. If we draw a line through these two points we will find a P_{CO_2} value of 31 mm. Hg, an $HHCO_3$ value of 0.93 m*M*./L. and an HCO_3^- value of 12.1 m*M*./L. All values are only approximate. (From Van Slyke D. D. and Sendroy, J. Biol. Chem., 79: 781, 1928. Reproduced with the permission of the author and The American Society of Biological Chemists, Inc.)

The validity of the results obtained by calculation depends on the accuracy of the two experimentally determined parameters. If only one of these is in error, the calculated figure must be in error as well. If we assume an acceptable experimental error for the pH determination in Example 1 of ± 0.02 and for the CO_2 content of ± 0.7 m*M*./L., the calculated "true" HCO_3^- concentration may be anywhere between 24.9 and 26.2 m*M*./L. and the "true" value for $HHCO_3$ may be between 1.08 and 1.25 m*M*./L.

The Henderson-Hasselbalch equation uses the average value of 6.1 for the pK′ of carbonic acid. Although it is true that few samples will differ significantly from this average value, it is now well established that deviations of as much as 0.2 units can occur.[39] In our own laboratory, we analyzed several thousand samples for P_{CO_2} by both methods (P_{CO_2} electrode and calculation from pH and total CO_2) and found a significant number of discrepancies; thus, we feel that the pH, total CO_2, and P_{CO_2} should be determined experimentally whenever possible. The calculation of these values should be carried out only when the experimental methods cannot be done or in addition to the experimental methods as a part of the quality control program.

THE DETERMINATION OF TOTAL CO_2 (CO_2 CONTENT)

The CO_2 content of blood, plasma, or serum may be determined by several different techniques, but the two most commonly employed methods in a modern clinical chemistry laboratory are the gasometric techniques[17,19] and the AutoAnalyzer procedure.[34]

Gasometric Measurement of CO_2 Content

The determination of the CO_2 content is most frequently performed on serum, but may also be performed on whole blood or plasma. As explained in Definitions of Terms, the technique determines the amount of physically dissolved CO_2 as well as that released from HCO_3^-, CO_3^{--}, $HHCO_3$, and carbamino compounds. The amount of CO_2 gas present is measured either volumetrically (volume of gas at atmospheric pressure = V) or manometrically (pressure of gas at a fixed volume = P). Initially, both of these techniques were carried out with the respective Van Slyke apparatus. Today, both have largely been replaced by the more convenient Natelson microgasometer, which measures the CO_2 content manometrically.

Principle of Manometric Determination of the CO_2 Content with the Natelson Microgasometer

An anaerobically collected sample of serum, plasma, or whole blood is introduced anaerobically into the pipet attached to the microgasometer, followed by lactic acid, an antifoam reagent, and deionized water. Lactic acid releases CO_2 from HCO_3^- and $HHCO_3$, the antifoam reagent (e.g., caprylic alcohol) prevents foaming, and the water washes the sample and reagents into the reaction chamber. The sample and each of the reagents is separated by a mercury button and the water is brought into the reaction chamber with the aid of mercury, which at the same time prevents the introduction of air. The reaction chamber is then sealed by closing the stopcock and a vacuum is applied. This will cause a transfer of the CO_2 gas and physically dissolved CO_2 from the liquid phase into the vacuum above the liquid. After 1 minute of agitation to assure complete release of the gas, the liquid level is advanced to a predetermined position and the pressure of the gas (which is now compressed into a fixed volume) is measured manometrically (P_1). The introduction of an alkali, such as NaOH, into the reaction chamber will cause a selective reabsorption of the CO_2 gas (formation of Na_2CO_3), allowing for measurement of the residual (non-CO_2) gas (P_2). The difference between P_1 and P_2 is a measure of the amount of CO_2 present.

The manometer of the Natelson microgasometer is calibrated in mm. Hg, but this figure has to be corrected for the temperature and must be converted to m*M*./L. by multiplication with the conversion factor supplied with the instrument. Correction for temperature is necessary since the pressure of a gas is directly related to the temperature if the volume is kept constant (see also the discussion on gasometry in Chapter 2).

REFERENCES

Natelson, S.: Am. J. Clin. Path., *21*:1153, 1951.
Natelson, S.: Microtechniques of Clinical Chemistry. 2nd ed. Springfield, Ill., Charles. C Thomas, Publisher, 1961. pp, 152 – 157.

Determination of CO_2 Content by AutoAnalyzer

The specimen is treated with sulfuric acid to release the CO_2 gas, and a fixed aliquot of this gas is reabsorbed in a weak alkaline carbonate-bicarbonate buffer solution containing phenolphthalein. As CO_2 is reabsorbed, the pH of the buffer solution decreases, resulting in a decrease of the red color. This change in the color of phenolphthalein is proportionate to the CO_2 content of the sample.

REFERENCE

Skeggs, L. T. Jr.: Am. J. Clin. Path., *33*:181, 1960.

CLINICAL SIGNIFICANCE

Knowledge of the CO_2 content of serum (plasma, blood), together with other clinical and laboratory information (pH, P_{CO_2}), is very useful in the evaluation of the acid-base balance; however, the determination of the CO_2 content alone, without additional information, is of limited value. A high CO_2 content may be observed in respiratory acidosis (retention of CO_2) as well as in metabolic alkalosis (increase in HCO_3^-). Low CO_2 content may be observed in respiratory alkalosis (loss of CO_2 due to hyperventilation) or in metabolic acidosis (decrease of HCO_3^- due to compensatory mechanisms). Additional laboratory determinations, such as pH and P_{CO_2}, will permit differentiation between metabolic and respiratory conditions. Examples of disorders associated with abnormal CO_2 content are described in Chapter 11.

NORMAL VALUES

Specimen	*Range in* m*M*./L.
Venous plasma (serum)	23–30
Capillary plasma (serum)	21–28
Venous (whole) blood	22–26
Arterial (whole) blood	19–24

The Determination of CO_2 Combining Power

The CO_2 combining power is an index of the amount of CO_2 that can be bound by plasma at a P_{CO_2} of 40 mm. Hg at 25°C. The test is performed by placing the sample (serum, plasma) into a suitable container such as a separatory funnel (to enlarge the surface of the specimen) and equilibrating it with a gas mixture containing CO_2 at a P_{CO_2} of 40 mm. Hg, the average CO_2 tension of alveolar air. This can be realized by a gas mixture containing 5.2 per cent (v/v) CO_2 in air at 760 mm. Hg (or 5.4 per cent CO_2 at 740 mm. Hg). In practice, it is customary to use expired air from a normal person, which approximates the desired P_{CO_2}. After equilibration, a portion of the sample is removed and analyzed for total CO_2 in the same way as outlined under CO_2 content. This method, as explained before, determines not only bicarbonate but also the $HHCO_3$. Thus, the value obtained must be corrected by subtracting 1.3 m*M*./L. if the equilibration was performed at 38°C. or 1.8 m*M*./L. if the equilibration was performed at 25°C. (The solubility of CO_2 increases with a decrease in temperature.) Although the test has some value in evaluating acid-base balance in metabolic disturbances, it is now being replaced by more accurate, more modern, and more convenient techniques. The value of the test is considerably less and results may even be misleading in respiratory disturbances since the sample is equilibrated with a P_{CO_2} that corresponds to the average *normal* alveolar air rather than to the condition existing in the patient under investigation, who may have a low P_{CO_2} (in respiratory alkalosis) or a high P_{CO_2} (in respiratory acidosis).

The blood specimen for this determination can be collected aerobically; although, at times, this may be a convenience, the technic also requires equilibration, which is inconvenient and also a potential source of error.

The Determination of Plasma Bicarbonate

The bicarbonate is the second largest fraction of the anions in plasma (21 to 28 m*M*./L.). Small amounts of carbamino compounds are present in plasma and are customarily included in this fraction. Bicarbonate has important functions as a component of the bicarbonate buffer system and it also serves as a transport form for CO_2 from the tissues to the lungs (see Chapter 11). Although bicarbonate can be determined by direct titration,[19,30] the test is rarely requested in clinical medicine since the total CO_2 content serves as an approximate measure of bicarbonate concentration (approximately within 1 m*M*./L.).

As indicated in the section, The Interrelation of Total CO_2, Bicarbonate, Carbonic Acid, P_{CO_2}, and pH in the Henderson-Hasselbalch Equation, the bicarbonate can also be calculated or it can be obtained with the aid of a nomogram. It can also be approximated by subtracting 1.3 m*M*./L. from the experimentally determined total CO_2 content.

COMMENTS

Specimens used for the determination of total CO_2 must be handled anaerobically at all times (see Collection of Blood for the Determination of pH, CO_2 content, and P_{CO_2}).

Determination of Blood, Plasma, or Serum pH

The determination of the blood (plasma, serum) pH has been a very difficult procedure; however, in recent years several pH meters have been made available that measure the blood pH with a sufficient degree of accuracy and with little effort. Today, it is not the pH meter but the way in which the specimen is collected that limits the accuracy of the pH determinations (see under collection of specimen).

Modern pH meters are generally accompanied by instruction manuals that contain detailed procedures for blood pH measurements; therefore, no procedure for pH measurement determinations is included in this chapter. There are certain factors, however, that apply to all pH measurements regardless of the instruments used; these will now be listed.

1. Specimens must be collected and, if not analyzed immediately, stored as outlined under collection of blood. If the test is performed on whole blood, the specimen must be remixed before analysis. The report must indicate the type of specimen used for the pH measurement, since normal values will differ slightly, depending on the type of specimen used.

2. The pH meter must be standardized with buffers that are accurate to 0.005 pH units. Some manufacturers, such as Instrumentation Laboratories, Inc. (Watertown, Mass.) and Radiometer, Inc. (Cleveland, Ohio), make buffers available that are standardized to 0.001 pH units, but such accuracy is not attainable, nor can solutions be maintained under routine laboratory conditions. As buffers are exposed to air their pH usually gradually decreases because of absorption of atmospheric CO_2. The National Bureau of Standards has also recently released instructions for the preparation of acceptable pH buffers.

pH meters should be standardized against at least two buffers to check the slope of the pH calibration curve and thus the linearity of pH measurements. The effect of deviations from such linearity can be minimized by choosing a buffer that is close (within approximately 0.2 pH) to the pH of the sample being measured.

3. During blood (plasma, serum) pH determinations it is important to note and record the temperature of the electrode. Although the pH of most phosphate buffers used for blood pH standardization decreases by about 0.001 pH per °C. rise, the coefficient for blood is about 0.015 pH per °C., and about 0.012 per °C. for serum. Thus, if the pH determination is expected to have an accuracy of 0.01 pH unit, the temperature variation should be less than ±0.5°C. For greater variations in temperature, a correction is to be applied.

$$\text{pH at } 38°\text{C.} = \text{pH at } t - (0.015^{*} \times (38 - t))$$

where t is the temperature at which the pH measurement was performed.

Thus, a blood pH of 7.40 at 25°C. becomes

$$7.40 - (0.015 \times 13) = 7.40 - 0.19 = 7.21 \text{ at } 38°\text{C.}$$

For variations in body temperature of the patient from whom the blood is drawn, one can apply a similar correction, in order to determine the pH at the temperature of the patient. Thus, the pH of a patient with a body temperature of 40°C., but which was measured at 38°C. and was found to be 7.30, can be corrected for temperature as follows:

$$\text{pH at } 40°\text{C.} - \text{pH at } 38°\text{C.}\ (-2 \times 0.015) = 7.33.$$

4. The pH electrode should be thoroughly rinsed with saline before, in between, and after the pH measurements. The electrode must be stored in accordance with the instructions of the manufacturer. This may be either with water, with saline, or with a cleaning solution containing HCl and pepsin.

CLINICAL CONDITIONS ASSOCIATED WITH CHANGES IN BLOOD pH

The pH of blood is controlled by the ratio of the concentration of bicarbonate: carbonic acid, as described in detail in Chapter 11. Any decrease in the ratio, and, therefore, of the blood pH, is referred to as acidosis. This condition may be caused by a primary decrease in bicarbonate as found in metabolic disorders (e.g., diabetes, renal failure) or by a primary increase in the carbonic acid as in respiratory disorders (e.g., decreased rate or depth of respiration, or both). If the ratio of bicarbonate:carbonic acid is decreased but the pH is still within normal limits, we refer to these

* The exact temperature factor given by Rosenthal (Freier, E. F., Clayson, K. J., and Benson, E. S.: Clin. Chim. Acta, 9:348, 1964.) is 0.0147, but most laboratories employ the factor 0.015.

conditions as compensated acidosis. If the pH is below 7.35, the term uncompensated acidosis is applied although some degree of compensation may already have taken place; however, the degree of compensation in these cases was not sufficient to return the pH into the normal range.

Any condition that causes an increase in the ratio of bicarbonate:carbonic acid is referred to as alkalosis. The increase may be due to a primary increase in the bicarbonate, as found in metabolic alkalosis, or due to a primary decrease in the carbonic acid, as found in respiratory alkalosis (hyperventilation). If the pH is outside the upper normal of 7.45, the term uncompensated alkalosis applies, although, here again, some degree of compensation may have taken place. In conditions in which the various compensatory mechanisms have succeeded in restoring normal body pH, we speak of compensated alkalosis. Examples for these various conditions are discussed in Chapter 11.

The determination of blood (plasma, serum) pH is the most valuable single factor in the evaluation of acid-base balance. The determination of the CO_2 content alone is often insufficient to evaluate existing conditions, especially when dealing with respiratory disorders or respiratory disorders superimposed on metabolic disorders. Therefore, it is recommended that a pH determination be performed when investigating the acid-base balance of a patient.

NORMAL VALUES

For serum and plasma that are collected under routine conditions, the normal range is generally considered to be 7.35 to 7.45, with an average of 7.40. This figure allows for the experimental error experienced in routine determinations. The true range of serum or plasma pH is probably narrower, with an upper limit of 7.42. The pH of whole blood is 0.01 to 0.03 pH units lower than that of plasma and serum because of an effect of the red cells on the liquid junction potential rather than because of an inherent difference ("suspension effect").[32]

COMMENTS

Special care must be taken not to expose the sample to air during collection or when transferring it from the collection tube to the pH meter (see also Collection of Blood for the Determination of pH, CO_2 Content, and P_{CO_2}).

P_{CO_2}

The pressure of a mixed gas, such as air, is the sum of the partial pressures of the individual gases (see Chapter 11, equation 15). That part of the pressure which is contributed by CO_2 is called the partial pressure of CO_2 or P_{CO_2} and is usually expressed in mm. Hg. The amount of CO_2 and, therefore, $HHCO_3$ dissolved in plasma is in direct relation to the P_{CO_2} and may be calculated from the latter figure by multiplying the P_{CO_2} in mm. Hg (at 38°C.) by 0.0301 (see also the discussion on $HHCO_3$).

The only place in the body where the blood plasma is in contact with a gas phase is in the lung alveoli (see Chapter 11). The P_{CO_2} of the blood plasma not in contact with the gas phase (e.g., arterial, capillary, and venous blood) refers to the P_{CO_2} in a hypothetical gas phase with which the blood plasma *would* be in equilibrium.

CLINICAL SIGNIFICANCE

The determination of P_{CO_2} in serum is one of the most valuable laboratory tools in the diagnosis of acid-base disturbances and particularly of respiratory disturbances. The reader is referred to the discussion in Chapter 11.

PROCEDURE

P_{CO_2} determinations in a modern hospital laboratory are generally performed on whole blood, plasma, or serum, using a P_{CO_2} electrode. The operating principle of this electrode is based on the fact that CO_2 from the sample crosses a thin Teflon membrane and diffuses into a $KHCO_3$ solution at a rate proportional to the P_{CO_2} value of the sample. The diffusion of CO_2 into the $KHCO_3$ solution causes a change in pH that can be measured directly by a glass electrode. The shift in pH of the $KHCO_3$ solution can be directly related to the P_{CO_2} of the sample.

The P_{CO_2} can also be calculated with the aid of the Henderson-Hasselbalch equation or with a nomogram as previously discussed, if the pH and either the total CO_2 content or the true bicarbonate concentration are known.

NORMAL VALUES

The normal range for P_{CO_2} in arterial blood is 35 to 46 mm. Hg (average 40) and that of venous blood is 38 to 49 mm. Hg (average 46). The average value for P_{CO_2} in alveolar air is 36 mm. Hg and that in tissue cells is 60 mm. Hg. The P_{CO_2} of plasma (serum) approximates the values given for whole blood.

COMMENTS

Specimens must not be exposed to air during collection or transfer to the P_{CO_2} electrode (see also Collection of Blood for the Determination of pH, CO_2 Content, and P_{CO_2}).

Carbonic Acid

The carbonic acid fraction of blood plasma or serum, by definition, includes the (undissociated) $HHCO_3$ and the physically dissolved CO_2.

$$[HHCO_3] = [\underset{\textit{Hydrated form}}{HHCO_3} + \underset{\textit{Dissolved gas}}{CO_2}]$$

The latter is present in approximately 700 to 1000 times larger quantity (m*M*./L.) than that of undissociated $HHCO_3$. The proportions of the two forms are governed by the mass law:

$$K \times [HHCO_3] = [CO_2] \times [H_2O]$$

The concentration of $HHCO_3$ is directly related to the P_{CO_2} and can be calculated by multiplying the P_{CO_2} value (in mm. Hg) by the proportionality constant (a) 0.0301 (see the following equation).

The proportionality constant (a) is derived from the solubility coefficient of CO_2 (α = ml. CO_2 dissolved per ml. solvent). The value for α for normal serum, at 38°C., is 0.510.[42] The value for α for pure water is slightly higher, namely, 0.545, because of the greater solubility of CO_2 in water as compared to serum. Salts and proteins in serum depress the solubility, but lipids increase its solubility. In fact, α for highly lipemic sera may be even higher than that for water. Because of the differences in composition of individual sera and the associated differences in the solubility of carbonic acid in this medium, the Henderson-Hasselbalch equation, as written earlier, is directly applicable only to sera with an average composition of solutes. It is apparent that any changes in the solubility coefficient and, therefore, the proportionality constant would change the product of $\frac{\text{total } CO_2 - (a \times P_{CO_2})}{a \times P_{CO_2}}$ and, thus, decrease the validity of results obtained by the Henderson-Hasselbalch equation, as generally written, unless a value for *a* applicable to the specific serum is used. The composition and concentration of solutes in serum also affects the pK′ as discussed under Interrelation of total CO_2, Bicarbonate, Carbonic Acid, P_{CO_2}, and pH in the Henderson-Hasselbalch equation.

In order to determine the amount of carbonic acid dissolved in 1 L. of plasma at 38°C., the following equation may be applied:

$$\text{m}M\text{. } HHCO_3\text{/L.} = \frac{\text{solubility coefficient } (\alpha) \text{ at } 38^{\circ}\text{C.} \times 1000 \text{ (ml./L.)}}{22.26 \text{ (ml./m}M\text{.)}} \times \frac{P_{CO_2}\text{(mm. Hg)}}{760 \text{ (mm. Hg)}}$$

$$= 0.0301 \times P_{CO_2} \text{ (at } 38^{\circ}\text{C. and 760 mm. Hg)}$$

NORMAL VALUES

Total Carbonic Acid

Sample	*Average* (in m*M*./L.)	*Range* (in m*M*./L.)
Venous blood	1.35	1.15–1.50
Arterial blood	1.30	1.05–1.45
Venous plasma	1.20	1.02–1.38

CLINICAL SIGNIFICANCE

Since the total carbonic acid is directly related to the P_{CO_2}, both values have the same clinical significance. For more detail see the section on P_{CO_2}.

CALCIUM AND PHOSPHORUS

The metabolisms of calcium and phosphorus are so closely related that it is justifiable to discuss these elements at the same time. More than 99 per cent of the body calcium and 80 per cent of the phosphorus are present in the bones as $Ca_3(PO_4)_2$; the remainder of the calcium and phosphorus have varied and significant functions in the body. For example, calcium ions decrease cell membrane permeability and neuromuscular excitability; they are engaged in the transmission of nerve impulses and participate in blood coagulation. Furthermore, calcium ions activate some enzymes, such as succinate dehydrogenase and adenosine triphosphatase. Phosphorus, on the other hand, is involved in the intermediary metabolism of carbohydrates (see Chapter 4) and is a component of other physiologically important substances, such as phospholipids, nucleic acids, nucleotides, and ATP.

Both calcium and phosphorus are absorbed in the upper small intestine, with an active absorption process taking place in the ileum. The absorption of both ions is favored at acid pH and is greatly decreased at alkaline pH, where both ions form insoluble compounds. Presence of vitamin D is essential for calcium absorption. Increased levels of this vitamin promote, and decreased levels reduce, calcium absorption. The normal values for serum calcium and phosphorus are listed below:

Normal Values for Calcium and Phosphorus in Serum

	Calcium		Phosphorus	
	(mg./100 ml)	(mEq./L.)	(mg./100 ml.)	(mEq./L.)
Adults (younger age group)	9.0–10.7	4.5–5.4	3.0–4.5	1.7–2.5
Elders	8.5–10.5	4.3–5.3	3.0–4.5	1.7–2.5
Children	9.0–11.0	4.5–5.5	4.5–6.5	2.5–3.6

All of the calcium of blood is present in the serum; however, phosphorus is present mainly in the cells as organic phosphate, with only a small but definite amount being in serum as inorganic phosphate.

Calcium is present in serum in two distinct forms, the *nondiffusible protein-bound form*, which constitutes approximately 40 to 50 per cent of the total serum calcium, and the *diffusible fraction*. The latter can be further subdivided into *complexed calcium* (by citrate and phosphate), which ranges from approximately 0.2 to 0.5 mg./100 ml., and *ionized calcium*, which is the physiologically active form (4.2 to 5.5 mg./100 ml.). Recent reports indicate that nonspecific salt effects may further decrease the ionized calcium so that the physiologically active calcium is only 20 per cent of the total serum calcium. A significant decrease in the ionized fraction, regardless of the total serum calcium level, results in tetany. Among the factors influencing the level of ionized calcium is the pH. Any increase in the blood pH decreases the level of ionized calcium without affecting the total serum calcium levels.

The *ionized calcium* can be determined by measuring the calcium levels before and after treatment with dextran gel,[29] or by ultrafiltration.[23] It has also been recommended that the level of total calcium in cerebrospinal fluid (CSF) be used as an index of ionized calcium in serum, since this fluid may be considered an ultrafiltrate of plasma.

Normal CSF calcium levels are 4.2 to 5.8 mg./100 ml. Results obtained on spinal fluid indeed correlate very closely with the ionized calcium, as determined by other methods. Recently ion specific electrodes have been made available that permit the determination of ionized calcium;[18] however, techniques, at present, are not advanced sufficiently to make them useful for routine clinical laboratories.

The approximate concentration of ionized calcium may also be calculated with a formula that is based on the nomogram by McLean and Hastings.[16]

$$\text{mg. Ca}^{++}/100\text{ ml.} = \frac{(6 \times \text{total serum calcium in mg./100 ml.}) - \dfrac{(\text{gm. serum protein/100 ml.})}{3}}{\text{gm. protein/100 ml.} + 6}$$

It should be realized that this formula is based on the relationship between calcium and protein at a serum pH of 7.35 at 25°C. There is also a significant difference in the ability of the various protein fractions to bind calcium, and cephalin is able to bind calcium in a manner similar to protein. Consequently, any differences in serum pH, in the proportions of serum proteins, or increase in phospholipids may affect the results obtained by this formula and limit its usefulness.

Factors Influencing Serum Calcium Levels

Of the many factors that are known to influence serum calcium levels, the following are the most important:

Parathyroid hormone. This hormone maintains a serum calcium level by mobilization of calcium from the bones. If the levels decrease, the mobilization of calcium from the bones will also decrease, resulting in low serum and urine calcium levels. The parathyroid hormone has also an effect on the tubular reabsorption of phosphorus, which increases with decreased parathyroid hormone levels and vice versa (see under Clinical Significance).

Plasma proteins. Since approximately 50 per cent of calcium is bound to protein, any decrease in serum proteins frequently results in a decrease of total serum calcium levels. This decrease, however, affects mainly the nondiffusible fraction and, therefore, tetany is rarely observed in these conditions. Similarly, an increase in protein, as seen in multiple myeloma, may increase the total serum calcium without significant change in the diffusible fraction.

Plasma phosphates. There appears to be a reciprocal relationship between calcium and phosphorus. Any increase in the serum phosphorus causes a decrease in serum calcium by a mechanism that is not clearly understood. The best example of such a situation is the increased serum phosphorus level in renal retention of phosphorus (e.g., in uremia), which results in decreased serum calcium levels. It has been speculated that the decreased calcium level in uremia is due to the acidosis, which causes a dissociation of some calcium from its protein bond. This portion of calcium then is most likely excreted by the kidneys to maintain a constant level of the ionized calcium fraction in plasma.

Vitamin D. This vitamin has an important role in the absorption of calcium; increased levels favor, decreased levels depress, the absorption of calcium and, consequently, the serum levels.

CLINICAL SIGNIFICANCE OF CALCIUM AND PHOSPHORUS MEASUREMENTS

Hypercalcemia (increased serum calcium level) is observed in hyperparathyroidism (together with decreased serum phosphorus levels and increased urine calcium and phosphorus levels); however, if calcium supplies are exhausted, serum calcium levels in this condition may be normal or even low. Hypervitaminosis, multiple myeloma, and some neoplastic diseases of bone may also be accompanied by increased serum calcium levels, but, unlike the situation in hyperparathyroidism, serum phosphorus levels are normal or even elevated.

Hypocalcemia may be observed in hypoparathyroidism, together with normal or increased serum phosphorus levels and decreased urinary calcium and phosphorus levels. The low serum calcium levels in this disease may lead to an increase in neuromuscular irritability and thus to tetany. Serum calcium levels may be low in steatorrhea (due to decreased absorption), in nephrosis (due to loss of protein), in nephritis (due to increase in serum phosphorus), and in pancreatitis (due to formation of calcium soaps).

Hyperphosphatemia (increased serum phosphorus levels) may be found in hypervitaminosis (vitamin D), hypoparathyroidism, and renal failure. *Hypophosphatemia* (low serum phosphorus levels) may be seen in ricketts (vitamin D deficiency), in hyperparathyroidism, and in the Fanconi syndrome, which is, among others, associated with a defect in reabsorption of phosphorus from the glomerular filtrate.

Determination of Serum Calcium by Oxalate Precipitation and Redox Titration

PRINCIPLE

A solution of ammonium oxalate is added to a diluted serum sample and calcium is precipitated as calcium oxalate (1). The precipitate is washed with dilute ammonium hydroxide to remove the excess of ammonium oxalate and is then dissolved in sulfuric acid (2). The oxalic acid thus formed from the calcium oxalate is titrated with a standardized solution of potassium permanganate (3). The manganese atom is reduced by oxalic acid from the +VII to the +II valent form. The titration end point is indicated by the first appearance of a purple color (excess potassium permanganate), which occurs when all oxalic acid has reacted with the $KMnO_4$:

$$Ca^{++} + C_2O_4^{--} \longrightarrow CaC_2O_4\downarrow \qquad (1)$$

$$CaC_2O_4 + H_2SO_4 \longrightarrow H_2C_2O_4 + CaSO_4 \qquad (2)$$

$$2KMn^{(VII)}O_4 + 5H_2C_2O_4 + 3H_2SO_4 \xrightarrow{70°C} K_2SO_4 + 2Mn^{(II)}SO_4 + 10CO_2 + 8H_2O \qquad (3)$$

SPECIMEN

Calcium determinations are generally performed on serum collected in the fasting state. Blood collected with oxalate or from patients receiving EDTA treatment is unsuitable for this analysis; the former will remove calcium from serum by precipitation as oxalate, and the latter will chelate calcium and thus make it unavailable for analysis by most procedures.

REAGENTS

1. Ammonium oxalate, 4 per cent (w/v). Dissolve 4.0 gm. of ammonium oxalate in water and dilute to 100 ml.; filter.

2. Sulfuric acid, 1.0 *N*. Into a 1 L. volumetric flask filled partially with distilled water, add carefully 29 ml. of concentrated H_2SO_4 and dilute to the mark with water. Mix.

3. Ammonium hydroxide, 2 per cent. Dilute 2 ml. of concentrated ammonium hydroxide to 100 ml.

4. Acid sodium oxalate, 0.100 *N* (stock). Place 6.700 gm. of sodium oxalate, A. R. (previously dried at 105°C. and cooled in a desiccator), into a 1 L. volumetric flask. Dissolve in water, add 30 ml. of concentrated sulfuric acid, cool, and dilute to the mark with deionized water. Mix thoroughly. It is stable, if well-stoppered.

5. Acid sodium oxalate, 0.0100 *N*. Pipet exactly 10 ml. of the stock solution into a 100 ml. volumetric flask and dilute to volume with deionized water. The solution is stable if kept well-stoppered.

6. Potassium permanganate, 0.100 *N* (stock). Dissolve 3.2 gm. of potassium permanganate, A. R., in water and dilute to 1000 ml. Let stand for at least 3 days in the dark to allow for stabilization. Filter through glass wool or a sintered glass filter to remove sediments of MnO_2 and standardize as follows: Pipet exactly 25 ml. of 0.100 *N* sodium oxalate into a titration vessel, add 1 ml. of concentrated sulfuric acid, warm to about 70°C., and titrate with the permanganate solution. The amount (ml.)

of permanganate solution needed to obtain the first permanent pink color = T. To prepare a solution with a normality of 0.100, add to a 1 L. volumetric flask 40 × (25 − T) ml. of water and dilute to the mark with the permanganate solution. Confirm titer by repeating the titration. Keep the solution in a brown bottle.

7. Potassium permanganate, 0.01 *N*. Pipet exactly 10 ml. of potassium permanganate stock solution into a 100 ml. volumetric flask and dilute to the volume with deionized water; mix. Determine the titer of this solution by titrating 2.0 ml. of 0.0100 *N* sodium oxalate solution heated to 70°C., making sure that this temperature is maintained until the end of the titration. The factor

$$F = \frac{2}{\text{ml. of } 0.010\ N \text{ permanganate solution}}$$

should be 0.95 to 1.05; if not, a new solution must be prepared.

8. Calcium standard (10.00 mg./100 ml. or 5.00 mEq./L.). Place 0.2497 gm. of dried Ca-carbonate, A. R., into a 1000 ml. volumetric flask. Add approximately 9 ml. deionized water and 1 ml. concentrated HCl. Shake until dissolved. Fill up to the mark with deionized water. Store in a Pyrex bottle. The solution is stable.

PROCEDURE

1. Into acid-cleaned 12 or 15 ml. conical centrifuge tubes, add 2 ml. of deionized water, followed by 2 ml. of serum or standard, respectively. Then add to all tubes 1 ml. of ammonium oxalate solution. Mix thoroughly, but do not invert.

2. Allow to stand for 4 hours at 37°C. or preferably overnight.

3. Centrifuge all tubes for at least 5 minutes at 2000 rpm and decant supernatant fluid carefully.

4. Invert tubes and allow them to drain on a pad of filter paper or gauze for approximately 3 minutes and then carefully wipe the mouth of the tube dry.

5. Add 1 drop of ammonium hydroxide solution and break up precipitate by tapping the tube with your fingers or, preferably, with the aid of a Vortex mixer. Add 3 ml. of 2 per cent ammonium hydroxide solution while rotating the tube to wash down all oxalate from the sides.

6. Centrifuge as in step 3, decant, add 1 drop of 1 *N* H_2SO_4, break up precipitate, and add 2 ml. of 1 *N* H_2SO_4 while rotating the tube as just described. Place all tubes in a boiling water bath for approximately 1 minute to dissolve the oxalate precipitate.

7. Titrate standards and unknowns with 0.010 *N* of permanganate solution to a faint pink end point. When approaching the end point, place the tubes back into the water bath to make sure that the solution has a temperature of at least 70°C. If this is not observed, false results will be obtained.

CALCULATIONS

$$\text{mg. Ca/100 ml.} = F \times T \times 0.2 \times \frac{100}{2}$$

or

$$\text{mg. Ca/100 ml.} = 10 \times F \times T$$

where F = the factor obtained in standardizing the 0.010 *N* permanganate solution (see reagents)

T = ml. of permanganate used to titrate the unknown

0.2 = mg. of calcium that is equivalent to 1 ml. of 0.0100 *N* permanganate solution.

COMMENTS ON THE METHOD

There is considerable controversy concerning the time needed for complete precipitation of the calcium oxalate. In the experience of this author, serum samples should be allowed to stand for at least 4 hours. Precipitation of calcium in a non-proteinaceous solution, such as standard solutions or urine, is complete within 1 hour. Washing the precipitate of calcium oxalate with ammonium hydroxide should be done only once. The second washing, as recommended by some authors, causes a slight loss of calcium; however, the washing step is critical, since failure to remove the excess oxalate would yield high results.

The pH of the sample at the time of precipitation should be between pH of 2.7 and 7.6. In the lower pH range, precipitation of calcium oxalate is incomplete, and in the higher pH range there is a coprecipitation of $Mg(OH)_2$ and $MgNH_4PO_4$. Under the conditions of the test, there is no interference from magnesium since magnesium oxalate, being soluble at pH less than 7.0, remains in the supernatant. pH adjustments for serum are not necessary since the combined pH of serum and reagent will be in the specified range.

REFERENCE

Kramer, B., and Tisdall, F. F.: J. Biol. Chem, *47*:475, 1921; modified by Clark, E. P., and Collip, J. B.: J. Biol. Chem., *63*:461, 1925.

The Determination of Serum Calcium by Precipitation with Chloranilic Acid

PRINCIPLE

Calcium is precipitated from the sample as calcium chloranilate by a saturated solution of sodium chloranilate. The precipitate is washed with isopropyl alcohol to remove the excess of chloranilic acid and is then treated with EDTA, which chelates with calcium and releases chloranilic acid. The latter is colored and can be measured photometrically.

$$Ca^{++} + \text{chloranilate} \rightarrow \text{Ca-chloranilate}\downarrow$$

$$\text{Ca-chloranilate} + \text{EDTA} \rightarrow \text{Ca-EDTA} + \text{Chloranilic acid (purple color)}$$

REAGENTS

1. Sodium chloranilate. Dissolve 6.13 gm. of sodium chloranilate in 500 ml. of deionized water. Shake well to saturate the solution and filter through filter paper.
2. Isopropyl alcohol, 50 per cent (v/v). Dilute 250 ml. isopropyl alcohol, A. R., to 500 ml. with deionized water.
3. EDTA. Dissolve 25 gm. tetrasodium ethylenediaminetetraacetate in deionized water and dilute to 500 ml.
4. Calcium standard. See preceding reagent section.

PROCEDURE

1. Pipet 2 ml. of serum, plasma (heparinized), or aqueous standard into a 15 ml. conical centrifuge tube (preferably in duplicate).
2. Add forcefully to all tubes 1 ml. of saturated sodium chloranilate. Mix tubes well; do not invert! (It is important to blow in the chloranilate to achieve immediate mixing; otherwise, proteins may be precipitated.)
3. Place tubes into a 37°C. water bath for at least 3 hours.
4. Centrifuge for 10 minutes at approximately 2000 rpm, decant the supernatant immediately, and drain tubes for approximately 2 minutes. Wipe the mouth of the tube with cotton gauze or filter paper to remove any excess of chloranilate.

5. Add 1 drop of isopropyl alcohol to all tubes and break up the precipitate by tapping the tube against your fingers or, preferably, with the aid of a Vortex mixer. Wash precipitate with 6 to 7 ml. of 50 per cent isopropyl alcohol. The alcohol is conveniently dispensed from a squirt bottle by directing the stream directly into the tip of the tube.

6. Centrifuge and drain tubes for approximately 2 minutes on cotton gauze or filter paper; wipe the mouth of the tube dry.

7. Add 1 drop of EDTA to the precipitate and break up the precipitate as outlined in step 5.

8. Add to all tubes exactly 6 ml. of EDTA.

9. Allow all tubes to stand for approximately 10 minutes and read at 520 nm. (or Klett filter No. 52) against an EDTA blank. Although the color should appear immediately, in the opinion of this author it is advantageous to allow the tubes to stand for the specified time to insure complete solution of the precipitate. The color is extremely stable.

CALCULATION

$$\frac{\text{absorbance of unknown}}{\text{absorbance of standard}} \times \frac{100}{2} \times 0.2$$

or

$$\frac{\text{absorbance of unknown}}{\text{absorbance of standard}} \times 10 = \text{calcium in mg./100 ml.}$$

If values are expressed in mEq./L. multiply by 5 instead of 10.

COMMENTS ON THE METHOD

Some samples tend to give a cloudy solution in the final step of the procedure (after the addition of EDTA). In some cases this is due to faulty washing of the precipitate with isopropyl alcohol; in other cases, it is thought to be due to lipids. In the latter instance, extraction of the final solution with ether may be helpful.

REFERENCE

Ferro, P. V., and Ham, A. B.: Am. J. Clin. Path., *28*:208, 1957, *28*:689, 1957.

Determination of Serum Calcium by EDTA Titration (Micromethod)

PRINCIPLE

A diluted serum sample is titrated with EDTA in the presence of calcein and at an alkaline pH (to avoid magnesium interference). The initial yellow-green fluorescence caused by the calcium-calcein complex changes to a nonfluorescent salmon-pink color (of free calcein) when all calcium present has been chelated by EDTA (Fig. 10-2).

REAGENTS

1. Calcium standard (10 mg./100 ml.). See the method by Clark and Collip.

2. Potassium hydroxide, 1.25 *N*. Dissolve 7.0 gm. of KOH, A. R., in deionized water, add 0.050 gm. of KCN, and dilute to 100 ml. with deionized water.

3. Calcein indicator. Dissolve 0.025 gm. of calcein in 100 ml. of 0.25 *N* NaOH. Store in polyethylene bottles and keep refrigerated. Replace the indicator if it turns greenish.

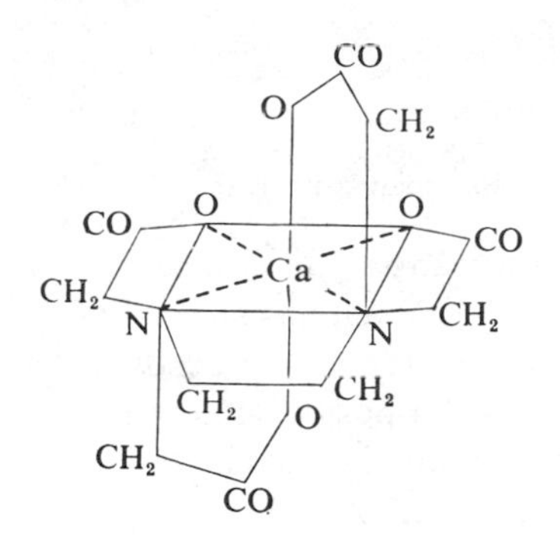

Figure 10-2. Chelate complex of EDTA with calcium.

4. EDTA, 0.020 *N*. Dissolve 0.372 gm. of disodium dihydrogen ethylenediaminetetraacetic acid in deionized water and dilute to 100 ml. Store in a small polyethylene bottle.

PROCEDURE

1. Add to microtitration cups (e.g., Beckman No. 314463) 40 μl. of water, standard, and sample, respectively. Add to all cups 200 μl. of KOH and 20 μl. of indicator. The indicated amounts may be added conveniently with a Beckman micropipet as described in Chapter 3. If such pipets are used, standards and unknowns should be pipetted with the same pipet.

2. Titrate blank, standards, and unknowns with 0.020 *N* EDTA solution, using one of the commercially available microtitrators as described in Chapter 3. Disappearance of the yellow-green fluorescence and the appearance of an orange-red color marks the end point. It has been found more convenient to carry out the titration in the dark (self-constructed dark box with viewing window or Oxford Titrator) and with the aid of an ultraviolet light source. It is much easier to detect the disappearance of the fluorescence in the absence of daylight than to detect the appearance of the salmon-pink color.

CALCULATION

$$\frac{\text{reading of unknown} - \text{reading of blank}}{\text{reading of standard} - \text{reading of blank}} \times 10 = \text{mg. Ca/100 ml.}$$

COMMENTS ON THE PROCEDURE

The sensitivity and speed of complexometric titration procedures are the main reasons for their use in the clinical laboratory. Unfortunately, all of these methods seem to have the great disadvantage that the titration end point is hard to detect, particularly in the presence of hemoglobin, jaundice, lipemia, and high phosphate concentration. In complexometric titrations of this type the slowly added complexing agent (EDTA) removes calcium little by little from the calcium-indicator-complex, resulting in a gradual color change from the yellow-green fluorescence to the non-fluorescent salmon-pink until all of the indicator is in the latter form.

A number of indicators have been used, e.g., murexide (ammonium purpurate), Cal-Red, and Eriochrome Black T. In some of these determinations (Eriochrome Black T) magnesium ions interfere; in others (Cal-Red and calcein) magnesium interference can be eliminated by titration at a strongly alkaline pH.

REFERENCES

Appleton, H. D., West, M., Mandel, M., and Sala, A.: Clin. Chem., *5*:36, 1959.
Diehl, H., and Ellingboe, J. L.: Anal. Chem., *28*:882, 1956.
Technical Bulletin U M-TB-007 E, Beckman Instruments, Inc., Fullerton. Calif.

Many attempts have been made to determine calcium in diluted serum samples by flame photometry. Although calcium has an arc line at 422.7 nm. and 2 molecular bands at 554 and 662 nm., flame photometric procedures have not been very successful. Among the difficulties are: (1) the positive interference by sodium and potassium, (2) the inhibition of calcium emission by phosphates and sulfates, and (3) the fact that calcium cannot be excited easily even in a "hot" flame. Attempts have been made to eliminate interference by phosphorus, by adding an excess of PO_4 ions to standard and samples, which minimizes the effect of phosphorus in the sample. This has solved one problem but created another, namely, a further decrease in sensitivity due to the addition of phosphates.

Isolation of calcium by precipitation with oxalate before analysis has resulted in the elimination of interfering substances; however, here again, oxalate depresses the emission. This is possibly due to the formation of degradation products with low excitation potential.

Normal values for flame photometric methods are generally slightly higher than those obtained with chemical methods.

REFERENCE

Margoshes, M. and Vallee, B. L.: Flame photometry and spectrometry, principles and applications. *In* Methods of Biochemical Analysis. New York, Interscience Publishers, Inc., 1956, vol. 3.

Determination of Calcium by Atomic Absorption Spectrophotometry

PRINCIPLE

Calcium compounds, when introduced into a flame, dissociate into free calcium atoms. In this form calcium absorbs light of characteristic wavelength (e.g., 4227 Å) produced by a hollow cathode lamp (see also Chapter 2, Analytical Procedures). Under the described conditions only a small fraction of calcium atoms (about 1 out of 1000) will be raised to a higher energy level and will emit light on returning to the ground state (see flame photometry).

Some anions, such as phosphates, bind with calcium to form highly refractory compounds in the flame, thus causing falsely low results. This interference is eliminated by the addition of La^{+++} (or Sr^{++}), which binds preferentially with phosphate and prevents the formation of calcium phosphates. Under the conditions of the following procedure, an 0.7 per cent solution of La^{+++} is capable of preventing phosphate interference up to concentrations of 1.6 gm. phosphate phosphorus/1000 ml. Proteins are precipitated by trichloroacetic acid to eliminate the depressing effect of proteins on the measurement and to increase reproducibility and accuracy.

REAGENTS

1. Stock standard (200 mg./100 ml.). Transfer quantitatively 2.500 gm. of dried $CaCO_3$ into a 500 ml. volumetric flask. Dissolve in approximately 10 ml. of deionized water and 5 ml. of concentrated HCl. After all the $CaCO_3$ has dissolved, dilute with deionized water to the 500 ml. mark.

2. Calcium working standards. Into six 100 ml. volumetric flasks place, respectively, 2, 3, 4, 5, 6, and 7 ml. of calcium stock standard and dilute with deionized water to the 100 ml. mark. The respective concentrations are 4, 6, 8, 10, 12, and 14 mg. Ca/100 ml.

3. Diluent (0.7 per cent La^{+++} in 4 per cent TCA). Place 17.8 gm. $LaCl_3 \cdot 6H_2O$ and 40 gm. of trichloroacetic acid into a 1000 ml. volumetric flask; dissolve in and dilute to the mark with deionized water.

INSTRUMENT

The following procedure has been developed for the Perkin-Elmer 303 Atomic Absorption Spectrophotometer, which was equipped with an Intensitron Hollow Cathode lamp and a Perkin-Elmer burner with a Boling head. The instrument was connected to a Perkin-Elmer recorder readout and a Texas Instrument recorder (Servo/Rider II). The procedure may be adapted to other instruments with sufficient sensitivity and stability.

Instrumental parameters. The Perkin-Elmer 303 Atomic Absorption Spectrophotometer is operated at a wavelength of 4227 Å, the slit is set to No. 4 (1 mm., 13 Å,) and the source is set to 10 mA. The air and acetylene supply should be adjusted to a flow meter reading of "10." At this setting the flame should be blue with narrow yellow streaks. The sample tubing should be approximately 300 mm. in length and 0.015 inch I.D. × 0.043 inch O.D. (Technicon Polyethylene tubing No. 4010). The recorder is set to scale expansion No. 1, noise suppression No. 3 and a speed of 0.5 or 1.0 inch/minute.

PROCEDURE

1. Pipet 1.00 ml. of each of the working standards, unknowns, and controls into appropriately labeled test tubes (duplicates if possible).

2. Add forcefully 9 ml. of diluent to each test tube. An automatic dispensing device may be used for this addition. Mix and allow to stand for 10 minutes. Centrifuge for approximately 10 minutes at 2000 rpm. The supernatant of some samples may be slightly turbid; however, this does not interfere with the test.

3. Insert the sample tubing into the supernatant, making sure that it is inserted into all tubes to the same level. The procedure may be semiautomated by aspirating the supernatant from an AutoAnalyzer Sampler II unit equipped with an adjustable cam (120/hr.) with a wash time of 8 seconds and an aspiration time of 22 seconds. In order to detect any possible drift it is recommended that standards be run frequently (e.g., after 10 samples).

CALCULATION

Prepare a standard curve and read the results either manually or with the aid of an AutoAnalyzer chart reader.

COMMENTS ON THE PROCEDURE AND SOURCES OF ERROR

Calcium values obtained with the preceding procedure are on an average approximately 0.2 mg./100 ml. higher than those obtained with the Ferro and Ham method. It is believed that the difference is due to incomplete precipitation of calcium and loss of small amounts of calcium when washing the precipitate in the manual procedure.

Some authors[40] have claimed that precipitation of proteins before analysis is not necessary, and serum dilutions (e.g., 1:50) containing 0.1 per cent La^{+++} may be analyzed directly. It has been our experience that such an approach will generally give results in good correlation with the classic methods, but that samples will occasionally differ by more than 5 per cent.

The method just presented may be applied to urine, provided that the urine has been acidified as outlined in connection with the manual methods that follow. No phosphorus interference is observed with levels up to 1.6 gm./1000 ml. Presence of phosphorus in concentrations of 3 gm./1000 ml. depresses calcium values by approximately 3 per cent. When aspirating the supernatant in step 3 of the procedure, the sample tubing should be held at approximately the same level for all determinations. Significant differences in the position of the sample tubing will produce a different flow rate of supernatant into the burner and thus will affect the results.

REFERENCE

Tietz, N. W., Fiereck, E. A., and Green, A.: unpublished.

Determination of Calcium in Urine by Oxalate Precipitation

PRINCIPLE

The basic principle of the procedure is identical to that for serum. Since urine may contain precipitated calcium salts, it is necessary to dissolve these by acidification to pH 1 (followed by adjustment of pH to above 4.0) before the precipitation of calcium with oxalate. The urine may also be collected in a bottle containing 10 ml. of 6 *N* HCl.

SPECIMEN

Values for urinary output of calcium are most meaningful if the patient has been on a low calcium, neutral ash diet for at least 3 days before the urine collection, and if a 24 hour urine specimen is collected.

PROCEDURE

1. Collect a 24 hour urine specimen, measure the volume, and acidify to a pH of approximately 1.0 by adding concentrated HCl drop by drop.
2. Heat the sample briefly in a 100°C. water bath, allow to come to room temperature, and centrifuge.
3. Pipet 5.0 ml. of this urine into a 15 ml. conical centrifuge tube and add 1.0 ml. of 4 per cent ammonium oxalate, 1 drop of 0.1 per cent methyl red indicator and 2 per cent NH_4OH until the appearance of an orange color (pH 4.5).
4. Prepare a standard (2.0 ml.) as outlined under serum calcium, add 1.0 ml. of ammonium oxalate to all tubes, and allow to stand for 4 hours at 37°C. or overnight at room temperature. Continue the procedure as outlined under serum calcium step 3.

CALCULATION

$$\text{mg. Ca/24 hr.} = T \times F \times 10 \times \frac{2}{5} \times \frac{\text{total urine volume}}{100}$$

The factor $\frac{2}{5}$ in the formula is included to correct for the use of 5.0 ml. urine instead of 2.0 ml. as is used for the standard.

NORMAL VALUES

The urinary excretion of calcium during a 24 hour period differs greatly with the intake of calcium.

Type of Diet	*Amount Excreted*
Low calcium	Less than 50 mg./24 hr.
Average calcium	50–150 mg./24 hr.
High calcium	100–300 mg./24 hr.

COMMENTS

The Sobel and Sobel[35,36] acidimetric oxalate precipitation method has certain advantages over the Clark and Collip method. In this method, urine calcium is precipitated as calcium oxalate at a pH of approximately 3.3. After centrifugation, the precipitate is washed and heated at 500°C. to convert calcium oxalate to calcium carbonate. The carbonate is then dissolved in boric acid, forming calcium borate; this is titrated with a stronger acid (0.01 *N* HCl), which replaces the borate. The end point (change of the indicator methyl purple to a purple color) is reached when all borate has been replaced by chloride. The method can also be adapted for serum calcium determinations.

The Sobel and Sobel method has the following advantages: the titration is carried out with a stable standard acid; the titration can be done at room temperature; and the use of an unstable redox reagent is not necessary.

REFERENCES

See under Serum Calcium.

Determination of Calcium in Urine by Chloranilate Precipitation

PRINCIPLE

Refer to the method for serum calcium by precipitation with chloranilic acid.

SPECIMEN

Refer to the method for urine calcium by oxalate precipitation.

PROCEDURE

1. Follow steps 1 and 2 of the procedure for urine by oxalate precipitation.
2. Pipet 2.0 ml. of urine into a 12 or 15 ml. conical centrifuge tube and add 1 ml. of sodium chloranilate as outlined under serum. If low calcium values are expected, use 4 ml. of urine and 2 ml. of chloranilate.

CALCULATION

Refer to the serum method. If 4.0 ml. of urine was used, divide the values by 2.

Semiquantitative Determination of Calcium in Urine (Sulkowitch Test)

PRINCIPLE

The Sulkowitch reagent contains oxalate and is buffered with acetate. When added to urine, it will produce a fine white precipitate of calcium oxalate without coprecipitation of other urine constituents. The amount of turbidity produced is the basis for the approximate quantitation.

REAGENTS

Sulkowitch reagent. Dissolve 2.5 gm. of oxalic acid and 2.5 gm. of ammonium oxalate in approximately 100 ml. of water, add 5 ml. of glacial acetic acid, and dilute to 150 ml. with deionized water.

PROCEDURE

1. If urine is turbid, it is filtered or centrifuged and the clear supernatant is used for analysis.
2. Add 5 ml. of protein-free urine to a test tube. Urine may be deproteinized by acidifying the sample with 10 per cent acetic acid and heating it to 100°C. for several minutes. Restore the original volume by adding distilled water, and filter.
3. Add 5 ml. of the Sulkowitch reagent.
4. Mix and let stand for 2 to 3 minutes.

INTERPRETATION

No precipitate of calcium oxalate:	−
Faint turbidity visible against black background:	+
Turbidity visible without black background:	++
Opaque cloud:	+++
Flocculent precipitate:	++++

NORMAL VALUES

Healthy individuals on a normal diet show results equal to one plus (+) or two plus (++). Absence of a precipitate indicates that no calcium was present in urine and that the serum calcium levels are probably below 8.5 mg./100 ml. A four plus (++++) reaction indicates increased rate of secretion and suggests that the serum values may be above 12 mg./100 ml.

The test is only a screening test and doubts have been raised whether the test is sufficiently accurate for clinical use. Its use is discouraged.

Determination of Inorganic Phosphate In Serum and Urine

PRINCIPLE

A trichloroacetic acid filtrate of serum or urine is treated with molybdate reagent, which reacts with phosphate to form ammonium molybdophosphate (ammonium phosphomolybdate). This is thought to have the formula $(NH_4)_3[PO_4(MoO_3)_{12}]$. The addition of a suitable reducing agent such as aminonaphtholsulfonic acid produces a blue color of heteropolymolybdenum blue. A mild reducing agent is employed in order to avoid reduction of the excess of molybdate present. Other reducing agents, such as stannous chloride, ascorbic acid, Elon (p-methyl amino phenol) and N-phenyl-p-phenylendiamine (Semidine) have been utilized in this reaction, but aminonaphtholsulfonic acid is still widely used.

REAGENTS

1. Trichloroacetic acid, 5 per cent (w/v). Place 50 gm. of trichloroacetic acid, A. R., into a 1000 ml. volumetric flask; dissolve in and fill to the mark with deionized water.

2. Sulfuric acid, 10 N. Slowly add 300 ml. of concentrated sulfuric acid, A. R., to 750 ml. of deionized water, mix well, and cool.

3. Molybdate reagent. Dissolve 25 gm. of ammonium molybdate, A. R., in about 200 ml. of deionized water. Into a 1 L. volumetric flask place 300 ml. of 10 N sulfuric acid, add the molybdate solution, dilute with washings to 1 L. with deionized water, and mix. The solution is stable indefinitely. Discard the reagent if blanks show a blue color.

4. Sodium bisulfite, 15 per cent (w/v). To 30 gm. of sodium bisulfite, A. R., in a beaker, add 200 ml. of deionized water from a graduate cylinder. Stir to dissolve and, if turbid, allow to stand well stoppered for several days and then filter. Keep reagent well stoppered.

5. Sodium sulfite, 20 per cent (w/v). Dissolve 20 gm. of sodium sulfite anhydrous, A. R., in deionized water and dilute to the 100 ml. mark. Filter if necessary. Keep well stoppered.

6. Aminonaphtholsulfonic acid reagent. Place 195 ml. of 15 per cent sodium bisulfite solution into a glass-stoppered cylinder or other suitable container. Add 0.5 gm. of 1, 2, 4-aminonaphtholsulfonic acid and 5 ml. of 20 per cent sodium sulfite. Stopper and shake until the powder is dissolved. If solution is not complete, add, with continuous shaking, 1 ml. of sodium sulfite at a time, until solution is complete. Avoid excess of sodium sulfite. Transfer the solution to a brown glass bottle and store in the cold. The solution is stable for about 1 month.

7. Stock standard (0.4 mg. P in 5 ml.). Place exactly 0.351 gm. of dry potassium dihydrogen phosphate, A. R., into a 1 L. volumetric flask, dissolve in deionized water, add 10 ml. of 10 N sulfuric acid, and dilute to the mark with deionized water.

8. Working standard (0.004 mg. P/ml.). Place 5.00 ml. of the stock phosphate standard into a 100 ml. volumetric flask and make up to the volume with 5 per cent trichloroacetic acid.

PROCEDURE

1. Place 0.5 ml. of serum into a 15 × 150 ml. test tube or a 10 ml. glass-stoppered cylinder.

2. Blow in 9.5 ml. of 5 per cent trichloroacetic acid, mix, and let stand for 5 minutes.

3. Centrifuge or filter through Whatman No. 42 filter paper.

4. Pipet 5 ml. of clear filtrate into a test tube or glass-stoppered cylinder graduated at 10 ml. Prepare a blank by using 5 ml. of 5 per cent trichloroacetic acid and a standard by using 5 ml. of working standard ($5 \times 0.004 = 0.02$ mg. P).

5. Add 1 ml. of molybdate reagent to all test tubes.

6. Add 0.4 ml. of aminonaphtholsulfonic acid reagent; mix.

7. Dilute to the 10 ml. mark with deionized water, mix, and allow to stand for 5 minutes.

8. Set blank at 100 per cent T or zero A and read standard and unknowns at 690 nm.

CALCULATIONS

Read results from a standard curve or calculate as follows:

$$\frac{A_U}{A_S} \times 0.02 \times \frac{10}{5} \times \frac{100}{0.5} = \text{mg. P/100 ml.}$$

or

$$\frac{A_U}{A_S} \times 8 = \text{mg. P/100 ml.}$$

COMMENTS ON THE PROCEDURE

1. If urine is to be analyzed, follow steps 1 to 3 of the serum procedure. In step 4 use two test tubes, one with 0.5 and one with 2.5 ml. of filtrate. Make up the missing volume with 5 per cent trichloroacetic acid. Calculate as for serum and multiply the results by 10 or 2, respectively, to correct for the smaller amount of filtrate used. In case of a 24 hour urine collection, express values in mg. P/24 hr., using the following formula:

$$\frac{\text{mg. P/100 ml.} \times \text{total 24 hr. urine volume}}{100}$$

The normal values for the excretion of urine phosphorus differ greatly with the dietary intake of phosphates, but average values most widely accepted are 0.9 to 1.3 gm./24 hr.

2. Phosphate in plasma (serum) of pH 7.4 is present as HPO_4^{--} and $H_2PO_4^{-}$ in a ratio of 80:20 or $\frac{0.8}{1}:\frac{0.2}{1}$. This ratio increases as the pH increases and decreases as the pH decreases. At a pH of 4.5, as it may be found in urine, the ratio of HPO_4^{--} to $H_2PO_4^{-}$ decreases to approximately 1:100. Thus, a biological sample has a different ratio of the two forms of phosphates, depending on the pH of the sample. On the other hand, the amount of phosphate phosphorus is unaffected by changes in pH. It is for this reason, that values are more commonly expressed in mg. phosphate-phosphorus/100 ml. rather than in terms of mEq. phosphate/L. as is customary for other electrolytes.

To convert mg. P/100 ml. serum to mEq. phosphate/L., the former is multiplied by an average factor of 0.58. This factor is derived from the following formula:

$$10\left(\frac{0.8}{1} \times \frac{2}{31}\right) + 10\left(\frac{0.2}{1} \times \frac{1}{31}\right) = 0.58$$

The figure 10 converts the volume of 100 ml. to 1 L.; the figures $\frac{0.8}{1}$ and $\frac{0.2}{1}$ indicate the ratio of HPO_4^{--} and $H_2PO_4^{-}$, respectively, at pH of 7.4; the figures 2 and 1 represent the valence of the two forms of phosphate; the number 31 represents the atomic weight of phosphorus.

3. Control of the pH during the color development is extremely important. At strongly acid pH, such as is provided by trichloroacetic acid, the reduction of the [P-Mo] complex is favored and reduction of the molybdate reagent itself is inhibited. At higher pH's, such as pH 2 or 3, reduction of the molybdate reagent takes place.

Labile organic phosphate esters, such as creatine phosphate or glucose-1-phosphate, are hydrolyzed at strongly acid pH. If inorganic phosphorus is to be measured in the presence of one of these esters, the pH must be adjusted to 4.5 to 5.0, and ascorbic acid must be used as a reducing agent. The latter is a mild reducing agent that will not reduce the molybdate reagent at this pH.

4. Lipid phosphorus and most of the phosphate from organic esters are not detected with this method.

REFERENCE

Fiske, C. H. and Subbarow, Y.: J. Biol. Chem., *66*:375, 1925.

MAGNESIUM

PHYSIOLOGY AND CLINICAL SIGNIFICANCE

Among the intracellular cations magnesium is second in quantity only to potassium. Like calcium, it is absorbed in the upper intestine, but unlike calcium there is no requirement for vitamin D or any similar factor. About 65 per cent of the magnesium of blood serum is diffusible; the remainder is bound to protein.

Little is known about the factors regulating magnesium levels in blood. A reciprocal relation between serum magnesium and serum calcium levels has been observed in some conditions and, in other conditions, between serum magnesium and serum phosphate; however, no details about the mechanism of these relationships are known.

Magnesium ions serve as activators for a number of important enzyme systems engaged in hydrolysis and transfer of phosphate groups[41] such as alkaline phosphatase (from red cells and bone), prostatic acid phosphatase, hexokinase, and creatine kinase.

A "magnesium deficiency tetany" has been described. It is characterized by low serum magnesium and normal serum calcium values and can be distinguished from "calcium deficiency tetany" in that it responds to magnesium but not to calcium administration. This is probably the most important application for magnesium determinations in serum. At the time of tetany, serum magnesium levels of 0.3 to 1.5 mEq./L. have been reported in the presence of normal serum calcium levels and normal pH. Treatment with magnesium sulfate resulted in all cases in a rise in the serum magnesium level and a concomitant disappearance of tetany and convulsions.[41] Serum levels of magnesium are, at times, a poor measure of cellular magnesium deficiency. Plasma levels do not decrease below 1 mEq./L. until at least 25 per cent of the cellular magnesium is lost. Thus, in these conditions, urine magnesium excretion is a better measure of cellular loss of magnesium.

Decreased serum magnesium levels (hypomagnesemia) have been found in the malabsorption syndrome, acute pancreatitis, chronic alcoholism and delirium tremens, chronic glomerulonephritis, aldosteronism, and excessive loss of magnesium in the urine. This last situation has been observed in renal tubular reabsorption defects, in patients with congestive heart failure treated with ammonium chloride and mercurial diuretics, and after treatment with chlorothiazides. With the exception of the last mentioned condition, the body has a great ability to preserve magnesium by tubular reabsorption. In states of magnesium deficiency, *urine* levels are extremely low.

Increased serum magnesium levels (hypermagnesemia) have been observed in dehydration, severe diabetic acidosis, and Addison's disease. Since magnesium is filtered through the glomeruli, any condition interfering with this process results in retention and thus elevation of serum magnesium levels. Some of the highest values have been observed in uremia. Hypermagnesemia leads to an increase in the atrio-ventricular conduction time of the electrocardiogram.

NORMAL VALUES

The normal concentration of magnesium in serum is 1.4 to 2.3 mEq./L. (1.7–2.8 mg./100 ml.). The concentration in cerebrospinal fluid is 2.4 to 3.0 mEq./L. Red cells contain about 4.5 to 6.0 mEq./L. The *urinary* excretion is 6 to 8.5 mEq./24 hr.

SPECIMENS

Serum magnesium determinations should be done on samples drawn in the fasting state. Since the magnesium concentration in red cells is substantially larger than that in serum, hemolyzed specimens are not suitable for analysis. Magnesium levels are stable for several days if the serum is stored in the refrigerator, separated from the red cells.

Determination of Serum Magnesium by the Titan Yellow Method

PRINCIPLE

A trichloroacetic acid filtrate of serum is treated with the dye titan yellow (methyl benzothiazide-1,3-4,4′-diazo aminobenzol-2,2′-disulfonic acid) in alkaline solution. The red lake that forms is thought to be dye adsorbed on the surface of colloidal particles of magnesium hydroxide, which are kept in solution with the aid of polyvinyl alcohol. This last reagent also increases the sensitivity of the method by a factor of approximately 2.

REAGENTS

1. Trichloroacetic acid, 5.0 per cent (w/v). Dissolve 50.0 gm. of trichloroacetic acid, A. R., in deionized water and make up to 1 L. Store in glass-stoppered borosilicate bottles.
2. Sodium hydroxide, 5.0 *N*. Dissolve 200.0 gm. of NaOH in deionized water and dilute to 1 L. The solution may also be prepared from a stronger solution by dilution.
3. Polyvinyl alcohol, 0.1 per cent (w/v). Suspend 1.0 gm. of polyvinyl alcohol (Elvanol, grade 70-05, E. I. Du Pont de Nemours & Co., Wilmington, Del.) in 40 to 50 ml. of 95 per cent ethanol and pour the mixture into 500 to 600 ml. of swirling deionized water. Warm on a hot plate until the solution is clear. Allow to cool and dilute to 1 L. Polyvinyl alcohol, dissolved directly in water, tends to form lumps; thus, the compound should be suspended in alcohol as outlined.
4. Titan yellow (stock solution). Dissolve 75 mg. of titan yellow in and make up to 100 ml. with 0.1 per cent polyvinyl alcohol; filter if not clear. The reagent is stable for 2 months if stored in a brown bottle at room temperature.
5. Titan yellow (working solution). Dilute 10 ml. of the stock titan yellow solution to 100 ml. with 0.1 per cent polyvinyl alcohol. The reagent is stable for 1 week if stored in a brown bottle.
6. Stock standard (20 mEq./L.). Dissolve 243.2 mg. of bright magnesium metal turnings, A. R., in a covered beaker with 50 ml. of deionized water and 2 ml. of concentrated hydrochloric acid. Avoid open flames since hydrogen is produced in this reaction. When the reaction subsides, add dropwise hydrochloric acid to dissolve the magnesium metal completely. Transfer the solution quantitatively to a 1 L. volumetric flask and make up the volume with deionized water. Accurately pipet 5, 10, 15, and 20 ml. aliquots of this stock standard into 100 ml. volumetric flasks and dilute to 100 ml. with deionized water. These standards contain 1.0, 2.0, 3.0, and 4.0 mEq./L. The stock standard is stable if well stoppered.

PROCEDURE

1. Add 1 ml. of the unknown serum into a 15 × 150 mm. test tube and blow in 5 ml. of 5 per cent trichloroacetic acid. Some prefer to use 7.5 per cent trichloroacetic acid.

2. Mix tubes gently but thoroughly, let stand for 5 minutes, and centrifuge for 5 minutes at 2000 rpm.

3. Transfer 3 ml. of the clear supernatant to a Coleman or other suitable cuvet.

4. Prepare a standard set by pipetting 0.5 ml. of each magnesium working standard into separate cuvets followed by 2.5 ml. of 5 per cent trichloroacetic acid. Prepare a reagent blank by substituting distilled water for the magnesium standard.

5. Add to all cuvets 2 ml. of titan yellow working solution and 1 ml. of 5.0 *N* NaOH. Mix tubes thoroughly and read after 5 minutes, but not later than 30 minutes, at a wavelength of 540 nm. with the reagent blank set to 100 per cent T or zero absorbance.

CALCULATION

Construct a standard curve or employ the following formula:

$$\frac{A_U}{A_S} \times C_S = \text{mEq. Mg/L.}$$

where A = absorbance readings

C_S = concentration of magnesium standard most nearly corresponding to the value of the unknown

COMMENTS ON THE PROCEDURE

Since the procedure follows Beer's Law only over a short range, it is necessary to run a standard set with each determination; only in narrow band spectrophotometers will the procedure follow Beer's Law over the range observed in serum.

It is critical that the test be read at the absorption peak; thus, the maximum absorbance should be determined for the instrument used in this determination.

The procedure just outlined is invalid for sera from patients receiving calcium gluconate or mercurial diuretics.

This method is simple and fast, but has the disadvantage that its accuracy is only within 10 per cent. The lack of greater accuracy is probably due to the "erratic and unsystematic variations presumably related to the colloidal nature of the material whose color is being measured."[37]

REFERENCE

Basinski, D. H.: *In* Standard Methods of Clinical Chemistry. S. Meites, Ed. New York, Academic Press, Inc., 1965, vol. 5.

Determination of Serum Magnesium by Fluorometric and Complexometric Techniques

Magnesium ions and 8-hydroxy-5-quinoline sulfonic acid form a chelate compound that fluoresces if exited at a wavelength of 380 to 410 nm. The peak fluorescence occurs at 510 nm.

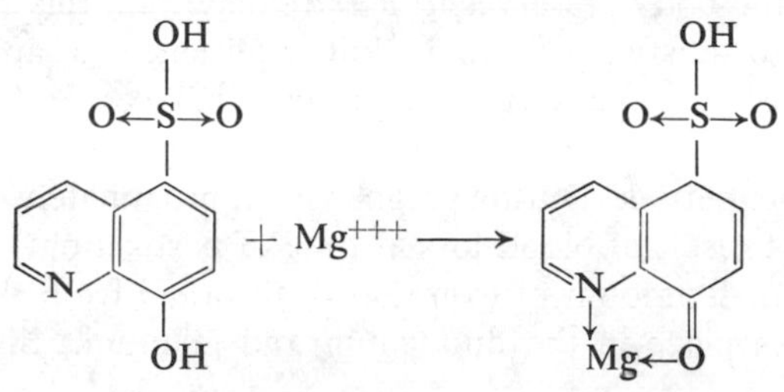

Other divalent ions will form similar complexes with this reagent. Because the binding capacity of the reagent for magnesium is greater than that for other ions and because the molar fluorescence of the magnesium compound is considerably greater, interferences by other ions in serum are negligible.

Although this method is extremely sensitive and simple, it is not widely used because of difficulties inherent in fluorometric methods. The enhancing or quenching effect of other compounds is the drawback that makes this method completely unsuitable for the determination of magnesium in urine.

Complexometric methods[1] employing Eriochrome, murexide or other dyes are also rarely used because of the difficulties inherent in such procedures (see Complexometric Determination of Serum Calcium).

Determination of Magnesium by Atomic Absorption Spectrophotometry

This technique has proved to be a fast, accurate, and reasonably simple method for determining magnesium in biological material.[1,38] Although difficulties similar to those discussed in calcium determinations by atomic absorption spectrophotometry have been observed, they have been overcome by the addition of lanthanum and strontium salts or EDTA.

Samples have also been analyzed by using aqueous dilutions in the presence of either Sr^{++}, La^{+++}, or EDTA or after protein precipitation with trichloroacetic acid.

REFERENCE

Thiers, R. E.: *In* Standard Methods of Clinical Chemistry. S. Meites, Ed. New York, Academic Press, Inc. 1965, vol. 5.

SERUM IRON AND IRON-BINDING CAPACITY

The total amount of iron in an adult is approximately 4 to 5 gm. About 70 to 75 per cent of this relatively small amount of iron has an active and vital physiological role, and the remaining 25 to 30 per cent is present in various storage forms that can readily be mobilized if needed (see Fig. 10-3). The physiologically active iron is mainly present in the form of oxygen carrying pigments, such as hemoglobin (approximately 65 per cent) and myoglobin (approximately 3 to 5 per cent), as well as in form of a number of enzymes involved in electron transfer reactions, such as the cytochromes, cytochrome oxidase, peroxidase, and catalase (less than 1 per cent). The main storage form for iron (15 to 20 per cent) is ferritin, which is made up of ferric hydroxide–ferric phosphate attached to a protein called apoferritin (molecular weight 460,000). If the amount of apoferritin is not sufficient to bind the iron, the excess of iron is deposited in form of small iron oxide granules, generally called hemosiderin (less than 0.1 per cent). Iron is stored mainly in the liver, spleen, and bone marrow.

The iron needs of the body are met by a dietary intake of approximately 5 to 20 mg./day. Since the body preserves iron extremely well, this rather minute intake of iron is sufficient to satisfy the normal adult requirement of approximately 12 mg./day. Less than 1 mg. is lost per day through the skin, feces, and urine of adult males and nonmenstruating females. Loss through normal menstruation, however, may be as much as 80 mg. per period. During pregnancy, approximately 400 mg. iron are lost to the fetus and as a result of blood loss at time of parturition.

Aside from a small amount of iron that is absorbed from the stomach, most of the absorption takes place in the duodenum and jejunum. Since only iron in its

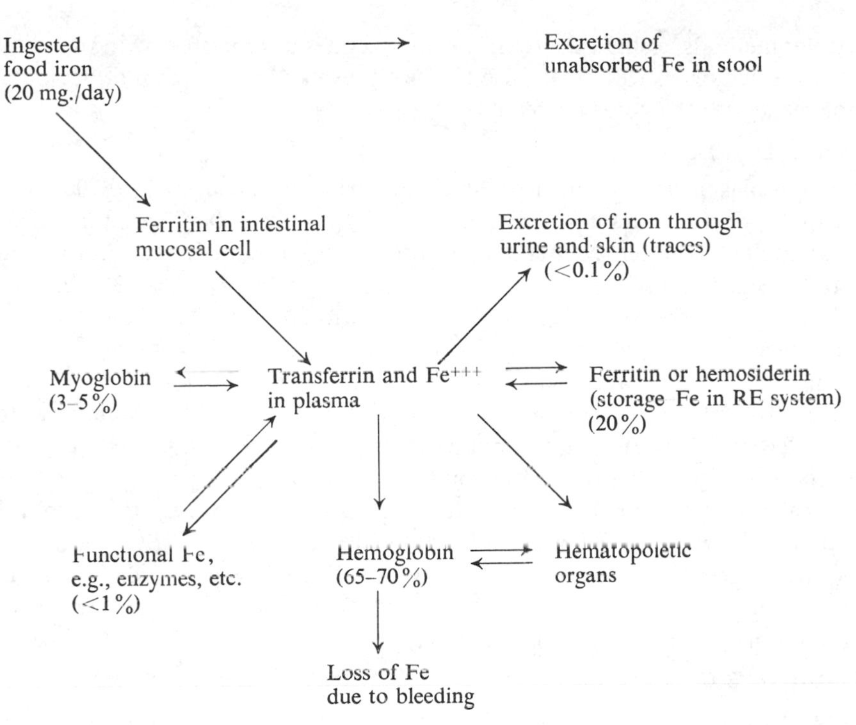

Figure 10-3. Pathways of iron.

ferrous (Fe^{++}) form can be absorbed, all dietary iron present in the Fe^{+++} form must first be reduced to the ferrous form. The rate of absorption is greatly dependent on pH. An acid pH, as provided by the gastric HCl, prevents precipitation of iron as phosphates, helps in solubilizing iron in food, and, finally, aids in the reduction of Fe^{+++} to Fe^{++}. The iron absorbed from the intestinal tract is immediately oxidized in the mucosal cells to the ferric state and is temporarily stored as ferritin. When the storage capacity of the mucosal cell for ferritin is exhausted, no further iron is absorbed. A largely unknown mechanism will, on demand, release iron from the mucosal cell into the blood, where it circulates mainly as ferri-transferrin, which in turn is in equilibrium with an extremely small amount of free Fe^{+++}. Transferrin (formerly called siderophilin) is the plasma iron transport protein, with a molecular weight of 90,000, that migrates electrophoretically with the β_1-globulin fraction and has the ability to form a complex with iron. The iron is then carried in this form to the various body storage areas such as the liver, and bone marrow, and, in smaller amounts, to most other tissues, where it is released from the protein thus providing a dynamic equilibrium between the iron stores. The total circulating amount of apo transferrin (protein capable of binding iron) is generally only 25 to 30 per cent saturated with iron. The additional amount of iron in μg./100 ml. that can potentially be adsorbed by apo-transferrin constitutes the latent or unsaturated iron-binding capacity (UIBC) of plasma. The total amount of iron circulating in the plasma plus the unsaturated (latent) iron-binding capacity constitutes the total iron-binding capacity (TIBC) of plasma.

SPECIMEN

Blood samples for serum or plasma iron and iron-binding capacity determinations should be drawn in the morning and in the fasting state since there appears to be a decrease in iron of approximately 30 per cent during the course of the day. In most of the newer methods, a slight degree of hemolysis does not interfere with the test, but any extensive hemolysis will cause false high iron levels. Iron in form of hemoglobin does not react with the color reagent.

NORMAL VALUES

Average plasma and serum iron levels at birth approach 200 μg./100 ml., but fall off rapidly within hours to average values below 50 μg./100 ml. and then increase to normal adult levels after the first 3 weeks of life. Adult males have a normal range of 60 to 150 μg./100 ml. and values for females range from 50 to 130 μg./100 ml. Serum iron levels decrease in elders to levels of 40 to 80 μg./100 ml.

Procedures for serum iron that do not require protein precipitation (see the section on serum iron determinations) give normal values that are approximately 10 to 20 μg./100 ml. higher than those just listed. It is felt that these values are closer to the true serum values and that the lower values observed with the other methods are due to loss of iron with the protein precipitate.

Normal values for the TIBC in healthy adults are 270 to 380 μg./100 ml., with some authors reporting normal ranges of 300 to 360 μg./100 ml. TIBC values tend to decrease with age and are an average of 70 μg./100 ml. lower in individuals above 70 years of age. Values for TIBC obtained with methods not requiring protein precipitation are 280 to 400 μg./100 ml.

CLINICAL SIGNIFICANCE

Alterations in the level of serum iron or iron-binding capacity have been observed in a number of conditions (see Table 10-1). In most instances, it is not the value of one, but the combination of both iron and iron-binding capacity that is of most clinical significance. In general, it can be stated that *increases in serum or plasma iron* may occur: (1) in conditions characterized by increased red cell destruction (hemolytic anemia, decreased survival time of red cells); (2) in cases of decreased utilization (decreased formation of blood as in lead poisoning or pyridoxine deficiency); (3) in situations in which increased release of iron from body stores occurs (e.g., release of ferritin in necrotic hepatitis); (4) in states in which the process of iron storage is defective (as in pernicious anemia); and (5) in conditions in which there is an increased rate of absorption (e.g., hemochromatosis, hemosiderosis). *Decreases in serum or plasma iron* levels are generally due to a deficiency in the total amount of iron present in the body, which, in turn, may be caused by lack of sufficient intake or absorption of iron, increased loss of iron (chronic blood loss or nephrosis), or increased demand on the body stores (pregnancy). Diminished iron levels may also be caused by a decreased release of iron from body stores (reticuloendothelial system) as seen in infections or with turpentine abscesses. *Increase in the total iron-binding capacity* of serum may be caused by increased production of the iron-binding protein transferrin, as found in the various states of chronic iron deficiency, or it may be caused by an increased release of ferritin, as in hepatocellular necrosis. *Decreases in the total iron-binding capacity* may be caused by a deficiency in ferritin, as found in cirrhosis and hemochromatosis, or as a result of an excessive loss of protein (transferrin), as occurs in nephrosis.

It has been recommended by some authors that the latent iron-binding capacity be expressed in terms of per cent unsaturated iron-binding capacity, which is calculated as follows:

$$\frac{\text{TIBC} - \text{serum Fe}}{\text{TIBC}} \times 100$$

TABLE 10-1. *Serum Iron and TIBC Values in Various Diseases*

Disease	*Serum Iron*	*TIBC*
Iron deficiency anemia (dietary, malabsorption, chronic hemorrhage, late pregnancy)	↓↓	↑
Anemia of chronic infections	↓→	↓→
Anemia of neoplastic disease	↓→	↓→
Hemolytic anemia	↑	↓→
Pernicious anemia	↑→	↓
Hemochromatosis	↑↑	↓
Hemosiderosis	↑↑	→
Hepatitis	↑↑	↑
Chronic liver disease	↓→	↓→
Obstructive jaundice	→	→
Polycythemia	↓→	↑
Nephrosis	↓	↓↓
Hodgkin's disease (terminal)	↑↑	

Discussion of Methods for the Determination of Iron in Serum

Most of the serum iron, as previously mentioned, is bound to protein; therefore, the first step in any iron procedure is the disruption of the iron-protein complex. An early approach was the wet digestion of the sample; however, hemoglobin iron (in plasma) was released as well, resulting in too high serum iron values. More recent methods rely on the release of iron at strongly acid pH, which is provided by addition of either hydrochloric, sulfuric, or trichloroacetic acids. The last reagent acts simultaneously to release the iron and to precipitate the serum proteins, but the use of hydrochloric or sulfuric acids requires the additional use of a protein precipitating agent. Henry[12] has indicated, however, that cold trichloroacetic acid gives incomplete recovery of iron and, therefore, he recommends the use of hot trichloroacetic acid.

The next step in most procedures is the reaction of the iron in the protein-free filtrate with one of the following chromogens, listed in order of increasing sensitivity: thiocyanate, which with Fe^{+++} yields a strongly colored ferric thiocyanate $Fe(SCN)_3$; α,α'-dipyridyl; 2,2',2''-terpyridine (2,2',2''-tripyridine); 1,10-phenanthroline (o-phenanthroline); 4,7-diphenyl-1,10-phenanthroline (bathophenanthroline) or its water soluble sulfonated form, and TPTZ (2,4,6 tripyridyl-s-triazine). The molar absorptivity for the iron complex with the preceding reagents is 7000, 8600, 11,000, 11,000, 22,400, and 22,600, respectively.

These reagents, except thiocyanate, react only with Fe^{++}, thus, requiring reduction of the Fe^{+++} with a suitable reducing agent. Hydrazine, hydrosulfite, sulfite, and especially hydroxylamine, ascorbic acid, and thioglycolic acid have been used for this purpose (see also Henry[12]).

Ramsay[21] combined the protein precipitating agent (acetate buffer, pH 5.0), the reducing agent (hydroxylamine HCl), and the color reagent (2,2',2'' terpyridine) and heated this mixture together with serum. Thus, protein precipitation, reduction of Fe^{+++}, and the color development are accomplished in one step. The procedure is extremely reliable and reproducible but, unfortunately, not very sensitive. The detailed method will be described later.

A unique approach has been used by Schade.[27] In this method iron is released from its protein complex by adjusting the serum to approximately pH 6.0 with phosphate buffer of pH 5.3. Under these conditions the serum proteins remain in the solution. The Fe^{+++} is then reduced by addition of ascorbic acid (stabilized with sodium metabisulfite) and sulfonated diphenylphenanthroline is added for color development:

$$\text{Ferri-transferrin} \xrightarrow{\text{pH } 6.0} Fe^{+++} + \text{transferrin}$$

$$\downarrow + \text{ascorbic acid and sodium metabisulphite}$$

$$Fe^{++} + \text{sulfonated diphenylphenanthroline} \longrightarrow \text{color complex}$$

Correction for the interference by the nonspecific colors or opacity of serum, or both, is made by preparing individual serum blanks, which contain all the reagents except diphenylphenanthroline.

Webster[43] used a similar approach but employed sodium dithionite as the reducing agent and added an anionic detergent, Teepol 710 or 610 (Norske Shell, A/S or Shell Chemical Company, New York, N.Y.). The latter aids in the release of iron from its protein complex and also renders lipemic sera clear. This method, as modified by Askevold and Vellar[2] will be described later in this chapter.

Iron has also been determined by atomic absorption spectrophotometry[44,46] either directly after a 1:1 dilution of serum or after chelation of the iron with bathophenanthroline and extraction of the complex into methyl isobutyl ketone (4-methyl-2-pentanone, isopropyl acetone, MIBK). Both methods, at this time, appear to be impractical, especially because of the relatively low sensitivity; however, it is conceivable that further advances in technique and instrument design will make these the methods of choice.

Discussion of Methods for the Determination of Serum Iron Binding Capacity

The apo-transferrin of serum is only partially saturated with iron (see under serum iron). The amount of iron that can be bound, in addition to that already bound, is called the unsaturated (or latent) iron-binding capacity (UIBC or LIBC). Both values together represent the total iron-binding capacity (TIBC).

$$\text{Serum Fe} + \text{UIBC} = \text{TIBC}$$

The TIBC may be determined by one of two major approaches. (1) Excess iron equivalent to 500 μg. Fe/100 ml. in the form of ferric ammonium citrate is added to serum that was previously adjusted to a pH of 8.0 or above with a buffer solution. At this pH the iron transferrin complex is stable, and apo-transferrin will bind iron until all apo-transferrin has been saturated with iron (transferrin). The excess iron is then removed either by an ion exchange resin[13,15] or by magnesium carbonate,[20] and the iron remaining in the supernatant (serum) is determined by one of the serum iron methods to be described. The value obtained represents the TIBC.

$$\text{UIBC} = \text{TIBC} - \text{Serum Fe}$$

(2) In the other possible approach,[27] excess iron is added as outlined under (1), but the iron in the supernatant is then determined at a pH above 7.5 at which only the

excess (free) iron will react, since only this fraction can be reduced to Fe^{++}. Iron bound as transferrin remains as Fe^{+++} and will give no color reaction. In this case:

$$\text{UIBC} = (\text{iron added}) - (\text{excess iron measured})$$

The Determination of Iron and Iron-Binding Capacity

METHOD OF RAMSEY, MODIFIED BY MANDEL AND NIESPODZRANY

REAGENTS

1. Acetate buffer, pH 5.0. Place 57.81 gm. of anhydrous sodium acetate or 95.598 gm. of sodium acetate $\cdot 3H_2O$ into a 2000 ml. volumetric flask, dissolve in deionized water, and add 17.2 ml. of concentrated acetic acid. Make up to volume with deionized water; check the pH and adjust if necessary.

2. Tripyridine reagent (2:2′:2″ tripyridine, 0.004 per cent, and hydroxylamine HCl, 0.05 per cent). Into a 1000 ml. volumetric flask place 40 mg. of 2:2′:2″ tripyridine and dissolve in approximately 2 ml. of 95 per cent ethanol. Add 0.50 gm. of hydroxylamine-HCl and then slowly add acetate buffer to the mark. The solution will first turn milky white and then will clear on further addition of acetate buffer.

3. Ferric ammonium citrate (0.05 mg. Fe/ml.). Place 10 ml. of $FeCl_3$ stock standard (see under stock standard) into a 50 ml. conical centrifuge tube and add 1 *N* NH_4OH dropwise until no further precipitate of $Fe(OH)_3$ occurs. Generally, about 2 ml. of NH_4OH is adequate. After precipitation is complete, centrifuge and discard the supernatant. Dissolve the precipitated $Fe(OH)_3$ by adding a few drops of water followed by crystals of citric acid. The clear contents of the tube are now quantitatively transferred with deionized water to a 200 ml. volumetric flask and the volume is made up to 200 ml. If necessary, the pH is adjusted to pH 6.5 to 6.6 with 1 *N* NH_4OH. The yellowish green, clear solution is stable at room temperature for 1 month.

4. Resin. Approximately 250 gm. of Amberlite IRA-410 (Rohm & Haas Co., Philadelphia) is suspended in approximately 3 *N* HCl overnight. The HCl is decanted and the resin washed thoroughly with water. The resin is then suspended in barbiturate buffer, pH 7.5. The pH is checked and adjusted with 1 *N* NaOH. The resin is then allowed to settle. The supernatant is filtered through a Buchner funnel and the resin is dried in a flat dish at 37°C.

5. Stock standard, 1 mg. Fe/ml. Dissolve 1 gm. of iron wire, A. R., in about 25 ml. of hot 6 *N* HCl in a small beaker; quantitatively transfer the solution to a 1000 ml. volumetric flask and make up to the mark with deionized water.

6. Dilute standard, 1 μg./ml. Daily, before use, dilute 1 ml. of stock standard to 1 L. with deionized water.

7. Barbiturate buffer, pH 7.5. Into a 1 L. volumetric flask place 2.3 gm. of sodium diethylbarbiturate and 6.0 gm. of diethylbarbituric acid and dissolve in about 500 ml. of 0.11 *M* NaCl solution. Upon complete clearing, which may be accelerated by mild heating, the NaCl solution is added to the 1000 ml. mark and the pH is adjusted, if necessary.

8. Sodium chloride, 0.11 *M*. Place 6.38 gm. of NaCl, A. R., into a 1000 ml. volumetric flask and dilute to the mark with deionized water.

PROCEDURE

Serum iron

1. Pipet 1 ml. of serum* into a 15 × 120 mm. test tube. The test should be run in duplicates, if possible.

2. Prepare a blank and a standard by adding into respective tubes 1 ml. of deionized water and 1 ml. of dilute standard.

3. Add 4 ml. of tripyridine reagent to all test tubes and mix thoroughly.

4. Cover all tubes with a glass marble and place into a 100°C. bath for 5 minutes. Incomplete protein precipitation and, thus, turbidity are observed if the heating process is not adequate. This may be due to water bath temperatures below 100°C. or the use of thick wall centrifuge tubes. If the procedure is followed as outlined, no turbidity will be observed in the supernatant.

5. Place all test tubes into cold water for 5 minutes and centrifuge at approximately 2500 rpm for 15 minutes.

6. Decant the supernatant carefully and *completely* into a suitable cuvet such as the Coleman 12 × 75 mm. cuvet. Complete transfer of the supernatant is necessary, since 1.8 ml. of liquid is needed to make an accurate reading in this type of cuvet.

7. Determine the absorbance of all tubes at 552 nm. with the reagent blank set to zero A (or 100% T).

CALCULATION

$$\frac{A \text{ unknown}}{A \text{ standard}} \times 100 = \mu\text{g. Fe/100 ml.}$$

If jaundiced or turbid serum was used:

$$\frac{A \text{ unknown} - A \text{ serum blank}}{A \text{ standard}} \times 100 = \mu\text{g. Fe/100 ml.}$$

Iron-binding capacity

1. Pipet 1 ml. of serum into a 15 × 120 mm. test tube. (Perform the test in duplicates if possible.)

2. Add 0.1 ml. of ferric ammonium citrate reagent, and mix; let stand for 10 minutes with occasional mixing.

3. Add approximately 0.5 ml. of dry, prepared resin. This may conveniently be done with the aid of a spoonula, which holds approximately the indicated amount of resin. Mix the serum with the resin and let stand for 15 minutes with occasional mixing.

4. Add 2 ml. of barbiturate buffer and mix with a glass rod, taking care that no resin granules cling to the side of the tube.

5. Centrifuge at approximately 2000 rpm for 5 minutes.

6. Transfer 2.0 ml. of supernatant to properly labeled 15 × 120 mm. test tubes.

7. Set up a blank consisting of 2 ml. deionized water and a standard consisting of 2 ml. dilute standard.

8. Follow steps 3 to 7 of the procedure for serum iron.

---

* If the serum is icteric or turbid, set up a serum blank consisting of 1 ml. of serum and 4 ml. of acetate buffer. Also, set up a buffer blank consisting of 1 ml. of deionized water and 4 ml. of acetate buffer. Follow steps 4 to 7 of procedure and read the serum blank against the buffer blank.

CALCULATION

$$\frac{A\ \text{unknown}}{A\ \text{standard}} \times \frac{100}{1} \times \frac{3.1}{2} \times 2 = \frac{A\ \text{unknown}}{A\ \text{standard}} \times 310 = \text{TIBC in } \mu\text{g./100 ml.}$$

If jaundiced or cloudy serum was used:

$$\frac{A\ \text{unknown} - 2/3 \text{ of } A \text{ serum blank}}{A\ \text{standard}} \times 310 = \text{TIBC in } \mu\text{g./100 ml.}$$

PROCEDURE NOTES

1. All glassware and pipets used for making reagents and performing the test must be acid washed in either diluted HCl (1:1) or nitric acid (1:3), and rinsed well with deionized water.

2. Specimens with barely visible hemolysis are suitable for this determination since iron bound to hemoglobin is not detected with this procedure. Specimens showing definite hemolysis (pink to red appearance) are unsuitable for this procedure.

3. Because of the low sensitivity of this procedure it is necessary to estimate the per cent T to the closest 0.25.

REFERENCES

Ramsay, W. N. M.: Biochem. J., *53*:227, 1953; modified by Mandel, E., and Niespodzrany, L.: unpublished.

The Determination of Iron and Iron-Binding Capacity

METHOD OF ASKEVOLD AND VELLAR (MODIFIED)

PRINCIPLE

See the previous general discussion.

REAGENTS

1. Sodium hydroxide, 3 per cent (w/v). Dissolve 30 gm. of NaOH, A. R., in deionized water and make up to 1000 ml. with deionized water.

2. Magnesium sulfate solution. Dissolve 10 gm. of $MgSO_4 \cdot 7H_2O$, A. R., in deionized water and make up to 1000 ml. with deionized water.

3. Teepol solution, iron-free. Dissolve 5.0 gm. of sodium dithionite ($Na_2S_2O_6$, A. R.) in 90 ml. of magnesium sulfate solution and add with mixing 14 gm. of Na_2HPO_4, A. R., 91 ml. of Teepol 710 or 110 ml. of Teepol 610. After thorough mixing, add 45 ml. of 3 per cent (w/v) sodium hydroxide, A. R. On adding the sodium hydroxide, the mixture will turn cloudy and should be allowed to stand at room temperature for 15 minutes. At the alkaline pH provided by the sodium hydroxide, magnesium and iron hydroxides will form and precipitate. The solution is then transferred into centrifuge tubes (250 ml. centrifuge tubes are convenient), centrifuged, and the clear supernatant is decanted. The pH is then adjusted to pH 5.5 to 5.6 by adding acetic acid, A. R., until the desired pH is reached. Approximately 1 ml. of acetic acid is required.

This solution is then transferred to small polyethylene bottles, which are completely filled without leaving any air space. The reagent is stored in the freezer and portions are removed as needed. The reagent may be kept up to 1 week at refrigerator temperatures.

Note: The reagent should not be permitted to stand long after precipitation of the hydroxides by sodium hydroxide, since these particles may partially redissolve. Standing may also cause oxidation of the sodium dithionite.

4. Bathophenanthroline reagent. Dissolve 0.36 gm. of disodium bathophenanthroline sulfonic acid in 10 ml. of deionized water and add 1 drop of the iron-free Teepol solution. The reagent may be slightly opaque or reddish.

5. Standard, 1 μg. Fe/ml. See the method by Mandel.

PROCEDURE

Serum iron

1. Into a series of respectively labeled 15 × 120 ml. test tubes place 1 ml. of water for the blank, 1 ml. of working standard for the standard, and 1 ml. of serum for each of the unknowns.

2. Add 3 ml. of Teepol reagent to all tubes, mix, and allow to stand at room temperature for 15 minutes.

3. Read the standard and unknown against the blank in a spectrophotometer at 540 nm.

4. Add 1 drop bathophenanthroline reagent to all test tubes, mix, and allow to stand for 10 minutes. Read the absorbance as in step 3.

CALCULATION

Subtract the absorbance reading obtained in step 3 from that obtained in step 4 (ΔA).

$$\frac{\Delta A \text{ of unknown}}{\Delta A \text{ of standard}} \times 100 = \mu\text{g. Fe/100 ml.}$$

Iron-binding capacity

1. Follow steps 1 through 5 of the procedure for iron-binding capacity by Mandel and Niespodzrany.

2. Transfer 1 ml. of supernatant to a properly labeled 15 × 120 mm. test tube.

3. Set up a blank, consisting of 1 ml. of deionized water, and a standard, consisting of 1 ml. of dilute standard.

4. Follow steps 1 to 4 of the procedure for serum iron by Askevold and Vellar.

CALCULATION

$$\frac{\Delta A \text{ unknown}}{\Delta A \text{ standard}} \times \frac{100}{1} \times \frac{3.1}{1} \times 1$$

or

$$\frac{\Delta A \text{ unknown}}{\Delta A \text{ standard}} \times 310 = \text{TIBC } \mu\text{g./100 ml.}$$

PROCEDURE NOTES

1. No correction for bilirubin or turbidity is necessary with this method since calculation of Δ absorbance based on readings before and after addition of the color reagent corrects for any absorbance contributed by these factors.

2. The excess iron may also be removed by adding 0.2 gm. of magnesium hydroxy carbonate (Merck Cat. No. 5827) and mixing the contents gently on an automatic shaker for 45 minutes. After centrifugation, 1 ml. of the supernatant (serum) may be used for analysis as just outlined above (steps 2 to 4).

REFERENCE

Askevold, R., and Vellar, O. D.: Scand. J. Clin. Lab. Invest., *20*:122, 1967 (modified).

COPPER

PHYSIOLOGY AND CLINICAL SIGNIFICANCE

Copper belongs to the group of essential trace elements. Deficiency causes severe derangements in growth and metabolism. More specifically, there appears to be a not clearly understood impairment of erythropoiesis, a decrease in erythrocyte survival time, impaired mitochondrial function, and a decrease in the catalytic action of a number of enzymes, of which copper is a part, e.g., tyrosinase, ceruloplasmin, and cytochrome oxidase.

Most of the copper that is taken in with the diet is lost through the stool; only a small portion is absorbed by the upper small intestine and reaches the blood. In plasma, most of it is first loosely bound to albumin, and then incorporated into a plasma α_2 globulin, *ceruloplasmin*, which is the copper transport protein. More than 90 per cent of the total serum copper is present in the form of this copper-containing enzyme, which has catalytic activities of an oxidase. A variable but small amount (less than 10 per cent) of copper is loosely bound to albumin and an even smaller quantity is dialyzable. The remainder of the body copper is mainly incorporated in various forms of copper proteins, such as erythrocuprein in the erythrocytes. The highest concentration is in the liver as hepatocuprein. Copper that has once been absorbed is retained and only very small amounts are excreted in urine.

The most significant clinical application of copper determinations is in the diagnosis of hepatolenticular degeneration (Wilson's disease). This disease is associated with a decrease in the synthesis of ceruloplasmin, which results in a low serum level of this enzyme. The amount of free and albumin bound copper, however, is greater than normal, and this is attributed to a greater and uncontrolled rate of absorption of copper. Despite this fact, the *total* serum copper concentration is generally decreased (because of low ceruloplasmin values). The amount of copper deposited in tissues (e.g., liver) is greatly increased and there is also an increased urinary excretion of copper (due to an increase in free serum copper?). Thus, the determination of serum ceruloplasmin, of total serum copper, and of the urinary excretion of copper are of great help in the diagnosis of this disease. Low serum copper levels have also been observed in a number of hypoproteinemias as a result of malnutrition, malabsorption, and the nephrotic syndrome. In the last condition an appreciable amount of ceruloplasmin reaches the urine, but in normal urine there is essentially no ceruloplasmin present.

Increased serum copper levels (hypercupremia) are found in a number of acute and chronic diseases, such as malignant diseases (including leukemia), hemochromatosis, cirrhosis, and various infections. Serum copper levels are also high after administration of estrogens and it is believed that the increase in these hormones during pregnancy is also responsible for the high copper and ceruloplasmin levels in this condition.

NORMAL VALUES

Normal adult *serum* copper levels range from 70 to 140 μg./100 ml. with only minor differences between males and females (mean 100 vs. 108, respectively).

Values in newborns appear to be considerably higher although there is no agreement in the literature as to the exact level; values of 80 to 280 μg./100 ml. have been suggested by different authors. *Urine* copper output in adults ranges from 0 to 30 μg./ 24 hr.

Determination of Copper in Serum

PRINCIPLE

Copper is released from its bond with serum protein by treatment with dilute hydrochloric acid. The proteins are then precipitated by trichloroacetic acid and an aliquot of the filtrate is reacted with biscyclohexanoneoxalyldihydrazone, which forms a stable blue-colored compound with cupric ions. The procedure follows Beer's Law for samples within the physiological range.

REAGENTS

1. Hydrochloric acid, 2 *N*. Add 166 ml. of HCl, A. R., to deionized water and dilute to 1 L.
2. Trichloroacetic acid, 20 per cent (w/v). Dissolve 20.0 gm. of trichloroacetic acid (sulfate free, iron free, Eastman No. 259) in deionized water and dilute to 100 ml.
3. Buffer solution. Add 35.7 ml. of saturated sodium pyrophosphate solution, 35.7 ml. of saturated sodium citrate solution, and 80.3 ml. of concentrated ammonium hydroxide into a 1000 ml. volumetric flask and dilute to the mark with deionized water.
4. Biscyclohexanoneoxalyldihydrazone (Cuprizone, G. Frederick Smith Chemical Co., Columbus, Ohio). Dissolve 0.5 gm. of the reagent in 100 ml. of 50 per cent (v/v) ethanol.
5. Standard copper solution, 0.10 mg./ml. Dissolve 0.3928 gm. of copper sulfate pentahydrate, A. R., in distilled water and dilute to the 1000 ml. mark with deionized water.
6. Working standard. Dilute 2 ml. of the stock standard to 100 ml. with deionized water. This solution contains 200 μg./100 ml.

PROCEDURE

1. Into 12 ml. conical centrifuge tubes pipet 1.0 ml. of water, 1.0 ml. of working standard, and 1.0 ml. of serum (or heparinized plasma), respectively.
2. Add 0.70 ml. of 2.0 *N* hydrochloric acid to all tubes, mix, and let stand at room temperature for 10 minutes.
3. Add 1.0 ml. of 20 per cent trichloroacetic acid, mix with a thin stirring rod, and allow to stand for 10 minutes.
4. Cover the tubes with parafilm, and centrifuge rapidly for 10 minutes at 2500 rpm.
5. Transfer 2.0 ml. of the clear supernatant to another test tube.
6. Add 2.8 ml. of buffer solution and 0.20 ml. of Cuprizone reagent. Mix and allow to stand for 20 minutes.
7. Read absorbance at 620 nm.

CALCULATION

$$\frac{A_{\text{unknown}}}{A_{\text{standard}}} \times 200 = \mu\text{g./100 ml. serum}$$

PROCEDURE NOTES

Syringes for the collection of blood as well as all glassware used in the test must be free of trace contamination. All reagents must be made up with chemicals of highest purity.

Some authors feel that protein precipitation by hot trichloroacetic acid extracts copper more quantitatively and thus gives better recovery results.

REFERENCE

Rice, E. W. : Principles and Methods of Clinical Chemistry. Springfield, Ill., Charles C Thomas, Publisher 1960, pp. 157–159

Alternate Methods for the Determination of Copper

In Henry's modification[12] of the method by Gubler et al.,[10] copper is dissociated from proteins by addition of hydrochloric acid and proteins are then precipitated by hot trichloroacetic acid. The copper in the protein-free filtrate is reacted with diethyldithiocarbamate in the presence of citrate, which prevents interference from iron by forming a soluble iron complex.

In the determination of copper in *urine*, a sample is digested by wet ashing and the color produced after the addition of diethyldithiocarbamate is extracted into isoamyl alcohol.

Copper in serum can also be determined by treating the sample with a hydrochloric acid solution of oxalyldihydrazide. Addition of trichloroacetic acid precipitates the proteins and the subsequent addition of ammonium hydroxide and acetaldehyde causes the formation of an intense lavender color as a result of the reaction of oxalyldihydrazide with $Cu^{(II)}$.[22] The molar absorption of the resulting color complex is approximately 22,000 at 542 nm. in contrast to the molar absorbance of 16,000 for the copper complex with Cuprizone and 8000 for the copper complex with diethyldithiocarbamate. Thus, the method offers the advantage of greater sensitivity, but it is slightly more complex than the preceding method. Copper in serum can also be determined successfully by atomic absorption spectrophotometry.

AMMONIA

CLINICAL SIGNIFICANCE

Determinations of blood ammonia (NH_4^+) are said to be of value in the detection of existing or impending hepatic coma, although many reports indicate that this value is very limited. These problems are discussed in Chapter 13 on liver function and the reader is referred there.

Determination of Ammonia in Blood, Plasma, and Serum

The ammonia content of freshly drawn normal blood rises rapidly on standing (due to enzymatic deamination of labile amides like glutamine) to two or three times its original value in the course of several hours at room temperature, less rapidly at refrigerator temperature, and remains constant for several days if kept in a deep freeze at −20°C. No change in ammonia content is noted if the blood is placed immediately in an ice bath and analyzed within 20 minutes. Some authors suggest immediate preparation of trichloracetic acid or tungstic acid filtrates since such filtrates show no changes in ammonia content on standing.

The problem of ammonia formation carries over into the procedure itself. Conway and Cooke[7] published a diffusion method in which the ammonia is released after addition of potassium carbonate and the released ammonia is reabsorbed in hydrochloric acid, containing a suitable acid-base indicator. The amount of ammonia is determined by titration. Other authors recommended that the amount of ammonia absorbed by the acid be determined by nesslerization or by the Berthelot reaction (see serum urea nitrogen method). Claims have been made that under the conditions of this test (pH of potassium carbonate-serum mixture = 12.5 to 13.0) ammonia formation continues. Thus, Faulkner and Britton[8] have used a solution of potassium bicarbonate and potassium carbonate, that gives a final pH of 9.9 at which apparently less ammonia formation takes place.

Forman[9] recommends the treatment of plasma with a strongly acidic cation exchange resin (sulfonated polystyrene cation exchanger, sodium form, 60 to 80 mesh), which captures the ammonium ion. Addition of sodium phenoxide in the presence of hypochlorite and nitroprusside (catalyst) will simultaneously elute the ammonium ion and colorimetrically react with it to produce a stable blue color. This procedure gives values lower than those obtained with other procedures and it is assumed that there is a loss of ammonia on the column.

A recent and possibly promising approach to the determination of plasma ammonia has been the use of an enzymatic method based on the reaction of ammonia and α-ketoglutaric acid in the presence of glutamic dehydrogenase:

$$\alpha\text{-ketoglutarate} + NH_4^+ + NADH \xrightarrow{\text{glutamic dehydrogenase}} \text{glutamic acid} + H_2O + NAD^+$$

Quantitation is obtained by measuring the decrease in fluorescence as NADH is converted into NAD.[26]

Neither the resin method nor the enzymatic method eliminate the problems related to collection of specimens and the enzymatic method is also affected by the liberation of ammonia during the test procedure.

In view of the many unresolved problems related to the determination of ammonia, and because of the questionable value of this test, no procedure is given here in detail. The interested reader is referred to the references given.

NORMAL VALUES

Enzymatic method: 40 to 80 μg./100 ml.
Resin method: 15 to 45 μg./100 ml.
Conway diffusion methods: 40 to 110 μg./100 ml.
Levels in impending or actual hepatic coma: up to 400 μg./100 ml.

Organic Acids

The organic acid fraction of the total anions of serum is approximately 5 to 6 mEq./L. (see Table 11-1, Chapter 11). It is composed of a great variety of individual acids, of which lactic acid, with about 1 mEq./L. (9 mg./100 ml.), is normally the most abundant. All others (fatty acids, amino acids, β-hydroxybutyric and acetoacetic acid as well as various other acidic products of metabolism) constitute collectively about 80 per cent of the total organic acid fraction, but individual acids are normally present in relatively insignificant amounts. In various disease states, however, some of these acids may accumulate to a rather significant extent. The levels of ketone acids (acetoacetic acid and β-hydroxybutyric acid) in diabetic acidosis, for example, may increase to more than 20 mEq./L. (about 200 mg./100 ml.). In severe renal disease, Seligson and associates[31] reported increases in the total organic acid fraction of as much as 26 mEq./L. In acute methyl alcohol intoxication quantities of formic acid (metabolite of methyl alcohol) may be as high as 15 mEq./L. (69 mg./100 ml.) and the levels of salicylates in salicylate intoxication may be up to 5 mEq./L. (70 mg./100 ml.). "Lactic acidosis" may cause increases in serum lactate concentration of up to 25 mEq./L. (225 mg./100 ml.). In any of these conditions there may occur a profound alteration in the relative proportions of anions. This must be kept in mind when checking electrolyte reports (see also the discussion on the production of metabolic acidosis in Chapter 11 and the calculations on page 618).

LACTIC ACID

Oxygen deprivation of tissues is followed by a blockage of aerobic oxidation of pyruvic acid in the tricarboxylic acid cycle, with accumulation of pyruvic acid and subsequent glycolytic reduction of this compound into lactic acid. This leads to a severe acidosis called "lactic acidosis," as just stated.

Accumulation of lactic acid has been observed in shock associated with the failure of effective peripheral blood flow. It has also been observed in diabetic patients after treatment with the hypoglycemic drug phenformin hydrochloride and in cases of "essential lactic acidosis" in which the underlying cause is not well known.

NORMAL VALUES

Normal values for lactic acid differ greatly with the method of analysis and especially the method of blood collection. Some authors report arterial blood values of 3.1 to 7 mg./100 ml. (0.34 to 0.78 m*M*./L.) in individuals after normal exercise and values up to 3 mg./100 ml. (0.33 m*M*./L.) in individuals at bed rest. Broder and Weil[6] reported normal values of 0.73 m*M*./L. (S.E. $\pm$ 0.356). Normal values for venous blood are 5 to 15 mg./100 ml. (0.55 to 1.15 m*M*./L.).

Methods for the Determination of Lactic Acid in Blood

Collection of a satisfactory specimen for lactic acid analysis requires a special procedure in order to avoid changes in blood lactic acid levels while the sample is being drawn. Either the puncture should be performed without applying a tourniquet or the withdrawal of the sample should be made 2 minutes after removal of the tourniquet. The first 3 to 4 ml. of blood should be discarded and the blood should then be drawn either directly into TCA or $HClO_4$ or it should be precipitated with these agents immediately after collection of the sample.

A wide variety of methods have been published and are reviewed in Henry.[12] The two most widely used methods are the colorimetric method by Barker and Summerson[5] and methods employing the enzyme lactic dehydrogenase. The colorimetric method is based on the conversion of lactic acid to acetaldehyde by treatment of a protein-free filtrate with concentrated sulfuric acid and heat. Acetaldehyde is subsequently reacted with p-hydroxydiphenyl in the presence of copper ions to form a purple colored compound.

The most promising approach is the enzymatic method based on the following reaction:

$$\text{Lactic acid} + \text{NAD} \rightleftharpoons \text{Pyruvic acid} + \text{NADH}_2$$

The reaction to the left is strongly favored; however, at a pH of 9.7, in the presence of excess NAD^+ and in the presence of a carbonyl binding substance, such as semicarbazide (to trap pyruvic acid), the reaction proceeds to the right. A method utilizing the last approach has recently been published by Rosenberg and Rush.[24] The quantity of lactic acid is determined by measuring the amount of NADH formed, which is proportional to ΔA_{340}.

KETONE BODIES

The metabolism of fatty acids results in the formation of a small amount of acetoacetic acid, which is subsequently metabolized in the peripheral tissues. In conditions in which there is carbohydrate deprivation (e.g., starvation) or decreased utilization of carbohydrates (e.g., diabetes mellitus) lipids are the main source of energy. This causes an increased production of acetoacetic acid, which may exceed the capacity of the peripheral tissues to metabolize this compound. Thus, the acetoacetic acid accumulates in the blood and is in part converted to acetone by spontaneous decarboxylation and in part converted to β-hydroxybutyric acid in accordance with the following reactions:

1. $$\underset{\textit{Acetoacetic acid}}{CH_3{-}CO{-}CH_2{-}COOH} \rightarrow \underset{\textit{Acetone}}{CH_3{-}CO{-}CH_3} + CO_2$$

2. $$\underset{\textit{Acetoacetic acid}}{CH_3{-}CO{-}CH_2{-}COOH} + NADH_2 \xrightleftharpoons[\text{(in liver)}]{\beta\text{-hydroxybutyric dehydrogenase}} \underset{\textit{β-hydroxybutyric acid}}{CH_3{-}\overset{\overset{OH}{|}}{\underset{\underset{H}{|}}{C}}{-}CH_2{-}COOH} + NAD$$

The relative proportions of the three ketone bodies in blood may differ; an average figure is 78 per cent β-hydroxybutyric acid, 20 per cent acetoacetic acid, and 2 per cent acetone. The most commonly used methods for the determination of ketone bodies in serum or urine do not react with all ketone bodies. Gerhardt's ferric chloride test reacts with acetoacetic acid only, and the various tests employing nitroprusside are 15 to 20 times more sensitive for acetoacetic acid than for acetone and give no reaction at all with β-hydroxybutyrate. Thus, these tests (to be described) essentially measure acetoacetic acid only. Tests for β-hydroxybutyric acid are indirect; they require brief boiling of the urine to remove acetone and acetoacetic acid by evaporation, which is followed by gentle oxidation of β-hydroxybutyric acid to acetoacetic acid and acetone by peroxide, ferric ions, or dichromate. The acetoacetic acid thus formed may be detected with Gerhardt's tests or one of the procedures employing nitroprusside (see procedure).

CLINICAL SIGNIFICANCE

Excessive formation of ketone bodies results in increased blood levels (ketonemia) and excessive excretion in the urine (ketonuria). This has been observed in conditions associated with a decreased supply of carbohydrates, such as starvation, digestive disturbances, dietary imbalance, frequent vomiting, and glycogen storage disease (Von Gierke's disease). Another, and possibly more frequent cause of increased production of ketone bodies is decreased utilization of carbohydrates, such as is found in diabetes mellitus. Alkalosis, owing to an obscure mechanism (decreased carbohydrate utilization in the liver?), may also result in excessive production of ketone bodies.

Determination of ketone bodies in blood is an extremely helpful guide (more so than the determination of these compounds in urine) in the treatment of ketonemia associated with diabetes. In fact, some authors (e.g., Lee and Duncan[14]) claim that the knowledge of the degree of ketonemia is the most valuable guide to insulin therapy and offers more information than the knowledge of the degree of ketonuria or of the blood sugar levels. Although this viewpoint is opposed by others, it still points out the great importance of this test.

Determination of Ketone Bodies in Serum

Although a number of quantitative and semiquantitative determinations for the estimation of ketone bodies have been devised,[12] it is generally agreed that the semiquantitative Acetest and Ketostix (Ames Co., Div. Miles Laboratories, Elkhart, Ind.) offer sufficient information for clinical purposes.

SPECIMEN

The sera should be free of visible hemolysis since discoloration of the tablet or reagent strip may occur if an excessive amount of hemoglobin is present. If there is any significant delay in performing the determination, the specimens should be kept well stoppered at refrigerator temperatures.

DETECTION OF KETONE BODIES BY ACETEST

PRINCIPLE

The Acetest tablets contain a mixture of aminoacetic acid (glycine), sodium nitroprusside, disodium phosphate, and lactose. Acetoacetic acid or acetone in the presence of glycine will form a complex of lavender-purple color with nitroprusside. The disodium phosphate provides an optimum pH for the reaction and lactose enhances the color.

PROCEDURE

A detailed procedure for the detection of ketone bodies by Acetest is supplied by the manufacturer with each package of tablets, and the reader is referred to these instructions.

The Acetest was mainly designed for the detection of ketone bodies in urine. If serum is used, the tablets should be crushed and a drop of serum should be added to the powder. Failure to do so will result in false low results.

A 1+ positive reaction (appearance of a purple-lavender color) indicates the presence of 5 to 10 mg. of ketone bodies per 100 ml. A color chart provided with the package may be used to estimate higher concentrations of ketone bodies. A 4+ reaction (deep lavender color) corresponds to approximately 40 to 50 mg./100 ml. If desirable, dilutions of serum with saline can be prepared and levels of ketone bodies above 40 mg. can be estimated.* Since a 4+ reaction in an undiluted sample corresponds to approximately 40 mg./100 ml., a 4+ reaction in a 1:4 dilution corresponds to approximately 160 mg./100 ml. Similar calculations can be performed if other dilutions are used.

DETECTION OF KETONE BODIES BY KETOSTIX

Ketostix is a modification of the nitroprusside test in which a reagent strip is used instead of tablets. There is some indication that this test is more satisfactory for the determinations of acetoacetic acid and acetone in serum than the tablet test. The Ketostix test gives a positive reaction with 5 to 10 mg. of acetoacetic acid per 100 ml. within 15 seconds. Approximate serum acetoacetic acid values assigned to the color blocks on the package are 10 mg./100 ml. for "small," 30 mg./100 ml. for "moderate," and 80 mg./100 ml. for "large." Acetone reacts also, but to a considerably lesser extent.

Determination of Ketone Bodies in Urine

Acetest and Ketostix are also suitable for the detection of ketone bodies in urine. The sensitivity and specificity of the tests are the same as outlined for serum. The original test by Rothera[25] has essentially been replaced by these two modifications.

Gerhardt's test is based on the reaction of ferric chloride with acetoacetic acid, resulting in the production of a wine red color. Other compounds such as salicylates, phenol, and antipyrine give a similar color; thus, a positive reaction merely indicates the *possible* presence of acetoacetic acid. To confirm its presence, urine is heated to drive off the volatile acetoacetic acid and the test is repeated. If the test is now negative, it can be assumed that the original color was due to acetoacetic acid. This test also has been replaced by the Ketostix and Acetest procedures.

REFERENCES

Fraser, J., Fetter, M. C., Mast, R. L., and Free, A. H.: Clin. Chim. Acta, *11*:372, 1965.
Free, A. H., and Free, H. M.: Am. J. Clin. Path., *30*:7, 1958.

PROTEINS

Plasma (serum) contains a number of electrolytes, including protein, that are normally not determined in connection with the customary "electrolyte pattern." The proteins contribute approximately 16 mEq./L. to the anions and thus are the third largest fraction among the anions (see Chapter 11, Table II-1). Therefore, major changes in the protein content of plasma (serum) can cause significant derangements in the relative anion composition.

The chemistry of proteins, the method for their determination, and a discussion of the clinical significance are given in Chapter 5.

---

* Since the reaction is affected by proteins, any dilution with saline introduces a certain error.

REFERENCES

1. Alcock, N. W., and MacIntyre, I.: Methods for estimating magnesium in biological materials. *In* Methods of Biochemical Analysis. D. Glick, Ed. New York, Interscience Publishers, Inc., 1966, vol. 14.
2. Askevold, R., and Vellar, O. D.: Scand. J. Clin. Lab. Invest., *20*:122, 1967.
3. Asper, S. P., Schales, J. R., and Schales, S. S.: J. Biol. Chem., *168*:779, 1947.
4. Barker, S. B.: Standard Methods of Clinical Chemistry, D. Seligson, Ed. New York, Academic Press, 1961, vol. 3.
5. Barker, S. B., and Summerson, W. H.: J. Biol. Chem., *138*:535, 1941.
6. Broder, G., and Weil, M. H.: Science, *143*:1457, 1964.
7. Conway, E. J., and Cooke, R.: Biochem. J., *33*:457, 1939.
8. Faulkner, W. R., and Britton, R. C.: Cleveland Clinic Quart., *27*:202, 1960.
9. Forman, D. T.: Clin. Chem., *10*:497, 1964.
10. Gubler, C. J., Labey, M. E., Ashenbrucher, H., Cartwright, G. E., and Wintrobe, M. M.: J. Biol. Chem., *196*:209, 1952.
11. Hastings, A. B., Sendroy, J., Jr., and Van Slyke, D. D.: J. Biol. Chem., *79*:183, 1928.
12. Henry, R. J.: Clinical Chemistry—Principles and Technics. New York, Harper & Row Publishers, 1964.
13. Henry, R. J., Sobel, C., and Chiamori, N.: Clin. Chim. Acta, *3*:523, 1958.
14. Lee, C. T., and Duncan, G. G.: Metabolism, *5*:144, 1956.
15. Mandel, E., and Niespodzrany, L.: Unpublished modification of method by W. Ramsey.[21]
16. McLean, F. C., and Hastings, A. B.: J. Biol. Chem., *108*:285, 1935.
17. Natelson, S.: Microtechniques of Clinical Chemistry for the Routine Laboratory. Springfield, Ill., Charles C Thomas, Publisher, 1957, p. 147.
18. Oreskes, I., Hirsch, C., Douglas, K. S., and Kupfer, S.: Clin. Chim. Acta, *21*:303, 1968.
19. Peters, J. P., and Van Slyke, D. D.: Quantitative Clinical Chemistry, Methods. Baltimore, The Williams & Wilkins Co., 1932, vol. 2, pp. 245, 283.
20. Ramsay, W. N. M.: *In* Advances in Clinical Chemistry. H. Sobotka and C. P. Stewart, Eds. New York, Academic Press, Inc. 1958, vol. 1, pp. 2–39.
21. Ramsay, W. N. M.: Biochem. J., *53*:227, 1953.
22. Rice, E. W.: Copper in serum. *In* Standard Methods of Clinical Chemistry. D. Seligson, Ed. New York, Academic Press, Inc. 1963, vol. 4.
23. Robertson, W. J., and Peacock, M.: Clin. Chim. Acta, *20*:315, 1968.
24. Rosenberg, J. C., and Rush, B. F.: Clin. Chem., *12*:299, 1966.
25. Rothera, A. C. H.: J. Physiol., *37*:491, 1908.
26. Rubin, M., and Knott, L.: Clin. Chim. Acta, *18*:409, 1967.
27. Schade, A. L., Oyama, J., Reinhart, R. W., and Miller, J. R.: Proc. Soc. Exp. Biol. Med., *87*:443, 1954.
28. Schales, O., and Schales, S. S.: J. Biol. Chem., *140*:879, 1941.
29. Schatz, B. C.: Thesis, Graduate School, University of Southern California, 1962.
30. Segal, M.: Am. J. Clin. Path., *25*:1212, 1955.
31. Seligson, D., Bruemle, L. W., Webster, G. D., and Senesky, D.: J. Clin. Invest., *38*:1042, 1959.
32. Severinghaus, J. W., Stupfel, M., and Bradley, A. F.: J. Appl. Physiol., *9*:189, 1956.
33. Skeggs, L. T., Jr.: Manual, Technicon Instrument Corp., Ardsley, N.Y.
34. Skeggs, L. T., Jr.: Am. J. Clin. Path., *33*:181, 1960.
35. Sobel, A. E., and Sklensky, S.: J. Biol. Chem., *122*:665, 1937–38.
36. Sobel, A. E., and Sobel, B. A.: J. Lab. Clin. Med., *26*:585, 1940.
37. Stewart, C. P., and Frazer, S. C.: *In* Advances in Clinical Chemistry. H. Sobotka and C. P. Stewart, Eds. New York, Academic Press, Inc. 1963, vol. 6.
38. Sunderman, F. W., and Carrol, J. E.: Am. J. Clin. Path., *43*:302, 1965.
39. Trenchard, D., Noble, M. I. M., and Guz, A., Clin. Sci., *32*:189, 1967.
40. Trudeau, D. L., and Freier, E. F.: Clin. Chem., *13*:101, 1967.
41. Ulmer, D. D.: Trace elements and clinical pathology. *In* Progress in Clinical Pathology. New York, Grune & Stratton, Inc., 1966.
42. Van Slyke, D. D., Sendroy, J., Jr., Hastings, A. B., and Neill, J. M.: J. Biol. Chem., *78*:765, 1928.
43. Webster, D.: J. Clin. Path., *13*:246, 1960.
44. Zaino, E. C.: Atomic Absorption Newsletter, *6*:93, 1967.
45. Zall, D. M., Fisher, D., and Garner, M. O.: Anal. Chem., *28*:1665, 1956.
46. Zettner, A., Sylvia, L. C., and Capacho-Delgado, L.: Am. J. Clin. Path., *45*:533, 1966.

ADDITIONAL READINGS

Cantarow, A. and Trumper, M.: Clinical Biochemistry. 6th ed., Philadelphia, W. B. Saunders Co., 1962.

Cotlove, E.: Determination of chloride in biological materials. Methods of Biochemical Analysis, D. Glick, Ed., New York, Interscience Publishers Inc., 1964, vol. 12.

Davenport, H. W.: The ABC of Acid-Base Chemistry. 4th ed. Chicago, University of Chicago Press, 1958.

Davidson, I., and Henry, J. B. (Eds.): Todd-Sanford Clinical Diagnosis by Laboratory Methods. 14th ed. Philadelphia, W. B. Saunders Co., 1969.

Gambino, S. R.: Workshop Manual on Blood pH, P_{CO_2}, Oxygen Saturation, and P_{O_2}. American Society of Clinical Pathologists, 1963.

Gambino, S. R.: pH and P_{CO_2}. Standard Methods of Clinical Chemistry. S. Meites, Ed., New York, Academic Press, Inc., 1965, vol. 5.

Gutman, A. B.: The biological significance of uric acid. The Harvey Lecture Series 60. New York, Academic Press, Inc. 1966.

MacIntyre, I. M., and Wootton, I. B. P.: Magnesium metabolism—analytical methods. Ann. Rev. Biochem., *29*:642, 1960.

Oreskes, I., Hirsch, C., Douglas, K. S., and Kupfer, S.: Measurement of ionized calcium in human plasma with a calcium selective electrode. Clin. Chim. Acta, *21*:303, 1968.

Peters, J. P., and Van Slyke, D. D.: Quantitative Clinical Chemistry, Interpretation. Baltimore, The Williams & Wilkins Co., 1932, vol. 1.

Peters, J. P., and Van Slyke, D. D.: Quantitative Clinical Chemistry, Methods. Baltimore, The Williams & Wilkins Co., 1932, vol. 2.

Ulmer, D. D.: Trace elements and clinical pathology. Prog. Clin. Path., *1*:176, 1966

Wacker, W. E. C.: Magnesium metabolism. Am. Diet. Assoc., *44*:362, 1964.

Weisberg, H. F.: Water, Electrolyte and Acid-Base Balance. 2nd ed. Baltimore, The Williams & Wilkins Co., 1962.

Chapter 11

ELECTROLYTE (CATION-ANION) BALANCE*

by Norbert W. Tietz, Ph.D.

Metabolic processes in the body result in the production of relatively large amounts of acids, such as carbonic acid, lactic acid, β-hydroxybutyric acid, sulfuric acid, and phosphoric acid. A person weighing 70 kg. disposes daily the equivalent of 360 L. of 0.1 *N* acid as carbon dioxide (through the lungs) and two additional liters of 0.1 *N* acid as nonvolatile acids (through the kidneys). These products of metabolism are transported to the excretory organs (lungs and kidneys) via the extracellular fluid without producing any appreciable changes in the body pH. This is accomplished by the combined functions of the buffer system of the blood, the respiratory system, and the renal mechanism. These three systems are also of great importance in maintaining the normal composition of cations and anions of the body. Table 11-1 gives the

TABLE 11-1. *Concentration of Cations and Anions in Serum (Expressed in* mEq./L.)

Cations		*Anions*	
Na^+	142	Cl^-	103
K^+	4	HCO_3^-	27
Ca^{++}	5	$HPO_4^=$	2
Mg^{++}	2	$SO_4^=$	1
Others (trace elements)	1	Organic acids$^-$	5
	154	Protein$^-$	16
			154

approximate composition of the cations and anions of blood serum. This fluid (or plasma) is easily accessible and, therefore, is generally used to evaluate the cation-anion composition of the body. Note that there is an exact equality of *total* anions and

* Several illustrations and portions of the text were taken from an article "Anion-cation balance" by N. W. Tietz (The Chicago Medical School Quarterly, *22*:156, 1962). Reproductions were made with the permission of the publisher.

total cations. This *electrical neutrality* is maintained at all times; any increase in one anion is accompanied either by a corresponding decrease in some other anions, or by an increase of one or more cations, or both, so that total electrical neutrality is invariably maintained. Similarly, any decrease in anions involves either a corresponding increase in other anions, or a decrease in cations, or both.

The term electrical neutrality is not to be confused with acid-base neutrality (pH = 7.0, where the number of H^+ equals the number of OH^-), nor does it indicate that the pH is normal (7.4); it indicates only that there is an equality between total anions and total cations. The actual pH of plasma will depend on the distribution of the various compounds and their degree of dissociation.

In clinical medicine it is still customary to use the term acid-base balance instead of cation-anion balance. This practice derives from the outdated concept that cations, such as Na^+, are bases and anions, such as Cl^-, are acids. According to Brønstedt's theory (1923), an acid is a substance that can donate protons (H^+) and a base is a substance that can accept protons (H^+).

$$\text{acid} \rightleftharpoons H^+ + \text{base}$$
$$HCl \rightleftharpoons H^+ + Cl^-$$
$$NH_4^+ \rightleftharpoons H^+ + NH_3$$
$$\text{glycine} \rightleftharpoons H^+ + \text{glycinate}^-$$
$$\text{glycinium}^+ \rightleftharpoons H^+ + \text{glycine (neutral)}$$

Thus, HCl, NH_4^+, and amine ions (glycinium$^+$) are acids, and NH_3, glycinate$^-$, and free (neutral) amines are bases. Some (hydrogen-containing) anions, such as HCO_3^- and HPO_4^{--}, are both acids and bases (called ampholytes or amphoteric substances) as are all amino acids (e.g., glycine) and all proteins. These definitions of acid and base will be very useful in discussions of anion-cation balance.

Electrolyte concentrations of body fluids are commonly expressed in milliequivalents per liter (mEq./L.). This is particularly important in studies on cation-anion balance or acid-base balance, when it is essential to be able to compare and relate to each other the equivalent concentrations of various constituents. For example, a 100 mEq./L. sodium chloride solution contains $(100/1000) \times (23 + 35.5) = 5.85$ gm. sodium chloride in 1 L., where the numbers 23 and 35.5 represent one equivalent weight of sodium and chloride, respectively. (See also the discussion on solutions in Chapter 2.)

Exchange of Ions Between Fluid Compartments

The body has two main fluid compartments, the intracellular and the extracellular, the latter being further divided into blood plasma (vascular compartment) and interstitial fluid. For simplicity, we will adhere to this division, although it should be realized that the previously named compartments can be divided further into subcompartments with varying composition in regard to water, electrolytes, and other components (e.g., bones).

Plasma, which is of main interest in our discussion, generally has a volume of 1300 to 1800 ml./sq. m. of body surface and constitutes approximately 5 per cent of the body weight. Its composition has been discussed earlier and is described in Table 11-1 and Figure 11-1. *Interstitial fluid* is essentially an ultrafiltrate of blood plasma and makes up approximately 15 per cent of the total body weight. The endothelial linings of the capillaries act as a semipermeable membrane and allow passage of water and

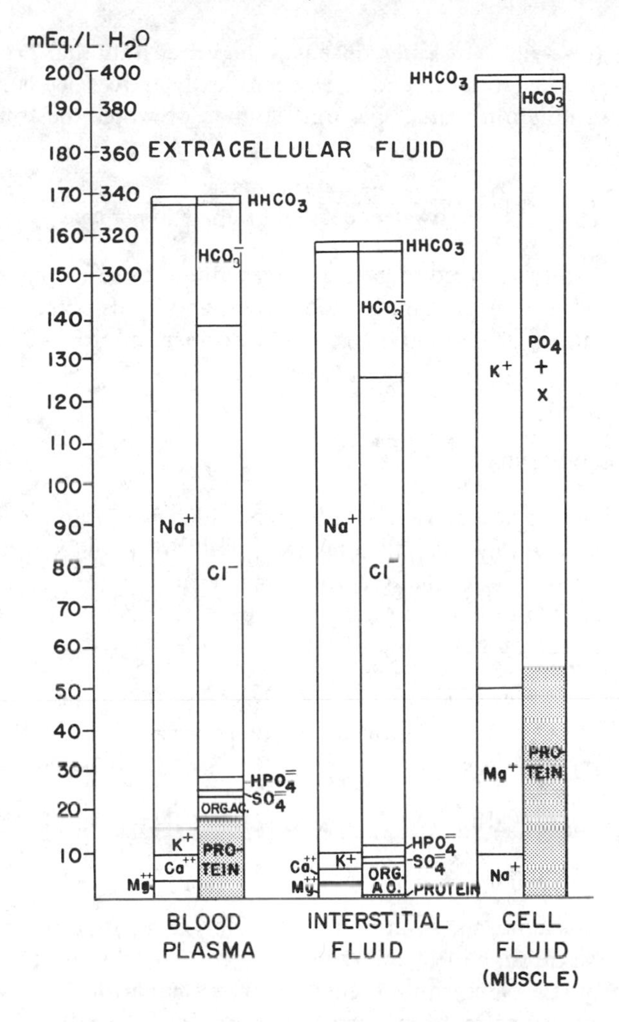

Figure 11-1. Electrolyte composition of blood plasma, interstitial fluid, and cell fluid. Values are given in mEq./L. of water. (After Gamble, J., Chemical Anatomy, Physiology, and Pathology of Extracellular Fluid, Cambridge, Harvard University Press, 1950.)

diffusible solutes, but not of compounds of large molecular weight such as proteins. The fact that this impermeability is not absolute is demonstrated by the varying content of protein in interstitial fluids. The exact composition of *intracellular fluid* is extremely hard to measure because of the relative unavailability of cells free of contamination. Although erythrocytes are easily accessible, it would be incorrect to make any generalizations based on the composition of these highly specialized cells. Data on cell composition (Fig. 11-1), therefore, are only considered approximations. Intracellular fluids contribute approximately 50 per cent of the total body weight.

Figure 11-1 illustrates the average composition of extracellular and intracellular fluids; note that the numbers are given in mEq./L. of water rather than in mEq./L. of plasma. In clinical medicine, as in most analytical chemistry, where samples are measured volumetrically, it is customary to express concentrations of solutes per unit volume of solution (e.g., mEq./L. or mg./100 ml.) since this is more easily measured. When comparing the concentration of a solute in two aqueous solutions of widely different water content, however, it is usually more meaningful to relate concentrations per unit volume (or weight) of *water* since it is the water only that provides the space in which solutes are free to move and to be osmotically active. To convert values per unit volume of sample into values per unit volume of water the following formula may be applied:

$$\text{mEq./L.} = \frac{\text{mEq./L. sample} \times 100}{\text{water content of sample in per cent}}$$

Serum is generally assumed to have a water content of 93 per cent; however, the actual value will depend on the amount of proteins and lipids present. Thus, a serum sodium concentration of 148 mEq./L. would correspond to $148 \times 100/93 = 159$ mEq./L. water.

Gibbs-Donnan Equilibrium

Two solutions separated by a semipermeable membrane will establish an equilibrium in such a way that all ions are equally distributed provided that the solution contains only solutes that can freely move through the membrane (equation 1). At the state of equilibrium the total ion concentration and, therefore, the osmotic pressure, are the same on both sides of the membrane.

Before steady state			*After steady state*		
Compartment I		Compartment II	Compartment I	Compartment II	
$[Na^+]$	$\rightarrow$	$[K^+]$	$[K^+]$	$[K^+]$	
$[Cl^-]$	$\rightarrow$	$[Cl^-]$	$[Cl^-]$	$[Cl^-]$	
$[Na^+]$	$\leftarrow$	$[K^+]$	$[Na^+]$	$[Na^+]$	(1)
$[Cl^-]$	$\leftarrow$	$[Cl^-]$	$[Cl^-]$	$[Cl^-]$	
	Membrane		Membrane		

If, on the other hand, the solution on one side of a membrane contains ions that cannot freely move through the membrane (e.g., proteins), distribution of the movable ions at the steady state is unequal (equation 2). In equation 2, C_1 and C_2 refer to the initial concentrations in Compartments I and II; R^- indicates a nondiffusible ion such as protein; X indicates the amount of diffusible ions that moved from Compartment II into Compartment I.

Before steady state				*After steady state*					
Compartment I		Compartment II		Compartment I			Compartment II		
$[Na^+]$	$[R^-]$ $\leftarrow$	$[Na^+]$	$[Cl^-]$	$[Na^+]$	$[Cl^-]$	$[R^-]$	$[Na^+]$	$[Cl^-]$	
C_1	C_1	C_2	C_2	$C_1 + X$	X	C_1	$C_2 - X$	$C_2 - X$	(2)
	Membrane					Membrane			

In order to preserve electrical neutrality, Na^+ and Cl^- have to diffuse in pairs across the membrane until a steady state is established. It can actually be shown experimentally and can be calculated from thermodynamic considerations that at the steady state the product of the concentrations of the diffusible ions on one side of the membrane will be the same as the product of the concentrations of the same ions on the other side of the membrane (Gibbs-Donnan law).

$$\underset{\text{Compartment I}}{[Na^+] \times [Cl^-]} = \underset{\text{Compartment II}}{[Na^+] \times [Cl^-]} \qquad (3)$$

Considering equations 2 and 3 we can state that

$$(C_1 + X) \times X = (C_2 - X)^2 \tag{4}$$

or

$$C_1X + X^2 = (C_2)^2 - 2C_2X + X^2$$

or

$$C_1X = (C_2)^2 - 2C_2X$$

or

$$C_1X + 2C_2X - (C_2)^2$$

or

$$(C_1 + 2C_2)X = (C_2)^2$$

or

$$X = \frac{(C_2)^2}{C_1 + 2C_2}$$

If, for example, we assume an initial concentration for C_1 and C_2 of 3.0 m*M*., then $X = \frac{3^2}{3 + (2 \times 3)} = \frac{9}{9} = 1$. If we enter this figure into equation 2, the following ion distribution in m*M*. results:

Before steady state				*After steady state*					
Compartment I		Compartment II		Compartment I			Compartment II		
$[Na^+]$	$[R^-]$	$[Na^+]$	$[Cl^-]$	$[Na^+]$	$[Cl^-]$	$[R^-]$	$[Na^+]$	$[Cl^-]$	(5)
3.0	3.0	3.0	3.0	3.0 + 1.0	1.0	3.0	3.0 − 1.0	3.0 − 1.0	

Note that after steady state has been established, solutions on both sides of the membrane have electrical neutrality and that the requirements of equation (3) for diffusible ions (Na^+, Cl^-) are fulfilled. (Compartment I: $(3.0 + 1.0) \times 1.0 = 4$; Compartment II: $(3.0 - 1.0) \times (3.0 - 1.0) = 4$.) At the same time, it should also be noted that the total ion concentration in Compartment I is greater than that of Compartment II. (Compartment I: $3.0 + 1.0 + 1.0 + 3.0 = 8.0$; Compartment II: $(3.0 - 1.0) + (3.0 - 1.0) = 4$.) Assuming that the size of the compartment is maintained constant, this results in an increased osmotic pressure in Compartment I. If the nondiffusible ion in Compartment I is a colloid, such as protein, this increased part of the osmotic pressure is referred to as *colloidal osmotic pressure*. Presence of protein in the intravascular compartment is an important factor in maintaining adequate osmotic pressure and preventing excessive water from moving into the tissues.

Distribution of Ions in Intracellular and Extracellular Compartments

Examination of Figure 11-1 reveals that the electrolyte composition of both blood plasma and interstitial fluid (both extracellular fluids) is similar, but their compositions differ markedly from that of intracellular fluid. The major extracellular ions are Na^+, Cl^-, and HCO_3^-; in intracellular fluids the main cations are K^+ and Mg^{++} and the major anions are organic phosphates and protein.

Until now, it has not been possible to explain adequately the unequal distribution of most electrolytes (or other constituents) between intracellular and extracellular fluids in terms of physicochemical theories of diffusion. The Gibbs-Donnan law (see previous section) has successfully explained the unequal distribution of chloride and bicarbonate between red cells and plasma. Van Slyke[5] has shown that the higher chloride and bicarbonate content of plasma is due to the presence of the negatively charged, nondiffusible hemoglobin inside the red cell. Although this relationship has been established, there appears to be no other proven application of the Gibbs-Donnan law to the distribution of other ions in blood or to the many cases of unequal distribution of ions in tissue compartments; however, other applications are suspected. Cell membranes cannot be considered simply as semipermeable membranes. It has been shown that some cell membranes have selective permeability for some molecules or ions; also, in many cases, active transport mechanisms across the cell membrane have been either suspected or demonstrated. These transport mechanisms have not been adequately explained in physicochemical terms, but it is believed that they are in some way intimately linked to metabolic processes of the cell. This concept recognizes that the living cell uses metabolic energy to control the unequal distribution of ions and other solutes on the two sides of a living membrane.

Osmolality

The osmolality of a solution is defined as the sum of the moles of all dissolved ions and undissociated molecules in 1 kg. of water; thus, molal* solutions of nonionized substances such as glucose (180 gm./1 kg. water) or urea (60 gm./1 kg. water) will have the same osmolal concentration (1 osmole/kg. or 1000 mOsm./kg.). On the other hand, a 0.1 molal solution of sodium chloride (5.85 gm./kg. water) that is ionized to give two ions, Na^+ and Cl^-, has an osmolal concentration of 200 mOsm./kg. water.† An 0.1 molal solution of calcium chloride (11.1 gm./kg. water) would be 300 mOsm./kg. since there are three ions per molecule of calcium chloride. The osmolal concentration of a solution containing a mixture of electrolytes and nonelectrolytes is equal to the sum of the individual osmolal concentrations of all of its constituents. For example, the osmolality of blood serum is about 305 mOsm. /kg. (295–315 mOsm./kg.) because of the presence of approximately 148 Na^+, 4 K^+, 2.5 Ca^{++}, 108 Cl^-, 27 HCO_3^-, 2 HPO_4^{--}, 1 SO_4^{--}, 5 organic acids, 5 glucose, and 2 urea (all concentrations in mOsm./kg.).

The significance of osmolality of body fluids is that it controls the distribution of water between various compartments. In general, water diffuses from a region of lower osmolal concentration to a region of higher concentration so that osmolal concentrations are equalized. Such movement can take place only if the fluid compartment with the higher osmolality can be expanded; however, if the volume of the compartment is fixed, it will undergo an increase in hydraulic pressure.

The Henderson-Hasselbalch Equation

The Henderson-Hasselbalch equation aids in the understanding and explanation of pH control of body fluids; this will become clearer in the later discussions of the compensatory mechanisms of the body. The equation applied to carbonic acid may be written as follows:

$$pH = pK'_{(HHCO_3)} + \log \frac{HCO_3^-}{HHCO_3} \tag{6}$$

* Values are generally expressed in osmolality (number of particles per kg. of water) rather than in osmolarity (number of particles in 1 L. solution).

† In fact, the osmolality of such a solution will be slightly less than 200 mOsm./kg. solution, since the effective dissociation of salts in other than highly diluted solutions is less than 100%.

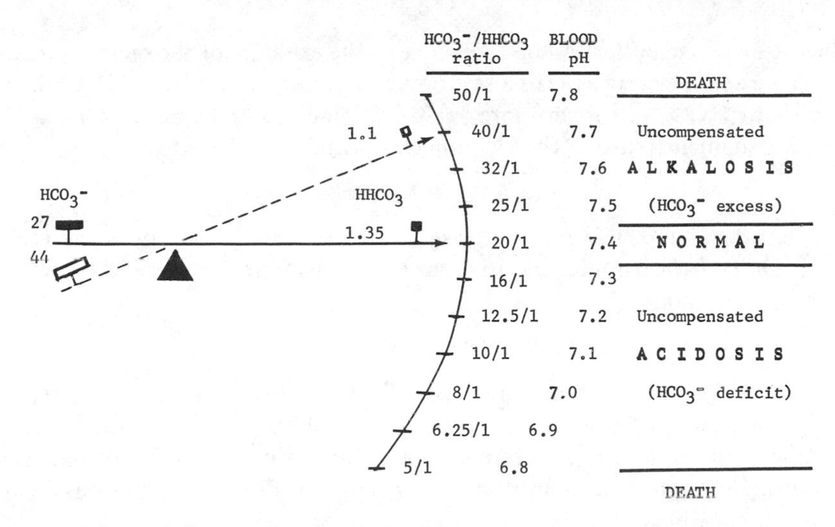

Figure 11-2. Scheme demonstrating the relation between pH and ratio of bicarbonate to carbonic acid. If the ratio in blood is 20:1 (e.g., 27 mEq. HCO_3^- and 1.35 mEq. of $HHCO_3$/L.), the resultant pH will be 7.4 as demonstrated by the solid beam. The dotted line shows a case of uncompensated alkalosis (bicarbonate excess) with a bicarbonate concentration of 44 mEq./L. and a carbonic acid concentration of 1.1 mEq./L. The ratio therefore is 40:1 and the resultant pH 7.7. In a case of uncompensated acidosis, the pointer of the balance would point to a pH between 6.8 and 7.35, depending on the HCO_3^-:$HHCO_3$ ratio. pH values below 6.8 or above 7.8 are incompatible with life. (After Weisberg, H. F.: Surg. Clin. N. Amer., *39*: 93, 1959; Snively, W. D. and Wessner, M.: J. Ind. State Med. Assn., *47*: 957, 1954.)

The pK′ of carbonic acid under the conditions of blood plasma is 6.1,* which is lower than the pK in the absence of other electrolytes, since pK values of all acids decrease as the ionic strength increases; thus, the symbol pK′ is used instead of pK. The average normal concentration of bicarbonate/carbonic acid in plasma is 27/1.35 mEq./L. = 20/1; the $\log_{10}$ of 20 is 1.3. Formula (6) applied to normal plasma can then be written:

$$pH = 6.1 + 1.3 = 7.4 \tag{7}$$

It follows that any change in the concentration of bicarbonate or carbonic acid and, therefore, in the ratio of $HCO_3^-/HHCO_3$ must be accompanied by a change in pH. Such changes in the ratio can occur through a change either in the numerator (HCO_3^-) or in the denominator ($HHCO_3$) as will be discussed later in this chapter.

Clinical conditions, as we shall see, are classified as primary disturbances in the HCO_3^- or in the $HHCO_3$ content. Various compensatory mechanisms that attempt to correct these primary disturbances are geared toward reestablishment of the normal ratio of $HCO_3^-/HHCO_3$. Here again the compensatory mechanisms may result in changes in the bicarbonate or the carbonic acid.

Snively and Wessner,[3] as well as Weisberg,[6] have brought the Henderson-Hasselbalch equation into a scheme that simplifies its explanation (see Fig. 11-2).

BUFFER SYSTEMS AND THEIR ROLE IN REGULATING BODY pH

Buffers are defined as solutes that minimize pH changes of a solution if an acid or a base is added to it. They consist of a weak acid and its salt with a strong base or of a weak base and its salt with a strong acid.

---

* Although the figure 6.1 is an average figure, it is very close for most sera. It has recently been reported,[4] however, that the pK′ of some abnormal sera may differ from this average value by as much as 0.2.

The action of the buffer can be explained by the example of the bicarbonate buffer system. If we add a strong acid to a solution containing HCO_3^- and $HHCO_3$, the H^+ will react with HCO_3^- to form more $HHCO_3$. The latter, being a weak acid, will cause only a slight increase in the H^+ concentration.

$$HCO_3^- + H^+ \rightarrow HHCO_3 \quad (8)$$

On the other hand, if we add a base to the same buffer solution, the OH^- will react with the H^+ of carbonic acid to form bicarbonate and water. The pH change, therefore, will be small.

$$HHCO_3 + OH^- \rightarrow HCO_3^- + H_2O \quad (9)$$

The buffer systems of immediate clinical interest in connection with regulation of body pH are those of plasma and erythrocytes. With the exceptions of the hemoglobin buffer and the serum protein buffer, all others are present in both plasma and erythrocytes, although their quantitative distribution is different. A discussion of the most important buffers follows.

The Bicarbonate:Carbonic Acid Buffer System

The most important buffer of plasma is the bicarbonate:carbonic acid buffer, which is also present in red cells, but at a lesser concentration. At the pH of blood this buffer is most effective in buffering H^+ as shown in equation (8), but it can also buffer OH^- as demonstrated in equation (9). The effectiveness of the bicarbonate buffer is based on its high concentration and the fact that $HHCO_3$ can readily be disposed of as carbon dioxide or increased by retention of carbon dioxide; both result from the function of the respiratory mechanism, to be explained later. Bicarbonate can be either eliminated or preserved by the action of the renal mechanism (see the Renal Compensatory Mechanisms). Thus, the lungs and the kidneys increase the efficiency of the bicarbonate buffer by supplying or removing $HHCO_3$ and HCO_3^-, respectively, as needed.

The Serum Protein Buffer System

The buffer capacity of proteins depends on the number of acidic and basic groups in the molecule, which varies with the pH. The system is involved in buffering acids as shown in the example of $HHCO_3$:

$$HHCO_3 + \text{Protein}^- \rightleftharpoons H \cdot \text{Protein} + HCO_3^- \quad (10)$$

This function is insignificant, however, when compared with that of the hemoglobin and the bicarbonate buffer systems.

The Phosphate Buffer System

At a plasma pH of 7.4, the ratio of $HPO_4^{--}/H_2PO_4^-$ is 80/20. The total amount of this buffer in both erythrocytes and plasma is less than that of other major buffer systems. The phosphate buffer reacts with acids and with bases as follows:

$$HPO_4^{--} + H^+ \rightleftharpoons H_2PO_4^- \quad (11)$$

$$H_2PO_4^- + OH^- \rightleftharpoons HPO_4^{--} + H_2O \quad (12)$$

This system has importance in the excretion of acids in the urine as will be explained in the section on renal mechanism.

The Hemoglobin Buffer System

Hemoglobin is present in blood as oxygenated hemoglobin (generally referred to as oxyhemoglobin or $HHbO_2$); as deoxygenated hemoglobin (HHb); and as their potassium salts ($KHbO_2$ and KHb). Hemoglobin has an effective buffering action because of its presence as free acid and its salt: $HbO_2^-/HHbO_2$ and Hb^-/HHb. The most important buffer action of hemoglobin, however, is the acceptance of protons (H^+) by the imidazole group of histidine in the hemoglobin molecule. The extent to which H^+ is accepted by this group is affected by the degree of oxygenation of the hemoglobin molecule. Removal of oxygen from the hemoglobin iron will result in an increased rate of acceptance of H^+ by the imidazole group (Fig. 11-3).

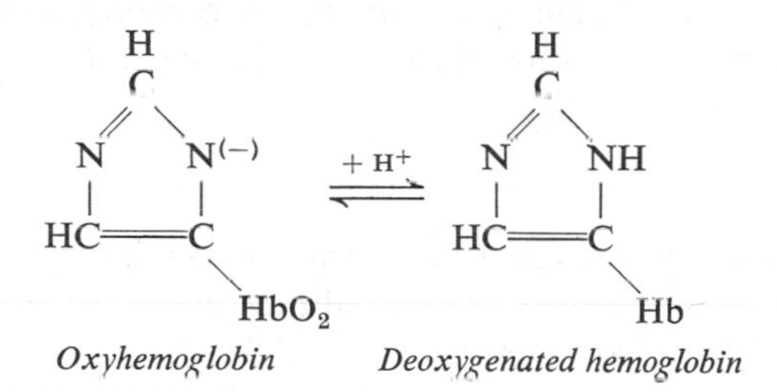

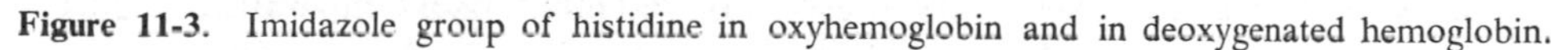
Figure 11-3. Imidazole group of histidine in oxyhemoglobin and in deoxygenated hemoglobin.

The transformation of oxygenated hemoglobin (stronger acid, pK 6.7) into deoxygenated hemoglobin (weaker acid, pK 7.9) results in a removal of H^+ from the solution. The importance of the system is enhanced by the fact that an increase in P_{CO_2} and a decrease in pH will favor the dissociation of oxygen from oxygenated hemoglobin, thus, permitting more hydrogen ions to be accepted by the imidazole group.

The hemoglobin buffer system mainly buffers carbonic acid produced during metabolic processes. For each millimole of oxygen that dissociates from hemoglobin, 0.7 m*M*. H^+ are removed from the solution according to the following equation:

$$HbO_2^- + HHCO_3 \rightleftharpoons HHb + O_2 + HCO_3^- \qquad (13)$$

This makes the hemoglobin buffer system the most effective single system in buffering $HHCO_3$.

RESPIRATION AND THE RESPIRATORY REGULATION OF CATION-ANION BALANCE

The respiratory mechanism is responsible for the adequate supply of oxygen to cells and, at the same time, for the removal of carbon dioxide produced during metabolic processes. Oxygen is transported from the lungs to the tissues, and carbon dioxide is brought from the place of production to the lungs with a minimum in pH change. In fact, respiration and the respiratory mechanism involved in the maintenance of body pH are so closely related that the discussion of both in the same chapter is justified.

The exchange of gases in the lungs takes place through the membranes of the alveoli. The average amount of oxygen that crosses the membrane and is absorbed by the blood is 250 ml./min. and the average amount of carbon dioxide eliminated is 200 ml./min. During exercise these figures increase significantly, possibly as much as tenfold. The gas exchange proceeds very rapidly, which is essential, since blood remains in the lungs only 0.75 second in resting subjects and 0.3 second during exercise. Some pathological conditions cause reduced flow of blood through the lungs or changes in the membranes of the alveoli. Such changes interfere with the gas exchanges and result in a decrease in both oxygenation and removal of carbon dioxide from the blood. This shall be discussed in more detail in connection with respiratory acidosis.

The exchange and transport of gases is governed by some of the following gas laws, which shall be mentioned briefly. *Henry's law* states that the solubility of a gas in a liquid is directly related to the partial pressure of the gas above the liquid. For example, this means that the amount of oxygen or carbon dioxide carried in the blood in physically dissolved form (not chemically bound) is directly proportional to the partial pressure of the respective gas.

$$\text{Dissolved } O_2 = a \times P_{O_2} \tag{14}$$

The proportionality constant a is dependent on the temperature and presence of other solutes, but independent of the partial pressure of other gases. It was found to be 0.023 ml./ml. of blood at 1 atmosphere and 38°C.

Example: Assume a barometric pressure of 760 mm. Hg, a temperature of 38°C., and a P_{O_2} of 100 mm. Hg. How much oxygen is dissolved in the blood?

$$\text{Dissolved } O_2 = 0.023 \times 100/760 = 0.00303 \text{ ml./ml. of blood}$$

If the figure is to be converted into volume per cent, then:

$$0.00303 \times 100 = 0.303 \text{ volume per cent}$$

Volume per cent is defined as the amount of gas in milliliters dissolved in 100 ml. of solution.

In gas mixtures such as atmospheric air, the total barometric pressure (P_B) equals the sum of the individual (partial) pressures (*Dalton's law*):

$$P_B = P_{O_2} + P_{CO_2} + P_{N_2} + P_{H_2O} \tag{15}$$

or

$$P_B - P_{H_2O} = P_{O_2} + P_{CO_2} + P_{N_2}$$

Dry atmospheric air has been found to have a composition of 20.93 per cent oxygen, 0.03 per cent carbon dioxide, and 79.04 per cent nitrogen. The nitrogen is generally measured by difference and the figure includes small amounts of other gases. In connection with respiration and clinical problems the gas concentration (partial pressure) is generally expressed in terms of mm. Hg rather than in volume %. The formula for the conversion of volume % into mm. Hg is shown for oxygen, but the same formula applies to other gases.

Example:

$$P_{O_2} = (P_B - P_{H_2O}) \times \frac{\% O_2}{100} = \text{mm. Hg} \tag{16}$$

$$P_{O_2} = (760 - 47) \times \frac{15}{100} = 107 \text{ mm. Hg.} \tag{17}$$

The number 47 is generally accepted as the value for PH_2O in alveolar air, 760 represents the existing barometric pressure, and 15 represents the concentration of oxygen (in per cent) in (dry) alveolar air.

Transport of Oxygen

Although a small amount of oxygen may be transported in blood as physically dissolved oxygen (see previous section), most of the oxygen is carried in the form of a bond with Fe^{++} of hemoglobin. One mole or 32 gm. of oxygen (in the form of oxygenated hemoglobin, oxyhemoglobin) can be bound by 16,570 gm. of hemoglobin; thus,

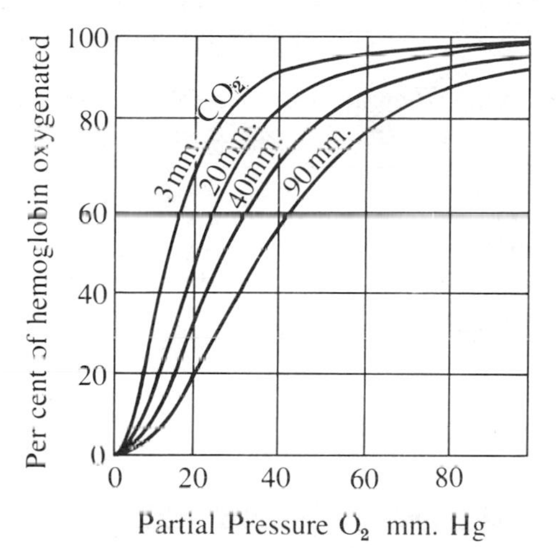

Figure 11-4. The dissociation curves of hemoglobin at partial pressures of carbon dioxide equal to 3, 20, 40, and 90 mm. Hg. (After Barcroft J.: The Respiratory Function of the Blood, London, Cambridge University Press, 1928.)

1 mole of hemoglobin (66,280 gm.) can bind 4 moles of oxygen. Since 1 mole of oxygen occupies 22,400 ml. at 0°C., 1 gm. of hemoglobin can combine with 22,400/16,570 = 1.36 ml. O_2/gm. Hb, and it follows that a blood sample with a hemoglobin concentration of 16 gm./100 ml. can carry 16 × 1.36 or 21.76 ml. O_2/100 ml. as oxyhemoglobin. (An additional small amount of oxygen is carried in plasma and blood as dissolved oxygen.) The binding of oxygen by hemoglobin is so constant that some methods of hemoglobin standardization take advantage of this fact and determine the amount of hemoglobin from the amount of oxygen that it binds.

The ability of hemoglobin to bind oxygen is influenced by a number of factors, such as PO_2, PCO_2, and pH. At the conditions of alveolar air (PO_2 = 100 mm. Hg, PCO_2 = 40 mm. Hg, and temperature = 38°C.), there is an oxygen saturation of hemoglobin of approximately 95 to 98 per cent. With decrease in PO_2, there is initially only a slight decrease in the saturation of hemoglobin; e.g., at PO_2 values of 70 mm. Hg, oxygen saturation is still more than 90 per cent. Below this level, however, decreases in the PO_2 will cause a significant drop in the degree of saturation, in accordance with the S-shaped curve shown in Figure 11-4. At PO_2 levels of 20 mm. Hg, as it exists frequently in tissues, the oxygen saturation of hemoglobin drops below 30 per cent. The high degree of dissociation of oxygen from hemoglobin at low PO_2 values (e.g., 20 to 70 mm. Hg) is an important factor in the supply of tissues with oxygen.

An increase in the P_{CO_2} or a decrease in the pH of blood favors the dissociation of oxygen from oxyhemoglobin. This effect is especially pronounced at low P_{O_2} levels. Again, the factors that influence the dissociation of oxygen from hemoglobin are advantageous in regard to oxygen supply of tissues where the P_{CO_2} values are increased and the pH values are decreased.

Transport of Carbon Dioxide in Blood

THE ISOHYDRIC AND CHLORIDE SHIFT

The carbon dioxide produced in metabolic processes is carried to the lungs in the blood (see Table 11-2). This is accomplished with a minimum of pH change of body

TABLE 11-2. *Average Distribution of Carbon Dioxide in 1 L. of Normal Blood Assuming a Hematocrit of 40%*

	Arterial		*Venous*		*Difference**	
Carbon Dioxide	m*M*	% of total	m*M*	% of total	m*M*	% of total
In plasma (600 ml.):						
as dissolved CO_2	0.72	3.32	0.81	3.47	0.09	5.38
as HCO_3^-†	15.27	70.56	16.26	69.75	0.99	59.28
In erythrocytes:						
as dissolved CO_2	0.36	1.66	0.40	1.71	0.04	2.39
as carbamino-CO_2	0.98	4.52	1.39	5.96	0.41	24.55
as HCO_3^-	4.31	19.91	4.45	19.09	0.14	8.38
Total	21.64		23.31		1.67	

* The difference between arterial and venous blood is considered to be that amount of carbon dioxide which is disposed of by the lungs.

† Plasma contains a small amount of carbamino-CO_2 (up to 0.5 m*M*), which is traditionally included in the HCO_3^- fraction.

fluids by a series of reactions generally referred to as the isohydric and chloride shift. Although most of these reactions are simultaneous, it is advantageous to discuss them step by step.

Owing to the continuous production of carbon dioxide within the tissue cells, there is a concentration differential for carbon dioxide from these cells to the plasma and the red cells, which causes a shift of physically dissolved carbon dioxide from the tissue cells into the plasma and the erythrocytes. A small portion of the carbon dioxide entering the plasma stays as dissolved carbon dioxide and another small portion reacts with water to form $HHCO_3$. The increased amount of H^+ is buffered by the plasma buffers, including the proteins (Fig. 11-5, reaction 1). Another small portion combines with the amino groups of proteins and forms carbamino compounds (Fig. 11-5, reaction 2). Up to a maximum of 0.5 m*M*. CO_2/L. may be carried in this form in plasma.

$$R—NH_2 + CO_2 \rightleftharpoons R—NHCOO^- + H^+$$

Most of the carbon dioxide entering the erythrocytes will react with water to form $HHCO_3$. This reaction is catalyzed by the enzyme carbonic anhydrase and, therefore, proceeds at a relatively high speed (Fig. 11-5, reaction 3).

The $HHCO_3$ formed in reaction 3 tends to lower the pH. This pH change will be fully or partially compensated by the release of oxygen from oxyhemoglobin,

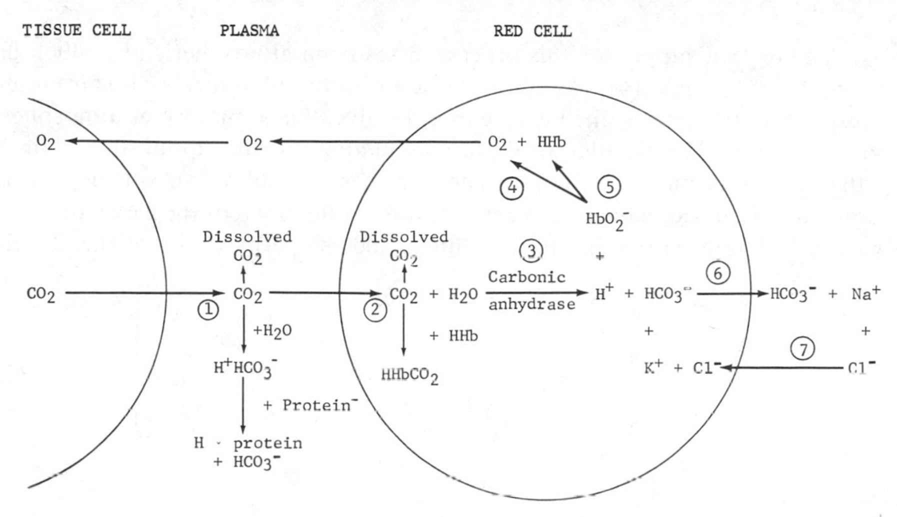

Figure 11-5. Scheme demonstrating the isohydric and chloride shift. The encircled numbers refer to the reactions described in the text.

which involves the conversion of the stronger acid ($HHbO_2$) into a weaker acid (HHb) and the imidazole group of HHb will accept the H^+ as outlined in the discussion of the hemoglobin buffer (see Fig. 11-3*A* and *B*). The oxygen released in this reaction will move from the red cells through the plasma into the tissue cells and supply these with the vital oxygen (see Fig. 11-5, reactions 3–5).

The removal of hydrogen ions from $HHCO_3$ in reaction 5 results in the formation of HCO_3^-, which is electrically balanced by K^+ previously released from oxyhemoglobin. The various transformations described so far (Fig. 11-5, reactions 1–5) are referred to as the *isohydric shift*.

The continuous formation of HCO_3^- within the red cells produces a concentration differential that results in the movement of HCO_3^- into the plasma, according to the law of membrane equilibrium.

Potassium ions will remain in the erythrocytes since they cannot readily diffuse through the cell wall. (Fig. 11-5, reaction 6.)

In order to maintain electrical neutrality in red cells and plasma, a corresponding number of Cl^- ions diffuse from the plasma into the red cells, a movement generally referred to as the *chloride shift*. The excess Na^+ formed by the shift of chloride from the plasma into the red cells is now balanced by the HCO_3^- entering the plasma. At the same time the K^+ released by the movement of bicarbonate will now be balanced by the Cl^- that entered the cell (Fig. 11-5, reaction 7).

The HCO_3^-, the carbamino compounds and the dissolved carbon dioxide are transported in the blood to the pulmonary capillaries and alveoli. The comparatively low P_{CO_2} in the alveoli will cause a shift of carbon dioxide from the erythrocytes and the plasma into the alveoli. On the other hand, the high P_{O_2} in the alveoli causes a shift of oxygen into the plasma and the erythrocytes. This exchange causes a reversal of reactions 1 to 7. The removal of carbon dioxide from the blood and the oxygenation of the blood are the major reactions which convert venous blood into arterial blood.

Respiration and its Role in Maintenance of Normal Body pH

The removal of carbon dioxide from the blood and the supply of tissues with oxygen as discussed in the previous paragraph, are the main purposes of the respiratory process. The oxygen supply for this process arises from atmospheric air, which has an average P_{O_2} of 159 mm. Hg. As atmospheric air is inhaled, it meets with air present in the respiratory tree; thus, the air reaching the alveoli is a mixture of atmospheric air and expired air. The alveolar air itself is in an approximate equilibrium with the blood that passes through the lungs. The degree of equilibration will depend on the factors just discussed, namely, the concentration differential of the gases, the speed with which the blood passes through the lungs, and the permeability of the alveolar

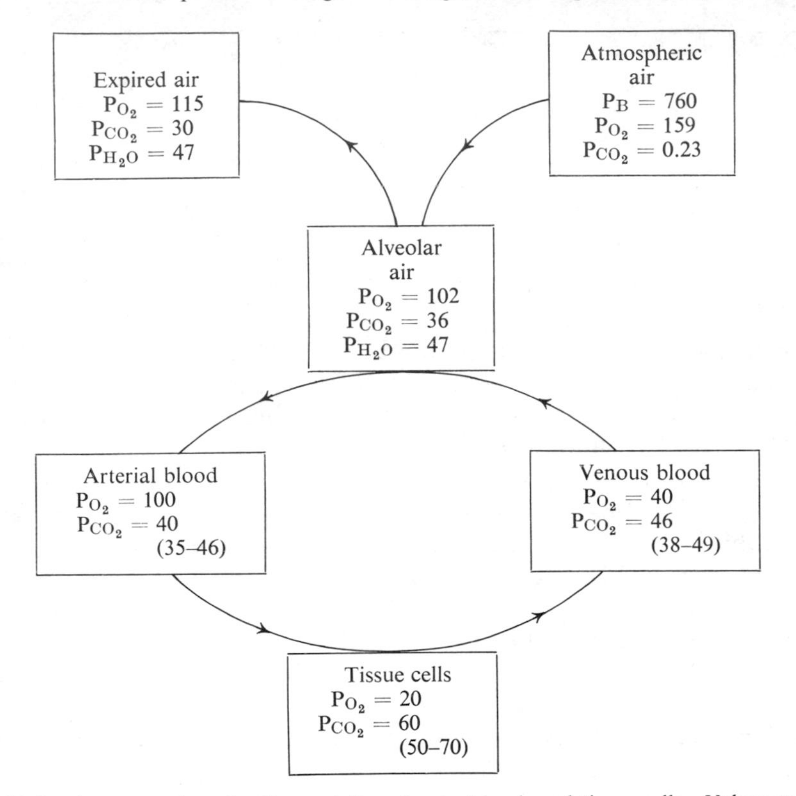

Figure 11-6. Average values for P_{O_2} and P_{CO_2} in air, blood, and tissue cells. Values are given in mm. Hg.

membranes for blood gases. In normal individuals the equilibration is close to 100 per cent and is shown by a comparison of the P_{O_2} of alveolar air with arterial blood (102 and 100 mm. Hg, respectively) and the P_{CO_2} in alveolar air with arterial blood (36 and 40 mm. Hg). As arterial blood passes through the tissues, oxygen diffuses from the blood into the cells and carbon dioxide is accepted by the blood, thus, converting arterial blood into venous blood. Venous blood is again converted to arterial blood in its passage through the lungs, and this completes the cycle. Figure 11-6 represents this process in schematic form and gives the respective average values for P_{O_2} and P_{CO_2}.

The *respiratory mechanism* is regulated by the medullary respiratory center and is affected by the P_{CO_2}, pH, P_{O_2}, and the temperature. Any decrease in the P_{O_2} of blood depresses the respiratory center, but at the same time causes a stimulation of the carotid and aortic chemoreceptors (chemoreflex drive). Since the latter reaction exceeds the former, decreased P_{O_2} values result in increased respiration. This mechanism is clinically important for the sufficient supply of oxygen to the blood during exercise, when gas exchange in the lung is inhibited by any pathological condition, or when the Po_2 of atmospheric air is decreased as, for example, at high altitudes. The increased respiratory rate at high altitudes must be taken into consideration when establishing normal values for Pco_2 and pH for individuals living in such geographical areas.

Decrease in the Pco_2 or increase in the pH depresses the respiratory center and results in *hypoventilation*; on the other hand, any increase in Pco_2 or decrease in pH stimulates the respiratory center through the chemoreceptors in the aortic arc and carotid sinus and causes *hyperventilation*. Increased temperatures also accelerate the respiratory rate.

The regulation of the rate of pulmonary ventilation in response to changes in Pco_2 and pH is the basis of the pulmonary compensatory mechanism in acidosis and alkalosis. In metabolic acidosis (primary HCO_3^- deficit) the pH is low, causing an increase in the respiratory rate (hyperventilation). This reduces the amount of $HHCO_3$ in blood and helps to restore the ratio of HCO_3 /$HHCO_3$, although the total carbon dioxide may be low.

Example: In a case of metabolic acidosis before compensation, the HCO_3 / $HHCO_3$ ratio may be 21.6/1.35 = 16/1, which represents a pH of 7.3 (see Fig. 11-2). An increased rate and depth of respiration causes an elimination of carbon dioxide and thus a reduction in plasma $HHCO_3$. This changes the HCO_3^-/$HHCO_3$ ratio, e.g., to 21.6/1.08 = 20/1, which results in a normal pH of 7.4. Such a condition is called *compensated metabolic acidosis*. The word compensated implies that the pH has returned to normal and acidosis implies that there is still a primary HCO_3^- deficit (HCO_3^- = 21.7 mEq./L). Those cases in which the HCO_3^-/$HHCO_3$ ratio stays below the normal of 18/1 to 22/1 after compensation are referred to as *uncompensated metabolic acidosis*.

In metabolic alkalosis (primary HCO_3^- excess), the pH increase induces hypo ventilation, which causes a retention of carbon dioxide and, therefore, an increase in plasma $HHCO_3$. This, again, aids in establishing the normal ratio of HCO_3^-/$HHCO_3$, resulting in normal blood pH or in a pH approaching the normal; however, the total carbon dioxide may be above normal.

Example: A patient in metabolic alkalosis, before compensation, has a ratio of HCO_3^-/$HHCO_3$ of 56/1.4 = 40/1, resulting in a pH of 7.7. Decreased rate of respiration results in retention of carbon dioxide and an increase in plasma $HHCO_3$. The HCO_3^-/$HHCO_3$ ratio changes to 56/1.75 — 32/1, resulting in a drop of pH to 7.6. Such a condition is called *uncompensated metabolic alkalosis*.* The term uncompensated implies that the pH, despite some compensatory measures, is not in the normal range and metabolic alkalosis implies that there is a primary excess of HCO_3^-; in this case, HCO_3^- = 56 mEq./L.

If the compensatory mechanisms had been successful in restoring the ratio of 20/1, we would have a (fully) *compensated metabolic alkalosis*.

The high speed with which this compensation takes place makes the respiratory mechanism an extremely important link in the effort of the body to maintain a normal pH and cation-anion balance.

Besides the respiratory mechanism described here, there are, of course, the following renal mechanisms that aid in maintaining or restoring the normal HCO_3^-/ $HHCO_3$ ratio of 20/1.

*A more appropriate term, although rarely used, is "partially compensated metabolic alkalosis."

THE RENAL COMPENSATORY MECHANISMS

The normal pH of plasma, as well as that of the glomerular filtrate, is 7.4; the urinary pH of fasting individuals is about 6.0. This drop in pH is brought about by the kidney, which excretes the nonvolatile (fixed) acids produced by metabolic processes. The various functions of the renal mechanism adjust to the specific requirements: in case of acidosis, there is an increased excretion of acids and a preservation of base; in alkalosis, there is an increased excretion of base and preservation of acids. The pH of the urine changes correspondingly and may vary in random specimens from pH 4.5 to 8.2 (normal values from 4.8 to 7.8). This ability to excrete variable amounts of acid or base is of utmost importance and makes the kidney the final defense mechanism against any drastic changes in body pH and cation-anion composition.

The various acids produced during metabolic processes are buffered in the extracellular fluid, although at the expense of HCO_3^- (see bicarbonate buffer system). Therefore, the supply of HCO_3^- would finally be exhausted if the kidneys did not excrete the acids and restore the HCO_3^-.

Excretion of Acids

None of the strong acids, such as sulphuric acid, hydrochloric acid, and most lactic acid, can be excreted in their free form by the kidneys, only in the form of their salts. This means that each of the anions will remove an equal number of cations to provide electrical balance; only some of the weak acids, such as acetoacetic acid, β-hydroxybutyric acid, and to a small extent citric acid, which are present in blood almost entirely ionized, may be partially excreted in the form of the free acid provided that the urine pH is acid.

$$Na^+ + \beta\text{-hydroxybutyrate}^- \xrightarrow{H^+} \beta\text{-hydroxybutyric acid} + Na^+ \qquad (18)$$

The H^+ are probably derived from metabolic processes within the tubular cell and are excreted into the tubular urine in exchange for sodium ions that return into the tubular cell (see Fig. 11-7, reaction 3a).

The phosphate ions in plasma and in glomerular filtrate of pH 7.4 are present as HPO_4^{--} and $H_2PO_4^-$ in a ratio of approximately 4:1. With increase in acidity, this ratio decreases gradually to 1:99 at a pH of 4.8 and, at a urine pH of 4.5, essentially all phosphate is present in the form of $H_2PO_4^-$ in a ratio of 1:100. This means that each HPO_4^{--} ion can take up to 1 H^+ ion and release in exchange one cation (mainly Na^+) according to the formula:

$$2Na^+ + HPO_4^{--} + H^+ \rightleftharpoons Na^+ + H_2PO_4^- + Na^+ \qquad (19)$$

Again, the H^+ are probably derived from metabolic processes within the tubular cells and are released in exchange for a cation, mainly Na^+. Since HPO_4^{--} and $H_2PO_4^-$ are present in unmodified glomerular filtrate in a ratio of 4:1, each five molecules of phosphate are capable of accepting and excreting 4 H^+ ions. At a pH of 4.5, nearly all phosphate (approximately 99 per cent) is present as $H_2PO_4^-$; thus, when this pH is attained, essentially no further H^+ can be accepted by this system.

The mechanism by which the Na^+ from the modified glomerular filtrate are exchanged for H^+ from the tubular cells (Na^+ — H^+ exchange) is not exactly known

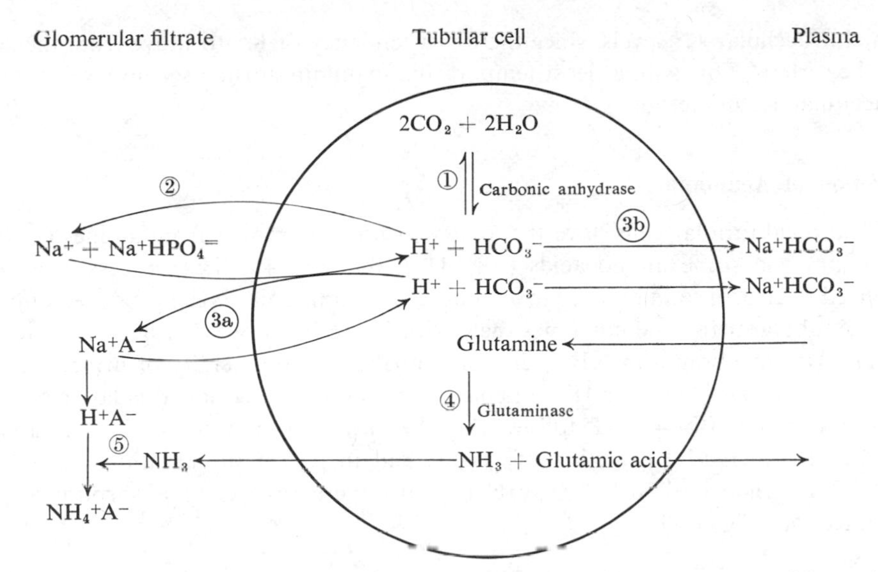

Figure 11-7. Scheme demonstrating acidification of urine by Na^+—H^+ exchange and formation of ammonia from amides. The encircled numbers refer to the reactions described in the text. (From Tietz, N. W.: The Chicago Medical School Quarterly, *22*:156, 1962.)

and the matter is strongly debated, but the following *ion exchange theory* by Pitts[2] is currently the most accepted one.

1. Metabolic processes in the tubular cells produce carbon dioxide, which reacts with water in the presence of carbonic anhydrase to form $HHCO_3$ (Fig. 11-7, reaction 1).

2. The H^+ passes into the lumen of the tubule, where it either forms $H_2PO_4^-$ from HPO_4 (Fig. 11-7, reaction 2) or one of the previously named weak acids (Fig. 11-7, reaction 3a).

3. Na^+ equal in number to H^+, which passed into the tubular lumen, move into the tubular cell where they combine with HCO_3^- formed in reaction 1, Figure 11-7. The $Na^+HCO_3^-$ then moves into the tubular blood, thus maintaining or restoring the HCO_3^- level. (It is now believed that there are a series of intermediary steps, but for the purpose of this discussion these will be omitted.)

The place where the Na^+ — H^+ exchange takes place was formerly thought to be the distal portion of the renal tubule, but recently some evidence has been presented that it may take place in a larger portion of the tubule (collecting ducts). In renal tubular acidosis, the Na^+ — H^+ exchange does not take place or is reduced; thus, the acids formed during metabolic processes are not effectively removed. This results in an increased accumulation of these acids in blood with the signs of metabolic acidosis.

Potassium ions compete in some way with the H^+ in the H^+ — Na^+ exchange. If the intracellular K^+ level (including those of renal tubular cells) is high, more K^+ and less H^+ are exchanged for Na^+; therefore, the urine becomes less acid and the acidity of body fluids increases. If there is K^+ depletion, more H^+ are exchanged for Na^+, the urine becomes more acid, and the body fluids more alkaline. Since the body mechanism against metabolic alkalosis is relatively ineffective, K^+ depletion frequently results in a metabolic alkalosis. Alkalosis caused by potassium depletion, therefore, can only be permanently corrected if the intracellular potassium levels are restored. It should be emphasized that serum K^+ levels are not always an accurate indication of the intracellular K^+ levels, since there is a tendency of K^+ to move from the cells into the serum. This will, at least temporarily, maintain normal serum levels despite intracellular K^+ depletion.

Excretion of Ammonia

The renal tubular cells have the ability to form ammonia from amides (mainly glutamine) and some amino acids (Fig. 11-7, reaction 4). The process is greatly enhanced in acidosis and reduced in alkalosis. The ammonia (NH_3) produced diffuses into the tubular urine and combines there with H^+ to form NH_4^+ (Fig. 11-7, reaction 5). The H^+ thus bound in NH_4^+ do not contribute to the acidity of urine, making further exchange of Na^+ for H^+ possible. (As mentioned earlier, this latter process stops at a urine pH of 4.5.) It follows that the formation of ammonia in the tubular cells permits increased excretion of H-ions and increased preservation of cations, mainly Na^+. Increased $Na^+ - H^+$ exchange also increases HCO_3^- absorption as just explained (see Fig. 11-7).

Reabsorption of Filtered Bicarbonate

The glomerular filtrate contains the same amount of HCO_3^- as blood plasma; however, with increasing acidification of the urine, the urine HCO_3^- content decreases and may even approach zero, and the P_{CO_2} increases. Many theories have been postulated, but the most accepted involves the following reactions.

1. With decrease in urinary pH (due to excretion of H^+ as just described), urinary HCO_3^- will be converted into $HHCO_3$, which in turn will be converted to water and carbon dioxide.

$$HCO_3^- + H^+ \rightarrow HHCO_3 \tag{19}$$

$$HHCO_3 \rightarrow H_2O + CO_2 \tag{20}$$

2. The increase in urinary P_{CO_2} causes carbon dioxide to diffuse across the tubule wall into the tubular cell. Therefore, the term "reabsorption of bicarbonate" is not quite correct since the bicarbonate is not reabsorbed in its original form but as carbon dioxide.

3. Along with the decrease of HCO_3^- in tubular urine, there is an increase in Cl^-. Both changes are accompanied by a simultaneous increase of HCO_3^- and decrease of Cl^- in blood.

The increased amount of bicarbonate reabsorbed helps restore the $HCO_3^-/HHCO_3$ ratio, which is low in acidosis. In alkalosis, the reabsorption of bicarbonate is decreased (excretion in urine increased), which helps lower the $HCO_3^-/HHCO_3$ ratio. Thus, it becomes apparent that the renal mechanism works with the other compensatory mechanisms either to retain or restore the normal bicarbonate:carbonic acid ratio and, thus, normal body pH.

CONDITIONS ASSOCIATED WITH ABNORMALITIES IN THE ANION-CATION PATTERN

Many pathological conditions are accompanied or caused by disturbances of the electrolyte composition of the body. These changes are usually reflected in the anion-cation pattern of extracellular fluid. For this reason the electrolyte composition of blood plasma or serum is often determined in clinical medicine and deviations from the normal are noted (see Figs. 11-8 to 11-10). It is important to realize that such results may not always be a true indication of the electrolyte composition of the whole body. Abnormalities in the anion-cation pattern are usually classified in one of the following four groups.

Metabolic Acidosis (Primary Bicarbonate Deficit)

This condition is usually caused by one or more of the following processes:

1. Production of organic acids, which exceeds the rate of elimination (e.g., production of acetoacetic acid and β-hydroxybutyric acid in diabetic acidosis). Acidosis may be accompanied, or augmented, or both, by the loss of cations that are excreted with the anions as explained earlier.
2. Reduced excretion of acids (e.g., renal failure, tubular acidosis).
3. Excessive loss of bicarbonate (base). This usually occurs if there is an excessive loss of duodenal fluid, as in diarrhea.

When any of these conditions exists, compensatory mechanisms act to restore the normal pH. If this restoration is complete (or nearly complete so that the pH remains between 7.35 and 7.45), the condition is called *compensated metabolic acidosis*. If, in spite of the compensatory mechanisms, the pH stays below 7.35, the condition is called *uncompensated metabolic acidosis* as explained by the example in the section on respiration. The ratio of HCO_3^- to $HHCO_3$ is decreased because of the primary decrease in the bicarbonate.

COMPENSATORY MECHANISMS IN METABOLIC ACIDOSIS

Buffer systems. The buffer systems of the blood, mainly the bicarbonate: carbonic acid buffer, tend to minimize changes in pH (see page 680). The bicarbonate decreases and the ratio of bicarbonate/carbonic acid will be less than 20/1. The respiratory mechanism and the renal mechanisms attempt to correct this ratio by increased excretion of $HHCO_3$ (as carbon dioxide) and by restoring the HCO_3^-.

Respiratory mechanism. The decrease in pH caused by any of the conditions just listed stimulates the respiratory mechanism and produces hyperventilation (Kussmaul respiration). The $HHCO_3$ diminishes and the ratio of $HCO_3^-/HHCO_3$ is regained or approaches its normal value of 20/1 (e.g., $HCO_3^-/HHCO_3$ before compensation = 15/1.2 or 12.5/1 = pH 7.2; after compensation = 15/0.95 or 16/1 = pH 7.3.)

Renal mechanism. The kidneys attempt to restore the original electrolyte composition and the pH by increased excretion of acid and preservation of base (increased rate of $Na^+ - H^+$ exchange, increased ammonia formation, and increased reabsorption of bicarbonate).

LABORATORY FINDINGS IN METABOLIC ACIDOSIS

The plasma bicarbonate and carbonic acid (P_{CO_2}) levels and also the total carbon dioxide content are decreased. The pH, in uncompensated cases, is decreased, the degree depending on the ratio of $HCO_3^-/HHCO_3$. The remaining components of the electrolyte pattern vary depending on the pathologic condition. In diabetic acidosis, for example, the fraction of organic acids is increased by the production of ketone bodies. The Na^+ and K^+ are decreased because of the associated polyuria and their excretion as salts of acetoacetic acid and β-hydroxybutyric acid. (Serum levels of K^+ may be normal or even high despite a severe K^+ depletion. The actual level is determined by the amount of K^+ lost through the urine, the amount of K^+ shifted from cells into extracellular fluid, and by the state of dehydration.) In renal failure,

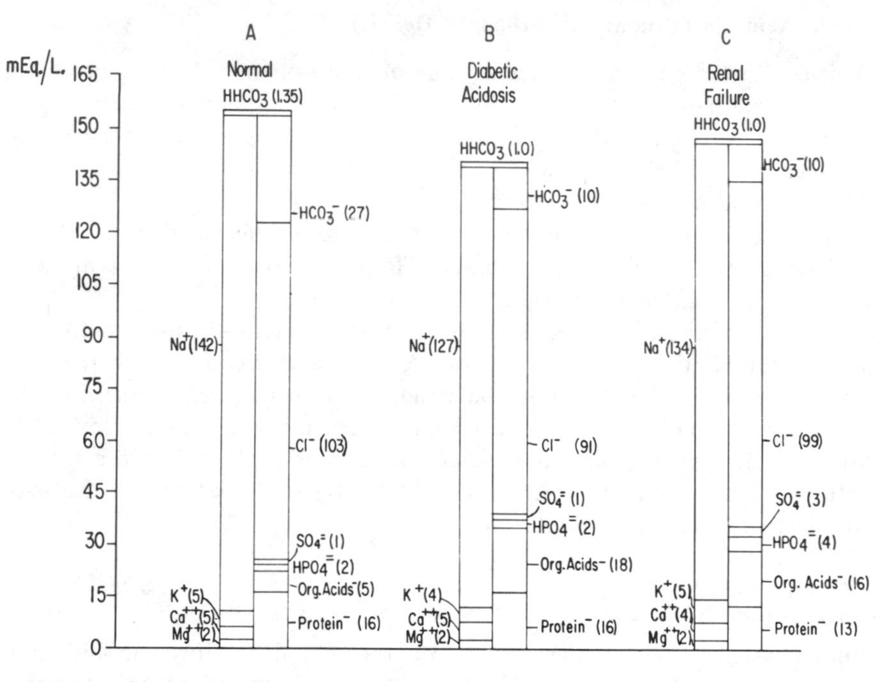

Figure 11-8. *A*, Gamblegram illustrating normal electrolyte composition of plasma. *B*, Example of anion-cation pattern as may be found in diabetic acidosis. Na^+ and Cl^- are decreased because of polyuria and excretion of Na^+ as salt of ketone acids. The organic acid fraction is increased because of excessive formation of ketone bodies. The ratio of $HCO_3^-/HHCO_3$ is 10/1; thus, the plasma pH must be 7.10. *C*, Example of anion-cation pattern as may be seen in renal failure. Organic acids as well as phosphates and sulfates are retained owing to the decreased renal functional efficiency. The ratio of $HCO_3^-/HHCO_3$ is 10/1, which results in a pH of 7.10. (From Tietz, N. W.: The Chicago Medical School Quarterly, *22*:156, 1962.)

organic acids as well as phosphate and sulfate ions are increased because of retention (Fig. 11-8).

Urinary acidity and urinary ammonia are increased provided that the renal mechanism is functioning.

Metabolic Alkalosis (Primary Bicarbonate Excess)

This condition is most frequently caused by one of the following processes, which in turn cause a bicarbonate excess:

1. Administration of excess alkali, especially $NaHCO_3$.
2. Excessive loss of hydrochloric acid from the stomach as seen after prolonged vomiting, for example, in pyloric or high intestinal obstruction.
3. Potassium depletion as seen, for example, in Cushing's syndrome, after administration of ACTH or adrenocortical hormones, and in aldosteronism.
4. Roentgen ray, radium, or ultraviolet treatment.

If one of these conditions exists, the compensatory mechanisms of the body act to restore the normal plasma pH. If compensation is complete, we have a state of compensated metabolic alkalosis with a pH value within the normal range. With progression of the disturbance, the compensatory mechanisms are not effective enough and the pH will increase. In such a case, the ratio of $HCO_3^-/HHCO_3$ is more than 20/1 because of primary increase in bicarbonate, e.g., 48/1.5 = 32/1, resulting in a pH of 7.6.

If the increase in pH is great enough, tetany may develop even in the presence of a normal serum calcium content. The cause for this is thought to be a decrease in the ionization of calcium caused by the pH increase.

COMPENSATORY MECHANISMS IN METABOLIC ALKALOSIS

Buffer systems. As a result of loss of acid (e.g., HCl), excess base will react with the carbonic acid of the $HCO_3^-/HHCO_3$ buffer system to form an increased amount of HCO_3^-, thereby minimizing pH changes.

Respiratory mechanism. The increase in pH depresses the respiratory center, causing a retention of carbon dioxide, which in turn causes an increase in $HHCO_3$. Thus, the ratio of $HCO_3^-/HHCO_3$, which was originally increased (see earlier section), approaches its normal value although the actual levels of both HCO_3^- and $HHCO_3$ are increased (e.g., the ratio of $HCO_3^-/HHCO_3$ before compensation was 48/1.5 or 32/1 = pH 7.6; the ratio after partial compensation is 48/1.8 or 25.5/1 = pH 7.51 and the total CO_2 is 48 + 1.8 = 49.8 m*M*./L.

Renal mechanism. The kidneys respond to the state of alkalosis by decreased Na^+–H^+ exchange, decreased formation of ammonia, and decreased reabsorption of bicarbonate.

LABORATORY FINDINGS IN METABOLIC ALKALOSIS

Blood plasma levels of HCO_3^-, $HHCO_3$, and P_{CO_2} and, therefore, the total plasma CO_2 content are increased with an increase in the ratio of $HCO_3^-/HHCO_3$ (if uncompensated). The pH is increased, and the remaining ions of the electrolyte pattern vary depending on the condition. In cases of prolonged vomiting, Cl^- and possibly K^+ levels tend to be low because of the loss of these ions through the vomitus. Protein values may be increased owing to dehydration and, if food intake is inadequate, formation of ketone bodies may increase the organic acid fraction. In case of excessive administration of $NaHCO_3$, Na^+ levels are increased. In K^+ depletion, decreased levels of Cl^- are very common, but serum K^+ levels are not necessarily low (Fig. 11-9).

Urinary pH values are increased because of the decreased excretion of acid and increased excretion of bicarbonate. Urinary ammonia values are decreased because of decreased formation of ammonia in the tubules.

Respiratory Acidosis (Primary $HHCO_3$ Excess)

Any condition that results in a decreased elimination of carbon dioxide through the lungs results in a primary $HHCO_3$ excess (respiratory acidosis). Inefficiency in carbon dioxide elimination may be mechanical, as in bronchopneumonia, pulmonary emphysema, and pulmonary fibrosis, or it may be caused by decreased circulation as

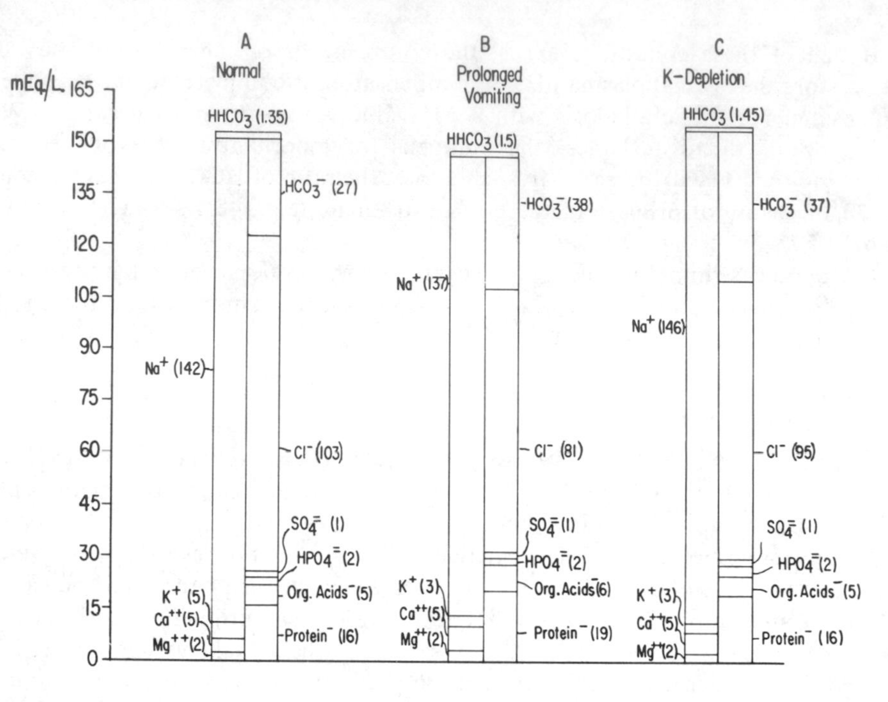

Figure 11-9. *A*, Electrolyte composition of normal plasma. *B*, Example of plasma electrolyte composition after prolonged vomiting, showing the decrease in K^+, Cl^- and the increase in protein$^-$, organic acids$^-$, HCO_3^-, and $HHCO_3$. The ratio of the last two is $38/1.5 = 25.3$. The pH is therefore 7.50. *C*, Example of typical plasma electrolyte composition in a patient with intracellular K^+ depletion. There is a decrease in K^+ and Cl^- and an increase in HCO_3^- and $HHCO_3$, resulting in a ratio of $37/1.45 = 25.4$ (pH = 7.50). (From Tietz, N. W.: The Chicago Medical School Quarterly, *22*:156, 1962.)

in cardiac disease. Rebreathing, or breathing of air high in CO_2 content, may also cause high P_{CO_2}. Increase in P_{CO_2} results in an increase of $HHCO_3$, which in turn causes a decrease in the $HCO_3^-/HHCO_3$ ratio (e.g., the ratio may be $27/1.8 = 15$, resulting in a pH of 7.28). In respiratory conditions, the pH rarely goes below 7.20.

COMPENSATORY MECHANISMS IN RESPIRATORY ACIDOSIS

Buffer system. Carbonic acid entering the blood in excess is to a great extent buffered by the hemoglobin and protein buffer systems. Some Cl^- will move from the plasma into the erythrocytes.

Respiratory mechanism. The increase in P_{CO_2} stimulates the respiratory center and results in increased pulmonary rate provided that the primary defect is not in a decreased activity of the respiratory center. The elimination of carbon dioxide through the lungs results in a decrease of $HHCO_3$ and the ratio of $HCO_3^-/HHCO_3$ approaches the normal value. This is accompanied by a pH change toward normal.

Renal mechanism. The kidneys respond to respiratory acidosis the same way that they do to metabolic acidosis, namely, increased Na^+–H^+ exchange, increased ammonia formation, and increased reabsorption of bicarbonate.

LABORATORY FINDINGS IN RESPIRATORY ACIDOSIS

Plasma levels of $HHCO_3$, P_{CO_2}, HCO_3^- and, therefore, the total carbon dioxide content are elevated. The ratio of $HCO_3^-/HHCO_3$ is decreased (due to an increase in $HHCO_3$), resulting in a low pH. The plasma chloride may be normal or slightly decreased, and urinary acidity and ammonia content are increased (see Fig. 11-10*B*).

Respiratory Alkalosis (Primary $HHCO_3$ Deficit)

A primary deficit in $HHCO_3$ occurs in all conditions that cause an increased rate, or depth of respiration, or both (e.g., fever, high external temperatures, hysteria, anoxic anoxia, and salicylate poisoning). The excessive elimination of carbon dioxide

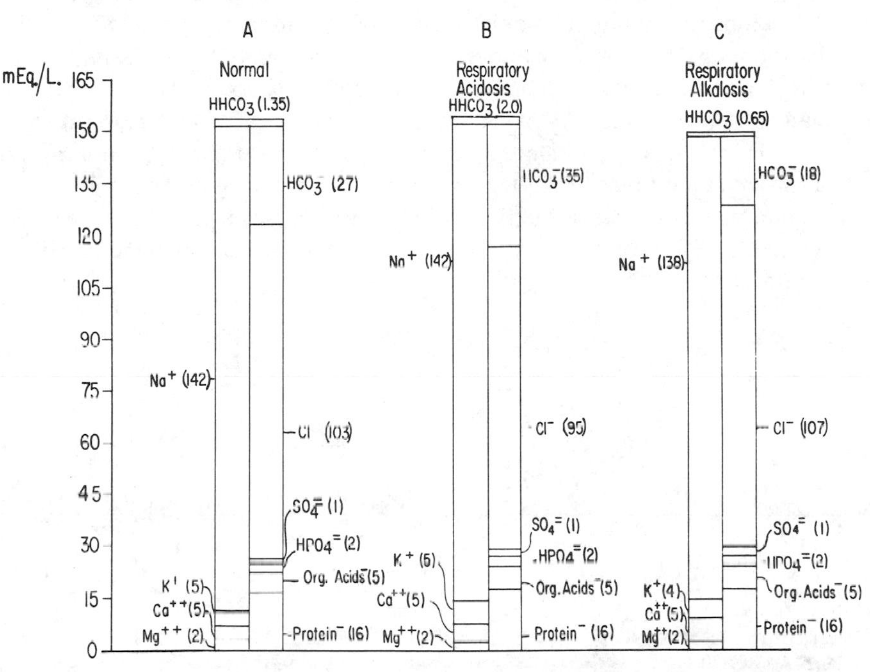

Figure 11-10. *A*, Electrolyte composition of normal plasma. *B*, Example of possible electrolyte pattern in a patient with respiratory acidosis. Note the increase in HCO_3^- and $HHCO_3$. The increase in the latter fraction is more pronounced than that of HCO_3^-. The ratio therefore is decreased, and the chloride fraction shows a decrease. *C*, Theoretical electrolyte pattern of patient in respiratory alkalosis. There is a decrease in the HCO_3^- and especially $HHCO_3$ fraction. Therefore, the ratio of $HCO_3^-/HHCO_3$ and the pH are increased. The Na^+ is at the lower limit of normal. (From Tietz, N. W.: The Chicago Medical School Quarterly, *22*:156, 1962.)

causes a decrease in $HHCO_3$ and, therefore, an increase in the $HCO_3^-/HHCO_3$ ratio associated with an increase in pH (see Fig. 11-10*C*).

COMPENSATORY MECHANISMS IN RESPIRATORY ALKALOSIS

The compensatory mechanisms of the body to respiratory alkalosis are functions mainly of the kidneys and these mechanisms correspond to those outlined in the discussion of metabolic alkalosis.

LABORATORY FINDINGS IN RESPIRATORY ALKALOSIS

In this condition, the $HHCO_3$, Pco_2, HCO_3^-, and, thus, the total CO_2 content are decreased. The ratio of $HCO_3^-/HHCO_3$ is increased, causing an increase in pH, which, however, rarely exceeds 7.60. In prolonged severe alkalosis there may be an increase of ketone bodies due to decreased carbohydrate utilization, and phosphate levels may be significantly decreased.

Combinations of Metabolic and Respiratory Conditions

The conditions just discussed are either respiratory or metabolic; however, clinicians are often confronted with cases in which a metabolic condition is superimposed on a respiratory condition or vice versa. For example, a patient in diabetic acidosis develops bronchopneumonia, resulting in impaired removal of carbon dioxide by the lungs. Laboratory findings in such a case may be confusing unless the exact electrolyte disturbance is recognized. The total CO_2, for example, may be much higher than one would normally expect at a given pH (e.g., before respiratory complication: pH = 7.1, CO_2 content = 8 m*M*./L.; with respiratory complication: pH 7.0, CO_2 content 12 m*M*./L.). The increase in carbon dioxide in such a case may indicate a bad prognosis, since the patient who already has a metabolic acidosis has now developed a respiratory acidosis. Figure 11-11 may be of help in recognizing such unusual cation-anion imbalances.

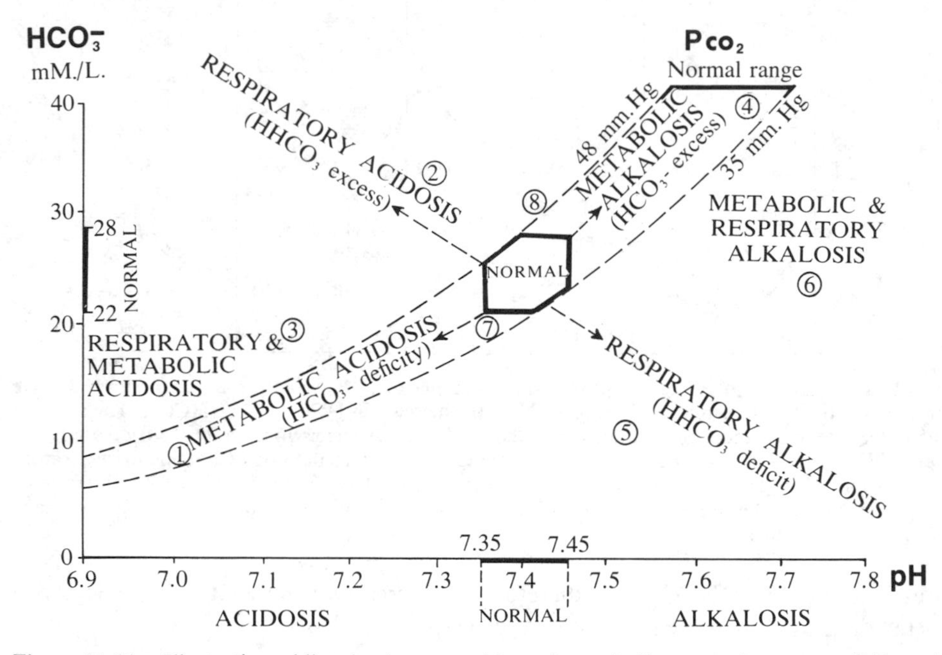

Figure 11-11. Illustration aiding in the recognition of metabolic or respiratory conditions. The hexagonal area in the center represents the range where all three, P_{CO_2}, HCO_3^-, and pH are normal. To use the diagram, draw a vertical line through the pH and a horizontal line through the HCO_3^- value (or total CO_2 value minus 1.0). The intersection of the two lines will then indicate the possible condition. Examples: No. 1 (pH 7.0; HCO_3^- 8 mEq./L.) represents an uncompensated metabolic acidosis. No. 2 (pH 7.28; HCO_3^- 34 mEq./L.; P_{CO_2} high) indicates a respiratory acidosis. No. 3 (pH 7.13; HCO_3^- 20 mEq./L.; P_{CO_2} above normal) is an example of a respiratory acidosis superimposed on a metabolic acidosis. No. 4 (pH 7.66; HCO_3^- 38 mEq./L.; P_{CO_2} normal) represents an uncompensated metabolic alkalosis. No. 5 (pH 7.52; HCO_2^- 10 mEq./L.; P_{CO_2} below normal) indicates a respiratory alkalosis. No. 6 (pH 7.72; HCO_3^- 24 mEq./L.; P_{CO_2} below normal) is an example of a respiratory and metabolic alkalosis. No. 7 is an example of compensated metabolic acidosis (pH normal, but HCO_3^- is still low). No. 8 is a typical example of a compensated metabolic alkalosis, the compensatory retention of CO_2 ($HHCO_3$) is superimposed on the metabolic condition resulting in a normal pH.

APPARENT IMBALANCE IN THE CATION-ANION PATTERN

When evaluating the acid-base balance, it is customary to determine the sodium, potassium, chloride, and CO_2 content, as well as the pH of plasma or serum. There are, however, some conditions in which all five of these determinations will not satisfactorily explain an existing electrolyte disturbance. In such cases the concentration of one of the *unmeasured ions* (Ca^{++}, Mg^{++}, HPO_4^{--}, SO_4^{--}, organic acids$^-$, proteins$^-$) may be changed. Most frequently occurring conditions of this kind are:

1. Impairment of kidney function, which results in a retention of acids, mainly amino acids, phosphates, and sulfates. Such a condition may be suggested by an increase in serum urea nitrogen. The electrolyte pattern is as shown in Figure 11-8.
2. An increased production and accumulation of ketone bodies as observed in diabetes mellitus and sometimes in starvation. These cases are accompanied by an elevated level of ketone bodies (organic acid fraction) in serum and urine. An example of the electrolyte pattern found in diabetic acidosis is shown in Figure 11-8.
3. Production of some unmeasured acids as a result of poisoning. Methyl alcohol poisoning, for example, brings about a high level of formic acid in the plasma.
4. Accumulation of some acids may result in severe acidosis (e.g., lactic acid acidosis).
5. Major changes in the plasma proteins, which are the third largest anion fraction (16 mEq./L.).

REFERENCES

1. Gamble, J.: Chemical Anatomy, Physiology, and Pathology of Extracellular Fluid. Cambridge, Harvard University Press, 1950.
2. Pitts, R. F.: Am. J. Med., *9*:356, 1950.
3. Snively, W. D., and Wessner, M.: J. Indiana State Med. Assn., *47*:957, 1954.
4. Trenchard, D., Noble, M. I. M., and Guz, A.: Clin. Sci., *32*:189, 1967.
5. Van Slyke, D. D., Wu, H., and McClean, F. C.: J. Biol. Chem., *56*:765, 1923.
6. Weisberg, H. F.: Surg. Clin. N. Amer., *39*:93, 1959.

ADDITIONAL READINGS

Cantarow, A., and Trumper, M.: Clinical Biochemistry. 6th ed. Philadelphia, W. B. Saunders Co., 1962.

Davenport, H. W.: The ABC of Acid-Base Chemistry. Chicago, The University of Chicago Press, 1958.

Weisberg, H. F.: Water, Electrolyte and Acid-Base Balance. Baltimore, The Williams & Wilkins Co., 1962.

Weyer, Ed. M. (Ed.): Ann. N.Y. Acad. Sci., *133*(Art. 1):1, 1966.

Chapter 12

RENAL FUNCTION TESTS

by Willard R. Faulkner, Ph.D. and John W. King, M.D., Ph.D.

The human kidneys are paired, bean-shaped structures situated in the posterior part of the abdominal cavity. Imbedded in the perirenal fat and other supporting structures, the kidneys are firmly fixed to the posterior wall of the abdomen at an area between the upper border of the twelfth thoracic vertebra and the lower border of the third lumbar vertebra. Because the liver occupies the right upper quadrant of the abdominal cavity, the right kidney, which lies behind and slightly below it, is located somewhat lower than the left kidney. The right kidney is in immediate contact with the capsule of the liver and the descending loop of the duodeneum. The left kidney lies behind and slightly below the spleen. In the adult human, each kidney weighs about 150 gm. and measures about 5 × 12 cm. The kidneys are slightly larger in the male than in the female. Their combined weight in proportion to the body weight is usually given as about 1:240. The mass of endocrine tissue on the superior pole of each kidney is the adrenal or suprarenal gland, which is not part of the renal system (see Fig. 12-1*A*, *B*).

On the medial aspect of each kidney is a region called the hilus. If one thinks of the kidney as bean-shaped, the hilus of the kidney corresponds in location to the scar on the bean that marks the area of attachment to the pod. Blood vessels and nerves pass into and out of the kidney at this point. The upper end of the ureter forms a pocket in the hilar area into which the urine that is formed almost continuously by the kidney is collected. Aided by the peristaltic action of the ureter the urine eventually reaches the bladder, which serves as a reservoir.

Malformations of the kidney occur. Sometimes only one kidney is present and functional, or the two kidneys are fused, usually at the superior pole, to form a horseshoe kidney. Double ureters or extra blood vessels are fairly common. Although many of these anomalies are compatible with life and health, this is not always the case, and many kidney problems, even those first manifesting themselves in later life, occur directly or indirectly as the result of congenital anatomic anomalies. One very serious developmental problem is the so-called polycystic kidney. In this condition, those parts of the kidney arising from the Wolffian duct and those parts arising from the metanephrogenic blastema fail to fuse normally. As a consequence, there is no

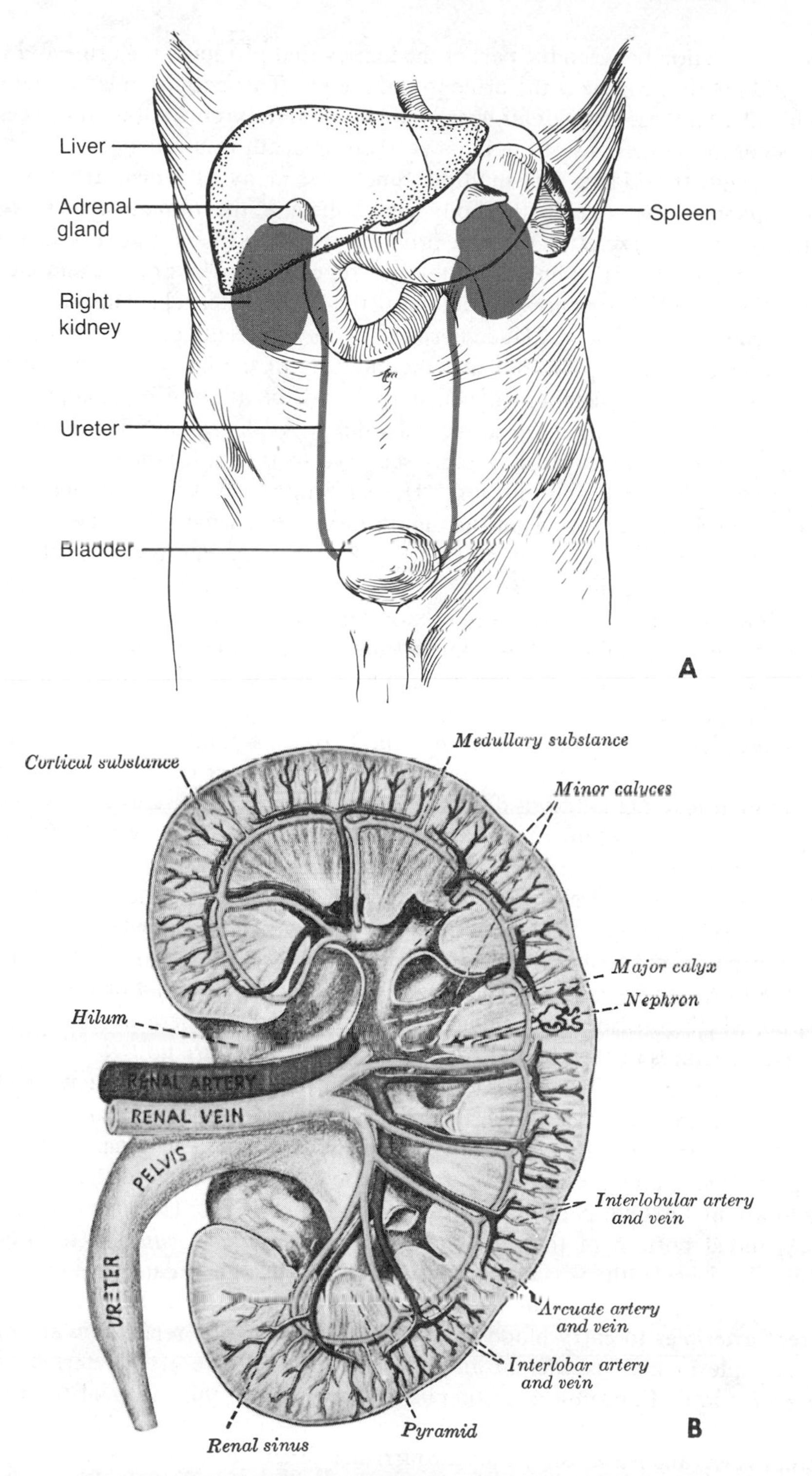

Figure 12-1. *A*, Schematic drawing of right and left kidneys with respect to liver, stomach, duodenal loop, spleen, adrenal gland, and bladder. *B*, Diagram of a vertical section through the kidney. Nephron and blood vessels greatly enlarged. (From Gray's Anatomy, 28th Edition, Edited by C. M. Goss, Lea & Febiger, 1966.)

proper connection between the part of the kidney that produces the urine and the part of the kidney that conducts the urine to the ureter. This condition is not compatible with health and these individuals usually die in renal failure. In other instances, when there is complete agenesis of the kidneys, there is death in utero.

Although the kidney has multiple functions, many of which are not yet well known, its principal role in the body metabolism is the *formation of urine*. This entails not only the excretion of waste products from the blood, but also the provision for the preservation of essential solutes and regulation of hydration and electrolyte balance. The vital importance of maintaining the body's internal environment has been recognized since the time of Claude Bernard. This nineteenth century French scientist was among the first to point out that the integrity of the body was dependent upon selective excretion of metabolites, which could not be allowed to accumulate within the body without causing harm to the individual. At the same time this excretion of metabolic products must be sufficiently selective so that substances that are utilized or required by the body are not lost. Homer Smith said, "The composition of the blood is determined not by what the mouth ingests but what the kidney keeps."

To accomplish its complex mission, the kidney must act first as a selective filter to remove water and filtrable solutes from the blood plasma. This process occurs in the glomeruli and the fluid formed is known as the glomerular filtrate. If the kidneys were to stop at this point and pass this filtrate to the bladder to be excreted, the individual could not replace the lost fluid and solutes and, therefore, life would be impossible. It is estimated that the volume of the glomerular filtrate is around 125 ml. per minute and may exceed 150 L. in a single day. Among the components of the glomerular filtrate are many substances that are necessary for normal function. To prevent their loss, the kidney is designed to resorb water and useful solutes selectively. By this process the glomerular filtrate is converted to urine, which contains end products of metabolism that might be injurious if accumulated; it also contains any excess of essential solutes that might be present. Of course, sufficient water to keep these solutes in solution is also lost, but this volume is so small that the individual can usually replace this easily.

The filtration apparatus of the kidney is built around a functional unit called the *nephron*, which consists of the *glomerulus* and the *tubule*. There are about 1.2 million of these structures in each kidney. All of these structures are not working at any one time, but their very presence gives the kidney considerable reserve capacity in the event of stress, disease, or injury. Each nephron is supplied by a small blood vessel called the *afferent arteriole*. It carries blood from a branch of the renal artery into the nephron. These afferent arterioles are estimated to carry blood to the nephrons at a rate of about 1200 ml. per minute (total renal blood flow). The arteriole enters into an expanded portion of the renal tubule called *Bowman's capsule* (see Fig. 12-2). Within the capsule, the vessel breaks up into a plexus of capillaries, which ultimately recombine to form an *efferent arteriole*. This efferent arteriole then joins with other efferent arterioles to carry blood from the nephrons to the renal tubular area. The capillary plexus and its afferent and efferent arterioles are often referred to as the *glomerular tuft*. Filtration is accomplished through the thin walls of the capillaries that make up the plexus. The blood flows into the plexus from the relatively large afferent vessel and leaves the plexus through an efferent vessel that has a smaller lumen. This difference in size of the two vessels produces an increase in hydrostatic pressure within the capillaries that has been reported to be about 75 mm. Hg. This pressure is almost twice as high as that found in capillaries in other parts of the body. In part

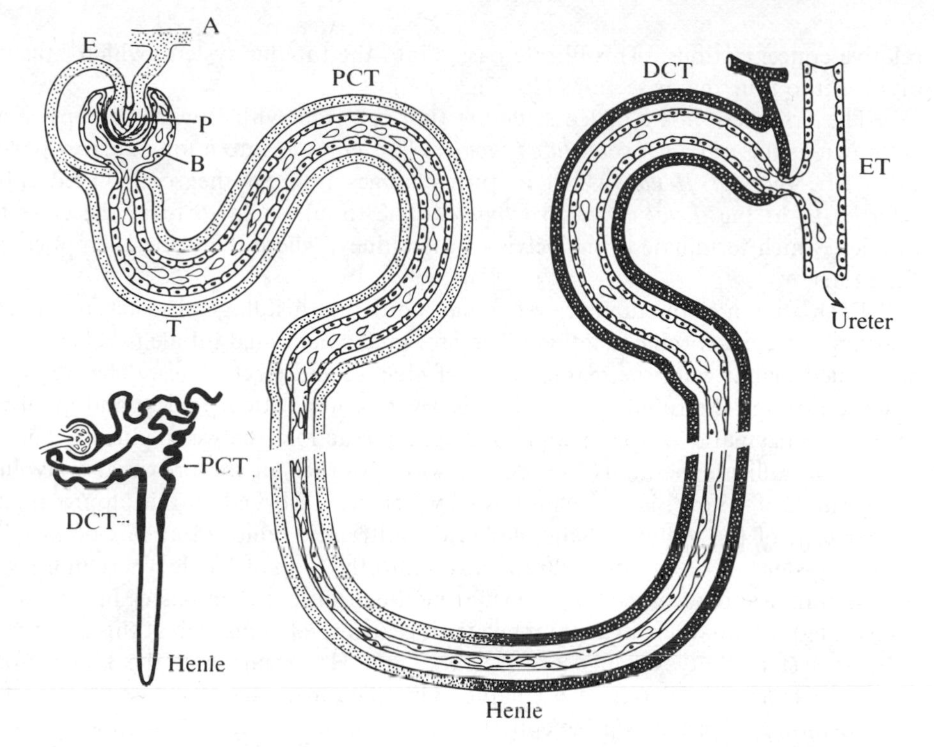

Figure 12-2. Schematic drawing of the glomerulus and the tubular system. *A*, Afferent arteriole; *E*, efferent arteriole; *P*, plexus of capillaries (glomerular tuft); *B*, Bowman's capsule; *T*, tubular blood supply; PCT proximal convoluted tubule; Henle, loop of Henle, DCT, distal convoluted tubule; ET, excretory tubule or duct. The blood capillaries shown along the tubular system (T) gradually change to venous capillaries as they pass down the tubular system.

because of the relatively high hydrostatic pressure, the filtrate is forced through the thin capillary epithelium and is caught in Bowman's capsule. As mentioned before, Bowman's capsule envelopes the glomerular tuft and is connected with the tubular system of the kidney where concentration and modification of the filtrate takes place. This is mainly an active process of selective secretion and reabsorption by the tubular epithelium of the kidney. By this process the kidney conserves solutes and metabolites that the body requires and, at the same time, disposes of molecules such as creatinine, urea, and acidic metabolites that are not needed by the body and that must be eliminated.

Some of the substances that are present in the glomerular filtrate are known as *threshold* substances and appear in the urine only after they have reached certain minimum concentrations in the blood. Glucose is such a substance and does not appear in the urine until plasma glucose levels reach about 180 mg./100 ml. Other substances, such as creatinine, may be excreted without appreciable reabsorption. A foreign substance that behaves similarly is the polysaccharide inulin. Therefore, both creatinine and inulin are useful in measuring the glomerular filtration rate (see the next section).

About 20 per cent of the volume of the plasma that passes through the glomerular tuft passes into Bowman's capsule as glomerular filtrate. This filtrate has a specific gravity of 1.010 $\pm$ 0.002, a pH of approximately 7.4, and is isoosmotic with plasma. It contains all the filtrable molecules present in plasma and at approximately the same relative concentrations. This filtrate passes into the tubular system, which is usually divided into four major sections (see Fig. 12-2).

The first section is a coiled structure that connects with Bowman's capsule and is known as the *proximal convoluted tubule*. This empties into a long narrow portion called the *loop of Henle*, which in turn changes into another convoluted tubule referred to as the *distal convoluted tubule*. The tubules join to form the collecting tubules, which terminate in the pelvis of the kidney, where the urine is emptied into the ureter.

Each section of the tubular system has a different histological structure and each appears to have a specific function. The proximal convoluted tubule is believed to be concerned mainly with the reabsorption of glucose and electrolytes. The passage of glucose through the tubular epithelium is a very complicated process and involves a series of enzymatic coupling and uncoupling reactions between glucose and the phosphate radical. Sodium chloride is also reabsorbed in the proximal convoluted tubule and, of course, large quantities of water are absorbed so that approximately 85 per cent of the volume of the glomerular filtrate is removed at this point of the tubular system. The modified filtrate passes into the loop of Henle where more water and sodium are reabsorbed. The further modified filtrate then passes into the distal convoluted tubule where again more water is reabsorbed. It is also at this point in the tubular system that exchange of Na^+, K^+, and H^+ occurs and the kidney forms ammonia from amides (see also Chapter 11). The urine has now assumed its final composition and passes, via the collecting tubules, to the pelvis of the kidney, to the ureter, and finally into the bladder. This remarkable process is subject to many types of disturbances and it is sometimes possible to diagnose the type of lesion present from a knowledge of renal physiology and the results of some of the kidney function tests. Problems in proximal convoluted tubular activity may be measured by the phenolsulfonphthalein (PSP) excretion or the para-aminohippurate excretion tests, both of which will be discussed later in this chapter. Distal tubular activity will affect the various concentration and dilution tests and when damage is severe enough it may be reflected in alteration in serum electrolyte levels.

Normal values for urine show a much wider range than do normal values for constituents of plasma and other body fluids. The composition of urine is more markedly affected than corresponding values for plasma by factors such as fluid intake, diet, and environmental temperature. Thus, a patient may have a large decrease in daily urine volume merely because of an increase in the room temperature or because of a personal dislike for the fluids being served. A patient on a voluntary restriction of fluids, associated with an Addis or Fishberg concentration test, may reduce his urinary output to a third of his normal output and at the same time raise his urinary specific gravity from his usual value to 1.025, 1.030, or higher. Similarly, the pH of the urine may be decreased by the administration of acidifying drugs or raised by the ingestion of foods that have an alkaline residue. When therapeutic changes in blood pH are desired, ammonium chloride may be administered to the patient to produce an acid urine and excess doses of sodium bicarbonate to produce an alkaline urine. Even after the urine has entered the bladder its pH may be altered because of bacterial action in the bladder. Further changes may occur in the specimen bottle. For this reason, interpretation of pH values must be made with caution and changes in urinary pH cannot be ascribed to metabolic activity unless the urine is fresh and contains few or no bacteria.

Normal kidney function depends upon an adequate *blood flow* to the organ. Any situation that interferes with renal blood supply will result in diminished kidney function and will alter the amount and composition of the urine that is produced and result in an accumulation of metabolic products in the blood. Patients in cardiac failure frequently have diminished kidney function that is not due to any intrinsic lesion in the kidney, but rather to an inadequate flow of blood to the kidney.

Intravascular changes that interfere with the flow of blood to the kidney may also produce increases in blood pressure. This condition is known as renal hypertension. Reduction of kidney function may also develop as a consequence of damage to the epithelium of the capillaries, which make up the glomerular tuft, as in glomerulonephritis. In this condition the endothelium is damaged to the extent that not only are water and other small molecules able to pass through the glomerular membrane, but also red blood cells and some plasma proteins that are not normally filtered. The relatively small molecules such as urea are subsequently (passively) reabsorbed by the tubules, but red cells and larger molecules (protein) cannot return to the blood and, therefore, are excreted with the urine. In chronic glomerulonephritis and other causes of the nephrotic syndrome, large amounts of protein are lost in urine.

Tubular as well as glomerular tuft damage may be the result of overloading the system with a toxic substance. One of the most common situations in which this phenomenon takes place is following massive hemolytic reactions such as occur after transfusion with incompatible blood. In this situation, massive amounts of hemoglobin are released into the plasma and exert a toxic effect on the tubules. In addition, there may be an agglutination of red cells, resulting in anoxia and, consequently, damage to the nephron. Both factors result in an impairment of kidney function and possibly complete renal failure.

Administration of various poisons to the body may selectively interfere with certain functions of the tubule. Actual mechanical trauma to the kidney or the development of neoplastic growth, which either destroys tissue directly or produces pressure on the kidney tissue, may also disrupt its normal functioning.

Sometimes solutes present in the urine are such that precipitation of these substances occurs either within the pelvis of the kidney or in the ureter or bladder. These precipitates are referred to as *calculi* and have an appearance similar to natural stones (see Chapter 17). They may produce injury by causing obstruction and subsequent back pressure, which reduces filtration rate and decreases urinary excretion. Pain may be produced as the ureter contracts against the stone and this pain is described as one of the most severe pains imaginable.

Occasionally patients are seen who, because of an inherited deficiency in certain enzyme systems, are unable to carry out one or more normal metabolic processes. Such disorders are called *inborn errors of metabolism.* They are of interest not only to the clinician in his care of the patient, but also to the biochemist because they can give considerable insight into fundamental body chemistry. One example of such a disorder is cystinuria. In this condition, the kidney tubule is unable to reabsorb cystine and several other amino acids. As a consequence, these amino acids occur in excess in the urine and their plasma levels are low.

Laboratory Tests Aiding in the Evaluation of Kidney Function

In an organ as complex as the kidney, it is obvious that there are many processes that may be impaired; therefore, for the physician it is quite helpful to know the location of the defect. The diagnosis of kidney disease, to a great extent, is made in the clinical chemistry laboratory. This is true because clinical signs and symptoms may be minimal or absent entirely and, even when present, will not always reflect the severity of the disease or the prognosis for the patient. It is indeed fortunate that the laboratory has a battery of kidney function tests available, which, when properly applied, can give valuable information about the status of kidney functions and frequently about the location of the defect. It must be remembered, however, that the kidney has a considerable functional reserve and kidney function tests may be normal even in the presence of relatively severe renal pathology.

Kidney function tests, in general, can be affected by *prerenal, renal, or postrenal* causes. Among the prerenal causes is dehydration, which may be found in pyloric and intestinal obstruction and in prolonged diarrhea. Conditions of shock and excessive loss of blood, as in severe intestinal bleeding or cardiac failure, are other prerenal causes for decreased kidney function. As discussed later in connection with renal blood flow, decreased kidney function in these conditions may occur either because of decreased plasma volume or decreased blood flow.

Among the renal causes for decreased kidney function are diseases affecting the glomerular filtration rate, the tubular function, or any changes in the renal vascular system that decrease the blood flow.

A postrenal cause for decrease in kidney function is obstruction of the urine flow, for example, caused by enlargement of the prostate, stones in the urinary tract, or tumors of the bladder. In these cases the decreased function is due to reduction in effective filtration pressure of the glomeruli.

The comments on the causes for decreased renal function as well as the initial discussion on the physiology of the kidney make it most practical to separate kidney function tests into three major groups: (1) those measuring glomerular filtration, (2) those measuring renal blood flow, and (3) those measuring tubular function. Tests measuring the retention of nonprotein nitrogenous compounds in plasma are listed in (1) although their plasma values may be affected by changes in any one of the three.

GENERAL CHARACTERISTICS OF URINE

Urine Volume

The volume of urine produced by a normal adult over a 24 hour period is usually between 800 and 1800 ml. These values are subject to considerable variation because many factors determine the exact amount of urine any given individual will produce on any given day. In essence, the urinary output is the difference between total fluid intake and loss of fluid by other means (e.g., lung, perspiration, stool). Thus, under some conditions the urinary output of normal individuals may be far outside of the range just given. For example, voluntary restriction of fluid intake, such as for an Addis or Fishberg concentration test, can reduce the daily urine volume to less than 500 ml. A moderate beer drinker can increase his daily output of urine well above the normal upper limit without much difficulty because not only is his urinary output affected by the volume of beer drunk, but also beer is believed to contain a diuretic factor, which further increases urine excretion. The prophylactic use of drugs for motion sickness may have the undesirable side effect of polyuria (increased urinary output). Nervousness, associated with stage fright, may be accompanied by polyuria that cannot be completely accounted for by increased intake of water.

Decrease of urinary output is known as *oliguria* and, in the absence of fluid deprivation, is usually an ominous sign. It may occur during the early development of edema or ascites or during shock. When occurring because of terminal uremia, it may be a prelude to anuria and death.

Pathologic states associated with *polyuria* are diabetes mellitus, diabetes insipidus, and some types of central nervous system injuries. In some kidney diseases that ultimately result in oliguria, the initial effect will be polyuria as the kidney, unable to concentrate properly, seeks to compensate by increasing the urine volume.

Care must be taken to distinguish between polyuria and frequency. The person with a bladder infection may have frequency without significant increase in the total volume of urine produced; however, both conditions may be present at the same time, as in diabetes.

Appearance of Urine

Urine is ordinarily clear when voided, but may become turbid upon standing. Precipitation of solutes occurs as the urine cools and as the pH changes as a result of bacterial action. This may have little or no clinical significance; however, cloudiness in a *freshly* voided urine may have clinical significance. The most common reasons for this may be presence of blood, pus, or bacteria, or all three, in the urine. Blood rarely occurs to the extent that the urine is actually red except following surgery or trauma. Usually hematuria (red cells in urine) is of the microscopic type and is said to produce a smoky urine, which some imaginative observers have described as looking like smoky Scotch whisky. Some foods may color the urine, although, for example, the amount of beets necessary to produce red urine is greater than most people will eat. Some of the B-vitamins, notably riboflavin, will produce a yellow urine. Many drugs, including several urinary tract antiseptics, change the color of the urine. The Diagnex Blue test, a technique for detecting the presence of hydrochloric acid in the stomach, may color the urine blue for several days. Other things being equal, concentrated urine of relatively high specific gravity will be of darker color than more dilute urine.

Odor of Urine

The odor of normal urine has been described in many ways, aromatic and nutty being two adjectives used; however, none have really described it properly. Let us simply state that fresh urine has a characteristic and not unpleasant odor. Abnormal odors can be produced by some foods (asparagus and garlic) and by various pathologic states. The odor of the urine of an uncontrolled diabetic is described as fruity because of the presence of acetone and acetoacetic acid. *Proteus* infections produce an ammoniacal urine and less characteristic odors are produced by infections due to other bacteria.

Although modern laboratory workers, for obvious reasons, do not make a practice of tasting the specimens sent to the laboratory, many significant observations in the past were made by urine tasters. The one that is perhaps the most well known is the distinction between diabetes mellitus and diabetes insipidus. In the first condition, the urine tastes sweet; in the latter, it is virtually tasteless.

Urine pH

The urine pH is approximately 6.0 (fasting 5.5 to 6.5, random 4.8 to 7.8). Night urine is generally more acid than day urine and this (besides being more concentrated) is one of the reasons that specimens collected for examination for casts and other formed elements are collected in the early morning. Casts, in particular, dissolve more quickly in alkaline than in acid urine specimens. Metabolism of fats produces more acid residues than the metabolism of carbohydrates. Starvation, with consequent utilization of stored body fat, will also produce ketosis, and thus acid urine. In the treatment of urinary infection, acid producing drugs are sometimes used to reduce the pH of the urine. One of the more commonly used drugs of this type, which is also used as a mild diuretic, is ammonium chloride. Bacterial infections can alter the pH in either direction, depending upon the end products of bacterial metabolism. Ammonia producing organisms obviously produce alkaline urine, but most other bacteria will cause the urine to become acid.

Formed Elements

Formed elements in urine are defined as objects in urine that may be observed by direct microscopic examination of the urine specimen. This term usually refers to white and red blood cells, casts, crystals, bacteria, ova, and parasites. A discussion of the clinical significance and the identification of these elements is beyond the scope of this text and the reader is referred to special texts.

The problem of identifying crystals does not seem nearly as critical to the laboratory workers of today as it was to their counterparts of a century ago. It is still important to recognize a sulfa crystal, but we see sulfa crystals much less often than we did a few years ago. We have much better ways of identifying amino-acidurias than by the microscope and if we do this procedure today, it is usually an exercise in hindsight. Many crystals are merely a reflection of the patient's diet; it has been said that the presence of oxalic acid crystals in a patient's urine merely means that he has had rhubarb for breakfast. This is an oversimplification, but it is safe to say that many urine specimens contain crystals and, in general, their presence and their identification have little clinical usefulness.

"Routine Urinalysis" Tests

Routine urinalysis procedures, as carried out in many hospitals, consist of determinations of urinary pH and specific gravity, tests for the detection of reducing sugars, protein, and ketone bodies, and, finally, a microscopic examination of the urinary sediment to detect the possible presence of red and white blood cells, casts, or any other formed elements. Some of these procedures are not tests of kidney function since they detect abnormalities that are reflections of pathology elsewhere in the body. An example of this is diabetes mellitus. In this disease, profound chemical abnormalities, such as presence of glucose and ketone bodies in the urine, may be noted as the kidney strives to correct abnormal physiological activities elsewhere in the body and to keep the internal environment within reasonable limits. The detailed laboratory procedures for urinary sugars, albumin, and ketone bodies are discussed in Chapters 4, 5, and 10, respectively.

Tests Measuring Glomerular Filtration Rate

CLEARANCE TESTS

The group of tests generally referred to as renal clearance tests are an extremely useful, effective, and sensitive way of measuring the actual excretory capacity of the kidney since they measure the amount of a substance excreted in the urine as compared to the concentration of the same substance in the plasma. Clearance tests, therefore, are considerably more sensitive and clinically more useful than the tests measuring retention of nonprotein nitrogenous compounds in serum, which may not be abnormal until there is a significant decrease in kidney function (below 50 per cent of normal). The amount of substance, under investigation, cleared by the kidney is generally expressed as that volume of plasma which contains the quantity of the substance excreted in the urine. This may be mathematically expressed as follows:

$$\text{ml. plasma cleared per minute} = \frac{U}{P} \times V$$

where U = the concentration of the substance in urine

P = the concentration of the substance in plasma

V = the urine volume per minute, expressed in ml.

The concentration of the substance in urine and plasma may be expressed in any convenient unit, but it is customary to use mg./100 ml. U and P must be expressed in the same unit. All other factors being equal, the clearance rate is roughly proportional to the size of the kidney and the body surface area of the individual. Therefore, the calculation for the clearance of any given substance should provide for correction for deviations from the average adult body surface. This is done by adding the factor $1.73/A$, where 1.73 is the generally accepted average body surface in square meters and A is the body surface of the patient under investigation. The formula for calculating the renal clearance, therefore, expands as follows:

$$\text{ml. cleared per minute} = \frac{U}{P} \times V \times \frac{1.73}{A}$$

The body surface area may be determined more conveniently from one of the available nomograms (see the Appendix) or it may be calculated from the weight and height of the patient by means of the following formula:

$$\log A = (0.425 \log W) + (0.725 \log H) - 2.144$$

where A = the body surface area in square meters

W = the weight of the patient in kg.

H = the height of the patient in cm.

To convert inches to centimeters multiply by 2.54.
To convert American lb. to kg. multiply by 0.45.

Example: Let us assume a creatinine clearance test has been performed on a child and a clearance (uncorrected) of 12 ml./min. was obtained. The patient has a weight of 4 kg. and a height of 35 cm.

Therefore:

$$\text{Log } A = (0.425 \times \log 4) + (0.725 \times \log 35) - 2.144$$
$$\text{Log } A = (0.425 \times 0.602) + (0.725 \times 1.544) - 2.144$$
$$\text{Log } A = 0.2559 + 1.1194 - 2.144$$
$$\text{Log } A = 1.3753 - 2.144$$
$$\text{Log } A = (-0.7687) \quad \text{or} \quad \begin{array}{r} 1.0000 \\ -0.7687 \\ \hline 0.2313 \end{array} - 1$$
$$A = \text{antilog } (0.2313) - 1 = 0.1703 \text{ sq. m.}$$

The clearance of 12 ml./min. corrected for the body surface of 0.1703 sq. m therefore, is:

$$12 \times \frac{1.73}{0.170} = 123 \text{ ml./min.}$$

Correction for body surface is absolutely mandatory if the body surface of the patient differs greatly from that of the average person. The error otherwise introduced is substantial and in infants may be up to several hundred per cent as shown by the previous example.

The selection of the type of clearance test to be used is made by taking into consideration the information desired by the physician, the mode of excretion of the substance to be tested, the safety and convenience to the patient, and the ease of detection and quantitation of the substance in the laboratory. Theoretically, a substance may be excreted by glomerular filtration alone, by filtration plus tubular excretion (depending on the substance, tubular excretion will more or less dominate), or be filtered first but then subsequently reabsorbed by the tubules. If only a single facet of renal physiology, the glomerular filtration rate, is to be studied, a substance

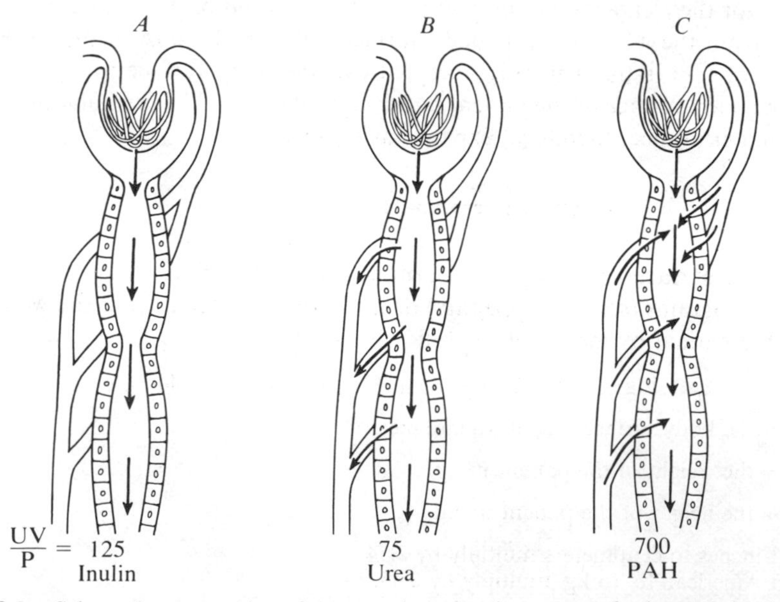

Figure 12-3. Schematic presentation of the excretion of various types of substances by the nephron. *A*, Inulin is excreted by glomerular filtration and passes through the tubular system without reabsorption. No inulin is removed by tubular secretion. Creatinine behaves very similarly to inulin. *B*, Urea is filtered through the glomerulus, but is subsequently partially reabsorbed by the tubular system. *C*, PAH (p-aminohippurate) is, to a limited extent, filtered by the glomerulus, but is mainly excreted by the tubules. Phenolsulfonphthalein behaves similarly. (After Cantarow and Trumper.)

should be selected that is excreted (filtered) either completely or predominantly by the glomeruli without being either excreted or reabsorbed by the tubules. Inulin (and mannitol) is such a substance; however, it is rarely employed because of the unavailability of procedures for the quantitation of inulin in most routine laboratories. Creatinine is a substance that behaves similarly to inulin. Since the determination for creatinine in plasma and urine is readily available in most routine laboratories, the creatinine clearance test has become one of the most popular tests for measuring the glomerular filtration rate. For the detailed discussion of the inulin and creatinine tests, the reader is referred to a later portion of this chapter.

Historically, the urea clearance test was the first clearance procedure commonly used. Urea, although cleared by the glomeruli, is subsequently partially reabsorbed by the tubules at an average rate of 40 per cent. Furthermore, the rate of reabsorption, which is a process of passive diffusion, varies with the amount of water reabsorbed. Thus, the urea clearance test is not a measure of the glomerular filtration rate, but is more a measure of overall renal function. For measurement of glomerular filtration, it has been widely replaced by the creatinine clearance test. Some of the tests measuring renal blood flow are also based on the clearance concept, but they are discussed in connection with tests measuring tubular blood flow.

The Inulin Clearance Test

A number of substances are eliminated by the kidney, wholly or predominantly by glomerular filtration. Examples are inulin, mannitol, thiosulfate and creatinine. The polysaccharide inulin, having a molecular weight of about 5100, obtained from dahlias and artichokes has become the substance of choice for precise investigative work because it is filtered freely by the glomerulus but is neither secreted nor absorbed by the tubules.

Although the use of inulin for measuring the glomerular filtration rate cannot be regarded as a routine laboratory test (a fact that should be emphasized at the outset), a brief description of the procedure is given. Since inulin is not normally present in the plasma, it must be introduced at a suitable concentration in order to allow its clearance by the kidneys to be measured. This may be done by giving a sufficient quantity as a priming dose (25 ml. of a 10 per cent solution of inulin) by intravenous injection to produce a satisfactory plasma level and then maintaining this level throughout the test period by a slow, continuous infusion of a less concentrated solution (500 ml. of a 1.5 per cent solution).

An adequate fluid intake (1000 ml.) is maintained during the hour before the test. It is not necessary for the patient to fast. A blood specimen is taken into a tube containing an anticoagulant and is used as the control. The patient empties his bladder and saves the sample of urine, which is also to be used as a control. The priming dose of inulin is then introduced slowly over a few minutes. This is followed by the maintenance solution, given at about 4 ml./min. After about one-half hour, urine is collected in three accurately timed specimens of about 20 minutes each. Blood specimens are withdrawn at the beginning and at the end of each 20 minute period. The mean of the two is used in the calculation. Inulin is measured in all blood and urine specimens.

Inulin clearance is calculated for the three timed specimens, using the appropriate blood sample in each case as follows:

$$\text{inulin clearance} = \frac{\text{mg. inulin/100 ml. urine}}{\text{mg. inulin/100 ml. plasma}} \times \text{ml. urine excreted/min.}$$

An average of the three clearance values is made; for a normal adult this is about 125 ml./min. or about 70 ml./sq. m. of body surface area. Goldring and Chasis reported a range of 110 to 152 ml./min. for men and 102 to 132 ml./min. for women.

CLINICAL SIGNIFICANCE

Inulin is almost completely cleared by glomerular filtration at an average rate of 125 ml./min. The amount of inulin, once filtered, is not reabsorbed by the tubules and, thus, is quantitatively excreted in the urine. The inulin clearance is the most accurate measure of glomerular filtration at the present time. Its clinical application is limited, however, by the fact that the test is cumbersome, expensive, time-consuming, and uncomfortable to the patient. The test requires an intravenous infusion and constant attendance by the physician during the test period. For these reasons, the use of the inulin clearance test is restricted to research institutions or institutions specializing in the study of kidney diseases. Routine laboratories generally perform the creatinine clearance, which is described later in this chapter.

REFERENCES

Dick, A., and Davies, C. E.: J. Clin. Path., *2*:67, 1949.

Goldring, W., and Chasis, H.: Hypertension and Hypertensive Disease. New York, The Commonwealth Fund, 1949.

Smith, H. W.: Kidney Structure and Function in Health and Disease. New York, Oxford University Press, 1951.

Moller E., McIntosh, J. F., and Van Slyke, D. D.: J. Clin. Invest., *6*:427, 1928.

Gary, A., and Discombe, G.: Clinical Pathology. Philadelphia, F. A. Davis Co., 1966.

Urea Clearance

CLINICAL SIGNIFICANCE

The urea clearance test has considerable historical significance; it was the first of the clearance tests to be used widely. The test is a reasonably reliable measure of the renal functional status if the rate of urinary excretion is 2 ml. or more per minute.

Urea is filtered by the glomerulus and subsequently partially reabsorbed by the tubules; therefore, results of this test are not a true measure of the glomerular filtration rate but of the overall renal function. Consequently, considerable popularity has been gained by other clearance tests that measure specific aspects of renal function.

Urea clearance values between 75 and 125 per cent of normal clearance are considered to be normal, and values below 70 per cent of normal clearance are suggestive of decreased renal function. Values below 50 per cent are usually accompanied by elevations of the nonprotein nitrogeneous (NPN) compounds in blood (serum) and clearance values below 10 per cent indicate severe renal impairment.

SPECIMEN

Two carefully timed 1 hour urine specimens and a whole blood sample are required for this test of renal function. Some laboratories prefer to collect one 2 hour specimen of urine, which is completely permissible and which may in some instances confer greater accuracy on the measurement.

Any of the common anticoagulants except ammonium oxalate* are suitable. Although serum or plasma yield more valid results, whole blood is more often used because of tradition.

---

* Ammonium oxalate may be used if the urea is measured by a method employing diacetyl monoxime as color reagent.

PRINCIPLE

The clearance of endogenously occurring urea is most often measured. Clearances for most substances are stated in terms of the milliliters of plasma cleared per minute. This terminology can also be applied to urea but because of certain factors peculiar to the excretion of urea, its clearance is usually expressed as "per cent of average normal clearance." This is a result of the fact that urea clearances are affected by the rate of urine flow. When the excretion is 2 ml. or more per minute, the average normal clearance is 75 (64 to 99) ml./min. and is known as the maximum clearance (C_m). When the urine flow is less than 2 ml./min., however, a larger portion of the urea in the glomerular filtrate passively diffuses into the tubules, thus lowering the clearance, which is then proportional to the square root of the excretory rate. The average normal is then 54 (41 to 68) ml./min., and is termed the standard clearance (C_s). The expression of urea clearances as per cent of normal is thus an attempt to make the two clearances (C_m and C_s) comparable in the same individual under the same physiologic or pathologic circumstances in spite of possible differences in the rate of urine flow and their partial dependence on this factor.

PROCEDURE

Preparation of the patient
(to be carried out by the nursing staff of the hospital ward)

1. Start the test in the morning after the patient has eaten a breakfast excluding coffee. The test may also be performed when the patient is in the fasting state. Do not allow the patient to exercise vigorously before or during the test.

2. Instruct the patient to void, emptying the bladder completely; discard the specimen. Record the exact time.

3. Have the patient drink 2 glasses of water and then void at the end of an hour; save the complete specimen in a clean container. It is not necessary for the time period to be exactly 1 hour as long as the exact length is known and recorded.

4. Have the patient drink another glass of water and at the end of the second hour void completely; save the entire specimen in a separate appropriately labeled container. Note and record the exact time interval.

5. Near the end of the first hour or at the beginning of the second, collect a blood specimen, using a suitable anticoagulant.

Laboratory procedure

1. Measure and record the volume of each of the two urine specimens.

2. Determine the urea nitrogen concentrations in the blood and in the two urine specimens according to the procedure for urea nitrogen.

CALCULATIONS

1. Calculate the number of milliliters of urine excreted per minute by dividing the urine volumes by the number of minutes over which they were collected.

2. Calculate urea clearance by substituting appropriate data in the following formula:

(a) When the urine volume is between 1.0 and 2.0 ml./min.

$$C_s = \frac{U\sqrt{V}}{B} \times 1.85 \times \frac{1.73}{A}$$

where C_s = standard clearance in per cent of normal
1.85 = (100/54) converts the clearance (in ml./min.) to per cent of normal clearance
54 = the average standard clearance in ml./min.
B = blood concentration of urea nitrogen in mg./100 ml.

U, V, and the factor $1.73/A$ have been defined in the preceding general discussion.

(b) When the urine volume is 2.0 ml. or more per minute.

$$C_m = \frac{U}{B} \times V \times 1.33 \times \frac{1.73}{A}$$

where C_m = maximum clearance in per cent of normal

1.33 = (100/75) converts the clearance (in ml./min.) to per cent of normal

75 = the average maximum clearance in ml./min.

NORMAL VALUES

	Mean	*Range*	*Range per cent of Normal*
Standard clearance	54 ml./min.	41 to 68 ml./min.	75 to 125
Maximum clearance	75 ml./min.	64 to 99 ml./min.	75 to 125

SOURCES OF ERROR

1. The most common source of error in this test is the inaccurate timing of the urine collection periods. Also, incomplete emptying of the bladder will lead to falsely low results.

2. Values are most accurate, provided all other factors are equal, when the urine flow is 2.0 ml. or more per minute; accuracy tends to decrease with decreasing urine volumes. Some laboratories regard samples representing urine flows of less than 1 ml./min. as unfit for analysis.

REFERENCES

Smith, H. W.: Kidney Structure and Function in Health and Disease. New York, Oxford University Press, 1951.

Austin, J. H., Stillman, E., and Van Slyke, D. D.: J. Biol. Chem., *46*:91, 1921.

Moller, E., McIntosh, J. F., and Van Slyke, D. D.: J. Clin. Invest., *6*:427, 1928.

Creatinine Clearance Test (Endogenous)

CLINICAL SIGNIFICANCE

The creatinine clearance test is a relatively accurate and useful measure of the glomerular filtration rate and it is rapidly replacing the less accurate urea clearance test. The reasons for the greater degree of accuracy of creatinine clearances are that creatinine is not reabsorbed by the tubules and that fluid intake and excretion affect the clearance of creatinine much less than that of urea. Also, blood creatinine values are relatively stable; hence, the blood specimen can be collected at any time during the urine collection.

When plasma creatinine levels increase considerably above the normal, creatinine is also secreted by the tubules and thus the creatinine clearance value may be greater than the actual glomerular filtration rate.

SPECIMEN

A precisely timed urine specimen and a serum sample are required for this test. The substance is subject to a slow equilibrium reaction with creatine, which is accelerated by hydrogen and hydroxyl ions. The best way to retard this process, as well as bacterial decomposition, is to store the specimen in a refrigerator until analyzed; this should be no longer than 1 working day after the collection.

PRINCIPLE

Creatinine occurring through metabolic production is eliminated from the plasma predominantly by glomerular filtration and, therefore, measurement of its rate of clearance affords a measure of this process.

The general principle of clearance as described for urea is also applicable in this situation but with two modifications. (1) The clearance of creatinine is much less subject to the rate of urine flow than urea and for this reason the distinction between standard and maximum clearance does not apply. (2) Creatinine clearance is stated only in terms of volume of plasma cleared per minute and not in per cent of normal. The test described here is an endogenous clearance test in which no creatinine is administered. In the exogenous clearance test, creatinine is administered either orally or intravenously. It is less reliable than the endogenous test because at high concentrations of creatinine some is eliminated by tubular secretion in addition to glomerular filtration.

REAGENTS

These are the same as those used for the determination of plasma and urinary creatinine.

PROCEDURE

Preparation of the patient

1. Hydrate the patient by administering a minimum of 600 ml. of water. Withhold tea, coffee, and drugs on the day of the test.
2. Have the patient void and discard the specimen.
3. Collect a 4, 12, or 24 hour specimen, and record the exact times of starting and completing the collection.
4. Collect a specimen of clotted blood during the urine collection period.

Laboratory procedure

1. Determine the creatinine in an aliquot of the well mixed urine specimen and in the plasma or serum according to the procedure to be described later.

CALCULATION

$$\frac{UV}{P} \times \frac{1.73}{A} = \text{ml. of plasma cleared/min.}$$

where U = concentration of creatinine in urine
V = volume of urine in ml./min.
A = body surface area in square meters.

NORMAL VALUES

If the creatinine method described in this chapter is used, the normal creatinine clearance values for males are 105 ± 20 ml./min. and for females 95 ± 20 ml./min. If a more specific method for the determination of creatinine is used (e.g., one employing Lloyd's reagent), the respective values for the clearance of creatinine by males and females are 117 ± 20 ml./min. and 108 ± 20 ml./min., respectively.

SOURCES OF ERROR

1. Faulty timing or improper collection of the urine specimen is the most common source of error.

2. Vigorous exercise during the test may cause alteration in the clearance rate.
3. Proper hydration of the patient insuring a urine flow of 2 ml./min. or more tends to eliminate retention of urine in the bladder as a source of negative error.

REFERENCES

Tobias, G. J., McLaughlin, R. F., and Hooper, J.: New England J. Med., *266*:317, 1962.
Edwards, K. D. G., and Whyte, H. M.: Aust. Ann. Med., *8*:218, 1959.

NONPROTEIN NITROGEN IN SERUM (BLOOD)

The nonprotein nitrogen (NPN) fraction of serum (and blood) is composed of all nitrogenous compounds other than protein. Its major component is urea nitrogen, which constitutes approximately 45 per cent of the total. The other compounds included in the NPN fraction, listed in order of their quantitative distribution, are amino acids, uric acid, creatinine, creatine, and ammonia. Other nitrogenous compounds that are generally not identified are grouped together as "undetermined nitrogen."

The total NPN may be determined as a group or the individual components may be measured in accordance with the procedures to be described.

The Determination of Nonprotein Nitrogen (NPN)

CLINICAL SIGNIFICANCE

Although the test for nonprotein nitrogen in whole blood and serum has for many years served as a test for kidney function, it has been replaced in recent years by the more accurate and more convenient test for urea nitrogen. The route of elimination of various nonprotein nitrogenous compounds of blood differ considerably (some are excreted by glomerular filtration only, some mainly by tubular excretion, and others are first excreted in the glomerular filtrate and are then partially absorbed by the tubules). Thus, the nonprotein nitrogen value is the result of many interacting factors. Increases in the NPN fraction are mainly a reflection of an increase in urea nitrogen, which makes up approximately 45 per cent of the total NPN.

The nonprotein nitrogen determinations do not offer any information in addition to that provided by the urea nitrogen determination. Although an exception to this may be the simultaneous determination of NPN and urea nitrogen in patients with hepatic failure in the presence of renal disease. Under such conditions the ratio of nonprotein nitrogen to urea nitrogen may be substantially higher than that normally found. This change in the ratio is due to the decreased ability of the liver to synthesize urea and to deaminate amino acids.

Serum NPN levels above 35 mg./100 ml. and blood levels above 50 mg./100 ml. suggest renal insufficiency. Serum NPN levels may increase to more than 400 mg./100 ml., but these values are mostly seen in the terminal stages of renal failure.

SPECIMENS

Nonprotein nitrogen may be determined in whole blood, plasma, serum, or other biological fluids. Any anticoagulant that does not introduce nitrogen into the sample may be used. No preservative is required since the NPN is quite stable as long as gross bacterial contamination does not occur. Since a major portion of the NPN is distributed uniformly throughout the body water, hemolysis causes little difficulty. The nonprotein nitrogen compounds in filtrates of biological fluids vary, depending upon the protein precipitating agents; however, the filtrate resulting from the widely used tungstic acid protein precipitation, employed in this procedure, contains such nitrogen compounds as amino acids, ammonia, creatine, creatinine, urea, uric acid, and minor amounts of other nitrogenous compounds. Normally about 45 per cent of the NPN is urea, but as the total rises, the proportion of the urea does also.

Since normal values differ, depending upon the specimen (blood or serum), it is important to indicate in any report, the type analyzed.

PRINCIPLE

The nitrogen in whatever form in a protein-free filtrate of the specimen is converted to the ammonium ion (NH_4^+) by digestion with hot concentrated sulfuric acid in the presence of a catalyst such as copper sulfate. Catalysts such as mercury and selenium are used in other modifications. Also, a few procedures dispense with catalysts and introduce hydrogen peroxide to complete the digestion. This step in the process is known as a Kjeldahl digestion. The ammonium formed is converted to ammonium hydroxide by the alkali of Nessler's solution; this in turn reacts with the double iodide, also contained in this reagent, to produce the colloidal dimercuric ammonium iodide, which appears yellow when nitrogen is low to medium in concentration and orange-brown in high concentration. For most procedures Beer's law is valid to concentrations of about 75 mg./100 ml. At very high levels (100 to 150 mg./100 ml.) the colloidal material forms large aggregates and precipitates.

The following chemical reactions are involved:

1. $\text{N-compounds} + H_2SO_4 \xrightarrow[\text{heat}]{CuSO_4} NH_4^+$
2. $NH_4^+ + H_2SO_4 \longrightarrow (NH_4)HSO_4$
3. $(NH_4)HSO_4 + 2NaOH \longrightarrow Na_2SO_4 + 2NH_3 + 2H_2O$

The final reaction of ammonia with the double iodide is not definitely known, but it has been postulated to be:

$$2HgI_2 \cdot 2KI + 2NH_3 \rightarrow NH_2Hg_2I_3 + 4KI + NH_4I$$

Other formulas that have been proposed for the Nessler complex, based on x-ray diffraction studies, are:

$$Hg_2NI \cdot H_2O \text{ and } HgOHNHHgI$$

REAGENTS

1. Ammonia-free water. Add 1 ml. of concentrated sulfuric acid, A. R., to 1 L. of deionized water and distill, collecting the distillate in a thoroughly cleaned flask. In order to minimize contamination with ammonia, introduce the outlet tube into the collecting flask through a hole in a parafilm or aluminum foil cover. Also carry out the distillation in a room not containing open urine specimens or other sources of ammonia vapors. Fresh deionized water may sometimes be suitable; however, it should be checked for a low reading by running a blank in the procedure.

2. Sulfuric acid, 2/3 *N* (0.66 *N*). Add 35 gm. (18.8 ml.) of concentrated sulfuric acid with mixing to about 500 ml. of ammonia-free water in a 1 L. volumetric flask. Dilute to the mark and mix thoroughly. Titrate with standard alkali and adjust the volume if necessary.

3. Sodium tungstate solution, 10 per cent. Dissolve 100 gm. of sodium tungstate ($Na_2WO_4 \cdot 2H_2O$) in about 500 ml. of ammonia-free water in a 1 L. volumetric flask, mix, and dilute to volume. Allow to stand for several days. Decant the clear supernatant solution from any precipitate that may have formed.

4. Digestion mixture. Mix 300 ml. of 85 per cent phosphoric acid (H_3PO_4) and 50 ml. of 5 per cent copper sulfate ($CuSO_4 \cdot 5H_2O$), A. R., in a 1 L. Erlenmeyer flask. Add 100 ml. of concentrated sulfuric acid and mix. Stopper the flask and allow to stand for a few days, permitting any calcium sulfate formed to settle. After measuring the volume of the clear supernatant fluid, pour it slowly with mixing into an equal volume of ammonia-free water and mix thoroughly.

5. Nessler's solution.[8] Dissolve 45.5 gm. of mercuric iodide, HgI_2, A. R., and 34.9 gm. of potassium iodide, A. R., in about 100 ml. of distilled water. In another container, dissolve 112 gm. of KOH in about 500 ml. of distilled water, mix, and cool to room temperature. Mix the two solutions and make up to 1 L. After allowing to stand for several days, decant the clear supernatant liquid. Store in an alkali-resistant bottle.

6. Standard ammonium sulfate (0.0500 mg. N/ml.). Dry several grams of ammonium sulfate $(NH_4)_2SO_4$, A. R., for several hours in an oven at 100°C. Place 0.236 gm. of the salt in a 1 L. volumetric flask and dissolve in water. Add a few drops of concentrated sulfuric acid and dilute to the mark.

7. Boiling beads.

PROCEDURE

Preparation of protein-free filtrate (1:10 dilution)

1. To 1.00 ml. of blood in a 50 ml. Erlenmeyer flask, add 7.00 ml. of water and mix. Add 1.00 ml. of 10 per cent sodium tungstate solution and mix. Add slowly with mixing 1.00 ml. of 0.66 *N* sulfuric acid: (If serum, plasma, or cerebrospinal fluid is the specimen, add half volumes of precipitating agents and 8 ml. of water instead of 7 ml.) Stopper the flask, shake vigorously, and allow to stand 10 minutes. If the protein has been properly precipitated, no foaming will occur and the color will change from red to dark brown. If precipitation is incomplete, the specimen may sometimes be salvaged by adding 1 to 2 drops of 10 per cent sulfuric acid with shaking until foaming ceases and the dark brown color is seen. If this does not suffice, it is advisable to repeat the deproteinization on another sample.

2. Filter the mixture through a Whatman No. 1 filter paper. This particular paper has been reported to contain negligible amounts of ammonia. The ammonia content may be tested by filtering 10 ml. of ammonia-free water and treating a portion of this filtered water as a protein-free filtrate.

Digestion of sample

1. Label oven dried NPN tubes (25 mm. × 200 mm. Pyrex tubes graduated at 35 and 50 ml.) and make the following additions:

	Blank	*Standard*	*Unknown*
Boiling beads (Pyrex or other comparable glass)	2	2	2
Standard ammonium sulfate solution	0	3.00 ml.	0
Protein-free filtrate of unknown	0	0	5.00 ml.
Digestion mixture	1.00 ml.	1.00 ml.	1.00 ml.

2. Place all tubes in a digestion rack.

3. Boil the contents slowly over a microburner until all water has been evaporated and dense white fumes of sulfur trioxide begin to form at the bottom of the tube. This should occur in 3 to 7 minutes and be accompanied by charring.

4. Reduce the size of the flame until boiling almost ceases. Cover all the tubes with a small watch glass or with a large marble and continue heating gently for about 2 minutes, timing from the moment the tube becomes filled with white fumes. The oxidation in the unknown tubes should be complete at the end of the 2 minute heating period, which will be indicated by the disappearance of all yellow or brown color. Heating should be continued until the solution becomes almost colorless.

5. Remove the flames and allow the tubes to cool.

6. Wash the watch glasses or the marbles with a few milliliters of water from a wash bottle allowing the water to run into each respective tube. Cool the tubes in cold running tap water and add sufficient cold deionized water to bring the volume within a few milliliters of the 35 ml. mark.

Nesslerization

1. Treat each tube individually as follows: Place 15 ml. of Nessler's solution in a graduated cylinder. While gently swirling the contents of the digested mixture, pour the Nessler solution with one continuous motion from the graduate to the tube. Immediately add sufficient water to bring to the 50 ml. mark. Place a clean rubber stopper in the tube or cover with Parafilm and mix the content thoroughly by inversion. Immediately treat the remaining tubes of the set in a similar manner.

2. Allow the tubes to stand 10 minutes for maximum color development to take place; then measure their respective absorbance (A) within the next 10 minutes at a wavelength of 500 nm., setting the blank to zero A.

CALCULATION

$$\text{NPN in mg./100 ml. of blood} = \frac{A \text{ unknown}}{A \text{ standard}} \times \frac{10.00}{5.00} \times 100 \times 3 \times 0.05$$

$$= \frac{A \text{ unknown}}{A \text{ standard}} \times 30$$

NORMAL VALUES

Plasma or serum	20 to 35 mg./100 ml.[3]
Whole blood	25 to 50 mg./100 ml.

SOURCES OF ERROR

1. If tubes have not been oven dried, excessive bumping and loss of fluid is often experienced.

2. Tubes must be heated slowly until water has been driven off; otherwise, liquid will boil above the 35 ml. mark and may not be digested later in the process, thereby resulting in erroneously low results.

3. Boiling after the solution has cleared is often the source of cloudiness during nesslerization.

4. The contents of the tubes must be cold (temperature of cold running tap water) before nesslerization; otherwise, turbidity is likely to develop.

5. Nesslerization must be accomplished quickly in one continuous motion to avoid erratic results.

REFERENCE

Folin, O., and Wu, H.: J. Biol. Chem., *38*:81, 1919.

UREA NITROGEN

Urea is synthesized in the liver from ammonia produced as a result of deamination of amino acids. This biosynthetic pathway is the chief means of excretion of surplus nitrogen by the body.

It is customary in most laboratories in the United States to express urea as urea nitrogen. This came about through the desire to compare the quantity of nitrogen in urea with that of other components included in the nonprotein (NPN) category. Since the NPN is seldom measured today, the expression of urea as urea nitrogen has little practical value or logic but continues to be used because of tradition.

The structure of urea is $NH_2{-}\overset{\overset{\displaystyle O}{\|}}{C}{-}NH_2$. Since its molecular weight is 60 and it contains 2 nitrogen atoms, with a combined weight of 28, a urea nitrogen value can be converted to urea by multiplying by 60/28 or 2.14.

Determination of Urea Nitrogen

CLINICAL SIGNIFICANCE

The determination of serum urea nitrogen is presently the most popular screening test for the evaluation of kidney function. The test is frequently requested along with the serum creatinine test since simultaneous determination of these two compounds appears to aid in the differential diagnosis of prerenal, renal, and postrenal hyperuremia (see the discussion on creatinine). Serum urea determinations are considerably less sensitive than urea clearance (and creatinine clearance) tests and levels may not be abnormal until the urea clearance has diminished to less than 50 per cent. As outlined in the preceding general discussion, increases in serum urea nitrogen may be due to prerenal causes (cardiac decompensation, water depletion due to decreased intake or excessive loss, increased protein catabolism). Among the renal causes are acute glomerulonephritis, in which only moderate increases are observed, chronic nephritis, polycystic kidney, nephrosclerosis, and tubular necrosis. Postrenal causes are any type of obstruction of the urinary tract (stone, enlarged prostate gland, tumors).

Urine urea nitrogen determinations are rarely done unless they are a part of the urea clearance test.

SPECIMEN

Urea nitrogen may be determined directly in plasma, serum, urine, and most other biological fluids with the method presented here. Whole blood, however, must be deproteinized to eliminate the colorimetric interference of hemoglobin. Sodium fluoride must not be used as an anticoagulant because it inhibits the action of urease employed in the assay.

Since urea may be lost through bacterial action, the specimen should be analyzed within several hours of collection or should be preserved by refrigeration. Urine is particularly susceptible to loss as a result of bacterial decomposition of urea; therefore, in addition to refrigeration, a crystal of thymol will help to reduce the loss of urea.

PRINCIPLE

Urea is hydrolyzed to ammonium carbonate by urease, and the ammonia that is released from the carbonate by alkali reacts with phenol and sodium hypochlorite in an alkaline medium to form the blue indophenol. Sodium nitroprusside serves as a catalyst. The intensity of the blue color is proportional to the quantity of urea in the specimen. Ammonia already present in urine specimens is removed by adsorption on Lloyd's reagent (sodium aluminum silicate) or on Permutit.

The three reactions can be represented as follows:

$$1.\quad H_2N{-}\overset{\overset{\displaystyle O}{\|}}{C}{-}NH_2 + 2H_2O + H^+ \xrightarrow{\text{urease}} (NH_4)_2CO_3 + H^+ \longrightarrow 2NH_4^+ + HCO_3^-$$

$$2.\quad NH_4^+ + OH^- \longrightarrow NH_3 + H_2O$$

$$3.\quad NH_3 + NaOCl + 2\,C_6H_5OH \xrightarrow[Na_2Fe(CN)_5NO]{NaOH} O{=}C_6H_4{=}N{-}C_6H_4{-}O^-$$

Indophenol
(blue in dissociated form)

REAGENTS

1. Ammonia-free water. Allow distilled water to pass through a mixed cation-anion exchange resin bed and collect it in a glass-stoppered bottle.

2. Phenol-nitroprusside solution. Place 10 gm. of "pink-white" phenol and 0.050 gm. of sodium nitroprusside, $Na_2Fe(CN)_5NO{\cdot}2H_2O$, A. R., in a 1 L. volumetric flask containing about 500 ml. of ammonia-free water. Dissolve the reagents, dilute to the mark, and mix thoroughly. Store this solution in a refrigerator at 5°C. and discard after 2 months.

3. Alkaline-hypochlorite solution. Place 5 gm. of sodium hydroxide in about 500 ml. of ammonia-free water in a 1 L. volumetric flask. Cool, add 0.42 gm. of sodium hypochlorite (commercial bleaches such as Clorox are satisfactory), dilute to the mark, and mix thoroughly. Store in an amber bottle in a refrigerator. Discard after 2 months.

4. Sodium ethylenediaminetetraacetate (EDTA), 1 per cent, pH 6.5. Dissolve 10 gm. of the disodium salt of EDTA in about 800 ml. of ammonia-free water. Adjust the pH to 6.5 with 1 *N* sodium hydroxide and dilute to 1 L. EDTA binds cations that might interfere with urease activity.

5. Urease stock solution (approximately 40 modified Sumner units/ml.). Suspend 0.2 gm. of urease in 10 ml. of water and add 10 ml. of glycerol. Type V urease containing 3500 to 4100 units/gm. (obtainable from the Sigma Chemical Co., St. Louis, Mo.) is suitable. Store in a refrigerator and discard after 4 months.

6. Urease working solution (0.4 units/ml.). Dilute 1 ml. of the urease stock to 100 ml. with the EDTA solution. Store in a refrigerator and discard after 3 weeks.

7. Urea nitrogen stock standard (500 mg. urea nitrogen/100 ml.). Dissolve 1.0717 gm. of dry urea, A. R., in about 50 ml. of ammonia-free water in a 100 ml.

volumetric flask. Add 0.1 gm. of sodium azide, dilute to the mark, and mix. Sodium azide serves as a preservative that does not inhibit urease action. Store this standard in a refrigerator and discard after 6 months. Ammonium sulfate, $(NH_4)_2SO_4$, is sometimes used as a standard; it has the advantage of being stable, but the disadvantage of not controlling the enzymatic steps in the procedure.

8. Urea nitrogen working standard (50 mg./100 ml.). Dilute 10 ml. of the stock standard to about 90 ml. with ammonia-free water in a 100 ml. volumetric flask. Add 0.1 gm. of sodium azide, dilute to mark, and mix thoroughly. Store in a refrigerator and discard after 6 months.

9. Permutit, according to Folin, 40 to 60 mesh.

PROCEDURE

For plasma or serum

1. Label three tubes, blank, standard, and unknown, respectively.
2. Pipet 1.0 ml. of urease working solution into each tube.
3. With a Kirk transfer or other micropipet add 10 μl. of unknown serum and working standard to the appropriate tubes. Mix and incubate all tubes for 15 minutes at 37°C.
4. Add rapidly and successively, mixing after each addition, 5.0 ml. of the phenol-nitroprusside solution and 5 ml. of the alkaline hypochlorite.
5. Place the tubes in a water bath at 37°C. for 20 minutes.
6. Measure the absorbances at 560 nm., using the blank as a reference.

For urine

1. Place about 0.5 gm. of Permutit in a 25 ml. volumetric cylinder and wash twice with water. Drain completely.
2. Add 1.0 ml. of urine and about 5 ml. of water. Mix by swirling for 5 minutes. Add water to the mark, mix, and allow the Permutit to settle.
3. Label 15 × 120 mm. tubes. Make the following addition and mix.

	Blank	*Standard*	*Unknown*
Urease working solution	1.0 ml.	1.0 ml.	1.0 ml.
Diluted urine from step 2	0	0	10 μl.
Urea working standard	0	10 μl.	0

4. Incubate all tubes at 37°C. for 15 minutes.
5. Add quickly, one after another, with mixing, 5 ml. of the phenol-nitroprusside reagent and 5 ml. of the alkaline hypochlorite reagent.
6. Incubate all tubes at 37°C. for 20 minutes.
7. Measure the absorbances at 560 nm. using the blank as a reference.

CALCULATIONS

For plasma or serum

$$\frac{A \text{ unknown}}{A \text{ standard}} \times 50 = \text{mg. urea nitrogen/100 ml.}$$

For urine

1. $$\frac{A \text{ unknown}}{A \text{ standard}} \times 1250 = \text{mg. urea nitrogen/100 ml.}$$

2. $$\text{Urea nitrogen/100 ml.} \times \frac{\text{24 hr. excretion (in ml.)}}{100 \times 1000} = \text{gm. urea nitrogen/24 hr.}$$

PROCEDURAL NOTES

The color produced is of such intensity that it is not practical to use the optimum wavelength of 628 nm. A wavelength of 560 nm. allows a much wider range of concentration to be measured.

NORMAL VALUES

Plasma or serum	7 to 18 mg. urea nitrogen/100 ml. (15 to 38 mg. urea/100 ml.)
Urine	12 to 20 gm. urea nitrogen/24 hr. (25 to 43 gm. urea/24 hr.)

SOURCES OF ERROR

1. Ammonia in any of the reagents or in the atmosphere of the room in which the procedure is carried out will result in falsely high values.

2. Lipemic sera cause turbidity in the final colored solution. This may be corrected by extracting the final solution with several milliliters of ether.

DISCUSSION

Methods based on principles other than those just described are also widely used. Two of the most prevalent are nesslerization after urease hydrolysis and the direct reaction of urea with diacetyl.

In the former method, a urease suspension is added to the specimen (blood, plasma, serum). After enzymatic action is complete, the specimen is deproteinized with tungstic acid. The protein-free filtrate is treated with Nessler's reagent as described for nonprotein nitrogen.[3]

The latter method is based on the reaction of urea with diacetyl to form a yellow compound. Because diacetyl is unstable, it is replaced in most methods by the more stable diacetyl monoxime. The color is intensified by pentavalent arsenic or other polyvalent ions.[2]

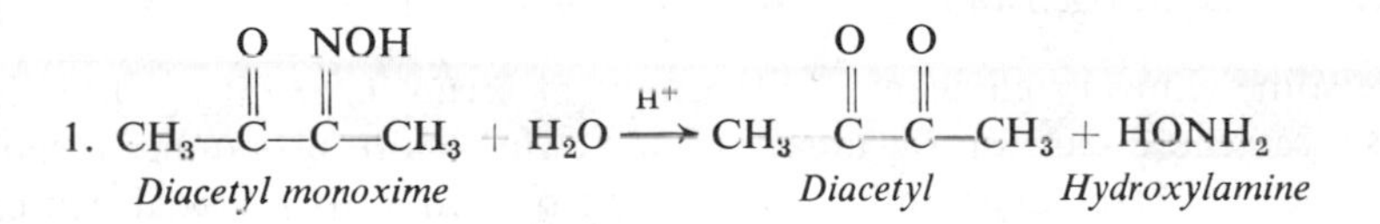

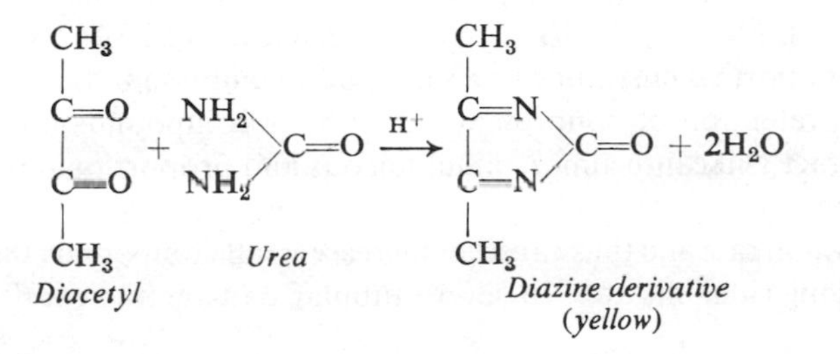

Since diacetyl reacts directly with urea and not with ammonia, the latter compound does not need to be removed from urine specimens.

In the Technicon AutoAnalyzer procedure, diacetyl reacts with urea in the presence of thiosemicarbazide, which intensifies the color of the reaction product.[6]

REFERENCES

Kaplan, A.: Urea nitrogen and urinary ammonia. *In* Standard Methods of Clinical Chemistry. S. Meites, Ed. New York, Academic Press, Inc. 1965, vol. 5, pp. 245–256.
Chaney, A. L., and Marbach, E. P.: Clin. Chem., *8*:130, 1962.

CREATINE AND CREATININE

Creatine phosphate acts as a reservoir of high energy, readily convertible to ATP in the muscles and other tissues. Creatine itself is synthesized in the liver and pancreas from three amino acids, arginine, glycine, and methionine. After synthesis, creatine diffuses into the vascular system and is thus supplied to many kinds of cells, particularly those of muscle, where it becomes phosphorylated. Creatine and creatine phosphate total about 400 mg./100 gm. of fresh muscle. Both compounds are spontaneously converted into creatinine at the rate of about 2 per cent per day. Creatinine is a waste product derived from creatine and is excreted by the kidneys.

Determination of Creatine and Creatinine

CLINICAL SIGNIFICANCE

Creatinine is removed from plasma by glomerular filtration and is then excreted in the urine without being reabsorbed by the tubules to any significant extent. This results in a relatively high clearance rate for creatinine, e.g., as compared with urea (125 vs. 70 ml./min.). In addition, when plasma levels increase above the normal, the kidney can also excrete creatinine through the tubules. Consequently, serum or blood creatinine levels in renal disease generally do not increase until renal function is substantially impaired. In the presence of normal renal blood flow, any increase in creatinine values above 2 to 4 mg./100 ml. is suggestive of moderate to severe kidney damage. This lack of sensitivity is in contrast to the creatinine clearance test, which is one of the most sensitive tests for measuring the glomerular filtration rate.

Simultaneous urea nitrogen and creatinine determinations $\left(\text{normal ratio } \frac{15\text{–}24}{1}\right)$ appear to have some clinical significance. Elevations of serum urea nitrogen levels in renal disease (see the discussion on urea) are somewhat more pronounced than those of creatinine. In cases of retention of urea nitrogen due to prerenal causes (especially severe intestinal bleeding), the ratio between urea nitrogen and creatinine levels will be even higher, up to 40/1. Urea nitrogen levels of 35 or even 40 mg./100 ml. in the presence of normal creatinine levels are not uncommon in these conditions. On the other hand, retention of nonprotein nitrogenous compounds due to obstruction of the urinary tract will cause almost simultaneous and proportional increases in both urea nitrogen and creatinine levels. These conditions will mechanically suppress the glomerular filtration rate and thus cause an increase in all compounds that are normally found in the glomerular filtrate. In severe tubular damage the ratio may be as low as 10/1.

Creatinine determinations have one advantage over urea determinations; they are not affected by a high protein diet as is the case for urea levels. Determination of *urine creatinine* levels is of little or no help in evaluating renal function unless it is done as a part of a creatinine clearance test. Since the excretion of creatinine in one given person is relatively constant, 24 hour urine creatinine levels are used as a check on the completeness of a urine collection. Determinations of the excretion ratio of another compound under investigation to that of creatinine is considered advantageous in some cases. Such an example is the excretion of vanillylmandelic acid (3-methoxy-4-hydroxymandelic acid), which can either be reported in terms of the complete 24 hour excretion or the amount excreted per milligrams of creatinine.

Creatine in serum represents a small part of the nonprotein nitrogen fraction. No significant variations of this compound have as yet been associated with kidney diseases so that its determination has little clinical value in these conditions. Diseases associated with extensive muscle destruction may result in elevated levels of serum creatine as well as creatinuria. Creatine tolerance tests, which measure the ability of the individual to retain a test dose of creatine, are almost diagnostic in such conditions, if found to be abnormal.

SPECIMEN

Creatine and creatinine may be determined on any biological fluid, but plasma, serum, and urine are the specimens most commonly employed. Plasma and serum are preferred to whole blood since considerable amounts of noncreatinine chromogens are present in red cells. If kept for a few days, specimens for creatine and creatinine are best stored at refrigerator temperatures; if kept for longer periods, they should be frozen.

Aqueous solutions of creatine and creatinine very slowly approach a state of equilibrium with respect to each other, as indicated in the following:

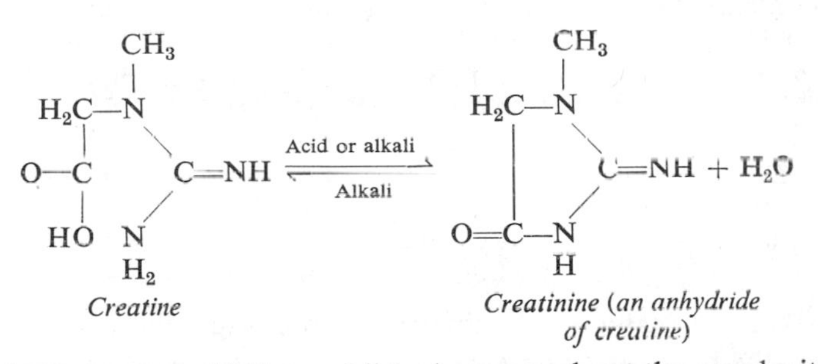

Creatine

Creatinine (an anhydride of creatine)

Although there are conflicting published reports about the speed with which this equilibrium is reached, it probably requires days or weeks at neutral pH. Creatinine, however, is formed rather quickly from creatine in either alkaline or acid solutions. Although this reversible reaction is catalyzed in both directions by hydroxyl ions, hydrogen ions promote the reaction only toward the right. Because of the lability of creatine and creatinine, it is advisable to carry out analysis for these two substances on fresh specimens. When this is not possible, adjustment of the pH to 7.0 or freezing, or both, may delay the change for indefinite periods.

Urine contains only small amounts of noncreatinine chromogens.

PRINCIPLE

Creatinine. This substance is determined in diluted urine or in a protein-free filtrate of plasma or serum after applying the Jaffe reaction. This results in the production of an amber colored substance of unestablished composition after the addition of an alkaline picrate solution.

Creatine. It is determined as the difference between the preformed creatinine and the total that results after the creatine present has been converted to creatinine by heating at an acid pH.

REAGENTS

1. Picric acid, 0.04 *M*. Dissolve about 9.3 gm. of picric acid, A.R., in about 500 ml. of water at 80°C. Cool to room temperature, dilute to 1 L. with water, and titrate with 0.1 *N* NaOH, using phenolphthalein as the indicator. Dilute as necessary to make 0.04 *N*.

2. Sodium hydroxide, 0.75 *N*. Dissolve 30 gm. of sodium hydroxide, A.R., in water and when cool, dilute to 1 L.

3. Creatinine stock standard, 1 mg./ml. Dissolve 0.100 gm. of creatinine, A.R., in 100 ml. of 0.1 *N* HCl. Store in a refrigerator.

4. Creatinine working standard, 20 μg./ml. Dilute 2 ml. of the stock solution to 100 ml. with water in a volumetric flask. Add a few drops of chloroform as a preservative.

5. Sulfuric acid, 2/3 (0.66) *N*. Add 18.8 ml. of concentrated H_2SO_4, A.R., to about 500 ml. of water. When cool, dilute to 1 L.

6. Sodium tungstate, 5 per cent. Dissolve 50 gm. of $Na_2WO_4 \cdot 2H_2O$, A.R., in water and dilute to 1 L.

STANDARDIZATION

1. Place the following in cuvets, mixing after each addition:

	Creatinine working standard (20 μg./ml.) (ml.)	*Water* (ml.)	*Picric acid* 0.04 *M* (ml.)	*Sodium hydroxide* (ml.)	*Equivalent to:* *for plasma or serum* (mg./100 ml.)	*for urine* (mg./100 ml.)
	0.25	3.75	1.0	1.0	1.0	100
	0.5	3.5	1.0	1.0	2.0	200
	1.0	3.0	1.0	1.0	4.0	400
	2.0	2.0	1.0	1.0	8.0	800
	3.0	1.0	1.0	1.0	12.0	1200
	4.0	0.0	1.0	1.0	16.0	1600
Blank	0	4.0	1.0	1.0	0	0

2. Allow tubes to stand for 15 minutes; then measure the absorbance of each at 500 nm., using the blank as a reference.

3. Construct an absorbance-concentration curve on rectangular coordinate paper, plotting absorbances as the ordinate.

PROCEDURE

Creatinine in plasma, serum, or urine

1. Add 1.0 ml. of 5 per cent sodium tungstate, 1.0 ml. of 2/3 *N* sulfuric acid and 1.0 ml. of water to 1.0 ml. of plasma or serum. Mix thoroughly and filter. If urine is analyzed, make a 1:400 dilution of urine with water.

2. Make the additions to labeled tubes as indicated below:

	Blank (ml.)	*Standard* (ml.)	*Unknown* (ml.)
Water	4.0	3.5	2.0
Creatinine working standard	0	0.5	0
Diluted urine (1: 400) or plasma (serum) filtrate	0	0	2.0
Picric acid, 0.04 *M*	1.0	1.0	1.0
Sodium hydroxide, 0.75 *N*	1.0	1.0	1.0

3. Allow to stand for 15 minutes and measure the absorbances at 500 nm.
4. Read the creatinine content from the preceding standard curve.

Note: The standard prepared in this run is equivalent to 2.0 mg. creatinine/100 ml. plasma (serum) or 200 mg. of creatinine/100 ml. of urine.

Creatine in plasma, serum or urine

1. Determine the preformed creatinine according to the procedure just described.
2. Measure the following into three 12 ml. graduated centrifuge tubes, mixing after each addition:

	Blank (ml.)	*Standard* (ml.)	*Unknown* (ml.)
Water	4.0	3.5	2.0
Creatinine standard	0	0.5	0
Protein-free filtrate of plasma or serum or diluted urine (1:400)	0	0	2.0
Picric acid, 0.04 *M*	1.0	1.0	1.0

3. Heat for 1 hour in a constant boiling water bath.
4. Cool and make up the volume to 5.0 ml.
5. Add 1 ml. of 0.75 *N* sodium hydroxide, mix, and allow to stand for 15 minutes.
6. Measure the absorbances at 500 nm. and read the values from the standard curve to obtain the total creatinine.
7. Subtract the preformed creatinine from the total and multiply the difference by 1.16 to obtain the concentration of creatine in mg./100 ml. of plasma, serum or urine. The factor 1.16 is the ratio of the molecular weight of creatine to creatinine. Some laboratories use a factor such as 1.25 to allow for incomplete conversion of creatine to creatinine.

NORMAL VALUES

	Men	*Women*
Plasma or serum		
Creatinine*	0.9–1.5 mg./100 ml.	0.8–1.2 mg./100 ml.
Creatinine	0.6–1.2 mg./100 ml.	0.5–1.0 mg./100 ml.
Creatine	0.17–0.50 mg./100 ml.	0.35–0.93 mg./100 ml.
Urine		
Creatinine	1.0–2.0 gm./24 hr.	0.8–1.8 gm./24 hr.
Creatine	0–40 mg./24 hr.	0–80 mg./24 hr.

DISCUSSION

The methods just presented, although not completely specific, are adequate for clinical purposes. A procedure for the isolation of creatinine from interfering substances has been presented by Owens and his associates.[7] After deproteinization of the specimen, creatinine is adsorbed from an acid medium onto Lloyd's reagent, an aluminum silicate, and subsequently desorbed in an alkaline solution. This type of method is highly specific but more time-consuming than that given here.

The optimum pH conditions for the hydrolytic conversion of creatine to creatinine are carefully controlled in this procedure by the amount and concentration of picric acid used.

* Includes unspecific chromogens.

The final color in the Jaffe reaction slowly fades and should be read within ½ hour.

Slight hemolysis, although it does not affect the values for creatinine, can cause appreciable positive errors for creatine.

REFERENCES

Bonsnes, R. W., and Taussky, H. H.: J. Biol. Chem., *158*:581, 1945.
Brod, J., and Sirota, J. M.: J. Clin. Invest., *27*:645, 1948.
Tierney, N. A., and Peters, J. P.: J. Clin. Invest., *22*:595, 1943.
Varley, H.: Practical Clinical Chemistry. New York, Interscience Publishers, Inc., 1967, pp. 197–200.
Henry, R. J.: Clinical Chemistry: *Principles and Technics.* New York, Hoeber Medical Div., Harper & Row, Publishers, 1964, p. 300.

URIC ACID

Uric acid has the following structure:

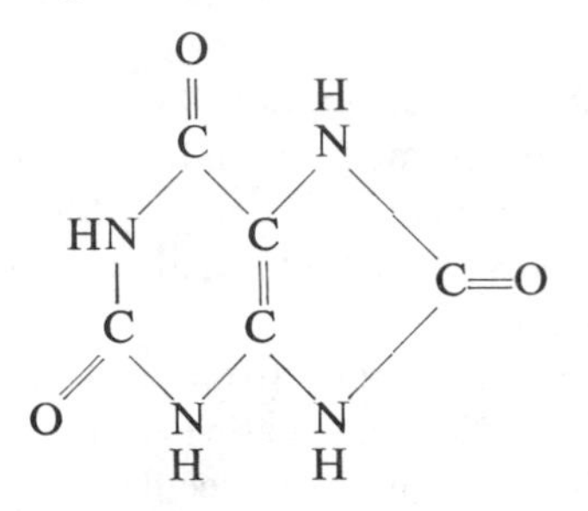

It is a waste product, derived from purines of the diet and those synthesized in the body. It has been shown that the healthy adult human body contains about 1.1 gm. of uric acid and that about one sixth of this is present in the blood, the remainder being in other tissues. Normally, about one half of the total uric acid is eliminated and replaced each day, partly by way of urinary excretion and partly through destruction in the intestinal tract by microorganisms. Uric acid is one of the components of the NPN fraction of plasma, which was discussed previously.

Determination of Uric Acid

CLINICAL SIGNIFICANCE

Plasma uric acid is filtered by the glomeruli and is subsequently reabsorbed to about 90 per cent by the tubules. It is the end product of purine metabolism in man, in the anthropoid ape, and in the Dalmatian dog. Other mammals are able to metabolize the uric acid molecule to the more soluble end product, allantoin. Uric acid concentrations in serum are greatly affected by extrarenal as well as renal factors.

Determination of serum uric acid levels are most helpful in the diagnosis of gout, in which serum levels are frequently between 6.5 and 10 mg./100 ml. Occasional normal blood (serum) levels are found in this disease, but it is believed that repeated determinations will reveal hyperuricemia at some point of the disease. Serum uric acid levels are also increased whenever there is increased metabolism of nucleoproteins, such as in leukemia and polycythemia or after the intake of food rich in nucleoproteins, for example, liver, kidney, or sweetbread. Increased serum uric acid levels are also a constant finding in familial idiopathic hyperuricemia, of which there seem to be at least two types. In one type there is an overproduction of uric acid in the presence of normal excretion, and in the other there is a decreased rate of excretion in the presence of normal uric acid production.

Uric acid levels are elevated in decreased renal function. In severe renal impairment values up to 20 to 35 mg./100 ml. have been observed, depending on the method employed (the lower values are observed with the method by Brown). Although any decrease in renal function is generally accompanied by increases in serum uric acid levels, this test is rarely used in this connection because of the great effect of extrarenal factors on serum levels.

Urine uric acid levels are generally a reflection of the endogenous nucleic acid breakdown and of the amount of dietary purines (see normal values). Unless hyperuricemia is due to decreased excretion of uric acid, it is generally accompanied by increased levels of uric acid in urine.

SPECIMEN

Uric acid is stable in serum for several days at room temperature and for longer periods if refrigerated, but since it is susceptible to destruction by bacterial action, the addition of thymol may increase the stability. Any of the common anticoagulants, except potassium oxalate, can be used. Potassium phosphotungstates, which are insoluble, are formed when potassium salts are introduced into the system, which results in turbidity.

Uric acid has been reported to be stable in urine for several days at room temperature. If urine specimens are refrigerated, urates may precipitate, which may be brought into solution by adjustment of the pH to 7.5 to 8.0 and by warming the specimen to approximately 50°C.

PRINCIPLE

Uric acid is oxidized to allantoin and carbon dioxide by a phosphotungstic acid reagent in alkaline solution. Phosphotungstic acid is reduced in this reaction to tungsten blue, which is measured at 710 nm.

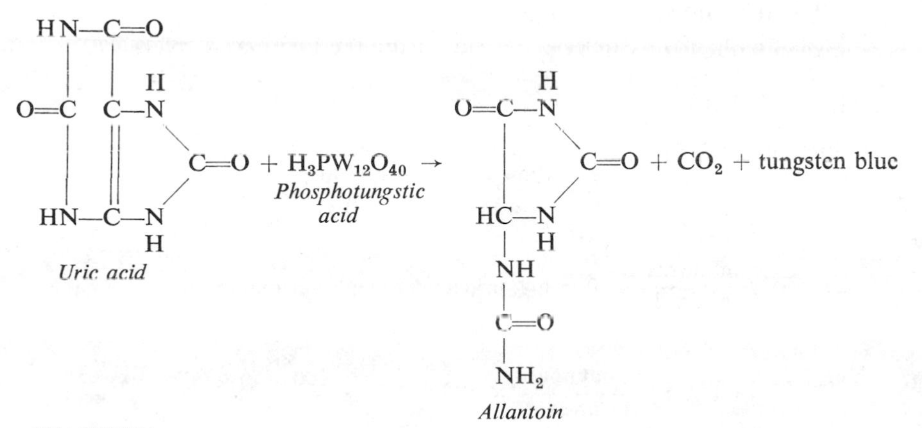

Uric acid

Allantoin

REAGENTS

1. Sodium tungstate, 10 per cent (w/v). Dissolve 100 gm. of $Na_2WO_4 \cdot 2H_2O$, A.R., in water and dilute to 1000 ml.

2. Sulfuric acid, 0.66 *N*. Slowly add 18.8 ml. of concentrated sulfuric acid, A.R., to about 500 ml. of water and dilute to 1000 ml.

3. Phosphotungstic acid reagent. Dissolve 40 gm. of sodium tungstate, A.R., in 300 ml. of water. Add 32 ml. of 82 per cent orthophosphoric acid and several glass beads. Reflux gently for 2 hours. Cool to room temperature and dilute to 1 L. Mix thoroughly. Dissolve 32 gm. of $Li_2SO_4 \cdot H_2O$, A.R., in the reagent and mix thoroughly. Store in a refrigerator.

4. Sodium carbonate, 14 per cent (w/v). Dissolve 70 gm. of anhydrous Na_2CO_3, A.R., in water and dilute to 500 ml. Store in a polyethylene bottle.

5. Uric acid stock standard, 1 mg./ml. Measure 100 mg. of uric acid, A.R., and 60 mg. of Li_2CO_3, A.R., into a 100 ml. volumetric flask. Add about 50 ml. of water and warm to about 60°C. to aid in the solution of the reagents. Cool to room temperature, dilute to the mark, and mix thoroughly. This reagent may be stable for several months in a refrigerator.

6. Uric acid working standards. Dilute 0.5, 1.0, and 1.5 ml. of the 1 mg./ml. stock standard to 100 ml. with water. (Equivalent to 5, 10, and 15 mg. of uric acid/100 ml. plasma and 50, 100, and 150 mg. of uric acid/100 ml. urine.)

PROCEDURE

1. For plasma or serum, prepare a tungstic acid protein-free filtrate by mixing 1.0 ml. of sample with 8.0 ml. of water, 0.5 ml. of 0.66 N H_2SO_4 and 0.5 ml. of 10 per cent sodium tungstate. Mix and filter.

2. Measure the following into tubes in the order shown. Mix after each addition.

	Blank (ml.)	*Standard* (ml.)	*Unknown* (ml.)
Protein-free filtrate (1:10) or urine (diluted 1:100 with water)	0	0	3.0
Water	3.0	0	0
Uric acid working standard	0	3.0	0
Sodium carbonate, 14 per cent	1.0	1.0	1.0
Phosphotungstic acid	1.0	1.0	1.0

3. Allow tubes to stand 15 minutes.

4. Measure the absorbances at 710 nm. using the blank as a reference.

CALCULATIONS

For plasma or serum

$$\frac{A \text{ unknown}}{A \text{ standard}} \times 0.03^* \times \frac{100}{0.3}$$

or:

$$\frac{A \text{ unknown}}{A \text{ standard}} \times 10 = \text{mg. uric acid/100 ml. plasma or serum}$$

For urine

$$\frac{A \text{ unknown}}{A \text{ standard}} \times 0.03^* \times 10 \times \frac{100}{0.3}$$

or:

$$\frac{A \text{ unknown}}{A \text{ standard}} \times 100^* = \text{mg. uric acid/100 ml. urine}$$

* This factor applies if the 0.01 mg./ml. standard is used.

NORMAL VALUES

Plasma or serum

men:	2.5 to 7.0 mg./100 ml.
women:	1.5 to 6.0 mg./100 ml.

Urine

average diet:	250 to 750 mg./24 hr.
low purine diet:	up to 450 mg./24 hr.
high purine diet:	up to 1 gm./24 hr.

SOURCES OF ERROR

This method has been reported to follow Beer's law up to an absorbance of 0.8 in a spectrophotometer; however, it is advisable to prepare several standards to check any deviation at higher concentrations.

DISCUSSION

Uric acid can be determined on the basis of another principle, that of differential spectrophotometry, in which uric acid is destroyed by the action of uricase. The decrease in absorbance after incubation with uricase is measured in the ultraviolet region (290 to 293 nm.) and is proportional to the uric acid initially present. This method has great specificity because, as far as is known, uricase acts only on uric acid. It is more time-consuming, however, and requires the use of an instrument capable of reading in the ultraviolet region; therefore, it is seldom used routinely.

In the Archibald modification[1] of the Kern and Stransky method, the specificity of the method is enhanced by pretreatment of the serum with sodium hydroxide, which causes an oxidative destruction of ascorbic acid and sulfhydryl compounds that otherwise would lead to false high values. Phosphotungstic acid serves both as a protein precipitating agent and a color agent. A glycerine silicate reagent increases the sensitivity, and also provides alkalinity for the reduction of phosphotungstic acid to tungsten blue. Sodium polyanetholsulfonate (Liquoid, Hoffmann-La Roche Inc., Nutley, N.J.) is added to prevent turbidity.

REFERENCES

Henry, R. J., Sobel, C., and Kim, J.: Am. J. Clin. Path., *28*:152, 1957.
Henry, R. J.: Clinical Chemistry: Principles and Technics. New York, Hoeber Medical Div., Harper & Row, Publishers, 1964, pp. 278–280.
Blauch, M. B., and Koch, F. C.: J. Biol. Chem., *130*:443, 1939.

AMMONIA IN BLOOD, SERUM, AND URINE

CLINICAL SIGNIFICANCE

Although ammonia is a part of the nonprotein nitrogen fraction of plasma and serum, it is present in both fluids only in small amounts and its determination is of very little or no value in the study of renal disease. Relatively significant increases in blood and serum ammonia are observed in impending and existing hepatic coma, which is discussed in Chapter 13, Liver Function Tests (see also Chapter 10).

The determination of ammonia in urine is used as an index of the ability of the kidney to produce ammonia. This process is discussed in more detail in Chapter 11, Electrolyte (Cation-Anion) Balance. The determination of ammonia in urine has been replaced by more modern and accurate tests for kidney function.

Urinary ammonium may be elevated as the result of the breakdown of urea by urease producing bacteria. Because of this, particular care must be taken in the collection and preservation of the urine specimens for urinary ammonium levels.

Determination of Ammonia in Urine

SPECIMEN

For the determinations of ammonia in urine, 24 hour specimens are obtained. In order to reduce potential positive errors resulting from the bacterial decomposition of urea, it is best to add a few milliliters of 1 *N* HCl to the container and to keep the specimen refrigerated until it is analyzed.

PRINCIPLE

The ammonia in an aliquot of a urine specimen reacts with phenol and sodium hypochlorite (Berthelot reaction) to form the blue indophenol described for the determination of urea.

REAGENTS

1. Ammonia-free water. This is the same as in the urea method.
2. Phenol-nitroprusside solution. This is the same as in the urea method.
3. Alkaline-hypochlorite solution. This is the same as for the urea method.
4. Ammonia-nitrogen stock standard (500 mg./100 ml). Dissolve 2.3581 gm. of dry ammonium sulfate, A.R., in ammonia-free water and dilute to 100 ml. in a volumetric flask. Add a few drops of concentrated H_2SO_4, A.R., to increase the stability.
5. Ammonia-nitrogen working standard (50 mg./100 ml). Dilute 10 ml. of the stock standard to 100 ml. with ammonia-free water. Store in a refrigerator.

PROCEDURE

1. Label three tubes, blank, standard, and unknown, respectively.
2. Pipet 1.0 ml. of ammonia-free water into each tube.
3. Dilute 1.0 ml. of the urine specimen with 9.0 ml. of ammonia-free water.
4. Add 10 μl. of the diluted urine specimen and 10 μl. of the working standard to the appropriate tubes. (Use a micropipet.)
5. Add rapidly and successively, mixing after each addition, 5.0 ml. of the phenol-nitroprusside solution and 5.0 ml. of alkaline-hypochlorite solution.
6. Place the tubes in a water bath at 37°C. for 20 minutes.
7. Measure the absorbance of the standard and the unknown at 560 nm., using the blank as a reference.

CALCULATION

$$\frac{A \text{ unknown}}{A \text{ standard}} \times 500 = \text{mg. ammonia-nitrogen/100 ml.}$$

NORMAL VALUES

500 to 1200 mg./24 hr. (36 to 86 mEq./24 hr.)

SOURCES OF ERROR

1. Ammonia in any of the reagents or in the atmosphere of the room in which the measurement is carried out may lead to erroneously high results.

2. Bacterial action in the specimen may cause values to be high, through hydrolysis of urea.

Note: For the determination of serum ammonia, see Chapter 10, Electrolytes.

REFERENCES

Goldstein, L., Bearle, R. R., and Dearborn, E. H.: J. Lab. Clin. Med., *48*:324, 1956.
Kaplan, A.: Urea nitrogen and urinary ammonia. *In* Standard Methods of Clinical Chemistry. S. Meites, Ed. New York, Academic Press, Inc. 1965, vol. 5, pp. 245–256.
Chaney, A. L., and Marbach, E. P.: Clin. Chem., *8*:130, 1962.

AMINO ACIDS

CLINICAL SIGNIFICANCE

Although amino acids are a part of the nonprotein nitrogen fraction of blood and serum, their quantitative determination finds clinical application only in some selected congenital renal disorders. Amino acids in plasma are filtered by the glomerulus and appear in the glomerular filtrate in the same proportion as they do in plasma. A portion of the amino acids is subsequently reabsorbed by the proximal convoluted tubules. In some congenital disorders there is a defect in the reabsorption of amino acids, resulting in aminoaciduria. An example of such a condition is cystinuria, in which there is a failure to reabsorb dibasic amino acids (cystine, lysine, arginine, and ornithine). In Fanconi's syndrome there is a failure to reabsorb a wide variety of amino acids.

Rather significant elevations in serum and urine amino acid nitrogen occur in conditions associated with parenchymal liver damage. The reader is referred to Chapter 13 on liver function and especially to Chapter 5, Proteins and Amino Acids, in which more detailed information regarding the clinical significance of amino acid determinations can be found. Chapter 5 also contains detailed methodology.

Tests Measuring Tubular Function

The renal tubules are engaged in a wide variety of activities and, consequently, there are several groups of tubular function tests that have been employed in the clinical laboratory.

The first group of these tests measures the excretory ability of the tubules. For this purpose substances are injected into the blood, which are cleared either exclusively or predominantly by the tubules. Such a compound is sodium p-aminohippurate, which is removed to about 90 per cent in a single passage through the kidney (see p-Aminohippurate Clearance Test). The clearance test employing this substance is not widely used, however, because of technical difficulties in performance. The phenolsulfonphthalein test (to be described) is the most widely used test to evaluate the excretory capacity of the kidney. Like other renal function tests, both procedures are affected by the renal blood flow.

The second group of tubular function tests is concerned with the concentrating ability of the tubules. In cases of tubular damage, this is generally the first function to be decreased. Commonly used tests in this group are the measurement of the specific gravity, the measurement of the osmolality, and the Fishberg concentration test (all to be described).

Other tubular function tests include the determination of ammonia in urine, which is a measure of the ability of the tubules to produce ammonia in states of acidosis. The test for "tubular reabsorption of phosphorus" is, strictly speaking, also a kidney function test; however, since this function of the tubule is greatly influenced by the parathyroid hormone, the test is mainly used to evaluate the status of the parathyroid gland.

p-AMINOHIPPURATE CLEARANCE TEST

Certain substances, foreign to the human body, in addition to being filtered through the glomerulus, are also removed by the kidney via the tubules. Within certain limits of plasma concentration, up to 90 per cent of some of these substances (e.g., p-aminohippurate and Diodrast) are removed from the plasma in a single passage through the kidney. Since renal function is dependent on the renal blood flow, a clearance value of p-aminohippurate will provide a measure of the effective renal plasma flow in the absence of tubular functional impairment and vice versa. The clearance values of para-aminohippurate and Diodrast are 600 to 700 ml./min. or 350 to 400 ml./min./sq. m. of body surface area. Since about 10 per cent of the blood circulating through the kidney does not come into contact with functional cells, the renal plasma flow, as measured by this technique, will be about 650 ml./min. or 390 ml./min./sq. m. of body surface area. The actual total renal blood flow is about 1200 ml./min. and the plasma flow about 750 ml./min.

The p-aminohippurate clearance is performed in essentially the same way as the inulin clearance and is also to be regarded as a strictly investigative procedure, not a routine clinical one. Para-aminohippurate is given to the subject in a priming dose by intravenous injection, which is followed by a slow continuous administration of solution of low concentration to maintain a constant level. Blood and urine specimens are collected before the drug is given and are used for controls. Three timed urine specimens are collected, as well as blood samples at the beginning and at the end of each urine collection period. The mean of the two is used in the calculation. An aliquot of each specimen is assayed for its p-aminohippurate concentration. The principle of this assay is a coupling of p-aminohippurate with diazotized N-(1-naphthyl)ethylenediamine dihydrochloride. The three clearance values are calculated and averaged. Values of plasma clearance are given in ml./min. and also in ml./min./sq. m. of body surface area.

REFERENCES

See the references under Glomerular Filtration and Inulin Clearance Test.

PHENOLSULFONPHTHALEIN (PSP) TEST

CLINICAL SIGNIFICANCE

Phenolsulfonphthalein is a dye that is removed from plasma to about 60 to 70 per cent during one passage through the kidney (renal clearance approximately 400 ml./min.). About 20 per cent is normally removed by the liver. Of that portion of the dye that is removed by the kidneys, about 6 per cent is excreted by glomerular filtration and about 94 per cent by tubular excretion. Thus, the test is mainly a measure of the secretory capacity of the tubules. In the absence of renal disease, the test may be used as a measure of renal blood flow.

Traditionally, the test has been performed as a 2 hour test. It has been well established, however, that the first specimen (15 minute specimen) is of greatest clinical value since the dye excretion during the first 15 minutes is affected to a lesser degree by extraneous factors such as the renal blood flow and the total urine flow. Therefore, many laboratories collect only one 15 minute specimen.

SPECIMEN

Urine specimens, collected 15, 30, 60, and 120 minutes after the injection of dye, are used. If a split renal function (differential) test is ordered, urine specimens, collected from each kidney, are tested. Some laboratories use different collection times, for example only one 15 or 20 minute sample.

PRINCIPLE

A standard dose of phenolsulfonphthalein (PSP) is injected intravenously into the patient.

The quantity of PSP excreted is measured colorimetrically after alkalizing the specimen to convert the dye to the colored form.

REAGENTS

1. Sterile 1 ml. vials of phenolsulfonphthalein (6 mg./ml.) (Hynson, Westcott and Dunning, Baltimore, Md.).
2. Sodium hydroxide (10 per cent). Dissolve 100 gm. of reagent grade sodium hydroxide in water and, when cool, dilute to 1000 ml.

PROCEDURE

Patient preparation (This is carried out by the nursing and clinical staffs on the hospital ward.)

1. Hydrate the patient by giving about 600 ml. of water to drink.
2. After 10 minutes, inject intravenously 1 ml. of sterile PSP solution (6 mg./ml.).
3. Collect separate urine samples at 15, 30, 60, and 120 minutes after injection.

Laboratory procedure

1. Transfer each specimen into a 1000 ml. graduated cylinder.
2. Add 10 ml. of 10 per cent NaOH to each cylinder. Filter, dilute to the 1 L. mark with water, and mix. If the color is pale, as it may be in the 60 and 120 minute samples, dilute to only 250 or 500 ml. and make corrections for this in the calculation.
3. Measure the per cent transmittance of each diluted specimen at 540 nm., using water as a reference.
4. Read the per cent of PSP excretion in each specimen from a calibration curve.

CALIBRATION CURVE

1. Pipet 1 ml. of PSP solution (6 mg./ml.) in a 1 L. volumetric flask containing 200 to 300 ml. of water. Add 10 ml. of 10 per cent NaOH, dilute to the mark, and mix. This represents the 100 per cent standard.
2. Place in separate tubes the volumes of the 100 per cent standard and water indicated in the following table and mix thoroughly.

100% *Standard* (ml.)	*Water* (ml.)	*Standard* (%)
10.0	0	100
7	3	70
5	5	50
3	7	30
1	9	10

3. Measure the per cent transmittance of each standard at 540 nm., using water as a reference.

4. Plot the per cent transmittances of the standards against the corresponding concentrations on semilogarithmic paper. (Alternatively, plot absorbance values against concentration on rectangular coordinate paper.)

SOURCES OF ERROR

1. Incomplete urine collection resulting from retention in the bladder is the most frequent source of error in this test.

2. Injection of an incorrect amount of dye is another very common error. The PSP vial contains more than 1 ml.; thus, not all the dye should be injected.

NORMAL VALUES

Specimen	*Per cent Excretion*
15 min.	25–50
30 min.	15–25
60 min.	10–15
120 min.	5
Total excretion	60–85

It should be noted that the results of the 15 and the 30 minute specimens are the most significant regardless of the outcome of the subsequent collections.

REFERENCES

Wells, B. B., and Halstead, J. A.: Clinical Pathology: Interpretation and Application. 4th Ed. Philadelphia, W. B. Saunders Co., 1967, p. 288.

Roundtree, L. G., and Geraghty, J. T. : Arch. Int. Med., *9*:284, 1912.

SPECIFIC GRAVITY

Specific gravity is the ratio of the weight of a substance to the weight of an equal volume of water at a specified temperature. It is a direct but not proportional function of the number of particles in the urine. Since the work done by the kidney in eliminating substances is directly related to the number of particles, specific gravity is one measure of the function of the kidney. Because each substance contributes differently to the specific gravity, this measure is not strictly a function of the number of particles as is the measurement of osmolality. Specific gravity is easily done, however, and furnishes information of considerable clinical value. It is still performed more frequently than the more informative osmolality measurement.

CLINICAL SIGNIFICANCE

The determination of the specific gravity of a urine sample is an important part of the routine urinalysis. Although there is a considerable fluctuation in the specific gravity values from day to day and also during the course of a day, the determination of the specific gravity of a randomly collected urine specimen has clinical value as a screening test to measure the concentrating ability of the renal tubules. In cases of renal tubular damage, this function is generally the first to be lost. Elevations in urine specific gravity in the absence of dehydration are most commonly seen in patients with uncontrolled diabetes with glycosuria; extremely high specific gravity values (above

1.050) may be seen in patients who have recently had urinary tract diagnostic studies that use mercurial X-ray contrast media.

Specific gravity measurements of urine are also a part of the various concentration tests (see Fishberg Concentration Test), which are more accurate measures of the concentrating ability of the kidney than a specific gravity measurement on a random urine specimen.

It has recently been felt that osmolality measurements should replace the specific gravity test for both routine urinalysis and the Fishberg concentration test. Although such a development would be desirable, the specific gravity test probably will still be used widely because of its simplicity and speed.

NORMAL VALUES

Specific gravity values observed in normal individuals vary greatly with fluid intake and the state of hydration. Thus, the normal values for a random specimen are from 1.002 to 1.030 and the normal values for a 24 hour specimen are usually considered to be from 1.015 to 1.025.

Determination of Specific Gravity with the Urinometer

SPECIMEN

The specific gravity is most often performed as a part of a routine urinalysis on a random urine specimen. Less frequently, however, specific gravity is measured on timed specimens after water restriction, in which case more exact information is derived.

PRINCIPLE

The urinometer is a hydrometer designed for the measurement of urinary specific gravity. When placed in the specimen contained in a cylinder, it sinks to the level characteristic of the specific gravity of the specimen. The value may then be read directly from the calibrations on the stem. It is important that the urinometer float freely without sticking to the walls of the container.

PROCEDURE

1. Pour the specimen into the urinometer tube until it is about three-fourths full. Sufficient space should be allowed so that it will not overflow when the urinometer is floated in the sample.

2. Place the urinometer in the specimen with a slight twisting motion so that it will spin and have less tendency to stick to the sides of the tube.

3. Read the scale on the stem where it is intersected by the lowest line of the meniscus.

4. For greatest accuracy, measure the temperature and make the following correction: Add 0.001 to the specific gravity for each 3°C. or 5.4°F. that the temperature is above the urinometer calibration temperature. Subtract 0.001 for each 3°C. or 5.4°F. that it is below the calibration.

NOTES

1. The urinometer should be calibrated against distilled water and should read 1.000 at its calibration temperature. It should also be calibrated at a high value. This may be done by testing in a mixture of 75 ml. of xylene and 28 ml. of bromobenzene, which has a specific gravity of 1.030.

2. If the urine contains significantly large quantities of protein, a correction should be applied to compensate for this factor. Subtract 0.003 from the reading for every gm. of protein/100 ml. of urine.

3. Subtract 0.004 for each gm./100 ml. of glucose present.

SOURCES OF ERROR

1. The urinometer must float freely, not adhering to the sides of the tube; otherwise, the reading may be in error. Also there should be no bubbles clinging to the stem.

2. Failure to compensate for temperature or for gross proteinuria and glycosuria.

REFERENCES

Wells, B. B., and Halsted, J. A.: Clinical Pathology: Interpretation and Application. 4th ed. Philadelphia, W. B. Saunders Co., 1967, p. 273.

Monroe, L., and Hopper, J., Jr.: J. Lab. Clin. Med., *31*:934, 1946.

Determination of Specific Gravity by Refractometry

SPECIMEN

See under Determination of Specific Gravity with the Urinometer.

PRINCIPLE

The refractive index and the specific gravity of a urine specimen are both related functions of the quantity and type of dissolved substance in the specimen. Each substance contributes differently to the refractive index and also to the specific gravity; however, because various urine specimens are likely to contain dissolved substances of similar types and proportions, the refractive index and the specific gravity may be correlated. Increased amounts of abnormal substances such as glucose and protein may partially invalidate the correlation and give specific values that are misleading.

The instrument most commonly used for this purpose is the TS (total solids) Meter (American Optical Corp., Scientific Instrument Div., Buffalo, N.Y.). This is a hand refractometer with two temperature compensated scales, which allows direct measurement of total solids of serum or the specific gravity of urine. Other refractometers have two scales, one calibrated in refractive index and the other in total serum protein concentration.

PROCEDURE

1. Place a small drop of sample on the lower glass plane surface of the TS Meter. Then bring the upper hinged surface down firmly on the drop so that the two glass planes are parallel.

2. Hold the meter toward a source of light so that the beam passes through the sample and the prisms.

3. Read the specific gravity from the proper scale at the sharp line of contrasting light and dark areas that falls across the scales.

NOTES

1. Accuracy is excellent as long as the protein content is low. Unfortunately there seems to be no published correction factor to compensate for the contribution of protein to the refractive index.

2. Wolf and his associates[9] describe a more accurate but more time-consuming method. The specific gravity is calculated from conversion tables provided with the instrument, through substitution of refractive index values of the untreated urine and of its supernatant after deproteinization with acetic acid.

REFERENCES

American Optical Co. Bull. No. SB 10400-1061.
Rubini, M. E., and Wolf, A. V.: J. Biol. Chem., *225*:869, 1957.

Determination of Osmolality of Serum and Urine

CLINICAL SIGNIFICANCE

In recent years there has been an increasing appreciation of the clinical value of both serum and urine osmolality determinations. This is partially due to the availability of more precise and convenient instrumentation, and also to the realization that the concentrating activity of the renal tubules is regulated by "osmoreceptors," which are sensitive to changes in solute concentration rather than to changes in the specific gravity. Consequently, the measurement of the urine osmolality, especially as part of the concentrations tests, should be a more accurate test of the concentrating ability of the kidney than the specific gravity measurement. Holmes performed a comparative study of specific gravity versus osmolality values in normal individuals and in unselected patients with renal disease. Although in the first group there appears to be a reasonably close relationship between specific gravity and osmolality, this relationship is maintained to a much lesser degree in patients with renal disease. This decrease in correlation can be explained at least partially by the fact that the presence of large molecules, such as proteins and glucose, or heavy ions, such as Hg^{++}, affect the specific gravity of urine substantially more than the osmolality.

Measurement of the osmolality of urine, although clinically useful, should only be considered a screening procedure; however, the simultaneous determination of

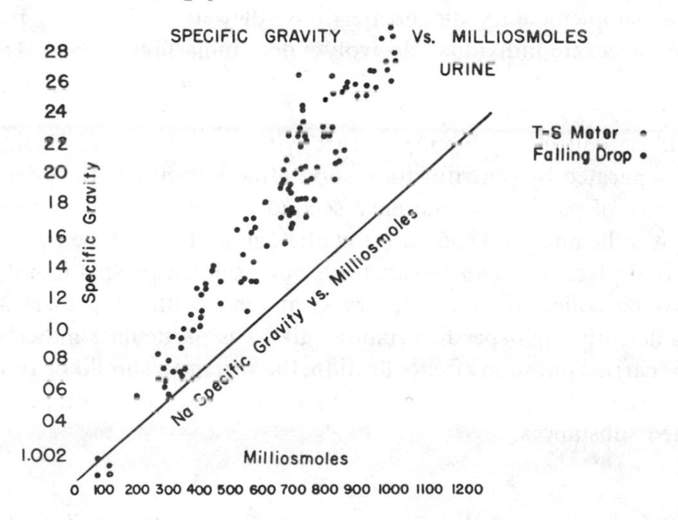

Figure 12-4. Comparison of the urinary specific gravity and urinary osmolality in a series of urines obtained from healthy medical students. The straight line represents comparative readings on various concentrations of sodium chloride solutions. (From Holmes, J. H.: Measurement of osmolality in serum, urine, and other biological fluids by the freezing point determination. *In:* Workshop on Urinalysis and Renal Function Studies. Commission on Continuing Education, American Society of Clinical Pathologists, 1962.)

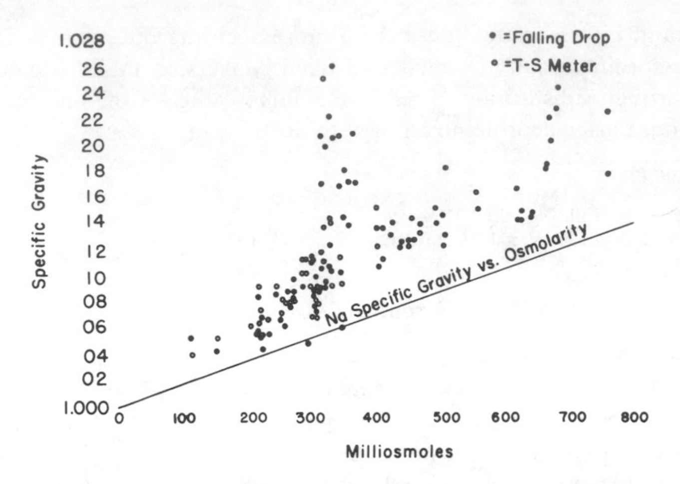

Figure 12-5. Comparison of the urinary specific gravity and urinary osmolality in a series of unselected urines obtained from patients on the renal service. The straight line represents comparative readings on various concentrations of sodium chloride solutions. (From Holmes, J. H.: Measurement of osmolality in serum, urine, and other biological fluids by the freezing point determination. *In:* Workshop on Urinalysis and Renal Function Studies. Commission on Continuing Education, American Society of Clinical Pathologists, 1962.)

serum and urine osmolality is a considerably more accurate way of measuring the concentrating ability of the tubules. Analogous to the clearance tests, here we measure the ratio of the concentration of the osmotically active particles in urine to serum. This ratio expresses the actual degree to which the kidney has concentrated the glomerular filtrate, which, in respect to osmolality, is very close to serum. The normal ratio of urine osmolality to serum osmolality is 3.0 or above.

Serum osmolality measurements have been suggested by some authors as a satisfactory replacement for electrolyte studies. Although such measurements may be very helpful as supplementary diagnostic tools, they are unlikely to replace the considerably more accurate individual electrolyte determinations.

SPECIMEN

Blood should be collected by venipuncture with a minimum of stasis, and the serum should be separated by centrifugation soon after collection. In order to lessen the possible presence of particulate matter, a second centrifugation is recommended. If the serum is not to be analyzed soon after centrifuging, it should be refrigerated or frozen. Heparinized plasma is also satisfactory, but oxalated plasma is not.

Urine should be collected in clean, dry containers without preservatives and centrifuged at sufficiently high speed to remove all gross particulate material. If the analysis cannot be carried out soon after collection, the specimen should be refrigerated. Before analysis, refrigerated specimens should be warmed to aid the complete solution of any precipitated substances.

PRINCIPLE

Osmolality is a measure of the total concentration of dissolved particles in a solution, without regard for the homogeneity or the nonhomogeneity of the molecular species, the molecular weights, the particle size, or the density.

Any substance dissolved in a solvent produces four mathematically interrelated physical characteristics known as colligative properties. These properties are:

1. Depression of the freezing point below that of the pure solvent.

2. Elevation of the boiling point above that of the pure solvent.
3. Decrease in the vapor pressure above that of the pure solvent.
4. Increase in the osmotic pressure above that of the pure solvent.

Theoretically, the measurement of any one of these properties will afford a means to calculate the other three and in turn to calculate the concentration of dissolved particles. Because of the chemical and physical nature of biological fluids, however, it is most practicable to measure the freezing point depression with any one of the commercially available osmometers. Briefly, the principle of operation of this instrument is to supercool the specimen a few degrees (7°C.) below the freezing point and then to initiate the freezing process. As freezing occurs, heat is liberated as ice crystals form, causing the temperature to rise and to approach the freezing point, the point of equilibrium, where the temperature is read. The measured freezing point then allows a calculation of the concentration of the dissolved particles. Although osmometers actually measure the freezing point, they are calibrated in terms of osmolality. Standard solutions of sodium chloride having known freezing points and corresponding osmotic pressures are used in the calibrating procedure.

One mole of any nonionic solute dissolved in a kilogram of water lowers the freezing point 1.86°C. and represents 1000 milliosmols (mOsm.).

Osmotic pressure is generally proportional to the molal concentration, which is defined as moles of solute per kilogram of *solvent* (H_2O). This differs from the less rigorous term molar concentration, which is the number of moles of solute per liter of *solution*. The osmolality of a solution may be calculated as follows:

$$\text{Osmols/kg. } H_2O = \phi \times N \times M$$

where ϕ = osmotic activity coefficient

N = the number of particles/molecule in solution

M = the concentration of the solution in mole/kg. H_2O

For a solution of sodium chloride of 1 mole/kg. H_2O, $\phi = 0.93$; $N = 2$ particles per molecule; $M = 1$.

Thus, the osmolality of the sodium chloride solution is:

$$(0.93) \times (2) \times (1) = 1.86$$

PROCEDURE

Although several different osmometers are now commercially available,* and although each one must be operated in a manner appropriate to the particular instrument, as described in its corresponding manual, the following procedure is general and may be applied in principle to any instrument.

1. Centrifuge the specimen twice to eliminate any gross particulate material.
2. Precool the specimen to several degrees (about 7°C.) above the freezing point by placing it in a cooling bath.
3. Place the proper volume of specimen into a sample tube and place the tube in position in the instrument.
4. Supercool the specimen (about −7°C.) and allow isothermalization to take place.
5. Initiate the freezing process and permit plateau to be established.
6. Measure the freezing point and read the value in terms of milliosmols.

---

* Fiske Osmometer, Fiske Associates Inc., Uxbridge, Mass. Osmette 2007, Aloe Scientific Div., Brunswick Corp., St. Louis, Mo. Osmometer, Advanced Instrument, Inc., Newton Highlands, Mass.

STANDARDIZATION

Osmometers are sometimes calibrated at the factory; however, it is necessary to check the calibration. This is usually accomplished by setting the instrument on the basis of standard sodium chloride solutions. The following are four such standards.*

gm. NaCl/kg. H_2O	mOsm./kg.	*Expected freezing point depression*, °C.
3.094	100	0.186
15.930	500	0.929
32.120	1000	1.86
44.980	1400	2.60

NORMAL VALUES†

Because controversy exists about normal values of osmolality, values from several sources are presented:

Serum	(mOsm./kg.)†	*Reference*
	289 ± 4 S.D.	Hendry
	289–308	Lindemann
	275–295	Lobdell
Urine		
Osmotic limits of renal dilution and concentration	40–1400	Wolf
During maximal urine concentration	967–1324	Lindemann
During maximal urine concentration	855–1335	Jacobson

PRECISION

It has been reported by Johnson and Hoch[5] that a reproducibility of ±1 mOsm./kg. with an equal accuracy can be obtained if standards are correct; however, under routine conditions an accuracy and precision of ±2 mOsm./kg. is quite acceptable.

SOURCES OF ERROR

Faulty standardization, faulty use of the osmometer, and the presence of particulate matter in the specimen are the most common sources of inaccurate results.

REFERENCES

Osmometer Manual. Fiske Associates, Inc., Uxbridge, Mass.
Lobdell, D. H.: St. Vincent's Hosp. Med. Bull. (Bridgeport, Conn.), *8*:7, 1966.
Warhol, R. M., Eichenholz, A., and Mulhausen, R. O.: Arch. Int. Med., *116*:743, 1965.
Hendry, E. B.: Clin. Chem., 7:156, 1961.

---

* Adapted from Johnson, R. B., and Hoch, H.: Osmolality of serum and urine. *In:* Standard Methods of Clinical Chemistry. S. Meites, Ed. New York, Academic Press, Inc., vol. 5.

† Please note that there is some inconsistency in the literature as to the units in which osmolality is expressed. Although the most accurate unit is mOsm./kg. H_2O, many investigators have used the term mOsm./L. plasma (serum). Assuming a water content of plasma of 93 per cent, the relation between mOsm./L. and mOsm./kg. H_2O is as follows:

$$\text{mOsm./kg. } H_2O = \text{mOsm./L. plasma} \times \frac{100}{93}$$

Thus, values expressed in mOsm./kg. H_2O are approximately 7 per cent higher than those expressed in mOsm./L. plasma. The values by Hendry and Lobdell are given in mOsm./L.

Lindemann, R. D., Van Buren, H. C., and Raisz, L. G.: New England J. Med., *262*:1306, 1960.
Wolf, A. V.: The Urinary Function of the Kidney. New York, Grune & Stratton, Inc., 1950.
Jacobson, M. H., Levy, S. E., Kaufman, R. M., Gallinek, W. E., and Donnelly, O. W.: Arch. Int. Med. *110*:83, 1962.

THE FISHBERG CONCENTRATION TEST

CLINICAL SIGNIFICANCE

Concentration tests of renal function are not used as frequently as they were several decades ago; then, such tests were used extensively in the evaluation of the kidney status of a patient. The tests were considered to be simple and readily adaptable to office practice and to be more sensitive than any other measure of renal function; however, more modern procedures have gradually replaced these tests. Of the many procedures available, the Fishberg test has perhaps had the widest application. Following the regimen described, the normal patient can concentrate his urine to a specific gravity of 1.025, often reaching 1.032. (The respective osmolality values are 850 and 1350 mOsm./L.) The ability to concentrate is lost to some extent with age so that older patients, whose kidney function is apparently normal, usually show values in the lower portion of the normal range. Tubular epithelial damage, as may occur with nephrotoxic drugs, severe alkalosis, shock syndrome, or impairment of tubular blood supply, may cause impairment of the concentrating ability. During recovery from severe damage, polyuria with urine of low specific gravity may occur, which, in this case, is not a true index of the concentrating ability of the tubules.

In severe functional impairment, levels of 1.010 to 1.020 are observed and most often the value seen is nearer the lower figure. When healing occurs, e.g., in acute nephritis, the concentrating power of the kidney is the last function to return to normal. This may reflect functional inadequacy of the newly regenerated tubular epithelial cells. Patients with edema who are receiving therapy and are losing their edema water may have low specific gravities, which have no connection with their renal status.

SPECIMEN

Three urine specimens are used in this test. They are the entire collections voided by the patient in the morning at 6 a.m., at 8 a.m., and at 10 a.m. If, because of fluid restriction, the patient cannot void at the specified time, collect three specimens whenever possible between 6 and 10 a.m.

PRINCIPLE

The ability of the kidneys to concentrate urine under conditions of water deprivation are tested by measuring the specific gravity of urine voided at intervals in the morning, following an overnight period of fluid restriction.

PROCEDURE

Preparation of the patient

1. Allow the patient no more than 200 ml. of total fluid intake during the evening meal on the day before the test. Allow the patient no fluid from 8 p.m. to 10 a.m.

2. Discard any urine voided by the patient during the night.

3. Collect the entire specimens, separately, voided by the patient at 6 a.m., at 8 a.m., and at 10 a.m., or up to 10 a.m.

For the laboratory

1. Measure the specific gravity or the osmolality of each of the three urine specimens.

NORMAL VALUES

The specific gravity of one or more of the specimens should have a value of 1.025 or higher. The osmolality should be 850 mOsm./L. or more. Values below this figure indicate a decrease in the concentrating ability of the kidney.

SOURCES OF ERROR

If the patient ingests more fluid than allowed, the results will not be valid.

REFERENCE

Fishberg, A. M.: Hypertension and Nephritis. 5th ed. Philadelphia, Lea & Febiger, 1954.

REFERENCES

1. Archibald, R. M.: Colorimetric measurement of uric acid. Clin. Chem., *3*:102, 1957.
2. Friedman, H. S.: Modification of the determination of urea by the diacetyl monoxime method, Anal. Chem., *25*:662, 1953.
3. Gentzkow, C. J.: An accurate method for determination of blood urea nitrogen by direct nesslerization. J. Biol. Chem., *143*:531, 1942.
4. Henry, R. J.: Clinical Chemistry: Principles and Technics. New York, Hoeber Medical Div., Harper & Row Publishers, 1964.
5. Johnson, R. B., and Hoch, H.: Osmolality of serum in urine. *In:* Standard Methods of Clinical Chemistry. S. Meites, Ed. New York, Academic Press, Inc., 1965, vol. 5, p. 159.
6. Marsh, W. H., Fingerhut, B., and Miller, H.: Automated and manual direct methods for the determination of blood urea. Clin. Chem., *11*:624, 1965.
7. Owens, J. K., Iggo, B., Scandrette, F. J., and Stewart, C. P.: The determination of creatinine in plasma or serum and in urine; a critical examination. Biochem. J., *58*:426, 1954.
8. Vanselow, A. P.: Preparation of Nessler's reagent. Industr. Eng. Chem., Analyt. Ed., *12*:516, 1940.
9. Wolf, A. V., Fuller, J. B., Goldman, E. J., and Mahony, T. D.: New refractometric methods for the determination of total protein in serum and urine. Clin. Chem., *8*:158, 1962.

ADDITIONAL READINGS

Abbrecht, P. H.: An outline of renal structure and function. Chem. Eng. Progr., Symp. Ser., *64*:1, 1968.

Cantarow, A., and Trumper, M.: Renal function. *In:* Clinical Biochemistry. 6th ed. Philadelphia, W. B. Saunders Co., 1962, pp. 373–446.

Caraway, W. T.: Uric acid. *In:* Standard Methods of Clinical Chemistry. New York, Academic Press, Inc., 1963, vol. 4, pp. 239–247.

Corcoran, A. C., Hines, D. C., and Page, I. H.: Kidney Function in Health; Kidney Function in Disease. Indianapolis, Lilly Laboratory for Clinical Research.

Josephson, B., and Eck, J.: The assessment of the tubular function of the kidneys. *In:* Advances in Clinical Chemistry. H. Sobotka and C. P. Stewart, Eds., New York, Academic Press, Inc. 1958, vol. 1.

Thurau, K., Valtin, H., and Schnermann, J.: Ann Rev. Physiol., *30*:441, 1968.

Chapter 13

LIVER FUNCTION TESTS

by Joseph I. Routh, Ph.D.

The liver is a large organ (1200 to 1600 gm.) shaped like a wedge with its base to the right. It has two major lobes, the larger right one being about six times the size of the left in the normal adult. The right lobe lies behind the rib cage with its upper border at about the level of the fifth rib. The smaller left lobe is tongue-shaped and its apex reaches to the dome of the left diaphragm (Fig. 13-1). The circulatory system of the liver is characterized by a dual blood supply. Approximately 80 per cent of the influent blood comes from the portal vein and 20 per cent from the hepatic artery. This rich flow of blood, amounting to about 1500 ml. per minute, brings oxygen and nutrient material to the liver. There is also an extensive system of lymph vessels that carry the interstitial fluid or liver lymph from that organ into the general circulation. In addition, a network of bile ducts transports substances secreted or excreted by the liver cells for storage in the gallbladder and subsequent use in the digestive process. The circulatory system will be examined in greater detail in connection with the function of liver cells and the biliary system.

The liver is composed of a multitude of functional units called *lobules*, which are small and hexahedral-shaped with vertical sides about twice the diameter of the horizontal cross section. A branch of the hepatic vein forms the central core of the lobule, and along the vertical sides in the portal tract or space lie branches of the portal vein, the hepatic artery, bile ducts, and lymph vessels (Fig. 13-2). The portal vein and hepatic artery provide blood that flows through special capillaries, called *sinusoids*, that empty into the central hepatic vein at right angles. The walls of the sinusoids are lined with special endothelial cells called *Küpffer cells*, which are phagocytic and are part of the reticuloendothelial system. Parenchymal liver cells in the form of plates or sheets one cell thick surround and separate the sinusoids and radiate out from the central core to the periphery of the lobule. Between the plates and the capillaries are tissue spaces (spaces of Disse) in which interstitial tissue fluid is formed and carried toward the periphery to the lymphatic system. Small bile canaliculi surround each liver cell and empty into capillaries that carry the bile to the periphery of the lobule.

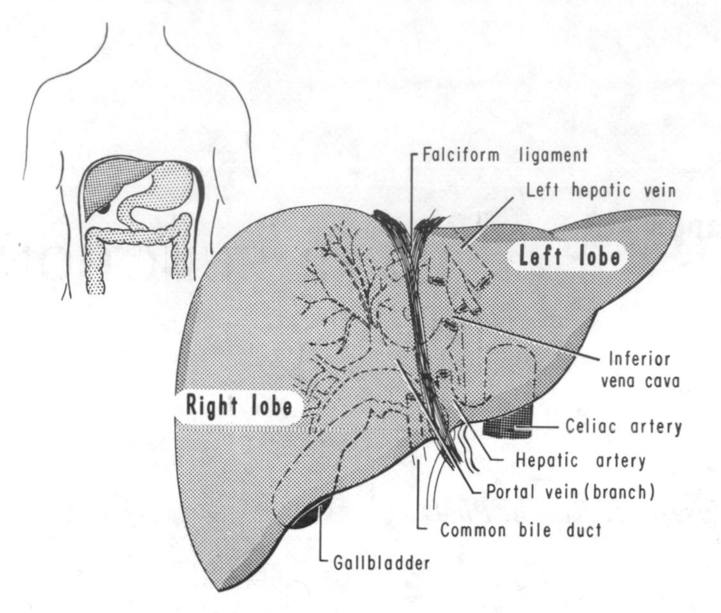

Figure 13-1. Anterior view of the liver showing major blood and bile vessels.

The gross appearance of a cut section of the liver would be a pattern of red dots representing the central vein of the lobule separated by greyish dots that represent the portal tracts, all against the brown background of liver tissue. The central veins are from 0.3 to 1.0 mm. apart and are a measure of the diameter of the lobule. The liver cells are approximately polyhedral in shape with a diameter of about 30 μ. Each gram of liver tissue contains about 200 million cells, of which 170 million are parenchymal and 30 million are sinusoidal epithelial cells (Kupffer cells).

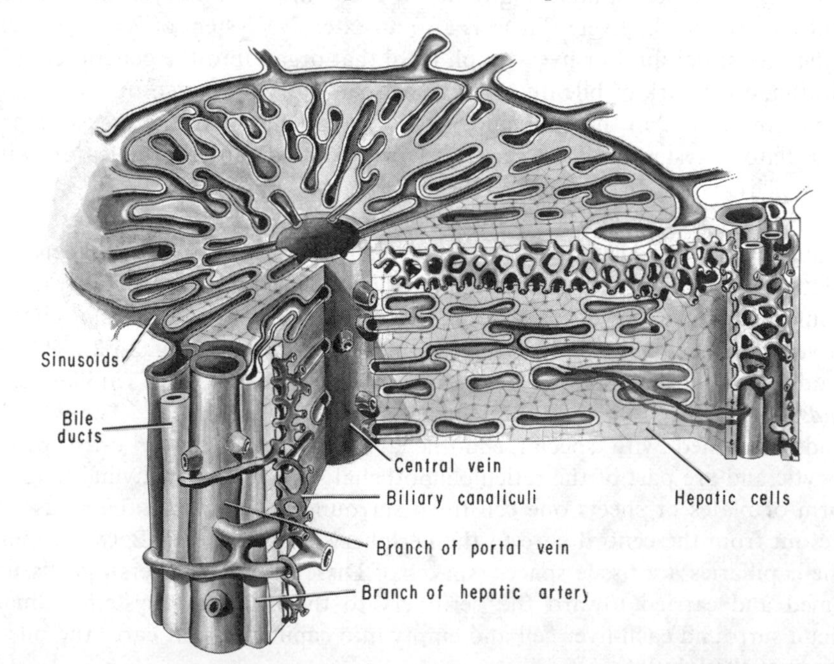

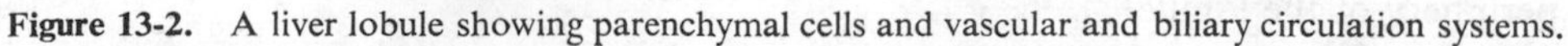

Figure 13-2. A liver lobule showing parenchymal cells and vascular and biliary circulation systems.

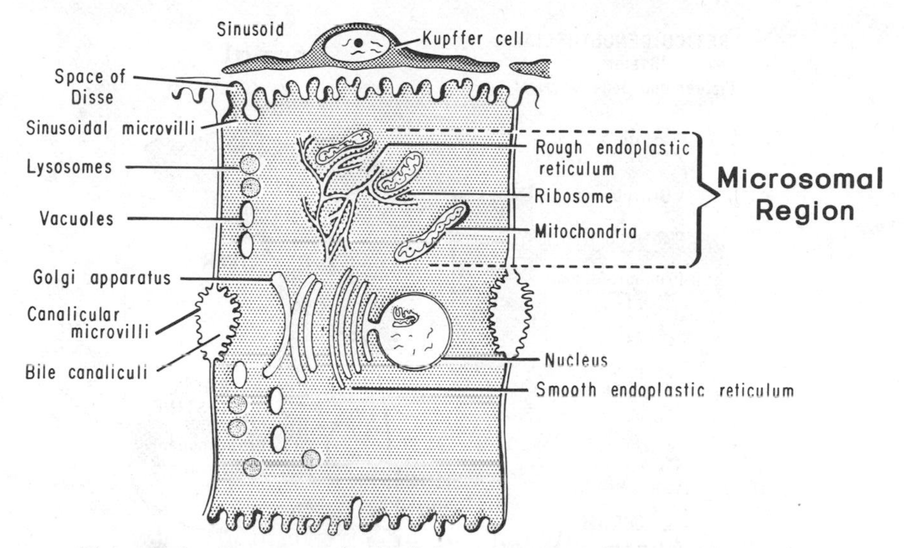

Figure 13-3. Structure and subcellular components of a normal liver cell.

In recent years, attention has been focused on the specific function of subcellular components; therefore, since the *parenchymal cells* of the liver are responsible for its functions, the lobule as a discrete unit may not be as important as had been thought. As is well known, the liver is the major site of intermediary metabolism and synthesis of many important compounds, the site of conjugation and detoxication of natural and potentially toxic foreign substances, and the site of storage of glycogen. It is also where synthesis and destruction of many of the factors involved in the blood clotting mechanism take place. These functions are carried out by subcellular components such as the nucleus, mitochondria, Golgi apparatus, endoplasmic reticulum, ribosomes, and the lysosomes (Fig. 13-3). The *mitochondria* are responsible for a large number of metabolic reactions, especially those that provide energy and involve oxidative phosphorylation. They contain many enzymes and coenzymes that are essential in the cycles and schemes of carbohydrate, protein, and fat metabolism. A very important function of the mitochondria is their ability to transform energy released from food into the energy bonds of adenosine triphosphate (ATP). The *ribosomes* exhibit related functions, largely synthetic and conjugating. The ribosomes are rich in ribonucleic acid (RNA) and are the site of protein synthesis within the cell. The *microsomal fraction*, obtained by centrifugation, is composed of free and bound ribosomes and pieces of rough and smooth endoplasmic reticulum that are involved in amino acid activation before protein synthesis, in cholesterol synthesis, in the conjugation of bilirubin, and in the detoxication of drugs. The *Golgi apparatus* is believed to function in the storage, transportation, and secretion of molecules such as proteins and bilirubin formed by the microsomal components. The *lysosomes* contain many hydrolytic enzymes and are involved in the metabolism of iron, bile pigments, and copper. The *Kupffer cells* contain a nucleus, mitochondria, and lysosomes, and sometimes phagocytosed material. They function as part of the reticuloendothelial system of the body.

Bilirubin Metabolism

The hemoglobin of aged or damaged red blood cells is converted by a complex series of reactions to the bile pigment, *bilirubin*. The average life of the red blood cell

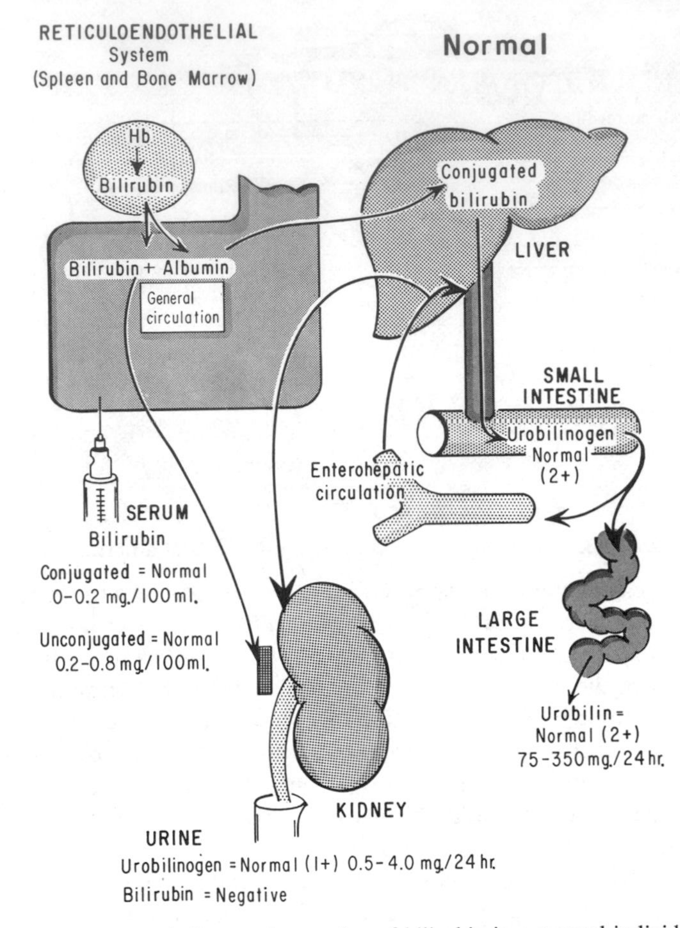

Figure 13-4. Metabolism and excretion of bilirubin in a normal individual.

in the body is 120 days and a total of about 6 gm. of hemoglobin is released each day from the disintegration of overaged cells. The cells of the reticuloendothelial system, especially in the spleen, liver, and bone marrow, first phagocytose the red cells and then convert the released hemoglobin into bilirubin. Hemoglobin consists of a protein, globin, attached to heme groups, which are porphyrin compounds containing iron. Each heme molecule is composed of four pyrrole rings joined by methene bridges and containing a central iron atom (see Chapter 6). One of the methene bridges of heme in hemoglobin appears to be split first by the reticuloendothelial cells to produce a biliverdin-iron-protein complex known as verdohemoglobin or choleglobin. The iron is then removed, the globin is released, and the biliverdin is reduced to form bilirubin (see Chart 13-1). The bilirubin is then transported from the extrahepatic sources as a *bilirubin-albumin complex* into the hepatic sinusoids. From the sinusoids, the bilirubin complex passes through the sinusoidal microvilli into the liver cell. The protein is separated from the complex, and bilirubin is converted into *bilirubin diglucuronide* by a reaction with uridine diphosphate glucuronate catalyzed by the enzyme system, *UDP-glucuronyl transferase.* It was concluded from the original experiments involving electrophoresis and chromatography of fresh bile that both mono- and diglucuronides of bilirubin were present in normal bile. Recent

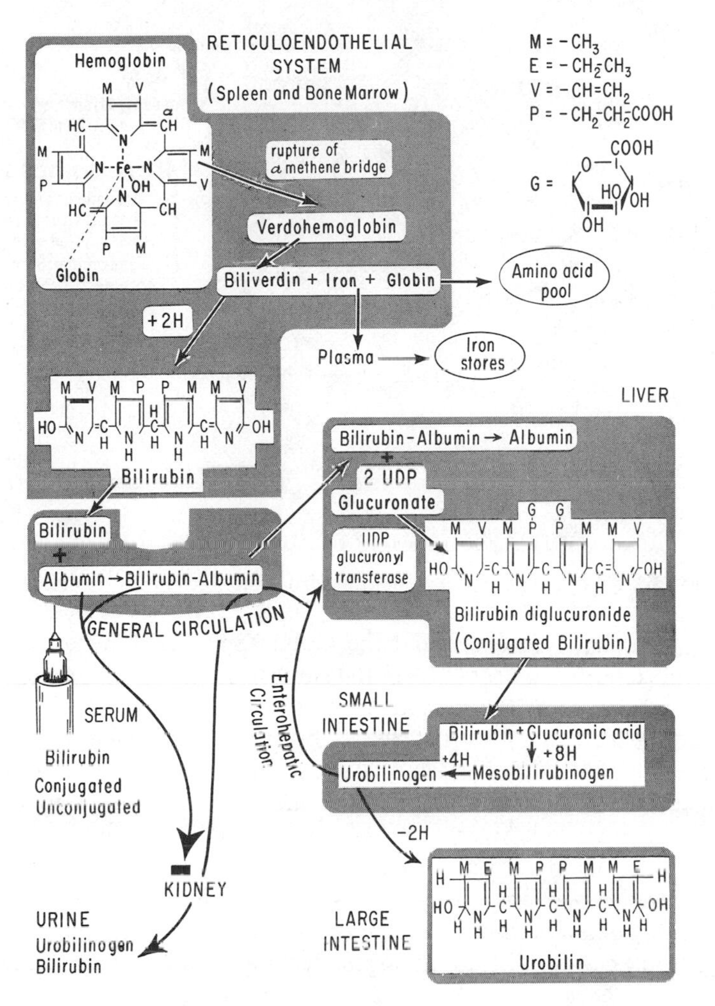

Chart 13-1. The normal metabolism of bilirubin.

studies have shown that "Pigment I," the monoglucuronide, is a complex of unconjugated bilirubin and bilirubin diglucuronide and that the monoglucuronide is not formed in the liver in the reaction catalyzed by UDP-glucuronyl transferase. Unconjugated bilirubin and bilirubin glucuronides are loosely bound to albumin and transported in this form in the blood (see Chart 13-1 and Figures 13-4, 13-5).

Conjugation to the glucuronide is a prerequisite for excretion of bilirubin into the bile and may be a limiting factor in transportation because conjugated bilirubin, when injected into the blood, is excreted in the bile much more rapidly than the unconjugated pigment. There are some clinical conditions in which excretion is abnormal despite a normal conjugating system which suggests that an active transport system may be required for excretion. From experimental work with fluorescent dyes similar to bilirubin, it has been postulated that after passing into the liver cell, bilirubin is first concentrated at the sinusoidal membrane surface and is then moved across the cell and concentrated at the canalicular membrane surface before excretion into the bile. The unidirectional nature of this process is vital; disturbances of these mechanisms produce leaks from the bile canaliculi back into the sinusoids, resulting in jaundice.

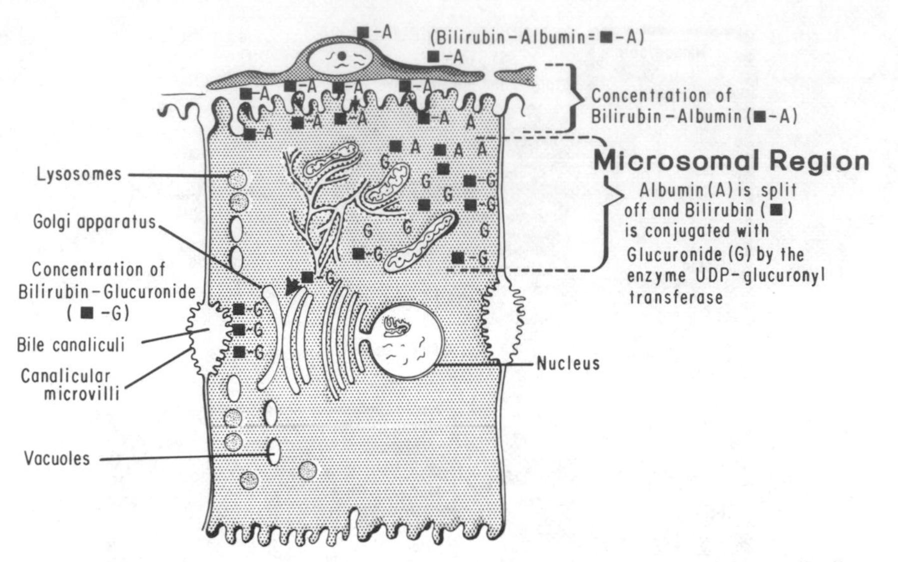

Figure 13-5. Normal transport of bilirubin from the sinusoids to the bile canaliculi.

Pinocytotic vacuoles, lysosomes, and the Golgi apparatus have all been implicated in the transportation of bilirubin from the sinusoids to the bile. A representation of this process is illustrated in Figure 13-5 (see also Figure 13-6).

The glucuronide, along with some free bilirubin, is excreted into the bile, passes into the small intestine, and is exposed to the reducing action of the enzymes of anaerobic bacteria. The reduction products, mesobilirubinogen, stercobilinogen, and urobilinogen, known collectively as *urobilinogen*, are first formed in the colon, then a portion of the urobilinogen is absorbed into the portal circulation and returned to the liver. The normal liver removes all but a small amount of the urobilinogen from the blood, probably oxidizes some of it to bilirubin, and excretes both into the bile for a return trip to the colon. The urobilinogens remaining in the colon are excreted in the

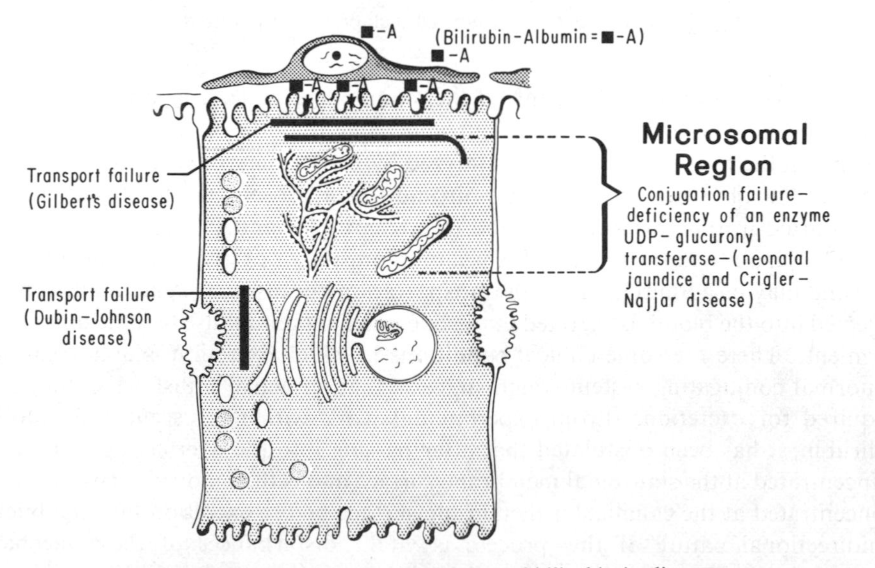

Figure 13-6. Impairment of transport of bilirubin in disease.

stool after being oxidized to form *urobilin* (stercobilin), an orange–brown-colored pigment. The scheme of bilirubin metabolism from the breakdown of hemoglobin to the formation of urobilin is shown in Chart 13-1.

JAUNDICE

Before consideration of the three major types of jaundice, the normal fate of bilirubin and urobilinogen should be illustrated. Figure 13-4 is a representation of the events that occur in the metabolism of bilirubin and the disposal of its compounds by the normal individual. The reticuloendothelial cells in the spleen and bone marrow convert the hemoglobin of red blood cells into bilirubin, which is loosely bound to albumin and carried to the liver, where the Kupffer cells may also produce more bilirubin from hemoglobin. After passage of the bilirubin-protein complex into the parenchymal liver cells, the protein is split from the complex and bilirubin is converted into the diglucuronide by enzyme systems in the microsomal region and is then passed into the bile canaliculi. The bile is transported to the common bile duct and then to the gallbladder, where it is concentrated and emptied into the small intestine. The bilirubin glucuronide is thus carried to the large intestine and, under the influence of the reducing enzymes of the bacteria in the colon, is changed into urobilinogen. Most of this urobilinogen is oxidized to D-, I-, and L-urobilin and excreted in the stool. The other part is carried back to the liver by the portal vein and may either be converted by the liver cells to bilirubin glucuronides or excreted into the bile canaliculi unchanged. The glucuronides and any urobilinogen that escapes transformation are carried back to the intestine by the bile and thus complete the enterohepatic circulation. The small amount of urobilinogen that is not taken up by the liver cells is carried in the general circulation to the kidney. Urobilinogen will pass through the kidney and normally 0.5 to 4.0 mg. per day is excreted. Unconjugated bilirubin is not excreted by the kidney and is absent from urine, whereas conjugated bilirubin, if present in abnormal concentration, is excreted by the kidney tubules.

The abnormal metabolism or retention of bilirubin usually results in *jaundice*, a condition that is characterized by an increase of bilirubin in the blood and a brownish-yellow pigmentation of the skin, sclera, and mucous membranes. One of several suggested schemes for the classification of jaundice is as follows:

Types of jaundice

A. Prehepatic jaundice: Hemolytic jaundice
   1. a. Acute hemolytic anemia
      b. Neonatal physiologic jaundice
   2. Chronic hemolytic anemia
B. Hepatic jaundice
   1. Conjugation failure
      a. Neonatal physiologic jaundice
      b. Crigler-Najjar disease*
   2. Bilirubin transport disturbances
      a. Preconjugation failure (Gilbert's disease)*
      b. Postconjugation failure (Dubin-Johnson disease)

---

* Some authors have classified conditions in which there is an inability to convert bilirubin to glucuronides as "prehepatic;" thus, the Crigler-Najjar disease and Gilbert's disease are, in some classifications, listed in this group.

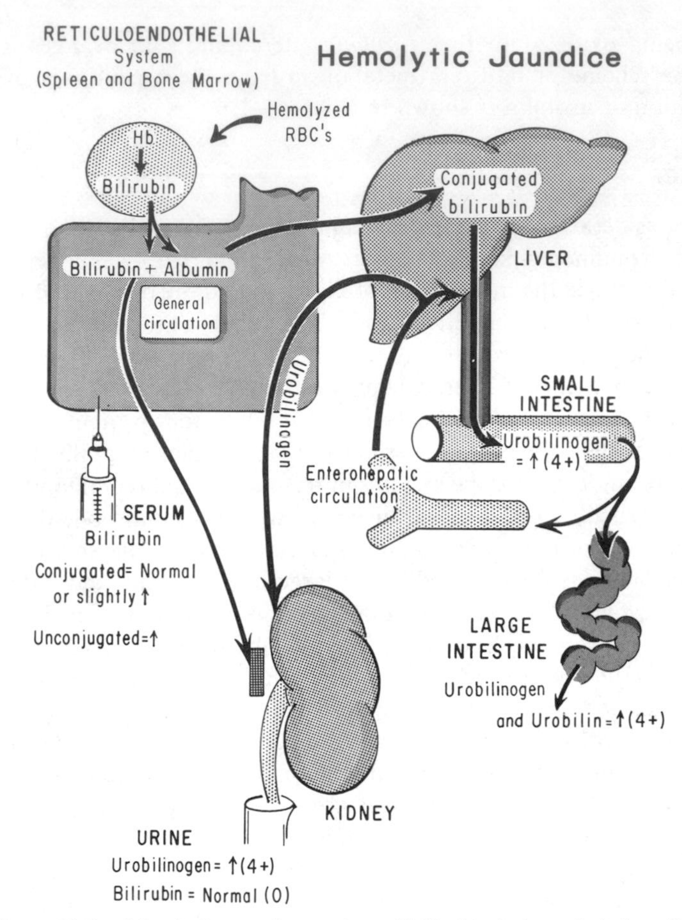

Figure 13-7. Metabolism and excretion of bilirubin in hemolytic jaundice.

3. Diffuse hepatocellular damage or necrosis
   a. Viral hepatitis
   b. Toxic hepatitis
   c. Cirrhosis
4. Intrahepatic obstruction (e.g., edema)

C. Posthepatic jaundice: obstruction of the common bile duct
   1. By stones
   2. By neoplasms
   3. By spasms or stricture

The *prehepatic type of jaundice* is commonly caused by hemolytic anemia. The increased destruction of red cells brings a larger load of bilirubin protein complex to the liver than the organ can handle (Fig. 13-7). When this bilirubin is converted to the glucuronide and excreted into the intestinal tract, an increased amount of urobilinogen is formed in the colon. A portion of this is excreted in the stool; the other portion is reabsorbed and returned to the liver. The liver cannot pick up the large quantities of urobilinogen; thus, increased amounts escape into the general circulation and are excreted into the urine by the kidneys. The increased breakdown of hemoglobin also causes an increase in the amount of unconjugated bilirubin in the blood without a corresponding increase in conjugated bilirubin.

The determination of urobilinogen in urine (and in stool) is a very useful clinical test in the diagnosis of *hemolytic anemia*. In this disorder, the 24 hour urine specimen may contain from 5 mg. to as high as 350 mg. of urobilinogen and the amount excreted in the feces may range from 350 to 1800 mg./24 hr. These values are considerably higher than the normal 0.5 to 4 mg. for urine and 75 to 400 mg./24 hr. for feces.

There are many cases of jaundice that involve the liver cell directly. There may be specific defects, as will be mentioned, or there may be diffuse damage such as in viral hepatitis, drug toxicity, and intrahepatic obstruction. The posthepatic type of jaundice usually involves the blockage of the common bile duct by a stone, stricture, or neoplasm.

Impairment of the liver cell mechanisms can result in several types of hepatic jaundice. As shown in Figure 13-6, the *transportation of bilirubin* or its protein complex from the sinusoidal membrane to the microsomal region may be impaired in the familial type of nonhemolytic jaundice known as *Gilbert's disease*. A deficiency of the enzyme system involving glucuronyl transferase is the cause of the hyperbilirubinemia of the *Crigler-Najjar disease*. This deficiency is also responsible for the lack of conjugation of bilirubin in the *physiologic jaundice* of the newborn, especially premature infants. This last mentioned type of jaundice is ordinarily short-lived, the deficient enzymes appear soon after the first few days of life, and there is also a concomitant decrease in hemoglobin destruction. Levels of unconjugated bilirubin

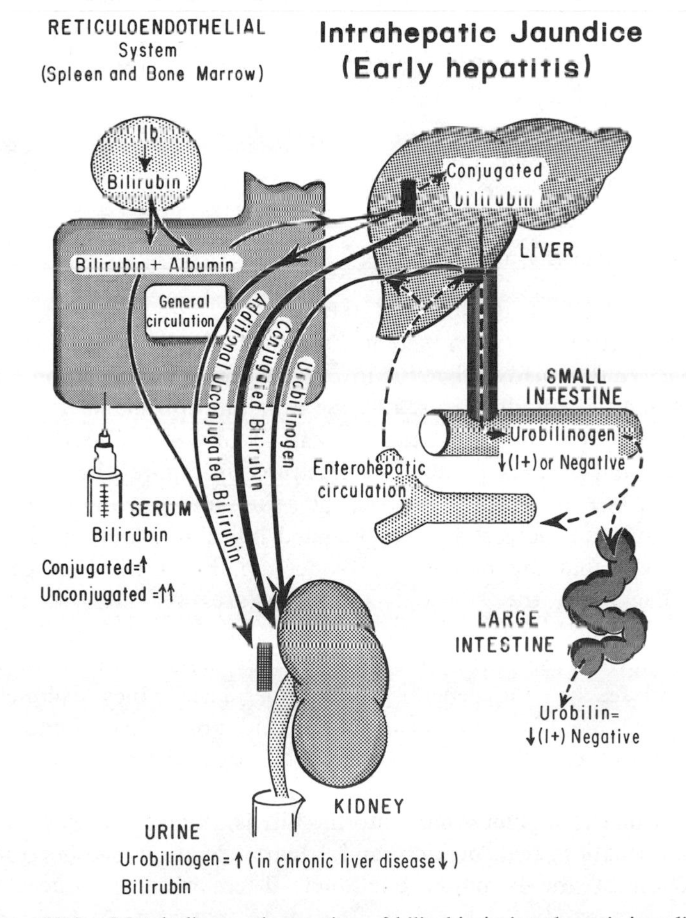

Figure 13-8. Metabolism and excretion of bilirubin in intrahepatic jaundice.

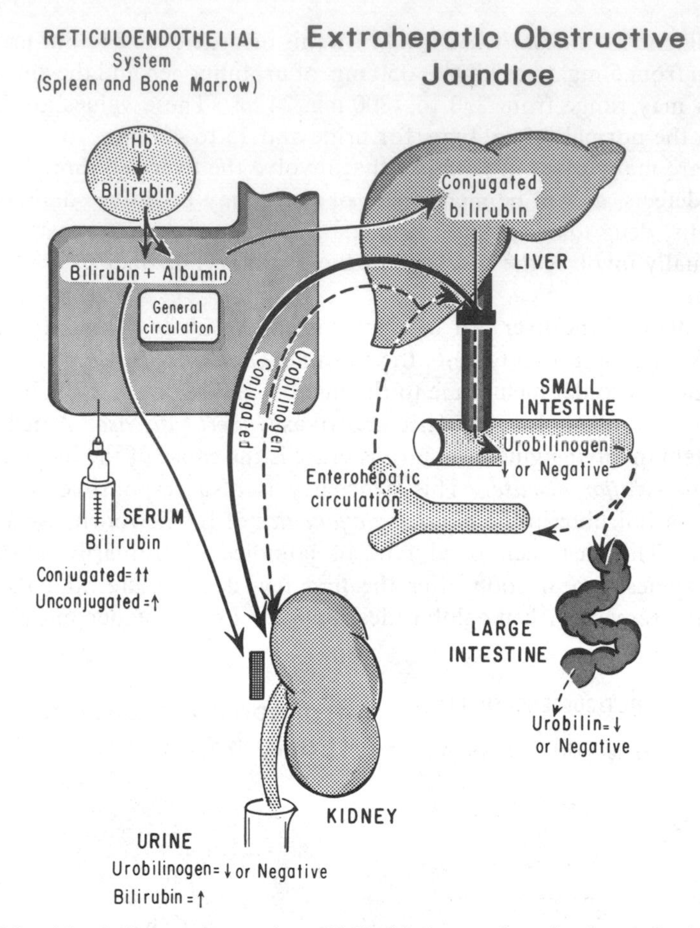

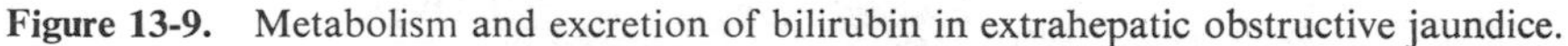

Figure 13-9. Metabolism and excretion of bilirubin in extrahepatic obstructive jaundice.

may reach 10 mg./100 ml. of serum at the height of the physiologic jaundice. A disturbance of the transportation system that carries the bilirubin glucuronides from the microsomes to the bile canaliculi is involved in the *Dubin-Johnson type* of hyperbilirubinemia and the jaundice caused by the drug chlorpromazine.

Viral hepatitis, *toxic hepatitis*, and *cirrhosis* involve overall damage and necrosis of the liver cells and produce jaundice as illustrated in Figure 13-8. The injured cells lose their ability to extract bilirubin from the serum and to conjugate bilirubin with glucuronate; thus, an increase in unconjugated bilirubin occurs in the serum. The parenchymal cells that are damaged provide a path for the leakage of bilirubin glucuronide back into the sinusoids, which causes an increase in conjugated bilirubin as well. Furthermore, the damaged cells are not capable of removing all of the urobilinogen from the portal blood in the enterohepatic circulation and more of this compound reaches the general circulation. The kidney excretes the excess urobilinogen and the bilirubin glucuronides, and appreciable quantities of these compounds are found in the urine. Urine urobilinogen determinations are useful in the detection of early viral hepatitis, whereas in later stages of the disease the formation and passage of bilirubin glucuronide into the bile is decreased to such an extent that urobilinogen formation, reabsorption, and urinary excretion also decrease.

Obstruction of the common bile duct, if complete, produces a condition illustrated in Figure 13-9. A gallstone, spasm, or neoplasm that blocks the passage of bile into the intestine causes regurgitation of bile into the sinusoids. The bilirubin glucuronides cannot reach the intestine and, therefore, no urobilinogen is produced for recirculation to the liver or excretion in the stool. Urobilin, the pigment of the stool, is lacking and the feces vary from a light brown to a chalky white color. The urine does not contain any urobilinogen, but does contain appreciable amounts of bilirubin glucuronides. Since there are conditions in which the posthepatic blockage is intermittent or incomplete, the stools may contain decreased amounts of urobilinogen, and small quantities of this compound may appear in the urine. The blood contains increased amounts of bilirubin and bilirubin glucuronides caused by regurgitation and impairment of liver cell function. In a severe case of *obstructive jaundice*, the serum may contain 12 mg./100 ml. of conjugated bilirubin and a total bilirubin level of 18 mg./100 ml. or higher.

Methods for the Evaluation of Liver Function

Icterus Index

A patient with jaundice is sometimes referred to as icteric. The *icterus index* was developed by Meulengracht in 1919 as a measure of the degree of icterus in a plasma or serum specimen in cases of jaundice. A serum sample (1 ml.) was diluted with saline until the color matched that of a 1:10,000 (w/v) solution of potassium dichromate and the dilution factor was reported as units. The method has been subsequently modified and under favorable conditions serves as an approximate measure of the concentration of bilirubin in the serum. The relation of icterus index units to milligrams of bilirubin is approximately 10 units = 1 mg. This relationship is better at higher than at lower levels of bilirubin because interfering pigments such as carotenes and hemoglobin contribute relatively more to the reading of samples with low icterus index than to samples with high icterus index, where the interference may be negligible. Since the icterus index only approximates the level of bilirubin in the serum, this method should not replace the quantitative estimation of bilirubin.

CLINICAL SIGNIFICANCE

Substances other than bilirubin in the serum may contribute to the icterus index, thus limiting its clinical usefulness. A value in the normal range is of diagnostic significance, since it excludes the possibility of the presence of increased amounts of bilirubin. In general, the greatest value of the method is in serial determinations of the index to follow the progression of established jaundice in liver disease or obstruction.

SPECIMEN

For an accurate determination of the icterus index, the plasma or serum sample used should be free from hemolysis and turbidity. A fasting sample is therefore preferred.

PRINCIPLE

The yellow color of a serum or plasma sample diluted with a sodium citrate solution is read at a wavelength of 460 nm. The results are expressed in units obtained by comparison with a standard potassium dichromate solution.

REAGENTS

1. Sodium citrate solution, 5 per cent (w/v). Dissolve 50 gm. of $Na_3C_6H_5O_7{\cdot}2H_2O$ in water and dilute to volume in a 1 L. volumetric flask.

2. Potassium dichromate standard. Dissolve 157 mg. of $K_2Cr_2O_7$ in water, add 1 drop of concentrated sulfuric acid and dilute with water to the volume in a 1 L. volumetric flask. The solution represents 1 icterus unit and is stable indefinitely at room temperature.

PROCEDURE

1. Dilute clear, nonhemolyzed serum 1:10 with the sodium citrate solution. The final volume should be adequate to read in the cuvets used in the laboratory. After mixing, read the absorbance at 460 nm. against a blank of sodium citrate solution.

2. Read the absorbance of the dichromate standard against a sodium citrate blank at 460 nm.

CALCULATIONS

$$\text{Units of icterus index} = \frac{\text{absorbance of specimen}}{\text{absorbance of standard}} \times 10$$

If the absorbance of the specimen is three to five times higher than that of the standard, make further dilutions of the specimen with sodium citrate solution until the absorbance values are in the same range. The new dilution factor should then be used instead of the (10) in the preceding equation.

NORMAL VALUES

Normal values range from 3 to 8 icterus units.

PROCEDURE NOTES

1. In the original method, visual matching against a 1:10,000 solution of $K_2Cr_2O_7$ (0.01 per cent) yielded icterus index units. Henry et al. (1953) modified the method for spectrophotometers and found at a wavelength of 460 nm. that 0.0157 per cent solutions of $K_2Cr_2O_7$ gave more accurate icterus index units.

2. Both bilirubin in serum and dichromate solutions follow Beer's law in a narrow band-pass spectrophotometer, but not in a filter photometer. When using the Klett colorimeter, the sample should be diluted until the reading will be less than 150 Klett units, unless a standard curve is constructed.

SOURCES OF ERROR

1. Slight turbidity or hemolysis does not markedly affect the values obtained by the method, but specimens with visible lipemia or hemolysis are unsatisfactory. Hemoglobin interference is minimized by reading at 460 nm. instead of 420 nm. as carried out in earlier modifications.

2. Lipochrome pigments such as carotene are detected with the method; therefore, foods such as carrots should be avoided the day before the test.

3. Since bilirubin is destroyed by light, the sample should be protected from strong lights.

REFERENCE

Henry, R. J., Golub, O. J., Berkman, S., and Segalove, M.: Am. J. Clin. Path., *23*:841, 1953.

SERUM BILIRUBIN

In 1883, Ehrlich described the coupling reaction of bilirubin with diazotized sulfanilic acid to form a blue pigment in strongly acid or alkaline solutions. Van den Bergh applied this color reaction to the quantitative determination of serum bilirubin and reported the effect of alcohol on the rate of the coupling reaction. Malloy and Evelyn in 1937 introduced the use of 50 per cent methanol in the serum determination, a concentration that avoided precipitation of the proteins. These authors developed the first useful quantitative method. For many years, results of bilirubin determinations in serum were reported as values for direct bilirubin (the fraction that produced a color in the van den Bergh method in aqueous solution) and indirect bilirubin (the fraction that produced a color only after alcohol was added). Since 1956, it has been known that the "*direct*" *reaction* is given by the diglucuronide of bilirubin, so-called conjugated bilirubin that is water soluble. The "*indirect*" *reaction*, on the other hand, is given by unconjugated bilirubin that is water insoluble, but dissolves in alcohol to couple with the diazotized sulfanilic acid.

The reactions involved in the coupling of bilirubin with Ehrlich's reagent are shown in the following scheme:

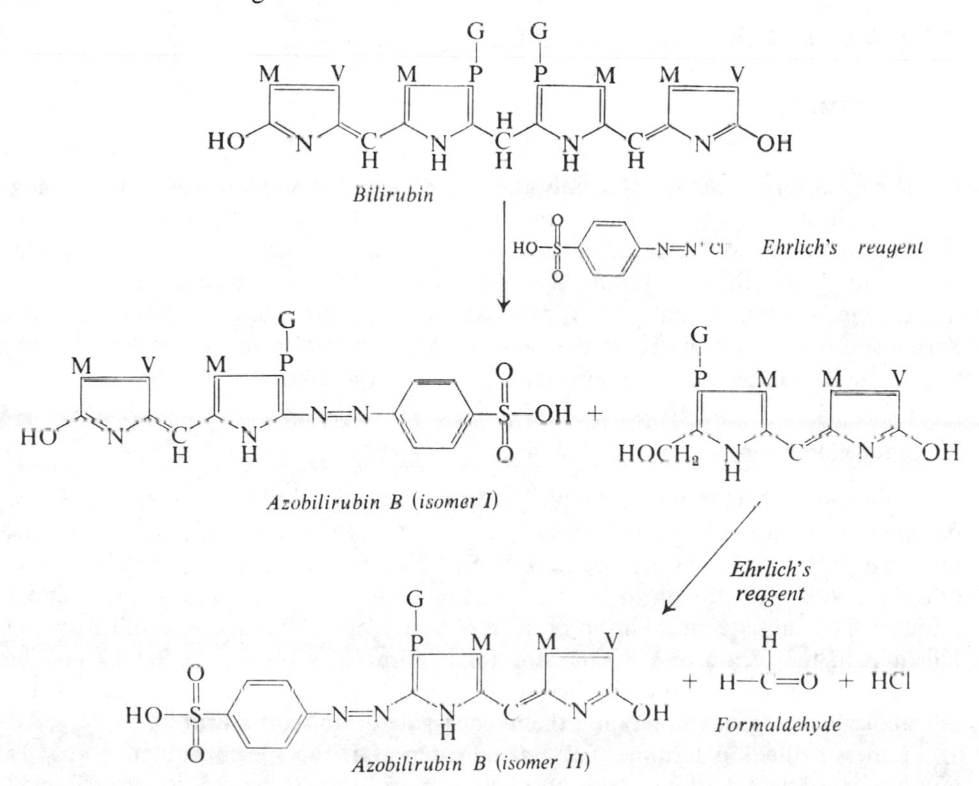

Chart 13-2. The formation of isomer I and II of azobilirubin B through the reaction of bilirubin glucuronide with Ehrlich's reagent. Unconjugated bilirubin reacts in the same way, resulting in isomer I and II of azobilirubin A.

The literature is filled with discussions and modifications of the van den Bergh method, especially that proposed by Malloy and Evelyn. Compounds other than methyl alcohol have been used to accelerate the color reactions of bilirubin, and considerable attention has been paid to the time period required to assure reaction of the glucuronide in aqueous solution. Also, it has been established that hemoglobin interferes with the diazo reaction and depresses the values for total bilirubin in hemolyzed serum specimens commonly obtained from newborn infants. Therefore, it is desirable to have available a method for total bilirubin that is not affected by hemolysis and can be carried out on the small specimens of serum obtained from newborns.

CLINICAL SIGNIFICANCE

As discussed earlier in the chapter, an increase primarily in the serum concentration of bilirubin glucuronides is seen in obstructive jaundice. An increase primarily in the "indirect" fraction, representing unconjugated bilirubin, is observed in hemolytic jaundice and in neonatal jaundice. In the newborn, jaundice may be caused by Rh, ABO, or other blood group incompatibilities, hepatic immaturity, or by hereditary defects in bilirubin conjugation (see Types of Jaundice). Both conjugated and unconjugated bilirubin are increased in the serum in hepatitis.

Determination of Direct and Total Bilirubin in Serum

SPECIMEN

To avoid lipemia, serum is preferably obtained from a blood specimen collected in the postabsorptive state. Hemolysis must be avoided since it causes falsely low results in the diazo method. Before analysis, the serum should be kept in the dark and the determination carried out as soon as possible and not later than 2 to 3 hours after clotting of the blood. Direct sunlight at room temperature may cause up to 50 per cent decrease in bilirubin within 1 hour, especially when serum is kept in capillary tubes. Serum specimens may be kept in the dark in a refrigerator for up to a week and in the freezer for 3 months without appreciable change in the bilirubin content.

PRINCIPLE

Bilirubin in the serum is coupled with diazotized sulfanilic acid to form *azobilirubin.* The intensity of the purple color that is formed is proportional to the bilirubin concentration in the serum. Bilirubin glucuronides, the conjugated or direct bilirubin, react with the diazo reagent in aqueous solution to form a color within 1 minute. The subsequent addition of alcohol accelerates the reaction of all forms of bilirubin in the serum and a value for total bilirubin is obtained after letting the specimen stand for 30 minutes. The total bilirubin value represents the sum of the bilirubin glucuronides (direct) and the unconjugated (indirect) bilirubin.

Other methods determine the total bilirubin after the specimen stands only 15 minutes, since the additional color production from 15 to 30 minutes is very minimal. No appreciable errors are introduced if the method is standardized under conditions of 15 minute standing.

It should be noted that the azobilirubin color has indicator properties and in strongly acid or strongly alkaline solution the color is blue. This blue color is more intense than the familiar purple color produced at moderately acid pH and is less affected by nonbilirubin pigments. Stoner and Weisberg developed a micromethod that employs a strongly acid pH; in the method of Jendrassik and Grof the diazotization reaction is accelerated by caffeine and sodium benzoate and the final blue color is developed at a strongly alkaline pH.

REAGENTS

1. Methyl alcohol, absolute, A.R.
2. Reagent A. Dissolve 5 gm. of sulfanilic acid in 60 ml. of concentrated hydrochloric acid in a 1 L. volumetric flask and dilute to the mark with water. The solution is stable indefinitely.
3. Sodium nitrite solution, 20 per cent (w/v) (stock). Dissolve 20 gm. of $NaNO_2$ in water and dilute to a volume of 100 ml. The solution is stable up to 4 weeks when stored in the refrigerator.
4. Reagent B. Sodium nitrite solution, 2 per cent. Prepare fresh daily by diluting the 20 per cent $NaNO_2$ solution 1:10.
5. Diazo reagent. Add 0.3 parts of Reagent B to 10 parts of Reagent A and mix. This reagent should be prepared within 30 minutes of the time of use.
6. Diazo blank. Dilute 60 ml. of concentrated hydrochloric acid to a volume of 1 L. with water. The reagent is stable indefinitely at room temperature.

PROCEDURE

1. Into a test tube labeled B pipet 0.5 ml. of clear unhemolyzed serum and 9.5 ml. of distilled water and mix by gentle inversion.
2. Transfer 5.0 ml. of the diluted serum to a second tube labeled X.
3. Add 1.0 ml. of diazo blank solution to tube B and 1.0 ml. of diazo reagent to tube X and mix immediately.
4. Exactly 1 minute after the addition of the diazo reagent, read the absorbance of tube X at 540 nm. against tube B, which was set at zero absorbance, or at 100 per cent T. When several specimens are run at the same time, the addition of the diazo reagent (step 3) should be spaced to enable readings to be carried out at exactly 1 minute.
5. Blow 6.0 ml. of methyl alcohol into each tube, mix by gentle inversion, let stand for 30 minutes, and read the absorbance of tube X against B.

CALCULATIONS

1. Values for total bilirubin may be obtained directly from the calibration curve for total bilirubin.
2. Values for bilirubin glucuronides, the direct or 1 minute bilirubin, may be obtained by dividing by 2 the values that are read from the calibration curve of absorbance versus concentration of total bilirubin.
3. The values for unconjugated (indirect) bilirubin are obtained by subtracting the 1 minute (direct) bilirubin from the total bilirubin.
4. In laboratories where microcuvets are used, the method may be scaled down by using one tenth of the volumes in the procedure. A special precaution is necessary for use with specimens from infants with neonatal jaundice: if hemolysis is apparent, low values will be obtained for total bilirubin.

CALIBRATION CURVE

It has recently been recommended by a committee composed of representatives from the American Academy of Pediatrics, the College of American Pathologists, the American Association of Clinical Chemists, and the National Institutes of Health, that an acceptable bilirubin for the preparation of calibration curves gives a 1 cm.

molar absorptivity between 59,100 and 62,300 at 453 nm. in chloroform at 25°C. The molar absorptivity is equal to the absorbance of a 1 *M* solution in a specified solvent at a specified wavelength and a 1 cm. light path. For example, the average acceptable molar absorptivity for bilirubin at 453 nm., in chloroform at 25°C., would be 60,700. This means that a 1 molar solution of bilirubin would yield an absorbance (*A*) reading of 60,700 or, for practical spectrophotometry, a 1/60,700 molar solution (9.52 mg./*L*.) would produce an *A* of approximately 1.0. In addition, the committee recommended the preparation of *bilirubin standards in serum* and standardization of the color produced from azobilirubin in the serum standards. Bilirubin standards in serum require an acceptable serum diluent, which is prepared by pooling approximately 150 ml. of nonhemolyzed, nonicteric, nonlipemic serum. The acceptability is checked by diluting 1 ml. of the pooled serum to 25 ml. with 0.85 per cent sodium chloride solution. The absorbance of this diluted serum read against a 0.85 per cent sodium chloride blank set at zero *A* must be less than 0.100 at 414 nm. and less than 0.040 at 460 nm. to be acceptable as a bilirubin diluent. The total bilirubin values to be used for the preparation of the calibration curve should be obtained as follows:

1. Weigh out exactly 20.0 mg. of an acceptable bilirubin, transfer to a 100 ml. volumetric flask, and dissolve by adding 2 ml. of 0.1 *M* Na_2CO_3 and 1.5 ml. of 0.1 *N* NaOH. If standards more concentrated than 20 mg./100 ml. are prepared, add 0.5 ml. of 0.1 *N* NaOH for each additional 5 mg. of bilirubin. The solution should be clear and red in color. Dilute this solution to 100 ml. with the acceptable pooled serum and mix well. Since the bilirubin is light sensitive, the flask should immediately be wrapped with aluminum foil and kept in the dark as much as possible.

2. Standard solutions for the preparation of the calibration curve are obtained as follows:

Total bilirubin (mg./100 ml.)	*Pooled serum* (ml.)	*20 mg./100 ml. standard* (ml.)
0	4	0
2	9	1
5	3	1
10	2	2
15	1	3
20	0	4

3. Determine the absorbance of each of the preceding solutions using the procedure for the analysis of serum. Subtract the absorbance value of the respective pooled serum blank from the absorbance of each of the five standard solutions. The corrected absorbance values may then be used to plot absorbance versus concentration to prepare the calibration curve for total bilirubin.

NORMAL VALUES

Normal values for serum bilirubin content of newborns, infants, and adults are shown in the following tabulation:

Age	*Premature* (mg./100 ml.)	*Full term* (mg./100 ml.)
Up to 24 hours	1–6	2–6
Up to 48 hours	6–8	6–7
Day 3 to 5	10–12	4–6
After 1 month	0–0.3 (conjugated)	
(all normal infants)	0.1–0.7 (total)	
Adults:		
Conjugated:	0–0.2	
Unconjugated:	0.2–0.8	
Total:	0.2–1.0	

REFERENCES

Jendrassik, L., and Grof, P.: Biochem. Z., *297*:81, 1938.
Malloy, H. T., and Evelyn, K. A.: J. Biol. Chem., *119*:481, 1937 (modified).
Meites, S., and Hogg, C. K.: Clin. Chem., *5*:470, 1959.
Standard Methods of Clinical Chemistry, *5*:75, 1965.
Stoner, R. A., and Weisberg, H. F.: Clin. Chem., *3*:22, 1957.

Direct Spectrophotometric Method for Total Bilirubin in Serum

SPECIMENS

Serum is obtained from capillary blood of infants, which is drawn from the heel or fingertip. Use of such specimens poses two major problems: the necessity of a micro- or ultramicromethod because of the small amount of blood that is drawn, and the fact that there is frequent hemolysis of the serum sample.

PRINCIPLE

The absorbance of bilirubin in the serum at 455 nm. is proportional to its concentration. The serum of newborn infants does not contain lipochromes, such as carotene, and other pigments that increase the absorbance at 455 nm; however, these pigments may be present in serum from adults. The absorbance of hemoglobin at 455 nm. is corrected by subtracting its absorbance at 575 nm.

REAGENTS

Phosphate buffer, pH 7.4. Weigh out 7.65 gm. of $Na_2HPO_4 \cdot 7H_2O$ and 1.74 gm. of anhydrous KH_2PO_4. Dissolve in water and dilute to a volume of 1000 ml. in a volumetric flask. Check the pH with a pH meter.

PROCEDURE

1. With a micropipet, add 20 μl. of serum to a microcuvet (10 mm. light path) containing 1 ml. of phosphate buffer. Rinse the pipet several times with the buffer.
2. Add 1 ml. of phosphate buffer to a microcuvet and, using this as a blank, set at zero absorbance; read the absorbancies of the diluted serum at 455 and 575 nm. To obtain the accurate values of absorbancies required in this method, a spectrophotometer capable of transmitting bandwidths of 10 nm. or less should be used.

CALCULATIONS

The absorbance values of the serum at 455 nm. or 575 nm. are due to both bilirubin and hemoglobin. Two equations are required to express the effect of both pigments.

$$A_{455} = (K_{b455} \times C_b) + (K_{h455} \times C_h) \quad (1)$$

$$A_{575} = (K_{b575} \times C_b) + (K_{h575} \times C_h) \quad (2)$$

In equations (1) and (2), C_b is concentration of bilirubin in the sample used, A_{455} and A_{575} are the absorbancies at 455 and 575 nm., and K_{b455}, K_{b575}, K_{h455}, and K_{h575} are the absorption constants of bilirubin and hemoglobin solutions at 455 and 575 nm., respectively. These two equations may be solved for C_b by multiplying (1) by K_{h575} and (2) by K_{h455}

$$A_{455} \times K_{h575} = (K_{b455} \times K_{h575} \times C_b) + (K_{h455} \times K_{h575} \times C_h) \quad (3)$$

$$A_{575} \times K_{h455} = (K_{b575} \times K_{h455} \times C_b) + (K_{h575} \times K_{h455} \times C_h) \quad (4)$$

Subtracting (4) from (3), the expression $K_{h455} \times K_{h575} \times C_h$ cancels and C_b may be expressed as

$$C_b = \frac{(A_{455} \times K_{h575}) - (A_{575} \times K_{h455})}{(K_{b455} \times K_{h575}) - (K_{b575} \times K_{h455})} \qquad (5)$$

The absorption constants for bilirubin and hemoglobin are determined with the same spectrophotometer that is used in the procedure with diluted serum specimens. Pure bilirubin dissolved in Na_2CO_3 solution or hemoglobin prepared from lysed erythrocytes (previously washed with saline) is added to a 5 per cent albumin solution or a hemoglobin-free, low bilirubin (<0.3 mg./100 ml.) pooled serum to prepare two series of standard solutions (bilirubin 2 to 20 mg. per cent, and hemoglobin 5 to 100 mg. per cent). These solutions are treated as serum specimens (see under Procedure) and the *A* value for each solution is determined. The absorption constants at 455 and 575 nm. are obtained by dividing *A* by the milligram per cent for each solution. The *K* values are averaged to obtain the four values required in equation (5). An example using reasonable *K* values will illustrate the simplification of equation (5).

$$K_{b455} = 0.780$$
$$K_{b575} = 0.0115$$
$$K_{h455} = 0.0103$$
$$K_{h575} = 0.0098$$

$$C_b = \frac{0.0098A_{455} - 0.0103A_{575}}{(0.780 \times 0.0098) - (0.0115 \times 0.0103)} = \frac{0.0098A_{455} - 0.0103A_{575}}{0.00753}$$
$$= 1.30A_{455} - 1.37A_{575}$$
$$C_b\text{, mg. per cent} = \text{dilution} \times (1.30A_{455} - 1.37A_{575})$$

REFERENCES

Meites, S., and Hogg, C. K.: Clin. Chem., *6*:421, 1960 (modified).
White, D., Haidar, G. A., and Reinhold, J. G.: Clin. Chem., *4*:211, 1958.

URINE BILIRUBIN

It has been recognized for many years that the presence of conjugated bilirubin in the urine suggests hepatocellular disease or obstructive jaundice. Generally, one is concerned only with whether bilirubin is present in urine, not with the exact amount; for this reason, many qualitative tests have been devised. These tests have been based on the observation of the color of the urine, the characteristic colors formed on oxidation of bilirubin, the addition of methylene blue until the yellow-brown color of the urine becomes blue, and diazotization of the bilirubin. The sensitivity and specificity of a qualitative method is of prime importance in the choice of a method. In general, methods based on diazotization are most satisfactory.

CLINICAL APPLICATION

If a method is available that consistently gives a negative test with normal urines, and yet is sensitive enough to detect slightly increased quantities of bilirubin, it is valuable in clinical diagnosis. In any form of hepatitis that involves impairment or destruction of liver cells, in transportation defects such as Dubin-Johnson syndrome, and in obstructive jaundice, conjugated bilirubin is excreted in the urine.

Methods for the Determination of Bilirubin in Urine

Although many attempts have been made to modify the various qualitative methods, as yet there has not been an acceptable quantitative method developed for the routine laboratory.

FOUCHET'S TEST

A test that is often used in the laboratory is based on the green color produced by the reaction of bilirubin with Fouchet's reagent (1 per cent ferric chloride in 25 per cent trichloroacetic acid solution). A common modification of the Fouchet's test is carried out as follows: A strip of thick filter paper, impregnated with a saturated solution of barium chloride and dried, is inserted for about half of its length into the urine specimen. The strip is removed and a few drops of Fouchet's reagent are added at the boundary between the wet and dry portion of the strip. A green color is produced if bilirubin is present in the urine; the intensity of color varies with the amount of bilirubin. The color response is usually graded 0 to 4+

ICTOTEST

A qualitative test employing diazotization that is both sensitive and specific has been developed by Free and Free. Semiquantitative results may also be obtained by this method.

SPECIMEN

A fresh specimen of urine is most satisfactory for this test. Bilirubin in the urine is unstable and will decrease in concentration in light and at room temperature. Specimens may be kept in the dark in a refrigerator for 1 day.

PRINCIPLE

Diazotization of bilirubin in the urine specimen under acid conditions produces a blue to purple color. The diazo agent is p-nitrobenzenediazonium p-toluenesulfonate, which produces a color within 30 seconds. The speed of color development and the intensity of the color are related to the amount of bilirubin in the urine.

REAGENTS

1. A test mat composed of equal amounts of asbestos and cellulose fibers is used to adsorb and concentrate the urine bilirubin and other bile pigments on its surface.
2. A powdered mixture composed of 0.6 part of p-nitrobenzenediazonium p-toluenesulfonate, 10 parts of $NaHCO_3$, 100 parts of sulfosalicylic acid, and 20 parts of boric acid may be used to carry out the diazotization reaction.

The proper test mats and tablets containing the diazo mixture are conveniently available under the trade name Ictotest (Ames Co., Div. Miles Laboratories, Elkhart, Ind.).

PROCEDURE

1. Place 5 drops of fresh urine in the center of one square of the test mat.
2. Place an Ames Ictotest reagent tablet in the center of the moistened area.
3. Add 2 drops of water onto the tablet, making sure that the water flows from the tablet onto the mat.

4. A positive test is indicated by a blue or purple color around the tablet within 30 seconds. A pink to red color or any color that develops after 30 seconds should be ignored.

5. Semiquantitation of the test can be carried out by making serial dilutions of positive reacting urine specimens. The highest dilution that still gives a positive test will have a bilirubin concentration of 0.1 mg./100 ml. A simple calculation (multiplication by the dilution factor) will yield the approximate bilirubin concentration of the original urine.

SOURCES OF ERROR

The specificity of many of the qualitative tests for bilirubin in urine is questionable. The Ictotest just outlined has a high degree of specificity; it seldom (see the following) yields false positive tests and is sensitive enough to pick up weak positive tests sometimes missed by other methods. The sensitivity of the test ranges from 0.05 to 0.1 mg. of bilirubin per 100 ml. of urine. High levels of indican, urobilin, or salicylate will produce a red color in the test; Pyridium and Serenium, drugs used in urinary tract infections, color the urine red and also yield an atypical color in the test.

REFERENCES

Free, A. H., and Free, H. M.: Gastroenterology, *24*:414, 1953.
Watson, C. V., and Hawkinson, V.: J. Lab. Clin. Med., *31*:914, 1946.

UROBILINOGEN IN URINE AND FECES

As already mentioned under Bilirubin Metabolism, urobilinogen is the name given to the end products of bilirubin metabolism: mesobilirubinogen, stercobilinogen, and urobilinogen. These compounds, known collectively as urobilinogen, are colorless reduction products of bilirubin, and are oxidized to the brown pigments D-, I- and L-urobilin (stercobilin).

In the normal individual, part of the urobilinogen that is formed by the reduction of bilirubin in the colon is excreted in the feces, and the remainder is reabsorbed into the portal blood and returned to the liver. A small amount escapes re-excretion into the intestine by the liver and is excreted in the urine.

CLINICAL SIGNIFICANCE

In various forms of hepatitis involving impairment or destruction of liver cells, the liver is unable to remove an appreciable fraction of the urobilinogen from the portal blood; thus, increasing amounts are excreted by the kidney. The determination of urine urobilinogen is, in fact, a very sensitive and useful test for detecting the early stages of hepatitis. In obstructive jaundice, bilirubin excretion into the intestine stops or is greatly decreased. This decreases the formation of urobilinogen; consequently, its excretion in the urine is decreased. Fecal excretion of urobilinogen in the presence of a normally functioning liver is dependent on the rate of breakdown of hemoglobin. Hemolytic anemia increases the amount of urobilinogen excreted, but anemias not related to destruction of red blood cells decrease it. Liver disease, in general, reduces the flow of bilirubin glucuronides to the intestine and thus decreases fecal excretion of urobilinogen. Complete obstruction of the bile duct reduces the urobilinogen of the stool to very low values. The "clay-colored" stool of obstructive jaundice reflects the exclusion of bile pigments from the intestine.

Determination of Urine Urobilinogen

SPECIMEN

A fresh urine specimen is collected over a 2 hour period from 2 p.m. to 4 p.m. (or 1 p.m. to 3 p.m.). The patient is asked to empty his bladder at 2 p.m. and this specimen is discarded. He is then given a glass of water. All urine specimens voided during the next 2 hours are collected and pooled. The specimen should be kept cool and protected from light. To prevent the oxidation of urobilinogen to urobilin, there must be no delay in running the analysis once the sample is collected. If a 24 hour sample is collected, it should be collected in a dark bottle that contains 5 gm. of sodium bicarbonate to minimize oxidation of urobilinogen and 100 ml. of toluene to minimize bacterial growth and to form a protective layer against oxygen from the air.

PRINCIPLE

The majority of the quantitative methods for urobilinogen are based on the reaction of this substance with p-dimethylaminobenzaldehyde to form a red color. This reaction was first described by Ehrlich in 1901, and methods involving the Ehrlich reagent have been modified over the years to improve their specificity. Major improvements were made in 1925 by Terwen, who used alkaline ferrous hydroxide to reduce urobilin to urobilinogen and added sodium acetate to eliminate interference from indole and skatole. In 1936, Watson introduced the use of petroleum ether rather than diethyl ether for the extraction of urobilinogen to assist in the removal of other interfering substances. Studies by Henry et al. in 1964 indicated that even the so-called quantitative method that involves extraction with petroleum ether is not capable of complete recovery of added urobilinogen from urine. These investigators suggest that the more rapid semiquantitative method would yield values of sufficient clinical significance.

The principle of the method chosen, therefore, involves the spectrophotometric determination of urobilinogen that has reacted with p-dimethylaminobenzaldehyde in a fresh urine specimen. Ascorbic acid is added as a reducing agent to maintain urobilinogen in a reduced state and prevent the reformation of urobilin. Sodium acetate is used to reduce the acidity after the reaction of urobilinogen with Ehrlich's reagent. The sodium acetate also inhibits color formation from indole and skatole, and intensifies the color with Ehrlich's reagent. The results of the method are expressed in Ehrlich units (E.U.), where 1 Ehrlich unit is equivalent to the color produced by 1 mg. of urobilinogen.

REAGENTS

1. Ehrlich's reagent, modified. Dissolve 0.7 gm. of p-dimethylaminobenzaldehyde in 150 ml. of concentrated HCl, A.R. grade; add to 100 ml. of distilled water and mix. This reagent is stable.

2. Saturated sodium acetate solution. Saturate 1 L. of distilled water with either A.R. grade anhydrous or triple-hydrated sodium acetate, maintaining extra crystals in the solution to insure saturation. Two to three pounds of sodium acetate will be required, depending on the form that is used. This reagent is stable at room temperature.

3. Ascorbic acid, powder, A.R.

4. Standard PSP dye solution. Dissolve 20.0 mg. of phenolsulfonphthalein (phenol red) in 100 ml. of 0.05 per cent NaOH. Use the acid form of the dye, which

may be obtained from Eastman Kodak Company (Rochester, N.Y.) and other suppliers.

5. Working standard. Dilute the stock solution (step 4) 1:100 with 0.05 per cent NaOH. This solution contains 0.2 mg./100 ml. and is equivalent to a solution of urobilinogen-aldehyde containing 0.346 mg. of urobilinogen per 100 ml. of final colored solution in the method. PSP salts or intravenous injection solutions should not be used to prepare the standard. The working standard should have an absorbance of 0.384 at 562 nm. in a cuvet with a 10 mm. light path on a high resolution spectrophotometer. Other methods use a mixture of Pontacyl Violet and Pontacyl Carmine 2B as the standard dye solution. This mixture has an absorption spectrum curve different from that of urobilinogen-aldehyde, and when it is employed as a standard one must be aware that the analytical results vary with the spectral resolution of the spectrophotometer used.

PROCEDURE

1. Measure the volume of the 2 hour urine sample.

2. Test the urine for bilirubin. If more than a trace is present, mix 2.0 ml. of 10 per cent $BaCl_2$ solution with 8.0 ml. of urine and filter. The final result must be multiplied by 1.25 to correct for this step.

3. Dissolve 100 mg. of ascorbic acid in 10 ml. of urine (centrifuge if cloudy) and place 1.5 ml. aliquots in each of two tubes labeled B for blank and X for unknown.

4. To the B tube add 4.5 ml. of a freshly prepared mixture of 1 volume of Ehrlich's reagent and 2 volumes of saturated sodium acetate; mix.

5. To the X tube add 1.5 ml. of Ehrlich's reagent, mix well, and immediately add 3.0 ml. of saturated sodium acetate; mix.

6. Measure the absorbance of X and B at 562 nm. against water set at zero within 5 minutes of step 5. Measure the absorbance of the PSP working standard against water at the same wavelength.

CALCULATION

$$\text{Ehrlich units/100 ml. urine} = \frac{A_x - A_b}{A_s} \times 0.346 \times \frac{6.0}{1.5} = \frac{A_x - A_b}{A_s} \times 1.38$$

and

$$\text{Ehrlich units/2 hr.} = \frac{A_x - A_b}{A_s} \times 0.0138 \times \text{urine volume in ml.}$$

Note: Multiply answer by 1.25 if bilirubin was removed by $BaCl_2$ (step 2).

NORMAL VALUES

The normal range by this method is 0.1 to 1.0 Ehrlich units per 2 hours.

SOURCES OF ERROR

1. The specimen should be fresh and the procedure should be carried out without delay in the absence of sunlight or bright fluorescent or other light.

2. Pure solutions of urobilinogen develop color immediately when mixed with Ehrlich's reagent. In the urine, Ehrlich's reagent produces color with nonurobilinogen substances on standing. For this reason it is important to stop these slower color reactions by the addition of saturated sodium acetate immediately after the urine is mixed with Ehrlich's reagent.

3. Since the urobilinogen-aldehyde color slowly decreases in intensity, the spectrophotometric readings should be made within 5 minutes after the production of the color.

4. Compounds other than urobilinogen that may be present in the urine, such as sulfonamides, procaine, and 5-hydroxyindoleacetic acid, may also react with Ehrlich's reagent. Bilirubin, when present, will form a green color and must, therefore, be removed before analyzing for urobilinogen.

REFERENCES

Henry, R. J., Jacobs, S. L., and Berkman, S.: Clin. Chem., *7*:231, 1961 (modified).
Henry, R. J., Fernandez, A. A., and Berkman, S.: Clin. Chem., *10*:440, 1964.
Watson, C. J., and Hawkinson, V.: Am. J. Clin. Path., *17*:108, 1947.
Watson, D. J., Schwartz, S., Sborov, V., and Bertie, E.: Am. J. Clin. Path., *14*:605, 1944.

Determination of Fecal Urobilinogen (Semiquantitative)

SPECIMEN

Any single fresh specimen may be used and the analysis should be carried out in the absence of direct sunlight or bright artificial light.

PRINCIPLE

This method involves the same principles described earlier for the urine procedure. It is carried out on an aqueous extract of fresh feces, and any urobilin present is reduced to urobilinogen by treatment with alkaline ferrous hydroxide before Ehrlich's reagent is added.

REAGENTS

1. Ferrous sulfate solution, 20 per cent (w/v). Prepare just before use.
2. NaOH solution, 10 per cent (w/v).
3. Ehrlich's reagent (for reagents 3, 4, and 6, see under urine urobilinogen).
4. Saturated sodium acetate solution.
5. Ascorbic acid powder.
6. Stock and working standard of PSP dye solution.

PROCEDURE

1. Transfer 10 gm. of a blended or homogenized sample of fresh feces to a large mortar. Add water to a 250 ml. graduated cylinder to the 190 ml. mark. Add 20 ml. of water from the cylinder to the mortar and grind the feces to a paste. Add 80 ml. more of water and grind again. To a 500 ml. Erlenmeyer flask, add 100 ml. of the 20 per cent ferrous sulfate solution and pour in the supernatant suspension of feces from the mortar. Add another 50 ml. of water to the residue in the mortar, grind, and transfer the supernatant material to the flask. Repeat this process with the remaining water in the cylinder. Slowly add 100 ml. of 10 per cent NaOH to the flask with swirling. Stopper the flask, shake, and allow to stand in the dark at room temperature for 1 to 3 hours.

2. Mix the contents of the flask and filter a portion of the contents. Dilute 5.0 ml. of the filtrate to 50 ml. with water. This solution should be nearly colorless. If the filtrate is highly colored, measure 50 ml. into a flask, add 25 ml. of the ferrous sulfate solution and 25 ml. of 10 per cent NaOH, and mix well. Stopper the flask and allow to stand in the dark for 1 to 3 hours. Mix the contents and filter as before;

dilute 10 ml. of the filtrate to 50 ml. with water before proceeding to step 3 to correct for the additional dilution.

3. Dissolve 100 mg. of ascorbic acid in 10 ml. of the diluted filtrate and place 1.5 ml. aliquots in each of two tubes labeled B for blank and X for unknown.

4. Carry the tubes through steps 4, 5, and 6 as described in the urine procedure.

CALCULATION

$$\text{Ehrlich units/100 gm. wet feces} = \frac{A_x - A_b}{A_s} \times 0.346 \times \frac{6.0}{1.5} \times \frac{50}{5} \times \frac{400}{10} = \frac{A_x - A_b}{A_s} \times 552$$

NORMAL VALUES

A range of 75 to 275 Ehrlich units/100 gm. of fresh feces, or 125 to 400 Ehrlich units/24 hr. specimens, is considered normal.

REFERENCES

See references under Determination of Urine Urobilinogen, page 765.

CARBOHYDRATE METABOLISM AND LIVER FUNCTION

It is well known that the liver is essential for normal carbohydrate metabolism. Monosaccharides such as fructose and galactose are converted to glucose and the glucose is stored as glycogen or may enter various metabolic pathways to be converted into amino acids or fatty acids or be broken down to carbon dioxide and water with the release of energy. The liver serves as a source of readily available glucose, either from its store of glycogen or by the process of gluconeogenesis. It plays an important role in the metabolism of lactic and pyruvic acids and produces acetone bodies in the process of oxidation of fatty acids (see Chapter 4, Carbohydrates). The liver is involved in so many essential metabolic reactions that one would think it would be a simple matter to devise tests to study abnormalities in these reactions. Although many investigators have attempted to provide such sensitive tests, in general, they have failed, since extensive impairment of liver cells or destruction of liver tissue is required before there is significant interference with metabolic function. For this reason, the clinical laboratory seldom carries out function tests based on impairment of the metabolic activities of the liver.

It is reasonable to assume that the glucose tolerance test (see Chapter 4) would be grossly abnormal in the presence of liver disease, but attempts at correlation of this test with various forms of liver disease have been disappointing. At times even normal curves are seen in serious liver disease and it must be concluded that this test is of limited value in liver function assessment. The most common abnormality in acute liver disease is the occurrence of a rapid sharp peak of about 200 mg./100 ml. in the blood glucose level within the first hour, followed by a return to normal or hypoglycemic level within 2 to 4 hours. Of all the tests of metabolic function that have been devised, the galactose tolerance test is probably the most valuable.

The Galactose Tolerance Test

The galactose that is carried to the liver cells from the intestinal tract is normally converted into glucose, which is then further converted into glycogen. If galactose is injected into the blood, the speed of removal of this sugar is related to the integrity of the liver and the normal functioning of its cells.

CLINICAL SIGNIFICANCE

Although the test is of some value in assessing liver function, its major limitation is its insensitivity. Abnormal tolerance is not evident until rather severe impairment or destruction of liver cells occurs. A progressive decrease of tolerance, however, does parallel a developing liver cell necrosis and may assist in distinguishing obstructive jaundice from hepatitis. Nevertheless, its use has been discarded by many clinicians.

PREPARATION OF THE PATIENT

The test is commonly carried out after oral ingestion of galactose, but many investigators strongly recommend the intravenous test to rule out the possibility of decreased absorption of the sugar.

In the oral test, the patient is given 40 gm. of galactose in about 200 ml. of water, which may be flavored with lemon juice. Blood samples (heparinized) are drawn at 30 and 60 minutes after giving the galactose. All urine specimens voided within the 5 hour period after ingestion are collected and combined.

In the intravenous test, 1 ml. of a 50 per cent galactose solution per kilogram of body weight (0.5 gm./kg. if less concentrated solutions are used) is injected by a physician under proper clinical conditions. A heparinized blood sample is drawn 60 minutes after the completion of the injection.

The Determination of Galactose in Blood

PRINCIPLE

Methods for the determination of galactose in the blood or urine ordinarily involve removal of glucose by fermentation with yeast or by treatment with glucose oxidase. The concentration of the remaining reducing sugar is determined and is expressed as galactose. Several modifications of such methods have been proposed and they all suffer from a lack of specificity. With the recent commercial availability of *galactose oxidase*, a more specific method can be presented. The enzymatic reaction is similar to that used for the determination of glucose with glucose oxidase.

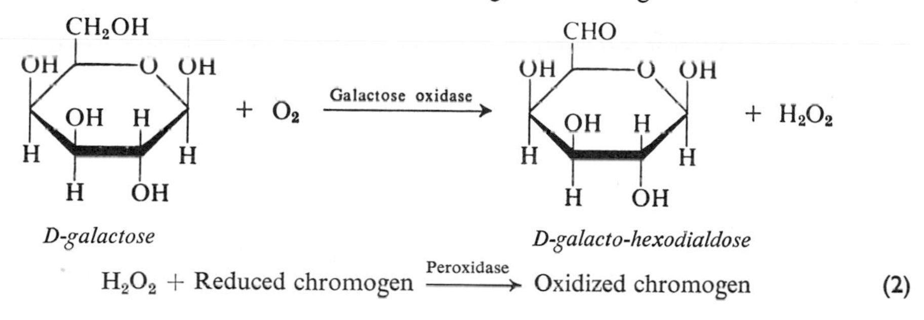

REAGENTS

1. Enzyme reagent. To a 50 ml. volumetric flask add 2.5 mg. of o-tolidine, dissolve in 0.5 ml. of methanol, and add 30 ml. of $M/15$ phosphate buffer, pH 7.0. Then add 2.5 mg. of peroxidase (horseradish peroxidase, Worthington Biochemical Corp., Freehold, N.J.) and 4.0 mg. of galactose oxidase (Worthington Biochemical Corp., or Sigma Chemical Co., St. Louis, Mo.). Mix until dissolved, add 5 ml. of 2 per cent Triton X-100 (Rohm & Haas Co., Philadelphia, Pa.) and dilute to mark with the phosphate buffer.

2. Phosphate buffer *M*/15, pH 7.0 at 25°C. Add 5.79 gm. of Na_2HPO_4, anhydrous, A.R. and 3.53 gm. of KH_2PO_4, anhydrous, A.R. to a 1 L. volumetric flask. Dissolve in water and dilute to the mark. Check the pH with a pH meter, adjusting to pH 7.0 with either 0.1 *N* HCl or 0.1 *N* NaOH if necessary.

3. Triton X-100, 2 per cent solution (v/v). Dilute 2 ml. of Triton X-100 to 100 ml. with water in a volumetric flask.

4. Glycine buffer, 0.25 *M*, pH 9.7 at 25°C. Add 12.22 gm. of glycine and 9.5 gm. of NaCl to a 1 L. volumetric flask. Then add about 300 ml. of water and 87.5 ml. of 1 *N* NaOH solution and mix. Dilute to the mark with water, and check the pH with a pH meter.

All of the preceding reagents are available commercially in a Galactostat kit prepared by the Worthington Biochemical Corp.

5. Zinc sulfate, 5 per cent (w/v) in water.

6. Barium hydroxide, 0.3 *N* solution.

7. Galactose standard, 50 mg./100 ml. Add 50 mg. of galactose to a 100 ml. volumetric flask. Dissolve in water and dilute to the mark. Prepare fresh.

8. Galactose standard, 25 mg./100 ml. Dilute 1 volume of 50 mg./100 ml. galactose solution (step 7) with 1 volume of water. Prepare fresh.

PROCEDURE

1. Pipet 0.5 ml. of water into three 12 ml. centrifuge tubes labeled X, S_1, and S_2. Pipet 1.0 ml. of water into a 12 ml. centrifuge tube labeled B for blank.

2. Add 0.5 ml. of blood or urine to the X tube and 0.5 ml. of each galactose standard to the S_1 and S_2 tubes, and mix.

3. Add 2.0 ml. of 0.3 *N* $Ba(OH)_2$ solution to each of the four tubes; mix and add 2.0 ml. of 5 per cent $ZnSO_4$ solution. Mix and let stand for 5 minutes.

4. Centrifuge the tubes for 5 minutes at 2500 rpm. A clear supernatant fluid should be obtained.

5. Transfer 2.0 ml. of the supernatant to appropriately labeled test tubes.

6. Add 2.0 ml. of the enzyme reagent to each tube and place the tubes in a water bath or a dry bath at 37°C.

7. After 1 hour, add 6.0 ml. of 0.25 *M* glycine buffer, pH 9.7, to stop the reaction and stabilize the color. Mix and read the absorbance of each tube against water at 425 nm. The color is stable for several hours.

CALCULATIONS

$$\text{mg. galactose/100 ml.} = \frac{A_x - A_b}{A_s - A_b} \times \text{concentration of galactose standard.}$$

For the calculation use the standard whose absorbance most closely matches that of the unknown.

NORMAL VALUES

In the oral test the blood galactose level should reach a peak of 40 to 60 mg./100 ml. in 30 to 60 minutes. The total galactose excreted in the urine in the 5 hour period should not exceed 3 gm.

In the intravenous test the blood level should not exceed 42 mg./100 ml. after 60 minutes.

PROCEDURE NOTES

1. The Triton X-100 is a surface active agent that is added to the enzyme reagent to prevent turbidity in the final colored solution.

2. A standard curve may be constructed by the simultaneous determination of a series of standard galactose solutions, e.g., 10, 25, 50, and 75 mg./100 ml.

3. If the blank solution cannot be set at zero absorbance at 425 nm., use the lowest wavelength that will permit the zero setting.

REFERENCES

Sempere, J. M., Gancedo, D., and Asensio, C.: Anal. Biochem., *12*:509, 1965.
Worthington Enzyme Manual, Galactostat Brochure. Freehold, N. J., Worthington Biochemical Corp., 1967.

PROTEIN METABOLISM AND LIVER FUNCTION

Normal functioning liver cells are essential for normal protein metabolism. Deamination and transamination of amino acids, urea formation, and synthesis of prothrombin and of many of the plasma proteins is dependent on normal liver function. In protein metabolism, as in carbohydrate metabolism, an extensive impairment or destruction of liver cells is required before abnormal function can be clearly demonstrated. In general, liver function tests that measure overall protein metabolism are less sensitive than those that depend, for example, on abnormal concentrations of specific plasma proteins such as albumin and gamma (γ) globulin.

Plasma Proteins

The plasma proteins such as albumin, fibrinogen, and the majority of the globulins, with the exception of γ-globulin, are synthesized by the liver. The decrease in albumin and increase in γ-globulin and lipoproteins such as β-globulin are characteristic of chronic liver disease. In obstructive jaundice, the albumin is only slightly decreased and the γ-globulin slightly increased, but a definite increase in the lipoprotein fractions, α_2- and β-globulins, is observed. In general, the plasma protein changes in parenchymal liver disease consist of a decrease in albumin and an overall increase of the globulin fractions. Total protein determinations alone are not particularly helpful, since normal values may be obtained by the combination of low albumin and high globulins. The A/G ratio is often reversed in liver disease, but this is not specific and indicates a need for further fractionation of the plasma proteins. Changes in the plasma proteins occur in chronic hepatitis and cirrhosis and are readily demonstrated by electrophoretic analysis (see Chapter 5, Proteins and Amino Acids). It should be emphasized that changes in the plasma proteins as measured by the methods discussed in this section are the result of prolonged or extensive liver cell impairment and are of little assistance in the detection of early liver disease.

The methods of electrophoresis, total protein determination, and A/G ratio are described in Chapter 5.

FIBRINOGEN

Fibrinogen is a plasma protein component that is essentially involved in the clotting of blood. It is synthesized only in the liver, a fact that suggests that a deficiency in the clotting mechanism may be caused by liver disease. Fortunately, although fibrinogen levels are sometimes lowered, they are ordinarily close to normal in hepatitis and cirrhosis as shown by electrophoretic studies of plasma proteins. Apparently, the liver has enough functional reserve to synthesize this vital clotting constituent. In severe liver injury (acute yellow atrophy), the plasma concentration of fibrinogen and other clotting factors (see next section) may be decreased to such an extent that hemorrhage may result. A method for the determination of fibrinogen is described in Chapter 5.

PROTHROMBIN

Prothrombin is a protein found in the plasma; like fibrinogen, it is involved in blood clotting. It is formed by the liver and requires adequate amounts of vitamin K for its synthesis. Decreased concentrations of prothrombin result in hemorrhagic tendencies, which may be corrected by injection of vitamin K if only vitamin K deficiency (malabsorption) is present. No response to vitamin K is noted in liver cell injury.

CLINICAL SIGNIFICANCE

It has been known for years that patients with jaundice exhibit a delayed coagulation of the blood. This may be observed in both obstructive jaundice and hepatitis. Since vitamin K absorption may be impaired in biliary obstruction, the injection of this vitamin may increase the concentration of prothrombin. An increase in prothrombin 24 hours after injection is suggestive of uncomplicated obstructive jaundice with adequate hepatic cell function. If the plasma prothrombin level is not changed by vitamin K administration, liver cell impairment or destruction is suggested and surgical procedures that involve the risk of bleeding should be avoided. Fortunately, it requires a great decrease in prothrombin levels, to about 10 per cent of normal, before bleeding tendencies occur. For this reason, the prothrombin level seldom drops low enough in obstructive jaundice or hepatitis to cause spontaneous bleeding; moreover, deficiency of prothrombin alone is seldom encountered and other clotting factors are usually involved.

The prothrombin concentration as measured by the prothrombin time is essential in following patients on dicumarol therapy. Dicumarol is now widely used in the prophylaxis of intravascular clotting, including thrombophlebitis, pulmonary embolism, and thrombotic or embolic occlusion of coronary and peripheral arteries. The drug is given orally and the dosage is adjusted to keep the prothrombin level at about 20 per cent of normal. This level corresponds to a prothrombin time of about 30 seconds when the normal time is 13 seconds. Dicumarol administration obviously requires frequent determinations of the prothrombin time of the patient since bleeding tendencies begin to show up when the prothrombin level drops to 10 per cent or lower.

The Determination of Prothrombin

The prothrombin concentration, as well as factors V, VII, and X in the plasma, is estimated by means of the prothrombin time. There are two types of methods, the one-stage and the two-stage procedures. The one-stage method was developed by Quick and is commonly used in clinical laboratories because it is relatively simple and reliable. It is presented here as an example of procedures measuring clotting mechanism. Further details and additional procedures may be obtained from texts such as Hougie (see References, p. 772). The two-stage method is more complicated and is used by research investigators in further studies of the clotting mechanism.

SPECIMEN

To obtain reproducible results in a system that is composed of many clotting factors it is essential to control carefully the collection and handling of the specimen, the addition of reagents, and the technical details of the procedure. Fresh plasma is required for the test. It is obtained by collecting exactly nine volumes of whole blood in a syringe that contains 1 volume of sterile 0.1 *M* sodium oxalate, mixing, transferring to a centrifuge tube, and centrifuging 5 minutes at 2000 rpm. Alternatively, 1 volume of 0.1 *M* sodium oxalate may be placed in a centrifuge tube and 9 volumes of whole blood added immediately after drawing. The plasma should be withdrawn from the tube without disturbing the red cells and should be used for the test within 10 minutes. If a delay is necessary, keep the plasma chilled and warm it to 37°C. just before carrying out the test. The plasma may be stored in the refrigerator (4°C.) for 6 to 8 hours if necessary, but the prothrombin time will decrease after overnight storage.

PRINCIPLE

Oxalated plasma is allowed to react with an excess of thromboplastin and an optimum concentration of Ca^{++}. The clotting time is determined and is expressed as the prothrombin time of the plasma. The ratio of the clotting time of the specimen to the clotting time of a normal control may be used to obtain a measure of the relative prothrombin activity of the patient's plasma, or the prothrombin may be expressed as a percentage of normal activity as determined by a standard curve.

REAGENTS

1. Sodium oxalate solution, 0.1 *M*. Add 1.34 gm. of sodium oxalate, anhydrous, to a 100 ml. volumetric flask. Dissolve in water and dilute to the mark. If a sterile solution is used in the collection of the specimen, 5 ml. ampules may be filled with the solution, sealed, and autoclaved. Prothrombin Vacutainer tubes that contain the correct amount of sodium oxalate may be purchased.
2. Calcium chloride solution, 0.0125 *M*. To a 100 ml. volumetric flask add 138.8 mg. of $CaCl_2$. Dissolve in water and dilute to the mark.
3. Thromboplastin. See Note.
4. Normal control plasma. See Note.

Note: Several preparations of rabbit brain thromboplastin, as well as normal control plasma, for prothrombin time are available commercially. Most clinical laboratories select a given product that yields the most satisfactory results and adhere to their choice.

PROCEDURE

1. Warm all the reagents and the plasma specimen to 37°C. before the test and carry out the test at 37°C.
2. Add 0.1 ml. of plasma to a small test tube.
3. Mix 1 part of a commercial thromboplastin solution and 1 part of the 0.0125 *M* $CaCl_2$ solution in a small test tube. Maintain the solution at 37°C.
4. With a 0.2 ml. micropipet, blow 0.2 ml. of the thromboplastin-$CaCl_2$ mixture into the 0.1 ml. of warm plasma specimen or warm control plasma. Mix rapidly and thoroughly, and start the timer with the addition of the plasma.
5. Stop the timer when the clot forms. The end point may be marked by several means: detection of a clot, visible while tilting the tube; drawing a loop of No. 20

nichrome wire through the mixture; or, instrumentally, by an electrode reaction or by a photoelectric recorder. Instruments employing end point detection by an electrode reaction use specially designed disposable plastic sample vessels that are prewarmed in heating wells of a thermostatically controlled chamber. Plasma and reagents are added to the vessels and a timer is started. An electrode dips into the vessel and when the fibrin gel forms across the electrodes, a circuit is completed and the instrument is stopped automatically.

CALCULATIONS

To obtain a measure of the prothrombin level of a patient's plasma, the clotting time obtained in the preceding procedure is compared with that of the normal plasma control. The results may be reported either in seconds or as a percentage of normal activity. Unfortunately, the increase in prothrombin time over the control, or the ratio of the two times, is only indirectly related to the prothrombin concentration. A hyperbolic curve results when prothrombin time in seconds is plotted against per cent of normal prothrombin concentration. The shape of the curve is mainly affected by the nature of the diluent, which may be either saline or adsorbed plasma (plasma treated with $BaSO_4$). Over the years, this difficulty in expression of prothrombin concentration from prothrombin time values has resulted in the reporting of only the patient's prothrombin time and the prothrombin time of the normal control plasma. These two values enable the physician to assess the relative prothrombin level of his patient's plasma and to follow his progress when serial determinations are requested.

PROCEDURE NOTES

1. The prothrombin time values obtained in the previous procedure are similar to, but not the same as, in vivo, since exogenous tissue thromboplastin may behave somewhat differently from endogenous thromboplastin.

2. The best source of tissue prothrombin is an acetone extract of desiccated rabbit brain.

REFERENCES

Quick, A. J.: Am. J. Clin. Path., *10*:222, 1940 (modified).
Hougie, C.: Fundamentals of Blood Coagulation in Clinical Medicine. New York, McGraw-Hill Book Co., Inc., 1963.

The Blood Clotting Process

The clotting of blood is a very complex process that involves several plasma protein components, many of which are synthesized by the liver. Investigators are still involved in establishing the exact mechanism of blood coagulation and many precursors, activators, and inhibitors have been proposed. The difficulties encountered in the isolation of pure, active complex protein clotting factors has led to considerable confusion. The essential steps involved in the clotting process had been originally outlined by Howell. In essence, the blood clot is formed by *fibrinogen*, which is present in a soluble form in the plasma and is transformed into an insoluble gel, fibrin, by the clotting mechanism. In Howell's basic theory, the fibrinogen is converted into fibrin by *thrombin*, which exists as *prothrombin* in the unclotted blood.

Prothrombin is converted to thrombin by the action of *thromboplastin* and *calcium*. The steps involved in the basic theory may be outlined as follows:

Step I Formation of plasma thromboplastin

Step II Prothrombin $\xrightarrow{\text{Thromboplastin} + \text{Ca}^{++}}$ Thrombin

Step III Fibrinogen $\xrightarrow{\text{Thrombin}}$ Fibrin

To aid in the understanding of the effect of the liver and liver disease on the blood clotting process, we will consider in greater detail each substance in the preceding mechanism.

THROMBOPLASTIN

The active factor of plasma thromboplastin is thought to be a lipoprotein that contains cephalin. In studies of patients with hemophilia and other coagulation defects, several of the individual blood clotting factors were discovered. One called *antihemophilic globulin* or AHG (Factor VIII) in the plasma reacts with tissue or platelet thromboplastin to initiate the clotting process. This component is lacking in the plasma of hemophilic patients. Other factors in the plasma are *plasma thromboplastin component* or PTC (Factor IX or Christmas factor), *plasma thromboplastin antecedent* or PTA (Factor XI), a labile factor called *accelerator globulin* or Ac-globulin (Factor V), the *Hageman factor* (Factor XII), and the *Stuart-Prower factor* (Factor X). The platelet thromboplastic factor is released when the platelets are disintegrated on contact with a rough surface. To account for the role of each of these factors, it has been proposed that active thromboplastin is formed in a series of reactions as follows:

PTC is activated by the Hageman factor and reacts with AHG + PTA + platelet factors → product.

The product + Ac-globulin + Stuart-Prower factor + Ca^{++} → active thromboplastin.

PROTHROMBIN

Prothrombin is a glycoprotein in the plasma with an electrophoretic mobility similar to α_2 globulin. In Step II of the clotting process, it is converted to thrombin. The series of reactions involved in the conversion require thromboplastin and Ca^{++}, the Stuart-Prower factor, and a labile as well as a stable factor. The labile factor, accelerator globulin (Factor V), is consumed during the clotting process and is not found in the serum. The stable factor, *serum prothrombin converting factor* or SPCA (Factor VII) is not consumed during the process and is stable to storage. A prolonged prothrombin time is characteristic of defects in any of the components required in Step II because each factor just mentioned is necessary for the conversion of prothrombin to thrombin. Since these factors are mainly produced by the liver, their level is an indicator of liver function.

THROMBIN

Thrombin is a glycoprotein that is formed from the prothrombin-cephalin-calcium complex in Step II. Purified preparations of thrombin exhibit all the properties of a proteolytic enzyme. When added to normal blood, clotting occurs almost immediately, suggesting a catalytic role in earlier steps in the clotting process. It is also accepted that thrombin increases the disintegration rate of the platelets and that the activation of thrombin is an autocatalytic process.

FIBRINOGEN

In Step III, fibrinogen, a plasma globulin, is changed into fibrin by the action of thrombin. This reaction involves the splitting off of two small polypeptides to produce a smaller fibrin monomer, which undergoes polymerization to form the fibrin clot. Although the liver is the sole source of fibrinogen, liver disease seldom leads to such low levels as to cause bleeding. Bleeding usually results from levels below 100 mg /100 ml. Hypofibrinogenemia may be observed, however, in severe liver injury, in cases of carcinoma of the prostate, and in complications of pregnancy, such as premature separation of the placenta and retention of a dead fetus. The condition is not generally caused by deficient production, but by increased clot production and breakdown by the action of fibrinolytic enzymes.

FIBRIN

Although, when first formed, the molecular weight of fibrin is less than that of fibrinogen, a polymerization reaction causes the production of a large insoluble fibrin clot. On standing, the gel-like clot contracts into a harder mass and extrudes the serum. This process is called *syneresis*, or clot retraction, and can be observed in the laboratory in specimen tubes containing clotted blood. In addition to the mechanisms outlined for the clotting mechanism, there is also a mechanism concerned with clot lysis. This system includes precursors such as profibrinolysin (plasminogen), and activators present in plasma, such as urokinase, that lead to the formation of *fibrinolysin* (*plasmin*), which in turn causes lysis of the fibrin clot. This process is of obvious importance for the dissolution of intravascular clots. Normally, the two systems of blood clotting and clot lysis are under a well-balanced control.

A slightly more elaborate scheme of the clotting mechanism, including all of the factors just discussed, may be outlined as follows:

Step IA. Blood is shed; contact with a rough surface releases platelet factors and activates the Hageman factor.
B. PTC activated by the Hageman factor + platelet factors + AHG + PTA → thromboplastin intermediate.
C. Thromboplastic intermediate + Ac-globulin + Stuart-Prower factor + Ca^{++} → thromboplastin.

Step IIA. Prothrombin + Ac-globulin + SPCA + Stuart-Prower factor + Ca^{++} + thromboplastin → prothrombin-cephalin-calcium complex.
B. Prothrombin-cephalin-calcium complex → thrombin.

Step IIIA. Fibrinogen + thrombin → fibrin.
B. Fibrin polymerization → fibrin clot.

CHANGES IN THE RELATION OF PROTHROMBIN PRODUCTION TO CLOTTING MECHANISM IN LIVER DISEASE

The condition of *hypoprothrombinemia* is not uncommon, although it is severe only when liver damage is so extensive that it causes a significant decrease in prothrombin synthesis. In obstructive jaundice the impairment of the absorption of vitamin K from the intestine may result in hypoprothrombinemia. The liver also produces Ac-globulin and SPCA and requires vitamin K for their synthesis and activity. A deficiency in either factor, vitamin K or the Stuart-Prower factor, will result in a prolonged prothrombin time. In acute liver cell damage, SPCA (Factor VII) is decreased first, then prothrombin and the Stuart-Prower factor (Factor X).

FLOCCULATION AND TURBIDITY TESTS

Both qualitative and quantitative alterations in the plasma protein fractions in disease have been used as the basis for diagnostic tests for many years. Although many tests were devised and some were applied to liver disease, at present the three most commonly used are the *cephalin-cholesterol flocculation*, *zinc sulfate turbidity*, and *thymol turbidity* tests. In general, the response to these tests depends on the state of balance between the stabilizing and precipitating factors in the serum. The precipitating factors include γ-globulins and lipoproteins such as the β-globulins. The stabilizing factors are albumin and α_1-globulin. In normal serum, the distribution of the protein components is such that the stabilizing factors prevent turbidity or flocculation when any of the three tests are carried out. In the design of each test, therefore, the concentration of the precipitating reagent is adjusted to produce minimal or negative response with normal serum, while exhibiting a reaction to changes in the precipitating and stabilizing factors. Such changes are found in liver disease.

CLINICAL SIGNIFICANCE

In varying degrees, the three tests are useful in distinguishing the several types of hepatitis from extrahepatic biliary obstruction. The *cephalin-cholesterol flocculation test* responds readily to serum specimens containing increased γ-globulin and decreased albumin. This condition is common in a high percentage of cases of viral hepatitis, cirrhosis, and hepatic necrosis and occurs less frequently in posthepatic obstruction. The test is very sensitive to *qualitative* changes in serum albumin, which may explain its rapid response to acute hepatitis. More specifically, a positive test is obtained under any of the following conditions: decrease in albumin, increase in γ-globulin, or production of an abnormal albumin with less stabilizing power. The last condition is frequently evident in patients with subclinical hepatitis and patients recovering from hepatitis.

The *zinc sulfate turbidity test* is more affected by changes in γ-globulin alone than are the other two tests; therefore, it is positive in any disease that produces an elevation of γ-globulin in the serum. The high values for γ-globulin characteristic of chronic hepatitis and cirrhosis are readily detected by the zinc sulfate turbidity test. One of the outstanding characteristics of this test is the low value obtained in extrahepatic obstructive jaundice; thus, the test is most useful in distinguishing between hepatitis and obstructive jaundice.

The *thymol turbidity* and *flocculation test* is affected by a decrease in albumin and increase in γ-globulin as are the other tests, but it is also affected by an increase in lipids and β-globulins. The test does not respond as rapidly to viral hepatitis as the cephalin-cholesterol flocculation, but the increase in thymol turbidity persists at times longer than an abnormal cephalin-cholesterol test. Hyperglobulinemic states and any condition that results in high serum lipoprotein levels, such as nephrosis, result in high values of thymol turbidity.

The Cephalin-Cholesterol Flocculation Test

SPECIMEN

Blood drawn from a patient in the fasting state is allowed to clot and the serum is separated by centrifugation. Fresh serum should be used and it should be protected from light and heat. Unpredictable stability and falsely positive tests have been observed when specimens are allowed to stand in the refrigerator or at room temperature for one or more days. Frozen specimens should not be used.

PRINCIPLE

Serum from patients with liver cell impairment will react with a cephalin-cholesterol suspension to produce a flocculant precipitate. This precipitate is probably an α and β *globulin-cholesterol complex.*

REAGENTS

1. Sodium chloride solution, 0.85 per cent (w/v). Dissolve 8.5 gm. of NaCl, C.P. or A.R., in water, and dilute to a volume of 1 L.
2. Ether, anesthesia grade (E. R. Squibb & Sons, New York, N.Y.).
3. Cephalin-cholesterol, stock solution. The cephalin-cholesterol mixture is available in dried form from commercial sources. Although the commercial preparations are often called antigens, this is a misnomer since an antigen-antibody reaction is not involved in the test. The stock solution is prepared by adding the proper amount of ether (usually 5 or 8 ml., respectively) to the dried material in a vial. To obtain complete solution of the material, the mixture should be allowed to stand several hours. The solution must be clear and is stable for months, if kept tightly stoppered in a refrigerator.
4. Cephalin-cholesterol suspension. Heat 35 ml. of freshly distilled water to 65 to 70°C., add slowly with constant stirring 1 ml. of the clear stock solution. Heat gently and allow to simmer until the volume is reduced to 30 ml. A stable milky emulsion should result and it should be cooled to room temperature before use. Stored in a refrigerator, the emulsion is stable for 2 weeks. Considerable care and precise technique should be exercised in the preparation of this suspension.

PROCEDURE

1. To a small tube, add 4 ml. of 0.85 per cent NaCl solution and 0.2 ml. of the patient's serum or a positive or negative control serum, and mix.
2. A reagent control tube containing 4 ml. of 0.85 per cent NaCl should be prepared in addition to the known positive and negative control serums.
3. Add to each tube 1 ml. of the cephalin-cholesterol emulsion, mix gently by inversion, and let stand in the dark at room temperature for 24 hours.
4. Examine the tubes and judge the reaction as follows:
   Negative—no flocculation or precipitation.
   1+ reaction—slight flocculation or precipitation.
   2+ reaction—definite flocculation or precipitation.
   3+ reaction—almost complete precipitation with a somewhat cloudy supernatant fluid.
   4+ reaction—complete precipitation with a clear supernatant fluid.

It is the practice in some laboratories to read the tubes again after 48 hours in the dark at room temperature. Although, in general, there is a slight increase in reaction (a 1+ may become a 2+), no advantage in clinical diagnosis is obtained from this 24 hour delay. To shorten the time required for the test, some laboratories read the tubes after a 4 hour incubation at 37°C.

NORMAL VALUES

Serum from normal individuals will give negative or 1+ reactions.

PROCEDURE NOTES

Positive control sera are not available commercially; however, many laboratories save positive specimens from patients for checking the method and all new batches of the cephalin-cholesterol emulsion.

SOURCES OF ERROR

There are several sources of error in this flocculation reaction and care must be exercised to keep the conditions of the test constant. One major problem involves alterations in the emulsion. Only milky smooth emulsions that are free of coarse particles and exhibit a homogeneous shimmering appearance should be used in the test. In addition, use carefully cleaned glassware, rinsed with distilled water; use freshly distilled water to prepare reagents; use fresh serum; and protect the reaction from heat and light.

REFERENCES

Hanger, F. J.: Trans. Assn. Am. Physicians, *53*:148, 1938; J. Clin. Invest. *18*:261, 1939 (modified).
Neefe, J. R., and Reinhold, J. G.: Science, *100*:83, 1944.

The Zinc Sulfate Turbidity Test

SPECIMEN

Serum is used and the blood should be drawn in the postabsorptive state. Heparin interferes with the test as does turbidity in serum. Bilirubin and hemoglobin do not affect the turbidity readings.

PRINCIPLE

Serum from patients with high γ globulins (liver disease) will produce varying degrees of turbidity when mixed with a dilute solution of zinc sulfate in a barbiturate buffer of pH 7.5.

REAGENTS

1. Zinc sulfate reagent. Add 24 mg. of $ZnSO_4 \cdot 7H_2O$, 302 mg. of barbital, and 190 mg. of sodium barbital to a 1 L. volumetric flask. Dissolve in distilled water that has been boiled to drive off CO_2 and dilute to the mark. The pH should be 7.5 ± 0.05 at a temperature of 25°C. ± 3°C., and this should be checked at intervals, especially if the bottle is opened to the air for any length of time.

2. Barium chloride solution, 0.0962 *N*. Dissolve 1.175 gm. of $BaCl_2 \cdot 2H_2O$ in water and dilute to 100 ml. in a volumetric flask.

3. Sulfuric acid, 0.2 *N*. Dilute 6 ml. of concentrated H_2SO_4 to 1 L. with water. Standardize against 0.1 *N* NaOH. This reagent may also be prepared by appropriate dilutions of previously standardized sulfuric acid solutions more concentrated than 0.2 *N*.

4. Turbidity standards. Dilute 3.0 ml. of 0.0962 N $BaCl_2$ to volume in a 100 ml. volumetric flask by the rapid addition of 0.2 N H_2SO_4 at 10°C. At this temperature, the particle size of the $BaSO_4$ that is formed results in a more permanent suspension; thus, both solutions should be brought to a temperature of 10°C. before they are mixed.

10 unit turbidity standard. Mix 2.7 ml. of the $BaSO_4$ suspension and 3.3 ml. of 0.2 N H_2SO_4.

20 unit turbidity standard. Mix 5.4 ml. of the $BaSO_4$ suspension and 0.6 ml. of 0.2 N H_2SO_4. Absorbances of these two standards are read against a distilled water blank at 650 nm. and the constant K is calculated.

$$K = \frac{\text{units of standard}}{A \text{ of standard}}$$

The K for each instrument should be determined, preferably using several different preparations of the turbidity standards. A standard curve may be prepared using proper dilutions of the turbidity standards, but the curve should be checked daily with commercial (glass) standards.

It is important to note that these standards are the *Shank and Hoagland standards* prepared with 0.0962 N $BaCl_2$ and accepted as turbidity unit standards by clinical chemists in the United States. For comparison, 1 *Maclagan unit* = 2 Shank-Hoagland units.

PROCEDURE

1. Add 0.1 ml. of serum from a pipet that measures between two marks or wash the contents of a 0.1 ml. TC pipet into 6.0 ml. of the zinc sulfate reagent. Stopper the tubes, mix thoroughly, and let stand 30 minutes at a temperature of 25°C. ± 3°C.
2. Mix again thoroughly and read the absorbance against a reagent blank at 650 nm.

CALCULATION

Zinc sulfate turbidity units = $A \times K$; or the values may be read from a standard curve.

NORMAL VALUES

The normal adult range is 2 to 12 units. This covers a mixture of races, since the normal range for Caucasians, 2 to 9, is lower than that for Negroes, 5 to 12.

REFERENCES

Kunkel, H. G.: Proc. Soc. Exp. Biol. Med., *66*:217, 1947 (modified).
MacLagen, N. F.: Brit. J. Exp. Path., *25*:334, 1944.

The Thymol Turbidity Test

SPECIMEN

Blood is drawn from a patient in the postabsorptive state; it is allowed to clot and the serum is used for the test the same day it is drawn. Heparin and lipemia interfere with the test, but it is not sensitive to bilirubin or hemoglobin.

PRINCIPLE

Serum specimens from patients with hepatitis will produce definite turbidity when mixed with a thymol solution in barbiturate buffer. The turbidity is caused by the precipitation of a *globulin-thymol-phospholipid complex*, and it is also observed with sera from patients with diseases that produce an increase in β globulin and with lipemic sera.

REAGENTS

1. Thymol reagent. Add 6.0 gm. of colorless U.S.P. grade or recrystallized thymol crystals to a 2 L. Erlenmeyer flask. Heat 1200 ml. of distilled water to boiling in another flask and boil for 5 minutes to remove CO_2. Weigh out 3.09 gm. of barbital and 1.69 gm. of sodium barbital. Measure 1 L. of the boiling water into a graduated cylinder and pour about 300 ml. over the thymol crystals. Add the barbital and sodium barbital and remaining 700 ml. of hot water to the flask containing the thymol. Stopper the flask and mix by swirling the contents. When cool, transfer the solution to a liter volumetric flask and dilute to the mark with CO_2-free water. Then return the solution to the original flask, add 1 gm. of thymol, and mix thoroughly. Allow the flask to stand overnight at 25°C., mix and filter through Whatman No. 1 filter paper. The pH must be between 7.50 and 7.60 and the flask should be tightly stoppered and stored at or near 25°C. It can be used until it becomes opalescent; then it should be discarded.

2. Turbidity standards. The 10 and 20 unit turbidity standards should be prepared as outlined under the zinc sulfate turbidity test.

Note: A turbidity standard prepared from a suspension of powdered glass is available commercially (Difco Laboratories, Detroit, Mich.). This standard may be used for both thymol turbidity and zinc sulfate turbidity methods.

PROCEDURE

1. Add 0.1 ml. of serum from a pipet that measures between two marks or wash the contents of a 0.1 ml. T.C. pipet into 6.0 ml. of the thymol reagent. Mix thoroughly and let stand 30 minutes at a temperature of 25°C. ± 3°C.

2. Mix again thoroughly and read the absorbance against a thymol-buffer blank at 650 nm.

CALCULATION

Thymol turbidity units $= A \times K$ (see zinc sulfate turbidity test).

NORMAL VALUES

The normal adult range is 0 to 5 Shank-Hoagland units.

SOURCES OF ERROR

It is essential that the thymol reagent is colorless and clear and that it is adequately mixed with the serum for the test. The original reagent of Maclagan had a pH of 7.8, but a reagent of pH 7.55 was found to be more sensitive in the test. The turbidity decreases with increasing temperature; therefore, a water bath at 25°C. should be used if the room temperature is 28°C. or above.

The test should not be performed on lipemic sera since such specimens will give false high results. In the screening of blood donors for occult hepatitis, postprandial hyperlipemia would interfere with the interpretation of the test. In these cases, a thymol flocculation test may be carried out. The serum-reagent mixture from Procedure, step 1, is transferred to a conical centrifuge tube and allowed to stand overnight at 25°C. ± 3°C. If the turbidity was due to lipemia, little or no precipitate will be observed; however, if the turbidity was due to a dysproteinemia, a flocculation similar to that in a cephalin flocculation test occurs.

REFERENCES

Reinhold, J. G., and Yonan, V. L.: Am. J. Clin. Path., *26*:669, 1956 (modified).
Reinhold, J. G.: Advances in Clinical Chemistry. New York, Academic Press, 1960, vol. 3, p. 84.

Mucoproteins

The mucoproteins of the serum are found in the α_1 globulin fraction and range in concentration from 80 to 200 mg./100 ml. in normal individuals. In viral hepatitis, toxic hepatitis, and cirrhosis, the level decreases, whereas a progressive increase above normal occurs in extrahepatic obstructive jaundice. The determination of mucoproteins in the serum has been suggested as a valuable diagnostic aid in the differentiation of jaundice in hepatitis from that in extrahepatic obstruction (see the discussion on jaundice and Chapter 5).

Amino Acids

The normal concentration of amino acid nitrogen in the serum is 3 to 6 mg./100 ml. Since the amino acids undergo deamination in the liver, an increase in the serum level could result from extensive impairment or destruction of liver cells. In cases of acute hepatic necrosis caused by chemical agents such as phosphorus or carbon tetrachloride and in the terminal stages of liver disease, the level may increase to 15 to 25 mg./100 ml. The changes are too slight in early hepatitis or mild cirrhosis to be of diagnostic assistance (see Chapter 5).

Ammonia

The level of circulating ammonia in the blood is extremely low in normal individuals and ranges from 10 to 70 μg./100 ml. This low concentration is surprising when one considers the continuous processes of oxidative deamination and transamination of dietary and tissue amino acids. Since any appreciable level of ammonia in the blood would adversely affect acid-base balance and brain function, a major mechanism for its removal is essential. Although a small amount of ammonia is used for the synthesis of *glutamine*, the synthesis of *urea* is mainly responsible for its removal from the blood. In the first reaction of urea synthesis in the liver, ammonia combines with carbon dioxide and water under the influence of carbamyl phosphate synthetase to form carbamyl phosphate.

$$NH_3 + CO_2 + H_2O + 2ATP \xrightarrow[\text{synthetase}]{\text{Carbamyl phosphate}} \underset{\textit{Carbamyl phosphate}}{H_2N{-}CO{-}OPO_3H_2} + 2ADP + PO_4 + H^+$$

This important reaction requires the expenditure of two molecules of ATP and is irreversible. The complete urea cycle involves the formation of *citrulline* from *ornithine* and carbamyl phosphate, the addition of another amino group from aspartic acid in the formation of *argininosuccinic acid*, a hydrolysis to form *arginine*, and the action of liver arginase, yielding urea and the starting material ornithine.

Glutamine is also synthesized in the brain, and this synthesis increases when the blood ammonia is elevated. In terminal liver disease, the ammonia not removed by the liver is used for glutamine synthesis by the brain, resulting in an elevated blood concentration of glutamine. Since glutamic acid is used in glutamine synthesis, this would cause a drain on the glutamic acid and on the citric acid cycle intermediates of the brain, decreasing oxidative metabolism. Respiration decreases and coma may result from a mechanism similar to the coma of hypoglycemia or hypoxia. The relationship between glutamine synthesis, glutamic acid, and the citric acid cycle intermediates in the brain may be represented as follows:

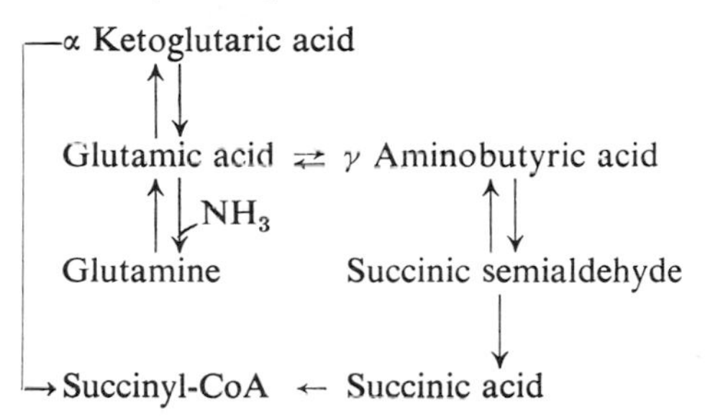

Hepatic coma and the terminal stages of cirrhosis are often marked by an increase in blood ammonia. In fact, the determination is most commonly requested to assist in establishing a diagnosis of impending or existing hepatic coma. The determination of ammonia is discussed in Chapter 10, Electrolytes.

LIPID METABOLISM AND LIVER FUNCTION

The liver plays a major role in lipid metabolism. It is involved in the complex transportation of lipid material between the blood and the bile. The liver is an important site of synthesis of fatty acids, bile acids, ketone bodies, cholesterol and cholesterol esters, phosphatides, and lipoproteins. Oxidation of fatty acids to produce energy and the removal of cholesterol and phosphatides from the plasma are carried out by the liver (see Chapter 7, The Lipids).

Cholesterol and Cholesterol Esters

Cholesterol is widely distributed in the tissues of the body and to some extent is synthesized by all tissues. Precursors for this sterol molecule have been shown to be small molecules such as acetate and acetyl-CoA. The liver is the major source of blood cholesterol and the main site of cholesterol synthesis, storage, and excretion. The serum concentration of *total cholesterol* of a normal individual is governed mainly by the rate of synthesis by the liver. The supply of dietary cholesterol and lipids and the serum and extrahepatic tissue concentration of the sterol control the rate of synthesis by the liver. Cholesterol is converted into several types of compounds, such as cholesterol esters and bile acids, in the liver. The bile acids and some of the cholesterol are excreted into the intestine by way of the bile and may be partially reabsorbed. The portion of the cholesterol and bile acid derivatives that is not reabsorbed is excreted in the feces.

Normal values for total cholesterol in the serum have been the subject of extended discussion. The increased interest in atherosclerosis in recent years has resulted in the determination of increasing numbers of serum levels by acceptable methods. These results indicate a level of 140 to 250 mg./100 ml. in both men and women at age 20.

The levels increase with age to a maximum of about 150 to 300 mg./100 ml. between 50 and 60 years, and then decrease slightly. Many clinicians feel that levels of over 250 mg./100 ml. are cause for concern, although the high fat diet in this country results in many apparently healthy individuals with cholesterol levels around 300 mg./100 ml. Esterified cholesterol is usually expressed as per cent of total cholesterol. The range for normal individuals is 65 to 75 per cent as esterified cholesterol.

As mentioned previously, the liver is the key organ in the synthesis and excretion of cholesterol; therefore, diseases of the liver or biliary tract would affect the plasma concentration of the free and ester forms. In *obstructive jaundice* there is a disproportionate increase in serum levels of free cholesterol, since free cholesterol and bile acids are normally excreted in the bile. Any type of obstruction, either intra- or extrahepatic, will cause an increase in total cholesterol levels in the serum, but the ratio of esterified to free form will always be less than that in normal serum. Since cholesterol esters are synthesized by normal parenchymal cells, the ester concentration will decrease in the presence of *parenchymal damage*. Acute obstruction may be characterized by serum levels of 300 to 400 mg./100 ml., whereas chronic biliary obstruction may result in levels around 800 mg./100 ml.

The esterification of cholesterol with polyunsaturated fatty acids by the liver cells depends on the presence of a transferase enzyme and involves a different mechanism than the synthesis of cholesterol. This is shown by the decrease in the serum level of cholesterol esters in liver disease, in which there is an impairment of the synthesis of the transferase. Intrahepatic obstruction tends to increase free cholesterol levels and an overall increase in the total level may be seen in early hepatitis. In chronic conditions, such as cirrhosis, that involve considerable destruction of liver cells, the cholesterol level eventually falls below normal since decreased synthesis is taking place. In cirrhosis the total cholesterol level may be below 100 mg./100 ml. and the ester level less than 20 mg./100 ml. In other conditions, such as hypothyroidism and nephrosis, an increase in total serum cholesterol is observed. The proportions of free and ester forms are normal in these diseases.

For details of the methods for cholesterol and cholesterol esters determination and their relation to other lipids in the body, see Chapter 7.

Bile Acids

Bile acids such as *cholic acid* are catabolic products of cholesterol formed in the liver. The acids are conjugated with either glycine or the sulfur-containing compound taurine to form bile salts that are called glycocholates or taurocholates. The bile salts are excreted in the bile and recirculate back to the liver through an enterohepatic pathway. About 0.8 gm. of the salts are excreted per day in the feces. The bile salts normally function in the emulsification of dietary fat, the activation of the lipases, and in the absorption of lipids through the intestinal mucosa.

CLINICAL SIGNIFICANCE

The normal level of bile acids in the serum is 0.3 to 3.0 mg./100 ml. Increased levels are commonly observed in hepatitis and obstructive jaundice with no clear-cut differentiation between intra- and extrahepatic obstruction. The increased levels in the serum are caused by a decrease in the conjugating mechanism with glycine and taurine in liver cell damage, and in an impairment of the enterohepatic circulation in obstruction. In the normal individual only small amounts of bile salts are excreted in the urine. As the level in the serum increases in liver disease, the urinary output increases.

Methods

Unfortunately, complex methods are required for the determination of bile acids in serum. These methods involve extraction with organic solvents, partition chromatography, and spectrophotometry or ultraviolet light absorption or fluorescence measurements in concentrated sulfuric acid. Investigators employing a recently developed complicated fluorescence method concluded that even when a sensitive quantitative method was available, the results added nothing to the diagnostic assistance gained from other tests for liver function.

Modifications of the Pettenkofer test may be used to detect the increased urinary excretion of bile acids in liver disease. The Pettenkofer reaction results in the development of a red-purple color when cholic acids react with fructose or furfural after the addition of sulfuric acid. The Mylius modification for the detection of bile salts in urine is carried out as follows: To 5 ml. of urine in a test tube, add 3 drops of 0.1 per cent aqueous furfural solution, then add 2 to 3 ml. of concentrated H_2SO_4 to form a lower layer. Cool under running water and shake carefully. The presence of appreciable amounts of bile acids will produce a red color.

Fatty Acids

Fatty acids are continually produced in the liver from acetyl-CoA, resulting from glucose metabolism. The acetyl-CoA is converted to malonyl-CoA to initiate this synthesis. The fatty acids are used in the synthesis of triglycerides, cholesterol esters, and phosphatides. A second source of fatty acids that are of more clinical interest are the unesterified, nonesterfied, or free fatty acids (UFA, NEFA, or FFA) released from adipose tissue and carried in the plasma to the liver and other tissues. They are carried in the blood in combination with albumin or lipoproteins and increase in concentration when glucose is not readily available for energy requirements in the tissues. The names NEFA or FFA are related to their lack of combination in glycerides. Several hormones, including epinephrine, norepinephrine, growth hormone, and ACTH, influence the release of additional quantities of UFA from the triglycerides in adipose tissue when energy is needed by muscle and other tissues. The rapid turnover rate of these fatty acids suggests that they represent a significant fraction of the daily caloric output.

CLINICAL SIGNIFICANCE

The normal plasma levels of UFA in the adult range from 0.4 to 0.9 mEq./L.; higher values are observed in children and obese individuals. An increase over the normal level occurs whenever the body tissues need energy or when carbohydrate utilization is impaired. In all forms of hepatitis and obstructive jaundice there is an increase in the plasma level, probably because there is an increased mobilization of lipids from adipose tissue.

Additional information concerning UFA and methods for their measurement may be found in Chapter 7.

Phosphatides

Phosphatides are conjugated lipids that contain a phosphate group and a nitrogenous base such as choline, ethanolamine, or sphingosine. An approximate distribution of the phosphatides in normal plasma would be 70 per cent lecithin, 20 per cent sphingomyelin, and 10 per cent cephalin. The phosphatides in the plasma occur in combination with globulins as lipoproteins. Although many cells in the body probably synthesize phosphatides for their own purposes, the plasma phosphatides are produced in the liver and apparently are also metabolized by that organ.

The normal range of phosphatides in the plasma, expressed as lecithin, is 175 to 275 mg./100 ml. and, expressed as phosphorus, 7 to 11 mg./100 ml. Since the liver is involved in both the synthesis and utilization of plasma phosphatides, the plasma level reflects the balance of the phosphatide metabolism in the liver. The levels are higher than normal in chronic obstructive jaundice compared with the lower levels in hepatitis, but the determination of phosphatides in the plasma is not commonly used to distinguish between the two conditions.

CONJUGATION AND DETOXICATION

In addition to the role played by the liver in bilirubin, carbohydrate, protein, and lipid metabolism, another essential function involves conjugation and detoxication. The liver by means of conjugation is able to convert many toxic substances into nontoxic compounds, change active drugs into inactive conjugates, and alter the solubility of metabolites by esterification or conjugation to assist in normal excretion. Phenols, menthol, camphor, salicylates, indole, hormones, bromsulfalein (BSP dye), and bilirubin are common examples of compounds that are detoxified or conjugated in the liver. Glycine, glucuronic acid, and sulfates are often used as the *conjugating agents*. Benzoic, nicotinic, and salicylic acids may be conjugated with glycine to form hippuric, nicotinuric, and salicyluric acids. Several conjugation mechanisms involve the formation of *glucuronides* of drugs such as salicylates and phenacetin, phenols, benzoic acid, and steroids. One of the essential steps in the metabolism of bilirubin is the formation of bilirubin diglucuronides by the liver cell before excretion in the bile. So many compounds are excreted in combination with glucuronic acid that the increase in glucuronides after the administration of a test substance has been proposed as a test of liver function. Although the conjugation processes of the liver fulfill many useful functions for the body, the presence of conjugated forms often complicates the determination of drugs and metabolites in the urine. For example, to measure the total excretion of salicylates or of steroids such as 17-hydroxycorticosteroids or 17-ketosteroids, the conjugated forms must be hydrolyzed before analysis.

The Hippuric Acid Test

Of all the tests that have been proposed to evaluate the ability of the liver to detoxify or conjugate, the hippuric acid test remains the most practical. Benzoic acid in the form of sodium benzoate is conjugated with glycine to form *hippuric acid* for excretion by the kidney. Since the test requires both the presence of the conjugating agent and the enzyme system involved in the conjugation, it should provide a measure of liver function unlike tests described earlier. Nevertheless, it is little used today in liver function testing.

CLINICAL SIGNIFICANCE

Functioning liver cells are essential for the proper rate of conjugation and excretion of a foreign substance when kidney function is normal. Decreased rates are observed in liver cell impairment or destruction, as found in hepatitis, cirrhosis, carcinoma of the liver, and hepatic necrosis. Uncomplicated or early extrahepatic obstruction does not decrease the rate of conjugation. In conditions involving renal impairment or disease, the results of the test are inconclusive. If abnormalities of intestinal absorption are suspected, the test substance must be administered intravenously.

PREPARATION OF PATIENT

For the oral test, the patient should be in the postabsorptive state. After emptying the bladder, he is given 6.0 gm. of sodium benzoate dissolved in about 200 ml. of water. The urine is collected for a 4 hour period.

For the intravenous test, a sterile solution containing 1.77 gm. of sodium benzoate in 20 ml. of water is slowly injected by a physician. Just before the injection, the patient empties his bladder and drinks a glass of water. The urine is collected for 1 hour after the injection with complete emptying of the bladder by the patient.

PRINCIPLE

Sodium benzoate is conjugated with glycine to form hippuric acid, which is excreted in the urine. Sodium chloride is added to decrease the solubility of hippuric acid and enhance its precipitation from acidified urine. Hippuric acid is then isolated, dissolved, and titrated with a standard solution of alkali using phenolphthalein as an indicator. The solubility of hippuric acid in the acidified urine is decreased from about 0.5 per cent to about 0.123 per cent by use of the NaCl.

REAGENTS

1. Sodium chloride, A.R., solid.
2. Sulfuric acid, concentrated, A.R.
3. Sodium chloride solution, 30 per cent (w/v). Store in a refrigerator and use the cold solution.
4. Standard sodium hydroxide solution, standardized near 0.1 *N*.

PROCEDURE

1. Measure and record the total volume of urine sample.
2. Transfer one tenth of the total volume of urine to a centrifuge tube of proper size. Add 3.0 gm. of solid NaCl for every 10 ml. of urine used and dissolve by warming if necessary.
3. Add 0.1 ml. of concentrated H_2SO_4 for every 10 ml. of urine used and mix. Place in refrigerator for 30 minutes. If precipitation does not occur, scratch the inside of the tube with a glass rod to initiate crystallization and place tube in the refrigerator for another 30 minutes.
4. Centrifuge and discard the supernatant fluid.
5. Wash the precipitate with 10 ml. of cold 30 per cent NaCl by rinsing the walls of the tube, mixing, and recentrifuging. Discard the supernatant fluid. Wash again with NaCl solution and discard the supernatant fluid.
6. Dissolve the precipitate in about 10 ml. of boiling hot water and transfer to a small flask. Titrate with the standardized 0.1 *N* NaOH, using phenolphthalein as an indicator.

CALCULATION

Several factors must be considered in the calculation of excreted hippuric acid. First, 1 ml. of 0.1 *N* NaOH is equivalent to 0.0179 gm. of hippuric acid; second, hippuric acid is slightly soluble, 0.123 per cent, in the acidified urine treated with NaCl and this solubility must be corrected for in the calculation. Since one tenth of the total urine specimen was used in the test, the calculation would be as follows:

$$\text{gm. hippuric acid} = 10[(0.0179 \times \text{ml. NaOH*}) + (0.00123 \times \text{ml. urine aliquot used})]$$
$$= (0.179 \times \text{ml. NaOH*}) + (0.0123 \times \text{ml. urine aliquot used})$$
$$\text{gm. hippuric acid} \times 0.68 = \text{gm. hippuric acid expressed as benzoic acid}$$

NORMAL VALUES

The normal excretion of hippuric acid expressed as benzoic acid in the oral test is 3.0 to 3.5 gm./4 hr.

The normal excretion of hippuric acid in the intravenous test is 0.6 to 0.9 gm./1 hr.

REFERENCES

Quick, A. J.: Am. J. Med. Sci., *185*:630, 1933.
Weichselbaum, T. E., and Probstein, J. G.: J. Lab. Clin. Med., *24*:636, 1939 (modified).

THE EXCRETORY FUNCTION OF THE LIVER

One of the major functions of the liver involves its excretory ability. Many substances, such as bilirubin, cholesterol, and alkaline phosphatase, are excreted by the liver cells into the bile. In the early stages of liver cell damage, the *excretory function* of the liver is diminished and thus a test that measures changes in the excretory power of the liver would be a valuable function test.

The Bromsulfophthalein Test

Phenolphthalein and several of its derivatives are excreted by the kidney and the liver. The brominated derivative of phenolsulfophthalein, called *bromsulfophthalein* or *BSP*, was found to be excreted almost entirely by the liver. The excretion of this dye has been used as a general liver function test since 1925. The rate of removal of BSP and its excretion into the bile depends on several factors: the blood level of the dye, the hepatic blood flow, the condition of the liver cells, and the patency of the bile ducts.

CLINICAL SIGNIFICANCE

In the absence of jaundice, the bromsulfophthalein test provides a simple, sensitive test of liver function that is capable of detecting early lesions of the liver cells. In the presence of jaundice from both intra- and posthepatic disorders including fatty liver, the test is elevated. In cirrhosis, in the absence of jaundice, an increased retention of 25 to 45 per cent of BSP is frequently observed.

SPECIMEN

The serum specimen should be free from hemolysis or lipemia.

---

* Based on a normality of NaOH of 0.100. If normality is different, apply appropriate correction factor.

PRINCIPLE

A serum specimen drawn after the injection of the dye is diluted with an alkaline buffer and the absorbance of the purple color is measured at 580 nm. An acid reagent is then added and the absorbance read again. The difference in absorbance is due to the BSP dye present in the serum. The strong anion, p-toluenesulfonate, is added to the alkaline buffer to release the BSP dye from the serum albumin.

REAGENTS

1. Alkaline buffer, pH 10.6 to 10.7. Dissolve 6.46 gm. of Na_2HPO_4, 1.77 gm. of $Na_3PO_4 \cdot 12H_2O$ and 3.2 gm. of sodium p-toluenesulfonate in water and dilute to volume in a 500 ml. volumetric flask. Check the pH and adjust if necessary with 1 *N* NaOH or 1 *N* HCl.

2. Acid reagent, 2 *M* NaH_2PO_4. Dissolve 27.6 gm. of $NaH_2PO_4 \cdot H_2O$ in water and dilute to volume in a 100 ml. volumetric flask.

3. BSP dye standard, 5 mg./100 ml.; equivalent to 50 per cent retention. The intravenous BSP solution, 50 mg./ml., is diluted 1:1000 (for example, 0.5 ml. of the dye diluted to 500 ml. with water). The diluted standard is stable for 1 week.

PROCEDURE

1. To avoid lipemia, the test is usually run in the morning on a patient in the fasting state. The dye, 5 mg./kg. of body weight, is injected by a physician under proper clinical conditions and a blood specimen is obtained 45 minutes after the injection of the dye.
2. To 1.0 ml. of the serum add 7.0 ml. of alkaline buffer and mix.
3. Read the absorbance (A_1) at 580 nm.
4. Add 0.2 ml. of the acid reagent, mix, and read absorbance (A_2) at 580 nm.
5. To 1.0 ml. of the BSP dye standard add 7.0 ml. of alkaline buffer, mix, and read the absorbance (A_3) at 580 nm.

CALCULATIONS

$$\text{per cent retention of BSP} = \frac{A_1 - A_2}{A_3} \times 50$$

NORMAL VALUES

With the dosage of 5 mg./kg., normal adults will show less than 6 per cent retention of BSP dye at 45 minutes. Very obese or markedly underweight people have different normal values, unless the amount of dye is corrected for the body surface.

PROCEDURE NOTES

A standard curve may be constructed by diluting the 50 mg./ml. BSP solution 1:500 to prepare a stock 10 mg./100 ml. solution representing 100 per cent retention. Dilute the stock solution to prepare 5 mg., 2.5 mg., 1 mg., and 0.5 mg./100 ml. standards, respectively representing 50, 25, 10, and 5 per cent retention. To 1.0 ml. of each of the five standard solutions, add 7.0 ml. of alkaline buffer, mix, and read the absorbance at 580 nm. against a water blank. Plot absorbance against per cent retention. Since the curve follows Beer's law, the 50 per cent retention standard may be used as in the preceding procedure.

REFERENCE

Seligson, D., Marino, J., and Dodson, D.: Clin. Chem., *3*:638, 1957 (modified).

SERUM ENZYMES RELATED TO LIVER FUNCTION

The search for useful liver function tests has naturally included a study of the activity of a number of enzymes in the serum. Many of these enzymes are involved in intermediary metabolism and are present in the liver cell in high concentration. When the cells are injured or disrupted, as occurs in acute liver disease, these enzymes are released into the serum and their increased levels are often of diagnostic significance. The level of certain enzymes, such as alkaline phosphatase, increases to a greater extent in obstructive jaundice; the serum concentrations of other enzymes, such as transaminase and isocitric dehydrogenase (ICD), are also increased in liver cell damage. Other enzymes, such as pseudocholinesterase, are decreased in active hepatitis. Detailed information on the various enzymes, as well as methods for their determination, is given in Chapter 8, Enzymes.

Alkaline Phosphatase

There are still unsettled questions concerning the formation and excretion or inactivation of alkaline phosphatase in the body. The normal serum level apparently is the result of two factors: the rate of release of alkaline phosphatase from tissues, mainly liver and bone; and the rate of inactivation of alkaline phosphatase, as well as its excretion. In recent investigations the alkaline phosphatase from intestine, placenta, liver, and bone have been differentiated and the isoenzyme originating in the liver was shown to be the main component of the serum of normal adults. In liver disease, the increased level of alkaline phosphatase results from increased liberation of the enzyme from the sinusoidal surface of the liver cell and from regurgitation of the biliary isoenzyme back into the serum. On the basis of extensive studies of the serum level in liver disease, it has been concluded that significant differences are found in obstructive jaundice versus hepatitis. An arbitrary diagnostic dividing line has been set to differentiate the two conditions. Using Bodansky units as an example, normal values are 1.5 to 4 units/100 ml.; values over 15 units are indicative of obstructive jaundice, or of intrahepatic cholestasis (hepatic cholestasis), whereas those between 4 and 15 units are more suggestive of various forms of hepatitis. It has been found that the serum alkaline phosphatase level is a more sensitive index of bile stasis than the serum bilirubin level.

Transaminases

Transaminases are found mainly in the liver, heart, kidney, and muscle tissue. These enzymes are involved in the exchange of amino groups from one amino acid to a keto acid to form a new amino acid. An acute destruction of tissue, such as in myocardial infarction or hepatocellular necrosis, results in a rapid release of the enzyme into the serum. For example, glutamic oxalacetic transaminase (GOT) levels in serum greatly increase after myocardial infarction. Another major enzyme, *glutamic pyruvic transaminase* (GPT) usually exhibits higher serum levels in acute viral hepatitis than does GOT. Normal serum values for both of these enzymes are less than 40 units/100 ml. In acute hepatitis, such as the early stages of viral hepatitis, the GPT levels usually range from 300 to 3000 units and may reach 4000 units/100 ml. The actual level found depends on the stage of the disease when the sample is withdrawn, and, generally, on the approximate extent of necrosis. The majority of cases of hepatitis have serum values around 2000 units and serial determinations are commonly used to follow the progress of the disease. Elevated values of 50 to 300 units may be found in cirrhosis and obstructive jaundice, but levels well above 400 are usually seen only in acute toxic or viral hepatitis.

The simultaneous determination of GOT and GPT often yields valuable diagnostic information. In myocardial infarction, GOT is higher than GPT, whereas in acute liver disease, such as hepatitis, GPT is usually higher. In uncomplicated obstructive jaundice, elevations are slight, but the GOT is generally higher than the GPT. When the obstruction is followed by secondary liver necrosis, a reversal of the relative concentrations may occur. In chronic liver diseases, such as cirrhosis or carcinoma, the GOT is higher.

Isocitric Dehydrogenase

ICD is found in many tissues, including the liver; it is an enzyme that catalyzes the conversion of isocitric acid into α-ketoglutarate in the Krebs cycle. The normal concentration in the serum ranges from 50 to 150 units/ml., using the millimicromole scale. The serum levels in obstructive jaundice and cirrhosis are only moderately elevated, and in myocardial infarction they are usually normal. In the early stages of viral hepatitis marked increases, up to 1500 units, are observed. High values are also seen in neoplastic disease with liver metastases. It is the feeling of some investigators that the determination of the level of ICD in the serum might be more valuable than that of GOT or GPT in the diagnosis of viral hepatitis. The procedure can be used to differentiate between myocardial infarction and hepatitis.

Lactic Dehydrogenase (LDH)

LDH is found in the liver, heart muscle, and many other tissues. It is a glycolytic enzyme involved in the reversible conversion of pyruvic acid to lactic acid. In methods based on the reduction of pyruvic acid by $NADH_2$(DPNH) and the measurement of the decrease in absorbance of $NADH_2$ at 340 nm., the normal serum level ranges from 200 to 450 units/ml. Moderate increases in the serum LDH values occur in hepatitis; values are usually higher in myocardial infarction, pulmonary embolism, leukemia, and hemolytic anemia. Fortunately, it is possible to distinguish between the LDH of heart muscle and that of liver tissue. By starch gel electrophoresis, five different *LDH isoenzymes* have been found in serum; they have been designated LDH_1 to LDH_5, with LDH_1 exhibiting the fastest mobility in electrophoresis. The slowest moving, LDH_5, is the liver LDH, whereas LDH_1 is of cardiac origin. Three main patterns of starch gel electrophoresis are observed in disease. High LDH_5 is characteristic of acute hepatic necrosis, high LDH_1 of acute myocardial infarction, and high $LDH_{2,3,4}$ of, e.g., malignancy. The myocardial LDH may also be distinguished from hepatic LDH by heating buffered serum at 65°C. for 30 minutes. The activity remaining after this treatment represents LDH_1; the hepatic LDH_5 is destroyed in the process. This technique is more practical than the electrophoretic one, which offers some technical difficulties.

Pseudocholinesterase

This enzyme is found in the serum and is capable of splitting acetylcholine and other esters. The concentration of the enzyme in normal serum ranges from 2.2 to

TABLE 13-1. *Summary of Liver Function Tests (with Normal Values)*

Normal values may change with the method employed. This is especially true for the enzyme tests where a wide range of different normal values exists, depending on the reaction selected and the type of unit employed.

Test	*Normal Values*	*Acute Hepatitis*	*Obstructive Jaundice*	*Chronic Hepatitis*
Serum Enzyme Determinations				
Aldolase	15–35 units/ml.	Marked increase	Normal or slight increase	Normal or slight increase
Alkaline phosphatase	1.5–4 units (Bodansky)	Mild increase <15 units	Marked increase >15 units	Mild increase 5–10 units
α-Hydroxybutyric dehydrogenase	114–290 units/ml.	Normal	Normal	Normal
Isocitric dehydrogenase	50–150 units	Marked increase up to 1500 units	Mild increase	Increased
Lactic dehydrogenase	200–450 units/ml.	Moderate increase	Normal or slight increase	Normal
Pseudocholinesterase	2.5–5.6 micromoles/ min./ml.	Decreased	Normal	Decreased
Sorbitol dehydrogenase	Zero	Marked increase	Near zero	Near zero
Transaminase GOT	8–40 units	Marked increase 300–3000 units	Mild increase 50–300 units	Mild increase 50–300 units
Transaminase GPT	8–40 units	Marked increase up to 4000 units	Mild increase 50–300 units	Mild increase 50–300 units
Flocculation Tests				
Cephalin-cholesterol	0 to 1+	3+ to 4+	0 to 1+	2+ to 3+
Thymol turbidity	0–5 units	Increased	Normal (often)	Increased
Zinc sulfate turbidity	2–7 (12) units	Increased	Normal	Increased
Miscellaneous Tests (tests measuring conjugation, detoxication, excretion, carbohydrate metabolism, and/or synthesis)				
Bilirubin, serum				
conjugated	0.2 mg./100 ml.	Increased	Increased	Variable
unconjugated	0.8 mg./100 ml.	Increased	Increased	Variable
Bilirubin, urine	Negative	3+	4+	2+
Bromsulfophthalein Retention	0–6 per cent retention (45 min.)	Increased (Test not carried out in presence of jaundice)	Increased	Increased
Cholesterol, total, serum	140–250 mg./100 ml. Age dependent	Normal	Increased	Normal or decreased
Cholesterol, ester, serum	65–75 per cent	Decreased	Normal	Decreased
Galactose tolerance	0–3 gm. (in urine)	Increased	Normal	Increased
Hippuric acid	3–3.5 gm./4 hrs.	Decreased	Normal	Decreased
Icterus index	3–8 units	Increased	Increased	Variable
Protein, total, serum	6.0–8.0 gm./100 ml.	Normal	Normal	Normal or decreased
Albumin	3.2–4.8 gm./100 ml.	Decreased	Normal	Decreased
Globulin	2.2–3.5 gm./100 ml.	Increased	Normal	Increased or normal
Prothrombin time	Normal	Increased	Increased	Increased
Urobilinogen, stool	75–275 E.U./100 gm. or 100–400 E.U./ 24 hr.	Slight decrease	Decreased	Slight decrease
Urobilinogen, Urine	0.1–1.0 units/2 hrs.	Increased (often)	Decreased	Increased

5.6 micromoles of substrate hydrolyzed per minute per ml. The determination of the serum level of cholinesterase appears to be a good test of liver injury, since lower than normal values are found in hepatitis and cirrhosis. No change in the level is observed in obstructive jaundice. The major drawback to the application of this determination is the lack of a control level for each patient before the liver disease study.

α-Hydroxybutyric Dehydrogenase (HBD)

HBD is closely associated with LDH and is found in serum and the tissues that contain LDH. It catalyzes the reduction of α-keto acids, especially α-ketobutyric acid. Of the five LDH isoenzymes, the two electrophoretically fastest components (LDH_1 and LDH_2), characteristic of heart muscle, exhibit the highest HBD activity. The normal serum level, expressed in the same units as LDH, ranges from 114 to 290 units/ml. Although LDH is increased in viral hepatitis, the serum level of HBD is essentially unchanged. The ratio of HBD/LDH in normal serum is about 0.72; however, in acute hepatitis, the ratio decreases markedly and differentiates between the LDH increase in hepatitis and myocardial infarction without the necessity of LDH isoenzyme determination.

Sorbitol Dehydrogenase (SDH)

SDH is found mainly in liver tissue and is involved in the conversion of sorbitol. into fructose. Normal serum contains little, if any, SDH, but the levels are markedly elevated in acute hepatitis. Little SDH activity has been found in sera from obstructive jaundice, cirrhosis, or acute myocardial infarction. The determination of SDH should, therefore, provide valuable diagnostic assistance in the differentiation between hepatitis on the one hand and obstruction and myocardial infarction on the other

Aldolase

This enzyme, like many of the dehydrogenases, is concerned with the metabolism of carbohydrates in tissues such as the liver. It splits fructose 1,6-diphosphate into dihydroxyacetone phosphate and glyceraldehyde 3-phosphate. The normal serum level, based on the decrease in absorbance of $NADH_2$ at 340 nm., ranges from 15 to 35 units/ml. An increase in the serum level of aldolase is observed in acute hepatitis, myocardial infarction, and muscular dystrophy. Since normal or mildly elevated levels are found in obstructive jaundice, the determination should be helpful in the diagnosis of viral hepatitis.

Table 13-1 compares the normal values obtained from the liver function tests described in this chapter with the changes observed in acute hepatitis, obstructive jaundice, and chronic hepatitis.

REFERENCES

1. Henry, R. J.: Clinical Chemistry: Principles and Technics. New York, Hoeber Medical Div., Harper & Row Publishers, 1964.
2. Leevy, C. M.: Evaluation of Liver Function. Indianapolis, Eli Lilly Research Laboratories, 1965.
3. Popper, H., and Schaffner, F.: Liver: Structure and Function. New York, McGraw-Hill Book Co., Inc., 1957.
4. Sherlock, S.: Diseases of the Liver and Biliary System. Springfield, Ill., Charles C Thomas, Publisher, 1965.

Chapter 14

GASTRIC ANALYSIS

by Norbert W. Tietz, Ph.D.

The human stomach consists of three major zones, the cardiac zone, the body, and the pyloric zone. (Fig. 14-1). The upper *cardiac zone* contains mucus-secreting *surface epithelial cells.* The *body* of the stomach contains cells or cell groups of four different types: (1) the surface epithelial cells, which secrete mucus; (2) the *parietal cells*, which are the main, and possibly only, source of hydrochloric acid; (3) the *chief* or *peptic cells*, which secrete a considerable amount of pepsinogen; and, finally, (4) the *neck chief cells* or *mucus cells*, which secrete mucus and pepsinogen. The third portion of the stomach, the *pyloric zone*, may be subdivided into the *antrum*, the *pyloric canal*, and the *sphincter*. Its cells secrete mucus, some pepsinogen, and gastrin. The pyloric zone plays a major rǫle in the process of emptying food into the duodenum by virtue of its strong musculature, which controls the flow of food into the duodenum by opening or closing the pylorus.

The functions of the stomach include acceptance, mixing, storing, and discharge of food into the duodenum. Its main functions, however, are the secretion of enzymes, intrinsic factor, and, especially, hydrochloric acid, which activates pepsinogen, provides the optimal pH for pepsin action, and destroys bacteria. During passage through the stomach, the proteinaceous foods are partially digested by the action of the enzyme *pepsin;* salivary *amylase* initiates the digestion of starch. The small amount of *lipase* that is secreted by the stomach contributes only slightly to the digestion of fats; only well emulsified fats, such as milk, are attacked to any significant degree.

An evaluation of these motor and secretory activities of the stomach is generally done by analyzing the *gastric residue* (content of the stomach after a fast of approximately 12 hours), which is obtained by aspiration through a gastric tube. The normal total volume is between 20 and 100 ml., usually being below 50 ml.; volumes above 100 ml. may be considered abnormal. Among the causes for increased volume are delayed emptying, as in *pyloric obstruction;* increase in gastric secretion, as in *duodenal ulcer;* and admixture of regurgitated material from the duodenum. In the last situation, the gastric residue generally contains an excessive amount of bile, which can be confirmed by chemical tests (e.g., Ictotest). The total volume of gastric secretion over 24 hours is about 2000 ml.; the volume of 12 hour nocturnal secretion varies between 150 and 1200 ml. The consistency is rather fluid, but, if thicker, there is

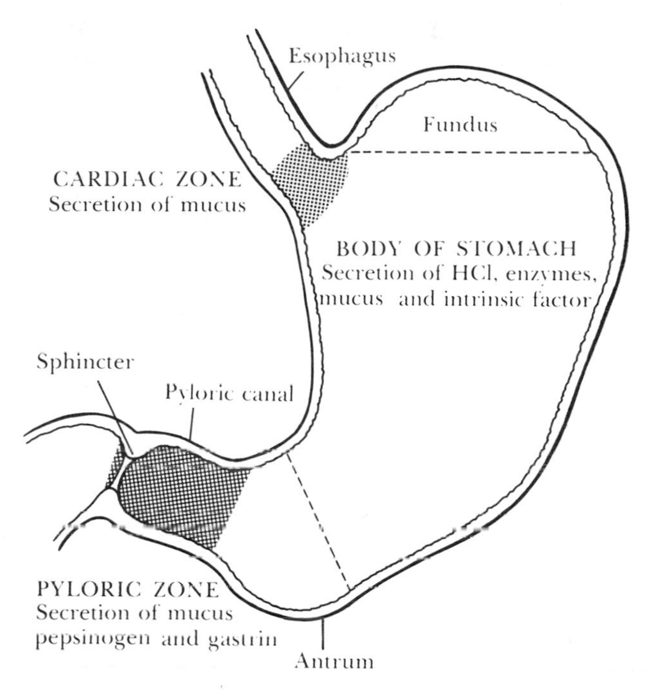

Figure 14-1. Schematic drawing of the stomach, with major zones.

probably an excessive amount of mucus present. The normal *odor* of gastric juice is sharply sour; a foul smelling gastric juice generally indicates fermentation. The residue is usually colorless, but its color may be slightly yellow or green due to admixture of regurgitated bile; this is the case in approximately 25 per cent of normal people and in the majority of patients after partial gastrectomy or gastroenterostomy. A red or brownish color in the gastric residue is usually due to blood, and simple tests are available to confirm this (e.g., Hematest).

MAJOR CONSTITUENTS OF GASTRIC RESIDUE

FREE HYDROCHLORIC ACID

Hydrochloric acid is secreted only by the parietal cells and at a constant concentration of approximately 155 mEq./L. (0.155 *N*). The hydrochloric acid mixes with the other gastric contents and it is believed that the variations in the concentration of hydrochloric acid found in gastric juice are due to changes in the proportion of hydrochloric acid relative to other secretions of the stomach. Also saliva, regurgitated materials, and ingested food can decrease the final concentration of free hydrochloric acid. The secretion of hydrochloric acid by the mucosa is continuous, but the volume fluctuates considerably. The flow increases by a neural mechanism mediated through the vagus (psychic or cephalic phase), and also by entrance of food into the stomach or duodenum, as well as by certain types of chemical stimulation of the gastric mucosa. The hormone *gastrin*, which is secreted by the antral mucosa, appears to have a significant role in stimulating the secretion of hydrochloric acid, pepsinogen, and the intrinsic factor, as well as pancreatic bicarbonate and especially pancreatic enzymes (see Test Meals and Other Stimuli of Gastric Secretion). Protein, polypeptides and alcohol in the antrum, in addition to antral distention, stimulate the release of gastrin into the blood. As the reaction at the antrum turns strongly acid, gastrin secretion ceases; this safeguards against overacidification of the intestinal tract. The gastrointestinal hormone *enterogastrone* is a further factor in the regulation of gastrin secretion. This hormone is released into the circulation on contact of the intestinal mucosa with fats and sugars, as well as by a strongly acid reaction. It not only inhibits the secretion of gastrin, but thus also the secretion of hydrochloric acid and pepsinogen; in addition, it decreases gastric motility. Emotional disturbances usually decrease output of hydrochloric acid.

The concentration of free hydrochloric acid in gastric residue of normal individuals is approximately 0 to 40 mEq./L., formerly often designated 0 to 40° of acidity.* Gastric secretion may also be measured as the absolute amount of HCl expressed in mEq. excreted per hour. The normal values are 0 to 5 mEq. for 1 hour basal output and 1 to 20 mEq. for a one hour period after stimulation with Histalog (1.7 mg./kg. body weight[5]).

Approximately 4 per cent of young normal individuals may have no free hydrochloric acid in the fasting stomach. The percentage of individuals without free hydrochloric acid increases to about 25 per cent at the age group of 60 years. Absence of free hydrochloric acid in gastric residue is considered abnormal only if there is also absence of hydrochloric acid secretion after proper stimulation with test meals, or preferably with Histalog. Therefore, it follows that test meals or gastric stimulants should always be tried in patients without free hydrochloric acid before the diagnosis of *achlorhydria* is made. *False achlorhydria* is a term used for those cases in which free hydrochloric acid is secreted, but is subsequently partially or fully neutralized by either saliva, food, or regurgitated materials. A determination of chloride, to be discussed, might help avoid misinterpretations.

The determination of the concentration of free hydrochloric acid is generally done by titration. Measurement of the pH of the gastric residue will also give an approximate indication of the amount of acid present, and is the preferred method according to some investigators[3] (see method that follows).

The pH of pure parietal secretion is approximately 0.9, but increases to pH 1.5 to 4.0 as the secretion mixes with the other gastric contents.

The hydrochloric acid secreted into the lumen of the stomach derives ultimately from the blood. Although the mechanism of this phenomenal process is still not understood, it is presently believed that there is an active transport of these ions across the membrane of the parietal cells and that the energy for this process is supplied by aerobic glycolysis. The secretion of gastric hydrochloric acid is accompanied by a decrease in blood chlorides and an increase in alkalinity of the blood and urine, the so-called "*alkaline tide*."

Several investigators have recommended the measurement of the total chloride content in gastric fluid as an indicator of the secretory activity of the gastric mucosa; such a test would not only measure the chloride present in free hydrochloric acid, but would also detect the chloride "bound" to protein and the hydrochloric acid neutralized by other gastric constituents, or duodenal material, or both. Very few clinicians request the assay, however. The total chloride content of gastric residue is considered to be 45 to 155 mEq./L.

---

* The degree of acidity is defined as milliliters of 0.10 *N* sodium hydroxide that are required to neutralize 100 ml. of gastric content. Degree of acidity per 100 ml. corresponds to mEq./L., and since mEq./L. is a better way of expressing concentrations, this term is now preferable.

TOTAL ACIDITY AND TOTAL TITRATABLE ACIDS

This fraction includes free hydrochloric acid, that portion of hydrochloric acid which is bound to protein, acid salts, e.g., $H_2PO_4^-$, mucus, and organic acids, such as lactic and butyric acid. These last two acids are normally not found in gastric juice except in some conditions (to be discussed later). The total acid is usually between 10 and 50 mEq./L., but occasionally may be slightly higher.

In recent years, some investigators[3] have advocated the determination of "total titratable acidity" by titrating a sample of gastric juice to pH 7.0 or 7.4 instead of the fractionation into free and combined acidity.

COMBINED ACIDITY

The difference between free and total acidity is due to acids other than free hydrochloric acid (see Total Acidity) and amounts to about 10 to 20 mEq./L. The term "*combined acidity*" stems from the belief that some of the acid is bound to other material, especially proteins. Although there has never been any definite proof that such binding of hydrochloric acid really occurs, the recent "cookie titrations" by Moore[3] indicate that binding of hydrochloric acid indeed may occur. Moore has also recommended that the term combined acidity be replaced by the term "nonionized acids." There is considerable doubt whether the determination of the "combined acidity" has any clinical significance and many investigators suggest that this test be discontinued.

ORGANIC ACIDS

Lactate and *butyrate* are formed in gastric content by bacterial action when food is retained in the stomach for long periods (over 6 hours) at neutral or slightly alkaline pH, a condition usually associated with carcinoma of the stomach or pyloric stenosis. If free hydrochloric acid is present, these organic acids are absent.

ENZYMES

The chief cells of the stomach store and secrete *pepsinogen*, which, in a medium below pH 6, is readily activated to *pepsin*. The activation process is autocatalytic. Pepsin seems to be always present when free hydrochloric acid is present. The optimum pH for the action of pepsin is pH 2.0, but 70 per cent of the activity is still present at pH 4.5. Gastric pH above 8.0 leads to denaturation of the enzyme. Pepsinogen is secreted not only into the lumen of the stomach, but also to some extent into the blood. It is then excreted in the urine as *uropepsinogen*, which is activated below pH 5.4 to *uropepsin*. Determinations of uropepsin in urine have been used in the attempt to determine the rate of pepsinogen secretion of the stomach. Uropepsinogen values are frequently found elevated in peptic ulcers, are essentially zero after total gastrectomy, and are very low in extensive atrophy of the gastric mucosa.

If both pepsin and free hydrochloric acid are absent from gastric juice, then *true achylia* (*achylia gastrica*) exists. This condition is constantly found in pernicious anemia. Gastric fluid also contains a small amount of other enzymes previously mentioned, but, at present, they do not appear to have any practical clinical significance.

MUCUS

Mucus is produced by the surface epithelial cells and the neck chief cells of the stomach and is present in only small amounts in normal gastric contents; however, increased amounts are sometimes found in gastric carcinoma, in gastritis, and in cases of mechanical irritation due to passage of a stomach tube. There is generally no need to analyze gastric content for mucus, which usually has a pH of approximately 7.4 to 8.2 and is chemically composed of mucopolysaccharides and protein moieties.

BLOOD

In some pathologic conditions, such as carcinoma of the stomach, peptic ulcer, and gastritis, blood may be present in gastric residue. Its appearance will vary with the pH of the gastric content. At the strongly acid pH in the stomach, acid hematin is formed; it has a brownish appearance, like coffee grounds. Fresh red blood is likely due to accidental trauma and does not generally signify any pathologic condition; however, in some instances underlying lesions, such as ulcer or carcinoma, may be responsible for the presence of fresh blood.

FOOD

Normal gastric residue does not contain any appreciable amount of food. Excessive accumulation of food particles indicates decreased motor activity of the stomach or pyloric obstruction.

MISCELLANEOUS MATERIALS

Gastric residue may contain *mucus cells*, *chief cells*, *parietal cells*, and regurgitated materials such as pancreatic juice, bile, and duodenal secretions. It also contains the *intrinsic factor*, which combines in some way with vitamin B_{12} and makes absorption of this vitamin possible. Lack of this factor causes vitamin B_{12} deficiency with resultant arrest in the development of red cells. It is now believed that the intrinsic factor is a mucoprotein with a molecular weight of above 50,000 and a terminal structure similar to that of specific blood group carbohydrates. It is secreted by the cells of the fundus and behaves electrophoretically as a β-globulin. The determination of the intrinsic factor is now feasible,[7] but it is rarely done except by the Schilling test (see Chapter 15, Pancreatic Function and Intestinal Absorption).

TEST MEALS AND OTHER STIMULI OF GASTRIC SECRETION

Since free hydrochloric acid may be absent from the gastric residue in a perfectly normal individual, it is imperative that one of the gastric stimulants, preferably Histalog, be used before making the diagnosis of achlorhydria.

EWALD MEAL

This stimulant meal, formerly quite popular, consists of two slices of unbuttered toast and 8 oz. of water or weak tea without cream or sugar. The main objection against this test meal is that it does not provide a very strong stimulus, and, in addition, technical difficulties in aspirating stomach content may result from food particles plugging up the tube. Therefore, these and other similar test meals have been largely replaced by the use of caffeine, alcohol, and mainly, histamine (Histalog) stimulation.

CAFFEINE STIMULATION

After withdrawing the gastric residue, 500 mg. of caffeine sodium benzoate in 200 ml. of warm water is admitted into the stomach through a gastric tube. Thirty minutes later, the entire gastric content is aspirated and specimens are collected for 90 minutes at 10 to 20 minute intervals. The test causes few side reactions. Caffeine is a stronger stimulant than the Ewald meal, although weaker than histamine and Histalog. It acts by stimulating gastrin liberation, which in turn stimulates hydrochloric acid secretion.

ALCOHOL STIMULATION

Again, the fasting content is withdrawn and 50 ml. of a 7 per cent (v/v) solution of ethanol is instilled into the stomach. Specimens are withdrawn at 10 to 20 minute intervals for 2 hours. Alcohol is a relatively potent stimulus for hydrochloric acid secretion and the danger of untoward reaction is small; however, the degree of stimulation is less than that given by histamine or Histalog. Alcohol acts by stimulating the secretion of gastrin, which in turn stimulates hydrochloric acid secretion.

HISTAMINE AND HISTALOG STIMULATION

Histamine and Histalog (Eli Lilly & Co., Indianapolis, Ind.) are the most powerful of the stimulants listed here. Approximately 40 per cent of patients who fail to show response to carbohydrate meals (such as the Ewald meal) do show response to histamine. Apparently histamine stimulates only the parietal cells (hydrochloric acid) and does not provoke the chief cells to secrete pepsin. In true achylia, there is no secretion of hydrochloric acid after histamine or Histalog stimulation.

Histamine has been found to produce allergic side effects in some patients and should not be given if such sensitivity on the part of the patient is known or suspected. It should always be administered by a physician, and, if the systolic blood pressure is below 110 mm., histamine is contraindicated. Some authors have recommended the administration of an antihistamine 30 minutes before the histamine injection, since antihistamines prevent the undesirable side effects of histamine without affecting hydrochloric acid secretion.

When performing the test, the stomach content is first aspirated and checked for free hydrochloric acid. If the test is positive, there is generally no need to give the histamine. If no free hydrochloric acid is present, a dose of 0.019 mg. of histamine phosphate/kg. body weight (equivalent to 0.01 mg. histamine base) is given subcutaneously or intramuscularly and further gastric specimens are aspirated at 15 to 20 minute intervals for 2 hours. The peak of hydrochloric acid secretion occurs 90 to 120 minutes after stimulation. Histalog (3-β-aminoethylpyrazol dihydrochloride), a histamine analog, is preferred by many investigators because it produces fewer side effects. The recommended dose is 0.5 mg. (I.M.)/kg. of body weight, although maximal stimulation is obtained by a dose of 2 mg./kg. body weight. For adults with average weight, a standard dose of 1.0 ml. of a solution containing 50 mg./ml. may be given.

GASTRIN STIMULATION

Gastrin I and II have recently been identified as polypeptides containing a chain of 17 amino acids, in which gastrin II differs from gastrin I only by the presence of a sulfate group on the tyrosine residue. Both are released by vagal stimulation, as well as by protein, polypeptides, and alcohol in the antrum of the stomach. Maximal stimulation of acid secretion by gastrin is obtained by a single subcutaneous injection of 2 μg. of gastrin per kg. of body weight. The extent of stimulation is 20 per cent above that produced by the maximal dose of histamine.

This stimulus has not yet been used routinely; however, since the compound is the natural stimulus of gastric hydrochloric acid secretion, and since this compound can now be synthesized, it is conceivable that gastrin will become the stimulus of choice in the future.

INSULIN STIMULATION

Insulin has been found to increase gastric hydrochloric acid secretion by stimulation of gastric mucosal cells through the vagus. This form of stimulus is very rarely used, although it is said to be helpful in evaluating the advisability of vagotomy. The recommended dose is 20 units of insulin, after which the stomach content is aspirated for 45 minutes.[5]

PROCEDURES FOR GASTRIC ANALYSIS

Determination of Free Hydrochloric Acid and Total Acidity

PREPARATION OF THE PATIENT

Routine gastric analysis is generally performed on gastric residue, which is defined as the content of the stomach after the patient has been fasting and has had no fluids for at least 12 hours. The patient is not permitted to smoke the morning of or during the test and should also avoid any form of exercise during this period.

PRINCIPLE

Toepfer's (Töpfer's) reagent (0.5 per cent dimethylaminoazobenzene) is an indicator that changes color in the pH range of 2.8 to 4.2. Two drops of this reagent are added to gastric fluid: a cherry red color denotes the presence of *free hydrochloric acid* and a yellow color the absence of it. If there is no free hydrochloric acid present, one of the available test meals or stimulants (see preceding section) should be given to check the response of the gastric mucosa to stimulation; however, there is generally no need to give a test meal if free hydrochloric acid is present in the fasting specimen. Other indicators, such as Congo red, thymol blue, and bromphenol blue, have been used for determining free hydrochloric acid, and phenol red has been used to determine total acidity. Although these indicators give satisfactory results, Toepfer's reagent and phenolphthalein are most commonly used.

A known amount (e.g., 5 or 10 ml.) of gastric juice containing 2 drops of Toepfer's reagent is titrated with 0.10 *N* NaOH until the indicator turns salmon pink, which occurs approximately at pH 2.8 to 3.0. It is believed that titration to this pH will detect free hydrochloric acid specifically and not any of the other acid gastric constituents (however, see Remarks and Sources of Error).[2] At a pH of 3.0, the concentration of hydrogen ions is only approximately 1 mEq./L., which represents essentially complete titration of all free acids. Some authors recommend titration to canary yellow (pH 4) since the yellow end point is easier to detect, but this approach is less desirable because titration values are slightly higher. The amount of 0.10 *N* NaOH used to reach the end point is an indication of the amount of free hydrochloric acid present (see Calculation). The *total acidity* of gastric fluid is determined by adding phenolphthalein to the same sample and continuing the titration until a pink color is obtained. The pH at this point is approximately 8.4. Titration to this pH will measure all acid gastric constituents, including mucus which has a pH range of approximately 7.4 to 8.2. *Combined acidity* is defined as the difference between total acidity and free hydrochloric acid.

REAGENTS

1. Toepfer's reagent. Dissolve 0.5 gm. of p-dimethylaminoazobenzene in 100 ml. of 95 per cent ethanol.

2. Phenolphthalein. Dissolve 1 gm. of phenolphthalein in 100 ml. of 95 per cent ethanol.*

3. 0.10 *N* NaOH, standardized.

PROCEDURE

1. Determine the pH of the gastric specimen. If the test involves a fractional meal, check the pH of each sample submitted. The pH may best be determined with a pH meter, but narrow range pH papers are usually satisfactory. If a timed (e.g., 12 hour) specimen is collected, the total volume must be measured and recorded.

2. Pipet whatever volume is conveniently available (e.g., 5.0 ml., but not more than 10 ml.) into a clean sputum jar or other appropriate container and record amount of gastric content used. If the gastric juice contains food particles, centrifuge the sample or filter it through gauze.

3. Add 2 drops of Toepfer's reagent. If there is a red color present, titrate with standard 0.10 *N* NaOH to the end point (salmon pink). The titration may also be carried out with a pH meter, in which case 0.10 *N* NaOH is added until a pH of 2.8 is reached. Calculate acidity by using the formula given under Calculation. If a determination of the total acidity is requested, continue as follows:

4. Add 1 drop phenolphthalein and titrate with standard 0.10 *N* NaOH to a pink color. (The color will change from the salmon pink of Toepfer's reagent to canary yellow and then again to the pink of phenolphthalein.)

5. Calculate total acidity by entering the entire amount of NaOH required (steps 3 and 4) into the following formula. The combined acidity may be calculated by taking the difference between total and free acidity.

CALCULATION

$$\text{free or total acidity in mEq./L.} = \frac{(\text{ml. of } 0.10\ N\ \text{NaOH}\dagger) \times 1000}{\text{ml. of gastric content used} \times 10}$$

or if 5 ml. of gastric content was used:

$$\text{mEq./L.} = \text{ml. of } 0.10\ N\ \text{NaOH} \times 20$$

NORMAL VALUES

There is considerable disagreement in the literature as to precise normal values, but the following are generally accepted.

Total volume: 20 to 100 ml. (generally below 50 ml.).

---

* Some investigators neutralize the phenolphthalein solution by adding sodium hydroxide until the first appearance of a pink color. The neutralized indicator does not contribute to the acidity of the gastric content; however, the error introduced by omitting this step is so small that most investigators use the phenolphthalein as outlined above.

† If the standardized NaOH solution is not exactly 0.10 *N*, multiply by appropriate normality factor.

Free HCl (without stimulation): 0 to 40 mEq./L.

Free HCl (after stimulation): 10 to 90 mEq./L. after alcohol and 10 to 130 mEq./L. after histamine or Histalog.

Combined acidity: 10 to 20 mEq./L.

Total acidity (without stimulation): 10 to 60 mEq./L.

REMARKS AND SOURCES OF ERROR

Normal gastric content is colorless, has a sour odor, and its total volume should not exceed 100 ml. If these criteria are not met, the appropriate observation should be noted on the requisition. The appearance of the specimen may suggest performance of additional tests such as the following: a test for blood in case of brownish looking gastric content, a test for bile if the gastric content appears yellowish or green, a test for lactic acid if a foul smell is observed.

Michaelis[2] established titration curves for gastric juice and for hydrochloric acid in water and found that both curves diverge at about pH 2.8. Therefore, he recommended that pH 2.8 should be used as an end point in the titrimetric determination of the hydrogen ion concentration; however, Moore[3] recently showed that this is an erroneous concept since the activity coefficients for hydrogen ions in the two solutions (hydrochloric acid and gastric juice) are different. Thus, the customary gastric titrations have an inherent source of error that can be reduced by performing pH determinations as discussed in the following section.

REFERENCES

Michaelis, L.: Harvey Lectures, 1926–1927. New York, Academic Press, Inc. 1928, pp. 59–89.
Cannon, D. C.: Examination of gastric and duodenal contents. *In* Todd-Sanford Clinical Diagnosis by Laboratory Methods. I. Davidsohn, and J. B. Henry, Eds. 14th ed. Philadelphia, W. B. Saunders Co., 1969, pp. 762–780.

Determination of pH of Gastric Juice

Intragastric pH measurements with electrodes, mainly glass electrodes, represent the most recent approach to pH measurement of gastric juice. Technical difficulties, such as proper placement of the electrode, and pH fluctuations due to swallowed air, limit this technique mainly to research work, and make it impractical for routine work.

Determinations of pH on aspirated gastric juice are best done with a pH meter, but many clinicians feel that narrow range pH papers are sufficiently accurate for routine use. It should be realized, however, that errors of more than 1 pH unit may be introduced when using the latter technique. There is no agreement on whether the gastric juice should be centrifuged before performance of the test. Many investigators recommend this technique, but others feel that some acid may be adsorbed to the sediment and thus is not being detected. Clear gastric content with little mucus may be analyzed without centrifugation.

Moore[3], as previously mentioned, stated that the hydrogen ion concentration of gastric juice cannot accurately be determined by either titration or direct pH measurement since the activity coefficient of acids is different in single and in mixed electrolyte solutions. He presented such an activity coefficient and recommended that single pH determinations be done followed by calculation of the true hydrogen ion concentration on the basis of the activity coefficient for mixed electrolytes such as gastric juice.

Tubeless Gastric Analysis (Diagnex Blue Test)

The passage of a stomach tube is an unpleasant experience for the patient and, in some conditions, it is even contraindicated; therefore, much effort has been expended to develop a tubeless form of gastric analysis.[4,6] Although the techniques now available have their limitations (see Comments on the Method and Sources of Error), they have become useful as screening procedures. The most popular of the various tests devised is the Diagnex Blue Test.

PRINCIPLE

After an overnight fast, the patient is given a stimulant, generally caffeine, which is followed 1 hour later by administration of an azure A resin. The azure A resin has been prepared by exchanging the hydrogen ions of a carboxylic cation exchange resin with azure A (3-amino-7-dimethylaminophenazathionium chloride). When this resin indicator compound comes in contact with free hydrochloric acid, the azure A indicator material is released as H^+ exchanges with it on the resin. This release begins at a pH below 3.0 and has its maximum at a pH of 1.5. The dye that is thus released is absorbed by the intestinal tract and excreted through the urine. The dye may be excreted in its blue form or as a colorless compound. It is not known whether the colorless compound is a reduced leucoform, or a conjugated form, or a combination of both, but it can be converted into the blue form by boiling with acid. The amount of dye excreted is determined either semiquantitatively with the aid of a comparator block or quantitatively by spectrophotometry.

PREPARATION OF THE PATIENT

The patient should be kept fasting overnight and during the test and should refrain from smoking on the morning of the test since nicotine stimulates adrenalin secretion, which in turn increases gastric secretion. Intake of water in limited quantity is permissible. On the morning of the test day, the patient is given 500 mg. of caffeine sodium benzoate with one glass of water. A Diagnex Blue test kit containing caffeine sodium benzoate and the resin is available commercially. Some investigators feel that caffeine is too weak a stimulant and have used histamine or Histalog instead. After 1 hour, the patient is asked to void and this urine specimen is discarded. The patient should now swallow the azure A resin suspended in water. From this time on, all urine voided for a 2 hour period is collected and the entire urine specimen is sent to the laboratory for examination.

REAGENTS

1. 6 *N* HCl. Place 50 ml. of concentrated HCl into a 100 ml. volumetric flask containing approximately 30 ml. of distilled water. Bring to mark with distilled water.
2. Diagnex Blue test kit (E. R. Squibb & Sons, New York, N.Y.).

PROCEDURE

1. Place the entire 2 hour urine sample into a graduated cylinder and dilute to 300 ml. with distilled water. If the total urine volume is greater than 300 ml., use the urine undiluted as is.

2. Place approximately 10 ml. of the diluted urine specimen into each of three test tubes. Label two tubes "control" and the other tube "test."

3. Add 30 mg. ascorbic acid (1 capsule supplied with kit) to both control tubes. The ascorbic acid will reduce the blue color due to azure A, but not most other pigments present in the urine.

4. Place the two controls into the slots of the Squibb Diagnex Comparator block that are marked 0.3 and 0.6 mg. The 0.3 and 0.6 mg. standard slots are prepared with a blue screen of a color intensity that is comparable to that of urine containing 0.3 or 0.6 mg. azure A dye, respectively. The tube marked test is placed into the middle slot labeled test.

5. The Diagnex Comparator block with the three test tubes is held against a suitable light source and the color intensity of the test is compared with those of the controls in the respective standard slots. If the color intensity of the test urine is equal to or exceeds that of the 0.6 mg. standard, the test is completed at this point and "presence of free gastric hydrochloric acid" may be presumed. The two control urines in the 0.3 and 0.6 mg. slot help prevent misinterpretations of the findings because of unspecific chromogens.

6. If the color intensity of the test is less than that of the 0.6 mg. standard, the test tube with urine (step 2) should be acidified with 1 drop of Squibb acid-copper sulphate solution or 2 drops of 6 *N* HCl and then placed into a 100°C. heating bath for 10 minutes.

7. Remove all tubes from the heating bath and allow to cool for 2 hours. This time is needed to bring out the full color of the azure A dye. Determine the color intensity of the test as outlined in step 5. A color intensity of the test which exceeds that of the 0.6 mg. standard suggests that "free hydrochloric acid is present." If the color of the test falls between the 0.6 and 0.3 mg. standard, there is "presumptive evidence for hypochlorhydria." A color intensity of the test less than that of the 0.3 mg. standard is "presumptive evidence for achlorhydria."

COMMENTS ON THE METHOD AND SOURCES OF ERROR

Although the tubeless gastric analysis eliminates the use of the stomach tube, it creates the need for a rather lengthy procedure for the patient and it has many potential sources of error (proper timing, complete urine collection, and so on).

The test gives only an indication of the presence or absence of hydrochloric acid. It provides no information as to the concentration of hydrochloric acid present, nor does it indicate the total volume of gastric residue.

If hydrochloric acid was indeed secreted by the gastric mucosa, but was subsequently neutralized by either regurgitated material, food particles, or saliva, a *false negative* result will be obtained. In a regular gastric analysis, one has a sample on hand for observation of the presence of food particles or bile.

The outcome of the tubeless gastric analysis depends not only on the presence of hydrochloric acid, but also on such factors as rate of emptying of the stomach, rate of intestinal absorption of the free azure A dye, and rate of excretion of the dye through the kidneys. Abnormalities in any of these steps could affect the amount of dye appearing in the urine and, thus, the test result; therefore, the test is unreliable in subtotal gastrectomy, pyloric obstruction, malabsorption, marked congestive heart failure, severe liver and kidney disorders, and urinary retention.

The replacement of the azure A dye from the resin is accomplished not only by hydrogen ions, but also by a number of other cations such as sodium, potassium, barium, and iron. Segal[4,5] states, however, that this results in the excretion of less than 0.35 mg. of azure A in the first 2 hours.

The control generally compensates for pigments that might be present in urine; however, false positive results may be obtained if dyes other than azure A are present, which decolorize after addition of ascorbic acid. In this case, the nonspecific colors contribute to the absorbance of the unknown without proper compensation for this color by the control. It is, therefore, best that patients to be tested receive no colored medication or colored foods (e.g., beets) for at least 24 hours before the test.

SUMMARY

The tubeless gastric analysis has provided useful information in selected cases but should be considered as a screening test only. The various sources of error just listed are responsible for approximately 5 to 10 per cent false negative and 2 to 5 per cent false positive results.

Stimulation of hydrochloric acid secretion by caffeine has been found ineffective in many patients; therefore, many investigators have replaced caffeine with Histalog or histamine.

REFERENCES

Segal, A. L., Miller, L. I., and Plumb, E. J.: Gastroenterology, *28*:402, 1955.
Diagnex Blue test kit package insert, E. R. Squibb & Sons, New York, N.Y.

Detection of Lactic Acid in Gastric Contents

Lactic acid in gastric contents (and also butyric and acetic acid) is most likely the product of bacterial action and fermentation as a result of food stagnation, or absence of free hydrochloric acid, or both. The determination of lactic acid is of relatively little clinical value, but the test is occasionally requested because the presence of lactic acid, together with gastric retention and hypochlorhydria, is a rather common finding in carcinoma of the stomach.

PRINCIPLE

A portion of gastric content is extracted with ether. A portion of this ether extract is treated with ferric chloride, which gives a yellow-greenish color with low concentrations of lactic acid and an intense yellow-green color with high concentrations.

REAGENTS

1. Ether, A.R.
2. Ferric Chloride, 10 per cent (w/v) $FeCl_3 \cdot 6H_2O$ in H_2O.

METHOD

1. Introduce 5 ml. of strained or centrifuged stomach contents into a separatory funnel.
2. Extract the lactic acid by addition of 20 ml. of ether and vigorous shaking for 1 minute. Permit the ether phase to separate.
3. Transfer 5 ml. of the upper ether layer to another separatory funnel, add 20 ml. of distilled water and 2 drops of 10 per cent ferric chloride. Mix gently.

INTERPRETATION

A slight yellow-greenish color is observed if lactic acid is present in concentrations of more than 50 mg./100 ml. of gastric content. Concentrations of more than 100 mg./100 ml. give an intense yellow-greenish color. Lack of color development indicates absence of lactic acid.

COMMENTS ON THE METHOD

There does not appear to be any need for a more accurate quantitation of lactic acid than that just presented.

Some authors have added the ferric chloride reagent directly to the gastric content (development of a yellow color indicates presence of lactic acid), but since there can be false readings because of interference from food particles, the procedure involving ether extraction is preferred.

REFERENCE

Strauss: *In* Hawk's Physiological Chemistry. B. L. Oser, Ed. 14th Ed. New York, McGraw-Hill Book Co., Inc., 1965, p. 486.

Conditions Associated with Abnormal Gastric Function

CARCINOMA OF THE STOMACH

Achlorhydria is found in about 50 per cent of cases; a large portion of the remainder of this group have hypochlorhydria, and only a small percentage have hyperacidity. In some cases, there is blood in the stomach. The lactic acid test is positive in a high percentage of cases. Also the lactobacillus of Boas-Oppler can frequently be demonstrated.

DUODENAL ULCERS

Hyperchlorhydria is observed in more than 70 per cent of cases; however, a small percentage of cases with hypochlorhydria have been reported. Histamine-fast achlorhydria excludes the diagnosis of duodenal ulcer. The volume of gastric secretion is usually increased and may be twice the normal volume.

GASTRIC ULCERS

In some cases, gastric acidity is high, but, in the majority of cases, it is normal. Hypoacidity, which is sometimes found, is thought to be due to chronic gastritis. Blood is constantly or intermittently found in many cases, and the same is true for occult blood in the stool. The diagnosis is made chiefly on the basis of the clinical and x-ray findings.

ATROPHY OF THE STOMACH (ATROPHIC GASTRITIS)

Achlorhydria and low volume is a rather constant finding.

PERNICIOUS ANEMIA

True achylia is a constant finding. The combined acid and the volume are both low and there is histamine-fast achlorhydria. Demonstration of free acid in a patient with megaloblastic anemia provides strong evidence that the diagnosis is *not* pernicious anemia.

PYLORIC OBSTRUCTION

If there is pyloric obstruction, the volume is usually large and an excessive amount of food can be demonstrated. Although hydrochloric acid may have been secreted, it is frequently neutralized by the food. Lactic and butyric acid, as well as yeast cells, can be demonstrated when pyloric obstruction is accompanied by gastric carcinoma.

ZOLLINGER-ELLISON SYNDROME

In many functioning pancreatic cell adenomas and carcinomas, there is a secretion of gastrin or a gastrin-like substance by the tumor cells. This substance causes increased and prolonged gastric acid secretion. The occurrence of ulcers in this syndrome is extremely high.

REFERENCES

1. Farrar, G. E., and Bower, R. J.: Ann. Rev. Physiol., *29*:141, 1967.
2. Michaelis, L.: Harvey Lectures 1926–1927. New York, Academic Press, Inc., 1928, pp. 59–89.
3. Moore, E. W.: Ann. N.Y. Acad. Sci., *140*:866, 1967.
4. Segal, H. L.: Ann. N.Y. Acad. Sci., *140*:896, 1967.
5. Segal, H. L.: J.A.M.A. *196*:655, 1966.
6. Segal, H. L., Miller, L. L., and Plumb, E. J.: Gastroenterology; *28*:402, 1955.
7. Yamaguchi, N., and Glass, G. B. J.: Ann. N.Y. Acad. Sci., *140*:924, 1967.

ADDITIONAL READINGS

Cantarow, A., and Trumper, M.: Clinical Biochemistry. 6th ed. Philadelphia, W. B. Saunders Co., 1962.

Clinical Symposia, Summit, N. J., Ciba Pharmaceutical Products, Inc., vol. 11, no. 1, Jan. Feb., 1959.

Henry, R. J.: Clinical Chemistry, Principles and Technics. New York, Hoeber Medical Div., Harper & Row, Publishers, 1964.

Chapter 15 /

PANCREATIC FUNCTION AND INTESTINAL ABSORPTION

by Norbert W. Tietz, Ph.D.

The pancreas is a long, narrow gland that lies across the posterior wall of the abdomen. The head is located in the duodenal curve and the body and tail are directed toward the left, extending to the spleen (see Fig. 15-1 *A*, *B*). Because of its position deep in the retroperitoneal space in the abdomen, the pancreas is inaccessible for direct physical examination. Radiologic examination, except in a case of calcification of the pancreas, is generally of little help. Clearly then, chemical tests are of paramount value in the diagnosis of disorders of this organ. Most of the laboratory tests listed in this chapter are not specific for pancreatic disease and allowance must be made for other disorders that give rise to similar manifestations. There is also a great overlap between results observed in normal individuals and those with pancreatic disorders. This is partly a result of the large functional reserve of the pancreas. For example, it has been estimated that until 50 per cent or more of the acinar cells have been destroyed, pancreatic insufficiency cannot be clearly demonstrated.

The functions of the pancreas can be separated into two major groups, endocrine and exocrine.

Endocrine Functions of the Pancreas

The principal endocrine function of the pancreas is production of insulin and glucagon. Both are produced in the islands (or islets) of Langerhans, which are groupings of cells surrounded by the acinar cells.

GLUCAGON

Glucagon is a hormone secreted by the alpha cells of the islets of Langerhans. Its physiological role is not precisely known, but it is thought to be in the control of blood glucose levels in the postabsorptive period, at which time glucose levels are

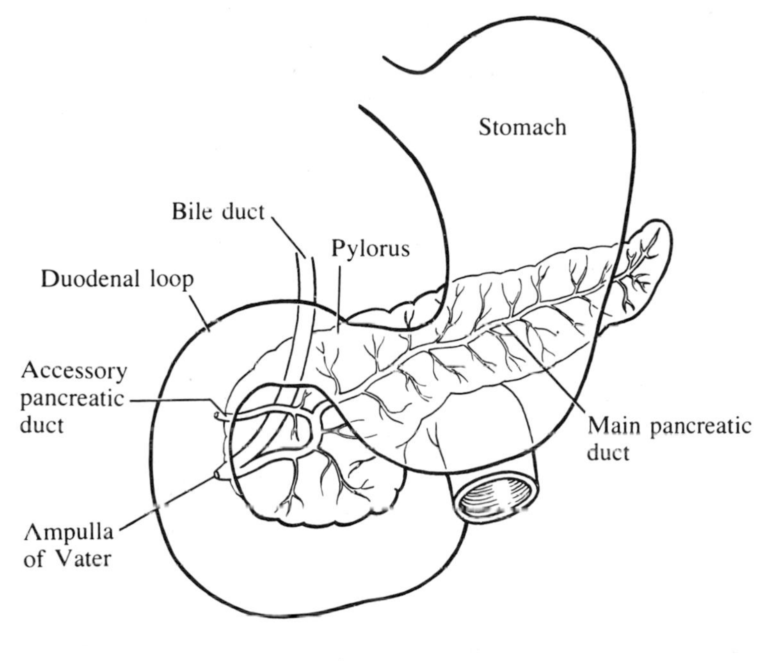

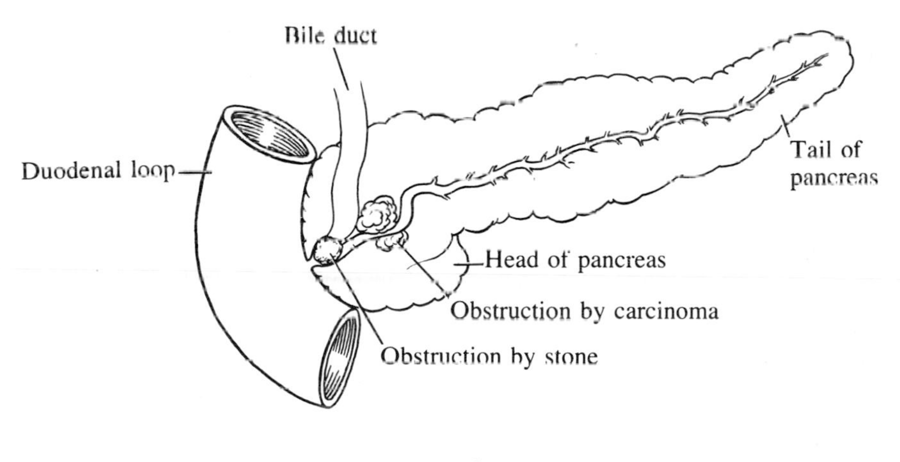

Figure 15-1. Cross sections through pancreas.

lowest. This hyperglycemic effect (increasing blood glucose levels) is most likely due to stimulation of hepatic glycogenolysis, in turn caused by activation of the phosphorylase system. The molecular weight of glucagon is of the order of 3485. No simple laboratory tests are available for its determination.

INSULIN

Insulin, a hormone secreted by the beta cells of the islets of Langerhans, decreases blood sugar levels. Although the mode of action is not precisely known, there is now evidence that insulin is primarily concerned with increasing the rate of entrance of glucose into the muscle and adipose tissue cells. This action is the result of either an increase in cell permeability or an acceleration of the hexokinase reaction, which converts the glucose in the cell into glucose-6-phosphate (see also Chapter 4, Carbohydrates, and Chapter 7, Lipids). Insulin is determined by bioassay or immunologically (see Chapter 9, Endocrinology).

LIPOCAIC

There is some evidence that the internal secretions of the pancreas contain a third compound (hormone?), which has been called lipocaic. The compound is produced by the alpha cells of the pancreas and its biological action is the prevention of fatty infiltrated liver and the activation of oxidation of fatty acids, as well as increase of the phospholipid exchange. Initially, there was considerable controversy about the existence of such an "anti–fatty liver factor," but numerous recent reports, mainly from Europe, indicate that such a substance indeed exists.

Exocrine Functions of the Pancreas

The exocrine functions of the pancreas are the production and secretion of pancreatic juice. The major solutes of pancreatic juice are sodium bicarbonate, amylase, lipase, and trypsinogen (see the following section). The pancreatic juice is produced in the acinar cells (see Fig. 15-2) and is secreted into the lumen of the acinus. From there it passes through the ductules into the pancreatic ducts and finally into the main pancreatic duct (duct of Wirsung). The main pancreatic duct empties into the ampulla of Vater, which is situated in the second (lower) part of the duodenum. Frequently, the bile duct joins the pancreatic duct just proximal to or at the ampulla of Vater (see Fig. 15-1*A*). The pancreas frequently has a second pancreatic duct (duct of Santorini), which may empty separately or jointly with the main duct into the duodenum. In the duodenum, the pancreatic juice mixes with the food material coming from the stomach. The combination of enzymes secreted by the pancreas, at the proper pH provided by the bicarbonate, can digest virtually any food material.

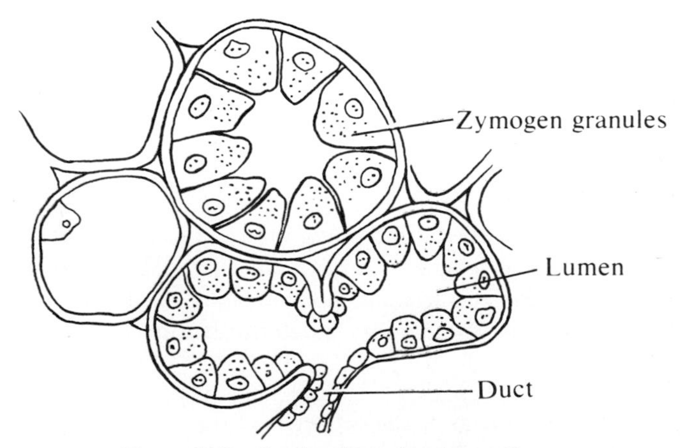

Figure 15-2. Section through acinar cells.

The secretion of pancreatic juice is regulated by several mechanisms. First of all, it is known that distention of the stomach induces secretion of pancreatic juice and there is some evidence that the hormone *gastrin* (secreted by the pyloric mucous membrane) stimulates the exocrine secretion of the pancreas. *Secretin* and *pancreozymin* are hormones produced by the upper intestine mucosa when it comes in contact with hydrochloric acid or gastric chyle. Secretin stimulates the secretion of an increased amount of pancreatic juice rich in bicarbonate; pancreozymin elicits a juice of greater enzyme concentration.

ENZYMES

The normal pancreas secretes a number of enzymes that pass almost entirely into the duodenum. Only a very small fraction reaches the blood directly, where small amounts can be demonstrated. There is some evidence that the blood enzymes such as amylase and lipase derive not only from the pancreas, but also from other sources.

In some disorders (obstruction due to stones or carcinoma and in pancreatitis), the flow of enzymes and bicarbonate into the duodenum is impeded. This results in a decreased secretion of pancreatic juice into the duodenum as demonstrated by direct examination of duodenal content (see secretin test), but, at the same time, an increased amount of enzyme can frequently be demonstrated in the blood serum. The mechanism by which the enzymes enter the circulation is not exactly known, but it is thought to be due to changes in the pressure in the pancreatic duct and ductules, changes in the permeability of acinar cells, or disruption of the acinar limiting membrane. The clinically most important enzymes in serum, in regard to evaluation of pancreatic function, are amylase and lipase.

Serum Amylase

Normally amylase is present in both serum and urine and its concentration shows wide variations. In pancreatitis and obstruction of the pancreatic duct, serum amylase usually increases rapidly on the first day and then decreases after the second or third day, but increases in urinary amylase often persist much longer. It follows that the diagnostic significance of serum amylase is limited to blood taken within the first few days after the beginning of an attack.

Amylase is an enzyme that hydrolyzes starch and glycogen to dextrin, and further to maltose. Its level in serum is generally determined by a saccharogenic or amyloclastic method (see Chapter 8, Enzymes). Amylase is usually absent in the serum of the newborn; it appears approximately at the age of two months and reaches normal adult levels at the end of the first year. The normal values for adults are 40 to 160 Somogyi units/100 ml. serum. The enzyme derived from the pancreas is an α-amylase.

CLINICAL SIGNIFICANCE

In acute pancreatitis or obstruction of the pancreatic duct, amylase values in serum are usually increased significantly. Values above 500 units are generally considered to be highly suggestive of pancreatitis; however, it should be emphasized that amylase elevations in pancreatitis may be less or may not occur at all, depending on the severity of the case, the timing of blood collection, and the amount of residual functioning pancreatic tissue. Serum amylase levels tend to increase rapidly after an attack and may be demonstrated as early as 6 to 8 hours after its onset. Levels stay elevated for 1 to 3 days and then generally return to normal. Amylase values that remain high for a longer period of time are frequently due to pancreatic pseudocysts.

In intra-abdominal diseases such as perforated peptic, gastric, or duodenal ulcers, intestinal obstruction, or acute peritonitis, elevated amylase values are frequently observed. Although values in these disorders rarely exceed 500 units, values of more than 1000 units have been observed in a few cases. Amylase elevations have also been observed in acute diseases of the salivary glands such as mumps, in renal disease, and after administration of drugs such as morphine, codeine, and meperidine (Demerol). These drugs cause a transient spasm of the duodenal musculature and Oddi's sphincter and thus produce an obstructive mechanism.

Low serum amylase values have been found in abscesses of the liver, acute hepatocellular damage, cirrhosis, cancer of the liver and bile duct, and cholecystitis. Jaundice alone and chronic diseases of the gallbladder do not influence amylase levels.

Urinary Amylase

Urinary amylase values in acute pancreatitis or obstruction of the pancreatic duct are sometimes more elevated than are serum amylase levels and the increased levels appear sooner and may persist over a longer period. In cases of pancreatic carcinoma with obstruction of the pancreatic duct, urinary amylase levels also may be occasionally elevated. One could speculate that a slight or moderate increase in serum amylase would result in increased excretion of amylase by the kidney and thus be more readily detectable in urine than in serum. Only after the rate of increase of serum amylase exceeds the renal excretion rate would there be a significant rise in serum concentration.

The urine amylase is most accurately determined by the Somogyi method (normal values up to 5000 units/24 hr.). The Wohlgemuth method has also been used successfully for this purpose (see Chapter 8, Enzymes).

Serum Lipase

Lipase is an enzyme that catalyzes the hydrolysis of esters of glycerol with long chain fatty acids. It acts at the interface of water and insoluble substrate present in an emulsion. The enzyme normally found in serum derives most likely from several sources, including the pancreas. Like amylase, pancreatic lipase is produced by the acinar cells and also reaches the blood in the same way (see Serum Amylase).

CLINICAL SIGNIFICANCE

In general, the changes in serum lipase levels have the same significance as those of amylase; however, at times, there are differences that make the lipase determination more important in the diagnosis of pancreatitis. Some reports indicate that lipase levels in acute pancreatitis decrease more gradually and may be found to be elevated for as long as 14 days. Our own studies[11] on 35 cases and studies of others confirm this for many but not all cases (see Fig. 15-3). Serum lipase levels in acute pancreatitis or obstruction of the pancreatic duct generally increase significantly and may reach levels of more than 10 units (normal: 0 to 1.0 unit). Serum lipase levels may be normal in advanced diseases of the pancreas if little or no functioning pancreatic tissue remains. Lipase is normal in acute diseases of the salivary glands.

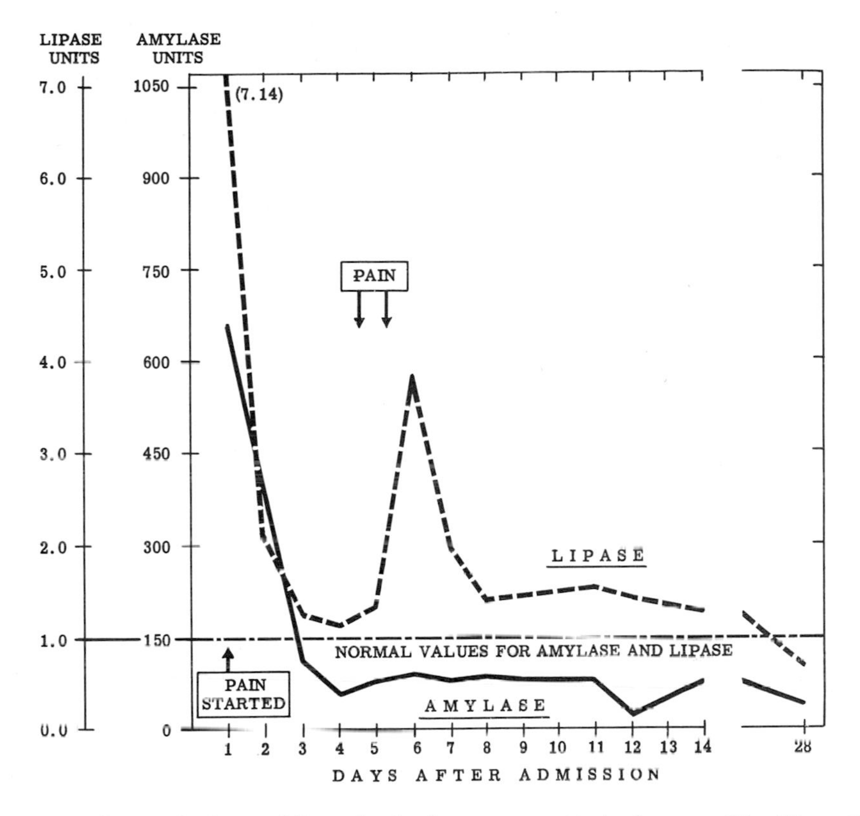

Figure 15-3. Serum amylase and lipase levels after an acute attack of pancreatitis. (From Tietz et al., Am. J. Clin. Path., *31*:148, 1959.)

The literature referring to the clinical significance of serum lipase determinations is very confusing and unsatisfactory, perhaps because of the use of nonspecific methods that measure esterase but not lipase activity.[12]

Urinary Lipase

Lipase is excreted by the kidneys and can be demonstrated in the urine. It has been found elevated in hemorrhagic pancreatitis and obstruction of the pancreatic duct.

The determination of lipase in the urine presents difficulties because of the presence of a dialyzable inhibitor, and presently available urinary lipase methods are either too tedious or not fully reliable.

Leucine Aminopeptidase

Leucine aminopeptidase (LAP) is found in almost all human tissues and also in serum, urine, and bile. Goldbarg and Rutenburg[3] developed a method for its assay in serum and urine and found significant elevations of this enzyme in all cases of carcinoma of the pancreas. The test seems to have little diagnostic value because it is nonspecific in as much as elevations have also been reported in obstructive jaundice, infectious hepatitis, metastatic cancer, acute pancreatitis, and other disorders. At one time, the test was believed to have clinical significance for excluding the diagnosis of carcinoma of the pancreas; however, several cases of nonmetastatic carcinoma of the pancreas showing no elevations in LAP have been reported. Its value, therefore, seems to be very limited.

Trypsin

Trypsin is the name commonly used for a group of protein-splitting enzymes formerly thought to be a single enzyme. It is now known that the proteolytic component of pancreatic juice consists of trypsinogen, chymotrypsinogen, and procarboxypeptidase. There is some indication that there is an even larger number of protein-splitting enzymes in pancreatic juice. *Trypsin* is formed from trypsinogen by the action of enterokinase. Subsequent activation proceeds autocatalytically (newly formed trypsin activates trypsinogen). Trypsin attacks almost all proteins and breaks them down to polypeptides and some amino acids. It is believed that the carboxyl group of arginine or lysine is the site of preferential action. Chymotrypsinogen is converted to *chymotrypsin* by trypsin. The enzyme is an endopeptidase that seems to attack peptide linkages involving the carboxyl group of tyrosine and phenylalanine. *Carboxypeptidase* is most likely formed from procarboxypeptidase by the action of trypsin. It is an exopeptidase that splits off terminal amino acids possessing a free carboxyl group (see also Chapter 8, Enzymes).

The activation of the three named proenzymes takes place in the intestines after secretion by the pancreas; therefore, the enzymes can be demonstrated either in duodenal contents or in stool. The semiquantitative determination of trypsin in stool has become a very popular screening test for cystic fibrosis.

The significance of trypsin determinations in stool is based on the assumption that excretion of trypsin is decreased or absent whenever the secretory function of the pancreas is decreased. This obviously is the case in obstruction of the pancreatic duct by stone or tumor and in cystic fibrosis, in which a large portion of the acinar tissue has been replaced by fibrous tissue.

Results of determinations of trypsin in stool should be interpreted carefully because of the possible presence of proteolytic bacteria in stool. A culture for proteolytic bacteria should always be done if the result is positive unless a control with antitrypsin is run at the same time (see the method in Chapter 8). This is especially true for adults, in whom the bacterial flora are generally well developed.

EXAMINATION OF PANCREATIC JUICE

Secretin Test

A direct method of determining pancreatic function is measurement of the volume and the quantitative analysis of pancreatic juice for bicarbonate and enzyme content. Stimulation of pancreatic function with secretin* (1 to 3 units/kg. of body weight) or secretin plus pancreozymin has made this test more reliable. Table 15-1 shows the normal values for duodenal contents after pancreatic stimulation with secretin.

Abnormally low concentrations of constituents of pancreatic juice after secretin stimulation have been reported in cases of chronic pancreatitis, pancreatic cysts, cystic fibrosis, calcification, carcinoma of the pancreas, and edema of the pancreas.

* Vitrum Co., Stockholm, Sweden. Preparations from the United States and England have been of inconsistent purity.

TABLE 15-1. *Normal Values of Secretin Response and Critical Values for Volume, Bicarbonate, and Amylase in the 60 Minute Specimen**

	Total Volume (ml.)	Total Volume (ml./kg. body wt.)	HCO_3^-, *Max. Concentration* (mEq./L.)	HCO_3^- (mEq. *in total 60 min. spec.*)	*Amylase (units in total 60 min. spec.)*
Observed range	91–270	1.6–4.8	88–137	13.2–23.7	204–1621
Mean	164.7	2.72	107.7	17.81	722.4
Critical value	102	1.7	91	14.1	173

* After Dreiling, D. A.: J. Mt. Sinai Hosp., *21*:363, 1954–1955.

Findings in these conditions are very similar and are of relatively little value in the differential diagnosis. The greatest value of the secretin test is probably in excluding pancreatic dysfunction, such as in fibrocystic disease, and in differentiating steatorrhea of pancreatic origin from that caused by sprue or celiac disease. In cases of decreased acinar function, earliest deficiencies are usually found in amylase and lipase secretion; decreases in trypsin, total volume, and bicarbonate follow with progression of the disease.

In general, it is sufficient to analyze the duodenal drainage for one of the enzymes only. Amylase is the enzyme of choice in adults, but, if the test is to be done on infants, trypsin should be determined. Amylase production of infants is extremely low and therefore contamination by salivary amylase is a serious source of error. Whenever trypsin is to be determined, duodenal fluid must be collected over ice since trypsin is readily inactivated at room temperature.

Simultaneous administration of secretin and of drugs producing an obstructive action (morphine sulphate, methacholine chloride, or Urecholine) has been used to determine the amount of remaining functioning acinar cell tissue. If one of the drugs mentioned is given, it will cause a constriction of the duodenal musculature and prevent the outflow of pancreatic juice into the duodenum. In that case, stimulation of pancreatic secretion with secretin will cause a significant increase in serum amylase and lipase levels, which are substantially greater in patients with pancreatitis as compared to normal individuals. Patients with substantially decreased amount of functioning acinar tissue, for example, patients with cystic fibrosis or chronic pancreatitis, show significantly lower serum amylase levels. A normal response to this combined secretin-morphine test results in serum amylase elevations of 390 units (average). Use of this test in pancreatitis may be contraindicated because of the possible risk involved.

TECHNIQUES OF THE SECRETIN TEST

Patients should be in the fasting state. A double-lumen gastroduodenal tube is introduced under fluoroscopic guidance in such a way that the shorter end of the tube lies in the stomach and the longer end in the duodenum distal to the ampulla of Vater. Constant suction is then applied and aspiration is continued until the duodenal fluid becomes clear and not contaminated with gastric juice. At this time, 1 unit of secretin/kg. of body weight is given intravenously (some investigators recommend 3 units/kg.), and specimens of duodenal and gastric content are collected at six 10 minute intervals. Each specimen is examined for pH (Hydrion paper). A sudden increase in pH of the gastric content indicates contamination by duodenal fluid; a sudden decrease in pH of the duodenal content indicates contamination by gastric juice. The 10 minute specimens from the duodenum are then analyzed for volume and occult blood.

The presence of blood, occult or overt, may be helpful in pinpointing gastrointestinal bleedings. After 60 minutes, all duodenal specimens are pooled and analyzed for volume and bicarbonate content, as well as for amylase, lipase, or trypsin activity. The pH can be determined with Hydrion paper or a pH meter and the bicarbonate concentrations can be determined with a gasometer (see Chapter 10, Electrolytes). The methods for amylase, lipase, and trypsin are discussed in Chapter 8.

ABSORPTION TESTS

The absorption tests are based on the theory that proper pancreatic function is essential for normal intestinal absorption of certain substances. Starch, fats, and proteins must be hydrolyzed before they can be absorbed and the pancreas is the main source of the enzymes amylase, lipase, and trypsin, which cause these hydrolyses. Although inadequate absorption may be the result of causes other than lack of pancreatic enzymes, normal absorption would rule out pancreatic deficiency. Careful interpretation of results is necessary, however, since the pancreas has a large functional reserve and absorption may be normal even in the presence of pancreatic disorder. Only if a significant amount of pancreatic tissue is destroyed, or, if the flow of pancreatic juice into the duodenum is greatly interfered with, will absorption tests be found abnormal. The greatest value of absorption tests seems to be in the diagnosis or exclusion of cystic fibrosis, a disease in which pancreatic function and thus intestinal absorption are decreased.

Vitamin A Tolerance Test

Serum vitamin A levels in children with cystic fibrosis of the pancreas and in adults with pancreatic insufficiency are usually below the normal range of 30 to 65 μg./100 ml. In these patients, oral administration of large amounts of vitamin A in oil causes only a slight increase in the serum vitamin A level. Persons with normal absorption show a peak increase from 200 to 600 μg. of vitamin A/100 ml. of serum in the 3 or 6 hour specimen, or both.

The test is based on the belief that vitamin A, a fat soluble vitamin, can be absorbed to any significant extent only if fats are hydrolyzed. Therefore, presence of a sufficient amount of pancreatic lipase is necessary for the absorption of both compounds. Absorption of lipids and vitamin A is also dependent on the secretion of a sufficient amount of bile by the liver, which is necessary for proper emulsification of fats before hydrolysis and absorption. A further limitation of the test is the fact that a malabsorption due to inadequate intestinal function, even in presence of normal pancreatic function, may be the cause for reduced vitamin A absorption. Consequently, the main value of this test is the exclusion of a diagnosis of pancreatic insufficiency such as cystic fibrosis.

Measurement of the serum carotene level has a similar clinical application although it is not as reliable. Values of 50 to 200 μg./100 ml. are considered normal, but, in severe malabsorption, values below 30 μg./100 ml. are observed.

SPECIMEN

A fasting blood specimen is collected and 5000 units of vitamin A (in oil) per kg. of body weight are given by mouth. Further blood specimens are drawn 3, 6, and sometimes 24 hours later. The patient may drink water or consume light meals during the test period. All specimens are analyzed for vitamin A content, as outlined later.

Structure and Biological Action of Vitamin A and Carotene

Beta-carotene is one of a group of naturally occurring pigments called carotenoids, and it is the major precursor of vitamin A. It is a symmetrical molecule containing two β-ionone rings, which are connected by an 18C hydrocarbon chain with 11 conjugated double bonds. Symmetrical oxidative scission of β-carotene (at the central double bond) results in the formation of two molecules of vitamin A_1. Since this oxidative process readily takes place in the organism, the β-carotene is frequently termed provitamin A_1. The other carotenoids such as α- and γ-carotene, as well as cryptoxanthin, contain only one β-ionone ring. Thus, oxidative scission will result in the formation of only one molecule of vitamin A (see Fig. 15-4).

β-carotene is widely distributed in both plant and animal tissues, but mainly in yellow-colored vegetables and fruits; however, *Vitamin A_1* is found only in animal tissues and animal products, such as milk and egg yolk. Since the liver of salt water fish contains a relatively high amount of vitamin A_1, fish liver oil concentrates are commonly used as vitamin A preparations for therapeutic purposes.

Both β-carotene and vitamin A are essentially insoluble in water, but very soluble in fats and fat solvents. Therefore, both compounds are absorbed only if the fat in which they are dissolved is hydrolyzed. This is the basis of the vitamin A absorption test discussed in the next section. The vitamin is present either in the form of the free alcohol or as an ester. It is esterified during the absorption process and consequently, immediately after absorption is present in serum mainly in the form of the ester. Blood, however, contains a factor capable of hydrolyzing the ester, which may explain why, in a fasting state, approximately 80 to 95 per cent of the total vitamin A is in the form of the free alcohol and only the remainder in the form of an ester. The vitamin A ester, as well as the β-carotene, is transported in the blood by low density lipoproteins; the vitamin A alcohol is transported by a protein of a density greater than 1.21. Both vitamin A and β-carotene are destroyed by oxidation and exposure to light; thus, if specimens or standards have to be kept for any period of time, they should be protected from light and stored under nitrogen.

An adequate body supply of vitamin A or the provitamin is needed to maintain the structure and function of a number of specialized epithelial and glandular tissues and vitamin A plays an essential role in the photochemical phase of the vision process. The impairment of dark adaptation is one of the earliest functional defects observed in vitamin A deficiency. Vitamin A is also required for normal growth and plays a role in the construction of bone.

Hypervitaminosis may lead to drowsiness, sluggishness, severe headache, vomiting, and skin defects. (For more details, consult one of the textbooks listed in the References.)

Determination of Vitamin A and Carotenoids in Serum

A variety of methods for the determination of vitamin A in serum have been proposed, but the reaction between vitamin A and antimony trichloride in chloroform,

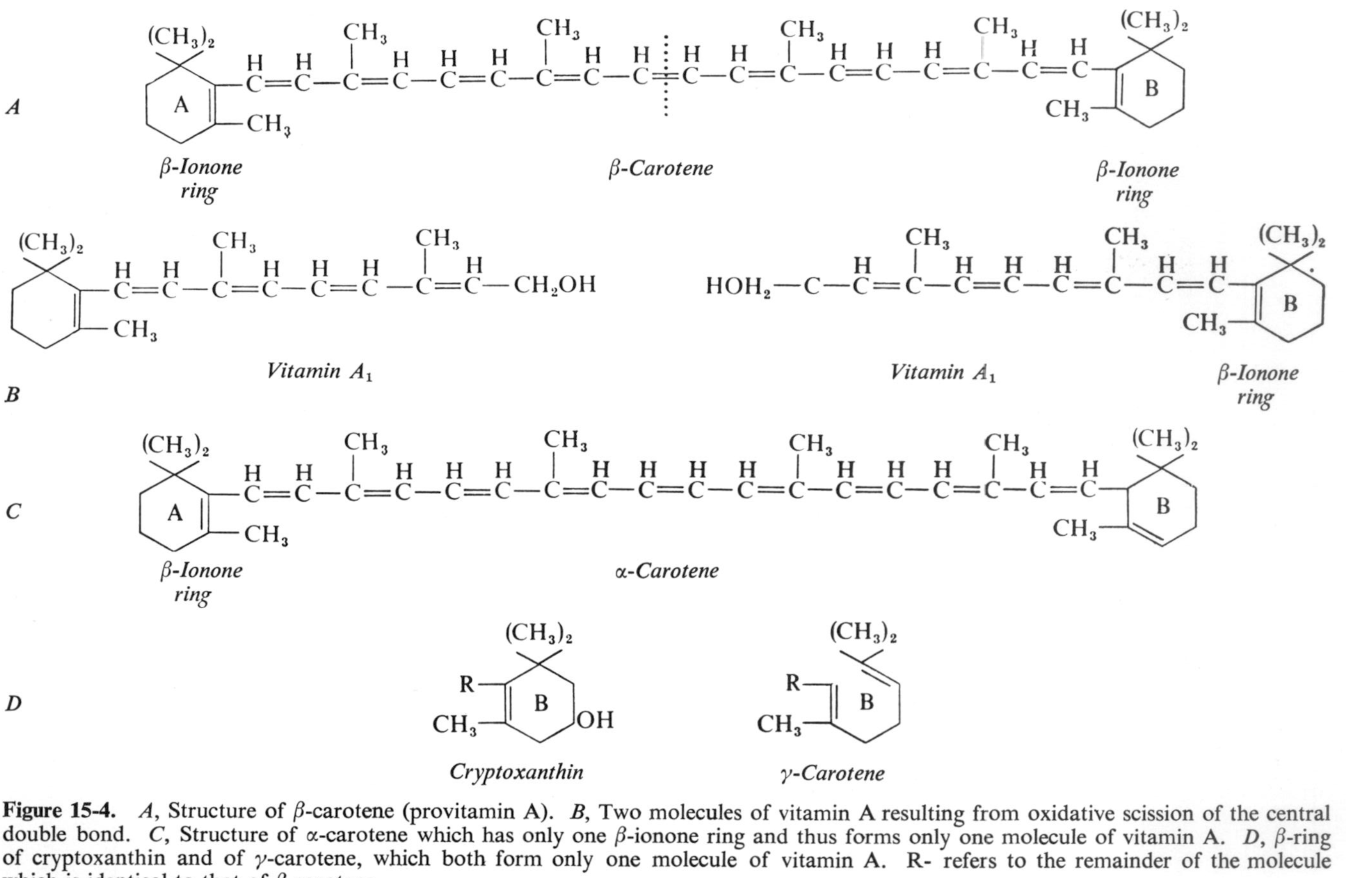

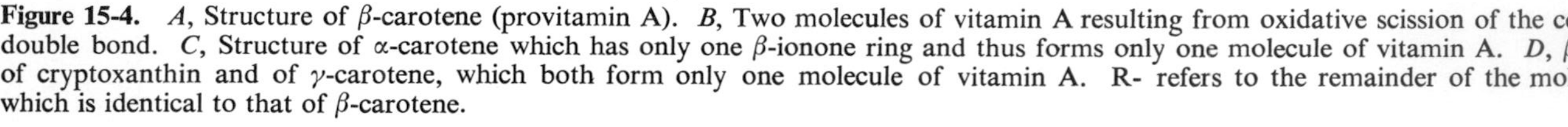

Figure 15-4. *A*, Structure of β-carotene (provitamin A). *B*, Two molecules of vitamin A resulting from oxidative scission of the central double bond. *C*, Structure of α-carotene which has only one β-ionone ring and thus forms only one molecule of vitamin A. *D*, β-ring of cryptoxanthin and of γ-carotene, which both form only one molecule of vitamin A. R- refers to the remainder of the molecule which is identical to that of β-carotene.

as introduced by Carr and Price[2], is most widely accepted in clinical laboratories. The method has the advantages of speed and simplicity and, in addition, the vitamin alcohol and the ester react equally in color production. The disadvantages of the method, namely, the instability of the color and the need for exclusion of moisture, are, in the opinion of this author, no real problem if the precautions listed later are observed.

Sobel and Snow[10] have applied the color reaction between glyceroldichlorohydrin (1,3-dichloro-2-propanol) and vitamin A to the assay of this compound in serum. Although the color is more stable and the test is unaffected by moisture, there is the distinct disadvantage that the molar absorptivity of the color is only one fifth of that obtained with the Carr and Price reaction and, in addition, a preliminary saponification of the vitamin A ester is necessary. This is especially significant in connection with vitamin A tolerance tests, where the ester fraction in serum may be excessively high.

Bessey and co-workers[1] and others[6,8] irradiated the specimen for vitamin A and carotene assay with ultraviolet light and measured the ΔA_{328} before and after irradiation of the specimen with ultraviolet light of a wavelength between 310 and 400 nm. Since vitamin A (and carotene) is destroyed under these conditions, ΔA may be taken as a measure of the vitamin A concentration. The method may also be applied to the determination of β-carotene, in which case ΔA is measured at 460 nm. This method also requires a preliminary saponification and appears to be unnecessarily long and involved for routine clinical purposes.

Beta-carotene is generally determined either by the method just described or by measuring the absorbance of the petroleum ether layer after addition of equal amounts of alcohol and petroleum ether to the sample (see detailed method). Vitamin A is also present in the petroleum ether layer, but its color is so weak that it essentially does not interfere with the β-carotene determination.

PRINCIPLE

Serum proteins are precipitated by an alcohol–petroleum ether mixture. Ethanol ruptures the lipoprotein complex with vitamin A and carotene and permits their extraction into the petroleum ether layer. Lutein and other xanthophylls, as well as bilirubin, will stay in the ethanol layer. The carotenoids are determined by measuring the absorption of the petroleum ether layer in a spectrophotometer at a wavelength of 440 nm.

Vitamin A content is determined by evaporating the petroleum ether and adding antimony trichloride to the (completely dry) residue. The blue color produced (Carr-Price reaction[1]) is determined colorimetrically. The vitamin A alcohol and the ester give equal color intensity in this reaction. Presence of moisture produces cloudiness and greatly interferes with the photometer reading. Carotenoids give a slight color in this reaction, and this is compensated for by applying an experimentally determined correction factor.

REAGENTS

1. Antimony trichloride, 22 per cent. Place 11.0 gm. of antimony trichloride, A.R., into a 50 ml. volumetric flask, dissolve in anhydrous chloroform, and fill to the mark. If necessary, the chloroform should be redistilled from anhydrous sodium carbonate. Keep the reagent in a tightly stoppered brown bottle. It is imperative that the reagent be kept free of water in order to avoid turbidity.

2. Ethanol, 95 per cent.
3. Petroleum ether, low boiling.
4. Beta-carotene standard. Transfer 30.0 mg. of β-carotene (Eastman Kodak Co., Rochester, N.Y.) to a 100 ml. volumetric flask, dissolve in approximately 4 ml. of chloroform, and dilute to volume with petroleum ether. Before use, dilute 1.0 ml. of stock standard to 100 ml. with petroleum ether. This diluted standard contains 3 μg. β-carotene/ml.

If the β-carotene is of insufficient purity, 100 mg. of the substance may be dissolved in 2 ml. of chloroform and reprecipitated with 20 ml. of methanol. The filtered precipitate should be washed with a few drops of methanol and dried in a vacuum desiccator.

5. Vitamin A stock standard (10 μg./ml.). Transfer 10 mg. of vitamin A alcohol (or 11.47 mg. vitamin A acetate) to a 1000 ml. volumetric flask and dilute to volume with petroleum ether. A standardized concentrate of vitamin A is also suitable for preparing this standard solution if the stock is diluted 1:10 with petroleum ether. Dilutions of the stock standard for preparation of the standard curve (see Calibrations) can be made with petroleum ether.

The stock standard is stable for approximately 4 weeks if kept well stoppered, protected from light, and at refrigerator temperature. Diluted standards are stable for a few days only, and must be stored in the same way as the stock standard.

PROCEDURE

1. Into each of two 5 ml. glass stoppered centrifuge tubes pipet exactly 1.0 ml. of serum, 2.0 ml. of 95 per cent ethanol, and 2.0 ml. of petroleum ether. Stopper tubes tightly and shake vigorously for 10 minutes (preferably on a shaker).
2. Centrifuge the stoppered tubes and transfer 1.0 ml. of the petroleum ether layer (top) into a small cuvet (for example, a 10 × 75 mm. Coleman cuvet).
3. Read immediately in a spectrophotometer at a wavelength of 440 nm. against a petroleum ether blank. Calculate carotenoid values from calibration curve or chart (see Calibration).
4. Evaporate the petroleum ether to dryness by placing the cuvets into a water bath of 40 to 45°C. and by gently blowing dry air (or nitrogen) into the cuvet. A stream of dry air can be obtained by passing the air first through a bottle containing a drying agent such as silica gel.
5. Set the spectrophotometer to a wavelength of 620 nm. and adjust to 100 per cent transmittance, using the antimony trichloride solution as blank.
6. Add exactly 1.0 ml. of the 22 per cent antimony trichloride solution, mix well and read within 3 to 5 seconds, since the blue color produced is extremely evanescent. Any cloudiness that might develop at this point is usually due to moisture in the cuvet or in the antimony trichloride solution. In such a case, the test must be repeated, taking precautions to avoid moisture. (Some authors add acetic anhydride to the reagent to remove moisture.)

CALCULATIONS

Carotenoids. Determine the amount of carotenoids per cuvet from the standard curve and carry out the following calculations:

$$\mu\text{g. carotenoids/100 ml. serum} = \frac{\mu\text{g. carotenoids/cuvet} \times \text{dilution*} \times 100}{\text{ml. serum used}}$$

* The factor for the dilution is 2 since the carotenoids from 1 ml. of serum are extracted into 2 ml. of petroleum ether.

If this procedure is followed with respect to quantities, the equation reduces to:

$$\mu g. \text{ carotenoids/100 ml. serum} = \mu g. \text{ carotenoids/cuvet} \times 200$$

Vitamin A. Determine the amount of vitamin A per cuvet from the standard curve or chart and carry out the following calculations:

$$\mu g. \text{ vitamin A/100 ml. serum} = \frac{\mu g. \text{ vitamin A/cuvet} \times \text{dilution} \times 100}{\text{ml. serum used}} - (0.075^* \times \mu g. \text{ carotenoids/100 ml.})$$

If the quantities used are those given in the procedure, the equation reduces to:

$$\mu g. \text{ vitamin A/100 ml. serum} = \mu g. \text{ vitamin A/cuvet} \times 200 - (0.075 \times \mu g. \text{ carotenoids/100 ml.})$$

CALIBRATION

Carotenoids. Prepare a calibration curve by using solutions of β-carotene in petroleum ether (see reagents) in concentrations of 0.5, 1.0, 1.5, 2.0, 2.5, and 3.0 μg./ml. Plot concentration against readings obtained with the spectrophotometer.

Vitamin A. The calibration curve for vitamin A is prepared by using aliquots of a solution of vitamin A alcohol in petroleum ether (see reagents), resulting in concentrations of 0.5, 1.0, 1.5, 2.0, 2.5, and 3.0 μg. vitamin A/cuvet. Proceed as outlined in steps 4 to 6, Procedure. Plot concentration against spectrophotometer reading.

NORMAL VALUES

The normal carotenoid level (mainly β-carotene and xanthophyll) is 60 to 200 μg./100 ml. serum, calculated as β-carotene. Vitamin A is normally found in serum in concentrations of 30 to 65 μg./100 ml. (approximately 100 to 200 U.S.P. units).

REMARKS AND SOURCES OF ERROR

The Carr-Price reaction gives a color with carotenoids as well as with vitamin A. The carotenoid value, therefore, must be determined before the vitamin A content can be calculated.

All reagents and all glassware used must be water-free, since moisture results in cloudiness and erroneous results.

The color produced in the Carr-Price reaction must be read within five seconds; any delay will result in falsely low results.

The cuvets, after use, should be rinsed in concentrated HCl to remove a white film of SbOCl, which is formed on contact of $SbCl_3$ with water.

FAT ABSORPTION TESTS

Normal absorption of fat can take place only in the presence of normal pancreatic function (in this case, secretion of lipase) since fats are mainly absorbed in the form of free fatty acids and monoesters with glycerol. This theory is the basis of fat

---

* Subtraction of μg. carotenoids $\times$ 0.075 is necessary to correct for the amount of color produced by the carotenoids. The factor of 0.075 may vary depending on the instrument and cuvet used. It should be confirmed by running a series of carotene standards through the vitamin A procedure.

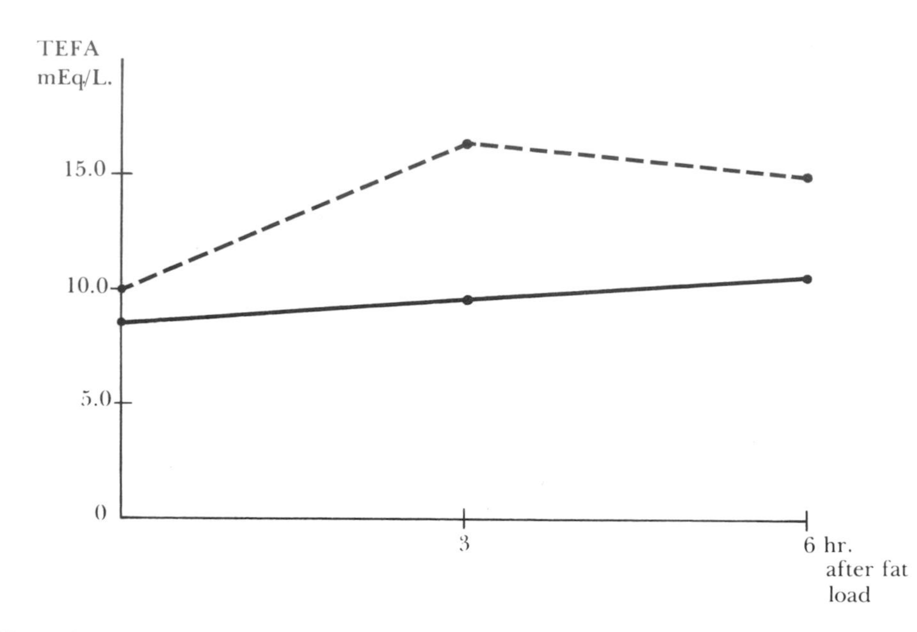

Figure 15-5. Example of fat absorption curve of patient with pancreatic insufficiency (———) as compared to an individual with normal fat absorption (- - - - - - -). TEFA represents total esterified fatty acids found in serum before and after fat load.

absorption tests. Several approaches have been used to measure the rate of fat absorption from the intestinal tract. The most commonly used test is the determination of the total fat in a fasting serum specimen and in specimens withdrawn 3 and 6 hours after administration of a high fat meal. Any method measuring total lipids or total esterified fatty acids may be used (see Chapter 7, The Lipids). The normal values vary somewhat with the procedure employed, but, in general, the lipid content in the 3 hour sample should be 60 to 100 per cent higher than the fasting level (9.0 to 16.0 mEq. fatty acids/L., or 300 to 500 mg. total lipids/100 ml.). Patients with malabsorption show a flat type of absorption curve with increases in the 3 hour sample of substantially less than 60 per cent (see Fig. 15-5).

The difference in absorbance of the serum before and after administration of a fat meal, determined at 640 to 700 nm., has also been used as a measure of fat absorption. This test is much simpler to perform, but is less accurate, since the degree of turbidity of serum does not always represent its true fat content. A method such as the one suggested by Schwarz and co-workers[7] may be used for such serum turbidity measurements.

Measurement of Absorption of I^{131}-Labeled Triolein and Oleic Acid

More recent procedures are based on the administration of I^{131}-labeled triolein and oleic acid on separate days and the determination of the per cent of I^{131} found in the blood. Normal absorption of I^{131}-labeled triolein indicates that pancreatic function is probably normal or not greatly impaired, since absorption of this compound to any significant extent must be preceded by lipolytic hydrolysis. Normal or near normal absorption of oleic acid in the presence of abnormal triolein uptake suggests impaired pancreatic function. Failure of absorption of both, triolein and oleic acid, indicates a malabsorption syndrome of nonpancreatic origin. The analysis of I^{131} in stool, in addition to that of blood, increases the accuracy of the test and therefore is preferred by some.[9]

PROCEDURE

The evening before the test the patient receives iodine, e.g., 10 drops Lugol's solution, to saturate the thyroid gland. On the test day, I^{131}-labeled triolein or oleic acid (50 μc.) is given with milk and the radioactivity of blood *plasma* is measured after 4 and 6 hours. Appearance of more than 1.7 per cent of the administered dose per liter of plasma is considered normal for both oleic acid and triolein. In the presence of decreased absorption of triolein, oleic acid levels may also be slightly decreased because of increased intestinal motility or other causes. In such a case, oleic acid values approaching the normal range are considered normal (see Fig. 15-6) as long as the oleic acid absorption is substantially greater than that of triolein. For more detail on the procedure consult Silver.[9]

Recently it has been suggested that triolein and oleic acid, each labeled with a different isotope (I^{101}, C^{14}, or H^{0}), be administered at the same time. The ratio of the absorbed triolein to the absorbed oleic acid is used to differentiate between malabsorption of pancreatic and nonpancreatic origin. A ratio of approximately 1:1 would indicate normal absorption, whereas a low ratio would indicate malabsorption of pancreatic origin.

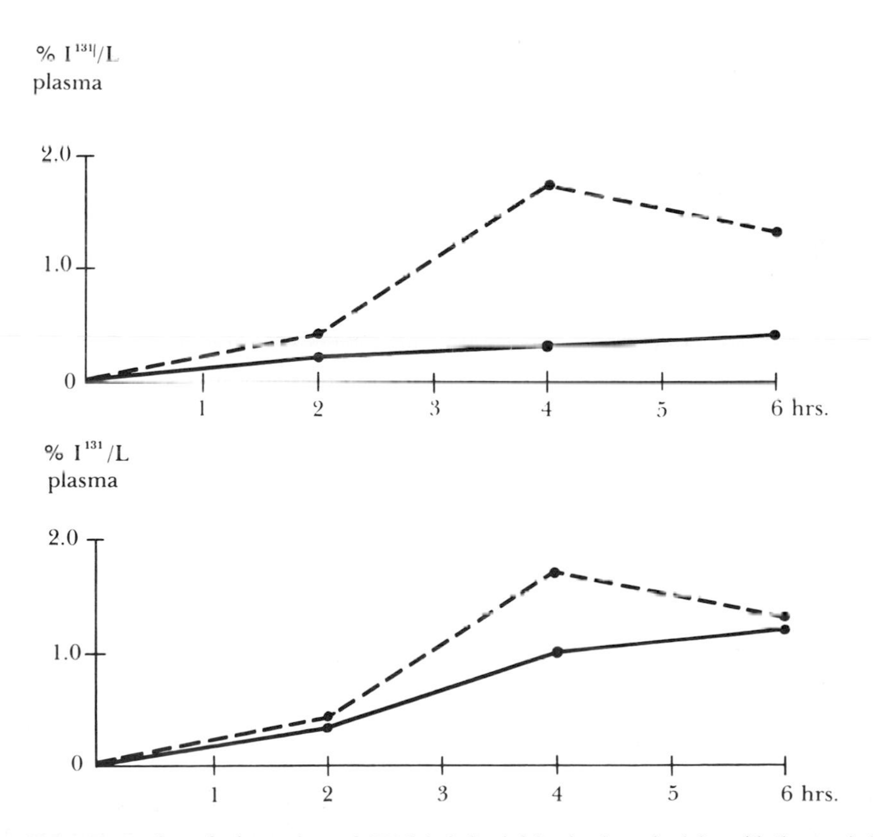

Figure 15-6. Examples of absorption of I^{131}-labeled triolein (top) and oleic acid (bottom) in a patient with pancreatic insufficiency; (- - - -) indicates lower limit of normal range.

In malabsorption of nonpancreatic origin, the ratio may be normal or slightly decreased, but the absolute amount of both triolein and oleic acid is significantly decreased.

Since both oleic acid and triolein are administered on the same day, errors due to daily fluctuation in the absorption of either one of these substances are eliminated.

The determination of I^{131} in *stool* after oral administration of I^{131}-labeled triolein and oleic acid is considered substantially more accurate than plasma determinations. The remaining radioactivity in a 72 hour stool specimen should be less than 5 per cent of the administered dose. In pancreatic insufficiency, triolein values may be as high as 20 to 30 per cent, but oleic acid values are normal or only slightly increased.

Measurement of Electrolyte Composition of Sweat and Saliva

It has been found that the sodium and chloride contents of sweat in patients affected with cystic fibrosis are significantly elevated, even in the absence of gastrointestinal or respiratory symptoms, and even when pancreatic insufficiency cannot be demonstrated by other tests. The sweat test is the most valuable single test in the diagnosis of fibrocystic disease of the pancreas. In most affected infants, it becomes positive between 3 and 5 weeks of age.

False normal results can be obtained in patients with pure salt depletion, which is common in the affected group of patients during hot weather periods. Additional electrolyte studies on serum help to avoid misinterpretation of results in such instances. Elevated levels of electrolytes in sweat have also been reported in meconium ileus, in adrenal insufficiency, and in some cases of renal disease.

COLLECTION OF SWEAT

The back of the child is carefully washed with distilled water, and a double layer of gauze sponges is placed on either side of the spine. The sponges should be approximately 2 inches square in size and must be cleaned with distilled or deionized water, dried, and weighed exactly in a labeled Petri dish or weighing bottle before use. A plastic sheet approximately 4 inches square is put on top of the gauze and tightly sealed with surgical tape. A plastic bag is then placed around the body of the child and tied around the neck. Sweating can be accelerated by wrapping the patient in an additional blanket, by using a heating lamp, or by using an electric blanket. After a sufficient amount of sweat is collected (usually after 15 to 60 minutes), the gauze is removed and sent to the laboratory in the same Petri dish. (Use forceps to prevent contamination and cover the Petri dish to prevent evaporation.)

Note: Avoid overheating the child, and do not prolong the sweating period beyond 60 minutes.

COLLECTION OF SWEAT BY IONTOPHORESIS

Principle. Pilocarpine is a drug that, when entered into the skin, induces sweating. The entrance of the pilocarpine can be enhanced with the aid of a current provided by a transformer and two electrodes. Such instruments are commercially available as iontophoresis kits. Pilocarpine, being positively charged, will move into the skin toward the negative electrode on the opposite side of the arm.

PROCEDURE

1. With the use of forceps, place two pre-washed 2 inch square gauzes (or two Whatman No. 42 filter papers, diameter 5.5 cm.) into a weighing bottle and determine the combined weight accurately. The weighing bottle should be handled with tissue or gauze to avoid direct contact with the fingers.

2. Place two other 2 inch square gauzes (not the preweighed gauze) on the anterior surface of the forearm and moisten them well with a freshly prepared 0.2 per cent solution of pilocarpine nitrate. Place two gauzes saturated with saline on the posterior (outside) surface of the arm. The positive electrode is placed over the gauze with pilocarpine and the negative electrode on the saline soaked gauze. Insure good contact and secure the electrode, e.g., with rubber bands. If the arm is too small to secure the electrode, as in small children, use the thigh of the patient.

3. Apply a current of 2.5 mamp. for 5 minutes. The current will tend to increase during this time interval and should be maintained at approximately 2.5 mamp. After 5 minutes, take off the electrodes and clean the area well with distilled water; dry the area.

If the patient complains about discomfort, discontinue the test. A tickling sensation at the site of the electrode is a common finding and should be disregarded. After the test, the skin may be somewhat reddish, but this will disappear within a few hours.

4. With the aid of forceps, place the preweighed gauze or filter paper over the skin area that was exposed to pilocarpine. Place an approximately 4 inch square plastic sheet over this area and seal with surgical tape. Allow the sweat to accumulate on the gauze or filter paper. This usually takes approximately 20 to 30 minutes, but the time of sweat collection may be extended as long as necessary. In general, the appearance of droplets on the plastic sheet indicates that enough sweat has accumulated.

5. Remove the gauze or filter paper with forceps, place it immediately into the weighing bottle and stopper. Send the weighing bottle to the laboratory.

RECOVERY OF SWEAT

Accurately determine the weight of the bottle or Petri dish containing the gauze or filter paper and calculate the amount of sweat by difference. One gram of sweat is assumed to be 1 ml. of sweat. (Generally 0.2 to 0.5 ml. of sweat is obtained.)

Many methods have been suggested to recover the sweat from the gauzes or filter paper. The method of choice will depend on the particular laboratory set-up. The sponges, for example, may be placed in a funnel and washed off with deionized water. The weighing container should also be rinsed with water and the washings should be combined and brought up to a predetermined volume (e.g., 50 ml.). To avoid such an elaborate procedure, it has been recommended recently that a golf tee be placed point down into a 15 ml. conical centrifuge tube and that the sponges be put on the top of the tee. The tube, tightly sealed with a rubber cap, is then centrifuged at moderate speed. The sweat will accumulate in the tip of the tube in its original concentration. If this procedure is used, weighing of the sponges is not necessary.

A detailed example of sweat recovery for a chloride determination by the Schales and Schales method is now outlined.

1. Add deionized water to the weighing container with the gauze so that the total volume of sweat plus water will be 5 ml. (e.g., if 0.3 ml. of sweat was collected, add 4.7 ml. of water). Stir well to elute the chloride from the paper or gauze.

2. Place 2.5 ml. of the extract into a suitable titration vessel and determine the chloride content as outlined in Chapter 10.

CALCULATION

$$\text{chloride in mEq./L.} = \frac{\text{reading of unknown} \times 0.2 \times 100 \times 5.0}{\text{reading of standard} \times \text{total amount sweat in ml.} \times 2.5}$$

The formula reduces to:

$$\text{chloride in mEq./L.} = \frac{\text{reading of unknown} \times 40}{\text{reading of standard} \times \text{total amount of sweat in ml.}}$$

Note: The figure 0.2 in the preceding formula indicates the amount of sample for which the chloride procedure has been set up; 100 is the factor for the concentration of the standard; 5.0 is the final volume of the dilution in the weighing container; 2.5 is the amount of extract used in the titration.

TABLE 15-2. *Sodium and Choride Values of Sweat in Normal and Abnormal Individuals in mEq./L.*

	Normal Value (Homozygotes)	*Normal Value (Heterozygotes)*	*Values in Cystic Fibrosis (Homozygotes)*
Na^+	10–40	40–70	80–190
Cl^-	5–35	35–60	60–160

REMARKS AND SOURCES OF ERROR

False high results can be obtained if the skin of the child is not cleaned properly or if contaminated sponges are used. Thus, gauzes have to be cleaned and dried carefully before use or a blank determination has to be done to assure that the gauze or filter paper is low in chloride content. Some laboratories have abandoned the use of sponges and instead have collected the sweat with a capillary tube directly from the bag and the skin of the patient. The amount of sweat collected with this procedure is larger, but the results are less reliable since evaporation can cause an increase in electrolyte concentration.

It has been found that determination of the chloride concentration is the most reliable index for diagnosing cystic fibrosis; therefore, many laboratories use only this test.

AGAR PLATE TEST

Since it is sometimes hard to obtain the amount of sweat necessary to perform the sweat test, modified procedures have been devised. These tests are based on the appearance of a white precipitate of silver chloride on an agar plate or on a commercially available paper prepared with silver salts that have been in contact with the skin of the patient. Such tests should be used only for screening purposes, since results are not as reliable as those obtained from direct measurement of sweat electrolyte contents.

ELECTROLYTE COMPOSITION OF SALIVA

Similar elevations of sodium, potassium, and chloride have been found in the saliva of patients with fibrocystic diseases of the pancreas,[4] but the results of these determinations are inconsistent and by no means as reliable as those obtained from the sweat tests.

Normal values for saliva without stimulation (resting) expressed in mEq./L. are: sodium 14.8 (6.5 to 21.7), potassium 22.1 (19 to 23), chloride up to 10. After stimulation of saliva production by chewing paraffin for 1 hour, the respective values for sodium, potassium, and chloride are: 44.6 (43 to 46), 18.3 (18 to 19), and 44 mEq./L.

STOOL EXAMINATIONS

The analysis of feces for various constituents was a test once widely used to measure pancreatic function. The simplest approach is the microscopic identification and demonstration of a large amount of undigested cell nuclei and meat fibers (creatorrhea), of increased fat (steatorrhea), and of increased starch (amylorrhea). Demonstration of these conditions is suggestive of impaired absorption (see the following section). In severe cases, the feces are usually pale, bulky, and unusually foul smelling. In steatorrhea of pancreatic origin, fat droplets usually appear in the feces on standing.

Unfortunately, these and many quantitative tests have no diagnostic value in mild cases of pancreatic disease and in those cases in which the acute phase has subsided. In addition, some patients with pancreatic insufficiency may have almost normal digestion owing to the activity of microorganisms and intestinal enzymes. On the other hand, patients with celiac disease or sprue may have stools similar to those of patients with pancreatic insufficiency because increased intestinal motility decreases the time that food remains in the intestinal tract, during which digestive enzymes can act. The quantitative assay of fat and nitrogen in the stool is clinically more helpful, but it also has its limitations.

Fat in stool. Ingested fat is normally split by pancreatic lipase into fatty acids, glycerol, and monoesters of glycerol, and the products of hydrolysis are absorbed by the intestinal tract. Therefore, the stool content of neutral fat, free fatty acids, and soaps is relatively low. The total content of fat in feces should not exceed 6 gm./24 hr. provided that the patient has been on a diet containing 60 to 150 gm. fat/day. In pancreatic insufficiency, the fat content of stool will increase rather substantially and may reach 20 gm./24 hr. or even more. Patients on a completely fat free diet will excrete 2 to 3 gm. fat/24 hr. The reason for this is not exactly known, but it is assumed that the source of this fat is the intestinal epithelium. There is also some secretion of phospholipids from blood into the intestinal lumen.

It was formerly believed that fractionation of the total lipids into neutral fat and free fatty acids might aid in the differential diagnosis of pancreatic insufficiency and malabsorption from other causes; however, today most investigators agree that such an approach gives very little additional information because of the extensive overlap in results. Enzymatic and spontaneous hydrolysis of fats during collection and storage time and incorrect collection of specimens add to the unreliability of the test.

Method for Determining Total Fat in Feces*

PRINCIPLE

The total lipids of stool (neutral fat, phospholipids, fatty acids, and soaps) are extracted with petroleum ether after acidification with hydrochloric acid and addition

* This method has been used successfully in our routine laboratory. A more complex research procedure is given in Chapter 7, The Lipids.

of ethanol. Acidification converts soaps into free fatty acids, which are soluble in organic solvents where most soaps are insoluble. Addition of alcohol prevents extraction of extraneous material such as amino acids into the petroleum ether. A portion of the petroleum ether extract is placed into a weighing container; it is evaporated and the residue is weighed. The per cent dry matter of stool may be determined by drying an aliquot of the stool and determining the difference between wet and dry weight of stool.

REAGENTS

1. Hydrochloric acid, A.R., concentrated.
2. Petroleum ether, low boiling.
3. Ethanol, 95 per cent.

PROCEDURE

1. Weigh the entire specimen (preferably a 72 hour specimen) and homogenize by grinding in a large mortar or a blender.

2. Weigh 5 gm. of wet feces (if the specimen is watery take 10 gm.) into a 100 ml. preweighed beaker, which is then placed under a heating lamp under a hood. When the specimen appears to be dry, place the beaker into a drying oven until a constant weight is reached. Record the final weight.

3. While the beaker is drying, weigh 5 gm. of wet feces and place in a 100 ml. ground glass-stoppered cylinder (if watery, weigh 10 gm.); bring the volume to 10 ml. with distilled water. Add 3 ml. of concentrated hydrochloric acid and mix well.

4. Put the cylinder with the specimen into a 100°C. bath for 10 minutes. Cool after incubation.

5. Add 10 ml. ethanol and 50 ml. petroleum ether. Stopper, shake vigorously, and allow the layers to separate.

6. Heat a 50 ml. beaker or other suitable container in an oven for about 15 minutes at 110°C. to insure that it is completely dry and allow to cool in a desiccator. Weigh the beaker accurately on an analytical balance and record the weight.

7. Transfer 25 ml. of the clear petroleum ether layer with a volumetric pipette into the preweighed beaker and evaporate to dryness on a hot plate at "low" temperature.

8. Place the beaker into a desiccator until a constant weight is obtained. Determine the amount of fat in the beaker by calculating the weight difference.

CALCULATIONS

$$\text{gm. of fat/total specimen} = \frac{\text{fat in beaker in gm.} \times 2 \times \text{total weight of feces}}{\text{gm. feces used}}$$

$$\text{per cent of dry matter} = \frac{\text{gm. of dry feces in beaker} \times 100}{\text{gm. of wet feces used}}$$

COMMENTS ON METHOD AND SOURCES OF ERROR

It has generally been found that the determination of total lipids is one of the most useful tests in detecting steatorrhea. The fractionation of the total lipids into fatty acids and neutral fats is of little additional help. If desired, the neutral fat alone can be determined by making the stool extract slightly alkaline with sodium hydroxide instead of acidifying it (Procedure, step 3). In an alkaline medium, fatty acids and soaps are not extracted by petroleum ether. The difference between total lipids and neutral fat represents the combined amount of free fatty acids and soaps.

The greatest source of error in this test is in the method of collection of the stool specimen. It has been suggested that markers such as charcoal or Congo red be used to identify the specimen produced during a given time interval; however, this approach is rarely used. Errors because of improper stool collection can be minimized by collecting a 72 hour sample rather than a 24 hour stool sample. During the collection period, the stool samples must be kept refrigerated and, at the end of the collection period, all stool specimens must be combined and homogenized as outlined in step 1 of Procedure. In the case of single stool specimens, it is advisable to determine the per cent fat in dry matter rather than the total amount of fat.

NORMAL VALUES

Normal adults and older children excrete up to 5 gm. of total fat/24 hr.; values above 6 gm. must be considered as abnormal. Children up to 6 years excrete up to 2 gm. total fat/24 hr. Free fatty acids and soaps normally constitute more than 40 per cent of the total fat fraction. Stool contains approximately 25 per cent dry solids (dry matter), and up to 25 per cent of this dry matter may be total lipids.

Miscellaneous Findings in Pancreatic Diseases

A significant number of patients with acute and chronic pancreatitis, as well as those with carcinoma of the body and tail of the pancreas, show a decreased glucose tolerance. Fasting hyperglycemia is observed in acute pancreatitis, but only in 15 to 30 per cent of cases of chronic pancreatitis. Glycosuria has been reported in 5 to 25 per cent of cases of acute pancreatitis.

Serum calcium levels are decreased to the range of 8.0 to 8.5 mg./100 ml. in the majority of cases of acute pancreatitis and are occasionally as low as 7.0 mg./100 ml. These changes have been observed from 24 to 72 hours after the attack and are attributed to a sudden withdrawal of calcium from extracellular fluids in the formation of calcium soaps. Some authors believe that the extent of the decrease of calcium levels parallels somewhat the severity of the condition.

TESTS MEASURING INTESTINAL ABSORPTION

D-Xylose Absorption Test*

Xylose is a pentose not normally present in significant amounts in blood. When given orally, it is absorbed in the intestine, passes unchanged through the liver, and is excreted by the kidneys. The xylose absorption test is considered a reliable index of intestinal carbohydrate absorption. Low absorption of xylose is observed in intestinal malabsorption, but not in malabsorption due to pancreatic insufficiency. In the latter condition, the absorption of xylose will be essentially normal provided that there is no significant increase in intestinal motility. Therefore, the test is of some help in distinguishing between these two types of malabsorption.

---

* Xylose has not as yet been released by the Food and Drug Administration for oral administration; thus, the test may be used only for investigative purposes.

PRINCIPLE

A 25 gm. dose of xylose is given to the patient and all of the urine voided over the next 5 hours is collected. A blood specimen is taken approximately 2 hours after the xylose is given (in children after 1 hour). The xylose of the 5 hour urine specimen and the blood specimen are determined by treating the diluted urine and a protein-free filtrate of the blood with p-bromoaniline in an acid medium. Xylose, when heated with acids, will form furfural, which in turn reacts with p-bromoaniline. A pink color is produced in this reaction. The concentration of xylose in blood and urine is calculated on the basis of a standard xylose solution.

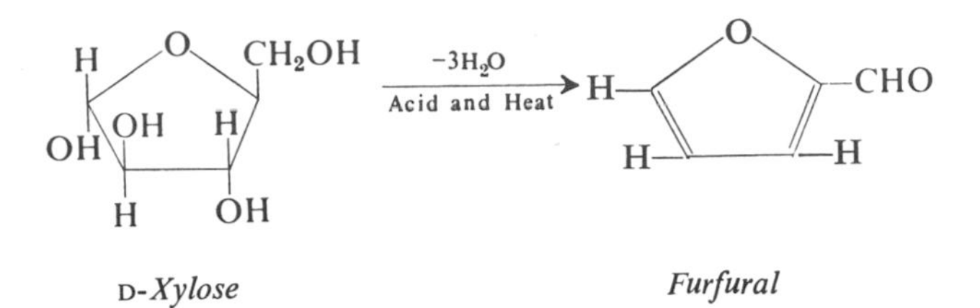

SAMPLE COLLECTION

Adults

1. Keep patient fasting overnight and during test period.

2. In the morning (e.g., at 7 a.m.) have the patient empty his bladder completely and discard this urine.*

3. Give 25 gm. of D(+)-xylose (or 5 gm. if so specified by physician) dissolved in 250 ml. water, followed immediately by an additional 250 ml. water.

4. Collect and pool all urine specimens voided during the next 5 hour period, including the 5 hour specimen.

5. After approximately 2 hours, collect a blood sample, preferably in a tube containing sodium fluoride.

Children

1. Keep patient fasting overnight and during test period.

2. In the morning (e.g., at 7 a.m.) have the patient empty his bladder completely.

3. Give 0.5 gm. of D(+)-xylose/lb. of body weight, up to 25 gm., cutting the amount of water according to the weight of the patient.

4. Collect and pool all urine specimens voided during the next 5 hour period.

5. Collect a blood sample approximately 60 minutes after xylose administration. The time of blood collection is different than in adults because the peak of absorption in children is reached sooner.

NORMAL VALUES

Adults, after a 25 gm. dose of xylose, should excrete more than 4.0 gm. of xylose/5 hr. urine specimen. In patients over 65 years of age, the minimum excretion is 3.5 gm./5 hr. Blood values are normally more than 25 mg. xylose/100 ml. blood. Normal children excrete 16 to 33 per cent of the ingested xylose and their blood xylose level is more than 30 mg./100 ml.

---

* Some investigators recommend checking a fasting urine and blood sample for nonspecific chromogens that react with p-bromoaniline; such interferences have been reported in some cases. Correction may be made by subtracting the base value from that found in the test specimen.

REAGENTS

1. Zinc sulfate ($ZnSO_4 \cdot 7H_2O$), 5 per cent (w/v) in water.
2. Barium hydroxide, 0.3 *N*. Dissolve 25 gm. of $Ba(OH)_2 \cdot 8H_2O$ in water and dilute to 500 ml. with water. Boil for a few minutes, stopper, cool, and filter.
3. p-Bromoaniline reagent. Prepare a saturated solution of thiourea in glacial acetic acid by shaking about 4 gm. of thiourea/100 ml. of glacial acetic acid and decant supernatant. Dissolve 2 gm. of p-bromoaniline in 100 ml. of the decanted thiourea solution.
4. Stock standard, 200 mg. of xylose/100 ml. Place 200 mg. D(+)-xylose into a 100 ml. volumetric flask and dissolve in and fill up to volume with 0.3 per cent benzoic acid.
5. Working standard (0.1 mg./ml.). Dilute stock standard 1:20 with saturated benzoic acid.
6. Working standard (0.2 mg./ml.). Dilute stock standard 1:10 with saturated benzoic acid.

PROCEDURE

1. Urine. If 5 to 15 gm. xylose is given, dilute the urine to 500 ml.; if 15 gm. or more xylose is given, dilute the urine to 1000 ml. (freeze a part of the first diluted urine for possible repeat). Further dilute a convenient portion of the urine 1:40.
2. Blood. Deproteinize the blood by using: 1 volume of blood, 7 volumes of H_2O, 1 volume of $ZnSO_4$ (5 per cent), and 1 volume of $Ba(OH)_2$ (0.3 *N*). Mix after each addition and centrifuge.
3. Add the following amounts (in ml.) of blood filtrate, urine, standard solutions, and p-bromoaniline to a set of test tubes as indicated in the following:

	Blood		*Urine*		*Standard (0.1 mg./ml.)*		*Standard (0.2 mg./ml.)*	
	Test	*Blank*	*Test*	*Blank*	*Test*	*Blank*	*Test*	*Blank*
Supernatant	0.5	0.5	—	—	—	—	—	—
Diluted urine	—	—	0.5	0.5	—	—	—	—
0.1 mg./ml. Xylose	—	—	—	—	0.5	0.5	—	—
0.2 mg./ml. Xylose	—	—	—	—	—	—	0.5	0.5
p-Bromoaniline	2.5	2.5	2.5	2.5	2.5	2.5	2.5	2.5

Place all tests into a 70°C. water bath for 10 minutes and all blanks in the dark at room temperature.

Cool the incubated tubes and place all test tubes in the dark. After 70 minutes, read the test against the blank, using a spectrophotometer or colorimeter at a wavelength of 520 nm. Alternatively, all tests and blanks may be read against a water blank.

CALCULATION

Blood

$$\text{mg. xylose/100 ml.} = \frac{A(\text{test})}{A(\text{Standard, 0.1 mg./ml.})} \times 100$$

If all test and blank tubes are read against water:

$$\text{mg. xylose/100 ml.} = \frac{A(\text{test}) - A(\text{blank})}{A(\text{standard}) - A(\text{blank})} \times 100$$

Urine

If urine was diluted to 1000 ml:

$$\text{gm. xylose excreted} = \frac{A(\text{test})}{A(\text{standard, 0.2 mg./ml.})} \times 8$$

If urine was diluted to 500 ml.:

$$\text{gm. xylose excreted} = \frac{A(\text{test})}{A(\text{standard, 0.2 gm./100 ml.})} \times 4$$

If all test and blank tubes were read against water, the absorbance of the blank must be subtracted from the absorbance of the test as indicated under blood.

COMMENTS ON METHOD AND SOURCES OF ERROR

The accuracy of the procedure depends not only on the rate of absorption, but on the rate of excretion of xylose by the kidneys. Thus, patients with renal insufficiency will excrete a decreased amount of xylose. In order to eliminate misinterpretations because of renal retention, a blood determination of xylose is carried out along with the determination of xylose in urine. Normal blood xylose level in the presence of decreased urine xylose levels would suggest renal retention, myxedema, or incomplete urine collection and invalidate the test.

The test findings are abnormal in 80 per cent of patients with malabsorption syndrome.

Some abdominal discomfort or slight diarrhea may be observed in some patients. These symptoms can be eased by use of the 5 gm. xylose dose; however, results in this case are not as reliable. Normal urinary excretion of xylose after a 5 gm. dose is more than 1.2 gm./5 hr.

Absorption of Co^{58} Labeled Vitamin B_{12}

SCHILLING TEST

The rate of absorption of orally administered Co^{57}, Co^{58}, or Co^{60}-labeled vitamin B_{12} can be measured by determining the radioactivity in feces, in urine, or in serum, or by externally scanning the liver uptake. The most frequently applied technique is measurement of radioactivity in a 24 hour urine sample, which is collected after oral administration of 0.5 or 1 μc. of radioactive cobalt-labeled vitamin B_{12} following an overnight fast. The patient also receives a 1 mg. dose of unlabeled vitamin B_{12} intramuscularly to saturate the binding capacity of plasma and liver for vitamin B_{12} and to assure adequate excretion of the vitamin. Urine specimens are collected for 24 hours and assayed for radioactivity. In normal individuals, more than 7.5 per cent of the dose administered will be excreted in the urine.

Initially, a Co^{60}-labeled vitamin B_{12} was used for this test, but Co^{57} or Co^{58} are now preferred, since they have a shorter half life and the counting efficiency is greater.

Absorption of vitamin B_{12} requires presence of the intrinsic factor (see Chapter 14, Gastric Analysis). Since this factor is missing in pernicious anemia, the Schilling test was initially designed to detect this disorder. Oral administration of the intrinsic factor simultaneously with the vitamin B_{12} will significantly increase the rate of absorption in individuals with pernicious anemia, but will not essentially affect the absorption rate in patients with malabsorption.

DISACCHARIDASE DEFICIENCY AND MALABSORPTION

The brush border region of the epithelial cells of the small intestine contains a series of enzymes that hydrolyze disaccharides such as lactose, sucrose, and maltose. Hydrolysis of these disaccharides by the respective enzymes is a prerequisite for their absorption as monosaccharides.

A deficiency in one or all of the disaccharidases has been observed in some newborns and, in rare instances, in adults.[5] Deficiency in the enzyme causes an intolerance in the patient for the respective disaccharide that frequently results in diarrhea, which may be accompanied by metabolic acidosis. Fatalities from this condition during the first year of life have been reported. Lactose intolerance has been observed most frequently. This appears to be quite important because of the relatively high intake of milk and therefore lactose by infants.

The most direct diagnostic test for the listed enzyme deficiencies is a biopsy and examination of the brush border region of the small intestine; however, this is difficult and is done only rarely. A more practical approach is the administration of one disaccharide at a time, followed by determinations of serum glucose levels. If the respective disaccharidase is present, the disaccharides will be hydrolyzed and absorbed as monosaccharides, and the glucose component can be determined in blood by employing one of the standard procedures for glucose. In cases of disaccharidase deficiency, serum glucose levels will increase only insignificantly. Changes of less than 20 mg./100 ml. should be considered as abnormal. The dose of disaccharides given is generally 50 gm./sq. m. body surface.

Appearance of the monosaccharide in serum obviously depends not only upon the presence of the enzyme, but also on the ability of the intestines to absorb monosaccharides. In order to eliminate misinterpretations because of this, a glucose tolerance test may be performed on a separate day. Appearance of this monosaccharide in serum in normal amounts would exclude this type of malabsorption. Another possibility is to give the respective enzyme (usually 500 mg.) simultaneously with the disaccharide. If the glucose appears in serum to a greater extent than after administration of the disaccharide alone, a deficiency of the respective enzyme may be suspected.

REFERENCES

1. Bessey, O. A., Lowry, O. H., Brock, M. J., and Lopez, J. A.: J. Biol. Chem., *166*:177, 1946.
2. Carr, F. H., and Price, E. A.: Colour reactions attributed to vitamin A. Biochem. J., *20*:498, 1926.
3. Goldbarg, J. A., and Rutenburg, A. M.: The colorimetric determination of leucine amino peptidase in urine and serum of normal subjects and patients with cancer and other diseases. Cancer, *11*:283, 1958.
4. Kaiser, E., Kunstadter, R. H., and Mendelsohn, R. S.: Electrolyte concentrations in sweat and saliva. A.M.A. J. Dis. Child., *92*:369, 1956.
5. Lifshitz, F., and Holman, G. H.: Disaccharidase deficiencies with steatorrhea. J. Ped., *64*:34, 1964.
6. Östlund, S. G., and Björlin, G.: Semi-micro method for determination of vitamin A and carotene in blood serum. Odontologisk Revy, *17*:208, 1966.
7. Schwarz, L., Woldow, A., and Dunsmore, R.: Determination of fat tolerance in patients with myocardial infarction. Method utilizing serum turbidity changes following a fat meal. J.A.M.A., *149*:364, 1952.
8. Sherman, B. S.: The use of ultraviolet irradiation in the estimation of retinol (vitamin A alcohol) and its derivatives. Clin. Chem., *13*:1039, 1967.

9. Silver, S.: Radioactive Isotopes in Medicine and Biology. Philadelphia, Lea & Febiger, 1962.
10. Sobel, A. E. and Snow, S. D.: Estimation of serum vitamin A with activated glycerol dichlorohydrin. J. Biol. Chem., *171*:617, 1947.
11. Tietz, N. W., Borden, T., and Stepleton, J. D.: An improved method for the determination of lipase in serum. Am. J. Clin. Path., *31*:148, 1959.
12. Tietz, N. W., and Fiereck, E. A.: A specific method for serum lipase determination. Clin. Chim. Acta, *13*:352, 1966.

ADDITIONAL READINGS

Cantarow, A. and Trumper, M.: Clinical Biochemistry. 6th ed. Philadelphia, W. B. Saunders Co., 1962.
Dreiling, D. A.: The technic of the secretin test: normal ranges. J. Mt. Sinai Hosp., *21*:363, 1954–1955.
Popper, H. L., and Necheles, H.: Pancreas function tests. Med. Clin. N. Am., *43*:401, 1959.
Silver, S.: Radioactive Isotopes in Medicine and Biology. Philadelphia, Lea & Febiger, 1962.

Chapter 16 /

TOXICOLOGY

by Robert V. Blanke, Ph.D.

The incidence of poisoning, both accidental and suicidal, has increased to an alarming level in recent years. In 1961, accidental poisoning was responsible for approximately 3000 deaths in the United States. In the same year 4500 suicides by poison occurred. Valid statistics on the number of nonfatal poisonings are not available, but the National Clearing House for Poison Control Centers estimated that 500,000 incidents of ingestion of poison occurred in 1964 in children under the age of 5 years.[5]

The high incidence of poisonings of all types has led most hospitals and clinical laboratories to recognize the need for laboratory services that give rapid and reliable information about the type and quantity of poison ingested by the patient.

The purpose of this chapter is to describe a number of laboratory procedures that can be carried out using equipment common to most clinical laboratories. These procedures require careful attention to detail, but not more than would be necessary in any other clinical test conducted by a well trained technologist. Some of the tests described are qualitative in nature, others enable an accurate, quantitative estimation of serum and urine levels to be made. Some of these tests are of the "screening" type; that is, a negative result indicates that the toxic substance is not present in significant amounts, and a positive result indicates that toxic substances may be present in significant quantities. Other procedures are very specific and yield positive results with only one or a limited group of substances.

The modern, toxicological specialist uses common, as well as other, more sophisticated, tools of science in isolating, identifying, and quantitating toxic substances in biological material. Activation analysis, x-ray diffraction, crystallography, optical rotatory dispersion, and other specialized or complex techniques may be required for specific problems. In addition, the toxicological specialist should be in a position to interpret the laboratory results, advise as to treatment of the patient, or give an opinion as to the effect, if any, of the toxic substance on the condition of the patient.

Every clinical laboratory director, whether he actually performs toxicological tests or not, should be informed about the special problems relating to toxicological tests. He should also know the closest toxicological specialist to whom he can turn for advice or to whom he can refer special problems.

In medicolegal cases, that is, cases in which the possibility of criminal poisoning exists or cases in which it is probable that legal action may be necessary in order to determine liability, the advice and services of an experienced toxicologist should be sought. In these cases, special precautions are necessary in collecting, transporting, and storing specimens to insure that loss, tampering, or contamination cannot occur. Also, the specimens and other evidence must be in the custody of a responsible individual at all times in order to establish a proper chain of custody and avoid legal objections during the presentation of evidence in court.

Ideally the analyst will collect the specimen and keep it in his custody until the analysis can be completed. If this is impractical, the individual collecting the specimen should place it in a proper container labeled with the name of the patient, the time and date it was collected, and the nature of the specimen. Many ingenious sealing devices have been described to make the container tamperproof. One of the most practical is a simple gummed-paper strip placed across the cap or stopper of the container so that the cap cannot be removed without tearing the strip. The paper strip should carry the signature or initials of the person who collected the specimen. The analyst can then note whether the seal was intact at the time he received the specimen, and he can also preserve the seal for later identification by the person who prepared it.

Other aspects of specimen handling of concern in legal cases are: the use of clean containers, the use of proper preservatives or anticoagulants, the procurement of receipts when specimens are transferred from one individual to another, the mailing of specimens by "registered mail," preparation of a written description of the container and its contents as received in the laboratory (photography is useful in this situation), and the preservation of unused specimens by freezing for duplicate analysis by an independent laboratory. These and other problems of a similar nature are considered in Stewart and Stolman's book.[46] It is best to obtain the advice of a toxicological specialist on these matters before collecting and sending him specimens.

The costly and complex laboratory procedures together with the legal problems and inconveniences of court testimony have had the result that most clinical laboratory directors are reluctant even to contemplate the performance of toxicological tests. Such considerations need not, and indeed should not, prevent clinical laboratories from carrying out those tests that are within the limitations of their personnel and facilities. The useful information gained can be rapidly communicated to the clinician, which results in more intelligent and faster treatment and possibly the saving of lives.

In the procedures to be described, an effort has been made to keep them as simple and economical as possible without sacrificing accuracy and specificity. With each procedure additional references are given to more specific or more rapid methods that require special equipment or training.

The analytical toxicologist, realizing that some overlapping of categories always occurs, tends to classify poisons according to the following groups: (1) gases, (2) volatile substances (i.e., substances, usually liquids, that are separable by steam distillation), (3) corrosives, (4) metals, (5) nonmetals (elemental or combined forms), and (6) nonvolatile organic substances.

No attempt has been made here to present a systematic approach to the "general unknown" type of problem, i.e., one in which the type of poison present, if any, is not known. Neither has any attempt been made to include procedures for the detection of all toxic substances. Space does not permit such an all inclusive approach. Only the most common toxic substances in each group are considered.

For those interested in other toxicological tests useful in the clinical laboratory, the excellent small book by Curry[11] or the comprehensive two volume work edited by Stewart and Stolman[46] is recommended. An excellent source of information concerning diagnosis and treatment is the book by Gleason, Gosselin, and Hodge[23] with its supplements.

Gases

CARBON MONOXIDE

Carbon monoxide, the product of incomplete combustion of organic substances, is the most common of the gaseous poisons. It is present in the free state in manufactured gas (coal gas), but not in natural gas. Both types of gas, when used as fuel

TABLE 16-1. *Carbon Monoxide Toxicity**

% (v/v) *in Air*	*Response*
0.01	Allowable for an exposure of several hours.
0.04–0.05	Can be inhaled for 1 hour without appreciable effect.
0.06–0.07	Causing a just noticeable effect after 1 hour's exposure.
0.1–0.12	Causing unpleasant but not dangerous symptoms after 1 hour's exposure.
0.15–0.20	Dangerous for exposure of 1 hour.
0.4 and above	Fatal in exposure of less than 1 hour.

* From Deichmann, W. B., and Gerarde, H. W.: Symptomatology and Therapy of Toxicological Emergencies. New York, Academic Press Inc., 1964. © Copyright Academic Press Inc.

for stoves, furnaces, and other appliances, release carbon monoxide as one of the combustion products. This is true, of course, of coal, oil, and other types of fuel as well. Malfunctioning or poorly ventilated heating appliances, therefore, are frequently a cause of carbon monoxide poisoning. Since this gas is also a component of exhaust fumes from internal combustion engines, accidental poisonings can occur when gasoline powered tools or outboard motors, as well as automobile engines, are used under conditions of poor ventilation or improper operation.

Carbon monoxide combines reversibly with hemoglobin in a manner almost identical to oxygen, but the bond is about 210 times as strong as that in oxygenated hemoglobin (oxyhemoglobin). As a result, carbon monoxide is not readily displaced from hemoglobin (except at high oxygen tension) and accidental poisonings can occur even at low levels of carbon monoxide in the atmosphere with prolonged exposure. (See Table 16-1.) In these instances the carbon monoxide level in blood builds up slowly until toxic levels are reached.

The detection and estimation of carbon monoxide in biological specimens can be approached in two general ways: (1) release of the gas from the hemoglobin complex with subsequent direct or indirect measurement of the gas, or (2) estimation of carboxyhemoglobin by its typical color or absorption bands. The first approach can be carried out by gasometric techniques,[55] gas chromatography,[24] microdiffusion,[18] or infrared spectrophotometry.[47] The second approach utilizes spectrophotometric[29] or spectrographic analysis or simple color comparison. An example of each approach will be described.

Regardless of the analytical method used, the specimen to be analyzed must contain hemoglobin. Relatively little carbon monoxide dissolves in the aqueous or lipid fractions of tissue, compared to that bound to hemoglobin.

This rather obvious statement is made since some clinicians request serum, spinal fluid, or even urine carbon monoxide levels. The most satisfactory specimen is whole blood. Clotted blood is less desirable and must be homogenized with a minimum exposure to air before analysis. This can be done using a Ten Broeck hand homogenizer. The clot is gently disintegrated, avoiding the introduction of air into the specimen or "whipping" the specimen. In those fatalities in which the victim is so badly burned or mutilated that blood is not available, tissue rich in hemoglobin, such as bone marrow or spleen, can be used. In these cases, estimation of carbon monoxide, rather than carboxyhemoglobin, is done.

Determination of Carbon Monoxide by Microdiffusion

The principle of diffusion can be used to separate a number of toxic substances from biological material. It is particularly applicable for separating gases and volatile substances. The specimen, containing a substance to be separated, and a "trapping" solution are placed in separate containers inside a third sealed container. The specimen and the trapping solution are in contact with the same atmosphere. The substance to be separated, because of its vapor pressure, leaves the specimen and enters the atmosphere from which it is absorbed by the trapping solution. Thus by gaseous diffusion the substance to be separated is transferred from the specimen to the trapping solution until an equilibrium is reached. If the trapping solution contains a reagent that converts the separated substance into a different compound, equilibrium does not occur and a quantitative transfer results. The entire operation can be carried out with a small specimen by using a Conway unit. This unit consists of two round, concentric chambers molded into a porcelain or glass dish that can be sealed by a glass plate. The trapping solution is placed in the center well and the specimen in the outer compartment (see Fig. 3-8, Chapter 3).

The time for completion of the diffusion process is variable, depending on the vapor pressure of the substance to be separated, the volume of the specimen solution, the nature of the trapping solution, and the temperature at which the process is carried out. In general, the diffusion time is shortened by higher temperatures, by small volumes of specimen solutions, and by substances of high vapor pressure. It should be emphasized that procedures worked out for a specific diffusion assembly may require different time periods for completion in an assembly of different dimensions. Also, if a diffusion is carried out at an elevated temperature, precautions must be taken to insure against loss of expanding gases, which tend to lift the glass plate. Even at room temperature this can occur if cold solutions are used. It is advisable to place a 10 ounce weight on the lid at room temperature and at elevated temperatures the lid should be clamped in place. For a mathematical consideration of these variables see Conway.[9]

PRINCIPLE

The microdiffusion technique can be utilized to detect carbon monoxide gas that has been released from hemoglobin by the action of dilute sulfuric acid or by lactic acid-ferricyanide solution. Carbon monoxide will reduce $Pd^{(II)}$ to metallic palladium, which appears as a silvery-black film on the surface of the reagent:

$$PdCl_2 + CO + H_2O \rightarrow Pd + CO_2 + 2HCl$$

The amount of carbon monoxide can be estimated indirectly by determining the amount of palladium reduced or the amount of hydrochloric acid produced in the reaction. The former is a colorimetric method, and the latter is titrimetric. We have found that microdiffusion is a very useful screening test, since the amount of metallic palladium produced is proportional to the amount of carbon monoxide in the specimen. Although this procedure can be carried out quantitatively, it is generally more convenient to use the spectrophotometric method, described later, for this purpose.

In the original method[18] on which this procedure is based, the authors recommend the use of 10 per cent sulfuric acid as an agent to liberate carbon monoxide from hemoglobin. We have found that some substances, particularly formic acid, may be converted to carbon monoxide by this treatment. In order to avoid such false positive results we recommend the use of lactic acid-ferricyanide solution as the liberating agent.

REAGENTS

1. Palladium chloride solution. Dissolve 0.22 gm. of $PdCl_2$ in 250 ml. of 0.01 *N* HCl. The solution is stable if protected from the carbon monoxide in the atmosphere.
2. Hemolyzing solution. Mix equal parts of 3.2 per cent $K_3Fe(CN)_6$ solution and 0.8 per cent (v/v) lactic acid solution.
3. Sealing compound. Either stopcock grease or petrolatum can be used.

PROCEDURE

Place 1.0 ml. of blood into the outer compartment of a Conway diffusion dish and 2.0 ml. of $PdCl_2$ solution into the center compartment. Add 2.0 ml. of hemolyzing solution into the outer compartment opposite the blood and cover the dish with the ground glass plate treated with sealing compound. Carefully mix the blood and hemolyzing solution by swirling the Conway dish gently. Allow diffusion to proceed for 1 hour at room temperature. In the presence of carbon monoxide, a mirror of metallic palladium will be noted on the surface of the palladium chloride solution in the inner compartment.

The only interference in this test is from the presence of sulfides in putrified blood. In the absence of such interference, a small spot of reduced palladium is consistent with the amount of carbon monoxide due to heavy smoking or subtoxic exposures. A bright mirror covering the entire compartment is typical of a lethal level of carbon monoxide in the blood.

For those interested in carrying out this procedure quantitatively, the original procedure should be consulted.[18] Much more dilute solutions are used in the titrimetric method, together with dilute, standardized acidic and basic solutions.

REFERENCE

Feldstein, M., and Klendshoj, N.: J. Forensic Sci., *2*:39, 1957.

Carbon Monoxide by Spectrophotometric Determination

Hemoglobin and its derivatives have characteristic absorption bands in the visible region that can be utilized to detect carboxyhemoglobin and to measure the quantity present. Oxygenated hemoglobin and carboxyhemoglobin have similar, double bands in alkaline solution. The absorption maxima for oxygenated hemoglobin are 576 to 578 and 540 to 542 nm.; for carboxyhemoglobin they are 568 to 572 and 538 to 540 nm. Deoxygenated hemoglobin has a single, broad band at 555 nm. (see Figs. 16-1 and 16-2).

If a weakly alkaline dilution of blood is treated with sodium hydrosulfite, oxygenated hemoglobin (and any methemoglobin present) is converted to deoxygenated hemoglobin. Carboxyhemoglobin is unaffected by such treatment.

$$\begin{matrix} HbO_2 \\ \text{or} \\ MetHb \end{matrix} + Na_2S_2O_4 \rightarrow Hb$$

$$HbCO + Na_2S_2O_4 \rightarrow \text{No reaction}$$

This is the basis of several methods for the determination of per cent saturation of hemoglobin by carbon monoxide. The method to be described[29] works satisfactorily with fresh, oxalated blood, but is not satisfactory with postmortem blood or specimens containing denatured hemoglobin. For these problem cases, the method of van Kampen and Klouwen[54] is more satisfactory, but it requires great care and excellent instrumentation.

REAGENTS

1. Ammonium hydroxide, 0.4 per cent. Dilute 15.9 ml. of concentrated NH_4OH to 1.0 L. The solution is stable.

2. Sodium hydrosulfite, powder. This is also called sodium dithionite. When added to 3 ml. of 0.4 per cent NH_4OH, 10 mg. of this powder should give a clear solution. If a turbid solution results, use a different lot of $Na_2S_2O_4$.

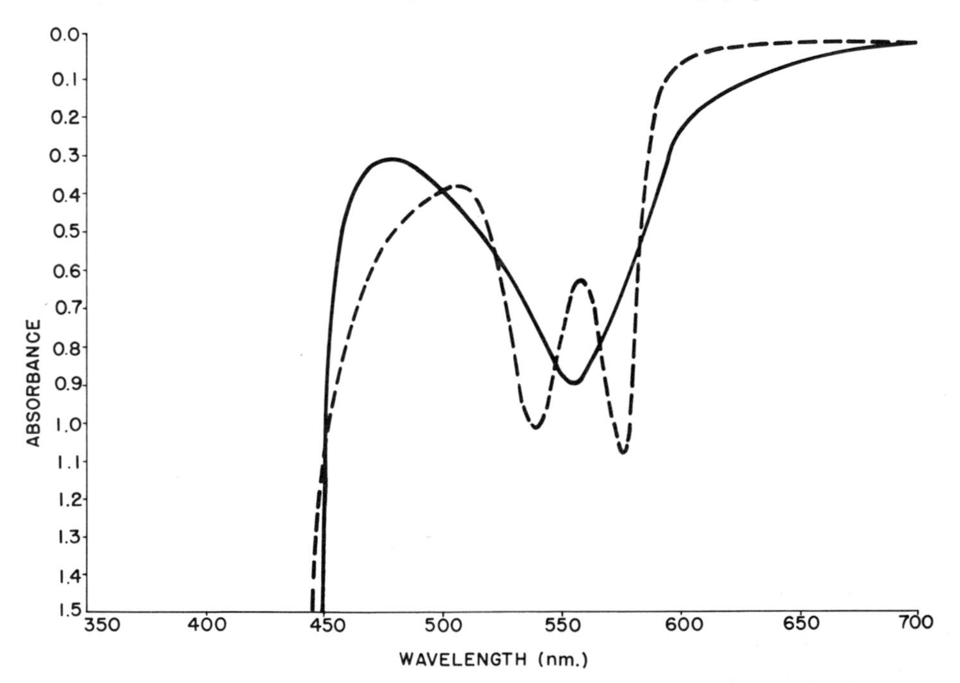

Figure 16-1. Spectral curve of normal blood, before and after treatment with $Na_2S_2O_4$. Dashed curve, oxyhemoglobin; solid curve, deoxygenated hemoglobin.

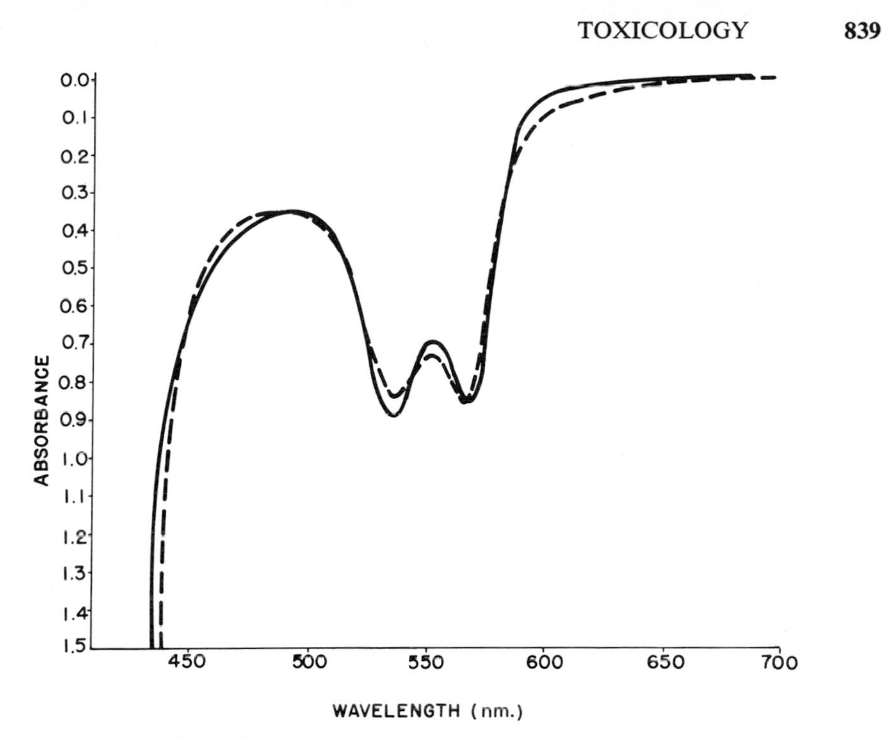

Figure 16-2. Spectral curve of blood containing a high level of carboxyhemoglobin, before and after treatment with $Na_2S_2O_4$. Note that the oxygenated curve (solid curve) is primarily carboxyhemoglobin with some oxyhemoglobin, and the deoxygenated curve (dashed curve) is primarily carboxyhemoglobin with some deoxygenated hemoglobin.

PROCEDURE

1. Dilute 0.1 ml. of oxalated blood with 19.9 ml. of 0.4 per cent NH_4OH. Mix well and let stand until clear. Occasionally it is necessary to centrifuge the diluted specimen until it is clear.

2. Transfer a portion of this dilution to a 1.0 cm. cuvet. Fill the reference cuvet with 0.4 per cent NH_4OH.

3. Add 10.0 mg. of $Na_2S_2O_4$ to both sample and reference cuvets; mix the solutions by inverting the cuvets 10 times. Allow cuvets to stand for 5 minutes. Then scan the solutions in the visible region, preferably with a recording spectrophotometer since the shape of the resulting spectrum is valuable for interpretation. For quantitative purposes, however, use only the absorbance readings at 555 and 480 nm. These can be made with a manual instrument with good resolution (band peak less than 2 nm.).

CALCULATION

$$\text{The ratio: } R = \frac{A_{555}}{A_{480}}$$

is calculated and the per cent saturation of hemoglobin by carbon monoxide is estimated from a table similar to Table 16-2.

The "R" values in Table 16-2 were calculated from standards made by mixing 100 per cent HbO_2 blood and 100 per cent HbCO blood in various proportions. Absorbance measurements were made with a Beckman DU spectrophotometer. Within limits, these "R" values should be similar in similar instruments but must be confirmed since deviations have been observed.

TABLE 16-2.

R	% *Sat.*	*R*	% *Sat.*	*R*	% *Sat.*
3.15	0	2.60	35	2.21	70
3.05	5	2.53	40	2.16	75
2.97	10	2.47	45	2.11	80
2.89	15	2.41	50	2.07	85
2.81	20	2.36	55	2.02	90
2.74	25	2.31	60	1.98	95
2.67	30	2.26	65	1.93	100

INTERPRETATION

Normal carbon monoxide levels depend on the degree of exposure to this gas, without signs and symptoms of poisoning being produced. For example, in smokers the following levels may occur:[6]

Smokers (one to two packs per day): up to 4 to 5 per cent saturation of hemoglobin with carbon monoxide.

Heavy smokers (more than two packs per day): up to 8 to 9 per cent saturation of hemoglobin with carbon monoxide.

Nonsmokers: 0.5 to 1.5 per cent saturation of hemoglobin with carbon monoxide.

In our experience, patients can survive brief periods of 70 to 75 per cent saturation. Prolonged periods at these high levels can, of course, be fatal. Interpretation of lethal levels must be related to other factors in each case, i.e., time of exposure, normal hemoglobin level of the patient, age and general health of the patient, degree of activity, and so on.

In patients treated with oxygen, carbon monoxide is fairly rapidly released from hemoglobin. Frequently, a patient is treated with oxygen while being transported to the hospital. By the time a blood sample is drawn and analyzed, the carbon monoxide level may be close to normal. These patients should be kept quiet with good oxygenation to insure that all tissue-bound carbon monoxide (e.g., with myoglobin and heme-containing enzymes) is dissipated before the patient is discharged.

TABLE 16-3. *Carbon Monoxide Toxicity**

Carboxyhemoglobin (%)	*Effect*
10	Shortness of breath on vigorous muscular exertion.
20	Shortness of breath on moderate exertion; slight headache.
30	Decided headache; irritation; ready fatigue; disturbance of judgment.
40–50	Headache, confusion, collapse, and fainting on exertion.
60–70	Unconsciousness; respiratory failure and death if exposure is long continued.
80	Rapidly fatal.
Over 80	Immediately fatal.

* From Deichmann, W. B., and Gerarde, H. W.: Symptomatology and Therapy of Toxicological Emergencies. New York, Academic Press Inc. 1964.

REFERENCE

Klendshoj, N. C., Feldstein, M., and Sprague, A. L.: J. Biol. Chem., *183*:297, 1955.

Volatile Substances

This group of toxic compounds consists mainly of liquids that have boiling points of 100°C. or lower. For this reason, they can be separated from biological specimens by steam distillation. Members of this group include almost all types of chemical compounds, and many are solvents commonly used in industry or in household products.

Steam distillation, although a useful procedure, necessitates assembly of glass apparatus and, particularly for some very volatile substances, great care in conducting the separation. The principle of microdiffusion can also be used to separate many members of this group (e.g., alcohols, aldehydes, ketones, chlorinated aliphatic hydrocarbons, and aromatic hydrocarbons[17]) with the advantages of small sample size, minimal equipment, and simplicity of operation. The use of the Conway dish with its various modifications greatly simplifies this type of separation.

Although only a few substances are considered in detail here, many other toxic, volatile substances can be determined using the Conway dish. For further details see Conway[9] or Feldstein.[17]

ETHANOL

Ethanol (ethyl alcohol) is the most common toxic substance involved in medico-legal cases. Not only is it lethal in its own right, but it is commonly a contributory factor in accidents of all types. In the case of a patient brought to the hospital in coma, the effect of alcohol, if any, must be ruled out in a differential diagnosis of the cause of coma.

There are probably more published methods for the determination of ethanol in blood than for any other toxic substance. In general they can be divided into those methods that are simple but nonspecific, and those that are specific but complex. The ideal method, as far as specificity and rapidity are concerned, is based on the use of gas chromatography.[24] Even this procedure is not absolutely specific by itself, although it is extremely unlikely that an interfering substance would be encountered in biological samples. A highly specific but time-consuming method is an enzymatic one utilizing alcohol dehydrogenase.[32] Besides the time factor, an ultraviolet spectrophotometer is necessary for this test; however, this method lends itself well to automation.

Two methods that are easily adaptable for use in most clinical laboratories use the principle of diffusion for separation of ethanol from the specimen. The Conway diffusion method has already been referred to and the method to be described here is a macrodiffusion method.[34]

Determination of Ethanol by Macrodiffusion

PRINCIPLE

The ethanol released from the sample is absorbed in an acid dichromate solution. Ethanol is oxidized by this solution to acetic acid. During this process, the yellow dichromate is reduced to the green chromic ion and this color change constitutes a rough qualitative indication of the progress of the reaction:

$$3C_2H_5OH + 2K_2Cr_2^{(VI)}O_7 + 8H_2SO_4 \rightarrow 3CH_3COOH + 2K_2SO_4 + 2Cr_2^{(III)}(SO_4)_3 + 11H_2O$$

The amount of unreacted dichromate is measured by reacting it with potassium iodide to form iodine, which is estimated by titration with thiosulfate:

$$2K_2Cr_2^{(VI)}O_7 + 6KI + 7H_2SO_4 \rightarrow 4K_2SO_4 + Cr_2^{(III)}(SO_4)_3 + 3I_2 + 7H_2O$$

$$I_2 + 2Na_2S_2O_3 \rightarrow 2NaI + Na_2S_4O_6$$

APPARATUS

1. Jars. Use wide-mouthed, squat, 16 oz. specimen jars that can be sealed from the atmosphere. The original article[34] described an apparatus that is commercially available (Aimer Products, Ltd., London, England); however, a clamp-on, glass-topped Mason jar with rubber gasket is suitable and is available at less cost. A screw-top Mason jar with rubber gasket is also suitable provided acid solution is kept away from the metal parts. The balance of this discussion assumes the use of a clamp-on, glass-topped Mason jar.
2. Beakers, Corning No. 1040, 100 ml. capacity (Corning Glass Works, Corning, N.Y.).
3. A 10 or 25 ml. buret with 0.02 ml. divisions.

REAGENTS

1. Potassium dichromate solution, 0.100 *N*. Accurately weigh 4.9035 gm. of reagent grade, dry $K_2Cr_2O_7$ and dissolve in 1000 ml. of cooled 50 per cent (v/v) H_2SO_4. Store in an all glass bottle. The solution is stable. (*Caution*—corrosive agent.)
2. Potassium iodide, A.R.
3. Sodium thiosulfate solution, 0.100 *N*. Dissolve 24.818 gm. of reagent grade $Na_2S_2O_3 \cdot 5H_2O$ in 1.0 L. of recently boiled water. Add 2 to 3 gm. of borax as a preservative. It is stable.
4. Starch solution. Prepare a 1 per cent solution of reagent grade, soluble starch in saturated sorbic acid solution. The solution is stable.
5. Alcohol standards. In this procedure the $K_2Cr_2O_7$ solution can be made up with a greater degree of accuracy than an alcohol standard. Nevertheless, it is desirable to prepare standards in order to maintain a proper quality control program.

Select 2 pints of outdated blood bank blood that, when analyzed by the following procedure, give negative results. Add 4.7 gm. of NaF to each pint in order to inhibit the production of alcohol by bacterial action. Combine the 2 pints.

Prepare a 2.00 gm./100 ml. alcohol solution by adding 2.53 ml. of anhydrous ethanol to about 90 ml. of blood while shaking the solution. Continue to shake the blood solution for 15 minutes. Make up to 100.0 ml. with blood. Absolute ethanol absorbs water from the air after it is opened. If there is doubt that the ethanol is anhydrous, measure its specific gravity, calculate the ethanol content from tables found in chemical handbooks, and use a corrected volume of ethanol to prepare this solution.

The following standards are made by adding the indicated volumes of the 2.00 gm./100 ml. ethanol blood standard solution to tared bottles (for standard solutions expressed in w/w) or to 100 ml. volumetric flasks (for standard solutions expressed in w/v).

Ethanol Standard (mg./100 ml. or mg./100 gm.)	*ml. of Ethanol Blood Standard Solution* (2.00 gm./100 ml.)
40	2.0
80	4.0
160	8.0
240	12.0
320	16.0
400	20.0

Make the solutions up to 100 gm. (w/w) or to 100 ml. (w/v) with ethanol-free blood. Store the standards in glass bottles, tightly stoppered, in the refrigerator. They are stable 2 months at refrigerator temperature.

PROCEDURE

Pipet accurately 10.00 ml. of 0.100 *N* $K_2Cr_2O_7$ acid solution into a jar. Weigh or pipet 2.00 gm. or ml. of blood into a 100 ml. beaker. Place the beaker containing the specimen into the jar containing the $K_2Cr_2O_7$ solution, position the rubber gasket, and clamp the top in place. In a similar manner prepare a blank, omitting only the sample. Place the blank and specimen jars in an incubator at 70°C. for a period of 1 hour.

After incubation, remove jars and cool. Remove the top, lift the beaker, and rinse the outside surface of the beaker with distilled water from a wash bottle, catching the rinsings in the jar. Dilute the contents of the jar with about 250 to 300 ml. of distilled water. Add about 2.0 gm. of solid KI and stir. Titrate with 0.100 *N* $Na_2S_2O_3$ until a faint yellow color remains. Add 1 to 2 ml. of starch solution and complete the titration to the disappearance of the blue-black starch-iodine color. Record the volume of titrant used. Repeat the same procedure with the blank.

CALCULATION

The difference between the volume of $Na_2S_2O_3$ solution used to titrate the blank and that used to titrate the sample, multiplied by the normality of the $Na_2S_2O_3$ solution, gives the milliequivalents of $K_2Cr_2O_7$ consumed in the oxidation of the same number of milliequivalents of alcohol (see the balanced equations under Principle).

Thus:

$$(\text{blank ml. } Na_2S_2O_3 - \text{sample ml. } Na_2S_2O_3) \times 0.100\,N\,Na_2S_2O_3 \times 11.5 \text{ mg. } C_2H_5OH/\text{mEq. } C_2H_5OH \times \frac{100}{2.0 \text{ gm. or ml. sample}} = \text{mg. } C_2H_5OH/100 \text{ gm. or ml.}$$

or

$$(\text{blank ml. } Na_2S_2O_3 - \text{sample ml. } Na_2S_2O_3) \times 57.5 = \text{mg. } C_2H_5OH/100 \text{ gm. or ml.}$$

Example:

Volume of titrant for blank	= 9.93 ml.
Volume of titrant for sample	= 4.19 ml.
Difference	5.74 ml.

5.74 ml. × 57.5 = 330 mg. C_2H_5OH/100 gm. or ml. of blood

INTERPRETATION

Any volatile reducing substance will interfere with this method. The possibility of interference by methanol, isopropanol, and aldehydes must be ruled out by separate tests for these substances. If the incubation is done at 37°C. or less, for a period of 8 hours, acetone does not interfere with this test.

Blood levels up to 10 mg. ethanol/100 gm. (or ml.) can be considered negative, since negative bloods may contain this amount of volatile reducing substances. With blood levels of 50 to 100 mg./100 gm. (or ml.), various signs of intoxication may be observed: flushing, loquaciousness, slowing of reflexes, impairment of visual acuity, and so on; however, there is much individual variation in this regard. Above 100 mg./100 gm. (or ml.), all individuals are under the influence of alcohol, and depression of the central nervous system is more apparent. Because of impairment of good judgment and visual acuity, as well as slowing of reflexes, driving a motor vehicle or operating machinery is hazardous when an individual is under the influence of alcohol.

With higher blood alcohol levels, central nervous system impairment is more pronounced and true coma may appear at levels of 300 mg./100 gm. (or ml.). Death may occur with levels above 400 mg./100 gm. (or ml.).

In many areas, state laws require that blood alcohol levels, when measured for legal purposes, be stated in terms of weight per cent. In these cases weighing the sample is mandatory. For clinical work, or when results in weight/volume per cent are permissible, pipetting the sample is more convenient. Interpretation of blood levels is the same in each case.

REFERENCE

Nickolls, L. C.: Analyst, *85*:840, 1960.

Ethanol Determination with Alcohol Dehydrogenase (ADH)

PRINCIPLE

Ethanol is oxidized in the presence of ADH to acetaldehyde. In the course of this reaction, NAD, a coenzyme, is reduced:

$$C_2H_5OH + NAD^+ \overset{ADH}{\rightleftharpoons} CH_3CHO + NADH + H^+$$

The increase in NADH can be measured by the increase in absorbance at its absorption maximum of 340 nm. The equilibrium for this reaction lies strongly to the left. At neutral pH and at normal NAD concentrations, less than 1 per cent of the ethanol present is oxidized to acetaldehyde. The reaction, however, can be driven almost completely to the right, by maintaining a high pH and removing the acetaldehyde as it is formed by reacting it with semicarbazide.

Both yeast and mammalian liver ADH can be used for this reaction. At a pH of 8, the Michaelis constant is 30 times greater for yeast than for liver ADH. The turnover number, however, is greater for yeast than for liver ADH. These differences are only of slight importance in analytical work when a large excess of the enzyme is used.

For details of the analytical procedure, the original reference should be consulted.[32]

REFERENCE

Lundquist, F.: *In* Methods of Biochemical Analysis. D. Glick, Ed. New York, Interscience Publishers Inc., 1959, vol. 7, p. 217.

METHANOL

Methanol (methyl or wood alcohol) is a widely used solvent in paints, varnishes, and paint removers. It is used alone as an antifreeze fluid and with ethanol and soap as a solid canned fuel. Poisonings are usually due to accidental ingestion by children or by alcoholics. In some areas, methanol may be a contaminant in "moonshine."

Determination of Methanol

Methanol can be determined by a variety of methods most of which involve measuring the color intensity after oxidation of methanol to formaldehyde, followed by the development of a color by reacting formaldehyde with chromotropic acid (CTA):

$$CH_3OH + MnO_4^- + 2H^+ \rightarrow CH_2O + MnO_2 + 2H_2O$$

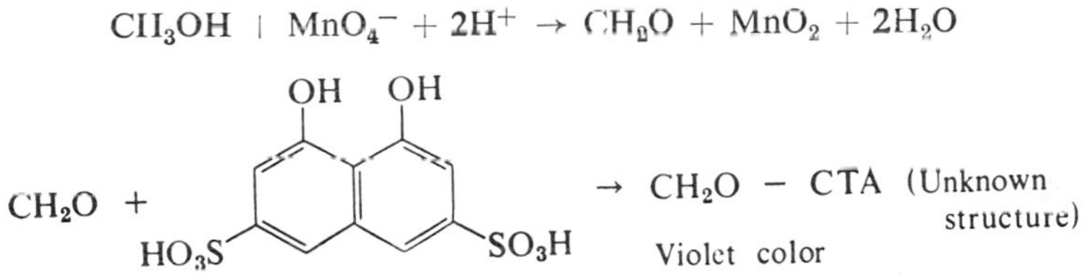

These methods work well since chromotropic acid is specific for formaldehyde and, hence, for methanol after oxidation. The microdiffusion method referred to earlier[17] is useful for the determination of methanol, and it also utilizes CTA for color development.

The CTA colorimetric procedure for methanol has two major drawbacks. First, methanol is not quantitatively oxidized to formaldehyde. It is readily apparent that after formation of formaldehyde by the oxidation reaction just noted, the formaldehyde itself can be oxidized to formic acid and further to carbon dioxide as follows:

$$CH_2O + MnO_4^- + H^+ \rightarrow CO_2 + MnO_2 + 2H_2O$$

This means that before a quantitative procedure can be devised, conditions must be chosen such that constant proportions of methanol are oxidized. Thus, the method is empirical and the set conditions must be established and adhered to rigidly before quantitative results can be achieved.

Second, the presence of reducing substances other than methanol will affect the system so that the procedure can no longer be applied quantitatively. The most common interference in cases of methanol poisoning is ethanol. It is not generally appreciated that the presence of ethanol invalidates a methanol procedure based on oxidation followed by CTA color development, if the calibration curve has been set up using pure methanol standards.

The following procedure, which is by Hindberg and Wieth[27] with modifications, obviates both drawbacks. First, the procedure must be carried out identically for standards and unknowns. Second, an excess of ethanol is added to *both* standards and unknowns. This results in a constant "interference" of a magnitude much greater than would ever be encountered in practice.

A third, but minor, drawback of any CTA procedure for determining methanol is the use of concentrated sulfuric acid for development of the final color. The dehydrating effect of concentrated sulfuric acid can produce formaldehyde from appropriate organic compounds and there will be false high results. We have encountered this interference occasionally in patients with severe acidosis. Apparently, some substances may appear in a trichloroacetic acid filtrate, such as glycolic acid, which reacts as follows:

$$H_2C(OH)COOH \xrightarrow[\text{Heat}]{H_2SO_4} HCHO + CO + H_2O$$

This type of interference can be detected by running a blank for comparison. A portion of filtrate is carried through the procedure except the oxidation step is omitted. Any color developed in the unoxidized specimen is due to formaldehyde contaminating the original specimen, or to glycolic acid.

Glycolic acid and its homologs can be eliminated as an interference by using gas chromatography, or by first separating the methanol by subjecting the specimen to steam distillation or microdiffusion in a Conway unit.

REAGENTS

1. Ethanol, 8.1 per cent (w/w). Add 10.0 ml. of 99 per cent ethanol to 90.0 ml. distilled water. The solution is stable if well stoppered.
2. Permanganate-phosphoric acid solution. Dilute 15.8 gm. of $KMnO_4$ and 14.0 ml. of 85 per cent H_3PO_4 to 500 ml. with distilled water. This is a 1.0 *N* (0.2 *M*) solution of $KMnO_4$. The solution is stable.
3. Bisulfite solution. Dissolve 5.0 gm. of $NaHSO_3$ in 100 ml. H_2O. Make fresh daily.
4. Phosphoric acid, 85 per cent, A.R.
5. Chromotropic acid solution (CTA). Dissolve 750 mg. of the sodium salt of 1,8-dihydroxynaphthalene-3,6-disulfonic acid in 50 ml. distilled H_2O. Make fresh daily.
6. Sulfuric acid, 95 per cent, A.R.
7. Methanol standard, stock solution. Dilute 10.00 ml. of absolute methanol, A.R., to 1.00 L. with distilled H_2O. This stock solution will contain approximately 785 mg./100 ml. methanol, depending on the quality of the methanol used.
8. Phosphoric acid, 2.8 per cent (v/v). Add 14.0 ml. of 85 per cent H_3PO_4 to distilled H_2O and dilute to 500 ml.

PREPARATION OF STANDARD CURVE

Using the stock solution, prepare five standard solutions as follows:

Stock Solution (ml.)	*Diluted to* (ml.)	CH_3OH *Concentration* (mg./100 ml.)
10.00	100.0	78.50
8.00	100.0	62.80
5.00	100.0	39.25
3.00	100.0	23.55
1.00	100.0	7.85

Each standard solution is run through the following procedure, including preparation of the filtrate or distillate. A straight line calibration curve should result with this group of standards.

Calculate the slope of the curve $\frac{(\text{mg. } CH_3OH/100 \text{ ml. sample})}{\text{Absorbance}}$ for each of the five points on the curve and compute the mean value of the slope of the calibration curve. This calibration constant is used for subsequent calculations.

PROCEDURE

1. Prepare a 1:10 tungstic acid filtrate or steam distill 5.0 ml. of specimen and collect 50.0 ml. of distillate.
2. Prepare three tubes labeled sample, sample blank, and reagent blank.
3. Pipet 1.0 ml. of filtrate or distillate into sample and sample blank tubes and 1.0 ml. of distilled water into reagent blank tube.
4. Into each tube pipet 1.0 ml. of the 8.1 per cent ethanol solution; mix.
5. Into sample and reagent blank tubes, pipet 5.0 ml. of the $KMnO_4$–H_3PO_4 solution. Into sample blank tube pipet 5.0 ml. of 2.8 per cent (v/v) H_3PO_4 solution. Mix all tubes thoroughly.
6. After 10 minutes at room temperature (time accurately), add 2.5 ml. of bisulfite solution to each tube, followed by 0.5 ml. of 85 per cent H_3PO_4. Mix all tubes until colorless.
7. Pipet 1.0 ml. of each reaction mixture into clean, labeled tubes.
8. Add 1.0 ml. of CTA solution and 8.0 ml. of concentrated H_2SO_4 to each tube. Mix carefully and place in a boiling water bath for 5 minutes.
9. Cool and read in a spectrophotometer at 570 nm. after setting the instrument at zero with the reagent blank.

CALCULATIONS

$$C = A \times F \times \frac{V_d}{W_s}$$

where C = Concentration of methanol in the sample in mg./100 ml.
A = Absorbance
F = Mean value of slope of calibration curve in mg./100 ml./Absorbance
V_d = Volume of total distillate or filtrate
W_s = Volume (or weight) of sample used

INTERPRETATION

Methanol poisoning is considerably more dangerous than that due to ethanol. Methyl alcohol is metabolized in man to formaldehyde and formic acid. The accumulation of formic, and other acids, severely reduces the alkali reserve, resulting in a metabolic acidosis (see Chapters 10 and 11). In addition, necrosis of the pancreas and serum amylase elevations have been demonstrated. Therefore, in addition to blood methanol levels, plasma carbon dioxide content, serum amylase determinations, and electrolyte studies are useful laboratory tests for determining the severity of the poisoning and following the progress of treatment.

Metabolites of methyl alcohol can damage the optic nerve, resulting in either temporary or permanent blindness. The mechanism of this effect is not well understood, nor is it a constant finding; nevertheless, prompt treatment of these cases may not only be lifesaving but may also preserve the eyesight.

As little as 2 teaspoonsful of methanol are considered toxic; fatal results have been reported with dosages between 2 and 8 ounces. A blood level greater than 80 mg./100 ml. is dangerous to life.

Treatment is twofold. First, the acidosis is treated, generally with sodium bicarbonate, both intravenously and orally. Second, it has also been proposed that in severe cases, ethyl alcohol be administered to saturate the alcohol dehydrogenase enzyme system. Since ethyl alcohol is the preferred substrate for this enzyme, this would prevent the conversion of methanol to its toxic metabolites.

REFERENCE

Hindberg, J., and Wieth, J. O.: J. Lab. Clin. Med., *61*:355, 1963.

CYANIDE

Cyanide inhibits cellular respiration because of its combination with important respiratory enzymes. This mechanism of action is the same whether cyanide is inhaled as the gas, hydrocyanic acid, or ingested as the potassium or sodium salt, or other combined form. Since death follows very quickly if sufficient cyanide is absorbed, the patient rarely survives long enough for treatment. Despite this fact, it is desirable to have a test for this poison available in order to confirm a suspected cyanide death.

Determination of Cyanide

PRINCIPLE

In the method to be described,[18] cyanide is separated from the specimen by microdiffusion, trapped in dilute alkali, and converted to cyanogen chloride:

$$CN^- + \text{chloramine-T} \rightarrow ClCN$$

The ClCN is then reacted with pyridine to form N-cyanopyridinium chloride (König reaction):

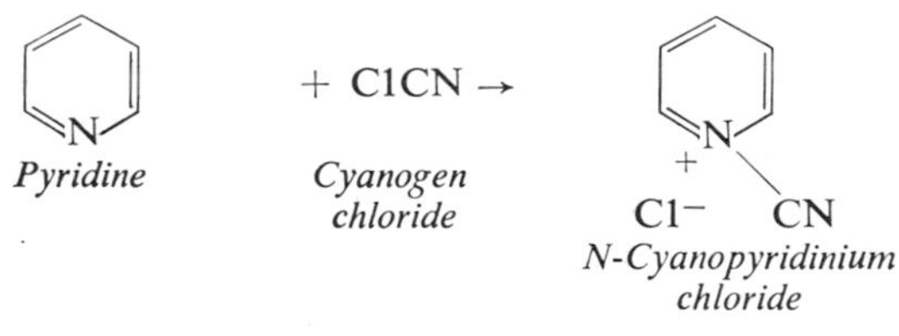

This is followed by a modified Aldrich reaction in which the N-cyanopyridinium chloride is cleaved to form an anil of glutaconic aldehyde. The aldehyde is coupled with barbituric acid to form a colored, highly resonant system:

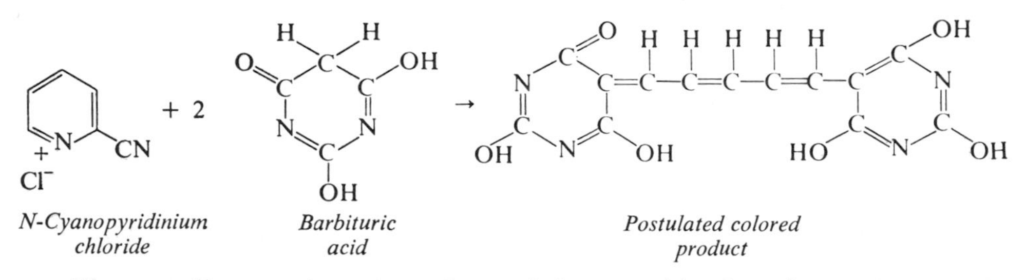

The preceding reaction cannot be carried out on blood or tissues that contain formalin since formaldehyde reacts with cyanide to form cyanohydrin, which is readily hydrolyzed to glycolic acid and ammonia:

$$HCN + HCHO \longrightarrow H_2C(OH)CN \xrightarrow{H_2O} H_2C(OH)COOH + NH_3$$

REAGENTS

1. Sulfuric acid, 10 per cent (v/v).
2. Sodium hydroxide, 0.10 *N*.
3. Sodium phosphate, monobasic, 1.0 *M*.

4. Chloramine-T, 0.25 per cent aqueous solution. This is stable when kept in the refrigerator.

5. Pyridine-barbituric acid reagent. Add 15 ml. of pyridine to 3.0 gm. of barbituric acid in a 50 ml. volumetric flask; mix. Add 3.0 ml. of concentrated HCl; mix. Dilute to volume with distilled water. Mix thoroughly since the ingredients dissolve slowly. Let stand for 30 minutes and filter if necessary. Prepare fresh as needed.

PROCEDURE

1. Place 4.0 ml. of blood or urine into the outer compartment of a Conway diffusion dish. If tissue is used, homogenize a portion in an equal weight of saline and use a measured quantity of homogenate. Place 2.0 ml. of 0.10 *N* NaOH in the center compartment and prepare the cover with silicone grease for a tight seal. Add 6 drops of 10 per cent H_2SO_4 to the outer compartment, seal the top quickly, and swirl gently to mix. Allow diffusion to proceed for 3 to 4 hours at room temperature.

2. After diffusion is complete, transfer 1.0 ml. of absorbing solution (center well) to a test tube.

3. Prepare a blank consisting of 1.0 ml. of 0.10 *N* NaOH in a second test tube.

4. To each tube add 2.0 ml. of NaH_2PO_4 solution and 1.0 ml. of chloramine-T solution. Mix and let stand 2 to 3 minutes.

5. To each tube add 3.0 ml. of pyridine-barbituric acid solution. Mix and allow to stand 10 minutes.

6. Observe the color, if any, or read the absorbance at 580 nm. in a spectrophotometer and compare with standards run through the same procedure.

CALCULATIONS

Prepare a series of standards ranging from 5 to 100 μg. cyanide/100 ml. of 0.10 *N* NaOH and carry these standards throughout the entire procedure, including the diffusion step. Construct a standard curve by plotting absorbance readings against concentration. The calibration curve will be a straight line up to values of 200 μg./100 ml.

INTERPRETATION

A red color is a positive test for cyanide in this procedure. Since chloramine-T can oxidize certain substances (e.g., glycine) to produce cyanide, care must be exercised to avoid mechanical contamination of the absorbing solution in the center well by trace amounts of specimen.

Although cyanide is very toxic, levels up to 15 μg./100 ml. blood can be found in adults without symptomatology. In cases of death due to ingestion of an overdose of a cyanide salt, levels of 1.0 mg./100 ml. or higher may be found. A lethal level is about 0.1 mg./100 ml.

REFERENCE

Feldstein, M., and Klendshoj, N.: J. Forensic Sci. *2*:39, 1957.

KEROSINE

Kerosine and other petroleum hydrocarbons are frequently ingested accidentally by children. Any hydrocarbon aspirated into the lungs or absorbed material excreted into the lungs can produce a dangerous chemical pneumonitis. Frequently, the odor of the breath is an indication that this material has been ingested.

Kerosine is used as a fuel in some areas, and this material is a common solvent for household products.

The aliphatic hydrocarbons are chemically inert. This makes the problem of testing for them exceedingly difficult. Generally, one can rely on the characteristic odor of this substance in gastric contents as a positive test. If a sufficient quantity of the hydrocarbon is present and can be physically separated, it can be identified by its physical constants. No simple laboratory tests for detecting the presence of kerosine are available.

Corrosives

This group includes those strong mineral acids or fixed alkalies that produce chemical burns on contact. There are no good tests that can be carried out on blood, serum, or urine by which the type of acid or alkali can be detected and the ingested quantity estimated.

The only specimen that can be examined profitably is gastric contents. Frequently, this specimen is not available unless the patient has vomited, since gastric lavage is contraindicated in this type of poisoning. If gastric contents are available, the pH should be measured. Ions such as Na^+, K^+, Cl^-, SO_4^{--}, and PO_4^{---} can be demonstrated by methods used in the laboratory for routine analysis. Obviously most of the common ions would be present normally in gastric contents. To be of significance in this type of case, a large excess must be present. Since many compounds in the group of corrosives can cause major disturbances in acid-base balance, it is advisable to perform electrolyte studies on blood. Usually the clinician has evidence from lesions in the mouth and esophagus that a corrosive substance has been ingested.

Metals

All metals are toxic if a sufficient quantity is absorbed. Generally they are not encountered in their toxic form in the elemental or free state, but rather in the form of salts. The degree of toxicity of a given metal is dependent on the solubility of the salt; the greater the solubility, the more likely it will be absorbed and the greater will be its toxicity. For example, barium chloride is soluble and extremely toxic, but barium sulfate is insoluble enough to be used as a radiopaque medium for the gastrointestinal tract.

In general, metals can be detected after burning away the organic material in the specimen and measuring the metal in the inorganic residue by some standard procedure. The combustion process may be either a wet digestion with strong, oxidizing acids, or a dry ashing procedure in a furnace.

Some metals and their salts are volatile at high temperatures (e.g., mercury) and special precautions must be taken to avoid loss during the ashing step. Some metals, such as sodium, potassium, calcium, magnesium, and iron, are commonly analyzed in clinical laboratories and thus will not be considered here. The so-called trace metals are present in biological material in only minute amounts, even after ingestion of toxic amounts. For this reason most metal determinations require analytical techniques used in trace analysis. These are difficult and require considerable experience if they are to be carried out validly. Instrumental analysis, particularly atomic absorption spectrophotometry, makes it possible to conduct trace metal analyses more simply.

ARSENIC AND RELATED METALS

Arsenic, despite its reputation, is not a common poison. It is still a favorite homicidal poison, but homicidal poisonings are rare. Since arsenic is an ingredient in some herbicides and insecticides, accidental poisonings, both acute and chronic, may still be encountered on occasion.

Clinically, the symptoms of both acute and chronic arsenic poisonings can easily be confused with a variety of other conditions. It is not uncommon, therefore, for a clinician to request that the presence of arsenic be ruled out as an aid in the differential diagnosis. The specimen of choice in this case is urine, even if 2 to 3 weeks have elapsed after ingestion of the poison. In long-term chronic cases, analysis of hair and nails may be informative, but this is subject to difficulties in interpretation (see under Interpretation).

In the older literature, arsenic is frequently described as a "protoplasmic poison." This term is as good as any for describing the mode of action of the metal. Arsenic combines readily with proteins because of its great affinity for sulfhydryl groups. This results in the precipitation of proteins, producing gastrointestinal irritation and irreversible inhibition of important enzyme systems, which are important toxic effects of arsenic. The great affinity of arsenic for tissue proteins is also responsible for the rapid removal of arsenic from the blood. Blood, therefore, is not a good specimen except in cases in which a large overdose of this substance has been ingested.

Detection of Arsenic

The test to be described is commonly referred to as the Reinsch test. It depends on the fact that metallic copper in the presence of acid will reduce arsenic to the elemental form. The arsenic deposits on the copper as a visible, dark film:

$$3Cu^0 + 2As^{+++} \xrightarrow{HCl} 3Cu^{++} + 2As^0$$

The oxidized forms of antimony, bismuth, mercury, and selenium can also be reduced by metallic copper under these conditions. Thus, the same test constitutes an exclusion test for these metals as well. The test was applied by Gettler[22] in a systematic way to biological material and its modification by Rieders[38] will be described.

REAGENTS

1. Hydrochloric acid, concentrated, A.R.
2. Copper spiral. Wind bright, clean copper wire around a 3 mm. glass rod about eight to ten times to make a tight spiral. A 1.0 cm. square copper foil may also be used.

PROCEDURE

To 100 ml. of urine in a shallow dish, add 10 ml. of concentrated HCl. Add a copper spiral and gently boil the solution until the volume is reduced to about 20 ml.

Remove the copper, rinse gently with distilled water, examine, and note any color change.

INTERPRETATION

If the copper is still bright, arsenic (25 μg./L. or more), mercury (50 μg./100 ml. or more) and selenium (50 μg./100 ml. or more) have been ruled out. In the presence of arsenic or selenium, the surface of the copper will be gray to black and in the presence of mercury, the film will be light gray to silvery and becomes shiny on rubbing. Some sulfur compounds, antimony, bismuth, or tellurium, also give gray to black deposits.

In the case of a positive test, the nature of the deposit on the copper must be verified by further tests.[22] Arsenic can be quantitated after wet digestion of another specimen, by an excellent colorimetric method.[35] Recently, the technique of atomic absorption spectroscopy has been used in a practical determination of arsenic in biological material. The principle of atomic absorption spectroscopy is discussed in Chapter 2.

Normal arsenic levels in urine are less than 50 μg./L. In cases of chronic poisoning, arsenic levels in urine will rise to 100 μg./L. and in acute poisoning, 1.0 mg./L. or more may be present.

Since arsenic is readily bound by sulfhydryl groups in protein, considerable arsenic is bound by keratin and subsequently deposited in hair and nails. This phenomenon has led to the analysis of hair and nails in an effort to determine whether a previous exposure to arsenic has occurred. Interpretation of these analyses is difficult because of the problem of differentiating between surface contamination of the hair and endogenous arsenic. If such an examination is required, a minimum of 1.0 gm. of clean hair (a large handful), clipped close to scalp, should be submitted to a toxicological specialist.

LEAD

Lead is still one of the most serious of the metallic poisons. In adults, inorganic and organic lead compounds may be encountered in industrial exposures. An increasing awareness of this danger has promoted the use of prophylactic measures. Education of workers about the hazards of lead intoxication has also been of help in minimizing industrial poisonings.

Unfortunately, children are particularly sensitive to lead poisoning and the exposure of children to lead-containing paint and plaster, particularly in lower class housing, has continued despite regulations, labeling laws, and attempts to educate the public. Severe poisoning in a child can cause lead encephalopathy and the mortality rate is high. Those children who survive frequently show evidence of permanent central nervous system damage.

The diagnosis of lead poisoning is difficult and the demonstration of an elevated lead level in blood or urine constitutes the most positive indication of absorption of a lead compound. Being an ubiquitous element, lead is normally present in trace amounts in biological material. Analytical procedures must be extremely sensitive and conducted with great care in order to achieve valid results.

These requirements generally make lead analyses the function of a special laboratory, particularly one where experience with trace metal analyses and their special problems is recognized. This can be illustrated by some facts relating to lead analysis. An average normal lead level in blood is 30 μg./100 ml., and an amount of 100 μg./100 ml. represents a toxic level. Thus, 5 ml. of the normal blood specimen contains 1.5 μg., and the abnormal sample contains 5 μg. It is obvious that any method used must not only be extremely sensitive, but it must also have an excellent accuracy and precision in order to discriminate between the 3.5 μg. separating the normal and toxic lead levels in blood. In addition, all of the glassware and reagents used in the analysis contain traces of lead. Thus, after careful selection of reagents and cleaning of glassware, the analyst must still exercise meticulous technique in order to keep blank values of lead low.

Detection of Lead or Lead Poisoning

The actual analysis may follow one of many techniques: colorimetric analysis with diphenylthiocarbazone,[20] polarography,[2] or atomic absorption spectroscopy,[62] to name some of the most reliable ones.

The clinical laboratory performs two very important functions that aid in the diagnosis of lead poisoning even if the lead analysis is done by others. First, the specimens to be analyzed must be collected in a valid way, that is, free of contamination. Second, other diagnostic tests can be done, for screening purposes or for confirmation. These tests are based on the effects of lead on erythropoiesis. Lead interferes in the biosynthesis of hemoglobin, which results in anemia. Although the precise mechanism of this interference is still not understood, one result is a buildup of precursors of hemoglobin. Two precursors that accumulate in lead poisoning are δ-aminolevulinic acid and coproporphyrin III, and urinary excretion of these substances increases markedly. Methods for the detection of these substances are discussed in Chapter 6.

Elevated urinary porphyrin levels can occur in conditions other than lead poisoning; however, after several hundred comparisons in the author's laboratory, it has been noted that in all cases of proven lead poisoning a corresponding elevation of urinary coproporphyrin levels occurred. Table 16-4 shows the correlation of urinary lead and elevated coproporphyrin levels in 140 cases of lead poisoning.

SPECIMEN COLLECTION

Collecting a proper specimen must be done with care. If the age and condition of the patient are suitable, a 24 hour urine specimen is the specimen of choice. The patient should void directly into a lead-free container (a borosilicate glass or polyethylene container from which surface lead has been removed by washing, then rinsing with hot 10 per cent nitric acid, and rinsing twice with metal-free water). A preservative should not be added because it might contaminate the specimen. The entire specimen, or a minimum of 100 ml. after noting the total volume, is submitted to the toxicological laboratory for analysis. Catheterized specimens should not be used unless it is unavoidable. In this case, the catheter should be cleansed (as just noted) to remove surface lead before sterilization. In some cases we have found that an indwelling catheter through which urine has been flowing freely for 24 to 48 hours is usually free from surface lead. The possibility of contamination should always be borne in mind when catheterized specimens are submitted for analysis. In an emergency, it may be necessary to analyze a random urine specimen rather than a 24 hour urine specimen. In such a case, the specimen must be collected with the same care as

TABLE 16-4. *Comparison of Urinary Lead and Coproporphyrin Levels in 140 Cases of Established Lead Poisoning*

	Number of Cases (total = 140)			
Urinary Lead Range μg./L.	*Test for Coproporphyrin Positive*	*Test for Coproporphyrin Doubtful*	*Test for Coproporphyrin Negative*	
			A*	B†
0–80 (Normal)	5	3	68	—
80–100	4	0	3	2
100–200	19	3	5	1
200–300	10	0	0	2
300–400	3	0	0	1
400–500	2	0	1	0
500–600	2	0	0	0
600–700	2	0	0	0
700–800	2	0	0	0
800–900	2	0	0	0

* A—Single specimen analyzed for lead.
† B—First specimen, Pb abnormal; other specimens, Pb normal.

just outlined. Interpretation of the result is subject to the same difficulty as discussed next in connection with blood specimens.

Blood specimens can be analyzed as readily as urine, but lead levels may fluctuate widely in different blood specimens from the same patient. We have had the experience of seeing normal lead levels in occasional blood specimens from lead-poisoned patients in well-documented cases. For this reason, properly collected 24 hour urine specimens are preferable. In very young or acutely ill patients, blood may be the only practical specimen available. In these cases it may be necessary to run several specimens before lead poisoning can be ruled out.

If the test is to be performed on blood, a minimum of 10 ml. should be collected. An anticoagulant or preservative should not be used unless the exact lead content of these agents is known so that proper correction can be made. The needle, syringe, test tube, and stopper should be of lead-free material, cleaned as previously described. Special tubes for blood-lead collection are commercially available. Since most of the lead is in the erythrocytes, a serum lead level is of little value.

As with any analysis of trace substances, the sensitivity of the analysis and the expected level of substance controls the amount of specimen to be collected. For example, if a lead method is used that is sensitive to 1 μg. of lead and in which the known reagent blank is also 1 μg. and the expected blood level is within normal limits, or about 30 μg./100 ml., then a minimum of 10.0 ml. of blood must be collected. This quantity of specimen would contain 3 μg. of lead, a level that can be differentiated from a blank with some degree of validity.

NORMAL VALUES

Normal lead levels are up to 80 μg./L. of urine or 80 μg./100 ml. of blood, with an average of 30 μg. Levels higher than normal indicate elevated absorption of lead compounds and levels greater than 100 μg./L. of urine, or 100 μg./100 ml. of blood, are usually associated with signs and symptoms of lead poisoning. Some clinicians prefer urine lead levels to be reported on a per diem basis. In the author's opinion, it is preferable to report these levels in μg./L. together with the total volume of the 24 hour specimen. This allows the clinician to correlate the 24 hour excretion of lead with other factors that may be related to an excessively high or low urinary output.

Colorimetric Determination of Lead

PRINCIPLE

Lead (as Pb^{++}) forms a red complex with diphenylthiocarbazone (dithizone) that is soluble in a number of organic solvents. Interference by other metal ions, such as zinc, cobalt, nickel, cadmium, silver, copper, mercury, stannous tin, bismuth, and thallous thallium, is eliminated by use of complexing agents and performance of extractions at controlled pH levels.

The reaction of lead with dithizone can be represented as follows:

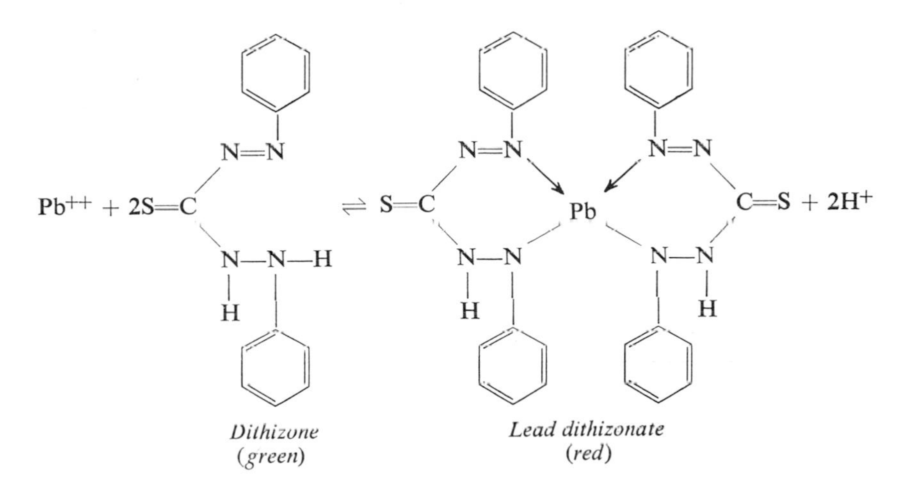

Dithizone (*green*) *Lead dithizonate* (*red*)

Approximately 10 ml. of clotted blood, or 100 ml. of urine, are weighed into a silica dish and are evaporated to dryness. The dish with its dried residue is placed into a muffle furnace and ashed overnight at 500°C. (Some authors prefer wet digestion with nitric acid.) After cooling, the dish is treated with a few milliliters of concentrated nitric acid, evaporated, and returned to the furnace at 500°C. for 30 minutes. This treatment converts the tin and thallium, if present, to the oxidized form, which does not interfere. The residue is treated with a few milliliters of concentrated HCl, evaporated to dryness, dissolved in dilute HCl and transferred to a separatory funnel. Citrate is added to complex the iron, the pH made alkaline with NH_4OH and the solution extracted with a CCl_4 solution of dithizone. Lead, zinc, copper, nickel, cobalt, cadmium, mercury, and bismuth are extracted quantitatively and silver partially. The dithizone solution is then shaken with dilute HCl, of pH 2. This step removes lead, zinc, and cadmium from the dithizone solution, leaving the other metals in the CCl_4-dithizone layer. Copper can be determined in the CCl_4 phase if desired. The dilute HCl solution is now treated with citrate, it is made alkaline with NH_4OH, cyanide is added to complex zinc and cadmium, and the solution is extracted with dithizone in toluene. Unreacted dithizone is extracted into the alkaline aqueous phase, leaving the red lead dithizonate in the toluene. After separating and filtering the toluene layer, the color intensity is read in a spectrophotometer at 520 nm. The absorbance is compared with standards carried through the same procedure. It is essential to run blank determinations and correct the final result for any trace quantities of lead present in the reagents and glassware. Zinc can also be determined in the dilute HCl solution of pH 2, by a modified extraction procedure. For more details of the method see the original article.

REFERENCE

Gant, V. A.: Industr. Med. 7:608, 1938.

Determination of Lead by Atomic Absorption Spectroscopy

PRINCIPLE

Bi^{+++} is added to urine that has been alkalinized with NH_4OH and this results in the precipitation of polyvalent ions, specifically lead. Na^+ and K^+ do not precipitate. This is useful since light scattering and plugging of the burner, as well as errors due to viscosity differences between specimens and standards, are minimized. Unlike some extraction procedures, this method is also suitable for partially decomposed or precipitated urines. Specimens kept at room temperature for 10 days gave an average recovery of 98.9 per cent lead with a standard deviation of ± 4.54.

APPARATUS

A Perkin-Elmer model 303 spectrophotometer with strip chart recorder and the following settings may be used:

Monochromator 2833 Å
Boling premix type burner
Air-acetylene flame at about 1:1 flow ratio
Sample flow 8 ml./min.
Aspiration time 15 seconds
Hollow cathode lamp current 6 mA.

REAGENTS

1. HCl and HNO_3, concentrated with low lead content.

2. Bismuth nitrate (10 per cent Bi w/w). Slowly and cautiously add excess of 10 *M* HNO_3 (about 50 ml.) to 10 gm. of bismuth pellets. Make up to 100 ml. with lead-free water. Metallic bismuth with a certified purity of 99.999 per cent is commercially available (Spex Industries Inc., Metuchen, N.J.).

3. Lead standard, stock. Dissolve 1.598 gm. of desiccated lead nitrate in 100 ml. of lead-free water. This standard is stored in an amber polyethylene bottle. The Pb concentration of this standard is 10.0 mg./ml. (For stability of standards see Comments on the Procedure.)

PROCEDURE

1. Transfer 25 ml. of a well mixed 24 hour urine specimen (collected in an acid washed, plastic container) to a conical polypropylene centrifuge tube.

2. Add 1.0 ml. of 10 per cent $Bi(NO_3)_3$ solution.

3. Adjust the pH to about 9 by adding 0.5 ml. of NH_4OH (about 28 per cent NH_3).

4. Centrifuge at 2000 to 2500 rpm for 10 minutes.

5. Discard supernatant, add 2 ml. of concentrated HCl, and stir with a thin Teflon rod or platinum wire until the precipitate dissolves.

6. Transfer to a 5 ml. volumetric flask and make up to volume with lead-free water. The lead from the original urine is now concentrated by a factor of 5. If a tenfold concentration is desired, use 50 ml. of urine and carry it through steps 2 to 6. (Since lead hollow cathode lamps vary in performance, it is sometimes desirable to use the higher concentration.)

7. Prepare fresh standard solutions containing 0, 50, 100, and 200 μg. Pb/L. Transfer 25 ml. portions of the standards to 50 ml. conical polypropylene centrifuge tubes and carry them through steps 2 to 6. Best results are obtained when the urine and standard solutions are prepared simultaneously and analyzed immediately after preparation. A blank containing all reagents is used to indicate cumulative lead contamination.

CALCULATION

Prepare a standard curve by plotting concentration against recorder readings obtained with the standard solutions and determine the concentration of the unknowns. The standard curve is linear up to 1 mg. Pb/L.

COMMENTS ON THE PROCEDURE

Without previous sample preparation, 200 μg. Pb/L. of urine can be detected by spraying the urine directly into the flame.

The amount of bismuth added in the sample preparation is sufficient to recover 10 to 20 times the normal lead level in urine. The optimal pH for coprecipitation of lead is 8.5 to 10.5. Bismuth does not interfere with the analysis and Na^+, K^+, Mg^{++}, Ca^{++}, and PO_4^{---} (which do not precipitate with Bi) are virtually removed with the supernatant.

The 2170 Å line is twice as sensitive but four times as noisy as the 2833 Å line. The latter is sufficiently sensitive for this study.

Lead standards deteriorate rapidly in polystyrene or borosilicate glass containers exposed to light. This is probably due to adsorption of lead cations on the glass or plastic surface. Aqueous solutions containing 0.5 or 1.0 μg. Pb/L. are stable several days in polystyrene or polyethylene containers wrapped in several layers of carbon paper. Upon exposure to light, a solution containing 0.5 μg. Pb/L. lost 20 per cent lead within 3 hours, 50 per cent within 6 hours and 90 per cent within 24 hours. More concentrated solutions saturate the container surface and after losing about 0.5 μg. Pb/L., remain constant for several days. Borosilicate adsorbed more lead than polystyrene. To prevent errors, solutions containing less than 0.2 μg. Pb/L. should be prepared in a darkened room and analyzed immediately.

No appreciable lead loss occurs in urine stored under conditions similar to those described for aqueous solutions. Urine stored without a preservative in clear polystyrene containers, exposed to light at room temperature, showed no significant lead change for a period of 10 days. These observations of Kopito and Schwachman emphasize the problems associated with trace metal analysis.[30]

For the determination of lead in blood by atomic absorption spectroscopy, reference should be made to the method of Pierce and Cholak.[37] These workers recommend dry ashing of clotted blood. The residue is dissolved in a small volume of 10 per cent HCl and this solution sprayed into the flame. The original article should be consulted for details.

REFERENCE

Kopito, L., and Schwachman, H.: J. Lab. Clin. Med., *70*:326, 1967.

THALLIUM

This metal is rarely encountered. Thallium salts are used as rodenticides, usually by professional exterminators. Cases have been reported of children eating fruit slices or other foods used as bait and treated with thallium salts, which had been carelessly scattered by exterminators intending to kill rats. Formerly, thallium salts were used, both internally and externally, as a depilatory. Fortunately, because of the high toxicity of these substances, this use has been virtually abandoned.

In many respects, the toxic effects of thallium are similar to those of lead. One characteristic feature of poisoning by this metal is loss of hair, or occasionally loss of nails and the skin of the feet. In some intoxications in children, loss of hair was the only sign. A lethal dose can be as little as 0.2 gm. of a soluble thallium salt. Since this metal is not present in biological material, except in extreme trace quantities, any amount demonstrated in the urine is significant.

The following procedure is simple and useful for the detection of this substance.[38]

Detection of Thallium

PRINCIPLE

Thallium is converted to its oxidized form by the action of bromine water. After destruction of excess bromine, the thallic ion is complexed with methyl violet to form a blue to violet compound of unknown structure that is soluble in benzene.

REAGENTS

1. Hydrochloric acid, concentrated, A.R.
2. Bromine water. Into a glass-stoppered bottle containing 50 ml. of water, add 2.0 ml. of liquid bromine (*caution:* do this in a fume hood! Avoid contact with skin!). Stopper the bottle and shake thoroughly. Some undissolved liquid bromine should remain in the bottom of the bottle. Store in the fume hood.
3. Sulfosalicylic acid, 20 per cent (w/v) solution in water.
4. Methyl violet, A.R., 20 per cent (w/v) solution in water.
5. Benzene, A.R.

PROCEDURE

1. Into a glass-stoppered tube, place 1.0 ml. of urine and 3 drops of concentrated HCl; mix.
2. Add 5 drops of bromine water; mix thoroughly.
3. Add 5 drops of 20 per cent sulfosalicylic acid to decolorize the bromine.
4. Add 1 drop of methyl violet solution and mix.
5. Add 1.0 ml. of benzene, stopper the tube, and shake thoroughly. After separation of the layers, decant or aspirate off the benzene and observe its color, if any.

INTERPRETATION

A colorless benzene layer rules out the presence of 0.8 μg. or more of thallium. A positive test imparts a blue to violet color to the benzene. No interference to the test is seen with levels up to 1.0 mg. of borate, oxalate, chlorate, nitrate, phosphate, sulfate, chloride, bromide, perchlorate, or EDTA. The color formation is inhibited by 1.0 mg. quantities of nitrite, sulfite, sulfide, thiosulfate, and thiocyanate; 0.2 mg. or more of iodide gives a false positive test. Of all the metals tested, the only one that gives a false positive is 0.01 mg. of $Hg^{(II)}$. Alkyl aryl sulfonate detergents give false positives, but these also give a color if the bromine and sulfosalicylic acid are omitted. This test can be made quantitative by reading the color at 610 nm.

Other procedures that can be used for the quantitative determination of thallium in blood, tissues, and other specimens involve emission spectrography or various colorimetric methods following digestion of the specimen. Atomic absorption spectroscopy (see under Lead) can also be applied for the determination of this metal.

REFERENCE

Rieders, F.: Ann. N.Y. Acad. Sci., *111*: 591, 1964.

Nonmetals

The toxic nonmetals are usually encountered as compounds with other elements, or as sodium and potassium salts. They are infrequently found in the free elemental form. Perhaps a more descriptive heading for this group would be toxic anions.

BORON

Boric acid and borate salts are commonly found in the home laundry or medicine cabinet. Accidental poisonings occur chiefly in children who ingest these preparations, or in infants treated with talcum powders containing borates, which may be absorbed through abraded or irritated skin.

Quantitative analysis of biological material for boron is a difficult problem. Not the least of the difficulties is that boron-free glassware must be used. The following simple test does not require special equipment and is convenient for screening purposes.[38]

Detection of Boron

PRINCIPLE

Turmeric or curcuma is a plant native to the East Indies and China. It is used as a condiment (curry powder) in the tropical East. From the root of the plant is obtained an orange-yellow coloring matter, which is curcumin or turmeric yellow. By an unknown reaction, this substance forms with borates a characteristic color. This has been used as a qualitative test for boron for many years and papers treated with turmeric are readily available.

REAGENTS

1. Hydrochloric acid, concentrated, A.R.
2. Turmeric paper (commercially available), A.R.
3. Ammonium hydroxide, concentrated, A.R.

PROCEDURE

1. Mix 5 drops of urine with 1 drop of concentrated HCl.
2. Place 1 drop of the acidified urine on turmeric paper. Observe the color.
3. Let the paper dry and again observe the color.
4. Hold the paper over concentrated NH_4OH and observe the color.

INTERPRETATION

A positive test is indicated by a brownish-red color of the acidified urine on the wet or dry paper. A green-black or blue color results after exposure to ammonia fumes.

If the wet acidified spot does not change color, the urine contains less than 10 mg. borate/L.

If, after drying, the paper is still negative, less than 5 mg. borate/L. is present. If, after exposure to ammonia, no color is produced, less than 3 mg. borate/L. is present.

Normally, less than 2 mg. borate/L. is present in urine. After a patient has been exposed to borate, the urine level increases sharply, but even at levels of 10 mg./L. no particular signs and symptoms of borate poisoning are seen. This test, although not sensitive to normal urinary borate levels, can detect elevated levels before the display of toxic effects.

REFERENCE

Rieders, F.: Ann. N.Y. Acad. Sci., *111*:591, 1964.

BROMIDES

Bromides are used in both organic and inorganic forms in medicine, chiefly for the purpose of sedation. These drugs are sometimes abused or may be taken in overdosage accidentally. The ready availability of nonprescription drugs containing bromide makes them easily available to the patient prone to drug abuse.

Determination of Bromide in Serum

PRINCIPLE

The procedure to be described[26] measures free Br^- only; thus, the bromine in most of the organic compounds is not detected. When organic bromides are ingested, however, they are metabolized eventually to inorganic bromide (see Interpretation).

The bromide anion readily displaces chloride from gold trichloride, forming gold tribromide:

$$AuCl_3 + 3Br^- \rightarrow AuBr_3 + 3Cl^-$$

The formation of gold tribromide may also be accompanied by the formation of $AuBrCl_2$ and $AuBr_2Cl$. The resulting brown color is very stable in acid solution and can be read quantitatively at 440 nm.

REAGENTS

1. Trichloroacetic acid, 10 per cent (w/v) aqueous solution.

2. Gold (auric) chloride solution. Wash the contents of a 1.0 gm. ampule of gold chloride into a 200 ml. volumetric flask and dilute to the mark with water. The solution is stable.

3. Trichloroacetic acid (10 per cent)–sodium chloride (0.06 per cent) mixture. Place 0.6 gm. of NaCl in a 1 L. volumetric flask and add 500 ml. of water. Add 100 gm. of trichloroacetic acid and dilute to volume with water.

4. Standards.

a. Stock, 10 mg./ml. Weigh exactly 1.000 gm. of NaBr, A.R., dissolve in water, and dilute to 100 ml.

b. Dilute standard, 0.5 mg./ml. Pipet 10.0 ml. of stock standard into a 200 ml. volumetric flask and dilute to volume with the trichloroacetic acid–NaCl mixture.

PROCEDURE

1. Prepare a 1:10 trichloroacetic acid filtrate of serum.
2. Pipet 5.0 ml. of clear filtrate (sample) into one tube and 5.0 ml. of 10 per cent trichloroacetic acid solution (blank) into a second tube.
3. Prepare standards as follows:

a. Pipet 0.5 ml. of dilute standard into a labeled tube and add 4.5 ml. of 10 per cent trichloroacetic acid–NaCl mixture. Mix well (corresponds to 50 mg. NaBr/100 ml.).

b. Pipet 2.0 ml. of dilute standard into a labeled tube and add 3 ml. of 10 per cent trichloroacetic acid–NaCl mixture. Mix well (corresponds to 200 mg. NaBr/100 ml.).

4. Add 0.5 ml. of 0.5 per cent $AuCl_3$ solution to all tubes. Mix well.
5. Read at 440 nm.

CALCULATION

$$\frac{\text{A unknown}}{\text{A standard*}} \times \text{concentration of standard*} = \text{mg. NaBr/100 ml.}$$

INTERPRETATION

Although normal bromide levels in serum are 0.8 to 1.5 mg./100 ml., this method may occasionally give results up to 5 mg./100 ml. even with normal serum. It has been suggested that this is due to a slight turbidity that may at times develop. Therapeutic levels may be in the order of 100 mg./100 ml. and toxic levels are usually greater than 150 mg./100 ml. With a single overdose of an organic bromide compound, serum levels of inorganic bromide do not rise above normal levels. After prolonged therapy with these drugs, serum levels of inorganic bromide may increase to more than 100 mg./100 ml. At these levels, mental disturbances may be elicited.

REFERENCE

Hepler, O. E.: Manual of Clinical Laboratory Methods. 4th ed. Springfield, Ill., Charles C Thomas, Publisher, 1963, p. 325.

FLUORIDE

This element is accessible to the public in the form of the sodium salt. Sodium fluoride is a common ingredient in roach and ant poisons and, as such, it is frequently kept around the house and even in the kitchen. For this reason, accidental poisonings have occurred, especially since the white crystalline material can be mistaken for ordinary salt or baking powder. In recent years a blue dye is usually added to these preparations to avoid this type of accident.

---

* Use the standard closest to the unknown.

The fatal dose of sodium fluoride is 5 to 10 gm. Once the compound reaches the stomach, the acidity of the gastric contents converts the salt to the free hydrofluoric acid, which produces a dark red corrosion of the mucous membrane. For this reason, inorganic fluorides could also be classified with the corrosives.

A number of organic fluoride compounds are extremely toxic, and one of these, sodium fluoroacetate (sometimes called "1080"), has been used as a rat poison. The toxicity of this substance is due to its competition with acetate in the tricarboxylic acid cycle with the eventual formation of fluorocitric acid. It is estimated that a lethal oral dose in man is about 50 mg.

Despite the marked toxicity of these substances, poisonings of this type have not been common in the past. This was fortunate for the analyst because the difficulty and length of time needed for fluoride analysis led him to avoid it, if at all possible. Now, with the aid of microdiffusion in plastic dishes, this analysis has been greatly simplified. Plastic containers are ideal for collecting specimens as well as for conducting the analysis, since silica reacts with fluoride to form a volatile product, resulting in loss of fluoride.

Rieders[38] has described a screening test using modified polypropylene Conway cells. The method takes about 1 hour for completion; however, the test to be described,[39] although somewhat lengthy, yields somewhat better quantitative results.

Determination of Fluoride in Biological Samples

PRINCIPLE

Fluoride is separated from the specimen by diffusing it into solid NaOH, at an elevated temperature for 20 hours. The separated fluoride is then estimated by developing a color with a cerium or lanthanum complex with alizarin complexone:

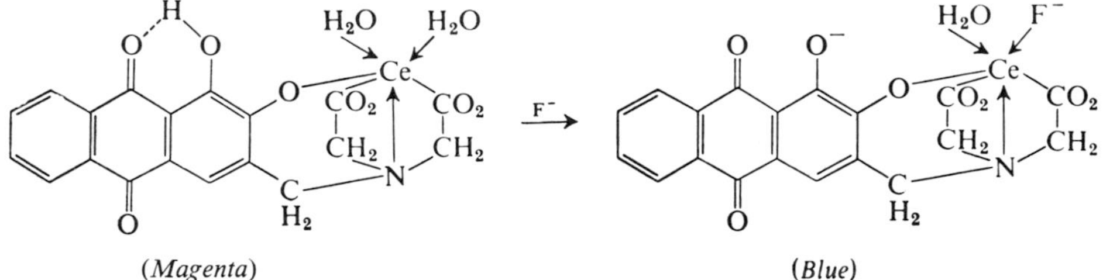

(*Magenta*) (*Blue*)

(*Cerium and lanthanum function similarly in these complexes*)

REAGENTS

1. Perchloric acid, 70 per cent, A.R.
2. Silver perchlorate, A.R.
3. Sodium hydroxide, 0.50 *N*. Dissolve 2.00 gm. of NaOH in 50 ml. of water and dilute to 100 ml. with 95 per cent ethanol.
4. Dye solution. Dissolve 10.0 gm. of Amadac-F (Burdick & Jackson Laboratories, Inc., Muskegon, Mich.) in 60 per cent isopropyl alcohol and make up to 100 ml. with the same solvent. Amadac-F contains all the necessary components of the dye reagent.
5. Fluoride standard. Dissolve 0.2210 gm. of sodium fluoride, A.R., in water and make up to 100 ml; 1.0 ml. of this solution contains 1.0 mg. of fluoride. A 1:100 dilution of this solution will result in a useful working standard in which 1.0 ml. contains 0.010 mg. of fluoride.

PROCEDURE

1. Place 0.10 ml. of 0.50 *N* NaOH in the center of the inside top of a plastic Petri dish (Millipore Filter Corp., Bedford, Mass., Cat. No. PD 10 047 00). Distribute the solution evenly and evaporate to dryness with a fan.
2. Transfer 1.0 ml. of blood, urine, or gastric contents to the bottom of the plastic Petri dish diffusion unit. Add about 0.2 gm. of silver perchlorate to the specimen.
3. Add 2.0 ml. of perchloric acid to the specimen and cover immediately with the prepared receiver top. Mix by gentle swirling.
4. Place the unit in an oven, previously heated to 50°C., and allow it to remain at this temperature for 20 hours.
5. Carefully remove the unit from the oven and allow to cool. Remove the receiver top and add 1.0 ml. of distilled water to the NaOH residue in the receiver top.
6. Gently stir with the tip of a dropping pipet, aspirate the solution, and transfer to a 5 ml. volumetric flask. Repeat last part of step 5 and first part of step 6 twice more.
7. Add 1.0 ml. of Amadac-F reagent and dilute the solution to 5.00 ml.
8. After 1 hour, read the color at 620 nm. and compare with a water blank and standard carried through the entire procedure.

INTERPRETATION

Normal blood fluoride levels range up to 0.050 mg./100 ml. In fatal cases, blood levels may be as high as 0.2 to 0.3 mg./100 ml. Excretion of fluoride in the urine is about 1.0 mg./24 hr. Even subtoxic doses of fluoride can result in sharp increases in urine levels, and in severe poisoning the urine concentration can be several milligrams/100 ml.

REFERENCE

Rowley, R. J., and Farrah, G. H.: Am. Industr. Hyg. Assn. J., *23*:314, 1962.

PHOSPHORUS

This is one nonmetal that is toxic in the form of the element itself. Phosphorus occurs in two forms, red phosphorus and yellow (or white) phosphorus. The red form is relatively nontoxic since it is extremely insoluble and hence is not absorbed. This is the form used in matches, fireworks, and other commercial applications.

Yellow phosphorus is very toxic and is still used in some types of rat poisons. It has a garlic-like odor and often can be detected in vomitus by its odor, combined with the phosphorescent effect that can be seen when the specimen is examined in a dark room.

If phosphorus is absorbed as the gas phosphine, PH_3, death can occur very rapidly because of cardiac collapse. Usually the elemental form is ingested accidentally or with suicidal intent. After absorption it remains in the blood for several days and is slowly oxidized to hypophosphorus and phosphorous acids. If the patient dies within one to three days, no significant changes are observed at autopsy. If the patient survives for a week or more, the effect of phosphorus on the parenchymatous cells of many organ systems becomes evident by extreme fatty changes. Both fat and protein metabolism are altered, the liver appears yellow with marked fatty degeneration, and severe jaundice is usually present.

Quantitative estimation of phosphorus is of little value in these cases. Since elemental phosphorus is not a normal constituent of biological material, any amount of this element that can be demonstrated is significant. The following test is useful for screening purposes only.

Detection of Phosphorus in Biological Samples

PRINCIPLE

Elemental phosphorus is converted to hypophosphorus acid and phosphine (step 1). These in turn reduce silver ions to metallic silver (step 2a) or form silver phosphide (step 2b), both of which impart a brown stain to $AgNO_3$-impregnated filter paper.

1. $P_4 + 6H_2O \rightarrow 3H_3PO_2 + PH_3$

2a. $H_3PO_2 + 2H_2O + 4AgNO_3 \rightarrow 4HNO_3 + H_3PO_4 + 4Ag$

2b. $PH_3 + 3AgNO_3 \rightarrow 3HNO_3 + Ag_3P$

REAGENTS

1. Silver nitrate, 0.25 *N*. Dissolve 4.3 gm. of silver nitrate in water and dilute to 100 ml. Store in a dark glass bottle, protected from light.
2. Lead acetate, 0.5 *N*. Dissolve 19 gm. of $Pb(C_2H_3O_2)_2 \cdot 3H_2O$ in water and dilute to 100 ml. Keep tightly stoppered.
3. Sulfuric acid solution, 10 per cent (v/v).

PROCEDURE

1. Into a 250 ml. Erlenmeyer flask, place 25 gm. of vomitus or tissue, finely chopped and suspended in water.
2. Prepare two strips of filter paper about 0.5 × 4.0 cm. and insert one end of each paper in a slit cut in the bottom of a cork stopper of proper size to fit the flask.
3. Moisten one filter paper strip with lead acetate solution and the other with silver nitrate solution.
4. Acidify the specimen with 10 per cent H_2SO_4 and suspend the treated paper strips over the solution by inserting the stopper in the flask. Carry out this procedure in a fume hood.
5. Place the flask over a steam bath and warm gently. Swirl the flask from time to time.
6. After 60 minutes examine the paper strips for the appearance of a dark stain of silver or silver phosphide.

INTERPRETATION

If both strips are not discolored, elemental phosphorus is not present in the specimen. If the silver nitrate treated strip is brown to black and the lead acetate treated strip is not affected, phosphorus may be present in the specimen. If both strips are discolored, phosphorus may be present, but sulfides are also present and interfere with the test.

This test, as described, is useful only when it is negative. To confirm the presence of silver phosphide on the paper, further tests are necessary.[11] The test may also be modified to differentiate between phosphorus, phosphine, and phosphides in the specimen.

As stated earlier, the determination of quantitative levels is not practical. The probable lethal oral dose of yellow phosphorus is about 60 mg.

REFERENCE

Scherer, J.: Ann. Chem. Phys., *112*:214, 1859.

Nonvolatile Organic Substances

This, the largest group of substances, includes most drugs and alkaloids. The problems associated with analysis of this group are quite complex. Extraction methods must usually be employed to separate the drug from the specimen. These are frequently not quantitative or may result in troublesome emulsions or may be pH dependent. Some drugs are rapidly metabolized, excreted, or bound to protein and this makes their detection difficult. Many drugs are chemically similar to naturally occurring substances and must be differentiated from them by purification steps or highly specific chemical tests.

Those drugs that are weak acids or bases are usually water soluble when they are in the form of salts. By reconverting the drug back to the free acid or base, it is made less water soluble but more soluble in solvents such as chloroform or ether. This property of organic acids and bases is used in separation and purification steps. By adjusting the pH of the aqueous phase and extracting with less polar immiscible solvents, separation is usually successful.

Those drugs that are neutral or amphoteric are more difficult to extract and purify. Chromatography, electrophoresis, sublimation, and countercurrent liquid-liquid extraction are some of the techniques that have been used.[46]

The various extracts can be rapidly screened for ultraviolet absorbing compounds by scanning with a recording spectrophotometer. Organic acids can be extracted from the immiscible solvent by a small quantity of 0.5 *N* NaOH. Organic bases are then extracted by a small quantity of 0.5 *N* HCl. The neutral compounds are recovered by evaporating the solvent to dryness and taking up the residue in alcohol. Both aqueous solutions and the alcohol solution are scanned, individually, from 220 to 350 nm. In general, aromatic compounds or compounds with conjugated double bonds can be detected in this manner if their extinction coefficient is great enough. Spectral curves thus obtained can be compared with similar curves obtained with known compounds.[51]

The advent of thin-layer chromatography has been a great help in detecting organic drugs in extracts of biological material. Although the extraction technique may not be quantitative, the speed and sensitivity of thin-layer chromatography enables the extract to be easily screened for the presence or absence of certain drugs or groups of drugs.

Only representative examples of some of the most common drugs in this group are given here. The determination of glutethimide is an example of thin-layer chromatography used in a semiquantitative way and utilizing improvised equipment and commercially available thin-layer plates. Identification of barbiturates and detection of amphetamine are further examples of thin-layer chromatography applications.

A number of books and articles discuss the technique more comprehensively and these should be consulted by anyone planning to use this extremely valuable tool more fully.[7,44,50] The technique of thin-layer chromatography in general is also discussed in Chapter 2, Analytical Procedures, and Chapter 7, The Lipids.

Once the substance has been extracted in more or less pure form, the next problem is to identify it and measure the quantity present. Obviously, the quantity isolated is usually quite small and cannot be handled, weighed, or visualized easily. Classically, color developing reagents, precipitating reagents, and microcrystallography have all been used for the identification of this group of poisons. Today, application of many of these techniques to extracts purified by thin-layer chromatography is much more reliable.

AMPHETAMINE

Amphetamine is a central nervous system stimulant. Both the dextrorotatory form and the less potent racemic mixture are used in medicine. This drug is prescribed for a variety of reasons, but since part of its effect is to inhibit sleep, it is sometimes abused by long-distance drivers. Although it is not truly an addicting drug, so-called "drug addicts" may misuse amphetamine, among other drugs. Prolonged abuse of this drug may produce bizarre effects on the behavior of the patient. In susceptible individuals gangrene of the extremities has been reported.

Chemically, amphetamine is an aromatic alkyl amine:

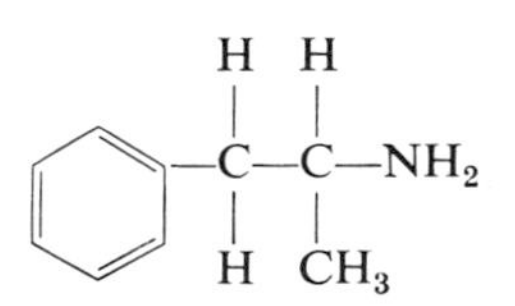

The asymmetric carbon atom gives rise to two optical isomers, but the test to be described measures both of them.

Detection of Amphetamine in Urine

PRINCIPLE

Amphetamine, in the form of the free base, can be separated from biological materials by steam distillation. Immiscible solvents can also be used to extract amphetamine, although some unwanted impurities may be carried along as well. The procedure presented here uses thin-layer chromatography to separate impurities from the drug prior to detection. Therefore, a simple extraction procedure is sufficient for isolating the drug.

Urine is extracted with chloroform under alkaline conditions and the chloroform extract is carefully evaporated. The residue, dissolved in methanol, is spotted on a thin-layer plate. After development, the amphetamine spot is visualized by spraying with ninhydrin.

REAGENTS

1. Hydrochloric acid, 10 per cent (v/v).
2. Sodium hydroxide, 5.0 *N*.

3. Chloroform, A.R.

4. Thin-layer plates. Coat 10 × 20 cm. glass plates with Silica Gel G to a thickness of 250 μ, dry at 110°C. for 1 hour, and store in a desiccator.

5. Developing solvent. Mix 85 ml. of ethyl acetate, 10 ml. of methyl alcohol, and 5 ml. of concentrated ammonium hydroxide. All solvents should be analytical, reagent grade.

6. Iodoplatinate solution. Mix 5.0 ml. of 5 per cent platinum trichloride, 45.0 ml. of 10 per cent potassium iodide, and 50.0 ml. of water. Store as a stock solution. Just before use, dilute with an equal volume of 2.0 *N* HCl.

7. Ninhydrin solution. Dissolve 400 mg. of triketohydrindene hydrate in 100 ml. acetone. Prepare fresh just before use.

PROCEDURE

1. Into a separatory funnel, place 50 ml. of urine (more can be used if available) and acidify with a few drops of HCl.

2. Add an equal volume of chloroform and extract the urine thoroughly. After separation of the layers, draw off and discard the chloroform layer. (This extraction is to remove organic acids; the chloroform will contain any acidic or neutral drugs, if present.)

3. To the aqueous phase, add 5.0 *N* NaOH until the pH is about 11. Cool, and check with a pH meter or indicator paper.

4. Add an equal volume of chloroform and again extract thoroughly. Allow the layers to separate. Centrifugation may be necessary.

5. Draw off the chloroform layer, which now contains the free base, and filter through Whatman No. 3 paper into an evaporating dish.

6. Evaporate the chloroform on a water bath to a volume of about 10 ml. (Do not let it go to dryness!) Transfer the solution to a conical tube, add 1 drop of glacial acetic acid, and continue the evaporation to dryness.

7. Take up the residue in 50 μl. of methyl alcohol and spot on a thin-layer plate, together with a standard for comparison.

8. Develop the plate in an equilibrated tank with the developing solvent (reagent 5) until the solvent front has moved about 10 cm. from the origin.

9. Allow the plate to dry and then spray it with the 0.4 per cent ninhydrin solution. Place the plate under an ultraviolet lamp for a 15 minute irradiation period. Amphetamine will show as a red spot. If iodoplatinate is used as a spray, a yellow-purple spot will develop.

INTERPRETATION

The ninhydrin spray is quite sensitive and will detect as little as 0.5 μg. of amphetamine. Iodoplatinate will show a color with about 4.0 μg. of the drug and will also react with other organic bases.

Blood or serum is not a good specimen for this test. The drug leaves the blood very rapidly and even large doses are difficult to detect in these specimens. Amphetamine is excreted in urine and the more urine that is extracted, the more likely the drug will be detected.

The ultraviolet absorption properties of amphetamine are weak and similar to many other substances. Thin-layer chromatography offers distinct advantages over other less specific techniques.

Fatalities due to overdose of this drug are rare in adults, but about 100 mg. (10 to 20 tablets) can be fatal in children. Fatalities have occurred when drivers of motor vehicles have ingested the drug, which resulted in erratic driving behavior and, therefore, accidents.

REFERENCE

Dole, V. P., Kim, W. K., and Eglitis, I.: J.A.M.A., *198*:349, 1966 (modified).

BARBITURATES

In cases associated with a drug overdosage, barbiturates are the leading offenders. These drugs are extremely useful in modern medicine and are commonly prescribed as treatment for a variety of conditions. Because of their availability, they are frequently the cause of accidental poisoning. Since they are hypnotics (sleep-producing drugs), they are commonly used for suicide or in suicide attempts. There are many individual drugs in this group, but all are chemically similar, and produce sleep and, upon overdosage, coma and death. They are chemically characterized by a pyrimidine ring with two substitutions on carbon atom 5 (see Table 16-5).

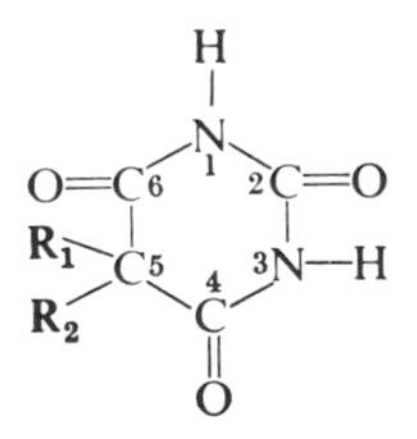

Some barbituric acid derivatives may have a methyl group substituted on one of the N atoms of the ring (e.g., mephobarbital and hexobarbital); others have an S atom instead of 0 at carbon atom 2 (e.g., thiopental and thiamylal). The sulfur derivatives are used as anesthetic agents and are not likely to be encountered in cases of overdosage.

Since these drugs can be readily extracted from blood, serum, or urine, and since they all have good ultraviolet absorption properties, the most reliable methods for determining barbiturates in biological materials require the availability of an ultraviolet spectrophotometer, preferably one that records automatically. The qualitative test for barbiturates that uses a color reaction produced by cobaltous acetate and an organic base under anhydrous conditions (the Koppanyi reaction) is unreliable. False positive results and, when improperly carried out, false negative results may be observed.

Recently, another colorimetric method has been reported. In this procedure a complex is formed between mercury$^{(II)}$ and barbituric acid derivatives.[1] The complex is soluble in organic solvents and the barbiturate can be estimated by using diphenylthiocarbazone to determine the amount of mercury present in the organic solvent. One atom of mercury combines with one molecule of a barbituric acid derivative that has two free >NH groups, resulting in a polymer, but one atom of mercury combines with two molecules of a barbituric acid derivative that has one free >NH group, resulting in a monomer. The exact structure of these complexes is not known. Serum barbiturate levels ranging from 0.1 to 2.0 mg./100 ml. can be determined in 1.0 ml.

TABLE 16-5. *Common Barbiturates and Their Toxic Blood Levels*

Generic Name	R_1	R_2	*Blood Level When Consciousness Regained* (mg./100 ml.)
Barbital (Veronal)	$-CH_2-CH_3$	$-CH_2-CH_3$	8.
Phenobarbital (Luminal)	$-CH_2-CH_3$	$-C_6H_5$	5.
Butabarbital (Butisol)	$-CH_2-CH_3$	$-C(H)(CH_3)CH_2-CH_3$	3.
Amobarbital (Amytal)	$-CH_2-CH_3$	$-CH_2-CH_2-C(CH_3)_2-H$	3.
Pentobarbital (Nembutal)	$-CH_2-CH_3$	$-C(H)(CH_3)CH_2-CH_2-CH_3$	1.
Secobarbital (Seconal)	$-CH_2-CH{=}CH_2$	$-C(H)(CH_3)CH_2-CH_2-CH_3$	1.

of serum. In the case of barbiturate overdosage, 0.2 ml. of serum is used and barbiturate levels of 0.5 to 12.5 mg./100 ml. can be measured.[63]

A similar method, which is rapid and does not require a colorimeter, has been described by Curry.[10,13] This method is useful for blood levels greater than 2.0 mg./100 ml.

Barbiturate Determination by Ultraviolet Spectrophotometry

PRINCIPLE

Ultraviolet spectrophotometric methods for the determination of 5,5-disubstituted barbiturates in biological material are based on the fact that these compounds exist in three forms in solution: a nonionized form in acid solution, with almost no absorption in the range 230 to 270 nm.; the first ionized form at pH 9.8 to 10.5 with an absorption maximum at 240 nm.; and the second ionized form at pH 13 to 14 with an absorption maximum at 252 to 255 nm. and a minimum at 234 to 237 nm. The three forms of the drug can be represented as follows:

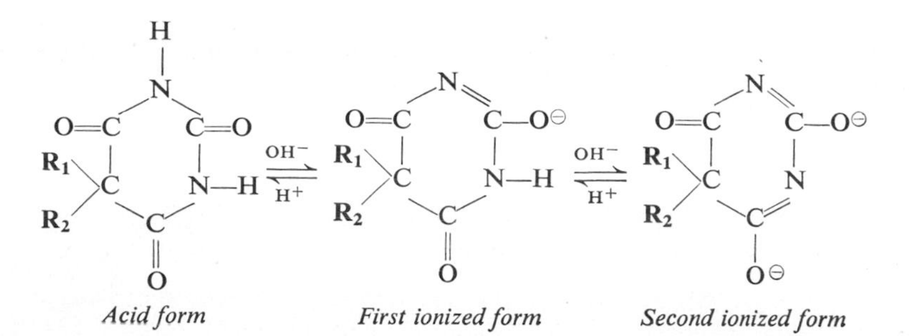

Acid form *First ionized form* *Second ionized form*

The 1,5,5-trisubstituted barbiturates exist only in two forms in solution since they lack one enolizable hydrogen:

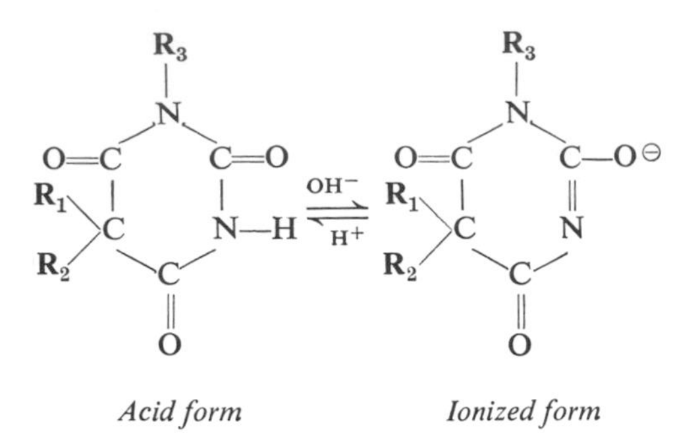

Acid form *Ionized form*

The nonionized form in acid solution has almost no absorption in the range of 230 to 270 nm.; the first and only ionized form at pH 9.8 to 14 has an absorption maximum at 245 nm.

In the acid form, barbiturates are relatively water insoluble, but they are soluble in organic solvents. In both the first and second ionized forms these drugs are very water soluble, but insoluble in organic solvents. Thus, they can be extracted from blood or serum at physiological pH values, or from acidified urine, by organic solvents. By washing the organic solvent with an aqueous phosphate buffer of pH 7.4, some interfering impurities, e.g., salicylates, can be removed, although some loss of barbiturates occurs. Shaking the organic solvent with dilute alkaline solution converts the free acid form of barbiturate into its salt, resulting in the transfer of the barbiturate into the aqueous phase. This aqueous extract is used for scanning in the ultraviolet spectrophotometer.

Proper interpretation of a blood or serum barbiturate level cannot be done unless the type of barbiturate present is known. For example, a 1 mg./100 ml. level of barbital is not too serious, but the same level of secobarbital is close to a lethal level (see Table 16-5). Also, 2 mg./100 ml. of phenobarbital in blood is not generally considered a dangerous level, but the same level of secobarbital can be lethal.

Since the prognosis in a given case of overdosage is influenced by the type of barbiturate involved, it is important to identify the drug or to determine the type of barbiturate present. This is done either by treatment of the barbiturate with hot alkaline solution or by thin-layer chromatography.* Both procedures will be outlined in more detail.

About 29 different barbituric acid derivatives are, or have been, used clinically. Of these, only about six or eight are commonly prescribed and available to the general public.

Pharmacologically, the barbiturates can be classified according to their duration of action. Four groups are commonly described: long, intermediate, short, and ultrashort acting.[23]

Long acting: barbital, phenobarbital, mephobarbital, diallylbarbituric acid.

Intermediate acting: amobarbital, aprobarbital, butabarbital, hexethal.

* E. F. Fiereck and N. W. Tietz recently reported a simple and fast gas chromatographic method suitable for routine use (National Meeting of the Society for Applied Spectroscopy, Chicago, Ill., May 17, 1968).

Short acting: cyclobarbital, pentobarbital, secobarbital.

Ultrashort acting: hexobarbital, thiamylal, thiopental.

The ultrashort acting barbiturates are used exclusively as anesthetic agents and consequently are rarely encountered in cases of accidental or intentional overdosage. The balance of this discussion is restricted to the classification of the first three groups: long acting, intermediate acting, and short acting barbiturates.

Classification of an unknown barbiturate into one of these major groups may be accomplished by treatment of the alkaline extract (step 9 of spectrophotometric procedure) with heat. Fast acting barbiturates are more stable to this treatment (alkaline hydrolysis at high temperatures) than the intermediate or slow acting barbiturates. Under this treatment barbiturate derivatives are hydrolyzed to malonic acid derivatives and urea. These compounds show little absorption in the ultraviolet region and, therefore, the decrease in ultraviolet absorption can be taken as a direct measure of the degree of hydrolysis of barbiturates (for more detail see under Procedure).

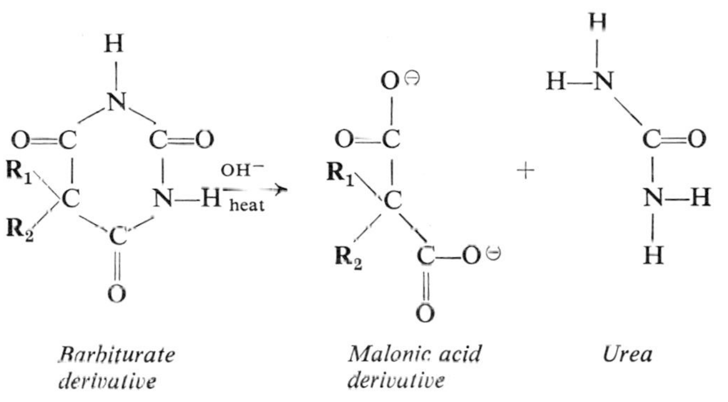

Barbiturate derivative *Malonic acid derivative* *Urea*

REAGENTS

1. Chloroform. Prepare a daily supply by washing sufficient U.S.P. chloroform with 1/10 volume of 1 *N* NaOH followed by two washings with 1/10 volume of distilled water. (Washing the $CHCl_3$ may be omitted if a blank determination shows the absence of interfering materials in the $CHCl_3$.)

2. Boric acid, 0.6 *M*, potassium chloride solution. Dissolve 37.1 gm. of H_3BO_3 and 44.7 gm. of KCl in distilled water and make up to the mark in a 1 L. volumetric flask. The solution is stable.

3. Sodium hydroxide, 0.45 *N*. Dissolve 18.0 gm. of NaOH in distilled water and make up to the mark in a 1 L. volumetric flask. This solution need not be standardized, but when equal volumes of 0.45 *N* NaOH are mixed with 0.6 *M* H_3BO_3–KCl, a pH of about 9.9 should result. Check with a pH meter. It may be necessary to add acid or alkali to the 0.45 *N* NaOH in order to get the correct pH on mixing with 0.6 *M* H_3BO_3–KCl. The solution is stable.

4. Potassium phosphate, monobasic, 0.5 *M*. Dissolve 17.0 gm. of KH_2PO_4 in distilled water and dilute to the mark in a 250 ml. volumetric flask. The solution is stable.

5. Sodium phosphate, dibasic, 0.5 *M*. Dissolve 179 gm. of $Na_2HPO_4 \cdot 12H_2O$ (or equivalent weight of anhydrous or other hydrated forms) in distilled water and dilute to the mark in a 1 L. volumetric flask. The solution is stable.

6. Phosphate buffer, pH 7.4. Mix 19.2 ml. of 0.5 *M* KH_2PO_4 (reagent 4) with 80.8 ml. of 0.5 *M* Na_2HPO_4 (reagent 5). Check the pH with a pH meter and adjust, if necessary, with reagent 4 or 5, respectively.

PROCEDURE

For the extractions use separatory funnels with Teflon stopcocks. Stopcock grease should not be used since some of the components of stopcock grease interfere with ultraviolet measurements.

1. Measure 10.0 ml. of blood, serum, urine, or gastric contents into a 125 ml. separatory funnel. With urine or gastric contents the pH should be checked with indicator paper and adjusted to a pH of 7 or less by the addition of dilute HCl. Volumes other than 10.0 ml. can be used with appropriate adjustment of the calculation. If the type of barbiturate need not be identified, 5.0 ml. of specimen are sufficient.
2. Add 30 ml. of chloroform (reagent 1) and extract the specimen by shaking for 1 minute.
3. Draw off the chloroform and filter through Whatman No. 3 filter paper into a second 125 ml. separatory funnel. Repeat the extraction two more times, using 30 ml. of chloroform each time. If an emulsion occurs at this point, add a few milliliters of chloroform in excess and gently invert the separatory funnel several times. The layers will separate easily with most specimens. If not, centrifuge at 2000 rpm for 5 minutes.
4. Wash the combined filtered chloroform extracts twice with 5 ml. of phosphate buffer (reagent 6). After the second phosphate wash, filter the chloroform through fresh Whatman No. 3 filter paper into a third 125 ml. separatory funnel.
5. Extract the chloroform by shaking for 3 minutes with 10.0 ml. of 0.45 *N* NaOH (reagent 3). Draw off the chloroform and discard it or save for analysis of other drugs.
6. Run the aqueous phase, together with any emulsion present, into a 15 ml. centrifuge tube. Centrifuge at 2000 rpm for 5 minutes. This is the "alkaline extract."
7. Prepare four test tubes as follows:

	Borate Blank	*Borate Sample*	NaOH *Blank*	NaOH *Sample*
0.6 *M* H_3BO_3–KCl	2.0 ml.	2.0 ml.	—	—
0.45 *N* NaOH	2.0 ml.	—	4.0 ml.	2.0 ml.
Alkaline extract	—	2.0 ml.	—	2.0 ml.

8. Scan these solutions in the ultraviolet spectrophotometer from 220 to 300 nm. as follows (if only a manual instrument is available, make readings at 5 nm. intervals): Read the borate sample against the borate blank as reference.* Read the NaOH sample against the NaOH blank as reference. If typical barbiturate curves are obtained (see Interpretation) and the type of barbiturate is known, proceed to Calculations. If the type of barbiturate is unknown, proceed as follows.
9. Pipet 5.0 ml. of alkaline extract from step 6 into a tube calibrated at 5.0 ml. and place into a boiling water bath for 15 minutes (time exactly!).
10. After 15 minutes, immediately transfer the tube to an ice-water bath. After cooling to room temperature, adjust the volume to 5.0 ml. with distilled water and mix thoroughly. This is the hydrolyzed alkaline extract.

---

* After completion of the reading of the borate sample, adjust the same recorder paper in such a way that the scan of the NaOH sample will exactly overlay the scan of the borate sample.

11. Prepare two test tubes as follows:

	NaOH-*Hydrolyzed Sample*	*Borate-Hydrolyzed Sample*
0.6 *M* H_3BO_3–KCl	—	2.0 ml.
0.45 *N* NaOH	2.0 ml.	—
Hydrolyzed alkaline extract	2.0 ml.	2.0 ml.

12. Scan these solutions in the ultraviolet spectrophotometer in the same manner as in Step 8, using the same blank solutions as reference solutions.

CALCULATIONS

Table 16-6 shows the calibration data for 5,5-disubstituted barbiturates commonly encountered. In this table, K_N and K_B are extinction coefficients of the particular

TABLE 16-6. *Calibration Data for 5,5-Disubstituted Barbiturates*

Barbiturate	*Duration of Action*	K_N	K_B	F	R %
Phenobarbital	Slow	0.0315	0.0086	43.7	31.8
Barbital	Slow	0.0329	0.0062	37.5	42.5
Butabarbital	Intermediate	0.0314	0.0056	38.8	49.6
Amobarbital	Intermediate	0.0292	0.0052	41.7	55.8
Pentobarbital	Fast	0.0268	0.0060	48.1	98.2
Secobarbital	Fast	0.0288	0.0078	47.5	97.6

barbiturate at a concentration of 1 mg./L. in 0.45*N* NaOH and in borate solution pH 9.9, respectively, at 260 nm. F is a calibration constant and is equal to $1/(K_N - K_B)$. R is the percentage of barbiturate remaining after alkaline hydrolysis for 15 minutes.

In the calculations to follow, these symbols are used:

A_1 = absorbance of NaOH sample at 260 nm.
A_2 = absorbance of borate sample at 260 nm.
A_3 = absorbance of NaOH hydrolyzed sample at 260 nm.
A_4 = absorbance of borate hydrolyzed sample at 260 nm.
C = concentration of barbiturate in sample.

$$(A_1 - A_2) \times F = C \text{ in } \mu\text{g./ml. of solution in cuvet} \quad \text{(a)}$$

or

$$(A_1 - A_2) \times F \times 0.2 = C \text{ in mg./100 ml. of specimen} \quad \text{(b)}$$

If specimen volumes other than 10.0 ml. are used, and the chloroform is extracted with volumes of 0.45 *N* NaOH other than 10.0 ml., then equation (c) is used:

$$(A_1 - A_2) \times F \times 0.2 \times \frac{\text{ml. 0.45 } N \text{ NaOH}}{\text{ml. sample}} = C \text{ in mg./100 ml. specimen} \quad \text{(c)}$$

If the barbiturate is known to be one of the six shown in Table 16-6, the appropriate F value is used. If the specific barbiturate is unknown, but it is found by hydrolysis to be slow acting, use an F value of 40.6; if intermediate acting, use F = 40.3 and if fast acting, use F = 47.8. If the type of barbiturate is unknown and insufficient specimen is available for hydrolysis, an approximate level can be estimated by using the mean F value of 42.9.

In order to determine the type of barbiturate present, the following calculation is made:

$$R = \frac{A_3 - A_4}{A_1 - A_2} \times 100$$

If: $R = 30$ to 45, a slow acting barbiturate is present.
$R = 45$ to 56, an intermediate acting barbiturate is present.
$R = 90$ to 98, a fast acting barbiturate is present.

INTERPRETATION

The criteria for the identification of barbiturates are as follows (see Figure 16-3):
In pH 9.9 solution—a maximum absorbance at 238 to 240 nm.
In pH 14 solution—a maximum absorbance at 252 to 255 nm.
In pH 14 solution—a minimum absorbance at 234 to 237 nm.
Isosbestic points*—227 to 230 nm. and 247 to 250 nm.

Dilute solutions of barbiturates may not give all of these characteristic points and the absorption peaks of salicylates and sulfonamides may obscure some of them; however, the phosphate wash should remove most of these interferences.

The value of R is valid only if a single barbiturate is present. For example, if a mixture of a slow and fast acting barbiturate is present, R may indicate an intermediate acting barbiturate.

Blood barbiturate levels must be interpreted cautiously for the following reasons: (1) there is variation in the response of different individuals to a given dose of any

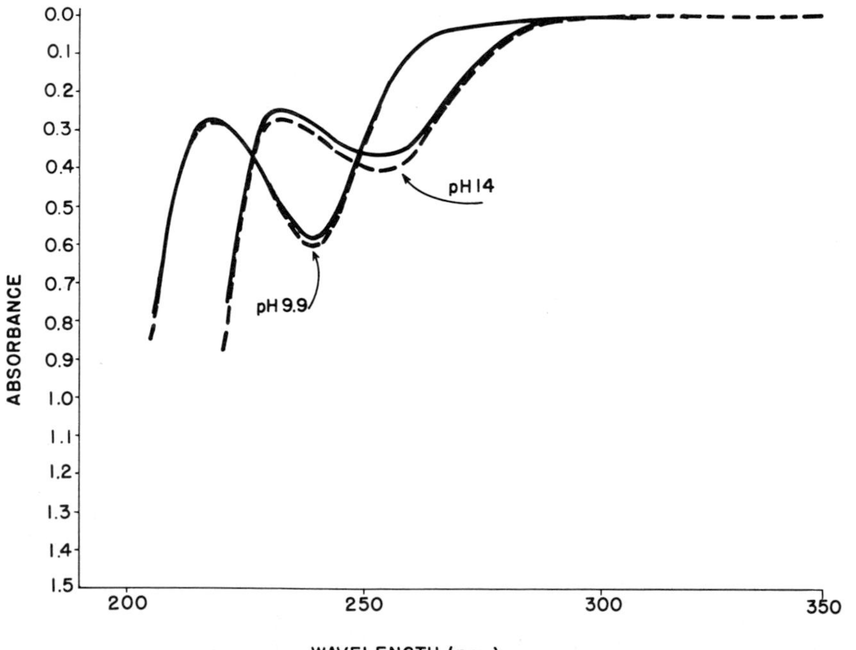

Figure 16-3. Typical ultraviolet absorption curve of secobarbital, a fast acting barbiturate. Dashed curve, alkaline extract; solid curve, hydrolyzed alkaline extract.

* Isosbestic points are points on a spectral curve at which two substances have equal absorbance at the same wavelength.

drug; (2) if the blood level is rising, a given concentration of barbiturate has a more profound effect than if the blood level is falling; (3) if other depressant drugs, particularly alcohol, are present in addition to the barbiturates, the effects observed in the patient will be more severe; (4) some individuals, who are tolerant or addicted to barbiturates, may have high blood levels without serious effects.

Table 16-5 shows approximate blood barbiturate levels at which coma disappears (falling blood levels). At levels higher than those indicated in Table 16-5, the prognosis is poor since fatalities have occurred at these levels. Except in epileptics treated with phenobarbital, normal therapeutic blood levels rarely rise higher than 0.1 to 0.2 mg./100 ml.

REFERENCE

Broughton, P. M. G.: Biochem. J., *63*:207, 1956.

OTHER METHODOLOGY

Numerous ultraviolet spectrophotometric methods have been published.[45,48] Barbiturates may be identified by infrared spectrophotometry,[53] gas chromatography,[36] paper chromatography,[11] and thin-layer chromatography.

Identification of Barbiturates by Thin-Layer Chromatography

The theory and principle of thin-layer chromatography are discussed in a number of books and articles.[7,44,50] Recently de Zeeuw[64] has shown that the use of unsaturated chambers results in better separations when multicomponent solvents are used. This is due to vapor adsorption on the dry adsorbent, which results in a gradient of the more polar vapor component, showing increasing polarity from the bottom to the top of the plate. This gradient results in better separation of the various barbiturates examined.

PROCEDURE

TLC apparatus

1. Tank chambers. Use rectangular glass jars, 21 × 21 × 9 cm. The top is a glass plate, sealed in place with a silicone grease.

2. Thin-layer plates. Coat 20 cm. square glass plates with a slurry prepared by making a smooth, homogenous mixture of 30 gm. Silica Gel GF 254 (E. Merck Co., c/o Brinkmann Instruments, Westbury, N.Y.) in 60 ml. distilled water. This is sufficient mixture to coat five plates. Apply the coating in a thickness of 250 μ, air dry for 15 minutes, heat at 110°C. for 30 minutes, cool and store in a desiccator.

REAGENTS

1. Standards. Prepare solutions of barbiturates in the acid (nonionized) form in chloroform at a concentration of 0.2 gm./100 ml. These standards are stable if stored in a well stoppered container, refrigerated, and protected from light.

2. Developing solvent. Mix 90 ml. of chloroform with 10 ml. of acetone. This solvent should be made fresh daily or after three runs.

3. Spray reagents.

a. Mercurous nitrate, 1 gm./100 ml. in distilled water.

b. Mercuric sulfate. Suspend 5.0 gm. of red HgO in 100 ml. distilled water. Carefully add, while mixing, 20 ml. of concentrated H_2SO_4. Cool and dilute to 250 ml. with distilled water.

c. Potassium permanganate, 0.2 per cent in distilled water.

d. Diphenylcarbazone. Dissolve 50 mg. of diphenylcarbazone in 500 ml. chloroform. Store in amber glass bottle in the refrigerator.

e. Alkaline fluorescein. Slowly add 20 gm. of NaOH pellets, with stirring, to 500 ml. distilled water. Add 20 mg. of sodium fluorescein. Mix well and store in glass-stoppered bottle.

PROCEDURE

1. Shake 5 to 10 ml. of oxalated blood with 30 ml. of $CHCl_3$ in a separatory funnel. Draw off the $CHCl_3$ layer and filter through a Whatman No. 3 dry filter paper into an evaporating dish. Repeat the extraction two more times.
2. Evaporate the combined $CHCl_3$ extracts to dryness. Redissolve the dry residue in 0.5 ml. of $CHCl_3$.
3. Apply 0.01 ml. of sample to a prepared and activated plate in four or five separate spots. Alternate several standards between the sample spots.
4. After drying, place the plate in the tank immediately after adding the developing solvent. Allow the solvent front to run 10 cm. (about 45 to 60 minutes).
5. Remove the plate, dry and spray a pair of sample and standard spots (by covering other spots with another plate) as follows:

a. Mercuric sulfate followed by diphenylcarbazone. First, spray with mercuric sulfate. After partially drying, spray with diphenylcarbazone. A violet color develops over the whole plate. This color fades gradually except for the areas containing the barbiturates, which remain violet for 5 to 10 minutes.

b. Mercurous nitrate. After spraying with this reagent, barbiturates appear as gray-white spots on a gray background.

c. Potassium permanganate. This spray detects thiobarbiturates or barbiturates with unsaturated substituents. Spots will show a yellow color against a purple background. This stain may be used alone or after mercurous nitrate.

d. Alkaline fluorescein. After spraying, view the plate in a dark room with a 254 nm. ultraviolet light. Barbiturates show as a purple spot against a yellow-green background. This may be followed by the $KMnO_4$ spray.

6. Mark the spots with a sharp writing instrument and measure the R_f values as compared with standards.

R_f Values

Barbiturate	*R_f*
Heptobarbital	0.21
Phenobarbital	0.31
Cyclobarbital	0.37
Allobarbital	0.39
Butobarbital	0.42
Pentobarbital	0.44
Secobarbital	0.52
Hexobarbital	0.62
Methylphenobarbital	0.73

Some of the barbiturates cannot be satisfactorily resolved by this solvent system, or other solvent systems that have been studied. Secobarbital and other barbiturates with unsaturated side chains can be differentiated by using the $KMnO_4$ spray.

REFERENCE

de Zeeuw, R. A.: Anal. Chem., *40*:915, 1968.

GLUTETHIMIDE (DORIDEN)

Glutethimide, a nonbarbiturate hypnotic, has been prescribed frequently in recent years. As a result, attempted and successful suicides, as well as abuse of this drug, have been reported with increasing frequency.

Chemically this drug is an imide and therefore is a very weak acid. Its acidic property is so weak, however, that it is usually extracted in the neutral fraction.

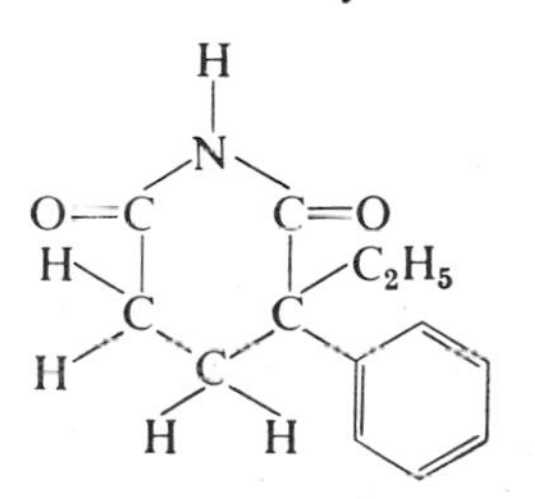

Glutethimide
2-ethyl-2-phenylglutarimide

Glutethimide has an absorption maximum in alkaline solution at 235 nm. Unfortunately, it is very unstable in alkaline solution and is easily hydrolyzed to the substituted monoamide of glutaric acid, thereby losing its ultraviolet absorbing property:

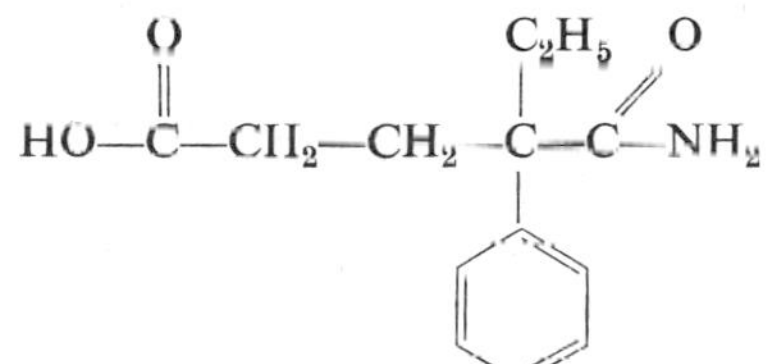

Goldbaum[25] utilized this property of glutethimide and, under rigidly controlled conditions, measured the rate of hydrolysis by plotting ultraviolet absorption at 235 nm. against time. By extrapolating back to zero time, it was possible to calculate the original concentration of the drug. This method, with modifications, was the only good procedure available for many years. It is somewhat time-consuming, however, and requires some experience in order to obtain reproducible results. Korzun et al.[31] reported a simple chromatographic procedure to identify glutethimide in blood. This procedure (to be described) is rapid, does not require any elaborate equipment, is sensitive, and uses 2.0 ml. of blood. It is capable of producing semiquantitative results by visually comparing the spot size of the unknown drug in the blood with the spot size of a known graded series of glutethimide standards.

Determination of Glutethimide by Thin-Layer Chromatography (TLC)

PRINCIPLE

Blood is extracted with chloroform and the residue, after evaporation of the solvent, is chromatographed by the thin-layer technique, which enables glutethimide to be detected.

REAGENTS

1. All solvents are analytical grade chemicals.
2. Developing solvent. Mix 95 ml. of benzene and 5 ml. of acetone.

3. Spray reagent. Prepare a 1 per cent aqueous solution of mercurous nitrate ($HgNO_3 \cdot H_2O$). Shake the solution well; all the $HgNO_3$ does not dissolve. The reagent must be freshly prepared.

4. Eastman Chromagrams, type K301R, 20 × 20 cm. (Eastman Kodak Co., Rochester, N.Y.).

5. Glutethimide standard. Dissolve 10.0 mg. of Doriden (Ciba Pharmaceutical Co., Summit, N.J.) in 25 ml. of chloroform. Make fresh daily.

PROCEDURE

1. Pipet 2.0 ml. of blood into a 60 ml. separatory funnel.

2. Add 10 ml. of chloroform and shake vigorously for 5 minutes.

3. Filter the chloroform layer (lower layer) into a 15 ml. centrifuge tube and wash the filter and funnel, each with 0.5 ml. of chloroform.

4. Evaporate the solvent from the centrifuge tube with the aid of a stream of nitrogen or air while warming the tube in a beaker of warm water (30 to 35°C.).

5. Take up the residue in 0.25 ml. chloroform.

6. Cut an Eastman Chromagram to a 10 × 20 cm. size and use without activation. Mark points for application of the samples with a pencil along the 20 cm. length, 1 cm. from the edge and 2 cm. apart. Mark the length of solvent travel by drawing a line across the top, 8.5 cm. from the point of the sample application. Nine samples may be spotted on one plate.

7. Before spotting, place 30 ml. of the developing solvent (benzene-acetone, 95:5, v/v) into a 600 ml. beaker and cover with Saran Wrap.

8. Spot 0.025 ml. of the prepared blood extract sample on the Chromagram.

9. Apply the glutethimide standard to five other points in volumes of 0.005, 0.01, 0.02, 0.03, and 0.04 ml. The size of the spot should be uniform and less than 1 cm. in diameter.

10. After spotting, scrape off enough adsorbent from the 10 cm. edges so that approximately 1 mm. of the film backing is exposed. Bend the Chromagram so that its uncoated surface is placed in contact with the inner surface of the developing beaker, with the coated side facing into the beaker, and the bottom edge dipping into the solvent. Cover the beaker with Saran Wrap and allow the solvent to ascend to the top line. Then remove the Chromagram and air dry it.

11. Pour the detecting reagent (mercurous nitrate) into a pie plate, large Petri dish, or other suitable vessel. With the adsorbent facing down, gently draw the Chromagram through the solution. Do not allow the Chromagram to remain in the solvent more than momentarily because this will dissolve the binder and the adsorbent will fall off.

12. After dipping, place the Chromagram face up on a sheet of white paper and examine. Glutethimide produces a grey-black spot with the reagent. Compare the blood extract with the glutethimide standards. If glutethimide is present, it will appear at the same distance from the point of application as the standard has migrated. If glutethimide is detected in the blood sample, compare it with the glutethimide standards for size, color, and intensity of the spots. Table 16-7 lists glutethimide standard applied and the amount it represents in mg./100 ml. of blood.

Figure 16-4 is an example of the completed chromatogram with glutethimide, and Figure 16-5 shows a separation of glutethimide, barbiturates, and other drugs.

Cyclohexane-triethylamine (9:1, v/v) is another solvent system used in the separation of some common drugs from each other. After the solvent has traveled to

TABLE 16-7.

Volume Applied	*Amount of Glutethimide*	*mg. of Glutethimide/ 100 ml. of blood*
0.005 ml.	0.002 mg.	1 mg./100 ml.
0.010 ml.	0.004 mg.	2 mg./100 ml.
0.020 ml.	0.008 mg.	4 mg./100 ml.
0.030 ml.	0.012 mg.	6 mg./100 ml.
0.040 ml.	0.016 mg.	8 mg./100 ml.

the top line, remove the Chromagram, air dry, and then replace into the chamber for a second development. After the second development, remove the Chromagram and dry in an oven (90°C.) for 30 minutes to remove the triethylamine. First view it under a short wave (254 nm.) ultraviolet lamp, then dip it into the mercurous nitrate reagent. The results are shown in Figure 16-6.

INTERPRETATION

Therapeutic blood levels of glutethimide range up to about 0.5 mg./100 ml. Levels higher than this usually indicate that an overdose has been ingested. Deep unconsciousness or coma may be produced at levels of 3.0 mg./100 ml., although

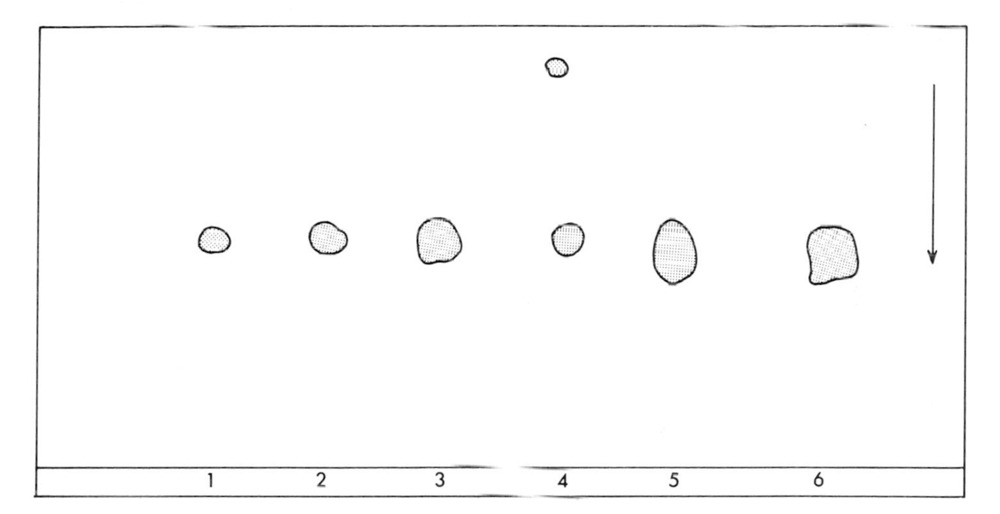

Figure 16-4. Example of glutethimide standards and blood extract on a chromatogram.

Position on Chromatogarm	*Type of Sample Applied*	*Amount Applied in μl.*	*Amount Applied in μg.*	*Corresponding Concentration*
1	Glutethimide	5 μl.	2 μg.	1 mg./100 ml. blood
2	Glutethimide	10 μl.	4 μg.	2 mg./100 ml. blood
3	Glutethimide	20 μl.	8 μg.	4 mg./100 ml. blood
4	Blood extract	25 μl.	2 μg.	approx. 1 mg./ 100 ml. blood
5	Glutethimide	30 μl.	12 μg.	6 mg./100 ml. blood
6	Glutethimide	40 μl.	16 μg.	8 mg./100 ml. blood

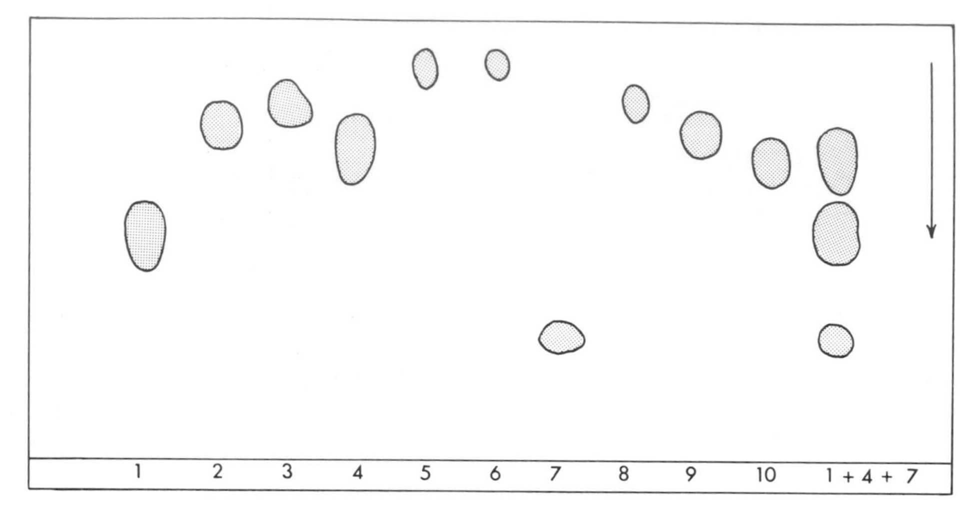

Figure 16-5. Chromatogram showing separation of glutethimide from other drugs and barbiturates: 1, glutethimide (Doriden); 2, Dial; 3, phenobarbital; 4, secobarbital; 5, meprobamate (Miltown); 6, chlordiazepoxide (Librium); 7, ethchlorvynol (Placidyl); 8, Noludar; 9, pentobarbital (Nembutal); 10, Megimide; 1 + 4 + 7—mixture of glutethimide, secobarbital and ethchlorvynol.

higher levels than this have been reported. Blood levels at death have been reported between 3 and 20 mg./100 ml.

Frequently, patients may have several central nervous system depressant drugs prescribed. The possibility of overdosage of drug combinations thus presents itself and interpretation of analytical results is very difficult in these situations. Combinations of glutethimide and alcohol or glutethimide and barbiturates are most common but almost every type of combination has been encountered. As a general rule, if the depth of coma cannot be reasonably explained by the analytical results, a drug combination should be suspected. If drugs available to the patient are not

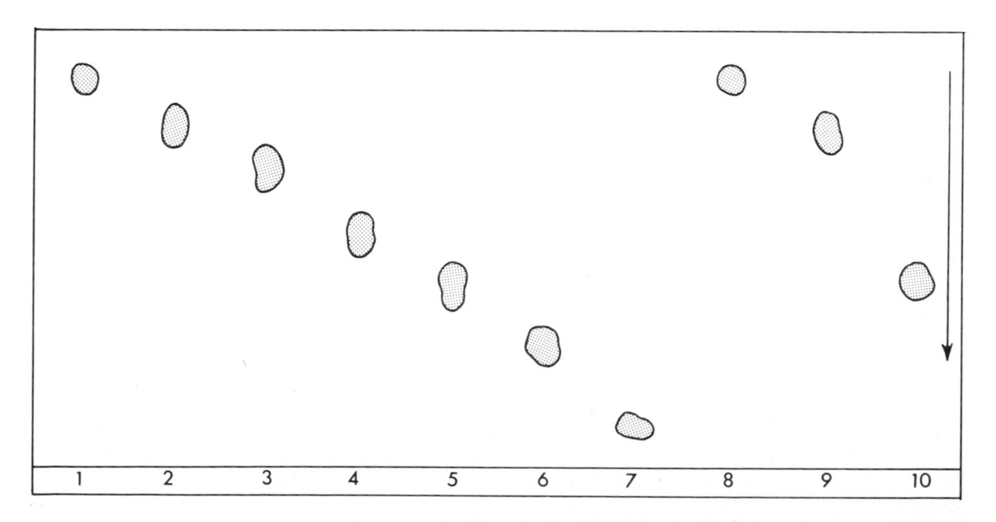

Figure 16-6. Chromatogram developed in cyclohexane-triethylamine (9:1) showing separation of the different drugs from each other: 1, meprobamate (Miltown); 2, phenobarbital; 3, Noludar; 4, Dial; 5, secobarbital; 6, glutethimide (Doriden); 7, ethchlorvynol (Placidyl); 8, chlordiazepoxide (Librium); 9, Megimide; 10, pentobarbital (Nembutal).

known, alcohol should be ruled out followed by barbiturates, chlordiazepoxide, meprobamate, and phenothiazine derivatives.

Although the procedure for glutethimide described here can be used on a semi-quantitative basis only, some of the other central nervous system depressant drugs may be detected at the same time as a further aid to the clinician. With some knowledge of possible drug combinations, a more rational approach to treatment of drug overdosage can be made.

REFERENCE

Korzun, B. P., Brody, S. M., Keegan, P. G., Luders, R. C., and Rehm, C. R.: J. Lab. Clin. Med., *68*:333, 1966.

PHENOTHIAZINE DERIVATIVES

Presently, there are a number of tranquilizers and antihistamines in clinical use that have in common the phenothiazine ring structure:

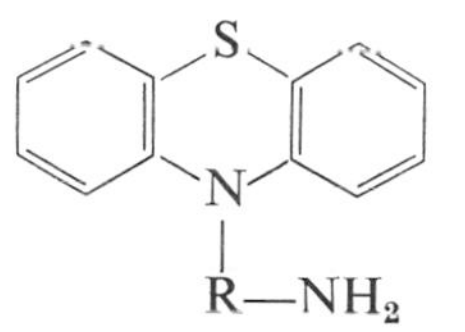

Detection of Phenothiazine Derivatives

These substances can be detected easily in urine by the use of "FPN (ferric-perchloric-nitric) reagent." Although not specific for this ring structure, the test is sensitive, and can detect the drug at therapeutic levels. It is a useful screening test since false negative results do not occur. Elevated urinary levels of bile metabolites, particularly urobilinogen, yield color reactions similar to the drugs. Phenylpyruvic acid, and p-aminosalicylic acid, as well as natural conjugated or synthetic estrogens, may give false positive results. These false positive reactions are generally of very weak intensity.

REAGENTS

1. Ferric chloride, 5 per cent (w/v). Not stable, use immediately to prepare FPN reagent.
2. Perchloric acid, 20 per cent (w/v). The solution is stable.
3. Nitric acid, 50 per cent (w/v). It is stable.
4. FPN reagent. Mix 5.0 ml. of 5 per cent ferric chloride, 45 ml. of 20 per cent perchloric acid and 50 ml. of 50 per cent nitric acid. The solution is stable.

PROCEDURE

Add 1.0 ml. of FPN reagent to 1.0 ml. of urine. The color is observed immediately. The stability of the color varies from 10 seconds at low dosages up to several minutes for high dosages.

INTERPRETATION

With a drug intake of 5 to 20 mg./day, a light pink-orange is observed. A drug intake of 25 to 70 mg./day produces more intense shades of pink. Doses of 75 to

120 mg./day produce gradually more intense shades of violet, and with doses above 125 mg./day deep purple colors are noted.

The value of this test is that a negative result effectively eliminates this group of drugs as a factor in diagnosing the condition of the patient. If the test is positive, several possibilities are suggested. First, the original author's interpretation (see preceding paragraph) is valid. Second, if considerable time has elapsed between possible drug ingestion and collection of the urine specimen (e.g., 24 to 48 hours), an overdosage is indicated since a therapeutic dose would have been excreted in this time period. Finally, the presence of an interfering substance is possible. To resolve these possibilities, one must use ultraviolet spectrophotometry[51] or thin-layer chromatography.[8]

SALICYLATES

Aspirin is responsible for more cases of accidental poisonings in children than any other substance. This extremely useful analgesic is so widely used and readily available (and carelessly handled) that children frequently ingest a toxic quantity by eating the flavored tablets like candy or mimicking adults. Toxic doses of salicylates initially produce a stimulation of the central nervous system. This may be reflected by hyperventilation, flushing, and fever. Unfortunately, an unrecognized case of salicylate poisoning may be thought to be a case of infection and further aspirin given in a vain attempt to control the fever. Central nervous system stimulation is followed by depression.

A complex disturbance of acid-base balance results from severe hyperventilation. Initially a respiratory alkalosis occurs, but this may be followed, especially in infants, by a metabolic acidosis. The net effect may be a decrease in the blood pH.

Determination of Salicylates in Biological Fluids

The procedure to be described for determining salicylate in urine, serum, or other specimen[52] is based on the formation of a violet colored complex between ferric iron and phenols.

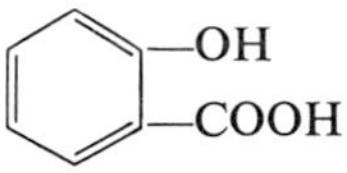

$$C_6H_4(OH)(COOH) + Fe^{+++} = \text{Violet color complex}$$

This test is not specific for salicylates but false negative results do not occur. The color developing solution contains acid and mercuric ions to precipitate protein.

REAGENTS

1. Color reagent. Dissolve 40 gm. of mercuric chloride, A.R., in 850 ml. of water by heating. Cool the solution and add 120 ml. of 1 *N* HCl and 40 gm. of ferric nitrate ($Fe(NO_3)_3 \cdot 9H_2O$). When all the ferric nitrate has dissolved, dilute the solution to 1 L. with water. It is stable indefinitely.

2. Salicylate standard, stock. Dissolve 580.0 mg. of sodium salicylate (500 mg. salicylic acid) in water and dilute to 250 ml. Add a few drops of chloroform as a preservative. This solution contains 2.0 mg. salicylic acid/ml. Store in refrigerator; it is stable for about 6 months.

3. Working standard. Dilute 25.0 ml. of stock salicylate solution to 100.0 ml. with water. Add a few drops of chloroform as a preservative. This solution contains 0.5 mg. of salicylic acid/ml. Store in refrigerator; the solution is stable for about 6 months.

PROCEDURE FOR SALICYLATE DETERMINATION IN CEREBROSPINAL FLUID, SERUM, OR WHOLE BLOOD

1. Pipet 1.0 ml. of specimen into a centrifuge tube and add 5.0 ml. of color reagent while shaking the tube. Continue shaking until the precipitate is finely dispersed.
2. Centrifuge at 2000 rpm for 2 minutes or filter through Whatman No. 42 filter paper.
3. Transfer the clear supernatant or filtrate to a 1.0 cm. cuvet and read at 540 nm. against a blank consisting of 1.0 ml. water mixed with 5.0 ml. of color reagent. The color is stable for one hour.
4. Prepare a series of standards as follows:

	A	*B*	*C*	*D*	*E*
ml. Working standard	0.2	0.4	0.6	0.8	1.0
ml. Water	0.8	0.6	0.4	0.2	0.0
mg. Salicylate/100 ml.	10.0	20.0	30.0	40.0	50.0

Run these standards as described above (steps 1 to 3), reading the absorbancies at 540 nm. against the reagent blank and plot a calibration curve. Beer's law is followed over this range.

5. If the unknown absorbance is greater than 0.7, repeat the analysis using a smaller portion of specimen diluted to 1.0 ml. with water.

PROCEDURE FOR SALICYLATE DETERMINATION IN URINE

1. Follow the procedure as outlined under serum. If the urine contains more than 50 mg./100 ml. salicylic acid, make an appropriate dilution of urine with distilled water and repeat the test.
2. After reading the absorbance of the unknown, obtain a sample blank reading by setting the instrument with water and reading the absorbance of a solution prepared by mixing 1.0 ml. of urine or diluted urine with 5.0 ml. of color reagent and 0.1 ml. of syrupy phosphoric acid (sp. gr. 1.75), using the same cuvets as before. Urine solutions may not require centrifuging. If it is necessary to clarify the solutions, centrifuge both unknown and urine sample blank.

CALCULATION

Urine salicylic acid (mg./100 ml.) = (mg./100 ml. in diluted unknown − mg./100 ml. in diluted sample blank) × dilution factor

INTERPRETATION

Blank values for serum, cerebrospinal fluid, and plasma are less than 1.1 mg./100 ml. as salicylic acid. For blood, the blank is less than 2.0 mg./100 ml. and for urine less than 4.5 mg./100 ml. Recoveries of added salicylate are quantitative and the following substances in the indicated concentrations do not interfere: phosphate (100 mg./100 ml.), bilirubin (20 mg./100 ml.), phenol (25 mg./100 ml.), heparin (10,000 I.U.), glucose (1000 mg./100 ml.), and urea (1000 mg./100 ml.). Acetoacetic acid forms a pink color with ferric iron, and at a level of 50 mg./100 ml. gives a value of 1 mg./100 ml. as salicylate. The procedure can easily be adapted to micro- or ultramicroscale, a useful feature in pediatric cases.

Therapeutic levels of salicylic acid rarely rise above 20 mg./100 ml. in blood or serum. Above 30 mg./100 ml., toxic symptoms such as headache, tinnitus, flushing, and hyperventilation may be seen. Serum electrolytes should be followed and any imbalance corrected. Lethal salicylate levels are usually greater than 60 mg./100 ml.

REFERENCE

Trinder, P.: Biochem. J., *57*:301, 1954.

SULFONAMIDES

This very useful group of compounds has been replaced in recent years by other antibiotics; nevertheless, for certain types of infections, sulfonamides are still useful and still occupy an important place in medicine. Although overdosage by these drugs is not common, their availability as well as the availability of a sensitive procedure for their detection makes it desirable to include a method for the determination of these drugs.

Determination of Sulfonamides

The procedure to be described uses the aromatic amino group of the sulfonamides for the production of a dye that can be measured colorimetrically. The amino group is diazotized with nitrous acid, the excess nitrous acid is destroyed by the addition of sulfamate, and the diazo compound coupled with N-(1-naphthyl)ethylenediamine to produce an intense color. This enables the original drug to be estimated in dilutions greater than 1 ppm.

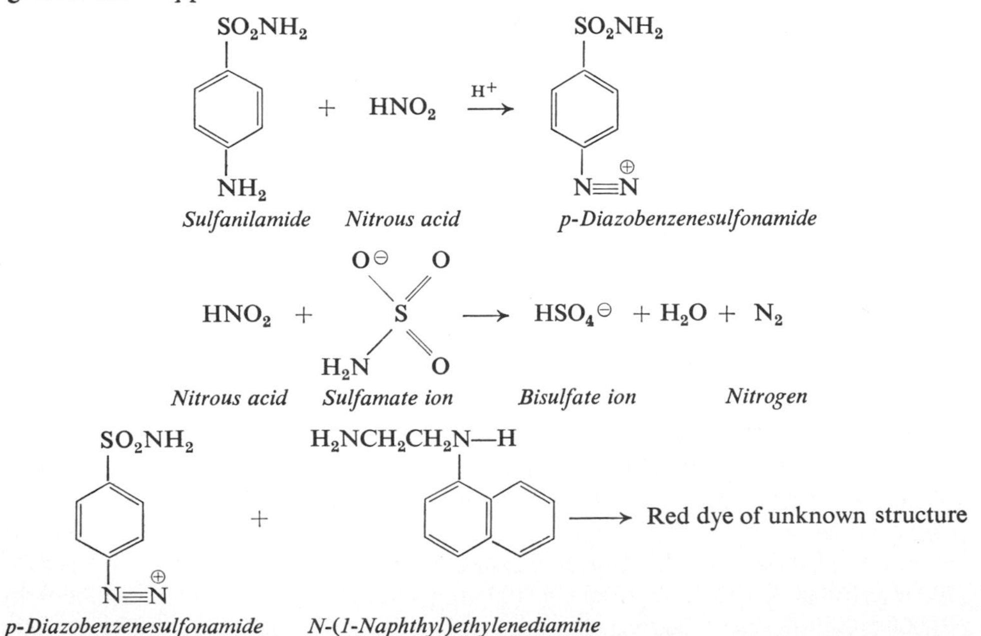

Metabolism of the sulfonamides results in acetylation of the free amino group. This prevents coupling of the compound to form the dye:

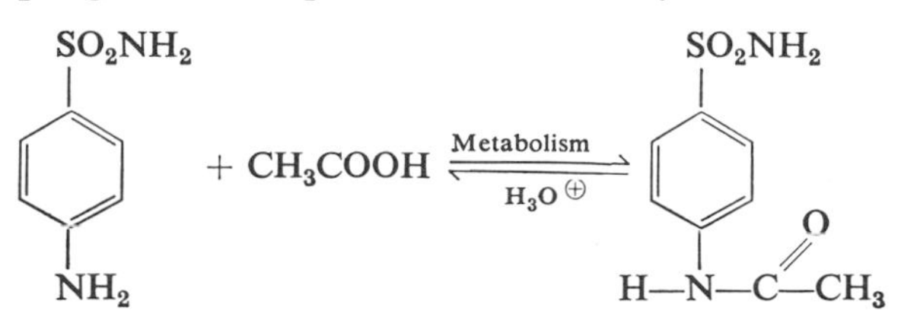

Thus, in order to estimate total sulfonamide, it is necessary to split the combined form by acid hydrolysis. This step can be introduced; however, the procedure to be described estimates only free sulfonamide, which is usually all that is necessary. It should be emphasized that any aromatic amine that can be diazotized may interfere with this test.

REAGENTS

1. Trichloroacetic acid, 15 per cent (w/v).
2. Sodium nitrite solution, stock. Prepare a 5 per cent solution (w/v) by dissolving 5.0 gm. of $NaNO_2$ in distilled water and make up to 100 ml. This solution is stable when stored in the refrigerator.
3. Sodium nitrite working solution. Just before use, dilute 1.0 ml. of stock solution to 50 ml. This 0.1 per cent solution is not stable.
4. Ammonium sulfamate, 0.5 per cent (w/v), stable.
5. N-(1-Naphthyl)ethylenediamine dihydrochloride, 0.1 per cent (w/v). For maximum stability, keep in dark bottle in refrigerator. Stable for 1 to 2 weeks.
6. Standard sulfanilamide solutions.

   a. Stock. Dissolve 20.0 mg. of sulfanilamide in and dilute to 100 ml. with water (1.0 ml. = 0.2 mg.).

   b. High working standard. Dilute 1.0 ml. of stock solution plus 20 ml. of 15 per cent trichloroacetic acid to 100 ml. (1.0 ml. = 0.002 mg.).

   c. Low working standard. Dilute 10.0 ml. of high working standard plus 20 ml. of 15 per cent trichloroacetic acid to 100 ml. with water (1.0 ml. = 0.0002 mg.).

PROCEDURE

1. To 15.9 ml. water, add 0.1 ml. of blood with a TC pipet. Mix to lake the cells.
2. Add 4 ml. of 15 per cent trichloroacetic acid, shake, and filter through Whatman No. 42 filter paper. (Filtrate must be clear.)
3. Prepare four tubes as follows:
   a. Blank: 8 ml. distilled water, 2 ml. 15 per cent trichloroacetic acid.
   b. Standard: 10 ml. high working standard.
   c. Standard: 10 ml. low working standard.
   d. Sample: 10 ml. filtrate (step 2).
4. To each tube add 1.0 ml. of 0.1 per cent $NaNO_2$ and shake. This is to diazotize the p-amino group. Let stand for 3 minutes.
5. Add 1.0 ml. of 0.5 per cent ammonium sulfamate to destroy excess nitrous acid. Shake and let stand 2 minutes.
6. Add 1.0 ml. of N-(1-naphthyl)ethylenediamine solution to form the dye. Mix and let stand 5 minutes. Read at 540 nm. against the reagent blank. The color is stable for 2 hours.

CALCULATION

Use the standard closest to the unknown.

High working standard:

$$\frac{A \text{ unknown}}{A \text{ standard}} \times 40 \times F = \text{mg./100 ml.}$$

Low working standard:

$$\frac{A \text{ unknown}}{A \text{ standard}} \times 4 \times F = \text{mg./100 ml.}$$

The factor F varies with the type of sulfonamide being measured as shown in the table below.

Drug	*Mol. Wt.*	*F*
Sulfanilamide	172	1.0
Sulfathiazole	255	1.48
Sulfadiazine	250	1.45
Sulfamerazine	264	1.54
Sulfasuxidine	355	2.06
Sulfathalidine	403	2.34
Sulfisoxazole	267	1.55
Sulfisomidine	281	1.63
Sulfaguanidine	218	1.27
Sulfapyridine	249	1.45

The standards may also be prepared using the sulfa drug being measured. In this case the factor is 1.0.

INTERPRETATION

Sulfonamides are normally not found in blood. Therapeutic levels are in the order of about 10 mg./100 ml.

Since this method will give positive results with any aromatic amine, other methods must be used to identify the drug more specifically. Ultraviolet spectrophotometry[51] and thin-layer chromatography[50] are convenient techniques for identifying unknown drugs.

COMMENTS

One side effect of this group of drugs is the production of methemoglobin under certain conditions. The mechanism of this effect and the measurement of methemoglobin is discussed in Chapter 6, Hemoglobin and Related Compounds. It should be emphasized, however, that a large number of chemical substances in addition to sulfonamides can produce this abnormal pigment.

REFERENCE

Bratton, A. C. and Marshall, E. K., Jr.: J. Biol. Chem. *128*:537, 1939.

DRUGS DETERMINED BY DERIVATIVE FORMATION

Some drugs require special techniques for their detection. It is beyond the scope of this chapter to list special techniques in detail, but an interesting approach to this problem has been the work of Wallace and co-workers.[60] This approach has been applied to drugs that can easily be extracted or separated from biological specimens, but which have weak ultraviolet absorption maxima and can not be readily detected. Wallace and his group found ways in which these drugs can be converted to derivatives that have strong, well defined absorption maxima. The drugs can therefore be determined indirectly by ultraviolet spectrophotometry of their derivatives.

In general, the procedures can be outlined as follows:

Separation of drug from specimen (extraction or steam distillation) → Derivative formation (usually an oxidation reaction)

Separation of derivative (extraction or steam distillation) → Concentration if necessary (vacuum evaporation) → Ultraviolet scan

Drugs for which this approach has been successfully used are: ethchlorvynol,[60] propoxyphene,[58] ephedrine,[56] amitriptyline,[59] and diphenylhydantoin.[57] The interested reader is referred to the original references for specific details regarding each procedure.

Miscellaneous Substances

There are always those substances that cannot be categorized. Many toxic substances (e.g., snake and insect venoms) are complex mixtures that have not been completely characterized chemically. Others (e.g., cholinesterase inhibitors) act by inhibiting enzyme systems and their effects can be more reliably estimated by determining enzyme activity. Many pesticides, some of the so-called "nerve gases," and some drugs fall into this category.

On the other hand, there are many substances that could be put into one or the other of the categories because of their chemical type, but for some other reason cannot be tested for in the usual way and must be sought by special, specific tests. Organic substances that are extremely inert or extremely insoluble are examples of this type.

Finally, the miscellaneous group includes the phenomenon of drug sensitivity or allergy. In this situation, an individual may succumb to a normal therapeutic dose of a drug, or less. In this case it is necessary to demonstrate that true sensitivity to a chemical agent did exist. For this the techniques and tools of serology are brought into play.

Fortunately, cases involving poisonings in this group are extremely rare. Considerable study and research are needed in this area since the difficulties encountered have discouraged many workers. The problems associated with members of the miscellaneous category are too numerous to be discussed here. The interested student is referred to more comprehensive texts.[46]

REFERENCES

1. Björling, C. O., Berggren, A., and Nygord, B.: Acta Chem. Scand., *16*:1481, 1962.
2. Blanke, R. V.: J. Forensic Sci., *1*:79, 1956.
3. Bratton, A. C., and Marshall, E. K., Jr.: J. Biol. Chem., *128*:537, 1939.
4. Broughton, P. M. G.: Biochem. J., *63*:207, 1956.
5. Bulletin: National Clearing House for Poison Control Centers. Jan.–Feb., 1965.
6. Burke, A.: Personal communication.
7. Cochin, J., and Daly, J. W.: J. Pharmacol. Exp. Ther., *139*:154, 1963.

8. Cochin, J., and Daly, J. W.: J. Pharmacol. Exp. Ther., *139*:160, 1963.
9. Conway, E. J.: Microdiffusion Analysis and Volumetric Error. 5th ed. London, Crosby Lockwood & Son Ltd., 1962.
10. Curry, A. S.: Brit. Med. J., *2*:1040, 1963.
11. Curry, A. S.: Poison Detection in Human Organs. 2nd ed. Springfield, Ill., Charles C Thomas, Publisher, 1969.
12. Curry, A. S.: *In* Toxicology—Mechanisms and Analytical Methods. C. P. Stewart and A. Stolman, Eds. New York, Academic Press, Inc., 1961, vol. 2, p. 185.
13. Curry, A. S.: Brit. Med. J., *1*:354, 1964.
14. Deichmann, W. B., and Gerarde, H. W.: Symptomatology and Therapy of Toxicological Emergencies. New York, Academic Press, Inc., 1964.
15. de Langen, C. D., and ten Berg, J. A. G.: Acta Med. Scand., *130*:37, 1948.
16. Dole, V. P., Kim, W. K., and Eglitis, I.: J.A.M.A., *198*:349, 1966.
17. Feldstein, M.: *In* Toxicology—Mechanisms and Analytical Methods. C. P. Stewart, and A. Stolman, Eds. New York, Academic Press, Inc., 1960, vol. 1, chap. 16.
18. Feldstein, M., and Klendshoj, N.: J. Forensic Sci., *2*:39, 1957.
19. Forrest, I. S., and Forrest, F. M.: Clin. Chem. *6*:11, 1960.
20. Gant, V.: Industr. Med., *7*:608, 1938; *7*:679, 1938.
21. Gerarde, H. W., and Skiba, P.: Clin. Chem. *6*:327, 1960.
22. Gettler, A. O., and Kaye, S.: J. Lab. Clin. Med., *35*:146, 1950.
23. Gleason, M. N., Gosselin, R. E., Hodge, H. C. and Smith, R. P. : Clinical Toxicology of Commercial Products. 3rd ed. Baltimore, The Williams & Wilkins Co., 1969.
24. Goldbaum, L. R., Schloegel, E. L., and Dominguez, A. M.: *In* Progress in Chemical Toxicology. A. Stolman, Ed. New York, Academic Press, Inc. 1963, vol. 1, chap. 1.
25. Goldbaum, L. R., Williams, M. D., and Koppanyi, T.: Anal. Chem., *32*:81, 1960.
26. Hepler, O. E.: Manual of Clinical Laboratory Methods. 4th ed. Springfield, Ill., Charles C Thomas, 1963, p. 325.
27. Hindberg, J., and Wieth, J. O.: J. Lab. Clin. Med., *61*:355, 1963.
28. Hirsh, J., Zander, H. L., and Drolette, B. M.: Arch. Environ Health, *3*:212, 1961.
29. Klendshoj, N. C., Feldstein, M., and Sprague, A. L.: J. Biol. Chem., *183*:297, 1950.
30. Kopito, L., and Schwachman, H.: J. Lab. Clin. Med., *70*:326, 1967.
31. Korzun, B. P., Brody, S. M., Keegan, P. G., Luders, R. C., and Rehm, C. R.: J. Lab. Clin. Med., *68*:333, 1966.
32. Lundquist, F.: *In* Methods of Biochemical Analysis. Glick, Ed. New York, Interscience Publishers Inc. 1959, vol. 7, p. 217.
33. McCord, C.: Industr. Med., *20*:185, 1951.
34. Nickolls, L. C.: Analyst, *85*:840, 1960.
35. Official Methods of Analysis of the Association of Official Agricultural Chemists. 10th ed. 1965, p. 357. American Association of Official Agricultural Chemists (Publ.).
36. Parker, K., Fontan, C. R., and Kirk, P. L.: Anal. Chem., *35*:418, 1963.
37. Pierce, J. O., and Cholak, J.: Arch. Environ. Health, *13*:208, 1966.
38. Rieders, F.: Ann. N.Y. Acad. Sci., *111*:591, 1964.
39. Rowley, R. J., and Farrah, G. H.: Am. Industr. Hyg. Assn. J., *23*:314, 1962.
40. Sainsbury, P.: Suicide in London: An Ecological Study. New York, Basic Books, Inc., Publishers, 1956.
41. Scherer, J.: Ann. Chem. Phys., *112*:214, 1859.
42. Schwartz, S., Zieve, L., and Watson, C. J.: J. Lab. Clin. Med., *37*:843, 1951.
43. Smith, H. and Lenihan, J. M. A.: *In* Methods of Forensic Science. A. S. Curry, Ed. New York, Interscience Publishers Inc. 1964, vol. 3.
44. Stahl, E.: Thin Layer Chromatography. Berlin, Springer-Verlag, 1965.
45. Stevenson, G. W.: Anal. Chem., *33*:1374, 1961.
46. Stewart, C. P. and Stolman, A. (Eds.): Toxicology—Mechanisms and Analytical Methods. New York, Academic Press, Inc., 1960, vol. 1; 1961, vol. 2.
47. Stewart, R. D., and Erley, D. S.: J. Forensic Sci., *8*:31, 1963.
48. Stokes, D. M., Camp, W. J. R., and Kirsch, E. R.: J. Pharmacol. Sci., *51*:379, 1962.
49. Stolman, A.: *In* Toxicology—Mechanisms and Analytical Methods. C. P. Stewart, and A. Stolman, Eds. New York, Academic Press, Inc., 1961, vol. 2, p. 787.
50. Sunshine, I.: Am. J. Clin. Path., *40*:576, 1963.
51. Sunshine, I., and Gerber, S. R.: Spectrophotometric Analysis of Drugs Including Atlas of Spectra. Springfield, Ill., Charles C Thomas, 1963.
52. Trinder, P.: Biochem. J. *57*:301, 1954.
53. Umberger, C. J., and Adams, G.: Anal. Chem., *24*:1309, 1952.
54. van Kampen, E. J., and Klouwen, H. M.: Rec. Trans. Chim. Paysbas, *73*:119, 1954.
55. Van Slyke, D. D., and Salvasen, H. A.: J. Biol. Chem., *40*:103, 1919.
56. Wallace, J. E.: Anal. Chem., *39*:531, 1967.

57. Wallace, J. E.: Anal. Chem., *40*:978, 1968.
58. Wallace, J. E., Biggs, J. D., and Dahl, E. V.: J. Forensic Sci., *10*:179, 1965.
59. Wallace, J. E., and Dahl, E. V.: J. Forensic Sci., *12*:484, 1967.
60. Wallace, J. E., Wilson, W. J., and Dahl, E. V.: J. Forensic Sci., *9*:342, 1964.
61. Watson, C. J., De Mello, R. P., Schwartz, S., Hawkinson, V. E., and Bossenmaier, I.: J. Lab. Clin. Med., *37*:831, 1951.
62. Willis, J. B.: Anal. Chem. *34*:614, 1962.
63. Zaar, B., and Gronwall, A.: Scand. J. Clin. Lab. Invest., *13*:225, 1961.
64. de Zeeuw, R. A.: Anal. Chem., *40*:915, 1968.
65. Zettner, A.: *In* Advances in Clinical Chemistry. New York, Academic Press, Inc. 1964, vol. 7, chap. 1.

Chapter 17

ANALYSIS OF CALCULI

by Ermalinda A. Fiereck, M.S.

Calculi are deposited chemicals in compact form. These concretions are frequently found in the urinary tract and gallbladder and less frequently in the salivary gland, pancreas, and prostate. The cause of calculus formation is not known; however, it is thought to be related to the solubility of the various crystalloids found in secretions or excretions.

URINARY CALCULI

Urinary calculi have received the most attention, perhaps because of their frequency and the variability of their composition. The occurrence of urinary "stones" dates back to antiquity; in recent explorations of Egyptian tombs dated from 8000 B.C. evidence of renal calculi has been found. In the impoverished areas of the world, urinary calculi formation in children is frequent; however, the incidence among European and American children has decreased in the past century. This decrease is thought to be related to an improvement in the dietary conditions among these populations. The majority of the patients requiring surgical removal of renal calculi in American hospitals are between the ages of 50 and 70.

Formation of Calculi

Winer[13] lists seven factors that may contribute to the formation of calculi: (1) metabolic disturbances such as cystinuria and gout; (2) endocrinopathies such as hyperparathyroidism; (3) urinary obstruction; (4) infections; (5) mucosal metaplasia, which occurs in vitamin A deficiency; (6) extrinsic conditions such as dehydration, dietary excess, drug excess, or chemotherapy; and (7) isohydruria, which is the loss of the normal acid-alkaline tides ("fixation of pH"). The presence of two or more of these findings is usually associated with stone formation, and, among the seven listed factors, isohydruria is found most often in association with one or more of the other findings. Fixation of pH may occur at almost any urinary pH within the physiological range. The type of calculi formed will depend upon the pH. Uric acid calculi are associated with a pH below 5.5, whereas calcium oxalate calculi occur at the pH range of 5.5 to 6.0 and calcium phosphate calculi occur at the alkaline pH range of 7.0 to 7.8. Prolonged isohydruria favors the formation of pure calculi, whereas calculi of mixed composition are found in patients showing a temporary and wide fluctuation of urinary pH.

Many calculi contain a clearly defined nucleus upon which the chemicals precipitate. Theories have been advanced as to how this is accomplished. Prien[10] states that some calculi have an indentation on one surface, with a structure in the bottom, which suggests an origin from a surface such as a renal papilla. The nuclei of stones that have a definite center are thought to arise by localized precipitation of salts from a supersaturated urine. Bacteria, blood clots, or epithelial cells may also serve as the nucleus for crystallization.

A matrix composed of mucoproteins and mucopolysaccharides is reported to be present throughout the calculus and constitutes approximately 3 per cent of its weight. According to Boyce and King,[2] this fibrous matrix surrounds the center of the calculus at a series of spaced intervals. The matrix is thought to be an essential factor in the initial phase of calculi formation and a determining factor in the position and composition of crystal deposits in calculi.

The growth of a calculus is dependent upon the pH of the urine, the solubility of each substance, and the availability of the various salts. The multiplicity of the crystalloids and colloids present and the interaction of these substances make it difficult, however, to predict the point of crystallization for the individual chemicals.

Frequently calculi have been found that contain distinct layers. Prien[10] states that calculi composed centrally of pure calcium oxalate, intermediary of calcium oxalate and apatite,* and peripherally of apatite and triple phosphate are rather frequently found. According to Prien,[10] the sequence that probably occurs is the precipitation of calcium oxalate in a sterile urine, which subsequently becomes infected with a nonurea-splitting organism. During this phase the intermediate layer is formed. After an infection with a urea-splitting organism, the urine becomes alkaline and the peripheral layer of magnesium ammonium phosphate is formed.

Urea-splitting bacteria are thought to function in stone formation by hydrolyzing urea to ammonium and carbonate ions. This raises the pH of the urine in the vicinity of the bacteria and favors the precipitation of magnesium ammonium phosphate. These bacteria or a small calculus, as previously mentioned, may serve as the nucleus upon which the precipitation of magnesium ammonium phosphate occurs. The increase in pH and additional magnesium and phosphate excretion are thought to lead to the rapid growth of the magnesium ammonium phosphate calculus. Successful treatment of the bacterial infection, with the reversal of the urinary pH to an acidic pH, may inhibit further growth of the stone. After surgical removal of the calculus, no recurrence will result if the cause was only bacterial; however, if the cause was metabolic, continued stone formation may occur.

Excessive excretion of some specific substances may contribute to the formation of urinary calculi. This excess may arise from a relative imbalance in the diet or as a consequence of a metabolic disorder, such as cystinuria. This complex metabolic defect is characterized by a decreased renal tubular reabsorption of certain amino acids, especially arginine, cystine, ornithine, and lysine, from the glomerular filtrate. Although greater quantities of lysine and arginine are excreted than cystine, cystine

* Apatite is a complex calcium phosphate sometimes containing carbonate.

calculi result, owing to the low solubility of cystine in urine. Cystine calculi occur in approximately 1 per cent of the calculi analyzed.

Uric acid calculi are rather frequently found and may occur in some cases in which there is an excess excretion of uric acid. At a pH below 5.7, uric acid is predominantly present as a free acid, which is insoluble in urine, and its presence favors the formation of uric acid calculi. Excessive excretion of uric acid may occur from a dietary excess or from tissue breakdown, such as occurs in severe systemic or neoplastic disease.

The average American diet contains considerably more calcium than is necessary for the maintenance of bones. This increased calcium load is generally not harmful and the excess calcium is excreted in the urine. Individuals who are predisposed to

TABLE 17-1. *Percentage Distribution of Chemical Constituents of Urinary Calculi*

Substance	*Prien and Frondel*[11]	*Herring*[5]	*Beeler et al.*[1]	*Kachmar*[6]
Calcium oxalate				
monohydrate	61.7	43.0	60.0	46.7% (includes $CaCO_3$)
dihydrate	46.0	60.1	37.0	
Apatite and tricalcium phosphate	20.8	61.8	46.0	88.8
Magnesium ammonium phosphate	17.0	15.7	15.0	18.7
Uric acid and ammonium urates	6.1	9.0	12.0	21.5
Cystine	3.8	0.89	5.0	0.93
Xanthine	0	0.04	1.0	0

urinary calculi formation may, because of this hypercalciuria, form calcium-containing calculi. Hypercalciuria may also arise from the endocrine disorder, hyperparathyroidism; therefore, a thorough study should be made to rule out this disorder as the cause of calcium-containing calculi. Analysis of calculi has shown that calcium is generally associated with oxalate, phosphate, carbonate, or mucoproteins. The anion associated with the calcium in the calculi varies, probably depending upon the pH of the urine during calculi formation. Prien[10] states that apatite occurs with calcium oxalate and uric acid in acid urine, and also with magnesium ammonium phosphate in alkaline urine and suggests that apatite may precipitate over the entire physiological urinary pH range.

Composition of Calculi

"Pure" calculi, those containing only one compound, are found in approximately 30 to 40 per cent of the calculi analyzed; the majority of the calculi are mixtures composed of various substances. Most calculi contain one or two principal constituents that are perhaps of most interest to the clinician. The composition is dependent to some degree on the location of the calculi. According to Lonsdale,[8] adult bladder stones are composed chiefly of uric acid, whereas kidney stones are generally composed of calcium oxalate, and apatite or triple phosphate, or all three. The composition of calculi is also related to the geographic location, diet, and metabolic state of the patient. This may explain the wide distribution of the constituents noted by the various authors shown in Table 17-1.

Urinary calculi that occur rarely include those composed of xanthine, sulfonamides, and fibrin. Fibrin calculi are thought to arise from blood clots and may also serve as the nucleus for other calculi.

Various other substances may be incorporated into the calculus during its formation. Trace amounts of many of the heavy metals have been found as well as larger amounts of fluoride, amino acids, mucopolysaccharides, and mucoproteins.

Nonbiological stones such as sand, gravel, plaster, and foreign material are occasionally encountered. The nonbiological stones may be submitted in urine by malingerers or by drug addicts in an attempt to obtain attention or a prescription for narcotics. These stones are generally extremely hard and nearly impossible to crush.

OTHER CALCULI

Prostatic calculi are composed of organic matter and inorganic constituents, such as carbonate, calcium, magnesium, ammonium, and phosphate. *Salivary calculi* are composed of "tartar," which contains magnesium, calcium, phosphate, and carbonate; some organic matter is included during the formation of these calculi.

Pancreatic calculi are extremely rare and, if present, are generally associated with gallstones. Calcium, carbonate, phosphate, organic matter, and small amounts of magnesium and oxalate have been detected in these calculi.

Gallstones (biliary calculi) occur very frequently. They are composed chiefly of cholesterol, bile pigments, calcium, phosphate, and carbonate. Trace amounts of iron, copper, magnesium, manganese, and organic matter are frequently found. The mechanism of gallstone formation, like that of the other calculi, is not fully known. The relative concentration of the various constituents in bile affect the solubility of these substances and may lead to calculus formation. An increase in the cholesterol level of the bile, a change in pH, stagnation of bile, infections, and a change in the ratio of the bile acids to cholesterol are thought to contribute to the formation of biliary calculi.

DISCUSSION OF METHODS FOR STONE ANALYSIS

Various methods have been utilized in the analysis of calculi. Beeler et al.[1] compared the methods of x-ray diffraction, infrared spectrophotometry, chemical analysis, optical analysis, electron microscopy, electron diffraction, and thermoluminescence. The last four methods were reported as unsuitable for stone analyses; the three remaining techniques, namely, x-ray diffraction, infrared spectrophotometry, and chemical analyses, showed fairly good agreement. Infrared spectrophotometry and x-ray diffraction methods have the advantage that the results can be quantitated and a permanent record is obtained, but there are also disadvantages. These instruments are not available in many clinical laboratories, the analyses are time-consuming, and much experience is needed in order to interpret the results correctly. Examples of infrared analysis of renal calculi containing a single constituent are shown in Figure 17-1. The spectra obtained on calculi containing numerous compounds, in contrast, are very complex and positive identification in such cases is extremely difficult, if not impossible. When infrared spectrophotometers and x-ray diffraction apparatus become readily available tools of clinical chemistry, these methods may

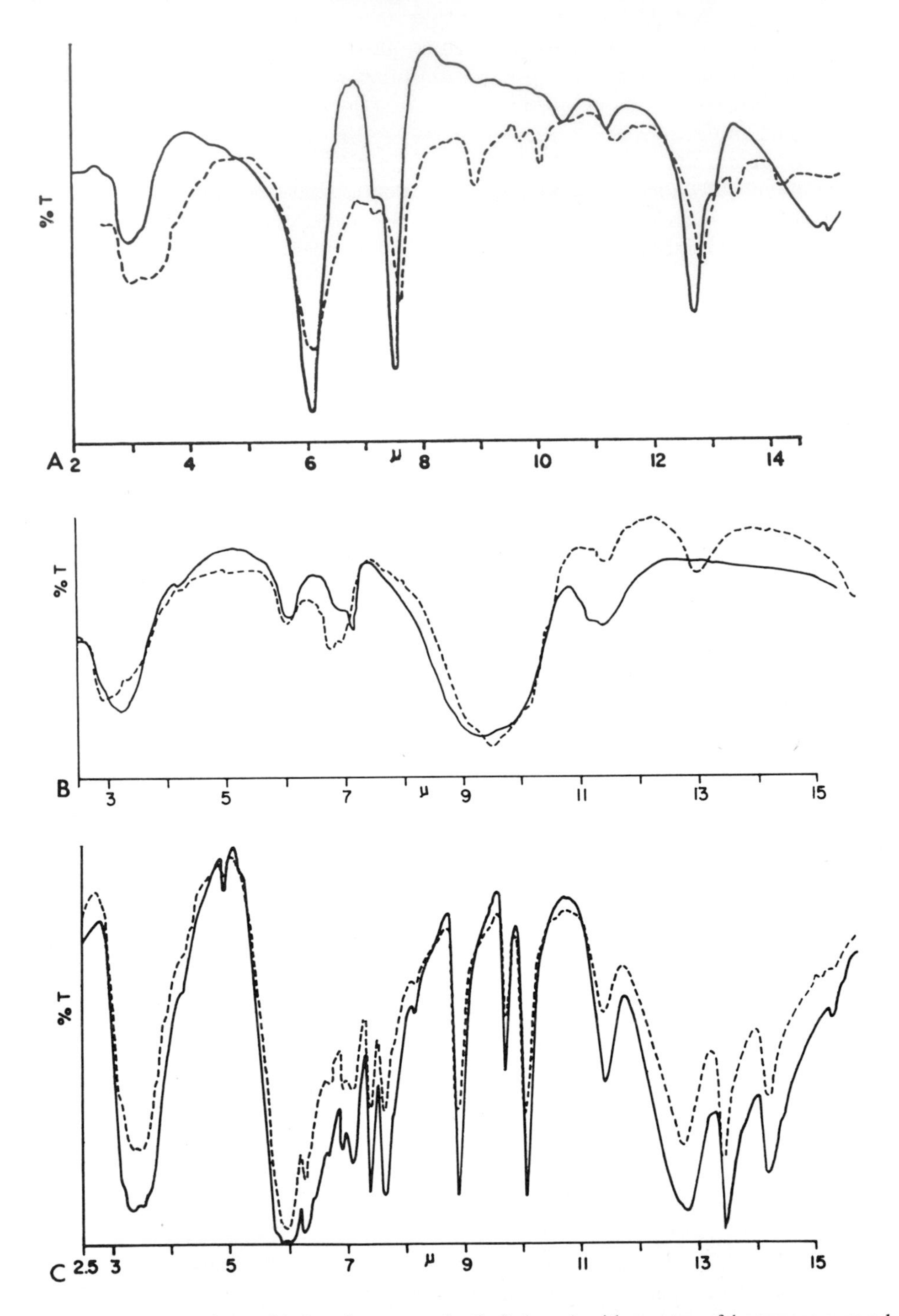

Figure 17-1. Comparison of infrared spectra of calculi (- - - -) with spectra of known compound (∼). *A*, Calcium oxalate; *B*, magnesium ammonium phosphate; *C*, uric acid. (From Weissman et al.: Anal. Chem., *31*:1335, 1959.)

supersede the chemical method for analysis of calculi. At the present time, however, the chemical method is still the method of choice since it is simple and accurate, requires no special equipment, and uses reagents that are readily available.

PROCEDURES FOR STONE ANALYSIS

Preliminary Examination of Calculi

The calculi submitted should be washed and dried. Included in the report should be a description stating the number of stones, the appearance, the weight, and measurements. If the stone is large, it should be cut, using a bone saw, and a portion of each layer should be crushed and used for separate analysis. A small stone is crushed and chemically analyzed in its entirety. The external appearance of calculi, the size, shape, and color, as well as the ashing of a portion of the calculus in an open flame, can be helpful in the identification of the constituents. For flame analysis, a small portion of the crushed calculus is heated in a platinum dish until it glows. If the original bulk is not appreciably changed by ashing and if there is little or no darkening, the calculus is mainly inorganic. If the sample chars and burns almost completely, the calculus is primarily organic. A description and the results of ashing of the frequently occurring calculi follow.

Calcium oxalate calculi are hard calculi that are difficult to crush. They vary from a small smooth type characterized as a hemp seed calculus to a large size possessing an extremely uneven surface. This rough surface may be due to the crystalline structure of calcium oxalate; a stone of this type is classed as a mulberry calculus. The color of the hemp seed type is usually brown, which is seen as a lighter colored powder when crushed. The mulberry type is usually light colored. After ashing a portion of a calcium oxalate stone, approximately two thirds of the bulk remains as a residue.

Calcium carbonate calculi are small and round. They vary from white to gray in color and have a hard, smooth texture. Ashing of the sample leaves a white residue of comparable size to the original sample.

Uric acid (urates) calculi are often small, round, smooth, dull stones and are frequently found in groups. Occasionally a single large urate stone with an irregular crater-like appearance is seen. Uric acid (urate) calculi are usually colored, yellow, brown, or reddish-brown and the crushed powder is yellow. A urate calculus, when ashed, blackens rapidly with the formation of oily brown rings that burn completely as they progress up the side of the crucible.

Phosphate calculi may have a smooth or rough, chalk-like appearance and may be white, gray, or yellow. The size and color of a sample after ashing remains unchanged.

Cystine calculi are waxy lustered, round stones, pale yellow or white in color and frequently they occur as multiple stones. Ashing of a portion of a cystine calculus results in a peculiar, sharp penetrating odor, similar to burning hair; the odor of sulfur dioxide is often noted as the sample begins to burn. Cystine burns completely and leaves the crucible empty.

Xanthine calculi have a waxy appearance and are generally white to brownish-yellow in color. A xanthine calculus, upon ashing, like uric acid, blackens rapidly and burns completely.

Fibrin calculi are small black lightweight stones that are insoluble in organic solvents, but soluble in potassium hydroxide in the presence of heat. When a portion of a fibrin calculus is ashed, the sample burns completely and an odor of burnt feathers is noted.

Qualitative Chemical Analysis

For a complete chemical analysis, place small portions of the powdered stone into three 10 × 75 mm. test tubes and two porcelain evaporating dishes, and proceed to check for the individual constituents according to the following scheme:*

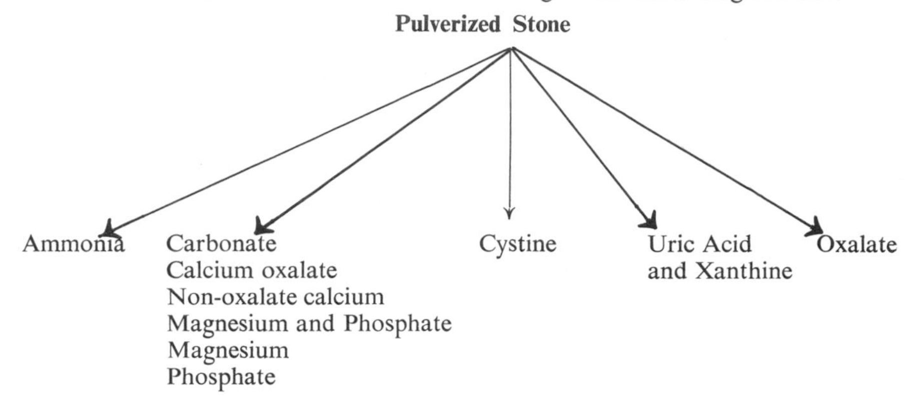

AMMONIA

To a small amount of the powdered calculus, in a 10 × 75 mm. test tube, add a few drops of H_2O. Carefully add 3 to 4 drops of 10 per cent (w/v) KOH into the bottom of the tube. Place a piece of H_2O moistened pink litmus paper over the mouth of the tube. (When adding the KOH to the tube be careful not to wet the top of the tube; contamination of the litmus with KOH will give a false positive result.) If the pink litmus changes to blue, *ammonia is present.* The volatile ammonia is released by the addition of KOH and combines with the indicator.

ALTERNATIVE PROCEDURE FOR AMMONIA

Heat a small amount of the powdered stone with 2 ml. of 5 per cent (v/v) HCl. Cool and neutralize the solution with 10 per cent (w/v) NaOH. Add 0.5 ml. of Nessler's reagent (see serum nonprotein nitrogen procedure). If an orange-brown precipitate forms, *ammonia is present.*

CARBONATE

To a small amount of the powdered calculus, in a 10 × 75 mm. test tube, add approximately 3 ml. of 5 per cent (v/v) HCl. If effervescence occurs, *carbonate is present.*

The addition of a strong acid (e.g., HCl) to a carbonate results in the release of CO_2, noted as small bubbles.

$$CO_3^{--} + 2H^+ \rightarrow H_2CO_3 \rightarrow H_2O + CO_2\uparrow$$

---

* Use only small portions of the stone; the use of large portions may lead to false positive results because of the precipitation of some constituents at their saturation point.

If the calculus submitted is very small, the stone can be analyzed according to the procedure given under Analysis of Small Calculi, included later in the chapter.

CALCIUM OXALATE

Heat the HCl solution from the carbonate procedure in order to solubilize the constituents. Cool and filter through Whatman No. 1 filter paper (5.5 cm. in diameter) into another tube. Add 0.5 ml. of saturated sodium acetate and adjust the pH to approximately 5 with 10 per cent (v/v) acetic acid. If a white precipitate forms, *calcium oxalate is present.*

NON-OXALATE CALCIUM

If a precipitate formed in the calcium oxalate procedure, filter using Whatman No. 1 filter paper; if no precipitate was noted, continue without filtering. Add 0.5 ml. of 5 per cent (w/v) potassium oxalate and adjust the pH to approximately 5.0 using 10 per cent (v/v) acetic acid. If a white precipitate forms, *non-oxalate calcium is present.*

$$Ca^{++} + C_2O_4^{--} \xrightarrow{\text{pH } 5.0} CaC_2O_4\downarrow$$

At a pH of 5, oxalate combines with calcium to form an insoluble white precipitate of calcium oxalate. It is important that the pH be approximately 5 because at an alkaline pH, phosphates, if present, will coprecipitate and at an acid pH, the precipitation of CaC_2O_4 is incomplete.*

MAGNESIUM AND PHOSPHATE

If a precipitate formed in the non-oxalate calcium procedure, filter; if no precipitate was noted, continue without filtering. Add NH_4OH until a pH of 8.0 is obtained. If a white precipitate forms, *magnesium and phosphate are present.* At a pH >8.0, the addition of ammonium ions to a solution containing magnesium and phosphate ions results in the precipitation of NH_4MgPO_4.

$$Mg^{++} + PO_4^{---} \xrightarrow[\text{pH } >8.0]{NH_4^+} NH_4MgPO_4\downarrow$$

If no precipitate forms, divide the solution in two portions and proceed as follows:

MAGNESIUM

To one portion, add 0.5 ml. of 5 per cent (w/v) Na_2HPO_4. If a white precipitate forms, *magnesium is present.* The addition of phosphate and ammonium ions to the solution containing magnesium ions, at a pH >8.0, results in the precipitation of NH_4MgPO_4.

PHOSPHATE

To the second portion add 0.5 ml. of 5 per cent (w/v) $MgSO_4$. If a white precipitate forms, *phosphate is present.* The addition of magnesium and ammonium ions to a solution containing phosphate ions, at a pH >8.0, results in the precipitation of NH_4MgPO_4.

ALTERNATIVE PROCEDURES FOR MAGNESIUM AND PHOSPHATE

Heat a small amount of the powdered stone with 4 ml. of 5 per cent (v/v) HCl. Cool and divide the solution into two parts.

Neutralize one portion with NH_4OH and add 1 ml. of 0.5 per cent (w/v) alcoholic p-nitrobenzeneazoresorcinol. If a blue color forms, *magnesium is present.*

Another approach is to add titan yellow and NaOH (see serum magnesium

* If only a small amount of calcium is present, a slowly forming cloudiness will be noted. Allow the test to stand 10 minutes before filtering to insure complete precipitation.

procedure) to a 5 per cent HCl solution of the calculus. If an orange-red color or precipitate forms, *magnesium is present.*

Neutralize the second portion with NaOH and add 0.5 ml. molybdate solution and 0.2 ml. of aminonaphtholsulfonic acid (see serum phosphorus procedures). If a blue color forms, *phosphate is present.*

CYSTINE

To a small amount of pulverized stone in an evaporating dish, add 1 to 2 drops of 10 per cent (w/v) NaOH, and heat. Add several drops of 10 per cent (w/v) lead acetate and heat again. If a black precipitate forms, *cystine is probably present.* Cystine decomposes in the presence of heat and NaOH to yield sodium sulfide, which combines with lead to form a black precipitate of lead sulfide. Sulfides from other sources such as organic matter give a positive lead sulfide test; therefore, the presence of cystine should be confirmed by the nitroprusside test or by a microscopic examination for crystals.

ALTERNATIVE PROCEDURE FOR CYSTINE

Boil a small amount of the powdered stone with 2 ml. of H_2O. Add 2 ml. of 5 per cent (w/v) NaCN. After 5 minutes, add 3 drops of a freshly prepared solution of sodium nitroprusside. If a red-wine color forms, *cystine is present.*

NaCN converts cystine to cysteine, which then reacts with sodium nitroprusside to form a red color.

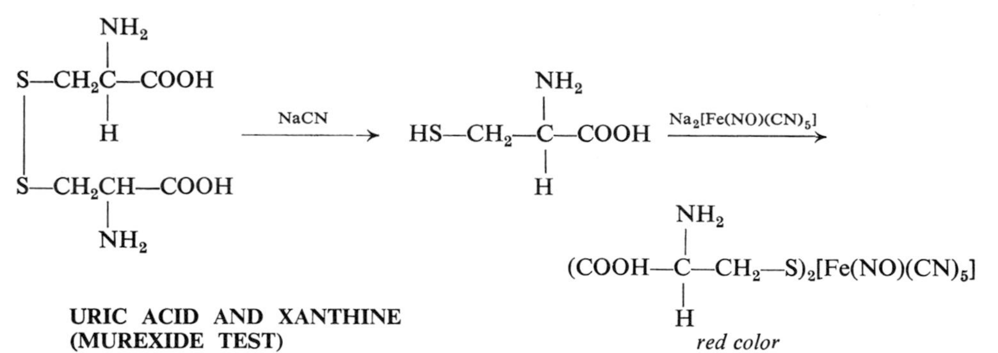

URIC ACID AND XANTHINE (MUREXIDE TEST)

To a small amount of pulverized stone in an evaporating dish, add 1 to 2 drops of concentrated HNO_3 and evaporate the solutions slowly just to dryness. Cool and add several drops of NH_4OH. If a purple color develops, *uric acid is present.* The reaction of uric acid with HNO_3 results in an oxidation of the uric acid, to dialuric acid and alloxan, which condense to form alloxantin. The addition of NH_4OH to alloxantin results in the formation of ammonium purpurate, murexide.

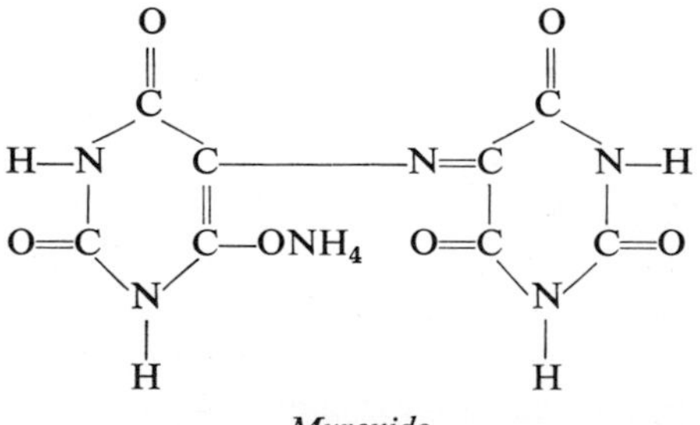

Murexide

If an orange color develops, which turns red with heat, *xanthine is present.*

TEST FOR DIFFERENTIATING BETWEEN XANTHINE AND URIC ACID

Although this test is not totally specific for xanthine, it can be used to distinguish xanthine from uric acid.

Boil a small amount of the powdered calculus in 2 ml. of freshly prepared Ehrlich's diazo reagent (see serum bilirubin procedure). Add 2 drops of 10 per cent (w/v) NaOH. Xanthine couples with the reagent to form a brilliant red-wine color; uric acid gives a faint yellow color or no color.

ALTERNATIVE PROCEDURE FOR URIC ACID

Heat a small amount of the powdered stone in 2 ml. of 5 per cent (v/v) HCl. Cool and neutralize with 10 per cent (w/v) NaOH. Add 1 ml. of phosphotungstic acid (see serum uric acid procedure). If a blue color forms, *uric acid is present*.

OXALATE

Heat a small amount of pulverized stone briefly in a porcelain dish. Cool and place the residue in a 12 × 75 mm. test tube. Add approximately 2 ml. of 5 per cent (v/v) HCl. If effervescence occurs, and there was *no* effervescence in the carbonate procedure, *oxalate is positive*. Moderate heat converts oxalate to carbonate, which upon acidification yields CO_2.

Prolonged intense heat will convert the oxalate to oxide, and invalidate this test.*

ALTERNATIVE PROCEDURE FOR OXALATE

Add 2 ml. of 5 per cent (v/v) HCl to a small portion of the powdered stone. Add a pinch of MnO_2. Do not shake the tube. MnO_2 oxidizes the oxalate to CO_2, which can be seen as tiny bubbles rising from the sediment. It may be necessary to heat the tube slightly to obtain the reaction if only trace quantities of oxalate are present.

Analysis of Small Calculi

Small stones, many less than 2 mm. in diameter, are often submitted for analysis. If the sample is not adequate for a complete stone analysis, the tests for the composition should be performed according to the frequency of their occurrence as noted in Table 17-1 or the selection of tests should be based upon clinical judgment (e.g., in a patient with a history of triple phosphate stones, check for $MgNH_4PO_4$; in patients with hyperparathyroidism, check for calcium; in patients with hyperuricemia, check for uric acid). When small amounts of sample are used, the volume of the reagents should be reduced accordingly, as false negative results may be obtained if the concentration of the constituent is too low.

An alternative approach for the analysis of small stones is to solubilize the crushed calculus in approximately 0.5 ml. of 5 per cent (v/v) HCl. Observe for effervescence when adding HCl. If there is effervescence, *carbonate is positive*. Heat the HCl solution to solubilize the constituents. Centrifuge. Using a microdropper,

* If effervescence was obtained in the carbonate procedure, effervescence would also occur here; therefore, oxalate cannot be determined by this method if carbonate is present.

place 2 drops of the supernatant into four wells of a spot plate. Use drop portions of the reagents and perform the test for calcium and the alternative tests for ammonia, magnesium, and phosphate already given. Place 1 to 2 drops of the HCl calculus solution on a microscopic slide, evaporate slightly, and examine microscopically for cystine crystals. Transfer the remaining HCl calculus solution, and any remaining undissolved calculus to an evaporating dish and perform the murexide test for uric acid.

Several kit procedures for stone analysis are commercially available. These procedures may be useful for the analysis of small stones since, in some kit methods, drop amounts of the reagents are used with very small portions of the calculus.

Microscopic Examination of Crystals

The microscopic examination of crystals has been used to a limited extent for the identification of some calculi constituents. This technique is perhaps most useful as a confirmatory test for cystine. Triple phosphate, calcium carbonate, and calcium oxalate crystals may sometimes be recognized; however, the various calcium salts are difficult to identify and may appear as amorphous material.

The method is as follows: a small portion of the powdered stone is dissolved in 6 *N* HCl on a microscopic slide. A portion of the liquid is evaporated and the residue is examined immediately under the low power of a microscope. The crystals are identified on the basis of a comparison to known crystalline preparations. When this technique is used as a confirmatory test for cystine, a small portion of the powdered calculus is dissolved in 10 per cent (w/v) NH_4OH on a microscopic slide. A portion of the liquid is evaporated and examined microscopically. Cystine crystals are seen as hexagonal plates (six-sided flat crystals).

BILIARY CALCULI

The following tests for cholesterol and bilirubin are performed only on gallstones.

CHOLESTEROL

Extract a portion of the stone with approximately 3 ml. of an absolute alcohol-ether mixture (1:1). Add 0.5 ml. of the sulfuric acid-acetic anhydride reagent (see serum cholesterol-ester procedure). If a green color develops, *cholesterol is present.* Cholesterol reacts with sulfuric acid and acetic anhydride to form a green colored compound.

BILIRUBIN

Extract a portion of the stone with several ml. of methanol. Add 0.5 ml. of diazotized sulfanilic acid (see serum bilirubin procedure). If a violet color develops, *bilirubin is present.* Bilirubin couples with diazotized sulfanilic acid to form azobilirubin.

REFERENCES

1. Beeler, M., Veith, D., Morriss, R., and Biskind, G.: Analysis of urinary calculus. Am. J. Clin. Path., *41*:553, 1964.

2. Boyce, W., and King, J. S.: Crystal-matrix interrelations in calculi. J. Urol., *81*:351, 1959.
3. Cantarow, A., and Trumper, M.: Clinical Biochemistry. 6th ed. Philadelphia, W. B. Saunders Co., 1962.
4. Henry, R. J.: Clinical Chemistry. Principles and Technics. New York, Harper & Row, Publishers, 1965.
5. Herring, L.: Observation on the analysis of ten thousand urinary calculi. J. Urol., *88*:545, 1962.
6. Kachmar, J.: personal communication.
7. Kolmer, J., Spaulding, E., and Robinson, H.: Approved Laboratory Technic. 5th ed. New York, Appleton-Century-Crofts, Inc., 1951.
8. Lonsdale, K.: Human stones. Science, *159*:3820, 1968.
9. Oser, B. L.: Hawk's Physiological Chemistry. 14th ed. New York, McGraw-Hill Book Co., Inc., 1965.
10. Prien, E. L.: Studies in urolithiasis: II Relationship between pathogenesis, structure and composition of calculi. J. of Urol., *61*:821, 1949.
11. Prien, E. L., and Frondel, C.: Studies in urolithiasis: Composition of urinary calculi. J. Urol., *57*:949, 1947.
12. Weissman, M., Klein, B., and Berkowitz, J.: Clinical applications of infrared spectroscopy, analysis of renal tract calculi. Anal. Chem., *31*:1334, 1959.
13. Winer, J.: Practical value of analysis of urinary calculi. J.A.M.A., *169*:1715, 1959.
14. Zinsser, H.: Urinary calculi. J.A.M.A., *174*:116, 1960.

Chapter 18

AMNIOTIC FLUID ANALYSIS

by Ermalinda A. Fiereck, M.S.

Neonatal or intrauterine deaths occur in approximately 20 per cent of pregnancies of Rh-negative, sensitized women. In the past many fatalities could possibly have been avoided by early induced delivery or by intrauterine blood transfusions if there had been an accurate means of assessing the status of the fetus *in utero*. Until recently, the determination of the Rh antibody titer, the obstetrical history, and radiographic analysis were the chief tools for evaluating the condition of the fetus in utero and for determining which patients could benefit most by preterm delivery. Deficiencies of these indirect means were frequently encountered. Many severely erythroblastotic infants who may have benefited by preterm delivery were not so managed, owing to low antibody titers and a previous normal obstetrical history in the mother. Conversely, in the presence of high antibody titers a normal or slightly affected infant was occasionally found, who would in all probability have progressed better had the pregnancy been allowed to continue. A more decisive means of evaluating the *in utero* condition of the fetus was needed since early induced delivery improves the prognosis of only the severely erythroblastotic fetuses, and the reliance on previous obstetrical history and antibody titers is not totally successful.

ABSORBANCE CURVE OF AMNIOTIC FLUID

In 1953, Bevis[1] observed that the amniotic fluid from some Rh incompatible pregnancies had a yellow-green coloration. This coloration was reported to be due to bilirubin or other intermediate products of heme metabolism. When the amniotic fluid absorbance curves of normal and erythroblastotic fetuses were compared, a distortion was noted in the absorption curve of the amniotic fluid obtained from the erythroblastotic fetuses. The degree of distortion at 450 nm. was used by Bevis[1] and Walker[8] to predict the severity of the hemolytic condition.

The absorbance curve of amniotic fluid from normal infants, at term, shows an approximately linear increase in absorbance,* with decrease in wavelength between

* The term absorbance will be used throughout this text, although in the medical literature related to amniotic fluid analysis, the older term, optical density (O.D.), is often used.

550 and 365 nm. (see Fig. 18-1*A*). This uniform increase in absorbance is probably due to the scattering of light by proteins. Any increase above this uniform increase can be determined by noting the extent to which the observed absorbance lies above a straight line, drawn between the observed absorbances at 550 nm. and 365 nm. (see Fig. 18-1*B*). Liley[4] used the differences between the observed absorbance at 450 nm. and that indicated by this straight line at 450 nm. to assess the extent of hemolytic disease. This difference is called "the absorbance peak at 450 nm."

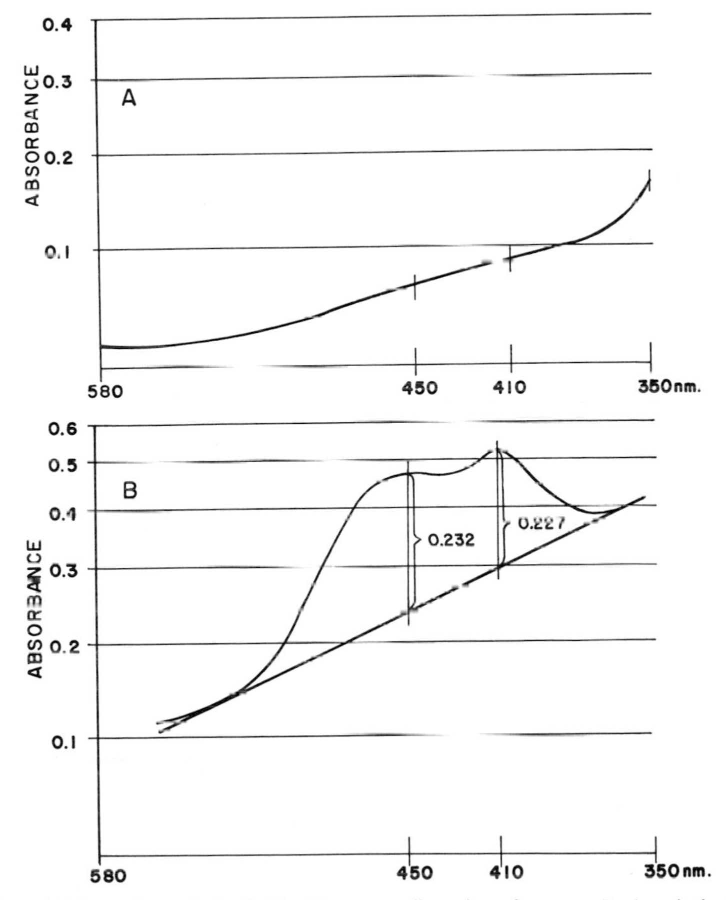

Figure 18-1. *A*, Normal amniotic fluid. Note near linearity of curve. *B*, Amniotic fluid showing bilirubin peak at 450 nm. and oxyhemoglobin peak, at approximately 410 nm. Note baseline drawn between linear parts of curve, from 550 to 365 nm.

RELATIONSHIP OF ABSORBANCE PEAK AT 450 nm. TO WEEKS OF GESTATION

Amniotic fluid from normal fetuses, or those with mild hemolytic disease, shows a moderate absorbance peak at 450 nm. early in pregnancy, which decreases with approaching maturity. This decrease is perhaps due to the continually increasing ability of the placenta to remove bilirubin.[7] At 28 weeks of gestation, the 450 nm. absorbance peak of amniotic fluid from normal fetuses or fetuses with mild hemolytic disease may be as high as 0.06 and at 40 weeks it may range from zero to 0.02. In severe hemolytic disease, the absorbance peak at 450 nm. is greatly increased and does not always decrease with approaching maturity; in fact, there is frequently an increase in absorbance. Generally the initial amniocentesis is performed between the 28th and 32nd weeks of gestation and, if indicated, amniocentesis is repeated every 2 weeks. For correct evaluation, the absorbance at 450 nm. must be related to the weeks of gestation. Liley[4] plotted the 450 nm. absorbance obtained on amniotic fluid specimens against the weeks of gestation. He noted a grouping of the results into zones indicative of the severity of the hemolytic disease. A modified graph of this type is shown in Figure 18-2.

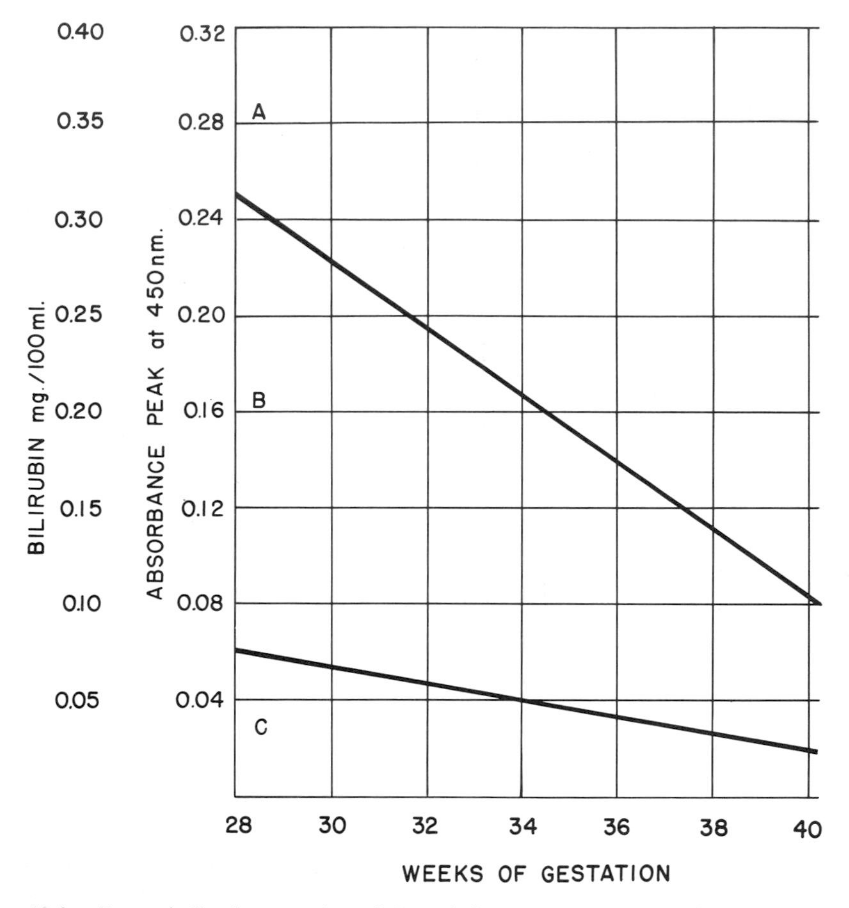

Figure 18-2. Zones indicating severity of hemolytic process. Zone *A*, Suggests a severe fetal hemolytic process. Zone *B*, Suggests a mild to severe hemolytic process. Zone *C*, Suggests the absence of hemolytic disease *or* the presence of a mild hemolytic process. (After Liley: Am. J. Obstet. & Gynec., *82*: 1359, 1961.)

RELATIONSHIP OF BILIRUBIN CONCENTRATION TO ABSORBANCE PEAK AT 450 nm.

The addition of bilirubin in known concentrations* to amniotic fluids with no absorbance peak at 450 nm., followed by the determination of the absorbance curve,

* The appropriate amounts of bilirubin are added in the form of an aqueous solution of bilirubin in Na_2CO_3 and NaOH as described in the procedure for serum bilirubin (Chapter 13).

has permitted us to relate absorbance at 450 nm. to bilirubin concentration. A direct linear relationship was obtained in which each 0.10 mg. of bilirubin/100 ml. corresponded to an absorbance peak of 0.08 at 450 nm. (In other words, absorbance peak $\times$ 1.25 $=$ mg. bilirubin/100 ml. amniotic fluid.)

A similar correlation was reported by Gambino and Freda,[2] using the chemical method of Jendrassik and Grof. The level of bilirubin in amniotic fluid can, therefore, be assessed by determining the absorbance peak at 450 nm. and expressing the results as mg. bilirubin/100 ml. amniotic fluid. The analysis of amniotic fluid by its absorbance curve avoids the difficulties of chemically measuring low concentrations of bilirubin and demonstrates not only the presence of bilirubin in amniotic fluid but also the possible presence of other interfering substances.

BILIRUBIN CONCENTRATIONS OF AMNIOTIC FLUID

Attempts have been made to determine bilirubin concentration of amniotic fluid chemically and to relate this to the severity of the hemolytic disease. Considerable disagreement in values exists in the literature, some perhaps due to difficulties in measuring the low bilirubin concentrations in amniotic fluid, which even in the presence of severe hemolytic disease may be as low as 0.3 mg./100 ml. Further disagreement was encountered when the bilirubin concentration was not related to the weeks of gestation.

Bilirubin concentrations reported by Watson et al.[9] at 32 to 33 weeks ranged from 0.01 to 2.74 mg./100 ml. and were, in general, directly related to the severity of the hemolytic condition. These workers reported that bilirubin glucuronide is not present in either normal amniotic fluid or in amniotic fluid obtained from erythroblastotic fetuses; however, the possibility of other bilirubin conjugates was not excluded.

Stewart and Taylor[6] chemically analyzed a number of amniotic fluid specimens that were submitted at the time of delivery (29 to 40 weeks). Bilirubin values less than 0.035 mg./100 ml. were obtained on normal or mildly affected fetuses who required no exchange transfusions; values from 0.035 to 0.060 mg./100 ml. were obtained on fetuses affected with hemolytic disease who survived but generally required exchange transfusions; and values above 0.060 mg./100 ml. were found only in the amniotic fluid of severely affected or stillborn fetuses.

Increased amniotic fluid bilirubin concentrations have been reported in two conditions other than hemolytic disease. Stewart and Taylor[6] reported a bilirubin value of 2.45 mg./100 ml. from a fetus whose mother had infectious hepatitis; Liley[3] reported an increased 450 nm. absorbance in the amniotic fluid of a fetus with duodenal atresia. The increased bilirubin concentration in the latter case possibly was due to the regurgitation of the bile into the amniotic fluid.

The *normal* amniotic fluid bilirubin values, based upon the relationship of weeks of gestation to the absorbance peak at 450 nm., as reported by Liley,[4] and the relationship of absorbance at 450 nm. to bilirubin concentration are shown in Figure 18-2. The bilirubin concentration of amniotic fluid from normal fetuses or those with mild hemolytic disease, at 28 weeks, may be as high as 0.075 mg./100 ml. and at 40 weeks from zero to 0.025 mg./100 ml.

CONTAMINATION OF AMNIOTIC FLUID BY BLOOD OR SERUM

Amniotic fluid may be contaminated with bilirubin from maternal or fetal blood as a result of trauma during aspiration. If the contamination is due to fetal blood from an erythroblastotic fetus, a substantial error may be introduced in the measurement of amniotic fluid bilirubin. For example, the inclusion of 0.05 ml. of fetal serum (approximately 0.1 ml. blood), with a bilirubin concentration of 10 mg./100 ml. in 5 ml. of amniotic fluid, would raise the amniotic fluid results by 0.1 mg. bilirubin/100 ml. amniotic fluid (or an absorbance of 0.08 at 450 nm.). This contamination would constitute a very serious error.

Contamination with whole blood is readily recognized because of the red color of hemoglobin, and an observant technician is warned that such a sample is unsuitable for analysis. If a specimen containing blood is centrifuged and only the clear portion submitted (e.g., a referral specimen), an error due to added bilirubin may not be suspected and incorrect results may be reported.

Repeated amniocentesis at short intervals (more than once a week) should be avoided since trauma may cause blood to enter the amniotic sac. The bilirubin formed from the hemoglobin would temporarily raise the bilirubin level in the amniotic fluid. It may require 2 or 3 weeks[3] for the elevated bilirubin value of the amniotic fluid, following such traumatic contamination, to return to its precontamination level.

OTHER CONSTITUENTS OF AMNIOTIC FLUID

Oxyhemoglobin

Oxyhemoglobin is frequently present in amniotic fluids. It may occasionally be a true constituent of the fluid, but generally results from the hemolysis of contaminating erythrocytes, which enter at the time of amniocentesis. The presence of oxyhemoglobin is noted by a peak in the region of 408 to 415 nm. (see Fig. 18-1*B*).[4,9] As a result of the proximity of the oxyhemoglobin peak to that of bilirubin (450 nm.), a positive error is introduced into the 450 nm. peak if a high concentration of hemoglobin is present. By experimental contamination Liley[4] has shown this error to be approximately 5 per cent; that is, 5 per cent of the observed oxyhemoglobin absorbance at approximately 410 nm. will appear as added absorbance at 450 nm. If oxyhemoglobin is present to a significant extent, noted by a deviation of absorbance at 410 nm. from the baseline, the 450 nm. absorbance peak, from which bilirubin is calculated, must be corrected (subtract 5 per cent of the 410 nm. absorbance from thc reading at 450 nm.). The method for applying this correction is illustrated in detail later in this chapter. This correction applies only to oxyhemoglobin and does not correct for the bilirubin derived from the serum portion of blood (see previous section).

Methemalbumin

The presence of methemalbumin, a pigment that originates in the fetal reticuloendothelial system, is a common finding in the amniotic fluid when there is intrauterine death.[4,9] According to Watson et al.,[9] methemalbumin probably results from a combination of albumin with the heme formed in this condition. Methemalbumin can be identified by its absorbance peak at 625 to 630 nm.; however, Liley[4] states that this finding adds but little information to that given by the absorbance at 450 nm.

Iron

Liley[4] determined the nonhem iron content of amniotic fluid and found a range of 8 to 30 μg./100 ml. in unaffected or mildly affected fetuses. A rapid increase to 200 or 300 μg./100 ml. is a dangerous sign, and is justification for prompt delivery of the fetus. According to Liley,[4] the nonhem iron determination is not as useful or as rapid to perform as is the amniotic fluid absorption curve and, therefore, it is not recommended.

Glucose, Urea, Protein, and Creatinine

The protein, glucose, and urea concentrations in amniotic fluid from normal and Rh-sensitized patients were compared by Watson et al.[9] No correlation was noted for urea or glucose concentrations with either the period of gestation or hemolytic disease. The glucose and urea concentrations ranged from 10 to 47 mg./100 ml. and 20 to 47 mg./100 ml. (9 to 22 mg. urea nitrogen/100 ml.), respectively. The protein concentration varied from 70 to 385 mg./100 ml.; high protein values tended to be associated with high bilirubin values.

Pitkin[5] reported that the amniotic fluid creatinine value may be used as a measure of the maturity of the fetus. Concentrations greater than 2 mg./100 ml., in the presence of normal maternal serum creatinine, were associated with mature fetuses.

METHOD FOR AMNIOTIC FLUID ANALYSIS

Collection of Specimen and Sample Preparation

Approximately 10 ml. of amniotic fluid is withdrawn by transabdominal amniocentesis. Since bilirubin is unstable in light, the specimen is immediately placed in a sterile tube and continually protected from light. (This can be accomplished conveniently by using foil-wrapped sterile tubes.) Liley[4] reported that the 450 nm. peak had a half-life of 10 hours in laboratory daylight and 12 to 18 minutes in winter sunlight. Storage of a sterile specimen in the dark preserves the specimen for 9 months in a refrigerator and 30 days at room temperature.[4]

Amniotic fluid specimens are generally turbid because of the presence of proteins and particulate matter. Immediate centrifugation, using a centrifuge such as the International Model (size 2, International Equipment Co., Needham, Mass.) at approximately 3000 rpm for 10 minutes, partially clears the sample by removing the larger particulate matter, such as erythrocytes and epithelial cells. Further clarification may be necessary if the sample is extremely turbid; this can be accomplished by centrifugation at 12,000 rpm, using a high speed centrifuge (e.g., Lourdes Model LRA, Lourdes Instrument Corp., Old Bethpage, N.Y.), at 0°C. for 30 minutes. Centrifugation does not eliminate the turbidity completely, but it usually clears the specimen adequately for spectral analysis.

Undiluted specimens are generally used; however, it may be necessary to dilute heavily pigmented specimens before analysis. If a dilution is necessary, 0.9 per cent (w/v) NaCl is used as the diluent.

Amniotic fluid specimens should be analyzed immediately. If there is a delay in analysis, the specimen should be centrifuged, protected from light, and stored in a refrigerator until analyzed.

The specimen may be analyzed either by determining the absorbance curve and noting the absorbance at 450 nm. or by chemically measuring the bilirubin present. The analysis of the amniotic fluid by its absorbance curve is preferred to the bilirubin method (see under Relationship of Bilirubin Concentration to Absorbance Peak at 450 nm.).

Bilirubin Measurement by Chemical Methods

A modified Evelyn and Malloy method[9] and the Jendrassik and Grof method[2] have been used for amniotic fluid bilirubin quantitation. No detailed chemical bilirubin method will be presented here since the analysis of amniotic fluid by its absorbance curve is generally preferred to that of chemical bilirubin measurement.

In cases in which a suitable spectrophotometer for the analysis of amniotic fluid by its absorbance at 450 nm. is not available, the Jendrassik and Grof procedure for bilirubin as presented by Gambino and Freda[2] could be successfully used. Excellent correlation between the absorbance at 450 nm. and the bilirubin concentration was found by these investigators.

Measurement of the 450 nm. Absorbance Peak

The instrument used for this analysis must have a narrow band-pass, preferably of 2 to 4 nm. and, in addition, the wavelength scale must correspond with the actual band of energy passed through the photometer, since inaccuracies in the spectral output may result in inaccurate values for the "bilirubin and oxyhemoglobin peaks."

The absorbance obtained from various instruments, even of the same model, may not be identical; therefore, serial samples should always be analyzed on the same instrument to insure correct interpretation of the results. The absorbance peak at 450 nm. obtained on amniotic fluids with known bilirubin concentration should correspond to the values given in Figure 18-2 or those of Gambino and Freda.[2]

Procedure

1a. If a recording spectrophotometer is employed, adjust the instrument and standardize as recommended by the manufacturer. Scan the amniotic fluid sample between 580 and 350 nm., using 0.9 per cent (w/v) NaCl in the reference cuvet.

1b. If a double beam spectrophotometer is used (without scanning attachment and recorder), take readings manually at 5 nm. intervals between 580 and 350 nm., using 0.9 per cent (w/v) NaCl in the reference cuvet.

1c. If a single beam spectrophotometer is used, set the instrument to zero absorbance with 0.9 per cent NaCl at each change in wavelength before measuring the absorbance of the sample. Take readings every 5 nm. between 580 and 350 nm.

2. Remove the plotted absorbance curve from the recorder. If the readings were taken manually, prepare the curve by plotting the absorbance against the specific wavelength.

3. Draw a line connecting the linear portion of the curve at approximately 550 and 365 nm. The absorbance peak at 450 nm. is calculated as the difference in absorbance between the top of the peak and the baseline (see Fig. 18-1*B*).

4. A peak at approximately 410 nm. indicates the presence of oxyhemoglobin, which causes about a 5 per cent error in the 450 nm. peak. This error can be corrected

by subtracting 5 per cent of the absorbance peak at 410 nm. from the bilirubin absorbance peak at 450 nm.[4]

Example (based on values from Fig. 18-1*B*):

A_{450} corrected for oxyhemoglobin $= A_{450} - (0.05 \times A_{410})$

where $A_{450} = 0.232$ (absorbance at 450 nm. measured from baseline to peak)

$A_{410} = 0.227$ (absorbance at approximately 410 nm. measured from baseline to peak)

$0.05 =$ correction factor (5 per cent) for the oxyhemoglobin error in the 450 nm. absorbance peak.

Substituting these values into the previous formula:

absorbance peak at 450 nm. corrected for oxyhemoglobin

$$= 0.232 - (0.05 \times 0.227)$$

$$= 0.221$$

5. The corrected absorbance at 450 nm. can be converted to mg. bilirubin/100 ml. amniotic fluid by multiplying the corrected 450 nm. absorbance by 1.25. In the example given, the 0.221 absorbance at 450 nm. is equivalent to 0.28 mg. bilirubin/100 ml.

6. If a dilution was necessary, the results must be multiplied by the dilution.

7. The absorbance or bilirubin concentration may be recorded in a graph such as that shown in Figure 18-2. This relates the absorbance or bilirubin concentrations, or both, to the weeks of gestation and suggests the severity of the hemolytic disease.

FACTORS AFFECTING THE ACCURACY OF THE ANALYSES

The following three factors decrease the accuracy of amniotic fluid analyses.

Contamination of the Sample

Amniotic fluid may be contaminated with fetal or maternal blood during aspiration.[3,7] If a large amount of blood is aspirated, the fluid is unsuitable for analysis; if a small amount is included, immediate centrifugation will remove most of the cellular contamination. Hemolysis of even a few erythrocytes results in the release of oxyhemoglobin into the sample. This contamination can be detected by the presence of an absorbance peak at approximately 410 nm. (see previous discussion of oxyhemoglobin).

If the blood serum contaminating the specimen has a high bilirubin concentration, as is the case with fetal blood in severe hemolytic disease, the results of the analysis are inaccurate (see previous discussion of bilirubin contamination of amniotic fluid).

Incorrect Fluid

The specimen submitted is assumed to be amniotic fluid; however, on occasions maternal urine, fluid from amniotic cysts, fetal ascitic fluid, or meconium have been submitted.[3,7] Any atypically shaped absorbance curve should be viewed with this in mind. A scan of a urine specimen from a normal pregnant woman is shown in Figure 18-3*B*.

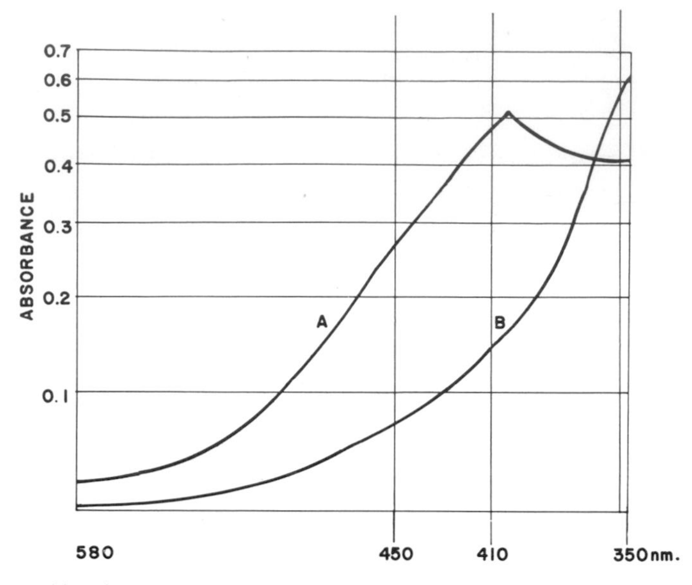

Figure 18-3. *A*, Absorbance curve of unidentified pigment, perhaps meconium.[7] *B*, Absorbance curve of urine from a normal pregnant woman.

A specimen of meconium (fetal intestinal contents) has a green-black color and is generally very turbid. If meconium is present as a contaminant of the amniotic fluid, the specimen may resemble an amniotic fluid of a fetus with severe hemolytic disease. Liley[4] states that a specimen containing meconium has a high 450 nm. absorbance and a higher peak at 408 to 415 nm., with no inflexion in the curve between these points. Figure 18-3*A* shows an absorbance curve of an amniotic fluid containing an unknown substance, perhaps meconium.[7]

Multiple Amniotic Sacs

The sample of amniotic fluid aspirated will assess only the condition of the fetus or fetuses included in the amniotic sac; therefore, if more than one amniotic sac is present, a sample from each should be submitted for analysis. Repeated samples but from different amniotic sacs could lead to misinterpretation of the results.[3]

INTERPRETATION OF THE RESULTS

The results of the amniotic fluid analysis used in conjunction with the previous obstetrical history and other clinical findings are useful in assessing the in utero condition of the fetus of an Rh-sensitized patient.

As stated previously, the bilirubin concentration (and the 450 absorbance peak) in amniotic fluid of normal or very mildly affected fetuses may be as high as 0.075 mg./100 ml. (0.06 absorbance) at 28 weeks and range from 0.00 to 0.025 mg./100 ml. (0.00 to 0.02 absorbance) at 40 weeks. This relationship of the bilirubin concentrations (or absorbance at 450 nm.) at specific weeks of gestation to the severity of the hemolytic disease is shown in Figure 18-2. Values falling in the *A* zone of the graph suggest a severe fetal hemolytic process, those falling in the *B* zone suggest a mild to severe fetal hemolytic disease, and values falling in the *C* zone suggest the absence of hemolytic disease or the presence of a mild hemolytic reaction.[4]

REFERENCES

1. Bevis, D.: Blood pigments in the hemolytic diseases of the newborn, J. Obstet. Gynaec. Brit. Emp., *63*:68, 1956.
2. Gambino, S. R., and Freda, V. J.: The measurement of amniotic fluid bilirubin by the method of Jendrassik and Grof. Am. J. Clin. Path., *46*:198, 1966.
3. Liley, A.: Errors in the assessment of hemolytic disease from amniotic fluid. Am. J. Obstet. & Gynec., *86*:485, 1963.
4. Liley, A.: Liquor amnii analysis in the management of the pregnancy complicated by Rhesus sensitization. Am. J. Obstet. & Gynec., *82*:1359, 1961.
5. Pitkin, R.: Prenatal estimation of fetal maturity. Abstract in American Medical Association Program, 1968.
6. Stewart, A., and Taylor, W.: Amniotic fluid analysis as an aid to ante-partum diagnosis of hemolytic disease. J. Obstet. & Gynaec. Brit. Comm., *71*:604, 1964.
7. Tietz, N. W.: Spectrophotometry-application. *In:* Workshop Manual on Instrumentation. Sponsored by Commission on Continuing Education, A.S.C.P./A.S.M.T., March 25–26, 1966.
8. Walker, A.: Liquor amnii studies in the prediction of hemolytic diseases of the newborn. Brit. Med. J., *2*:376, 1957.
9. Watson, D., Mackay, E., and Trevella, W.: Amniotic fluid analysis and foetal erythroblastosis. Clin. Chim. Acta, *12*:500, 1965.

APPENDIX

Compiled by Ermalinda A. Fiereck, M.S.

METRIC UNITS

Prefix Name	*Prefix Symbol*	*Equivalent*	*Units of Length (meters)*	*Units of Mass (grams)*	*Units of Capacity (liters)*
kilo-	k	10^3	kilometers (km.)	kilograms (kg.)	kiloliters (kl.)
		1	meters (m.)	grams (gm.)	liters (l.)
deci-	d	10^{-1}	decimeters (dm.)	decigrams (dg.)	deciliters (dl.)
centi-	c	10^{-2}	centimeters (cm.)	centigrams (cg.)	centiliters (cl.)
milli-	m	10^{-3}	millimeters (mm.)	milligrams (mg.)	milliliters (ml.)
micro-	μ	10^{-6}	micrometers (μm.)	micrograms (μg.)	microliters (μl.)
nano-	n	10^{-9}	nanometers (nm.)	nanograms (ng.)	nanoliters (nl.)
	Å	10^{-10}	Angstroms (Å)		
pico-	p	10^{-12}	picometers (pm.)	picograms (pg.)	picoliters (pl.)

CONVERSION CHARTS

1. Units of Length

Kilometers km.	*Meters* m.	*Decimeters* dm.	*Centimeters* cm.	*Millimeters* mm.	*Micrometers* μm.	*Nanometers* nm.	*Angstroms* Å	*Picometers* pm.	*Inches* in.
1	10^3	10^4	10^5	10^6	10^9	10^{12}	10^{13}	10^{15}	39.37×10^3
10^{-3}	1	10	10^2	10^3	10^6	10^9	10^{10}	10^{12}	39.37
10^{-4}	10^{-1}	1	10	10^2	10^5	10^8	10^9	10^{11}	39.37×10^{-1}
10^{-5}	10^{-2}	10^{-1}	1	10	10^4	10^7	10^8	10^{10}	39.37×10^{-2}
10^{-6}	10^{-3}	10^{-2}	10^{-1}	1	10^3	10^6	10^7	10^9	39.37×10^{-3}
10^{-9}	10^{-6}	10^{-5}	10^{-4}	10^{-3}	1	10^3	10^4	10^6	39.37×10^{-6}
10^{-12}	10^{-9}	10^{-8}	10^{-7}	10^{-6}	10^{-3}	1	10	10^3	39.37×10^{-9}
10^{-13}	10^{-10}	10^{-9}	10^{-8}	10^{-7}	10^{-4}	10^{-1}	1	10^2	39.37×10^{-10}
10^{-15}	10^{-12}	10^{-11}	10^{-10}	10^{-9}	10^{-6}	10^{-3}	10^{-2}	1	39.37×10^{-12}
2.54×10^{-5}	2.54×10^{-2}	2.54×10^{-1}	2.54	2.54×10	2.54×10^4	2.54×10^7	2.54×10^8	2.54×10^{10}	1

2. Units of Mass

Kilograms kg.	*Grams* gm.	*Decigrams* dg.	*Centigrams* cg.	*Milligrams* mg.	*Micrograms* μg	*Nanograms* ng.	*Picograms* pg.	*Ounces, Av.* oz.	*Pounds, Av.* lb.
1	10^{3}	10^{4}	10^{5}	10^{6}	10^{9}	10^{12}	10^{15}	35.27	2.2
10^{-3}	1	10	10^{2}	10^{3}	10^{6}	10^{9}	10^{12}	35.27×10^{-3}	2.2×10^{-3}
10^{-4}	10^{-1}	1	10	10^{2}	10^{5}	10^{8}	10^{11}	35.27×10^{-4}	2.2×10^{-4}
10^{-5}	10^{-2}	10^{-1}	1	10	10^{4}	10^{7}	10^{10}	35.27×10^{-5}	2.2×10^{-5}
10^{-6}	10^{-3}	10^{-2}	10^{-1}	1	10^{3}	10^{6}	10^{9}	35.27×10^{-6}	2.2×10^{-6}
10^{-9}	10^{-6}	10^{-5}	10^{-4}	10^{-3}	1	10^{3}	10^{6}	35.27×10^{-9}	2.2×10^{-9}
10^{-12}	10^{-9}	10^{-8}	10^{-7}	10^{-6}	10^{-3}	1	10^{3}	35.27×10^{-12}	2.2×10^{-12}
10^{-15}	10^{-12}	10^{-11}	10^{-10}	10^{-9}	10^{-6}	10^{-3}	1	35.27×10^{-15}	2.2×10^{-15}
28.35×10^{-3}	28.35	28.35×10	28.35×10^{2}	28.35×10^{3}	28.35×10^{6}	28.35×10^{9}	28.35×10^{12}	1	0.0625
0.454	454	454×10	454×10^{2}	454×10^{3}	454×10^{6}	454×10^{9}	454×10^{12}	16	1

3. Units of Capacity

Kiloliters kl.	*Liters* l.	*Deciliters* dl.	*Centiliters* cl.	*Milliliters* ml.	*Microliters* μl.	*Nanoliters* nl.	*Picoliters* pl.	*Ounces* oz.	*Quarts* qt.
1	10^{3}	10^{4}	10^{5}	10^{6}	10^{9}	10^{12}	10^{15}	33.81×10^{3}	1.06×10^{3}
10^{-3}	1	10	10^{2}	10^{3}	10^{6}	10^{9}	10^{12}	33.81	1.06
10^{-4}	10^{-1}	1	10	10^{2}	10^{5}	10^{8}	10^{11}	33.81×10^{-1}	1.06×10^{-1}
10^{-5}	10^{-2}	10^{-1}	1	10	10^{4}	10^{7}	10^{10}	33.81×10^{-2}	1.06×10^{-2}
10^{-6}	10^{-3}	10^{-2}	10^{-1}	1	10^{3}	10^{6}	10^{9}	33.81×10^{-3}	1.06×10^{-3}
10^{-9}	10^{-6}	10^{-5}	10^{-4}	10^{-3}	1	10^{3}	10^{6}	33.81×10^{-6}	1.06×10^{-6}
10^{-12}	10^{-9}	10^{-8}	10^{-7}	10^{-6}	10^{-3}	1	10^{3}	33.81×10^{-9}	1.06×10^{-9}
10^{-15}	10^{-12}	10^{-11}	10^{-10}	10^{-9}	10^{-6}	10^{-3}	1	33.81×10^{-12}	1.06×10^{-12}
29.57×10^{-6}	29.57×10^{-3}	29.57×10^{-2}	29.57×10^{-1}	29.57	29.57×10^{3}	29.57×10^{6}	29.57×10^{9}	1	3.125×10^{-2}
0.946×10^{-3}	0.946	0.946×10	0.946×10^{2}	0.946×10^{3}	0.946×10^{6}	0.946×10^{9}	0.946×10^{12}	32	1

CONVERSION BETWEEN mg. AND mEq.

$$\text{mg./100 ml.} = \frac{\text{mEq./L.}}{10} \times \text{equivalent weight}$$

$$\text{mEq./L.} = \frac{\text{mg./100 ml.} \times 10}{\text{equivalent weight}}$$

Ion	*Ionic Weight (gm.)*	*Equivalent Weight (gm.)*	*Conversion Factors (mEq./L.)*	*Conversion Factors (mg./100 ml.)*
Na^{+}	23.0	23.0	mg./100 ml. × 0.435	mEq./L. × 2.30
K^{+}	39.1	39.1	mg./100 ml. × 0.256	mEq./L. × 3.91
Ca^{++}	40.1	20.0	mg./100 ml. × 0.498	mEq./L. × 2.00
Mg^{++}	24.3	12.2	mg./100 ml. × 0.823	mEq./L. × 1.21
Cl^{-}	35.5	35.5	mg./100 ml. × 0.282	mEq./L. × 3.55
HCO_3^{-}	61.0	61.0	vol. % (CO_2) × 0.45*	m*M*./L. × 2.22 (vol. %)
HPO_4^{--}	96.0	53.3†	(P) mg./100 ml. × 0.580	mEq./L. × 1.72 (P)
SO_4^{--}	96.1	48.0	(S) mg./100 ml. × 0.613	mEq./L. × 1.60 (S)

* CO_2 is converted from vol. % to m*M*./L. At the normal pH of blood, HCO_3^- is the chief form, therefore, for practical purposes 1 m*M*./L. = 1 mEq./L.

† For phosphorus: at the normal pH of blood, 20% of the phosphorus is present as $H_2PO_4^-$ and 80% as HPO_4^{--}. The average valence, therefore, is $(1 \times 0.2) + (2 \times 0.8)$ or 1.8 and the equivalent weight is $96 \div 1.8$ or 53.3.

TEMPERATURE CONVERSIONS
(Centigrade-Fahrenheit)

$$°C. = 5/9 \times (°F. - 32)$$
$$°F. = (9/5 \times °C.) + 32$$

Temp. °C.	0	1	2	3	4	5	6	7	8	9
0	32.0	33.8	35.6	37.4	39.2	41.0	42.8	44.6	46.4	48.2
10	50.0	51.8	53.6	55.4	57.2	59.0	60.8	62.6	64.4	66.2
20	68.0	69.8	71.6	73.4	75.2	77.0	78.8	80.6	82.4	84.2
30	86.0	87.8	89.6	91.4	93.2	95.0	96.8	98.6	100.4	102.2
40	104.0	105.8	107.6	109.4	111.2	113.0	114.8	116.6	118.4	120.2
50	122.0	123.8	125.6	127.4	129.2	131.0	132.8	134.6	136.4	138.2
60	140.0	141.8	143.6	145.4	147.2	149.0	150.8	152.6	154.4	156.2
70	158.0	159.8	161.6	163.4	165.2	167.0	168.8	170.6	172.4	174.2
80	176.0	177.8	179.6	181.4	183.2	185.0	186.8	188.6	190.4	192.2
90	194.0	195.8	197.6	199.4	201.2	203.0	204.8	206.6	208.4	210.2
100	212.0	213.8	215.6	217.4	219.2	221.0	222.8	224.6	226.4	228.2

Name	Symbol	International Atomic Weight	Oxidative States
Aluminum	Al	26.98	+3
Antimony, stibium	Sb	121.8	+3, +5, −3
Argon	Ar	39.95	0
Arsenic	As	74.92	+3, +5, −3
Barium	Ba	137.3	+2
Beryllium	Be	9.012	+2
Bismuth	Bi	209.0	+3, +5
Boron	B	10.81	+3
Bromine	Br	79.91	+1, +5, −1
Cadmium	Cd	112.4	+2
Calcium	Ca	40.08	+2
Carbon	C	12.01	+2, +4, −4
Cerium	Ce	140.1	+3, +4
Cesium	Cs	132.9	+1
Chlorine	Cl	35.45	+1, +5, +7, −1
Chromium	Cr	52.00	+2, +3, +6
Cobalt	Co	58.93	+2, +3
Copper	Cu	63.54	+1, +2
Fluorine	F	19.00	−1
Gold, aurum	Au	197.0	+1, +3
Helium	He	4.003	0
Hydrogen	H	1.008	+1, −1
Iodine	I	126.9	+1, +5, +7, −1
Iron, ferrum	Fe	55.85	+2, +3
Lanthanum	La	138.9	+3
Lead, plumbum	Pb	207.2	+2, +4
Lithium	Li	6.939	+1
Magnesium	Mg	24.31	+2

Name	Symbol	International Atomic Weight	Oxidative States
Manganese	Mn	54.94	+2, +3, +4, +7
Mercury	Hg	200.6	+1, +2
Molybdenum	Mo	95.94	+6
Neon	Ne	20.18	0
Nickel	Ni	58.71	+2, +3
Nitrogen	N	14.01	+1, +2, +3, +4, +5, −1, −2, −3
Oxygen	O	16.00	−2
Palladium	Pd	106.4	+2, +4
Phosphorus	P	30.97	+3, +5, −3
Platinum	Pt	195.1	+2, +4
Potassium, kalium	K	39.10	+1
Selenium	Se	78.96	+4, +6, −2
Silicon	Si	28.09	+2, +4, −4
Silver, argentum	Ag	107.9	+1
Sodium, natrium	Na	22.99	+1
Strontium	Sr	87.62	+2
Sulfur	S	32.06	+4, +6, −2
Tellurium	Te	127.6	+4, +6, −2
Thallium	Tl	204.4	+1, +3
Thorium	Th	232.0	+4
Tin, stantium	Sn	118.7	+2, +4
Titanium	Ti	47.90	+2, +3, +4
Tungsten, wolfram	W	183.8	+6
Uranium	U	238.0	+3, +4, +5, +6
Vanadium	V	50.94	+2, +3, +4, +5
Xenon	Xe	131.3	0
Zinc	Zn	65.37	+2

* Values as of 1963, based on carbon-12 and rounded off to four significant figures.

BOILING POINTS OF COMMONLY USED SOLVENTS

Name	*Additional Names*	*Molecular Weight*	*Boiling Point**
Acetic acid	Ethanoic acid	60.05	118.5^{760}
Acetoacetic acid	3-Oxobutanoic acid	102.09	$<100°$d†
Acetone	2-Propanone	58.08	56.2
Aniline	Aminobenzene	93.13	184.3^{760}
Benzene		78.11	80.1
n-Butanol	1-Butanol	74.12	117.5^{760}
Carbon disulfide		76.14	45^{760}
Carbon tetrachloride	Tetrachloromethane	153.82	76.8^{760}
Chloroform	Trichloromethane	119.38	61.2^{760}
Ethanol	Ethyl alcohol	46.07	78.5
Ethyl acetate	Acetic acid ethyl ester	88.11	77.1^{760}
Ethyl ether	Diethyl ether	74.12	34.6
Ethylene dichloride	1,2-Dichloroethane	98.96	84^{760}
Heptane		100.21	98.4
Isoamyl acetate	Acetic acid 3-methylbutyl ester	130.2	142
Isoamyl alcohol	3-Methyl-1-butanol	88.15	131^{760}
Isobutyl alcohol	2-Methyl-1-propanol	74.12	108.4
Isopropyl alcohol	2-Propanol	60.09	82.4
Methanol	Carbinol; Methyl alcohol	32.04	65.0^{760}
Methyl isobutyl ketone	4-Methyl-2-propanone	100.16	116.9
Methylene chloride	Dichloromethane	84.93	40
Nitrobenzene		123.11	210.8^{760}
Petroleum ether		Varies with fraction	Appr. 40 to 120. Varies with fraction
Pyridine		79.10	115.5
Toluene	Methylbenzene	92.13	110.6
p-Xylene	1,4-Dimethylbenzene	106.16	138
m-Xylene	1,3-Dimethylbenzene	106.16	139
o-Xylene	1,2-Dimethylbenzene	106.16	144

* Superscript indicates the barometric pressure at which the boiling point was measured. If no figure is given, the barometric pressure was measured at approximately 1 atmosphere.

† d = decomposes.

PRIMARY AND SECONDARY STANDARDS

Primary Standards

	Formula	*Molecular Weight*	*Equivalent Weight (Quantity Needed for 1 L. of 1 N. solution)*
Sodium carbonate	Na_2CO_3	105.989	52.994 gm.
Sodium oxalate	$Na_2C_2O_4$	134.000	67.000 gm.
Sodium chloride	$NaCl$	58.443	58.443 gm.
Potassium iodate	KIO_3	214.005	35.67 gm.
Potassium dichromate	$K_2Cr_2O_7$	294.192	49.04 gm.
Potassium hydrogen phthalate	$KHC_8H_4O_4$	204.229	204.229 gm.
Succinic acid	$HOOC—(CH_2)_2—COOH$	118.090	59.045 gm.
Sodium tetraborate, decahydrate	$Na_2B_4O_7 \cdot 10H_2O$	381.373	190.686 gm.

Secondary Standards

	Formula	*Molecular Weight*	*Equivalent Weight (Quantity Needed for 1 L. of 1 N. solution)*
Oxalic acid	$H_2C_2O_4$	90.035	45.018 gm.
Oxalic acid, dihydrate	$H_2C_2O_4 \cdot 2H_2O$	126.067	63.038 gm.
Nitric acid	HNO_3	63.013	*
Hydrochloric acid	HCl	36.461	*
Sulfuric acid	H_2SO_4	98.077	*

* Calculate from specific gravity and assay of acid and establish exact concentration by titration against primary standard.

INDICATORS

Common Name	*Chemical Name and Formula*	pH *Range*	*Color Change*	*Commonly Used Concentration*
Bromcresol green	3,3′,5,5′-tetrabromo-m-cresolsulfonphthalein 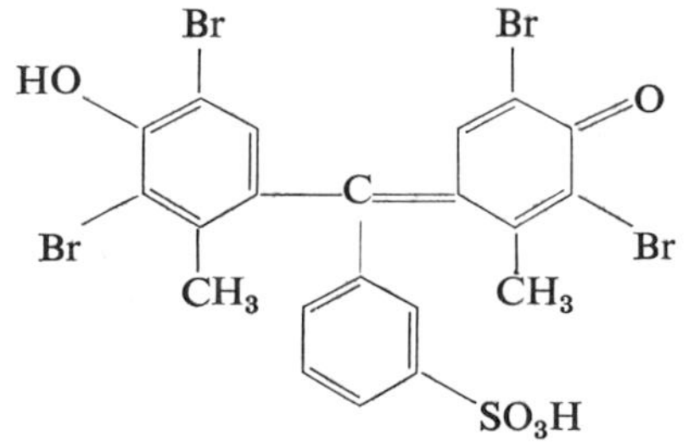	3.8–5.4	yellow to green	0.04% in 0.0006 *N*. NaOH
Bromphenol blue	3,3′,5,5′-tetrabromophenolsulfonphthalein 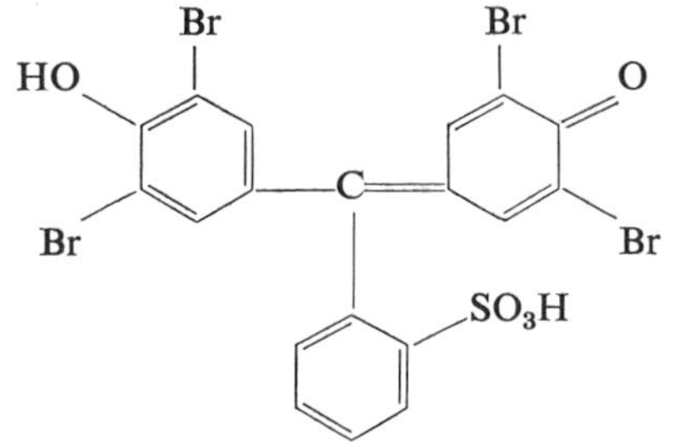	3.0–4.6	yellow to blue	0.04% in 0.0006 *N*. NaOH

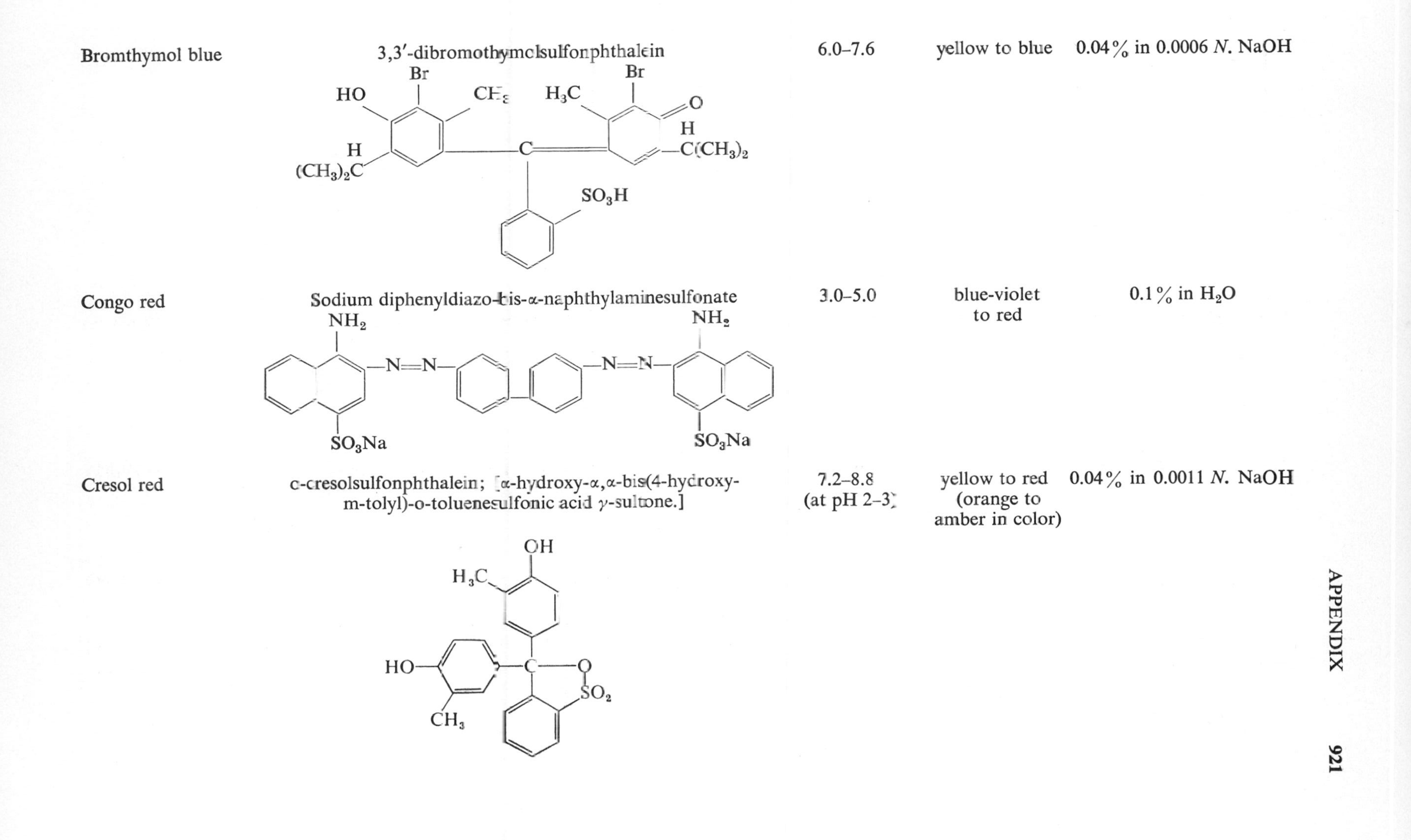

Bromthymol blue	3,3′-dibromothymolsulfonphthalein	6.0–7.6	yellow to blue	0.04% in 0.0006 *N.* NaOH
Congo red	Sodium diphenyldiazo-bis-α-naphthylaminesulfonate	3.0–5.0	blue-violet to red	0.1% in H_2O
Cresol red	o-cresolsulfonphthalein; [α-hydroxy-α,α-bis(4-hydroxy-m-tolyl)-o-toluenesulfonic acid γ-sultone.]	7.2–8.8 (at pH 2–3)	yellow to red (orange to amber in color)	0.04% in 0.0011 *N.* NaOH

INDICATORS (Continued)

Common Name	*Chemical Name and Formula*	pH *Range*	*Color Range*	*Commonly Used Concentration*
Litmus	Lacmus; tournesol; turnsole; lacca musica; lacca coerulea. (Blue coloring matter of various lichens.)	4.5–8.3	red to blue	
Methyl orange	Sodium 4′-dimethylaminoazobenzene 4-sulfonate $(CH_3)_2N—C_6H_4—N{=}N—C_6H_4—SO_3Na$	3.0–4.4	red to yellow	0.1% in H_2O
Methyl red	4′-dimethylaminoazobenzene 2-carboxylic acid $(CH_3)_2N—C_6H_4—N{=}N—C_6H_4—(HO—C{=}O)$	4.2–6.3	red to yellow	0.05% in 50% ethanol.
Phenol red	Phenolsulfonphthalein; α-hydroxy-α,α-bis (p-hydroxyphenyl)-o-toluenesulfonic acid γ-sultone SO_2, O, OH, OH	6.8–8.4	yellow to red	0.04% in 0.0011 *N*. NaOH

Phenolphthalein	3,3-bis(p-hydroxyphenyl)phthalide 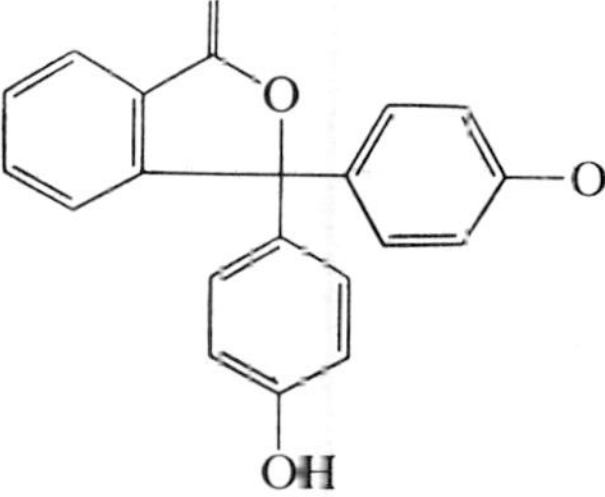 	8.3–10.0	colorless to red	0.1% in 95% ethanol.
Thymol blue	Thymolsulfonephthalein; α-hydroxy-α,α-bis(5-hydroxy-carvacryl)-o-toluenesulfonic acid γ-sultone	Acid Range: 1.2–2.8	red to yellow	If sodium salt is used: 0.04% in 95% ethanol.
		Alkaline range: 8.0–9.6	yellow to blue	If acid form is used: 0.04% in 0.001 *N*. NaOH

INDICATORS (Continued)

Common Name	*Chemical Name and Formula*	pH *Range*	*Color Change*	*Commonly Used Concentration*
Thymolphthalein	5′,5″-diisopropyl-2′,2″-dimethylphenolphthalein 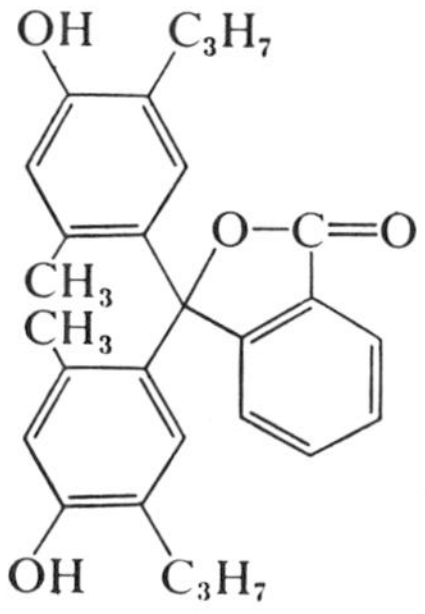	9.3–10.5	colorless to blue	0.1% in 95% ethanol.
Toepfer's reagent	Dimethylaminoazobenzene 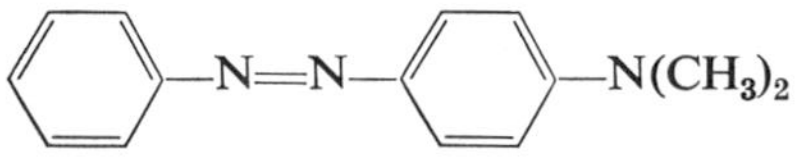 $C_6H_5{-}N{=}N{-}C_6H_4{-}N(CH_3)_2$	2.9–4.0	red to yellow	0.5% in 95% ethanol.

IONIZATION CONSTANTS (K) FOR COMMON ACIDS AND BASES IN WATER*

	K	pK_a
Acetic acid	1.75×10^{-5}	4.76
Acetoacetic acid	2.62×10^{-4}	3.58 (18°C.)
Ammonia	5.6×10^{-10}	9.25
Boric acid†	6.4×10^{-10}	9.19
Carbonic acid	4.47×10^{-7}	6.35
	4.68×10^{-11}	10.34
Citric acid	7.4×10^{-4}	3.13
	1.7×10^{-5}	4.77
	4.0×10^{-7}	6.40
Diethylbarbituric acid (Veronal)	3.7×10^{-8}	7.43
Ethylenediamine	1.4×10^{-7}	6.85
	1.12×10^{-10}	9.93
Ethylenediamine tetraacetate	1.00×10^{-2}	2.00
	2.16×10^{-3}	2.67
	6.92×10^{-7}	6.16
	5.50×10^{-11}	10.26
Formic acid	1.76×10^{-4}	3.75
Glycine	4.5×10^{-3}	2.35
	1.7×10^{-10}	9.77
Glycylglycine	7.24×10^{-4}	3.14
	5.62×10^{-9}	8.25
Hydroxylamine	9.1×10^{-9}	8.04
Imidiazole	1.01×10^{-7}	6.95
Isocitric acid	5.13×10^{-4}	3.29
	1.99×10^{-5}	4.70
	3.98×10^{-7}	6.40
p-Nitrophenol	7×10^{-8}	7.15
Oxalacetic acid	2.75×10^{-3}	2.56
	4.27×10^{-5}	4.37
Oxalic acid	6.5×10^{-2}	1.19
	6.1×10^{-5}	4.21
Phosphoric acid	7.5×10^{-3}	2.12
	6.2×10^{-8}	7.21
	4.8×10^{-13}	12.32
Phosphorous acid	5×10^{-2}	1.30
	2.6×10^{-7}	6.59
Pyruvic acid	3.23×10^{-3}	2.49
Succinic acid	6.2×10^{-5}	4.21
	2.3×10^{-6}	5.64
Sulfuric acid	$\gg 1$	—
	1.2×10^{-2}	1.92
Tartaric acid	1.1×10^{-3}	2.96
	6.9×10^{-5}	4.16
Triethanolamine	1.26×10^{-8}	7.90
Tris(hydroxymethyl)-amino methane	8.32×10^{-9}	8.08

* Temperature at or near room temperature (25°C.) unless otherwise indicated.
† Boric acid acts as a monotropic acid in aqueous solution

DATA ON COMPRESSED GASES*

	Relative Specific Gravity (Air = 1)	*State of Gas in Cylinder*	*Approximate Pressure in Cylinder (psi)*	*Purity of Best Grade in Cylinder*	*Flammability and Toxicity*	*Flammability Limit in Air by Volume*	*Special Notes*	*Leak Detection*
Acetylene	0.9073	Dissolved in Acetone	250 (21°C.)	>99.6%	Highly flammable. Nontoxic but asphyxiant and anesthetic	2.5–81%	1. Use only at pressures less than 30 psi. 2. Never use with unalloyed Cu, Ag, and Hg. 3. Store upright—acetone is included in cylinder	Soap solution
Air	1.000	Non liquefied	2200–2400 (21°C.)	Mixture	Nonflammable. Nontoxic		Dry air is inert to metals and plastics at ambient temperatures	Soap solution
Ammonia	0.5870	Liquefied	114 (21°C.)	>99.95%	Flammable. Toxic in concentrations greater than 100 ppm. for 8 hrs.	15–28%	1. Cu, Sn, Zn, and their alloys are attacked by moist ammonia 2. Never use mercury; ammonia can combine with Hg to form explosive compounds	1. Open bottle of HCl 2. Wet red litmus paper 3. Wet phenol-phthalein paper
Argon	1.38	Non liquefied	2200 (21°C.)	>99.9995 vol. %	Inert. Nontoxic but acts as a simple asphyxiant		Inert	Soap solution
Butane	2.076 (15°C., 1 atm.)	Liquefied	16.3 (21°C.)	>99.99 mole %	Extremely flammable. Nontoxic but acts as a simple asphyxiant and has anesthetic effect	1.9–8.5%	Noncorrosive	Soap solution

Carbon dioxide	1.5289	Liquefied	830 (21°C.)	>99.995 mole %	Nonflammable. Toxic in concentrations greater than 5000 ppm. for 8 hr.		Dry gas is relatively inert	Soap solution Aqueous ammonia
Carbon monoxide	0.9678 (21°C., 1 atm.)	Non liquefied	1500 (21°C.)	>99.8 mole %	Extremely flammable. Extremely toxic; maximum allowable concentration for 8 hr. exposure is 100 ppm.	12.5–74%	Corrosion by pure CO at low pressure can be considered negligible. At high pressures it will react with Fe, Ni, and other metals forming carbonyls	Soap solution
Helium	0.137	Non liquefied	2200 (21°C.)	>99.9995 mole %	Inert. Nontoxic but acts as a simple asphyxiant		Inert	Soap solution
Hydrogen	0.06952	Non liquefied	2000 (21°C.)	>99.9995 mole %	Extremely flammable. Nontoxic but can act as asphyxiant	4.0–75%	Noncorrosive	Soap solution
Hydrogen chloride	1.268 (gas, 0°C.)	Gas over liquid	613 (21°C.)	>99.0%	Non flammable. Highly toxic; 0.13–0.2% lethal in a few minutes. Maximum accepted concentration for 8 hr. is 5 ppm.		1. Hydrogen chloride is essentially inert and does not attack metals; however, in the presence of moisture, hydrogen chloride will corrode most metals except silver, platinum, and tantalum 2. *Always* shut off hydrogen chloride from the use end, backward to the cylinder	1. Large leaks evident by dense white fumes on contact with the atmosphere 2. Open bottle of conc. NH_4OH 3. Wet blue litmus paper

* Compiled from *Matheson Gas Data Book*, 4th Ed. The Matheson Company Inc., Herst Litho Inc., New York, 1966.

DATA ON COMPRESSED GASES* (Continued)

	Relative Specific Gravity (Air = 1)	*State of Gas in Cylinder*	*Approximate Pressure in Cylinder (psi)*	*Purity of Best Grade in Cylinder*	*Flammability and Toxicity*	*Flammability Limit in Air by Volume*	*Special Notes*	*Leak Detection*
Hydrogen sulfide	1.1895 (gas, 15°C.)	Liquefied	250 (21°C.)	>99.5%	Highly flammable. Highly toxic. Maximum allowable concentration for 8 hr. is 20 ppm.	4.3–45%		1. Soap solution 2. Cadmium chloride solution (turns yellow upon contact with H_2S) 3. Moist lead acetate paper (turns black upon contact with H_2S)
Methane	0.5549 (15°C. 1 atm.)	Non liquefied	2265	>99.993 mole %	Extremely flammable. Nontoxic but acts as a simple asphyxiant	5.3–14%	Noncorrosive	Soap solution
Neon	0.6964 (gas, 21°C., 1 atm.)	Non liquefied	1800	>99.995 mole %	Inert. Nontoxic but acts as an asphyxiant		Inert	Soap solution
Nitrogen	0.9670	Non liquefied	2200	>99.9999 mole %	Inert. Nontoxic but acts as an asphyxiant		Inert	Soap solution
Nitrous oxide	1.530 (Gas, 15°C.)	Liquefied	745 (21°C.)	>98.0%	Non flammable. Nontoxic but acts as asphyxiant and anesthetic		Non corrosive	Soap solution

Oxygen	1.1053	Non liquefied	2200 (21°C.)	>99.99 mole %	Nonflammable but may cause explosive oxidation of organic substances as oil, grease etc. Nontoxic		Non corrosive. Extreme caution should be taken to avoid contact with oil, grease, or other readily combustible substances as explosion may occur	Soap solution
Propane	1.5503 (15°C., 1 atm.)	Liquefied	110 (21°C.)	>99.99 mole %	Extremely flammable. Nontoxic but acts as an asphyxiant and in high concentration has an anesthetic action	2.2–9.5%	Non corrosive	Soap solution

* Compiled from *Matheson Gas Data Book*, 4th Ed., The Matheson Company Inc., Herst Litho Inc., New York, 1966.

DESIRABLE WEIGHTS FOR MEN AND WOMEN*

According to Height and Frame. Ages 25 and Over

Height (In Shoes)	*Weight in Pounds (In Indoor Clothing)*		
	Small Frame	*Medium Frame*	*Large Frame*
		Men	
5′ 2″	112–120	118–129	126–141
3″	115–123	121–133	129–144
4″	118–126	124–136	132–148
5″	121–129	127–139	135–152
6″	124–133	130–143	138–156
7″	128–137	134–147	142–161
8″	132–141	138–152	147–166
9″	136–145	142–156	151–170
10″	140–150	146–160	155–174
11″	144–154	150–165	159–179
6′ 0″	148–158	154–170	164–184
1″	152–162	158–175	168–189
2″	156–167	162–180	173–194
3″	160–171	167–185	178–199
4″	164–175	172–190	182–204
		Women	
4′ 10″	92–98	96–107	104–119
11″	94–101	98–110	106–122
5′ 0″	96–104	101–113	109–125
1″	99–107	104–116	112–128
2″	102–110	107–119	115–131
3″	105–113	110–122	118–134
4″	108–116	113–126	121–138
5″	111–119	116–130	125–142
6″	114–123	120–135	129–146
7″	118–127	124–139	133–150
8″	122–131	128–143	137–154
9″	126–135	132–147	141–158
10″	130–140	136–151	145–163
11″	134–144	140–155	149–168
6′ 0″	138–148	144–159	153–173

* Prepared by the Metropolitan Life Insurance Company. Derived primarily from data of the *Build and Blood Pressure Study, 1959*, Society of Actuaries. Reproduced with permission.

BROMSULPHALEIN DOSAGE SCHEDULE FOR BSP LIVER FUNCTION TEST

Vial concentration: 50 mg./ml.

Dose: 5 mg./kg. (5 mg./2.2 lb.)

Calculation:

$$\text{ml. needed for injection} = \frac{1\text{ ml.}}{50\text{ mg.}} \times \frac{5\text{ mg.}}{2.2\text{ lb.}} \times \text{wt. of patient in lb.}$$

or $0.0454 \times$ wt. of patient in lb.

Weight in lbs.	*ml. of Dye Solution*
60	2.7
70	3.2
80	3.6
90	4.1
100	4.5
110	5.0
120	5.5
130	5.9
140	6.4
150	6.8
160	7.3
170	7.7
180	8.2
190	8.6
200	9.1
210	9.5
220	10.0

NOMOGRAM FOR THE DETERMINATION OF BODY SURFACE AREA OF CHILDREN*

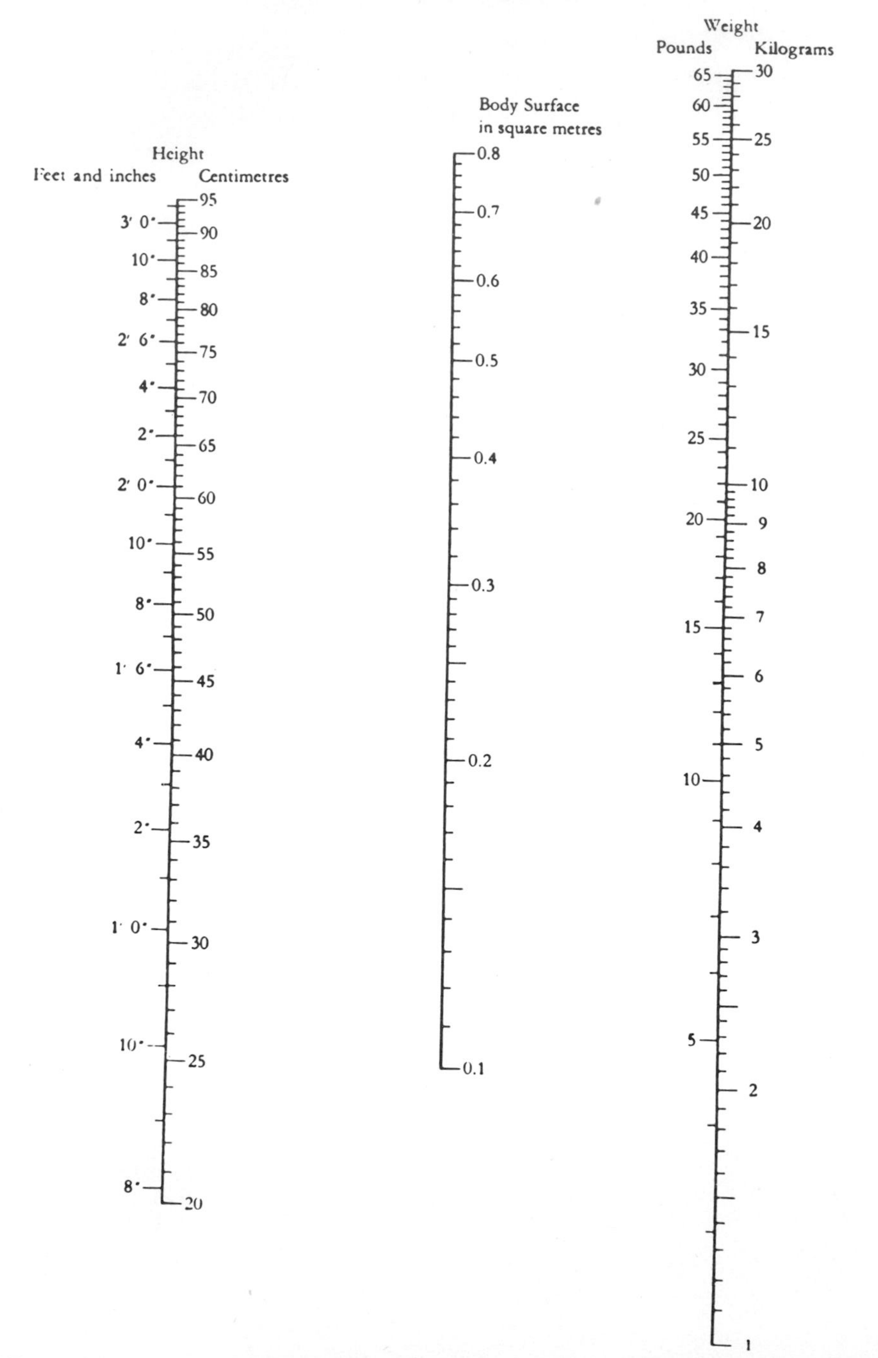

* From DuBois, E. F.: *Basal Metabolism in Health and Disease*, Philadelphia, Lea & Febiger, 1936.

NOMOGRAM FOR THE DETERMINATION OF BODY SURFACE OF CHILDREN AND ADULTS*

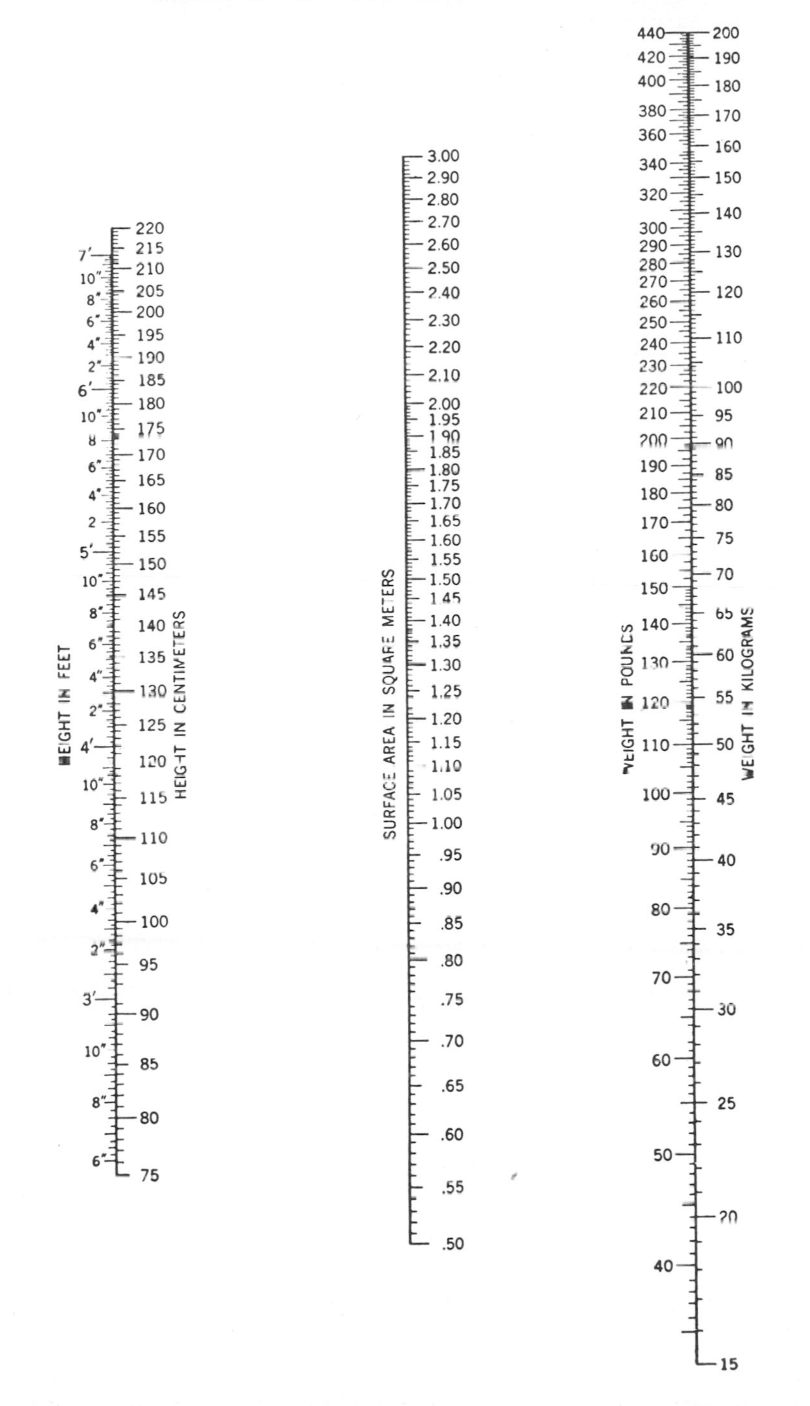

* From Boothby, W. M., and Sandiford, R. B.: Boston M. & S.J. *185*:337, 1921.

NORMAL VALUES*

Test	*Specimen*	*Value*	*Special Instructions and Interfering Substances*
Acetoacetic acid, qualitative	Serum or urine	Negative	
Acetone, qualitative	Serum or urine	Negative	
quantitative	Serum or plasma	0.3–2.0 mg./100 ml.	
quantitative (acetone and acetoacetic acid)	Serum or plasma	0.5–3.0 mg./100 ml.	
Albumin/globulin (A/G) ratio	Serum	1.1–1.8 by salt fractionation (27%) or electrophoresis	
	Spinal fluid	1.6–2.2	Presence of blood invalidates the test.
Albumin, quantitative	Serum	3.5–4.7 gm./100 ml. by salt fractionation (27%) or electrophoresis 3.8–5.0 gm./100 ml. by dye-binding methods	
	Spinal fluid	10–30 mg./100 ml.	Presence of blood invalidates the test
Alcohol	Serum or plasma	see Ethanol and Methanol	
Aldolase	Serum	Adults: 2–8 U./100 ml. at 37°C. (Sibley and Lehninger) 1.5–6.0 mI.U./ml. at 37°C. (Sibley and Lehninger) 0.8–3.0 mI.U./ml. at 30°C. (Sibley and Lehninger) Newborn: approx. 4 × adult level Children: approx. 2 × adult level	Keep at room temperature. Avoid hemolysis. Values lower at bed rest
Aldosterone	Urine, 24 hr.	3–32 μg./24 hr.	Refrigerate during collection. Acidify to pH 4–5 with HCl, acetic or boric acid
Alkapton bodies	Urine, random	Negative	
α-Amino acid nitrogen	Plasma	3–6 mg./100 ml.	
	Urine, 24 hr.	Adults: 50–200 mg./24 hr. (by chromatography) 200–700 mg./24 hr. (with less specific technique) Older children: 66–204 μM/kg./24 hr.	Preserve with thymol or HCl. Keep refrigerated

* The normal values listed, in this table and normal values in general, are not to be taken as absolute but only as guidelines. The normal values vary with the particular procedure employed and may indeed vary from laboratory to laboratory. Thus, ideally, normal values should be established for each constituent under the particular method and conditions of the laboratory (see Chapter 1). Values for various blood constituents, in general, are those obtained in the fasting state. The values listed in this table were compiled chiefly from the values listed in this text and supplemented by values obtained from the sources[2,3,4] listed at the end of this table. Other references[1,5,6] for normal values are also noted. Enzyme values are given in conventional units (U.) and in international units (I.U.). The names given indicate the author of the method.

TABLE OF NORMAL VALUES (Continued)

Test	*Specimen*	*Value*	*Special Instructions and Interfering Substances*
α-Amino acid nitrogen *cont'd.*		Premature newborns: approx. 6 × value for older children Full-term newborns: approx. 3 × value for older children	
p-Amino hippurate clearance test	Urine and plasma	600 ml./min./average body surface	
δ-Aminolevulinic acid	Serum	0.01–0.03 mg./100 ml.	
	Urine, fresh random	Adult: 0.1–0.6 mg./100 ml. Under 15 years: <0.50 mg./100 ml.	
	Urine, 24 hr.	1.5–7.5 mg./24 hr.	Collect 24 hr. specimen in bottle containing 10 ml. glacial acetic acid. Refrigerate
Ammonia nitrogen	Serum or plasma	Enzymatic method: 40–80 μg./100 ml. Resin method: 15–45 μg./100 ml. Conway diffusion method: 40–110 μg./100 ml. Newborn: 90–150 μg./100 ml. (higher in premature or jaundiced infants) Older children: 40–80 μg./100 ml.	For plasma, collect with sodium heparinate. Analyze immediately, as concentration increases rapidly on standing (see Chapter 10)
	Urine, 24 hr.	500–1200 mg./24 hr. or: 36–85 mEq./24 hr.	Avoid contamination with ammonia and bacteria
Amniotic fluid analysis	Amniotic fluid	*28 weeks* / *40 weeks*	Protect from light. Centrifuge and analyze immediately. If analysis is delayed, refrigerate
Absorbance at 450 nm.		0–0.048 / 0–0.02	
Bilirubin		0–0.075 mg./100 ml. / 0–0.025 mg./100 ml.	
Amylase	Serum	40–160 Somogyi units/100 ml. (Saccharogenic method)	Inhibited by oxalate and citrate
	Urine, 24 hr.	<5000 Somogyi units/24 hr. (Saccharogenic Method)	Refrigerate during collection
	Duodenal contents	50,000–80,000 Somogyi units/100 ml.	
Arsenic	Whole blood	less than 3 μg./100 ml.	Collect in acid-cleaned tube
	Urine, 24 hr.	less than 50 μg./L.	Collect in acid-cleaned container
Ascorbic acid	Plasma	0.5–1.5 mg./100 ml.	Unstable; analyze immediately. Stable in acid
	Whole blood	0.7–2.0 mg./100 ml.	

TABLE OF NORMAL VALUES (Continued)

Test	*Specimen*	*Value*	*Special Instructions and Interfering Substances*
Ascorbic acid *cont'd.*	Red cells White cells and platelets Urine, random Urine, 24 hr.	Approx. 2 × plasma conc. Approx. 20–40 × plasma conc. 1–7 mg./100 ml. >50 mg./24 hr.	solution. Collect with oxalate. Collect in metaphosphoric acid or TCA.
Barbiturate	Whole blood Urine, random	Negative Therapeutic levels: Short acting: <0.3 mg./100 ml. Long acting: 1–3 mg./100 ml. Negative	Collect with sodium heparinate
Bence-Jones Protein	Urine, random first morning	Negative	Analyze fresh specimen
Beryllium	Urine, 24 hr.	Less than 0.05 μg./day	Collect in acid-washed container
Bicarbonate	Plasma	21–28 m*M*/L.	Collect anaerobically
Bile, unaltered, qualitative	Feces, fresh random	Negative	
Bile acids	Serum	0.3–3.0 mg./100 ml.	Analyze immediately or freeze
Bile salts	Feces, 24 hr.	Approx. 0.8 gm./24 hr.	
Bilirubin	Serum	Total: 0.2–1.0 mg./100 ml. Conjugated: 0–0.2 mg./100 ml. Unconjugated: 0.2–0.8 mg./100 ml. Infants (total): 1–12 mg./100 ml.	Avoid exposure to direct light See Chapter 13 for details
Bilirubin, qualitative	Urine, fresh random	Negative	
Blood, occult	Urine, random Feces, random	Negative Negative	 Place patient on meat-free diet for 2 days
Blood volume	Whole blood Plasma	Adults: 60–90 ml./kg. Adults: 40–50 ml./kg.	Varies with hematocrit and sex
Borate	Urine, 24 hr.	Less than 2 mg./L.	
Bromide	Serum	Adult: <1.5 mg./100 ml. (as NaBr) Children <1.0 mg./100 ml. (as NaBr) Therapeutic Level: 100–200 mg./100 ml. (as NaBr)	

TABLE OF NORMAL VALUES (Continued)

Test	*Specimen*	*Value*	*Special Instructions and Interfering Substances*
Bromide *cont'd.*		Toxic level: above 200 mg./100 ml. (as NaBr)	
BSP (Bromsulfophthalein)	Serum	<10% after 30 minutes <6% after 45 minutes	Dose: 5 mg./kg.
Calcium	Feces, 24 hr.	Average: 0.64 gm./day (Varies with diet)	
	Serum	Ionized: 4.2–5.5 mg./100 ml. Total: 8.7–10.7 mg./100 ml. (4.4–5.4 mEq./L.)	
	Spinal fluid	4.2–5.8 mg./100 ml.	
	Urine, 24 hr.	Average: 50–150 mg./24 hr. (varies with intake)	Acidify during collection
qualitative (Sulkowitch test)	Urine, random	1+, 2+	
Carbon dioxide	Whole blood, arterial	Children: (1–3 years) 12–20 m*M*./L. (Increases slowly to adult level) Adult: 19–24 m*M*./L.	Collect anaerobically
	Plasma or serum, capillary	21–28 m*M*./L.	
	Plasma or serum, venous	23–30 m*M*./L.	
	Whole blood, venous	22–26 m*M*./L.	
Carbon dioxide pressure (P_{CO_2})	Alveolar air	average 36 mm. Hg	
	Whole blood, arterial	35–46 mm. Hg	Collect with heparinized syringe, and seal. Handle anaerobically
	Whole blood, venous	38–50 mm. Hg	
Carbon monoxide hemoglobin (carboxyhemoglobin)	Whole blood	Nonsmokers: 0.5–1.5% saturation of Hb Average smokers: <5% saturation Heavy smokers: 5–9% saturation	Avoid exposure to light and air
Carbonic acid	Whole blood, arterial	1.05–1.45 m*M*./L.	
	Whole blood, venous	1.15–1.50 m*M*./L.	
	Plasma, venous	1.02–1.38 m*M*./L.	
β-Carotene	Serum	Adult: 60–200 μg./100 ml. Newborn: Approx. 70 μg./100 ml.	

TABLE OF NORMAL VALUES (Continued)

Test	*Specimen*	*Value*	*Special Instructions and Interfering Substances*
Catecholamines, total	Urine, random	0–14 μg./100 ml.	Refrigerate and acidify collection
	Urine, 24 hr.	Less than 150 μg./24 hr. Borderline: 150–230 μg./24 hr. (Varies with muscular activity)	
Cephalin cholesterol	Serum	Negative to 1+	
Ceruloplasmin	Serum	250–570 U/100 ml. at 37°C. (Henry) 35–63 mI.U./ml. at 37°C. (Rice) 25–43 mg./100 ml. (Sass-Kortsak)	
Chloride	Gastric residue	45–155 mEq./L.	
	Saliva	Without stimulation: up to 10 mEq./L. With stimulation: up to 44 mEq./L.	
	Serum or plasma	Adult: 98–108 mEq./L. Children: 98–106 mEq./L.	
	Spinal Fluid	Adult: 118–132 mEq./L. Children: 120–128 mEq./L.	
	Sweat	Normal homozygotes: 5–35 mEq./L. Heterozygotes: 35–60 mEq./L.	
	Urine, 24 hr.	110–250 mEq./24 hr. (Varies with intake)	
Cholesterol, total	Serum	Adult: 130–250 mg./100 ml. (Varies with diet and age) Lower in children Infants: 45–170 mg./100 ml. At birth: <10 mg./100 ml.	
Cholesterol esters	Serum	Adults: 65–75% Infants: 50–70%	
Cholesterol/phosphatide ratio	Serum	0.55–1.05	
Cholinesterase, pseudo	Serum or plasma	130–310 U./ml. at 37°C. (de la Huerga) 2.3–5.2 mI.U./ml. at 37°C. (de la Huerga)	Refrigerate. Collect plasma with sodium heparinate
Colloidal gold test	Spinal fluid	000 111 100 0	Presence of blood invalidates the test
Concentration test (Fishberg)	Urine, after fluid restriction	Specific gravity: 1.025 or higher Osmolality: 850 mOsm./kg. or higher	

TABLE OF NORMAL VALUES (Continued)

Test	*Specimen*	*Value*	*Special Instructions and Interfering Substances*
Congo red test	Plasma	After 1 hr.: >65%	See Chapter 5
Copper	Serum	Males: 70–140 μg./100 ml.	Avoid hemolysis. Use acid-cleaned collection tubes
		Females: 80–155 μg./100 ml. Higher in children	
	Urine, 24 hr.	0–30 μg./24 hr.	Use acid-cleaned collection vessel
Coproporphyrins, quantitative	Feces, 24 hr.	Approx. 30 μg./gm. dry wt.	
qualitative	Urine, fresh random	Negative, trace	Avoid exposure to direct light. Collect in a dark bottle
quantitative	Urine, fresh random	3–15 μg./100 ml.	Avoid exposure to direct light
	Urine, 24 hr.	60–180 μg./24 hr. (Lower in children)	Collect with 5 gm. Na_2CO_3 to maintain pH 6.5 - 9.0. Refrigerate during collection Protect from direct light Collect in a dark bottle
Corticosteroids, 17–OH	Urine, 24 hr.	Adult male: 3–10 mg./24 hr. Adult female: 2–8 mg./24 hr.	Refrigerate during collection. Use 1 gm. boric acid/L. of urine For values after stimulation or suppression tests and drug interference, see Chapter 9
Cortisol	Plasma	9–10 a.m.: 6.5–26.3 μg./100 ml. 4 p.m.: 2–18 μg./100 ml.	For values after stimulation or suppression tests and drug interference see Chapter 9
Creatine	Serum or plasma	Males: 0.2–0.5 mg./100 ml. Females: 0.4–0.9 mg./100 ml.	
	Urine, 24 hr.	Males: 0–40 mg./24 hr. Females: 0–80(100) mg./24 hr. Higher in pregnancy (>12% of creatinine) Children: >30% of creatinine	Refrigerate during collection
Creatine phosphokinase	Serum	10–100 mI.U./ml. at 37°C. 7–60 mI.U./ml. at 30°C. (Rosalki) Values lower in females and for patients at bed rest	Some isoenzymes are unstable, therefore analyze as soon as possible
Creatinine	Serum or plasma	Using nonspecific method: Males: 0.9–1.5 mg./100 ml. Females: 0.8–1.2 mg./100 ml. Using specific method: Males: 0.6–1.2 mg./100 ml. Females: 0.5–1.0 mg./100 ml.	Refrigerate or freeze
	Urine, 24 hr.	Males: 1.0–2.0 gm./24 hr.	Refrigerate
		Females: 0.8–1.8 gm./24 hr.	

TABLE OF NORMAL VALUES (Continued)

Test	*Specimen*	*Value*	*Special Instructions and Interfering Substances*
Creatinine clearance (endogenous)	Urine and serum	Using nonspecific method: Males: 105 ± 20 ml./min. Females: 95 ± 20 ml./min. Using specific method: Males: 117 ± 20 ml./min. Females: 108 ± 20 ml./min.	Refrigerate or freeze. Correct for body surface, see Chapter 12
Creatinine coefficient	Urine, 24 hr.	Males: 20–26 mg./kg./24 hr. Females: 14–22 mg./kg./24 hr. Newborn: 7–12 mg./kg./24 hr. 1.5–22 mos.: 5–15 mg./kg./24 hr. 2.5–3.5 years: 12 mg./kg./24 hr. 4–4.5 years: 15–20 mg./kg./24 hr. 9–9.5 years: 18–25 mg./kg./24 hr.	
Cryoglobulins	Serum	Negative	Keep specimen at 37°C.
Cyanide	Blood	Negative. Lethal level: <0.1 mg./100 ml.	
Cystine, qualitative	Urine	Negative or trace	
Cystine and cysteine	Urine, 24 hr.	10–100 mg./24 hr.	
Diagnex (tubeless gastric analysis)	Urine, 2 hr.	Free HCl present	Do in postabsorptive state
Diodrast clearance	Urine and plasma	600–700 ml./min. (average body surface)	
Duodenal drainage	Duodenal juice	For secretin test, see Chapter 15	
Electrophoresis, hemoglobin	Whole blood	Chiefly HbA_1 and 1.8–3.5% HbA_2	
Electrophoresis, protein	Serum	% — gm./100 ml. alb. 53–65 — 3.5–4.7 α_1 2.5–5.0 — 0.17–0.33 α_2 7.0–13.0 — 0.42–0.87 β 8.0–14.0 — 0.52–1.05 γ 12.0–22.0 — 0.71–1.45	Hemolysis interferes
	Spinal fluid	*Average %* pre-alb. 4.1 ± 1.2 alb. 62.4 ± 5.6 α_1 5.3 ± 1.2 α_2 8.2 ± 2.0 β 12.8 ± 2.0 γ 7.2 ± 1.1	Concentrate specimen prior to analysis
	Urine, 24 hr.	*Average %* alb. 37.9 α_1 27.3 α_2 19.5 β 8.8 γ 3.3	Concentrate specimen prior to analysis

TABLE OF NORMAL VALUES (Continued)

Test	*Specimen*	*Value*	*Special Instructions and Interfering Substances*
Epinephrine	Urine, 24 hr.	0–20 μg./24 hr.	
Epinephrine tolerance test	Plasma	Fasting: normal glucose; 40–60 min.: increase of 35–45 mg./100 ml. 2 hr.: return to approx. fasting level	Dose: 1 ml. of 1:1000 solution of epinephrine HCl
Estrogens, total	Urine, 24 hr.	Males: 5–18 μg./24 hr.	Keep refrigerated. Preserve with boric acid
		Females: Postmenopausal: 14–19.6 μg./24 hr. Pregnant: up to 45,000 μg./24 hr. Nonpregnant: onset of menstruation: 4–25 μg./24 hr. ovulation peak: 28–99 μg./24 hr. luteal peak: 22–105 μg./24 hr.	
Estrogens	Urine, 24 hr.	Nonpregnant females (mid-cycle):	
Estradiol		0–10 μg./24 hr.	
Estriol		2–30 μg./24 hr.	
Estrone		2–25 μg./24 hr.	
Ethanol	Whole blood	Negative. (up to 10 mg./100 ml. of volatile reducing substances considered negative)	Avoid exposure to air. Keep container tightly stoppered and refrigerate
Fat	Feces, 72 hr.	Older children and adults: <5 gm./24 hr. (5–6 gm./24 hr., borderline) Children up to 6 years: <2 gm./24 hr. Total: less than 25% of dry weight. (Free fatty acids and soaps constitute more than 40% of total fat.)	Diet: 60–150 gm. of fat/day
Fat, neutral (triglycerides)	Serum	30–200 mg./100 ml.	Use fasting specimen. Remove serum from clot soon after drawing. Hemolysis interferes
Fat absorption test	Serum	Increase of 60–100% 3 hr. after a high fat meal. Value at 6 hr. below the 3 hr. value.	
Fatty acids esterified	Serum	9–15 m*M*/L.	Patient should be fasting. Remove serum from clot soon after drawing. Analyze promptly
"free" (NEFA)	Plasma	Newborn: 905 ± 470 μEq./L. 4–10 mos.: 699 ± 199 μEq./L. (14 hr. fast) 986 ± 235 μEq./L. (19 hr. fast)	(same as above)

TABLE OF NORMAL VALUES (Continued)

Test	*Specimen*	*Value*	*Special Instructions and Interfering Substances*
Fatty acids "free" *cont'd*		Adults: 448 ± 140 μEq./L. (14 hr. fast) 560 ± 157 μEq./L. (19 hr. fast)	
Fibrinogen	Plasma	0.2–0.4 gm./100 ml.	Use oxalate or citrate as anticoagulant
Figlu test (N-Formiminoglutamic acid)	Urine, 24 hr.	<3 mg./24 hr. after 15 gm. L-histidine: appr. 4 mg./8 hr.	
Fluoride	Whole blood Urine, 24 hr.	less than 0.05 mg./100 ml. less than 1 mg./24 hr.	
Follicle stimulating hormone (FSH) (measured as gonadotrophins)	Urine, 24 hr.	Adults: 10–50 Mouse uterine units/24 hr. Children, prepuberty: <10 M.U.U./24 hr. Postmenopausal: >50 M.U.U./24 hr.	Preserve with 2 ml. of 1% thymol in glacial acetic acid or adjust pH to 5–6.5 with glacial acetic acid
Fructose	Serum Urine, 24 hr.	<7.5 mg./100 ml. Approx. 60 mg./24 hr.	
Galactose	Blood	Children: up to 20 mg./100 ml.	
Galactose tolerance	Plasma	*I.V.*: 60 min.: <42 mg./100 ml.	Dose: 0.5 gm./kg. Collect blood with potassium oxalate and sodium fluoride or analyze without delay.
	Plasma and urine	*Oral:* 30–60 min.: 40–60 mg./100 ml. plasma 5 hr. urine: <3 gm./5 hr.	Dose: 40 gm. in 200 ml. H_2O. Collect blood with potassium oxalate and sodium fluoride or analyze without delay
Gamma globulin	Serum	Caucasians: 0.55–1.3 gm./100 ml. by salt fractionation. 0.75–1.4 gm./100 ml. by electrophoresis Negroes: 0.88–1.6 gm./100 ml. by salt fractionation	
Gastric analysis	Gastric residue		Fasting specimen
Acidity:			
Free		0–40 mEq./L.	
Combined		10–20 mEq./L.	
Total		10–50 mEq./L.	
pH		1.5–4.0	
Volume		20–100 ml.	
Gastric analysis, tubeless (Diagnex)	Urine, 2 hr.	Free HCl present	Do in postabsorptive state

TABLE OF NORMAL VALUES (Continued)

Test	*Specimen*	*Value*	*Special Instructions and Interfering Substances*
Globulins, total	Serum	2.3–3.5 gm./100 ml. by salt fractionation (27%) or by electrophoresis 2.3–3.2 gm./100 ml. by dye-binding methods	Presence of blood causes false high values
qualitative (Pandy)	Spinal fluid	Negative or trace	
quantitative	Spinal fluid	6–16 mg./100 ml.	
Glucose	Serum or plasma	70–105 mg./100 ml. ("True" glucose method) 75–110 mg./100 ml. (AutoAnalyzer method) 80–120 mg./100 ml. (Folin-Wu Method)	Collect with potassium oxalate and sodium fluoride, or analyze without delay. Serum or plasma glucose has greater stability than whole blood glucose
	Whole blood	65–95 mg./100 ml. ("True" glucose method) Capillary blood higher. See chapter 3.	
2 hr. postprandial	Plasma	less than 120 mg./100 ml.	See Glucose
	Spinal fluid	40–70 mg./100 ml. ("True" glucose method) 45–75 mg./100 ml. (AutoAnalyzer method)	
qualitative	Urine, random	<30 mg./100 ml. (Clinitest negative)	
quantitative	Urine, 24 hr.	0.5–1.5 gm./24 hr. (Total reducing substances) <0.5 gm./24 hr. as glucose (average 0.13 gm./24 hr.)	
Glucose tolerance	Plasma	*I.V.:* Fasting: Normal glucose level. 5 min.: maximum of 250 mg./100 ml. 60 min.: significant decrease 120 min.: <120 mg./100 ml. 180 min.: fasting level	Dose: *I.V.:* 0.33 gm./kg. Collect specimens with potassium oxalate and sodium fluoride or analyze without delay
		Oral: Fasting: normal glucose level. 30–60 min.: peak <60% above fasting level but should not exceed 170 mg./100 ml. 2 hr.: <120 mg./100 ml.	Dose: *Oral:* Adult: 100 gm. Children (below 12 yrs.): 1.75 gm./kg. Collect specimens with potassium oxalate and sodium fluoride or analyze without delay
Glucose-6-phosphate dehydrogenase (G-6-P-D)	Red cells	1200–2000 mI.U./ml. packed RBC's at 30°C (Zinkham)	
Glutethimide	Serum or whole blood	Negative Therapeutic level: approx. 0.3 mg./100 ml. Lethal level: >3.0 mg./100 ml.	

TABLE OF NORMAL VALUES (Continued)

Test	*Specimen*	*Value*	*Special Instructions and Interfering Substances*
Gonadotrophin, pituitary (FSH & LH)	Urine, 24 hr.	Adults: approx. 10–50 mouse uterine units/24 hr. Children, prepuberty: $<$10 M.U.U./24 hr. Post menopausal: $>$50 M.U.U./24 hr.	Preserve with 2 ml. of 1% thymol in glacial acetic acid or adjust pH to 5–6.5 with glacial acetic acid
Haptoglobin	Serum or plasma	70–140 mg./100 ml. (as hemoglobin binding capacity)	
Hematocrit	Whole blood	Males: 42–50 vol./100 ml. Females: 37–47 vol./100 ml. Higher at birth	
Hemoglobin	Whole blood	Males: 13–18 gm./100 ml. Females: 11–16 gm./100 ml. Newborn: 16–20 gm./100 ml. Increased values obtained in high altitudes	Collect with EDTA
electrophoresis	Whole blood	Normally only HbA_1 and 1.8–3.5% HbA_2	
free, qualitative	Serum or plasma Urine, random	Negative (by spectroscopy) Negative (by spectroscopy)	Hemolysis invalidates this test
free, quantitative	Plasma	0.5–2.5 mg./100 ml. $<$1 mg./100 ml. when special blood collection technique is used	Hemolysis invalidates this test
Hemoglobin A_2	Whole blood	1.8–3.5% of Hgb	Collect with EDTA
Hemoglobin solubility test	Whole blood	Genotype A/A: 90–105% of total hemoglobin	
Homogentisic acid	Urine, random	Negative	
Homovanillic acid (HVA)	Urine, 24 hr.	$<$15 mg./24 hr.	Use boric acid as preservative
α-Hydroxybutyric dehydrogenase	Serum	114–290 units/ml. (Wroblewski)	
17-Hydroxycortico-steroids (as cortisol)	Plasma	9–10 A.M.: 6.5–26.3 μg./100 ml. 4 P.M.: 2–18 μg./100 ml.	For values after stimulation or suppression tests and drug interference, see Chapter 9

TABLE OF NORMAL VALUES (Continued)

Test	*Specimen*	*Value*	*Special Instructions and Interfering Substances*
17-Hydroxycortico-steroids	Urine, 24 hr.	Adult males: 3–10 mg./24 hr. Adult females: 2–8 mg./24 hr. Lower in children	Refrigerate during collection. Preserve with 1 gm. boric acid/L. of urine For values after stimulation or suppression tests and drug interference see Chapter 9
5-Hydroxy-indole acetic acid	Urine, random Urine, 24 hr.	Negative 2–8 mg./24 hr.	Some drugs and fruits interfere, see Chapter 9
Icterus index	Serum	3–8 units	Hemolysis and turbidity interfere
Indican	Urine, 24 hr.	10–20 mg./24 hr.	
Insulin	Plasma	Fasting: 11 240 μI.U./ml. (bioassay) 0–60 μI.U./ml. (Radio immunoassay)	
Insulin tolerance test	Plasma	Fasting: normal plasma glucose 30 min.: decrease to 50% of fasting level 90–120 min.: approaches fasting level	Diet containing 300 gm. carbohydrate 2–3 days prior to test. Dosage: 0.1 unit insulin/kg. Collect specimens with potassium oxalate and sodium fluoride or analyze immediately. Hypoglycemic reaction may occur
Inulin clearance	Urine and whole blood	Males: 110–152 ml./min. Females: 102–132 ml./min.	Dose: 25 ml. of a 10% inulin solution (see Chapter 12)
Iodine, butanol extractable (BEI)	Serum	3.2–6.4 μg./100 ml.	Test not reliable if iodine-containing drugs or x-ray contrast media were given prior to test; see Chapter 9
Iodine, radioactive uptake (RAI)		15–45%	
Iodine, protein bound (PBI)	Serum	(3.5)4.0 8.0 μg./100 ml.	Test not reliable if iodine-containing drugs or x-ray contrast media were given prior to test; see Chapter 9
Iron, total	Serum	Using protein precipitation method: Males: 60–150 μg./100 ml. Females: 50–130 μg./100 ml. Elders: 40–80 μg./100 ml. Birth: 100–170 μg./100 ml. Children: 50–180 μg./100 ml. Without protein precipitation: Values 10–20 μg./100 ml. higher than those above	Avoid hemolysis. For diurnal variation see Chapter 10
Iron-binding capacity, total	Serum	Using protein precipitation methods: Adults: 270–380 μg./100 ml.	Avoid hemolysis

TABLE OF NORMAL VALUES (Continued)

Test	*Specimen*	*Value*	*Special Instructions and Interfering Substances*
Iron-binding capacity, total *cont'd.*	Serum	>70 yrs.: approx. 70 μg./100 ml. lower Children: 180–575 μg./100 ml. Without protein precipitation: Adults: 280–400 μg./100 ml.	
Isocitric dehydrogenase	Serum	50–250 units/ml. at 32°C. or 0.8–4.2 mI.U./ml. at 32°C. (Wolfson & Williams-Ashman)	Separate serum from clot soon after drawing
17-Ketogenic steroids, total	Urine, 24 hr.	Males: 5–23 mg./24 hr. Females: 3–15 mg./24 hr.	Refrigerate during collection. Preserve with 1 gm. boric acid/L. of urine. For values after stimulation or suppression tests and drug interference see Chapter 9
Ketone bodies, qualitative	Serum Urine, random	Negative Negative	
quantitative	Serum	0.5–3.0 mg./100 ml.	
17-Ketosteroids	Urine, 24 hr.	Males: 8–20 mg./24 hr. Young male: 9–22 mg./24 hr. Females: (5)6–15 mg./24 hr. Decline after 60 yr. Children: up to 1 yr.: <1 mg./24 hr. 1–4 yr.: <2 mg./24 hr. 5–8 yr.: <3 mg./24 hr. 8–12 yr.: 3–10 mg./24 hr. 13–16 yr.: 5–12 mg./24 hr.	Refrigerate during collection. Preserve with 1 gm. boric acid/L. of urine. For values after stimulation or suppression tests and drug interference, see Chapter 9
17-Ketosteroids (Gas chromatographic fractionation)	Urine, 24 hr.	Total: Males: 5–12 mg./24 hr. Females: 3–10 mg./24 hr. Androsterone M: 2.0–5.0 mg./24 hr. F: 0.5–3.0 mg./24 hr. Etiocholanolone M: 1.4–5.0 mg./24 hr. F: 0.8–4.0 mg./24 hr. Dehydroepiandrosterone M: 0.2–2.0 mg./24 hr. F: 0.2–1.8 mg./24 hr. 11-Ketoandrosterone M: 0.2–1.0 mg./24 hr. F: 0.2–0.8 mg./24 hr.	Refrigerate during collection. For values after stimulation and suppression tests, see Chapter 9

TABLE OF NORMAL VALUES (Continued)

Test	*Specimen*	*Value*	*Special Instructions and Interfering Substances*
17-Ketosteroids (Gas chromatographic fractionation) *cont'd.*	Urine, 24 hr.	11-Ketoetiocholanolone M: 0.2–1.0 mg./24 hr. F: 0.2–0.8 mg./24 hr. 11-Hydroxyandrosterone M: 0.1–0.8 mg./24 hr. F: 0.0–0.5 mg./24 hr. 11-Hydroxyetiocholanolone M: 0.2–0.6 mg./24 hr. F: 0.1–1.1 mg./24 hr.	For values after stimulation or suppression tests, see Chapter 9
17-Ketosteroids			
Alpha/beta ratio	Urine, 24 hr.	Above 5 (average 9)	
Beta/alpha ratio	Urine, 24 hr.	below 0.2	
Lactic acid	Whole blood, arterial	3.1–7.0 mg./100 ml. (0.34–0.78 m*M*/L.)	Collect blood without tourniquet. Use iodoacetate as a preservative or analyze immediately.
	Whole blood, venous	5.0–15.0 mg./100 ml. (0.55–1.15 m*M*/L.)	Patient must be fasting and at complete bed rest
Lactic dehydrogenase	Serum	150–450 Wroblewski units/ml. at 32°C. (Henry Modification)	Hemolysis causes elevated results. Store at room temperature
	Spinal Fluid	15–60 Wroblewski units/ml. at 32°C. (Henry Modification)	
Lactic dehydrogenase isoenzymes	Serum	1: 28–29% 2: 39–49% 3: 13–19% 4: 3 7% 5: 0–2%	Values for each isoenzyme are averages of the values reported in Table 5, Chapter 8
Lactose	Plasma Urine, random Urine, 24 hr.	<0.5 mg./100 ml. Children: <1.5 mg./100 ml. Adult: 12–40 mg./24 hr.	
Lactose tolerance	Plasma	Results similar to glucose tolerance curve of patient	Oral adult dose: 100 gm.
Lead	Urine, 24 hr.	<100 μg./24 hr.	Use lead free container for collection.
	Whole blood	<50 μg./100 ml. 50–80 μg./100 ml., borderline. Lower in children	Use glassware and anticoagulant (heparin) that are lead free
Leucine amino peptidase (LAP)	Serum	Males: 84–200 G-R U./ml. at 37°C. Females: 76–184 G-R U./ml. at 37°C.	

TABLE OF NORMAL VALUES (Continued)

Test	*Specimen*	*Value*	*Special Instructions and Interfering Substances*
Lipase	Serum	0.05–1.0 U./ml. at 37°C. 14–280 mI. U./ml. at 37°C. (Tietz and Fiereck)	
Lipids	Feces, 72 hr.	Older children and adults: <5 gm./24 hr. (5–6 gm./24 hr. borderline) Children up to 6 yr.: <2 gm./24 hr. Total: less than 25% of dry weight (Free fatty acids and soaps constitute more than 40% of total fat) Average values: Total fat 17.5% of wet weight Neutral fat: 7.3% of wet weight Free fatty acid: 5.6% of wet weight Soaps: 4.6% of wet weight Sterols: 1.8% of wet weight Total lipids: 19.3% of wet weight	Diet: 60–150 gm. of fat/day
	Serum or plasma	Total: 400–700 mg./100 ml.	Draw blood after a 12 hr. fast
Lipid phosphorus	Serum	7–11 mg./100 ml.	See also phospholipids and phosphatides. Hemolysis interferes
Lipoproteins, beta/alpha ratio	Serum	2.5–3.0	
Lithium	Serum	Therapeutic level: 0.5–1.5 mEq./L. Toxic level: >2.0 mEq./L.	
Long-acting thyroid stimulator (LATS)	Serum	None detected	
Luteinizing hormone (LH)	Plasma	Males: <11 mI.U./ml. Females: Premenopausal: <25 mI.U./ml. Midcycle: >3 × base level Postmenopausal: >25 mI.U./ml.	
Macroglobulins	Serum	0.07–0.43 gm./100 ml.	
Magnesium	Serum	Adults: 1.4–2.3 mEq./L. (1.7–2.8 mg./100 ml.) Children: 1.2–1.6 mEq./L. (1.4–1.9 mg./100 ml.)	Avoid hemolysis

TABLE OF NORMAL VALUES (Continued)

Test	*Specimen*	*Value*	*Special Instructions and Interfering Substances*
Magnesium *cont'd.*	Spinal fluid	2.4–3.0 mEq./L.	Presence of blood invalidates the test
	Urine, 24 hr.	6–8.5 mEq./24 hr.	
Melanin, qualitative	Urine, random	Negative	
Mercury	Urine, random	<50 μg./100 ml.	Collect in acid-cleaned glass container
Methanol	Serum	Negative	
Methemoglobin	Whole Blood	0–0.25 gm./100 ml. or 0.4–1.5% of total hemoglobin	Analyze immediately
3-Methoxy-4-hydroxy-mandelic acid (VMA)	Urine, 24 hr.	Adults: 1.8–8.0 mg./24 hr. (1.5–7.0 μg./mg. creatinine) Infants: <83 μg./kg./24 hr.	No coffee, desserts, or fruit two days prior to test. Refrigerate and acidify specimen during collection
Mucin	Urine, 24 hr.	100–150 mg./24 hr.	
Mucoproteins	Serum	75–135 mg./100 ml. (80–200 mg./100 ml.—by less specific technique)	
Myoglobin	Urine, 24 hr.	<1.6 mg./L.	
Nitrogen, total	Feces, 24 hr.	10% of intake or 1–2 gm./24 hr.	
Nonprotein nitrogen	Serum or plasma	20–35 mg./100 ml.	
	Whole blood	25–50 mg./100 ml.	
5-Nucleotidase	Serum	2–15 mI.U./ml. (Campbell)	
Occult blood	Urine, random	Negative	
	Feces, random	Negative, if patient on meat-free diet for two days	
Oleic acid-I^{131} absorption test	Plasma	1.7% of administered dose/L. after 4–6 hr.	Dose: 50 μc. in milk
	Feces, 72 hr.	Less than 5% of administered dose in 72 hr. specimen	
Ornithine carbamyl transferase	Serum	8–20 mI.U./ml. (Brown and Grisolia)	
Osmolarity	Serum	275–295 mOsm./L.	
	Urine, random	390–1090 mOsm./L. >850 mOsm./L. after fluid restriction	
Oxygen capacity	Whole blood, arterial	16–24 vol. % or 1.34 ml./gm. Hgb.	
content	Whole blood, arterial	15–23 vol. %	

TABLE OF NORMAL VALUES (Continued)

Test	*Specimen*	*Value*	*Special Instructions and Interfering Substances*
Oxygen pressure, P_{O_2}	Whole blood, arterial	Average 100 mm. Hg	Collect in heparinized syringe. Store on ice. Handle anaerobically
Oxygen saturation	Whole blood, arterial	90–95%	
	Whole blood, venous	60–85%	
P_{CO_2}(Pressure of carbon dioxide)	Alveolar air	Average 36 mm. Hg	
	Whole blood, arterial	35–46 mm.Hg.	Collect with heparinized syringe and seal. Handle anaerobically
	Whole blood, venous	38–49 mm. Hg	
P_{O_2}(Pressure of oxygen)	Whole blood, arterial	Average 100 mm. Hg	Collect with heparinized syringe. Store on ice. Handle anaerobically
Pandy test	Spinal fluid	Negative	Presence of blood causes false high results
Pentoses, total	Urine, 24 hr.	Adult: 2–5 mg./kg./24 hr. on fruit-free diet	
pH	Gastric contents	1.5–4.0	
	Serum or plasma, venous	7.35–7.45	Handle anaerobically. See Chapter 10 for blood collection
	Whole blood, venous	7.33–7.43	
	Whole blood, capillary	7.35–7.45	
	Urine, fasting	5.5–6.5	
	Urine, random	4.8–7.8	
Phenolsulfonphthalein	Urine	*% Excretion*	Dose: 6 mg., I.V.
	15 min.	25–50	
	30 min.	15–25	
	60 min.	10–15	
	120 min.	5	
		60–85 (Total)	
Phenylalanine	Serum	Adults: 0.8–1.8 mg./100 ml. Full-term, normal weight newborns: 1.2–3.4 mg./100 ml. Premature or low weight newborns: 2.0–7.5 mg./100 ml. (the results will fall into the normal range within 7–20 days.)	

TABLE OF NORMAL VALUES (Continued)

Test	*Specimen*	*Value*	*Special Instructions and Interfering Substances*
Phenylpyruvic acid, qualitative	Urine, fresh random	Negative by $FeCl_3$ Test	
Phosphatase, acid	Serum	Bessey, Lowry, and Brock method: Total: Males: 0.15–0.70 U./ml. at 37°C. (2.5–12.0 mI.U./ml.) Females: 0.02–0.55 U./ml. Prostatic fraction: Males: 0.01–0.30 U./ml. Females: 0–0.05 U./ml. Bodansky method: Total: 0–1.0 U./100 ml. at 37°C. Gutman-Gutman: Total: 0.6–3.1 U./100 ml. at 37°C. or 1.4–5.5 mI.U./ml. Shinowara, Jones and Reinhart: Total: 0–1.5 U./100 ml. at 37°C or 0.0–6.0 mI.U./ml. at 37°C.	Avoid hemolysis. Perform test without delay. Fluoride and oxalate inhibit the reaction. (see Chapter 8)
Phosphatase, alkaline	Serum	Adult: Bessey, Lowry, and Brock: 0.7–2.7 U./ml. at 37°C. or 13–38 mI.U./ml. Bodansky: 1.5–4.0 U./100 ml. at 37°C. or 8.0–22 mI.U./ml. at 37°C. Shinowara, Jones and Reinhart: 2.2–6.5 U./100 ml. at 37°C. or 15–35 mI.U./ml. at 37°C. King and Armstrong (Kind and King): 3.5–13 U./100 ml. at 37°C. or 25–92 mI.U./ml. Children: Bessey, Lowry, and Brock: 3.4–9.0 U./ml. Bodansky: 5–14 U./100 ml. King and Armstrong: (Kind & King) 10–30 U./100 ml. at 37°C.	
Phosphatides	Plasma	Birth: 61 ± 32 mg./100 ml. (as lecithin) Females, nonpregnant: 195 ± 37 mg./100 ml. Females, pregnant: 248 ± 43 mg./100 ml.	Avoid hemolysis

TABLE OF NORMAL VALUES (Continued)

Test	*Specimen*	*Value*	*Special Instructions and Interfering Substances*
Phosphatides *cont'd.*		Males (and females over 65 yrs.): 281 ± 85 mg./100 ml.	
Phospholipid-P	Serum	7–11 mg./100 ml.	Avoid hemolysis
Phosphorus, inorganic	Serum	Adults: 3.0–4.5 mg./100 ml. (1.7–2.5 mEq./L.) Children: 4.5–6.5 mg./100 ml. (2.5–3.6 mEq./L.)	Avoid hemolysis. Separate serum from cells promptly
	Urine, 24 hr.	0.9–1.3 gm./24 hr. (Varies greatly with intake)	
Porphobilinogen, qualitative	Urine, random	Negative	Refrigerate and avoid exposure to direct light
quantitative	Urine, 24 hr.	0–2.0 mg./24 hr.	Acidify urine to pH 4–6. Refrigerate and avoid exposure to light
Potassium	Serum	Adults: 3.5–5.3 mEq./L. Newborn: 4.0–5.9 mEq./L.	Avoid hemolysis
	Saliva	Without stimulation: 19–23 mEq./L. With stimulation: 18–19 mEq./L.	
	Sweat	5–17 mEq./L.	
	Urine, 24 hr.	Average diet: 30–90 mEq./24 hr. (Varies with diet)	
Pregnanediol	Urine, 24 hr.	Males: 0–1.0 mg./24 hr. Females: Proliferative phase: 0.1–1.3 mg./24 hr. Luteal phase: 0.2–9.5 mg./24 hr. Pregnant: 27–47 mg./24 hr. at peak (28–32 wk.)	Refrigerate during collection
Pregnanetriol	Urine, 24 hr.	0.5–2.0 mg./24 hr. Children: <0.5 mg./24 hr.	Refrigerate during collection
Protein	Serum	6.0–8.2 gm./100 ml. (Using the protein/nitrogen factor of 6.25)	
	Spinal fluid	15–45 mg./100 ml.	
	Ventricular fluid	10–15 mg./100 ml.	
	Lumbar fluid	20–45 mg./100 ml.	
	Urine, 24 hr.	50–100 mg./24 hr.	

TABLE OF NORMAL VALUES (Continued)

Test	*Specimen*	*Value*	*Special Instructions and Interfering Substances*
Protein *cont'd.*	Urine, random	<10 mg./100 ml.	
	Urine, first morning	15–20 mg./100 ml.	
Protein fractionation	Serum	Salt fractionation method (27%): Albumin: 3.8–4.7 gm./100 ml. Globulin: 2.3–3.5 gm./100 ml. Electrophoresis: % — gm./100 ml. Albumin 53–65 — 3.5–4.7 α_1 2.5–5.0 — 0.17–0.33 α_2 7.0–13.0 — 0.42–0.87 β 8.0–14.0 — 0.52–1.05 γ 12.0–22.0 — 0.71–1.45	Hemolysis interferes
Protein-bound iodine (PBI)	Serum	(3.5) 4.0–8.0 μg./100 ml.	Test not reliable if iodine-containing drugs or x-ray contrast media were given prior to test
Reducing substances, total	Urine, 24 hr.	0.5–1.5 gm./24 hr. <150 mg./100 ml. (as glucose)	
Renal blood flow	Whole blood	Approx. 1200 ml./min.	
	Plasma	Approx. 650 ml./min. (or 390 ml./min./m² body surface)	
Salicylates	Serum	Negative (<2.0 mg./100 ml. considered negative) Therapeutic: <30 mg./100 ml. Toxic: >40 mg./100 ml. Lethal: >70 mg./100 ml.	
	Urine, random	Negative (<4.5 mg./100 ml. considered negative)	
Secretin test	Duodenal contents	See Chapter 15	
Sodium	Saliva	Without stimulation: 6.5–21.7 mEq./L. After stimulation: 43–46 mEq./L.	
	Serum	135–148 mEq./L.	
	Spinal fluid	138–150 mEq./L.	
	Sweat	Normal homozygotes: 10–40 mEq./L.	
	Urine, 24 hr.	Varies with diet. Normal diet: 40–220 mEq./24 hr. (average = 120 mEq./24 hr.)	
Solids, total	Urine, 24 hr.	45–70 gm./24 hr.	Refrigerate during collection

TABLE OF NORMAL VALUES (Continued)

Test	*Specimen*	*Value*	*Special Instructions and Interfering Substances*
Specific gravity	Urine, 24 hr. Urine, random	1.015–1.025 1.002–1.030	
Stool analysis	Feces, 24 hr.		
Dry matter (See also individual constituent requested)		Up to 25% of weight	
Sulfhemoglobin	Whole blood	Negative	
Sulfonamides	Serum or whole blood	Negative Therapeutic levels: Approx. 10 mg./100 ml.	False positive caused by other aromatic amines
T_3 test, (resin)	Whole blood	25.5–37.5% resin uptake	
Testosterone	Plasma	Males: 400–1200 ng./100 ml. Females: 30–150 ng./100 ml.	Collect with heparin. Analyze immediately or freeze
Thymol flocculation	Serum	Negative	
Thymol turbidity	Serum	0–5 units (Shank and Hoagland)	Lipemia interferes
Thyroid stimulating hormone, TSH (Bioassay)	Plasma	$<$0.2 Mouse Units/ml.	
Thyroid uptake of I^{131}		15–45%	
Thyroxine, free	Serum	1.0–2.1 ng./100 ml., as iodine. 2.9–5.1 ng./100 ml., as T_4.	
Thyroxine (T_4) by Column	Serum	3.2–6.4 μg./100 ml., as iodine.	Organic iodine containing drugs and contrast media interfere
Thyroxine (Murphy-Pattee Method)	Serum	2.9–6.4 μg./100 ml., as iodine.	No interference by organic iodine containing compounds except excessive T_3, diiodothyronine and diphenylhydantoin
Thyroxine binding globulin	Serum	10–26 ng./100 ml., as thyroxine	
Titratable acidity	Urine, 24 hr.	10–60 mEq./24 hr.	Refrigerate and preserve with toluene
Tolbutamide (Orinase) tolerance test	Plasma	Fasting: Glucose normal; 30 min.: Decrease of 50% then return to normal	Dose: I.V. 1 gm. in 20 ml. H_2O Collect specimens with potassium oxalate and sodium fluoride. Severe hypoglycemic reaction may occur, therefore glucose for I.V. administration should be readily available

TABLE OF NORMAL VALUES (Continued)

Test	*Specimen*	*Value*	*Special Instructions and Interfering Substances*
Transaminase, glutamate oxalacetic, GOT (asparate aminotransferase)	Serum	Karmen (Henry Modification): 15–40 U./ml. at 32°C. 7–19 mI.U./ml. at 32°C. Reitman-Frankel: 18–40 U./ml. at 37°C.	
	Spinal fluid	Karmen (Henry Modification): 7–49 U./ml. at 32°C.	
Transaminase, glutamate pyruvate, GPT (alanine aminotransferase)	Serum	Karmen (Henry Modification): 6–35(40) U./ml. at 32°C. 3–17(19) mI.U./ml. at 32°C. Reitman-Frankel: 5–35 U./ml. at 37°C.	
	Spinal fluid	Karmen (Henry Modification): 0–5 U./ml. at 32°C.	
Triglycerides	Serum	30–200 mg./100 ml.	Hemolysis interferes. Use fasting specimen. Remove serum from clot soon after drawing
Triolein-I^{131}, absorption test	Plasma	1.7% of administered dose/L. after 4–6 hr.	Dose: 50 μc. in milk
	Feces, 72 hr.	Less than 5% of administered dose in 72 hr. specimen	
Trypsin	Feces, fresh random	Infants: Positive in 1:80 dilution. (Semiquantitative method using x-ray film.)	Patient must be off enzyme preparations 3 days prior to test
Tyrosine	Serum	Adult: 0.8–1.3 mg./100 ml. Full-term, normal weight newborns: 1.6–3.7 mg./100 ml. Premature and low weight full-term newborns: 7.0–24 mg./100 ml. (falls into the normal range in 7–20 days)	
Urea clearance	Urine and blood	Standard clearance: 41–68 ml./min. Maximum clearance: 64–99 ml./min. or: 75–125% of normal clearance	Correct for body surface
Urea nitrogen	Serum or plasma	7–18 mg./100 ml. (15–38.5 mg. urea/100 ml.) 1–2 yrs.: 5–15 mg.100 ml. (11–32 mg. urea/100 ml.)	
	Urine, 24 hr.	12–20 gm./24 hr. (26–43 gm. urea/24 hr.)	Refrigerate during collection

TABLE OF NORMAL VALUES (Continued)

Test	*Specimen*	*Value*	*Special Instructions and Interfering Substances*
Uric acid	Serum or plasma	Henry, Sobel, and Kim method: Males: 2.5–7.0 mg./100 ml. Females: 1.5–6.0 mg./100 ml. Archibald; Kern and Stransky; Caraway; and Uricase methods: Males: 3.8–7.1 mg./100 ml. Females: 2.6–5.6 mg./100 ml. Children: 2.0–5.5 mg./100 ml.	Separate serum from cells immediately
	Urine, 24 hr.	250–750 mg./24 hr. (varies with diet) Low purine diet: $<$450 mg./24 hr. High purine diet: $<$1.0 gm./24 hr.	Refrigerate during collection
Urobilinogen, quantitative	Feces, 24 hr.	100–400 Ehrlich units/24 hr.	Refrigerate during collection
	Feces, random	75–275 Ehrlich units/100 gm.	
	Urine, 2 hr.	0.1–1.0 Ehrlich units/2 hr.	
	Urine, 24 hr.	0.5–3.5 mg./24 hr. or 0.5–4.0 Ehrlich units/24 hr.	Refrigerate during collection. Preserve with 5 gm. Na_2HCO_3. Protect from light
Uropepsin	Urine, random	15–45 U./hr. at 37°C. (West)	Preserve with toluene
	Urine, 24 hr.	1500–5000 U./24 hr. at 25°C (Anson)	
Uroporphyrins, quantitative	Feces, 24 hr.	10–40 μg./24 hr. Children only trace	
qualitative	Urine, random	Negative	Refrigerate during collection. Protect from light
quantitative	Urine, 24 hr.	0–35 μg./24 hr.	
Vanillyl mandelic acid (VMA) (3-methoxy-4-hydroxy-mandelic acid)	Urine, 24 hr.	Adults: 1.8–8.0 mg./24 hr. (1.5–7.0 μg./mg. creatinine) Infants: $<$83 μg./kg./24 hr. (2–12 μg./mg. creatinine)	No coffee, desserts, or fruit two days prior to test. Refrigerate and acidify during collection
Vitamin A test	Serum	30–65 μg./100 ml. (lower at birth)	
Vitamin A tolerance	Serum	Fasting: 30–65 μg./100 ml. 3 or 6 hrs.: Increase to 200–600 μg./100 ml.	Dose: 5000 U.S.P. Units Vit. A (in oil)/kg.
Vitamin C (Ascorbic acid)	Plasma	0.5–1.5 mg./100 ml.	Analyze immediately, unstable. Stable in acid solution. Collect with oxalate
	Whole blood	0.7–2.0 mg./100 ml.	

TABLE OF NORMAL VALUES (Continued)

Test	*Specimen*	*Value*	*Special Instructions and Interfering Substances*
Vitamin C (Ascorbic acid) *cont'd.*	Red cells	Approx. 2 × plasma conc.	
	White cells and platelets	Approx. 20–40 × plasma conc.	
	Urine, random	1–7 mg./100 ml. or >50 mg./24 hr.	
Volume	Urine, 24 hr.	Males: 800–1800 ml./24 hr. Females: 600–1600 ml./24 hr.	
Xanthochromia	Spinal fluid	Absent	
Xylose	Urine, 24 hr.	Average: 49 mg./24 hr.	
Xylose absorption test	Whole blood and urine	Adults: Blood: >25 mg./100 ml. after 2 hr. Urine: >4 gm./5 hr. (after 65 yr. > 3.5 gm./5 hr.) Children: Blood: 30 mg./100 ml. after 1 hr. Urine: 16–33% of ingested xylose	Dose: Adult = 25 gm. Children = 0.5 gm./lb. Analyze urine immediately after collection. Collect blood with potassium oxalate and sodium fluoride
Zinc turbidity	Serum	Caucasians: 2–9 units Negroes: 5–12 units	Lipemia interferes

REFERENCES FOR NORMAL VALUE TABLE

1. Davidsohn, I., and Henry, J. B., Eds.: Todd–Sandford Clinical Diagnosis by Laboratory Methods. 14th ed. Philadelphia, W. B. Saunders Co., 1969.
2. Handbook of Specialized Diagnostic Laboratory Tests. 8th ed. Van Nuys, Calif., Bioscience Laboratories, 1968.
3. Henry, R. J.: Clinical Chemistry. Principles and Technics. New York, Hoeber Medical Div., Harper & Row Publishers, 1964.
4. O'Brien, D., Ibbott, F. A., and Rodgerson, D. O.: Laboratory Manual of Pediatric Micro-Biochemical Techniques. 4th ed. New York, Hoeber Medical Div., Harper & Row, 1968.
5. Page, L. B., and Culver, P. J.: A Syllabus of Laboratory Examinations in Clinical Diagnosis. Revised Ed. Cambridge, Mass., Harvard University Press, 1960.
6. Robinson, H. W.: Appendix, normal blood values. *In* Textbook of Pediatrics. W. E. Nelson, Ed., 8th ed. Philadelphia, W. B. Saunders Co., 1964, pp. 1583–1587.

INDEX

Page numbers in *italics* refer to illustrations; (t) refers to tables.

TABLE OF FOUR

	0	1	2	3	4	5	6	7	8	9
1.0	.0000	.0043	.0086	.0128	.0170	.0212	.0253	.0294	.0334	.0374
1.1	.0414	.0453	.0492	.0531	.0569	.0607	.0645	.0682	.0719	.0755
1.2	.0792	.0828	.0864	.0899	.0934	.0969	.1004	.1038	.1072	.1106
1.3	.1139	.1173	.1206	.1239	.1271	.1303	.1335	.1367	.1399	.1430
1.4	.1461	.1492	.1523	.1553	.1584	.1614	.1644	.1673	.1703	.1732
1.5	.1761	.1790	.1818	.1847	.1875	.1903	.1931	.1959	.1987	.2014
1.6	.2041	.2068	.2095	.2122	.2148	.2175	.2201	.2227	.2253	.2279
1.7	.2304	.2330	.2355	.2380	.2405	.2430	.2455	.2480	.2504	.2529
1.8	.2553	.2577	.2601	.2625	.2648	.2672	.2695	.2718	.2742	.2765
1.9	.2788	.2810	.2833	.2856	.2878	.2900	.2923	.2945	.2967	.2989
2.0	.3010	.3032	.3054	.3075	.3096	.3118	.3139	.3160	.3181	.3201
2.1	.3222	.3243	.3263	.3284	.3304	.3324	.3345	.3365	.3385	.3404
2.2	.3424	.3444	.3464	.3483	.3502	.3522	.3541	.3560	.3579	.3598
2.3	.3617	.3636	.3655	.3674	.3692	.3711	.3729	.3747	.3766	.3784
2.4	.3802	.3820	.3838	.3856	.3874	.3892	.3909	.3927	.3945	.3962
2.5	.3979	.3997	.4014	.4031	.4048	.4065	.4082	.4099	.4116	.4133
2.6	.4150	.4166	.4183	.4200	.4216	.4232	.4249	.4265	.4281	.4298
2.7	.4314	.4330	.4346	.4362	.4378	.4393	.4409	.4425	.4440	.4456
2.8	.4472	.4487	.4502	.4518	.4533	.4548	.4564	.4579	.4594	.4609
2.9	.4624	.4639	.4654	.4669	.4683	.4698	.4713	.4728	.4742	.4757
3.0	.4771	.4786	.4800	.4814	.4829	.4843	.4857	.4871	.4886	.4900
3.1	.4914	.4928	.4942	.4955	.4969	.4983	.4997	.5011	.5024	.5038
3.2	.5051	.5065	.5079	.5092	.5105	.5119	.5132	.5145	.5159	.5172
3.3	.5185	.5198	.5211	.5224	.5237	.5250	.5263	.5276	.5289	.5302
3.4	.5315	.5328	.5340	.5353	.5366	.5378	.5391	.5403	.5416	.5428
3.5	.5441	.5453	.5465	.5478	.5490	.5502	.5514	.5527	.5539	.5551
3.6	.5563	.5575	.5587	.5599	.5611	.5623	.5635	.5647	.5658	.5670
3.7	.5682	.5694	.5705	.5717	.5729	.5740	.5752	.5763	.5775	.5786
3.8	.5798	.5809	.5821	.5832	.5843	.5855	.5866	.5877	.5888	.5899
3.9	.5911	.5922	.5933	.5944	.5955	.5966	.5977	.5988	.5999	.6010
4.0	.6021	.6031	.6042	.6053	.6064	.6075	.6085	.6096	.6107	.6117
4.1	.6128	.6138	.6149	.6160	.6170	.6180	.6191	.6201	.6212	.6222
4.2	.6232	.6243	.6253	.6263	.6274	.6284	.6294	.6304	.6314	.6325
4.3	.6335	.6345	.6355	.6365	.6375	.6385	.6395	.6405	.6415	.6425
4.4	.6435	.6444	.6454	.6464	.6474	.6484	.6493	.6503	.6513	.6522
4.5	.6532	.6542	.6551	.6561	.6571	.6580	.6590	.6599	.6609	.6618
4.6	.6628	.6637	.6646	.6656	.6665	.6675	.6684	.6693	.6702	.6712
4.7	.6721	.6730	.6739	.6749	.6758	.6767	.6776	.6785	.6794	.6803
4.8	.6812	.6821	.6830	.6839	.6848	.6857	.6866	.6875	.6884	.6893
4.9	.6902	.6911	.6920	.6928	.6937	.6946	.6955	.6964	.6972	.6981
5.0	.6990	.6998	.7007	.7016	.7024	.7033	.7042	.7050	.7059	.7067
5.1	.7076	.7084	.7093	.7101	.7110	.7118	.7126	.7135	.7143	.7152
5.2	.7160	.7168	.7177	.7185	.7193	.7202	.7210	.7218	.7226	.7235
5.3	.7243	.7251	.7259	.7267	.7275	.7284	.7292	.7300	.7308	.7316
5.4	.7324	.7332	.7340	.7348	.7356	.7364	.7372	.7380	.7388	.7396
5.5	.7404	.7412	.7419	.7427	.7435	.7443	.7451	.7459	.7466	.7474
5.6	.7482	.7490	.7497	.7505	.7513	.7520	.7528	.7536	.7543	.7551
5.7	.7559	.7566	.7574	.7582	.7589	.7597	.7604	.7612	.7619	.7627
5.8	.7634	.7642	.7649	.7657	.7664	.7672	.7679	.7686	.7694	.7701
5.9	.7709	.7716	.7723	.7731	.7738	.7745	.7752	.7760	.7767	.7774

PLACE LOGARITHMS

	0	1	2	3	4	5	6	7	8	9
6.0	.7782	.7789	.7796	.7803	.7810	.7818	.7825	.7832	.7839	.7846
6.1	.7853	.7860	.7868	.7875	.7882	.7889	.7896	.7903	.7910	.7917
6.2	.7924	.7931	.7938	.7945	.7952	.7959	.7966	.7973	.7980	.7987
6.3	.7993	.8000	.8007	.8014	.8021	.8028	.8035	.8041	.8048	.8055
6.4	.8062	.8069	.8075	.8082	.8089	.8096	.8102	.8109	.8116	.8122
6.5	.8129	.8136	.8142	.8149	.8156	.8162	.8169	.8176	.8182	.8189
6.6	.8195	.8202	.8209	.8215	.8222	.8228	.8235	.8241	.8248	.8254
6.7	.8261	.8267	.8274	.8280	.8287	.8293	.8299	.8306	.8312	.8319
6.8	.8325	.8331	.8338	.8344	.8351	.8357	.8363	.8370	.8376	.8382
6.9	.8388	.8395	.8401	.8407	.8414	.8420	.8426	.8432	.8439	.8445
7.0	.8451	.8457	.8463	.8470	.8476	.8482	.8488	.8494	.8500	.8506
7.1	.8513	.8519	.8525	.8531	.8537	.8543	.8549	.8555	.8561	.8567
7.2	.8573	.8579	.8585	.8591	.8597	.8603	.8609	.8615	.8621	.8627
7.3	.8633	.8639	.8645	.8651	.8657	.8663	.8669	.8675	.8681	.8686
7.4	.8692	.8698	.8704	.8710	.8716	.8722	.8727	.8733	.8739	.8745
7.5	.8751	.8756	.8762	.8768	.8774	.8779	.8785	.8791	.8797	.8802
7.6	.8808	.8814	.8820	.8825	.8831	.8837	.8842	.8848	.8854	.8859
7.7	.8865	.8871	.8876	.8882	.8887	.8893	.8899	.8904	.8910	.8915
7.8	.8921	.8927	.8932	.8938	.8943	.8949	.8954	.8960	.8965	.8971
7.9	.8976	.8982	.8987	.8993	.8998	.9004	.9009	.9015	.9020	.9026
8.0	.9031	.9036	.9042	.9047	.9053	.9058	.9063	.9069	.9074	.9079
8.1	.9085	.9090	.9096	.9101	.9106	.9112	.9117	.9122	.9128	.9133
8.2	.9138	.9143	.9149	.9154	.9159	.9165	.9170	.9175	.9180	.9186
8.3	.9191	.9196	.9201	.9206	.9212	.9217	.9222	.9227	.9232	.9238
8.4	.9243	.9248	.9253	.9258	.9263	.9269	.9274	.9279	.9284	.9289
8.5	.9294	.9299	.9304	.9309	.9315	.9320	.9325	.9330	.9335	.9340
8.6	.9345	.9350	.9355	.9360	.9365	.9370	.9375	.9380	.9385	.9390
8.7	.9395	.9400	.9405	.9410	.9415	.9420	.9425	.9430	.9435	.9440
8.8	.9445	.9450	.9455	.9460	.9465	.9469	.9474	.9479	.9484	.9489
8.9	.9494	.9499	.9504	.9509	.9513	.9518	.9523	.9528	.9533	.9538
9.0	.9542	.9547	.9552	.9557	.9562	.9566	.9571	.9576	.9581	.9586
9.1	.9590	.9595	.9600	.9605	.9609	.9614	.9619	.9624	.9628	.9633
9.2	.9638	.9643	.9647	.9652	.9657	.9661	.9666	.9671	.9675	.9680
9.3	.9685	.9689	.9694	.9699	.9703	.9708	.9713	.9717	.9722	.9727
9.4	.9731	.9736	.9741	.9745	.9750	.9754	.9759	.9763	.9768	.9773
9.5	.9777	.9782	.9786	.9791	.9795	.9800	.9805	.9809	.9814	.9818
9.6	.9823	.9827	.9832	.9836	.9841	.9845	.9850	.9854	.9859	.9863
9.7	.9868	.9872	.9877	.9881	.9886	.9890	.9894	.9899	.9903	.9908
9.8	.9912	.9917	.9921	.9926	.9930	.9934	.9939	.9943	.9948	.9952
9.9	.9956	.9961	.9965	.9969	.9974	.9978	.9983	.9987	.9991	.9996

% TRANSMISSION – ABSORBANCE CONVERSION CHART

% T	A	% T	A	% T	A	% T	A
1	2.000	1.5	1.824	51	.2924	51.5	.2882
2	1.699	2.5	1.602	52	.2840	52.5	.2798
3	1.523	3.5	1.456	53	.2756	53.5	.2716
4	1.398	4.5	1.347	54	.2676	54.5	.2636
5	1.301	5.5	1.260	55	.2596	55.5	.2557
6	1.222	6.5	1.187	56	.2518	56.5	.2480
7	1.155	7.5	1.126	57	.2441	57.5	.2403
8	1.097	8.5	1.071	58	.2366	58.5	.2328
9	1.046	9.5	1.022	59	.2291	59.5	.2255
10	1.000	10.5	.979	60	.2218	60.5	.2182
11	.959	11.5	.939	61	.2147	61.5	.2111
12	.921	12.5	.903	62	.2076	62.5	.2041
13	.886	13.5	.870	63	.2007	63.5	.1973
14	.854	14.5	.838	64	.1939	64.5	.1905
15	.824	15.5	.810	65	.1871	65.5	.1838
16	.796	16.5	.782	66	.1805	66.5	.1772
17	.770	17.5	.757	67	.1739	67.5	.1707
18	.745	18.5	.733	68	.1675	68.5	.1643
19	.721	19.5	.710	69	.1612	69.5	.1580
20	.699	20.5	.688	70	.1549	70.5	.1518
21	.678	21.5	.668	71	.1487	71.5	.1457
22	.658	22.5	.648	72	.1427	72.5	.1397
23	.638	23.5	.629	73	.1367	73.5	.1337
24	.620	24.5	.611	74	.1308	74.5	.1278
25	.602	25.5	.594	75	.1249	75.5	.1221
26	.585	26.5	.577	76	.1192	76.5	.1163
27	.569	27.5	.561	77	.1135	77.5	.1107
28	.553	28.5	.545	78	.1079	78.5	.1051
29	.538	29.5	.530	79	.1024	79.5	.0996
30	.523	30.5	.516	80	.0969	80.5	.0942
31	.509	31.5	.502	81	.0915	81.5	.0888
32	.495	32.5	.488	82	.0862	82.5	.0835
33	.482	33.5	.475	83	.0809	83.5	.0783
34	.469	34.5	.462	84	.0757	84.5	.0731
35	.456	35.5	.450	85	.0706	85.5	.0680
36	.444	36.5	.438	86	.0655	86.5	.0630
37	.432	37.5	.426	87	.0605	87.5	.0580
38	.420	38.5	.414	88	.0555	88.5	.0531
39	.409	39.5	.403	89	.0505	89.5	.0482
40	.398	40.5	.392	90	.0458	90.5	.0434
41	.387	41.5	.382	91	.0410	91.5	.0386
42	.377	42.5	.372	92	.0362	92.5	.0339
43	.367	43.5	.362	93	.0315	93.5	.0292
44	.357	44.5	.352	94	.0269	94.5	.0246
45	.347	45.5	.342	95	.0223	95.5	.0200
46	.337	46.5	.332	96	.0177	96.5	.0155
47	.328	47.5	.323	97	.0132	97.5	.0110
48	.319	48.5	.314	98	.0088	98.5	.0066
49	.310	49.5	.305	99	.0044	99.5	.0022
50	.301	50.5	.297	100	.0000		.0000